Horizons

EXPLORING THE UNIVERSE

About the Author

Mike Seeds is Professor of Astronomy at Franklin and Marshall College, where he has taught astronomy since 1970. His research interests have focused on peculiar variable stars and the automation of astronomical telescopes. He is continuing his research by serving as Principal Astronomer in charge of the Phoenix 10, the first fully robotic telescope, located in southern Arizona. In 1989, he received the Christian R. and Mary F. Lindback Award for Distinguished Teaching. In addition to teaching, writing, and research, Mike has published educational tools for use in computer-smart classrooms. His interest in the history of astronomy led him to offer upper-level courses "Archaeoastronomy" and "Changing Concepts of the Universe," which he continues to develop. He has also published educational software for preliterate toddlers. Mike was Senior Consultant in the creation of the 26-episode telecourse to accompany this text. He is the author of *Astronomy: The Solar System and Beyond,* Fourth Edition (2005), and *Foundations of Astronomy,* Eighth Edition (2005), published by Brooks/Cole.

9

NINTH EDITION

Horizons

EXPLORING THE UNIVERSE

Michael A. Seeds

Joseph R. Grundy Observatory,
Franklin and Marshall College

Australia • Canada • Mexico • Singapore • Spain
United Kingdom • United States

For Janet and Kate

THOMSON

BROOKS/COLE

Acquisitions Editors: *Keith Dodson, Chris Hall*
Development Editor: *Alyssa White*
Assistant Editor: *Carol Benedict*
Editorial Assistant: *Kyra Engelberg*
Technology Project Manager: *Samuel Subity*
Executive Marketing Manager: *Julie Conover*
Marketing Assistant: *Leyla Jowza*
Marketing Communications Manager: *Stacey Purviance*
Project Manager, Editorial Production: *Lisa Weber*
Art Director: *Vernon Boes*
Print Buyer: *Karen Hunt*

Permissions Editor: *Kiely Sisk*
Production Service: *Heckman & Pinette*
Text Designer: *Linda Beaupré/Stone House Art*
Development Artist: *Lisa Torri*
Photo Researchers: *Mike Seeds and Kathleen Olson*
Copy Editor: *Margaret Pinette*
Illustrator: *Precision Graphics*
Cover Designer: *Irene Morris*
Cover Image: *NASA Hubble Heritage Team, AURA/STScI*
Cover and Text Printer: *Transcontinental Printing/Interglobe*
Compositor: *Thompson Type*

COPYRIGHT © 2006 Brooks/Cole, a division of Thomson Learning, Inc. Thomson Learning™ is a trademark used herein under license.

ALL RIGHTS RESERVED. No part of this work covered by the copyright hereon may be reproduced or used in any form or by any means—graphic, electronic, or mechanical, including but not limited to photocopying, recording, taping, Web distribution, information networks, or information storage and retrieval systems—without the written permission of the publisher.

Printed in Canada
1 2 3 4 5 6 7 09 08 07 06 05

For more information about our products, contact us at:
Thomson Learning Academic Resource Center
1-800-423-0563

For permission to use material from this text or product, submit a request online at
http://www.thomsonrights.com

Any additional questions about permissions can be submitted by email to
thomsonrights@thomson.com

Library of Congress Control Number: 2004117359

Student Edition: ISBN 0-495-01003-0
Instructor's Edition: ISBN 0-495-01004-9
International Student Edition: ISBN 0-495-01503-2
(Not for sale in the United States)

Thomson Brooks/Cole
10 Davis Drive
Belmont, CA 94002
USA

Asia
Thomson Learning
5 Shenton Way #01-01
UIC Building
Singapore 068808

Australia / New Zealand
Thomson Learning
102 Dodds Street
Southbank, Victoria 3006
Australia

Canada
Nelson
1120 Birchmount Road
Toronto, Ontario M1K 5G4
Canada

Europe / Middle East / Africa
Thomson Learning
High Holborn House
50/51 Bedford Row
London WC1R 4LR
United Kingdom

Latin America
Thomson Learning
Seneca, 53
Colonia Polanco
11560 Mexico D.F.
Mexico

Spain/Portugal
Paraninfo
Calle Magallanes, 25
28015 Madrid, Spain

Brief Contents

Part 1: The Sky

Part 2: The Stars

Part 3: The Universe of Galaxies

Part 4: The Solar System

Part 5: Life

Contents

Part 1: The Sky

Windows on Science

Concept Art Portfolios

Part 2: The Stars

Windows on Science

Concept Art Portfolios

Part 3: The Universe of Galaxies

Windows on Science

Concept Art Portfolios

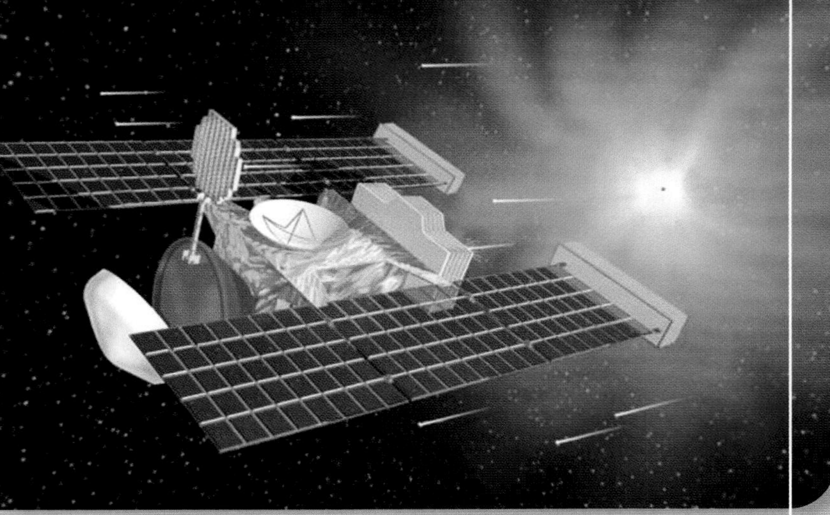

Part 4: The Solar System

Windows on Science

Concept Art Portfolios

Part 5: Life

Windows on Science

Concept Art Portfolios

A Note to the Student

From Mike Seeds

ASTRONOMY IS ABOUT US. Although an astronomy course covers planets and stars and galaxies, it is really about you and me. Astronomy helps us answer the ultimate question of human existence: What are we? Great minds like Plato, Beethoven, and Hemingway have tried to give their personal answers, but we must each find our own answer. Astronomy will help you understand the meaning of your own existence.

Of course, part of your personal answer must include philosophical and cultural issues. We define ourselves by what we create, what we worship, what we admire, and what we expect of each other. But a major part of your answer is embedded in the physical universe. You cannot hope to find a complete answer until you understand how the birth of the universe, the origin of galaxies, the evolution of stars, and the formation of planets created and brought together the atoms of which you are made. A basic understanding of astronomy is fundamental to knowing what you are. Astronomy is exciting because it is about you.

Part of the excitement of astronomy is the discovery of new things. Astronomers expect to be astonished. That excitement is reflected in this book in the form of new images, new discoveries, and new understandings that take you, in an introductory course, to the frontier of human knowledge. Here you will explore the new evidence of water on Mars and a vast ocean on Jupiter's moon Europa. You will study the newest images of dying stars, and you will struggle to understand new evidence that the expansion of the universe is speeding up. The Hubble Space Telescope, spacecraft orbiting Mars and Saturn, the Chandra X-Ray Observatory, and giant Earth-based telescopes on remote mountaintops provide us with a daily dose of excitement that goes far beyond sensationalism. These new discoveries in astronomy are exciting because they are about us—about you and me. They tell us more and more about what we are.

I hope you will share in the excitement and satisfaction that scientists feel as they puzzle out nature's secrets. That is why I have created 32 Concept Art Portfolios located throughout this book. Each presents a carefully selected subject visually, not only to make it exciting but to challenge you to synthesize your own undersanding. Rather than memorize facts and predigested descriptions of nature, you can create your own understanding through the guided-discovery design of these special features. The understanding we create ourselves is the most exciting and the most durable learning that we will ever experience.

In addition, I have used the principles of guided discovery to create dozens of new figures that will guide you to discover for yourself some of nature's most interesting secrets.

As a teacher, my quest is simple. I want you to understand your place in the universe—not just your location in space, but your role in the unfolding history of the physical universe. Not only do I want you to know where you are and what you are in the universe, but I want you to understand how we know. By the end of this book, I want you to know that the universe is very big, but that it is described by a small set of rules and that we have found a way to figure out the rules—a method called science.

The most important concept in introductory astronomy is the process of science, the process by which scientists ask questions of nature and gradually puzzle out the beautiful secrets of the physical world. You should not memorize facts or believe principles because some astronomer says it is so. Of course not! To be a participant in the adventure and not a mere observer, you need to understand how evidence and hypotheses interact to create new understanding. Those who understand how science works are part of the adventure.

Another reason you need to understand how science works is that you live in a technological world that is changing rapidly through the practical application of new scientific discoveries. To control your life in such a world, to guide and care for others, you need to understand the power and the limitation of scientific study.

Science is based on the interplay of evidence and hypothesis, and that interplay is the principal organizing theme for this book at every level—from the individual sentences to the order of the chapters. We will deal with each topic by surveying basic observational facts, synthesizing hypotheses, and then discussing further observations as evidence that supports or contradicts those hypotheses.

A good example is how a survey of the solar system in Chapter 16 leads to theories to explain the origin of the solar system, which is followed by further evidence in Chapters 17 through 19 that helps test and generalize the theories. At a different level, the paragraphs that discuss Jupiter's rings show how basic observations allow astronomers to propose theories for the origin of the ring particles.

One big advantage of emphasizing evidence and hypothesis is that it unifies the mass of facts that discourage so many non-science readers. Even the ordering of the chapters follows this theme, as you can see with Chapter 16, "The Origin of the Solar System." Furthermore, reading the solar system section after the stars and galaxies sections allows you to see how the solar nebula hypothesis develops from a modern understanding of stellar evolution, star formation, and nucleosynthesis. I want you to understand logical arguments. My approach makes the facts easier to remember, but, more importantly, the facts make sense because they fit into a logical framework. Understanding nature is the heart of science, and the interplay of evidence and hypothesis is the tool science uses to unlock the secrets of the universe. Once you begin to understand that exciting story, the facts will fall into place.

To help you grasp the logic of astronomy, I have created a number of features within the book.

- **Concept Art Portfolios,** mentioned earlier, provide an opportunity for you to create your own understanding and share in the satisfaction that scientists feel as they uncover the secrets of natural processes.
- **Guided discovery figures** illustrate important ideas visually and guide students to understand relationships and contrasts interactively.
- **Windows on Science** provide a parallel commentary on how all of science works. For example, the Windows point out where we are using statistical evidence, where we are reasoning by analogy, and where we are building a scientific model rather than a scientific hypothesis.
- **Guideposts** on the opening page of each chapter help you see the organization of the book. The Guidepost connects the chapter with preceding and following chapters to provide an overall organizational guide.
- **Inquire Review Analyze** at the end of each text section is a carefully designed question to help you review and synthesize concepts and understand the scientific procedures used in the section. A short answer follows to show how scientists construct logical arguments (from observations, evidence, theories, and natural laws) that lead to a conclusion. A further question then gives you a chance to construct your own argument on a related issue.
- **End-of-Chapter Review Questions** are designed to help you review and test your understanding of the material.
- **End-of-Chapter Discussion Questions** go beyond the text and invite you to think critically and creatively about scientific questions. You can think about these questions yourself or discuss them in class.
- **Critical Inquiries for the Web** conclude each chapter by challenging you to use the World Wide Web to explore further and to think creatively and analytically about astronomy.
- **Exploring *TheSky*™** are experiments you can perform with the software *TheSky,* experiments that will allow you to see the sky and celestial bodies in new ways on your own computer screen.

As additional aids, you should note that this book also includes, free of charge, the following electronic enhancements:

- **Virtual Astronomy Laboratories.** This set of 20 online labs is free with a passcode included with every new copy of this textbook. The labs cover topics from helioseismology to dark matter and allow you to submit your results electronically to your instructor or print them out to hand in. The first page of each chapter in this textbook notes which labs correlate to that chapter.
- **AceAstronomy (http://astronomy.brookscole.com/sh9e).** Take charge of your learning with the first assessment-centered student learning tool for astronomy. Access AceAstronomy free via the Web with the access code card bound into this book and begin to maximize your study time with a host of interactive tutorials and quizzes that help you focus on what you need to learn to master astronomy.
- ***TheSky* Student Edition CD-ROM.** With this CD-ROM, a personal computer becomes a powerful personal planetarium. Loaded with data on 118,000 stars and 13,000 deep-sky objects with images, it allows you to view the universe at any point in time from 4000 years ago to 8000 years in the future, to see the sky in motion, to view constellations, to print star charts, and much more.
- **The Brooks/Cole Astronomy Resource Center (www.brookscole.com/astronomy).** Thousands of students and instructors visit this free Web site every week. The Astronomy Resource Center offers—among other things—online study aids, virtual field trips, and links to other sites.

Do not be humble. Astronomy tells us that the universe is vast and powerful, but it also tells us that we are astonishing creatures. We humans are the parts of the universe that think. The human brain is the most complex piece of matter known, so as you explore the universe, remember that it is your human brain that is capable of understanding the depth and beauty of the cosmos.

To appreciate your role in this beautiful universe, you must learn more than just the facts of astronomy. You must understand what we are and how we know. Every page of this book reflects that ideal.

Mike Seeds
mike.seeds@fandm.edu

Acknowledgments

In 1973, I started writing astronomy textbooks, and over the years I have had the guidance of a great many people who care about astronomy and teaching. I would like to thank all of the students and teachers who have responded so enthusiastically to *Horizons: Exploring the Universe.* Their comments and suggestions have been very helpful in shaping the book over the years. I would especially like to thank the many reviewers whose careful analysis and thoughtful suggestions have been invaluable in completing this most recent edition.

The following institutions have provided the many images and diagrams for this book: Anglo-Australian Telescope Board; Arecibo Observatory; Arizona State University Computer Science and Geography Departments; Associated Universities for Research in Astronomy; Astronomical Journal; Astrophysical Journal; AT&T Archives; BOOMERANG Collaboration; Brookhaven National Laboratory; Calar Alto Observatory; California Astronomy Research Association; Caltech; Celestron International; Chandra X-Ray Observatory; European Southern Observatory; Cornell University; European Space Administration; Gemini Observatory; Granger Collection; High Altitude Observatory; Hubble Heritage Team; Hughes Aircraft Company; IPAC; JPL; Johns Hopkins Applied Physics Laboratory; Keck Observatory; Large Binocular Telescope Project; Lick Observatory; Lowell Observatory; Malin Space Science Systems; Max-Planck Institute for Extraterrestrial Physics; National Aeronautics and Space Administration; National Geophysical Data Center; National Museum of Natural History; National Oceanic and Atmospheric Administration; National Optical Astronomy Observatory; National Radio Astronomy Observatory; National Science Foundation; National Solar Observatory; New England Meteoritical; Observatories of the Carnegie Institution of Washington; Sac Peak Observatory; SETI Institute; Sloan Digital Sky Survey; Smithsonian Astrophysical Observatory; Solar and Heliospheric Observatory; Space Telescope Science Institute; Sudbury Neutrino Observatory; Swedish Vacuum Solar Telescope; Transition Region and Coronal Explorer; Two Micron All Sky Survey; United States Geological Survey; University of Hawaii; WIYN Consortium; Yerkes Observatory; and Yohkoh Solar Observatory.

I would especially like to thank the following individuals, who went out of their way to help me locate materials for *Horizons:* H. A. Abt, Dana Backman, Michael Bailey, Samuel Barden, Allen Beechel, Charles Blue, David Bradstreet, Jack Burns, Linda Bustos, Chip Clark, Jan Curtis, Don Davis, Frank Drake, R. J. Dufour, Michael Evans, Michelangelo Fazio, Robert Fosbury, Sidney Fox, Henry Freudenreich, Dan Gezari, Daniel Good, Randall Grubbs, John Grula, Calvin J. Hamilton, F. R. Harnden, William Hartmann, Phil Harvey, D. Hathaway, John Hill, Helen Horstman, George Jacoby, Paul Kalas, Victoria Kaspi, William Keel, Russell Kempton, Dean Ketelson, Stephen Larson, Tod Lauer, Christopher Lawton, Deborah Levine, Kwok-Yung Lo, John W. Mackenty, David Malin, Jack B. Marling, Stanley Miller, Richard Mundt, Harold Nations, P. J. E. Peebles, C. M. Pieters, Carolyn Porco, Ron Reese, Vera Rubin, Rudolph E. Schild, Maarten Schmidt, J. William Schopf, François Schweizer, Nigel Sharp, Virgil L. Sharpton, Seth Shostak, L. D. Simon, Mike Skrutskie, Steve Snowden, Joachim Trümper, Anthony Tyson, Mike Werner, Michael J. West, Gart Westerhout, Simon D. M. White, and Robin Witmore.

Certain diagrams in Chapters 9 and 10 are based on figures I designed that first appeared in my article "Stellar Evolution," *Astronomy,* February 1979.

I am happy to acknowledge the use of images and data from a number of important programs. In preparing materials for this book I used NASA's Sky View facility, located at NASA Goddard Space Flight Center. I have used atlas images and mosaics obtained as part of the Two Micron All Sky Survey (2MASS), a joint project of the University of Massachusetts and the Infrared Processing and Analysis Center/California Institute of Technology, funded by the National Aeronautics and Space Administration and the National Science Foundation. A number of solar images are used by the courtesy of the SOHO consortium, a project of international cooperation between ESA and NASA. NOAO images are copyrighted by the Association of Universities for Research in Astronomy, Inc. (AURA), all rights reserved.

It is always a pleasure to work with the Thomson Learning team at Brooks/Cole and Wadsworth. Special thanks go to all of the people who have contributed to this project, including Lisa Weber, Vernon Boes, Lisa Torri, Samuel Subity, Carol Benedict, Julie Conover, Stacey Purviance, Karen Hunt, Kiely Sisk, Linda Beaupré, Irene Morris, Leyla Jowza, and Kyra Engelberg.

I have enjoyed working on production with Margaret Pinette and Bill Heckman of Heckman & Pinette, and I appreciate their understanding and goodwill. Development editor Alyssa White has suggested many ways to improve this edition ranging from overall art design to the use of pronouns, and the book is much better for her innovations. I want especially to thank my editors Keith Dodson and Chris Hall for their initiative, encouragement, and support with this project.

Most of all, I would like to thank my wife, Janet, and my daughter, Kate, for their patience with "the books." They know all too well that textbooks are made of time.

Mike Seeds

Manuscript Reviewers

For the Eighth Edition

Nadine Barlow, University of Central Florida
Chris Churchill, Pennsylvania State University
John J. Cowan, University of Oklahoma
Manfred Cuntz, University of Texas–Arlington
Scott Dodelson, University of Chicago/Fermilab
Boyd Edwards, University of West Virginia
Doug Franklin, Western Illinois University
Daniel "Thor" Garber, Pensacola Junior College
Manoj Kaplinghat, University of California, Davis
Cynthia Peterson, University of Connecticut
Wayne Wooten, Pensacola Junior College
Gareth Wynn-Williams, University of Hawaii

For the Ninth Edition

Gene Byrd, University of Alabama
John J. Cowan, University of Oklahoma
D. Michael Crenshaw, Georgia State University
David Falk, Los Angeles Valley College
Marc Gagné, West Chester University
Stephen Gottesman, University of Florida
Scott T. Miller, Pennsylvania State University
Melvyn J. Oremland, Pace University
Bradley M. Peterson, Ohio State University
Peter Shull, Oklahoma State University
Gordon J. Stacey, Cornell University
G. Roger Stanley, San Antonio College

Horizons

EXPLORING THE UNIVERSE

1 The Scale of the Cosmos

The longest journey

begins with a single step.

CONFUCIUS

Visual-wavelength image

YOU ARE ABOUT to embark on a voyage to the end of the known universe. Marco Polo journeyed east, Columbus west, but you will travel outward, away from your home on Earth, past the moon and the sun and the other planets of our solar system, past the stars you see in the evening sky, and beyond a billion more stars that can only be seen with the aid of the largest telescopes. You will journey through great whirlpools of stars to visit the most distant galaxies visible from Earth, and then you will continue on, carried only by experience and imagination—searching for the structure of the universe itself. ▌ Your imagination is the key to discovery; it will be your scientific time machine transporting you into the past and into the future. Go back to watch the birth of the universe, the formation of the first stars, and ultimately the

So distant that light has taken 2,500 years to reach Earth, the Veil nebula was produced by the explosion of a star 15,000 years ago. It now drifts in front of a more distant star not related to the nebula.
(T. Rector, University of Alaska, and WIYN/NOAO/AURA/NSF)

Guidepost

Getting Started

You are already an expert on astronomy. You have enjoyed sunsets and moonrises. You have admired the stars and may know a few constellations. You have probably read about Mars and eclipses, and you may have read about distant galaxies. That is more than most Earthlings know about astronomy.

But you are a planetwalker, and you should understand what it means to live on a planet that whirls around a star drifting through a universe of stars and galaxies. You owe it to yourself to know where you are.

This Chapter

Here you will take a quick trip, a cosmic zoom, from objects you know up to the largest things in the universe. That quick survey will answer a few essential questions:

Where is Earth in relation to the sun, the planets, and the galaxies?

How do astronomers express distances?

Which objects are big, and which are small?

Are there other worlds like Earth?

This chapter will give you a sense of the scale of the universe that will carry you to new discoveries.

Looking Ahead

It is easy to learn a few facts, but it is the relationships among facts that are important. This chapter will give you the sense of scale that you need to understand where you are in the universe. The remaining chapters in this book will fill in the details, cite evidence and theories, and illustrate the wonderful intricacy and beauty of the universe. That journey begins here.

AceAstronomy™ The AceAstronomy icon throughout the text indicates an opportunity for you to test yourself on key concepts and to explore animations and interactions on the AceAstronomy website at: http://astronomy.brookscole.com/sh9e

origin of the sun and our Earth. Then rush into the future to see what will happen when the sun dies and Earth withers.

Although you will discover a beginning to the universe, you will not find an edge or an end in space. No matter how far you voyage, you will not run into a wall, but you will discover evidence that our universe may be infinite; that is, it may extend in all directions without limit. Such vastness may dwarf our earthly dimensions, but not our curiosity and imagination.

Astronomy is more than the study of stars and planets—it is the study of the universe in which we live. Although we are confined to a small planet circling a small star, the study of astronomy can take us beyond these boundaries and help us not only see where we are in the universe but understand what we really are.

Do not be humble. Although we are very small compared to stars and galaxies, we are intelligent creatures, and we can understand the universe in which we live.

Astronomy will introduce you to sizes, distances, and times far beyond your usual experience on Earth. Your task in this chapter is to grasp the meaning of these unfamiliar sizes, distances, and times. Believe it or not, the solution lies in a single word—scale. In this chapter, you will compare objects of different sizes in order to comprehend the scale of the universe.

Let's begin with something familiar, like the size of ourselves and our surroundings. ■ Figure 1-1 shows a region about 52 feet across occupied by a human being, a sidewalk, and a few trees—all objects whose size you can relate to. Each successive picture in this sequence of images will show you a region of the universe that is 100 times wider than the preceding picture. In ■ Figure 1-2 your field of vision has increased by a factor of 100, and you can now see an area 1 mile in diameter. Look at the arrow that points to the scene shown in the preceding photograph. People, trees, and sidewalks have vanished, but now you can see a college campus and the surrounding streets and houses. The dimensions of houses and streets are familiar. This is the world you know and can relate to the scale of your body.

You started this adventure using feet and miles, but you should use the metric system of units. Not only is it used by all scientists around the world, but it makes calculations much easier. If you are not already familiar with the metric system, or would like a review, study Appendix A before reading on.

The photo in Figure 1-2 is 1 mile in diameter. A mile equals 1.609 kilometers, so you can see that a kilometer is about two-thirds of a mile—a short walk across a neighborhood.

The view in ■ Figure 1-3 spans 160 kilometers. In this infrared photo, the green foliage shows up as various shades of red.

■ **Figure 1-1**

(Michael A. Seeds)

■ **Figure 1-2**

(USGS)

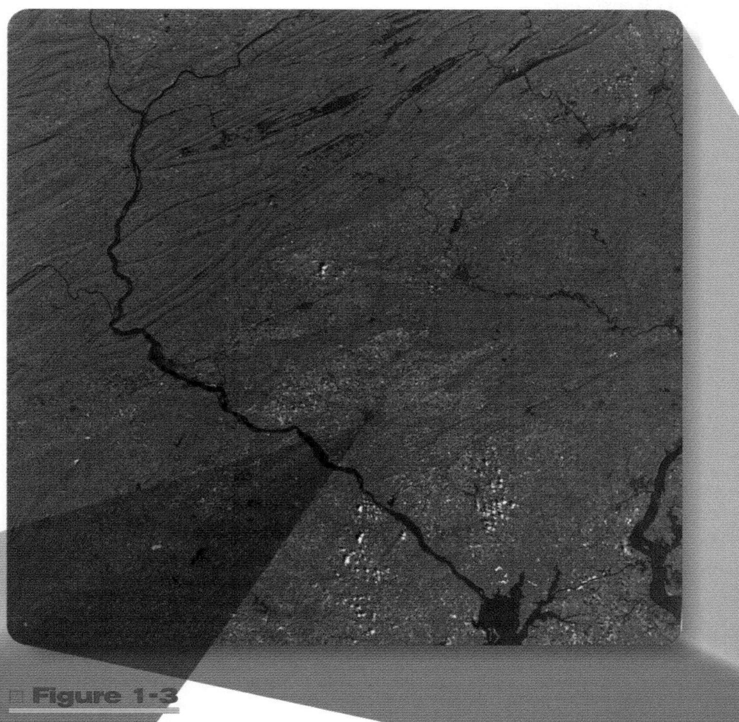

■ **Figure 1-3**

(NASA infrared photograph)

■ **Figure 1-4**

(NASA)

The college campus is now invisible, and the patches of gray are small towns and cities, including the suburbs of Philadelphia, which are visible in the lower right corner. At this scale, you can see the natural features of Earth's surface. Look at the Allegheny Mountains of southern Pennsylvania, which cross the image in the upper left, and the Susquehanna River, which flows southeast into the Chesapeake Bay. What look like white bumps are actually puffs of clouds.

These features are a reminder that you live on the surface of a changing planet. Forces in Earth's crust pushed the mountain ranges up into parallel folds, like a rug wrinkled by sliding on a smooth floor. The clouds tell you that Earth's atmosphere is rich in water, which falls as rain and then erodes the mountains, washing material down river and into the sea. Mountains and valleys are only temporary features on Earth, because they too are constantly changing. As you explore the universe, you will come to see that, like Earth's surface, it is always evolving.

Take a closer look at Figure 1-3 and notice the red color. This is an infrared photograph in which healthy green leaves and crops show up as red. Our eyes are sensitive to only a narrow range of colors. As you explore the universe, you will need to learn to use a wide range of "colors," from X rays to radio waves, to reveal sights invisible to our unaided eyes.

In the next step of this journey, you will see our entire planet (■ Figure 1-4), which is 12,756 km in diameter. Earth rotates on its axis once a day, exposing half of its surface to daylight at any particular moment. This photo shows most of the daylight side of the planet. However, the blurriness at the extreme right is the sunset line. The rotation of Earth carries us eastward, and as we cross the sunset line into darkness, we say that the sun has set. It is the rotation of our planet that causes the cycle of day and night.

Earth's interior is made mostly of iron and nickel, and its crust is mostly silicate rock. Only a thin layer of water makes up the oceans, and the atmosphere is only a few hundred kilometers deep. At the scale of this photograph, the atmosphere, on which our lives depend, is thinner than a strand of thread.

Enlarge your field of view again by a factor of 100, and you see a region 1,600,000 km wide (■ Figure 1-5). Earth is the small blue dot in the center, and the moon, whose diameter is only one-fourth that of Earth, is an even smaller dot along its orbit 380,000 km from Earth.

These numbers are so large that it is inconvenient to write them out. Astronomy is the science of big numbers, and you will use numbers much larger than these to discuss the universe. Rather than writing out these numbers as in the previous paragraph, it is more convenient to write them in **scientific notation.** This is nothing more than a simple way to write numbers without writing a great many zeros. In scientific notation, we write 380,000 as 3.8×10^5. If you are not familiar with scientific notation, read the section on "Powers of 10 Notation" in Appendix A. The universe is too big to discuss without using scientific notation.

When you once again enlarge your field of view by a factor of 100 (■ Figure 1-6), Earth, its moon, and the moon's orbit all lie in the small red box at lower left. But now you see the sun and two other planets that are part of our solar system. Our **solar system** consists of the sun, its family of planets, and some smaller bodies, such as moons and comets.

■ Figure 1-5

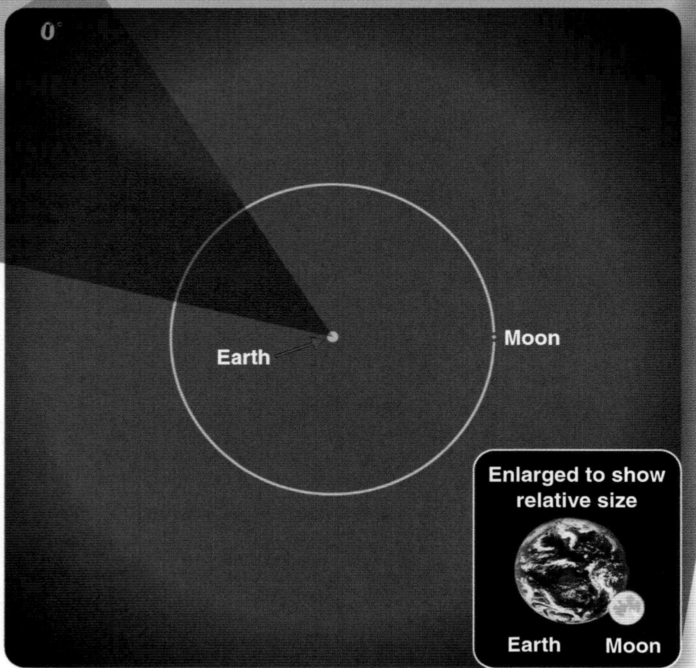

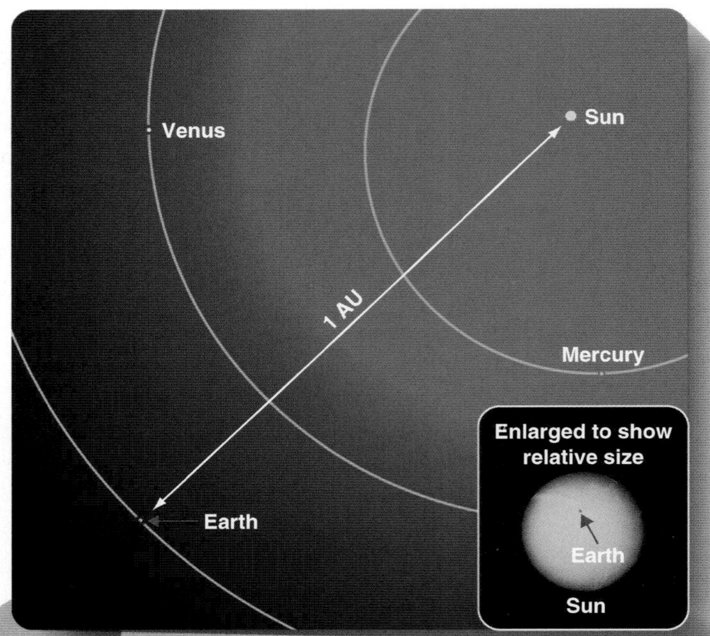

■ Figure 1-6

Like Earth, Venus and Mercury are **planets,** small, nonluminous bodies that shine by reflected light. Venus is about the size of Earth, and Mercury is a bit larger than our moon. On this diagram, they are both too small to be seen as anything but tiny dots. The sun is a **star,** a self-luminous ball of hot gas that generates its own energy. The sun is 109 times larger in diameter than Earth (inset), but it too is nothing more than a dot in this diagram.

This diagram has a diameter of 1.6×10^8 km. One way astronomers deal with large numbers is to define new units. The average distance from Earth to the sun is a unit of distance called the **astronomical unit (AU),** a distance of 1.5×10^{11} m. Using this unit you can say that the average distance from Venus to the sun is about 0.7 AU. The average distance from Mercury to the sun is about 0.39 AU.

The orbits of the planets are not perfect circles, and this is particularly apparent for Mercury. Its orbit carries it as close to the sun as 0.307 AU and as far away as 0.467 AU. Earth's orbit is more circular, and its distance from the sun varies by only a few percent.

Your first field of view was only 52 feet (about 16 m) in width. After only six steps of enlarging by a factor of 100, you now see the entire solar system (■ Figure 1-7). Your field of view is 1 trillion (10^{12}) times wider than in your first view. The details of the preceding figure are now lost in the red square at the center of this diagram. You see only the brighter, more widely separated objects as you back away. The sun, Mercury, Venus, and Earth lie so close together that you cannot separate them at this scale.

Mars, the next outward planet, lies only 1.5 AU from the sun. In contrast, Jupiter, Saturn, Uranus, Neptune, and Pluto are so far from the sun that they are easy to place in this diagram. These are cold worlds far from the sun's warmth. Light from the sun reaches Earth in only 8 minutes, but it takes over 4 hours to reach Neptune. Notice that Pluto's orbit is so elliptical that Pluto can come closer to the sun than Neptune does, as Pluto did between 1979 and 1999.

When you again enlarge your field of view by a factor of 100, the solar system vanishes (■ Figure 1-8). The sun is only a point of light, and all the planets and their orbits are now crowded into the small red square at the center. The planets are too small and reflect too little light to be visible so near the brilliance of the sun.

Nor are any stars visible except for the sun. The sun is a fairly typical star, and it seems to be located in a fairly average neighborhood in the universe. Although there are many billions of stars like the sun, none is close enough to be visible in this diagram, which shows an area only 11,000 AU in diameter. The stars are typically separated by distances about 10 times larger than the distance represented in this diagram. You will see stars in your next field of view, but, except for the sun at the center, this diagram is empty.

It is difficult to grasp the isolation of the stars. If the sun were represented by a golf ball in New York City, the nearest star would be another golf ball in Chicago. Except for the widely scattered stars and a few atoms of gas drifting between the stars, the universe is nearly empty.

In ■ Figure 1-9, your field of view has expanded to a diameter a bit over 1 million AU. The sun is at the center, and you see a few of the nearest stars. These stars are so distant that it is not reasonable to give their distances in astronomical units. We must define a new unit of distance, the light-year. One **light-year (ly)** is the distance that light travels in 1 year, roughly 10^{13} km or 63,000 AU. The diameter of your field of view is, in our new unit, 17 ly. The nearest star to the sun, Proxima Centauri, is 4.2 ly from Earth. In other words, light from Proxima Centauri takes 4.2 years to reach Earth.

Although these stars are roughly the same size as the sun, they are so far away that you cannot see them as anything but points of light. Even with the largest telescopes on Earth, you still see only points of light when you look at stars, and any planets that might circle those stars are much too small and too faint to be visible.

■ **Figure 1-7**

■ **Figure 1-8**

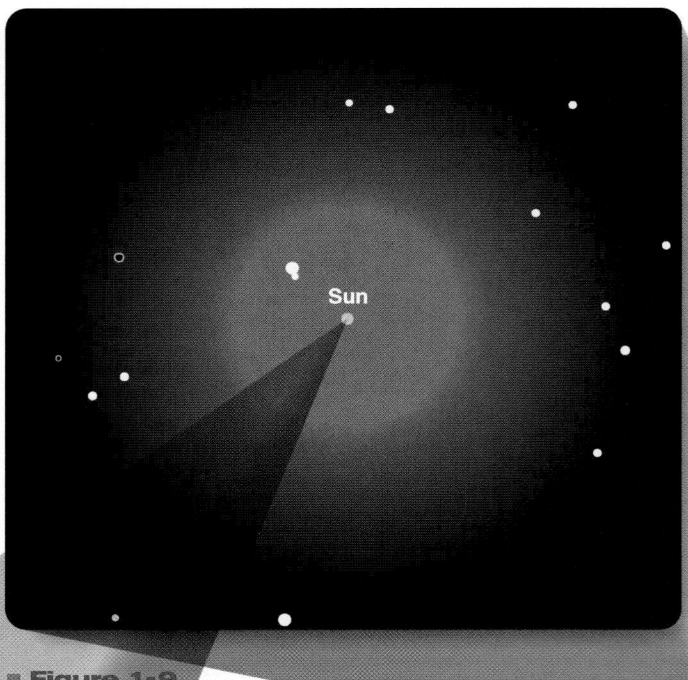

■ Figure 1-9

In Figure 1-9, the sizes of the dots represent not the size of the stars but their brightness. This is the custom in astronomical diagrams, and it is also how star images are recorded in photographs. Bright stars make larger spots on a photograph than faint stars. The size of a star image in a photograph tells you not how big the star is but only how bright it looks.

In ■ Figure 1-10, you expand your field of view by another factor of 100, and the sun and its neighboring stars vanish into the background of thousands of stars. The field of view is now 1700 ly in diameter. Of course, no one has ever journeyed thousands of light-years from Earth to look back and photograph the solar neighborhood, so this is a representative photo of the sky. The sun is a relatively faint star that would not be easily located in a photo at this scale.

What you do not see is critically important. You do not see the thin gas that fills the spaces between the stars. Although those clouds of gas are thinner than the best vacuum on Earth, it is those clouds that give birth to new stars. Our sun formed from such a cloud about 5 billion years ago. You will see more star formation in our next view.

If you expand your field of view by a factor of 100, you see our galaxy (■ Figure 1-11). A **galaxy** is a great cloud of stars, gas, and dust bound together by the combined gravity of all the matter. Galaxies range from 1500 to over 300,000 ly in diameter and can contain over 100 billion stars. In the night sky, you see our galaxy as a great, cloudy wheel of stars ringing the sky as the **Milky Way,** and we refer to our galaxy as the **Milky Way Galaxy.** Of course, no one can journey far enough into space to look back and photograph our galaxy, so the photo in Figure 1-11 shows a galaxy similar to our own. Our sun would be invisible in such a photo, but if you could see it, you would find it about two-thirds of the way from the center to the edge.

Our galaxy, like many others, has graceful **spiral arms** winding outward through the disk. You will discover that stars are born in great clouds of gas and dust as they pass through the spiral arms.

Ours is a fairly large galaxy with a visible disk roughly 75,000 ly in diameter. Only a century ago astronomers thought it was the entire universe—an island universe of stars in an otherwise empty vastness. Now we know that our galaxy is not unique. Indeed, ours is only one of many billions of galaxies scattered throughout the universe.

■ Figure 1-10

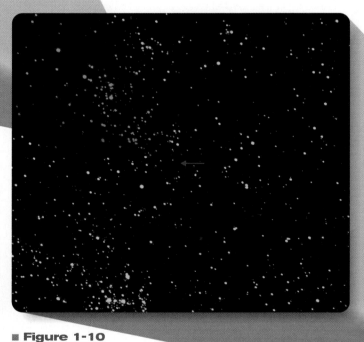

This box ■ represents the relative size of the previous frame. (NOAO)

As you expand your field of view by another factor of 100, our galaxy appears as a tiny luminous speck surrounded by other specks (■ Figure 1-12). This diagram includes a region 17 million ly in diameter, and each of the dots represents a galaxy. Notice that our galaxy is part of a cluster of a few dozen galaxies. You will find that galaxies are commonly grouped together in clusters. Some of these galaxies have beautiful spiral patterns like our own galaxy, but others do not. Some are strangely distorted. One of the mysteries we will try to solve is what produces these differences among the galaxies.

If you again expand your field of view, you see that the clusters of galaxies are connected in a vast network (■ Figure 1-13). Clusters are grouped into superclusters—clusters of clusters—and the superclusters are linked to form long filaments and walls outlining voids that seem nearly empty of galaxies. These appear to be the largest structures in the universe. Were you to expand your field of view another time, you would probably see a uniform fog of filaments and voids. When you puzzle over the origin of these structures, you are at the frontier of human knowledge.

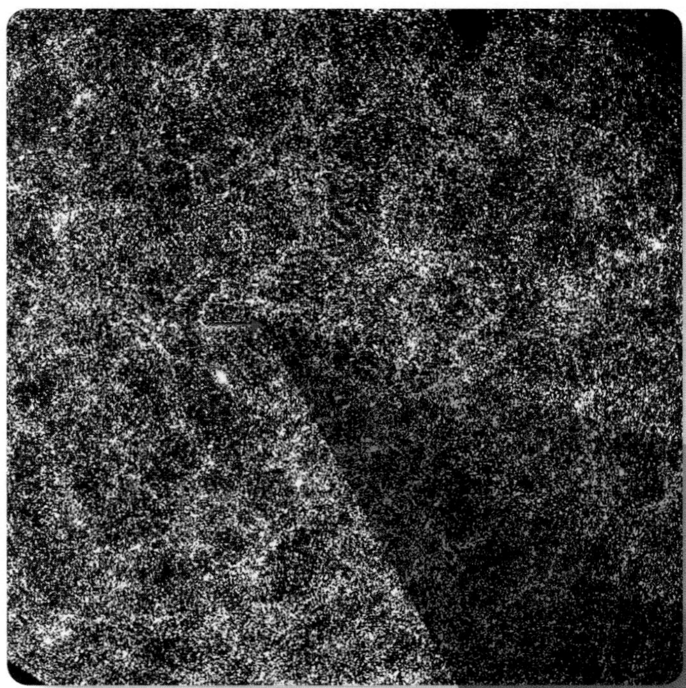

■ Figure 1-13

This box ▪ represents the relative size of the previous frame. (Detail from galaxy map from M. Seldner, B. L. Siebers, E. J. Groth, and P. J. E. Peebles, *Astronomical Journal 82* [1977])

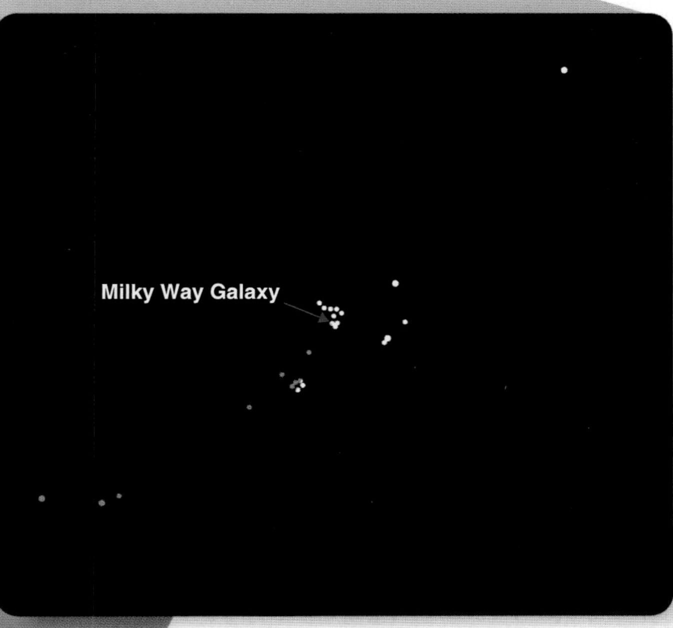

■ Figure 1-12

■ Figure 1-11

(© Anglo-Australian Telescope Board)

A problem in studying astronomy is keeping a proper sense of scale. Remember that each of the billions of galaxies contains billions of stars. Most of those stars probably have families of planets like our solar system, and on some of those billions of planets liquid-water oceans and protective atmospheres may have sheltered the spark of life. It is possible that some other planets are inhabited by intelligent creatures who share our curiosity and our wonder at the scale of the cosmos.

Summary

Your goal in this chapter was to preview the scale of astronomical objects. To do so you journeyed outward from a familiar campus scene by expanding your field of view by factors of 100. Only 12 such steps took you to the largest structures in the universe.

Where is Earth in relation to the sun, the planets, and the galaxies?

- We live on planet Earth, which orbits our star, the sun, once a year. As Earth rotates once a day, we see the sun rise and set.
- The other planets in our solar system, Mercury, Venus, Mars, Jupiter, Saturn, Uranus, Neptune, and Pluto, orbit the sun in ellipses that are nearly circular.
- The sun is just one out of the billions of stars that fill our home galaxy, the Milky Way spiral galaxy.
- Our galaxy is just one of billions of galaxies that fill the universe in great clusters, clouds, and filaments.

How do astronomers express distances?

- Astronomers use the metric system because it simplifies calculations and scientific notation for very large or very small numbers.
- The astronomical unit (AU) is the average distance from Earth to the sun. Mars, for example, orbits 1.5 AU from the sun. The light-year (ly) is the distance light can travel in one year. The nearest star is 4.2 ly from the sun.

Which objects are big, and which are small?

- The moon is only a fourth the diameter of Earth, but the sun is 109 times larger than Earth—a typical size for a star.
- Galaxies contain many billions of stars. Our galaxy is about 75,000 ly in diameter and contains over 100 billion stars.
- The largest things in the universe are the long filaments containing many clusters of galaxies.

Are there other worlds like Earth?

- Many stars are believed to have planets, but such small, distant worlds are difficult to detect. Only about 100 have been found so far.

New Terms

scientific notation (p. 6)
solar system (p. 6)
planet (p. 6)
star (p. 6)
astronomical unit (AU) (p. 6)
light-year (ly) (p. 7)
galaxy (p. 8)
Milky Way (p. 8)
Milky Way Galaxy (p. 8)
spiral arm (p. 8)

Review Questions

AceAstronomy™ Assess your understanding of this chapter's topics with additional quizzing and animations at **http://astronomy.brookscole.com/sh9e**

1. What is the largest dimension you have personal knowledge of? Have you run a mile? Hiked 10 miles? Run a marathon?
2. In Figure 1-4, the division between daylight and darkness is at the right on the globe of Earth. How do you know this is the sunset line and not the sunrise line?
3. What is the difference between our solar system, our galaxy, and the universe?
4. Look at Figure 1-6. How can you tell that Mercury follows an elliptical orbit? Can you detect the elliptical shape of any other orbits in this figure or the next?
5. Which is the outermost planet in our solar system? Why does that change?
6. Why are light-years more convenient than miles, kilometers, or astronomical units for measuring certain distances?
7. Why is it difficult to detect planets orbiting other stars?
8. What does the size of the star image in a photograph tell us?
9. What is the difference between the Milky Way and the Milky Way Galaxy?
10. What are the largest known structures in the universe?
11. Of the objects listed here, which would be contained inside the object shown at right? Which would contain the object? Stars, planets, galaxy clusters, filaments, spiral arms.

(Bill Schoening/NOAO/AURA/NSF)

Problems

1. The diameter of Earth is 7928 mi. What is its diameter in inches? In yards?
2. If a mile equals 1.609 km and the moon is 2160 mi in diameter, what is its diameter in kilometers?
3. One astronomical unit is about 1.5×10^8 km. Explain why this is the same as 150×10^6 km.
4. Venus orbits 0.7 AU from the sun. What is that distance in kilometers?
5. Light from the sun takes 8 minutes to reach Earth. How long does it take to reach Mars?
6. The sun is almost 400 times farther from Earth than is the moon. How long does light from the moon take to reach Earth?
7. If the speed of light is 3×10^5 km/s, how many kilometers are in a light-year? How many meters?
8. How long does it take light to cross the diameter of our Milky Way Galaxy?
9. The nearest large galaxy to our own is about 2 million light-years away. How many meters is that?
10. How many galaxies like our own would it take laid edge to edge to reach the nearest galaxy? (*Hint:* See Problem 9.)

Media Cluster

Critical Inquiries for the Web

1. Locate photographs of Earth taken from space. What do cities look like? Can you see highways? Is the presence of our civilization detectable from space?
2. Locate photographs of nearby galaxies and compare them with photos of very distant galaxies. What kind of detail is invisible for distant galaxies?
3. One of the biggest clusters of galaxies is the Virgo cluster. Find out how many and what kind of galaxies are in the cluster. How far away is the Virgo cluster?

Exploring *TheSky*

1. Locate and center one example of each of three different types of objects:
   a. A planet, such as Saturn. Find its rising and setting time. Such objects have distances measured in astronomical units (AU).
      How to proceed: Decide on the object you want to locate. Then find and center the object by pressing the **Find** button on the **Object Toolbar.** The second method is to press the **F** key. The third is to click **Edit,** then **Find.** Once you have the **Object Information** window, press the **center** button.
   b. A star. All stars in *TheSky* belong to our Milky Way Galaxy. Give the star's name, its magnitude, and its distance in light-years.
      How to proceed: Click on any star, which brings up an **Object Information** window containing a variety of information about the star.
   c. A galaxy; give its name and/or its designation.
      How to proceed: To show galaxies, click on the **Galaxies** button in the **Object Toolbar,** then click on any galaxy. Distances to galaxies are on the order of millions and billions of light-years.
2. Look at the solar system from beyond Pluto by clicking on **View** and then on **3D Solar System Mode.** Tip the solar system edge on and then face on. Zoom in to see the inner planets. Under **Tools,** set the **Time Skip Increment** to 1 day and **then go forward** in time to watch the planets move.
3. Identify some of the brightest constellations located along the Milky Way. (*Hint:* See **View, Reference Lines.**)

- Go to the Brooks/Cole Astronomy Resource Center **(http://astronomy.brookscole.com)** for critical thinking exercises, articles, and additional readings from InfoTrac College Edition, Brooks/Cole's online student library.

2 The Sky

The Southern Cross I saw every night abeam. The sun every morning came up astern; every evening it went down ahead. I wished for no other compass to guide me, for these were true.

CAPTAIN JOSHUA SLOCUM
Sailing Alone around the World

Visual-wavelength image

THE NIGHT SKY is the rest of the universe as seen from our planet. When you look up at the stars, you look out through a layer of air only a few hundred kilometers deep. Beyond that, space is nearly empty, with the stars scattered light-years apart. We begin our search for the natural laws that govern the universe by trying to understand what that universe looks like. ▌ As you read this chapter, keep in mind that you live on a planet. The stars are scattered into the void all around you, most very distant and some closer. Our planet rotates on its axis once a day, so from our viewpoint the sky appears to rotate around us each day. Not only does the sun rise in the east and set in the west, but so also do the stars. That apparent daily motion is caused by the rotation of our planet.

The sky above mountain top observatories far from city lights is the same sky you see from your window. The stars above you are other suns scattered through the universe. (Kris Koenig/Coast Learning Systems)

Guidepost

Looking Back

Astronomy is about us. As we learn about astronomy, we learn about ourselves. We search for an answer to the question, "What are we?" The cosmic zoom in the previous chapter showed you Earth in relation to other objects in the universe. Even in that quick preview, as you learned about stars and galaxies, you were also learning about yourself.

This Chapter

Now it is time to return to Earth and look closely at the sky. You have noticed stars and may know a few constellations and star names, but now you need to refine your terms and answer a few essential questions:

How do astronomers refer to stars?

How can you compare the brightness of the stars?

How does the sky move as Earth rotates?

This chapter will give you a tour of the sky and a vocabulary to describe it.

Looking Ahead

The remaining chapters of this book will fill in details and help you understand the objects in the sky. The next chapter discusses the motions of the sun and moon and will help you understand what the universe looks like seen from the surface of our spinning planet. But appearances can be deceiving. Chapter 4 will show how humanity finally understood that we all live on a planet. In fact, you will discover that modern science was born when people tried to understand the appearance of the sky.

Ace Astronomy™ The AceAstronomy icon throughout the text indicates an opportunity for you to test yourself on key concepts and to explore animations and interactions on the AceAstronomy website at: http://astronomy.brookscole.com/sh9e

2-1 The Stars

ON A DARK NIGHT far from city lights, you can see a few thousand stars in the sky. Like ancient astronomers, we will try to organize what we see by naming groups of stars and individual stars and by specifying the brightness of individual stars.

Constellations

All around the world, ancient cultures celebrated heroes, gods, and mythical beasts by naming groups of stars—**constellations** (■ Figure 2-1). You should not be surprised that the star patterns do not look like the creatures they represent any more than Columbus, Ohio, looks like Christopher Columbus. The constellations "celebrate" the most important mythical figures in each culture. The constellations named within western culture originated in Mesopotamia over 5000 years ago, with other constellations added by Babylonian, Egyptian, and Greek astronomers during the classical age. Of these ancient constellations, 48 are still in use.

■ **Figure 2-1**

The constellations are an ancient heritage handed down for thousands of years as celebrations of great heroes and mythical creatures. Here Sagittarius and Scorpius hang above the southern horizon.

To the ancients, a constellation was a loose grouping of stars. Many of the fainter stars were not included in any constellation, and regions of the southern sky not visible to the ancient astronomers of northern latitudes were not identified with constellations. Constellation boundaries, when they were defined at all, were only approximate (■ Figure 2-2a), so a star like Alpheratz could be thought of as part of Pegasus or part of Andromeda. In recent centuries astrono-

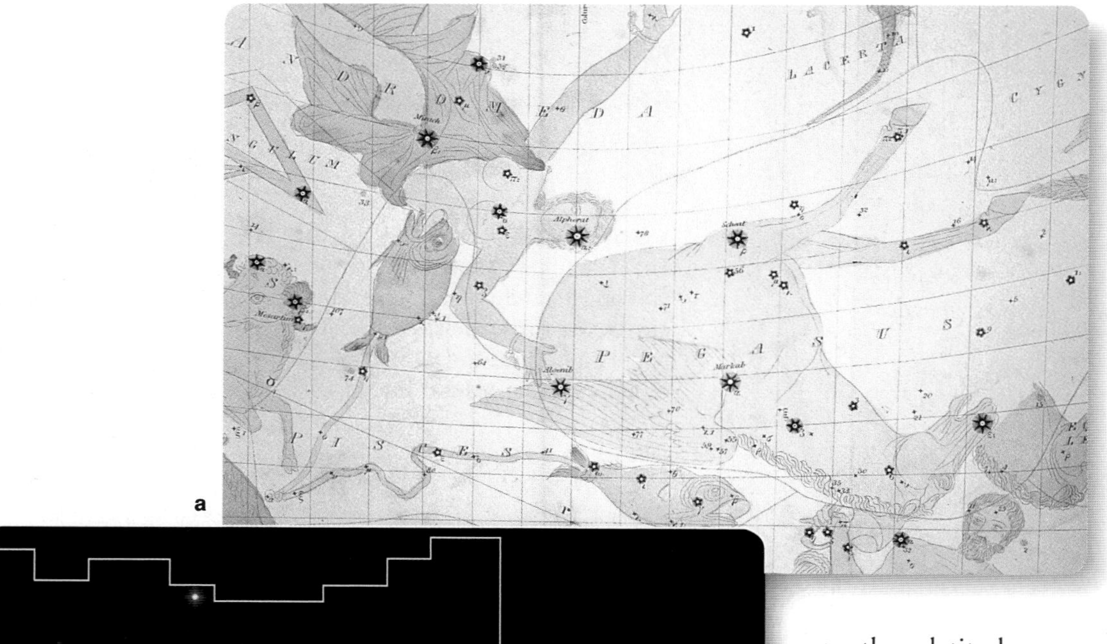

■ **Figure 2-2**

(a) In antiquity, constellation boundaries were poorly defined, as shown on this map by the curving dotted lines that separate Pegasus from Andromeda. (From Duncan Bradford, *Wonders of the Heavens*, Boston: John B. Russell, 1837) (b) Modern constellation boundaries are precisely defined by international agreement.

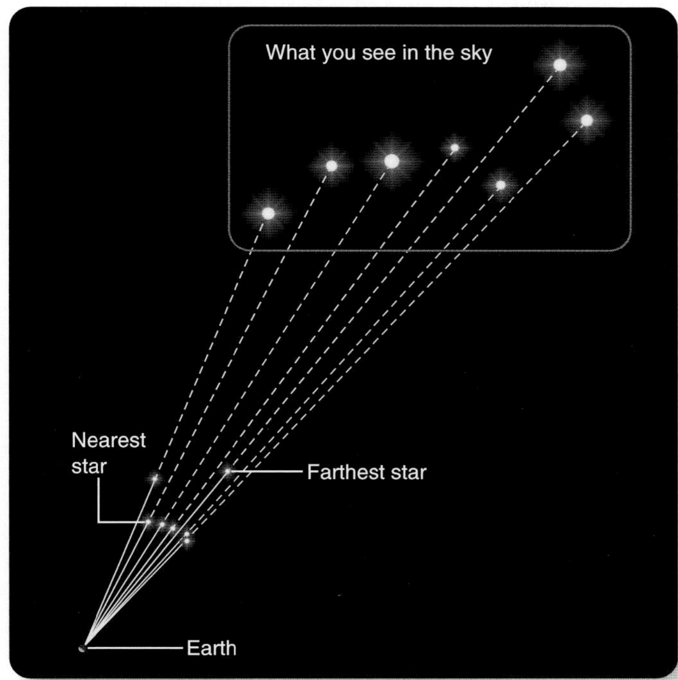

■ **Figure 2-3**

The stars we see in the Big Dipper—the brighter stars of the constellation Ursa Major, the Great Bear—are not at the same distance from Earth. You see the stars in a group in the sky because they lie in the same general direction as seen from Earth. The sizes of the star dots in the star chart represent the apparent brightness of the stars.

mers have added 40 modern constellations to fill gaps, and in 1928 the International Astronomical Union established 88 official constellations with clearly defined boundaries (Figure 2-2b). Thus a constellation now represents not a group of stars but an area of the sky, and any star within the region belongs to one and only one constellation.

In addition to the 88 official constellations, the sky contains a number of less formally defined groupings called **asterisms.** The Big Dipper, for example, is a well-known asterism that is part of the constellation Ursa Major (the Great Bear). Another asterism is the Great Square of Pegasus (Figure 2-2b), which includes three stars from Pegasus and one from Andromeda. The star charts at the end of this book will introduce you to the brighter constellations and asterisms.

Although we name constellations and asterisms, most are made up of stars that are not physically associated with one another. Some stars may be many times farther away than others and moving through space in different directions. The only thing they have in common is that they lie in approximately the same direction from Earth (■ Figure 2-3).

The Names of the Stars

In addition to naming groups of stars, ancient astronomers named the brighter stars, and modern astronomers still use many of those names. Whereas the names of the constellations are in Latin, the common language of science in Renaissance Europe, most star names derive from ancient Arabic, though much altered by the passing centuries. The name of Betelgeuse, the bright red star in Orion, for example, comes from the Arabic *yad al-jawza,* meaning "armpit of Jawza [Orion]." Aldebaran, the bright red eye of Taurus the bull, comes from the Arabic *al-dabar an,* meaning "the follower."

Naming individual stars is not very helpful because you can see thousands of them, and names do not help you locate stars in the sky. Another way to identify stars is to assign Greek letters (see Appendix A) to the bright stars in a constellation in approximate order of brightness. Thus the brightest star is usually designated α (alpha), the second brightest β (beta), and so on. For many constellations, the letters follow the order of brightness, but some constellations, by tradition, mistake, or the personal preferences of early chartmakers, are exceptions (■ Figure 2-4).

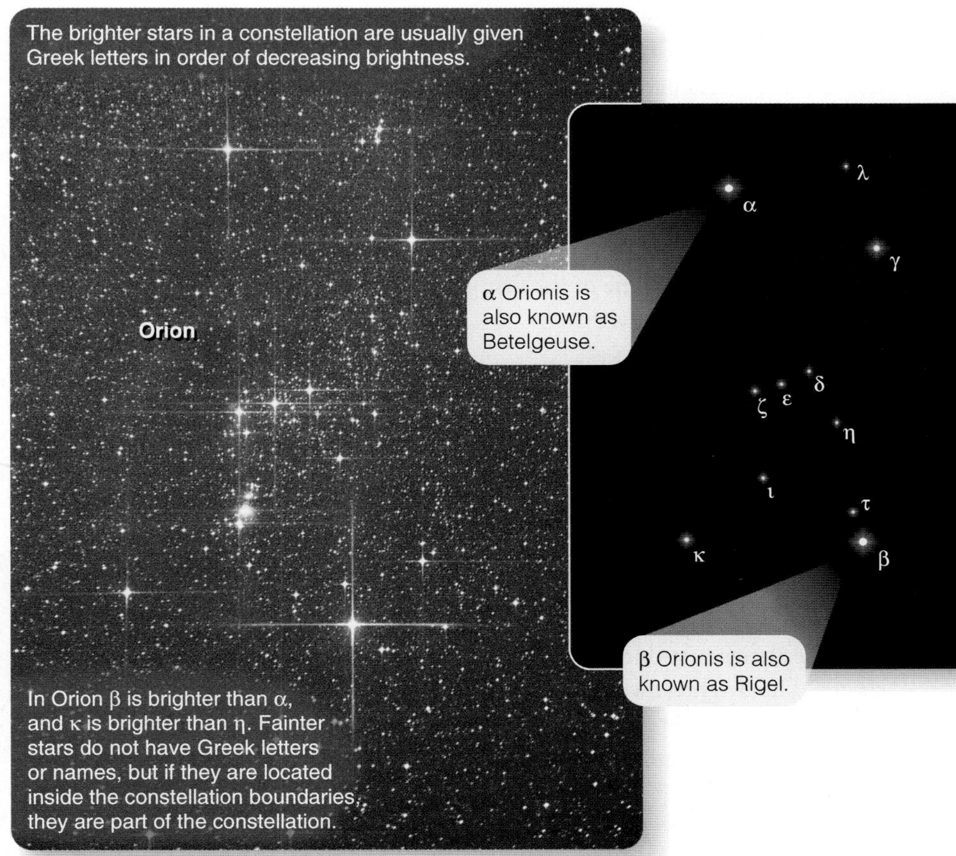

■ **Figure 2-4**

Stars in a constellation can be identified by Greek letters and by names derived from Arabic. The spikes on the star images in the photograph were produced by the optics in the camera. (William Hartmann)

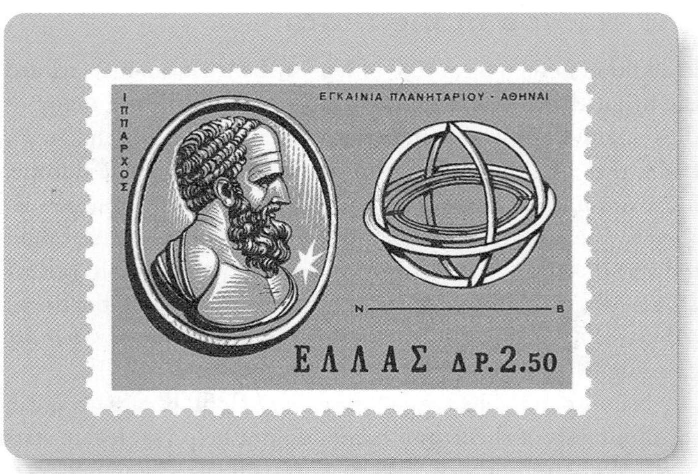

■ Figure 2-5

Hipparchus (2nd century BC) was the first great observational astronomer. Among other things, he constructed a catalog listing 1080 of the brightest stars. He is honored here on a Greek stamp that also shows one of his observing instruments.

To identify a star by its Greek-letter designation, give the Greek letter followed by the genitive (possessive) form of the constellation name; for example, the brightest star in the constellation Canis Major is α Canis Majoris. This both identifies the star and the constellation and gives a clue to the relative brightness of the star. Compare this with the ancient name for this star, Sirius, which tells you nothing about location or brightness.

This method of identifying a star's brightness is only approximate. In order to discuss the sky with precision, you must have an accurate way of referring to the brightness of stars, and for that you must consult one of the first great astronomers.

The Brightness of Stars

Astronomers measure the brightness of stars using the **magnitude scale,** a system that first appeared in the writings of the ancient astronomer Claudius Ptolemy about 140 AD. The system may have originated earlier than Ptolemy, and most astronomers attribute it to the Greek astronomer Hipparchus (160–127 BC) (■ Figure 2-5).

The ancient astronomers divided the stars into six classes. The brightest were called first-magnitude stars and those that were fainter, second-magnitude. The scale continued downward to sixth-magnitude stars, the faintest visible to the human eye. Thus, the larger the magnitude number, the fainter the star. This makes sense if you think of the bright stars as first-class stars and the faintest stars visible as sixth-class stars.

Hipparchus is believed to have compiled the first star catalog, and he may have used the magnitude system in that catalog. Almost 300 years later Ptolemy used the magnitude system in his own catalog, and successive generations of astronomers have continued to use the system.

Modern astronomers can measure the brightness of stars to high precision, so instead of saying that the star known by the charming name Chort (Theta Leonis) is third magnitude, they can say its magnitude is 3.34. If you measure magnitudes, you discover that some stars are brighter than 1.0. For example, Vega (α Lyrae) is so bright that its magnitude, 0.04, is almost zero. A few are so bright the magnitude scale must extend into negative numbers (■ Figure 2-6). On this scale, Sirius, the brightest star in the sky, has a magnitude of −1.47. If you use a telescope to search for very faint stars, you can find stars much fainter than the limit for the unaided eye. Thus the magnitude system has been extended to numbers larger than sixth magnitude to include fainter stars.

These numbers are known as **apparent visual magnitudes** **(m_v)**, and they describe how the stars look to human eyes observing from Earth. Although some stars emit large amounts of infrared or ultraviolet light, humans can't see it, and it is not included in the apparent visual magnitude. The subscript "v" reminds you that you are including only light you can see. Another problem is the distance to the stars. Very distant stars look fainter, and nearby stars look brighter. Apparent visual magnitude ignores the effect of distance and tells only how bright the star looks as seen from Earth.

Brightness is quite subjective, depending on both the physiology of human eyes and the psychology of perception. To be accurate you should refer to intensity—a measure of the light energy from a star that hits one square meter in one second. A simple relationship connects apparent visual magnitudes and the intensity of starlight (**Reasoning with Numbers 2-1**). Thus modern astronomers can measure the brightness of stars to high

■ Figure 2-6

The scale of apparent visual magnitudes extends into negative numbers to represent the brightest objects and to positive numbers larger than 6 to represent objects fainter than the human eye can see.

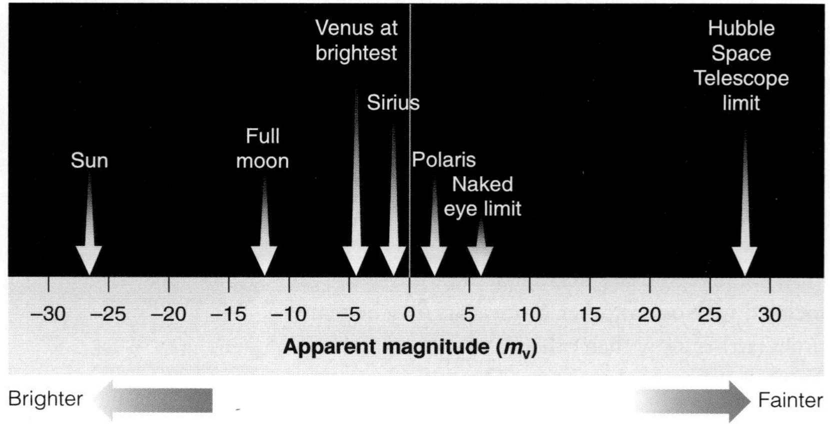

Reasoning with Numbers | 2-1

Magnitudes

Astronomers use a simple formula to convert between magnitudes and intensities. If two stars have intensities I_A and I_B, then the ratio of their intensities is I_A/I_B. Modern astronomers have defined the magnitude scale so that two stars that differ by 5 magnitudes have an intensity ratio of exactly 100. Then two stars that differ by 1 magnitude must have an intensity ratio of the fifth root of 100, $\sqrt[5]{100}$, which equals 2.512—that is, the light of one star must be 2.512 times more intense. Two stars that differ by 2 magnitudes will have an intensity ratio of 2.512×2.512, or about 6.3, and so on (■ Table 2-1).

Example A: Suppose star C is third magnitude and star D is ninth magnitude. What is the intensity ratio? **Solution:** The magnitude difference is 6 magnitudes, and the table shows the intensity ratio is 250. Therefore light from star C is 250 times more intense than light from star D.

A table is convenient, but for more precision you can express the relationship as a simple formula. The intensity ratio I_A/I_B is equal to 2.512 raised to the power of the magnitude difference $m_B - m_A$:

$$\frac{I_A}{I_B} = (2.512)^{(m_B - m_A)}$$

Example B: If the magnitude difference is 6.32 magnitudes, what is the intensity ratio? **Solution:** The intensity ratio must be $2.512^{6.32}$. A pocket calculator tells you the answer: 337.

■ Table 2-1 | Magnitude and Intensity

Magnitude Difference	Intensity Ratio
0	1
1	2.5
2	6.3
3	16
4	40
5	100
6	250
7	630
8	1600
9	4000
10	10,000
⋮	⋮
15	1,000,000
20	100,000,000
25	10,000,000,000
⋮	⋮

When you know the intensity ratio and want to find the magnitude difference, it is convenient to solve the formula for the magnitude difference:

$$m_A - m_B = 2.5 \log(I_B/I_A)$$

Example C: The light from Sirius is 24.2 times more intense than light from Polaris. What is the magnitude difference? **Solution:** The magnitude difference is 2.5 log(24.2). Your pocket calculator tells you the logarithm of 24.2 is 1.38, so the magnitude difference is $2.5 \times 1.38 = 3.4$ magnitudes.

precision while still making comparisons to observations of apparent visual magnitude that go back to the time of Hipparchus.

Inquire | Review | Analyze

Nonastronomers sometimes complain that the magnitude scale is awkward. Why would they think it is awkward, and how did it get that way? Two things might make the magnitude scale seem awkward. First, it is backwards; the greater the magnitude number, the fainter the star. Of course, that arose because ancient astronomers were not measuring the brightness of stars but rather classifying them, and first-class stars would be brighter than second-class stars. The second awkward feature of the magnitude scale is its mathematical relation to intensity. If two stars differ by 1 magnitude, one is about 2.5 times brighter than the other. But if they differ by 2 magnitudes, one is 2.5×2.5 times brighter. This mathematical relationship arises because of the way we perceive brightness as ratios of intensity.

If the magnitude scale is so awkward, why do you suppose astronomers have used it for over two millennia?

■ ■ ■

Connections: In this section, you have found a way to identify stars by name and by brightness. Next we will look at the sky as a whole and notice its motion.

2-2 The Sky and Its Motion

THE SKY ABOVE US seems to be a great blue dome in the daytime and a sparkling ceiling at night. Learning to look at the sky requires that we begin thousands of years ago.

The Celestial Sphere

Ancient astronomers believed the sky was a great sphere surrounding Earth with the stars stuck on the inside like thumbtacks in a ceiling. Modern astronomers know that the stars are scattered through space at different distances, but it is still convenient to think of the sky as a great starry sphere enclosing Earth.

Window on Science 2-1

Frameworks for Thinking about Nature: Scientific Models

In everyday language, we use the word *model* in various ways—fashion model, model airplane, model student—but scientists use the word in a very specific way. A **scientific model** is a carefully devised mental conception of how something works, a framework that helps scientists think about some aspect of nature. For example, astronomers use the celestial sphere as a way to think about the motions of the sky, sun, moon, and stars.

Although a scientific model is a mental conception, it can take many forms. Some models are imprecise—the psychologist's model of how the human mind processes visual information into images, for instance. But other models are so specific that they can be expressed as a set of mathematical equations. For example, an astronomer might use a set of equations to describe in detail how gas falls into a black hole. You could refer to such a calculation as a model. Of course, you could use metal and plastic to build a celestial globe, but the thing you build wouldn't really be the model any more than the equations are a model. A scientific model is a mental conception, an idea, that helps you think about nature.

On the other hand, a model is not meant to be a statement of truth. The celestial sphere is not real; you know the stars are scattered through space at various distances, but you can imagine a celestial sphere and use it to help you think about the sky. A scientific model does not have to be true to be useful. Chemists, for example, think about the atoms in molecules by visualizing them as balls joined together by rods. This model of a molecule is not fully correct, but it is a helpful way to think about molecules; it gives chemists a framework within which to organize their ideas.

Because scientific models are not meant to be totally correct, you must always remember the assumptions on which they are based. If you begin to think a model is true, it can mislead you instead of helping you. The celestial sphere, for instance, can help you think about the sky, but you must remember that it is only a mental crutch. The universe is much larger and much more interesting than this ancient scientific model of the heavens.

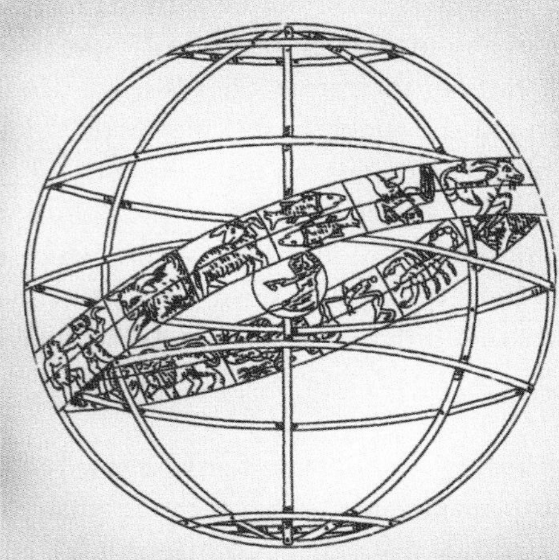

The ancient celestial sphere is a useful model of the sky.

As you study **The Sky around Us** on pages 20 and 21, notice two important points. First, the sky appears to rotate westward around Earth each day, but that is a consequence of the eastward rotation of Earth. Second, what you can see of the sky depends on where you are on Earth. Australians see many constellations and asterisms invisible from North America, but they never see the Big Dipper.

The celestial sphere is an example of a scientific model, a common feature of scientific thought (**Window on Science 2-1**). Notice that a scientific model does not have to be true to be useful. We will discuss many scientific models in the chapters that follow.

In addition to the daily motion of the sky, Earth's rotation adds a second motion to the sky that can be detected only over centuries.

Precession

Over 2000 years ago, Hipparchus compared a few of his star positions with those made nearly two centuries before and realized that the celestial poles and equator were slowly moving across the sky. Later astronomers understood that this motion is caused by the toplike motion of Earth.

If you have ever played with a gyroscope or top, you have seen how the spinning mass resists any change in the direction of its axis of rotation. The more massive the top and the more rapidly it spins, the more difficult it is to change the direction of its axis of rotation. But you probably recall that the axis of even the most rapidly spinning top sweeps around in a conical motion. The weight of the top tends to make it tip, and this combines with its rapid rotation to make its axis sweep around in a conical motion called **precession** (■ Figure 2-7a).

Earth spins like a giant top, but it does not spin upright in its orbit; it is tipped 23.5° from vertical. Earth's large mass and rapid rotation keep its axis of rotation pointed toward a spot near the star Polaris, and the axis would not wander if Earth were a perfect sphere. However, Earth, because of its rotation, has a slight bulge around its middle, and the gravity of the sun and moon pulls on this bulge, tending to twist Earth upright in its orbit. The combination of these forces and Earth's rotation causes Earth's axis to precess in a conical motion, taking about 26,000 years for one cycle (Figure 2-7b).

Because the celestial poles and equator are defined by Earth's rotational axis, precession moves these reference marks. We notice no change at all from night to night or year to year, but precise measurements reveal the precessional motion of the celestial poles and equator.

Over centuries, precession has dramatic effects. Egyptian records show that 4800 years ago the north celestial pole was near

Window on Science 2-2

Understanding versus Naming: The True Goal of Science

One of the fascinations of science is that it can reveal things you normally do not sense, such as a slow drift in the direction of Earth's axis. Science can take your imagination into realms beyond your experience, from the inside of an atom to the inside of a star. Because these experiences are so unfamiliar, you need a vocabulary of technical terms just to talk about them, and that leads to a common confusion between naming things and understanding things.

The first step in understanding something is naming the parts, but it is only the beginning. The scientist's true goal is not naming but understanding. Thus, biologists studying plants go beyond memorizing the names of the flowers' colors. They search for evidence to test their hypothesis that specific plants have developed specific flower colors to attract specific kinds of insects for efficient pollination. Naming the colors is only the beginning of an exciting insight into nature.

You must know the proper words, but real understanding comes from being able to use that vocabulary to discuss the way nature works. You must be able to tell the stories that scientists call theories and hypotheses. You must be able to use technical terms precisely to cite the evidence that makes us think those stories are true. That goes far beyond merely memorizing terminology; that is real understanding.

In folklore, naming a thing gives you power over it—recall the story of Rumpelstiltskin. In science, naming things is only the first step toward understanding.

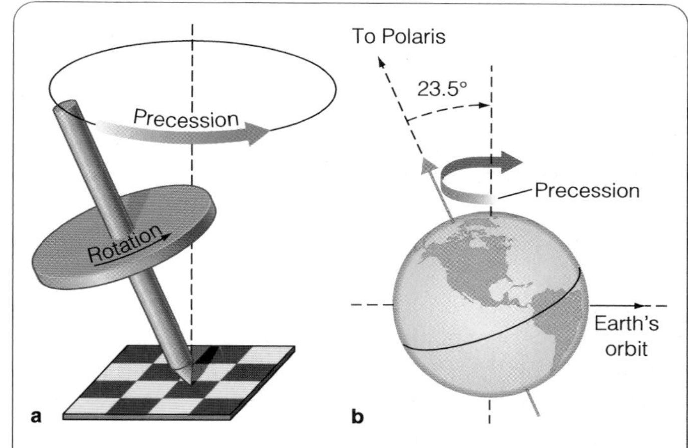

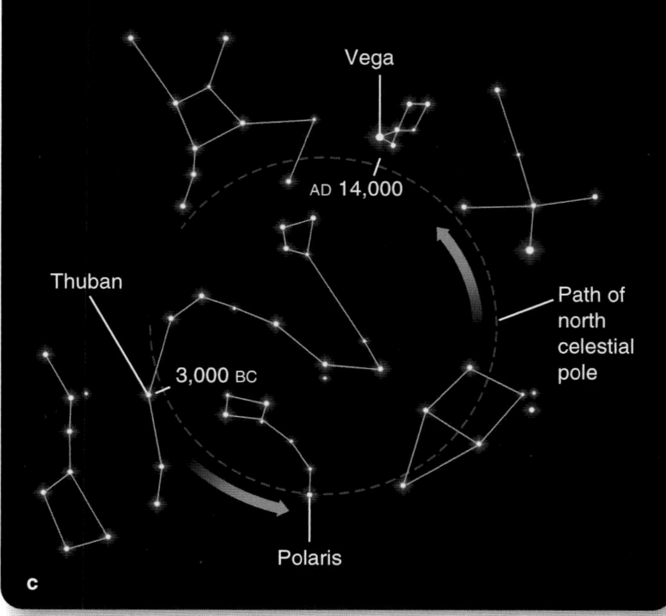

Figure 2-7

Precession. (a) A spinning top precesses in a conical motion around the perpendicular to the floor because its weight tends to make it fall over. (b) Earth precesses around the perpendicular to its orbit because the gravity of the sun and moon tend to twist it upright. (c) Precession causes the north celestial pole to drift among the stars, completing a circle in 26,000 years.

the star Thuban (α Draconis). The pole is now approaching Polaris and will be closest to it in about 2100. In about 12,000 years, the pole will have moved to within 5° of Vega (α Lyrae). Figure 2-7c shows the path followed by the north celestial pole.

As you study astronomy, notice the special terms used to describe such things as precession and the celestial sphere. You need to know those terms, but science is about understanding nature, not about naming its parts (**Window on Science 2-2**). Keep in mind that science is more than just vocabulary.

Inquire I Review I Analyze

Does everyone see the same circumpolar constellations?
Here you can use the celestial sphere as a convenient model of the sky. A circumpolar constellation is one that does not set or rise. Which constellations are circumpolar depends on your latitude. If you live on Earth's equator, you see all the constellations rising and setting, and there are no circumpolar constellations at all. If you live at Earth's North Pole, all the constellations north of the celestial equator never set, and all the constellations south of the celestial equator never rise. In that case, every constellation is circumpolar. At intermediate latitudes, the circumpolar regions are caps whose angular radius equals the latitude of the observer. If you live in Iceland, the caps are very large, and if you live in Egypt, near the equator, the caps are much smaller.

Locate Ursa Major and Orion on the star charts at the end of this book. For people in Canada, Ursa Major is circumpolar, but people in Mexico see most of this constellation slip below the horizon. From much of the United States, some of the stars of Ursa Major set, and some do

The Sky around Us

The sky above appears to be part of a sphere, the celestial sphere, with Earth at its center. From any location on Earth you see only half of the celestial sphere, the half above the horizon. The zenith marks the top of the sky above your head, and the nadir marks the bottom of the sky directly under your feet. What you see in the sky depends on where you are on Earth. The drawing at right shows the view for an observer in North America. An observer in South America would have a dramatically different horizon, zenith, and nadir.

Ace Astronomy™ Log into AceAstronomy and select this chapter to see Active Figure "Celestial Sphere." Notice how each location on Earth has its unique horizon.

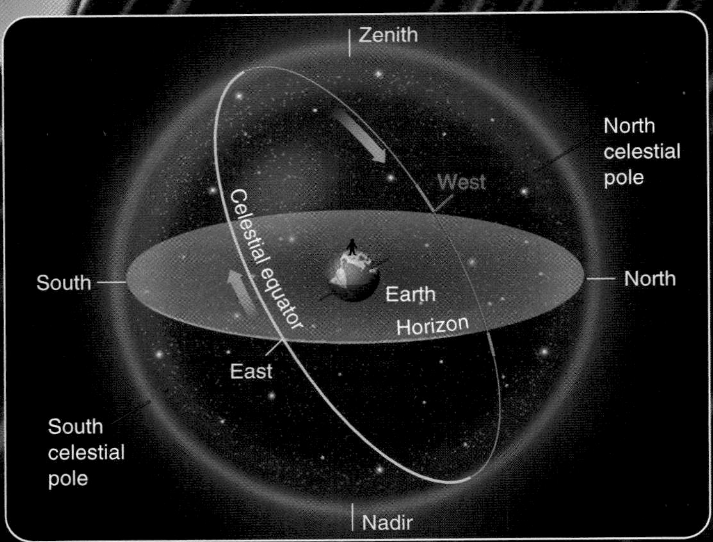

The eastward rotation of Earth causes the sun, moon, and stars to move westward in the sky as if the celestial sphere were rotating westward around Earth. The apparent pivot points are the **north celestial pole** and the **south celestial pole** located directly above Earth's north and south poles. Halfway between the celestial poles lies the **celestial equator.** Earth's rotation defines the directions you use every day. The **north point** and **south point** are the points on the horizon closest to the celestial poles. The **east point** and the **west point** lie halfway between the north and south points. The celestial equator always touches the horizon at the east and west points.

AURA/NOAO/NSF

This time exposure of about 30 minutes shows stars as streaks, called star trails, rising behind an observatory dome. The camera was facing northeast to take this photo. The motion you see in the sky depends on which direction you look, as shown at right. Looking north, you see the star Polaris, the North Star, located near the north celestial pole. As the sky appears to rotate westward, Polaris hardly moves, but other stars circle the celestial pole. Looking south from a location in North America, you can see stars circling the south celestial pole, which is invisible below the southern horizon.

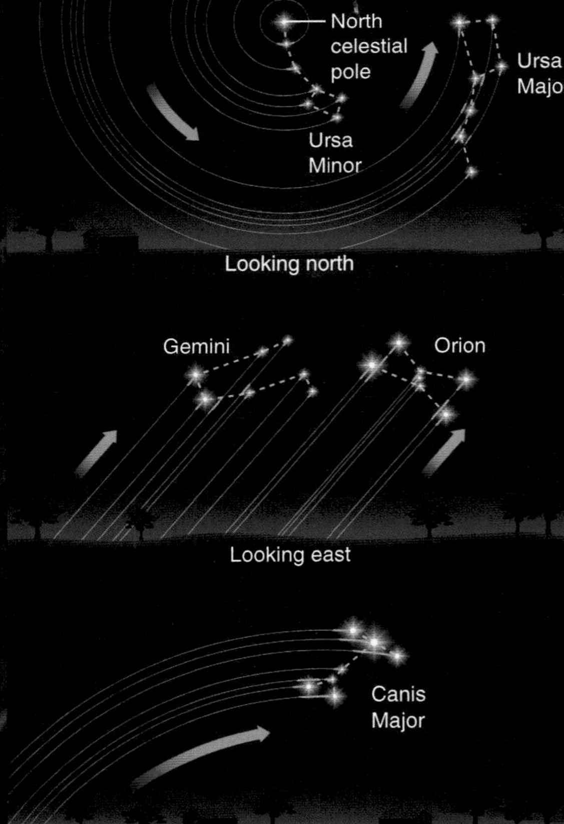

Ace Astronomy™ Log into AceAstronomy and select this chapter to see Active Figure "Rotation of the Sky." Look in different directions and compare the motions of the stars.

Astronomers measure distance across the sky as angles.

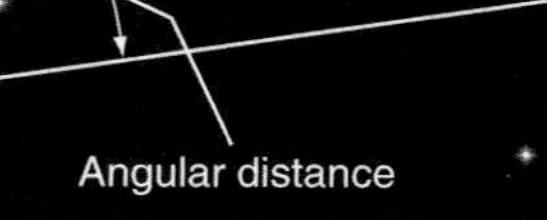

Astronomers might say, "The star was only 2 degrees from the moon." Of course, the stars are much farther away than the moon, but when you think of the celestial sphere, you can measure distances *on the sky* as **angular distances** in degrees, minutes of arc, and seconds of arc. A **minute of arc** is 1/60th of a degree, and a **second of arc** is 1/60th of a minute of arc. Then the **angular diameter** of an object is the angular distance from one edge to the other. The sun and moon are each about half a degree in diameter, and the bowl of the Big Dipper is about 10° wide.

Circumpolar constellations are those that never rise or set. From mid-northern latitudes, as shown below, you see a number of familiar constellations circling Polaris and never dipping below the horizon. As the sky rotates, the pointer stars at the front of the Big Dipper always point toward Polaris. Circumpolar constellations near the south celestial pole never rise as seen from mid-northern latitudes. From a high latitude such as Norway, you would have more circumpolar constellations, and from Quito, Ecuador, located on Earth's equator, you would have no circumpolar constellations at all.

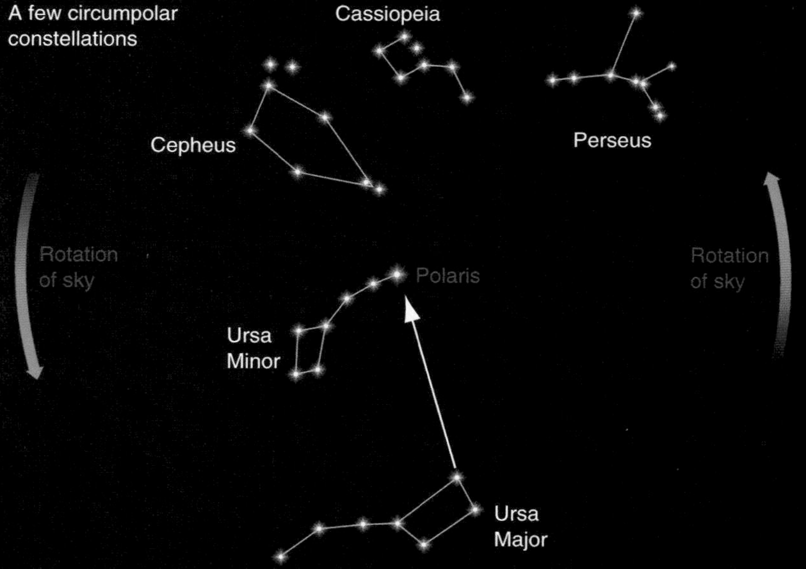

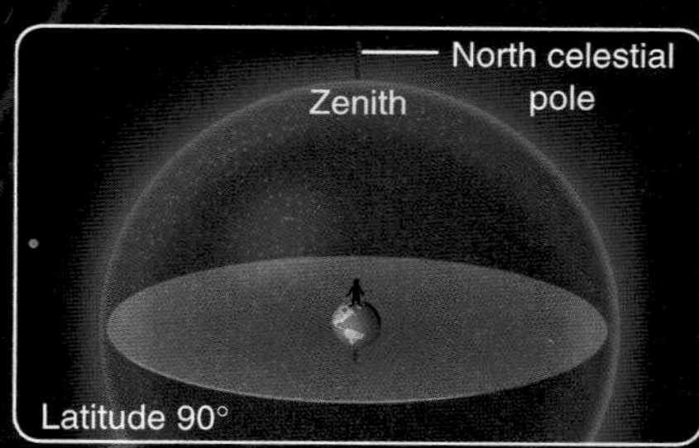

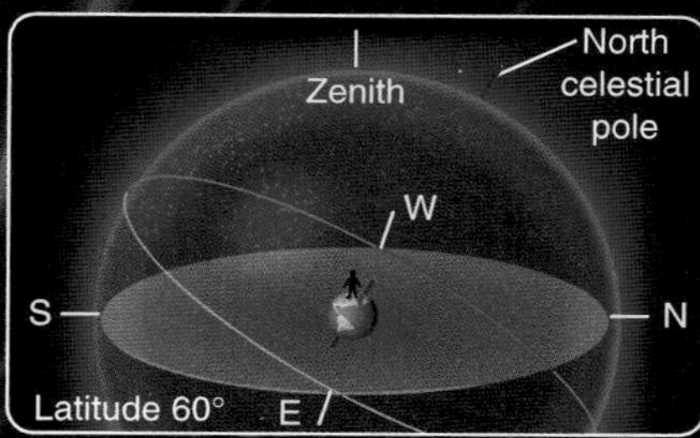

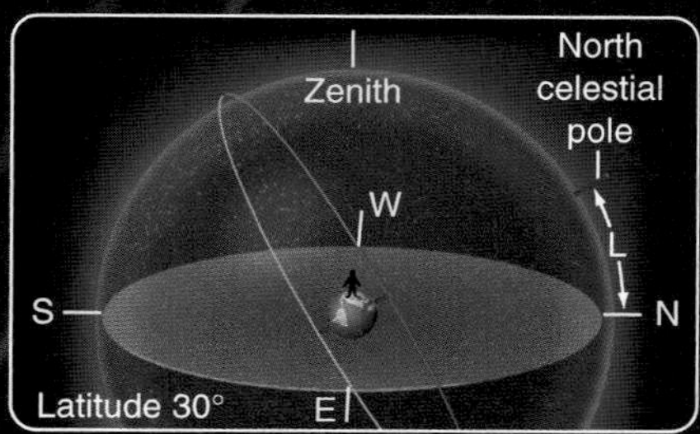

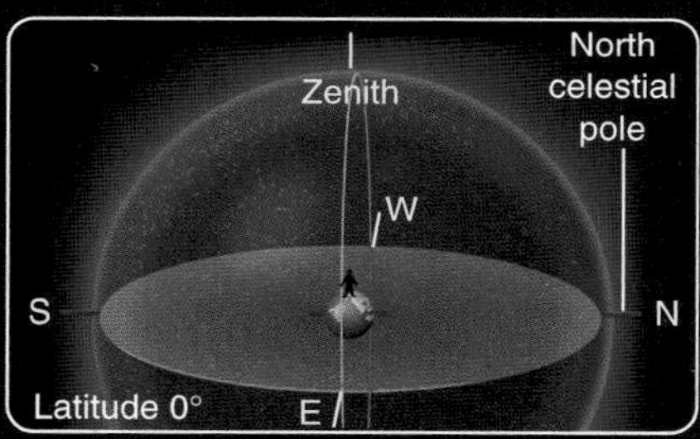

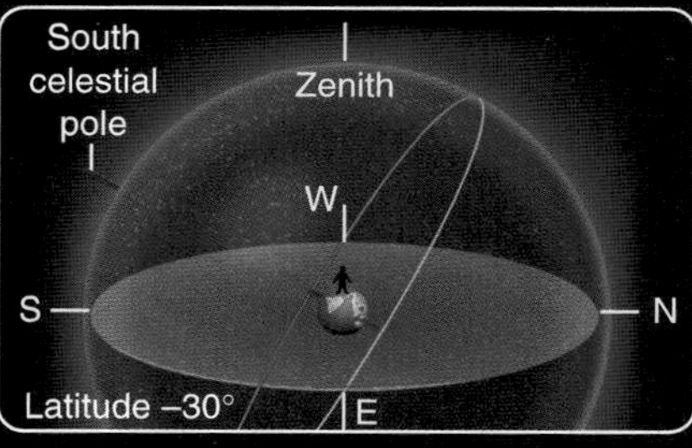

What you see in the sky depends on your latitude as shown above. Imagine that you begin a journey in the ice and snow at Earth's North Pole with the north celestial pole directly overhead. As you walk southward, the celestial pole moves toward the horizon, and you can see further into the southern sky. The angular distance from the horizon to the north celestial pole always equals your latitude (L)—the basis for celestial navigation. As you cross Earth's equator, the celestial equator would pass through your zenith, and the north celestial pole would sink below your northern horizon.

Log into AceAstronomy and select this chapter to see Active Figure "Constellations from Different Latitudes." Ace Astronomy™

not. In contrast, Orion rises and sets as seen from nearly everywhere on Earth. Explorers at Earth's poles, however, never see Orion rise or set. How would you improve the definition of a circumpolar constellation to clarify the status of Ursa Major? Would your definition help in the case of Orion?

■ ■ ■

Connections: You now have a model of the sky—the celestial sphere—complete with reference marks and angular measurements to guide you. You can locate constellations, name the stars, and express their brightness numerically. You are now ready to discuss the principal motions in the sky—the motions of the moon and sun.

Study and Review Tools

Summary

2-1 | The Stars

How do astronomers refer to stars?

- Astronomers divide the sky into 88 constellations. Although the constellations originated in Greek and Middle Eastern mythology, the names are Latin. Even the modern constellations, added to fill in the spaces between the ancient figures, have Latin names.
- The names of stars usually come from ancient Arabic, though modern astronomers often refer to a star by its constellation and a Greek letter assigned according to its brightness within the constellation.

How can we compare the brightness of stars?

- The magnitude system is the astronomer's brightness scale. First-magnitude stars are brighter than second-magnitude stars, which are brighter than third-magnitude stars, and so on. The magnitude you see when you look at a star in the sky is its apparent visual magnitude, which does not take into account its distance from Earth.

2-2 | The Sky and Its Motion

How does the sky move as Earth rotates?

- The celestial sphere is a model of the sky, carrying the celestial objects around Earth. Because Earth rotates eastward, the celestial sphere appears to rotate westward on its axis. The northern and southern celestial poles are the pivots on which the sky appears to rotate.
- The celestial equator, an imaginary line around the sky above Earth's equator, divides the sky in half.
- Astronomers often refer to angles "on" the sky as if the stars, sun, moon, and planets were equivalent to spots painted on a plaster ceiling. These angular distances are unrelated to the true distance between the objects in light-years.
- What you see of the celestial sphere depends on your latitude. Much of the southern hemisphere of the sky is not visible from northern latitudes. To see that part of the sky, you would have to travel southward over Earth's surface.
- The angular distance from the horizon to the north celestial pole always equals your latitude. This is the basis for celestial navigation.
- The gravitational forces of the moon and sun act on the spinning Earth and cause it to precess like a top. Earth's axis of rotation sweeps around in a conical motion with a period of 26,000 years, and consequently the celestial poles and celestial equator move slowly against the background of the stars.

New Terms

constellation (p. 14)
asterism (p. 15)
magnitude scale (p. 16)
apparent visual magnitude (m_v) (p. 16)
scientific model (p. 18)
precession (p. 18)
celestial sphere (p. 20)
horizon (p. 20)
zenith (p. 20)
nadir (p. 20)
north and south celestial poles (p. 20)
celestial equator (p. 20)
north, south, east, and west points (p. 20)
angular distance (p. 21)
minute of arc (p. 21)
second of arc (p. 21)
angular diameter (p. 21)
circumpolar constellation (p. 21)

Review Questions

Ace Astronomy™ Assess your understanding of this chapter's topics with additional quizzing and animations at http://astronomy.brookscole.com/sh9e

1. Why have astronomers added modern constellations to the sky?
2. What is the difference between an asterism and a constellation? Give some examples.
3. What characteristic do stars in a constellation or asterism share?
4. Do people from other cultures on Earth see the same stars, constellations, and asterisms that you see?
5. How does the Greek-letter designation of a star give us a clue to its brightness?
6. Why do some people say the magnitude system is backwards? How did that arise?
7. What does the word *apparent* mean in *apparent visual magnitude?*
8. In what ways is the celestial sphere a scientific model?
9. Why do astronomers use the word *on* to describe angles *on* the sky rather than angles *in* the sky?
10. If Earth did not rotate, could we define the celestial poles and celestial equator?
11. Where would you go on Earth if you wanted to be able to see both the north celestial pole and the south celestial pole at the same time?
12. Where would you go on Earth to place a celestial pole at your zenith?
13. Explain how to make a simple astronomical observation that would determine your latitude.

14. Why does the number of circumpolar constellations depend on the latitude of the observer?
15. How could we detect Earth's precession by examining star charts from ancient Egypt?
16. The stamp at right shows the constellation Orion. Explain why this looks odd to residents of the northern hemisphere.

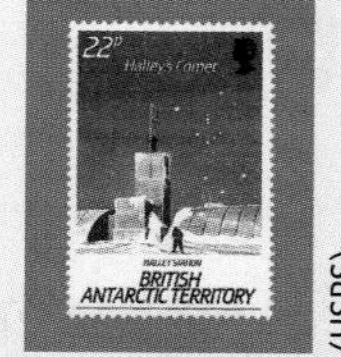

(USPS)

Discussion Questions

1. All cultures on Earth named constellations. Why do you suppose this was such a common practice?
2. If you were lost at sea, you could find your approximate latitude by measuring the altitude of Polaris. But Polaris isn't exactly at the celestial pole. What else would you need to know to measure your latitude more accurately?

Problems

1. If light from one star is 40 times more intense than light from another star, what is their difference in magnitudes?
2. If two stars differ by 8.6 magnitudes, what is their intensity ratio?
3. Star A has a magnitude of 2.5; Star B, 5.5; and Star C, 9.5. Which is brightest? Which are visible to the unaided eye? Which pair of stars has an intensity ratio of 16?
4. By what factor is sunlight more intense than moonlight? (*Hint:* See Figure 2-6.)
5. If you are at a latitude of 35 degrees north of Earth's equator, what is the angular distance from the northern horizon up to the north celestial pole? from the southern horizon down to the south celestial pole?

Media Cluster

ACTIVE FIGURES

Ace Astronomy™ To access the resources in the Media Cluster, log into AceAstronomy at **http://astronomy.brookscole.com/sh9e** and select Chapter 2.

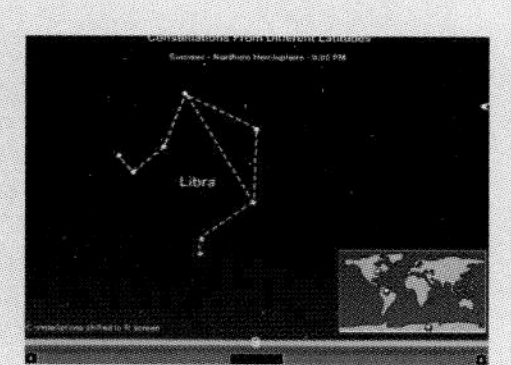

Constellations from Different Latitudes
Use this animation to observe how a change in latitude provides a different view of the constellations.

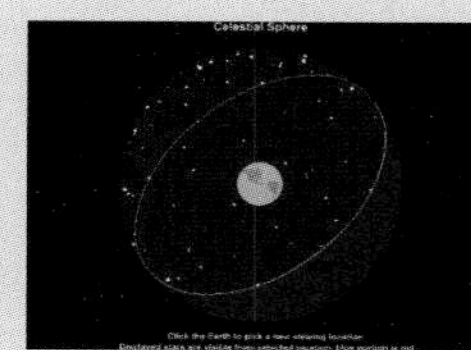

The Celestial Sphere
Observers on Earth see the sky as a celestial sphere. This animation lets you select your vantage point and visualize that location's half-dome of the sky.

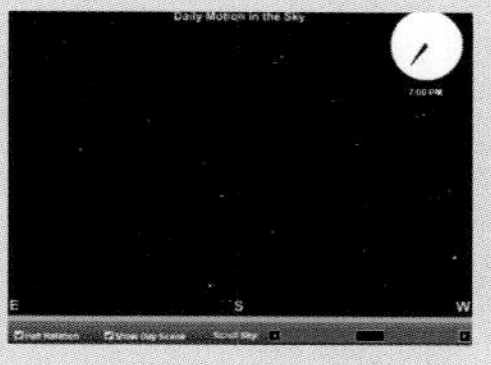

Rotation of the Sky
This animation shows how the stars (including the sun) appear to rotate in the sky every 24 hours, as seen by northern hemisphere observers.

Critical Inquiries for the Web

1. You've heard of the "North Star," but is there a "South Star" for southern hemisphere observers? As Earth precesses on its axis, Polaris will cease to be the pole star and others will eventually hold that title. Search for star maps available online and for information on precession and determine what bright southern hemisphere stars might be future pole stars for people living south of the equator.
2. Would the stars of a familiar constellation like Pegasus or Orion look similar to your view from Earth if you lived on a planet orbiting another star? Find a table of distance data for the bright stars in a familiar constellation and construct a diagram that visualizes the positions of these stars in space (similar to Figure 2-3).
3. Most cultures around the world developed their own constellations. Search the web to find constellation mythology from other cultures. Native Americans have a rich mythology of the sky, as do Polynesian, Asian, and African peoples.

Exploring *TheSky*

1. Describe the apparent rotation of the sky in the north, south, east, and west sky from your location. (*Hint:* Use **Site Information** under the **Data** menu to set location and date. Use **Look** under the **Orientation** menu and use the time skip arrows in the toolbar to go forward in time.)
2. What constellations are visible in your evening sky? (*Hint:* Use **Reference Lines** and **Labels** under the **View** menu to turn on the constellations and their labels. Use **Site Information** under the **Data** menu to set your location, date, and time.)
3. What would the sky look like if you could not see stars fainter than second magnitude? (*Hint:* Use **Filters** under the **View** menu to change the magnitude limits.)
4. Locate the constellation Orion as seen from your location, from Earth's north pole, and from Earth's south pole.

Go to the Brooks/Cole Astronomy Resource Center **(http://astronomy.brookscole.com)** for critical thinking exercises, articles, and additional readings from InfoTrac College Edition, Brooks/Cole's online student library.

3 Cycles of the Sky

Even a man who is
pure in heart
and says his prayers by night
may become a wolf when
the wolfbane blooms
and the moon shines
full and bright.

PROVERB FROM OLD WOLFMAN MOVIES

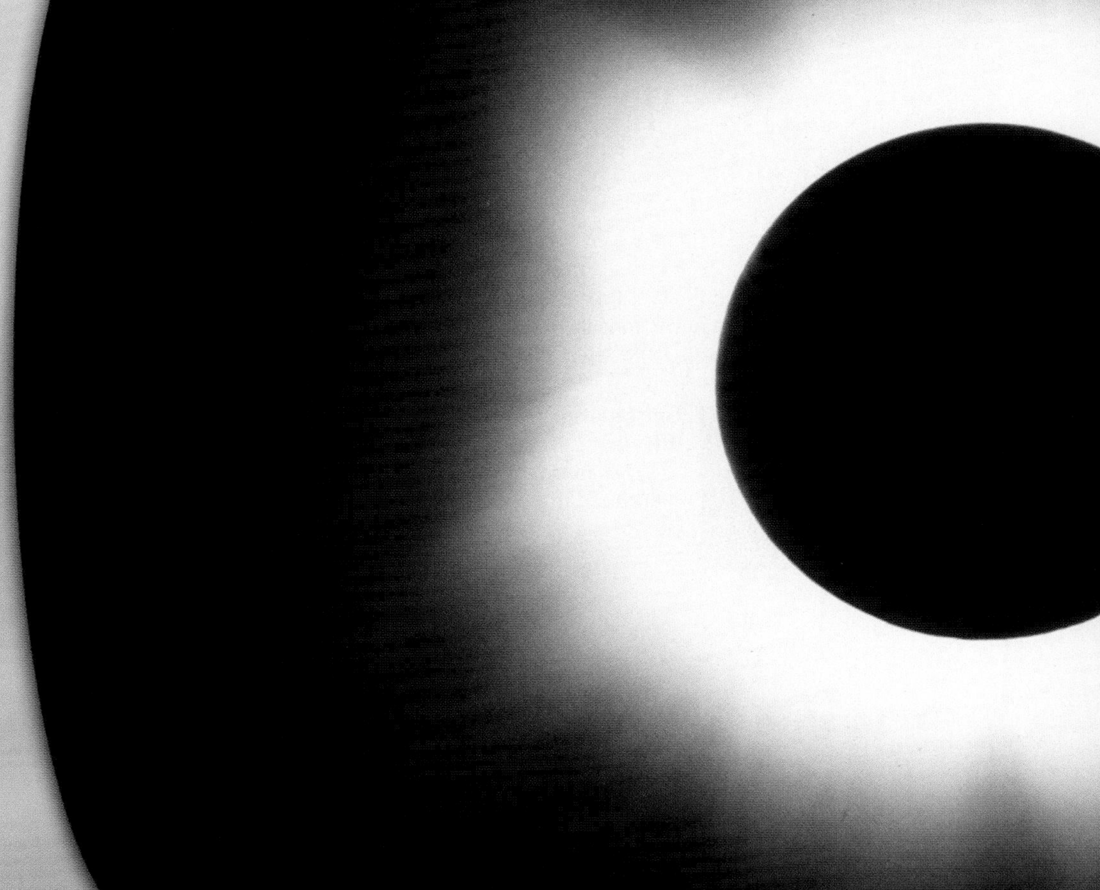

Visual-wavelength image

YOUR ALARM CLOCK is an astronomical instrument. So is your calendar. Our lives are shaped by the cycles of the sky. The rotation of Earth on its axis defines the day, and its revolution around its orbit defines the year. The orbital motion of the moon across our sky defines the month and divides it into four weeks. ▌ English defines **rotation** as the turning of a body on its axis. **Revolution** means the motion of a body around a point outside the body. Thus we are careful to say Earth rotates once a day on its axis and revolves once a year around the sun. Furthermore, we say the moon revolves once a month along its orbit around Earth. ▌ By thinking carefully about the motions you see in the sky, you will learn a great deal about the planet you live on.

A total solar eclipse occurs when the moon crosses in front of the sun and hides its brilliant surface. Then we can see the sun's extended atmosphere. (NSO/AURA/NSF)

Guidepost

Looking Back

In the preceding chapter you watched Earth turning on its axis and making the sun rise and set. That daily cycle is probably the most important cycle in your life, but there are others, and now you are ready to watch the sun and moon move against the background of the stars.

This Chapter

The motions of the sun and moon have direct influences on your life and produce dramatic sights in the sky. As you explore these you will find answers to five essential questions:

What causes the seasons?

Why does the moon go through phases?

What is a lunar eclipse?

What is a solar eclipse?

How do astronomical cycles affect Earth's climate?

This chapter also illustrates a powerful way to analyze certain kinds of problems. If a process repeats, you can analyze it by searching for cycles. The motions of the sun and moon become simple and elegant when you see them as cycles.

Looking Ahead

Once you have a 21st century understanding of your world and its motions, you will be ready to read the next chapter, where you will see how Renaissance astronomers analyzed what they saw in the sky and came to a revolutionary conclusion—that we live on a planet.

AceAstronomy™ The AceAstronomy icon throughout the text indicates an opportunity for you to test yourself on key concepts and to explore animations and interactions on the AceAstronomy website at: http://astronomy.brookscole.com/sh9e

3-1 The Cycle of the Sun

THE SUN RISES AND SETS because Earth rotates on its axis. That is only the most obvious of the solar cycles.

The Annual Motion of the Sun

Even in the daytime, the sky is filled with stars, but the glare of sunlight fills our atmosphere with scattered light, and you can only see the brilliant sun. If the sun were fainter and you could see the stars, you would notice that the sun was moving slowly eastward against the background of stars. This motion is caused by the orbital motion of Earth around the sun. In January, you would see the sun in front of the constellation Sagittarius (■ Figure 3-1). As Earth moves along its circular orbit (red), the sun appears to move eastward among the stars. By March, it is in front of Aquarius.

Through the year, the sun moves eastward among the stars following a line called the **ecliptic,** the apparent path of the sun among the stars. If the sky were a great screen, the ecliptic would be the shadow cast by Earth's orbit. That is why the ecliptic is often called the projection of Earth's orbit on the sky.

Earth circles the sun in 365.25 days, and consequently the sun appears to circle the sky in the same period. That means the sun, traveling 360° around the ecliptic in 365.25 days, travels about 1° eastward each day, about twice its angular diameter. You don't notice this motion because you cannot see the stars in the daytime, but the motion of the sun has an important consequence that you do notice—the seasons.

The Seasons

The seasons are caused by a simple fact: Earth's axis of rotation is tipped 23.5° from the perpendicular to its orbit. As you study **The Cycle of the Seasons** on pages 28 and 29, notice two important principles. First, the seasons are not caused by any variation in the distance from Earth to the sun. Earth's orbit is nearly circular, so it is always about the same distance from the sun. Second, the seasons are caused by the changes in solar energy that Earth's northern and southern hemispheres receive at different times of the year.

■ Active Figure 3-1

The motion of Earth around the sun makes the sun appear to move against the background of the stars. The circular orbit of Earth is thus projected on the sky as the circular path of the sun, the ecliptic. If you could see the stars in the daytime, you would notice the sun crossing in front of the distant constellations as Earth moves along its orbit.

AceAstronomy™ Log into AceAstronomy and select this chapter to see the Active Figure called "Constellations in Different Seasons." Notice how constellations in the morning sky move into the evening sky as the year passes.

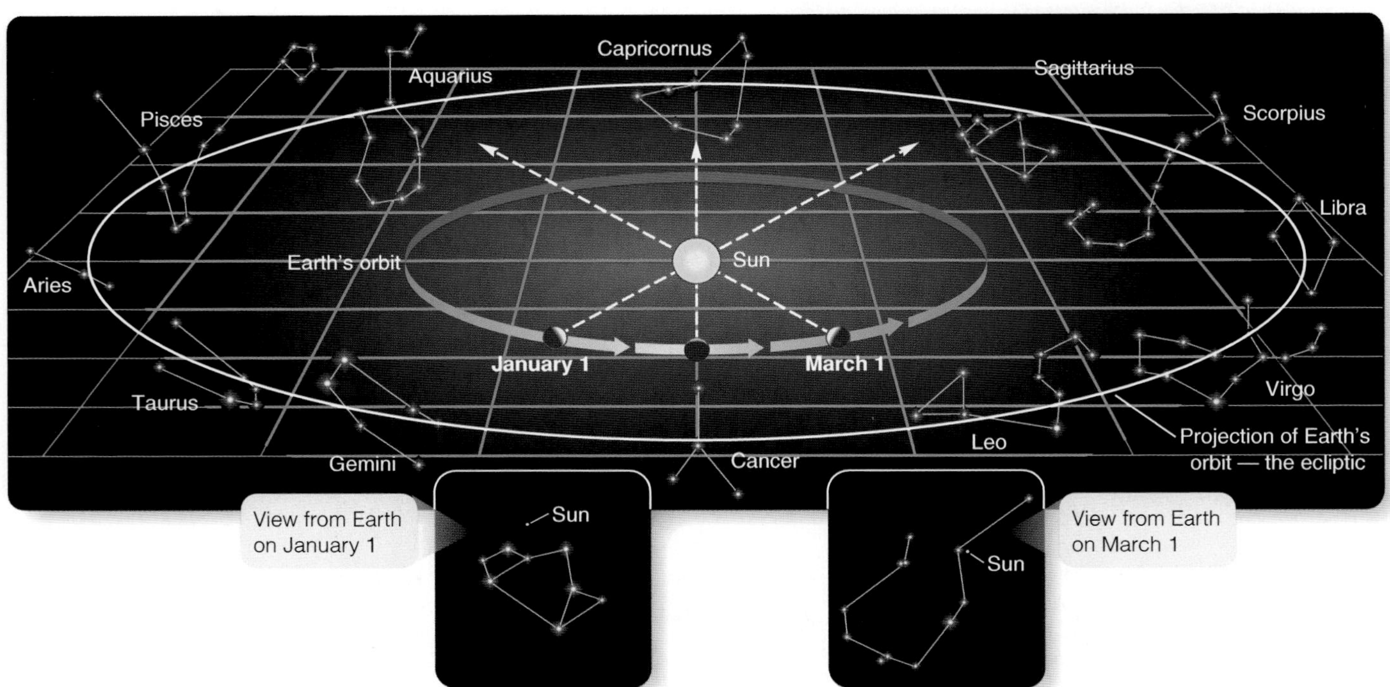

Window on Science 3-1

Astrology and Pseudoscience: Misusing the Rules of Science

Astronomers have a low opinion of astrology not so much because it is groundless but because it pretends to be a science. It is a pseudoscience, from the prefix *pseudo,* meaning "false." There are many examples of pseudoscience, and it is illuminating to consider the difference between pseudoscience and science.

A pseudoscience is a set of beliefs that appear to be based on scientific ideas but that fail to obey the most basic rules of science. For example, some years ago a fad arose in which people placed objects under pyramids made of paper, plastic, wire, and so on. The claim was that the pyramidal shape would focus cosmic forces on anything inside and so preserve fruit, sharpen razor blades, and do other miraculous things. Many books promoted this idea, but simple experiments showed that any shape would protect a piece of fruit from airborne spores and allow it to dry without rotting. Likewise, any shape would allow oxidation, thus improving the cutting edge of a razor blade. In short, experimental evidence contradicted the claim. Nevertheless, supporters of the theory declined to abandon or revise their claims. Thus, the fad of pyramid power was a pseudoscience.

One characteristic of a pseudoscience is that it appeals to our needs and desires. Thus, some pseudoscientific claims are self-fulfilling. For example, some people bought pyramidal tents to put over their beds and thus improve their rest. While there is no logical mechanism by which such a tent could affect a sleeper, people slept more soundly because they wanted and expected the claim to be true. Many pseudoscientific claims involve medical cures, ranging from copper or magnetic bracelets to crystals that focus spiritual power to astonishingly expensive and illegal treatments for cancer. Logic is a stranger to pseudoscience, but human fears and needs are not.

Astrology is a pseudoscience. Over the centuries, astrology has been tested repeatedly, and no correlation between the movement of the heavens and human lives has been found. But it survives, and its supporters disregard any evidence that it doesn't work. Like all pseudosciences, astrology is not open to revision in the face of contradictory evidence. Furthermore, astrology fulfills our human need to believe that there is order and meaning to our lives. It may comfort us to believe that our sweetheart has rejected us because of the motions of the sun, moon, and planets along the zodiac rather than to admit that we behaved badly on our last date. Comfort aside, astrology is a poor basis for life decisions.

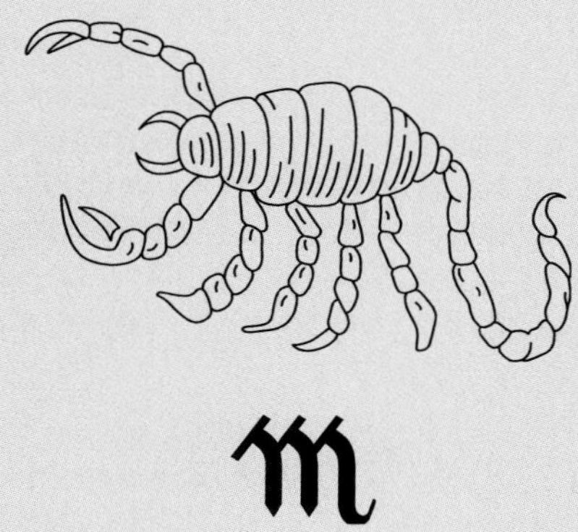

Astrology may be the oldest pseudoscience.

Human nature and human needs probably ensure that pseudoscientific beliefs will continue to plague us like emotional viruses propagating from person to person. If we recognize them for what they are, we can more easily guide our lives by rational principles and not assign the stars blame for our failures and credit for our successes.

The seasons are so important as a cycle of growth, harvest, death, and rebirth, that it is not surprising to realize that ancient peoples saw the motion of the sun around the ecliptic as a powerful influence on their daily lives. The ancient superstition of astrology is based on the cycle of the sun. The ecliptic marks the center line of the **zodiac,** a band 18° wide that encircles the sky. The signs of the zodiac take their names from the 12 principal constellations along the ecliptic. Astrology was once an important part of astronomy, but the two are now almost exact opposites—astronomy is a science that depends on evidence, and astrology is a superstition that survives in spite of evidence (**Window on Science 3-1**). Thus the signs of the zodiac are no longer important in astronomy. The zodiac itself is of interest only because it is on the path followed by the planets as they move around the sky.

AceAstronomy™ Log into AceAstronomy and select this chapter to see Astronomy Exercise "Sunrise through the Seasons" to watch the sun moving.

AceAstronomy™ Log into AceAstronomy and select this chapter to see Astronomy Exercise "The Seasons" and see how the sun's altitude changes.

The Motion of the Planets

The planets of our solar system produce no visible light of their own; we see them by reflected sunlight. Mercury, Venus, Mars, Jupiter, and Saturn are all easily visible to the naked eye; but Uranus is usually too faint to be seen, and Neptune is never bright enough. Pluto is even fainter, and we need a large telescope to find it.

All the planets of the solar system move in nearly circular orbits around the sun. If you were looking down on the solar system from the north celestial pole, you would see the planets moving in the same counterclockwise direction around their orbits, with the planets farthest from the sun moving the slowest.

When you look for planets in the sky, you always find them near the ecliptic because their orbits lie in nearly the same plane

The Cycle of the Seasons

You can use the celestial sphere to help you think about the seasons. The celestial equator is the projection of Earth's equator on the sky, and the ecliptic is the projection of Earth's orbit on the sky. Because Earth is tipped in its orbit, the ecliptic and equator are inclined to each other by 23.5°. As the sun moves eastward around the sky, it spends half the year in the southern half of the sky and half of the year in the northern half. That causes the seasons.

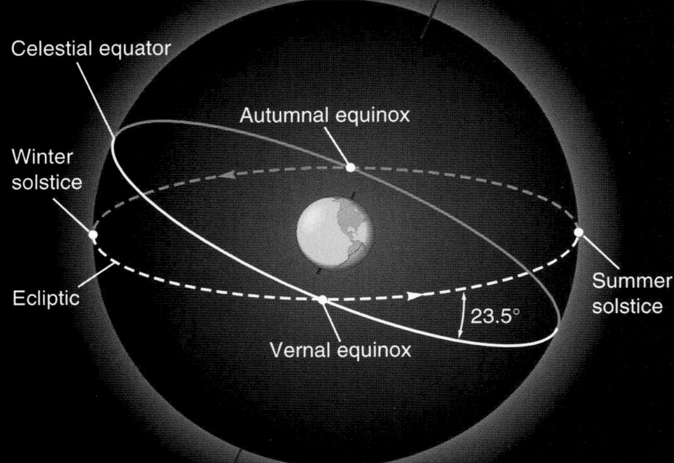

The sun crosses the celestial equator going northward at the point called the **vernal equinox.** The sun is at its farthest north at the point called the **summer solstice.** It crosses the celestial equator going southward at the **autumnal equinox** and reaches its most southern point at the **winter solstice.**

The seasons are defined by the dates when the sun crosses these four points, as shown in the table at the right. *Equinox* comes from the word for "equal"; the day of an equinox has equal amounts of daylight and darkness. *Solstice* comes from the words meaning "sun" and "stationary." *Vernal* comes from the word for "green." The "green" equinox marks the beginning of spring.

Event	Date*	Season
Vernal equinox	March 20	Spring begins
Summer solstice	June 22	Summer begins
Autumnal equinox	September 22	Autumn begins
Winter solstice	December 22	Winter begins

* Give or take a day due to leap year and other factors.

Ace Astronomy™

Log into AceAstronomy and select this chapter to see Active Figure "Seasons" and watch Earth orbiting the sun.

23.5°

To Polaris

40° N latitude

Equator

40° S latitude

Sunlight nearly direct on northern latitudes

To sun

Sunlight spread out on southern latitudes

On the day of the summer solstice in late June, Earth's northern hemisphere is inclined toward the sun, and sunlight shines almost straight down at northern latitudes. At southern latitudes, sunlight strikes the ground at an angle and spreads out. North America has warm weather, and South America has cool weather.

Earth's axis of rotation points toward Polaris, and, like a top, the spinning Earth holds its axis fixed as it orbits the sun. On one side of the sun, Earth's northern hemisphere leans toward the sun; on the other side of its orbit, it leans away. However, the direction of the axis of rotation does not change.

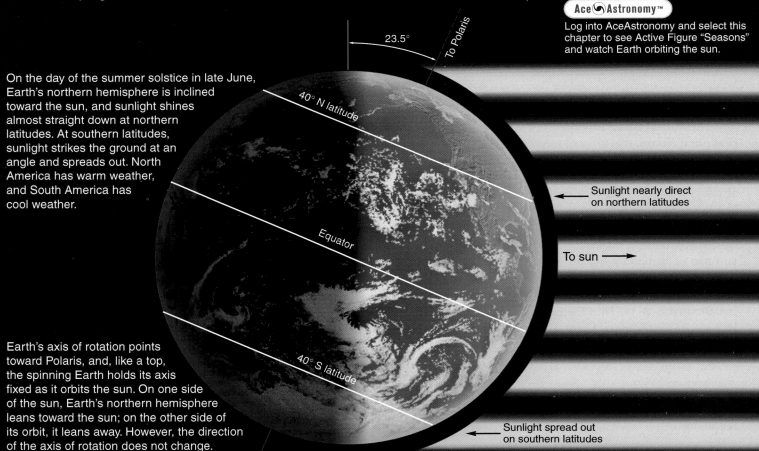

Earth at summer solstice

Light striking the ground at a steep angle spreads out less than light striking the ground at a shallow angle. Light from the summer-solstice sun strikes northern latitudes from nearly overhead and is concentrated. Light from the winter-solstice sun strikes northern latitudes at a much steeper angle and spreads out. The same amount of energy is spread over a larger area, so the ground receives less energy from the winter sun.

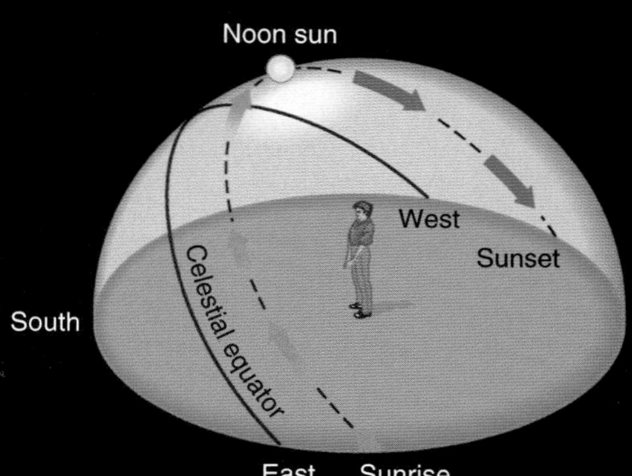

At summer solstice

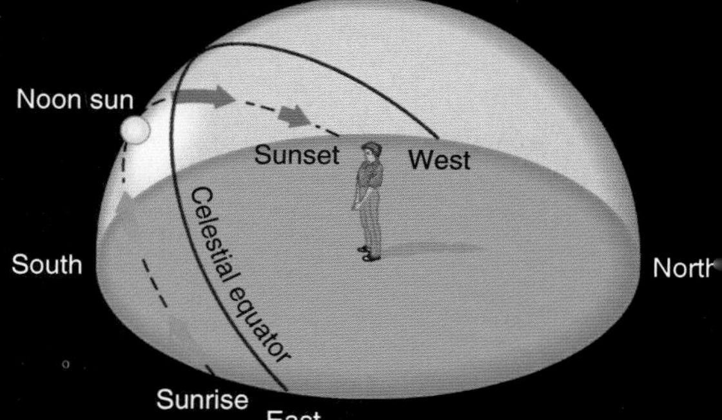

At winter solstice

The two causes of the seasons are shown at right for someone in the northern hemisphere. First, the noon summer sun is higher in the sky and the winter sun is lower, as shown by the longer winter shadows. Thus winter sunlight is more spread out. Second, the summer sun rises in the northeast and sets in the northwest, spending more than 12 hours in the sky. The winter sun rises in the southeast and sets in the southwest, spending less than 12 hours in the sky. Both of these effects mean that northern latitudes receive more energy from the summer sun, and summer days are warmer than winter days.

Ace Astronomy™

Log into AceAstronomy and select this chapter to see Active Figure "Path of the Sun" and see this figure from the inside.

23.5°

To Polaris

Sunlight spread out on northern latitudes

40° N latitude

Equator

To sun

Sunlight nearly direct on southern latitudes

40° S latitude

On the day of the winter solstice in late December, Earth's northern hemisphere is inclined away from the sun, and sunlight strikes the ground at an angle and spreads out. At southern latitudes, sunlight shines almost straight down and does not spread out. North America has cool weather and South America has warm weather.

Earth's orbit is only very slightly elliptical. About January 4, Earth is at **perihelion,** its closest point to the sun, when it is only 1.7 percent closer than average. About July 4, Earth is at **aphelion,** its most distant point from the sun, when it is only 1.7 percent farther than average. This small variation does not significantly affect the seasons.

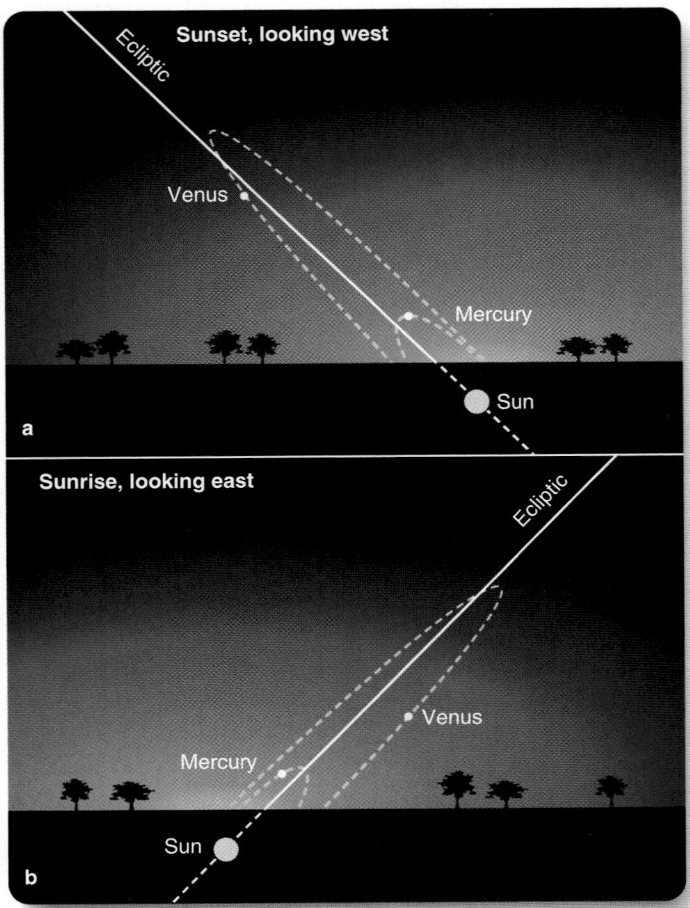

Figure 3-2

Mercury and Venus follow orbits that keep them near the sun, and they are visible only soon after sunset or before sunrise. Venus takes 584 days to move from morning sky to evening sky and back again, but Mercury zips around in only 116 days.

as Earth's orbit. As they orbit the sun, they appear to move eastward along the ecliptic.* Mars moves completely around the ecliptic in slightly less than 2 years, but Saturn, being farther from the sun, takes nearly 30 years.

As seen from Earth, Venus and Mercury can never be seen far from the sun because their orbits are inside Earth's orbit. They appear at times above the western horizon just after sunset or above the eastern horizon just before sunrise. Venus is easier to locate because its larger orbit carries it higher above the horizon than does Mercury's (■ Figure 3-2). Mercury's orbit is so small that it can never get farther than 27°50′ from the sun. Consequently, it is hard to see against the sun's glare and is often hidden in the clouds and haze near the horizon.

By tradition, any planet visible in the evening sky is called an **evening star,** even though planets are not stars. Similarly, any planet visible in the sky shortly before sunrise is called a **morning star.** Perhaps the most beautiful is Venus, which can become as bright as magnitude −4.7. As Venus moves around its orbit, it can dominate the western sky each evening for many weeks, but eventually its orbit carries it back toward the sun, and it is lost in the haze near the horizon. In a few weeks, it reappears in the dawn sky, a brilliant morning star.

*We will discuss occasional exceptions to this eastward motion in Chapter 4.

Inquire | Review | Analyze

How does the elliptical shape of Earth's orbit affect the seasons?
In the critical analysis of an idea it can be helpful to exaggerate the importance of a single factor. Doing so not only reveals the effect of that factor but can also reveal the inner workings of the process itself. In this case, you might try to imagine what life on Earth would be like if Earth's orbit were more elliptical than it really is. Earth now passes perihelion about January 4, and on that date we are slightly closer to the sun than average. If Earth's orbit were more elliptical, we would be significantly closer. This would make winters in the northern hemisphere much warmer than they are now; and the weather in the southern hemisphere, normally warm in January, would be dreadfully hot. Six months later, about July 4, Earth reaches aphelion, its farthest point from the sun. If Earth had a more elliptical orbit, it would be significantly farther from the sun on this date. Then the northern hemisphere would experience a much cooler summer than it now does; and in the southern hemisphere, the cool season would be bitterly cold.

Of course, Earth's orbit is nearly circular, so its distance from the sun varies by only about 3.4 percent from nearest to farthest. That should make the northern seasons a little milder and the southern seasons a little more extreme, an effect that is hardly noticeable. Now use exaggeration to analyze the true cause of the seasons. Try to imagine what the seasons would be like if Earth's axis were inclined more than 23.5° or less than 23.5° from the perpendicular to its orbit.

■ ■ ■

Connections: The cycle of the sun around the ecliptic dominates our lives through the seasons, but there is another moving light in the sky. What the cycles of our moon lack in intensity, they make up in beauty and elegance.

3-2 The Cycles of the Moon

THE MOON ORBITS EASTWARD around Earth once a month. Starting this evening, look for the moon in the sky. If it is a cloudy night or if the moon is in the wrong part of its orbit, you may not see it; but keep trying on successive evenings, and within a week or two you will see the moon. Then watch for the moon on following evenings, and you will see it following its orbit around Earth and cycling through its phases as it has done for billions of years.

The Motion of the Moon

If you watch the moon night after night, you will notice two things about its motion. First, you will see it moving eastward against the

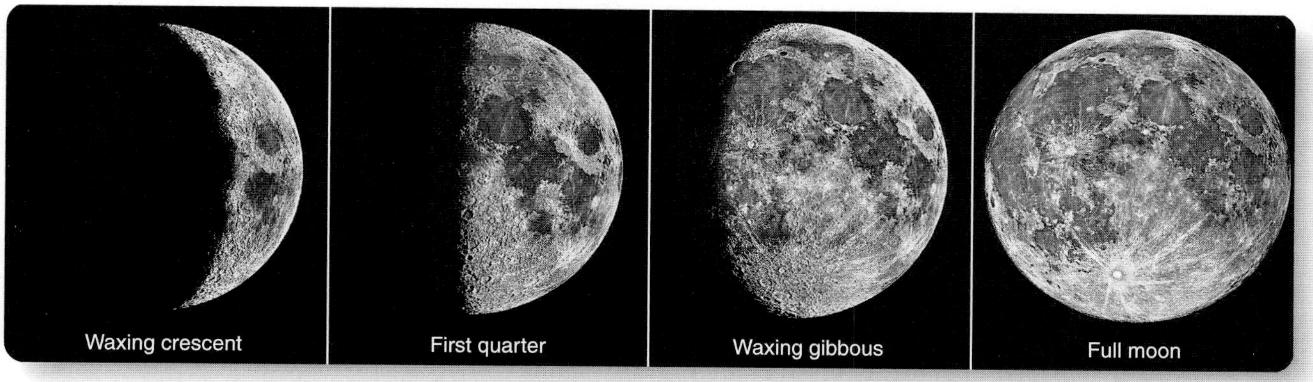

■ **Figure 3-3**

In this sequence of the waxing moon, you see the same face of the moon, the same mountains, and craters, and plains, but the changing direction of sunlight produces the lunar phases. (UCO/Lick Observatory)

background of stars; second, you will notice that the markings on its face don't change. These two observations will help you understand the motion of the moon and the origin of the moon's phases.

The moon moves rapidly among the constellations. If you watch the moon for just an hour, you can see it move eastward against the background of stars by slightly more than its angular diameter. In the previous chapter, you discovered that the moon is about 0.5° in angular diameter, so it moves eastward a bit more than 0.5° per hour. In 24 hours, it moves 13°. Each night when you look at the moon, you see it about 13° eastward of its location the night before. This eastward movement is the result of the motion of the moon along its orbit around Earth.

The Cycle of Phases

The changing shape of the moon as it orbits Earth is one of the most easily observed phenomena in astronomy. We have all noticed the full moon rising dramatically or a thin crescent moon hanging in the evening sky. Study **The Phases of the Moon** on pages 32 and 33 and notice three important points. First, the moon always keeps the same side facing Earth, and you never see the far side of the moon. "The man in the moon" is produced by the familiar features on the moon's near side. Second, the changing shape of the moon as it passes through its cycle of phases is produced by sunlight illuminating different parts of the side of the moon you can see. The third thing to notice is the difference between the orbital period of the moon around Earth and the length of the lunar phase cycle. That difference is a good illustration of how your view from Earth is produced by the combined motions of Earth and other heavenly bodies such as the sun and moon.

You can make a moon-phase dial from the middle diagram on page 32 by covering the lower half of the moon's orbit with a sheet of paper, aligning the edge of the paper to pass through the word "Full" at the left and the word "New" at the right. Push a pin through the edge of the paper at Earth's North Pole to make a pivot and, under the word "Full," write on the paper "Eastern Horizon." Under the word "New," write "Western Horizon." The paper now represents the horizon you see facing south.

You can set your moon-phase dial for a given time by rotating the diagram behind the horizon-paper. Set the dial to sunset by turning the diagram until the human figure labeled "sunset" is standing at the top of the Earth globe; the dial shows, for example, that the full moon at sunset would be at the eastern horizon.

We always see the same side of the moon looking down on us, but the changing shadows make the man in the moon shift his moods as the moon cycles through its phases (■ Figure 3-3). Occasionally something peculiar happens, and the moon darkens and turns copper-red in a lunar eclipse.

AceAstronomy™ Log into AceAstronomy and select this chapter to see Astronomy Exercise "Phases of the Moon" to take control of the phases.

AceAstronomy™ Log into AceAstronomy and select this chapter to see Astronomy Exercise "Moon Calendar" to find the phases for any month.

Lunar Eclipses

A **lunar eclipse** occurs at full moon when the moon moves through the shadow of Earth. Because the moon shines only by reflected sunlight, you see the moon gradually darken as it enters the shadow.

Earth's shadow consists of two parts (■ Figure 3-4). The **umbra** is the region of total shadow. In the umbra of Earth's shadow, you would see no portion of the sun. If you moved into the **penumbra,** however, you would be in partial shadow and would see part of the sun peeking around the edge of Earth. Thus, in the penumbra the sunlight is dimmed but not extinguished.

If the orbit of the moon carries it through the umbra, you see a total lunar eclipse (■ Figure 3-5). As you watch the moon in the sky, it first moves into the penumbra and dims slightly;

The Phases of the Moon

The moon goes through a month-long cycle of phases. This cycle is caused by the play of light and shadow on the moon as it follows its orbit around Earth.

As the moon orbits Earth, it rotates to keep the same side facing Earth. Consequently you always see the same features on the moon, and you never see the far side of the moon. A mountain on the moon that points at Earth will always point at Earth as the moon revolves and rotates.

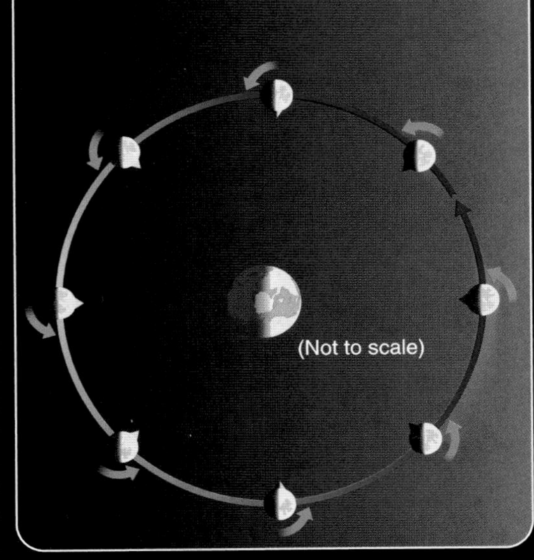

(Not to scale)

First quarter
Waxing gibbous
Waxing crescent
Sunset
North Pole
Full
Midnight
Noon
New
Sunlight
Earth's rotation
Sunrise
Waning gibbous
Waning crescent
Third quarter

Sunlight always illuminates half of the moon. Because you see different amounts of this sunlit side, you see the moon cycle through phases. At the phase called "new moon," sunlight illuminates the far side of the moon, and the side you see is in darkness. At new moon you see no moon at all. At full moon, the side you see is fully lit, and the far side is in darkness. How much you see depends on where the moon is in its orbit.

Notice that there is no such thing as the "dark side of the moon." All parts of the moon experience day and night in a month-long cycle.

In the diagram at the left, you see that the new moon is close to the sun in the sky, and the full moon is opposite the sun. The time of day depends on the observer's location on Earth.

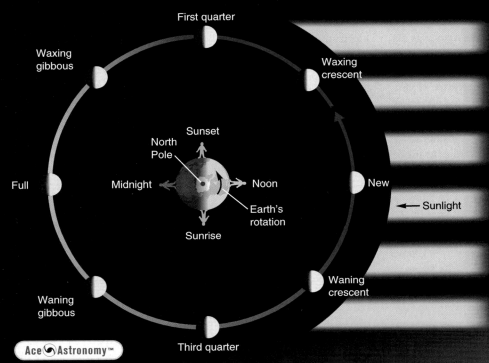

Ace Astronomy™

Log into AceAstronomy and select this chapter to see Active Figure "Lunar Phases" and take control of this diagram.

Gibbous comes from the Latin word for humpbacked.

The first quarter moon is one week through its 4-week cycle.
The first two weeks of the cycle of the moon are shown below by its position at sunset on 14 successive evenings. As the moon grows fatter from new to full, it is said to wax.
The full moon is two weeks through its 4-week cycle.
Waxing gibbous
Waxing crescent
THE SKY AT SUNSET
Full moon rises at sunset
New moon is invisible near the sun
East
South
West

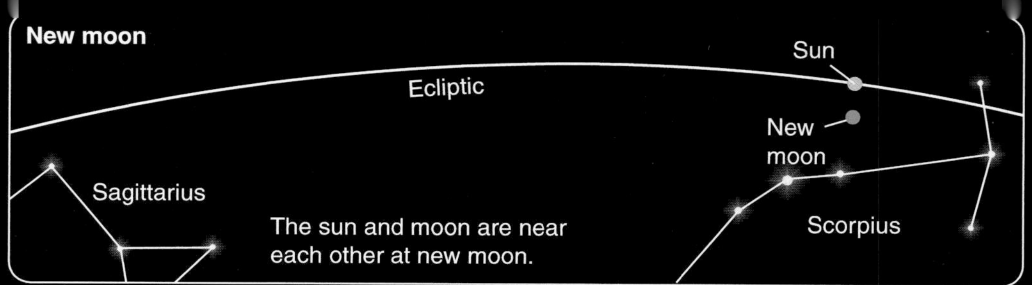
New moon
Sun
Ecliptic
New moon
Sagittarius
Scorpius
The sun and moon are near each other at new moon.

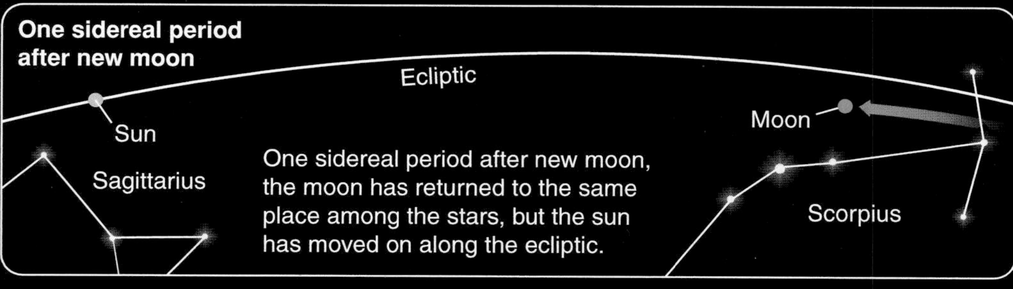
One sidereal period after new moon
Ecliptic
Sun
Moon
Sagittarius
Scorpius
One sidereal period after new moon, the moon has returned to the same place among the stars, but the sun has moved on along the ecliptic.

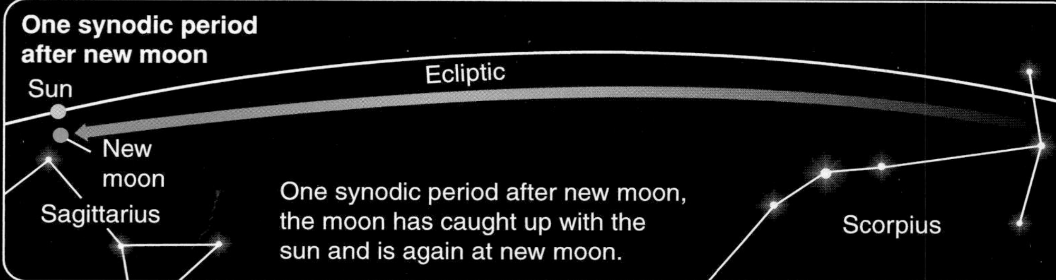
One synodic period after new moon
Sun
Ecliptic
New moon
Sagittarius
Scorpius
One synodic period after new moon, the moon has caught up with the sun and is again at new moon.

The moon orbits eastward around Earth in 27.32 days, its **sidereal period.** This is how long the moon takes to circle the sky once and return to the same position among the stars.

A complete cycle of lunar phases takes 29.53 days, the moon's **synodic period.** (Synodic comes from the Greek words for "together" and "path.")

To see why the synodic period is longer than the sidereal period, study the star charts at the left.

Although you think of the lunar cycle as being about 4 weeks long, it is actually 1.53 days longer than 4 weeks. The calendar divides the year into 30-day periods called months (literally "moonths") in recognition of the 29.53 day synodic cycle of the moon.

You can use the diagram on the opposite page to determine when the moon rises and sets at different phases.

TIMES OF MOONRISE AND MOONSET

Phase	Moonrise	Moonset
New	Dawn	Sunset
First quarter	Noon	Midnight
Full	Sunset	Dawn
Third quarter	Midnight	Noon

Earth Moon

This diagram is to scale. The moon's orbit is about 30 Earth diameters in radius. Because the orbit is slightly elliptical, the moon's angular diameter in the sky can vary by plus or minus 6 percent. Because the orbit is tipped a bit over 5 degrees, the moon does not follow the ecliptic exactly.

The third quarter moon is 3 weeks through its 4-week cycle.

The last two weeks of the cycle of the moon are shown below by its position at sunrise on 14 successive mornings. As the moon shrinks from full to new, it is said to wane.

Waning crescent

Waning gibbous

New moon is invisible near the sun

Full moon sets at sunrise

THE SKY AT SUNRISE

East South West

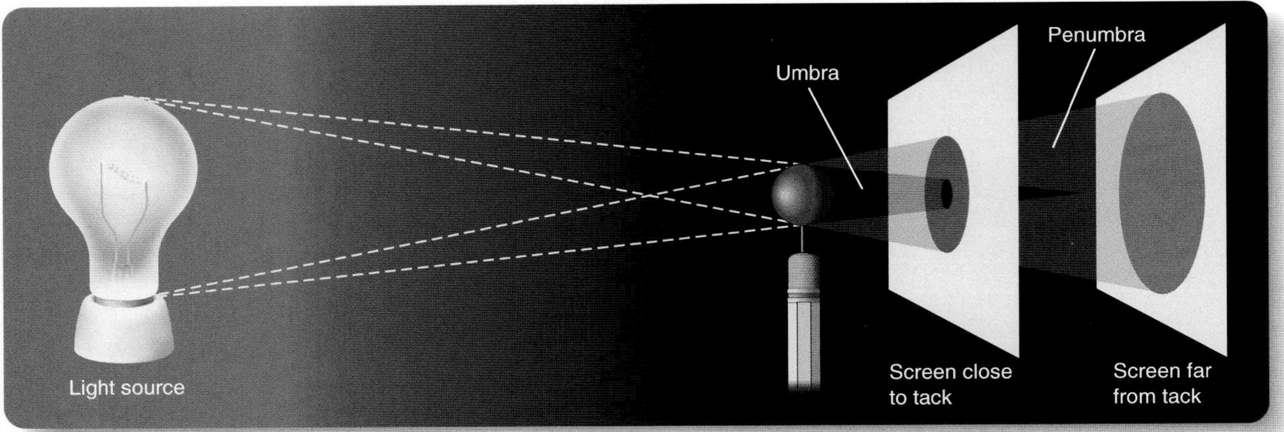
Penumbra
Umbra
Light source
Screen close to tack
Screen far from tack

Figure 3-4

The shadows cast by a map tack resemble those of Earth and the moon. The umbra is the region of total shadow; the penumbra is the region of partial shadow.

the deeper it moves into the penumbra, the more it dims. In about an hour, the moon reaches the umbra, and you see the umbral shadow darken part of the moon. It takes about an hour for the moon to enter the umbra completely and become totally eclipsed. Totality, the period of total eclipse, may last as long as 1 hour 45 minutes, though the timing of the eclipse depends on where the moon crosses the shadow.

Figure 3-5

During a total lunar eclipse the moon passes through Earth's shadow, as shown in this multiple-exposure photograph. A longer exposure was used to record the moon while it was totally eclipsed. The moon's path appears curved in the photo because of photographic effects. (© 1982 Dr. Jack B. Marling)

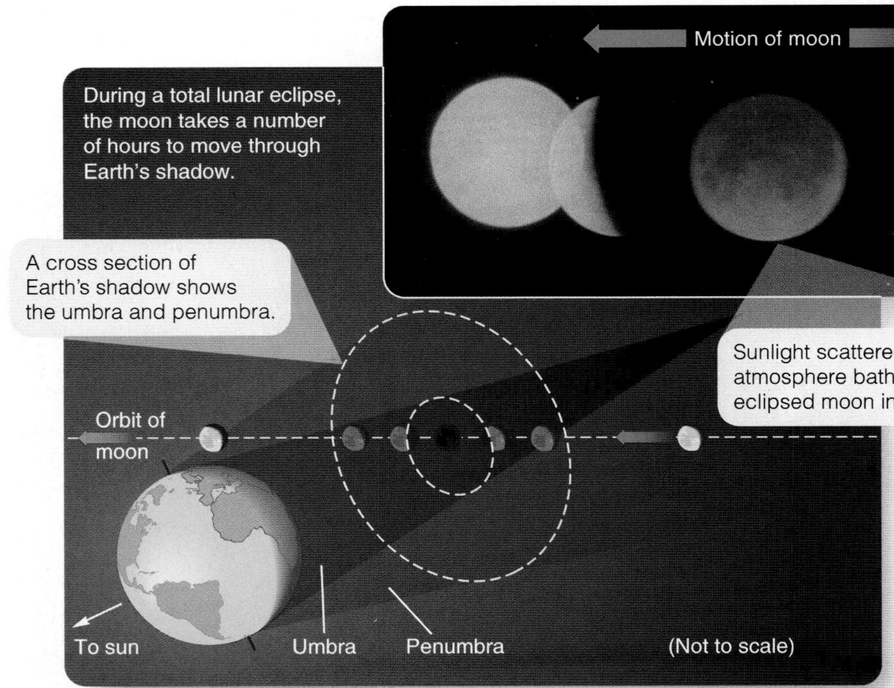
Motion of moon
During a total lunar eclipse, the moon takes a number of hours to move through Earth's shadow.
A cross section of Earth's shadow shows the umbra and penumbra.
Sunlight scattered from Earth's atmosphere bathes the totally eclipsed moon in a coppery glow.
Orbit of moon
To sun
Umbra
Penumbra
(Not to scale)

When the moon is totally eclipsed, it does not disappear completely. Although it receives no direct sunlight, the moon in the umbra does receive some sunlight refracted (bent) through Earth's atmosphere. If you were on the moon during totality, you would not see any part of the sun because it would be entirely hidden behind Earth. However, you would be able to see Earth's atmosphere illuminated from behind by the sun. The red glow from this "sunset" illuminates the moon during totality and makes it glow coppery red, as shown in Figure 3-5.

If the moon passes a bit too far north or south, it may only partially enter the umbra, and you see a partial lunar eclipse. The part of the moon that remains outside the umbra in the penumbra receives some direct sunlight, and the glare is usually great enough to prevent your seeing the faint coppery glow of the part of the moon in the umbra.

A penumbral lunar eclipse occurs when the moon passes through the penumbra but misses the umbra entirely. Because the penumbra is a region of partial shadow, the moon is only partially dimmed. A penumbral eclipse is not very impressive.

Although there are usually no more than one or two lunar eclipses each year, it is not difficult to see one. You need only be on the dark side of Earth when the moon passes through Earth's shadow. That is, the eclipse must occur between sunset and sunrise at your location.

Reasoning with Numbers | 3-1

The Small-Angle Formula

■ Figure 3-6 shows the angular diameter of an object, its linear diameter, and its distance. *Linear diameter* is the distance between an object's opposite sides. The linear diameter of the moon, for instance, is 3476 km. Recall that the *angular diameter* of an object is the angle formed by two lines extending from opposite sides of the object and meeting at our eye. Clearly, the farther away an object is, the smaller its angular diameter. You can use this formula to find any of these three quantities if you know the other two.

In the small-angle formula, you always express angular diameter in seconds of arc, and you always use the same units for distance and linear diameter:

$$\frac{\text{angular diameter}}{206{,}265^*} = \frac{\text{linear diameter}}{\text{distance}}$$

Example: The moon has a linear diameter of 3476 km and is about 384,000 km away. What is its angular diameter? **Solution:** You can leave linear diameter and distance in kilometers and find the angular diameter in seconds of arc:

$$\frac{\text{angular diameter}}{206{,}265} = \frac{3476 \cancel{\text{km}}}{384{,}000 \cancel{\text{km}}}$$

The angular diameter is 1870 seconds, which equals 31 minutes, of arc—about 0.5°.

*The number 206,265 is the number of seconds of arc in a radian. When you divide by 206,265, you convert the angle from seconds of arc into radians.

■ Table 3-1 will allow you to determine which upcoming total and partial lunar eclipses will be visible from your location.

Solar Eclipses

We who live on planet Earth can see a phenomenon that is not visible from most planets. It happens that the sun is 400 times larger than our moon and, on the average, 390 times farther away, so the sun and moon have nearly equal angular diameters—0.5°. (See **Reasoning with Numbers 3-1**.) Thus, the moon is just the right size to cover the bright disk of the sun and cause a **solar eclipse.** If the moon covers the entire disk of the sun, we see a total eclipse. If it covers only part of the sun, we see a partial eclipse.

Whether we see a total or partial eclipse depends on whether we are in the umbra or the penumbra of the moon's shadow. The umbra of the moon's shadow barely reaches Earth and casts a small circular shadow (■ Figure 3-7a). The shadow is never larger than 270 km (168 miles) in diameter. Standing in that umbral spot, you would be in total shadow, unable to see any part of the sun's surface, and the eclipse would be total (Figure 3-7b). But if

■ Table 3-1 | Total and Partial Eclipses of the Moon, 2005–2012

Date	Time* of Mideclipse (GMT)	Length of Totality (Min)	Length of Eclipse (Hr:Min)
2005 Oct. 17	12:04	Partial	0:56
2006 Sept. 7	18:52	Partial	1:30
2007 Mar. 3	23:22	74	3:40
2007 Aug. 28	10:38	90	3:32
2008 Feb. 21	3:27	50	3:24
2008 Aug. 16	21:11	Partial	3:08
2009 Dec. 31	19:24	Partial	1:00
2010 June 26	11:40	Partial	2:42
2010 Dec. 21	8:18	72	3:28
2011 June 15	20:13	100	3:38
2011 Dec. 10	14:33	50	3:32
2012 June 4	11:04	Partial	2:06

*Times are Greenwich Mean Time. Subtract 5 hours for Eastern Standard Time, 6 hours for Central Standard Time, 7 hours for Mountain Standard Time, and 8 hours for Pacific Standard Time. From your time zone, lunar eclipses that occur between sunset and sunrise will be visible, and those at midnight will be best placed.

■ Active Figure 3-6

The three quantities related by the small-angle formula. Angular diameter is given in seconds of arc in the formula. Distance and linear diameter must be expressed in the same units—both in meters, both in light-years, and so on.

AceAstronomy™ Log into AceAstronomy and select this chapter to see the Active Figure called "Small-Angle Formula." Notice that both distance and linear diameter affect an object's angular diameter.

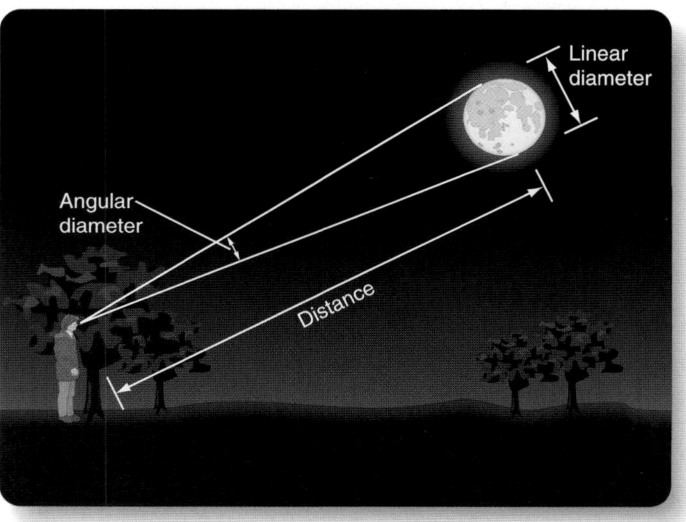

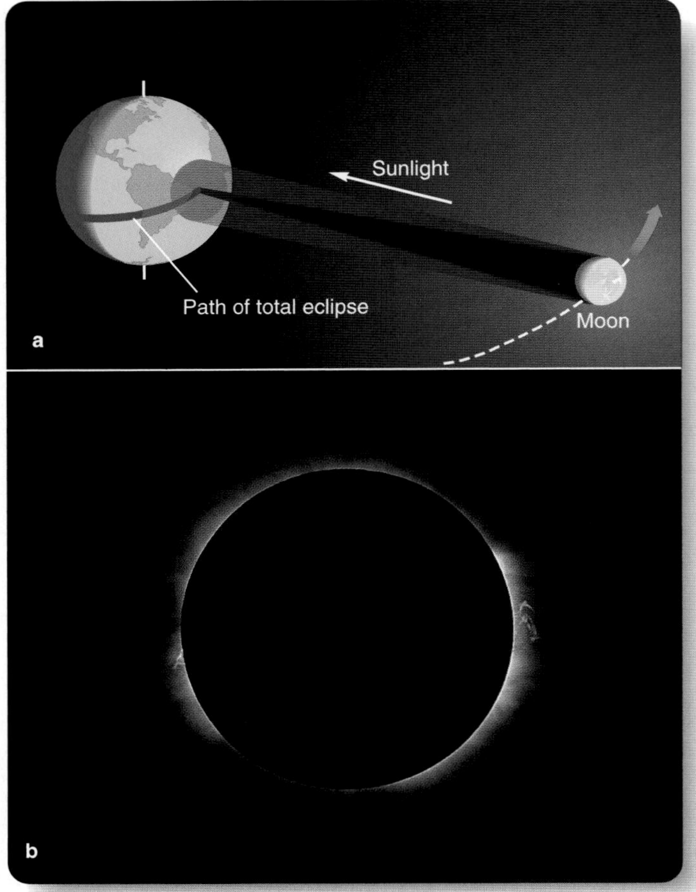
Sunlight
Path of total eclipse
Moon
a
b

■ Figure 3-7

(a) The umbral shadow of the moon sweeps over a narrow strip of Earth. (b) From a location inside the umbral shadow, you would see the moon cover the bright surface of the sun in a total solar eclipse. (Daniel Good)

A Total Solar Eclipse
The moon moving from the right just begins to cross in front of the sun.
The disk of the moon gradually covers the disk of the sun.
Sunlight begins to dim as more of the sun's disk is covered.
During totality, pink prominences are often visible.
A longer-exposure photograph during totality shows the fainter corona.

■ Figure 3-8

This sequence of photos shows the first half of a total solar eclipse. (Daniel Good)

you were located outside the umbra, in the penumbra, you would see part of the sun peeking around the edge of the moon and the eclipse would be partial. Of course, if you are outside the penumbra, you see no eclipse at all.

Because of the orbital motion of the moon and the rotation of Earth, the moon's shadow sweeps rapidly across Earth in a long, narrow path of totality. If you want to see a total solar eclipse, you must be in the path of totality. When the umbra of the moon's shadow sweeps over you, you see one of the most dramatic sights in the sky—the totally eclipsed sun.

The eclipse begins as the moon slowly crosses in front of the sun. It takes about an hour for the moon to cover the solar disk, but as the last sliver of sun disappears, dark falls in a few seconds. Automatic street lights come on, drivers of cars turn on their headlights, and birds go to roost. The sky becomes so dark you can even see the brighter stars.

The darkness lasts only a few minutes because the umbra is never more than 270 km (168 miles) in diameter and sweeps across Earth's surface at over 1600 km/hr (1000 mph). The sun cannot remain totally eclipsed for more than 7.5 minutes, and the average period of totality lasts only 2 or 3 minutes.

The brilliant surface of the sun is called the **photosphere.** When the moon covers the photosphere, you can see the fainter **chromosphere,** the higher layers of the sun's atmosphere, glowing a bright pink. Above the chromosphere you see the **corona,** the sun's outer atmosphere (■ Figure 3-8). The corona is a low-density, hot gas that glows with a pale white color. Streamers caused by the solar magnetic field streak the corona, as may be seen in the last frame of Figure 3-8. The chromosphere is often marked by eruptions on the solar surface called **prominences** (see Figure 3-7b). The corona, chromosphere, and prominences are visible only when the brilliant photosphere is covered. As soon as part of the photosphere reappears, the fainter corona, chromosphere, and prominences vanish in the glare, and totality is over. The moon moves on in its orbit, and in an hour the sun is completely visible again.

Just as totality begins or ends, a small part of the photosphere can peek out from behind the moon through a valley at the edge of the lunar disk. Although it is intensely bright, such a small part of the photosphere does not completely drown out the fainter corona, which forms a silvery ring of light with the brilliant spot of photosphere gleaming like a diamond (■ Figure 3-9). This **diamond-ring effect** is one of the most spectacular of astronomical sights, but it is not visible during every solar eclipse. Its occurrence depends on the exact orientation and motion of the moon.

Sometimes when the moon crosses in front of the sun, it is too small to fully cover the sun, and we see an **annular eclipse,** a solar eclipse in which a ring (or annulus) of the photosphere is visible around the disk of the moon (■ Figure 3-10). With a portion of the photosphere visible, the eclipse never becomes total; it never quite gets dark, and we can't see the prominences, chromosphere, and corona. Annular eclipses occur because the moon follows a slightly elliptical orbit around Earth, and thus its angular diameter can vary (■ Figure 3-11). When it is at **perigee,** its point of closest approach to Earth, it looks significantly larger than when it is at **apogee,** the most distant point in its orbit. Furthermore, Earth's orbit is slightly elliptical, so the Earth–sun distance varies slightly, and thus the diameter of the solar disk varies slightly. If the moon is in the farther part of its orbit during totality, its angular diameter will be less than the angular diameter of the sun, and thus we see an annular eclipse.

■ Figure 3-9

The diamond-ring effect can sometimes occur momentarily at the beginning or end of totality if a small segment of the photosphere peeks out through a valley at the edge of the lunar disk. (NOAO)

■ Figure 3-10

(a) The annular eclipse of 1994. A bright ring of photosphere remains visible around the moon, and the corona and prominences are not visible. (Daniel Good) (b) An annular eclipse occurs when the moon is in the farther part of its orbit and its umbral shadow does not reach Earth. From Earth, we see an annular eclipse because the moon's angular diameter is smaller than the angular diameter of the sun.

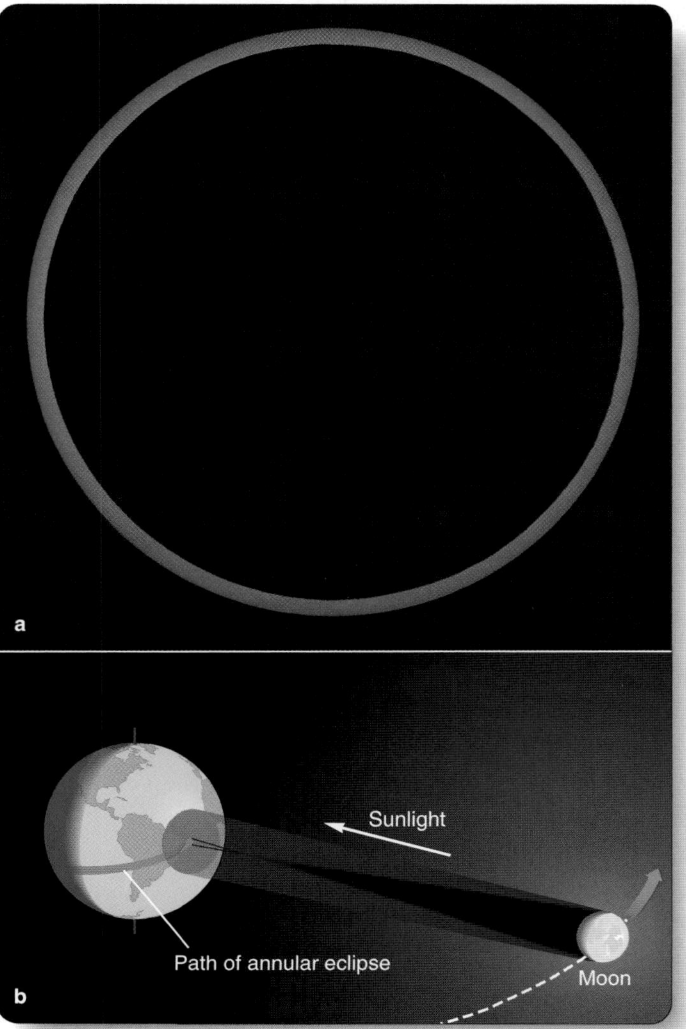

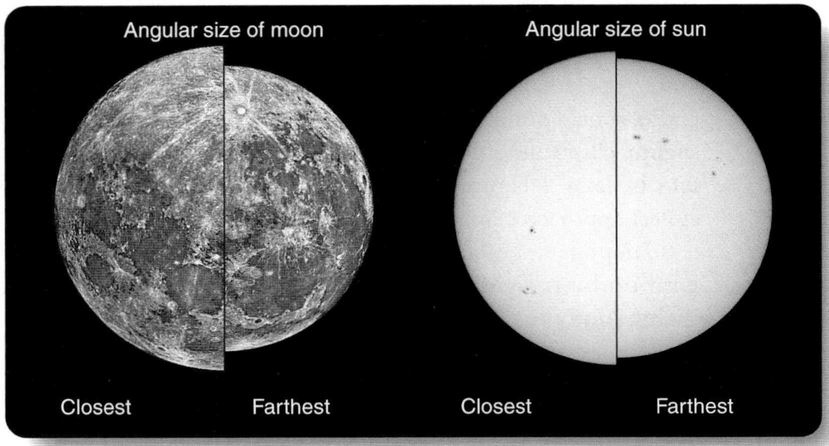

■ **Figure 3-11**

The angular diameter of the moon (left) varies dramatically because its orbit is elliptical and its distance varies from perigee (closest) to apogee (farthest). The angular diameter of the sun (right) varies by a smaller amount because Earth's orbit is slightly elliptical. (Moon: UCR/Lick Observatory; Sun: Daniel Good)

A list of future total and annular eclipses of the sun is given in ■ Table 3-2. If you plan to observe a solar eclipse, remember that the sun is bright enough to burn your eyes and cause permanent damage if you look at it directly. This is true whether there is an eclipse or not. Solar eclipses can tempt us to look at the sun in spite of its brilliance and thus can tempt us to risk our eyesight. During totality, the brilliant photosphere is hidden, and it is safe to look at the eclipse, but the partial phases can be dangerous. See ■ Figure 3-12 for a safe way to observe the partially eclipsed sun.

Predicting Eclipses

Predicting lunar or solar eclipses seems quite complex, and precise predictions require sophisticated calculations. But you can make general eclipse predictions by thinking about the geometry of an eclipse and the cyclic motions of the sun and moon.

You see a solar eclipse when the moon passes between Earth and the sun, that is, when the lunar phase is new moon. You see a lunar eclipse at full moon. However, you don't see solar eclipses at every new moon or lunar eclipses at every full moon. Why not? ■ Figure 3-13 is a drawing to scale of the umbral shadows of Earth and the moon. Notice that they are extremely long and narrow. Earth, moon, and sun must line up almost exactly, or the shadows miss their mark, and there is no eclipse.

■ **Table 3-2 | Total and Annular Eclipses of the Sun, 2005–2016**

Date	Total/Annular (T/A)	Time of Mideclipse* (GMT)	Maximum Length of Total or Annular Phase (Min:Sec)	Area of Visibility
2005 Apr. 8	AT	21^h	0:42	Pacific, N. of S. America
2005 Oct. 3	A	11^h	4:32	Atlantic, Spain, Africa
2006 Mar. 29	T	10^h	4:07	Atlantic, Africa, Turkey
2006 Sept. 22	A	12^h	7:09	N.E. of S. America, Atlantic
2008 Feb. 7	A	4^h	2:14	S. Pacific, Antarctica
2008 Aug. 1	T	10^h	2:28	Canada, Arctic, Siberia
2009 Jan. 26	A	8^h	7:56	S. Atlantic, Indian Ocean
2009 July 22	T	3^h	6:40	Asia, Pacific
2010 Jan. 15	A	7^h	11:10	Africa, Indian Ocean
2010 July 11	T	20^h	5:20	Pacific, S. America
2012 May 20	A	23^h	5:46	Japan, N. Pacific, W. US
2012 Nov. 13	T	22^h	4:02	Australia, S. Pacific
2013 May 10	A	0^h	6:04	Australia, Pacific
2013 Nov. 3	AT	13^h	1:40	Atlantic, Africa
2015 March 20	T	10^h	2:47	N. Atlantic, Arctic
2016 March 9	T	2^h	4:10	Borneo, Pacific
2016 Sept. 1	A	9^h	3:06	Atlantic, Africa, Indian Oc.

The next major total solar eclipse visible from the United States will occur on August 21, 2017.

*Times are Greenwich Mean Time. Subtract 5 hours for Eastern Standard Time, 6 hours for Central Standard Time, 7 hours for Mountain Standard Time, and 8 hours for Pacific Standard Time.

^{h}hours.

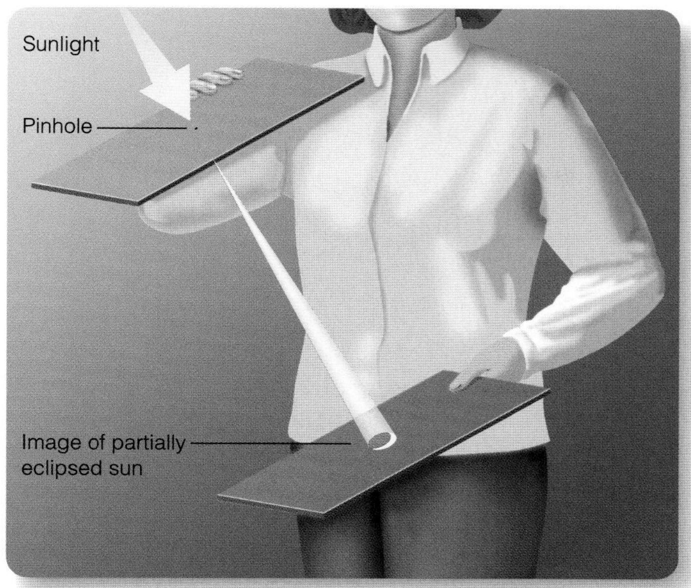

■ Figure 3-12

A safe way to view the partial phases of a solar eclipse. Use a pinhole in a card to project an image of the sun on a second card. The greater the distance between the cards, the larger (and fainter) the image will be.

To be eclipsed, the moon must enter Earth's shadow. However, because the moon's orbit is tipped relative to that of Earth, the full moon often misses the shadow, passing north or south of it, and there is no lunar eclipse.

To produce a solar eclipse, the moon's shadow must sweep over Earth. The inclination of the moon's orbit, however, is such that often, when the moon is new, its shadow passes north or south of Earth, and there is no solar eclipse.

For an eclipse to occur, the moon must reach full or new moon just as it passes through the plane of Earth's orbit; otherwise, the shadows miss. The points where it passes through the plane of Earth's orbit are called the **nodes** of the moon's orbit, and the line connecting these is called the *line of nodes.* In other words, the planes of the two orbits intersect along the line of nodes. Twice a year, this line of nodes points toward the sun, and for a few weeks eclipses are possible at new moons and full moons (■ Figure 3-14). These intervals when eclipses are possible are called eclipse seasons, and they occur about six months apart.

If the moon's orbit were fixed in space, the eclipse seasons would always occur at the same time each year. The moon's orbit precesses, however, because of the gravitational pull of the sun on the moon, and the precession slowly changes the direction of the line of nodes. The line turns westward, making one complete rotation in 18.61 years. As a result, the eclipse seasons occur about three weeks earlier each year. Many ancient peoples noticed this pattern and could guess which full and new moons were likely to produce eclipses.

Another way the ancients predicted eclipses was to notice that the pattern of eclipses repeats every 6585.3 days—the **Saros cycle.** After one Saros, the sun, moon, and nodes have circled the sky many times and finally returned to the same arrangement they occupied when the Saros began. Then the cycle of eclipses begins to repeat. One Saros equals 18 years $11\frac{1}{3}$ days. Because of the extra third of a day, an eclipse visible in North America will recur after one Saros, but it will be visible one-third of the way around the world in the North Pacific. Once ancient astronomers recognized the Saros cycle, they could predict eclipses from records of previous eclipses.

Inquire | Review | Analyze

What would solar eclipses be like if the moon's orbit were not tipped to the plane of Earth's orbit?

One way to analyze a complex process is to change one thing and imagine what would happen. In this case, you can review your understanding of real eclipses by trying to imagine a geometry that does not exist. If the moon's orbit were not inclined, then the moon would follow the ecliptic, and it would cross in front of the sun at every new moon. There would be a solar eclipse every month at new moon. Because the ecliptic is inclined 23.5° to the celestial equator, the sun can never be farther than 23.5° from the celestial equator. That means the moon's shadow would sweep over Earth from west to east somewhere in a band about 23.5° north or south of Earth's equator. The shadow would never sweep over anyone who lived too far north of Earth's equator. They would see the moon pass south of the sun at new moon, and they would never see a solar eclipse. If you lived too far south, the moon would pass north of the sun, and again, you would see no eclipse.

■ Figure 3-13

Umbral shadows of Earth and the moon. Notice how easy it is for the shadows to miss their mark at full moon and at new moon and fail to produce eclipses. That is, it is easy for the moon to reach full phase and not enter Earth's shadow. It is also easy for the moon to reach new phase and not cast its shadow on Earth. (The diameters of Earth and the moon are exaggerated by a factor of 2 for clarity.)

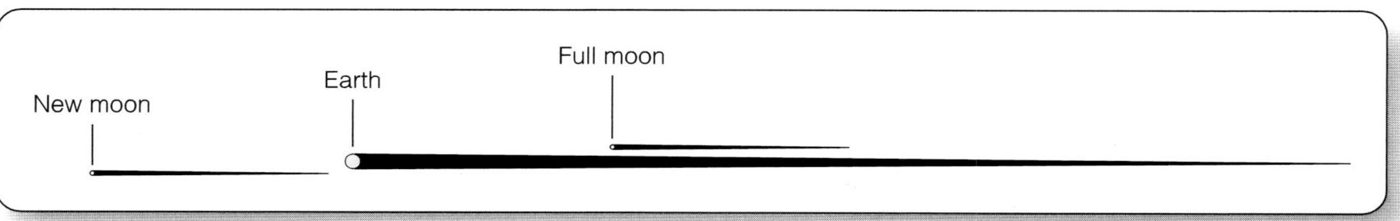

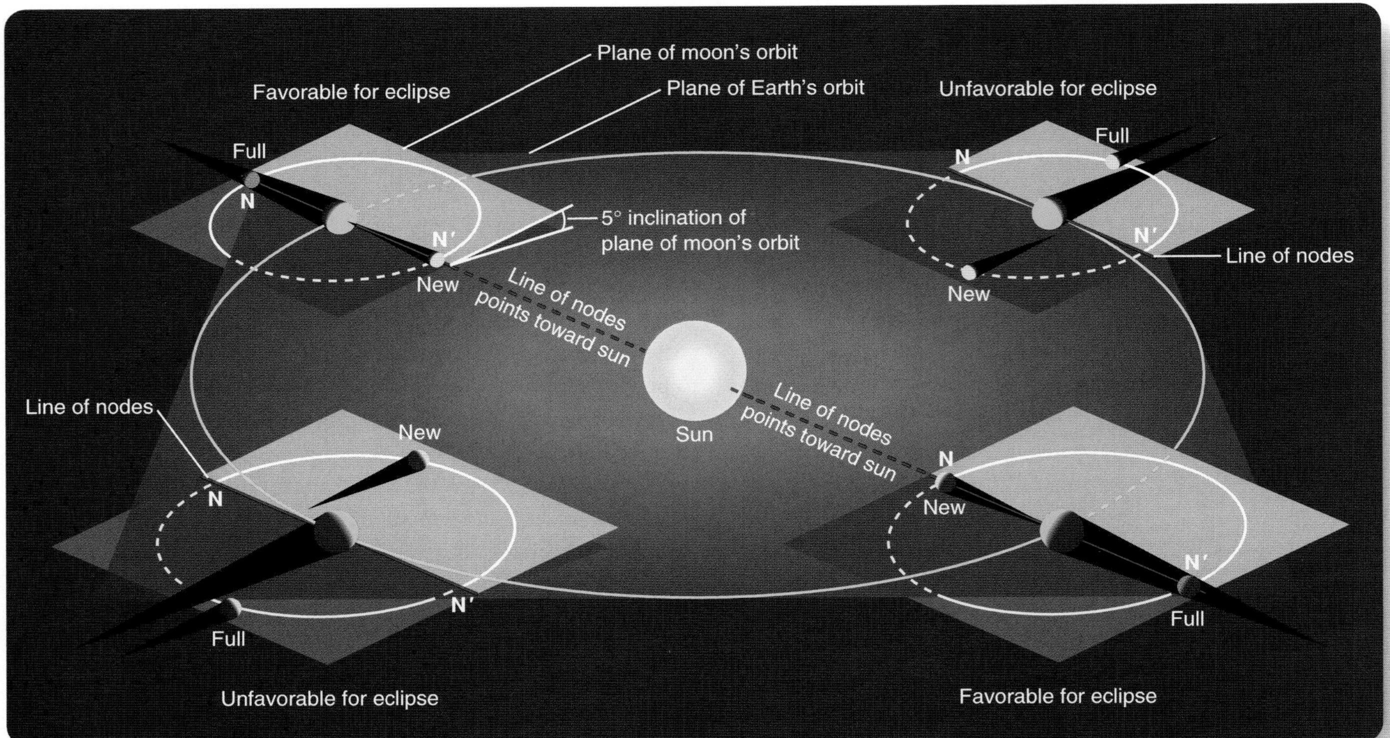

Figure 3-14

The moon's orbit is tipped about 5° to Earth's orbit. The nodes N and N′ are the points where the moon passes through the plane of Earth's orbit. If the line of nodes does not point at the sun, the shadows miss, and there are no eclipses at new moon and full moon. At those parts of Earth's orbit where the line of nodes points toward the sun, eclipses are possible at new moon and full moon.

Analyzing a problem with one parameter slightly altered is a good way to study a complicated process. You could repeat this analysis for lunar eclipses, but it will be more interesting to change a different factor. What would lunar eclipses look like if Earth had no atmosphere?

■ ■ ■

Connections: The elegant cycles of the sun and moon produce beautiful phenomena to decorate Earth's sky. Many cultures have romantic myths about the dance of the sun and moon. The Greek myth of the handsome Hippomenes racing for the hand of the beautiful Atalante is really about the sun racing the moon around the ecliptic.

The cycles in the sky are a rich part of our culture, but those same cycles may be affecting our planet in much more dramatic ways. They may be affecting the ice ages.

3-3 Astronomical Influences on Earth's Climate

WEATHER IS WHAT HAPPENS TODAY; climate is the average of what happens over decades and centuries. We know that Earth has gone through past episodes, called ice ages, when the worldwide climate was cooler and dryer and thick layers of ice covered northern latitudes. The earliest known ice age occurred about 570 million years ago, and the next about 280 million years ago. The most recent ice age began only about 3 million years ago and is still going on. We are living in one of the periodic episodes when the glaciers melt and Earth grows slightly warmer. The current warm period began about 12,000 years ago.

Ice ages seem to occur with a period of roughly 250 million years, and cycles of glaciation within ice ages occur with a period of about 40,000 years. Scientists now believe that these cyclic changes have an astronomical origin.

The Hypothesis

Sometimes a theory or hypothesis is proposed long before scientists can find the critical evidence to test it. That situation happened in 1920 when Yugoslavian meteorologist Milutin Milankovitch proposed what became known as the **Milankovitch hypothesis**—that small changes in Earth's orbit, in precession, and in inclination affect Earth's climate and trigger ice ages. We will examine each of these three motions in turn.

First, astronomers know that the shape of Earth's orbit varies slightly over a period of about 100,000 years. At present, Earth's orbit carries it 1.7 percent closer than average to the sun during northern hemisphere winters and 1.7 percent farther away in northern hemisphere summers. This makes the northern climate very slightly warmer, and that is critical—most of the land mass where ice can accumulate is in the northern hemisphere. If Earth's orbit became more elliptical, for example, northern summers might be too cool to melt all of the snow and ice from the previous winter. That would make glaciers grow larger.

A second factor is also at work. Precession causes Earth's axis to sweep around a cone with a period of about 26,000 years, and that changes the location of the seasons around Earth's orbit. Northern summers now occur when Earth is 1.7 percent farther from the sun, but in 13,000 years northern summers will occur on the other side of Earth's orbit where Earth is 1.7 percent closer to the sun. Northern summers will be warmer, which could melt all of the previous winter's snow and ice and prevent the growth of glaciers.

The third factor is the inclination of Earth's equator to its orbit. Currently at 23.5°, this angle varies from 22° to 24°, with a period of roughly 41,000 years. When the inclination is greater, seasons are more severe.

In 1920, Milankovitch proposed that these three factors cycle against each other to produce complex periodic variations in Earth's climate and the advance and retreat of glaciers (■ Figure 3-15a). But no evidence was available to test the theory in 1920, and scientists treated it with skepticism. Many thought it was laughable.

The Evidence

By the middle 1970s, Earth scientists could collect the data that Milankovitch had lacked. Oceanographers could drill deep into the seafloor and collect samples, and geologists could determine the age of the samples from the natural radioactive atoms they contained. From all this, scientists constructed a history of ocean temperatures that convincingly matched the predictions of the Milankovitch hypothesis (Figure 3-15b).

The evidence seemed very strong, and by the 1980s, the Milankovitch hypothesis was widely discussed as the leading hypoth-

■ Figure 3-15

(a) Mathematical models of the Milankovitch effect can be used to predict temperatures on Earth over time. Here, cool temperatures are represented by violet and blue, and warm temperatures by yellow and red. These globes show the warming that occurred beginning 25,000 years ago, which ended the last ice age. (Courtesy Arizona State University, Computer Science and Geography Departments) (b) Over the last 400,000 years, changes in ocean temperatures measured from fossils found in sediment layers from the seabed match calculated changes in solar heating. (Adapted from Cesare Emiliani)

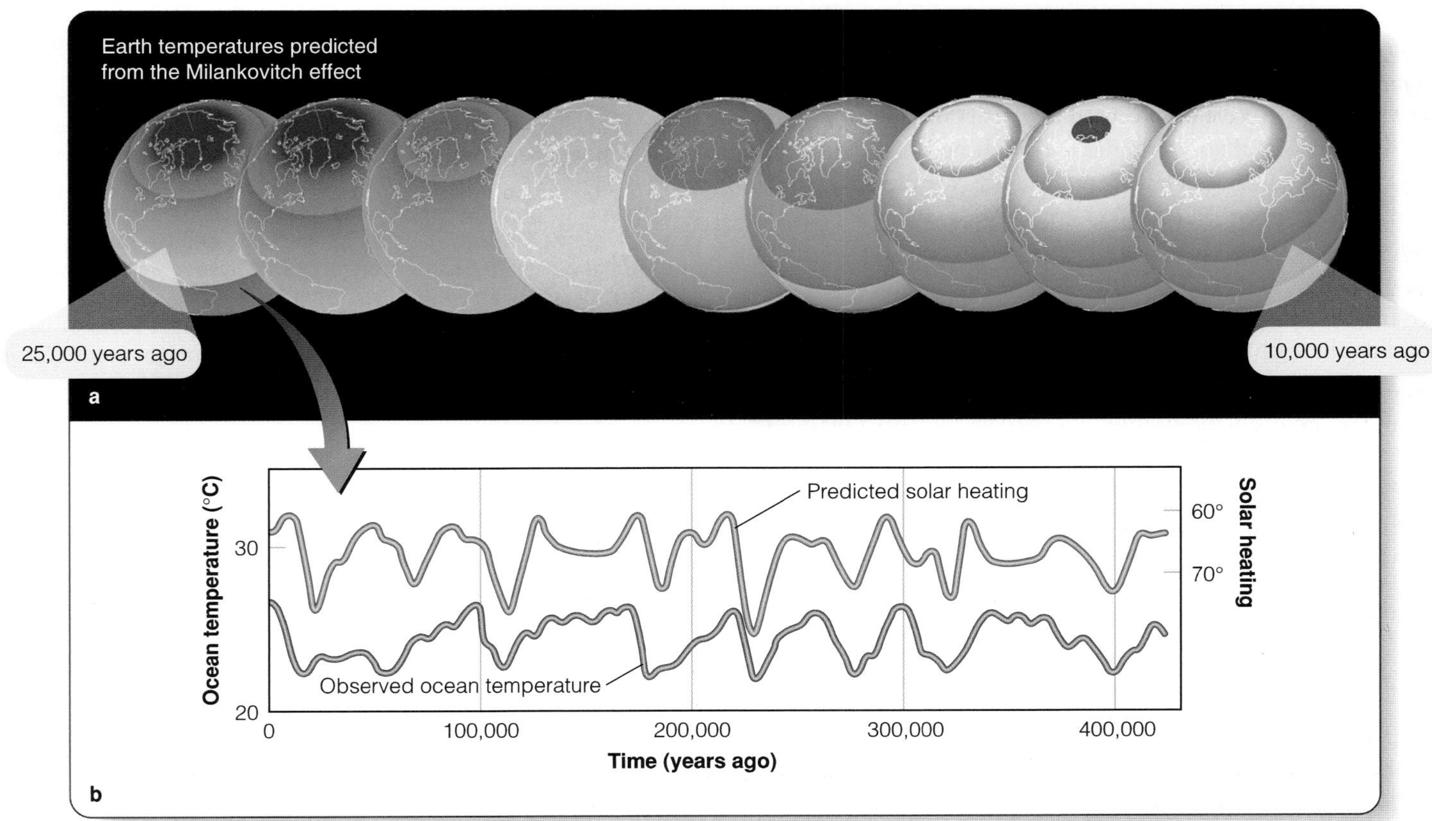

The Foundation of Science: Evidence

Science is based on evidence. Every theory and conclusion must be supported by evidence obtained from experiments or from observation. If a theory is supported by many pieces of evidence but is clearly contradicted by a single experiment or observation, scientists quickly abandon it. For a theory to be true, there can be no contradictory evidence.

Of course, scientists argue about the significance of particular evidence and often disagree on the interpretation of evidence. Some observations may seem significant at first glance, but upon closer examination you may find that the procedure was flawed, and so the piece of evidence is not important. Or you might conclude that the observational fact is correct but is being misinterpreted. The observation may not mean what it seems. Some of the most famous disagreements in science, such as those surrounding Galileo and Darwin, have arisen over the interpretation of well-established factual evidence.

Furthermore, scientists are not allowed to be selective in considering evidence. A lawyer in court can call a certain witness and intentionally fail to ask a critical question that would reveal evidence harmful to the lawyer's case. But a scientist may not ignore any known evidence. The difference in their methods is revealing. The lawyer is attempting to prove only one side of the case and rightly may ignore contradictory evidence. The scientist, however, is searching for the truth and so must test any theory against all available evidence. In a sense, the scientist, in dealing with evidence, must act as both the prosecution and the defense.

As you read about any science, look for the evidence in the form of measurements or observations. Every theory or conclusion should have supporting evidence. If you can find and understand the evidence, the science will make sense. All scientists, from astronomers to zoologists, demand evidence. You should, too.

esis. But science follows a mostly unstated set of rules that holds that a hypothesis must be tested over and over against all available evidence (**Window on Science 3-2**). In 1988, scientists discovered contradictory evidence.

While scuba diving in a water-filled crack in Nevada called Devil's Hole, scientists drilled out samples of calcite, a mineral that contains oxygen atoms. For 500,000 years, layers of calcite have built up in Devil's Hole, recording in their oxygen atoms the temperature of the atmosphere when rain fell there. Finding the ages of the mineral samples was difficult, but the results seemed to show that the previous ice age ended thousands of years too early to have been caused by Earth's motions.

These contradictory findings are irritating because we naturally prefer certainty, but such circumstances are common in science. The disagreement between ocean floor samples and Devil's Hole samples triggered a scramble to understand the problem. Were the ages of one or the other set of samples wrong? Were the ancient temperatures wrong? Or were scientists misunderstanding the significance of the evidence?

In 1997, a new study of the ages of the samples confirmed that those from the ocean floor are correctly dated. This seems to give scientists renewed confidence in the Milankovitch hypothesis. But the same study found that the ages of the Devil's Hole samples are also correct. Some now believe the temperatures at Devil's Hole tell us about local climate changes in the region that became the southwestern United States. The ocean floor samples seem to tell us about global climate. Other scientists argue that the data warn us that Earth's climate may depend on additional factors such as small variations in the sun's energy production or atmospheric dust produced by episodes of volcanism. Thus Milutin Milankovitch's hypothesis, first proposed in 1920, is still being tested as we try to understand the world we live on.

Inquire | Review | Analyze

How do precession and the shape of Earth's orbit interact to affect Earth's climate?

Here, as in an earlier section of this chapter, exaggeration is a useful analytical tool. If you exaggerate the variation in the shape of Earth's orbit, you can see dramatically the influence of precession. At present, Earth reaches perihelion during winter in the northern hemisphere and aphelion during summer. The variation in distance is only about 1.7 percent, and that difference doesn't cause much change in the severity of the seasons. But if Earth's orbit were much more elliptical, then winter in the northern hemisphere would be much warmer, and summer would be much cooler.

Now you can see the importance of precession. As Earth's axis precesses, it points gradually in different directions, and the seasons occur at different places in Earth's orbit. In 13,000 years, northern winter will occur at aphelion, and if Earth's orbit were highly elliptical, northern winter would be terrible. Similarly, summer would occur at perihelion, and the heat would be awful. Such extremes might deposit large amounts of ice in the winter but then melt it away in the hot summer, thus preventing the accumulation of glaciers.

Continue this analysis by exaggeration. What effect would precession have if Earth's orbit were more circular?

■ ■ ■

Connections: In this chapter, you have studied the cycles of the sun and moon and in passing have seen Earth spinning on its axis and whirling around the sun. Knowledge of such movements is only a few centuries old. The next chapter will tell the story of how the people of Earth discovered that they live on a moving planet.

Summary

3-1 | The Cycle of the Sun

What causes the seasons?

- Because Earth orbits the sun, the sun appears to move eastward along the ecliptic through the constellations. It circles the entire zodiac in a year.
- Because the ecliptic is tipped 23.5° to the celestial equator, the sun spends half the year in the northern celestial hemisphere and half in the southern celestial hemisphere.
- In the summer, the sun is above the horizon longer and shines more directly down on the ground. Both effects cause warmer weather.

3-2 | The Cycles of the Moon

Why does the moon go through phases?

- Because you see the moon by reflected sunlight, its shape appears to change as it orbits Earth.
- The lunar phases wax from new moon to first quarter to full moon and wane from full moon to third quarter to new moon.
- A complete cycle of lunar phases takes 29.53 days.

What is a lunar eclipse?

- If a full moon passes through Earth's shadow, sunlight is cut off, and the moon darkens in a lunar eclipse. If the moon only grazes the shadow, the eclipse is partial or penumbral and not total.
- The totally eclipsed moon looks copper-red because of sunlight scattered through Earth's atmosphere.

What is a solar eclipse?

- A total solar eclipse occurs if a new moon passes directly between the sun and Earth and the moon's shadow sweeps over Earth's surface. Observers outside the path of totality see a partial eclipse if the penumbra sweeps over their location.
- During a total eclipse, the bright photosphere of the sun is covered, and the fainter corona, chromosphere, and prominences become visible.
- If the moon is in the farther part of its orbit during a solar eclipse, it does not look large enough to cover the photosphere, and you see only an annular eclipse.

3-3 | Astronomical Influences on Earth's Climate

How do cycles of the sky affect Earth's climate?

- Changes in the shape of Earth's orbit, in its precession, and in its axial tilt can alter the planet's heat balance and seem to be at least partly responsible for the ice ages and glacial periods.

New Terms

rotation (p. 24)
revolution (p. 24)
ecliptic (p. 26)
zodiac (p. 27)
vernal equinox (p. 28)
summer solstice (p. 28)
autumnal equinox (p. 28)
winter solstice (p. 28)
perihelion (p. 29)
aphelion (p. 29)
evening star (p. 30)
morning star (p. 30)
lunar eclipse (p. 31)
umbra (p. 31)
penumbra (p. 31)
sidereal period (p. 33)
synodic period (p. 33)
solar eclipse (p. 35)
photosphere (p. 36)
chromosphere (p. 36)
corona (p. 37)
prominence (p. 37)
diamond-ring effect (p. 37)
annular eclipse (p. 37)
perigee (p. 37)
apogee (p. 37)
node (p. 39)
Saros cycle (p. 39)
Milankovitch hypothesis (p. 40)

Review Questions

Ace Astronomy™ Assess your understanding of this chapter's topics with additional quizzing and animations at **http://astronomy.brookscole.com/sh9e**

1. What is the difference between the daily and annual motions of the sun?
2. If Earth did not rotate, could we still define the ecliptic? Why or why not?
3. What would our seasons be like if Earth were tipped 35° instead of 23.5°? What would they be like if Earth's axis were perpendicular to its orbit?
4. Why are the seasons reversed in the southern hemisphere?
5. Do the phases of the moon look the same from every place on Earth, or is the moon full at different times as seen from different locations?
6. What phase would Earth be in if you were on the moon when the moon was full? at first quarter? at waning crescent?
7. Why have most people seen a total lunar eclipse, while few have seen a total solar eclipse?
8. Why isn't there an eclipse at every new moon and at every full moon?
9. Why is the moon red during a total lunar eclipse?
10. Why should the eccentricity of Earth's orbit make winter in the northern hemisphere different from winter in the southern hemisphere?
11. How could small changes in the inclination of Earth's axis affect world climate?
12. The stamp at right shows a crescent moon. Explain why the moon could never look this way.
13. The photo at right shows the annular eclipse of May 30, 1984. How is it different from the annular eclipse shown in Figure 3-10?

(Laurence Marschall)

Discussion Questions

1. Do planets orbiting other stars have ecliptics? Could they have seasons?
2. Why would it be difficult to see prominences if you were on the moon during a total lunar eclipse?

Problems

1. If Earth is about 4.6 billion (4.6×10^9) years old, how many precessional cycles have occurred?
2. Identify the phases of the moon if on March 21 the moon were located at (a) the vernal equinox, (b) the autumnal equinox, (c) the summer solstice, (d) the winter solstice.

3. Identify the phases of the moon if at sunset the moon were (a) near the eastern horizon, (b) high in the south, (c) in the southeast, (d) in the southwest.
4. About how many days must elapse between first-quarter moon and third-quarter moon?
5. Draw a diagram showing Earth, the moon, and shadows during (a) a total solar eclipse, (b) a total lunar eclipse, (c) a partial lunar eclipse, (d) an annular eclipse.
6. Phobos, one of the moons of Mars, is 20 km in diameter and orbits 5982 km above the surface of the planet. What is the angular diameter of Phobos as seen from Mars? (*Hint:* See Reasoning with Numbers 3-1.)
7. A total eclipse of the sun was visible from Canada on July 10, 1972. When did this eclipse occur next? From what part of Earth was it total?
8. When will the eclipse described in Problem 7 next be total as seen from Canada?

Media Cluster

ACTIVE FIGURES

Ace Astronomy™ To access the resources in the Media Cluster, log into AceAstronomy at **http://astronomy.brookscole.com/sh9e** and select Chapter 3.

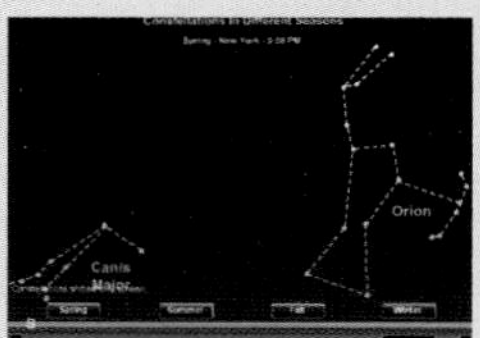

Constellations in Different Seasons
This animation lets you observe how the constellations change with the seasons, as Earth moves through its orbit of the sun.

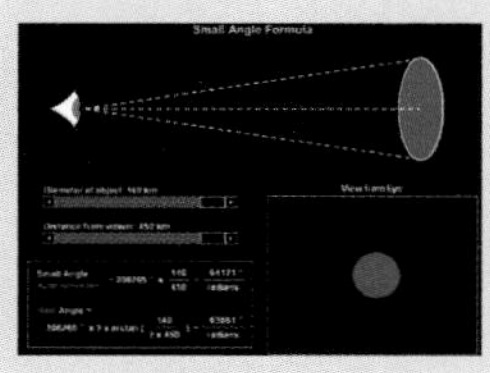

Small-Angle Formula
The small-angle formula animation illustrates how the size and distance of an object affect its angular size and how well this simple approximation works.

ASTRONOMY EXERCISES

Sunrise through the Seasons
In this animation, you can observe how the position of the sunrise changes throughout the year.

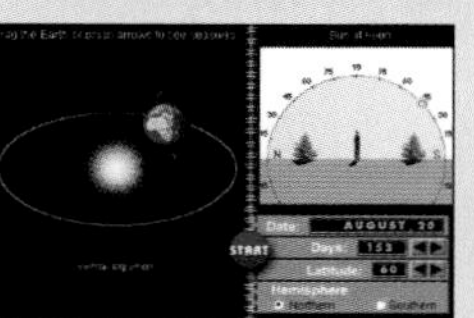

The Seasons
Earlier in this chapter, you saw how the angle of the sun in the sky brings about warmer days in the summer and colder days in the winter for the northern hemisphere. In this animation you can select the latitude and hemisphere to compare the angle of the sun for various points around the world.

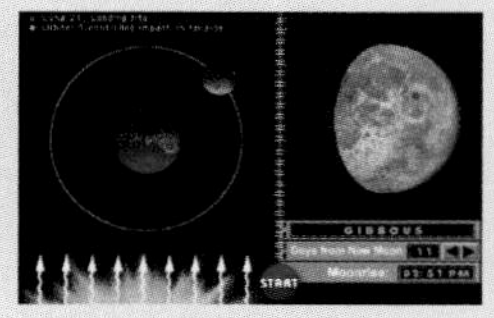

Phases of the Moon
Sunlight always illuminates half of the moon. Because we see different amounts of this sunlit side, we see the moon cycle through phases. Explore this relationship further with this interactive exercise.

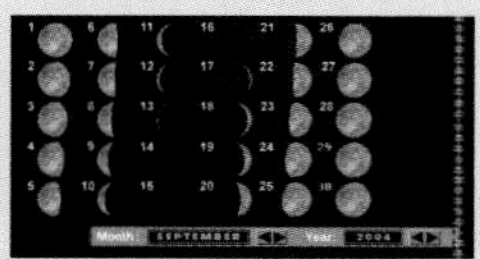

Moon Calendar
This exercise lets you select any month or year in the past, present, or future and see the moon phases for that month.

VIRTUAL ASTRONOMY LAB

Lab 6: Tides and Tidal Forces
This lab investigates the origin of tidal forces and compares the tidal forces on Earth caused by the moon and sun. It also investigates the role played by tidal forces in the formation of astronomical bodies by gravitational accretion.

Critical Inquiries for the Web

1. There are many websites on astrology, but nearly all accept the old superstition as real. Can you find websites that analyze astrology logically? Begin at the website for the Committee for the Scientific Investigation of Claims of the Paranormal.
2. Find photographs of recent lunar and solar eclipses. Where did the observers have to go to see the eclipse?
3. What is the latest news concerning the Milankovitch hypothesis? Is there new evidence to test the theory?

Exploring *TheSky*

1. How long is daylight on the day of the summer solstice and on the day of the winter solstice for your location? Use **Site Information** under the **Data** menu to set your location and the date. Find the time of sunrise and sunset.
2. How many days must elapse between new moons? Use the **Moon Phase Calendar** under the **Tools** menu to locate new moons.
3. Where would you have to go to see a total solar eclipse in the next year or two? Use the **Eclipse Finder** under the **Tools** menu to select a total solar eclipse. Check **Show Path of Totality** to see a map of Earth.
4. View the total solar eclipse you located in Activity 3. In the **Eclipse Finder,** click **View** to see the sun and moon from your present location. Then use **Site Information** under the **Data** menu to change your location to a spot in the path of totality. Set the **Time Step** to 5 minutes and click the **arrows** to move the moon across the sun.
5. Locate and observe an annular and a total lunar eclipse.

Go to the Brooks/Cole Astronomy Resource Center **(http://astronomy.brookscole.com)** for critical thinking exercises, articles, and additional readings from InfoTrac College Edition, Brooks/Cole's online student library.

4 The Origin of Modern Astronomy

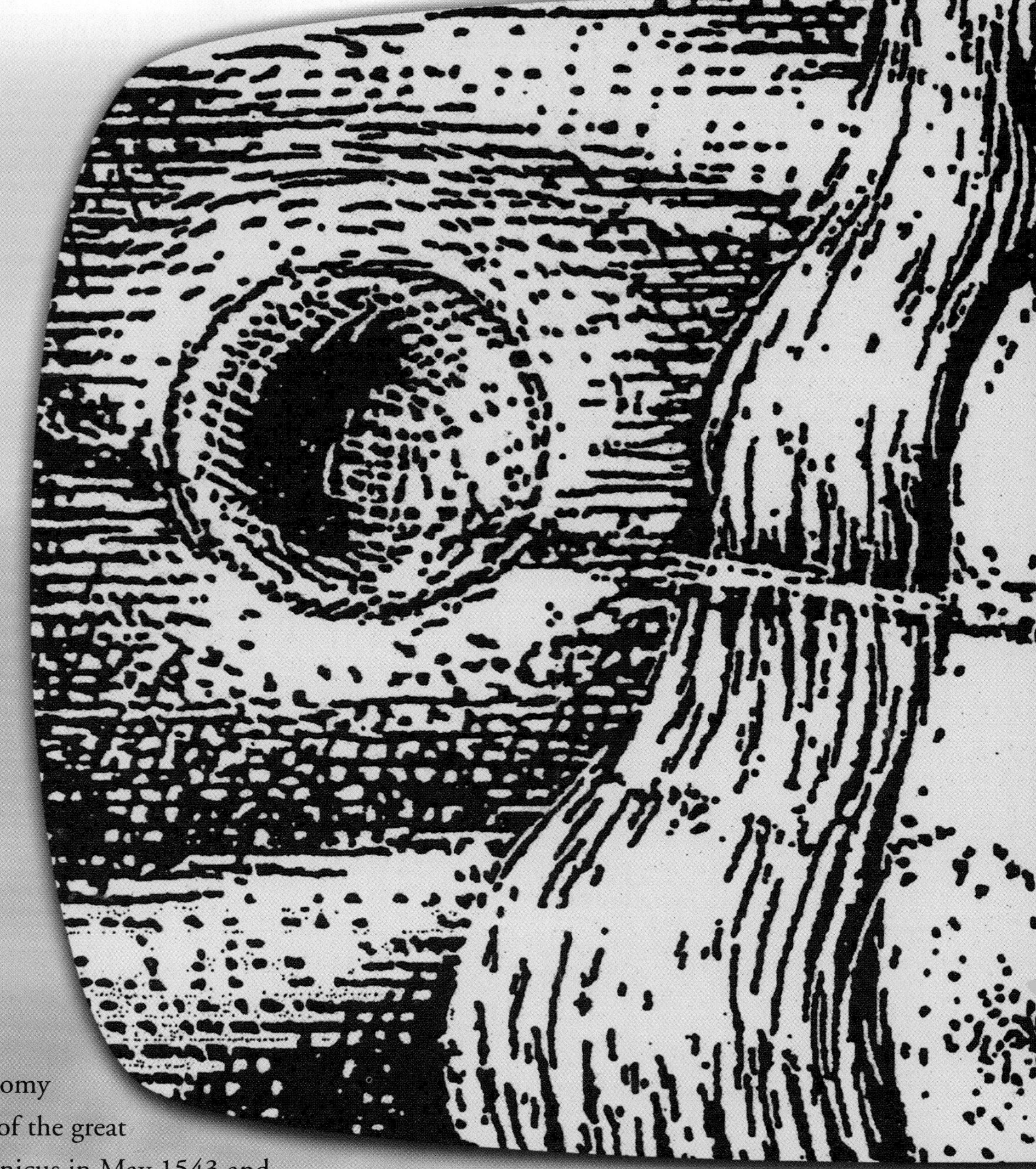

The passions of astronomy are no less profound because they are not noisy.

JOHN STEINBECK
The Short Reign of Pippin IV

THE STORY OF modern astronomy begins with an ending—the death of the great Polish astronomer Nicolaus Copernicus in May 1543 and the almost simultaneous publication of his model of the universe. That model revolutionized not only astronomy but all science, and it inspired a new consideration of humanity's place in nature. This chapter will trace this story over the 99 years from 1543 and Copernicus's death to 1642, the year that saw both the death of another great astronomer, Galileo, and the birth of one of the greatest scientists in history, Isaac Newton. ▌ Two themes run through this story: Earth's place in the cosmos and the character of planetary motion. The debate over Earth's place involved theological questions and eventually brought Galileo Galilei before the Inquisition. The understanding of planetary

Aristotle, Copernicus, Galileo, and many other astronomers tried to understand the place of the Earth in the universe but could never fully succeed until Isaac Newton understood gravity and orbital motion.

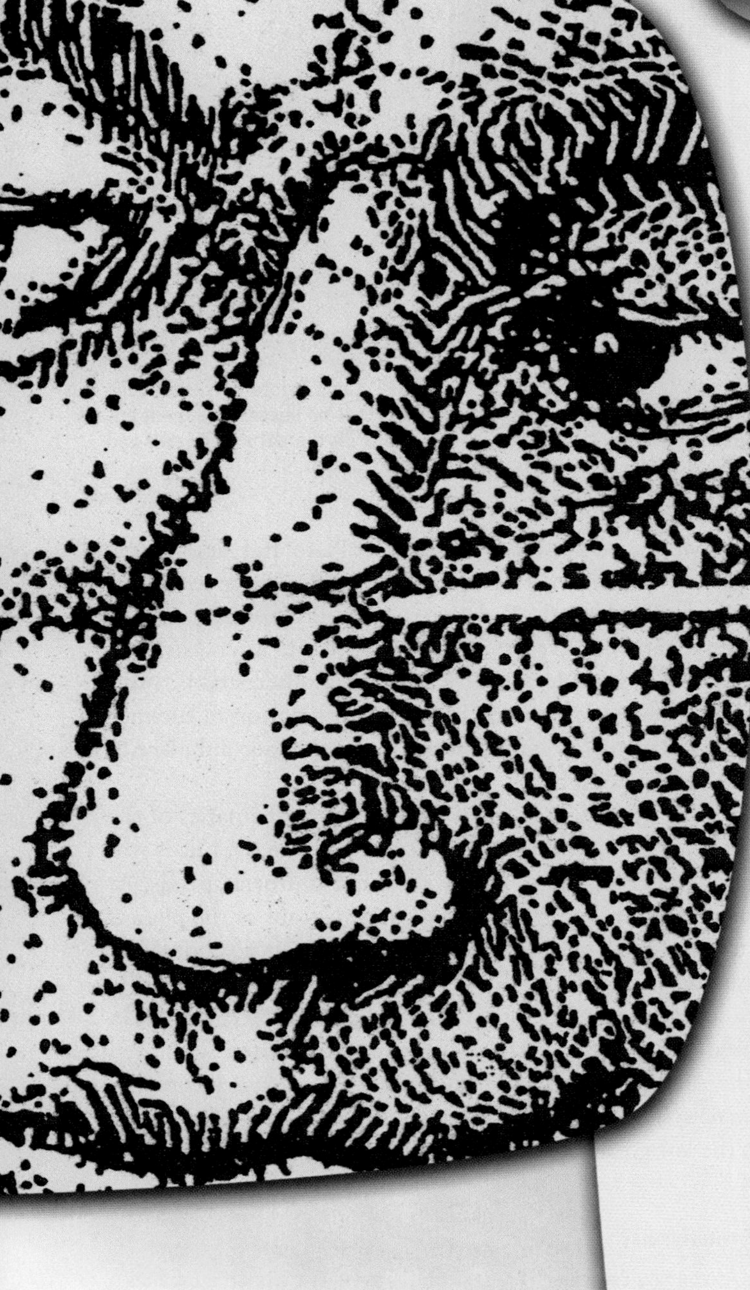

Guidepost

Looking Back

The previous chapters gave you a modern view of the heavens. You are now familiar with what you see in the sky, and you also can imagine how the Earth, moon, and sun move through space. Now you are ready to understand one of the most sweeping revolutions in human thought: the realization that we live on a planet.

This Chapter

By the 16th century, many astronomers were uncomfortable with the ancient theory that Earth sat at the center of a spherical universe. In this chapter, you will discover how an astronomer named Copernicus changed the ancient theory, how Galileo Galilei changed the rules of debate, and how Isaac Newton changed humanity's concept of nature. Here you will find answers to five essential questions:

How did the ancients describe Earth's place in the universe?

How did Copernicus revise that ancient theory?

Why was the Copernican model gradually accepted?

Why was Galileo condemned by the Inquisition?

How did Isaac Newton change humanity's view of nature?

This chapter is not just about the history of astronomy. As they struggled to understand Earth and the heavens, the astronomers of the Renaissance invented a new way of understanding nature—a way of thinking that is now called science.

Looking Ahead

This is a book of science, and every chapter that follows will use the methods that were invented when Copernicus tried to repair that ancient theory that Earth was the center of the universe. As you think about stars, galaxies, and planets through the rest of this book, you will be using the tools that Galileo, Newton, and many others struggled to devise.

Ace Astronomy™ The AceAstronomy icon throughout the text indicates an opportunity for you to test yourself on key concepts and to explore animations and interactions on the AceAstronomy website at: http://astronomy.brookscole.com/sh9e

motion involved the most basic physics of motion and eventually led Isaac Newton to understand gravitation and motion. It all started with the Copernican model of the universe and led to the birth of modern science.

To understand why the Copernican model was so important, we must backtrack to ancient Greece and meet the two great authorities of ancient astronomy, Aristotle and Ptolemy. These Greek astronomers struggled to understand Earth's place and the nature of planetary motion.

4-1 Pre-Copernican Astronomy

THE HISTORY OF ASTRONOMY EXTENDS back over millennia to ancient Babylon and beyond, but any discussion of ancient astronomy is dominated by a brilliant man, his teacher, and a follower of his principles who lived five centuries later.

The Aristotelian Universe

Philosophers of the ancient world attempted to deduce the true structure of the universe by reasoning from first principles. A first principle was something that was obviously true. This may strike you as peculiar; modern thinkers tend to observe how things work and make deductions from that evidence. However, reasoning from evidence, what you might call scientific thinking, had not been invented, and, in fact, the story of this chapter is the story of the invention of science as a new way to know about the physical world.

Study **The Ancient Universe** on pages 50 and 51 and notice three important ideas. First, ancient philosophers and astronomers accepted as a first principle that Earth was located at the center of the universe. Although a few writers mentioned the possibility that Earth might move, they did so mostly to point out the obvious objections.

Second, ancient philosophers and astronomers accepted without question that the heavens were perfect and that Earth, at the center, was imperfect. The lowest of the celestial spheres, the sphere of the moon, was the dividing line; everything below the sphere of the moon was imperfect and changeable, and everything above was perfect and unchanging. The perfection of the heavens was a first principle of ancient astronomy. The Greek philosopher Plato (427?–347 BC) argued that the sphere and circle were the only perfect geometrical figures, so all motion in the perfect heavens must be caused by the uniform rotation of spheres carrying objects around in circles.

Plato's student Aristotle (■ Figure 4-1) lived in Greece from 384 to 322 BC. He adopted many of his teacher's ideas. He became so famous that he was known throughout the Middle Ages as "The Philosopher," and the geocentric universe of nested spheres that he devised dominated astronomy.

The third critical idea involves a mathematical model of the universe. Although Claudius Ptolemy was a follower of Aristotle's ideas, he weakened the first principles of Plato and Aristotle by moving Earth slightly off-center and adjusting the speed of the planets using a device called the equant. Ptolemy, a brilliant mathematician who lived roughly five centuries after Aristotle, believed in the Aristotelian universe, but was interested in a different problem: creating a mathematical description of the motion of the planets. For him, first principles took second place to mathematical precision.

Aristotle's universe, as embodied in the mathematics of the Ptolemaic universe, dominated ancient astronomy, but it was wrong. The planets don't follow circles at uniform speeds. At first the Ptolemaic system predicted the positions of the planets with fair accuracy, but as centuries passed, errors accumulated, and Islamic and later European astronomers had to update the system, computing new constants and adjusting the epicycles. In the middle of the 13th century, a team of astronomers supported by King Alfonso X of Castile worked for 10 years analyzing the Ptolemaic system and publishing the result as the *Alfonsine Tables.* It was the last great adjustment of the Ptolemaic system.

■ Figure 4-1

Aristotle (384–322 BC), honored on this Greek stamp, wrote on such a wide variety of subjects and with such deep insight that he became the great authority on all matters of learning. His opinions on the nature of Earth and the sky were widely accepted for almost two millennia.

AceAstronomy™ Log into AceAstronomy and select this chapter to see Astronomy Exercise "Parallax" and experiment with the parallax of a spacecraft.

Inquire | Review | Analyze

Why did classical astronomers conclude the heavens were made up of spheres?

Classical astronomers did not use evidence and hypotheses the way modern scientists do. Rather they argued from first principles, and Plato argued that the perfect geometrical figure was a sphere. Then the heavens, which everyone agreed were perfect, must be made up of spheres. The natural motion of a sphere is rotation, and the only perfect motion is uniform motion, so the heavenly spheres were thought to move in uniform circular motion. Thus, classical philosophers argued that the daily motion of the heavens around Earth and the motions of the seven planets (the sun and moon were counted as planets) against the background of the stars had to be produced by the combination of uniformly rotating spheres carrying objects around in perfect circles.

Although ancient astronomers didn't use evidence as modern scientists do, they did observe the world around them. What observations led them to conclude that Earth didn't move?

■ ■ ■

Connections: Pre-Copernican astronomy consists of the elaboration of Aristotle's philosophy of the universe as expressed by Ptolemy in a mathematical model. For nearly 1500 years, astronomers believed in this geocentric universe. Earth was finally set in motion by a Polish church official with a talent for mathematics.

4-2 Copernicus

NICOLAUS COPERNICUS WAS BORN IN 1473 (■ Figure 4-2) in what is now Poland. At the time of his birth—and throughout his life—astronomy was based on the Ptolemaic system, even though, in spite of many revisions, it was still a poor predictor of planetary positions. Yet because of the authority of Aristotle, it was the officially accepted theory of the universe.

The Copernican Model

According to the Aristotelian universe, the most perfect region was in the heavens, and the most imperfect at Earth's center. Thus the classical geocentric universe matched the commonly held Christian geometry of heaven and hell, and anyone who criticized the geometry of the Aristotelian universe was challenging belief in heaven and hell and thus risking at least criticism and perhaps a charge of heresy.

Throughout his life, Copernicus was associated with the Church. His uncle, by whom he was raised and educated, was an important bishop in Poland; and, after studying canon law and medicine in some of the major universities in Europe, Copernicus became a canon of the Church at the unusually young age of 24. He served as secretary and personal physician to his powerful uncle for 15 years. When his uncle died, Copernicus went to live in quarters adjoining the cathedral in Frauenburg. Because of this long association with the Church and his fear of persecution, he hesitated to publish his revolutionary ideas.

What was this astonishing idea? Copernicus believed that the sun and not Earth was the center of the universe. That idea was shocking because it challenged the traditional belief in Earth's place at the center of all creation. Notice that the U.S. stamp in Figure 4-2 shows Copernicus holding a model of a sun-centered or **heliocentric universe.**

In fact, his idea was already being discussed long before he published it. His interest in astronomy had begun during his college days, and he apparently doubted the Ptolemaic system even then. Sometime before 1514, he wrote a short pamphlet that discussed the motion of the sky and outlined his hypothesis that the sun, not Earth, was the center of the universe and that Earth rotated on its axis and revolved around the sun. He distributed his summary in handwritten form without a title and, in some cases, anonymously. Perhaps his caution grew from a sense of modesty, but he may also have feared criticism for challenging the Ptolemaic system and the place of Earth in the universe.

■ Figure 4-2

Nicolaus Copernicus (1473–1543) pursued a lifetime career in the Church, but he was also a talented mathematician and astronomer. His work triggered a revolution in human thought. These stamps were issued in 1973 to mark the 500th anniversary of his birth.

The Ancient Universe

For 2000 years, the minds of astronomers were shackled by a pair of ideas. The Greek philosopher Plato argued that the heavens were perfect. Because the only perfect geometrical shape is a circle and the only perfect motion is uniform motion, Plato concluded that all motion in the heavens must be made up of combinations of circles turning at uniform rates. This was called uniform circular motion.

Plato's student Aristotle argued that Earth was imperfect and lay at the center of the universe. That is, he argued for a geocentric universe. He devised a model universe with 55 spheres turning at different rates and at different angles to carry the seven known planets (the moon, Mercury, Venus, the sun, Mars, Jupiter, and Saturn) across the sky.

Aristotle was known as the greatest philosopher in the ancient world, and his authority, lasting for 2000 years, chained the minds of astronomers and forced them to expect the universe to be geocentric and to move in uniform circular motion.

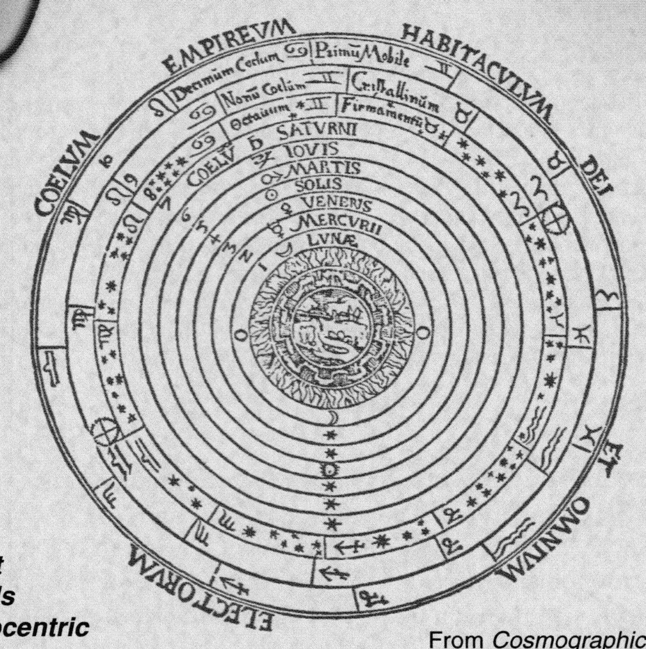

From *Cosmographica* by Peter Apian (1539).

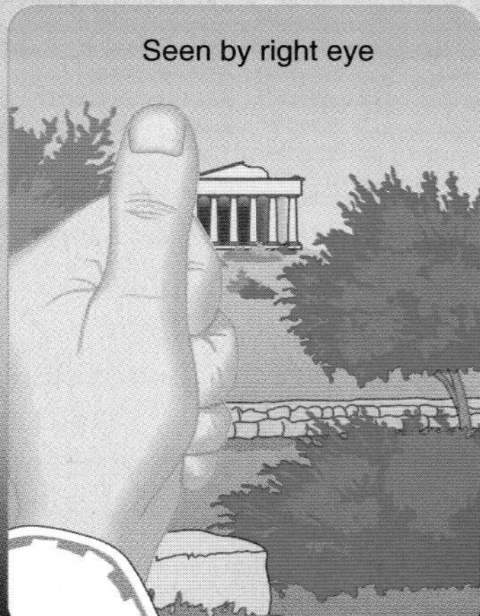

Ancient astronomers believed that Earth did not move because they saw no **parallax,** the apparent motion of an object because of the motion of the observer. To demonstrate parallax, close one eye and cover a distant object with your thumb held at arm's length. Switch eyes, and your thumb appears to shift position. If Earth moves, ancient astronomers reasoned, you should see the sky from different locations at different times of the year, and you should see parallax distorting the shapes of the constellations. They saw no parallax, so they concluded Earth could not move. Actually, the parallax of the stars is too small to see with the unaided eye.

Planetary motion was a big problem for ancient astronomers. In fact, the word *planet* comes from the Greek word for "wanderer," referring to the eastward motion of the planets against the background of the fixed stars. The planets did not, however, move at a constant rate, and they could occasionally stop and move westward for a few months before resuming their eastward motion. This backward motion is called **retrograde motion.**

East

Ecliptic

Oct. 3, 2005

Dec. 12, 2005

Taurus

West

Orion

Position of Mars at 5-day intervals

Every 2.14 years, Mars passes through a retrograde loop. Shown here is its path through Orion and Taurus during the fall of 2005.

Simple uniform circular motion centered on Earth could not explain retrograde motion, so ancient astronomers combined uniformly rotating circles much like gears in a machine to try to reproduce the motion of the planets.

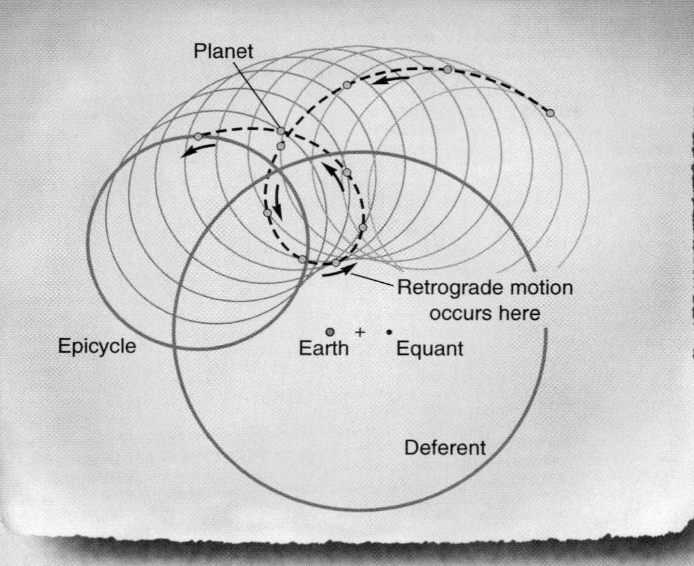

Uniformly rotating circles were key elements of ancient astronomy. Claudius Ptolemy created a mathematical model of the Aristotelian universe in which the planet followed a small circle called the **epicycle** that slid around a larger circle called the **deferent.** By adjusting the size and rate of rotation of the circles, he could approximate the retrograde motion of a planet.

To adjust the speed of the planet, Ptolemy supposed that Earth was slightly off center and that the center of the epicycle moved such that it appeared to move at a constant rate as seen from the point called the **equant.**

To further adjust his model, Ptolemy added small epicycles riding on top of larger epicycles (not shown here), producing a highly complex model.

Ace Astronomy™

Log into AceAstronomy and select this chapter to see Active Figure "Epicycles." Notice how the counterclockwise rotation of the epicycle produces retrograde motion.

Ptolemy's great book *Mathematical Syntaxis* (c. AD 140) contained the details of his model. Islamic astronomers preserved and studied the book through the Middle Ages, and they called it *Al Magisti* (The Greatest). When the book was found and translated from Arabic to Latin in the 12th century, it became known as *Almagest*.

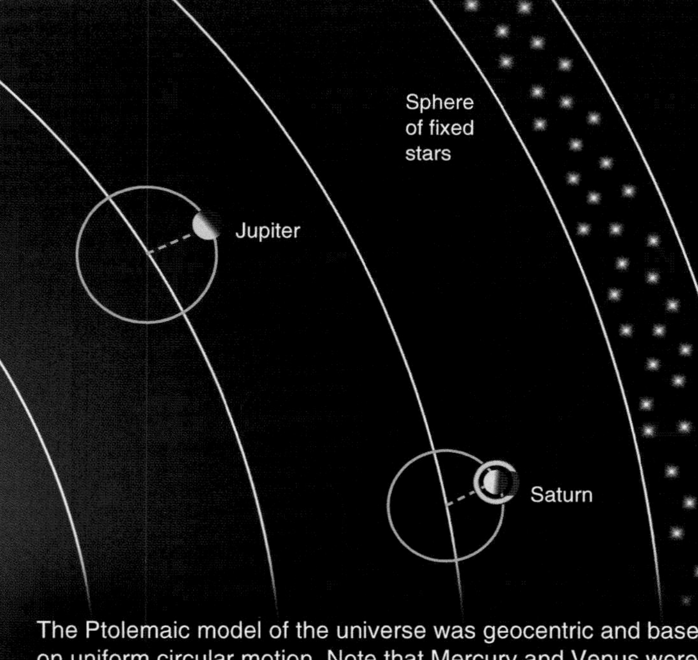

The Ptolemaic model of the universe was geocentric and based on uniform circular motion. Note that Mercury and Venus were treated differently from the rest of the planets. The centers of the epicycles of Mercury and Venus had to remain on the Earth–Sun line as the sun circled Earth through the year.

Equants and smaller epicycles are not shown here. Some versions contained nearly 100 epicycles as generations of astronomers tried to fine-tune the model to better reproduce the motion of the planets.

Notice that this modern illustration shows rings around Saturn and sunlight illuminating the globes of the planets, features that could not be known before the invention of the telescope.

De Revolutionibus

Copernicus worked on his book *De Revolutionibus Orbium Coelestium* over a period of many years and was essentially finished by about 1530, yet he hesitated to publish it, even though other astronomers knew of his theories. Church officials concerned about the reform of the calendar sought his advice and looked forward to the publication of his book.

One reason he hesitated was that the idea of a heliocentric universe was highly controversial. This was a time of rebellion in the Church—Martin Luther was speaking harshly about fundamental Church teachings, and others, both scholars and scoundrels, were questioning the authority of the Church. Even matters as abstract as astronomy could stir controversy. Remember, too, that Earth's place in astronomical theory was linked to the geometry of heaven and hell, so moving Earth from its central place was a controversial and perhaps heretical idea. Another reason Copernicus may have hesitated to publish was that his work was incomplete. His model could not accurately predict planetary positions. Copernicus was clearly concerned about how his ideas would be received, but in 1540 he allowed the visiting astronomer Joachim Rheticus (1514–1576) to publish an account of the Copernican universe in Rheticus's book *Prima Narratio (First Narrative)*. In 1542, Copernicus sent the manuscript for *De Revolutionibus* off to be printed. He died in the spring of 1543 before the printing was completed.

The most important idea in the book was placing the sun at the center of the universe. That single innovation had an astonishing consequence—the retrograde motion of the planets was immediately explained in a straightforward way without the large epicycles that Ptolemy used. In the Copernican system, Earth moves faster along its orbit than the planets that lie farther from the sun. Consequently, Earth periodically overtakes and passes these planets. Imagine that you are in a race car, driving rapidly along the inside lane of a circular race track. As you pass slower cars driving in the outer lanes, they fall behind, and if you did not know you were moving, it would seem that the cars in the outer lanes occasionally slowed to a stop and then backed up for a short interval. The same thing happens as Earth passes a planet such as Mars. Although Mars moves steadily along its orbit, as seen from Earth it appears to slow to a stop and move westward (retrograde) as Earth passes it (■ Figure 4-3). Because the planetary orbits do not lie in precisely the same plane, a planet does not resume its eastward motion in precisely the same path it followed earlier. Consequently, it describes a loop whose shape depends on the angle between the orbital planes.

Copernicus could explain retrograde motion without epicycles, and that was impressive. The Copernican system was elegant and simple compared with the whirling epicycles and off-center equants of the Ptolemaic system. However, *De Revolutionibus* failed to disprove the geocentric model for one critical reason—the Copernican theory could not predict the positions of the planets any more accurately than the Ptolemaic system could. To understand why it failed this critical test, you must understand Copernicus and his world.

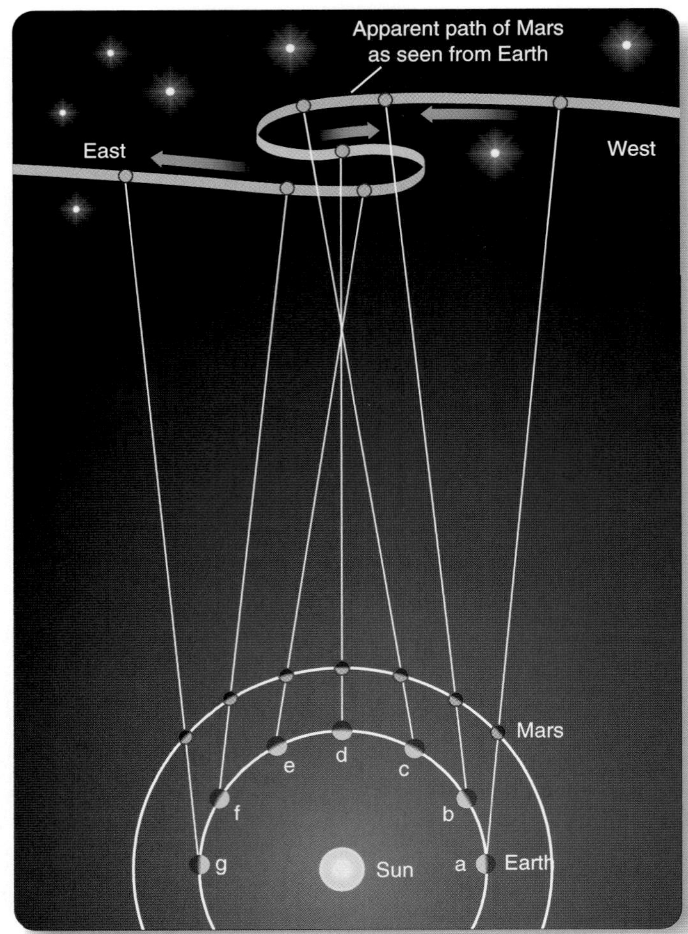

■ Figure 4-3

The Copernican explanation of retrograde motion. As Earth overtakes Mars (a–c), Mars appears to slow its eastward motion. As Earth passes Mars (d), Mars appears to move westward. As Earth draws ahead of Mars (e–g), Mars resumes its eastward motion against the background stars. The positions of Earth and Mars are shown at equal intervals of 1 month.

Although Copernicus proposed a revolutionary idea in making the planetary system heliocentric, he was a classical astronomer with tremendous respect for the old concept of uniform circular motion. In fact, Copernicus objected strongly to Ptolemy's use of the equant. It seemed arbitrary to Copernicus, a direct violation of the elegance of Aristotle's philosophy of the heavens. Copernicus called equants "monstrous" in that they violated both geocentrism and uniform circular motion. In devising his model, Copernicus returned to a strong belief in uniform circular motion. Although he did not need epicycles to explain retrograde motion, Copernicus discovered that the sun, moon, and planets suffered other smaller variations in their motions that he could not explain with uniform circular motion centered on the sun. Today astronomers recognize those variations as typical of objects following elliptical orbits, but Copernicus held firmly with uniform circular motion, so he had to introduce small epicycles to reproduce these minor variations in the motions of the sun, moon, and planets.

Because Copernicus imposed uniform circular motion on his model, it could not accurately predict the motions of the planets. The *Prutenic Tables* (1551) were based on the Copernican model, and they were not significantly more accurate than the 13th century *Alfonsine Tables* that were based on Ptolemy's model. Both could be in error by as much as 2°, which is four times the angular diameter of the full moon.

The Copernican *model* is inaccurate. It includes uniform circular motion and thus does not precisely describe the motions of the planets. But the Copernican *hypothesis* that the universe is heliocentric was correct, given how little astronomers of the time knew of other stars and galaxies. The planets circle the sun, not Earth, so their universe was heliocentric. Why that hypothesis gradually won acceptance in spite of the inaccuracy of the epicycles and deferents is a question historians still debate. There are probably a number of reasons, including the revolutionary temper of the times, but the most important factor may be the elegance of the idea. Placing the sun at the center of the universe produced a symmetry among the motions of the planets that is pleasing to the eye as well as to the intellect (■ Figure 4-4). In the Ptolemaic model, Mercury and Venus were treated differ-

■ Figure 4-4

(a) The Copernican universe as reproduced in *De Revolutionibus*. Earth and all the known planets revolve in separate circular orbits about the sun (Sol) at the center. The outermost sphere carries the immobile stars of the celestial sphere. Notice the orbit of the moon around Earth (Terra). (Courtesy Yerkes Observatory) (b) The model is elegant not only in the arrangement of the planets but also in their motions. Orbital velocities (blue arrows) decrease from that of Mercury, the fastest, to that of Saturn, the slowest. Compare the elegance of this model with the complexity of the Ptolemaic model as shown on page 51.

NICOLAI COPERNICI

net, in quo terram cum orbe lunari tanquam epicyclo contineri diximus. Quinto loco Venus nono menſe reducitur. Sextum denique locum Mercurius tenet, octuaginta dierum ſpacio circū currens. In medio uero omnium reſidet Sol. Quis enim in hoc

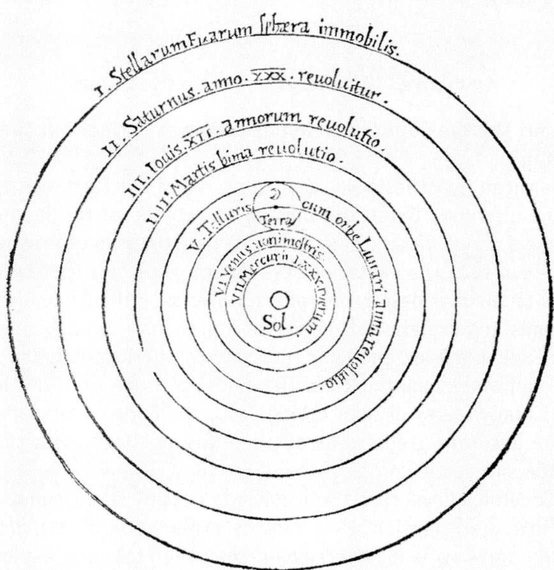

pulcherrimo templo lampadem hanc in alio uel meliori loco poneret, quàm unde totum ſimul poſsit illuminare? Siquidem non inepte quidam lucernam mundi, alij mentem, alij rectorem uocant. Trimegiſtus uiſibilem Deum, Sophoclis Electra intuentē omnia. Ita profecto tanquam in ſolio regali Sol reſidens circum agentem gubernat Aſtrorum familiam. Tellus quoque minime fraudatur lunari miniſterio, ſed ut Ariſtoteles de animalibus ait, maximam Luna cum terra cognationē habet. Cōcipit interea à Sole terra, & impregnatur annno partu. Inuenimus igitur ſub hac

a

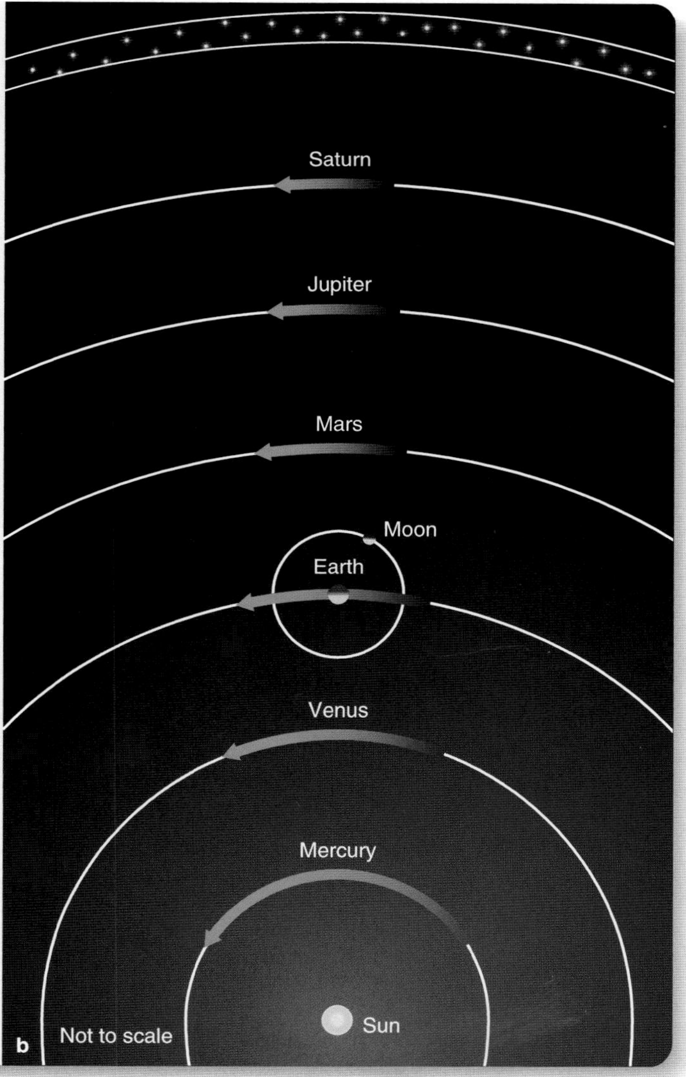

Window on Science 4-1

Creating New Ways to See Nature: Scientific Revolutions

The Copernican revolution is often cited as the perfect example of a scientific revolution. Over a few decades, astronomers abandoned a way of thinking about the universe that was almost 2000 years old and adopted a new set of ideas and assumptions. The American philosopher of science Thomas Kuhn has referred to a commonly accepted set of scientific ideas and assumptions as a scientific **paradigm.** Thus, the pre-Copernican astronomers had a geocentric paradigm that included uniform circular motion and the perfection of the heavens. That paradigm survived for many centuries until a new generation of astronomers was able to overthrow the old paradigm and establish a new paradigm that included heliocentrism and the motion of Earth.

A scientific paradigm is powerful because it shapes perceptions. It determines what are judged to be important questions and what is judged to be significant evidence. Thus, it is often difficult to recognize how paradigms limit what you can understand. For example, the geocentric paradigm contained problems that seem obvious to a modern mind, but because astronomers before Copernicus lived and worked inside that paradigm, they had difficulty seeing the problems. Overthrowing an outdated paradigm is not easy because you must learn to see nature in an entirely new way. Galileo, Kepler, and Newton saw nature from a new paradigm that would have been almost incomprehensible to astronomers of earlier centuries.

You can find examples of scientific revolutions in many fields, including biology, geology, genetics, and psychology. These scientific revolutions have been difficult and controversial because they have involved the overthrow of accepted paradigms. But that is why scientific revolutions are exciting. They give you not just a new idea or a new theory, but an entirely new insight into how nature works—a new way of seeing the world.

The ancients believed the stars were attached to a starry sphere. (NOAO and Nigel Sharp)

ently from the rest of the planets; their epicycles had to remain centered on the Earth–sun line. In the Copernican model, all of the planets were treated the same. They all followed orbits that circled the sun at the center.

The most astonishing consequence of the Copernican hypothesis was not what it said about the sun but what it said about Earth. By placing the sun at the center, Copernicus made Earth move along an orbit like the other planets. By making Earth a planet, Copernicus revolutionized humanity's view of its place in the universe and triggered a controversy that would eventually bring the astronomer Galileo Galilei before the Inquisition, a controversy over the nature of scientific truth that continues even today.

Although astronomers throughout Europe read and admired *De Revolutionibus,* they did not usually accept the Copernican hypothesis. The mathematics was elegant, and the astronomical observations and calculations were of tremendous value; but few astronomers believed, at first, that the sun actually was the center of the planetary system and that Earth moved. How the Copernican hypothesis was gradually recognized as correct has been called the Copernican revolution, because it was not just the adoption of a new idea but a total change in the way astronomers think about the place of the Earth (**Window on Science 4-1**).

Inquire | Review | Analyze

Why do we say the Copernican hypothesis was correct but the model was wrong?

The Copernican hypothesis was that the sun and not Earth was the center of the universe. Given the limited knowledge of the Renaissance astronomers about distant stars and galaxies, that hypothesis was correct. The sun is at the center of our planetary system. That hypothesis produces an elegant description of the universe without large epicycles and equants and explains retrograde motion in a simple way. The Copernican model, however, included not only the heliocentric hypothesis but also uniform circular motion. The model is wrong because the planets don't really follow circular orbits. Thus, the Copernican model could not make accurate predictions of planetary position even though it placed the sun at the center of the universe.

The Copernican hypothesis won converts because it is elegant and can explain retrograde motion. How does its explanation of retrograde motion work, and how is it more elegant than the Ptolemaic explanation?

■ ■ ■

Connections: The central issue in the Copernican hypothesis is the place of the Earth, but that was linked to a second issue, planetary motion. Thus, the story of the Copernican revolution shifts from Copernicus, a mathematician and theorist, to one of the first great observational astronomers, a man who measured the daily positions of the sun, moon, and planets.

4-3 Tycho Brahe

THE GREAT OBSERVATIONAL ASTRONOMER of this 99-year story was a Danish nobleman, Tycho Brahe (■ Figure 4-5), born December 14, 1546, only three years after the publication of *De Revolutionibus.* Tycho was not a churchman, like Copernicus, but rather a nobleman from an important Danish family. He was well known for his vanity and lordly manners.

The New Star

During his university days, Tycho was officially studying law, but he made it clear to his family that his real passion was mathematics and astronomy. Yet his life was shaped by two events outside his studies—a duel and a new star in the sky.

While at the university, Tycho fought a duel over some supposed insult and received a wound that disfigured his nose. For the rest of his life, he wore false noses made of gold and silver and stuck on with wax (see Figure 4-5). He was known as a proud and haughty nobleman, and the injury did little to improve his disposition.

Early in his university days, Tycho showed an interest in astronomy and began measuring the positions of the planets in the sky. In 1563, Jupiter and Saturn passed very near each other in the sky, nearly merging into a single point on the night of August 24. Tycho found that the *Alfonsine Tables* were a full month in error and that the *Prutenic Tables* were in error by a number of days.

■ Figure 4-5

Tycho Brahe (1546–1601) was, during his lifetime, the most famous astronomer in the world. Proud of his noble rank, he wears the elephant medal awarded him by the king of Denmark. His artificial nose is suggested in this engraving.

Then in 1572 a "new star" (now called Tycho's supernova) appeared in the sky, shining more brightly than Venus. Tycho carefully measured its position time after time through the night and found that it displayed no parallax. To understand the significance of this observation, note that Tycho believed the heavens rotated westward around Earth. The new star, according to classical astronomy, represented a change in the heavens and therefore had to lie below the sphere of the moon. In that case, Tycho reasoned, the new star would appear slightly too far east as it rose and slightly too far west as it set. That is parallax. Tycho saw no parallax in the position of the new star, so he concluded that it must lie above the sphere of the moon and was probably on the starry sphere itself. This contradicted Aristotle's belief that the starry sphere was perfect and unchanging.

No one before Tycho could have made this discovery because no one had ever measured the positions of stars so accurately. Tycho had great confidence in the precision of his measurements, so when he failed to detect parallax for the new star, he knew it was important evidence against the Ptolemaic theory. He announced his discovery in a small book, *De Stella Nova (The New Star),* published in 1573.

The book attracted the attention of astronomers throughout Europe, and soon Tycho was summoned to the court of the Danish King Frederik II and offered funds to build an observatory on the island of Hveen just off the Danish coast. Tycho also received a steady source of income as landlord of a coastal district from which he collected rents. (He was not a popular landlord.) On Hveen, Tycho constructed a luxurious home with six towers especially equipped for astronomy and populated it with servants, assistants, and a dwarf to act as jester. Soon Hveen was an international center of astronomical study.

Tycho Brahe's Legacy

Tycho made no direct contribution to astronomical theory. Because he could measure no parallax for the stars, he concluded that Earth had to be stationary, thus rejecting the Copernican hypothesis. However, he also rejected the Ptolemaic model because of its inaccurate predictions. Instead, he devised a complex model in which Earth was the immobile center of the universe around which the sun and moon moved. The other planets circled the sun. This model thus incorporated part of the Copernican model, but Earth—rather than the sun—was stationary. In this way, Tycho preserved the central, immobile Earth described in Scripture (■ Figure 4-6). Although Tycho's model was very popular at first, the Copernican model replaced it within a century.

The true value of Tycho's work was observational. Because he was able to devise new and better instruments, he was able to make highly accurate observations of the positions of the stars, sun, moon, and planets. Tycho had no telescopes—they were not invented until the next century—so his observations were made by the naked eye peering along sights. All of his instruments were

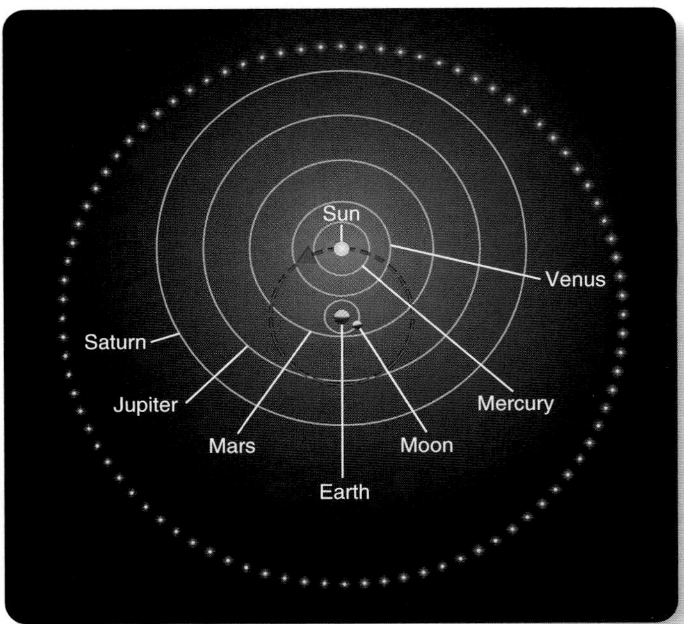

Figure 4-6

Tycho Brahe's model of the universe retained the first principles of classical astronomy; it was geocentric and the heavenly bodies moved on circular paths. The moon and sun circled Earth, but the rest of the planets circled the sun.

Figure 4-7

Much of Tycho's success was due to his skill in designing large, accurate instruments such as this one, known as the mural quadrant. In this engraving, the figure of Tycho at the center, his dog, and the scene in the background are a mural painted on the wall within the arc of the quadrant. The observer at the extreme right (also Tycho) peers through a sight out the loophole in the wall at the upper left and thus measures a celestial body's angular distance above the horizon. (The Granger Collection, New York)

designed to measure angles in the sky. For example, his quadrant could measure angles up to 90° above the horizon (■ Figure 4-7). By designing and building large instruments with great care, he was able to measure angles to high precision. He measured the positions of 777 stars to better than 4 minutes of arc and regularly measured the positions of the sun, moon, and planets for the 20 years he stayed on Hveen.

Unhappily for Tycho, King Frederik II died in 1588, and his young son took the throne. Suddenly Tycho's temper, vanity, and noble presumptions threw him out of favor. In 1596, taking most of his instruments and books of observations, he went to Prague, the capital of Bohemia, and became imperial mathematician to the Holy Roman Emperor Rudolph II. His goal was to revise the *Alfonsine Tables* and publish the revision as a monument to his new patron. It would be called the *Rudolphine Tables.*

Tycho did not intend to base the *Rudolphine Tables* on the Ptolemaic system but rather on his own Tychonic system, proving once and for all the validity of his hypothesis. To assist him, he hired a few mathematicians and astronomers, including one Johannes Kepler. Then in November 1601, Tycho collapsed at a nobleman's home. Before he died, 11 days later, he asked Rudolph II to make Kepler imperial mathematician. Thus the newcomer, Kepler, became Tycho's replacement (though at one-sixth Tycho's salary).

Inquire | Review | Analyze

Why did Tycho conclude that the new star of 1572 contradicted the Ptolemaic universe?

According to Aristotle, the heavens were perfect and therefore unchanging. Any change in the heavens had to occur below the moon, which was attached to the lowest of the crystalline spheres. When Tycho saw the new star of 1572 in the sky, he could detect no parallax. If the star lay below the moon's sphere, then it should appear to be east of its average position when it was rising in the east, and it should appear west of its average position when it was setting in the west. That is, it should show parallax. Tycho could measure no parallax at all, which meant the new star had to be at least above the moon's sphere and was probably part of the starry sphere itself. Thus, the new star was a change in the supposedly unchanging heavenly realm.

Tycho didn't believe in the Ptolemaic universe, and he produced his own model to explain planetary motion. In what ways did his model include ideas from classical astronomy and from the Copernican model?

■ ■ ■

Connections: Tycho Brahe tried to resolve the problem of the place of the Earth by assuming the universe was geocentric, but his attempts to understand planetary motion were cut short by his death. The newly appointed imperial mathematician, Johannes Kepler, used Tycho's observations to finally solve the second great problem of the age, the motion of the planets.

4-4 Johannes Kepler

An Astronomer of Humble Origins

NO ONE COULD HAVE BEEN MORE DIFFERENT from Tycho Brahe than Johannes Kepler (■ Figure 4-8). He was born on December 27, 1571, to a poor family in a region now included in southwest Germany. His father was unreliable and shiftless, principally employed as a mercenary soldier fighting for whoever paid enough. He finally failed to return from a military expedition, either because he was killed or because he found circumstances more to his liking elsewhere. Kepler's mother was apparently an unpleasant and unpopular woman. She was accused of witchcraft in later years, and Kepler had to defend her in a trial that dragged on for three years. She was finally acquitted but died the following year.

Kepler was the oldest of six children, and his childhood was no doubt unhappy. Not only was the family poor and often lacking a father, but Kepler was never healthy, even as a child, so it is surprising that he did well in school, winning promotion to a Latin school and eventually a scholarship to the university at Tübingen, where he studied to become a Lutheran pastor.

During his last year of study, Kepler accepted a job in Graz teaching mathematics and astronomy, a job he resented because he knew little about the subjects. Evidently he was not a good teacher either—he had few students his first year, and none at all his second. His superiors put him to work teaching a few introductory courses and preparing an annual almanac that contained astronomical, astrological, and weather predictions. Through good luck, in 1595 some of his weather predictions were fulfilled, and he gained a reputation as an astrologer and seer. Even in later life he earned money from his almanacs.

While still a college student, Kepler had become a believer in the Copernican hypothesis, and at Graz he used his extensive spare time to study astronomy. By 1596, the same year Tycho arrived in Prague, Kepler was ready to solve the mystery of the universe. That year he published a book called *The Forerunner of Dissertations on the Universe, Containing the Mystery of the Universe.* The book, like nearly all scientific works of that age, was written in Latin and is now known as *Mysterium Cosmographicum.*

By modern standards, the book contains almost nothing of value. It begins with a long appreciation of Copernicanism and then goes on to speculate on the reasons for the spacing of the planetary orbits. Kepler felt he had found the underlying architecture of the universe in the five regular solids*—the cube, tetrahedron, dodecahedron, icosahedron, and octahedron. Because these five solids were the only regular solids, he supposed they were the "spacers" between the planetary orbits (■ Figure 4-9).

*A regular solid is a three-dimensional body, each of whose faces is the same. A cube is a regular solid, each of whose faces is square.

■ Figure 4-8

Johannes Kepler (1571–1630) derived three laws of planetary motion from Tycho Brahe's observations of the positions of the planets. This Romanian stamp commemorates the 400th anniversary of Kepler's birth. Ironically, it contains an astronomical error—the horns of the crescent moon should be inclined upward from the horizon.

■ Figure 4-9

In this diagram based on one drawn by Kepler, the sizes of the celestial spheres carrying the outer three planets, Saturn, Jupiter, and Mars, are determined by spacers (blue) consisting of a cube and a tetrahedron. Inside the sphere of Mars, the remaining regular solids (not shown here) separate the spheres of Earth, Venus, and Mercury. The sun lay at the very center of this Copernican universe based on geometrical spacers.

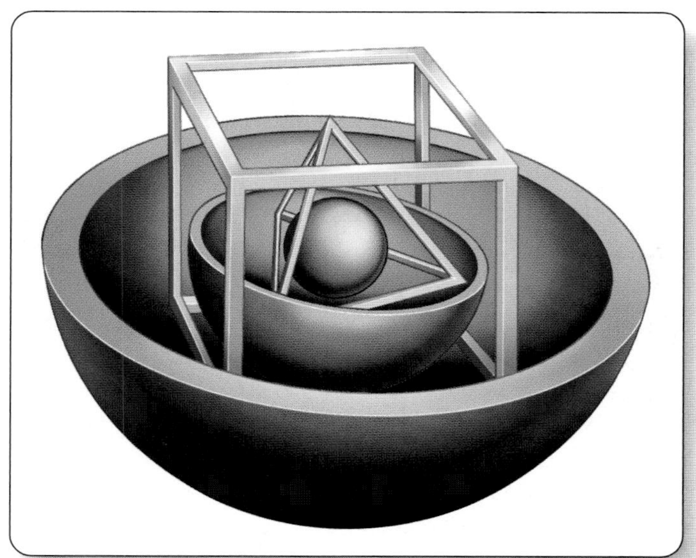

In fact, Kepler concluded that there could be only six planets (Mercury, Venus, Earth, Mars, Jupiter, and Saturn) because there were only five regular solids to act as spacers between their spheres. He advanced astrological, numerological, and even musical arguments for his theory.

The second half of the book is no better than the first, but it has one virtue—as Kepler tried to fit the five solids to the planetary orbits, he demonstrated that he was a talented mathematician and that he was well versed in astronomy. He sent copies to Tycho and to Galileo in Rome, and both recognized his talent in spite of the mystical content of the book.

Joining Tycho

Life was unsettled for Kepler because of the persecution of Protestants in the region, so when Tycho Brahe invited him to Prague in 1600, Kepler went readily, eager to work with the famous Danish astronomer. Tycho's sudden death in 1601 left Kepler in a position to use the observations from Hveen to analyze the motions of the planets and complete the *Rudolphine Tables.* Tycho's family, recognizing that Kepler was a Copernican and guessing that he would not follow the Tychonic system in completing the *Rudolphine Tables,* sued to recover the instruments and books of observations. The legal wrangle went on for years. Tycho's family did get back the instruments Tycho had brought to Prague, but Kepler had the books and he kept them.

Whether Kepler had any legal right to Tycho's records is debatable, but he put them to good use. He began by studying the motion of Mars, trying to deduce from the observations how the planet moved. By 1606, he had solved the mystery, this time correctly. The orbit of Mars is an ellipse and not a circle. Thus he abandoned the 2000-year-old belief in the circular motion of the planets. But the mystery was even more complex. The planets do not move at a uniform speed along their elliptical orbits. Kepler recognized that they move faster when close to the sun and slower when farther away. Thus Kepler abandoned both uniform motion and circular motion and finally solved the problem of planetary motion.

Kepler published his results in 1609 in a book called *Astronomia Nova (New Astronomy).* Like Copernicus's book, *Astronomia Nova* was written in Latin for other scientists and was highly mathematical. In some ways the book was surprisingly advanced.

In spite of the abdication of Rudolph II in 1611, Kepler continued his astronomical studies. He wrote about a supernova that had appeared in 1604 (now known as Kepler's supernova) and about comets, and he wrote a textbook about Copernican astronomy. In 1619, he published *Harmonice Mundi (The Harmony of the World)* in which he returned to the cosmic mysteries of *Mysterium Cosmographicum.* The only thing of note in *Harmonice Mundi* is his discovery that the radii of the planetary orbits are related to the planet's orbital periods. That and his two previous discoveries have become known as the three most fundamental rules of orbital motion.

Kepler's Three Laws of Planetary Motion

Although Kepler dabbled in the philosophical arguments of the day, he was a mathematician, and his triumph was the solution of the problem of the motion of the planets. The key to his solution was the ellipse.

An **ellipse** is a figure drawn around two points, called the *foci,* in such a way that the distance from one focus to any point on the ellipse and back to the other focus equals a constant. This makes it easy to draw ellipses with two thumbtacks and a loop of string. Press the thumbtacks into a board, loop the string about the tacks, and place a pencil in the loop. If you keep the string taut as you move the pencil, it traces out an ellipse (■ Figure 4-10a).

The geometry of an ellipse is described by two simple numbers. The **semimajor axis, *a*,** is half of the longest diameter (Fig-

■ Figure 4-10

The geometry of elliptical orbits. (a) Drawing an ellipse with two tacks and a loop of string. (b) The semimajor axis, *a,* is half of the longest diameter. (c) Kepler's second law is demonstrated by a planet that moves from *A* to *B* in 1 month and from *A′* to *B′* in the same amount of time. The two blue segments have the same area.

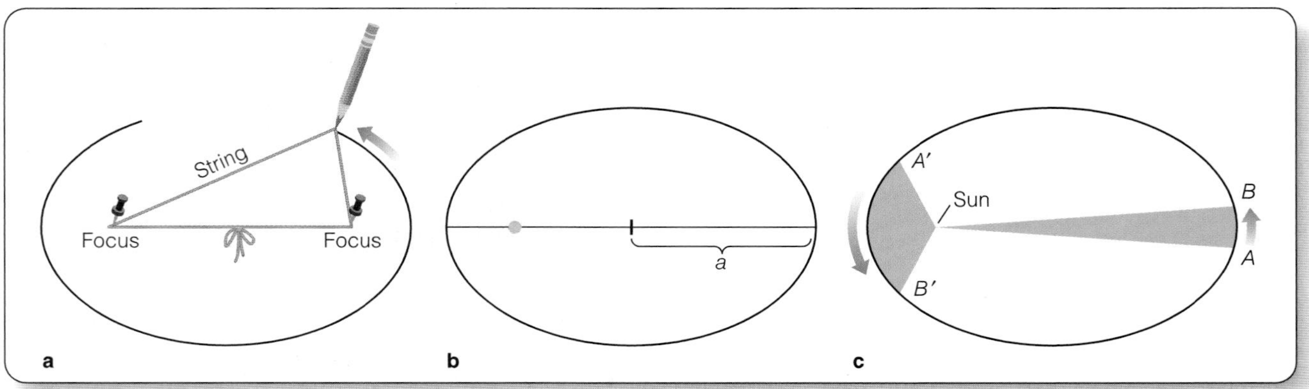

Hypothesis, Theory, and Law: Levels of Confidence

Even scientists misuse the words *hypothesis, theory,* and *law.* You must try to distinguish these terms from one another because they are key elements in science.

A **hypothesis** is a single assertion or conjecture that must be tested. It could be true or false. "All Texans love chili" is a hypothesis. To know whether it is true or false, you need to test it against reality by making observations or performing experiments. Copernicus asserted that the universe was heliocentric; his assertion was a hypothesis subject to testing.

In Chapter 2, you saw that a model (Window on Science 2-1) is a description of some natural phenomenon; it can't be right or wrong. A model is not a conjecture of truth but merely a convenient way to think about a natural phenomenon. Consequently, a model such as the celestial sphere cannot be a hypothesis. Copernicus used his hypothesis to build a model, but they are not the same thing.

A **theory** is a system of rules and principles that can be applied to a wide variety of circumstances. A theory may have begun as one or more hypotheses, but it has been tested, expanded, and generalized. Many textbooks refer to the "Copernican theory," but some historians argue that it was not complete and had not been tested enough to be a theory. It is probably better to call it the Copernican hypothesis.

A **natural law** is a theory that has been refined, tested, and confirmed so often that scientists have great confidence in it. Natural laws are the most fundamental principles of scientific knowledge. Kepler's laws are good examples.

Confidence is the key to understanding these terms. Scientists have more confidence in a theory than in a hypothesis and great confidence in a natural law. Nevertheless, scientists are not always consistent about these words. For example, Einstein's theory of relativity is much more accurate than Newton's laws of gravity and motion, but tradition dictates that you do not refer to Einstein's laws and Newton's theories. Darwin's theory of evolution has been tested many times, and scientists have great confidence in it, but no one refers to Darwin's law. These distinctions are subtle and sometimes depend more on custom than on levels of confidence.

ure 4-10b). The **eccentricity** of an ellipse, ***e*,** is half the distance between the foci divided by the semimajor axis. The eccentricity of an ellipse tells us its shape; if *e* is nearly equal to one, the ellipse is very elongated. If *e* is closer to zero, the ellipse is more circular. To draw a circle with the string and tacks shown in Figure 4-10a, you would move the two thumbtacks together, which shows that a circle is really just an ellipse with eccentricity equal to zero. As you move the thumbtacks farther apart, the ellipse becomes flatter, and the eccentricity moves closer to 1.

Kepler used ellipses to describe the motion of the planets in three fundamental rules that have been tested and confirmed so many times that astronomers now refer to them as natural laws (**Window on Science 4-2**). They are commonly called Kepler's laws of planetary motion (■ Table 4-1).

Kepler's first law states that the orbits of the planets around the sun are ellipses with the sun at one focus. Thanks to the precision of Tycho's observations and the sophistication of Kepler's mathematics, Kepler was able to recognize the elliptical shape of the orbits even though they are nearly circular. Of the planets known to Kepler, Mercury has the most elliptical orbit, but even it deviates only slightly from a circle (■ Figure 4-11).

Kepler's second law states that a line from the planet to the sun sweeps over equal areas in equal intervals of time. This means that when the planet is closer to the sun and the line connecting it to the sun is shorter, the planet moves more rapidly to sweep over the same area that is swept over when the planet is farther from the sun. Thus the planet in Figure 4-10c would move from point *A′*

■ Table 4-1 | Kepler's Laws of Planetary Motion

I. The orbits of the planets are ellipses with the sun at one focus.
II. A line from a planet to the sun sweeps over equal areas in equal intervals of time.
III. A planet's orbital period squared is proportional to its average distance from the sun cubed:

$$P_y^2 = a_{AU}^3$$

■ Figure 4-11

The orbits of the planets are nearly circular. Measure the horizontal and vertical diameters of this orbit to detect its elliptical shape.

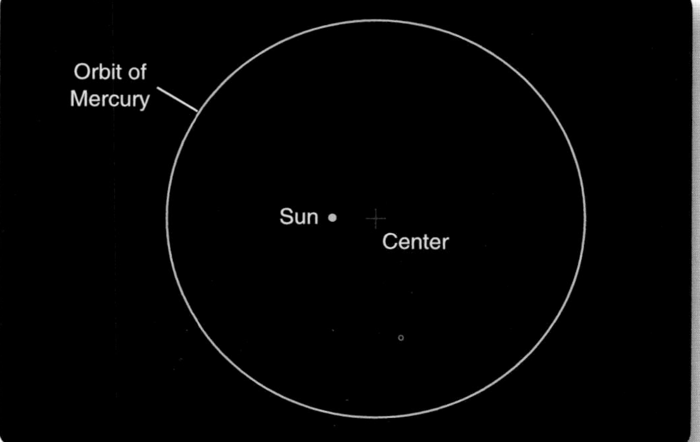

to point B' in 1 month, sweeping over the area shown. But when the planet is farther from the sun, 1 month's motion would be shorter, from A to B.

The time that a planet takes to travel around the sun once is its orbital period, P, and its average distance from the sun equals the semimajor axis of its orbit, a. Kepler's third law tells us that these two quantities are related; the orbital period squared is proportional to the semimajor axis cubed. Measuring P in years and a in astronomical units, you can summarize the third law as

$$P_y^2 = a_{AU}^3$$

For example, Jupiter's average distance from the sun is roughly 5.2 AU. The semimajor axis cubed would be about 140.6, so the period must be the square root of 140.6, about 11.8 years.

It is important to notice that Kepler's three laws are empirical. That is, they describe a phenomenon without explaining why it occurs. Kepler derived them from Tycho's extensive observations, not from any first principle, fundamental assumption, or theory. In fact, Kepler never knew what held the planets in their orbits or why they continued to move around the sun. His books are a fascinating blend of careful observation, mathematical analysis, and mystical theory.

The *Rudolphine Tables*

In spite of Kepler's recurrent involvement with astrology and numerology, he continued to work on the *Rudolphine Tables.* At last, in 1627, they were ready, and he financed their printing himself, dedicating them to the memory of Tycho Brahe. In fact, Tycho's name appears in larger type on the title page than Kepler's own. This is especially surprising when you recall that the tables were based on the heliocentric model of Copernicus and the elliptical orbits of Kepler and not on the Tychonic system. The reason for Kepler's evident deference was Tycho's family, still powerful and still intent on protecting the memory of Tycho. They even demanded a share of the profits and the right to censor the book before publication, though they changed nothing but a few words on the title page and added an elaborate dedication to the emperor.

The *Rudolphine Tables* were Kepler's masterpiece. They could predict the positions of the planets 10 to 100 times more accurately than previous tables. Kepler's tables were the precise model of planetary motion that Copernicus had sought but failed to find. The accuracy of the *Rudolphine Tables* was strong evidence that both Kepler's model for planetary motion and the Copernican hypothesis for the place of the Earth were correct. Copernicus would have been pleased.

Kepler died on November 15, 1630. He had solved the problem of planetary motion, and his *Rudolphine Tables* demonstrated his solution. Although he did not understand why the planets moved or why they followed ellipses, insights that had to wait half a century for Isaac Newton, Kepler's three rules worked. In science the only test of a theory is, "Does it describe reality?" Kepler's laws have been used for almost four centuries as a true description of orbital motion.

Inquire | Review | Analyze

What principles of classical astronomy made it difficult for Kepler to understand planetary motion?

Critical analysis is more than careful thinking. You can think carefully about what Kepler did; but, until you analyze the paradigm (the network of assumptions and principles) within which he worked, you cannot understand how impressive his achievement was. Kepler lived at a time when every astronomer believed in uniform circular motion. The heavens were assumed to be perfect; and, according to classical philosophers such as Plato and Aristotle, circular motion at a constant rate is the only perfect motion. Before Kepler could adopt elliptical orbits and variable planetary speeds, he had to step outside that classical paradigm and see nature in a new way. Only then could he recognize how the planets really moved.

This might make you wonder what false paradigms surround you and prevent you from understanding nature better. Kepler was never confused by believing in geocentrism. From his days in college, he accepted that the sun was the center of the universe. Of course, his first law partially violates a strict rule of heliocentrism. How?

■ ■ ■

Connections: Kepler was a Copernican, and he wrote a number of books that proclaimed his belief, but he was never seriously persecuted because he lived in northern Europe, beyond the sway of the Inquisition. Others were not so lucky. The next astronomer in this story was born in southern Europe under the rule of the Inquisition.

4-5 Galileo Galilei

MOST PEOPLE KNOW TWO FACTS ABOUT GALILEO, and both are wrong. Galileo did not invent the telescope, and he was not condemned by the Inquisition for believing that Earth moved around the sun. Then why is Galileo so important that in 1979, almost 400 years after his trial, the Vatican reopened his case? As you learn about Galileo, you will discover that his trial concerned not just the place of the Earth and the motion of the planets but also a new method of understanding nature, a method called *science.*

Telescopic Observations

Galileo Galilei (■ Figure 4-12) was born in Pisa (in what is now Italy) in 1564, and he studied medicine at the university there. His true love, however, was mathematics, and although he had to leave school early for financial reasons, he returned only four years later as a professor of mathematics. Three years after that he became professor of mathematics at the university at Padua, where he remained for 18 years.

■ **Figure 4-12**

Galileo Galilei (1564–1642), remembered as the great defender of Copernicanism, also made important discoveries in the physics of motion. He is honored here on an Italian 2000-lira note. The reverse side shows one of his telescopes at lower right and a modern observatory above it.

During this time, Galileo seems to have adopted the Copernican model, although he admitted in a 1597 letter to Kepler that he did not support Copernicanism publicly because of the criticism such a declaration would bring. It was the telescope that drove Galileo to publicly defend the heliocentric model.

Galileo did not invent the telescope. It was apparently invented around 1608 by lens makers in Holland. Galileo, hearing descriptions in the fall of 1609, was able to build working telescopes in his workshop. In fact, Galileo was not the first person to look at the sky through a telescope, but he was the first person to observe the sky systematically and apply his observations to the theoretical problem of the day—the place of the Earth (■ Figure 4-13).

What Galileo saw through his telescopes was so amazing he rushed a small book into print. *Sidereus Nuncius (The Sidereal Messenger)* reported three major discoveries. First, the moon was not perfect. It had mountains and valleys on its surface, and Galileo used the shadows to calculate the height of the mountains. Aristotle's philosophy held that the moon was perfect, but Galileo showed that it was not only imperfect but was also a world like Earth.

The second discovery reported in the book was that the Milky Way was made up of a myriad of stars too faint to see with the unaided eye. While intriguing, this discovery could not match the third discovery. Galileo's telescope revealed four new "planets" circling Jupiter, planets that we know today as the Galilean moons of Jupiter (■ Figure 4-14).

The moons of Jupiter supported the Copernican model over the Ptolemaic model. Critics of Copernicus had said Earth could not move because the moon would be left behind; but Jupiter moved, yet kept its satellites, so Galileo's discovery proved that Earth, too, could move and keep its moon. Also, Aristotle's philosophy included the belief that all heavenly motion was centered

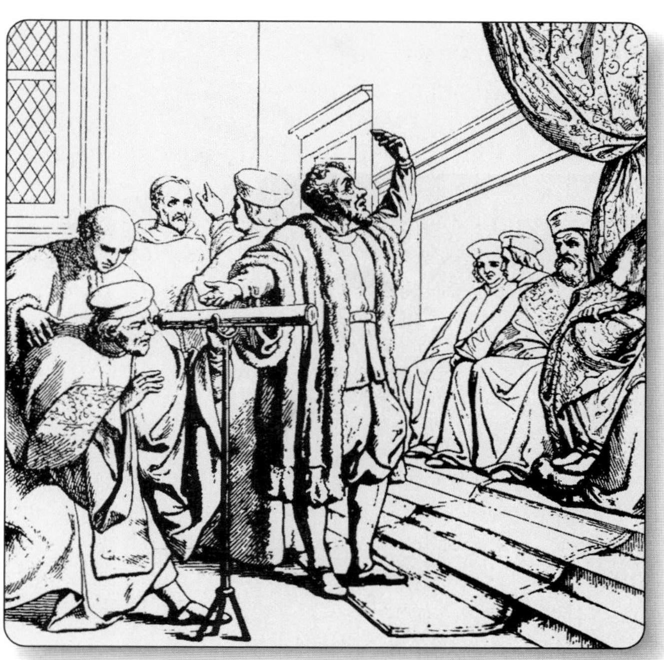

■ **Figure 4-13**

Galileo's telescopic discoveries generated intense interest and controversy. Some critics refused to look through a telescope lest it deceive them. (Courtesy Yerkes Observatory)

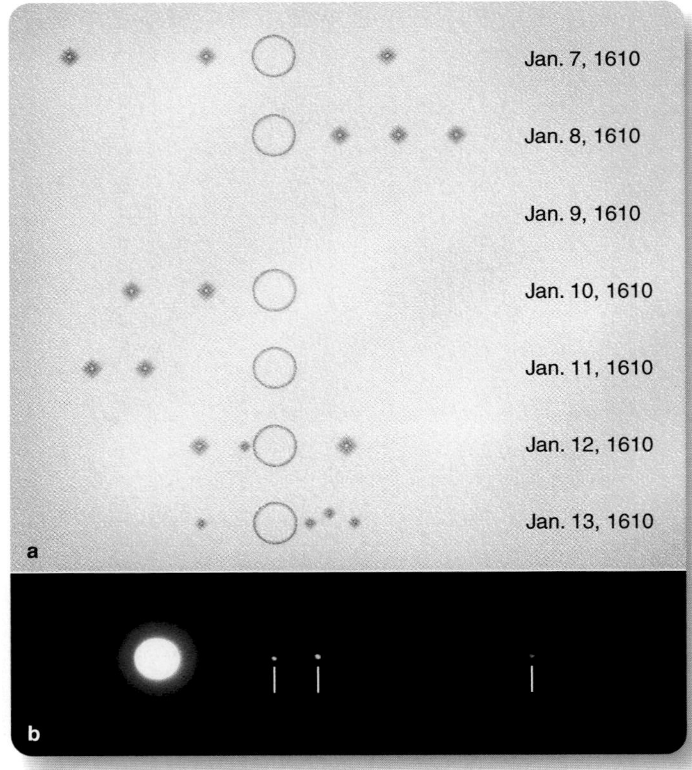

■ Figure 4-14

(a) On the night of January 7, 1610, Galileo saw three small "stars" near the bright disk of Jupiter and sketched them in his notebook. On subsequent nights (excepting January 9, which was cloudy), he saw that the stars were actually four moons orbiting Jupiter. (b) This photograph taken through a modern telescope shows the overexposed disk of Jupiter and three of the four Galilean moons. (Grundy Observatory)

on Earth. Galileo showed that Jupiter's moons revolve around Jupiter, so there could be other centers of motion.

Some time after *Sidereus Nuncius* was published, Galileo noticed something else that made Jupiter's moons even stronger evidence supporting the Copernican universe. When he was able to measure the orbital periods of the four moons, he found that the innermost moon had the shortest period and that the moons further from Jupiter had proportionally longer periods. In this way, Jupiter's moons made up a harmonious system ruled by Jupiter, just as the planets in the Copernican universe were a harmonious system ruled by the sun. (See Figure 4-4b.) The similarity isn't proof, but Galileo saw it as an argument that the solar system was sun-centered and not Earth-centered.

In the years following publication of *Sidereus Nuncius,* Galileo made two additional discoveries. When he observed the sun, he discovered sunspots, raising the suspicion that the sun was less than perfect. Further, by noting the movement of the spots, he concluded that the sun was a sphere and that it rotated on its axis. When he observed Venus, Galileo saw that it was going through phases like those of the moon. In the Ptolemaic model, Venus moves around an epicycle centered on a line between Earth and the sun. Thus, it would always be seen as a crescent (■ Figure 4-15a). But Galileo saw Venus go through a complete set of phases, which proved that it did indeed revolve around the sun (Figure 4-15b).

Sidereus Nuncius was very popular and made Galileo famous. He became chief mathematician and philosopher to the Grand Duke of Tuscany in Florence. In 1611, Galileo visited Rome and was treated with great respect. He had long, friendly discussions with the powerful Cardinal Barberini, but he also made enemies. Personally, Galileo was outspoken, forceful, and sometimes tactless. He enjoyed debate, but most of all he enjoyed being right. Thus,

■ Figure 4-15

(a) If Venus moved in an epicycle centered on the Earth–sun line, it would always appear as a crescent. (b) Galileo's telescope showed that Venus goes through a full set of phases, proving that it must orbit the sun.

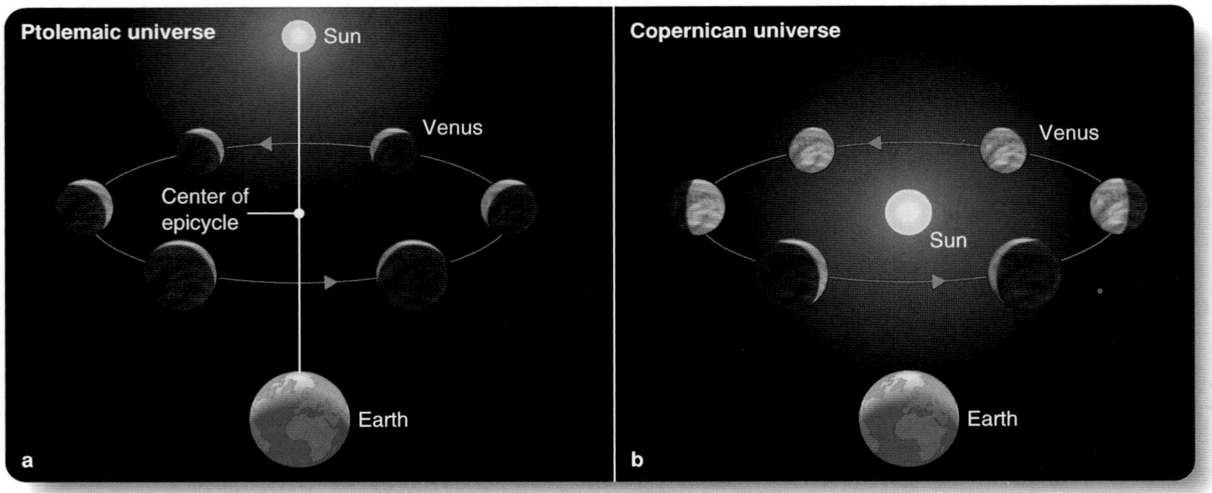

in lectures, debates, and letters he offended important people who questioned his telescopic discoveries.

By 1616, Galileo was the center of a storm of controversy. Some critics said he was wrong, and others said he was lying. Some refused to look through a telescope lest it mislead them, and others looked and claimed to see nothing (hardly surprising given the awkwardness of those first telescopes). Pope Paul V decided to end the disruption, so when Galileo visited Rome in 1616 Cardinal Bellarmine interviewed him privately and ordered him to cease debate. There is some controversy today about the nature of Galileo's instructions, but he did not pursue astronomy for some years after the interview. Books relevant to Copernicanism were banned, although *De Revolutionibus,* recognized as an important and useful book in astronomy, was only suspended pending revision. Everyone who owned a copy of the book was required to cross out certain statements and add handwritten corrections stating that Earth's motion and the central location of the sun were only theories and not facts.

Dialogo and Trial

In 1621 Pope Paul V died, and his successor, Pope Gregory XV, died in 1623. The next pope was Galileo's friend Cardinal Barberini, who took the name Urban VIII. Galileo rushed to Rome hoping to have the prohibition of 1616 lifted; and, although the new pope did not revoke the orders, he did encourage Galileo. Thus, Galileo began to write his great defense of the Copernican model, finally completing it on December 24, 1629. After some delay, the book was approved by both the local censor in Florence and the head censor of the Vatican in Rome. It was printed in February 1632.

Called *Dialogo Dei Due Massimi Sistemi (Dialogue Concerning the Two Chief World Systems),* it confronts the ancient astronomy of Aristotle and Ptolemy with the Copernican model. Galileo wrote the book as a debate among three friends. Salviati, a swift-tongued defender of Copernicus, dominates the book; Sagredo is intelligent but largely uninformed. Simplicio is the dismal defender of Ptolemy. In fact, he does not seem very bright.

The publication of *Dialogo* created a storm of controversy, and it was sold out by August 1632, when the Inquisition ordered sales stopped. The book was a clear defense of Copernicus, and, either intentionally or unintentionally, Galileo exposed the pope's authority to ridicule. Urban VIII was fond of arguing that, as God was omnipotent, He could construct the universe in any form while making it appear to humans to have a different form, and thus its true nature could not be deduced by mere observation. Galileo placed the pope's argument in the mouth of Simplicio, and Galileo's enemies showed the passage to the pope as an example of Galileo's disrespect. The pope thereupon ordered Galileo to face the Inquisition.

Galileo was interrogated by the Inquisition four times and was threatened with torture. He must have thought often of Giordano Bruno, tried, condemned, and burned at the stake in Rome in 1600. One of Bruno's offenses had been Copernicanism. However, Galileo's trial did not center on his belief in Copernicanism. *Dialogo* had been approved by two censors. Rather, the trial centered on the instructions given Galileo in 1616. From his file in the Vatican, his accusers produced a record of the meeting between Galileo and Cardinal Bellarmine that included the statement that Galileo was "not to hold, teach, or defend in any way" the principles of Copernicus. Many historians believe that this document, which was signed neither by Galileo nor by Bellarmine nor by a legal secretary, was a forgery. Others suspect it may be a draft that was never used. In any case, although it is possible that Galileo's true instructions were much less restrictive, Bellarmine was dead and could not testify at Galileo's trial.

The Inquisition condemned Galileo not for heresy but for disobeying the orders given him in 1616. On June 22, 1633, at the age of 70, kneeling before the Inquisition, Galileo read a recantation admitting his errors. Tradition has it that as he rose he whispered *"E pur si muove"* ("Still it moves"), referring to Earth. Although he was sentenced to life imprisonment, he was actually confined at his villa for the next 10 years, perhaps through the intervention of the pope. He died there on January 8, 1642, 99 years after the death of Copernicus.

Galileo was not condemned for heresy, nor was the Inquisition interested when he tried to defend Copernicanism. He was tried and condemned on a charge you might call a technicality. Then why is his trial so important that historians have studied it for almost four centuries? Why have some of the world's greatest authors, including Bertolt Brecht, written about Galileo's trial? Why in 1979 did Pope John Paul II create a commission to reexamine the case against Galileo?

To understand the trial, you must recognize that it was the result of a conflict between two ways of understanding the universe. Since the Middle Ages, scholars had taught that the only path to true understanding was through religious faith. St. Augustine (AD 354–430) wrote *"Credo ut intelligame,"* which can be translated as "Believe in order to understand." Galileo and other scientists of the Renaissance, however, used their own observations to try to understand the universe, and when their observations contradicted Scripture, they assumed their observations of reality were correct. Galileo paraphrased Cardinal Baronius in saying, "The Bible tells us how to go to heaven, not how the heavens go."

The trial of Galileo was not about the place of the Earth in the universe. It was not about Copernicanism. It wasn't really about the instructions Galileo received in 1616. It was about the birth of modern science as a rational way to understand the universe (■ Figure 4-16). The commission appointed by John Paul II in 1979, reporting its conclusions in October 1992, said of Galileo's inquisitors, "This subjective error of judgment, so clear to us today, led them to a disciplinary measure from which Galileo 'had much to suffer.'" Galileo was not found innocent in 1992 so much as the Inquisition was forgiven for having charged him in the first place.

■ **Figure 4-16**

Although he did not invent it, Galileo will always be remembered along with the telescope, as on these postage stamps. By depending on direct observation instead of the first principles of philosophy and theology, Galileo led the way to the invention of modern astronomy and modern science as a way to know about the natural world.

Inquire | Review | Analyze

How were Galileo's observations of the moons of Jupiter evidence against the Ptolemaic model?

Galileo presented his arguments in the form of evidence and conclusions, and the moons of Jupiter were key evidence. Ptolemaic astronomers argued that Earth could not move or it would lose its moon, but even in the Ptolemaic universe Jupiter moved, and the telescope showed that it had moons and kept them. Evidently, Earth could move and not leave its moon behind. Furthermore, moons circling Jupiter did not fit the classical belief that all motion was centered on Earth. Obviously there could be other centers of motion. Finally, the orbital periods of the moon were related to their distance from Jupiter, just as the orbital periods of the planets were, in the Copernican system, related to their distance from the sun. This similarity suggested that the sun rules its harmonious family of planets just as Jupiter rules its harmonious family of moons.

Of all of Galileo's telescopic observations, the moons of Jupiter caused the most debate, but the craters on the moon and the phases of Venus were also critical evidence. How did they argue against the Ptolemaic model?

■ ■ ■

Connections: Galileo faced the Inquisition because of a conflict between two ways of knowing and understanding the world. The Church taught faith and understanding through revelation, but the scientists of the age were inventing a new way to understand nature that relied on the evidence of observation and measurement. Thus, the problems of the place of the Earth and the motion of the planets triggered the birth of modern science. The final episode in that adventure has at its center a man who wondered what makes an apple fall.

4-6 Isaac Newton and Orbital Motion

THE BIRTH OF MODERN ASTRONOMY and of modern science date from the 99 years between the deaths of Copernicus and Galileo. The Renaissance is commonly taken to be the period between 1350 and 1600, and thus the 99 years of this story lie at the culmination of the reawakening of learning in all fields (■ Figure 4-17). Not only did the world adopt a new model of the universe, but it also adopted a new way of understanding humanity's place in nature.

The problem of the place of the Earth was resolved by the Copernican Revolution, but the problem of planetary motion was only partly solved by Kepler's laws. For the last 10 years of his life, Galileo studied the nature of motion, especially the accelerated motion of falling bodies. Although he made some important progress, he was not able to relate his discoveries about motion to the heavens. That final step fell to Isaac Newton.

Isaac Newton

Galileo died in January 1642. Some 11 months later, on Christmas day 1642,* a baby was born in the English village of Woolsthorpe. His name was Isaac Newton (■ Figure 4-18), and his life represented the first flower of the seeds planted by the four astronomers in this story.

Newton was a quiet child from a farming family, but his work at school was so impressive that his uncle financed his education at Trinity College, where he studied mathematics and physics. In 1665, plague swept through England, and the colleges were closed. During 1665 and 1666, Newton spent his time in Woolsthorpe, thinking and studying. It was during these years that he made most of his discoveries in optics, mechanics, and mathematics. Among other things, he studied optics, developed three laws of motion, divined the nature of gravity, and invented differential calculus. The publication of his work in his book *Principia* in 1687 placed science on a firm analytical base.

*Because England had not yet reformed its calendar, December 25, 1642, in England was January 4, 1643, in Europe. It is only a small deception to use the English date and thus include Newton's birth in our 99-year history.

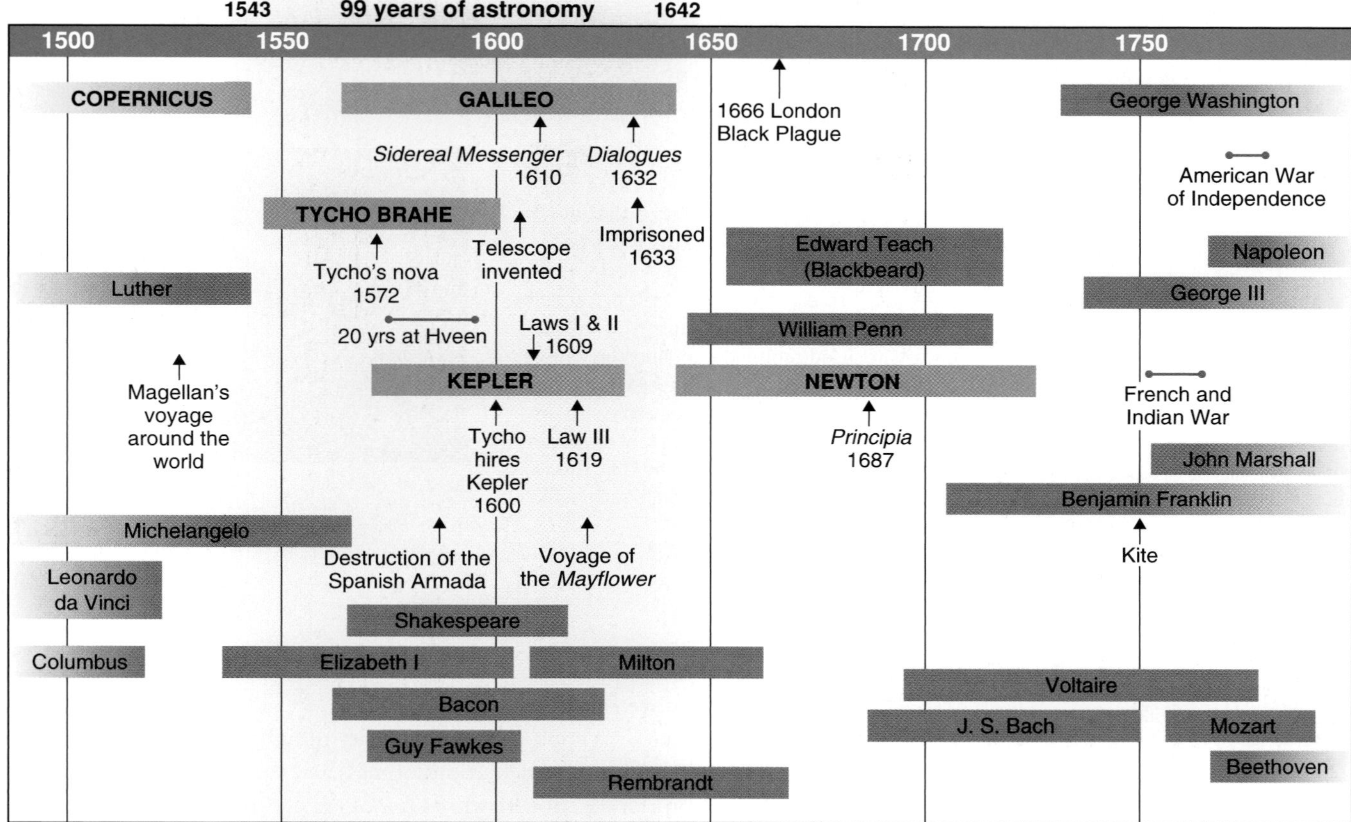

■ Figure 4-17

The 99 years between the death of Copernicus in 1543 and the birth of Newton in 1642 marked the transition from the ancient astronomy of Ptolemy and Aristotle to the revolutionary theory of Copernicus. This period saw the birth of modern scientific astronomy.

It is beyond the scope of this book to analyze all of Newton's work, but his laws of motion and gravity had an important impact on the future of astronomy. From his study of the work of Galileo, Kepler, and others, Newton extracted three laws that relate the motion of a body to the forces acting on it (■ Table 4-2). These laws made it possible to predict exactly how a body would move if the forces were known (**Window on Science 4-3**).

When Newton thought carefully about motion, he realized that some force must pull the moon toward Earth's center. If there were no such force altering the moon's motion, it would continue moving in a straight line and leave Earth forever. It can circle Earth only if Earth attracts it. Newton's insight was to recognize that the force that holds the moon in its orbit is the same as the force that makes apples fall from trees.

Newtonian gravitation is sometimes called universal mutual gravitation. Newton's third law points out that forces occur in pairs. If one body attracts another, the second body must also

■ Figure 4-18

Isaac Newton (1642–1727) worked from the discoveries of Galileo and Kepler to study motion and gravitation. He and some of his discoveries are honored on this English 1-pound note.

Window on Science 4-3

The Unstated Assumption of Science: Cause and Effect

One of the most often used and least often stated principles of science is cause and effect, and you could argue that Newton's second law of motion was the first clear statement of the principle. Ancient philosophers such as Aristotle argued that objects moved because of tendencies. They said that earth and water, and objects made of earth and water, had a natural tendency to move toward the center of the universe. This natural motion had no cause but was inherent in the nature of the objects. But Newton's second law says $F = ma$. If an object of mass m changes its motion by an acceleration a, then it must be acted on by a force F.

The principle of cause and effect goes far beyond motion. The principle of cause and effect gives scientists confidence that every effect has a cause. Hearing loss in certain laboratory rats, color changes in certain chemical dyes, and explosions on certain stars are all effects that must have causes. All of science is focused on understanding the causes of the effects seen. If the universe were not rational, then you could never expect to discover causes. Newton's second law of motion was arguably the first statement that the behavior of the universe depends rationally on causes.

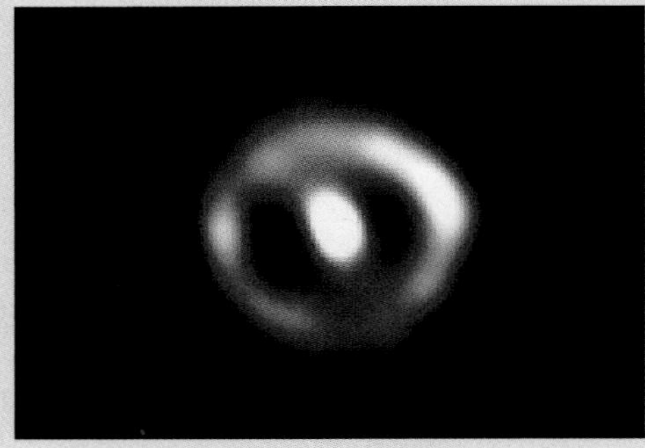

Cause and effect: Why did this star explode in 1992? There must have been a cause. (ESA/STScI and NASA)

Table 4-2 | Newton's Three Laws of Motion

I. A body continues at rest or in uniform motion in a straight line unless acted upon by some force.

II. A body's change of motion is proportional to the force acting on it and is in the direction of the force.

III. When one body exerts a force on a second body, the second body exerts an equal and opposite force back on the first body.

attract the first. Thus gravitation must be mutual. Furthermore, gravity must be universal. That is, all objects that contain mass must attract all other masses in the universe. The force between two bodies depends on the masses of the bodies and the distance between them.

The **mass** of an object is a measure of the amount of matter in the object, usually expressed in kilograms. Mass is not the same as weight. An object's *weight* is the force that Earth's gravity exerts on the object. Thus an object in space far from Earth might have no weight, but it would contain the same amount of matter and would thus have the same mass that it has on Earth.

Newton also realized that the distance between the objects is important. He recognized that the force of gravity decreases as the square of the distance between the objects increases. Specifically, if the distance from, say, Earth to the moon were doubled, the gravitational force between them would decrease by a factor of 2^2, or 4. If the distance were tripled, the force would decrease by a factor of 3^2, or 9. This relationship is known as the **inverse square relation.** (This is discussed in more detail in Chapter 8 where it is applied to the intensity of light.)

With these definitions of mass and the inverse square relation, you can describe Newton's law of gravity in a simple equation:

$$F = -G\frac{Mm}{r^2}$$

Here M and m are the masses of two objects, r is the distance between their centers, G is the gravitational constant, and F is the force of gravity acting between the two objects. The minus sign reminds you that the force is attractive, tending to make r decrease. To summarize, the force of gravity attracting two objects to each other equals the gravitational constant times the product of their masses divided by the square of the distance between the objects.

Orbital Motion

Newton's laws of motion and gravitation make it possible to understand why the moon orbits Earth and how the planets move along their orbits. You can even discover why Kepler's laws work.

To understand how an object can orbit another object, you must see orbital motion as Newton did. Objects in orbit are falling. Study **Orbiting Earth** on pages 68 and 69 and notice three important ideas. First, an object orbiting Earth is actually falling (being accelerated) toward Earth's center. The object continuously misses Earth because of its orbital velocity. Second, notice that objects orbiting each other actually revolve around their center of mass. Finally, notice the difference between closed orbits and open orbits. If you want to leave Earth never to return, you must give your spaceship a high enough velocity so it will follow an open orbit.

Window on Science 4-4

Universal Gravitation

Isn't it weird that Isaac Newton is said to have "discovered" gravity in the late 17th century—as if people didn't have gravity before that, as if people in the 16th century floated around holding onto tree branches? Newton's accomplishment was that he realized that gravity was universal. The same force that makes an apple fall holds the moon in its orbit, guides the planets around the sun, and shapes the motion of the universe. Newton's discovery made gravity the master of the universe and changed the way people thought about nature.

Gravity, Newton said, is a characteristic of matter, and any object made of matter exerts gravity on other objects. Earth, moon, sun, planets, and even the stars have their gravitational fields, and they move under the combined influence of those gravitational tugs. Before Newton, most people thought the heavens were governed by a different set of rules; the heavens were separate from humanity and un-Earthly. Universal gravitation linked everything together and made humans part of the universe. The Copernican revolution made Earth into a planet, and the Newtonian revolution made everyone part of nature.

The universe is filled with gravitational forces, and our planet, our star, and our galaxy move through space under the combined influence of every other object made of matter. The universe is a swirling waltz of matter danced to the music of gravity.

Gravity distorts two galaxies as they whirl past each other. (NASA, STScI, and ESA)

When the captain of the starship *Enterprise* says, "Put us into a standard orbit," the ship's computers must quickly calculate the velocity needed to achieve a circular orbit. That circular velocity depends only on the mass of the planet and the distance from the center of the planet (**Reasoning with Numbers 4-1**). Once the engines fire and the ship reaches circular velocity, the engines can shut down. The *Enterprise* is in orbit and will fall around the planet forever so long as it is above the atmosphere where there is no friction. No further effort is needed to maintain orbit, thanks to Newton's laws.

AceAstronomy™ Log into AceAstronomy and select this chapter to see Astronomy Exercise "Falling Bodies" to try one of Galileo's famous experiments.

AceAstronomy™ Log into AceAstronomy and select this chapter to see Astronomy Exercise "Orbital Motion" to experiment with an object in orbit.

AceAstronomy™ Log into AceAstronomy and select this chapter to see Astronomy Exercise "Escape Velocity" and try to escape from a planet.

Tides

Newton understood that gravity was mutual—Earth attracts the moon and the moon attracts Earth—and that meant the moon's gravity can explain the ocean tides. But Newton also realized that gravitation is universal (**Window on Science 4-4**), and that means there is much more to tides than just Earth's oceans.

Reasoning with Numbers 4-1

Circular Velocity

Circular velocity is the velocity a satellite must have to remain in a circular orbit around a larger body. If the mass of the satellite is small compared with the central body, then the circular velocity is given by

$$V_c = \sqrt{\frac{GM}{r}}$$

In this formula, M is the mass of the central body in kilograms, r is the radius of the orbit in meters, and G is the gravitational constant, 6.67×10^{-11} m^3/s^{2}kg. This formula is all you need to calculate how fast an object must travel to stay in a circular orbit.

For example, how fast does the moon travel in its orbit? The mass of Earth is 5.98×10^{24} kg, and the radius of the moon's orbit is 3.84×10^8 m. The moon's velocity is

$$V_c = \sqrt{\frac{6.67 \times 10^{-11} \times 5.98 \times 10^{24}}{3.84 \times 10^8}}$$

$$= \sqrt{\frac{39.9 \times 10^{13}}{3.84 \times 10^8}}$$

$$= \sqrt{1.04 \times 10^6} = 1020 \text{ m/s}$$

This calculation shows that the moon travels 1.02 km along its orbit each second.

Orbiting Earth

You can understand orbital motion by thinking of a cannon ball falling around Earth in a circular path.

Imagine a cannon on a high mountain aimed horizontally. A little gunpowder gives the cannonball a low velocity, and it doesn't travel very far before falling to Earth.

More gunpowder gives the cannonball a higher velocity, and it travels farther.

With enough gunpowder, the cannonball travels so fast it never strikes the ground. Earth's gravity pulls it toward Earth's center, but Earth's surface curves away from it at the same rate it falls. It is in orbit.

A satellite above Earth's atmosphere feels no friction and will fall around Earth indefinitely.

North Pole

The velocity needed to stay in a circular orbit is called the **circular velocity.** Just above Earth's atmosphere, circular velocity is 7790 m/s or about 17,400 miles per hour, and the orbital period is about 90 minutes.

Earth satellites eventually fall back to Earth if they orbit too low and experience friction with the upper atmosphere.

AceAstronomy™
Log into AceAstronomy and select this chapter to see Active Figure "Newton's Cannon" and fire your own version of Newton's cannon.

A Geosynchronous Satellite

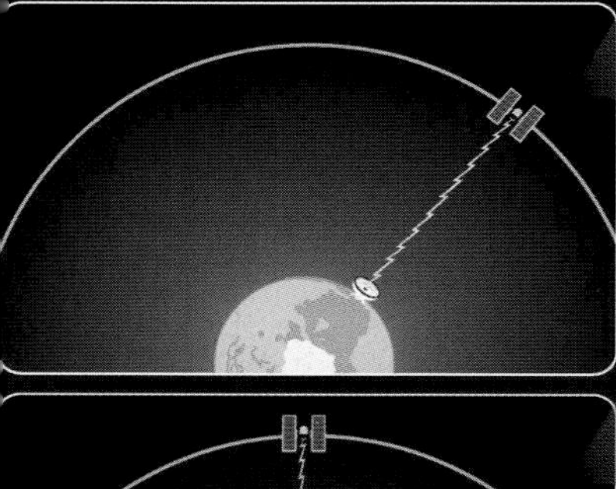

At a distance of 42,250 km (26,260 miles) from Earth's center, a satellite orbits with a period of 24 hours.

The satellite orbits eastward, and Earth rotates eastward under the moving satellite.

The satellite remains fixed above a spot on Earth's equator.

A **geosynchronous satellite** orbits eastward with the rotation of Earth and remains above a fixed spot — ideal for communications and weather satellites.

AceAstronomy™
Log into AceAstronomy and select this chapter to see Active Figure "Geosynchronous Orbit" and place your own satellite into geosynchronous orbit.

According to Newton's first law of motion, the moon should follow a straight line and leave Earth forever. Because it follows a curve, Newton knew that some force must continuously accelerate it toward Earth — gravity. Each second the moon moves 1020 m (3350 ft) eastward and falls about 1.6 mm (about 1/16 inch) toward Earth. The combination of these motions produces the moon's curved orbit. The moon is falling.

NASA

Astronauts in orbit around Earth feel weightless, but they are not "beyond Earth's ravity," to use a term from old science fiction movies. Like the moon, the astronauts are accelerated toward Earth by Earth's gravity, but they travel fast enough along their orbits that they continually "miss the Earth." They are literally falling around Earth. Inside or outside a spacecraft, astronauts feel weightless because they and heir spacecraft are falling at the same rate. Rather than saying they are weightless, you should more accurately say they are in free fall.

To be precise you should not say that an object orbits Earth. Rather the two objects orbit each other. Gravitation is mutual, and if Earth pulls on the moon, the moon pulls on Earth. The two bodies revolve around their common **center of mass,** the balance point of the system.

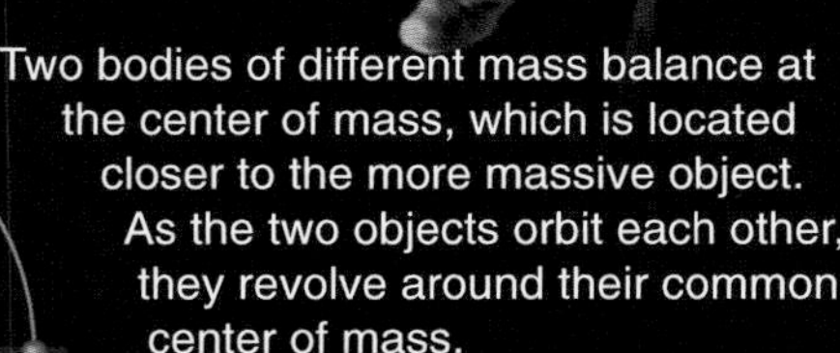

Two bodies of different mass balance at the center of mass, which is located closer to the more massive object. As the two objects orbit each other, they revolve around their common center of mass.

The center of mass of the Earth–moon system lies only 4708 km (2926 miles) from the center of Earth — inside the Earth. As the moon orbits the center of mass on one side, the Earth swings around the center of mass on the opposite side.

Ace Astronomy™ Log into AceAstronomy and select this chapter to see Active Figure "Center of Mass." Change the mass ratio to move the center of mass.

Closed orbits return the orbiting object to its starting point. The moon and artificial satellites orbit Earth in closed orbits. Below, the cannonball could follow an elliptical or a circular closed orbit.

f the cannonball travels as fast as **escape velocity,** he velocity needed to leave a body, it will enter an open orbit. An **open orbit** does not return he cannonball to Earth. t will escape.

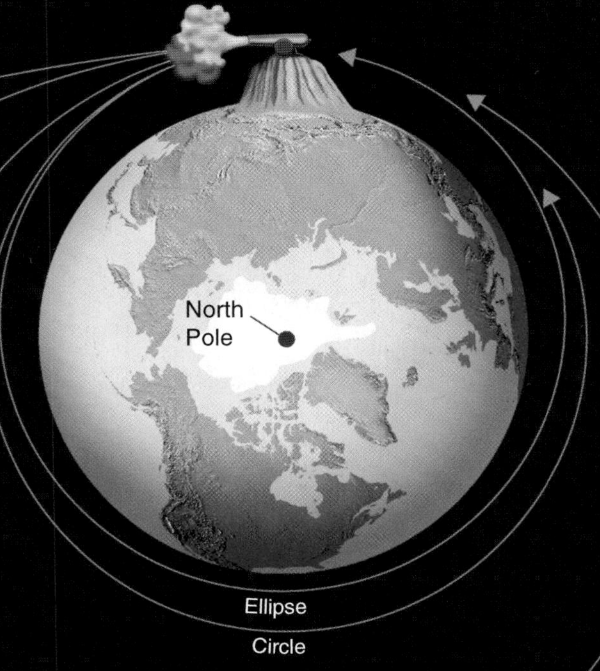

A cannonball with a velocity greater than escape velocity will follow a hyperbola and escape from Earth.

A cannonball with escape velocity will follow a parabola and escape.

As described by Kepler's Second Law, an object in an elliptical orbit has its lowest velocity when it is farthest from Earth (apogee), and its highest velocity when it is closest to Earth (perigee). Perigee must be above Earth's atmosphere, or friction will rob the satellite of energy and it will

Tides are caused by small gravitational forces. For example, Earth's gravity attracts your body downward with a force equal to your weight. The moon is less massive and more distant, so the moon attracts your body with a force equal to roughly 0.0003 percent of your weight. You don't notice that tiny force, but Earth's oceans respond dramatically.

The side of Earth facing the moon is about 4000 miles closer to the moon than is the center of Earth. Consequently, the moon's gravity, tiny though it is at the distance of Earth, is just a bit stronger when it acts on the near side of Earth than on the center. It pulls on the oceans on the near side of Earth a bit more strongly than on Earth's center, and the oceans respond by flowing into a bulge of water on the side of Earth facing the moon. There is also a bulge on the side away from the moon, which develops because the moon pulls more strongly on Earth's center than on the far side. Thus the moon pulls Earth away from the oceans, which flow into a bulge on the far side as shown at the top of ■ Figure 4-19.

You might wonder: If Earth and moon accelerate toward each other, why don't they smash together? The answer, of course, is that they are orbiting around their common center of gravity. The ocean tides are caused by the accelerations Earth and the oceans feel as they move around that center of gravity.

The rocky bulk of Earth also responds to these tidal forces. Although you don't notice, Earth flexes, and the mountains and plains rise and fall by a few centimeters in response to the moon's gravitational acceleration.

You can see dramatic evidence of this effect if you watch the ocean shore for a few hours. Though Earth rotates on its axis, the tidal bulges remain fixed with respect to the moon. As the turning Earth car-

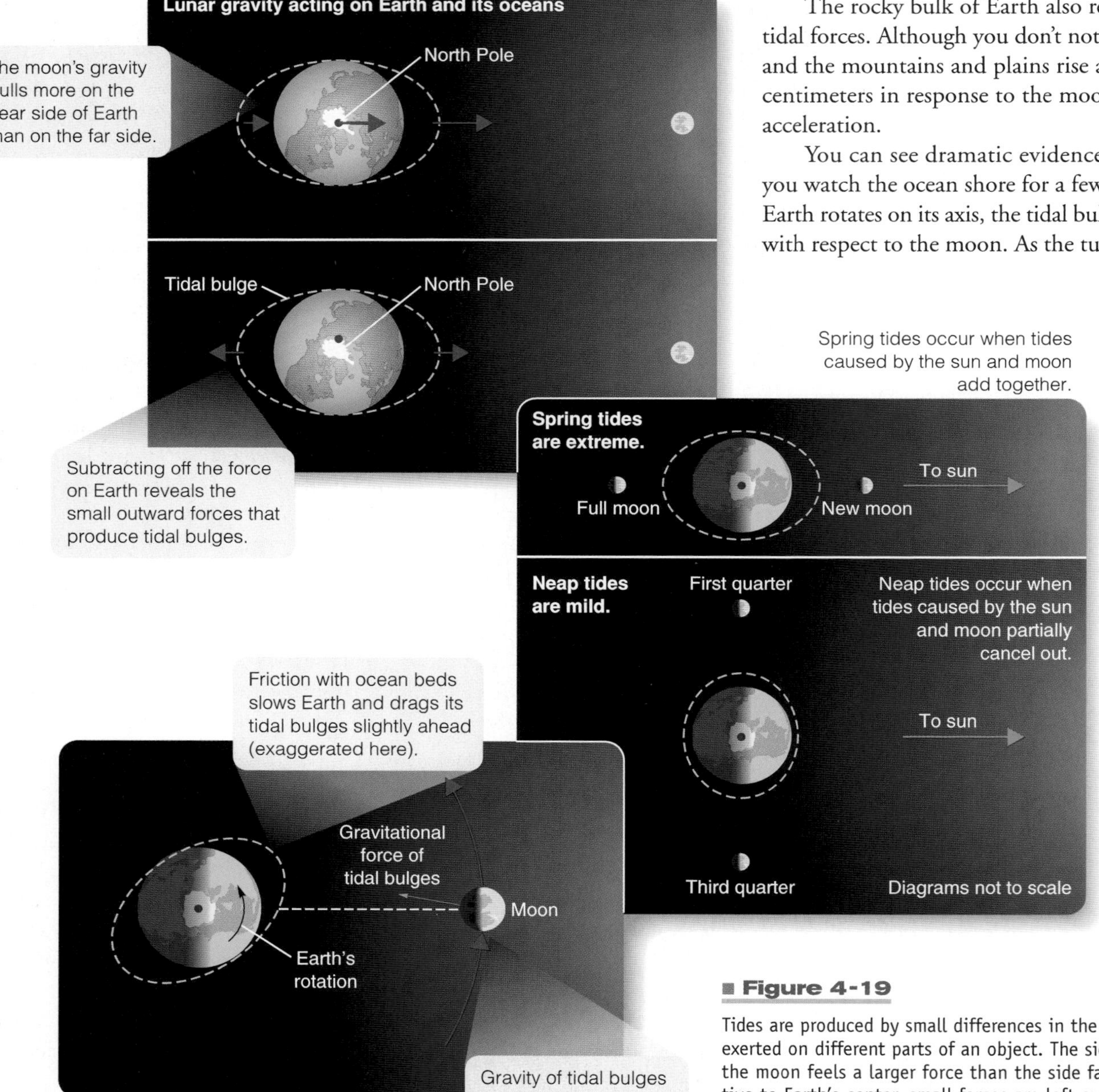

■ Figure 4-19

Tides are produced by small differences in the gravitational force exerted on different parts of an object. The side of Earth nearest the moon feels a larger force than the side farthest away. Relative to Earth's center, small forces are left over, and they cause the tides. Both the moon and the sun produce tides on Earth. Tides can alter both an object's rotation and orbital motion.

Testing a Theory by Prediction

When you read about any science, you should notice that scientific theories face in two directions. They look back into the past and explain previously observed phenomena. For example, Newton's laws of motion and gravity explained how the planets moved. But theories also face forward in that they enable you to make predictions about what you should find as you explore further. Thus, Newton's laws allowed astronomers to calculate the orbits of comets, predict their return, and eventually understand their origin.

Scientific predictions are important in two ways. First, if a theory leads to a prediction and scientists later discover the prediction was true, the theory is confirmed, and scientists gain confidence that it is a true description of nature. But predictions are important in science in a second way. Using an existing theory to make a prediction may lead you into an unexplored avenue of knowledge. Thus, the first theories of genetics made predictions that confirmed the genetic theory of inheritance, but those predictions also created a new understanding of how living creatures evolve.

As you read about any scientific theory, think about both what it can explain and what it can predict.

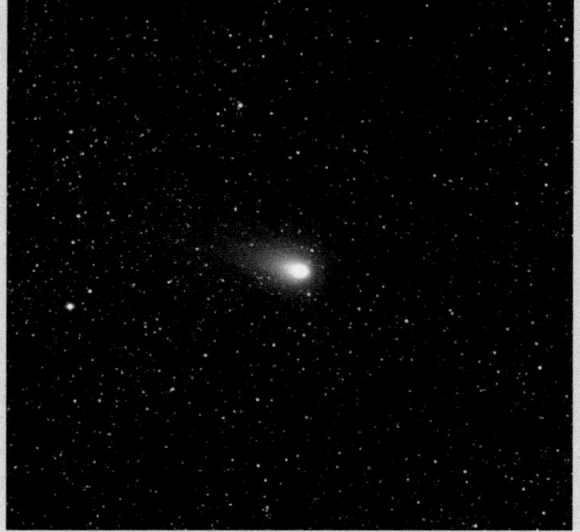

The motion of comets can be predicted using Newton's laws. (Nigel Sharp/NOAO/AURA/NSF)

ries you and your beach into a tidal bulge, the ocean water deepens, and the tide crawls up the sand. Later, when Earth's rotation carries you out of the bulge, the ocean becomes shallower, and the tide falls. Because there are two bulges on opposite sides of Earth, the tides should rise and fall twice a day on an ideal coast.

In reality, the tidal cycle at any given location can be quite complex because of the latitude of the site, shape of the shore, winds, and so on. Tides in the Bay of Fundy (New Brunswick, Canada), for example, occur twice a day and can exceed 40 feet, whereas the northern coast of the Gulf of Mexico has only one tidal cycle a day of roughly 1 foot.

Gravity is universal, so the sun, too, produces tides on Earth. The sun is roughly 27 million times more massive than the moon, but it lies almost 400 times farther from Earth. Consequently, tides on Earth caused by the sun are less than half those caused by the moon. At new moon and at full moon, the moon and sun produce tidal bulges that add together and produce extreme tidal changes; high tide is very high, and low tide is very low. Such tides are called **spring tides,** even though they occur at every new and full moon and not just in the spring. **Neap tides** occur at first- and third-quarter moons, when the moon and sun pull at right angles to each other, as shown in Figure 4-19. Then the tides caused by the sun reduce the tides caused by the moon, and the rise and fall of the ocean is less extreme than usual.

Galileo tried to understand tides, but not until Newton described gravity could astronomers analyze tidal forces and recognize their surprising effects. For example, the friction of the ocean waters with the ocean beds slows Earth's rotation and makes the length of a day grow by 0.0023 seconds per century. Fossils of marine animals and tidal sediments confirm that only 900 million years ago Earth's day was 18 hours long. In addition, Earth's gravitation exerts tidal forces on the moon, and although there are no bodies of water on the moon, friction within the flexing rock has slowed the moon's rotation to the point that it now always keeps the same face toward Earth.

Tidal forces can also affect orbital motion. Friction with the ocean beds drags the tidal bulges eastward out of a direct Earth–moon line. These tidal bulges contain a large amount of mass, and their gravitational field pulls the moon forward in its orbit, as shown at the bottom of Figure 4-19. As a result, the moon's orbit is growing larger by about 4 cm a year, an effect that astronomers can measure by bouncing laser beams off reflectors left on the lunar surface by the Apollo astronauts.

Newton's gravitation is much more than just the force that makes apples fall. In later chapters, you will see how tides can pull gas away from stars, rip galaxies apart, and melt the interiors of small moons orbiting near massive planets. Tidal forces produce some of the most surprising and dramatic processes in the universe.

The Newtonian Universe

Newton's insight gave the world a new conception of nature. His laws of motion and gravity were *general* laws that described the motions of all bodies under the action of external forces. In addition, the laws were *productive* because they made possible specific calculations that could be tested by observation. For example, Newton's laws of motion could be used to derive Kepler's third law from the law of gravity.

Newton's discoveries remade astronomy into an analytical science in which astronomers could measure the positions and motions of celestial bodies, calculate the gravitational forces acting on them, and predict their future motion (**Window on Science 4-5**).

Were you to trace the history of astronomy after Newton, you would find scientists predicting the motion of comets, the gravitational interaction of the planets, the motions of double stars, and so on. Astronomers built on the discoveries of Newton, just as he had built on the discoveries of Copernicus, Tycho, Kepler, and Galileo. It is the nature of science to build on the discoveries of the past, and Newton was thinking of that when he wrote, "If I have seen farther than other men, it is because I stood upon the shoulders of giants."

Inquire | Review | Analyze

What do the words *universal* and *mutual* mean in the term *universal mutual gravitation?*

Newton realized that the force that makes an apple accelerate downward is the same as the force that accelerates the moon and holds it in its orbit. You can learn more by thinking about Newton's third law of motion, which says that forces always occur in pairs. If Earth attracts the moon, then the moon must attract Earth. That is, gravitation is *mutual* between any two objects.

Furthermore, if Earth's gravity attracts the apple and the moon, then it must attract the sun, and the third law tells you that the sun must attract Earth. But if the sun attracts Earth, then it must also attract the other planets and even distant stars, which, in turn, must attract the sun and each other. Step by step, Newton's third law of motion leads you to conclude that gravitation must apply to all masses in the universe. That is, gravitation must be *universal.*

Newton's explanation of orbital motion was powerful because he explained it in terms of common processes known on Earth's surface. For example, why is it correct to say that the moon is falling?

■ ■ ■

Connections: Newton said he stood on the shoulders of giants, and you too have that vantage point. You can understand what you see in the sky because you understand the work of Copernicus, Kepler, and Newton—not just their discoveries, but also their creation of a new way to understand nature—by the evidence of your senses. In their search for evidence, modern astronomers have extended their senses with powerful instruments, the subject of the next chapter.

Study and Review Tools

Summary

4-1 | Pre-Copernican Astronomy

How did the ancients describe Earth's place in the universe?

- Ancient philosophers accepted as a first principle that Earth was the unmoving center of the universe. Another first principle was that the heavens were perfect, so philosophers such as Plato argued that because the circle was the only perfect geometrical form, the heavens must move in uniform circular motion.
- The geocentric universe became part of the teachings of the great philosopher Aristotle, who argued that the sun, moon, and stars were carried around Earth on rotating crystalline spheres.
- About 140 AD, Ptolemy gave mathematical form to Aristotle's model in the *Almagest.* Ptolemy preserved the principles of geocentrism and uniform circular motion, but he added epicycles, deferents, and equants to better predict the motions of the planets.

4-2 | Copernicus

How did Copernicus revise that ancient theory?

- Copernicus devised a heliocentric model. He preserved the principle of uniform circular motion, but he argued that Earth rotates on its axis and circles the sun once a year. His theory was controversial because it contradicted Church teaching.
- Copernicus published his theory in his book *De Revolutionibus* in 1543, the same year he died.

Why was the Copernican model gradually accepted?

- Because Copernicus kept uniform circular motion, his model did not predict the motions of the planets well, but it did offer a simple explanation of retrograde motion without using big epicycles.
- The Copernican model was also more elegant. Venus and Mercury were treated the same as all the other planets, and the velocity of each planet was related to its distance from the sun.

4-3 | Tycho Brahe

- Tycho Brahe developed his own model in which the sun and moon circled Earth and the planets circled the sun.
- Tycho's great contribution was to compile detailed observations over a period of 20 years, observations that were later used by Kepler.

4-4 | Johannes Kepler

- Kepler inherited Tycho's books of observations in 1601 and used them to uncover three laws of planetary motion. He found that the planets follow ellipses with the sun at one focus, that they move faster when near the sun, and that a planet's orbital period squared is proportional to its orbital radius cubed.
- Kepler's final book, the *Rudolphine Tables* (1627), combined heliocentrism with elliptical orbits and predicted the positions of the planets well.

4-5 | Galileo Galilei

Why was Galileo condemned by the Inquisition?

- Galileo used the newly invented telescope to observe the heavens, and he recognized the significance of what he saw there. His discoveries of the phases of Venus, the satellites of Jupiter, the mountains of the moon, and other phenomena helped undermine the Ptolemaic universe.
- Galileo based his analysis on observational evidence. In 1633, he was condemned before the Inquisition for refusing to halt his defense of Copernicanism.

4-6 | Isaac Newton and Orbital Motion

How did Isaac Newton change humanity's view of nature?

- Newton used the work of Kepler and Galileo to discover three laws of motion and the law of gravity. These laws made it possible to understand such phenomena as orbital motion and the tides.
- Newton's laws gave scientists a unified way to think about nature. Every effect has a cause, and science is the search for those causes.
- The 99 years from the death of Copernicus to the birth of Newton marked the beginning of modern science. From that time on, science depended on evidence to test theories and relied on the analytic methods first demonstrated by Kepler and Newton.

New Terms

heliocentric universe (p. 49)
uniform circular motion (p. 50)
geocentric universe (p. 50)
parallax (p. 50)
retrograde motion (p. 50)
epicycle (p. 51)
deferent (p. 51)
equant (p. 51)
paradigm (p. 54)
ellipse (p. 58)
semimajor axis, *a* (p. 58)
eccentricity, *e* (p. 59)
hypothesis (p. 59)
theory (p. 59)
natural law (p. 59)
mass (p. 66)
inverse square relation (p. 66)
circular velocity (p. 68)
geosynchronous satellite (p. 68)
center of mass (p. 69)
closed orbit (p. 69)
escape velocity (p. 69)
open orbit (p. 69)
spring tide (p. 71)
neap tide (p. 71)

Review Questions

Ace Astronomy™ Assess your understanding of this chapter's topics with additional quizzing and animations at **http://astronomy.brookscole.com/sh9e**

1. Why did Greek astronomers conclude that the heavens were made up of perfect crystalline spheres moving at constant speeds?
2. Why did classical astronomers conclude that Earth had to be motionless?
3. How did the Ptolemaic model explain retrograde motion?
4. In what ways were the models of Ptolemy and Copernicus similar?
5. Why did the Copernican hypothesis win gradual acceptance?
6. Why is it difficult for scientists to replace an old paradigm with a new paradigm?
7. Why did Tycho Brahe expect the new star of 1572 to show parallax? Why was the lack of parallax evidence against the Ptolemaic model?
8. How was Tycho's model of the universe similar to the Ptolemaic model? How did it resemble the Copernican model?
9. Explain how Kepler's laws contradict uniform circular motion.
10. What is the difference between a hypothesis, a theory, and a law?
11. How did the *Alfonsine Tables,* the *Prutenic Tables,* and the *Rudolphine Tables* differ?
12. Review Galileo's telescope discoveries and explain why they supported the Copernican model and contradicted the Ptolemaic model.
13. Galileo was condemned by the Inquisition, but Kepler, also a Copernican, was not. Why not?
14. Why did Newton conclude that gravitation had to be universal?
15. Explain why we might describe the orbital motion of the moon with the statement, "The moon is falling."
16. What three astronomical objects are represented here? What are the two rings?

17. Whose observatory is shown here? Why are there no telescopes?

Discussion Questions

1. Historian of science Thomas Kuhn has said that *De Revolutionibus* was a revolution-making book but not a revolutionary book. How was it classical?
2. Why might Tycho Brahe have hesitated to hire Kepler? Why do you suppose he appointed Kepler his scientific heir?
3. How does the modern controversy over creationism and evolution reflect two ways of knowing about the physical world?

Problems

1. If you lived on Mars, which planets would describe retrograde loops? Which would never be visible as crescent phases?
2. Galileo's telescope showed him that Venus has a large angular diameter (61 seconds of arc) when it is a crescent and a small angular diameter (10 seconds of arc) when it is nearly full. Use the small-angle formula to find the ratio of its maximum distance to its minimum distance. Is this ratio compatible with the Ptolemaic universe shown on page 51?
3. Galileo's telescopes were not of high quality by modern standards. He was able to see the moons of Jupiter, but he never reported seeing features on Mars. Use the small-angle formula to find the angular diameter of Mars when it is closest to Earth. How does that compare with the maximum diameter of Jupiter?
4. If a planet had an average distance from the sun of 10 AU, what would its orbital period be?
5. If a space probe were sent into an orbit around the sun that brought it as close as 0.5 AU to the sun and as far away as 5.5 AU, what would its orbital period be?
6. Pluto orbits the sun with a period of 247.7 years. What is its average distance from the sun?
7. Calculate the circular velocity of Venus and Saturn around the sun. (*Hint:* The mass of the sun is 2×10^{30} kg.)
8. The circular velocity of Earth around the sun is about 30 km/s. Are the arrows for Venus and Saturn correct in Figure 4-4b? (*Hint:* See Problem 7.)
9. What is the orbital velocity of an Earth satellite 42,250 km from Earth? How long does it take to circle its orbit once?

Media Cluster

Ace Astronomy™ To access the resources in the Media Cluster, log into AceAstronomy at **http://astronomy.brookscole.com/sh9e** and select Chapter 4.

ASTRONOMY EXERCISES

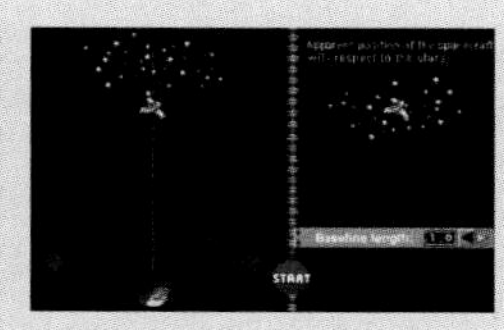

Parallax
As you learned in this chapter, parallax is the apparent motion of an object because of the motion of the observer. In this animation, see how a spaceship appears to move against a background of stars due to Earth's motion.

Falling Bodies
According to tradition, Galileo demonstrated that the acceleration of a falling body is independent of its weight by dropping various objects from the Leaning Tower of Pisa. You can repeat his experiment with this interactive exercise.

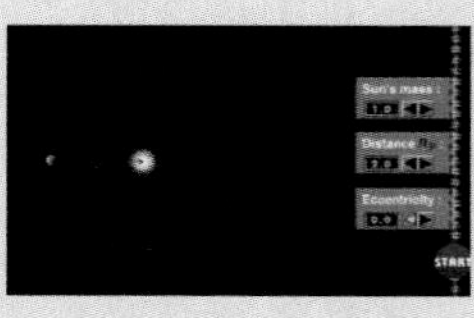

Orbital Motion
Explore how the orbit of a hypothetical planet is affected by changing the mass of the sun, distance to the sun, and eccentricity of orbit in this animation.

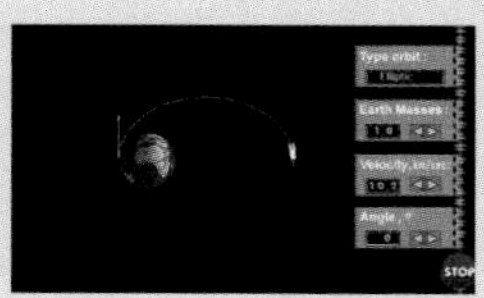

Escape Velocity
Escape velocity is the initial velocity an object needs to escape from the surface of a celestial body. See if you can determine the correct escape velocity for the rocket in this animation.

Critical Inquiries for the Web

1. The trial of Galileo is an important event in the history of science. Scientists now know, and the Church now recognizes, that Galileo's view was correct, but what were the arguments on both sides of the issue as it was unfolding? Research the Internet for documents chronicling the trial, Galileo's observations and publications, and the position of the Church. Use this information to outline a case for and against Galileo in the context of the times in which the trial occurred.
2. It's hard to imagine that an observatory could exist before the invention of the telescope, but Tycho Brahe's observatory at Hveen was a great astronomical center of its day. Search the web for sites on Tycho and his instruments and describe what an observing session at Hveen might have involved.

Exploring *TheSky*

1. Observe Mars going through its retrograde motion. (*Hint:* Use **Reference Lines** under the **View** menu to turn on the ecliptic. Be sure you are in **Free Rotation** under the **Orientation** menu. Locate Mars and use the time skip arrows to watch it move.)
2. Compare the size of the retrograde loops made by Mars, Jupiter, and Saturn.
3. Can you recognize the effects of Kepler's second law in the orbital motion of any of the planets? (*Hint:* Use **3D Solar System Mode** under the **View** menu.)
4. Can you recognize the effects of Kepler's third law in the orbital motion of the planets?

- Go to the Brooks/Cole Astronomy Resource Center **(http://astronomy.brookscole.com)** for critical thinking exercises, articles, and additional readings from InfoTrac College Edition, Brooks/Cole's online student library.

5 Astronomical Tools

He burned his house down for the fire insurance and spent the proceeds on a telescope.

ROBERT FROST
The Star-Splitter

STARLIGHT IS GOING to waste. Every night, light from the stars falls on trees, oceans, roofs, and empty parking lots, and it is all wasted. To an astronomer, nothing is so precious as starlight. It is our only link to the sky, and the astronomer's quest is to gather as much starlight as possible and extract from it the secrets of the stars. The telescope is the emblematic tool of the astronomer because its purpose is to gather and concentrate light for analysis. Nearly all of the interesting objects in the sky are faint sources of light, so modern astronomers are driven to build the largest possible telescopes to gather the maximum amount of light (■ Figure 5-1). Thus, any discussion of astronomical tools is concentrated on large telescopes and the specialized tools used to analyze light.

Sunset atop a Hawaiian mountain as astronomers prepare for a night's work. (NOAO/AURA/NSF)

Guidepost

Looking Back

In the early chapters of this book, you looked at the sky the way ancient astronomers did, with the unaided eye. In the last chapter, you got a glimpse through Galileo's telescope, and it revealed astonishing things about Jupiter, the moon, and Venus. Now it is time to examine the tools of the modern astronomer.

This Chapter

Begin by noting that these tools gather and focus light and its related forms of radiation. For that reason, the first of the essential questions is about mysterious light:

What is light?

How do telescopes work, and how are they limited?

What kind of instruments do astronomers use to record and analyze light?

Why do astronomers use radio telescopes?

Why must some telescopes go into orbit?

As you answer these questions, you will discover how important it is to understand light in all its forms.

Looking Ahead

Astronomy is almost entirely an observational science. Astronomers cannot visit distant galaxies and far-off worlds, so they must observe using astronomical telescopes. Fifteen chapters remain, and every one will discuss information gathered by telescopes.

AceAstronomy™ The AceAstronomy icon throughout the text indicates an opportunity for you to test yourself on key concepts and to explore animations and interactions on the AceAstronomy website at: http://astronomy.brookscole.com/sh9e

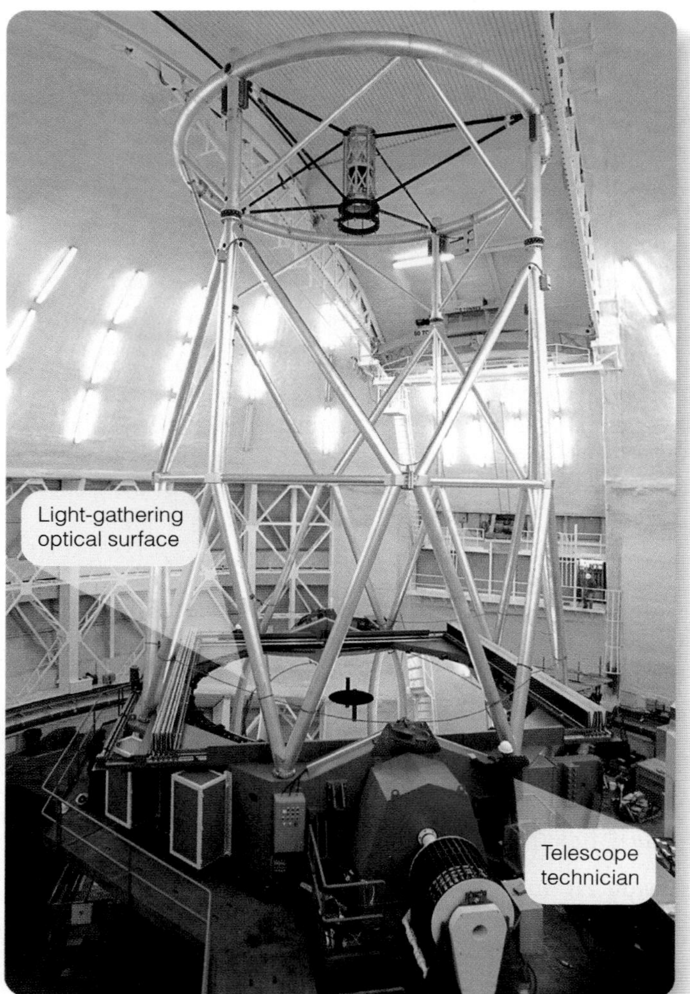

Figure 5-1

Astronomical telescopes are often very large to gather large amounts of starlight. The northern Gemini telescope stands over 19 m (60 feet) high when pointed straight up. Its optical mirror is 8.1 m (26.5 feet) in diameter—larger than some classrooms. The dome of this telescope is shown at the left of the photo on the opening page of this chapter. Note the telescope technician in the white hard hat for scale. (NOAO/AURA/NSF)

If you wish to gather visible light, a normal telescope will do; but, as you will see in this chapter, visible light is only one kind of radiation. You can extract information from other forms of radiation by using specialized telescopes. Radio telescopes, for example, give you an entirely different view of the sky. Some of these specialized telescopes can be used from Earth's surface, but some must go into orbit above Earth's atmosphere. Telescopes that observe X rays, for instance, must be placed in orbit.

Astronomers no longer study the sky by mapping constellations or charting the phases of the moon. Modern astronomers use the most sophisticated telescopes and instruments to analyze starlight. Consequently, this chapter begins with a discussion of the nature of light.

5-1 Radiation: Information from Space

JUST AS A BOOK ON BREAD BAKING might begin with a discussion of flour, this chapter on telescopes begins with a discussion of light—not just visible light, but the entire range of radiation from the sky.

Light as a Wave and a Particle

If you have admired the colors in a soap bubble, you have seen light behave as a wave. But when that light enters the light meter on a camera, it behaves as a particle. How it behaves depends on how you observe it—it is both wave and particle.

You experience waves whenever you hear sound. Sound waves are a mechanical disturbance that travels through the air from source to ear. Sound requires a medium; so, on the moon, where there is no air, there can be no sound. In contrast, light is made up of electric and magnetic waves that can travel through empty space. Unlike sound, light does not require a medium and thus can travel through a perfect vacuum.

Because light is made up of both electric and magnetic fields, it is referred to as **electromagnetic radiation.** As you are about to see, visible light is only one form of electromagnetic radiation. These oscillating electric and magnetic fields travel through space at about 300,000 km/s (186,000 mi/s). This is commonly referred to as the speed of light, c, but it is in fact the speed of all such radiation.

It may seem odd to use the word *radiation* when you speak of light. The word can be used to refer to high-energy particles emitted from radioactive atoms, and thus you have learned to be a little bit concerned when you see this word. But it really refers to anything that spreads outward from a source. Light radiates from a source, so you can correctly refer to light as a form of radiation.

Electromagnetic radiation is a wave phenomenon—that is, it is associated with a periodically repeating disturbance, a wave. You are familiar with waves in water: If you disturb a pool of water, waves spread across the surface. Imagine that you use a meter stick to measure the distance between the successive peaks of a wave. This distance is the **wavelength,** usually represented by the Greek letter lambda (λ).

Although light does behave as a wave, it also behaves as a particle. A particle of light is called a **photon,** and you can think of a photon as a bundle of waves.

The amount of energy a photon carries depends inversely on its wavelength. That is, shorter-wavelength photons carry more energy, and longer-wavelength photons carry less. We can express this relationship in a simple formula:

$$E = \frac{hc}{\lambda}$$

Here h is Planck's constant (6.6262×10^{-34} joule second), and c is the speed of light (3×10^8 m/s). We will not use this formula for a calculation. The important point is the inverse relationship between the energy E and the wavelength λ. As λ gets smaller, E gets larger. Thus a photon of visible light carries a very small amount of energy, but a photon with a wavelength much shorter than that of visible light can carry much more energy.

The Electromagnetic Spectrum

A spectrum is an array of electromagnetic radiation in order of wavelength. You are most familiar with the spectrum of visible light, which you see in rainbows. The colors of the spectrum differ in wavelength, with red having the longest wavelength and violet the shortest, as shown in the visible spectrum at the top of ■ Figure 5-2.

The average wavelength of visible light is about 0.0005 mm. You could put 50 light waves end to end across the thickness of a sheet of household plastic wrap. It is too awkward to measure such short distances in millimeters, so we will measure the wavelength of light using the **nanometer (nm),** one-billionth of a meter (10^{-9} m). Another unit that astronomers commonly use is called the **angstrom (Å)** (named after the Swedish astronomer Anders Jonas Ångström). One angstrom is 10^{-10} m. The wavelength of visible light ranges between 400 nm and 700 nm or between 4000 Å and 7000 Å. Radio astronomers often refer to wavelength using meters, centimeters, or millimeters.

Figure 5-2 shows how the visible spectrum makes up only a small part of the electromagnetic spectrum. Beyond the red end of the visible spectrum lies infrared radiation, where wavelengths range from 700 nm to about 1 mm. Your eyes are not sensitive to this radiation, but your skin senses it as heat. A heat lamp is nothing more than a bulb that gives off large amounts of infrared radiation.

Beyond the infrared part of the electromagnetic spectrum lie radio waves. The radio radiation used for AM radio trans-

■ Figure 5-2

The spectrum of visible light, extending from red to violet, is only part of the electromagnetic spectrum. Most radiation is absorbed in Earth's atmosphere, and only radiation in the visual window and the radio window can reach Earth's surface.

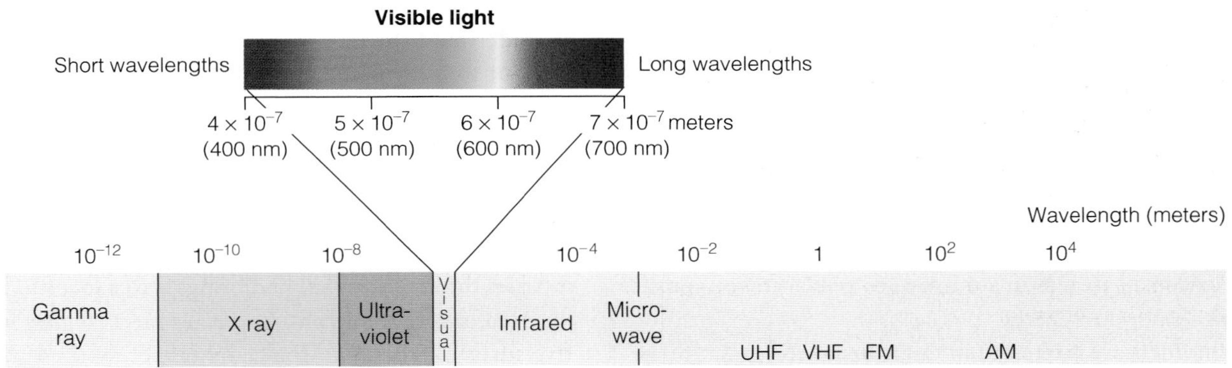

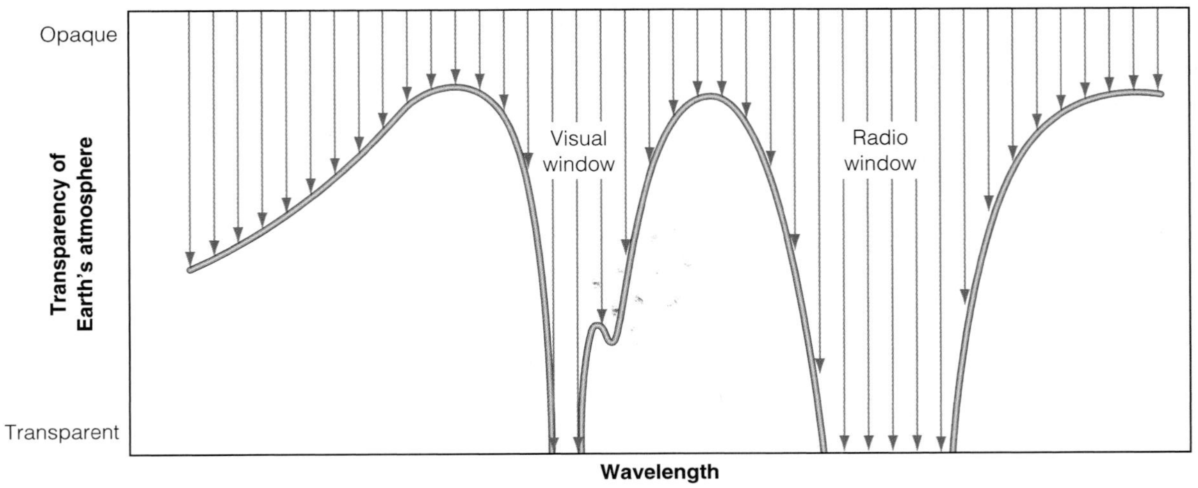

missions has wavelengths of a few kilometers down to a few hundred meters, while FM, television, and military, governmental, and ham radio transmissions have wavelengths that range down to a few tens of centimeters. Microwave transmissions, used for radar and long-distance telephone communications, for instance, have wavelengths from a few centimeters down to about 1 mm.

You may not think of radio waves in terms of wavelength because radio dials are marked in units of frequency, the number of waves that pass a stationary point in 1 second. To calculate the wavelength of a radio wave, divide the speed of light by the frequency. Thus, when you tune in your favorite FM station at 89.5 MHz (million cycles per second), you are adjusting your radio to detect radio photons with a wavelength of 335 cm.

The distinction between the wavelength ranges is not sharp. Long-wavelength infrared radiation and the shortest microwave radio waves are the same. Similarly, there is no clear division between the short-wavelength infrared and the long-wavelength part of the visible spectrum. It is all electromagnetic radiation.

Look once again at the electromagnetic spectrum in Figure 5-2 and notice that electromagnetic waves shorter than violet are called *ultraviolet.* Electromagnetic waves even shorter are called *X rays,* and the shortest are gamma rays. Again, the boundaries between these wavelength ranges are not clearly defined.

X rays and gamma rays can be dangerous, and even ultraviolet photons have enough energy to do you harm. Small doses produce a suntan; larger doses can cause sunburn, and extreme doses might produce skin cancers. Contrast this to the lower-energy infrared photons. Individually, they have too little energy to affect skin pigment, a fact that explains why you can't get a tan from a heat lamp. Only by concentrating many low-energy photons in a small area, as in a microwave oven, can you transfer significant amounts of energy.

Astronomers are interested in electromagnetic radiation because it carries clues to the nature of stars, planets, and other celestial objects. Earth's atmosphere is opaque to most electromagnetic radiation, as shown by the graph at the bottom of Figure 5-2. Gamma rays, X rays, and some radio waves are absorbed high in Earth's atmosphere, and a layer of ozone (O_3) at an altitude of about 30 km absorbs ultraviolet radiation. Water vapor in the lower atmosphere absorbs the longer-wavelength infrared radiation. Only visible light, some shorter-wavelength infrared, and some radio waves reach Earth's surface through two wavelength regions called **atmospheric windows.** Obviously, if you wish to study the sky from Earth's surface, you must look out through one of these windows.

AceAstronomy™ Log into AceAstronomy and select this chapter to see Astronomy Exercise "The Electromagnetic Spectrum" and explore different wavelength regions.

Inquire | Review | Analyze

What could you see if your eyes were sensitive only to X rays? Sometimes the critical analysis of an idea is easier if you try to imagine a totally new situation. In this case, you might at first expect to be able to see through walls, but remember that your eyes detect only light that already exists. There are almost no X rays bouncing around at Earth's surface, so if you had X-ray eyes, you would be in the dark and would be unable to see anything. Even when you looked up at the sky, you would see nothing, because Earth's atmosphere is not transparent to X rays. If Superman can see through walls, it is not because his eyes can detect X rays.

But suppose your eyes were sensitive only to radio waves. Would you be in the dark?

■ ■ ■

Connections: Now that you know something about electromagnetic radiation, you can study the tools astronomers use to gather and analyze that radiation.

5-2 Optical Telescopes

ASTRONOMERS BUILD OPTICAL TELESCOPES to gather light and focus it into sharp images. This requires sophisticated optical and mechanical designs, and it leads astronomers to build gigantic telescopes on the tops of high mountains. To begin, you need to understand the terminology of telescopes, but it is more important to understand how different kinds of telescopes work and why some are better than others.

Two Kinds of Telescopes

Astronomical telescopes focus light into an image in one of two ways, as shown in ■ Figure 5-3. A lens bends (refracts) the light as it passes through the glass and brings it to a focus to form a small inverted image. A mirror—a concave piece of glass with a reflective surface—forms an image by reflecting the light. In either case, the **focal length** is the distance from the lens or mirror to the image formed of a distant light source, such as a star (■ Figure 5-4). Short-focal-length lenses and mirrors must be strongly curved, and long-focal-length lenses and mirrors are less strongly curved. Grinding the proper optical shapes is an expensive process. The surfaces of lenses and mirrors must be shaped and polished to accuracies less than the wavelength of light. Creating the optics for a large telescope can take months or years; involve huge, precision machinery; and employ expert optical engineers and scientists.

The main lens in a refracting telescope is called the **primary lens,** and the main mirror in a reflecting telescope is called the **primary mirror.** These are also called the **objective lens** and **mirror.** Both kinds of telescopes form a very small, inverted image that is difficult to observe directly, so astronomers use a small lens called the **eyepiece** to magnify the image and make it convenient to view.

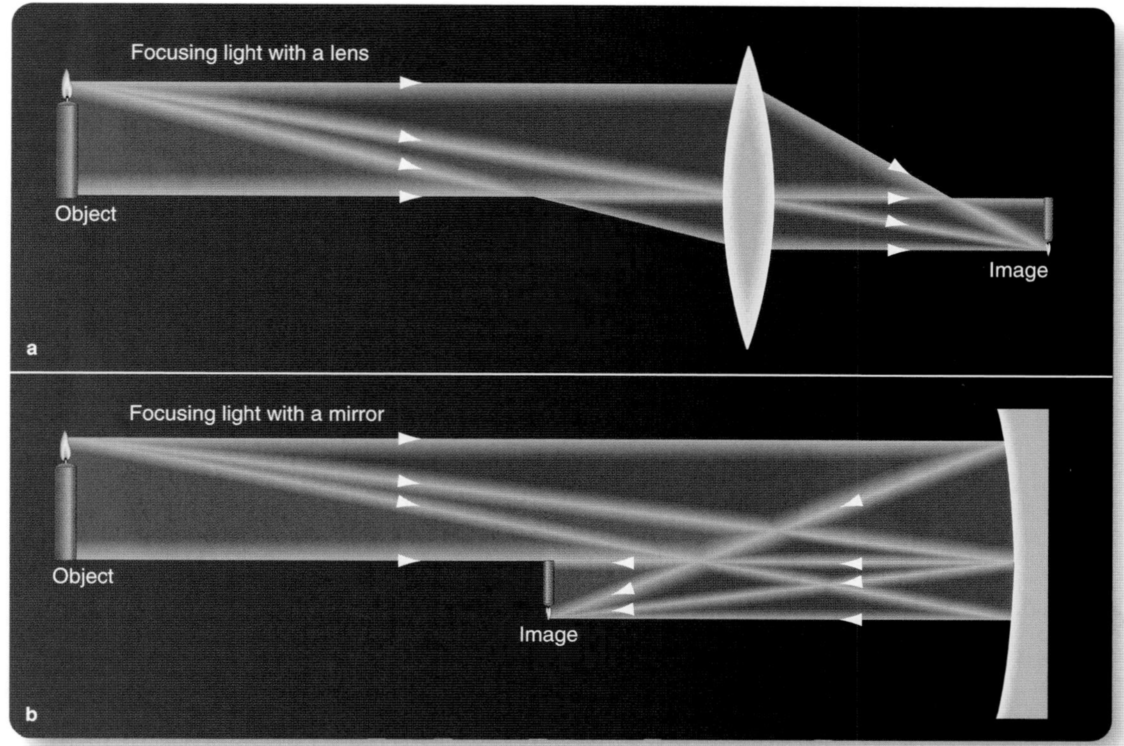

■ **Figure 5-3**

(a) To see how a lens can focus light, trace light rays from the flame and base of a candle through a lens, where they are refracted to form an inverted image. (b) A mirror forms an image by reflection from a concave surface. Notice that the light is reflected from the reflective front surface of the mirror and does not enter the glass.

Because there are two ways to focus light, there are two kinds of astronomical telescopes. **Refracting telescopes** use a large lens to gather and focus the light, whereas **reflecting telescopes** use a concave mirror (■ Figure 5-5). The advantages of the reflecting telescope have made it the preferred design for modern observatories.

Refracting telescopes suffer from a serious optical distortion that limits their usefulness. When light is refracted through glass, shorter wavelengths bend more than longer wavelengths, and blue light comes to a focus closer to the lens than does red light (■ Figure 5-6a). If you focus the eyepiece on the blue image, the red light is out of focus, and you see a red blur around the image. If you focus on the red image, the blue light blurs. The color separation is called **chromatic aberration.** Telescope designers can grind a telescope lens of two components made of different kinds of glass and so bring two different wavelengths to the same focus

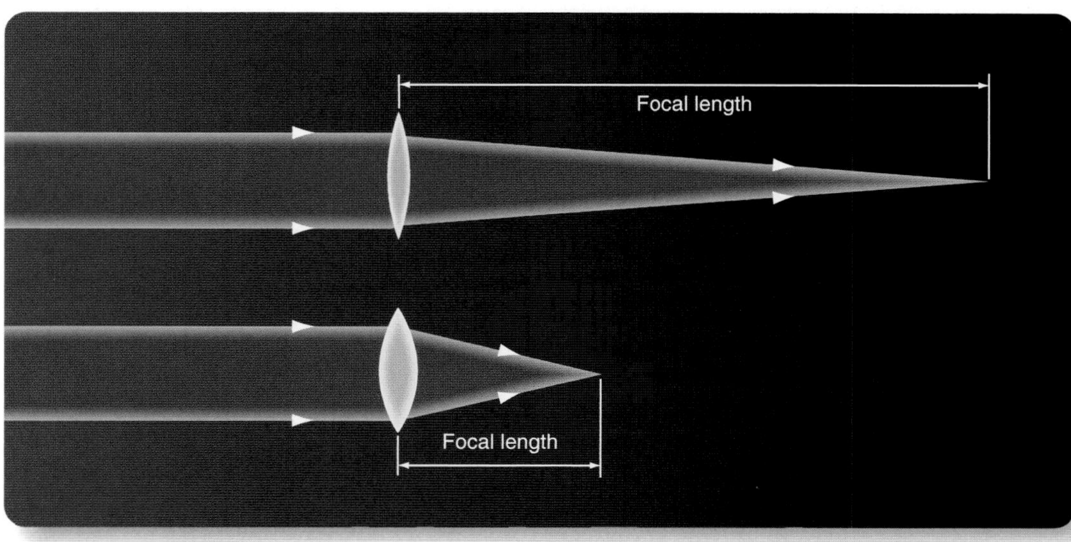

■ **Figure 5-4**

The focal length of a lens is the distance from the lens to the point where parallel rays of light come to a focus. The lens at the top has a longer focal length than does the lens at the bottom.

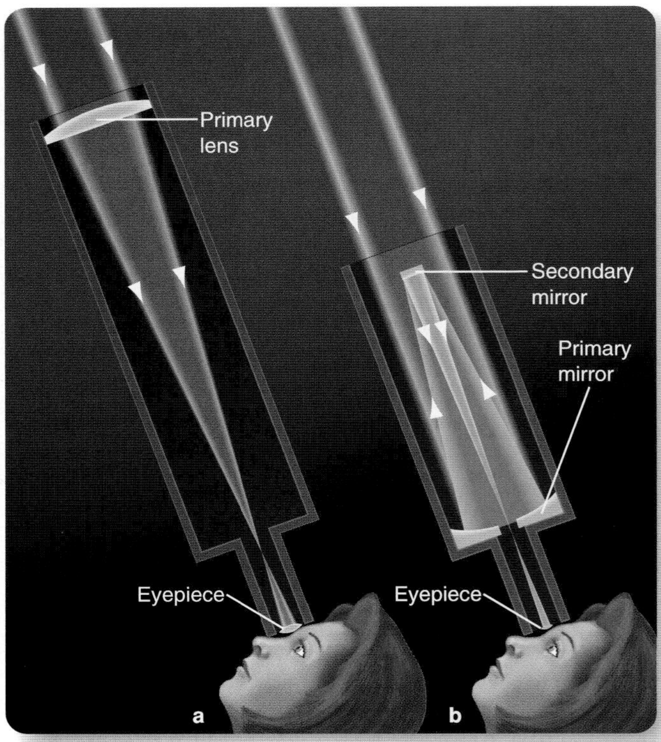

Active Figure 5-5

(a) A refracting telescope uses a primary lens to focus starlight into an image that is magnified by a lens called an eyepiece. The primary lens has a long focal length, and the eyepiece has a short focal length. (b) A reflecting telescope uses a primary mirror to focus the light by reflection. A small secondary mirror reflects the starlight back down through a hole in the middle of the primary mirror to the eyepiece.

Ace Astronomy™ Log into AceAstronomy and select this chapter to see the Active Figure called "Refractors and Reflectors." Watch light pass through the optics.

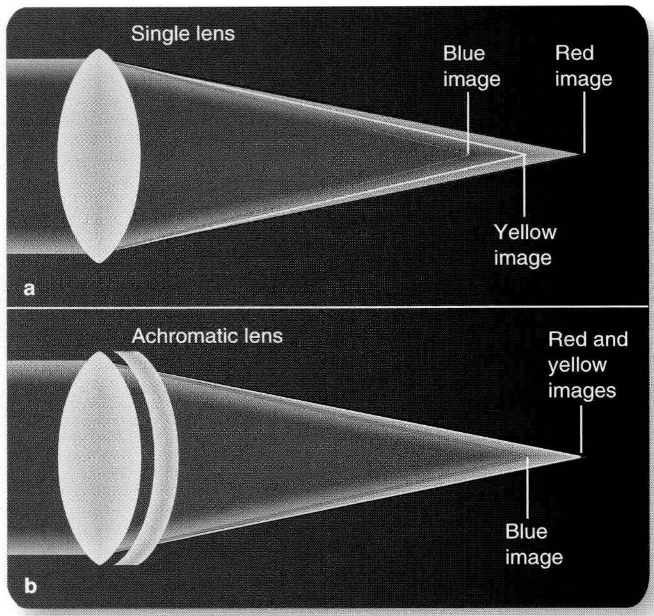

Figure 5-6

(a) A normal lens suffers from chromatic aberration because short wavelengths bend more than long wavelengths. (b) An achromatic lens, made in two parts, can bring any two colors to the same focus, but other colors remain slightly out of focus.

(Figure 5-6b). This does improve the image, but these **achromatic lenses** are not totally free of chromatic aberration, because other wavelengths still blur. Telescopes made with such lenses were popular until the end of the 19th century.

The primary lens of a refracting telescope is very expensive to make because it must be achromatic, and the glass must be pure and flawless because the light passes through the lens. The four surfaces must be ground precisely, and the lens can be supported only along its edge. The largest refracting telescope in the world was completed in 1897 at Yerkes Observatory in Wisconsin. Its lens, 1 m (40 inches) in diameter, weighs half a ton. Although modern glass would make it possible to build slightly larger refracting telescopes, reflecting telescopes have important advantages.

Reflecting telescopes are much less expensive because the light reflects from the front surface of the mirror. Consequently only the front surface need be ground to a precise shape. This front surface is coated with a highly reflective surface of aluminum alloy, and the light reflects off the surface without entering the glass. Consequently, the glass of the mirror need not be perfectly transparent, and the mirror can be supported over its back surface to reduce sagging. Most important, reflecting telescopes do not suffer from chromatic aberration, because the light is reflected toward the focus before it can enter the glass. For these reasons, every large astronomical telescope built since the beginning of the 20th century has been a reflecting telescope.

Ace Astronomy™ Log into AceAstronomy and select this chapter to see Astronomy Exercise "Lenses—Focal Length" to adjust the shape of a lens.

Ace Astronomy™ Log into AceAstronomy and select this chapter to see Astronomy Exercise "Telescopes: Objective Lens and Eyepiece," and you can build your own telescope.

The Powers of a Telescope

Astronomers struggle to build large telescopes because a telescope can help our eyes in three important ways—the three powers of a telescope. The most important of these depends on the diameter of the telescope.

Most interesting celestial objects are faint sources of light, so you need a telescope that can gather large amounts of light to produce a bright image. **Light-gathering power** refers to the ability of a telescope to collect light. Catching light in a telescope is like catching rain in a bucket—the bigger the bucket, the more

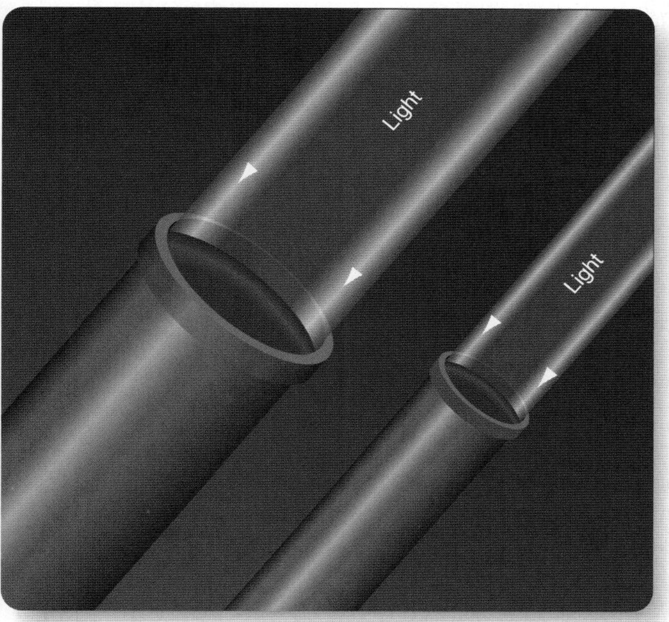

■ **Figure 5-7**

Gathering light is like catching rain in a bucket. A large-diameter telescope gathers more light and has a brighter image than does a smaller telescope of the same focal length.

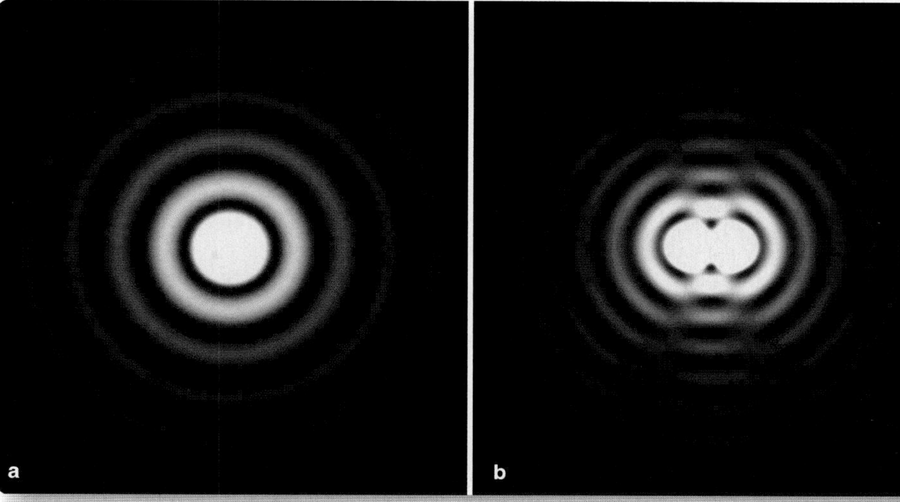

■ **Figure 5-8**

(a) Stars are so far away that their images are points, but the wave nature of light surrounds each star image with diffraction fringes (much magnified in this computer model). (b) Two stars close to each other have overlapping diffraction fringes and become impossible to detect separately. (Computer model by M. A. Seeds)

rain it catches (■ Figure 5-7). This is why astronomers use large telescopes and why they refer to telescopes by diameter.

The second power, **resolving power,** refers to the ability of the telescope to reveal fine detail. Because light acts as a wave, it produces a small **diffraction fringe** around every point of light in the image, and you cannot see any detail smaller than the fringe (■ Figure 5-8). Astronomers can't eliminate diffraction fringes, but the larger a telescope is in diameter, the smaller the diffraction fringes are. Thus the larger the telescope, the better its resolving power.

Two other factors—optical quality and atmospheric conditions—limit resolving power. A telescope must contain high-quality optics to achieve its full potential resolving power. Even a large telescope shows little detail if its optics are marred with imperfections. In addition, when you look through a telescope, you look through miles of turbulent air in Earth's atmosphere, which makes the image dance and blur, a condition called **seeing** (■ Figure 5-9). A related phenomenon is the twinkling of a star. The twinkles are caused by turbulence in Earth's atmosphere, and a star near the horizon, where you look through more air, will twinkle more than a star overhead.

On a night when the atmosphere is unsteady, the stars twinkle, the images are blurred, and the seeing is bad. Even under good seeing conditions, the detail visible through a large telescope is limited, not by its diffraction fringes, but by the air through which the observer must look. A telescope performs best on a high mountaintop, where the air is thin and steady,

■ **Active Figure 5-9**

The left half of this photograph of a galaxy is from an image recorded on a night of poor seeing. Small details are blurred. The right half of the figure is from an image recorded on a night when Earth's atmosphere above the telescope was steady and the seeing was better. Much more detail is visible under good seeing conditions. (Courtesy William Keel)

AceAstronomy™ **Log into AceAstronomy and select this chapter to see the Active Figure called "Resolution and Telescopes." You can control telescope diameter and watch resolution change.**

Resolving Power and the Accuracy of a Measurement

Have you ever seen a movie in which the hero magnifies a newspaper photo and reads some tiny detail? It isn't really possible, because newspaper photos are made up of tiny dots of ink, and no detail smaller than a single dot will be visible no matter how much you magnify the photo. In fact, all images are made up of elements of some sort. On a computer screen, the picture elements are called pixels. That means there is a limit to the amount of detail you can see in the image. In an image formed by a telescope on Earth's surface, the size of the picture element is set by the blurring called seeing. It would be foolish to try to resolve (detect) any detail smaller than this limit.

This limitation is true of all measurements in science. A zoologist might be trying to measure the length of a live snake, or a sociologist might be trying to measure the attitudes of people toward drunk driving, but both face limits to the resolution of their measurements. The zoologist might specify that the snake was 43.28932 cm long, and the sociologist might say that 98.2491 percent of people oppose drunk driving, but a critic might point out that it isn't possible to make these measurements that accurately. The resolution of the techniques does not justify the accuracy implied.

Science is based on measurement, and whenever you make a measurement you should ask yourself how accurate that measurement can be. The accuracy of the measurement is limited by the resolution of the measurement technique, just as the amount of detail in a photograph is limited by the resolution of the photo.

A high-resolution image of Mars reveals details such as mountains, craters, and the southern polar cap. (NASA)

but, even then, atmospheric turbulence spreads star images into blobs 1 to 0.5 seconds of arc in diameter. No smaller detail is visible.

This limitation on the amount of information in an image is related to the limitation on the accuracy of a measurement. All measurements have some built-in uncertainty (**Window on Science 5-1**), and scientists must learn to work within those limitations.

The third and least important power of a telescope is **magnifying power,** the ability to make the image bigger. Because the amount of detail you can see is limited by the seeing conditions and the resolving power, very high magnification does not necessarily show you more detail. Also, you can change the magnification by changing the eyepiece, but you cannot alter the telescope's light-gathering or resolving power.

Compare an astronomer's telescope with a biologist's microscope. A microscope is designed primarily to magnify and thus show you things too small to see. An astronomer's telescope solves a different problem. Its primary function is to gather light and show you things too faint to see.

Nearly all major observatories are located far from big cities and usually on high mountains. Astronomers avoid cities because **light pollution,** the brightening of the night sky by light scattered from artificial outdoor lighting, can make it impossible to see faint objects (■ Figure 5-10). In fact, many residents of cities are unfamiliar with the beauty of the night sky because they can see only the brightest stars. Astronomers prefer to place their telescopes on carefully selected high mountains. The air there is thin and more transparent, but, most important, astronomers select mountains where the air flows smoothly and is not turbulent. This produces the best seeing. Building an observatory on top of a high mountain far from civilization is difficult and expensive, but the dark sky and steady seeing make it worth the effort (■ Figure 5-11).

When you compare telescopes, you should consider their powers. **Reasoning with Numbers 5-1** shows how to calculate the powers of a telescope. This will be useful if you decide to buy a telescope of your own.

Ace Astronomy™ Log into AceAstronomy and select this chapter to see Astronomy Exercise "Telescopes and Resolution I" and experiment to discover what affects resolution.

Ace Astronomy™ Log into AceAstronomy and select this chapter to see Astronomy Exercise "Telescopes and Resolution II," and you can try different wavelengths of light to get the best resolution.

Ace Astronomy™ Log into AceAstronomy and select this chapter to see Astronomy Exercise "Particulate" and see what happens when Earth's atmosphere is polluted by dust.

Ace Astronomy™ Log into AceAstronomy and select this chapter to see Astronomy Exercise "Heat" and find out what happens when the air above your telescope is stirred by rising heat currents.

Ace Astronomy™ Log into AceAstronomy and select this chapter to see Astronomy Exercise "Light Pollution," and you can see why astronomers build observatories far from cities.

Figure 5-10

This satellite view of the continental United States at night shows the light pollution produced by outdoor lighting. Not only does the glare drown out the fainter stars and interfere with astronomy, but it wastes electrical power. Many astronomers work with city governments to enact laws that improve lighting on the ground and reduce light scattered into the night sky. (NOAA)

Buying a Telescope

Thinking about how you should shop for a new telescope will not only help you if you decide to buy one but will also illustrate some important points about astronomical telescopes.

Assuming you have a fixed budget, you should buy the highest-quality optics and the largest-diameter telescope you can afford. Of the two things that limit what you see, optical quality is under your control. You can't make the atmosphere less turbulent, but you should buy good optics. If you buy a telescope from a toy store and it has plastic lenses, you shouldn't expect to see very much. Also, you want to maximize the light-gathering power of your telescope, so you want to purchase the largest-diameter telescope you can afford. Given a fixed budget, that means you should buy a reflecting telescope rather than a refracting telescope. Not only will you get more diameter per dollar, but your telescope will not suffer from chromatic aberration.

You can safely ignore magnification. Department stores and camera stores may advertise telescopes by quoting their magnifica-

Figure 5-11

The domes of four giant telescopes are visible at upper left at Paranal Observatory, built by the European Southern Observatory at an altitude of 2635 m (8640 ft) atop the mountain Cerro Paranal in the Atacama desert of northern Chile. Located 120 km (75 mi) from the nearest city, the mountaintop was chosen for its steady air and isolation from lights. It is believed to be the driest area on Earth. (ESO)

Reasoning with Numbers | 5-1

The Powers of a Telescope

Light-gathering power is proportional to the area of the telescope objective. A lens or mirror with a large area gathers a large amount of light. The area of a circular lens or mirror of diameter D is $\pi(D/2)^2$. To compare the relative light-gathering powers (LGP) of two telescopes A and B, you can calculate the ratio of the areas of their objectives, which reduces to the ratio of their diameters (D) squared:

$$\frac{LGP_A}{LGP_B} = \left(\frac{D_A}{D_B}\right)^2$$

Example A: Suppose you compare a 4-cm telescope with a 24-cm telescope. How much more light will the large telescope gather? **Solution:**

$$\frac{LGP_{24}}{LGP_4} = \left(\frac{24}{4}\right)^2 = 6^2 = 36 \text{ times more light}$$

Example B: Your eye acts like a telescope with a diameter of about 0.8 cm, the maximum diameter of the pupil. How much more light can you gather if you use a 24-cm telescope? **Solution:**

$$\frac{LGP_{24}}{LGP_{eye}} = \left(\frac{24}{0.8}\right)^2 = (30)^2 = 900 \text{ times more light}$$

The resolving power of a telescope is the angular distance between two stars that are just barely visible through the telescope as two separate images. The resolving power α, in seconds of arc, equals 11.6 divided by the diameter of the telescope in centimeters:

$$\alpha = \frac{11.6}{D}$$

Example C: What is the resolving power of a 10-cm telescope? **Solution:**

$$\alpha = \frac{11.6}{10} = 1.16 \text{ seconds of arc}$$

If the lenses are of good quality, and if the seeing is good, you should be able to distinguish as separate points of light any pair of stars farther apart than 1.16 seconds of arc. If the stars are any closer together, diffraction fringes blur the stars together into a single image.

The magnification M of a telescope is the ratio of the focal length of the objective lens or mirror F_o divided by the focal length of the eyepiece F_e:

$$M = \frac{F_o}{F_e}$$

Example D: What is the magnification of a telescope whose focal length is 80 cm used with an eyepiece whose focal length is 0.5 cm? **Solution:** The magnification is 80 divided by 0.5, or 160 times.

tion, but it is not an important number. What you can see is fixed by light-gathering power, optical quality, and Earth's atmosphere. Besides, you can change the magnification by changing eyepieces.

Other things being equal, you should choose a telescope with a solid mounting that will hold the telescope steady and allow you to point at objects easily. Computer-controlled pointing systems are available for a price on many small telescopes. A good telescope on a poor mounting is almost useless.

You might be buying a telescope to put in your backyard, but you must think about the same issues astronomers consider when they design giant telescopes to go on mountaintops. In fact, some of the newest telescopes solve these traditional problems in new ways.

New-Generation Telescopes

For most of the 20th century, astronomers faced a serious limitation on the size of astronomical telescopes. Traditional telescope mirrors were made thick to avoid sagging that would distort the reflecting surface, but those thick mirrors were heavy. The 5-m (200-in.) mirror on Mount Palomar weighs 14.5 tons. These traditional telescopes were big, heavy, and expensive.

Modern astronomers have solved these problems in a number of ways. Study **Modern Astronomical Telescopes** on pages 88 and 89 and notice three important advances in telescope design made possible by high-speed computers. First, astronomers can now build simpler, stronger telescope mountings and depend on computers to move the telescope and follow the westward motion of the stars as Earth rotates.

Second, computer control of the shape of telescope mirrors allows the use of thin, lightweight mirrors—either "floppy" mirrors or segmented mirrors. Lowering the weight of the mirror lowers the weight of the rest of the telescope and makes it stronger and less expensive.

The third advance is the way astronomers use high-speed computers to reduce seeing distortion caused by Earth's atmosphere. Only a few decades ago, many astronomers argued that it wasn't worth building more large telescopes on Earth's surface because of the limitations set by seeing. Now a number of new giant telescopes have been built, and more are in development.

An international collaboration of astronomers built the Gemini telescopes with 8.1-m thin mirrors (■ Figure 5-12). One is located in the northern hemisphere and one in the southern hemisphere to cover the entire sky. The European Southern Observatory has built the Very Large Telescope (VLT) high in the remote Andes of northern Chile (Figure 5-12). The VLT consists of four telescopes with computer-controlled mirrors 8.2 m

The LBT has the light-gathering power of a 11.8-m telescope and the resolving power of a 22.8-m telescope.
The mirrors in the VLT telescopes are 8.2 m in diameter.
The two Gemini telescopes have 8.1-m mirrors with active optics.

■ **Figure 5-12**

Astronomers have begun building multiple telescopes to solve different problems. Two Gemini telescopes have been built, one in Hawaii and one in Chile, to observe the entire sky. The Large Binocular Telescope (LBT) carries two 8.4-m mirrors that combine their light. The four telescopes of the VLT are housed in separate domes, but they can combine to work as a single very large telescope. (Gemini: NOAO/AURA/NSF; LBT: Large Binocular Telescope Project and European Industrial Engineering; VLT: ESO)

in diameter and only 17.5 cm (6.9 in.) thick. The four telescopes can work singly or can combine their light to work as one large telescope. Italian, American, and German astronomers are building the Large Binocular Telescope, which carries a pair of 8.4-m mirrors on a single mounting, as shown in Figure 5-12. Around the world astronomers are drawing plans for large telescopes, including truly gigantic instruments with segmented mirrors 50 m and even 100 m in diameter. Computer control of the optics makes such huge telescopes worth considering.

The days when astronomers worked beside their telescopes through long, dark, cold nights are nearly gone. The complexity and sophistication of telescopes require a battery of computers, and almost all research telescopes are run from control rooms that astronomers call warm rooms. Astronomers don't need to be kept warm, but computers demand comfortable working conditions (■ Figure 5-13).

Interferometry

One of the reasons astronomers build big telescopes is to increase resolving power, and astronomers have been able to achieve very high resolution by connecting multiple telescopes together to work as if they were a single telescope. This method of synthesizing a larger telescope is known as **interferometry** (■ Figure 5-14).

To work as an interferometer, the separate small telescopes must combine their light through a network of mirrors, and the path that each light beam travels must be controlled so that it does not vary by more than some small fraction of the wavelength. Turbulence in Earth's atmosphere constantly distorts the light, and high-speed computers must continuously adjust the light paths. Recall that the wavelength of light is very short, roughly 0.0005 mm, so building optical interferometers is one of the most difficult technical problems that astronomers face. Infrared- and radio-wavelength interferometers are slightly easier to build because the wavelengths are longer. In fact, as you will

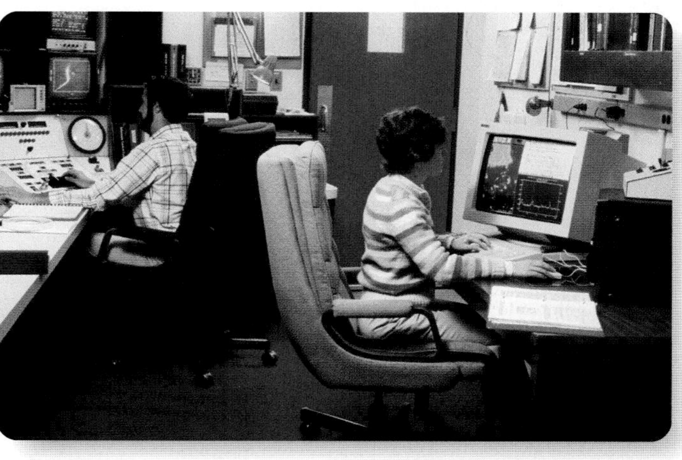

■ **Figure 5-13**

In the control room of the 4-meter telescope atop Kitt Peak National Observatory, the telescope operator at left manages the operation and safety of the telescope. The astronomer at right operates the instruments, records data, and makes decisions on the observing program. Astronomers work through the night controlling the computers that control the telescope and its instruments. (NOAO/AURA/NSF)

Modern Astronomical Telescopes

The traditional telescopes described on this page are limited by complexity, weight, and Earth's atmosphere. Modern solutions are shown on the opposite page.

In larger telescopes the light can be focused to a prime focus position high in the telescope tube. Although it is a good place to image faint objects, the prime focus is inconvenient for large instruments. A secondary mirror can reflect the light through a hole in the primary mirror to a Cassegrain focus. This focal arrangement may be the most common form of astronomical telescope.

Secondary mirror

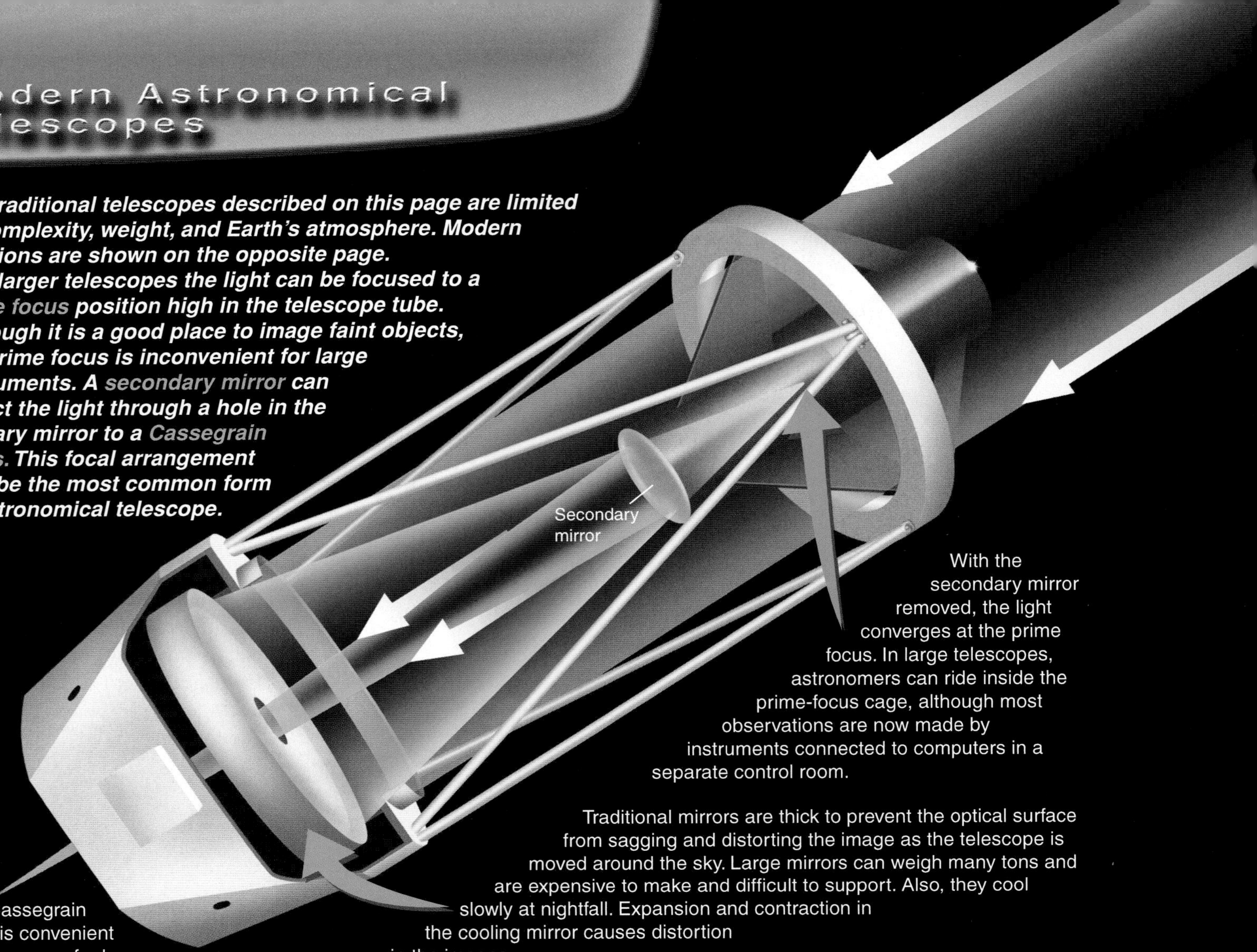

With the secondary mirror removed, the light converges at the prime focus. In large telescopes, astronomers can ride inside the prime-focus cage, although most observations are now made by instruments connected to computers in a separate control room.

Traditional mirrors are thick to prevent the optical surface from sagging and distorting the image as the telescope is moved around the sky. Large mirrors can weigh many tons and are expensive to make and difficult to support. Also, they cool slowly at nightfall. Expansion and contraction in the cooling mirror causes distortion in the images.

The Cassegrain focus is convenient and has room for large instruments.

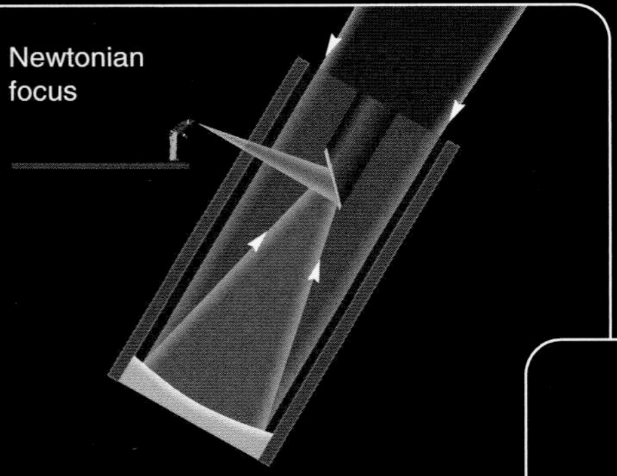

Newtonian focus

Smaller telescopes are often found with a **Newtonian focus,** the arrangement that Isaac Newton used in his first reflecting telescope. The Newtonian focus is inconvenient for large telescopes.

Many small telescopes use a **Schmidt-Cassegrain focus.** A thin correcting plate improves the image but is too slightly curved to introduce serious chromatic aberration.

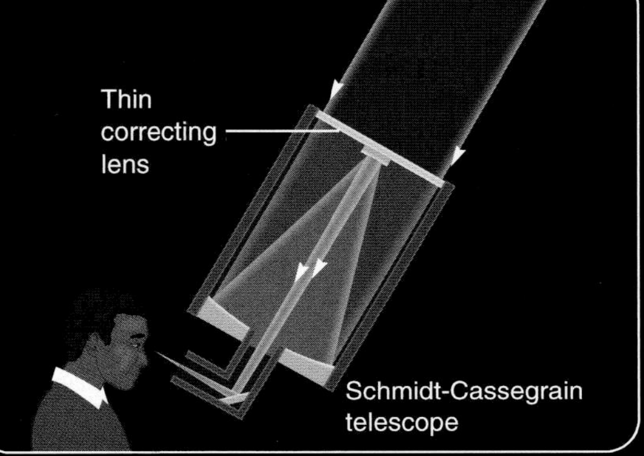

The 4-meter Mayall Telescope at Kitt Peak National Observatory in Arizona can be used at either the prime focus or the Cassegrain focus. Note the human figure at lower right.

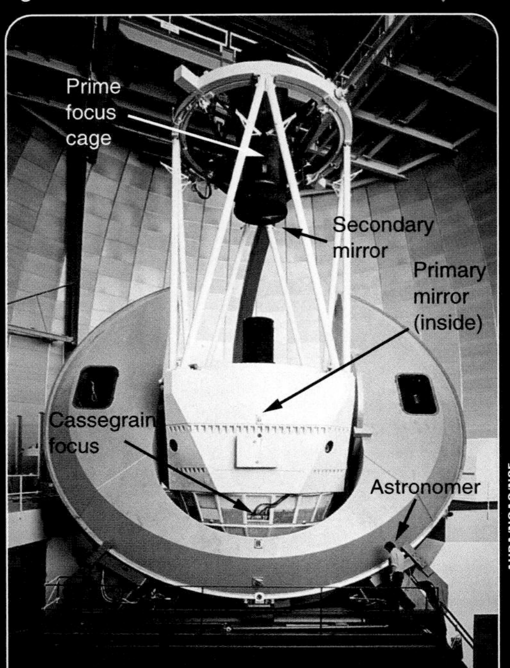

AURA/NOAO/NSF

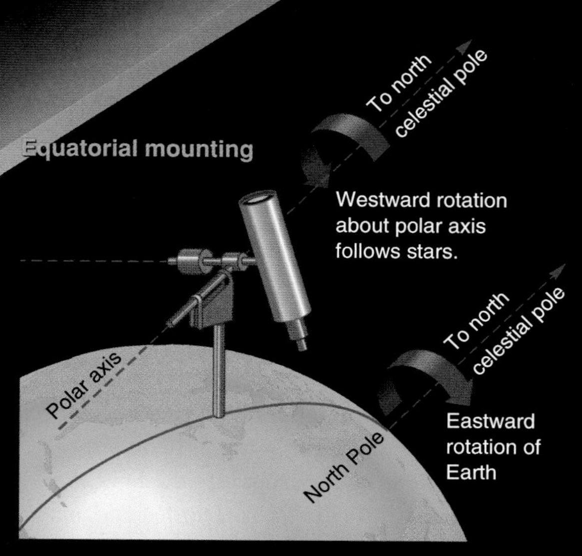

Alt-azimuth mounting

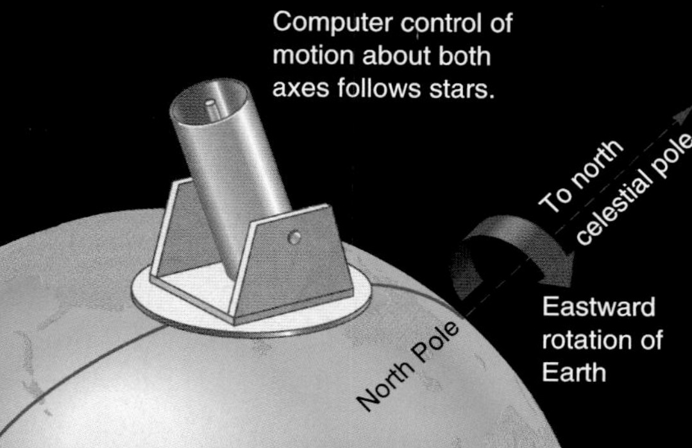

Telescope mountings must contain a **sidereal drive** to move smoothly westward and counter the eastward rotation of Earth. The traditional **equatorial mounting** has a **polar axis** parallel to Earth's axis, but the modern **alt-azimuth mounting** moves like a cannon — up and down and left to right. Such mountings are simpler to build but need computer control to follow the stars.

Unlike traditional thick mirrors, thin mirrors, sometimes called floppy mirrors, weigh less and require less massive support structures. Also, they cool rapidly at nightfall and there is less distortion from uneven expansion and contraction.

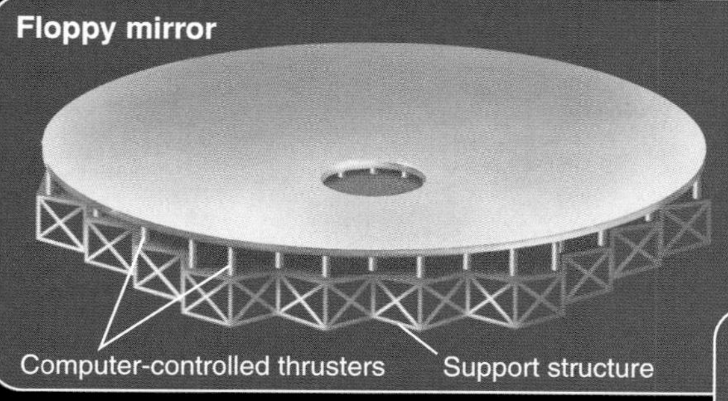

Grinding a large mirror may remove tons of glass and take months, but new techniques speed the process. Some large mirrors are cast in a rotating oven that causes the molten glass to flow to form a concave upper surface. Grinding and polishing such a preformed mirror is much less time consuming.

Mirrors made of segments are economical because the segments can be made separately. The resulting mirror weighs less and cools rapidly.

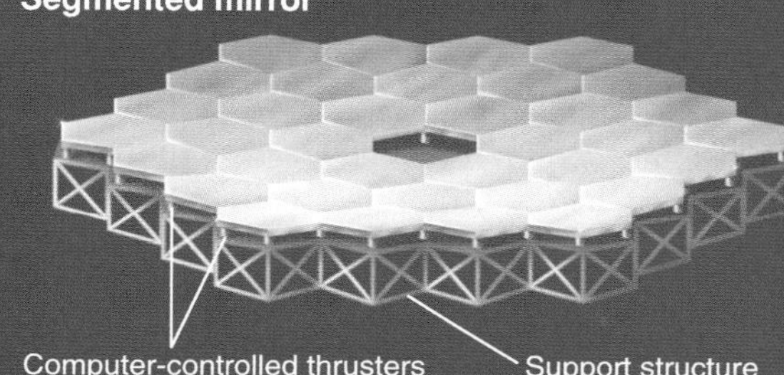

Both floppy mirrors and segmented mirrors sag under their own weight. Their optical shape must be controlled by computer-driven thrusters under the mirror in what is called **active optics.**

The two largest telescopes in the world, the Keck I and Keck II telescopes in Hawaii, contain segmented mirrors 10 m in diameter.

The thrusters are located behind the mirror segments in this photo of the Keck I mirror. The technician is sitting in the front of the light baffle over the Cassegrain hole in the center of the mirror.

Edge of mirror

W. M. Keck Observatory

Adaptive optics in Telescopes

Adaptive optics uses high-speed computers to monitor the image distortion caused by Earth's atmosphere and adjust the optics many times a second to compensate. This can reduce the blurring due to seeing and dramatically improve image quality in Earth-based telescopes.

Adaptive optics off

Object appears to be a single star.

Adaptive optics on

Object revealed as a pair of stars.

1 second of arc

Paul Kalas

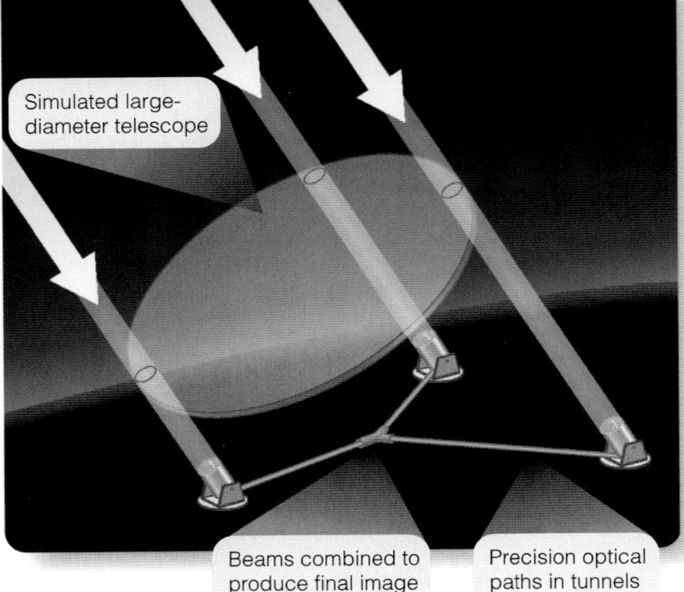

Figure 5-14

In an astronomical interferometer, smaller telescopes can combine their light through specially designed optical tunnels to simulate a larger telescope with a resolution set by the separation of the smaller telescopes.

discover later in this chapter, the first astronomical interferometers worked at radio wavelengths.

The VLT shown in Figure 5-12 consists of four 8.2-m telescopes that can operate separately, but they can be linked together through underground tunnels with three 1.8-m telescopes on the same mountaintop. The resulting optical interferometer provides the resolution of a telescope 200 meters in diameter.

Other telescopes can work as interferometers. The two Keck 10-m telescopes can be used as an interferometer. The Navy Prototype Optical Interferometer built near Flagstaff, Arizona, has small telescopes located along three arms up to 250 m in length. It is being used to study technical aspects of optical interferometry and to make high-precision measures of star positions. The CHARA array on Mt. Wilson combines six 1-meter telescopes to create the equivalent of a telescope one-fifth of a mile in diameter. The Large Binocular Telescope shown in Figure 5-12 can be used as an interferometer.

Although turbulence in Earth's atmosphere can be partially averaged out in an interferometer, plans are being made to put interferometers in space. The Space Interferometry Mission, for example, will work at optical wavelengths and study everything from the cores of erupting galaxies to planets orbiting nearby stars.

Inquire | Review | Analyze

Why do astronomers build observatories at the tops of mountains? Astronomers have joked that the hardest part of building a new observatory is constructing the road to the top of the mountain. It certainly isn't easy to build a large, delicate telescope at the top of a high mountain, but it is worth the effort. A telescope on top of a high mountain is above the thickest part of Earth's atmosphere. There is less air to dim the light, and there is less water vapor to absorb infrared radiation. Even more important, the thin air on a mountaintop causes less disturbance to the image, and thus the seeing is better. A large telescope with modern, high-quality optics is capable of detecting very small details. It is the seeing that limits the detail that astronomers can detect. It really is worth the trouble to build telescopes atop high mountains where the air is thin, dry, and steady.

Astronomers not only build telescopes on mountaintops; they also build gigantic telescopes many meters in diameter. What are the problems and advantages in building such giant telescopes?

■ ■ ■

Connections: Astronomers sometimes refer to a telescope that produces distorted images as a "light bucket." In a sense, all astronomical telescopes are light buckets, because the light they focus into images tells you very little until it is recorded and analyzed by special instruments attached to the telescopes.

5-3 Special Instruments

JUST LOOKING THROUGH A TELESCOPE doesn't tell you much. To use an astronomical telescope to learn about stars, you must be able to analyze the light the telescope gathers. Special instruments attached to the telescope make that possible.

Imaging Systems

The original imaging device in astronomy was the photographic plate. It could record faint objects in long exposures and could be stored for later analysis, but photographic plates have been almost entirely replaced in astronomy by electronic imaging systems.

Most modern astronomers use **charge-coupled devices (CCDs)** to record images. A CCD is a specialized computer chip containing roughly a million microscopic light detectors arranged in an array about the size of a postage stamp. These devices can be used like a small photographic plate, but they have dramatic advantages. They can detect both bright and faint objects in a single exposure, are much more sensitive than a photographic plate, and can be read directly into computer memory for later analysis. Although CCDs for astronomy are extremely sensitive and therefore expensive, less sophisticated CCDs are used in video cameras and digital cameras.

The image from a CCD is stored as numbers in computer memory, so it is easy to manipulate the image to bring out details that would not otherwise be visible. For example, astronomical images are often reproduced as negatives, with the sky white and the stars dark. This makes the faint parts of the image easier to see (■ Figure 5-15). Astronomers also manipulate images to produce **false-color images** in which the colors represent different levels of intensity and are not related to the true colors of the object (Figure 5-15).

Measurements of intensity and color were made in the past using a photometer, a highly sensitive light meter attached to a telescope. Today, however, most such measurements are made on

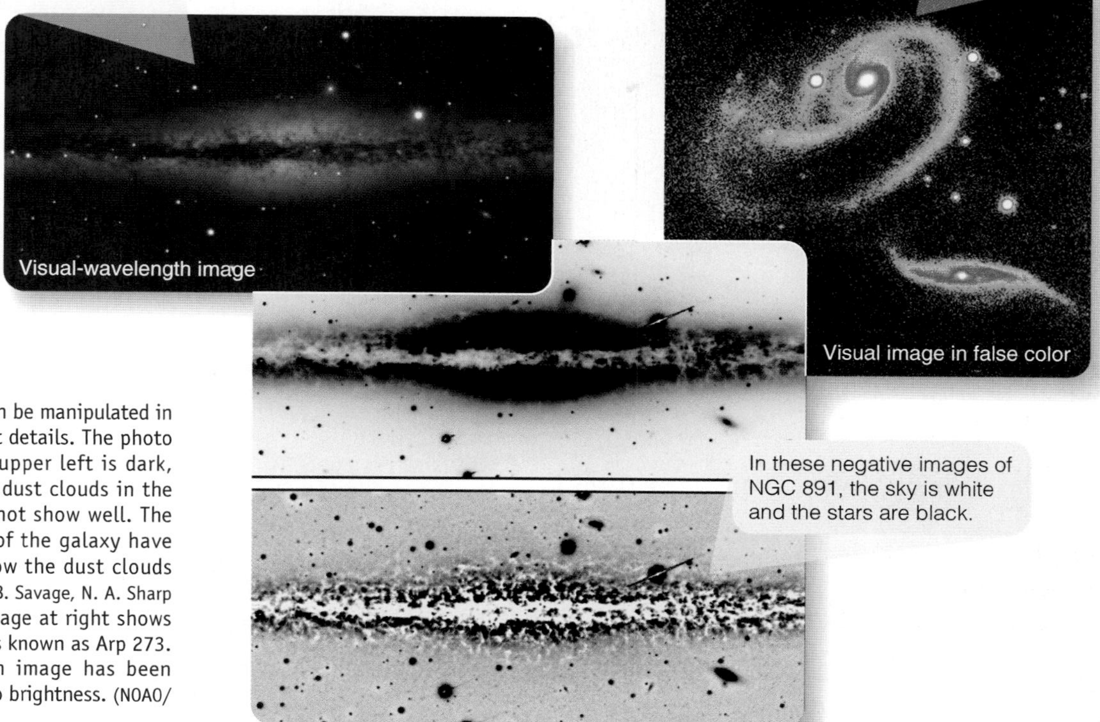
Galaxy NGC 891 as it would look to your eyes. It is edge-on and contains thick dust clouds.
In this image, color shows brightness. White and red are brightest, and yellow and green are dimmer.
Visual-wavelength image
Visual image in false color
In these negative images of NGC 891, the sky is white and the stars are black.
Visual-wavelength negative images

■ **Figure 5-15**

Astronomical images can be manipulated in many ways to bring out details. The photo of the disk galaxy at upper left is dark, and the details of the dust clouds in the disk of the galaxy do not show well. The two negative images of the galaxy have been processed to show the dust clouds more clearly. (C. Howk, B. Savage, N. A. Sharp NOAO/WIYN/NSF) The image at right shows two interacting galaxies known as Arp 273. The visual-wavelength image has been given color according to brightness. (NOAO/WIYN/NSF)

CCD images. Because the CCD image is easily digitized, brightness and color can be measured to high precision.

The Spectrograph

To analyze light in detail, you need to spread the light out according to wavelength into a spectrum, a task performed by a **spectrograph.** You can understand how this works if you reproduce an experiment performed by Isaac Newton in 1666. Boring a hole in his window shutter, Newton admitted a thin beam of sunlight into his darkened bedroom. When he placed a prism in the beam, the sunlight spread into a beautiful spectrum on the far wall. From this Newton concluded that white light was made of a mixture of all the colors.

Newton didn't think in terms of wavelength, but you can use that modern concept to see that the light passing through the prism is bent at an angle that depends on the wavelength. Violet (short wavelength) bends most, and red (long wavelength) least. Thus, the white light entering the prism is spread into a spectrum (■ Figure 5-16a). A typical prism spectrograph contains more than one prism to spread the light farther and lenses to guide the light into the prism and to focus the light onto a photographic plate.

Nearly all modern spectrographs use a grating in place of a prism. A **grating** is a piece of glass with thousands of microscopic parallel lines scribed onto its surface. Different wavelengths of light reflect from the grating at slightly different angles, so white light is spread into a spectrum and can be recorded, often by a CCD camera.

Recording the spectrum of a faint star or galaxy can require a long time exposure, so astronomers have developed multiobject spectrographs that can record the spectra of as many as 100 objects simultaneously. Fiber optic strands collect the light from many objects in the field of view and pipe the light to a single spectrograph. In some cases, a robotic arm can rapidly place the fibers in the right place to collect light from many galaxies in the telescope's field of view. Such multiobject spectrographs automated by computers have made possible large surveys of many thousands of stars or galaxies.

The spectrum of an astronomical object can contain hundreds of spectral lines produced by the atoms in the object. Because astronomers must measure the wavelength of the lines in a spectrum, they use a **comparison spectrum** as a calibration of their spectrograph. Special bulbs built into the spectrograph produce bright lines given off by such atoms as thorium and argon

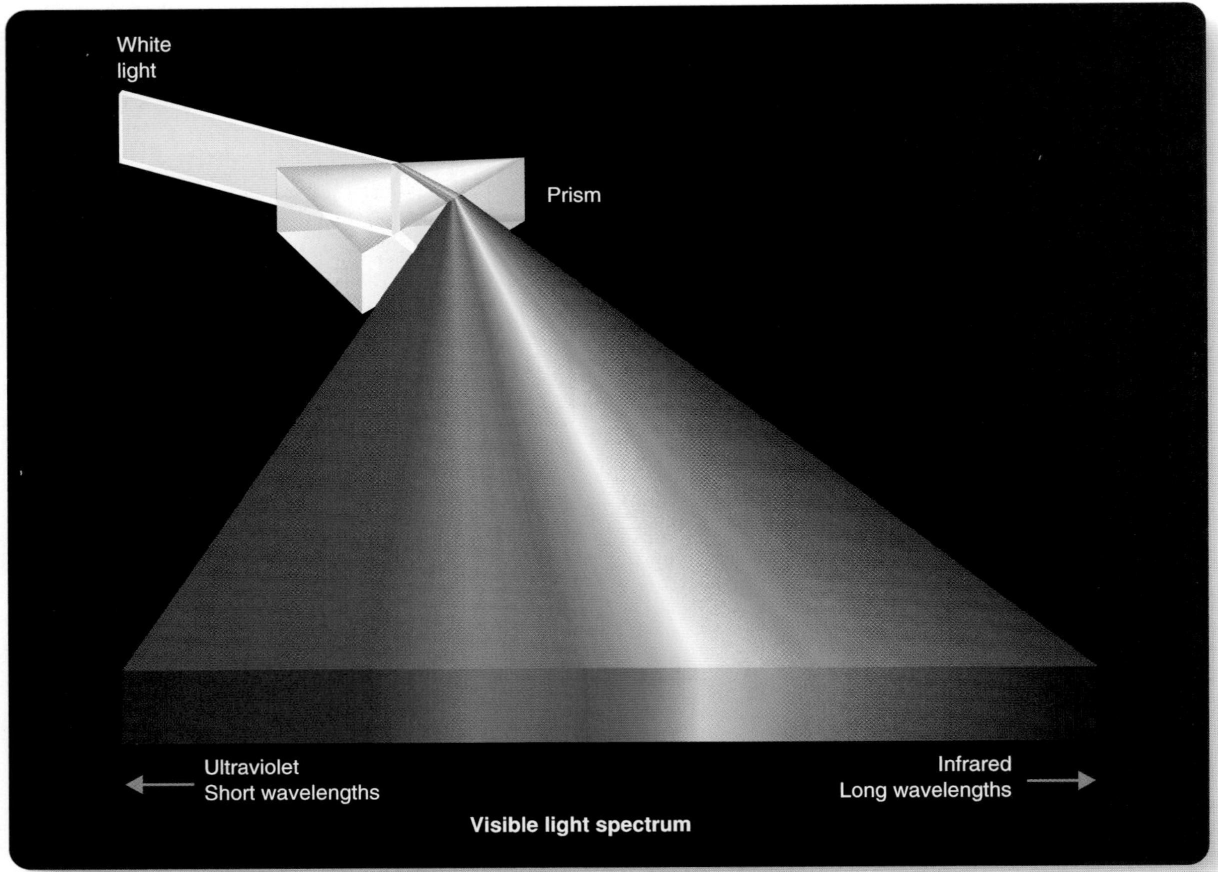

■ **Figure 5-16**

A prism bends light by an angle that depends on the wavelength of the light. Short wavelengths bend most and long wavelengths least. Thus, white light passing through a prism is spread into a spectrum.

or neon. The wavelengths of these spectral lines have been measured to high precision in the laboratory, so astronomers can use spectra of these light sources like roadmaps to measure wavelengths and identify spectral lines in the spectrum of a star, galaxy, or planet.

Because astronomers understand how light interacts with matter, a spectrum carries a tremendous amount of information (as you will see in the next chapter), and that makes a spectrograph the astronomer's most powerful instrument. An astronomer recently remarked, "We don't know anything about an object 'til we get a spectrum," and that is only a slight exaggeration.

Inquire | Review | Analyze

What is the difference between light going through a lens and light passing through a prism?

A refracting telescope producing chromatic aberration and a prism dispersing light into a spectrum are two examples of the same thing, but one is bad and one is good. When light passes through the curved surfaces of a lens, different wavelengths are bent by slightly different amounts, and the different colors of light come to focus at different focal lengths. This produces the color fringes in an image called chromatic aberration, and that's bad. But the surfaces of a prism are made to be precisely flat, so all of the light enters the prism at the same angle, and any given wavelength is bent by the same amount wherever it meets the prism. Thus, white light is dispersed into a spectrum. You could call the dispersion of light by a prism "controlled chromatic aberration," and that's good.

CCDs have been very good for astronomy, and they are now widely used. Explain why they are more useful than photographic plates.

■ ■ ■

Connections: So far, our discussion has been limited to visual wavelengths. Now it is time to consider the rest of the electromagnetic spectrum.

5-4 Radio Telescopes

Operation of a Radio Telescope

A RADIO TELESCOPE usually consists of four parts: a dish reflector, an antenna, an amplifier, and a recorder (■ Figure 5-17). The components, working together, make it possible for astronomers to detect radio radiation from celestial objects.

■ Figure 5-17

In most radio telescopes, a dish reflector concentrates the radio signal on the antenna. The signal is then amplified and recorded. For all but the shortest radio waves, wire mesh is an adequate reflector (above). (Courtesy Seth Shostak/SETI Institute)

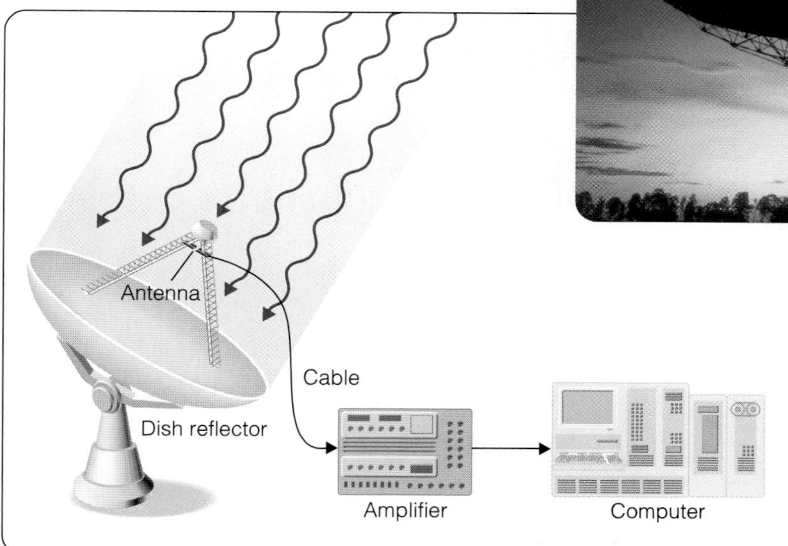

The dish reflector of a radio telescope, like the mirror of a reflecting telescope, collects and focuses radiation. Because radio waves are much longer than light waves, the dish need not be as smooth as a mirror. In some radio telescopes, the reflector may not even be dish shaped, or the telescope may contain no reflector at all.

Though a radio telescope's dish may be many meters in diameter, the antenna may be as small as your hand. Like the antenna on a TV set, its only function is to absorb the radio energy and direct it along a cable to an amplifier. After amplification, the signal goes to some kind of recording instrument. Most radio observatories record data into computer memory. However it is recorded, an observation with a radio telescope measures the amount of radio energy coming from a specific point on the sky.

Because humans can't see radio waves, astronomers must convert them into something perceptible. One way is to measure the strength of the radio signal at various places in the sky and draw a map in which contours mark areas of uniform radio intensity. You might compare such a map to a seating diagram for a baseball stadium in which the contours mark areas in which the seats have the same price (■ Figure 5-18a). Contour maps are very common in radio astronomy and are often reproduced using false colors (Figure 5-18b).

Limitations of the Radio Telescope

A radio astronomer works under three handicaps: poor resolution, low intensity, and interference. You saw that the resolving power of an optical telescope depends on the diameter of the objective lens or mirror. It also depends on the wavelength of the radiation. At very long wavelengths, like those of radio waves, images become fuzzy because of the large diffraction fringes. As with an optical telescope, the only way to improve the resolving power is to build a bigger telescope. Consequently, radio telescopes must be quite large.

Even so, the resolving power of a radio telescope is not good. A dish 30 m in diameter receiving radiation with a wavelength of 21 cm has a resolving power of about 0.5°. Such a radio telescope would be unable to show you any details in the sky smaller than the moon. Fortunately, radio astronomers can combine two or more radio telescopes to form a **radio interferometer** capable of much higher resolution. For example, the Very Large Array (VLA) consists of 27 dish antennas spread in a Y-shape across the New Mexico desert (■ Figure 5-19). In combination, they have the resolving power of a radio telescope 36 km (22 mi) in diameter. The VLA can resolve details smaller than 1 second of arc. Eight new dish antennas being added across New Mexico will give the VLA 10 times better resolving power. Another large radio interferometer, the Very Long Baseline Array (VLBA), consists of matched radio dishes spread from Hawaii to the Virgin Islands and has an effective diameter almost as large as Earth.

The second handicap radio astronomers face is the low intensity of the radio signals. You saw earlier that the energy of a photon depends on its wavelength. Photons of radio energy have such long wavelengths that their individual energies are quite low. In order to get strong signals focused on the antenna, the radio astronomer must build large collecting dishes.

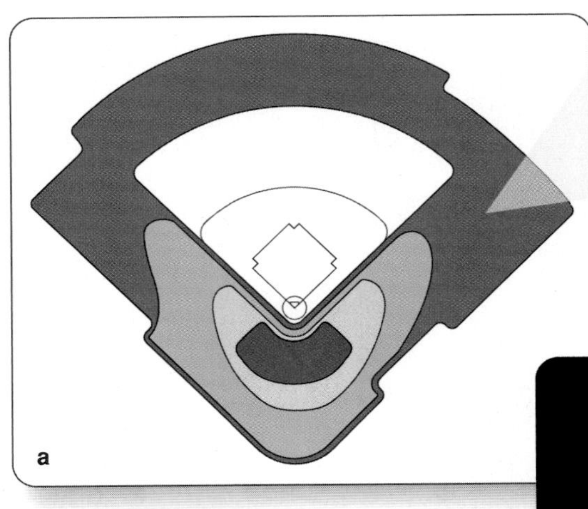

■ **Figure 5-18**

(a) A contour map of a baseball stadium shows regions of similar admission prices. The most expensive seats are those behind home plate. (b) A false-color radio map of Tycho's supernova remnant, the expanding shell of gas produced by the explosion of a star in 1572. The radio contour map has been color-coded to show intensity. (Courtesy NRAO)

The largest fully steerable radio telescope in the world is at the National Radio Astronomy Observatory in Green Bank, West Virginia (■ Figure 5-20a). The telescope has a reflecting surface 100 m in diameter, big enough to hold an entire football field, and can be pointed anywhere in the sky. Its surface consists of 2004 computer-controlled panels that adjust to maintain the shape of the reflecting surface.

The largest radio dish in the world is 300 m (1000 ft) in diameter. So large a dish can't be supported in the usual way, so it is built into a mountain valley in Arecibo, Puerto Rico. The reflecting dish is a thin metallic surface supported above the valley floor by cables attached near the rim, and the antenna hangs above the dish on cables from three towers built on three mountain peaks that surround the valley (Figure 5-20b). Although this telescope can look only overhead, the operators can change its aim slightly by moving the antenna above the dish and waiting for Earth's rotation to point the telescope in the proper direction. This may sound clumsy, but the telescope's ability to detect weak radio sources, together with its good resolution, makes it one of the most important radio observatories in the world.

The third handicap the radio astronomer faces is interference. A radio telescope is an extremely sensitive radio receiver listening to radio signals thousands of times weaker than artificial radio and TV transmissions. Such weak signals are easily drowned out by interference. Sources of such interference include everything from poorly designed transmitters in Earth satellites to automobiles with faulty ignition systems. To avoid this kind of interference, radio astronomers locate their telescopes as far from civilization as possible. Hidden deep in mountain valleys, they are able to listen to the sky protected from human-made radio noise.

■ **Figure 5-19**

The Very Large Array uses 27 radio dishes that can be moved to different positions along a Y-shaped set of tracks across the New Mexico desert. They are shown here in the most compact arrangement. Signals from the dishes are combined to create very high resolution radio maps of celestial objects. (NRAO/AUI/NSF)

Advantages of Radio Telescopes

Building large radio telescopes in isolated locations is expensive, but three factors make it all worthwhile. First, and most important, a radio telescope can show where clouds of

■ **Figure 5-20**

(a) The largest steerable radio telescope in the world is the GBT located in Green Bank, West Virginia. With a diameter of 100 m, it stands higher than the Statue of Liberty. (Mike Bailey: NRAO/AUII) (b) The 300-m (1000-ft) radio telescope in Arecibo, Puerto Rico, hangs from cables over a mountain valley. The Arecibo Observatory is part of the National Astronomy and Ionosphere Center, operated by Cornell University and the National Science Foundation. (David Parker/Science Photo Library)

cool hydrogen are located between the stars. Because 90 percent of the atoms in the universe are hydrogen, that is important information. Large clouds of cool hydrogen are completely invisible to normal telescopes because they produce no visible light of their own and reflect too little to be detected on photographs. However, cool hydrogen emits a radio signal at the specific wavelength of 21 cm. (You will see how the hydrogen produces this radiation when we discuss our galaxy in Chapter 12.) The only way to detect these clouds of gas is with a radio telescope that receives the 21-cm radiation. These hydrogen clouds are the places where stars are born, and that is one reason that radio telescopes are so important.

Nevertheless, there is a second reason. Because radio signals have relatively long wavelengths, they can penetrate the vast clouds of dust that obscure the view at visual wavelengths. Light waves are short, and they interact with tiny dust grains floating in space; thus, the light is scattered and never penetrates the dust to reach optical telescopes on Earth. However, radio signals from far across the galaxy pass unhindered through the dust, giving us an unobscured view.

Finally, a radio telescope can detect objects that are more luminous at radio wavelengths than at visible wavelengths. This includes everything from the coldest clouds of gas to the hottest stars. Some of the most distant objects in the universe, for instance, are detectable only at radio wavelengths.

Inquire | Review | Analyze

Why do optical astronomers build big telescopes, while radio astronomers build groups of widely separated smaller telescopes?

Comparison of seemingly similar concepts is a powerful tool in critical analysis because it often reveals subtle differences. In this case, optical astronomers are trying to maximize light-gathering power, but radio astronomers try to maximize resolving power. Because radio waves are so much longer than light waves, a single radio telescope can't see details in the sky much smaller than the moon. By linking radio telescopes miles apart, radio astronomers build a radio interferometer that can simulate a radio telescope miles in diameter and thus increase the resolving power.

This difference between the wavelengths of light and radio waves makes a big difference in building the best telescopes. But why don't radio astronomers build their telescopes on mountaintops as optical astronomers do?

■ ■ ■

Connections: Earth's atmosphere causes trouble for astronomers in two ways. It distorts images, and it absorbs many wavelengths. The only way to avoid the limitations completely is to send telescopes above the atmosphere, into space.

5-5 Astronomy from Space

YOU HAVE LEARNED about the observations that ground-based telescopes can make through the two atmospheric windows in the visible and radio parts of the electromagnetic spectrum. Most of the rest of the electromagnetic radiation—infrared, ultraviolet, X ray, and gamma ray—never reaches Earth's surface. To observe at these wavelengths, telescopes must fly above the atmosphere in high-flying aircraft, rockets, balloons, and satellites. The only exceptions are observations that can be made in the near-infrared and the near-ultraviolet.

The Great Observatories in Space

Over two decades ago, astronomers developed a plan to place a series of great observatories in space. Each observatory would cover a segment of the electromagnetic spectrum, and together they would paint a picture of the universe at a broad range of wavelengths. Those space telescopes have revolutionized our understanding of what we are and where we are in the universe.

Visual

Hubble image of Mars and its polar cap.

Visual-wavelength image

Hubble image of a nebula around an aging star.

The Hubble Space Telescope was carried into orbit by the Space Shuttle in 1990. The telescope contains a 2.4-m (96-inch) mirror and can observe from the near-infrared to the near-ultraviolet.

Orbiting above Earth's blurring atmosphere, Hubble is limited only by diffraction in its optics. It can detect details 10 times smaller than Earth-based telescopes.

The telescope, as big as a large bus, has been visited twice by astronauts, who repaired equipment and installed new instruments.

Named after Edwin Hubble, the astronomer who discovered the expansion of the universe, the telescope has been tremendously productive observing everything from the weather on Mars to the most distant galaxies visible in the universe.

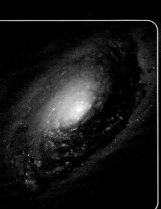

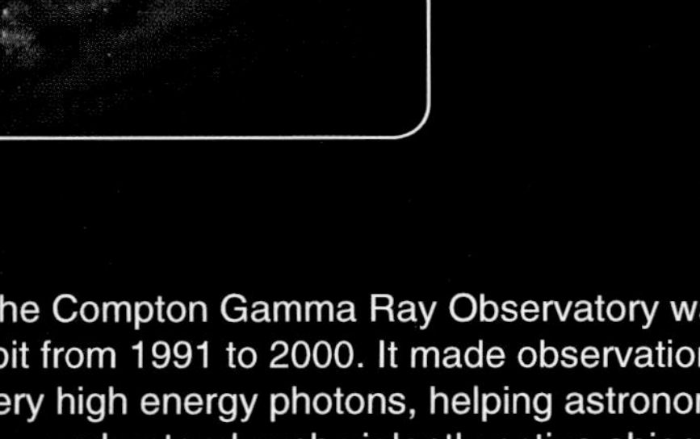

Hubble image of a dust-filled galaxy.

The Compton Gamma Ray Observatory was in orbit from 1991 to 2000. It made observations of very high energy photons, helping astronomers understand such violently active objects as neutron stars and black holes.

The Chandra X-Ray Observatory was placed in orbit 1/3 of the way to the moon in 1999. It is nearly 14 m (45 feet) long and carries highly precise mirrors 1.2 m (47 inches) in diameter. X rays would penetrate into regular mirrors, so Chandra's mirrors are designed as cylinders polished on the inside so that X rays just graze the surface and are focused onto detectors. The telescope was named after the late Indian-American Nobel Laureate Subrahmanyan Chandrasekhar who was a pioneer in many branches of theoretical astronomy.

Chandra can detect X-ray-emitting objects 50 times fainter and resolve details 10 times smaller than any previous X-ray telescope.

Chandra X-ray images

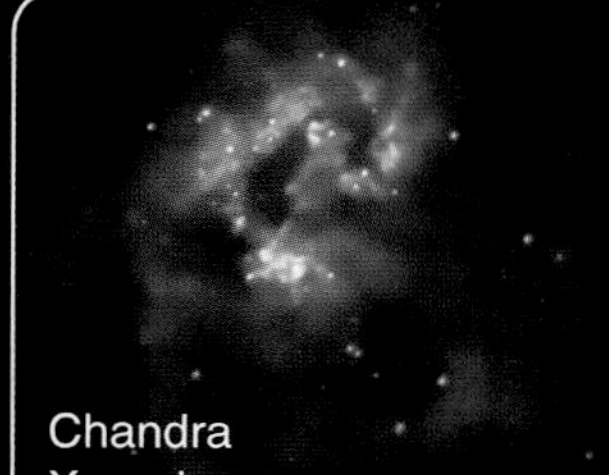

Two galaxies collide and trigger the birth of stars.

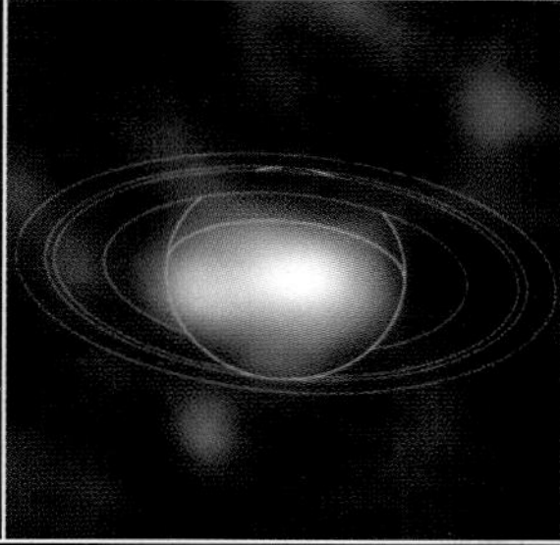

Saturn emits x-rays from near its equator.

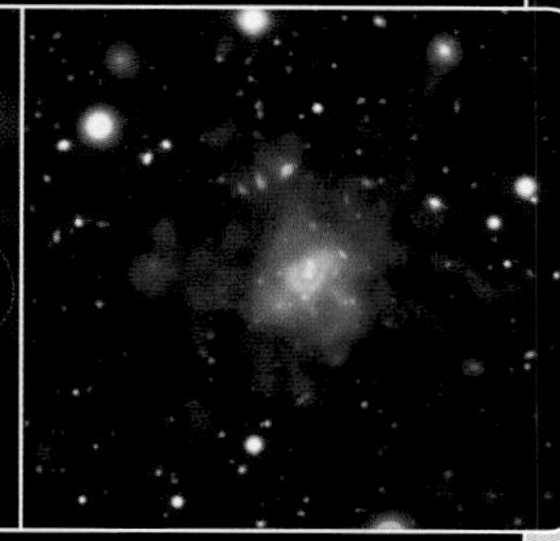

Very hot gas is trapped in a cluster of galaxies.

The Spitzer Space Telescope observes in the infrared. It was launched in 2003 and named in honor of the late astronomer Lyman Spitzer, a leader in space astronomy. The telescope is cooled to –273 °C (–459 °F) so it cannot orbit the warm Earth. Instead it is in an orbit around the sun and will drift slowly away from Earth during its lifetime. Protected from sunlight by a heat screen, it can observe a wide range of astronomical objects.

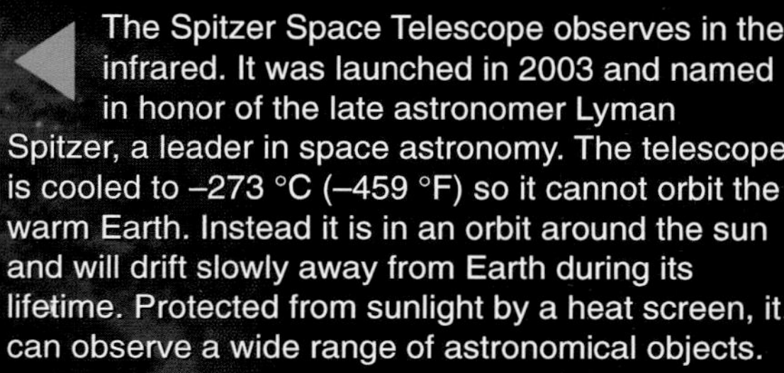

Spitzer infrared image

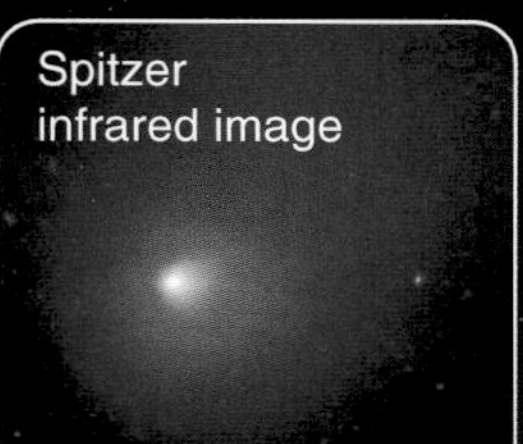

A comet glows in the infrared.

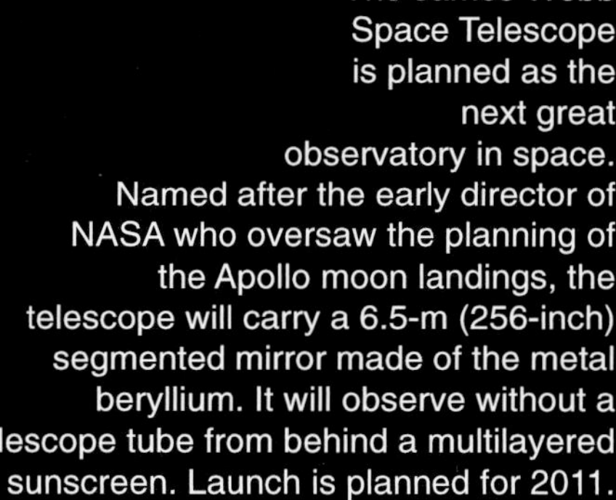

Spitzer infrared image

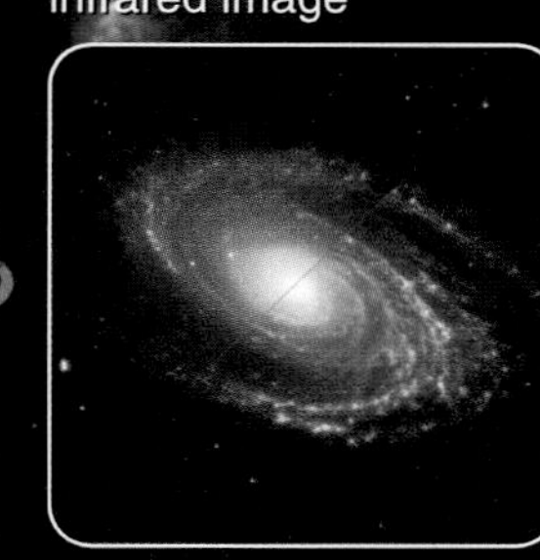

Dust warmed by hot young stars glows in the disk of this spiral galaxy.

Spitzer infrared image

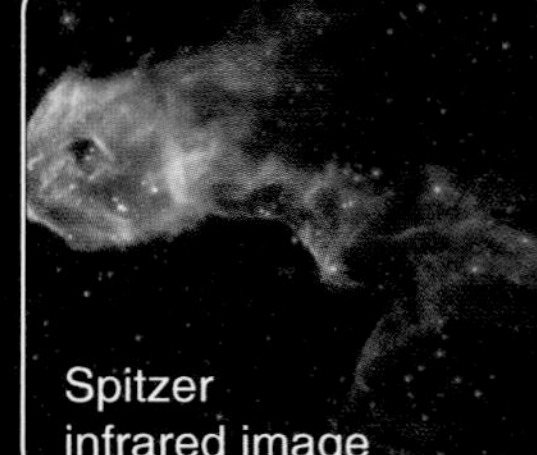

An infrared image penetrates a dusty nebula to reveal newborn stars just beginning to shine.

The James Webb Space Telescope is planned as the next great observatory in space. Named after the early director of NASA who oversaw the planning of the Apollo moon landings, the telescope will carry a 6.5-m (256-inch) segmented mirror made of the metal beryllium. It will observe without a telescope tube from behind a multilayered sunscreen. Launch is planned for 2011.

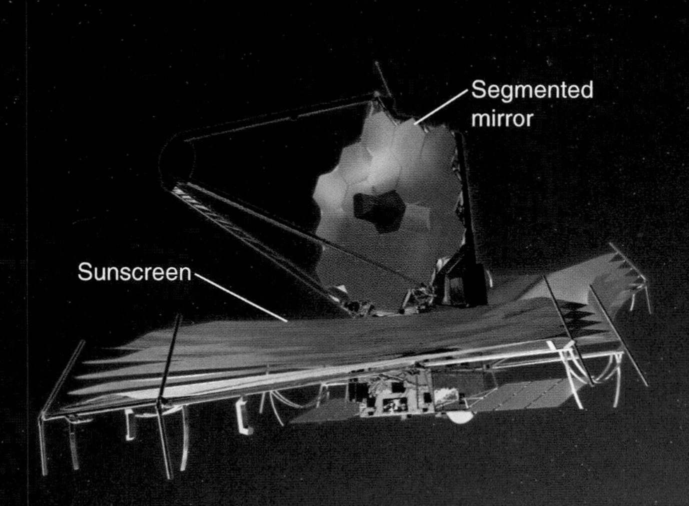

The Ends of the Visual Spectrum

Astronomers can observe in the near-infrared just beyond the red end of the visible spectrum. Some of this infrared radiation leaks through the atmosphere in narrow, partially open atmospheric windows scattered from 1200 nm to about 40,000 nm. Infrared astronomers usually measure wavelength in micrometers (10^{-6} meters), so they refer to this wavelength range as 1.2 to 40 micrometers or microns. In this range, much of the radiation is absorbed by water vapor, carbon dioxide, and oxygen molecules in Earth's atmosphere, so it is an advantage to place telescopes on mountains where the air is thin and dry. A number of important infrared telescopes, for example, observe from the 4150-m (13,600-ft) summit of Mauna Kea in Hawaii. At this altitude, they are above much of the water vapor, which is the main absorber of infrared (■ Figure 5-21).

The far-infrared range, which includes wavelengths longer than 40 micrometers, can tell us about planets, comets, forming stars, and other cool objects, but these wavelengths are absorbed high in the atmosphere. To observe in the far-infrared, telescopes must venture to high altitudes. Remotely operated infrared telescopes suspended under balloons have reached altitudes as high as 41 km (25 miles). For many years, a NASA jet transport carried a 91-cm infrared telescope and a crew of astronomers to altitudes of 12,000 m (40,000 ft) to get above 99 percent of the water vapor in Earth's atmosphere. Now retired from service, that airborne observatory will soon be replaced with the Stratospheric Observatory for Infrared Astronomy (SOFIA), a Boeing 747 that will carry a 2.5-m telescope to the fringes of the atmosphere.

If a telescope observes at far-infrared wavelengths, then it must be cooled. Infrared radiation is emitted by heated objects, and if the telescope is warm it will emit many times more infrared radiation than that coming from a distant object. Imagine trying to look at a dim, moonlit scene through binoculars that are glowing brightly. In a telescope observing near-infrared wavelengths, only the detector, the element on which the infrared radiation is focused, must be cooled. To observe in the far-infrared, however, the entire telescope must be cooled.

At the short wavelength end of the spectrum, astronomers can observe in the near-ultraviolet. Human eyes do not detect this radiation, but it can be recorded by photographic plates and CCDs. Wavelengths shorter than about 290 nm, the far-ultraviolet, are completely absorbed by the ozone layer extending from 20 km to about 40 km above Earth's surface. No mountain is that high, and no balloon or airplane can fly that high, so astronomers cannot observe in the far-ultraviolet without going into space.

Telescopes in Space

To observe far beyond the ends of the visible spectrum, astronomical telescopes must go above Earth's atmosphere into space. This is very expensive and difficult, but it is the only way to study some processes. Stars are born inside clouds of gas and dust, and visible wavelengths cannot escape from these dust clouds. Only observations in the infrared can reveal the secrets of star formation. Black holes are small and hard to detect, but matter flowing into a black hole emits X rays. Telescopes in space can explore these exciting processes that are invisible from Earth's surface.

One of the most successful space telescopes was the International Ultraviolet Explorer (IUE), launched in 1978. It carried a telescope only 45 cm (18 in.) in diameter and was expected to last only a year or two, but it became the little telescope that could. It made many observations and exciting discoveries until it finally failed in 1996.

Many space telescopes are small satellites designed to make specific observations for a short period, but some are large general-purpose telescopes. Study **The Great Observatories in Space** on pages 96 and 97 and notice three points. Not only can a telescope in space observe at a wide range of wavelengths, but it is above the atmospheric blurring called seeing. The Hubble Space Telescope observes mostly at visual wavelengths,

■ Figure 5-21

Comet Hale–Bopp hangs in the sky over the 3-meter NASA Infrared Telescope atop Mauna Kea. The air at high altitudes is steady and so dry that it is transparent to shorter infrared photons. Infrared astronomers can observe with the lights on in the telescope dome. Their instruments are usually insensitive to visible light. (Courtesy William Keel)

but it has the advantage of sharp images undistorted by seeing. The second thing to notice is how telescopes must be specialized for their wavelength range. The Spitzer Space Telescope must have a heat screen to protect it from sunlight, and the Chandra space telescope must have special optics. Finally, notice the different kinds of processes that these telescopes can observe. Warm dust and star formation are visible in the infrared, but it takes X-ray telescopes to detect the high-temperature gas clouds in some galaxy clusters.

These great observatories in space are controlled from research centers on Earth and are open to proposals from any astronomer with a good idea; but competition is fierce, and only the most worthy projects win approval.

Inquire | Review | Analyze

Why can infrared astronomers observe from high mountaintops, while X-ray astronomers must observe from space?

In this analysis, you find similar consequences springing from different causes. Although both infrared and X-ray telescopes have been put into orbit, they differ dramatically. Infrared radiation is absorbed mainly by water vapor in Earth's atmosphere, which is confined to the lower atmospheric layers. If you built an infrared telescope on top of a high mountain, you would be above most of the water vapor, and you could collect some infrared radiation from the stars. The longer-wavelength infrared is absorbed much higher in the atmosphere, so telescopes to observe in the far-infrared must go into space. X rays are absorbed in the uppermost layers of the atmosphere, and no mountain is high enough to extend above those layers. To observe the stars at X-ray wavelengths, a telescope must go into space.

X-ray and far-infrared telescopes must observe from space, but the Hubble Space Telescope observes in the visual wavelength range. Then why must it observe from orbit?

■ ■ ■

Connections: The tools of the astronomer are designed to gather radiation from the sky and extract information. Perhaps no tool is as important as the spectrograph, because no form of observation is as loaded with information as a spectrum. In the next chapter, you will see how astronomers can harvest the information in a star's spectrum.

Study and Review Tools

Summary

5-1 | Radiation: Information from Space

What is light?

- Light is the visible form of electromagnetic radiation, an electric and magnetic disturbance that transports energy at the speed of light. The electromagnetic spectrum includes gamma rays, X rays, ultraviolet radiation, visible light, infrared radiation, and radio waves.
- You can think of a particle of light, a photon, as a bundle of waves that acts sometimes as a particle and sometimes as a wave.
- The energy a photon carries depends on its wavelength. The wavelength of visible light, usually measured in nanometers (10^{-9}m), ranges from 400 nm to 700 nm. Infrared and radio photons have longer wavelengths and carry less energy. Ultraviolet, X-ray, and gamma-ray photons have shorter wavelengths and carry more energy.

5-2 | Optical Telescopes

How do telescopes work, and how are they limited?

- Astronomers use telescopes to gather light, see fine detail, and magnify the image. The first two of these three powers of the telescope depend on the telescope's diameter. Consequently, astronomers strive to build telescopes with large diameters.
- A refracting telescope uses a lens to bend the light and focus it into an image. Because of chromatic aberration, refracting telescopes cannot bring all colors to the same focus, resulting in color fringes around the images. An achromatic lens partially corrects for this, but such lenses are expensive and cannot be made much larger than about 1 m in diameter.
- Reflecting telescopes use a mirror to focus the light and are less expensive than refracting telescopes of the same diameter. Also, reflecting telescopes do not suffer from chromatic aberration. Most recently built large telescopes are reflectors.
- Astronomers build observatories on high mountains for two reasons. Turbulence in Earth's atmosphere blurs the image in an astronomical telescope, a phenomenon that astronomers refer to as seeing. Atop a mountain, the air is steady, and the seeing is better. The air on a mountaintop is also thin and dry and is more transparent, especially in the infrared.
- Sometimes astronomical telescopes can be linked together to form an interferometer, which has a resolution equivalent of a telescope as large in diameter as the separation between the telescopes.

5-3 | Special Instruments

What kind of instruments do astronomers use to record and analyze light?

- For many decades, astronomers used photographic plates to record images at the telescope, but modern electronic systems such as CCDs have replaced photographic plates in most applications.
- Spectrographs using prisms or a grating spread starlight out according to wavelength to form a spectrum revealing hundreds of spectral lines produced by atoms in the object being studied.

5-4 | Radio Telescopes

Why do astronomers use radio telescopes?

- Astronomers use radio telescopes for three reasons: They can detect cool hydrogen in space; they can see through dust clouds that block visible light; and they can detect certain objects invisible at other wavelengths.

- Most radio telescopes consist of a dish reflector, an antenna, an amplifier, and a data recorder. Such a telescope can record the intensity of the radio energy coming from a spot on the sky. Scans of small regions are used to produce radio maps.
- Because of the long wavelength, radio telescopes have very poor resolution, and astronomers often link separate radio telescopes together to form a radio interferometer capable of detecting much finer detail.

5-5 | Astronomy from Space

Why must some telescopes go into orbit?

- Earth's atmosphere is transparent in two wavelength ranges or windows, the visual window and the radio window. At other wavelengths, the atmosphere absorbs radiation. To observe at other wavelengths, telescopes must go into space.
- Earth's atmosphere distorts and blurs images. Telescopes in orbit are above this seeing distortion and are limited only by diffraction in their optics.

New Terms

electromagnetic radiation (p. 78)
wavelength (p. 78)
photon (p. 78)
nanometer (nm) (p. 79)
angstrom (Å) (p. 79)
atmospheric window (p. 80)
focal length (p. 80)
primary lens, mirror (p. 80)
objective lens, mirror (p. 80)
eyepiece (p. 80)
refracting telescope (p. 81)
reflecting telescope (p. 81)
chromatic aberration (p. 81)
achromatic lens (p. 82)
light-gathering power (p. 82)
resolving power (p. 83)
diffraction fringe (p. 83)
seeing (p. 83)
magnifying power (p. 84)
light pollution (p. 84)
interferometry (p. 87)
prime focus (p. 88)
secondary mirror (p. 88)
Cassegrain focus (p. 88)
Newtonian focus (p. 88)
Schmidt-Cassegrain focus (p. 88)
sidereal drive (p. 89)
equatorial mounting (p. 89)
polar axis (p. 89)
alt-azimuth mounting (p. 89)
active optics (p. 89)
adaptive optics (p. 89)
charge-coupled device (CCD) (p. 90)
false-color image (p. 90)
spectrograph (p. 91)
grating (p. 91)
comparison spectrum (p. 91)
radio interferometer (p. 93)

Review Questions

Ace Astronomy™ Assess your understanding of this chapter's topics with additional quizzing and animations at **http://astronomy.brookscole.com/sh9e**

1. Why would you not plot sound waves in the electromagnetic spectrum?
2. If you had limited funds to build a large telescope, which type would you choose, a refractor or a reflector? Why?
3. Why do nocturnal animals usually have large pupils in their eyes? How is that related to astronomical telescopes?
4. Why do optical astronomers sometimes put their telescopes at the tops of mountains, while radio astronomers sometimes put their telescopes in deep valleys?
5. Optical and radio astronomers both try to build large telescopes but for different reasons. How do these goals differ?
6. What are the advantages of making a telescope mirror thin? What problems does this cause?
7. Small telescopes are often advertised as "200 power" or "magnifies 200 times." As someone knowledgeable about astronomical telescopes, how would you improve such advertisements?
8. An astronomer recently said, "Some people think I should give up photographic plates." Why might she change to something else?
9. What purpose do the colors in a false-color image or false-color radio map serve?
10. How is chromatic aberration related to a prism spectrograph?
11. Why would radio astronomers build identical radio telescopes in many different places around the world?
12. Why do radio telescopes have poor resolving power?
13. Why must telescopes observing in the far-infrared be cooled to low temperature?
14. What might we detect with an X-ray telescope that we could not detect with an infrared telescope?
15. If the Hubble Space Telescope observes at visual wavelengths, why must it observe from space?
16. The two images at right show a star before and after an adaptive optics system was switched on. What causes the distortion in the first image, and how does adaptive optics correct the image?

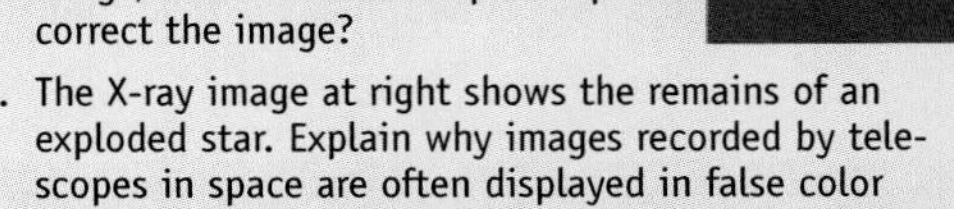
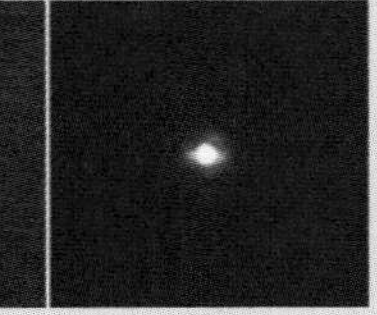

(ESO)

17. The X-ray image at right shows the remains of an exploded star. Explain why images recorded by telescopes in space are often displayed in false color rather than in the "colors" received by the telescopes.

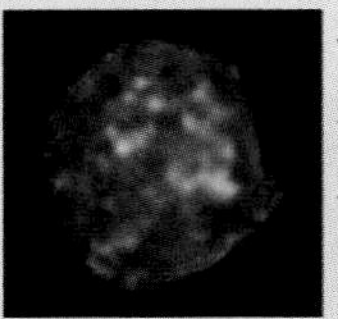

(NASA/CXC/PSU/S.Park)

Discussion Questions

1. Why does the wavelength response of the human eye match so well the visual window of Earth's atmosphere?
2. Basic research in chemistry, physics, biology, and similar sciences is supported in part by industry. How is astronomy different? Who funds the major observatories?

Problems

1. The plastic in plastic bags is about 0.001 mm thick. How many wavelengths of red light is this?
2. What is the wavelength of radio waves transmitted by a radio station with a frequency of 100 MHz?
3. Compare the light-gathering powers of a 5-m telescope and a 0.5-m telescope.
4. How does the light-gathering power of one of the Keck telescopes compare with that of the human eye? (*Hint:* Assume that the pupil of your eye can open to about 0.8 cm.)
5. What is the resolving power of a 25-cm telescope? What do two stars 1.5 seconds of arc apart look like through this telescope?
6. Most of Galileo's telescopes were only about 2 cm in diameter. Should he have been able to resolve the two stars mentioned in Problem 5?
7. How does the resolving power of the 5-m telescope compare with that of the Hubble Space Telescope? Why does the Hubble Space Telescope outperform the 5-m telescope?
8. If you build a telescope with a focal length of 1.3 m, what focal length should the eyepiece have to give a magnification of 100 times?
9. Astronauts observing from a space station need a telescope with a light-gathering power 15,000 times that of the human eye, capable of resolving detail as small as 0.1 second of arc, and having a magnifying power of 250. Design a telescope to meet their needs. Could you test your design by observing stars from Earth?
10. A spy satellite orbiting 400 km above Earth is supposedly capable of counting individual people in a crowd. What minimum-diameter telescope must the satellite carry? (*Hint:* Use the small-angle formula.)

Media Cluster

ACTIVE FIGURES

Ace Astronomy™ To access the resources in the Media Cluster, log into AceAstronomy at http://astronomy.brookscole.com/sh9e and select Chapter 5.

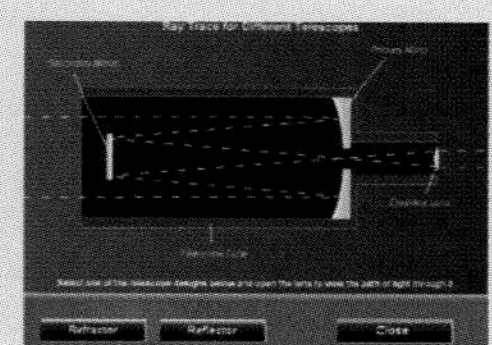

Refractors and Reflectors
Compare the two basic types of telescope designs and the paths light takes through each with this animation.

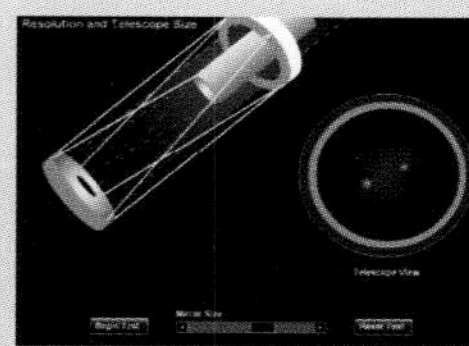

Resolution and Telescopes
This animation allows you to contrast the high resolution of larger-aperture telescopes with the lower resolution of smaller ones.

ASTRONOMY EXERCISES

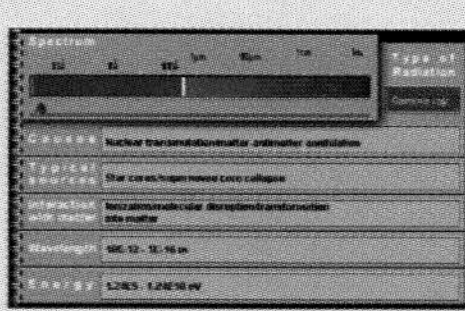

The Electromagnetic Spectrum
In this exercise, study the properties such as wavelength and sources of various parts of the electromagnetic spectrum.

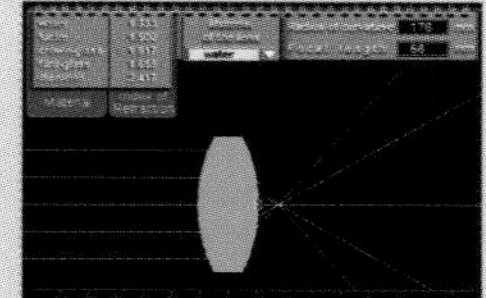

Lenses: Focal Length
Astronomers use lenses to focus starlight into an image. You can change the shape and material of the lens in this animation to study how these parameters affect the focal length.

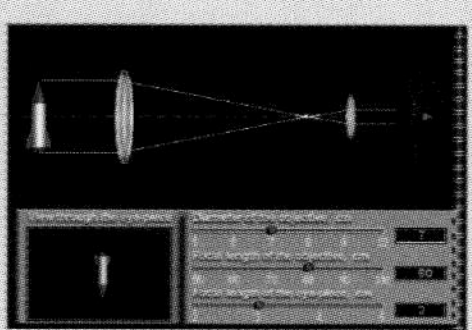

Telescopes: Objective Lens and Eyepiece
Use this animation to study how the objective lens and eyepiece work together to form an image in a telescope. You can change parameters for the diameter of the lens and the focal lengths of the objective lens and eyepiece.

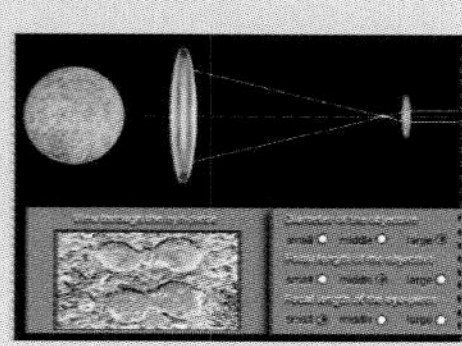

Telescopes and Resolution I
In this simulation, you can vary the focal length of the objective lens, the focal length of the eyepiece lens, and the diameter of the objective lens. Try them all and see which factor (or factors) determines how sharp the image is.

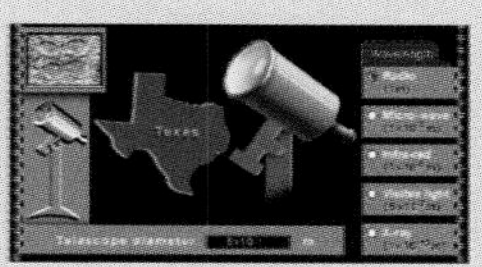

Telescopes and Resolution II
Get a feel for how radiation wavelength affects the resolution of a telescope by comparing the lens diameter needed to observe craters on the moon with a visible-light telescope versus other types such as radio and microwave telescopes.

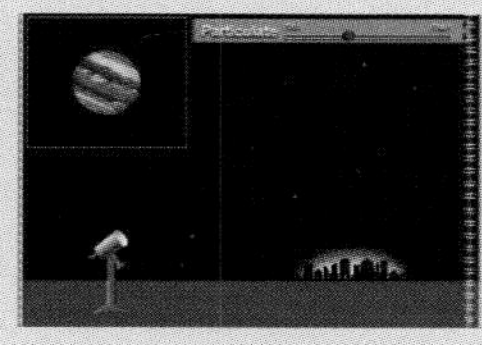

Particulate, Heat, and Light Pollution
These three animations let you study how environmental variables can affect viewing quality. How does this fit in with what you learned in this chapter about the locations of major astronomical telescopes?

VIRTUAL ASTRONOMY LAB

Lab 4: Solar Wind and Cosmic Rays
This lab begins with an overview of the properties of the sun's atmosphere and how energetic particles escape and travel through the solar system. The lab ends with a discussion of cosmic rays.

Critical Inquiries for the Web

1. How do professional astronomers go about making observations at major astronomical facilities? Visit several observatory websites to determine the process an astronomer would go through to secure observing time and make observations at the facility.
2. What can you learn about the current status of the first great observatory in space, the Hubble Space Telescope, and the most recently launched, the Spitzer Space Telescope? Can you find out more about the James Webb Space Telescope?

Exploring *TheSky*

1. Astronomical telescopes using equatorial mountings must be aligned precisely with the north celestial pole. Locate Polaris and determine how far it is from the north celestial pole. (*Hint:* Use **Reference Lines** under the **View** menu and check **Grid** under Equatorial. Be sure the spacing is set to auto/fine. Then locate the Little Dipper and zoom in on Polaris.)

- Go to the Brooks/Cole Astronomy Resource Center **(http://astronomy.brookscole.com)** for critical thinking exercises, articles, and additional readings from InfoTrac College Edition, Brooks/Cole's online student library.

6 Starlight and Atoms

Awake! for Morning
in the Bowl of Night
Has flung the stone
that put the Stars to Flight!
And lo! the Hunter
of the East has caught
The Sultan's Turret
in a Noose of Light.

EDWARD FITZGERALD, TRANSLATOR
The Rubáiyát of Omar Khayyám

Visual-wavelength image

NO LABORATORY JAR on Earth holds a sample labeled "star stuff," and no instrument has ever probed inside a star. The stars are far beyond human reach, and the only information you can obtain about them comes hidden in light (■ Figure 6-1). Whatever you want to know about the stars you must catch in a noose of light. ▌ Earthbound humans knew almost nothing about stars until the early 19th century, when the Munich optician Joseph von Fraunhofer studied the solar spectrum and found it interrupted by some 600 dark lines. As scientists realized that the lines were related to the various atoms in the sun and found that stellar spectra had similar patterns of lines, the door to an understanding of stars finally opened.

Hot stars and cool stars, clouds of glowing gas, and obscuring dust fill this region of the southern constellation Ara, the Altar. (ESO)

Guidepost

Looking Back

In the last chapter you read how hard astronomers work to gather light from the stars. You also read how spectrographs are used to spread light out into spectra. Now you are ready to see what all the fuss is about.

This Chapter

This chapter explains how light interacts with matter and how astronomers must understand that interaction to understand stars. Here you will find answers to four essential questions:

What is an atom?

How do atoms interact with light?

What kind of spectra do you see when you look at celestial objects?

What can you learn from a star's spectrum?

This chapter marks a change in the way you will look at nature. Up to this point, you have been thinking about what you can see with your eyes alone or aided by telescopes. In this chapter, you begin using modern astrophysics, the application of physics to study the sky. Now you can search out secrets of the stars that lie beyond what you can see.

Looking Ahead

The analysis of spectra is a powerful tool, and in the next chapter you will use that tool to study the sun. In the chapters that follow, you will study other suns—the stars.

Ace Astronomy™ The AceAstronomy icon throughout the text indicates an opportunity for you to test yourself on key concepts and to explore animations and interactions on the AceAstronomy website at: http://astronomy.brookscole.com/sh9e

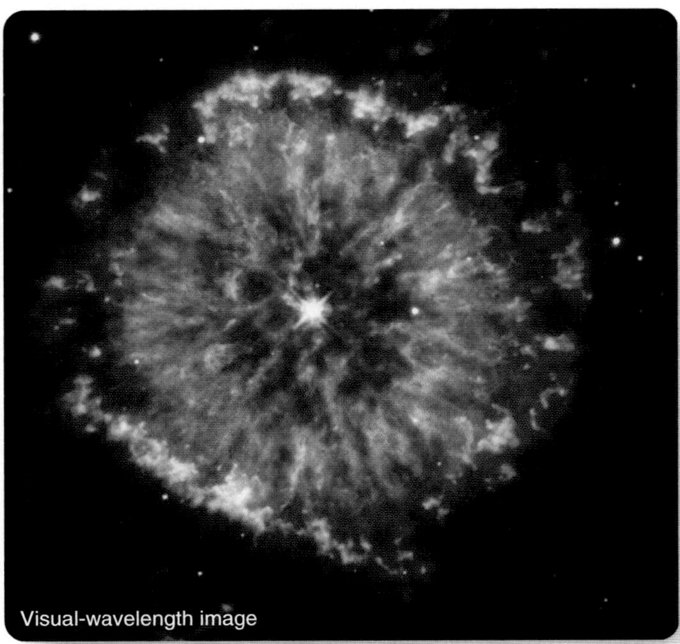

Figure 6-1

What's going on here? The sky is filled with beautiful and mysterious objects that lie far beyond your reach—in the case of the nebula NGC6751, about 6500 ly beyond your reach. An analysis of the light received from this object reveals it is a dying star surrounded by the expanding shell of gas that it ejected a few thousand years ago. This phenomenon is discussed in Chapter 10. (NASA Hubble Heritage Team/STScI/AURA)

In this chapter, you will go through that door by considering how atoms interact with light to produce **spectral lines**—dark or bright lines that cross spectra at specific positions. The chapter begins with the hydrogen atom because it is the most common atom in the universe, as well as the simplest. Other atoms are larger and more complicated, but in many ways their properties resemble those of hydrogen.

Once you understand how an atom's structure can interact with light to produce spectral lines, you will recognize certain patterns in stellar spectra. By classifying the spectra according to these patterns, you can arrange the stars in a sequence according to temperature. One of the most important pieces of information revealed in a star's spectrum is its temperature.

But, properly analyzed, a stellar spectrum can tell you much more. The spectrum gives you information about the chemical composition of the star and the star's motion relative to Earth.

6-1 Atoms

A Model Atom

TO THINK ABOUT ATOMS and how they can interact with light, you need a working model of an atom. In Chapter 2, you used a working model of the sky, the celestial sphere. You identified and named the important parts and described how they were located and how they interacted. We will begin our study of atoms by creating a model of an atom.

Our model atom has a positively charged **nucleus** at the center; this nucleus consists of two kinds of particles. **Protons** carry a positive electrical charge, and **neutrons** have no charge. Thus, the nucleus has a net positive charge.

The nucleus in this model atom is surrounded by a whirling cloud of orbiting **electrons,** low-mass particles with a negative charge. In a normal atom, the number of electrons equals the number of protons, and the positive and negative charges balance to produce a neutral atom. Because protons and neutrons each have a mass 1836 times greater than that of an electron, most of the mass of the atom lies in the nucleus. The hydrogen atom is the simplest of all atoms. The nucleus is a single proton orbited by a single electron, with a total mass of only 1.67×10^{-27} kg, about a trillionth of a trillionth of a gram.

An atom is mostly empty space. To see this, imagine constructing a simple scale model. The nucleus of a hydrogen atom is a proton with a diameter of about 0.0000016 nm, or 1.6×10^{-15} m. If you multiply this by 1 trillion (10^{12}) you can represent the nucleus of your model atom with a grape seed, which is about 0.16 cm in diameter. The region of a hydrogen atom containing the whirling electron has a diameter of about 0.4 nm, or 4×10^{-10} m. Multiplying by a trillion magnifies the diameter to about 400 m, or about 4.5 football fields laid end to end (■ Figure 6-2). When you imagine a grape seed in the midst of a sphere 4.5 football fields in diameter, you can see that an atom is mostly empty space.

Figure 6-2

Magnifying a hydrogen atom by 10^{12} makes the nucleus the size of a grape seed and the diameter of the electron cloud about 4.5 times longer than a football field. The electron itself is still too small to see.

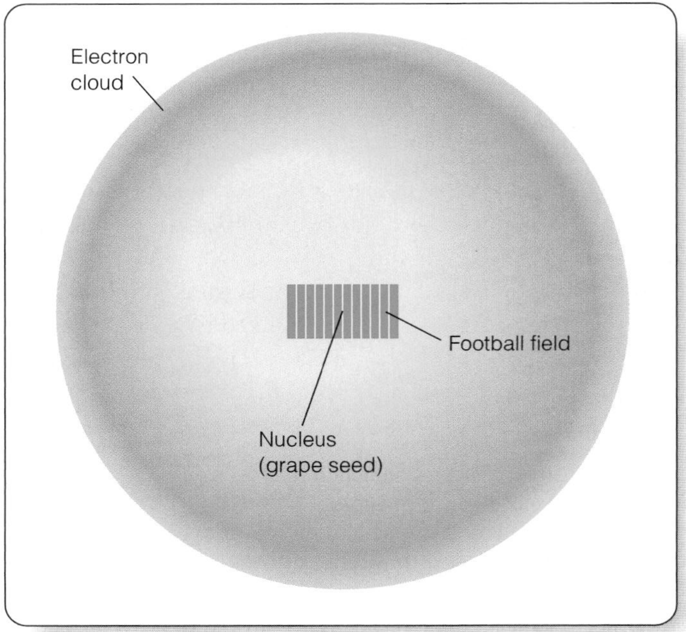

Window on Science 6-1

Quantum Mechanics: The World of the Very Small

Quantum mechanics is the set of rules that describe how atoms and subatomic particles behave. When you think about large objects, such as stars, planets, aircraft carriers, and hummingbirds, you don't have to think about quantum mechanics; but on the atomic scale, particles behave in ways that seem unfamiliar.

One of the principles of quantum mechanics specifies that you cannot know simultaneously the exact location and motion of a particle. This is why physicists refer to the electrons in an atom as if they were a cloud of negative charge surrounding the nucleus. Because you can't know the position and motion of the electron, you can't really describe it as a small particle following an orbit. You can use that image as a model to help your imagination, but the reality is much more interesting, and describing the electrons as a charge cloud gives you a better and more sophisticated model of an atom.

This raises some serious questions about reality. Is an electron really a particle at all? Quantum mechanics describes particles as waves and waves as particles. If you can't know simultaneously the position and motion of a specific particle, how can you know how it will react to a collision with a photon or another particle? The answer is that you can't know, and that seems to violate the principle of cause and effect (Window on Science 4-3).

Needless to say, these discrepancies can't be explored here. Scientists and philosophers of science continue to struggle with the meaning of reality on the quantum-mechanical level. Here you should note that the reality you see on the scale of stars and hummingbirds is only part of nature. Scientists have constructed some models to help them think about nature on the scale of atoms, but the truth is much more interesting and much more exciting than those simple models. Although you can use models of atoms to study stars, there is still much to learn about the atoms themselves.

Different Kinds of Atoms

There are over a hundred kinds of atoms, called *chemical elements.* Which element an atom is depends only on the number of protons in the nucleus. For example, carbon has six protons in its nucleus. An atom with one more proton than this is nitrogen, and an atom with one fewer proton is boron.

Although the number of protons in an atom of an element is fixed, you can change the number of neutrons in an atom's nucleus without changing the atom significantly. For instance, if you add a neutron to the carbon nucleus, you still have carbon, but it is slightly heavier than normal carbon. Atoms that have the same number of protons but a different number of neutrons are **isotopes.** Carbon has two stable isotopes. One form contains six protons and six neutrons for a total of 12 particles and is thus called carbon-12. Carbon-13 has six protons and seven neutrons in its nucleus.

Protons and neutrons are bound tightly into the nucleus, but the electrons are held loosely in the electron cloud. Running a comb through your hair creates a static charge by removing a few electrons from their atoms. This process is called **ionization,** and the atom that has lost one or more electrons is an **ion.** A carbon atom is neutral if it has six electrons to balance the positive charge of the six protons in its nucleus. If you ionize the atom by removing one or more electrons, the atom is left with a net positive charge. Under some circumstances, an atom may capture one or more extra electrons, giving it more negative charges than positive. Such a negatively charged atom is also considered an ion.

Atoms that collide may form bonds with each other by exchanging or sharing electrons. Two or more atoms bonded together form a **molecule.** Atoms do collide in stars, but the high temperatures cause violent collisions that are unfavorable for chemical bonding. Only in the coolest stars are the collisions gentle enough to permit the formation of chemical bonds. You will see later that the presence of molecules such as titanium oxide (TiO) in a star is a clue that the star is cool. In later chapters, you will see that molecules can form in cool gas clouds in space and in the atmospheres of planets.

Electron Shells

We mentioned the whirling cloud of orbiting electrons in a general way, but the specific way electrons behave within the cloud is very important in astronomy.

The electrons are bound to the atom by the attraction between their negative charge and the positive charge on the nucleus. This attraction is known as the **Coulomb force,** after the French physicist Charles-Augustin de Coulomb (1736–1806). To ionize an atom, you need a certain amount of energy to pull an electron away from its nucleus. This energy is the electron's **binding energy,** the energy that holds it to the atom.

An electron may orbit the nucleus at various distances. If the orbit is small, the electron is close to the nucleus, and a large amount of energy is needed to pull it away. Therefore, its binding energy is large. An electron orbiting farther from the nucleus is held more loosely, and less energy will pull it away. It therefore has less binding energy. The size of an electron's orbit is related to the energy that binds it to the atom.

Nature permits atoms only certain amounts (quanta) of binding energy, and the laws that describe how atoms behave are called the laws of **quantum mechanics** (Window on Science 6-1). Much of our discussion of atoms is based on the laws of quantum mechanics.

Because atoms can have only certain amounts of binding energy, your model atoms can have orbits of only certain sizes, called **permitted orbits.** These are like steps in a staircase: You can stand on the number-one step or the number-two step, but not on the number-one-and-one-quarter step. The electron can occupy any permitted orbit but not orbits in between.

The arrangement of permitted orbits depends primarily on the charge of the nucleus, which in turn depends on the number of protons. Thus, each kind of element has its own pattern of permitted orbits (■ Figure 6-3). Isotopes of the same elements have nearly the same pattern because they have the same number of protons. However, ionized atoms have orbital patterns that differ from their un-ionized forms. Thus the arrangement of permitted orbits differs for every kind of atom and ion.

Inquire | Review | Analyze

How many hydrogen atoms would it take to cross the head of a pin? This is not a frivolous question. In answering it, you will discover how small atoms really are, and you will see how powerful physics can be as a way to understand nature. First, assume that the head of a pin is about 1 mm in diameter. That is 0.001 m. The size of a hydrogen atom is represented by the diameter of the electron cloud; assume that the electron in your atom is in the second orbit. Then the diameter of the electron cloud is about 0.4 nm. Because 1 nm equals 10^{-9} m, you multiply and discover that 0.4 nm equals 4×10^{-10} m. To find out how many atoms would stretch 0.001 m, you divide the diameter of the pinhead by the diameter of an atom. That is, you divide 0.001 m by 4×10^{-10} m, and you get 2.5×10^{6}. It would thus take 2.5 million hydrogen atoms lined up side by side to cross the head of a pin.

This shows how tiny an atom is and also how powerful basic physics is. A bit of arithmetic gives you a view of nature beyond the capability of your eyes. Now use another bit of arithmetic to calculate how many hydrogen atoms you would need to equal the mass of a paper clip (1 g).

■ ■ ■

Connections: The energy levels in atoms are familiar territory to astronomers because the electrons in those levels can interact with light. Such interactions fill starlight with clues to the nature of the stars.

■ **Figure 6-3**

The electrons in atoms may occupy only certain, permitted orbits. Because different elements have different charges on their nuclei, the elements have different, unique patterns of permitted orbits.

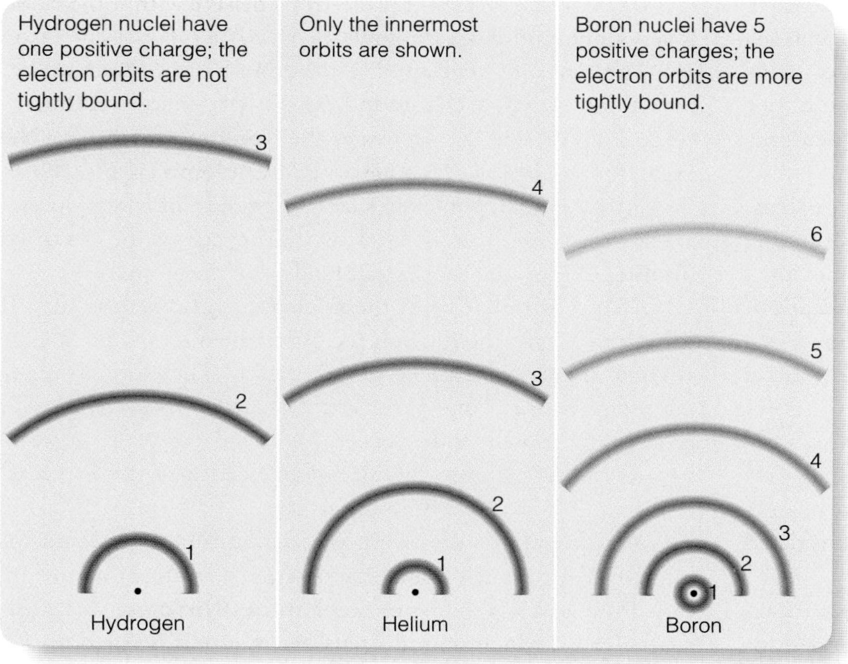

6-2 The Interaction of Light and Matter

LET'S BEGIN OUR STUDY of light and matter by considering the hydrogen atom. As we noted earlier, hydrogen is both simple and common. Roughly 90 percent of all atoms in the universe are hydrogen.

The Excitation of Atoms

Each orbit in an atom represents a specific amount of binding energy, so physicists commonly refer to the orbits as **energy levels.** Using this terminology, you can say that an electron in its smallest and most tightly bound orbit is in its lowest permitted energy level. You can move the electron from one energy level to another by supplying enough energy to make up the difference between the two energy levels. It is like moving a flowerpot from a low shelf to a high shelf; the greater the distance between the shelves, the more energy you need to raise the pot. The amount of energy needed to move the electron is the energy difference between the two energy levels.

If you move the electron from a low energy level to a higher energy level, we say the atom is an **excited atom.** That is, you have added energy to the atom in moving its electron. If the electron falls back to the lower energy level, that energy is released.

An atom can become excited by collision. If two atoms collide, one or both may have electrons knocked into a higher energy level. This happens very commonly in hot gas, where the atoms move rapidly and collide often.

Another way an atom can get the energy that moves an electron to a higher energy level is to absorb a photon. Only a photon with exactly the right amount of energy can move the electron from one level to another. If the photon has too much or too little energy, the atom cannot absorb it. Because the energy of a photon depends on its wavelength, only photons of certain wavelengths can be absorbed by a

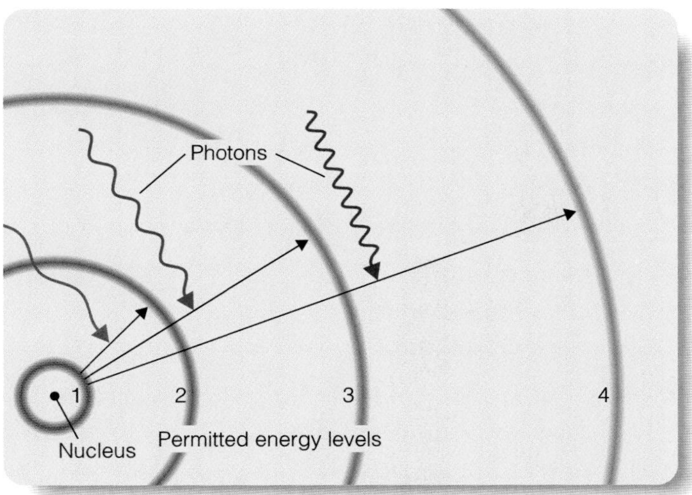

■ Figure 6-4

A hydrogen atom can absorb only those photons that move the atom's electron to one of the higher-energy orbits. Here three different photons are shown, along with the change they would produce if they were absorbed.

given kind of atom. ■ Figure 6-4 shows the lowest four energy levels of the hydrogen atom along with three photons the atom could absorb. The longest-wavelength photon only has enough energy to excite the electron to the second energy level, but the shorter-wavelength photons can excite the electron to higher levels. A photon with too much or too little energy cannot be absorbed. Because the hydrogen atom has many more energy levels than shown in Figure 6-4, it can absorb photons of many different wavelengths.

Atoms, like humans, cannot exist in an excited state forever. The excited atom is unstable and must eventually (usually within 10^{-6} to 10^{-9} s) give up the energy it has absorbed and return its electron to the lowest energy level. Because the electrons eventually tumble down to this bottom level, physicists call it the **ground state.**

When the electron drops from a higher to a lower energy level, it moves from a loosely bound level to one more tightly bound. The atom then has a surplus of energy—the energy difference between the levels—that it can emit as a photon. Study the sequence of events in ■ Figure 6-5 to see how an atom can absorb and emit photons. Because each type of atom or ion has its unique set of energy levels, each type absorbs and emits photons with a unique set of wavelengths. Thus you can identify the elements in a gas by studying the characteristic wavelengths of light absorbed or emitted.

The process of excitation and emission is a common sight in urban areas at night. A neon sign glows when atoms of the neon gas in the tube are excited by electricity flowing through the tube. As the electrons in the electric current flow through the gas, they collide with the neon atoms and excite them. As you have seen, immediately after an atom is excited, its electron drops back to a lower energy level, emitting the surplus energy as a photon of a certain wavelength. The photons emitted by excited neon produce a reddish-orange glow. Signs of other colors, erroneously called "neon," contain other gases or mixtures of gases instead of pure neon.

Radiation from a Heated Object

To begin our discussion of the interaction of light and matter, we must consider a simple but very important phenomenon. When you heat an object up, it begins to glow. Think of a blacksmith heating a horseshoe in a forge. The hot iron glows bright yellow-orange. What produces this light?

Recall that light is a changing electric and magnetic field. Whenever you produce a changing electric field, you produce electromagnetic radiation. If you run a comb through your hair, you disturb electrons in both hair and comb, producing static electricity. Because each electron is surrounded by an electric field, any sudden change in the electron's motion gives rise to electromagnetic radiation. Running a comb through your hair while standing near an AM radio produces radio static. This illustrates an important principle: Whenever you change the motion of an electron, you generate electromagnetic waves.

To see what this has to do with a heated object, think of what you mean when you say an object is hot. The molecules and atoms in an object are in constant motion, and in a hot object they are more agitated than in a cool object. You can refer to this agitation as *thermal energy.* When you touch a hot object, you feel **heat** as that thermal energy flows into your fingers. **Temperature,** on the other hand, refers to the average speed of the particles. Hot cheese and hot green beans can have the same temperature, but the cheese can contain more thermal energy and can burn your tongue. Thus, *heat* refers to the flow of thermal energy, and *temperature* refers to the intensity of the atomic motion.

When we refer to the temperature of astronomical objects such as stars, we will use the **Kelvin temperature scale.** On this

■ Figure 6-5

An atom can absorb a photon only if the photon has the correct amount of energy. The excited atom is unstable and within a fraction of a second returns to a lower energy level, reradiating the photon in a random direction.

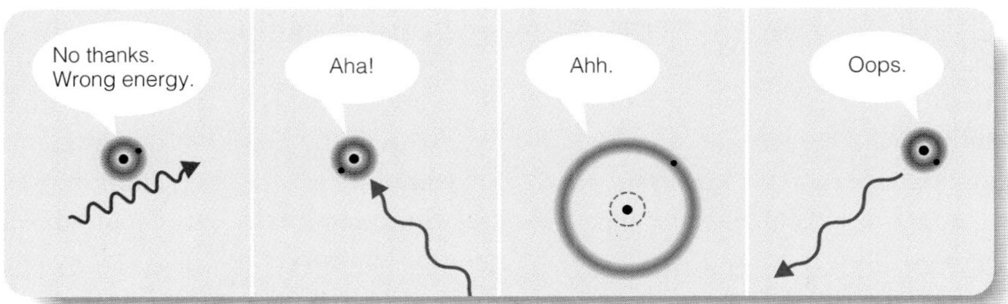

scale, zero degrees Kelvin (written 0 K) is **absolute zero** (−459.7°F), the temperature at which an object contains no heat energy that can be extracted. Water freezes at 273 K and boils at 373 K. The Kelvin temperature scale is useful in astronomy because it is based on absolute zero and thus is related to the motion of the particles in an object.

Now you can understand why a hot object glows. The hotter an object is, the more motion among its particles. The agitated particles collide with electrons, and when electrons are accelerated, part of the energy is carried away as a photon. The radiation emitted by a heated object is called **black body radiation,** a name that refers to the way a perfect emitter of radiation would behave. A perfect emitter would also be a perfect absorber and at room temperature would look black. Thus the term *black body* is often used to refer to objects that glow brightly.

Black body radiation is quite common. In fact, it is responsible for the light emitted by an incandescent lightbulb. Electricity flowing through the filament of the lightbulb heats it to high temperature, and it glows. You should also recognize the light emitted by a heated horseshoe in the blacksmith's forge as black body radiation. Many objects in astronomy, including stars, emit radiation approximately as if they were black bodies.

Hot objects emit black body radiation, but so do cold objects. Ice cubes are cold, but their temperature is higher than absolute zero, so they contain some thermal energy and must emit some black body radiation. The coldest gas drifting in space has a temperature only a few degrees above absolute zero, but it too emits black body radiation.

There are two important features of black body radiation. First, the hotter an object is, the more black body radiation it emits. Hot objects emit more radiation because their agitated particles collide more often and more violently with electrons. Thus you expect a glowing coal from a fire to emit more total energy than an ice cube of the same size.

The second feature is the relationship between the temperature of the object and the wavelengths of the photons it emits. The wavelength of a photon emitted when a particle collides with an electron depends on the violence of the collision. Only a violent collision can produce a short-wavelength (high-energy) photon. Because extremely violent collisions don't occur very often, short-wavelength photons are rare. Similarly, most collisions are not extremely gentle, so long-wavelength (low-energy) photons are also rare. Consequently, black body radiation is made up of photons with a distribution of wavelengths, and very short and very long wavelengths are rare. The **wavelength of maximum intensity (λ_{max}),** the wavelength at which the object emits the most radiation, occurs at some intermediate wavelength. (Make special note that λ_{max} does not refer to the maximum wavelength but to the wavelength of maximum.)

■ Figure 6-6 shows the intensity of radiation versus wavelength for three objects of different temperatures. The curves are high in the middle and low at either end, telling you that these objects emit most intensely at intermediate wavelengths. The total area under each curve is proportional to the total energy emitted, and you can see that the hotter object emits more total energy than the cooler objects. Closer examination of the curves reveals that the wavelength of maximum intensity depends on temperature. The hotter the object, the shorter the wavelength of maximum intensity. Notice in this figure how temperature determines the color of a glowing black body. The hotter object emits more blue light than red and thus looks blue, and the cooler object emits more red than blue and consequently looks red. You can see this among stars. Hot stars look blue and cool

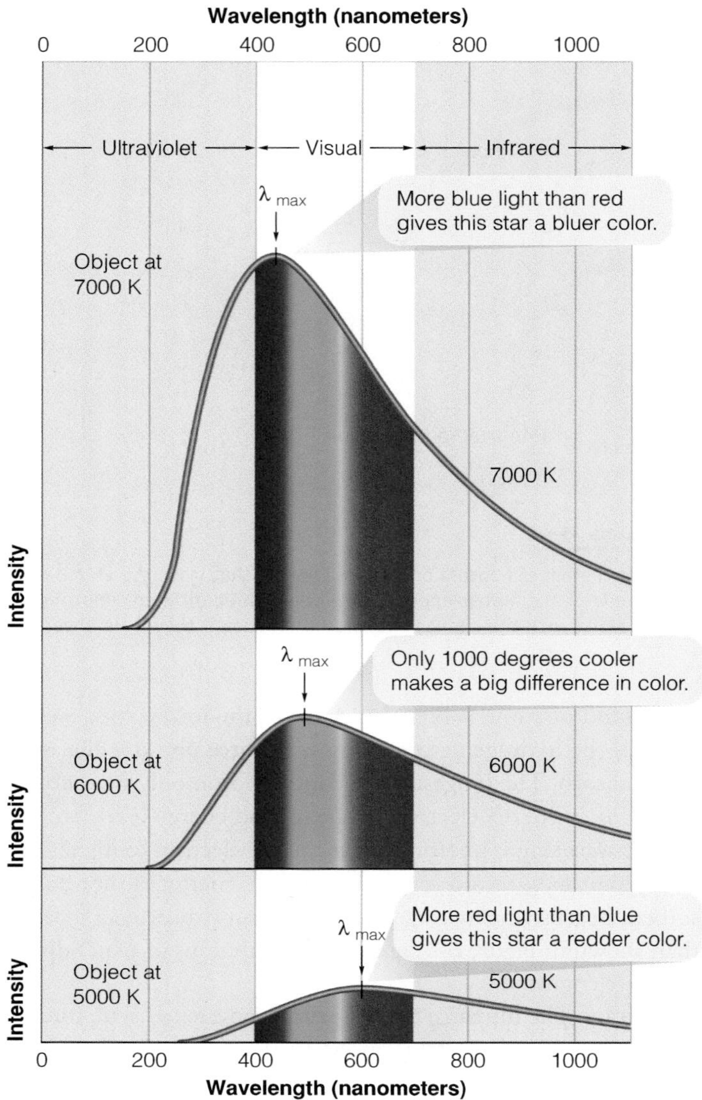

■ **Figure 6-6**

Black body radiation from three bodies at different temperatures demonstrates that a hot body radiates more total energy and that the wavelength of maximum intensity is shorter for hotter objects.

Reasoning with Numbers | 6-1

Black Body Radiation

Black body radiation is described by two simple laws. So many objects in astronomy behave like black bodies that these two laws are powerful tools for the analysis of light from the sky.

Wien's law expresses quantitatively the idea that the higher the temperature, the shorter the wavelength. According to this law, the wavelength of maximum intensity in nanometers, λ_{max}, equals 3,000,000 divided by the temperature in degrees Kelvin:

$$\lambda_{max} = \frac{3{,}000{,}000}{T}$$

For example, a cool star with a temperature of 3000 K will emit most intensely at a wavelength of 1000 nm, which is in the infrared part of the spectrum. A hot star with a temperature of 30,000 K, for example, will emit most intensely at a wavelength of 100 nm, which is in the ultraviolet.

The Stefan–Boltzmann law expresses the idea that the higher the temperature, the greater the radiated energy. According to this law, the total energy radiated in 1 second from 1 square meter of an object equals a constant times the temperature raised to the fourth power.*

$$E = \sigma T^4 \text{ (J/s/m}^2\text{)}$$

Here the temperature is expressed in degrees Kelvin and the energy in units called *joules.* One **joule (J)** is about the energy of an apple falling from a table to the floor. This law shows how strongly the energy radiated depends on temperature. If you doubled an object's temperature, for instance, it would radiate not 2 times, but rather 2^4, or 16, times more energy per second from each square meter of its surface. A small change in temperature can make a big difference to the brightness of a star.

*For the sake of completeness, note that the constant σ equals 5.67×10^{-8} J/m^2 s degree^4.

stars red. The properties of black body radiation are described in **Reasoning with Numbers 6-1.**

Notice that cool objects may emit little visible radiation but are still producing black body radiation. For example, the human body has a temperature of 310 K and emits black body radiation mostly in the infrared part of the spectrum. Infrared security cameras can detect burglars by the radiation they emit. Humans emit very few visible-wavelength photons and almost never emit X-ray or gamma-ray photons. Humans' wavelength of maximum intensity lies in the infrared part of the spectrum.

AceAstronomy™ Log into AceAstronomy and select this chapter to see Astronomy Exercise "Black Body"; you can change the temperature of an object and see how its black body curve changes.

AceAstronomy™ Log into AceAstronomy and select this chapter to see Astronomy Exercise "Stefan–Boltzmann Law" to watch the luminosity of a star change as you adjust its temperature.

Inquire | Review | Analyze

How could a doctor measure someone's temperature without touching him or her?

Doctors and nurses use a handheld device to measure body temperature by observing the infrared radiation emerging from a patient's ear. You might suspect the device depends on the Stefan–Boltzmann law and measures the intensity of the infrared radiation. A person with a fever will emit more energy than a healthy person. However, a healthy person with a large ear canal would emit more than a person with a small ear canal, so measuring intensity would not be accurate. The device actually depends on Wien's law in that it measures the "color" of the infrared radiation. A patient with a fever will emit at a shorter wavelength of maximum intensity, and the infrared radiation emerging from his or her ear will be a tiny bit "bluer" than normal.

Astronomers can measure the temperatures of stars the same way. Use Figure 6-6 to explain how that works.

■ ■ ■

Connections: The spectrum of a star is bursting with information. To interpret that information, astronomers must think carefully about how atoms and light interact at the surfaces of stars.

6-3 Stellar Spectra

SCIENCE IS A WAY OF UNDERSTANDING nature, and the spectrum of a star tells you a great deal about such things as temperature, motion, and composition. In later chapters, you will use spectra to study galaxies and planets, but let's begin by studying the spectra of stars, including the sun. Stellar spectra are the easiest to understand, and the nature of stars is central to your study of all celestial objects.

The Formation of a Spectrum

The spectrum of a star is formed as light passes outward through the gases near its surface. Study **Atomic Spectra** on pages 110 and 111. Notice three important properties of spectra.

Atomic Spectra

To understand how to analyze a spectrum, begin with a simple incandescent lightbulb.

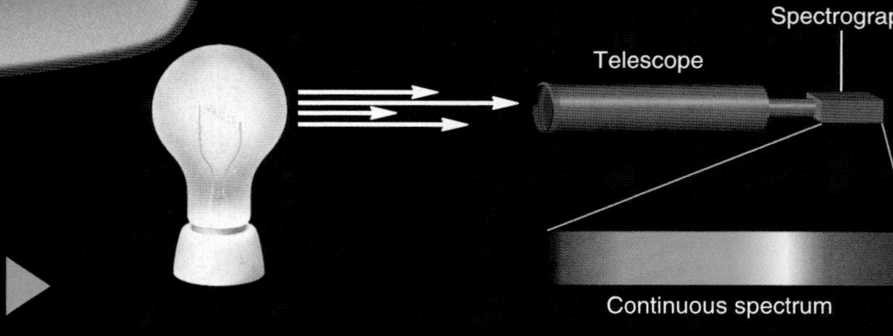

The hot filament emits black body radiation, which forms a **continuous spectrum.**

An **absorption spectrum** results when radiation passes through a cool gas. In this case you can imagine that the lightbulb is surrounded by a cool cloud of gas. Atoms in the gas absorb photons of certain wavelengths, which are missing from the spectrum, and you see their positions as dark **absorption lines.** Such spectra are sometimes called **dark-line spectra.**

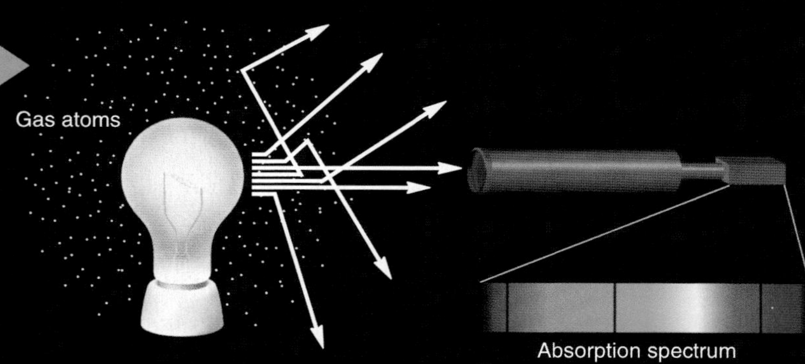

An **emission spectrum** is produced by photons emitted by an excited gas. You could see **emission lines** by turning your telescope aside so that photons from the bright bulb did not enter the telescope. The photons you would see would be those emitted by the excited atoms near the bulb. Such spectra are also called **bright-line spectra.**

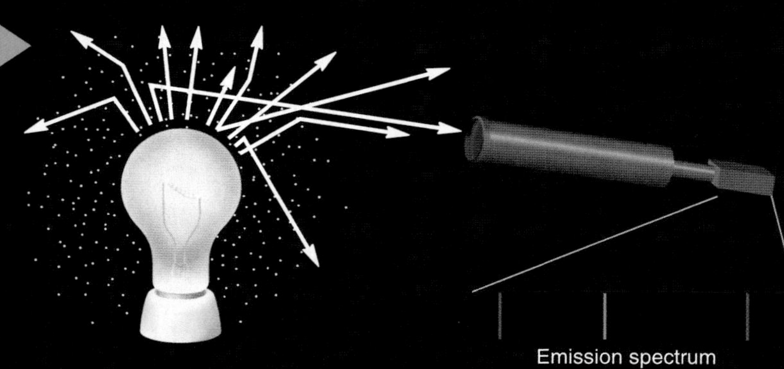

The spectrum of a star is an absorption spectrum. The denser layers of the photosphere emit black body radiation. Gases in the atmosphere of the star absorb their specific wavelengths and form dark absorption lines in the spectrum.

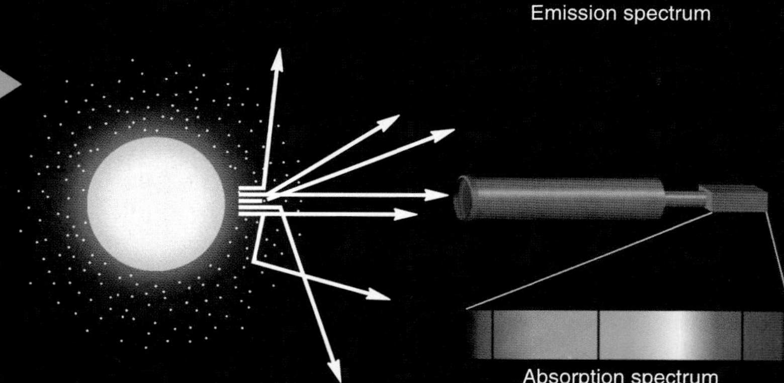

In 1859, long before scientists understood atoms and energy levels, the German scientist Gustav Kirchhoff formulated three rules, now known as **Kirchhoff's laws,** that describe the three types of spectra.

KIRCHHOFF'S LAWS

Law I: The Continuous Spectrum

A solid, liquid, or dense gas excited to emit light will radiate at all wavelengths and thus produce a continuous spectrum.

Law II: The Emission Spectrum

A low-density gas excited to emit light will do so at specific wavelengths and thus produce an emission spectrum.

Law III: The Absorption Spectrum

If light comprising a continuous spectrum passes through a cool, low-density gas, the result will be an absorption spectrum.

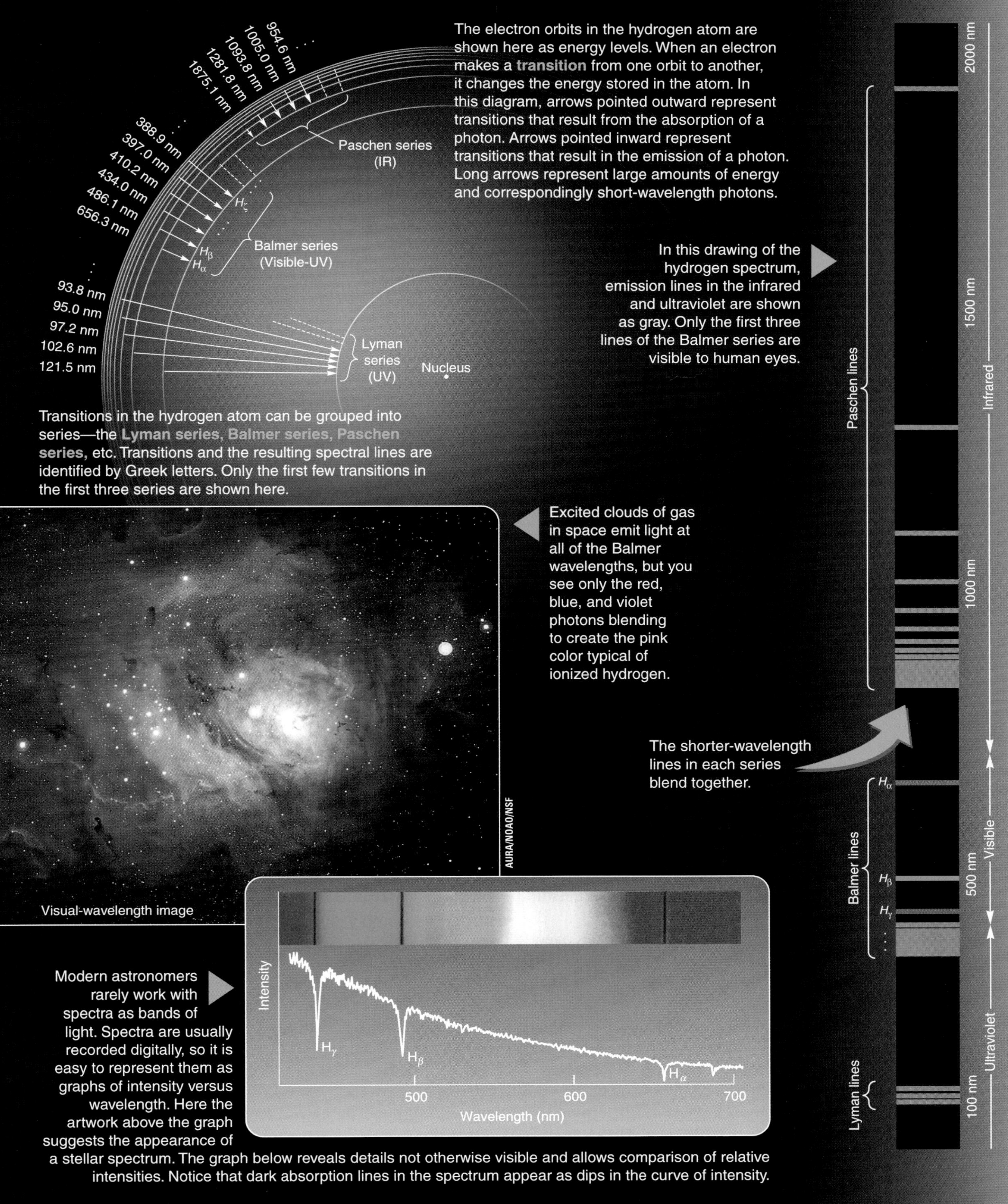

The electron orbits in the hydrogen atom are shown here as energy levels. When an electron makes a **transition** from one orbit to another, it changes the energy stored in the atom. In this diagram, arrows pointed outward represent transitions that result from the absorption of a photon. Arrows pointed inward represent transitions that result in the emission of a photon. Long arrows represent large amounts of energy and correspondingly short-wavelength photons.

Transitions in the hydrogen atom can be grouped into series—the **Lyman series, Balmer series, Paschen series,** etc. Transitions and the resulting spectral lines are identified by Greek letters. Only the first few transitions in the first three series are shown here.

In this drawing of the hydrogen spectrum, emission lines in the infrared and ultraviolet are shown as gray. Only the first three lines of the Balmer series are visible to human eyes.

Excited clouds of gas in space emit light at all of the Balmer wavelengths, but you see only the red, blue, and violet photons blending to create the pink color typical of ionized hydrogen.

The shorter-wavelength lines in each series blend together.

Modern astronomers rarely work with spectra as bands of light. Spectra are usually recorded digitally, so it is easy to represent them as graphs of intensity versus wavelength. Here the artwork above the graph suggests the appearance of a stellar spectrum. The graph below reveals details not otherwise visible and allows comparison of relative intensities. Notice that dark absorption lines in the spectrum appear as dips in the curve of intensity.

First, there are three kinds of spectra. When you see one of these types of spectra, you can recognize the kind of matter that emitted the light.

Notice also that the wavelengths of the photons that are absorbed or emitted in a spectrum are determined by the atomic energy levels in the atoms. The emitted photons coming from a hot cloud of hydrogen gas have the same wavelengths as the photons absorbed by hydrogen atoms in the gases of a star. Whether an atom is emitting a photon or absorbing a photon, the energy levels in the atom are the same, and the wavelengths are the same. Different atoms, however, have different energy levels, so their spectra contain different patterns of spectral lines.

Finally, notice that the hydrogen atom produces many spectral lines, from the ultraviolet to the infrared, but only three are visible to human eyes. These hydrogen lines are prominent in the spectra of stars and can tell you the temperatures of the stars.

Ace Astronomy™ Log into AceAstronomy and select this chapter to see Astronomy Exercise "Emission and Absorption Spectra." This exercise will give you an inside look at atoms as they interact with photons.

The Balmer Thermometer

You can use the Balmer lines as a thermometer to find the temperatures of stellar surfaces. From the discussion of black body radiation, you know that you can estimate stellar temperatures from color—red stars are cool, and blue stars are hot. The Balmer lines, however, give you much greater accuracy.

Astronomers use the Kelvin temperature scale when referring to the stellar temperatures. Notice that the temperature of a star refers to its surface temperature—typically 40,000 to 2000 K. The centers of stars are much hotter—many millions of degrees—but the spectra tell you about the surface layers from which the light originates.

The Balmer thermometer works because the Balmer absorption lines are produced by hydrogen atoms whose electrons are in the second energy level. If the surface of a star is as cool as the sun or cooler, there are few violent collisions between atoms to excite the electrons, and most atoms have their electrons in the ground state. These atoms can't absorb photons in the Balmer series. As a result, you should expect to find weak Balmer lines in the spectra of cool stars.

In the surface layers of stars hotter than about 20,000 K, however, there are many violent collisions between atoms, exciting electrons to high energy levels or knocking the electrons completely out of most atoms. That is, the atoms become ionized. Therefore, few hydrogen atoms have electrons in the second energy level to form Balmer absorption lines, and you should expect hot stars, like cool stars, to have weak Balmer absorption lines.

At an intermediate temperature, roughly 10,000 K, the collisions have the correct amount of energy to excite large numbers of electrons into the second energy level. With many atoms excited to the second level, the gas absorbs Balmer-wavelength photons well and thus produces strong Balmer lines.

To summarize, the strength of the Balmer lines depends on the temperature of the star's surface layers. Both hot and cool stars have weak Balmer lines, but medium-temperature stars have strong Balmer lines.

Theoretical calculations can predict just how strong the Balmer lines should be for stars of various temperatures. Such calculations are the key to finding temperatures from stellar spectra. The curve in ■ Figure 6-7a shows the calculated strength of the Balmer lines for various stellar temperatures. You could use this as a temperature indicator, except that the curve gives two answers. A star with Balmer lines of a certain strength might have either of two temperatures, one high and one low. How do you know the true temperature of the star? You must examine other spectral lines to arrive at the correct temperature.

You have seen how the strength of the Balmer lines depends on temperature. Temperature has a similar effect on the spectral lines of other elements, such as ionized calcium (Figure 6-7b). That is, their lines are weak at high and low temperatures and strong at some intermediate temperature. But the temperature at which their lines reach maximum strength is different for each element. If you add a number of these elements to your graph, you get a powerful tool for finding the temperature of stars (Figure 6-7c).

How do you use this tool? You determine a star's temperature by comparing the strengths of its spectral lines with your graph. For instance, if you recorded a spectrum of a star and found medium-strength Balmer lines and strong helium lines, you could conclude it had a temperature of about 20,000 K. But if the star had weak hydrogen lines and strong lines of ionized iron, you would assign it a temperature of about 5800 K, similar to that of the sun.

The spectra of stars cooler than about 3000 K contain dark bands produced by molecules such as titanium oxide (TiO). Because of their structure, molecules can absorb photons at many wavelengths, producing numerous, closely spaced spectral lines that blend together to form bands. These molecular bands appear only in the spectra of the coolest stars because, as mentioned before, molecules in cool stars are not subject to the violent collisions that would break up molecules in hotter stars. Thus, the presence of dark bands in a star's spectrum indicates that the star is very cool.

From stellar spectra, astronomers have found that the hottest stars have surface temperatures above 40,000 K and the coolest about 2000 K. Compare these with the surface temperature of the sun, which is about 5800 K.

Spectral Classification

You have seen that the strengths of spectral lines depend on the surface temperature of the star. From this you can predict that all stars of a given temperature should have similar spectra. If you learn to recognize the pattern of spectral lines produced by a 6000-K star, for instance, you need not use Figure 6-7c every time you see that kind of spectrum. You can save time by classifying stellar spectra rather than analyzing each one individually.

■ **Figure 6-7**

The strength of spectral lines can tell you the temperature of a star. (a) Balmer hydrogen lines of a certain strength could be produced by a hotter star or a cooler star. (b) Adding another atom to the diagram helps, and (c) adding many atoms and molecules to the diagram gives you a precise tool to find the temperatures of stars.

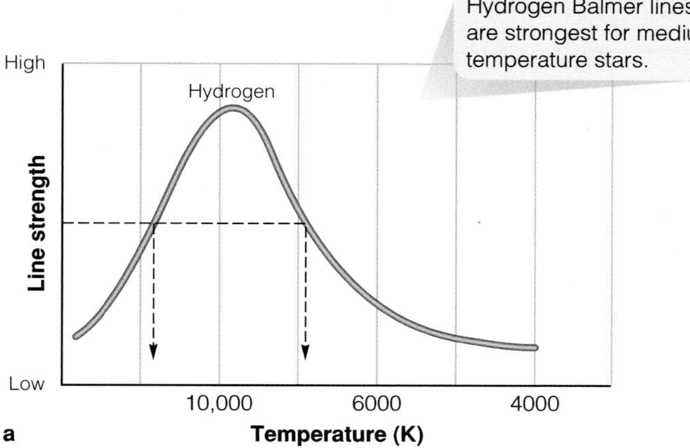

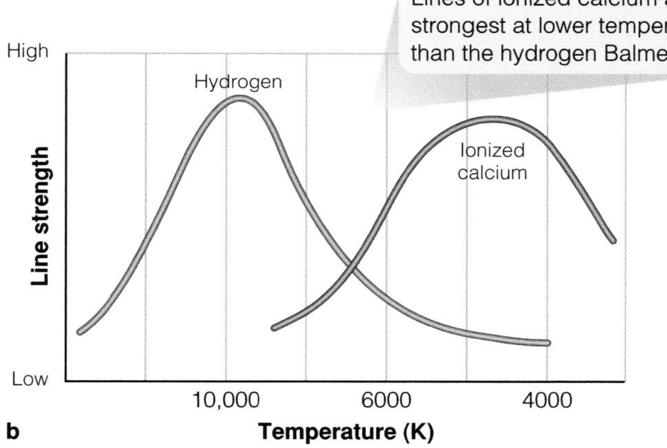

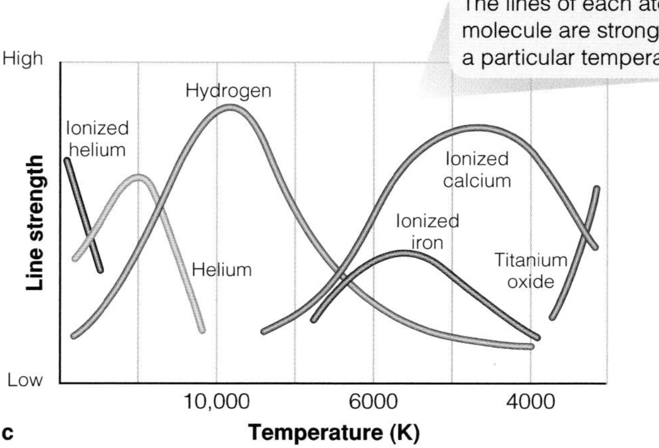

The first widely used classification system was devised by astronomers at Harvard during the 1890s and 1900s. One of them, Annie J. Cannon, personally inspected and classified the spectra of over 250,000 stars. The spectra were first classified in groups labeled A through Q, but some groups were later dropped, merged with others, or reordered. The final classification includes the seven **spectral classes,** or **types,** still used today: O, B, A, F, G, K, M.*

This sequence of spectral types, called the **spectral sequence,** is important because it is a temperature sequence. The O stars are the hottest, and the temperature continues to decrease down to the M stars, the coolest. For maximum precision, astronomers divide each spectral class into 10 subclasses. For example, spectral class A consists of the subclasses A0, A1, A2, . . . A8, A9. Next comes F0, F1, F2, and so on. This finer division defines a star's temperature to an accuracy within about 5 percent. The sun, for example, is not just a G star, but a G2 star, with a temperature of 5800 K.

Astronomers classify a star by the lines and bands in its spectrum, as shown in ■ Table 6-1. For example, if it has weak Balmer lines and lines of ionized helium, it must be an O star. This table is based on the same information used in Figure 6-7c.

■ Figure 6-8 shows 13 stellar spectra ranging from the hottest at the top to the coolest at the bottom. Notice how spectral features change gradually from hot to cool stars. Although these spectra are attractive, astronomers rarely work with spectra as color images. Rather, they display spectra as graphs of intensity versus wavelength with dark absorption lines as dips in the graph

* Generations of astronomy students have remembered the spectral sequence using the mnemonic "Oh, Be A Fine Girl (Guy), Kiss Me." More recent suggestions from students include, "Oh Boy, An F Grade Kills Me," and "Only Bad Astronomers Forget Generally Known Mnemonics."

■ **Table 6-1** | **Spectral Classes**

Spectral Class	Approximate Temperature (K)	Hydrogen Balmer Lines	Other Spectral Features
O	40,000	Weak	Ionized helium
B	20,000	Medium	Neutral helium
A	10,000	Strong	Ionized calcium weak
F	7500	Medium	Ionized calcium weak
G	5500	Weak	Ionized calcium medium
K	4500	Very weak	Ionized calcium strong
M	3000	Very weak	Titanium oxide strong

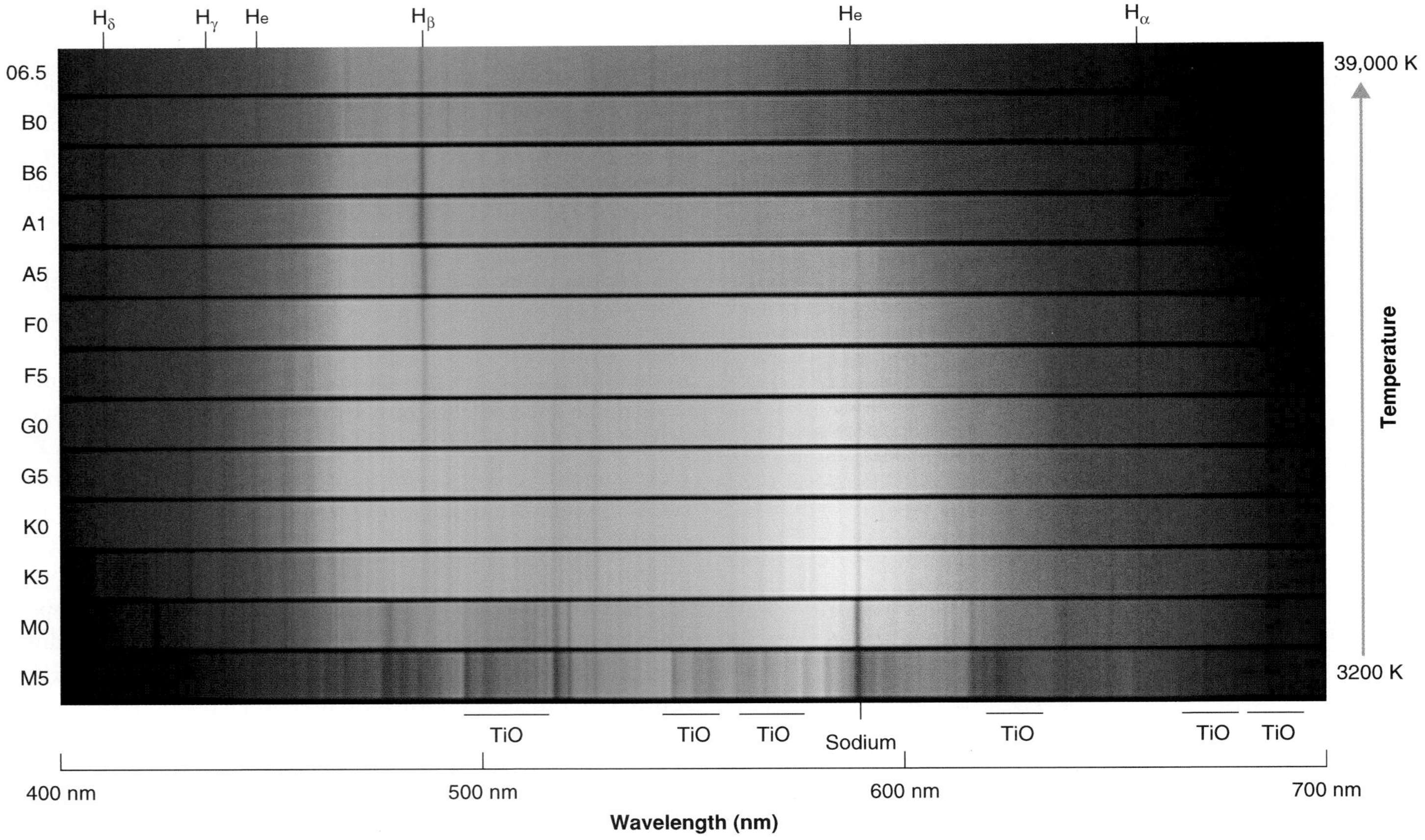

■ **Figure 6-8**

These spectra show stars from hot O stars at the top to cool M stars at the bottom. The Balmer lines of hydrogen are strongest about A0, but the two closely spaced lines of sodium in the yellow are strongest for very cool stars. Helium lines appear only in the spectra of the hottest stars. Notice that the helium line visible in the top spectrum has nearly but not exactly the same wavelength as the sodium lines visible in cooler stars. Bands produced by the molecule titanium oxide are strong in the spectra of the coolest stars. (AURA/NOAO/NSF)

(■ Figure 6-9). Such graphs show more detail than photographs. Notice also that the overall curves are similar to black body curves. The wavelength of maximum is in the infrared for the coolest stars and in the ultraviolet for the hottest stars.

Compare Figures 6-8 and 6-9 and notice how the strength of spectral lines depends on temperature. Note that the Balmer lines are strongest in A stars, where the temperature is moderate but still high enough to excite the electrons in hydrogen atoms to the second energy level, where they can absorb Balmer-wavelength photons. In the hotter stars (O and B), the Balmer lines are weak because the higher temperature excites the electrons to energy levels above the second level or ionizes the atoms. The Balmer lines in cooler stars (F through M) are also weak but for a different reason. The lower temperature cannot excite many electrons to the second energy level, so few hydrogen atoms are capable of absorbing Balmer-wavelength photons.

The spectral lines of other atoms also change from class to class. Helium is visible only in the spectra of the hottest classes, and titanium oxide bands are visible only in the coolest. Two lines of ionized calcium increase in strength from A to K and then decrease from K to M. Because the strength of these spectral lines depends on temperature, it requires only moments to study a star's spectrum and determine its temperature.

The study of spectral types is a century old, but astronomers continue to discover new types of stars. The **L dwarfs,** found in 1998, are cooler and fainter than M stars. The L dwarfs are clearly a different type of star. The spectra of M stars contain bands produced by metal oxides such as titanium oxide, but L dwarf spectra contain bands produced by chromium hydride and iron hydride (■ Figure 6-10). The newly discovered **T dwarfs** are even cooler and fainter than L dwarfs. The development of giant telescopes and highly sensitive infrared cameras and spectrographs is allowing astronomers to find and study these faintest of stars.

Not only can a star's spectrum tell you the surface temperature of the star, but it can also tell you how the star is moving through space.

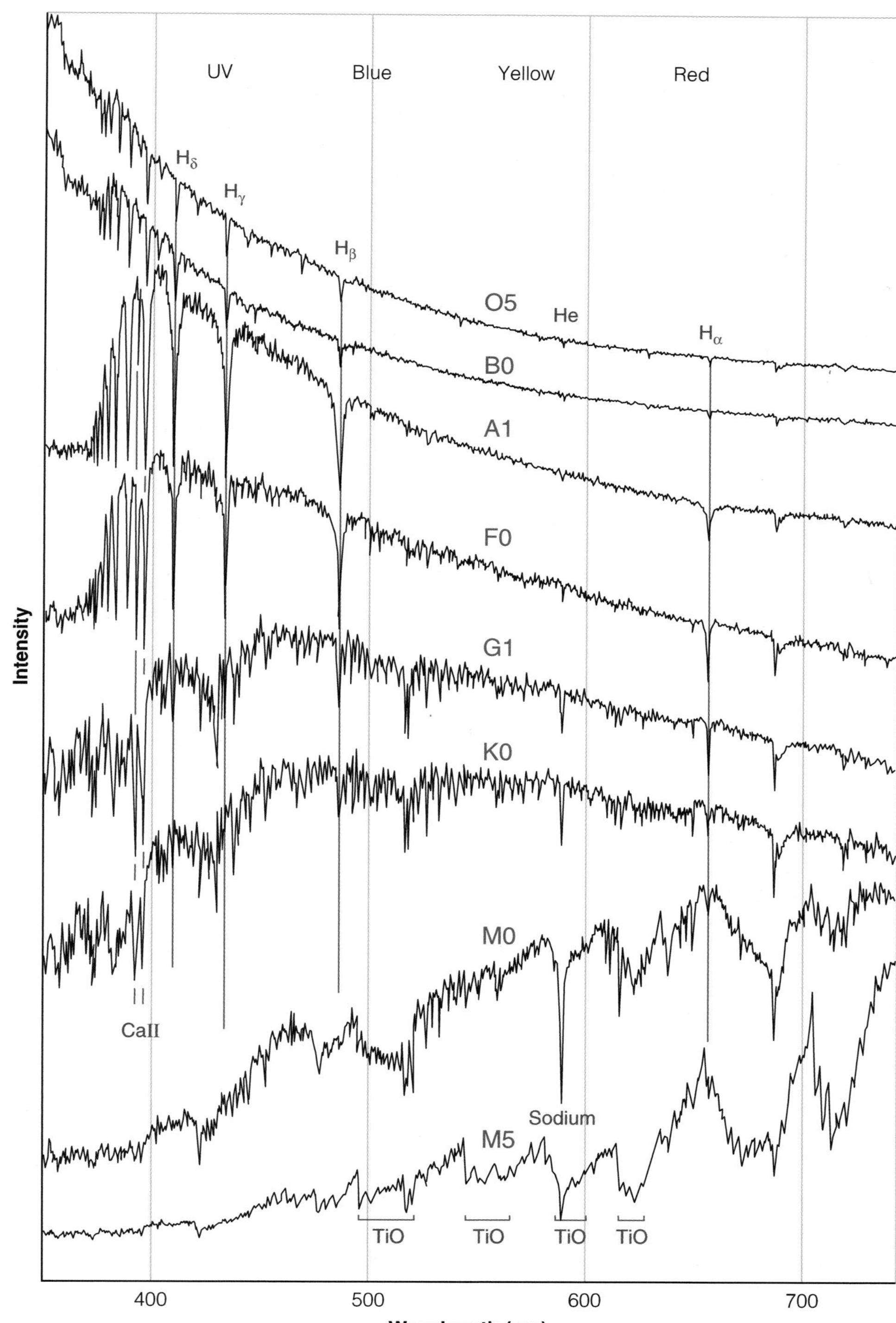

■ **Figure 6-9**

Modern digital spectra show how stellar spectra depend on spectral class. Here spectra are represented by graphs of intensity versus wavelength, and dark absorption lines appear as sharp dips in intensity. Hydrogen Balmer lines are strongest about A0, while lines caused by ionized calcium (CaII) are strong in K stars. Bands produced by TiO molecules are strong in the coolest stars. Compare these spectra with Figures 6-7c and 6-8. (Courtesy NOAO, G. Jacoby, D. Hunter, and C. Christian)

The Doppler Effect

Astronomers can measure the wavelengths of lines in a star's spectrum and find the velocity of the star. The **Doppler effect** is the apparent change in the wavelength of radiation caused by the motion of the source.

If a star is moving toward Earth, the lines in its spectrum will be shifted slightly toward shorter wavelengths. That is, they are shifted toward the blue end of the spectrum—a **blueshift.** If a star is moving away from Earth, the lines are shifted slightly toward the red end of the spectrum—a **redshift.** These Doppler shifts are small and don't change the color of the star. But it is easy to detect these changes in wavelength in a star's spectrum.

■ Figure 6-11a shows the Doppler effect in two spectra of the star Arcturus. Lines in the top spectrum are slightly blue-

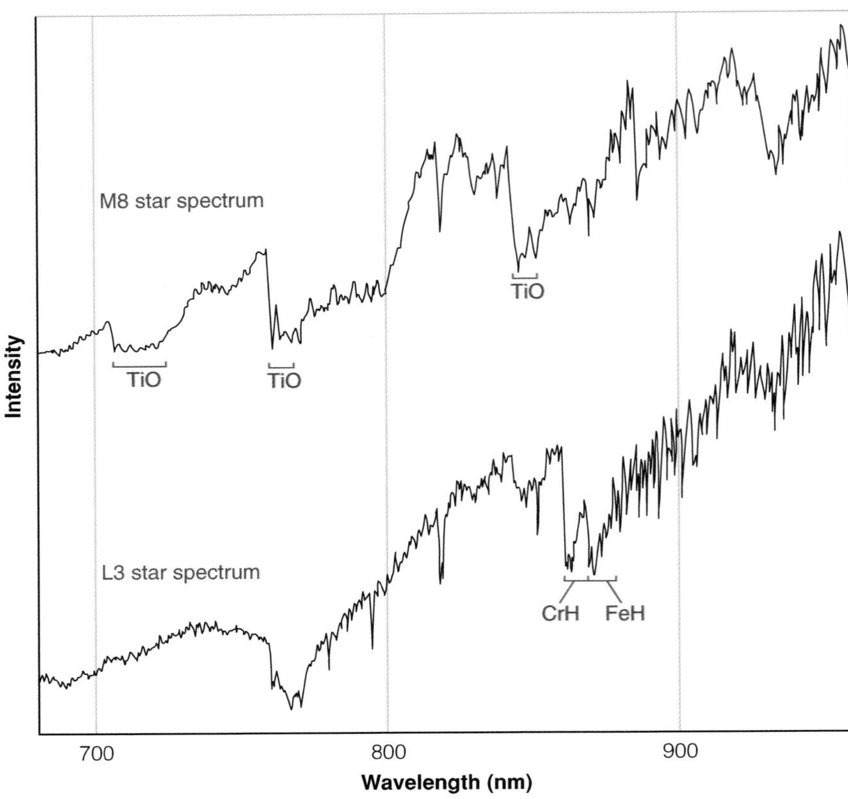

Figure 6-10

The spectra of an M star and an L star compared: M-star spectra contain strong bands produced by the titanium oxide molecule (TiO), but in L-star spectra TiO bands are weak or absent. Notice the missing TiO band near 710 nm in the L-star spectrum. Instead, the spectra of L stars contain bands produced by chromium hydride (CrH) and iron hydride (FeH). (Adapted from data by J. Davy Kirkpatrick)

shifted because the spectrum was recorded when Earth, following its orbit, was moving toward Arcturus. Lines in the bottom spectrum are redshifted because it was recorded six months later, when Earth was moving away from Arcturus.

The Doppler effect tells you how rapidly the distance between you and the source of light is increasing or decreasing. It does not matter whether you are moving or the star is moving. Only the relative velocity is important. Also, the Doppler shift is sensitive only to the part of the velocity directed away from you or toward you. This part of the velocity is called the **radial velocity (V_r).** You cannot use the Doppler effect to detect any part of the velocity that is perpendicular to your line of sight.

The Doppler effect occurs for any kind of radiation, not just light. Radio astronomers, for instance, observe the Doppler effect when they measure the wavelength of radio waves coming from objects moving toward or away from Earth. Whatever wavelength range you consider—radio, X rays, gamma rays—you should still refer to a lengthening of the observed wavelength as a redshift and a shortening of the observed wavelength as a blueshift.

You can even detect the Doppler shift in sound. Sound is not electromagnetic radiation, of course; it is a mechanical wave transmitted through air. Because it is a wave, however, it is subject to the Doppler effect. Sounds with long wavelengths have low pitches, and sounds with short wavelengths have higher pitches. You hear a Doppler shift as a car or truck passes and its pitch drops. Its sound is shifted to shorter wavelengths and higher pitches (a "blue" shift) while it is approaching and is shifted to longer wavelengths and lower pitches (a "red" shift) after it passes.

You can understand how the Doppler shift works by thinking about a fire truck approaching with a bell clanging once a second. When the bell clangs, the sound travels ahead of the truck to reach your ears. One second later, the bell clangs again, but not at the same place. During that 1 second, the fire truck moved closer to you, so the bell is closer at its second clang. Now the sound has a shorter distance to travel and reaches your ears a little sooner than it would have if the fire truck were not approaching. The third time the bell clangs, it is even closer. By timing the bell, you could observe that the clangs are slightly less than 1 second apart, all because the fire truck is approaching. If the fire truck were moving away from you, you would hear the clangs sounding more than 1 second apart, because each successive clang of the bell occurs farther from you.

Figure 6-11b shows a fire truck moving toward one observer and away from another observer. The position of the bell at each clang is shown by the small black bells, with the sound spreading outward as black circles. You can see that the clangs are squeezed together ahead of the fire truck and stretched apart behind.

Now you can substitute a source of light waves for the clanging of the bell. If the source is approaching, then each time the source emits the peak of a wave it will be slightly closer to you, and you will observe a shorter wavelength. If it is moving away, you will observe a longer wavelength. This is shown in Figure 6-11c, where the peaks of the waves appear compressed in front of the moving light source and stretched out behind the source.

Of course, how great the change in wavelength is depends on the velocity. Just as a slowly moving car has a smaller Doppler effect in sound than does a high-speed airplane, a slowly moving star has a smaller Doppler effect in light than does a high-speed star. Thus, you can measure the velocity by measuring the amount by which the spectral lines are shifted in wavelength. A large Doppler shift means a high radial velocity. This is a common tool in law enforcement. Police transmit radio signals toward passing autos and measure speed by measuring the wavelength shifts in the reflected signal. **Reasoning with Numbers 6-2** shows how you could do a Doppler effect calculation.

Stellar spectra are filled to bursting with information about stars and their motion, but the astronomer must know some physics to interpret a spectrum. For example, you have now learned

Figure 6-11

The Doppler effect. (a) Absorption lines in the spectrum of the bright star Arcturus are shifted to the blue in winter, when Earth's orbital motion carries it toward the star, and to the red in summer, when Earth moves away from the star. (The Observatories of the Carnegie Institution of Washington) (b) The clanging bell on a moving fire truck produces sounds that move outward (black circles). An observer ahead of the truck hears the clangs closer together, while an observer behind the truck hears them farther apart. (c) A moving source of light emits waves that move outward (black circles). An observer in front of the light source observes a shorter wavelength (a blueshift), and an observer behind the light source observes a longer wavelength (a redshift).

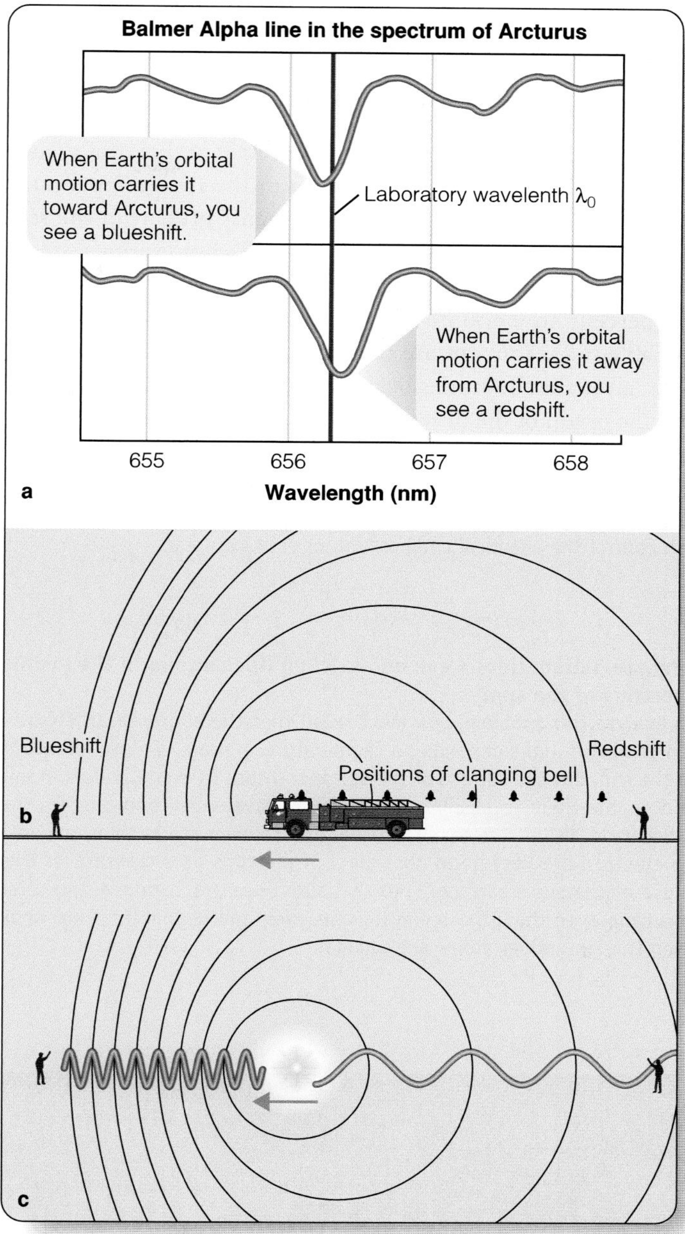

enough about the physics of stellar spectra to estimate the chemical composition of a star from its spectrum.

Ace Astronomy™ Log into AceAstronomy and select this chapter to see Astronomy Exercise "Doppler Shift." You can compare the Doppler shift for a source of sound with that for a source of light.

Ace Astronomy™ Log into AceAstronomy and select this chapter to see Astronomy Exercise "Stellar Rotation" and see an interesting application of the Doppler effect.

Chemical Composition

Identifying the elements in a star by identifying the lines in the star's spectrum is a relatively straightforward procedure. For example, two dark absorption lines appear in the yellow region of the solar spectrum at the wavelengths 589 nm and 589.6 nm. The only atom that can produce this pair of lines is sodium, so the sun must contain sodium. Over 90 elements in the sun have been identified this way.

However, just because the spectral lines characteristic of an element are missing, you cannot conclude that the element itself is absent. For example, the hydrogen Balmer lines are weak in the sun's spectrum, yet 90 percent of the atoms in the sun are hydrogen. The reason for this apparent paradox is that the sun is too cool to produce strong Balmer lines. Similarly, an element's

Reasoning with Numbers | 6-2

The Doppler Formula

Astronomers can measure radial velocity by using the Doppler effect. The laboratory wavelength λ_0 is the wavelength a certain spectral line would have in a laboratory where the source of the light is not moving. In the spectrum of a star, this spectral line is shifted by some small amount $\Delta\lambda$. If the wavelength is increased (a red shift), $\Delta\lambda$ is positive; if the wavelength is decreased (a blue shift), $\Delta\lambda$ is negative. The radial velocity, V_r, of the star is given by the Doppler formula:

$$\frac{V_r}{c} = \frac{\Delta\lambda}{\lambda_0}$$

That is, the radial velocity divided by the speed of light, c, is equal to $\Delta\lambda$ divided by λ_0. In astronomy, radial velocities are almost always given in kilometers per second, so c is expressed as 300,000 km/s.

For example, suppose the laboratory wavelength of a certain spectral line is 600.00 nm, and the line is observed in a star's spectrum at a wavelength of 600.10 nm. Then $\Delta\lambda$ is +0.10 nm, and the velocity is 0.10/600 multiplied by the speed of light. The radial velocity equals 50 km/s. Because $\Delta\lambda$ is positive, you know the star is receding from you.

spectral lines may be absent from a star's spectrum because the star is too hot or too cool to excite those atoms to the energy levels that produce visible spectral lines.

To derive accurate chemical abundances, you must use the physics that describes the interaction of light and matter to analyze a star's spectrum, take into account the star's temperature, and calculate the amounts of the elements present in the star. Such results show that nearly all stars have compositions similar to the sun's—about 91 percent of the atoms are hydrogen, and 8.9 percent are helium, with small traces of heavier elements (■ Table 6-2). You will use these results in later chapters when we analyze the life stories of the stars, the history of our galaxy, and the origin of the universe.

Ace Astronomy™ Log into AceAstronomy and select this chapter to see Astronomy Exercise "Stellar Atomic Absorption Lines," where you can control the chemical composition of your own star.

■ **Table 6-2 I The Most Abundant Elements in the Sun**

Element	Percentage by Number of Atoms	Percentage by Mass
Hydrogen	91.0	70.9
Helium	8.9	27.4
Carbon	0.03	0.3
Nitrogen	0.008	0.1
Oxygen	0.07	0.8
Neon	0.01	0.2
Magnesium	0.003	0.06
Silicon	0.003	0.07
Sulfur	0.002	0.04
Iron	0.003	0.1

Inquire I Review I Analyze

Why are helium lines weak and calcium lines strong in the visible spectrum of the sun?

To analyze this problem, you must recall that the ability of an atom or ion to absorb light depends on temperature. Helium is quite abundant in the sun, but the surface of the sun is too cool to excite helium atoms and enable them to easily absorb visible-wavelength photons. On the other hand, calcium is easily ionized, and calcium atoms that have lost an electron are very good absorbers of photons at temperatures like those at the sun's surface. Thus, calcium lines are strong in the solar spectrum even though calcium ions are rare, and helium lines are weak even though helium atoms are common.

If you are to analyze a star's spectrum and consider the strength of spectral lines, you must take into account the temperature of the star. Why don't you need to take temperature into account when you use the Doppler effect to measure a star's radial velocity?

■ ■ ■

Connections: The spectra of the stars are filled with clues about the gas emitting the light. Much of astronomy is based on unraveling these clues. We will begin in the following chapter by studying the nearest star—the sun.

Study and Review Tools

Summary

6-1 I Atoms

What is an atom?

- An atom consists of a nucleus surrounded by a cloud of electrons. The nucleus is made up of positively charged protons and uncharged neutrons.
- The number of protons in an atom determines which element it is. Atoms of the same element (that is, having the same number of protons) with different numbers of neutrons are called isotopes.
- A neutral atom is surrounded by a number of negatively charged electrons equal to the number of protons in the nucleus. An atom that has lost or gained an electron is called an ion.
- The electrons in an atom may occupy various permitted orbits around the nucleus but not orbits in between.

6-2 I The Interaction of Light and Matter

How do atoms interact with light?

- The size of an electron's orbit depends on the energy stored in the motion of the electron.
- An electron may be excited to a higher orbit during a collision between atoms, or it may move from one orbit to another by absorbing or emitting a photon of the proper energy.
- The motion among the particles in a solid, a liquid, or a dense gas causes the emission of black body radiation. This produces a continuous spectrum—a continuous rainbow of color without dark or bright lines.
- The hotter an object is, the more it radiates and the shorter is its wavelength of maximum intensity, λ_{max}. This allows astronomers to estimate the temperatures of stars from their colors.

6-3 | Stellar Spectra

What kind of spectra do you see when you look at celestial objects?

- Because orbits of only certain energies are permitted in an atom, photons of only certain wavelengths can be absorbed or emitted. Each kind of atom has its own characteristic set of spectral lines. The hydrogen atom has the Lyman series of lines in the ultraviolet, the Balmer series in the visible, and the Paschen series (and others) in the infrared.
- If light passes through a low-density gas on its way to your telescope, the gas can absorb photons of certain wavelengths, and you see dark lines in the spectrum at those positions. Such a spectrum is called an absorption spectrum.
- If you look at a low-density gas that is excited to emit photons, you see bright lines in the spectrum. Such a spectrum is called an emission spectrum.

What can you learn from a star's spectrum?

- Nearly all stellar spectra are absorption spectra, and the hydrogen lines you see there are Balmer lines. In cool stars, the Balmer lines are weak because atoms are not excited out of the ground state. In hot stars, the Balmer lines are weak because atoms are excited to higher orbits or are ionized. Only at medium temperatures are the Balmer lines strong.
- The strength of spectral lines in a star's spectrum can tell you its temperature. In its simplest form, this amounts to classifying the star's spectrum in the spectral sequence: O, B, A, F, G, K, M.
- Long after the spectral sequence was created, astronomers found the L and T stars at temperatures even cooler than the M stars.
- The wavelengths of spectral lines provide clues to the motions of the stars. When a star is approaching, you observe slightly shorter wavelengths (a blueshift), and when it is receding, you observe slightly longer wavelengths (a redshift). This Doppler effect reveals a star's radial velocity, that part of its velocity directed toward or away from Earth.
- A spectrum can tell you the chemical composition of the stars. The presence of spectral lines of a certain element shows that that element must be present in the star. But you must proceed with care. Lines of a certain element may be weak or absent if the star is too hot or too cool.

New Terms

spectral line (p. 104)
nucleus (p. 104)
proton (p. 104)
neutron (p. 104)
electron (p. 104)
isotope (p. 105)
ionization (p. 105)
ion (p. 105)
molecule (p. 105)
Coulomb force (p. 105)
binding energy (p. 105)
quantum mechanics (p. 105)
permitted orbit (p. 106)
energy level (p. 106)
excited atom (p. 106)
ground state (p. 107)
heat (p. 107)
temperature (p. 107)
Kelvin temperature scale (p. 107)
absolute zero (p. 108)
black body radiation (p. 108)
wavelength of maximum intensity (λ_{max}) (p. 108)
joule (J) (p. 109)
continuous spectrum (p. 110)
absorption spectrum (dark-line spectrum) (p. 110)
absorption line (p. 110)
emission spectrum (bright-line spectrum) (p. 110)
emission line (p. 110)
Kirchhoff's laws (p. 110)
transition (p. 111)
Lyman series (p. 111)
Balmer series (p. 111)
Paschen series (p. 111)
spectral class or type (p. 113)
spectral sequence (p. 113)
L dwarf (p. 114)
T dwarf (p. 114)
Doppler effect (p. 115)
blueshift (p. 115)
redshift (p. 115)
radial velocity (V_r) (p. 116)

Review Questions

Ace Astronomy™ Assess your understanding of this chapter's topics with additional quizzing and animations at **http://astronomy.brookscole.com/sh9e**

1. Why might you say that atoms are mostly empty space?
2. What is the difference between an isotope and an ion?
3. Why is the binding energy of an electron related to the size of its orbit?
4. Explain why ionized calcium can form absorption lines, but ionized hydrogen cannot.
5. Describe two ways an atom can become excited.
6. Why do different atoms have different lines in their spectra?
7. Why does the amount of black body radiation emitted depend on the temperature of the object?
8. Why do hot stars look bluer than cool stars?
9. What kind of spectrum does a neon sign produce?
10. Why are Balmer lines strong in the spectra of medium-temperature stars and weak in the spectra of hot and cool stars?
11. Why are titanium oxide features visible in the spectra of only the coolest stars?
12. Explain the similarities among Table 6-1, Figure 6-7c, Figure 6-8, and Figure 6-9.
13. Why does the Doppler effect detect only radial velocity?
14. How can the Doppler effect explain shifts in both light and sound?
15. Explain why the presence of spectral lines of a given element in the solar spectrum tells you that element is present in the sun, but the absence of the lines would not mean the element was absent from the sun?
16. The nebula shown at right contains mostly hydrogen excited to emit photons. What kind of spectrum would you expect this nebula to produce?
17. If the nebula in the image at right crosses in front of the star and the nebula and star have different radial velocities, what might the spectrum of the star look like?

T. Rector, University of Alaska, and WIYN/NURO/AURA/NSF

Discussion Questions

1. In what ways is the model of an atom a scientific model? In what ways is it incorrect?
2. Can you think of classification systems used to simplify what would otherwise be complex measurements? Consider foods, movies, cars, grades, and clothes.

Problems

1. Human body temperature is about 310 K (98.6°F). At what wavelength do humans radiate the most energy? What kind of radiation do we emit?
2. If a star has a surface temperature of 20,000 K, at what wavelength will it radiate the most energy?

3. Infrared observations of a star show that it is most intense at a wavelength of 2000 nm. What is the temperature of the star's surface?
4. If you double the temperature of a black body, by what factor will the total energy radiated per second per square meter increase?
5. If one star has a temperature of 6000 K and another star has a temperature of 7000 K, how much more energy per second will the hotter star radiate from each square meter of its surface?
6. Transition *A* produces light with a wavelength of 500 nm. Transition *B* involves twice as much energy as *A*. What wavelength light does it produce?
7. Determine the temperatures of the following stars based on their spectra. Use Figures 6-7 and 6-9.
   a. medium-strength Balmer lines, strong helium lines
   b. medium-strength Balmer lines, weak ionized-calcium lines
   c. strong TiO bands
   d. very weak Balmer lines, strong ionized-calcium lines
8. To which spectral classes do the stars in Problem 7 belong?
9. In a laboratory, the Balmer beta line has a wavelength of 486.1 nm. If the line appears in a star's spectrum at 486.3 nm, what is the star's radial velocity? Is it approaching or receding?
10. The highest-velocity stars an astronomer might observe have velocities of about 400 km/s. What change in wavelength would this cause in the Balmer gamma line? (*Hint:* Wavelengths are given on page 111)

Media Cluster

ACTIVE FIGURE

Ace Astronomy™ To access the resources in the Media Cluster, log into AceAstronomy at http://astronomy.brookscole.com/sh9e and select Chapter 6.

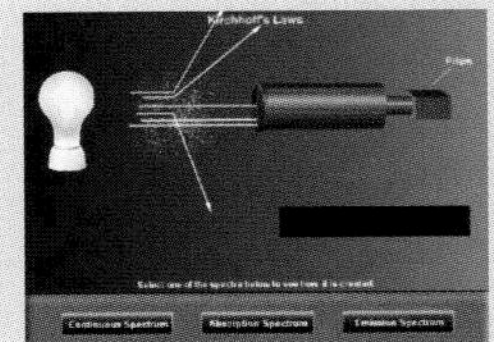

Kirchhoff's Laws
Kirchhoff studied and identified three different spectra in his experiments. Use this animation to study the conditions for producing each of the types of spectra.

ASTRONOMY EXERCISES

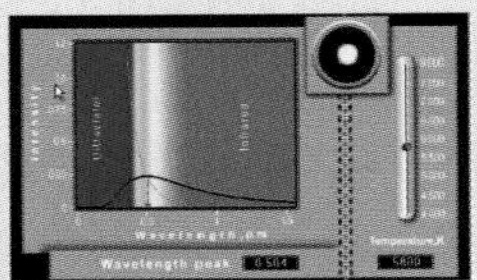

Black Body
In this animation, you can change the temperature of an object and see how its black body curve changes.

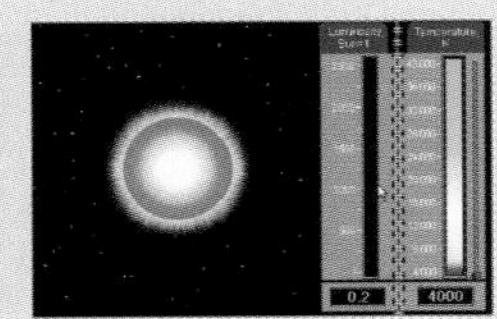

Stefan–Boltzmann Law
In this introduction to the Stefan–Boltzmann law, watch the luminosity of a star change as you adjust its temperature.

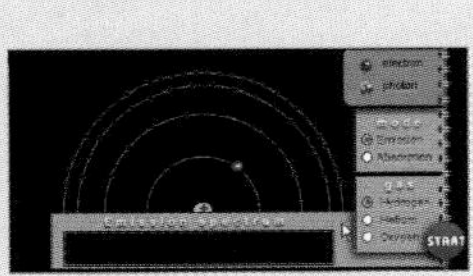

Emission and Absorption Spectra
This exercise will give you an inside look at atoms as they interact with photons. The photons collide with electrons inside the atom, sending the atoms to a higher energy level.

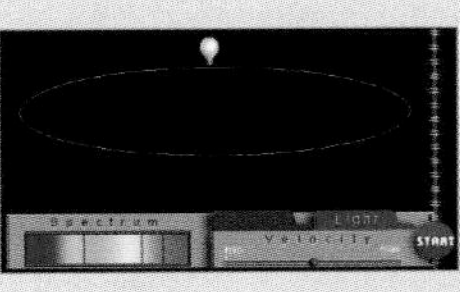

Doppler Shift
Recall that the Doppler effect is the change in wavelength of radiation due to relative radial motion of source and observer. You can compare the Doppler shift for a source of sound with that for a source of light in this animation.

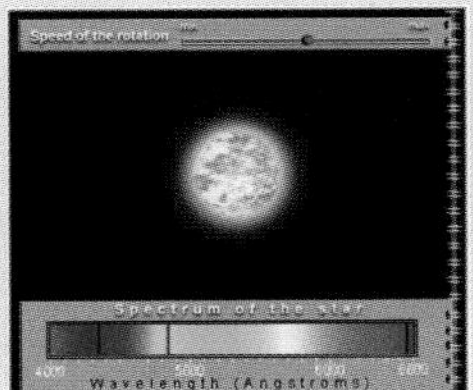

Stellar Rotation
This animation lets you increase or decrease the rotation speed of a star and see the effect on the star's spectrum.

Stellar Atomic Absorption Lines
In this exercise, you can essentially build your own star. Select from the type and ratio of element mixes in your new star.

VIRTUAL ASTRONOMY LABS

Lab 2: Properties of Light and Its Interaction with Matter
This lab examines the wave properties of electromagnetic radiation and how different regions of the electromagnetic spectrum are related. It ends by looking at the interaction between matter and radiation on the atomic scale.

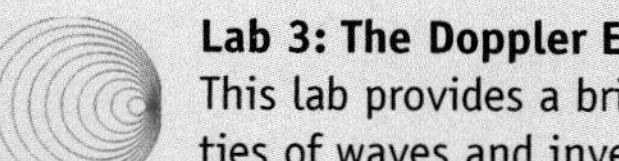

Lab 3: The Doppler Effect
This lab provides a brief review of some basic properties of waves and investigates the shift in observed wavelengths of sound and light waves caused by an emitting source being in motion with respect to an observer.

Lab 11: The Spectral Sequence and the H-R Diagram
This lab introduces the tools astronomers use to identify the stars of a main sequence and how that information is used to estimate the distance to the stars and the age of a star cluster.

Critical Inquiries for the Web

1. The name for the element helium has astronomical roots. Search the Internet for information on the discovery of helium. How and when was it discovered, and how did it get its name? Why do you suppose it took so long for helium to be recognized?
2. How was the model of the atom presented in the text you read developed? Search the web for information on historical models of the atom and compile a time line of important developments leading to our current understanding.

Exploring *TheSky*

1. Locate the following stars, click on them, and determine their spectral types: Antares in Scorpius, Betelgeuse in Orion, Aldebaran in Taurus, Sirius in Canis Major, Rigel in Orion.
2. How are spectral types correlated with the colors of stars? (*Hint:* Locate Orion and choose **Spectral Colors** under the **View** menu.)

- Go to the Brooks/Cole Astronomy Resource Center **(http://astronomy.brookscole.com)** for critical thinking exercises, articles, and additional readings from InfoTrac College Edition, Brooks/Cole's online student library.

7 The Sun—Our Star

All cannot live on the piazza,
but everyone may enjoy
the sun.

ITALIAN PROVERB

UV image

A WIT ONCE remarked that solar astronomers would know a lot more about the sun if it were farther away. This comment contains a grain of truth; the sun is only a humdrum star, and there are billions like it in the sky, but the sun is the only one close enough to show surface detail. Solar astronomers can see so much detail in the swirling currents of gas and arching bridges of magnetic force that present theories seem inadequate to describe it. Yet the sun is not an unusual object. It is just a star. In their general properties, stars are very simple. They are great balls of hot gas held together by their own gravity. Their gravity would make them collapse into small, dense bodies were they not so hot. The tremendously hot gas inside stars has such a high pressure that the stars would surely explode were it not for their own gravity. Like

This far-ultraviolet image of the sun reveals complex structures and eruptions undetected at visual wavelengths. (SOHO)

Guidepost

Looking Back

The interaction of light and matter, which you studied in Chapter 6, is a powerful tool. It can reveal the secrets of the stars. The next step is to apply that tool to the sun.

This Chapter

In this chapter, you will discover how little astronomers could know about the sun were it not for the analysis of its spectrum. Just a bit of spectrographic ingenuity reveals that the brilliance of the sun hides a complex atmosphere of hot gases churned by powerful storms and powered by nuclear reactions in its center.

This chapter will help you answer four essential questions:

What is the sun's surface made of?

What are the dark sunspots and other forms of solar activity?

Why does the sun go through a cycle of activity?

How does the sun make its energy?

Perhaps the most important question answered here is more general: How can astronomers learn about the sun and stars?

Looking Ahead

This is the first chapter that applies the methods of science to understand a celestial body. The tools developed in this chapter will help you explore stars, galaxies, and planets through the rest of this book.

Ace Astronomy™ The AceAstronomy icon throughout the text indicates an opportunity for you to test yourself on key concepts and to explore animations and interactions on the AceAstronomy website at: http://astronomy.brookscole.com/sh9e

soap bubbles, stars are simple structures balanced between opposing forces that individually would destroy them. Thus, our sun is a close-up example of a star.

Another reason to study the sun is that life on Earth depends critically on the sun. Earth gets nearly all its energy from the sun—oil and coal are merely stored sunlight. Also, Earth's pleasant climate is maintained by energy from the sun. Very small changes in the sun's luminosity can alter that climate, and a slightly larger change in the sun might make Earth uninhabitable. Furthermore, the sun's atmosphere of very thin gas reaches out past Earth's orbit, and any change in the sun, such as an eruption or a magnetic storm, can have a direct effect on Earth.

Finally, we study the sun because it is beautiful. Analysis of sunlight will reveal that the sun is both powerful and delicate. We study the sun not only because it is *a star,* not only because it is *our star,* but because it is the sun.

7-1 The Solar Atmosphere

THE SUN LOOKS LIKE A BRILLIANT, YELLOW DISK in the daytime sky (**Celestial Profile 1**); but, during a total solar eclipse when the moon hides the photosphere, you can see the outer layers of its atmosphere (see Chapter 3), the narrow chromosphere and extended corona. These are the layers of the solar atmosphere.

The sun is 109 times Earth's diameter and 333,000 times its mass, but it is gas from its center to its surface. Let's begin by looking at its hot surface.

Heat Flow in the Sun

Simple logic tells you that energy in the form of heat is flowing outward from the sun's interior. The solar spectrum reveals that the temperature of the sun's surface is about 5800 K. At that temperature every square centimeter of the sun's surface must be radiating more energy than a 6000-watt lightbulb. With all that energy radiating into space, the sun's surface would cool rapidly if energy did not flow up from the interior to keep the surface hot.

Not until the 1930s did astronomers understand how the sun makes its energy. Nuclear reactions occur in the core of the sun and generate energy that flows outward and keeps the surface hot. These nuclear reactions are discussed in detail later in this chapter, but here you should notice the importance of the energy flowing outward through the sun's surface.

As you study the atmosphere of the sun, you will find many phenomena that are driven by the energy flowing outward. Like a pot of boiling soup on a hot stove, the surface of the sun is in constant activity as the heat flows up from below.

Our discussion of the solar atmosphere begins with the quiescent, unchanging state of the photosphere, chromosphere, and corona. Later we will explore the violent solar activity.

The Photosphere

The visible surface of the sun looks like a smooth layer of gas marked by a few dark **sunspots.** Although the photosphere seems to be a distinct surface, it is not a solid surface. In fact, the sun is gaseous from its outer atmosphere right down to its center. The photosphere is the thin layer of gas from which Earth receives most of the sun's light. It is less than 500 km deep and has an average temperature of about 5800 K. If the sun magically shrank to the size of a bowling ball, the photosphere would be no thicker than a layer of tissue paper wrapped around the ball. For comparison, the chromosphere lies above the photosphere and is only a few times thicker in extent, but the corona, beginning above the chromosphere, extends far from the sun (■ Figure 7-1).

■ Figure 7-1

(a) A cross section at the edge of the sun shows the relative thickness of the photosphere and chromosphere. Earth is shown for scale. On this scale, the disk of the sun would be more than 1.5 m (5 ft) in diameter. (b) The corona extends from the top of the chromosphere to great height above the photosphere. This photograph, made during a total solar eclipse, shows only the inner part of the corona. (Daniel Good)

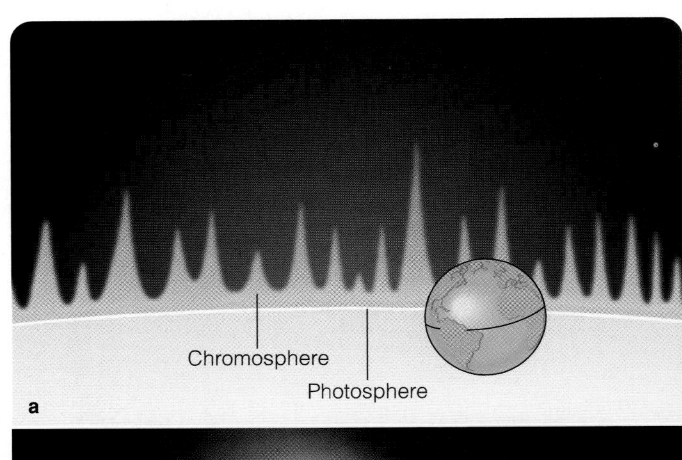

Below the photosphere, the gas is denser and hotter and therefore radiates plenty of light, but that light cannot escape from the sun because of the outer layers of gas. Thus, you cannot detect light from these deeper layers. Above the photosphere, the gas is less dense and so is unable to radiate or absorb much light. The photosphere is the layer in the sun's atmosphere that is dense enough to emit plenty of light but not so dense that the light can't escape.

Although the photosphere appears to be substantial, it is really a very low-density gas. Even in the deepest and densest layers visible, the photosphere is 3400 times less dense than the air you breathe. To find gases as dense as the air you breathe, you would have to descend about 7×10^4 km below the photosphere, about 10 percent of the way to the sun's center. With a fantastically efficient insulation system, you could fly a spaceship right through the photosphere.

The spectrum of the sun is an absorption spectrum, and that can tell you a great deal about the photosphere. You know from Kirchhoff's third law that an absorption spectrum is produced when a source of a continuous spectrum is viewed through a gas. In the case of the photosphere, the deeper layers are dense enough to produce a continuous spectrum, but atoms in the photosphere absorb photons of specific wavelengths, producing the absorption lines you see.

In good photographs, the photosphere has a mottled appearance because it is made up of dark-edged regions. The regions are called *granules,* and the visual pattern is called **granulation** (■ Figure 7-2a). A typical granule is about the size of Texas and lasts for only 10 to 20 minutes before fading away. Faded granules are continuously replaced by new granules. Spectra of these granules show that the centers are a few hundred degrees hotter than the edges, and Doppler shifts reveal that the centers are rising and the edges are sinking at speeds of about 0.4 km/s.

From this evidence, astronomers recognize granulation as the surface effect of convection just below the photosphere. **Convection** occurs when hot fluid rises and cool fluid sinks, as when, for example, a convection current of hot gas rises above a candle flame. You can observe convection in a liquid by adding a bit of cool nondairy creamer to an unstirred cup of hot coffee. The cool creamer sinks, warms, rises, cools, sinks again, and so on, creating small regions on the surface of the coffee that mark the tops of convection currents. Viewed from above, these regions look much like solar granules.

In the sun, the tops of rising currents of hot gas are brighter than their surroundings. As the gas cools slightly, it is pushed aside by rising gas from below. The cooler gas, sinking at the edge of the granule, is slightly dimmer. Consequently, granules have bright centers and dimmer edges (Figure 7-2b). The presence of granulation is clear evidence that energy is flowing upward through the photosphere.

The Chromosphere

Above the photosphere lies the chromosphere. Solar astronomers define the lower edge of the chromosphere as lying just above

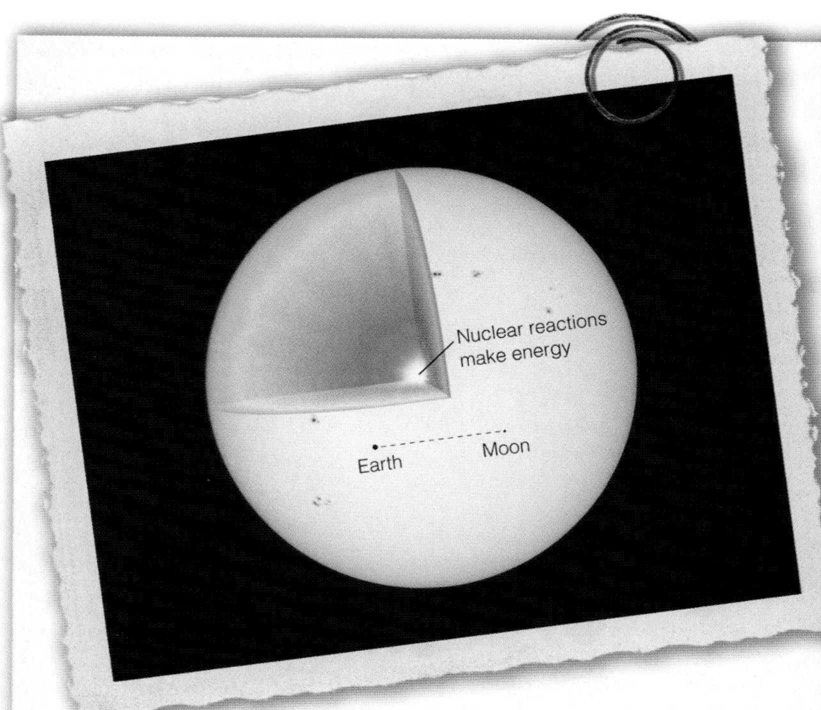

An image of the sun in visible light shows a few sunspots. Nuclear reactions occur near the sun's center and the energy flows outward. The Earth–moon system is added for scale. (Daniel Good)

Celestial Profile 1: The Sun

From Earth:

Average distance from Earth	1.00 AU (1.495979×10^8 km)
Maximum distance from Earth	1.0167 AU (1.5210×10^8 km)
Minimum distance from Earth	0.9833 AU (1.4710×10^8 km)
Average angular diameter seen from Earth	0.53° (32 minutes of arc)
Period of rotation	25.38 days at equator
Absolute visual magnitude	4.83

Characteristics:

Radius	6.9599×10^5 km
Mass	1.989×10^{30} kg
Average density	1.409 g/cm^3
Escape velocity at surface	617.7 km/s
Luminosity	3.826×10^{26} J/s
Surface temperature	5800 K
Central temperature	15×10^6 K
Spectral type	G2 V
Apparent visual magnitude	−26.74

Personality Point:

In Greek mythology, the sun was carried across the sky in a golden chariot pulled by powerful horses and guided by the sun-god Helios. When Phaeton, son of Helios, drove the chariot one day, he lost control of the horses, and Earth was nearly set ablaze before Zeus smote Phaeton from the sky. Even in classical times, people understood that life on Earth depends critically on the sun.

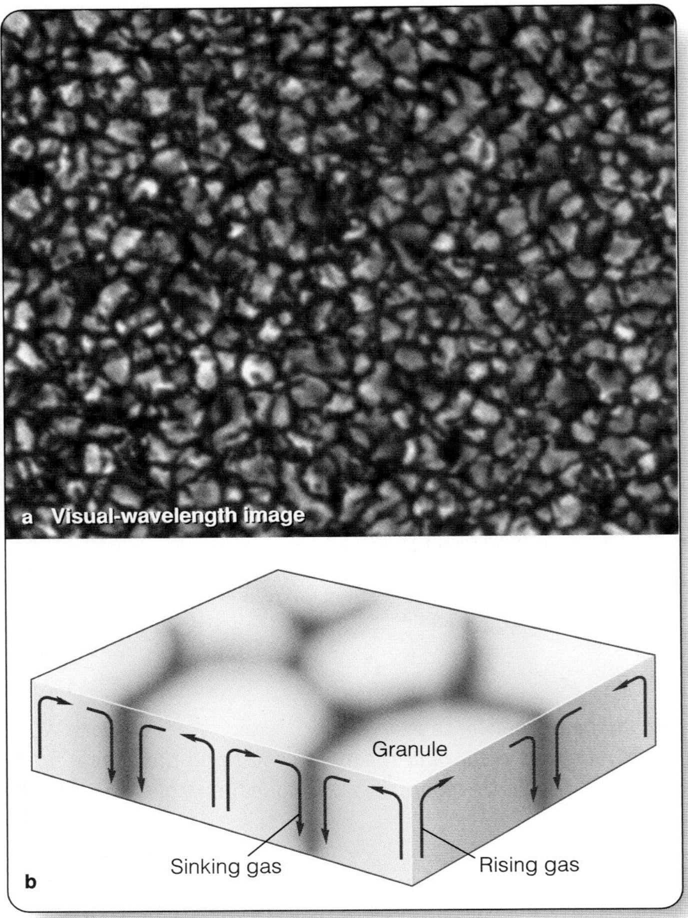

■ **Figure 7-2**

(a) This ultra-high-resolution image of the photosphere shows granulation. The largest granules here are about the size of Texas. (P. N. Brandt, G. Scharmer, G. W. Simon, Swedish Vacuum Solar Telescope, La Palma) (b) This model explains granulation as the tops of rising convection currents just below the photosphere. Heat flows upward as rising currents of hot gas and sinking currents of cool gas. The rising currents heat the solar surface in small regions that are seen as granules.

the visible surface of the sun, with its upper regions blending gradually with the corona. We can think of the chromosphere as being an irregular layer with a depth on average less than Earth's diameter (see Figure 7-1). Because the chromosphere is roughly 1000 times fainter than the photosphere, you can see it with your unaided eyes only during a total solar eclipse when the moon covers the brilliant photosphere. Then, the chromosphere flashes into view as a thin line of pink just above the photosphere. The word *chromosphere* comes from the Greek word *chroma,* meaning "color." The pink color is produced by the combined light of the red, blue, and violet Balmer emission lines of hydrogen.

Astronomers know a great deal about the chromosphere from its spectrum. The chromosphere produces an emission spectrum, and Kirchhoff's second law tells you the chromosphere must be an excited, low-density gas. The density is about 10^8 times less dense than the air you breathe.

Atoms in the lower chromosphere are ionized, and atoms in the higher layers of the chromosphere are even more highly ionized. That is, they have lost more electrons. From this, astronomers can find the temperature in different parts of the chromosphere. Just above the photosphere, the temperature falls to a minimum of about 4500 K and then rises rapidly (■ Figure 7-3) to the extremely high temperatures of the corona.

Solar astronomers can take advantage of some elegant physics to study the chromosphere. The gases of the chromosphere are transparent to nearly all visible light, but atoms in the gas are very good at absorbing photons of specific wavelengths. This produces certain dark absorption lines in the absorption spectrum of the photosphere. A photon at one of those wavelengths is very unlikely to escape from deeper layers. A **filtergram** is a photograph made using light in one of these dark absorption lines. Those photons can only have escaped from higher in the atmosphere. In this way filtergrams reveal detail in the upper layers of the chromosphere.

■ Figure 7-4 shows a filtergram made at the wavelength of the H_α Balmer line. This image reveals complex structure in the chromosphere including long dark **filaments** silhouetted against the bright surface. **Spicules** are flamelike jets of gas rising upward into the corona and lasting 5 to 15 minutes. The spicules appear to be cooler gas from the lower chromosphere extending

■ **Figure 7-3**

The chromosphere. If you could place thermometers in the sun's atmosphere, you would discover that the temperature increases from 5800 K at the photosphere to 10^6 K at the top of the chromosphere.

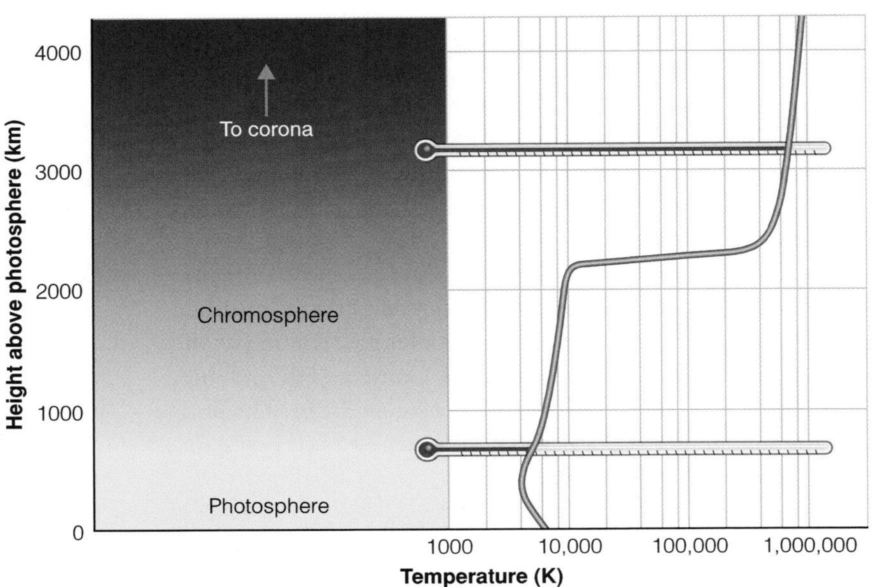

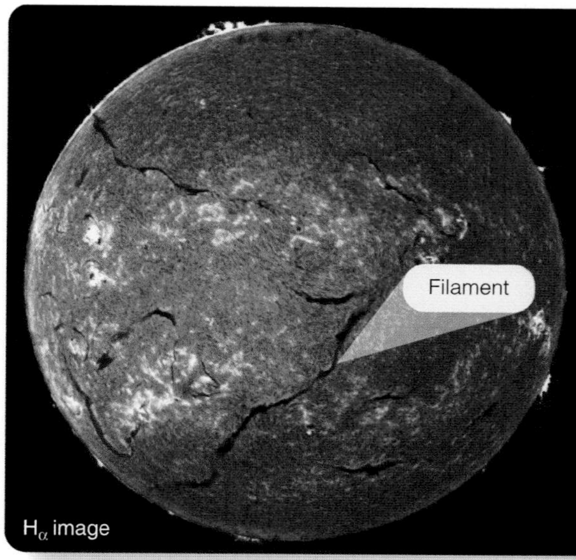

■ **Figure 7-4**

H_α filtergrams reveal complex structure in the chromosphere, including long, dark filaments and spicules springing from the edges of supergranules twice the diameter of Earth. (NOAA/SEL/USAF; © 1971 NOAO/NSO)

upward into hotter regions. Seen at the limb (edge) of the sun's disk, these spicules blend together and look like flames covering a burning prairie. Filtergrams of spicules located at the center of the solar disk show that they spring up around the edge of regions called **supergranules.** Over twice the diameter of Earth, supergranules are caused by larger convection currents circulating deeper below the photosphere.

Spectroscopic analysis of the chromosphere alerts us that it is a low-density gas in constant motion where the temperature increases rapidly with height. Just above the chromosphere lies even hotter gas.

The Solar Corona

The outermost part of the sun's atmosphere is called the *corona,* after the Greek word for "crown." The corona is so dim that it is not visible in the daytime sky because of the glare of the brilliant photosphere. During a total solar eclipse, the moon covers the photosphere, and the corona shines with a pearly glow not even as bright as the full moon (see Figure 7-1b). Observations with special telescopes on Earth and in space can block the light from the photosphere and image the corona out beyond 20 solar radii, almost 10 percent of the way to Earth. Such images show that magnetic fields link the sunspots with features in the chromosphere and corona (■ Figure 7-5).

The spectrum of the corona can tell you a great deal about the coronal gases and simultaneously illustrate how astronomers can analyze a spectrum. Some of the light from the outer corona produces a spectrum with absorption lines the same as the sun's spectrum. This light is just sunlight reflected from dust particles in the corona. In contrast, some of the light from the corona produces a continuous spectrum that lacks absorption lines, and that happens when sunlight from the photosphere is scattered off free electrons in the ionized coronal gas. Because the coronal gas has a temperature over 1 million K and the electrons travel very fast, the reflected photons suffer large, random Doppler shifts that smear out solar absorption lines to produce a continuous spectrum.

Superimposed on the corona's continuous spectrum are emission lines of highly ionized gases. In the lower corona, the atoms are not as highly ionized as they are at higher altitudes; this tells you that the temperature of the corona rises with altitude. Just above the chromosphere, the temperature is about 500,000 K, but in the outer corona the temperature can be as high as 2 million K or more.

The spectrum of the corona reveals that it is exceedingly hot gas, but it is not very bright. Its density is very low, only 10^6 atoms/cm^3 in its lower regions. That is about a trillion times less dense than the air you breath. In its outer layers the corona contains only 1 to 10 atoms/ cm^3, better than the best vacuum on Earth. Because of this low density, the hot gas does not emit much radiation.

Astronomers have wondered for years how the corona and chromosphere can be so hot. Heat flows from hot regions to cool regions, so how can the heat from the photosphere, with a temperature of only 5800 K, flow out into the much hotter chromosphere and corona? Observations made by the SOHO satellite have mapped a **magnetic carpet** of looped magnetic fields extending up through the photosphere. Turbulence below the surface may be whipping these fields about and churning the low density gases of the chromosphere and corona. That could heat the gas. In this instance, energy appears to flow outward as the agitation of the magnetic fields.

Gas follows the magnetic fields pointing outward and flows away from the sun in a breeze of ionized atoms called the **solar wind.** Like an extension of the corona, the low-density gases of the solar wind blow past Earth at 300 to 800 km/s, with gusts as high as 1000 km/s. Earth is bathed in the corona's hot breath.

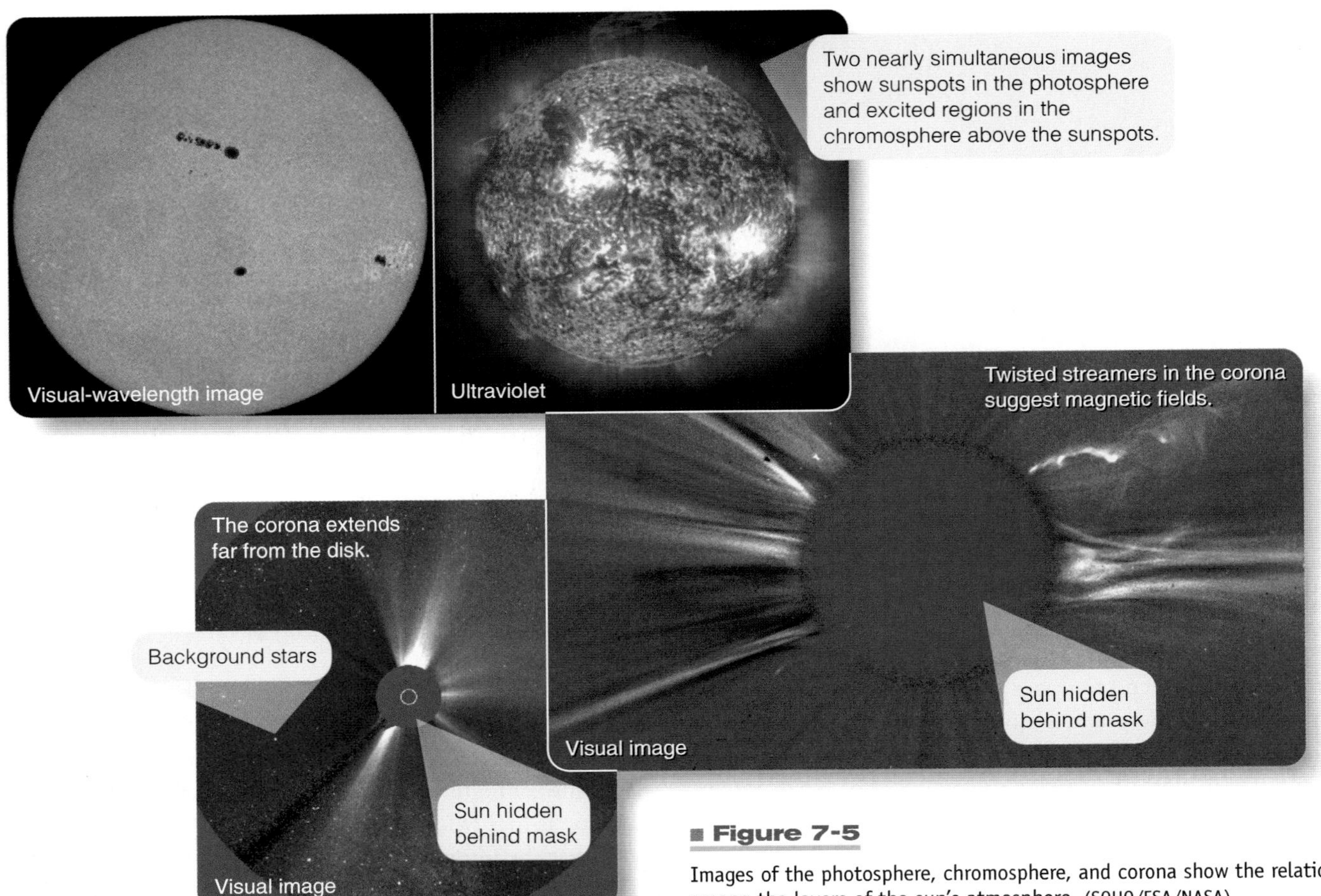
Two nearly simultaneous images show sunspots in the photosphere and excited regions in the chromosphere above the sunspots.
Visual-wavelength image
Ultraviolet
Twisted streamers in the corona suggest magnetic fields.
The corona extends far from the disk.
Background stars
Sun hidden behind mask
Visual image
Sun hidden behind mask
Visual image

■ Figure 7-5

Images of the photosphere, chromosphere, and corona show the relationships among the layers of the sun's atmosphere. (SOHO/ESA/NASA)

Because of the solar wind, the sun is slowly losing mass, but this is a minor loss, amounting to only about 10^7 tons per year, only about 10^{-14} of a solar mass per year. Later in life, the sun, like many other stars, will lose mass rapidly. You will see in later chapters how this affects stars.

Do other stars have chromospheres, coronae, and stellar winds like the sun? Ultraviolet and X-ray observations suggest the answer is yes. The spectra of many stars contain emission lines in the far ultraviolet that could only have formed in the low-density, high-temperature gases of a chromosphere and corona. Also, many stars are sources of X-rays, which appear to have been produced by the coronae. This observational evidence gives astronomers good reason to believe that the sun, for all its complexity, is a typical star.

Helioseismology

Almost no light emerges from below the photosphere, so you can't see details in the solar interior. However, solar astronomers can use the vibrations in the sun to explore its depths in a process called **helioseismology.** Random motions in the sun constantly produce vibrations, and some of those vibrations resonate like sound waves in organ pipes. These vibrations are just sound waves of very long period. A vibration with a period of 5 minutes is strongest, but the periods range from about 3 to 20 minutes.

Astronomers can detect these vibrations by observing Doppler shifts in the solar surface (■ Figure 7-6). As a vibrational wave travels down into the sun, the increasing density and temperature curve its path and it returns to the surface, where it makes the photosphere heave up and down. By observing these motions, astronomers can determine which vibrations resonate the strongest. Just as geologists can study Earth's interior by analyzing vibrations from earthquakes, so solar astronomers can use helioseismology to explore the sun's interior.

To understand the process better, think of a duck pond. If you stood on the shore and looked down at the water, you would see ripples arriving from all parts of the pond. Each duck produces ripples, so, in principle, you could analyze the ripples at your feet and draw a map showing the position and velocity of every duck on the pond. Of course, it would be difficult to untangle all the ripples, so you would need lots of data and a big computer.

Solar astronomers need large amounts of data for helioseismology. The Global Oscillation Network Group (GONG) uses a network of telescopes spread around the world to observe the

■ Figure 7-6

Helioseismology is the study of the modes of vibration of the sun. This computer image shows one of nearly 10 million possible modes of oscillation. Red regions are receding, and blue approaching. Helioseismological studies can reveal details about the sun's interior. (NOAO)

sun continuously as Earth turns. Spacecraft such as the Solar and Heliospheric Observatory (SOHO) can also observe the sun uninterrupted by the cycle of day and night. Analysis of such data has allowed astronomers to map the temperature, density, and rate of rotation inside the sun. Helioseismology has been able to detect great currents of gas flowing below the photosphere and to detect the emergence of a sunspot before it appeared in the photosphere. Helioseismology can even locate sunspots on the far side of the sun, sunspots that are not visible from Earth.

Inquire | Review | Analyze

How deeply into the sun can astronomers see?

This is a simple question, but it has a very interesting answer. To analyze this question you must avoid being misled by the words we use. When you look into the layers of the sun, your sight does not really penetrate *into* the sun. Rather, your eyes record photons that have escaped *from* the sun and traveled outward through the layers of the sun's atmosphere. If you observe at a wavelength at the center of a dark absorption line, then the photosphere and lower chromosphere are opaque, photons can't escape to your eyes, and the only photons you can see come from the upper chromosphere. What you see are the details of the upper chromosphere—a filtergram. On the other hand, if you observe at a wavelength that is not easily absorbed (a wavelength between spectral lines), the atmosphere is more transparent, and photons from deeper inside the photosphere can escape to your eyes.

By choosing the proper wavelength, solar astronomers can observe structures in the sun's chromosphere. But the corona is so thin and the gas below the photosphere so dense that this method doesn't work in these regions. How could you observe the corona and the deeper layers of the sun?

■ ■ ■

Connections: So far we have thought of the sun as a static, unchanging ball of gas with energy flowing outward from the interior and through the atmospheric layers. In fact, the sun is a highly variable body whose appearance is constantly changing. It is now time to think of the active sun.

7-2 Solar Activity

SOLAR ACTIVITY REFERS TO FEATURES on the sun that change over minutes, days, or years. As we use spectroscopic analysis to explore this aspect of solar astronomy, you will discover that solar activity is shaped by magnetic fields and driven by the powerful flow of energy rising from the sun's interior.

Observing the Sun

Solar activity is often visible with even a small telescope, but you should exercise great caution in observing the sun. Sunlight is very intense, and when it enters your eye it is absorbed and converted into heat. Equally dangerous is the infrared radiation in sunlight. Your eyes can't detect the infrared, but it is converted to heat in your eyes and can burn and scar the retina.

It is not safe to look directly at the sun, and it is even more dangerous to look at the sun through any optical instrument such as a telescope, binoculars, or even the viewfinder of a camera. The light-gathering power of such an optical system concentrates the sunlight and can cause severe injury. Never look at the sun with any optical instrument unless you are certain it is safe. ■ Figure 7-7a illustrates a safe way to observe the sun with a small telescope.

In the early 17th century, Galileo observed the sun and saw spots on its surface; day by day he saw the spots moving across the sun's disk. He rightly concluded that the sun was rotating. You could repeat his observations, and you would probably see something that looks like Figure 7-7b. You would see sunspots.

Sunspots

The dark spots that you see at visible wavelengths only hint at the complex processes that go on in the sun's atmosphere. To explore those processes, you must turn to the analysis of spectra at a wide range of wavelengths.

Study **Sunspots and the Sunspot Cycle** on pages 130 and 131 and notice three important ideas. First, sunspots are cool spots on the sun's surface caused by strong magnetic fields. Also notice that sunspots follow an 11-year cycle, which may subside and strengthen in ways that affect Earth's climate. Finally, notice the clear evidence that sunspots are part of a larger magnetic process that involves all layers of the sun's atmosphere.

The sunspot groups are merely the visible traces of the active regions. But what causes this magnetic activity? The answer

Sunspots and the Sunspot Cycle

The dark spots that appear on the sun are only the visible traces of complex regions of activity. Observations over many years and at a range of wavelengths tell you that sunspots are clearly linked to the sun's magnetic field.

A typical sunspot is about twice the size of Earth, but there is a wide range of sizes. They appear, last a few weeks to as long as two months, and then shrink away. Usually, sunspots occur in pairs or complex groups.

Size of Earth

Royal Swedish Academy of Sciences

Spectra show that sunspots are cooler than the photosphere with a temperature of about 4240 K. The photosphere has a temperature of about 5800 K. Because the total amount of energy radiated by a surface depends on its temperature raised to the fourth power, sunspots look dark in comparison. Actually, a sunspot emits quite a bit of radiation. If the sun were removed and only an average-size sunspot were left behind, it would be brighter than the full moon.

Umbra

Penumbra

Sunspots are not shadows, but astronomers refer to the dark core of a sunspot as its umbra and the outer, lighter region as the penumbra.

The number of spots visible on the sun varies in a cycle with a period of 11 years. At maximum, there are often over 100 spots visible. At minimum, there are very few.

Early in the cycle, spots appear at high latitudes north and south of the sun's equator. Later in the cycle, the spots appear closer to the sun's equator. If you plot the latitude of sunspots versus time, the graph looks like butterfly wings, as shown in this **Maunder butterfly diagram,** named after E. Walter Maunder of Greenwich Observatory.

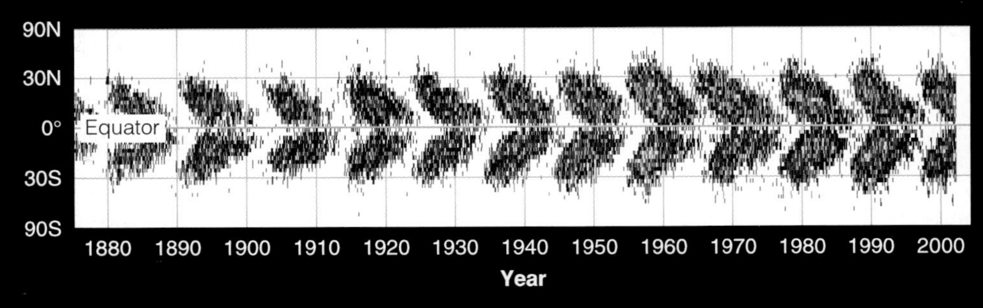

Astronomers can measure magnetic fields on the sun using the **Zeeman effect.** When an atom is in a magnetic field, the electron orbits are altered, and the atom is able to absorb a number of different wavelength photons even though it was originally limited to a single wavelength. In the spectrum, you see single lines split into multiple components, with the separation between the components proportional to the strength of the magnetic field.

Sunspot groups

Magnetic fields around sunspot groups

Ultraviolet filtergram

Magnetic image

Simultaneous images

Images of the sun show that sunspots contain magnetic fields a few thousan times stronger than Earth's. The strong fields are believed to inhibit gas moti below the photosphere; consequently, convection is reduced below the sunspot, and the surface there is cooler. Heat prevented from emerging through the sunspot is deflected and emerges around the sunspot, which can be detected in infrared image

Historical records show that there were very few sunspots fro about 1645 to 1715, a phenomenon known as the **Maunder minimum.** This coincides with a period called the "little ice age," a period of unusually cool weather in Europe and North America from about 1500 to about 1850. Other such periods of cooler climate are known. The evidence suggests that there is a lin between solar activity and the amount of solar energy Earth receives. This link has been confirmed by measurements made by spacecraft above Earth's atmosphere.

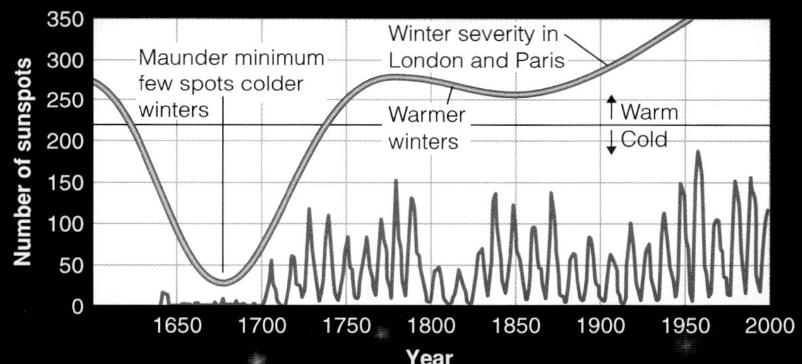

Magnetic fields can reveal themselves by their shape. For example, iron filings sprinkled over a bar magnet revea an arched shape.

Simultaneous images

The complexity of an active region becomes visible at short wavelengths.

Visual-wavelength image

Far-UV image

NASA/TRACE

Far-UV image

Observations at nonvisible wavelengths reveal that the chromosphere and corona above sunspots are violently disturbed in what astronomers call **active regions.** Spectrographic observations show that active regions contain powerful magnetic fields.

Arched structures above an active region are evidence of gas trapped in magnetic fields.

SOHO/EIT, ESA and NASA

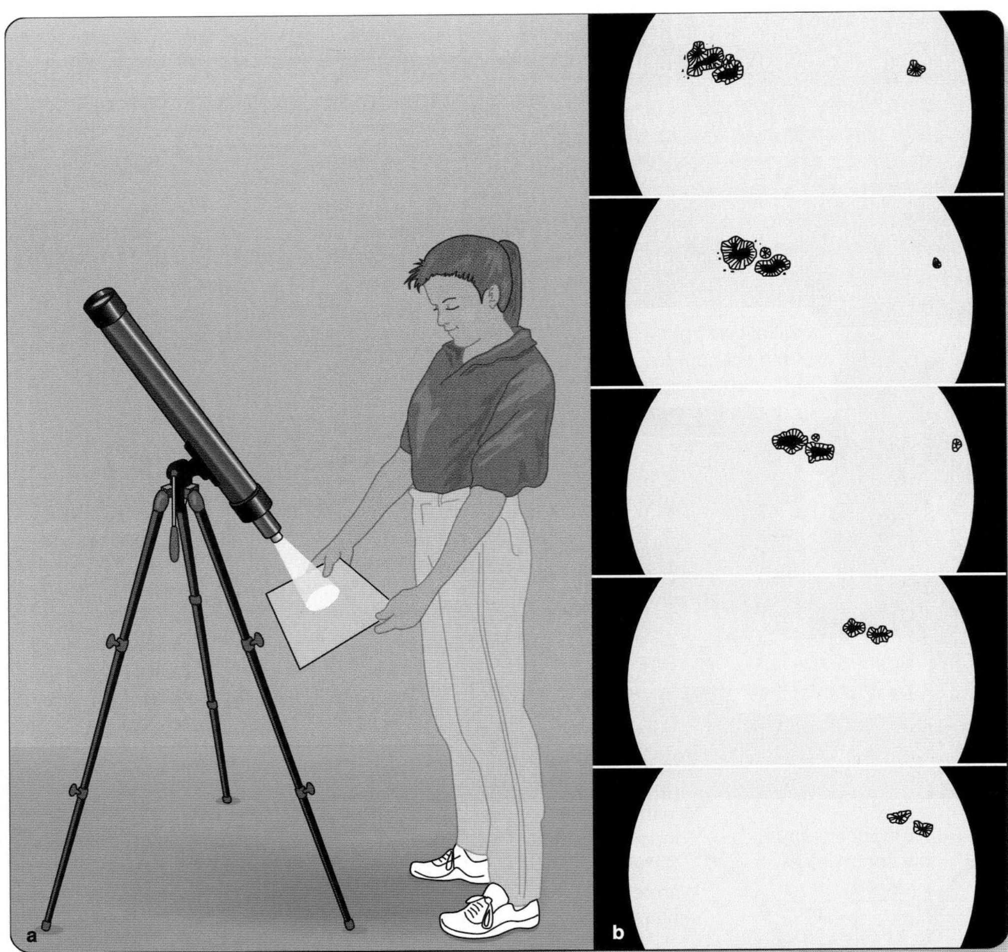

■ **Figure 7-7**

(a) Looking at the sun through a telescope can burn your eyes, but you can view the sun safely with a small telescope by projecting the image on a white screen. (b) If you sketch sunspots for a few days, you can see them develop or shrink. The sun's rotation carries sunspots across the disk over a period of two weeks.

appears to be linked to the waxing and waning of the sun's magnetic field.

AceAstronomy™ Log into AceAstronomy and select this chapter to see Astronomy Exercise "Zeeman Effect" and see how astronomers measure magnetic fields.

AceAstronomy™ Log into AceAstronomy and select this chapter to see Astronomy Exercise "Sunspot Cycle I"; you can explore the link between sunspots and the sun's magnetic field.

AceAstronomy™ Log into AceAstronomy and select this chapter to see Astronomy Exercise "Sunspot Cycle II." You can observe sunspots over the centuries.

The Sun's Magnetic Cycle

Sunspots are magnetic phenomena, so the 11-year cycle of sunspots must be caused by cyclical changes in the sun's magnetic field. To explore that idea, we must begin with the sun's rotation.

The sun does not rotate as a rigid body. It is a gas from its outermost layers down to its center, and some parts of the sun rotate faster than other parts. The equatorial region of the photosphere rotates faster than do regions at higher latitudes (■ Figure 7-8). At the equator, the photosphere rotates once every 25 days, but at a latitude of 45° one rotation takes 27.8 days. Helioseismology shows that deeper layers of gas rotate at different speeds. Different parts of an object rotating at different rates is called **differential rotation.** The magnetic cycle in the sun is linked to its differential rotation.

The sun's magnetic field appears to be powered by the energy flowing outward through the moving currents of gas. The gas is highly ionized, so it is a very good conductor of electricity. The **dynamo effect** occurs when a rapidly rotating conductor is stirred by convection to produce a magnetic field. This process is believed to produce Earth's magnetic field, and helioseismology provides evidence that the process also produces the sun's magnetic field at the bottom of the convection currents that lie below the sun's photosphere. The details of this process are still poorly understood, but the sun's magnetic cycle is clearly related to the creation of its magnetic field.

The magnetic behavior of sunspots gives an insight into how the magnetic cycle works. Sunspots tend to occur in groups or pairs. The magnetic field around such a sunspot pair resembles the magnetic field created by a magnet. That is, one end of the

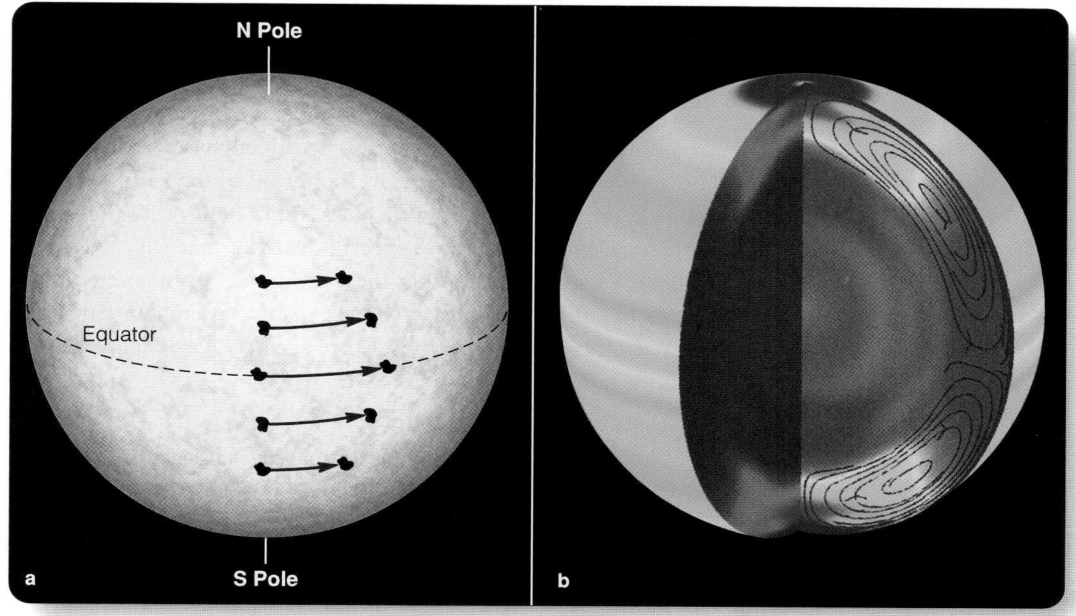

■ **Figure 7-8**

(a) In general, the photosphere of the sun rotates faster at the equator than at the higher latitudes. If you started five sunspots in a row, they could not stay aligned as the sun rotates. (b) Detailed analysis of the sun's rotation from helioseismology reveals regions of slow rotation (blue) and rapid rotation (red). Such studies show that the interior of the sun rotates slower than the surface and that currents similar to the trade winds in Earth's atmosphere flow through the sun. (NASA/SOI)

field is magnetic north and one end is magnetic south. At any one time, sunspot pairs south of the sun's equator have reversed polarity compared with those north of the sun's equator. ■ Figure 7-9 illustrates this by showing sunspot pairs south of the sun's equator with magnetic south poles leading and sunspots north of the sun's equator with magnetic north poles leading. At the beginning of the next 11-year sunspot cycle, the new spots appear with reversed magnetic polarity.

This magnetic cycle is not fully understood, but the **Babcock model** (named for its inventor) explains the magnetic cycle as a progressive tangling of the solar magnetic field. Because the electrons in an ionized gas are free to move, the gas is a very good conductor of electricity, and any magnetic field in the gas is "frozen" into the gas. If the gas moves, the magnetic field must move with it. Thus the sun's magnetic field is frozen into its gases, and the differential rotation wraps this field around the sun like a long string caught on a hubcap. Rising and sinking gas currents twist the field into ropelike tubes, which tend to float upward. Where these magnetic tubes burst through the sun's surface, sunspot pairs occur (■ Figure 7-10).

The Babcock model explains the reversal of the sun's magnetic field from cycle to cycle. As the magnetic field becomes tangled, adjacent regions of the sun's surface are dominated by magnetic fields that point in different directions. After about 11 years of tangling, the field becomes so complex that adjacent regions of the solar surface begin changing their magnetic fields to agree with neighboring regions. Quickly the entire field rearranges itself into a simpler pattern, and differential rotation begins winding it up to start a new cycle. But the newly organized field is reversed, and the next sunspot cycle begins with magnetic north replaced by magnetic south. Thus the complete magnetic cycle is 22 years long, and the sunspot cycle is 11 years long.

Notice the power of a scientific model. The Babcock model may in fact be incorrect in some details, but it gives you a framework on which to organize all of the complex solar activity. Even though the models of the sky (see Chapter 2) and the atom (see Chapter 6) were only partially correct, they served as organizing themes to guide your thinking. Similarly, although the details of

■ **Figure 7-9**

In sunspot groups, here simplified into pairs of major spots, the leading spot and the trailing spot have opposite magnetic polarity. Spots in the southern hemisphere have reversed polarity from those in the northern hemisphere.

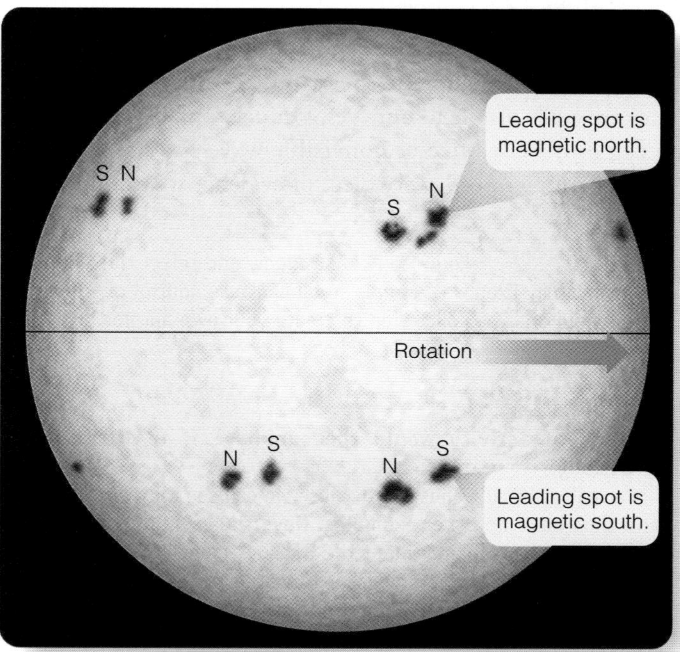

The Solar Magnetic Cycle

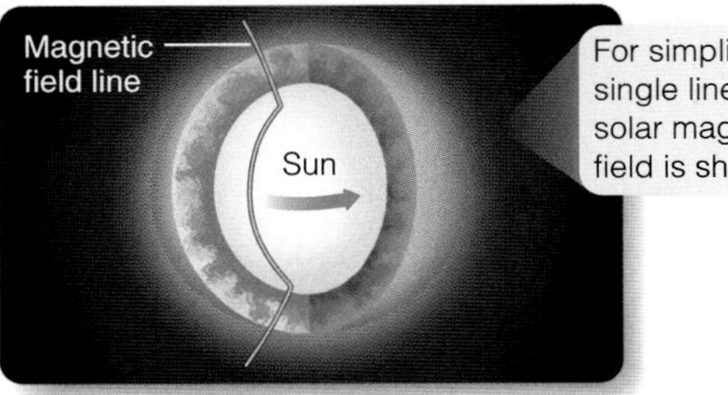
Magnetic field line
Sun
For simplicity, a single line of the solar magnetic field is shown.

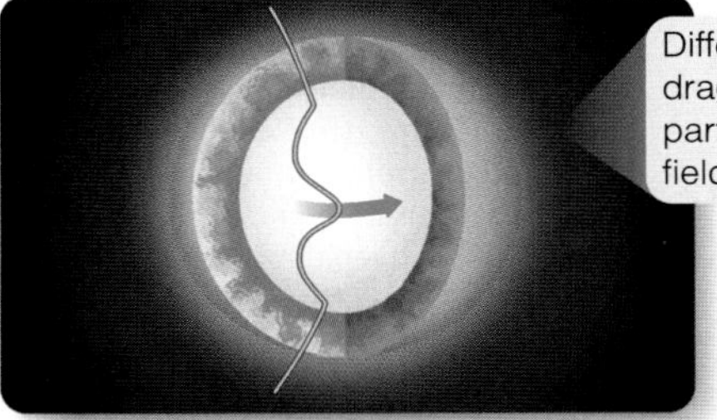
Differential rotation drags the equatorial part of the magnetic field ahead.

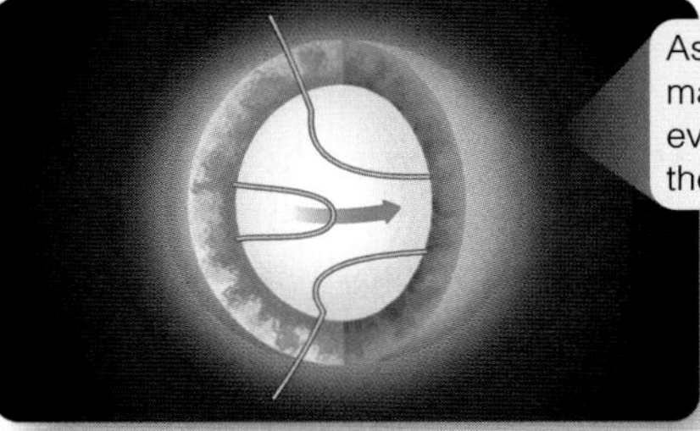
As the sun rotates, the magnetic field is eventually dragged all the way around.

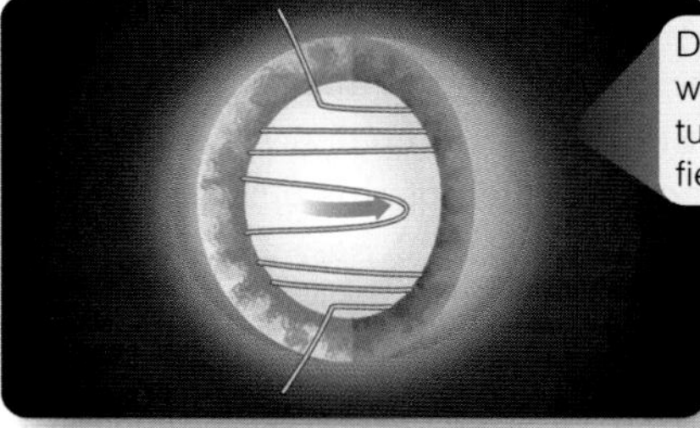
Differential rotation wraps the sun in many turns of its magnetic field.

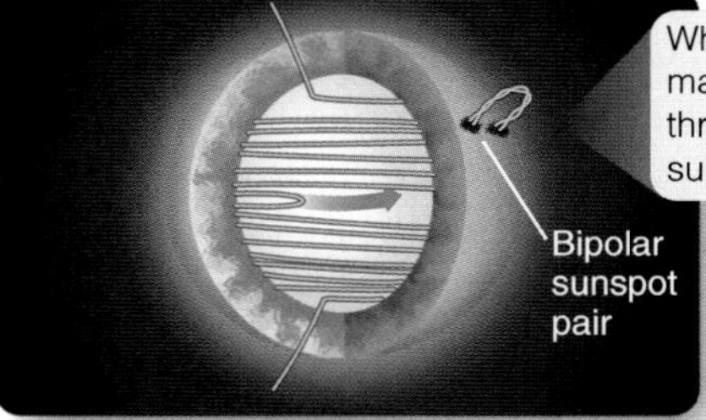
Where loops of tangled magnetic field rise through the surface, sunspots occur.
Bipolar sunspot pair

Figure 7-10

The Babcock model explains the sunspot cycle as a consequence of the sun's differential rotation gradually winding up the magnetic field near the base of the sun's outer, convective layer.

the solar magnetic cycle are not yet understood, the Babcock model gives you a general picture of the behavior of the sun's magnetic field (**Window on Science 7-1**). Further observations will help astronomers refine the model and make it a better description of what really happens in the sun.

If the sun is truly a representative star, you might expect to find similar magnetic cycles on other stars, but stars other than the sun are too distant for spots to be directly visible. Some stars, however, vary in brightness over a period of days in a way that reveals they are marked with dark spots believed to resemble sunspots. Other stars have spectral features that vary over periods of years, suggesting that they are subject to magnetic cycles much like the sun's. In fact, some stars display sudden flares that may resemble solar eruptions. Once again, the evidence tells you that the sun is a normal star.

AceAstronomy™ Log into AceAstronomy and select this chapter to see Astronomy Exercise "Convection and Magnetic Fields." You can adjust the sun's central temperature and change its magnetic field.

Chromospheric and Coronal Activity

The solar magnetic field extends high into the chromosphere and corona, where it produces beautiful and powerful phenomena. Study **Magnetic Solar Phenomena** on pages 136 and 137 and notice three important points.

First, all solar activity is magnetic. There are no such events on Earth because Earth's magnetic field is weak. Earth's atmosphere is not ionized and thus is free to move independent of the magnetic field. On the sun, however, the weather is a magnetic phenomenon.

Second, tremendous energy can be stored in arches of magnetic field. You can see these near the limb of the sun as prominences, and, seen from above, in filtergrams, you can see them as filaments (Figure 7-4a). When that stored energy is released, it can trigger powerful eruptions; and, although these flares occur far from Earth, they can affect us in dramatic ways.

The third thing to notice is how solar astronomers use every part of the electromagnetic spectrum to study the sun. Some things are easily visible to our eyes, but many phenomena on the sun can only be studied at nonvisible wavelengths. To understand our sun, we need every piece of evidence we can find.

AceAstronomy™ Log into AceAstronomy and select this chapter to see Astronomy Exercise "Auroras"; you can take control of Earth's magnetic field and the solar wind to create your own auroras.

Inquire | Review | Analyze

What kind of activity would the sun have if it didn't rotate differentially?

This is a really difficult question because we can see only one star close up and thus have no other examples. Nevertheless, you can make an educated guess by thinking about the Babcock model. If the sun didn't rotate differentially, with its equator traveling faster than higher latitudes, then the magnetic field might not get twisted up and there might

Building Confidence by Confirmation and Consolidation

While many textbooks describe science as the process of testing hypotheses by observation and experiment, you should not think that every astronomer approaches the telescope expecting to make an observation that will disprove long-held beliefs and trigger a revolution in science. Then what is the daily grind of science really about?

First, many observations and experiments merely confirm already-tested hypotheses. The biologist knows that all worker bees in a hive are sisters, but a careful study of the DNA from different workers further confirms that hypothesis. By repeatedly confirming a hypothesis, scientists build confidence in the hypothesis and may be able to extend it to a wider application. Of course, there is always the chance that a new observation or experiment will disprove the hypothesis, but that is usually very unlikely. Much of the daily grind of science is confirmation.

Another aspect of routine science is consolidation, the linking of a hypothesis to other well-studied phenomena. Chemists may understand certain kinds of carbon molecules shaped like rings, but by repeated study they find a carbon molecule shaped like a hollow sphere. To consolidate their findings, they must show that the chemical bonding in the two molecules follows the same rules and that the molecules have certain properties in common. No hypothesis is overthrown, but the chemists consolidate their knowledge and understand carbon molecules better.

The Babcock model of the solar magnetic cycle is an astronomical example of the scientific process. Solar astronomers know that the model explains some solar features but has shortcomings. Although most astronomers don't expect to discard the entire model, they work through confirmation and consolidation to better understand how the solar magnetic cycle works and how it is related to cycles in other stars.

not be a solar cycle. Twisted tubes of magnetic field might not form and rise through the photosphere to produce prominences and flares, although convection might tangle the magnetic field and produce some activity. Is the magnetic activity that heats the chromosphere and corona driven by differential rotation or by convection? It is hard to guess, but without differential rotation, the sun might not have a strong magnetic field and high-temperature gas above its photosphere.

This is very speculative, but sometimes in the critical analysis of ideas it helps to imagine a change in a single important factor and try to understand what might happen. For example, what do you think the sun would be like if it had no convection inside?

■ ■ ■

Connections: The sun is beautiful and complex, but the evidence tells us that all of the powerful activity on its surface is driven by the energy flowing up from its interior. It is time to descend into the sun and ask how that energy is made.

7-3 Nuclear Fusion in the Sun

ASTRONOMERS OFTEN USE THE WRONG WORDS to describe energy generation in the sun and stars. Astronomers will say, "The star ignites hydrogen burning." We use the word *ignite* to mean "catch on fire" and we use *burn* to mean "on fire." What goes on inside stars isn't really burning in the usual sense.

The sun is a star, and it is powered by nuclear reactions that occur near its center. The energy produced keeps the interior hot, and the gas is totally ionized. That is, the electrons are not attached to the atomic nuclei, and the gas is an atomic soup of rapidly moving particles colliding with each other at high velocity. When we discuss nuclear reactions inside stars, we refer to atomic nuclei and not to atoms.

How exactly can the nucleus of an atom yield energy? The answer lies in the forces that hold the nuclei together.

Nuclear Binding Energy

The sun generates its energy by breaking and reconnecting the bonds between the particles *inside* atomic nuclei. This is quite different from the way we generate energy by burning wood in a fireplace. The process of burning wood extracts energy by breaking and reconnecting chemical bonds between atoms in the wood. Chemical bonds are formed by the electrons in atoms, and you saw in Chapter 6 that the electrons are bound to the atoms by the electromagnetic force. Thus chemical energy originates in the electromagnetic force.

There are only four known forces in nature: the force of gravity, the electromagnetic force, the **weak force,** and the **strong force.** The weak force is involved in the radioactive decay of certain kinds of nuclear particles, and the strong force binds together atomic nuclei. Thus nuclear energy comes from the strong force.

Nuclear power plants on Earth generate energy through **nuclear fission** reactions that split uranium nuclei into less massive fragments. A uranium nucleus contains a total of 235 protons and neutrons, and it splits into a range of fragments containing roughly half as many particles. Because the fragments produced are more tightly bound than the uranium nuclei, energy is released during uranium fission (■ Figure 7-11).

Stars make energy in **nuclear fusion** reactions that combine light nuclei into heavier nuclei. The most common reaction,

Magnetic Solar Phenomena

Magnetic phenomena in the chromosphere and corona, like magnetic weather, result as constantly changing magnetic fields on the sun trap ionized gas to produce beautiful arches and powerful outbursts. Some of this solar activity can affect Earth's magnetic field and atmosphere.

A **prominence** is composed of ionized gas trapped in a magnetic arch rising up through the photosphere and chromosphere into the lower corona. Seen during total solar eclipses at the edge of the solar disk, prominences look pink because of the three Balmer emission lines. The image below shows the arch shape suggestive of magnetic fields. Seen from above against the sun's bright surface, prominences form dark filaments.

Sacramento Peak Observatory

H-alpha filtergram

This ultraviolet image of the solar surface was made by the NASA TRACE spacecraft. It shows hot gas trapped in magnetic arches extending above active regions. At visual wavelengths, you would see sunspot groups in these active regions.

Quiescent prominences may hang in the lower corona for many days, whereas eruptive prominences burst upward in hours. The eruptive prominences below are many Earth diameters long.

Far-UV image

SOHO, EIT, ESA and NASA

The gas in prominences may be 60,000 to 80,000 K, quite cold compared with the low-density gas in the corona, which may be as hot as a million Kelvin.

Trace/NASA

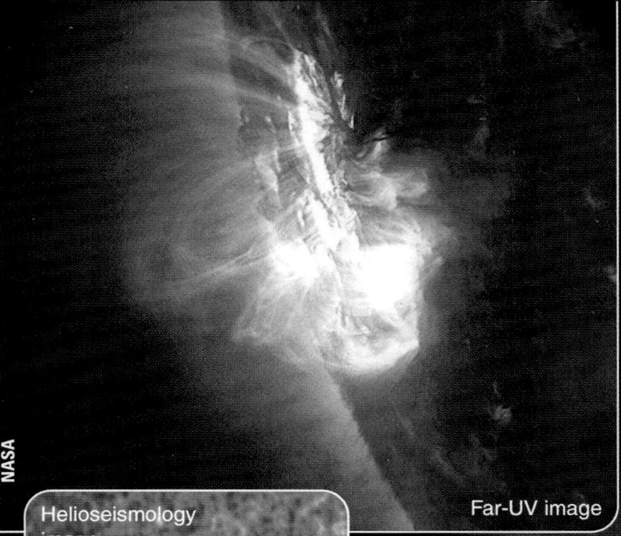

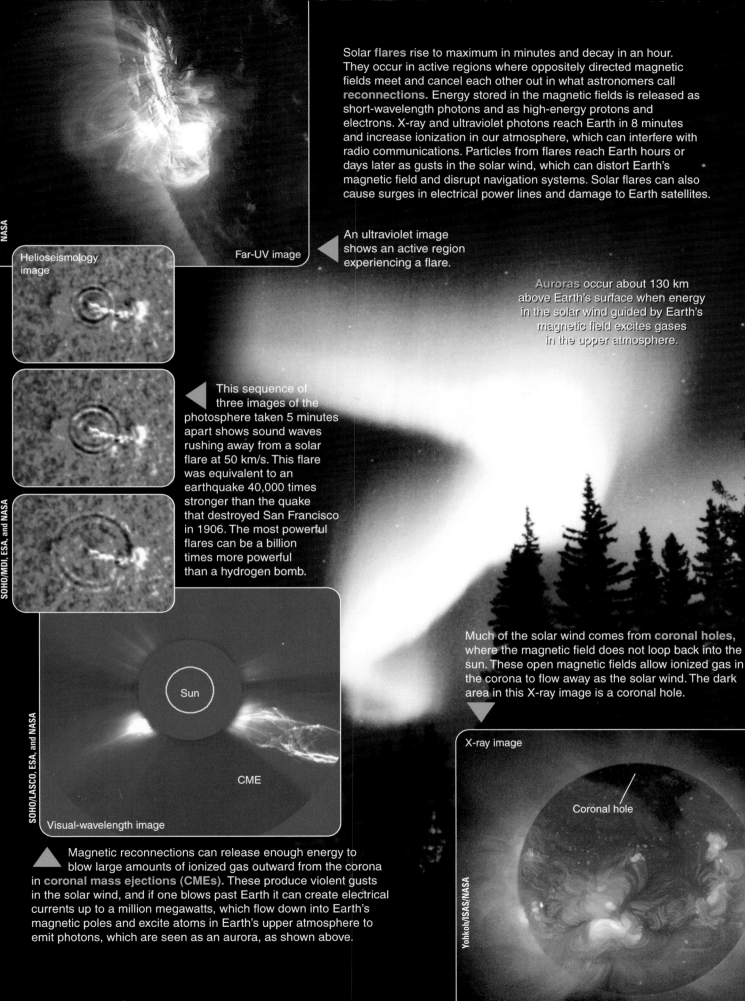

Solar **flares** rise to maximum in minutes and decay in an hour. They occur in active regions where oppositely directed magnetic fields meet and cancel each other out in what astronomers call **reconnections.** Energy stored in the magnetic fields is released as short-wavelength photons and as high-energy protons and electrons. X-ray and ultraviolet photons reach Earth in 8 minutes and increase ionization in our atmosphere, which can interfere with radio communications. Particles from flares reach Earth hours or days later as gusts in the solar wind, which can distort Earth's magnetic field and disrupt navigation systems. Solar flares can also cause surges in electrical power lines and damage to Earth satellites.

NASA

Far-UV image

An ultraviolet image shows an active region experiencing a flare.

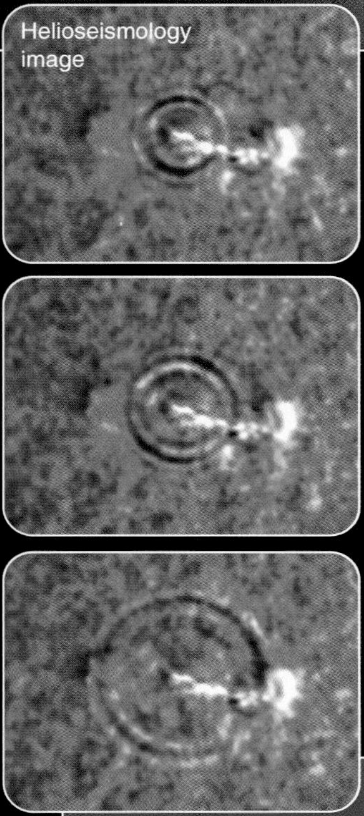

Auroras occur about 130 km above Earth's surface when energy in the solar wind guided by Earth's magnetic field excites gases in the upper atmosphere.

Helioseismology image

This sequence of three images of the photosphere taken 5 minutes apart shows sound waves rushing away from a solar flare at 50 km/s. This flare was equivalent to an earthquake 40,000 times stronger than the quake that destroyed San Francisco in 1906. The most powerful flares can be a billion times more powerful than a hydrogen bomb.

SOHO/MDI, ESA, and NASA

Copyright Jan Curtis

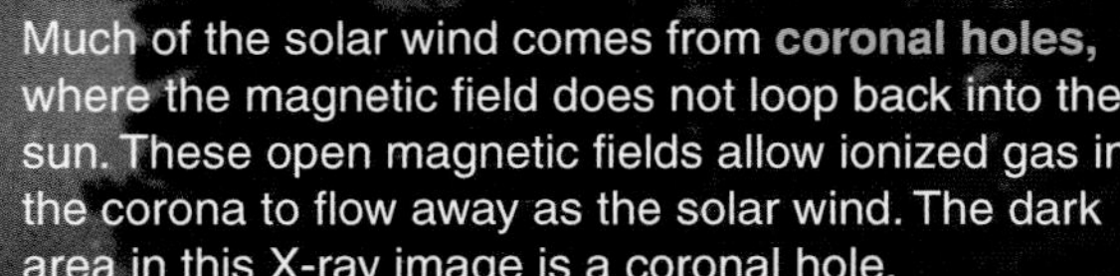

Much of the solar wind comes from **coronal holes,** where the magnetic field does not loop back into the sun. These open magnetic fields allow ionized gas in the corona to flow away as the solar wind. The dark area in this X-ray image is a coronal hole.

Sun

CME

Visual-wavelength image

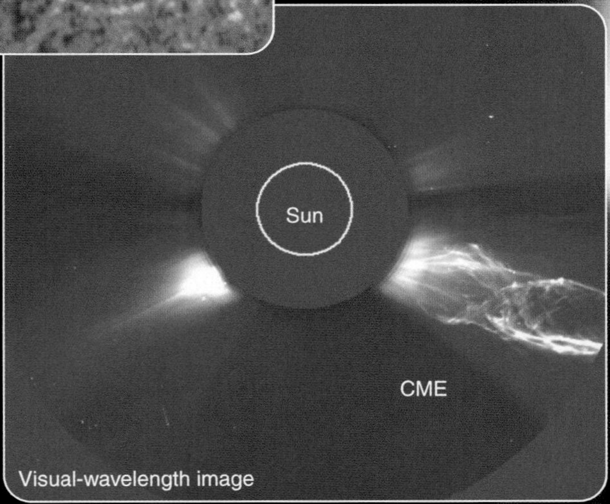

SOHO/LASCO, ESA, and NASA

Magnetic reconnections can release enough energy to blow large amounts of ionized gas outward from the corona in **coronal mass ejections (CMEs).** These produce violent gusts in the solar wind, and if one blows past Earth it can create electrical currents up to a million megawatts, which flow down into Earth's magnetic poles and excite atoms in Earth's upper atmosphere to emit photons, which are seen as an aurora, as shown above.

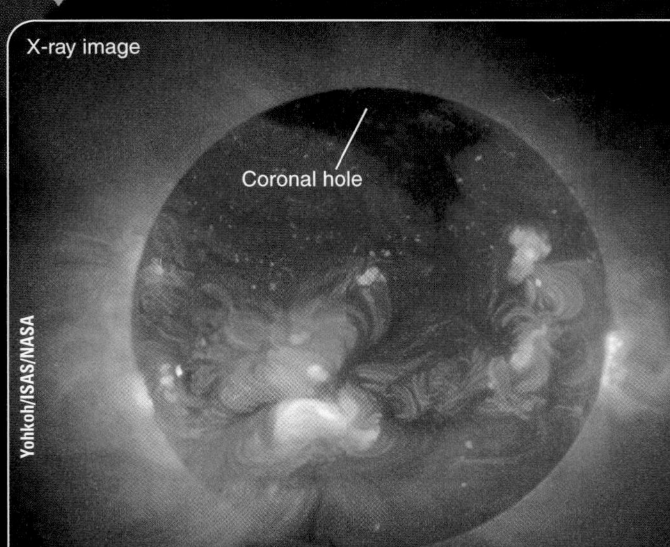

Yohkoh/ISAS/NASA

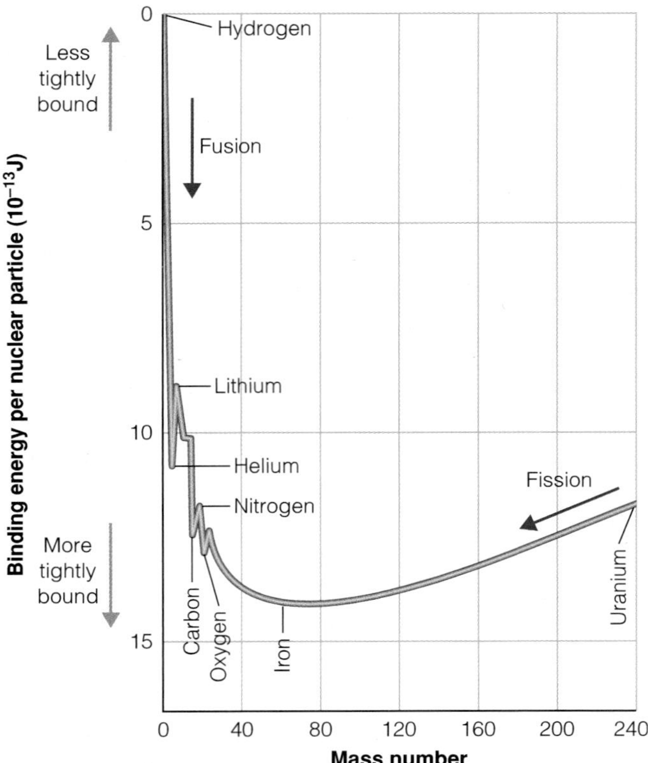

■ **Figure 7-11**

The red line in this graph shows the binding energy (the energy that holds an atomic nucleus together) for all the different atoms plotted by atomic mass (the number of protons and neutrons in their nucleus). Both fission and fusion nuclear reactions move downward in the diagram (arrows) toward more tightly bound nuclei. Iron has the most tightly bound nucleus, so no nuclear reactions can begin with iron and release energy.

Reasoning With Numbers | 7-1

Hydrogen Fusion

When four hydrogen nuclei fuse to make one helium nucleus, a small amount of matter seems to disappear:

$$\begin{array}{rl} \text{4 hydrogen nuclei} = & 6.693 \times 10^{-27}\ \text{kg} \\ -\ \text{1 helium nucleus} = & 6.645 \times 10^{-27}\ \text{kg} \\ \hline \text{difference in mass} = & 0.048 \times 10^{-27}\ \text{kg} \end{array}$$

That mass is converted to energy according to Einstein's equation:

$$E = mc^2$$
$$= (0.048 \times 10^{-27}\ \text{kg}) \times (3 \times 10^8\ \text{m/s})^2$$
$$= 0.43 \times 10^{-11}\ \text{J}$$

Recall that one joule (J) is roughly equal to the energy of an apple falling from a table to the floor.

including that in the sun, fuses hydrogen nuclei (single protons) into helium nuclei (two protons and two neutrons). Because the nuclei produced are more tightly bound than the original nuclei, energy is released. Notice in Figure 7-11 that both fusion and fission reactions move downward in the diagram toward more tightly bound nuclei. Thus both produce energy by releasing the binding energy of atomic nuclei.

Hydrogen Fusion

The sun fuses four hydrogen nuclei to make one helium nucleus. Because one helium nucleus contains 0.7 percent less mass than four hydrogen nuclei, it seems that some mass vanishes in the process. In fact, that mass is converted to energy, and you could figure out how much by using Einstein's famous equation, $E = mc^2$ (**Reasoning with Numbers 7-1**).

Creating one helium nucleus makes only a small amount of energy, hardly enough to raise a housefly 0.001 inch into the air. Only by concentrating many reactions in a small area can nature produce significant results. A single kilogram (2.2 lb) of hydrogen, for instance, converted entirely to energy, would produce enough power to raise an average-sized mountain 10 km (6 miles) into the air! The sun has a voracious energy appetite and needs 10^{38} reactions per second, transforming 5 million tons of mass into energy every second, just to balance its own gravity. This might sound as if the sun is losing mass at a furious rate, but in its entire 10-billion-year lifetime, the sun will convert less than 0.07 percent of its mass into energy.

It seems from this that nuclear fusion is very powerful, especially if you calculate that the fusion of a milligram of hydrogen (roughly the mass of a match head) produces as much energy as burning 30 gallons of gasoline. However, the nuclear reactions in the sun are spread through a large volume in its core, and any single gram of matter produces little energy. A person of normal mass eating a normal diet produces about 4000 times more heat per gram than the matter in the core of the sun. The sun produces a lot of energy because it contains a lot of grams of matter in its core.

Fusion reactions can occur only when the nuclei of two atoms get very close to each other. Because atomic nuclei carry positive charges, they repel each other with an electrostatic force called the Coulomb force. Physicists commonly refer to this repulsion between nuclei as the **Coulomb barrier.** To overcome this barrier, atomic nuclei must collide violently. Violent collisions are rare unless the gas is very hot, in which case the nuclei move at high speeds and collide violently. (Remember, an object's temperature is just a measure of the speed with which its particles move.)

Thus nuclear reactions in the sun take place only near the center, where the gas is hot and dense. A high temperature ensures that collisions between nuclei are violent enough to overcome the Coulomb barrier, and a high density ensures that there are enough collisions, and thus enough reactions, to meet the sun's energy needs.

You can symbolize this process with a simple nuclear reaction:

$$4\,^{1}\text{H} \rightarrow {}^{4}\text{He} + \text{energy}$$

In this equation, ^{1}H represents a proton, the nucleus of the hydrogen atom, and ^{4}He represents the nucleus of a helium atom. The superscripts indicate the approximate weight of the nuclei. The actual steps in the process are more complicated than this convenient summary suggests. Instead of waiting for four hydrogen nuclei to collide simultaneously, a highly unlikely event, the process can proceed step by step in a chain of reactions called the proton–proton chain.

The **proton–proton chain** is a series of three nuclear reactions that builds a helium nucleus by adding protons one at a time. This process is efficient at temperatures above 10,000,000 K. The sun, for example, manufactures its energy in this way.

The three reactions in the proton–proton chain are:

$$^{1}\text{H} + {}^{1}\text{H} \rightarrow {}^{2}\text{H} + e^{+} + \nu$$
$$^{2}\text{H} + {}^{1}\text{H} \rightarrow {}^{3}\text{He} + \gamma$$
$$^{3}\text{He} + {}^{3}\text{He} \rightarrow {}^{4}\text{He} + {}^{1}\text{H} + {}^{1}\text{H}$$

In the first reaction, two hydrogen nuclei (two protons) combine to form a heavy hydrogen nucleus called **deuterium,** emitting a particle called a *positron* (a positively charged electron, symbolized e^{+}) and another called a **neutrino** (ν). In the second reaction, the heavy hydrogen nucleus absorbs another proton and, with the emission of a gamma ray (γ), becomes a lightweight helium nucleus (^{3}He). Finally, two light helium nuclei combine to form a normal helium nucleus and two hydrogen nuclei. Because the last reaction needs two ^{3}He nuclei, the first and second reactions must each occur twice for each ^{4}He produced (■ Figure 7-12). The net result of this chain reaction is the transformation of four hydrogen nuclei into one helium nucleus plus energy.

The energy appears in the form of gamma rays, positrons, and neutrinos. The gamma rays are photons that are absorbed by the surrounding gas before they can travel more than a few centimeters. This heats the gas. The positrons produced in the first reaction combine with free electrons, and both particles vanish, converting their mass into gamma rays. Thus, the positrons also help keep the center of the star hot. The neutrinos, however, are particles that travel at nearly the speed of light and almost never interact with other particles. The average neutrino could pass unhindered through a lead wall 1 ly thick. Consequently, the neutrinos do not help heat the gas but race out of the star, carrying away roughly 2 percent of the energy produced.

It is time to ask the critical question that lies at the heart of science. What is the evidence to support this theoretical explanation of how the sun makes its energy? The search for that evidence will introduce you to one of the great problems of modern astronomy.

AceAstronomy™ Log into AceAstronomy and select this chapter to see Astronomy Exercise "Nuclear Fusion" and take control of fusion in the sun's core.

Neutrinos from the Sun's Core

The center of a star seems forever hidden, but the sun is transparent to neutrinos because these subatomic particles almost never interact with normal matter. Nuclear reactions in the sun's core produce floods of neutrinos that rush out of the sun and off into space. If you could detect these neutrinos, you could probe the sun's interior.

Because neutrinos almost never interact with atoms, you never feel the flood of over 10^{12} solar neutrinos that flows through your body every second. Even at night, neutrinos from the sun rush through Earth as if it weren't there, up through your bed, through you, and onward into space. Obviously you are lucky to be transparent to neutrinos, but it means that neutrinos are extremely hard to

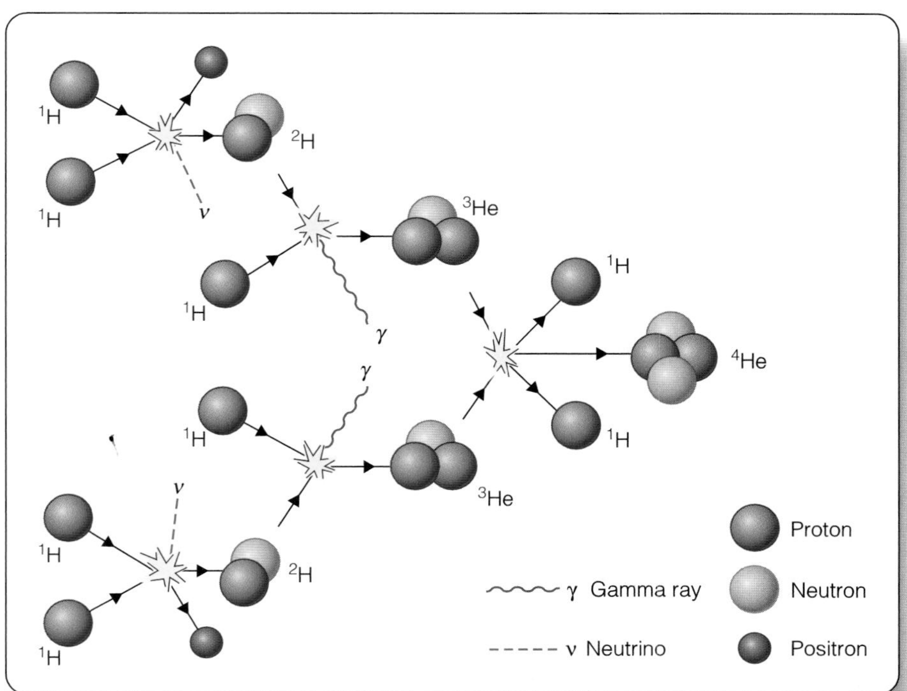

■ **Figure 7-12**

The proton–proton chain combines four protons (at left) to produce one helium nucleus (at right). Energy appears as gamma rays and as positrons, which combine with electrons to convert their mass into energy. Neutrinos escape, carrying away about 2 percent of the energy.

Avoiding Hasty Judgments: Scientific Faith

Scientists like to claim that every scientific belief is based on evidence, that every theory has been tested, and that the moment a theory fails a test, it is discarded or revised. The truth is much more complicated than that, and the solar neutrino problem is a good illustration. If the detection of solar neutrinos contradicted the theory of solar energy production, why wasn't the theory abandoned?

While scientists do indeed have tremendous respect for evidence, they also have a faith in theories that have been tested successfully many times. If a theory has been tested and confirmed over and over, they may even begin to call it a natural law, and that means scientists have great faith in its truth. That is, they have confidence that the theory or law is a good description of how nature works.

Nevertheless, it is not unusual for an experiment or an observation to contradict well-established theories. In many cases, the experiments and observations are simple mistakes, or they have not been interpreted correctly. Scientists resist abandoning a well-tested theory even when an observation continues to contradict it. If confidence in the theory is stronger than the evidence, scientists begin testing the evidence. Can it be right? Do they understand it correctly? Of course, if the evidence cannot be impeached and it continues to contradict the theory, scientists must eventually abandon or modify the theory no matter how many times it has previously been tested and confirmed. This ultimate reliance on evidence is the distinguishing characteristic of science.

It is only human nature to hang onto the principles you have come to trust, and that confidence in well-tested scientific principles helps scientists avoid rushing to faulty judgments. For example, claims for perpetual-motion machines occasionally crop up in the news, but the world's scientists don't instantly abandon the laws of energy and motion pending an analysis of the latest claim. Of course, if such a claim did prove true, the entire structure of scientific knowledge would come crashing down, but because the known laws of energy and motion have been well tested and no perpetual-motion machine has ever been successful, scientists know which way to bet. Like the keel on a ship, confidence in well-tested theories and laws keeps the scientific boat from rocking before every little breeze.

Some forms of faith must be absolute and unshakable—religious faith, for example—but scientific faith may be better described as scientific confidence because it must be open to change. If, ultimately, a single experiment or observation conclusively contradicts our most cherished law of nature, scientists must abandon that law and find a new way to understand nature. You can see this scientific confidence at work in many controversies, from the origin of the human race to the meaning of IQ measurements; but one of the best examples is the solar neutrino problem. While astronomers struggled to understand the origin of solar neutrinos, they continued to have confidence that they did indeed understand how stars make energy.

detect. Certain nuclear reactions, however, can be triggered by a neutrino of the right energy; and in the late 1960s, chemist Raymond Davis, Jr., began using such a reaction to detect solar neutrinos.

Davis filled a 100,000-gallon tank with the cleaning fluid perchloroethylene (C_2Cl_4). Theory predicts that about once a day, a solar neutrino will convert a chlorine atom in the tank into radioactive argon, which can be detected later by its radioactive decay. To protect the detector from cosmic rays from space, the tank was buried nearly a mile deep in a South Dakota gold mine (■ Figure 7-13a). Of course, the mile of rock overhead has no effect on the neutrinos.

The result of the Davis experiment startled astronomers. The cleaning fluid detected too few neutrinos—not one neutrino per day as predicted by models of the sun, but about one every three days. The experiment was refined, tested, and calibrated for over three decades, but it did not find the missing neutrinos. Other detectors were built, and they too counted too few neutrinos coming from the sun.

The missing solar neutrinos were one of the great mysteries of modern astronomy. Some scientists argued that astronomers didn't correctly understand how the sun and stars make their energy, but other scientists wondered if there was something about neutrinos that could explain the problem. Astronomers have great confidence in their theories of the sun's interior, and helioseismology confirmed those theories, so astronomers did not abandon their theories immediately (**Window on Science 7-2**).

As the 21st century began, scientists were able to solve the mystery. Physicists know of three kinds of neutrinos, which they call "flavors." The Davis experiment could detect (or taste) only one flavor, electron-neutrinos. Theory hinted that the electron-neutrinos produced in the core of the sun might oscillate among the three flavors as they rushed out through the sun and across space to Earth. Observations begun in 2000 confirm this theory (Figure 7-13b). Some of the electron neutrinos produced in the sun transform into tau- and muon-neutrinos, which most detectors cannot count.

This solution to the solar neutrino problem is exciting because neutrinos can't oscillate unless they have mass. Neutrinos were long thought to be massless, but if they have even a small mass, they are so common their gravity could affect the evolution of the universe as a whole—an idea you will meet in Chapter 15. The detection of neutrino oscillation excites astronomers for another reason. It confirms the theories that describe the interior of the sun and stars.

Inquire | Review | Analyze

Why does nuclear fusion require that the gas be very hot?
Only under certain conditions do the nuclei of atoms fuse together to form a new nucleus. Inside a star, the gas is ionized, which means the

Figure 7-13

(a) The Davis solar neutrino experiment used cleaning fluid and could detect only one of the three flavors of neutrinos. (Brookhaven National Laboratory) (b) The Sudbury Neutrino Observatory is a globe 12 meters in diameter containing water rich in deuterium in place of hydrogen. Buried 6800 feet deep in an Ontario mine, it can detect all three flavors of neutrinos and confirms that neutrinos oscillate. (Photo courtesy of SNO)

electrons have been stripped off the atoms, and the nuclei are bare and have a positive charge. For hydrogen fusion, the nuclei are single protons. These atomic nuclei repel each other because of their positive charges, so they must collide with each other violently to overcome that repulsion and get close enough together to fuse. If the atoms in a gas are moving rapidly, then it has a high temperature, and so nuclear fusion requires that the gas have a very high temperature. If the gas is cooler than about 10 million K, hydrogen can't fuse because the protons don't collide violently enough to overcome the repulsion of their positive charges.

The sun requires high temperature to make energy by nuclear fusion. Why does it need high density?

■ ■ ■

Connections: The sun is a star, and you now understand how it makes its energy and how that energy, flowing outward, drives solar activity. Now you can turn to a larger problem. What are other stars like?

Study and Review Tools

Summary

7-1 | The Solar Atmosphere

What is the sun's surface made of?

- The solar atmosphere consists of three layers of hot, low-density gas: the photosphere, chromosphere, and corona. The photosphere, or visible surface, is the level in the sun from which visible photons most easily escape.
- The granulation of the photosphere is produced by convection currents of gas rising from below.
- The chromosphere is most easily visible during total solar eclipses, when it flashes into view for a few seconds. It is a thin, hot layer of gas just above the photosphere, and its pink color is caused by the Balmer emission lines in its spectrum.
- Filtergrams of the chromosphere reveal spicules and filaments.
- The corona is the sun's outermost atmospheric layer. It is composed of a very low-density, very hot gas extending many solar radii from the visible sun. Its high temperature—up to 3×10^6 K—is believed to be maintained by the magnetic field and the rotation of the sun.
- Parts of the corona give rise to the solar wind, a breeze of low-density ionized gas streaming away from the sun.
- Solar astronomers can study the motion, density, and temperature of gases inside the sun by analyzing the way the solar surface oscillates. Known as helioseismology, this process requires large amounts of data and extensive computer analysis.

7-2 | Solar Activity

What are the dark sunspots and other forms of solar activity?

- Sunspots seem dark because they are slightly cooler than the rest of the photosphere. The average sunspot is about twice the size of Earth and contains magnetic fields a few thousand times stronger than Earth's.

Sunspots are thought to form because the magnetic field inhibits rising currents of hot gas and allows the surface to cool.

- Astronomers can use the Zeeman effect to measure magnetic fields on the sun.

Why does the sun go through a cycle of activity?

- The average number of sunspots varies over a period of about 11 years and appears to be related to a magnetic cycle. The sunspot cycle does not repeat exactly each cycle, and the decades from 1645 to 1715, known as the Maunder minimum, seem to have been a time when solar activity was very low and Earth's climate was slightly colder.
- Alternate sunspot cycles have reversed magnetic polarity, which has been explained by the Babcock model, in which the differential rotation of the sun winds up the magnetic field. Tangles in the field rise to the surface and cause active regions visible to our eyes as sunspot pairs. When the field becomes strongly tangled, it reorders itself into a simpler but reversed field, and the cycle starts over.
- Prominences and flares are other examples of solar activity. Prominences occur in the chromosphere; their arched shapes show that they are formed of ionized gas trapped in the magnetic field. You can see prominences in filtergrams as dark filaments silhouetted against the bright chromosphere.
- Flares are sudden eruptions of X-ray, ultraviolet, and visible radiation plus high-energy atomic particles produced when magnetic fields on the sun interact and reconnect. Flares are important because they can have dramatic effects on Earth such as communications blackouts and auroras.
- Spacecraft images show long streamers extending from the corona out into space. Coronal mass ejections occur when magnetic fields on the surface of the sun eject bursts of ionized gas that flow outward in the solar wind. Such bursts can produce auroras and other phenomena as they strike Earth.

7-3 | Nuclear Fusion in the Sun

How does the sun make its energy?

- The sun generates its energy near its center, where the temperature and density are high enough for nuclear fusion reactions to combine hydrogen nuclei to make helium nuclei.
- Observations of too few neutrinos coming from the sun's core are now explained by the oscillation of neutrinos among three different types. The neutrinos confirm that the sun makes it energy by hydrogen fusion.

New Terms

sunspot (p. 124)
granulation (p. 125)
convection (p. 125)
filtergram (p. 126)
filament (p. 126)
spicule (p. 126)
supergranule (p. 127)
magnetic carpet (p. 127)
solar wind (p. 127)
helioseismology (p. 128)
Maunder butterfly diagram (p. 130)
Zeeman effect (p. 131)
Maunder minimum (p. 131)
active region (p. 131)
differential rotation (p. 132)
dynamo effect (p. 132)
Babcock model (p. 133)
weak force (p. 135)
strong force (p. 135)
nuclear fission (p. 135)
nuclear fusion (p. 135)
prominence (p. 136)
flare (p. 137)
reconnection (p. 137)
aurora (p. 137)
coronal hole (p. 137)
coronal mass ejection (CME) (p. 137)
Coulomb barrier (p. 138)
proton–proton chain (p. 139)
deuterium (p. 139)
neutrino (p. 139)

Review Questions

Ace Astronomy™ Assess your understanding of this chapter's topics with additional quizzing and animations at **http://astronomy.brookscole.com/sh9e**

1. Why can't we see deeper than the photosphere?
2. What evidence can you cite that granulation is caused by convection?
3. How are granules and supergranules related? How do they differ?
4. How can a filtergram reveal structure in the chromosphere?
5. What evidence can you cite that the corona has a very high temperature?
6. What heats the chromosphere and corona to high temperature?
7. How are astronomers able to explore the layers of the sun below the photosphere?
8. What evidence can you cite that sunspots are magnetic?
9. How does the Babcock model explain the sunspot cycle?
10. What does the spectrum of a prominence tell you? What does its shape tell you?
11. How can solar flares affect Earth?
12. Why does nuclear fusion require high temperatures?
13. Why does nuclear fusion in the sun occur only near the center?
14. How can astronomers detect neutrinos from the sun?
15. How can neutrino oscillation explain the solar neutrino problem?
16. The upper two images here show two solar phenomena. What are they, and how are they related? How do they differ?
17. The lower image was recorded in the extreme ultraviolet by the SOHO spacecraft. Explain the features you see.

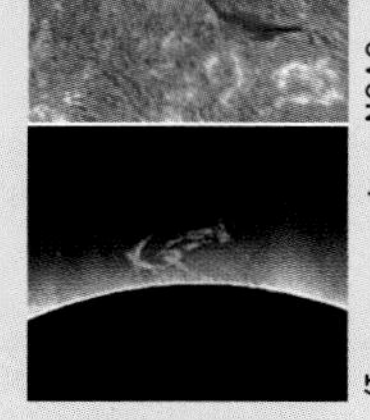

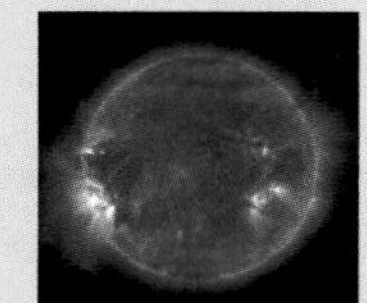

(Images courtesy NOAO and Daniel Good)

(NASA/SOHO)

Discussion Questions

1. What energy sources on Earth cannot be thought of as stored sunlight?
2. What would the spectrum of an auroral display look like? Why?
3. What observations would you make if you were ordered to set up a system that could warn astronauts in orbit of dangerous solar flares? Such a system exists.

Problems

1. The radius of the sun is 0.7 million km. What percentage of the radius is taken up by the chromosphere?
2. The smallest detail visible with ground-based solar telescopes is about 1 second of arc. How large a region does this represent on the sun? (*Hint:* Use the small-angle formula.)
3. What is the angular diameter of a star like the sun located 5 ly from Earth? Is the Hubble Space Telescope able to resolve detail on the surface of such a star?
4. If a sunspot has a temperature of 4200 K and the solar surface has a temperature of 5800 K, how many times brighter is the surface compared with the sunspot? (*Hint:* Use the Stefan–Boltzmann law, Reasoning with Numbers 6-1.)
5. A solar flare can release 10^{25} J. How many megatons of TNT would be equivalent? (*Hint:* A 1-megaton bomb produces about 4×10^{15} J.)
6. The United States consumes about 2.5×10^{19} J of energy in all forms in a year. How many years could the United States run on the energy released by the solar flare in Problem 5?
7. Neglecting energy absorbed or reflected by Earth's atmosphere, the solar energy hitting 1 square meter of Earth's surface is 1360 J/s (the solar constant). How long does it take a baseball diamond (90 ft on a side) to receive 1 megaton of solar energy? (*Hint:* See Problem 5.)
8. How much energy is produced when the sun converts 1 kg of mass into energy?
9. How much energy is produced when the sun converts 1 kg of hydrogen into helium? (*Hint:* How does this problem differ from Problem 8?)
10. A 1-megaton nuclear weapon produces about 4×10^{15} J of energy. How much mass must vanish when a 5-megaton weapon explodes?

Media Cluster

ACTIVE FIGURE

Ace Astronomy™ To access the resources in the Media Cluster, log into AceAstronomy at **http://astronomy.brookscole.com/sh9e** and select Chapter 7.

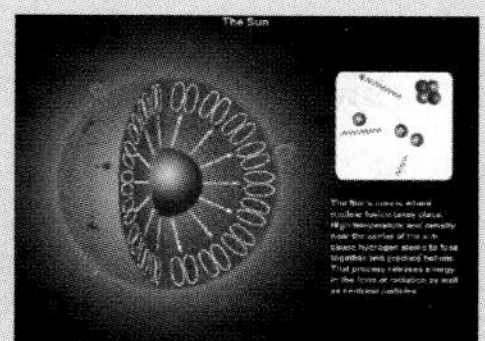

The Sun
Take a look inside the sun in this animation. You can click on various parts of the sun to see things like an animation of nuclear fusion and a movie of a solar flare.

ASTRONOMY EXERCISES

Zeeman Effect
The phenomenon known as the Zeeman effect allows astronomers to measure magnetic fields. In this animation you can adjust the strength of the magnetic field around a star to study the effects on the star's spectrum.

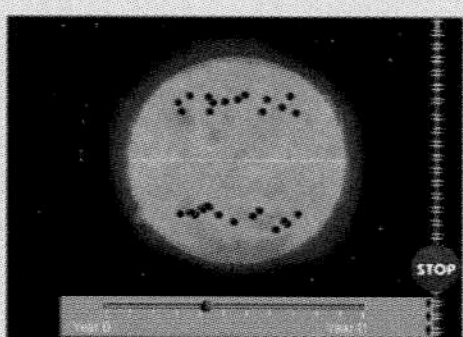

Sunspot Cycle I
Sunspots are relatively dark spots on the sun that contain intense magnetic fields. This animation lets you study the number and location of sunspots occurring over an 11-year cycle.

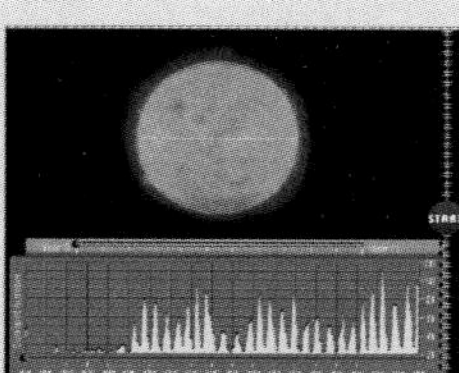

Sunspot Cycle II
In this animation, you can track sunspot activity over the past 300 years.

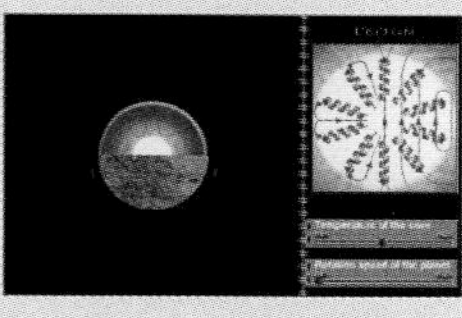

Convection and Magnetic Fields
You can vary the temperature of the sun's core in this animation. As you do so, notice how the amount of convection changes, as well as what happens to the magnetic field around the sun.

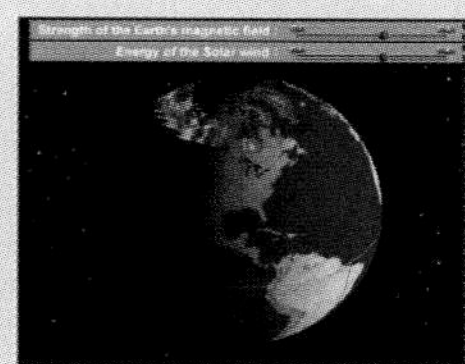

Auroras
This animation lets you vary the strength of Earth's magnetic field as well as the speed and number of particles in the solar wind. See if you can determine what factors cause the most and least intense auroras.

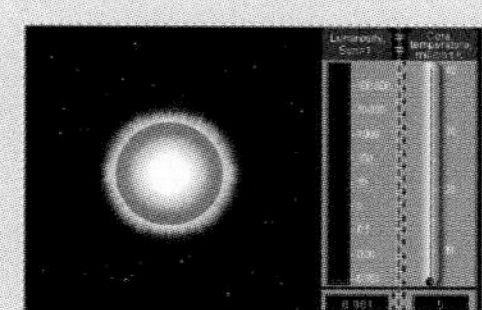

Nuclear Fusion
Nuclear fusion is a reaction that joins the nuclei of atoms to form more massive nuclei. Vary the temperature of the star in this animation to see how this affects the rate of nuclear fusion and luminosity.

VIRTUAL ASTRONOMY LABS

Lab 4: Solar Wind and Cosmic Rays
This lab begins with an overview of the properties of the sun's atmosphere and how energetic particles escape and travel through the solar system. It ends with a discussion of cosmic rays.

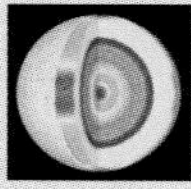

Lab 10: Helioseismology
This lab examines how helioseismology, the study of the sun's vibrations, allows us to obtain detailed information about its interior. Such information allows us to test our understanding of the sun very precisely.

Critical Inquiries for the Web

1. Do disturbances in one layer of the solar atmosphere produce effects in other layers? You have seen that filtergrams are useful in identifying the layers of the solar atmosphere and the structures within them. Visit a website that provides daily solar images, choose today's date (or one near it), and examine the sun in several wavelengths to explore the relation between disturbances in various layers.
2. Explore the web to find out how auroral activity is affected as solar activity rises and falls through the solar cycle. What changes in auroral visibility occur during this cycle? In what other ways can the increased activity associated with a solar maximum affect Earth?
3. What can you find on the web about Earth-based efforts to generate energy through nuclear fusion? How do nuclear fusion power experiments attempt to trigger and control nuclear fusion? So-called cold fusion has been largely abandoned as a false trail. How did it resemble nuclear fusion?

Exploring *TheSky*

1. Locate the six photos of the sun provided in *TheSky* and attempt to draw in the sun's equator in each photo. (*Hint:* In the sun's information box, choose **More Information** and then **Multimedia.** What features are visible in these images that help you recognize the orientation of the sun's equator?)

- Go to the Brooks/Cole Astronomy Resource Center **(http://astronomy.brookscole.com)** for critical thinking exercises, articles, and additional readings from InfoTrac College Edition, Brooks/Cole's online student library.

8 The Family of Stars

When he shall die,
Take him and cut him out
in little stars,
And he will make the face
of heaven so fine
That all the world will be
in love with night,
And pay no worship
to the garish sun.

SHAKESPEARE
Romeo and Juliet III,ii,21

Visual-wavelength image

JULIET SO LOVED Romeo that she compared him to the beauty of the stars. Had she known what stars are really like, she would have made the comparison even stronger. Stars are beautiful in the night sky, but they are also powerful lamps that nature uses to generate energy and light the universe. To understand what stars are and eventually to understand the structure of the universe, the origin of our Earth, and the nature of our existence, we must find a way to discover what Juliet did not know—the true nature of the stars. ▌ Unfortunately, finding out what a star is like is quite difficult. When you look at a star through a telescope or in a photograph (■ Figure 8-1), You see only a point of light. Just looking tells you almost nothing about the star's energy production, diameter, or mass. Rather than just looking at stars, you must

The Eagle Nebula is the site of active star formation, but intense radiation from nearby hot stars to the upper right is blowing the nebula apart. (Mark McCaughrean and Morten Andersen of the Astrophysical Institute, Potsdam, and the European Southern Observatory)

Guidepost

Looking Back

Science is based on measurement, but measurement in astronomy is very difficult. In Chapter 5 you studied the complex telescopes and instruments that astronomers use, and in Chapter 6 you read about the information that can be found in spectra. Chapter 7 showed how astronomers use their understanding of the interaction of light and matter to understand a star, the sun.

This Chapter

Here you will discover how those same tools—telescopes, instruments, and basic physics—can tell us what stars are like. Here you will find answers to five essential questions about stars:

How far away are the stars?

How much energy do stars make?

How big are stars?

How much matter do stars contain?

What is the typical star like?

With this chapter you leave our sun behind and begin your study of the billions of stars that dot the sky. In a sense, the star is the basic building block of the universe. If you hope to understand what the universe is, what our sun is, what our Earth is, and what we are, you must understand stars.

Looking Ahead

Once you know how to find the basic properties of stars, you will be ready to trace the history of the stars from birth to death, a story that begins in the next chapter.

Ace Astronomy™ The AceAstronomy icon throughout the text indicates an opportunity for you to test yourself on key concepts and to explore animations and interactions on the AceAstronomy website at: http://astronomy.brookscole.com/sh9e

■ **Figure 8-1**

A family portrait. Each point of light in this photo of the Trifid Nebula is a star, but nothing about the appearance of a star tells you its energy output, its diameter, or its mass. By puzzling out these properties of stars, astronomers can understand this photo as a family portrait. Stars of different sizes and masses fill the image, as newborn stars are forming at the center of the nebula. (AURA/NOAO/NSF)

analyze starlight with great care. This chapter concentrates on finding out three things about stars—how much energy they emit, how big they are, and how much mass they contain. These three properties, combined with stellar temperatures—already discussed in Chapter 6—will give you an overview of stars.

Although you begin this chapter with three things to learn about stars, you immediately meet a short detour. To find out how much energy a star emits, you must know how far away it is. If at night you see lights approaching on the highway, you cannot tell whether you are looking at the bright headlights of a distant truck or the faint lights of a pair of nearby bicycles. Only when you know the distance to the lights can you judge their intrinsic brightness. In the same way, to find the intrinsic brightness of a star—or, more precisely, the amount of energy it emits—you must know its distance from you. A short detour will provide you with a method of measuring stellar distances.

After you reach your three goals, you will put your data together to find out what the average star is like. How dense are stars? Which types of stars are most common? Which are rare? By the time you finish this chapter, you will know what the family of stars is like.

8-1 Measuring the Distances to Stars

DISTANCE IS THE MOST IMPORTANT and the most difficult measurement in astronomy, and astronomers have found many different ways to estimate the distance to stars. Yet each of those ways depends on a direct geometrical method that is much like the method surveyors would use to measure the distance across a river they cannot cross. Let's begin by reviewing this method and then apply it to stars.

The Surveyor's Triangulation Method

To measure the distance across a river, a team of surveyors begins by driving two stakes into the ground. The distance between the stakes is the baseline of the measurement. The surveyors then choose a landmark on the opposite side of the river, a tree perhaps, thus establishing a large triangle marked by the two stakes and the tree. Using their surveyor's instruments, they sight the tree from the two ends of the baseline and measure the two angles on their side of the river.

Knowing two angles of this large triangle and the length of the side between them, the surveyors can then find the distance across the river by simple trigonometry. Another way to find the distance is to construct a scale drawing. For example, if the baseline is 50 m and the angles are 66° and 71°, you can draw a line 50 mm long to represent the baseline. Using a protractor, you can construct angles of 66° and 71° at each end of the baseline, and then, as shown in ■ Figure 8-2, extend the two sides until they meet at *C*. Point *C* on your drawing is the location of the tree. If you measure the height of your triangle, you would find it to be 64 mm and thus conclude that the distance from the baseline to the tree is 64 m.

Modern surveyors use computers for these problems, not simple drawings. The point is not how the problem is solved, but that it can be solved. Surveyors use simple triangulation to find the distance across a river.

The Astronomer's Triangulation Method

To find the distance to a star, you must use a very long baseline, the diameter of Earth's orbit. If you take a photograph of a nearby star and then wait 6 months, Earth will have moved halfway around its orbit. You can then take another photograph of the star. This second photograph is taken at a point in space 2 AU (astronomical units) from the point where the first photograph was taken. Thus your baseline equals the diameter of Earth's orbit, or 2 AU.

You then have two photographs of the same part of the sky taken from slightly different locations in space. When you examine the photographs, you will discover that the star is not in exactly the same place in the two photographs. This apparent shift in the position of the star is called *parallax.*

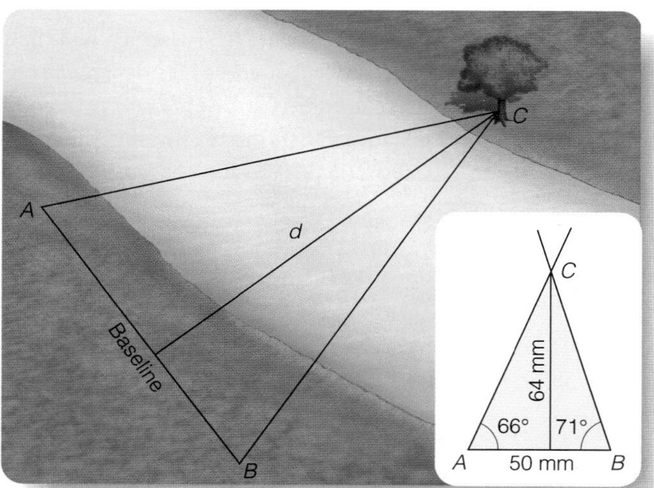

■ **Figure 8-2**

You can find the distance *d* across a river by measuring the baseline and the angles *A* and *B* and then constructing a scale drawing of the triangle.

Parallax is the apparent change in the position of an object due to a change in the location of the observer. In Chapter 4, you saw that parallax is an everyday experience. Your thumb, held at arm's length, appears to shift position against a distant background when you look with first one eye and then with the other (see page 50). In this case, the baseline is the distance between your eyes, and the parallax is the angle through which your thumb appears to move when you change eyes. The farther away you hold your thumb, the smaller the parallax. If you know the length of the baseline and measure the parallax, you can calculate the distance from your eyes to your thumb.

Because the stars are so distant, their parallaxes are very small angles, usually expressed in seconds of arc. The quantity that astronomers call **stellar parallax (*p*)** is half the total shift of the star, as shown in ■ Figure 8-3. Astronomers measure the parallax and surveyors measure the angles at the ends of the baseline, but both measurements tell the same thing—the shape of the triangle and thus the distance to the object in question.

Measuring the parallax *p* is very difficult because it is such a small angle. The nearest star, α Centauri, has a parallax of only 0.76 second of arc, and the more distant stars have even smaller parallaxes. To see how small these angles are, hold a piece of paper edgewise at arm's length. The thickness of the paper covers an angle of about 30 seconds of arc.

■ **Figure 8-3**

You can measure the parallax of a nearby star by photographing it from two points along Earth's orbit. For example, you might photograph it now and again in six months. Half of the star's total change in position from one photograph to the other is its stellar parallax, *p*.

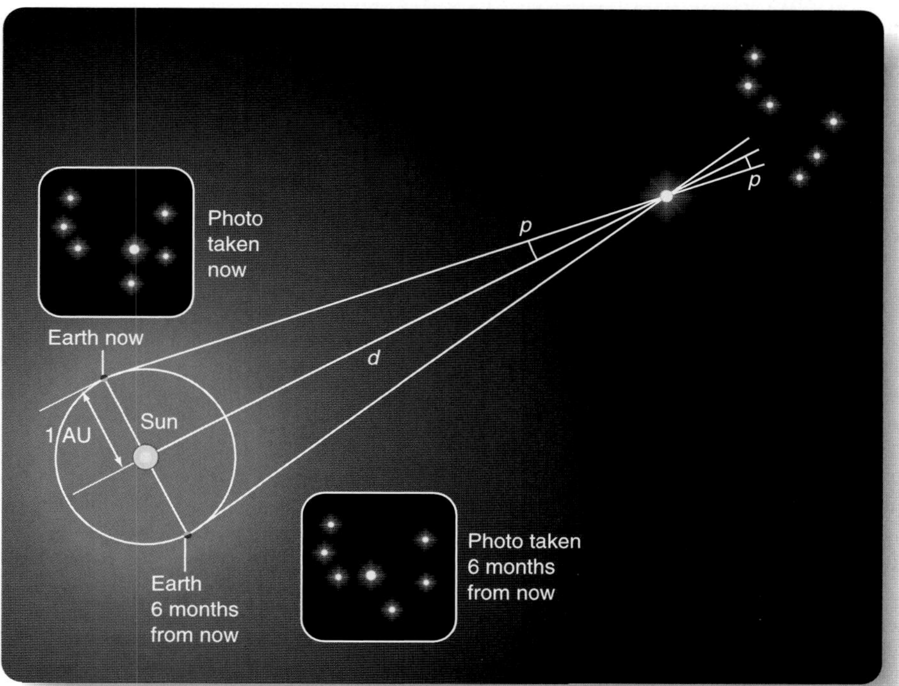

You cannot use scale drawings to find the distances to stars because the angles are so small and the distances are so large. Even for the nearest star, the triangle would have to be 300,000 times longer than it was wide. If the baseline in your drawing were 1 cm, the triangle would have to be about 3 km long. **Reasoning with Numbers 8-1** describes how you could find the distance from the parallax without drawing scale triangles.

The distances to the stars are so large that it is not convenient to use astronomical units. As Reasoning with Numbers 8-1 explains, when you measure distance via parallax, it is convenient to use the unit of distance called a **parsec (pc).** The word *parsec* was created by combining *par*allax and *sec*ond of arc. One parsec equals the distance to an imaginary star that has a parallax of 1 second of arc. A parsec is 206,265 AU, which equals roughly 3.26 ly (light-years).*

The blurring caused by Earth's atmosphere makes star images about 1 second of arc in diameter, and that makes it difficult to measure parallax. Even when they average together many observations, astronomers on Earth cannot measure parallax with an uncertainty smaller than about 0.002 second of arc. If you measure a parallax of 0.02 second of arc, the uncertainty is

* The parsec is used throughout astronomy because it simplifies the calculation of distance. However, there are instances in which the light-year is also convenient. Consequently, the chapters that follow use either parsecs or light-years as convenience and custom dictate.

Reasoning with Numbers 8-1

Parallax and Distance

To find the distance to a star from its measured parallax, imagine that you observe Earth from the star. Figure 8-3 shows that the angular distance you observe between the sun and Earth equals the star's parallax p. To find the distance, recall that the small-angle formula (see Reasoning with Numbers 3-1) relates an object's angular diameter, its linear diameter, and its distance. In this case, the angular diameter is p and the linear diameter is 1 AU. Then the small-angle formula, rearranged slightly, tells you that the distance to the star in AU is equal to 206,265 divided by the parallax in seconds of arc:

$$d = \frac{206{,}265}{p}$$

Because the parallaxes of even the nearest stars are less than 1 second of arc, the distances in AU are inconveniently large numbers. To keep the numbers manageable, astronomers have defined the parsec as their unit of distance in a way that simplifies the arithmetic. One parsec equals 206,265 AU, so the equation becomes

$$d = \frac{1}{p}$$

Thus, a parsec is the distance to an imaginary star whose parallax is 1 second of arc.

Example: The star Altair has a parallax of 0.20 second of arc. How far away is it? *Solution:* The distance in parsecs equals 1 divided by 0.2, or 5 pc:

$$d = \frac{1}{0.2} = 5 \text{ pc}$$

Because 1 pc equals about 3.26 ly, Altair is about 16.3 ly away.

about 10 percent. That means that ground-based astronomers can't measure accurate parallaxes smaller than about 0.02 second of arc, which corresponds to a distance of 50 pc. Consequently, ground-based parallax measurements are limited to only the closest stars. Since the first stellar parallax was measured in 1838, ground-based astronomers have been able to measure accurate parallaxes for only about 10,000 stars.

In 1989, the European Space Agency launched the satellite Hipparcos to measure stellar parallaxes from orbit above the blurring effects of Earth's atmosphere. The little satellite observed for four years, and the data were reduced by highly sophisticated computers to produce two parallax catalogs in 1997. One catalog contains 120,000 stars with parallaxes 20 times more accurate than ground-based measurements. The other catalog contains over a million stars with parallaxes as accurate as ground-based parallaxes. The Hipparcos data have given astronomers new insights into the nature of stars.

Inquire | Review | Analyze

Why are parallax measurements made from space better than parallax measurements made from Earth?

At first you might suppose that a satellite in orbit can measure the parallax of the stars better because the satellite is closer to the stars, but when you recall the immense distances to the stars, you see that being in space doesn't really put the satellite significantly closer to the stars. Rather, the satellite is above Earth's atmosphere. When you try to measure parallax, the turbulence in Earth's atmosphere blurs the star images and smears them out into blobs roughly 1 second of arc in diameter. It isn't possible to measure the position of these fuzzy images accurately. You can't measure parallax smaller than about 0.02 second of arc. A parallax of 0.02 second of arc corresponds to a distance of 50 pc, so you can't measure parallax accurately beyond that distance.

A satellite in orbit, however, is above Earth's atmosphere, so the only blurring in the star images is that produced by diffraction in the optics. In other words, the star images are very sharp, and a satellite in orbit can measure the positions of stars and thus their parallaxes to high accuracy.

If a satellite can measure parallaxes as small as 0.001 second of arc, then how far are the most distant stars it can measure? To put this distance into scale, express it as a percent of the diameter of our galaxy (see Chapter 1).

■ ■ ■

Connections: Distance is the key to the secrets of the stars. Now that you know how to find distances, you are ready to think about stellar brightness.

AceAstronomy™ Log into AceAstronomy and select this chapter to see Astronomy Exercise "Parallax I" and "Parallax II" to experiment with parallax.

8-2 Intrinsic Brightness

YOUR EYES TELL YOU that some stars look brighter than others, and in Chapter 2 you used the apparent magnitude scale to refer to stellar brightness. The faintest stars you can see with the naked eye are about sixth magnitude, with brighter stars having smaller magnitudes. In fact, the brightest stars you see in the sky have negative magnitudes, such as Sirius, whose apparent magnitude is −1.47.

The scale of apparent magnitudes only tells you how bright stars look, however, and you need to know their true, or intrinsic, brightness. *Intrinsic* means "belonging to the thing." When we refer to the intrinsic brightness of a star, we mean a measure of the total amount of light the star emits. An intrinsically very bright star might appear faint if it were far away. Thus to find out the true brightness of a star, you must correct for the influence of its distance.

Brightness and Distance

When you look at a bright light, your eyes respond to the visual-wavelength energy falling on the eye's retina, which tells you how bright the object looks. Thus, brightness is related to the flux of energy entering your eye. **Flux** is the energy in joules (J) per second falling on 1 square meter. Recall that a joule is about as much energy as is released when an apple falls from a table onto the floor. We commonly refer to 1 joule per second as 1 watt.

If you placed a screen 1 meter square near a lightbulb, a certain amount of flux would fall on the screen. If you moved the screen twice as far from the bulb, the light that previously fell on the screen would be spread to cover an area four times larger, and the screen would receive only one-fourth as much light. If you tripled the distance to the screen, it would receive only one-ninth as much light. Thus, the flux you receive from a light source is inversely proportional to the square of the distance to the source. This is known as the inverse square relation (■ Figure 8-4). (You first encountered the inverse square relation in Chapter 4, where it was applied to the strength of gravity.)

Now you can understand how the brightness of a star depends on its distance. If you know the apparent magnitude of a star and its distance from Earth, you can use the inverse square law to correct for distance and learn the intrinsic brightness of the star. Astronomers do that using a special kind of magnitude scale as described in the next section.

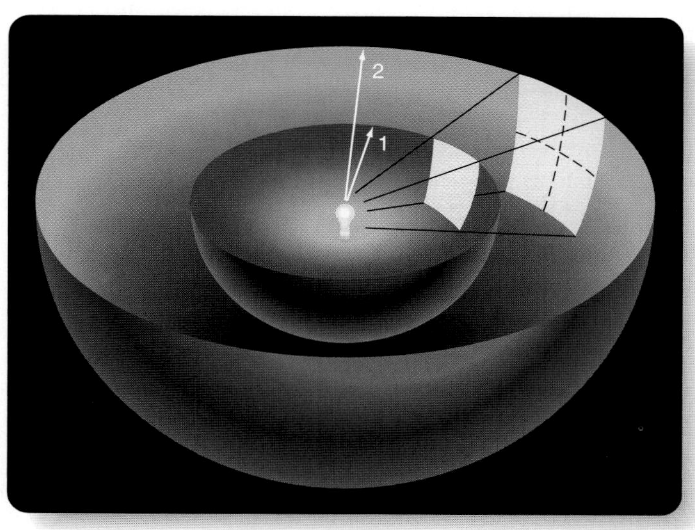

■ **Active Figure 8-4**

The inverse square relation. A light source is surrounded by spheres with radii of 1 unit and 2 units. The light falling on an area of 1 m^2 on the inner sphere spreads to illuminate an area of 4 m^2 on the outer sphere. Thus, the brightness of the light source is inversely proportional to the square of the distance.

AceAstronomy™ Log into AceAstronomy and select this chapter to see the Active Figure called "Brightness and Distance." Notice how distant sources of light appear fainter than nearby sources.

Absolute Visual Magnitude

If all the stars were the same distance away, you could compare one with another and decide which was emitting more light and which less. Of course, the stars are scattered at different distances, and you can't shove them around to line them up for comparison. If, however, you know the distance to a star, you can use the inverse square relation to calculate the brightness the star would have at some standard distance. Astronomers take 10 pc as the standard distance and refer to the intrinsic brightness of the star as its **absolute visual magnitude (M_v),** the apparent visual magnitude the star would have if it were 10 pc away.

The symbol for absolute visual magnitude is a capital M with a subscript v. The subscript tells you it is a visual magnitude based only on the wavelengths of light you can see. Other magnitude systems are based on other parts of the electromagnetic spectrum, such as the infrared and ultraviolet.

It is not difficult to find the absolute magnitude of a nearby star. You begin by measuring the apparent magnitude, which is an easy task in astronomy. Then you find the distance to the star. If the star is nearby, you can measure its parallax and from that find the distance. Once you know the distance, you can use a simple formula to correct the apparent magnitude for the distance and find the absolute magnitude (**Reasoning with Numbers 8-2**).

Astronomers can find the sun's absolute magnitude because they know the distance to the sun and can measure its apparent magnitude. The absolute magnitude of the sun is about 4.78. If the sun were only 10 pc from Earth (not a great distance in astronomy), it would look no brighter than the faintest star in the handle of the Little Dipper.

This detour to find the distance to stars has led you to absolute magnitude, a measure of the true brightness of the stars. You are now ready to reach the first of your three goals in this chapter.

AceAstronomy™ Log into AceAstronomy and select this chapter to see Astronomy Exercise "Apparent Brightness and Distance." You can change the distance to a star and see its apparent brightness change.

Luminosity

The first of your three goals for this chapter was to find out how much energy the stars emit. With the absolute magnitudes of the stars in hand, you can now compare other stars with our sun.

The intrinsically brightest stars known have absolute magnitudes of about −8, which means that if such a star were 10 parsecs away from Earth, you would see it nearly as bright as the moon. Such stars are 13 magnitudes brighter than the sun, so they must be emitting over 100,000 times more light than the sun. Yet the intrinsically faintest stars have absolute magnitudes of +15 or fainter. They are 10 magnitudes fainter than the sun, meaning they are emitting 10,000 times less light than the sun.

Your goal is to discover the **luminosities** (L) of the stars. That is, you want to know how much energy per second each star is emitting. You can use absolute magnitude to compare a star with the sun, but a small correction is necessary. Absolute visual magnitude refers to light, but you want to know the luminosity,

Reasoning with Numbers | 8-2

Absolute Magnitude

Apparent magnitude tells you how bright a star looks (see Reasoning with Numbers 2-1), but absolute magnitude tells you how bright the star really is. The absolute visual magnitude M_v of a star is the apparent visual magnitude of the star if it were 10 pc away. If you know a star's apparent magnitude and its distance, you can calculate its absolute magnitude. The equation that allows this calculation relates apparent magnitude m_v, distance in parsecs d, and absolute magnitude M_v:

$$m_v - M_v = -5 + 5 \log_{10}(d)$$

Sometimes it is convenient to rearrange the equation and write it in this form:

$$d = 10^{(m_v - M_v + 5)/5}$$

It is the same equation, so you can use whichever form is most convenient in a given problem. If you know the distance, the first form of the equation is convenient, but if you are trying to find the distance, the second form of the equation is best.

Example: Deneb is 490 pc from Earth and has an apparent magnitude of 1.26. What is its absolute magnitude? *Solution:* A pocket calculator tells you that $\log_{10}(490)$ equals 2.69, so you substitute into the first equation to get

$$1.26 - M_v = -5 + 5(2.69)$$

Solving for M_v tells you that the absolute magnitude of Deneb is −7.19. If it were only 10 pc from Earth, it would dominate the night sky.

which includes all energy. Hot stars emit a great deal of ultraviolet radiation that you can't see, and cool stars emit infrared. To add in the energy you can't see, astronomers make a small correction that depends on the temperature of the star. With that correction, the absolute magnitudes can tell you the luminosities of the stars.

Astronomers know the luminosity of the sun because they can send satellites above Earth's atmosphere and measure the amount of energy Earth receives from the sun in 1 second. The luminosity of the sun is about 4×10^{26} J/s.

You can express a star's luminosity in two ways. For example, you can say that the star Capella, which is 100 times more luminous than the sun, has a luminosity of 100 solar luminosities. You can also express this in real energy units by multiplying by the luminosity of the sun. The luminosity of Capella is 4×10^{28} J/s.

When you look at the night sky, the stars look much the same, yet your study of distances and luminosities has revealed an astonishing fact. Some stars are a million times more luminous than the sun, and some are nearly a million times fainter. Clearly, the family of stars is filled with interesting characters.

Inquire | Review | Analyze

Why do extremely cool stars look fainter than you might expect from their luminosities and distances?

Stars like the sun radiate most of their energy at visible wavelengths, and you can see those wavelengths, so they look about right for their luminosities and distances. But the coolest stars radiate the vast majority of their photons in the infrared, which you can't see. The luminosity includes all radiated energy. If you use that luminosity and a cool star's distance to predict how bright it should look, you might be disappointed. Most of the energy that cool stars radiate is invisible to your eyes, so they will look fainter than you might expect. Of course, the same thing happens for very hot stars, which radiate most of their energy in the ultraviolet part of the spectrum.

The luminosity is a very important piece of information about any star, but you must remember all the details that go into the measurement of luminosity. Explain exactly how errors in measuring the positions of blurred star images could introduce errors into your measurements of luminosity.

■ ■ ■

Connections: Although you detoured to find the distances to the stars, you have now reached your first goal. You have found the luminosities of the stars. When you reach the next goal, you will know even more about the stars.

8-3 The Diameters of Stars

YOUR SECOND GOAL in this chapter is to find the diameters of stars. You know little about stars until you know their diameters. Are they all the same size as the sun, or are some larger and some smaller? You certainly can't see their diameters through a telescope; the images of the stars are much too small for you to resolve their disks and measure their diameters. But there is a way to find out how big stars really are. If you know their temperatures and luminosities, you can find their diameters. That relationship will introduce you to the most important diagram in astronomy, where you will discover more family relations among the stars.

Luminosity, Radius, and Temperature

The luminosity and temperature of a star can tell you its diameter if you understand the two factors that affect a star's luminosity: surface area and temperature. For example, you can eat dinner by candlelight because the candle flame has a small surface area, and, although it is very hot, it cannot radiate much heat; it has a low luminosity. However, if the candle flame were 12 ft tall, it would have a very large surface area from which to radiate, and, although it might be no hotter than a normal candle flame, its luminosity would drive you from the table.

In a similar way, a star's luminosity is proportional to its surface area. A hot star may not be very luminous if it has a small surface area, although it could be highly luminous if it were larger. Even a cool star could be luminous if it had a large surface area. (See **Reasoning with Numbers 8-3.**) Because of this dependence on both temperature and surface area, you can use stellar luminosities to determine the diameters of stars if you can separate the effects of temperature and surface area.

The **Hertzsprung–Russell (H–R) diagram,** named after its originators, Ejnar Hertzsprung and Henry Norris Russell, is a graph that separates the effects of temperature and surface area on stellar luminosities and enables astronomers to sort the stars according to their diameters. Before discussing the details of the H–R diagram, let's look at a similar diagram you might use to sort automobiles.

You can plot a diagram such as ■ Figure 8-5 to show horsepower versus weight for various makes of cars. And in so doing, you will find that in general the more a car weighs, the more horsepower it has. Most cars fall somewhere along the sequence of cars, running from heavy, high-powered cars to light, low-powered models. You might call this the main sequence of cars. But some cars have much more horsepower than normal for their weight—the sport or racing models—and the economy models have less power than normal for cars of the same weight. Just as this diagram helps you understand the different kinds of autos, so the H–R diagram can help you understand the kinds of stars.

The H–R diagram is a graph with luminosity on the vertical axis and temperature on the horizontal axis. A star is represented by a point on the graph that tells you the luminosity of the star and its temperature. The H–R diagram in ■ Figure 8-6 also contains a scale of spectral type across the top. Because a star's spectral type is determined by its temperature, you could use either spectral type or temperature on the horizontal axis.

Reasoning with Numbers | 8-3

Luminosity, Radius, and Temperature

The luminosity L of a star depends on two things—its size and its temperature. If the star has a large surface area from which to radiate, it can radiate a great deal. Recall from our discussion of black body radiation in Reasoning with Numbers 6-1 that the amount of energy emitted per second from each square meter of the star's surface is σT^4. Thus, the star's luminosity can be written as its surface area in square meters times the amount it radiates per square meter:

$$L = \text{area} \times \sigma T^4$$

Because a star is a sphere, you can use the formula area $= 4\pi R^2$. Then the luminosity is

$$L = 4\pi R^2 \sigma T^4$$

This seems complicated, but if you express luminosity, radius, and temperature in terms of the sun, you get a much simpler form:*

$$\frac{L}{L_\odot} = \left(\frac{R}{R_\odot}\right)^2 \left(\frac{T}{T_\odot}\right)^4$$

Example A: Suppose you want to find the luminosity of a particular star. This star is 10 times the sun's radius but only half as hot. How luminous is it? *Solution:*

$$\frac{L}{L_\odot} = \left(\frac{10}{1}\right)^2 \left(\frac{1}{2}\right)^4 = \frac{100}{1} \times \frac{1}{16} = 6.25$$

The star has 6.25 times the sun's luminosity.

You can also use this formula to find diameters.

Example B: Suppose you found a star whose absolute magnitude is +1 and whose spectrum shows it is twice the sun's temperature. What is the diameter of the star? *Solution:* The star's absolute magnitude is 4 magnitudes brighter than the sun, and you recall from Reasoning with Numbers 2-1 that 4 magnitudes is a factor of 2.512^4, or about 40. The star's luminosity is therefore about 40 $L_\odot$. With the luminosity and temperature, you can find the radius:

$$\frac{40}{1} = \left(\frac{R}{R_\odot}\right)^2 \left(\frac{2}{1}\right)^4$$

Solving for the radius you get:

$$\left(\frac{R}{R_\odot}\right)^2 = \frac{40}{2^4} = \frac{40}{16} = 2.5$$

So the radius is

$$\frac{R}{R_\odot} = \sqrt{2.5} = 1.58$$

The star is 58 percent larger in radius than the sun.

* In astronomy the symbols ⊙ and ⊕ refer respectively to the sun and Earth. Thus $L_\odot$ refers to the luminosity of the sun, $T_\odot$ refers to the temperature of the sun, and so on.

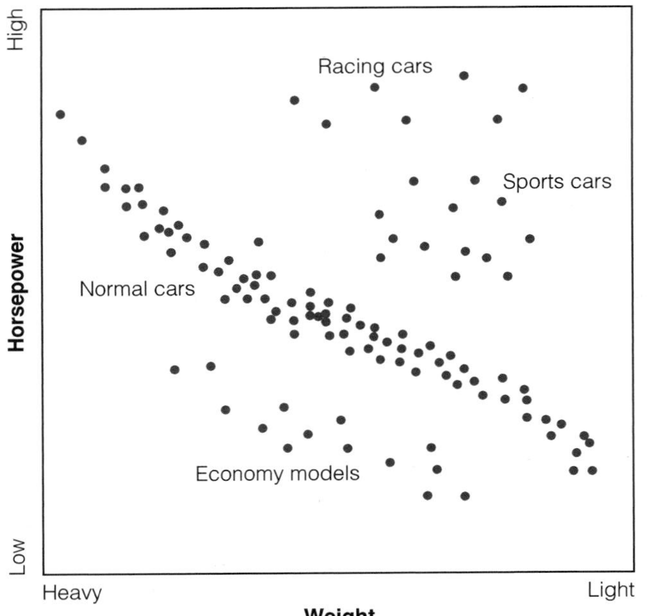

■ **Figure 8-5**

You could analyze automobiles by plotting their horsepower versus their weight and thus reveal relationships between various models. Most would lie somewhere along the main sequence of "normal" cars.

In an H–R diagram, the location of a point representing a star tells you a great deal about the star. Points near the top of the diagram represent very luminous stars, and points near the bottom represent very low-luminosity stars. Also, points near the right edge of the diagram represent very cool stars, and points near the left edge of the diagram represent very hot stars. Notice in Figure 8-6 how the artist has used color to represent temperature. Cool stars are red, and hot stars are blue (■ Figure 8-7).

Astronomers use H–R diagrams so often that they usually skip the words "the point that represents the star." Rather, they will say that a star is located in a certain place in the diagram. Of course, they mean the point that represents the luminosity and temperature of the star and not the star itself. The location of a star in the H–R diagram has nothing to do with the location of the star in space. Furthermore, a star may move in the H–R diagram as it ages and its luminosity and temperature change, but such motion in the diagram has nothing to do with the star's motion in space.

The **main sequence** is the region of the H–R diagram running from upper left to lower right. It includes roughly 90 percent of all normal stars. In Figure 8-6, the main sequence is represented by a curved line with dots for stars plotted along it. As you might expect, the hot main-sequence stars are more luminous than the cool main-sequence stars.

In addition to temperature, size is important in determining the luminosity of a star. Notice in the H–R diagram that some cool stars lie above the main sequence. Although they are cool, they are luminous, and that must mean they are larger and have more surface area than main sequence stars of the same temperature. These are called **giant** stars, and they are roughly 10 to 100 times larger than the sun. The **supergiant** stars, at the top of the H–R diagram, are as large as 1000 times the sun's diameter.

At the bottom of the H–R diagram lie the economy models, stars that are very low in luminosity

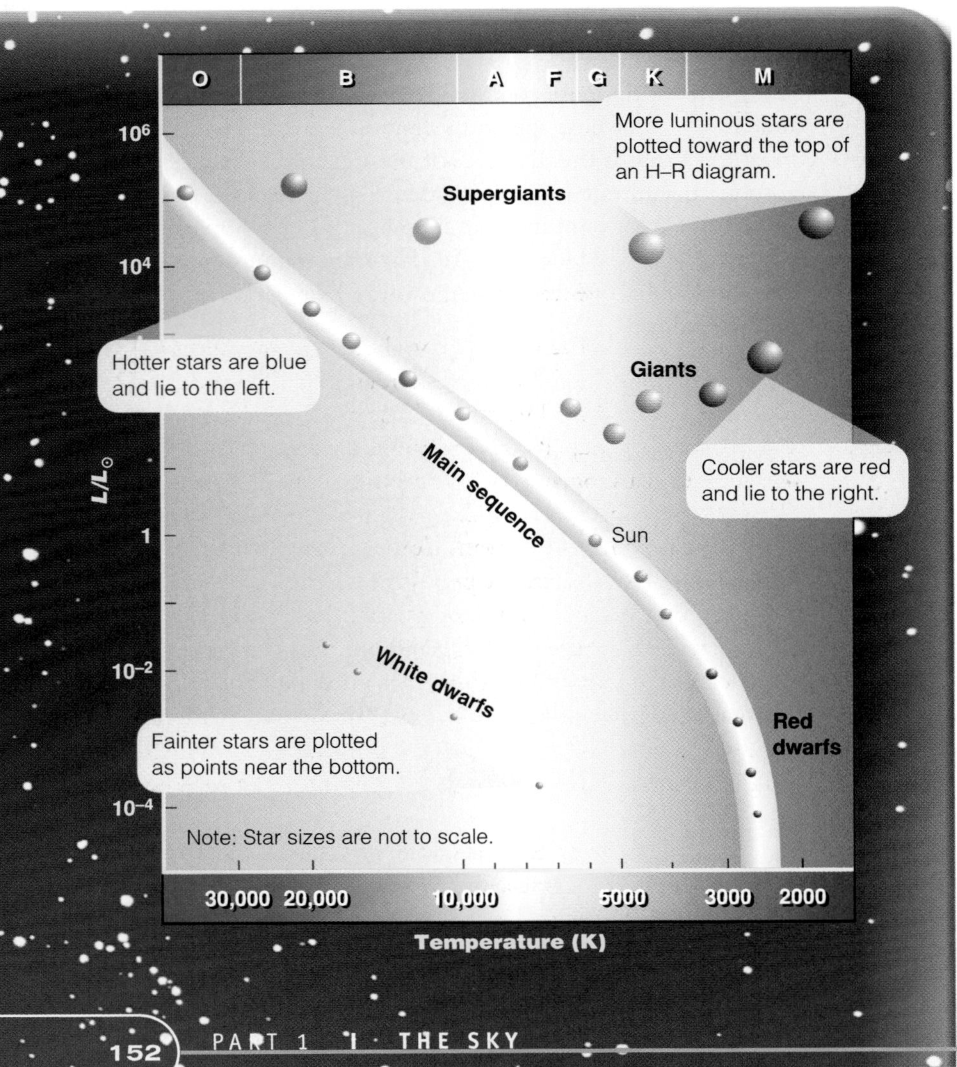

■ **Figure 8-6**

In an H–R diagram, a star is represented by a dot that shows the luminosity and temperature of the star. The background color in this diagram indicates the temperature of the stars. The sun is a yellow-white G2 star. Most stars fall along a sequence running from hot luminous stars at upper left to cool low luminosity stars at lower right. The exceptions—giants, supergiants, and white dwarfs—are discussed in the text.

■ **Figure 8-7**

Notice the colors of the stars in the small star cluster M39. The brightest stars are either hot and blue or cool and red. Compare with the most luminous stars in Figure 8-6. (Heidi Schweiker /NOAO/AURA/NSF)

because they are very small. At the bottom end of the main sequence, the **red dwarfs** are not only small, but they are also cool, and that gives them low luminosities. In contrast, the **white dwarfs** lie in the lower left of the H–R diagram, and although some white dwarfs are among the hottest stars known, they are so small they have very little surface area from which to radiate and that limits them to low luminosities.

The equation in Reasoning with Numbers 8-3 can be used to draw precise lines of constant radius across the H–R diagram. ■ Figure 8-8 shows the accurate luminosities and temperatures of a number of well-known stars along with lines of constant radius. For example, locate the line labeled $1 R_{\odot}$ (1 solar radius) and notice that it passes through the point representing the sun. Any star whose point is located along this line has a radius equal to the sun's. Notice also that the lines of constant radius slope downward to the right, because cooler stars are always fainter than hotter stars of the same size. These lines of constant radius show dramatically that the supergiants and giants are large stars. In the H–R diagram, the white dwarfs fall along the line labeled $0.01\ R_{\odot}$. They are all about the same radius—about the size of Earth.

Notice the great range of sizes among stars. The largest stars are 100,000 times larger than the tiny white dwarfs. If the sun were a tennis ball, the white dwarfs would be grains of sand, and the largest supergiants would be as big as football fields.

Ace Astronomy™ Log into AceAstronomy and select this chapter to see Astronomy Exercise "Stefan–Boltzmann Law II." You can control the size and temperature of a star and watch its luminosity change.

Luminosity Classification

A star's spectrum contains clues whether it is a main-sequence star, a giant, or a supergiant. The larger a star is, the less dense its atmosphere is, and the widths of spectral lines are partially determined by the density of the gas.

Some spectral lines are particularly sensitive to density. If the atoms collide often in a dense gas, their energy levels become distorted and the spectral lines are broadened. Hydrogen Balmer lines are an example. In the spectrum of a main-sequence star, the Balmer lines are broad because the star's atmosphere is dense and the hydrogen atoms collide often. In the spectrum of a giant star, the lines are narrower (■ Figure 8-9), because the giant star's atmosphere is less dense and the hydrogen atoms collide less often. In the spectrum of a supergiant star, the Balmer lines are very narrow.

Thus you can look at a star's spectrum and tell roughly how big it is. These are called **luminosity classes,** because the size of the star is the dominating factor in determining luminosity. Supergiants, for example, are very luminous because they are very large.

The luminosity classes are represented by the roman numerals I through V, with supergiants further subdivided into types Ia and Ib, as follows:

Luminosity Classes

Ia	Bright supergiant
Ib	Supergiant
II	Bright giant
III	Giant
IV	Subgiant
V	Main-sequence star

You can distinguish between the bright supergiants (Ia) such as Rigel (β Orionis) and the regular supergiants (Ib) such as Polaris, the North Star. The star Adhara (ε Canis Majoris) is a bright giant (II), Capella (α Aurigae) is a giant (III), and Altair (α Aquilae) is a subgiant (IV). The sun is a main-sequence star (V). The luminosity class appears after the spectral type, as in G2 V for the sun. White dwarfs don't enter into this classification, because their spectra are peculiar.

The positions of the luminosity classes on the H–R diagram are shown in ■ Figure 8-10. Remember that these are rather broad classifications and that the lines on the diagram are only approximate. A star of luminosity class III may lie slightly above or below the line labeled III. Notice that a scale of absolute magnitude has been added to the right edge of this H–R diagram for easy reference.

Luminosity classification is subtle and not too accurate, but it is an important tool in modern astronomy. As you will see in the next section, luminosity classification provides a way to find the distance to stars that are too far away to have measurable parallaxes.

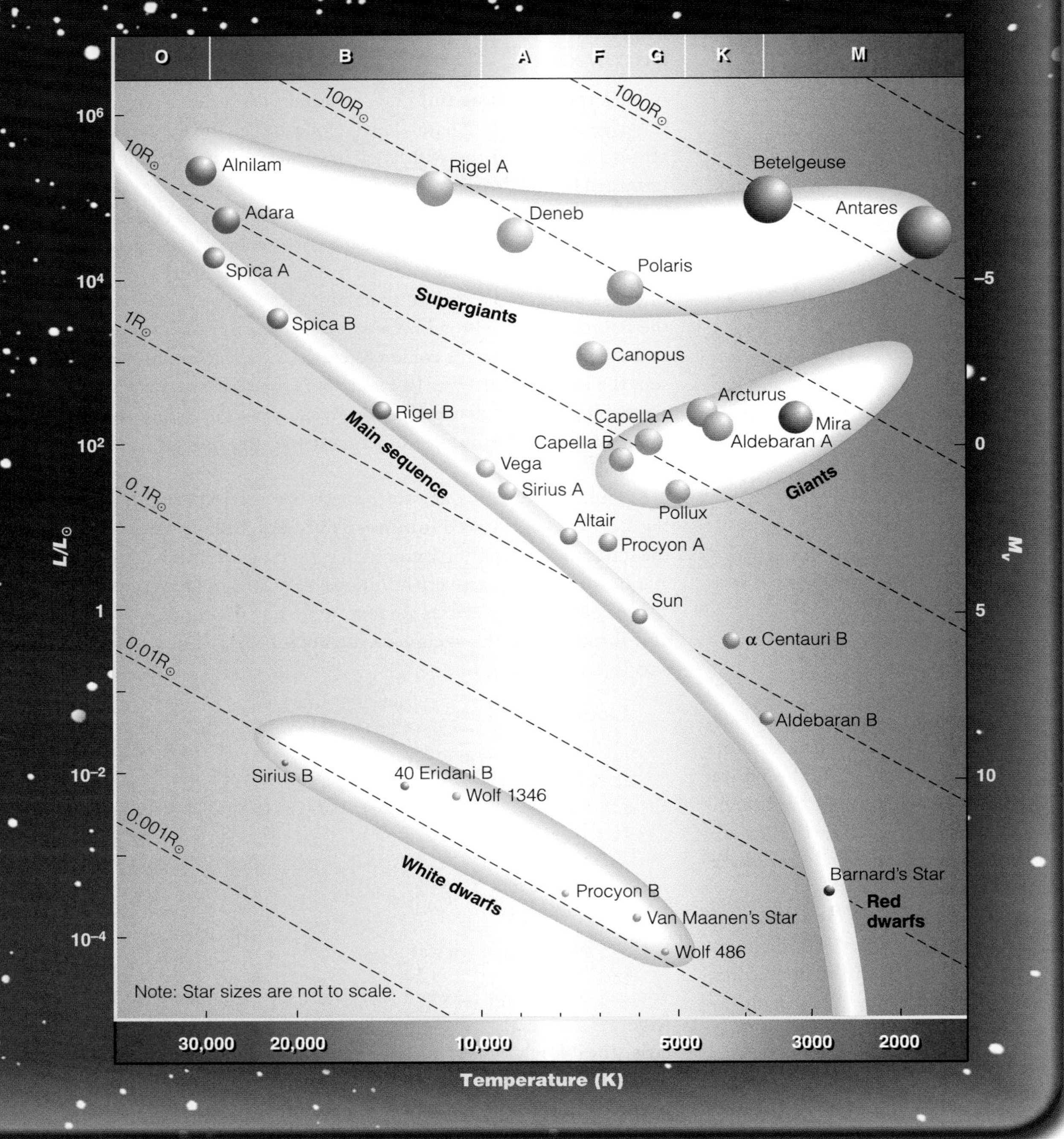

Figure 8-8

An H–R diagram showing the luminosity and temperature of many well-known stars. The dashed lines are lines of constant radius. The star sizes on this diagram are not to scale; try to sketch in the correct sizes for supergiants and white dwarfs using the size of the sun as a guide. (Individual stars that orbit each other are designated A and B, as in Spica A and Spica B.)

Spectroscopic Parallax

Astronomers can measure the stellar parallax of nearby stars, but most stars are too distant to have measurable parallaxes. They can find the distances to these stars if they can record the stars' spectra and determine their luminosity classes in a process called **spectroscopic parallax**—the estimation of the distance to a star from its spectral type, luminosity class, and apparent magnitude. Spectroscopic parallax is not an actual measure of parallax, but it does tell you the distance to the star.

The method of spectroscopic parallax depends on the H–R diagram. If you record the spectrum of a star, you can determine its spectral class, and that tells you its horizontal location in the

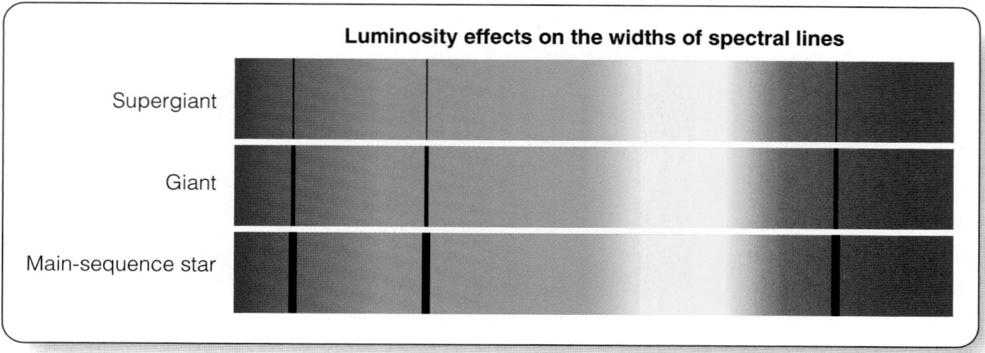

■ Figure 8-9

These schematic spectra show how the widths of spectral lines reveal a star's luminosity classification. Supergiants have very narrow spectral lines, and main-sequence stars have broad lines. In addition, certain spectral lines are more sensitive to this effect than others, so an experienced astronomer can inspect a star's spectrum and determine its luminosity classification.

H–R diagram. You can also determine its luminosity class by looking at the widths of its spectral lines, and that tells you the star's vertical location in the diagram. Once you plot the point that represents the star in the H–R diagram, you can read off its absolute magnitude. As you have seen earlier in this chapter, you can find the distance to a star by comparing its apparent and absolute magnitudes.

For example, Spica is classified B1 V, and its apparent magnitude is +1. You can plot this star in an H–R diagram such as Figure 8-10, where you would find it should have an absolute magnitude of about −4. Using the apparent and absolute magnitudes, you can find the distance from the equation in Reasoning with Numbers 8-2. Your use of spectroscopic parallax says Spica should be about 100 pc from Earth. The Hipparcos satellite finds the distance to be 80 pc, so the method of spectroscopic parallax has given you a good estimate of the distance, but a direct measurement of the parallax is better whenever possible.

Inquire | Review | Analyze

What evidence can you cite that giant stars really are bigger than the sun?

You can find stars of the same spectral type as the sun that are clearly more luminous than the sun. Capella, for example, is a G star with an absolute magnitude of 0. Because it is a G star, it must have about the same temperature as the sun, but its absolute magnitude is 5 magnitudes brighter than the sun's. A magnitude difference of 5 magnitudes corresponds to an intensity ratio of 100, so Capella must be about 100 times more luminous than the sun. If it has the same surface temperature as the sun but is 100 times more luminous, then it must have a surface area 100 times greater than the sun's. Because the surface area of a sphere is proportional to the square of the radius, Capella must be 10 times larger in radius. That is clear observational evidence that Capella is a giant star.

In Figure 8-8, you can see that Procyon B is a white dwarf only slightly warmer than the sun but about 10,000 times less luminous than the sun. Use the same logical procedure followed above to demonstrate that the white dwarfs must be small.

■ ■ ■

Connections: The first two goals of this chapter have been achieved: You know how to find the luminosities and diameters of stars. The next step is to find their masses.

■ Figure 8-10

The approximate location of the luminosity classes on the H–R diagram.

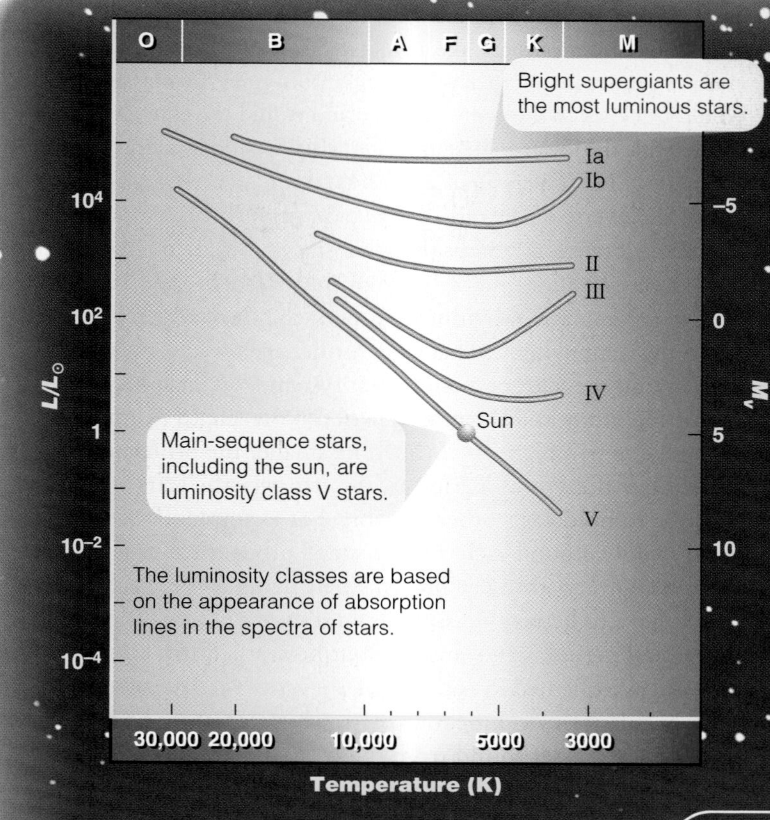

8-4 The Masses of Stars

YOUR THIRD GOAL is to find out how much matter stars contain, that is, to know their mass. Do they all contain about the same mass as the sun, or are some more massive and others less? Unfortunately, it's difficult to determine the mass of a star. Looking through a telescope at a star, you see only a point of light that tells nothing about the star. Gravity is the key. Matter produces a gravitational field, and you can figure out how much matter a star contains if you watch an object move through the star's gravitational field. To find the masses of stars, you must study **binary stars,** pairs of stars that orbit each other.

Binary Stars in General

The key to finding the mass of a binary star is understanding orbital motion, which provides clues to mass.

Chapter 4, illustrated orbital motion by imagining a cannonball fired from a high mountain (see page 68). If Earth's gravity didn't act on the cannonball, it would follow a straight-line path and leave Earth forever. Because Earth's gravity pulls it away from its straight-line path, the cannonball follows a curved path around Earth—an orbit. When two stars orbit each other, their mutual gravitation pulls them away from straight-line paths and makes them follow closed orbits around a point between the stars.

Each star in a binary system moves in its own orbit around the system's center of mass, the balance point of the system. If the stars were connected by a massless rod and placed in a uniform gravitational field such as that near Earth's surface, the system would balance at its center of mass like a child's seesaw (see page 69). If one star were more massive than its companion, then the massive star would be closer to the center of mass and would travel in a smaller orbit, while the lower-mass star would whip around in a larger orbit (■ Figure 8-11). The ratio of the masses of the stars M_A/M_B equals r_B/r_A, the inverse of the ratio of the radii of the orbits. If one star has an orbit twice as large as the other star's orbit, then it must be half as massive. Getting the ratio of the masses is easy, but that doesn't tell you the individual masses of the stars, which is what you really want to know. That takes further analysis.

To find the mass of a binary star system, you must know the size of the orbits and the orbital period—the length of time the stars take to complete one orbit. The smaller the orbits are and the shorter the orbital period is, the stronger the stars' gravity must be to hold each other in orbit. For example, if two stars whirl rapidly around each other in small orbits, then their gravity must be very strong to prevent their flying apart. Such stars would have to be very massive. From the size of the orbits and the orbital period, you can figure out how much mass the stars contain, as explained in **Reasoning with Numbers 8-4.** Such calculations yield the total mass, which, combined with the ratio of the masses found from the relative sizes of the orbits, can tell you the individual masses of the stars.

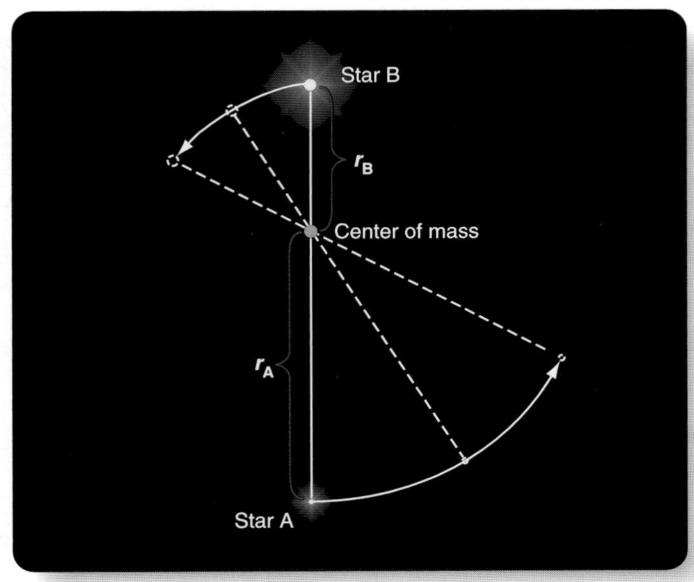

■ Active Figure 8-11

As stars in a binary star system revolve around each other, the line connecting them always passes through the center of mass, and the more massive star is always closer to the center of mass.

Ace Astronomy™ Log into AceAstronomy and select this chapter to see the Active Figure called "Center of Mass." Experiment to see how the ratio of the masses affects the relative size of the orbits.

Actually, figuring out the mass of a binary star system is not as easy as it might seem from this discussion. The orbits of the two stars may be elliptical; and, although the orbits lie in the same plane, that plane can be tipped at an unknown angle to your line of sight, further distorting the shapes of the orbits. Astronomers must find ways to correct for these distortions. In addition, astronomers analyzing binary systems must find the distances to the stars so they can estimate the true size of the orbits in astronomical units. Notice that finding the masses of binary stars requires a number of steps to get from what you can observe to what you really want to know, the masses. Constructing such sequences of steps is an important part of science (**Window on Science 8-1**).

Although there are many different kinds of binary stars, three types are especially important for determining stellar masses. These are discussed separately in the next sections.

Visual Binary Systems

In a **visual binary system,** the two stars are separately visible in the telescope. Only a pair of stars with large orbits can be separated visually, for if the orbits are small, the star images blend together in the telescope and you see only a single point of light. Because visual binary systems must have large orbits, they also have long orbital periods. Some take hundreds or even thousands of years to complete a single orbit.

Learning about Nature through Chains of Inference

Scientists can rarely observe the things they really want to know, so they must construct chains of inference. You can't observe the mass of stars directly, so you must find a way to use what you can observe—orbital period and angular separation—and step by step figure out the parameters you need to reach your goal. In this case, the goal is the masses of the two stars. You follow the chain of inference from the observable parameters to the unobservable quantities you want to know.

Chains of inference can be mathematical, as when geologists use earthquakes to calculate the temperature and density of Earth's interior. There is no way to drill a hole to Earth's center and lower a thermometer or recover a sample, so the internal temperature and density are not directly observable. Nevertheless, the speed of vibrations from distant earthquakes depends on the temperature and density of the rock they pass through. Geologists can't measure the speed of the vibrations deep inside Earth, but they can measure the delay in the arrival time at different locations on the surface, and that allows them to work their way back to the speed and, finally, the temperature and density.

Chains of inference can also be nonmathematical. Biologists studying the migration of whales can't follow individual whales for years at a time, but they can observe feeding and mating in different locations; take into consideration food sources, ocean currents, and water temperatures; and construct a chain of inference that leads back to the seasonal migration pattern for whales.

This chapter contains a number of chains of inference because it describes how astronomers know what stars are like. Almost all sciences use chains of inference. If you can link the observable parameters step by step to the final conclusions, you have gained a strong insight into the nature of science.

San Andreas fault: A chain of inference connects earthquakes to conditions inside Earth. (USGS)

Astronomers study visual binary systems by measuring the position of the two stars directly at the telescope or in images. In either case, the astronomers need measurements over many years to map the orbits. The first frame of ■ Figure 8-12 shows a photograph of the bright star Sirius, which is a visual binary system made up of the bright star Sirius A and its white dwarf companion Sirius B. The photo was taken in 1960. Successive frames in Figure 8-12 show the motion of the two stars as observed since 1960 and the orbits the stars follow. The orbital period is 50 years.

Binary systems are common; more than half of all stars are members of binary star systems. Few, however, can be analyzed completely. Many are so far apart that their periods are much too long for practical mapping of their orbits. Others are so close together they are not visible as separate stars.

Reasoning with Numbers | 8-4

The Masses of Binary Stars

Johannes Kepler's third law of orbital motion applied only to planets in our solar system, but Isaac Newton realized that mass was involved. Newton's version of the third law applies to any pair of objects that orbit each other. The total mass of the two objects is related to the average distance a between them and their orbital period P. If the masses are M_A and M_B, then:

$$M_A + M_B = \frac{a^3}{P^2}$$

In this formula, a is expressed in AU, P in years, and the mass in solar masses.

Notice that this formula is related to Kepler's third law of planetary motion (see Table 4-1). Almost all of the mass of the solar system is in the sun. If you apply this formula to any planet in our solar system, the total mass is 1 solar mass. Then the formula becomes $P^2 = a^3$, which is Kepler's third law.

In other star systems, the total mass is not necessarily 1 solar mass, and this gives you a way to find the masses of binary stars. If you can find the average distance in AU between the two stars and their orbital period in years, the sum of the masses of the two stars is just a^3/P^2.

Example A: If you observe a binary system with a period of 32 years and an average separation of 16 AU, what is the total mass? *Solution:* The total mass equals $16^3/32^2$, which equals 4 solar masses.

Example B: Let's call the two stars in the previous example A and B. Suppose star A is 12 AU away from the center of mass, and star B is 4 AU away. What are the individual masses? *Solution:* The ratio of the masses must be 12:4, which is the same as a ratio of 3:1. What two numbers add up to 4 and have the ratio 3:1? Star B must be 3 solar masses, and star A must be 1 solar mass.

A Visual Binary Star System

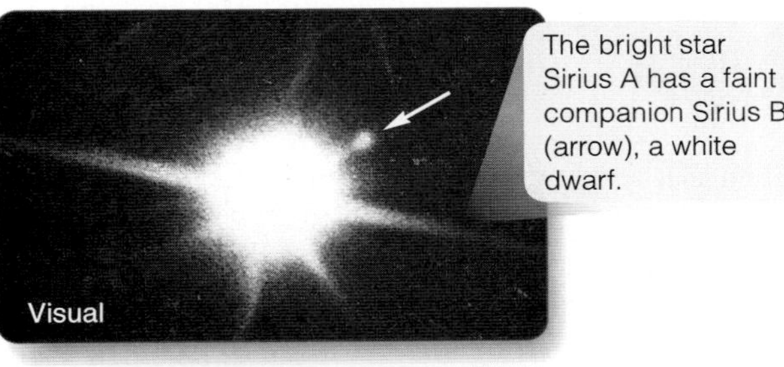

The bright star Sirius A has a faint companion Sirius B (arrow), a white dwarf.
Visual

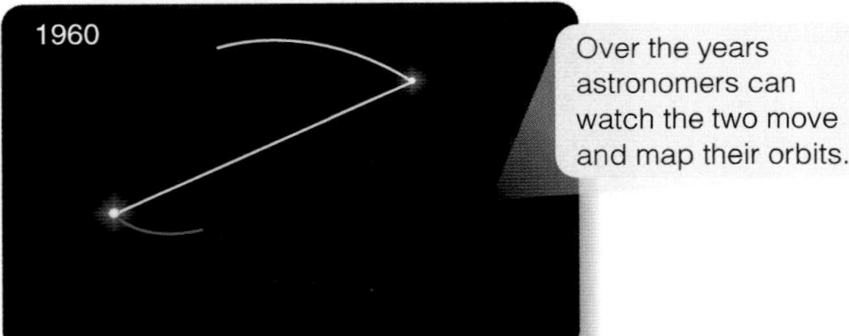

1960
Over the years astronomers can watch the two move and map their orbits.

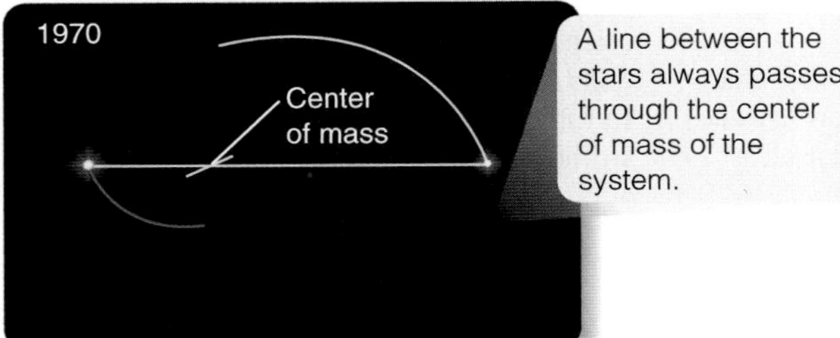

1970
Center of mass
A line between the stars always passes through the center of mass of the system.

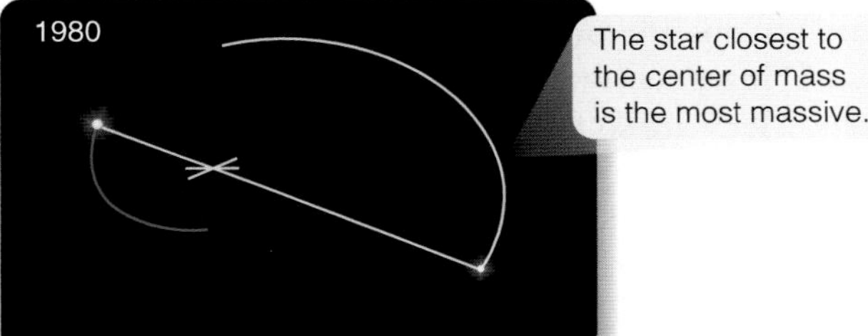

1980
The star closest to the center of mass is the most massive.

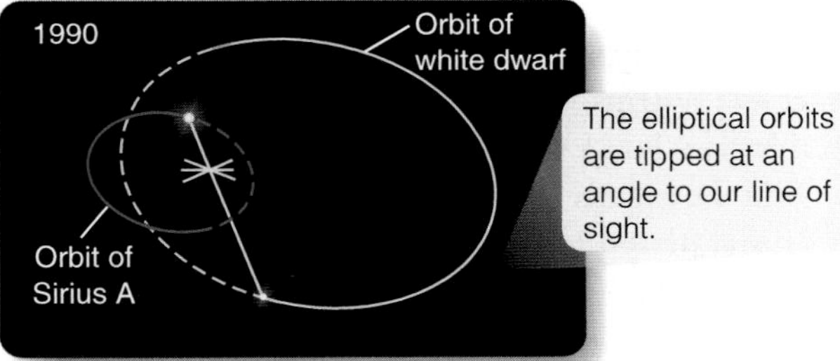

■ **Figure 8-12**

The orbital motion of Sirius A and Sirius B can reveal their individual masses. (Photo: Lick Observatory)

Spectroscopic Binary Systems

If the stars in a binary system are close together, the telescope, limited by diffraction and by the atmospheric blurring called seeing, shows a single point of light. Only by looking at a spectrum, which is formed by light from both stars and contains spectral lines from both, can astronomers tell that there are two stars present and not one. Such a system is a **spectroscopic binary system.**

■ Figure 8-13 shows a pair of stars orbiting each other in identical circular orbits. From the diagram you can guess that the two stars have equal masses, and you notice that the circular orbit appears elliptical because you see it nearly edge-on. Of course, if this were a spectroscopic binary system, you would not see the separate stars. Nevertheless, the Doppler shift would tell you there were two stars orbiting each other. In the first frame of the figure, star A is approaching while star B recedes. In the spectrum, you see a spectral line from star A blueshifted while the same spectral line from star B is redshifted. As you watch the two stars revolve around their orbits, they alternately approach and recede, and you see their spectral lines Doppler shifted first toward the blue and then toward the red. In a real spectroscopic binary, you can't see the individual stars, but the sight of pairs of spectral lines moving back and forth across each other would alert you that you were observing a spectroscopic binary system (■ Figure 8-14).

You can find the orbital period of a spectroscopic binary system by waiting to see how long it takes for the spectral lines to return to their starting positions. You can measure the size of the Doppler shifts to find the orbital velocities of the two stars. If you multiply velocity times orbital period, you can find the circumference of the orbit, and from that you can find the radius of the orbit. Of course, if you know the orbital period and the size of the orbit you can calculate the mass. One important detail is missing, however. You don't know how much the orbits are inclined to your line of sight.

You can find the inclination of a visual binary system because you can see the shape of the orbits. In a spectroscopic binary system, however, you cannot see the individual stars, so you can't find the inclination or untip the orbits. Recall that the Doppler effect only reveals the radial velocity, the part of the velocity directed toward or away from the observer. Because you cannot find the inclination, you cannot correct these radial velocities to find the true orbital velocities. Consequently, you cannot find the true masses. All you can find from a spectroscopic binary system is a lower limit to the masses.

More than half of all stars are in binary systems, and most of those are spectroscopic binary systems. Many of the familiar stars in the sky are actually pairs of stars orbiting each other (■ Figure 8-15).

You might wonder what happens when the orbits of a spectroscopic binary system lie exactly edge-on to Earth. The result is the most useful kind of binary system.

AceAstronomy™ Log into AceAstronomy and select this chapter to see Astronomy Exercise "Spectroscopic Binaries." You can see how the motions of the two stars affect their shared spectrum.

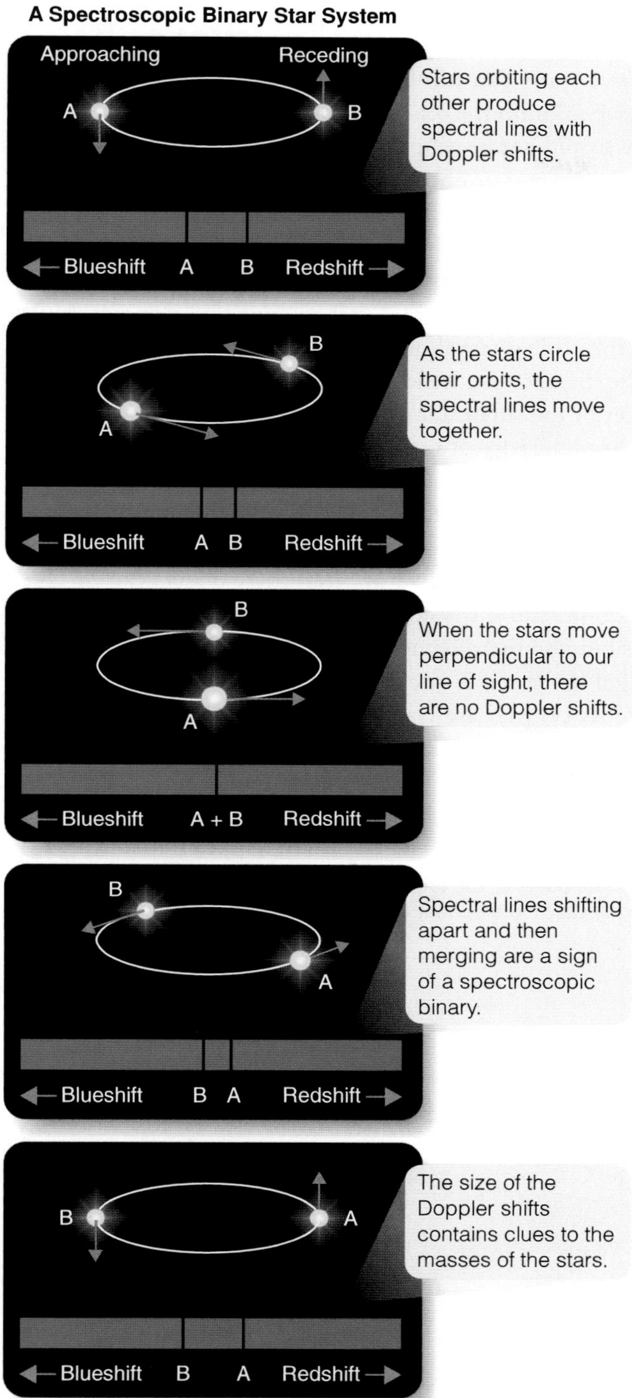

■ **Figure 8-13**

From Earth, a spectroscopic binary looks like a single point of light, but the Doppler shifts in its spectrum reveal the orbital motion of the two stars.

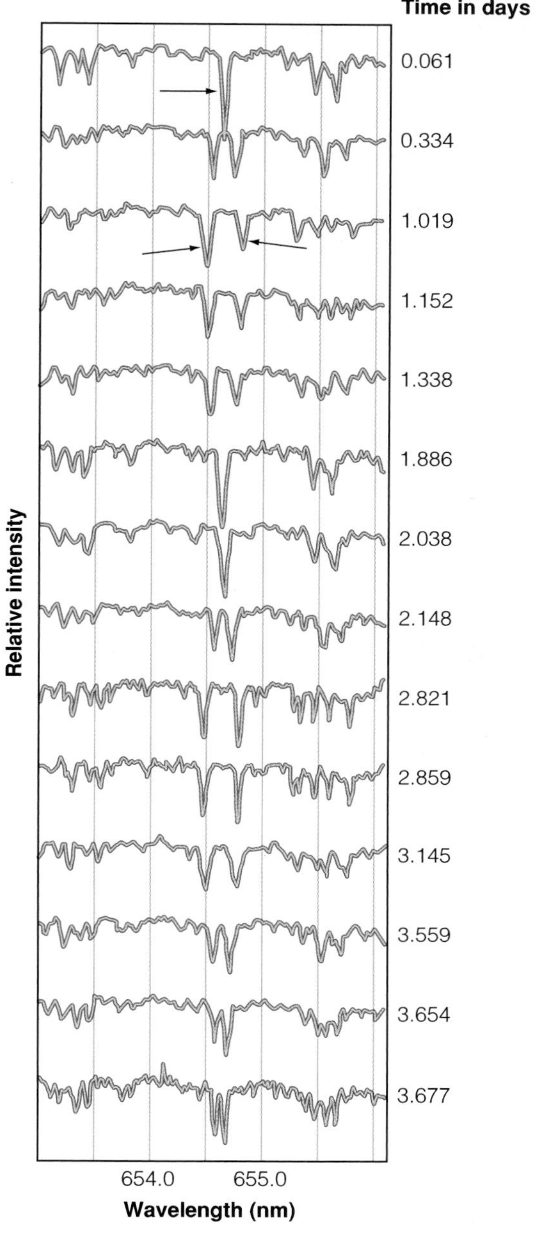

■ **Figure 8-14**

Fourteen spectra of the star HD80715 are shown here as graphs of intensity versus wavelength. A single spectral line (arrow in top spectrum) splits into a pair of spectral lines (arrows in third spectrum), which then merge and split apart again. These changing Doppler shifts reveal that HD80715 is a spectroscopic binary. (Adapted from data courtesy of Samuel C. Barden and Harold L. Nations)

Eclipsing Binary Systems

As we mentioned earlier, the orbits of the two stars in a binary system always lie in a single plane. If that plane is nearly edge-on to Earth, then the stars can cross in front of each other as seen from Earth. Imagine a model of a binary star system in which a cardboard disk represents the orbital plane and balls represent the stars, as in ■ Figure 8-16. If you see the model from the edge, then the balls that represent the stars can move in front of each other as they follow their orbits. The small star crosses in front of the large star, and then, half an orbit later, the large star crosses

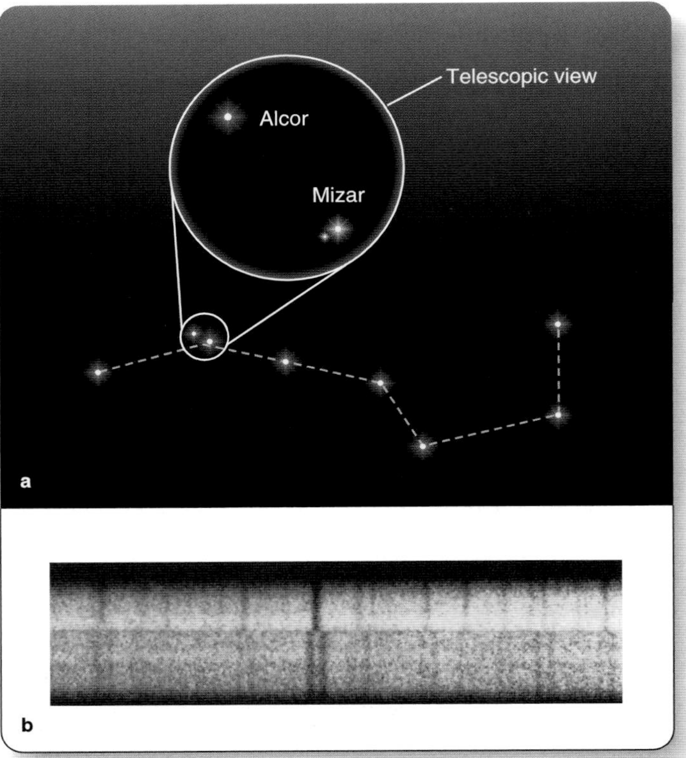

■ **Figure 8-15**

(a) At the bend of the handle of the Big Dipper lies a pair of stars, Mizar and Alcor. Through a telescope you can discover that Mizar has a fainter companion and so is a member of a visual binary system. (b) Spectra of Mizar recorded at different times show that it is itself a spectroscopic binary system rather than a single star. In fact, both the faint companion to Mizar and the nearby star Alcor are also spectroscopic binary systems. (The Observatories of the Carnegie Institution of Washington)

in front of the small star. When one star moves in front of the other, it blocks some of the light, and the star is eclipsed. Such a system is called an **eclipsing binary system.**

Seen from Earth, the two stars are not visible separately. The system looks like a single point of light. But when one star moves in front of the other star, part of the light is blocked, and the total brightness of the point of light decreases. ■ Figure 8-17 shows a smaller star moving in an orbit around a larger star, first eclipsing the larger star and then being eclipsed as it moves behind. The resulting variation in the brightness of the system is shown as a graph of brightness versus time, a **light curve.**

The light curves of eclipsing binary systems contain tremendous amounts of information about the stars, but the curves can be difficult to analyze. Figure 8-17 shows an idealized system. ■ Figure 8-18 shows the light curve of a real system in which the stars have dark spots on their surfaces and are so close to each other that their shapes are distorted.

Once the light curve of an eclipsing binary system has been accurately observed, you can construct a chain of inference that leads to the masses of the two stars. You can find the orbital period easily, and you can get spectra showing the Doppler shifts of the two stars. You can find the orbital velocity because you don't have to untip the orbits; you know they are nearly edge-on or there would not be eclipses. Then you can find the size of the orbits and the masses of the stars.

Earlier luminosity and temperature were used to calculate the radii of stars, but eclipsing binary systems give a way to measure the sizes of stars directly. From the light curve you can tell how long it took for the small star to cross the large star. Multiplying this time interval by the orbital velocity of the small star gives the diameter of the larger star. You can also determine the diameter of the small star by noting how long it took to disappear behind the edge of the large star. For example, if it took 300 seconds for the small star to disappear while traveling 500 km/s relative to the large star, then it must be 150,000 km in diameter.

Of course, there are complications due to the inclination and eccentricity of orbits, but often these effects can be taken into account, and the system can tell you not only the masses of its stars but also their diameters.

Algol (β Persei) is one of the best-known eclipsing binaries because its eclipses are visible to the naked eye. Normally, its magnitude is about 2.15, but its brightness drops to 3.4 in

■ **Figure 8-16**

Imagine a model of a binary system with balls for stars and a disk of cardboard for the plane of the orbits. Only if you view the system edge-on do you see the stars cross in front of each other.

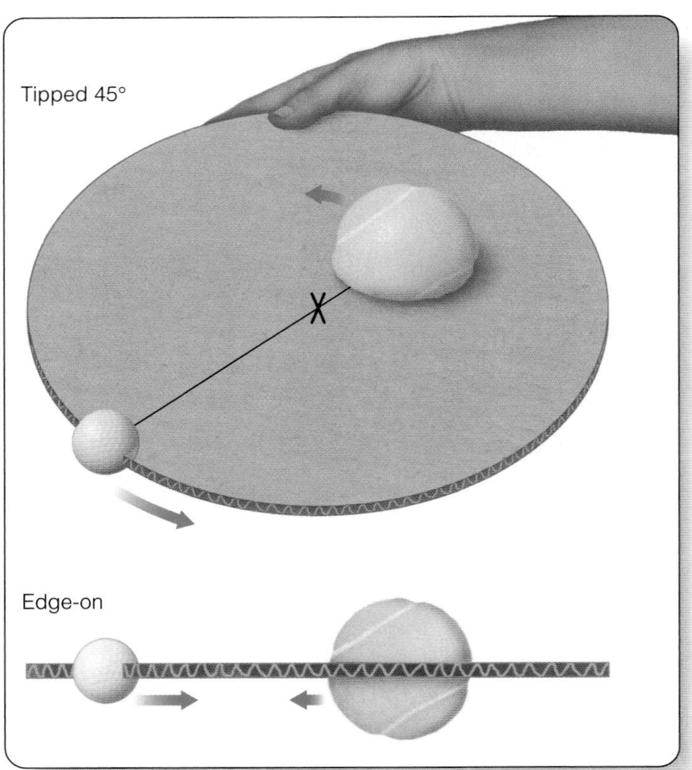

An Eclipsing Binary Star System

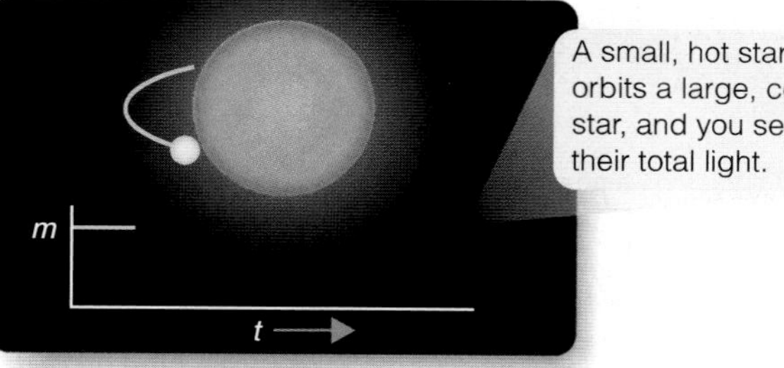

A small, hot star orbits a large, cool star, and you see their total light.
m
t

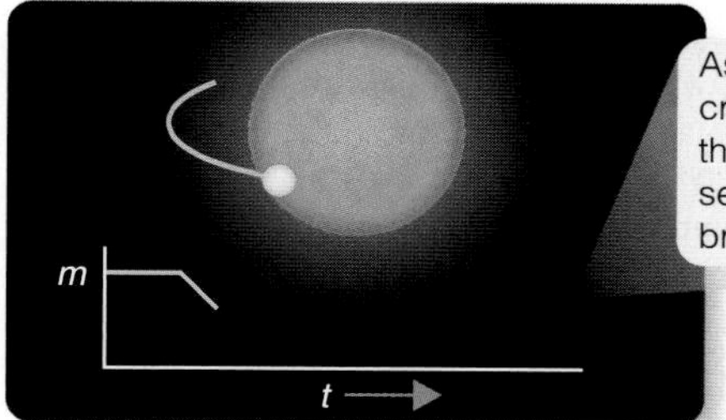

As the hot star crosses in front of the cool star, you see a decrease in brightness.
m
t

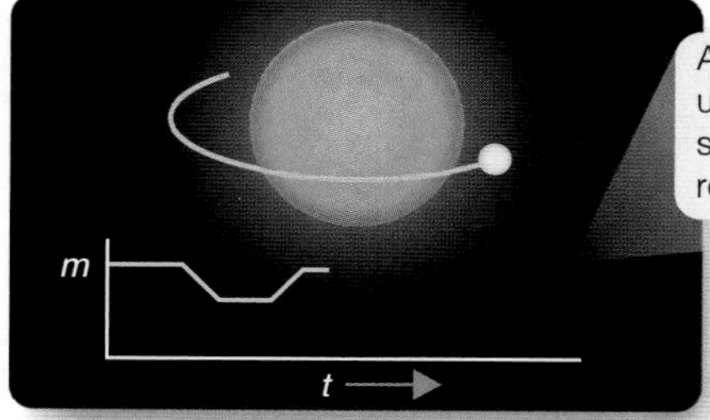

As the hot star uncovers the cool star, the brightness returns to normal.
m
t

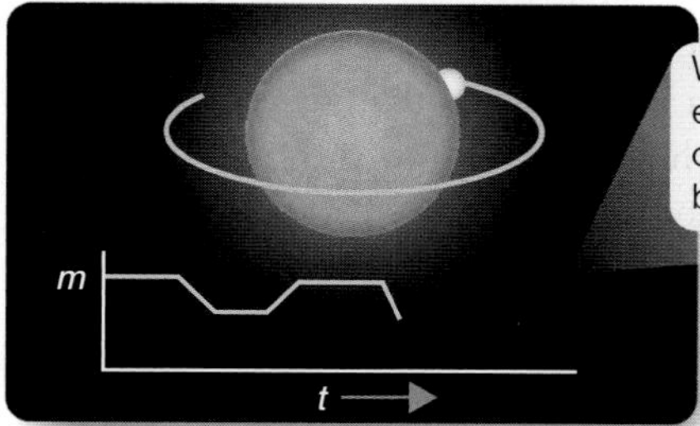

When the hot star is eclipsed behind the cool star, the brightness drops.
m
t

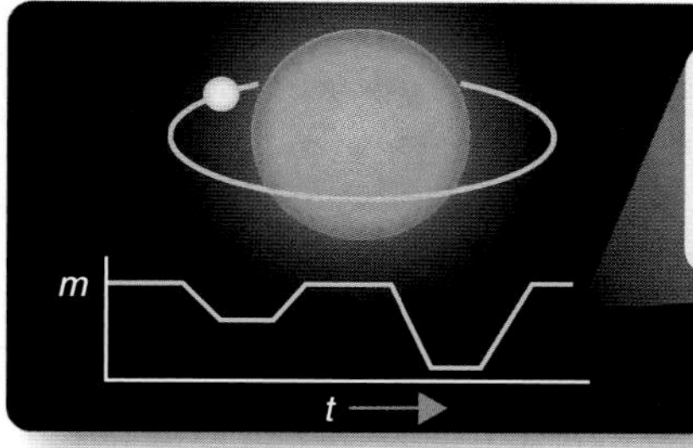

The depth of the eclipses depends on the surface temperatures of the stars.
m
t

■ Figure 8-17

From Earth, an eclipsing binary looks like a single point of light, but changes in brightness reveal that two stars are eclipsing each other. Doppler shifts in the spectrum combined with the light curve, shown here as magnitude versus time, can reveal the size and mass of the individual stars.

eclipses that occur every 68.8 hours. Although the nature of the star was not recognized until 1783, its periodic dimming was probably known to the ancients. *Algol* comes from the Arabic for "the demon's head," and it is associated in constellation mythology with the severed head of Medusa, the sight of whose serpentine locks turned mortals to stone (■ Figure 8-19). Indeed, in some accounts, Algol is the winking eye of the demon.

From the study of binary stars, astronomers have found that the masses of stars range from roughly 0.1 solar mass at the low end to nearly 100 solar masses at the high end. The most massive stars ever found in a binary system have masses of 83 and 82 solar masses. A few other stars are believed to be more massive, 100 solar masses to 150 solar masses, but they do not lie in binary systems, so astronomers must estimate their mass.

AceAstronomy™ Log into AceAstronomy and select this chapter to see Astronomy Exercise "Eclipsing Binaries." You can take control of a binary star system.

Inquire | Review | Analyze

When you look at the light curve for an eclipsing binary system with total eclipses, how can you tell which star is hotter?

If you assume that the two stars in an eclipsing binary system are not the same size, then you can refer to them as the larger star and the smaller star. When the smaller star moves behind the larger star, you lose the light coming from the total area of the small star. And when the smaller star moves in front of the larger star, it blocks off light from the same amount of area on the larger star. In both cases, the same amount of area, the same number of square meters, is hidden from your sight. Then the amount of light lost during an eclipse depends only on the temperature of the hidden surface, because temperature is what determines how much a single square meter can radiate per second. When the surface of the hotter star is hidden, the brightness will fall dramatically, but when the surface of the cooler star is hidden, the brightness will not fall as much. So you can look at the light curve and point to the deeper of the two eclipses and say, "That is where the hotter star is behind the cooler star."

But eclipsing binaries also reveal the diameters of the stars. How could you look at the light curve of an eclipsing binary with total eclipses and find the ratio of the diameters?

■ ■ ■

Connections: Binary stars give a powerful tool with which to discover the masses of the stars. With that information, you are now ready to assemble your data and begin trying to understand the secrets of the family of stars.

8-5 A Survey of the Stars

YOU HAVE ACHIEVED the three goals set at the start of this chapter. You know how to find the luminosities, diameters, and masses of stars. Now you can put that data (**Window on Science 8-2**) together to paint a family portrait of the stars; like most family portraits, both similarities and differences are important clues to the history of the family. As you begin trying to understand how stars are born and how they die, ask a simple

Information about Nature: Basic Scientific Data

In a simple sense, science is the process by which you look at data and search for relationships that show how nature works. But, as a result, science sometimes requires large amounts of data. For example, astronomers need to know the masses and luminosities of many stars before they can begin looking for relationships.

Compiling basic data is one of the common forms of scientific study. This work may not seem very exciting to an outsider, but scientists often love their work not so much because they want to know nature's secrets but because they love the process of studying nature. Using a microscope or a telescope is fun. Gathering plants in the rain forest or geological samples from a cliff face can be tremendously exciting and satisfying. It is sometimes the process of science that is most rewarding, and enjoyment of the process can lead scientists to gather significant amounts of information.

Solving a single binary star system to find the masses of the stars does not tell an astronomer a great deal about nature, but solving a binary is like solving a puzzle. It is fun and it is satisfying. Over the years, many astronomers have added their results to the growing data file on stellar masses. They can now analyze that data to search for relationships between the masses of stars and other parameters such as diameter and luminosity.

The history of science is filled with stories of hardworking scientists who compiled large amounts of data that later scientists used to make important discoveries. For example, Tycho Brahe spent 20 years recording the positions of the stars and planets (Chapter 4). He must have loved his work to spend 20 years on his windy island, but he did not live to use his data. It was his successor, Johannes Kepler, who used Tycho's data to discover the laws of planetary motion.

Whatever science you study, you will encounter accumulations of measurements and observations that have been compiled over the years, including everything from the hardness of rocks to the attention span of infants. Determining these basic scientific data is as much a part of science as is testing a hypothesis.

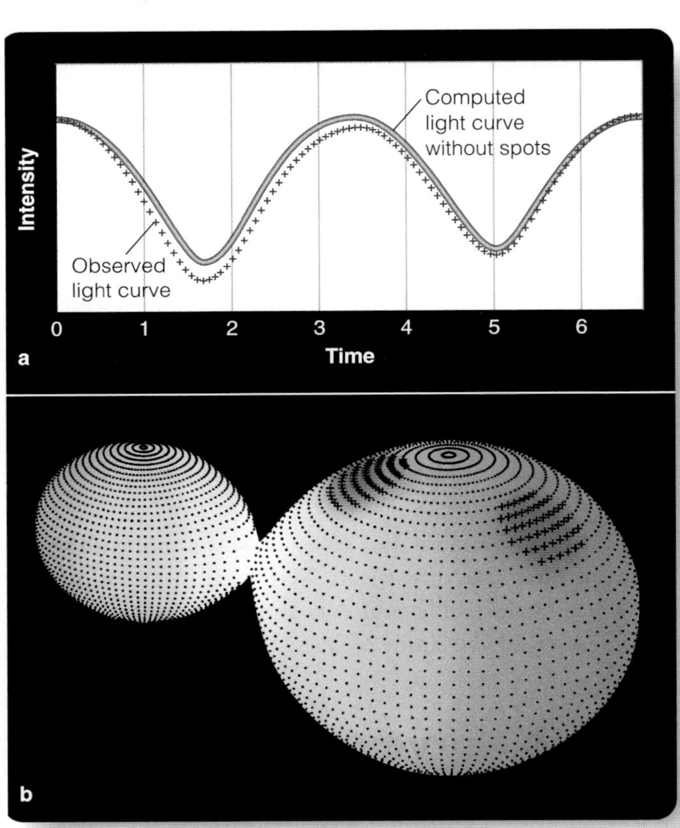

■ Figure 8-18

The observed light curve of the binary star VW Cephei (lower curve) shows that the two stars are so close together their gravity distorts their shapes. Slight distortions in the light curve reveal the presence of dark spots at specific places on the star's surface. The upper curve shows what the light curve would look like if there were no spots. (Graphics created with Binary Maker 2.0)

question: What is the average star like? Answering that question is both challenging and illuminating.

Mass, Luminosity, and Density

The H–R diagram is filled with patterns that give you clues as to how stars are born, how they age, and how they die. When you add your data, you see traces of those patterns.

If you label an H–R diagram with the masses of the plotted stars, as is done in ■ Figure 8-20, you will discover that the main-sequence stars are ordered by mass. The most massive main-sequence stars are the hot stars. As you run your eye down the main sequence, you will find lower-mass stars, and the lowest-mass stars are the coolest, faintest main-sequence stars.

Stars that do not lie on the main sequence are not in order according to mass. Some giants and supergiants are massive, while others are no more massive than the sun. All white dwarfs have about the same mass, somewhere in the narrow range of 0.5 to about 1 solar mass.

Because of the systematic ordering of mass along the main sequence, these main-sequence stars obey a **mass–luminosity relation**—the more massive a star is, the more luminous it is (■ Figure 8-21). In fact, the mass–luminosity relation can be expressed as a simple formula. (See **Reasoning with Numbers 8-5**.) Giants and supergiants do not follow the mass–luminosity relation very closely, and white dwarfs not at all. In the next chapters, the mass–luminosity relation will help you understand how stars generate their energy.

Though mass alone does not reveal any pattern among giants, supergiants, and white dwarfs, density does. Once you know

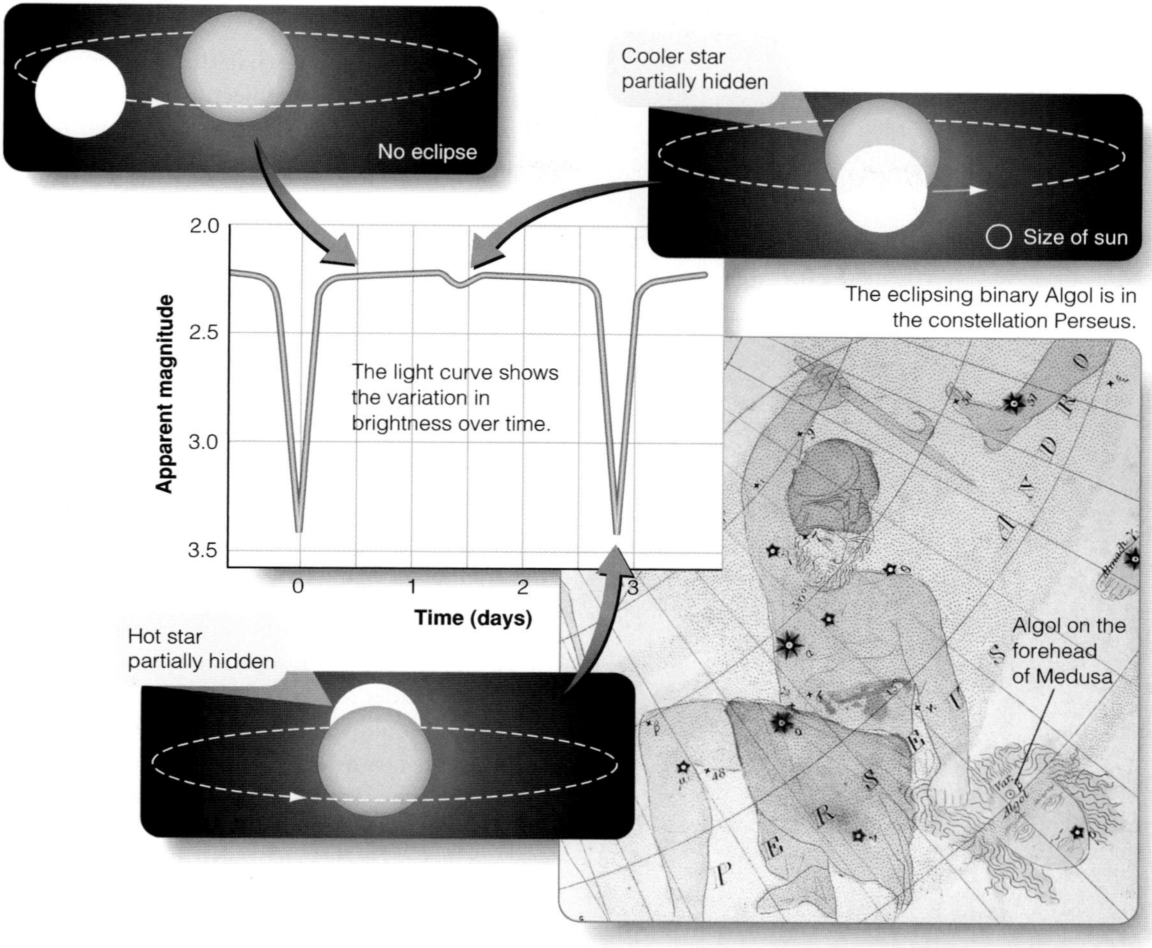

■ Figure 8-19

The eclipsing binary Algol consists of a hot B star and a cooler G or K star. The eclipses are partial, meaning that neither star is completely hidden during eclipses. The orbit here is drawn as if the cooler star were stationary.

a star's mass and diameter, you can calculate its average density by dividing its mass by its volume. Stars are not uniform in density but are most dense at their centers and least dense near their surface. The center of the sun, for instance, is about 100 times as dense as water; its density near the visible surface is about 3400 times less dense than Earth's atmosphere at sea level. A star's average density is intermediate between its central and surface densities. The sun's average density is approximately 1 g/cm^3—about the density of water.

Main-sequence stars have average densities similar to the sun's, but giant stars, being large, have low average densities, ranging from 0.1 to 0.01 g/cm^3. The enormous supergiants have still lower densities, ranging from 0.001 to 0.000001 g/cm^3. These densities are thinner than the air we breathe, and if you could insulate yourself from the heat, you could fly an airplane through these stars. Only near the center would you be in any danger, for there the material is very dense—about 3,000,000 g/cm^3.

The white dwarfs have masses about equal to the sun's but are very small, only about the size of Earth. Thus, the matter is compressed to densities of 3,000,000 g/cm^3 or more. On Earth, a teaspoonful of this material would weigh about 15 tons.

Density divides stars into three groups. Most stars are main-sequence stars with densities like the sun's. Giants and supergiants are very low-density stars, and white dwarfs are high-density stars. You will see in later chapters that these densities reflect different stages in the evolution of stars.

AceAstronomy™ Log into AceAstronomy and select this chapter to see Astronomy Exercise "Mass-Luminosity Relation." You can discover your own relation from real stars.

AceAstronomy™ Log into AceAstronomy and select this chapter to see Astronomy Exercise "H-R/Mass-Luminosity 3D Graph." You can explore this 3D graph and see how mass is related to position in the H-R diagram.

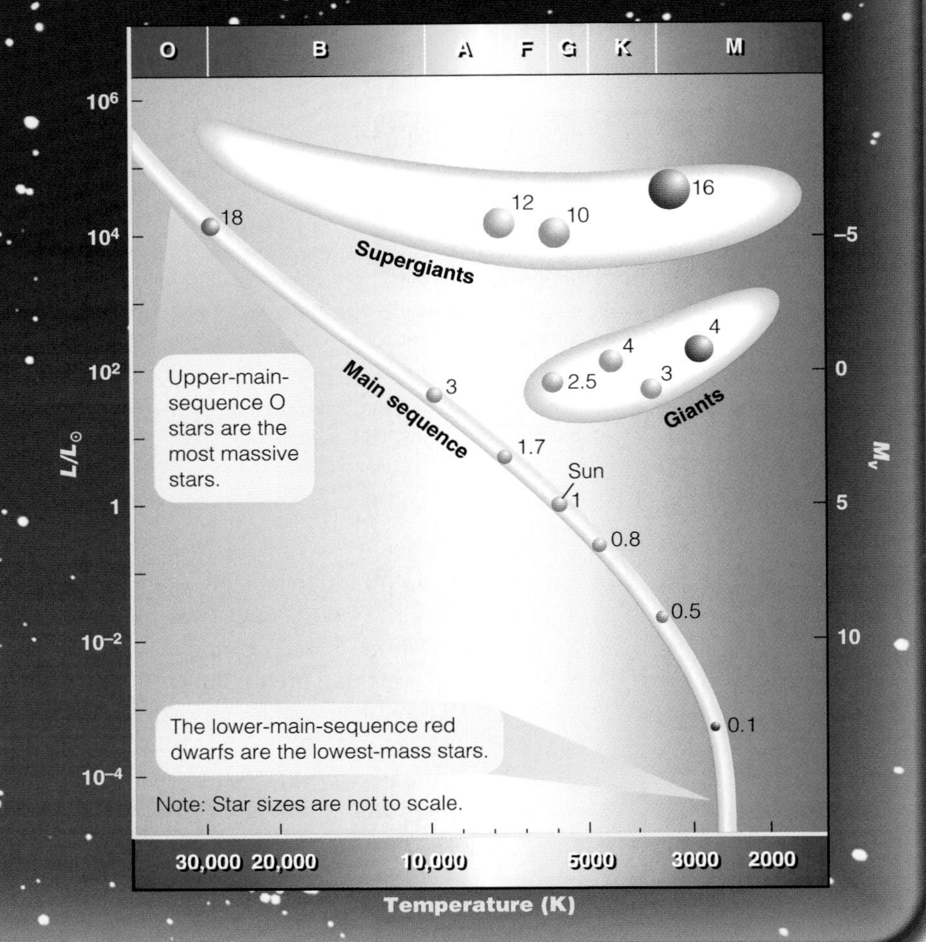

■ **Figure 8-20**

The masses of the plotted stars are labeled on this H–R diagram. Notice that the masses of main-sequence stars decrease from top to bottom but that masses of giants and supergiants are not arranged in any ordered pattern.

Surveying the Stars

If you want to know what the average person thinks about a certain subject, you take a survey. If you want to know what the average star is like, you must survey the stars. Such surveys reveal important relationships among the family of stars.

As the 21st century begins, astronomers are deeply involved in massive surveys. Remember the earlier mention of the Hipparchos satellite that surveyed the entire sky measuring the parallax of over a million stars. Powerful computers to control instruments and analyze data make such immense surveys possible. For example, the Sloan Digital Sky Survey will eventually map a quarter of the sky, measuring the position and brightness of 100 million stars and galaxies. Also, the Two Micron All Sky Survey (2MASS) has mapped the entire sky at three wavelengths in the near-infrared. A number of other sky surveys are underway. Astronomers will mine these mountains of data for decades to come.

What could you learn about stars from a survey of the stars near the sun? Because the sun is believed to be in a typical place in the universe, such a survey could reveal general characteristics of the stars. Study **The Family of Stars** on pages 166 and 167 and notice three important points.

First, taking a survey is difficult because you must be sure to get an honest sample. If you don't survey enough stars or don't notice the faintest stars, you can get biased results. The second important point is that most stars are faint, and the most luminous stars are rare. The most common kinds of stars are the lower-main-sequence red dwarfs and white dwarfs. Also notice that luminous stars, although they are rare, are easily visi-

Reasoning With Numbers | 8-5

The Mass–Luminosity Relation

You can calculate the approximate luminosity of a star from a simple equation. A star's luminosity in terms of the sun's luminosity equals its mass in solar masses raised to the 3.5 power:

$$L = M^{3.5}$$

This is the mathematical form of the mass–luminosity relation. It is only an approximation, as shown by the red line in Figure 8-21, but it applies to most stars over a wide range of stellar masses.

You can do simple calculations with this equation if you remember that raising a number to the 3.5 power is the same as cubing it and then multiplying by its square root.

Example: What is the luminosity of a star four times the mass of the sun? *Solution:* The star must be 128 times more luminous than the sun because

$$L = M^{3.5} = 4^{3.5} = 4 \cdot 4 \cdot 4\sqrt{4} = 64 \cdot 2 = 128$$

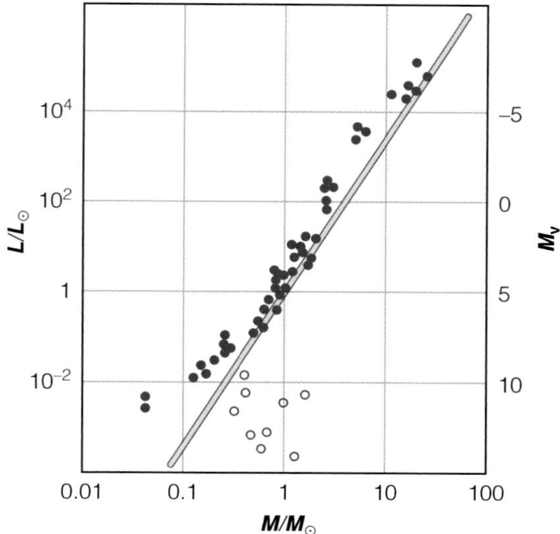

Figure 8-21

The mass–luminosity relation shows that the more massive a main-sequence star is, the more luminous it is. The open circles represent white dwarfs, which do not obey the relation. The red line represents the equation in Reasoning with Numbers 8-5.

ble even at great distances. Many of the brighter stars in the sky are highly luminous stars that you see even though they lie far away.

The night sky is a beautiful carpet of stars, but they are not all the same. Some are giants and supergiants, and some are dwarfs. The family of the stars is rich in its diversity.

Inquire I Review I Analyze

What kind of stars do you see if you look at a few of the brightest stars in the sky?

When you look at the night sky, the brightest stars are mostly giants and supergiants. Most of the bright stars in Canis Major, for instance, are supergiants. Sirius, the brightest star in the sky, is in Canis Major, but it is a main-sequence star; it looks bright because it is very nearby, not because it is very luminous. In general, the supergiants and giants are so luminous that they stand out and look bright, even though they are not nearby. When you look at a bright star in our sky, you are probably looking at a highly luminous star—a supergiant or a giant.

What kind of star do you see if you look at a few of the nearest stars to the sun? You can check these answers by consulting the tables of the brightest and nearest stars in the Appendix.

■ ■ ■

Connections: This chapter set out to find the basic properties of stars. Once you found the distance to the stars, you were able to find their luminosities, diameters, and masses—rather mundane data. But you have now discovered a puzzling situation. The largest and most luminous stars are so rare you might joke that they hardly exist, and the average stars are such dinky low-mass things they are hard to see even if they are near Earth in space. Why does nature make stars in this peculiar way? To answer that question, you must explore the birth, life, and death of stars. That quest begins in the next chapter.

The Family of Stars

What is the most common kind of star? Are some rare? Are some common? To answer those questions you must survey the stars. To do so you must know their spectral class, their luminosity class, and their distance. Your census of the family of stars produces some surprising demographic results.

You could survey the stars by observing every star within 62 pc of Earth. Such a survey would tell you how many stars of each type are found within a volume of a million cubic parsecs.

A sphere 62 pc in radius encloses a million cubic parsecs.

Your survey faces two problems.

1. The most luminous stars are so rare you find few in your survey region. There are no O stars at all within 62 pc of Earth.

2. Lower-main-sequence M stars, called red dwarfs, and white dwarfs are so faint they are hard to locate even when they are only a few parsecs from Earth. Finding every one of these stars in your survey sphere is a difficult task.

In this chart showing most of the constellation Canis Major, stars are represented as dots with colors assigned according to spectral class. The brightest stars in the sky tend to be the rare, highly luminous stars, which look bright even though they are far away. Most stars are of very low luminosity, so nearby stars tend to be very faint red dwarfs.

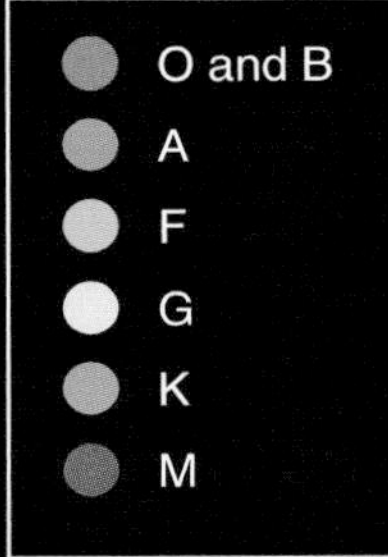

Sirius A (α Canis Majoris) is the brightest star in the sky. With a spectral type of A1V, it is not a very luminous star. It looks bright because it is only 2.6 pc away.

Sirius B is a white dwarf that orbits Sirius A. Although Sirius B is not very far away, it is much too faint to see with the unaided eye.

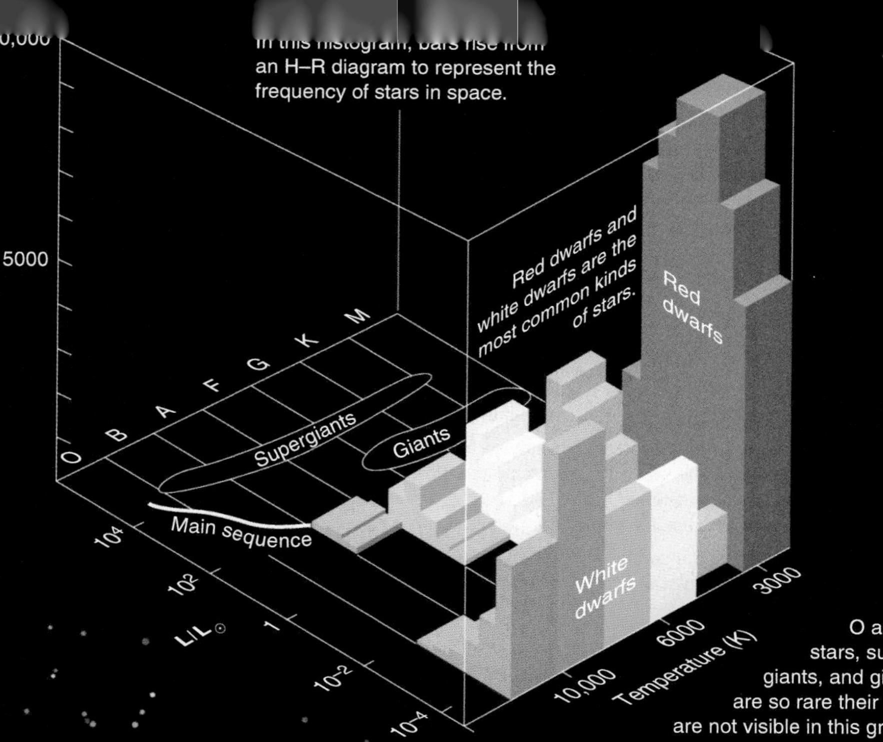

In this histogram, bars rise from an H–R diagram to represent the frequency of stars in space.
10,000
5000
Stars per 10^6 pc^3
O B A F G K M
Supergiants
Giants
Main sequence
Red dwarfs and white dwarfs are the most common kinds of stars.
Red dwarfs
White dwarfs
10^4
10^2
$L/L_\odot$
1
10^{-2}
10^{-4}
10,000
6000
3000
Temperature (K)
O and B stars, supergiants, and giants are so rare their bars are not visible in this graph

The universe is filled with uncountable stars, but their family relationships tell you how they were born and how they will die.

Visual-wavelength image

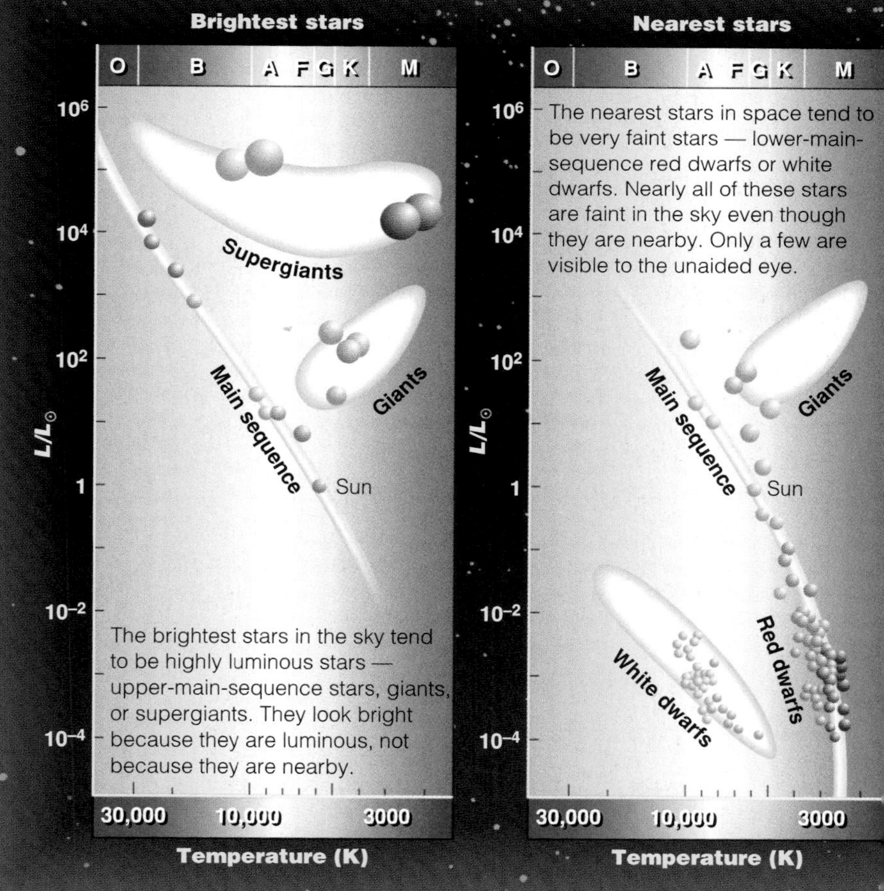

Brightest stars
O B A F G K M
10^6
10^4
10^2
1
10^{-2}
10^{-4}
$L/L_\odot$
Supergiants
Giants
Main sequence
Sun
The brightest stars in the sky tend to be highly luminous stars — upper-main-sequence stars, giants, or supergiants. They look bright because they are luminous, not because they are nearby.
30,000
10,000
3000
Temperature (K)
Nearest stars
O B A F G K M
The nearest stars in space tend to be very faint stars — lower-main-sequence red dwarfs or white dwarfs. Nearly all of these stars are faint in the sky even though they are nearby. Only a few are visible to the unaided eye.
Main sequence
Giants
Sun
Red dwarfs
White dwarfs

Although white dwarfs and red dwarfs are very common, not a single one is bright enough to be visible to the unaided eye

Summary

8-1 | Measuring the Distances to Stars

How far away are the stars?

- Distance is critical in astronomy. Your goal in this chapter was to characterize the stars by finding their luminosities, diameters, and masses. Before you could begin, you needed to find their distances. Only by first knowing the distance to a star could you find its other properties.
- Astronomers can measure the distance to nearer stars by observing their parallaxes. The most distant stars are so far away that their parallaxes are unmeasurably small. To find the distances to these stars, astronomers must use spectroscopic parallax.
- Stellar distances are commonly expressed in parsecs. One parsec is 206,265 AU—the distance to an imaginary star whose parallax is 1 second of arc.

8-2 | Intrinsic Brightness

How much energy do stars make?

- Once you know the distance to a star, you can find its intrinsic brightness expressed as its absolute magnitude. The absolute magnitude of a star equals the apparent magnitude it would have if it were 10 pc away.
- The luminosity of a star, found from its absolute magnitude, is a measure of the total energy radiated by the star in one second. Luminosity is often expressed in terms of the luminosity of the sun.

8-3 | The Diameters of Stars

How big are stars?

- The H–R diagram is a plot of luminosity versus surface temperature. It is an important graph in astronomy because it sorts the stars into categories by size.
- Roughly 90 percent of normal stars, including the sun, fall on the main sequence, with the more massive stars being hotter and more luminous. The giants and supergiants, however, are much larger and lie above the main sequence. Some of the white dwarfs are hot stars, but they fall below the main sequence because they are so small.
- The large size of the giants and supergiants means their atmospheres have low densities and their spectra have sharper spectral lines than the spectra of main-sequence stars. In fact, it is possible to assign stars to luminosity classes by the widths of their spectral lines. Class V stars are main-sequence stars. Giant stars, class III, have sharper lines, and supergiants, class I, have extremely sharp spectral lines.

8-4 | The Masses of Stars

How much matter do stars contain?

- The only direct way you can find the mass of a star is by studying binary stars. When two stars orbit a common center of mass, astronomers find their masses by observing the period and sizes of their orbits.
- Given the mass and diameter of a star, you can find its average density. On the main sequence, the stars are about as dense as the sun, but the giants and supergiants are very low-density stars. Some are much thinner than air. The white dwarfs, lying below the main sequence, are tremendously dense.
- The mass–luminosity relation says that the more massive a star is, the more luminous it is. Main-sequence stars follow this rule closely, the most massive being the upper-main-sequence stars and the least massive the lower-main-sequence stars. Giants and supergiants do not follow the relation precisely, and white dwarfs not at all.

8-5 | A Survey of the Stars

What is the typical star like?

- A survey in the neighborhood of the sun shows that lower-main-sequence stars are the most common type. Giants and supergiants are rare, but white dwarfs are quite common, although they are faint and hard to find.

New Terms

stellar parallax (*p*)(p. 147)
parsec (pc) (p. 147)
flux (p. 149)
absolute visual magnitude (M_v) (p. 149)
luminosity (*L*) (p. 149)
Hertzsprung–Russell (H–R) diagram (p. 151)
main sequence (p. 152)
giant (p. 152)
supergiant (p. 152)
red dwarf (p. 153)
white dwarf (p. 153)
luminosity class (p. 153)
spectroscopic parallax (p. 154)
binary stars (p. 156)
visual binary system (p. 156)
spectroscopic binary system (p. 158)
eclipsing binary system (p. 160)
light curve (p. 160)
mass–luminosity relation (p. 162)

Review Questions

Ace Astronomy™ Assess your understanding of this chapter's topics with additional quizzing and animations at **http://astronomy.brookscole.com/sh9e**

1. Why are Earth-based parallax measurements limited to the nearest stars?
2. Why was the Hipparcos satellite able to make more accurate parallax measurements than ground-based telescopes?
3. What do the words *absolute* and *visual* mean in the definition of absolute visual magnitude?
4. What does luminosity measure that is different from what absolute visual magnitude measures?
5. Why does the luminosity of a star depend on both its radius and its temperature?
6. How can you be sure that giant stars really are larger than main-sequence stars?
7. Why do astronomers conclude that white dwarfs must be very small?
8. What observations would you make to classify a star according to its luminosity? Why does that method work?
9. Why does the orbital period of a binary star depend on its mass?
10. What observations would you make to study an eclipsing binary star?
11. Why don't astronomers know the inclination of a spectroscopic binary? How do they know the inclination of an eclipsing binary?
12. How do the masses of stars along the main sequence illustrate the mass–luminosity relation?
13. Why is it difficult to find out how common the most luminous stars are? The least luminous stars?

14. What is the most common kind of star?
15. If you look only at the brightest stars in the night sky, what kind of star are you likely to be observing? Why?
16. If all of the stars in the photo here are members of the same star cluster, then they all have about the same distance. Then why are three of the brightest much redder than the rest? What kind of star are they?

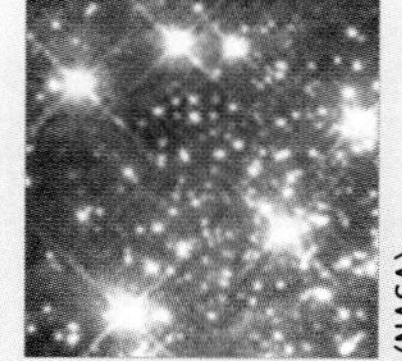
(NASA)

Discussion Questions

1. If someone asked you to compile a list of the nearest stars to the sun based on your own observations, what measurements would you make, and how would you analyze them to detect nearby stars?
2. The sun is sometimes described as an average star. Is that true? What is the average star really like?

Problems

1. If a star has a parallax of 0.050 second of arc, what is its distance in pc? in ly? in AU?
2. If you place a screen of area 1 m^2 at a distance of 2.8 m from a 100-watt lightbulb, the light flux falling on the screen will be 1 J/s. To what distance must you move the screen to make the flux striking it equal 0.01 J/s? (This assumes the lightbulb emits all of its energy as light.)
3. If a star has a parallax of 0.016 second of arc and an apparent magnitude of 6, how far away is it, and what is its absolute magnitude?
4. Complete the following table:

m	M_v	*d* (pc)	*P* (seconds of arc)
____	7	10	____
11	____	1000	____
____	−2	____	0.025
4	____	____	0.040

5. The unaided human eye can see stars no fainter than those with an apparent magnitude of 6. If you can see a bright firefly blinking up to 0.5 km away, what is the absolute magnitude of the firefly? (*Hint:* Convert the distance to parsecs and use the formula in Reasoning with Numbers 8-2.)
6. If a main-sequence star has a luminosity of 100 $L_\odot$, what is its spectral type? (*Hint:* See Figure 8-8.)
7. If a star is 10 times the radius of the sun and half as hot, what will its luminosity be? (*Hint:* See Reasoning with Numbers 8-3.)
8. A B0 V star has an apparent magnitude of +1. Use the method of spectroscopic parallax to find the distance to the star. Why might this distance be inaccurate?
9. Find the luminosity and spectral type of a 5-$M_\odot$ main-sequence star.
10. In the following table, which star is brightest in apparent magnitude? most luminous? largest? least dense? farthest away?

Star	Spectral Type	*m*
a	G2 V	5
b	B1 V	8
c	G2 Ib	10
d	M5 III	19
e	White dwarf	15

11. If two stars orbit each other with a period of 6 years and a separation of 4 AU, what is their total mass? (*Hint:* See Reasoning with Numbers 8-4.)
12. If the eclipsing binary in Figure 8-17 has a period of 32 days, an orbital velocity of 153 km/s, and an orbit that is nearly edge-on, what is the circumference of the orbit? the radius of the orbit? the mass of the system?
13. If the orbital velocity of the eclipsing binary in Figure 8-17 is 153 km/s and the smaller star becomes completely eclipsed in 2.5 hours, what is its diameter?
14. What is the luminosity of a 4-solar-mass star? of a 9-solar-mass star? of a 7-solar-mass star?

Media Cluster

Ace Astronomy™ To access the resources in the Media Cluster, log into AceAstronomy at **http://astronomy.brookscole.com/sh9e** and select Chapter 8.

ACTIVE FIGURES

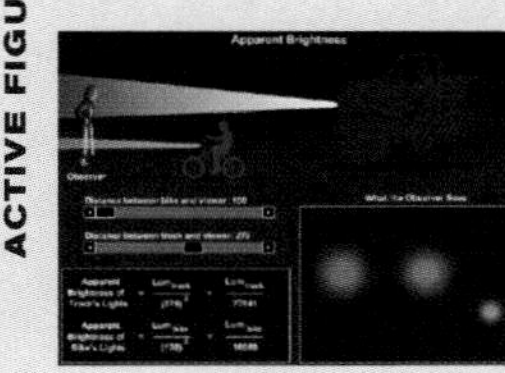

Brightness and Distance
In this animation, you can change the position of both the truck and bicyclist in the figure and observe the effect on the intensity of light seen by the observer. Notice how distant sources of light appear fainter than nearby sources.

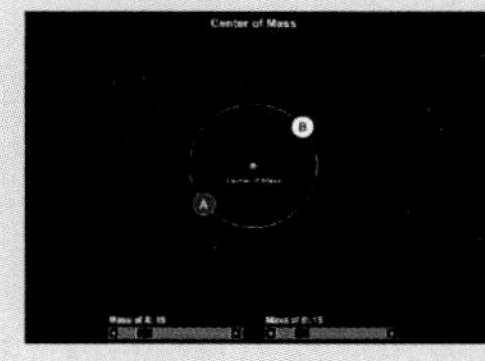

Center of Mass
We often think in terms of the moon orbiting Earth; but, more precisely speaking, both bodies orbit each other around a common center of mass. Experiment with this animation to see how the ratio of the masses affects the relative size of the orbits.

ASTRONOMY EXERCISES

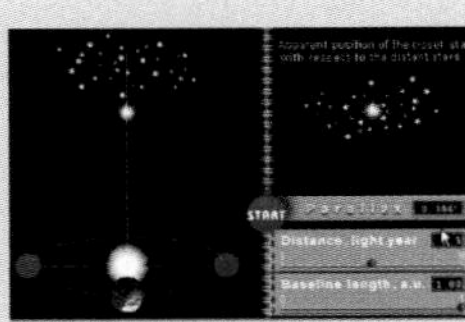

Parallax I and II
The apparent change in position of an object due to a change in the location of the observer is known as *parallax*. Use these two animations to study the concept of parallax in astronomy.

Apparent Brightness and Distance
The distance to stars affects how bright they look to us on Earth. You can change the distance to a star in this animation and see its apparent brightness change.

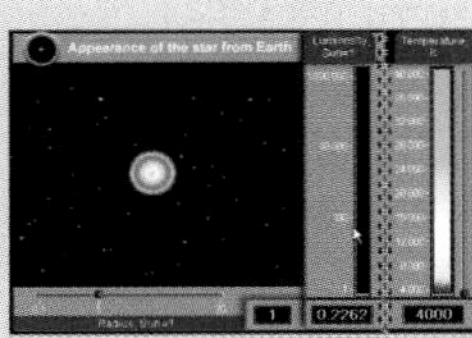

Stefan–Boltzmann Law II
In this exercise, you can vary the temperature of a star and its radius. As you change the temperature and radius, notice what happens to the total energy output of the star as indicated by the luminosity meter.

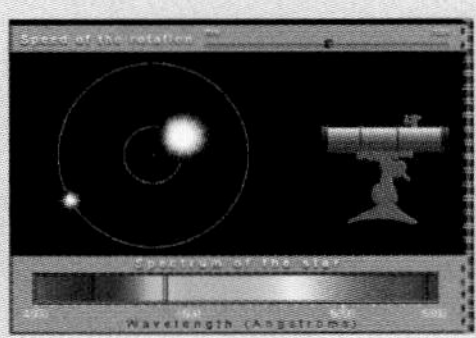

Spectroscopic Binaries
This animation shows a close double star system where each star orbits a common center of gravity. As the stars change position, notice what happens to the absorption lines in the spectrum of the star system.

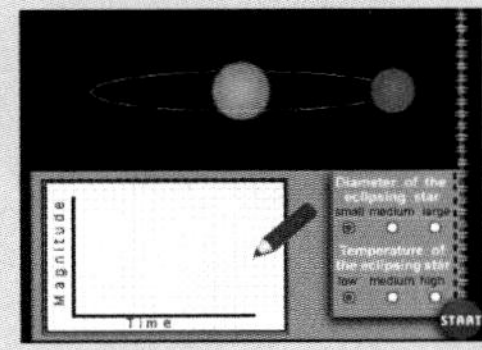

Eclipsing Binaries
Study an eclipsing binary system with this animation. As the stars move and periodically pass in front of one another, the total light output from the system is charted on a graph.

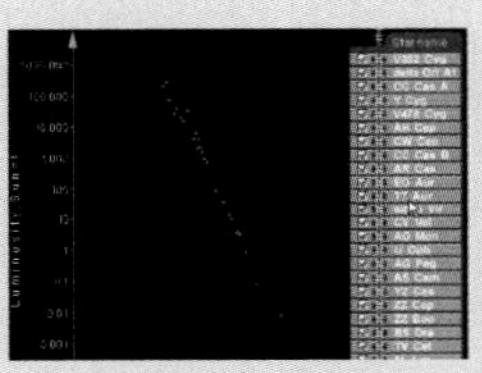

Mass–Luminosity Relation
This exercise provides a list of stars whose masses and luminosities have been measured. Click on each star to plot its point on the mass vs. luminosity axes.

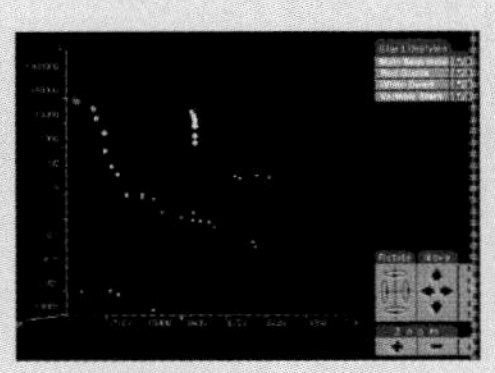

H–R/Mass–Luminosity 3D Graph
Explore this 3D graph and see how mass is related to position in the H–R diagram. You can select from main-sequence stars, red giants, white dwarfs, and variable stars.

VIRTUAL ASTRONOMY LABS

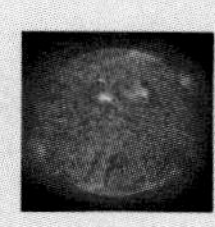

Lab 11: The Spectral Sequence and the H–R Diagram This lab introduces the tools astronomers use to identify the stars of the main sequence and how that information is used to estimate the distance to the stars and the age of a star cluster.

Lab 12: Binary Stars This lab investigates some things that can be learned from different types of binary stars. Newton's form of Kepler's third law is used to determine stellar masses. You also see how the Doppler effect can be used in the study of binary systems.

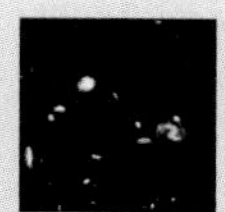

Lab 16: Astronomical Distance Scales In this lab you explore some methods for determining distances in astronomy.

Critical Inquiries for the Web

1. Hertzsprung–Russell diagrams allow astronomers to represent the distribution of stellar properties. Examine H–R diagrams at various websites and determine which types of stars are the most (and least) numerous in our galaxy.
2. What if Algol were oriented such that its stars did not eclipse each other as seen from Earth? Use the Internet to find information about Algol and binary stars in general and discuss how astronomers could still determine that it is a multiple-star system.

Exploring *TheSky*

1. Locate the following stars and determine their apparent magnitude, parallax, distance in parsecs and in light-years, and spectral classification: Sirius, Aldebaran, Vega, Deneb, Betelgeuse, Antares, and Altair. (*Hint:* To center on an object, use **Find** under the **Edit** menu and type the object's name followed by a period.)
2. Use the spectral type and parallax of the stars above to estimate their distance from Earth. Compare with distances given in *TheSky*.
3. Take a survey of the stars. Center on Orion and adjust the field until it is about 100° wide. Click on the ten brightest stars and record the spectral types. Now zoom in until only a few dozen stars are in the frame. Click on the 10 faintest stars and record their spectral types. Is there a difference between the brightest and faintest stars? Is the result what you expected? Are certain kinds of stars missing from the computer data base?
4. Repeat exercise 3 for a region centered on the Big Dipper.

- Go to the Brooks/Cole Astronomy Resource Center **(http://astronomy.brookscole.com)** for critical thinking exercises, articles, and additional readings from InfoTrac College Edition, Brooks/Cole's online student library.

9 The Formation and Structure of Stars

Jim he allowed [the stars] was made, but I allowed they happened. Jim said the moon could'a laid them; well, that looked kind of reasonable, so I didn't say nothing against it, because I've seen a frog lay most as many, so of course it could be done.

MARK TWAIN
The Adventures of Huckleberry Finn

Visual + infrared image

STARS EXIST BECAUSE of gravity. They form because gravity makes clouds of gas contract. Once they form, stars spend long, stable lives generating nuclear energy at their centers and balancing their own gravity with the pressure of the hot gas in their interiors. In the end, stars die because they exhaust their fuel supply and can no longer withstand the force of their own gravity. ▌ A star can remain stable only by maintaining great gas pressure in its interior. Gravity tries to make the stellar gas contract, but, if the internal temperature is high enough, the pressure of the gas pushes outward just enough to balance gravity. Thus, a star is a battlefield where gas pressure and gravity struggle for dominance. If gas pressure wins, the star must expand; but, if gravity wins, the star must contract.

The Great Nebula in Orion is filled with bright, blue, newborn stars; but, at infrared wavelengths, we see a cooler cloud of gas (red) where more stars are being born. (ESO)

Guidepost

Looking Back

Throughout the preceding chapters you have been building a general understanding of the sky and a facility with the tools of astronomy. The last chapter provided a detailed family portrait of the stars. You can now describe the basic properties of different kinds of stars.

This Chapter

This is the place where you will really begin to see how the universe works. Here you will begin putting together observations and theories to understand how nature makes stars. That will answer five essential questions about stars:

How are stars born?

How do stars make energy?

How do stars maintain their stability?

How long can a star survive?

What evidence do astronomers have that theories of star formation are correct?

Looking Ahead

The most important of the questions above may be the last. Testing theories against evidence is the basic skill required of all scientists, and you will use it over and over in the 11 chapters that follow.

AceAstronomy™ The AceAstronomy icon throughout the text indicates an opportunity for you to test yourself on key concepts and to explore animations and interactions on the AceAstronomy website at: http://astronomy.brookscole.com/sh9e

Only by generating tremendous amounts of nuclear energy can a star keep its interior hot enough to maintain the gravity–pressure balance. The sun, for example, generates enough energy in 1 second to power Earth's civilization at its current level for 2 million years.

To understand how a star forms, you must understand how a star generates its energy and how that energy eventually escapes from the star. Thus, you must explore the nuclear reactions that generate energy in the core of a star, and you must explore the star's internal structure, which resists the inward pull of the star's gravity. These explorations will allow you to imagine the slow evolution of stable stars over billions of years.

How can astronomers know what stars are like when they can't see inside them and don't live long enough to see them evolve? The answer lies in the methods of science. By constructing theories that describe how nature works and then testing those theories against evidence from observations, astronomers can unravel some of nature's greatest secrets. They can even understand the origin and structure of the stars.

9-1 The Birth of Stars

THE KEY TO UNDERSTANDING star formation is the correlation between young stars and clouds of gas and dust. Where you find the youngest groups of stars, you also find large clouds of gas illuminated by the hottest and brightest of the new stars (■ Figure 9-1). This should lead you to suspect that stars form from such clouds, just as raindrops condense from the water vapor in a thundercloud. To study the formation of stars, you must examine these cold clouds of gas that float between the stars and understand how these clouds contract, heat up, and become stars.

The Interstellar Medium

Astronomers can describe the gas and dust between the stars, the **interstellar medium,** in some detail. About 75 percent of the mass of the gas is hydrogen, and 25 percent is helium; there are also traces of carbon, nitrogen, oxygen, calcium, sodium, and heavier

■ Figure 9-1

Young stars are found in clouds of gas and dust from which they have been born. The nebula N44 is 170,000 ly from Earth in a nearby galaxy, and the Horsehead nebula is only about 1500 ly distant in our own galaxy. Gas in both nebulae is excited to glow by hot, young stars, and dust is visible as dark, twisting clouds seen against the bright background gas. (N44: ESO; Horsehead Nebula: NOAO and Nigel Sharp)

atoms. Roughly 1 percent of the mass is made up of microscopic dust about the size of the particles in cigarette smoke. The dust seems to be made mostly of carbon and silicates (rocklike minerals) mixed with or coated with frozen water. The average distance between dust grains is about 150 m.

This interstellar material is not uniformly distributed through space; it consists of a complex tangle of cool, dense clouds pushed and twisted by currents of hot, low-density gas. Although the cool clouds contain only 10 to 1000 atoms/cm^3 (fewer than any vacuum on Earth), astronomers refer to them as dense clouds in contrast with the hot, low-density gas that fills the spaces between clouds. That thin gas contains only about 0.1 atom/cm^3.

The preceding two paragraphs describe the interstellar medium, but you should not accept these facts blindly. Science is based on evidence, so you should demand to know what observational evidence supports these facts. How do astronomers know there is an interstellar medium, and how do they know its properties?

In some cases, the interstellar medium is easily visible as clouds of gas and dust. Astronomers call such a cloud a **nebula** from the Latin word for cloud. Such nebulae (plural) give clear evidence of an interstellar medium. Study **Three Kinds of Nebulae** on pages 176 and 177 and notice three important points. First, very hot stars can excite clouds of gas and dust to emit light, and the spectra of such nebulae show that the gas is mostly hydrogen at low densities. Second, where slightly cooler stars illuminate clouds, light is reflected from dust in a way that reveals that the dust particles are very small. The third thing to notice is that some dense clouds of gas and dust are detectable where they are silhouetted against background regions filled with stars or bright nebulae.

If a cloud is less dense, the starlight may be able to penetrate it, and stars can be seen through the cloud; but the stars look dimmer because the dust in the cloud scatters some of the light. Because shorter wavelengths are scattered more easily than longer wavelengths, the redder photons are more likely to make it through the cloud, and the stars look slightly redder than they should—an effect called **interstellar reddening** (■ Figure 9-2). (This is the same process that makes the setting sun look redder.) Distant stars are dimmed and reddened by intervening gas and dust, clear evidence of an interstellar medium. At near-infrared wavelengths, stars are more easily seen through the dusty interstellar medium because those longer wavelengths are scattered less often (Figure 9-2).

The dust in space is called **interstellar dust**, and it makes up roughly 1 percent of the mass of the interstellar medium. From the way the dust scatters different wavelengths, astronomers can determine that it is made up of very small particles of silicates, carbon, and other heavy elements such as iron with coatings of ice.

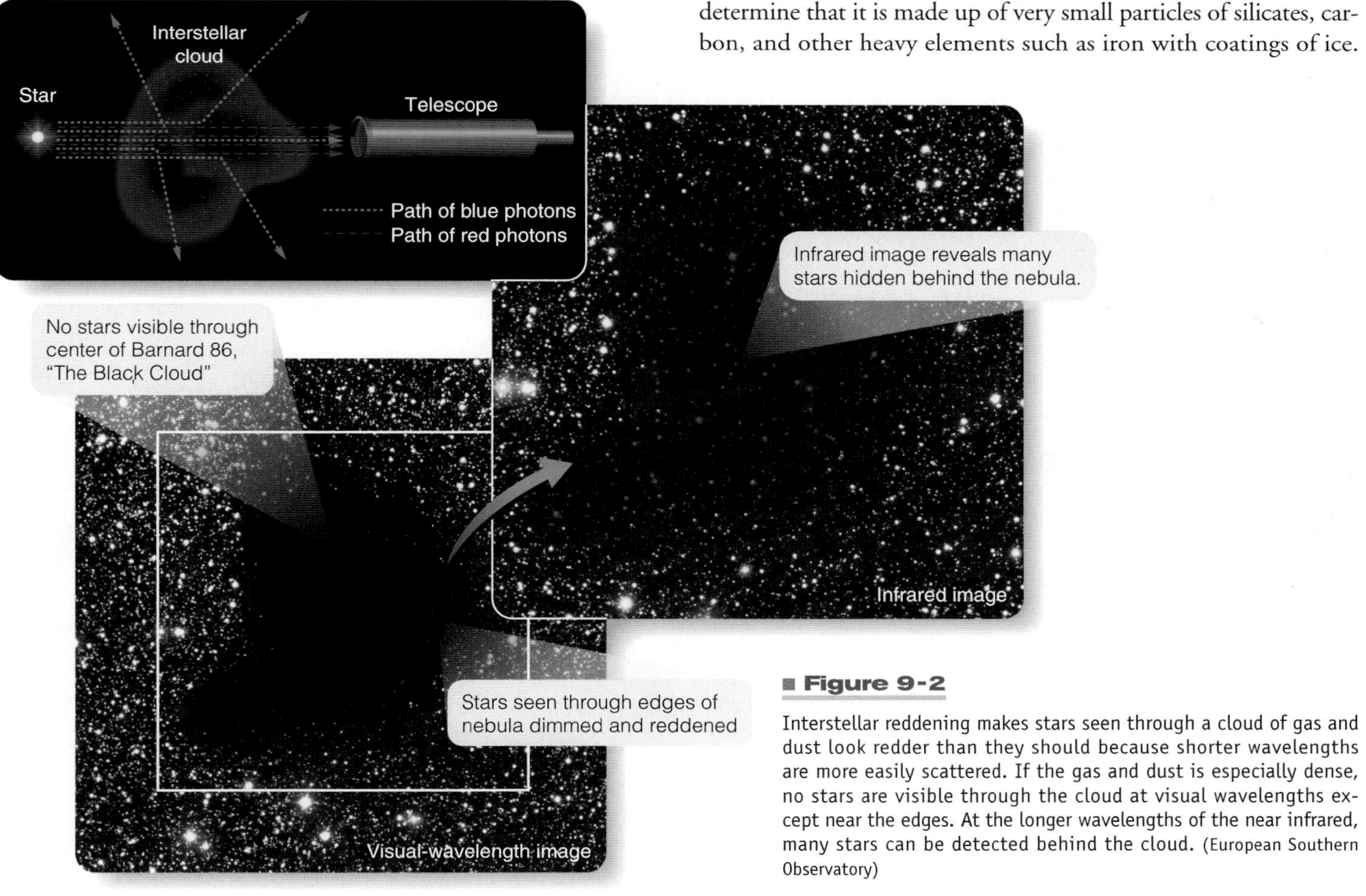

■ **Figure 9-2**

Interstellar reddening makes stars seen through a cloud of gas and dust look redder than they should because shorter wavelengths are more easily scattered. If the gas and dust is especially dense, no stars are visible through the cloud at visual wavelengths except near the edges. At the longer wavelengths of the near infrared, many stars can be detected behind the cloud. (European Southern Observatory)

Three Kinds of Nebulae

At visual wavelengths, astronomers see three kinds of nebulae. The differences between these clouds of gas and dust provide insights into the interstellar medium.

Emission nebulae are produced when a hot star excites the gas near it to produce an emission spectrum. The star must be hotter than about B1 (25,000 K). Cooler stars do not emit enough ultraviolet radiation to ionize the gas. Emission nebulae have a distinctive pink color produced by the blending of the red, blue, and violet Balmer lines.

Emission nebulae are also called **HII regions**, following the custom of naming gas with a roman numeral to show its state of ionization. HI is neutral hydrogen, and HII is ionized.

In an HII region, the ionized nuclei and free electrons are mixed. When a nucleus captures an electron, the electron falls down through the atomic energy levels, emitting photons at specific wavelengths. Spectra indicate that the nebulae have compositions much that like of the sun – mostly hydrogen.

Emission nebulae have densities of 100 to 1000 atoms per cubic centimeter, better than the best vacuums produced in laboratories on Earth.

Visual-wavelength image

European Southern Observatory

A **reflection nebula** is produced when starlight scatters from a dusty nebula. Consequently, the spectrum of a reflection nebula is just the reflected absorption spectrum of starlight. Gas is surely present in a reflection nebula, but it is not excited to emit photons.

Reflection nebulae look blue for the same reason the sky looks blue. Short wavelengths scatter more easily than long wavelengths.

Ace Astronomy™

Log into AceAstronomy and select this chapter to see Active Figure "Scattering." Watch photons scatter through Earth's atmosphere.

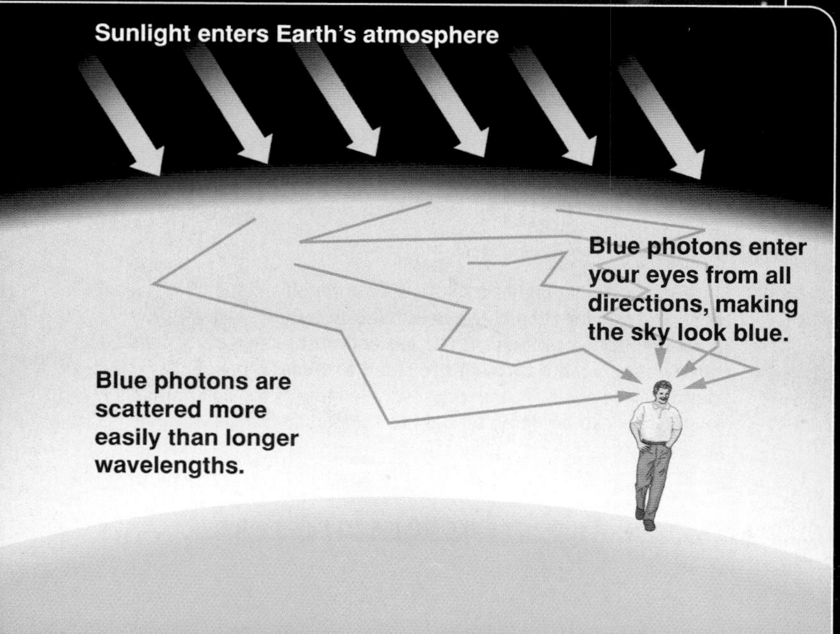

Sunlight enters Earth's atmosphere

Blue photons enter your eyes from all directions, making the sky look blue.

Blue photons are scattered more easily than longer wavelengths.

Reflection nebulae NGC 1973, 1975, and 1977 lie just north of the Orion nebula. The pink tints are produced by ionized gases deep in the nebulae.

Visual-wavelength image

AATB

The blue color of reflection nebulae shows that the dust particles must be very small in order to preferentially scatter the blue photons. Interstellar dust grains must have diameters ranging from 0.001 mm down to 1 nm or so.

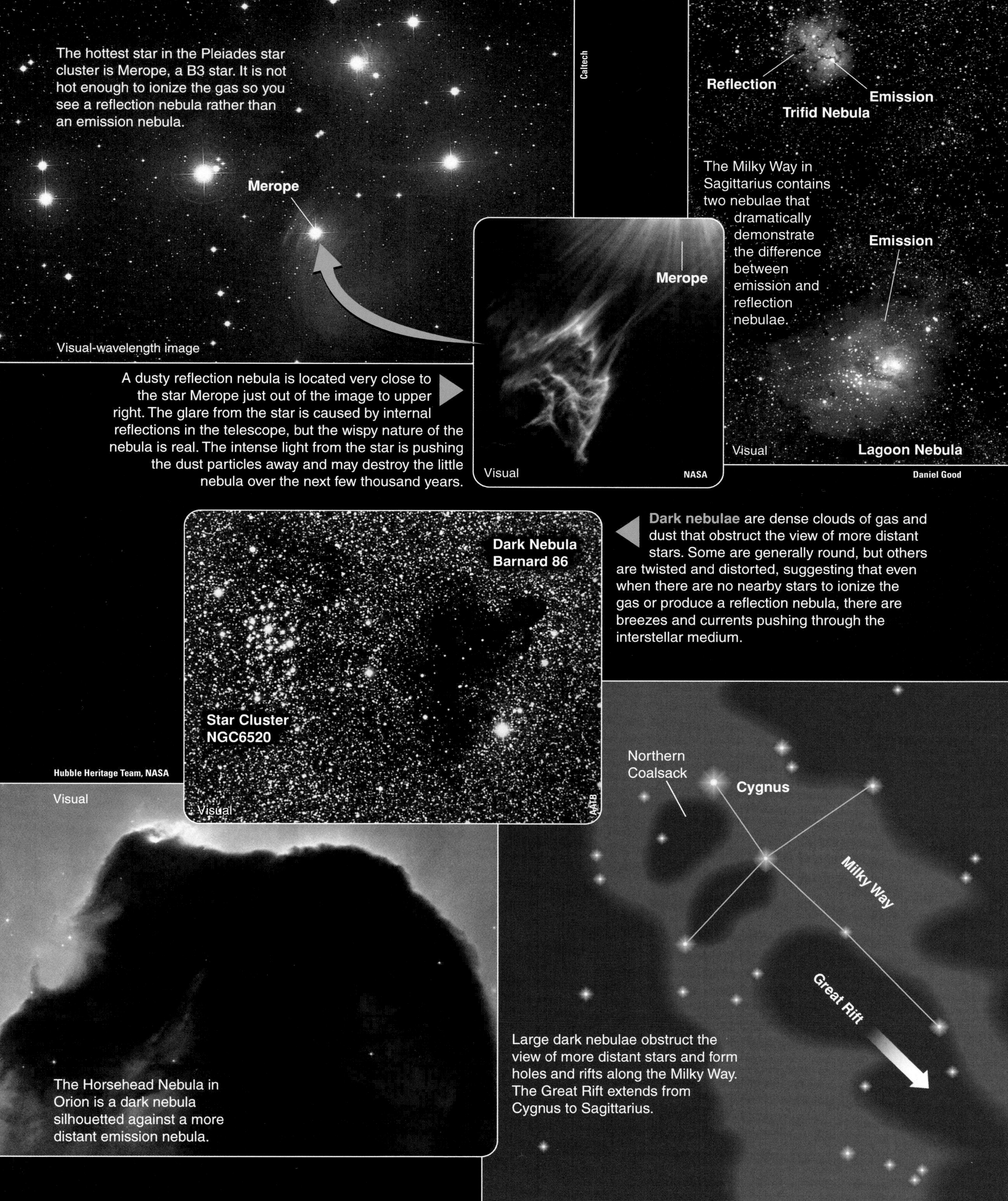

The hottest star in the Pleiades star cluster is Merope, a B3 star. It is not hot enough to ionize the gas so you see a reflection nebula rather than an emission nebula.

The Milky Way in Sagittarius contains two nebulae that dramatically demonstrate the difference between emission and reflection nebulae.

A dusty reflection nebula is located very close to the star Merope just out of the image to upper right. The glare from the star is caused by internal reflections in the telescope, but the wispy nature of the nebula is real. The intense light from the star is pushing the dust particles away and may destroy the little nebula over the next few thousand years.

Dark nebulae are dense clouds of gas and dust that obstruct the view of more distant stars. Some are generally round, but others are twisted and distorted, suggesting that even when there are no nearby stars to ionize the gas or produce a reflection nebula, there are breezes and currents pushing through the interstellar medium.

The Horsehead Nebula in Orion is a dark nebula silhouetted against a more distant emission nebula.

Large dark nebulae obstruct the view of more distant stars and form holes and rifts along the Milky Way. The Great Rift extends from Cygnus to Sagittarius.

Although you have been thinking of individual nebulae, the thin gas and dust of the interstellar medium fills the spaces between the nebulae. You can see evidence of that because the gas forms interstellar absorption lines in the spectra of distant stars (■ Figure 9-3a). As starlight travels through the interstellar medium, gas atoms of elements such as calcium and sodium absorb photons of certain wavelengths, producing narrow interstellar absorption lines. You can be sure these lines originate in the interstellar medium because they appear in the spectra of O and B stars—stars that are too hot to form calcium and sodium absorption lines in their own atmospheres. Also, the narrowness of the interstellar lines indicates they could not have been formed in the hot atmospheres of the stars. In a hot gas, the atoms move rapidly, and the Doppler shifts of the different atoms smear the spectral line, making it broad. The interstellar medium is very cold, 10 to 50 K, so the atoms are not moving very rapidly at all, and they form extremely sharp spectral lines. In fact, the narrow widths of the spectral lines are clear evidence of the extremely low temperature of the interstellar gas. Often multiple interstellar lines appear because the light from the star passed through a number of clouds on its way to Earth (Figure 9-3b).

Observations at nonvisible wavelengths provide valuable evidence about the interstellar medium. Although the interstellar dust is very cold and makes up only about 1 percent of the interstellar medium, the dust can radiate in the far-infrared. A single grain, cold and tiny, can radiate only a small amount of infrared radiation, but the interstellar medium contains vast quantities of such dust and can be studied in the far-infrared. For example, the Infrared Astronomy Satellite discovered a faint, wispy network of dusty clouds covering the entire sky now known as the **infrared cirrus** (■ Figure 9-4a). In contrast, X-ray observations can detect

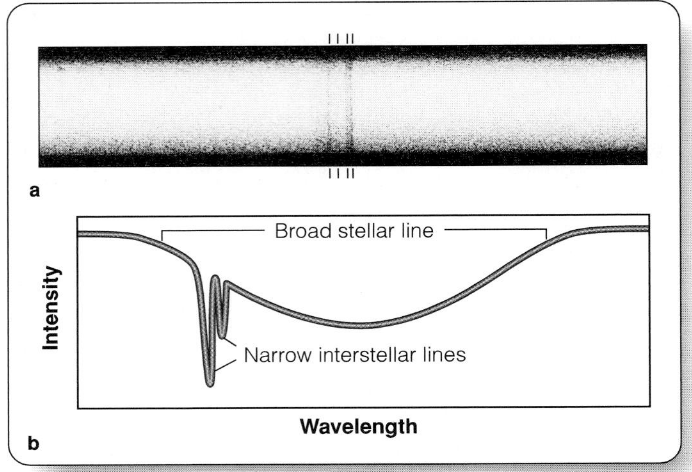

■ **Figure 9-3**

Interstellar absorption lines can be recognized in two ways. (a) The B0 supergiant ε Orionis is much too hot to show spectral lines of once-ionized calcium (CaII), yet this short segment of its spectrum reveals narrow, multiple lines of CaII (tic marks) that must have been produced in the interstellar medium. (The Observatories of the Carnegie Institution of Washington) (b) Spectral lines produced in the atmospheres of stars are much broader than the spectral lines produced in the interstellar medium. (Adapted from a diagram by Binnendijk) In both (a) and (b) the multiple interstellar lines are produced by separate interstellar clouds with slightly different radial velocities.

■ **Figure 9-4**

Infrared and X-ray observations tell about different components of the interstellar medium. (a) This far-infrared image shows the infrared cirrus, wispy clouds of very cold dusty gas spread all across the sky. (NASA/IPAC, Courtesy Deborah Levine) (b) This ROSAT X-ray image shows the Cygnus Superbubble, a cavity filled with gas from exploding stars. The gas is heated and emits X rays where the expanding bubble pushes into the surrounding gas. (Courtesy Steve Snowden and Max-Planck Institute for Extraterrestrial Physics, Germany)

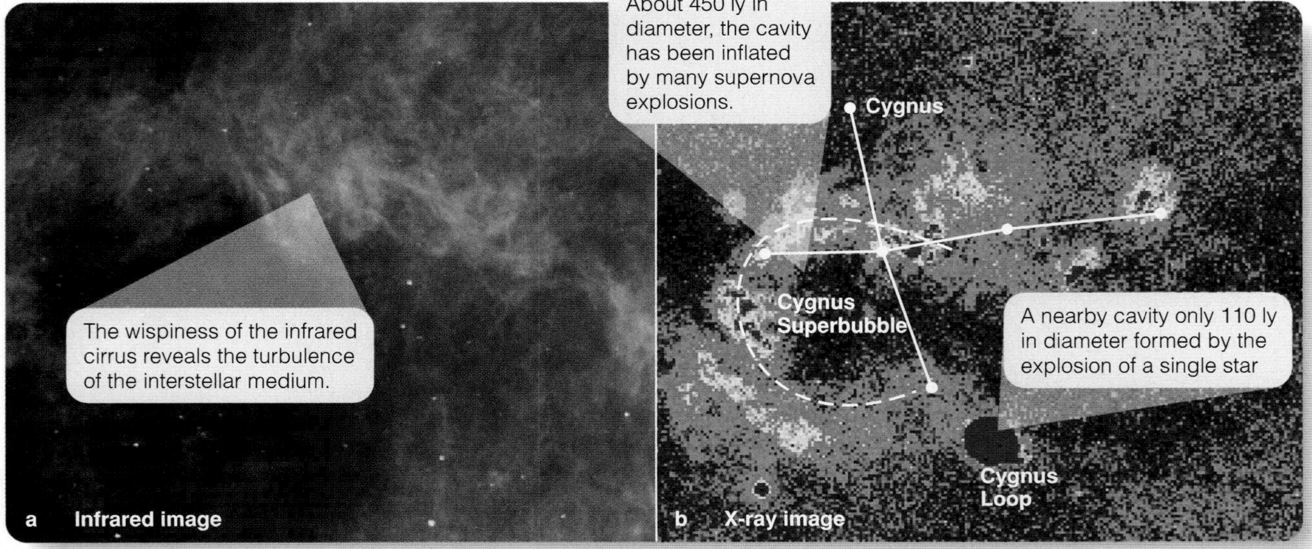

regions of very hot gas apparently produced by exploding stars (Figure 9-4b). Ultraviolet observations have been able to map the distribution of hydrogen in space and reveal that the sun is located just inside a cavity filled with hot gas. Furthermore, radio astronomers can study the radio emissions of specific molecules in the interstellar medium—the equivalent of emission lines in visible light. Such studies show that some of the atoms in space have linked together to form molecules.

There is no shortage of evidence concerning the interstellar medium. Clearly, space is not empty, and from the correlation between young stars and clouds of gas and dust you can suppose that stars form from these clouds. But how can the icy cold interstellar clouds create hot stars? The answer is gravity.

AceAstronomy™ Log into AceAstronomy and select this chapter to see Astronomy Exercise "Why is the Sky Blue?" The sky is blue and the setting sun is red for the same reason that reflection nebulae are blue and distant stars are reddened. You can explore that process here.

The Formation of Stars from the Interstellar Medium

To study the formation of stars, you must continue comparing theory with evidence. The theory of gravity predicts that the combined gravitational attraction of the atoms in a cloud of gas will squeeze the cloud, pulling every atom toward the center. Thus, you might expect that every cloud would eventually collapse and become a star; however, the heat in the cloud resists collapse. Temperature is a measure of the motion of the atoms or molecules in a material—in a hot gas, the atoms move more rapidly than do those in a cool gas. The interstellar clouds are very cold, but even at a temperature of only 10 K, the average hydrogen atom moves about 0.5 km/s (1100 mph). This thermal motion would make the cloud drift apart if gravity were too weak to hold it together.

The densest interstellar clouds contain from 10^3 to 10^5 atoms/cm^3, include from a few hundred thousand to a few million solar masses, and have temperatures as low as 10 K. In such clouds hydrogen can exist as molecules (H_2) rather than as atoms. These dense clouds are called **molecular clouds,** and the largest are called giant molecular clouds. Although hydrogen molecules cannot be detected by radio telescopes, the clouds can be mapped by the emission of carbon monoxide molecules (CO) present in small amounts in the gas. Stars form in these clouds when the densest parts of the clouds become unstable and contract under the influence of their own gravity.

Most clouds do not appear to be gravitationally unstable, but a stable cloud colliding with a **shock wave** (the astronomical equivalent of a sonic boom) can be compressed and disrupted into fragments. Theoretical calculations show that some of these fragments can become dense enough to collapse and form stars (■ Figure 9-5).

Thus, most interstellar clouds may not collapse and form stars until they are triggered by the compression of a shock wave. Happily for this hypothesis, space is filled with shock waves. Supernova explosions (exploding stars described in the next chapter) produce shock waves that compress the interstellar medium, and recent observations show young stars forming at the edges of such shock waves (■ Figure 9-6). Another source of shock waves may

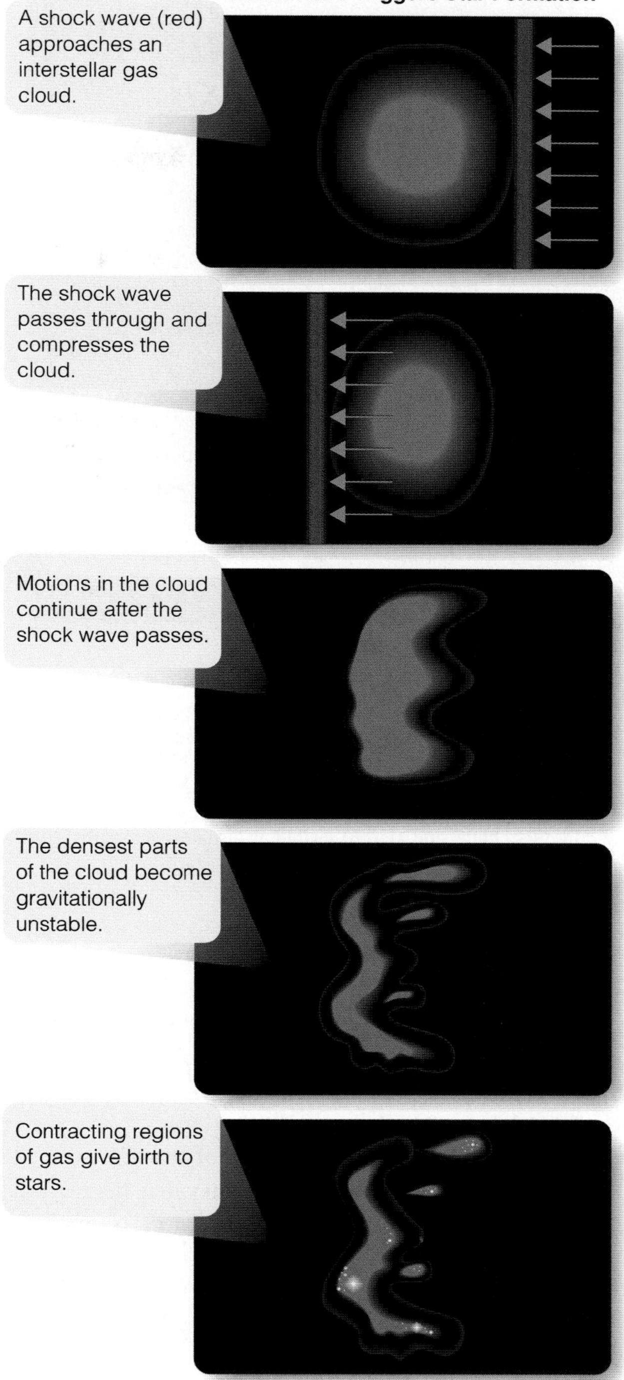

Shock Wave Triggers Star Formation
A shock wave (red) approaches an interstellar gas cloud.
The shock wave passes through and compresses the cloud.
Motions in the cloud continue after the shock wave passes.
The densest parts of the cloud become gravitationally unstable.
Contracting regions of gas give birth to stars.

■ **Figure 9-5**

In this summary of a computer model, an interstellar gas cloud is triggered into star formation by a passing shock wave. The events summarized here might span about 6 million years.

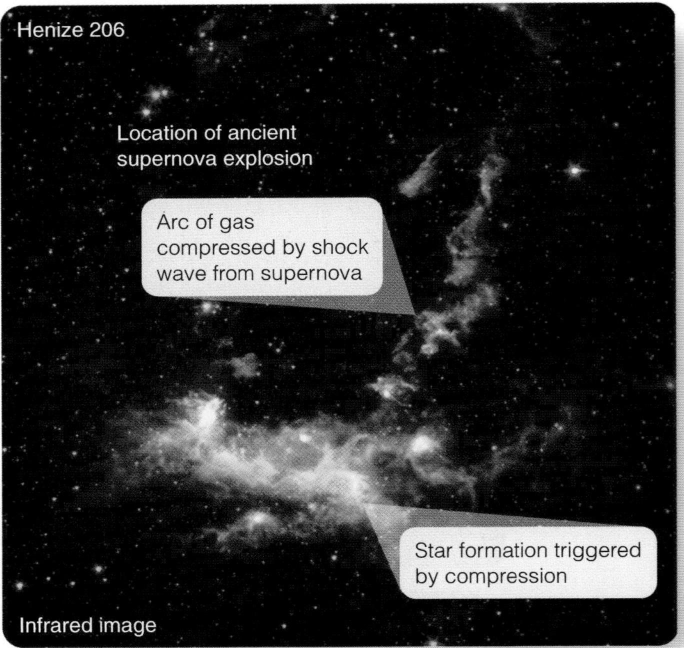

■ **Figure 9-6**

A few million years ago, a massive star exploded as a supernova in the upper left quarter of this image. The expanding shock wave compressed a nearby cloud of gas and triggered the birth of new stars. In this infrared image, you can see the arc of compressed gas left behind as the shock wave rushed through the gas cloud. Most of the bright, young stars in this nebula are hidden deep in dust clouds and are not visible at visual wavelengths. (NASA/JPL-Caltech/V. Gorjian, JPL & NOAO)

be the birth of very hot stars. A massive star is so luminous and hot that it emits vast amounts of ultraviolet photons. When such a star is born, the sudden blast of radiation can ionize and drive away nearby gas, forming a shock wave that could trigger further star formation. Even the collision of two interstellar clouds can produce a shock wave and trigger star formation.

Although these are important sources of shock waves, the dominant trigger of star formation in our galaxy may be the spiral pattern itself. In Chapter 1, you learned that our galaxy contains spiral arms; many astronomers believe the arms are long, curved shock waves traveling through the interstellar medium. If they are, then the collision of an interstellar cloud with a spiral arm could trigger the formation of new stars.

Astronomers have found a number of giant molecular clouds in which stars are forming in a repeating cycle. Both high-mass and low-mass stars form in such a cloud, but when the massive stars form, their intense radiation or eventual supernovae explosions push back the surrounding gas and compress it. This compression in turn can trigger the formation of more stars, some of which will be massive. Thus, a few massive stars can drive a continuing cycle of star formation in a giant molecular cloud.

Although low-mass stars do form in such clouds along with massive stars, low-mass stars also form in smaller clouds of gas and dust. Because lower-mass stars have lower luminosities and do not develop quickly into supernovae explosions, low-mass stars alone cannot drive a continuing cycle of star formation.

Once part of a cloud is triggered to collapse, gravity draws each atom toward the center. The in-falling atoms pick up speed as they fall until, by the time the gas becomes dense enough for the atoms to collide often, they are traveling very fast. Remember that temperature is a measure of the speed of the atoms in a gas, so as the atoms move faster, the temperature of the gas increases. Thus, the contraction of an interstellar cloud heats the gas by converting gravitational energy into thermal energy.

Such a collapsing cloud of gas does not form a single object; because of instabilities, it fragments, producing 10 to 1000 stars. Stars held in a stable group by their combined gravity are called a *star cluster.* An **association** is a group of stars that are not gravitationally bound to one another. Thus an association drifts apart in a few million years. The youngest associations are rich in young stars, including O and B stars.

The Formation of Protostars

To follow the story of star formation further, let's concentrate on a single fragment of a collapsing cloud as it forms a star. The initial collapse of the material is almost unopposed by any internal pressure. The material falls inward, picking up speed and increasing in density and temperature. This free-fall contraction transforms the cold gas into a warm protostar buried deep in the dusty gas. A **protostar** is an object that will eventually become a star.

When a star or protostar changes, it is said to evolve, and you can follow that evolution in the H–R diagram. As a star's temperature and luminosity change, its location in the H–R diagram shifts as well. You can draw an arrow or line called an **evolutionary track** to show how the star moves in the diagram. Of course, the evolutionary motion of the star in the diagram is not related in any way to its motion through space.

Stars begin contracting from the interstellar gas, which is very cold. Thus, to trace the contraction of a protostar, you must extend the H–R diagram to the right to include very low temperatures. ■ Figure 9-7, which includes temperatures below 100 K, shows the initial collapse of a 1-solar-mass cloud fragment as its temperature rises to more than 1000 K. The exact process is poorly understood because of the complex effect of the dust in the cloud. The dusty cloud also hides the protostar from sight during its contraction. If you could see it, it would be a luminous red object a few thousand times larger than the sun, and you would plot it in the red-giant region of the H–R diagram.

The protostar's original free-fall contraction is slowed by the increasing density and internal pressure. The hot gas inside the protostar resists gravity, and the star can continue to contract only as fast as it can radiate energy into space. Although this contraction is much slower than the free-fall contraction, the star must continue to contract because its interior is not hot enough to generate nuclear energy.

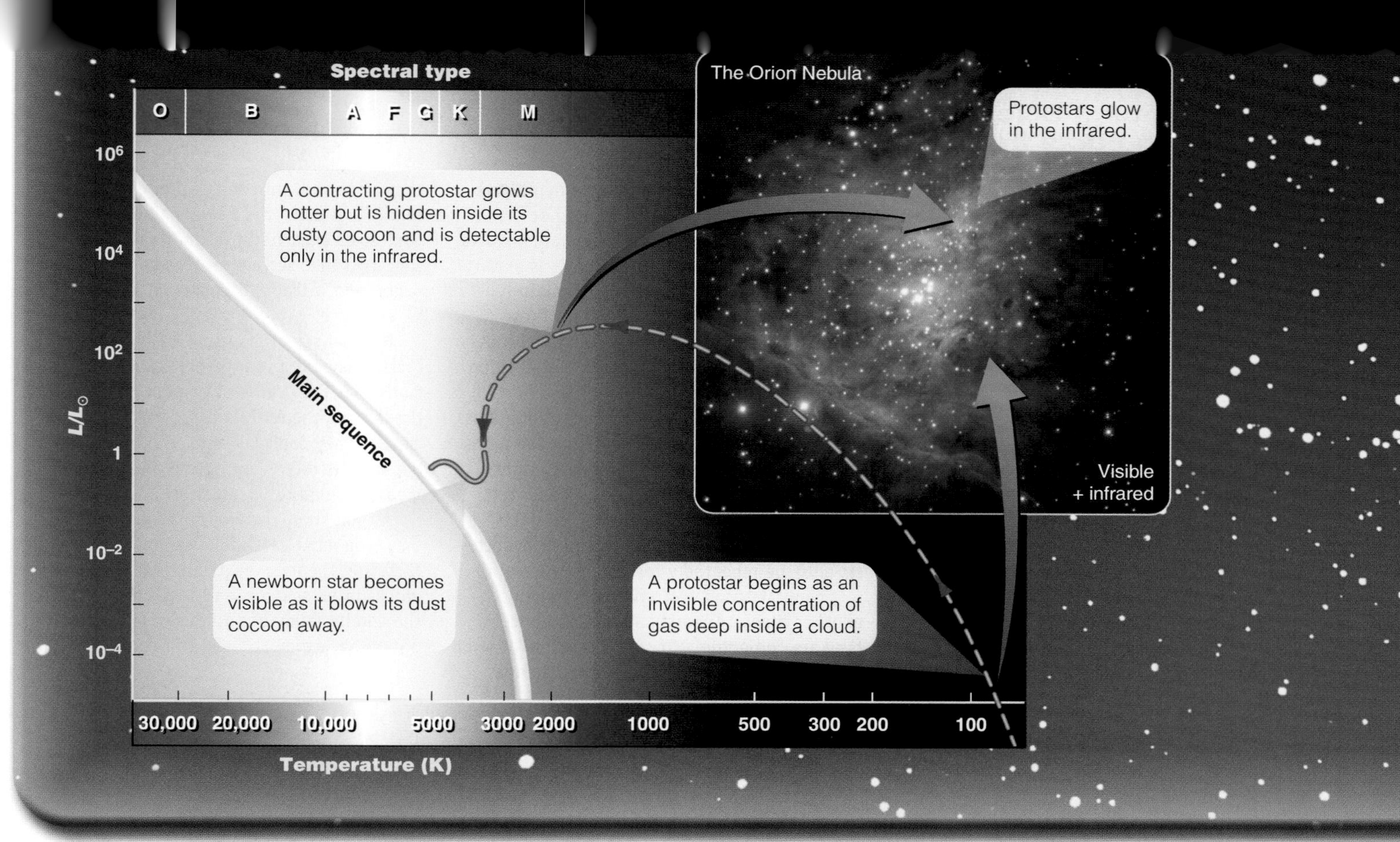

■ **Figure 9-7**

This H–R diagram has been extended to very low temperatures to show schematically the contraction of a dim, cool protostar. At visual wavelengths, protostars are invisible because they are deep inside dusty clouds of gas, but they are detectable at infrared wavelengths. The Orion Nebula contains both protostars and newborn stars that are just blowing their dust cocoons away. (ESO)

Throughout its contraction, the protostar converts its gravitational energy into thermal energy. Half of this thermal energy radiates into space, but the remaining half raises the internal temperature. As the internal temperature climbs, the gas becomes ionized, becoming a mixture of positively charged atomic nuclei and free electrons. When the center gets hot enough, nuclear reactions begin generating energy, the protostar halts its contraction, and, having blown away its cocoon of gas and dust, it becomes a stable, main-sequence star.

The time a protostar takes to contract from a cool interstellar gas cloud to a main-sequence star depends on its mass. The more massive the star, the stronger its gravity and the faster it contracts (■ Figure 9-8). The sun took about 30 million years to reach the main sequence, but a 15-solar-mass star can contract in only 160,000 years. Conversely, a star of 0.2 solar mass takes 1 billion years to reach the main sequence.

Thus far the story of the formation of a star from the interstellar medium has been based on theory. By understanding what the interstellar medium is like and by understanding how the laws of physics work, astronomers have been able to tell the story of how stars must form. But you can't accept a scientific theory without testing it, and that means you must compare the theory with the evidence. That constant checking of theories against evidence is the distinguishing characteristic of science, so it is time to carefully separate theory from evidence (**Window on Science 9-1**) and ask how much of this star formation story can be observed.

Observations of Star Formation

In astronomy, evidence means observations. Thus, astronomers must ask what observations confirm their theories of star formation. Unfortunately, a protostar is not easy to observe. The protostar stage is less than 0.1 percent of a star's total lifetime, so, although that is a long time in human terms, you cannot expect to find many stars in the protostar stage. Furthermore, protostars form deep inside clouds of dusty gas that absorb any light the protostar might emit. Only when the protostar is hot enough to drive away its enveloping cloud of gas and dust do you see it.

Understanding Science: Separating Facts from Theories

The fundamental work of science is the testing of theories by comparison with facts. As you think about science, you need to clearly distinguish between facts and theories. The facts are the evidence against which you test the theories.

Scientific facts are those observations or experimental results of which scientists are confident. An astronomer makes observations of stars, a botanist collects samples of related plants, and a chemist performs an experiment to measure the rate of a chemical reaction. A fact could be a precise measurement, such as the mass of a star expressed as a specific number, or a fact could be a simple observation, such as that a certain butterfly no longer visits a certain mountain valley. In each case, the scientist is gathering facts.

A theory, however, is a conjecture as to how nature works. If scientists are uncertain of the theory, they might call it a hypothesis. In any case, these conjectures are not facts; they are attempts to explain how nature works. In a sense, a theory or a hypothesis is a story that scientists have made up to explain how nature works in some specific case. These stories can be wonderfully detailed and ingenious, but without evidence they are nothing more than hunches.

When one of these stories is tested against the facts and confirmed, scientists have more confidence that the story is probably right. The more a story is tested successfully, the more confidence they have in it. The facts represent reality, and every theory or hypothesis must be repeatedly tested against reality.

Scientists can't test one theory against another theory. Theories are not evidence; they are conjectures. If scientists were allowed to test one theory against another theory, they might fall into the trap of circular reasoning. "Elves make the flowers bloom. I know that elves exist because the flowers bloom." Here two theories are used to confirm each other, and it leads to nonsense. Only facts can be evidence.

Nor can scientists use a theory to deduce facts. They can use a theory to make predictions that they can test against facts, but the predictions themselves can never be certainties, so they can't be facts. The only way to get facts is to consult nature and make direct measurements or observations.

As you study different problems in astronomy, you must carefully distinguish between facts and theories. Facts are the basic building blocks of all science, and they come only from the careful study of nature. To use the words of Galileo, we must "read the book of nature."

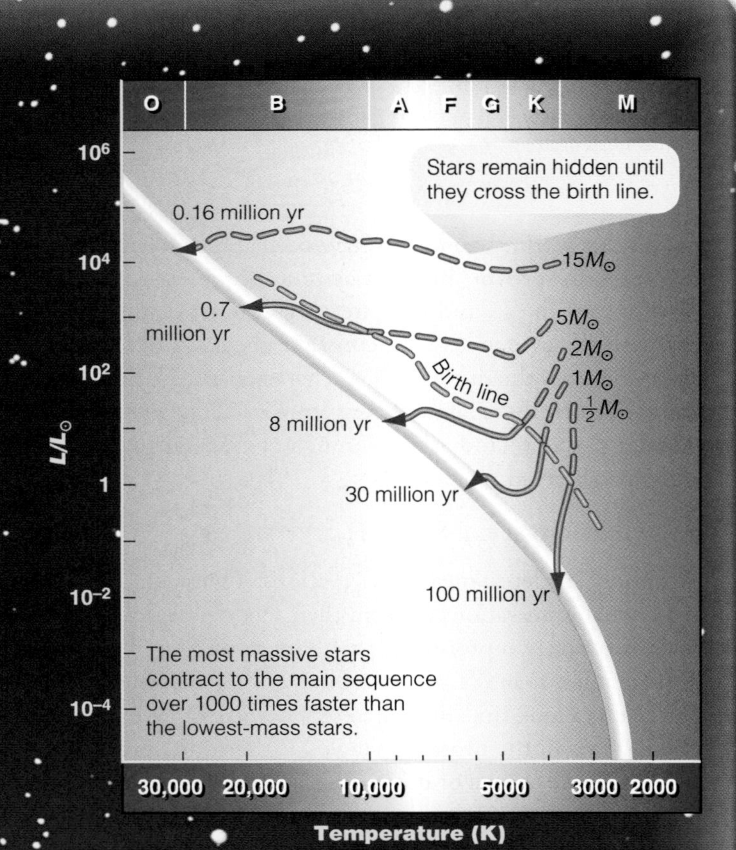

■ **Figure 9-8**

The more massive a protostar is, the faster it contracts. A 1-$M_\odot$ star requires 30 million years to reach the main sequence. (Recall that $M_\odot$ means "solar mass.") The dashed line is the birth line, where contracting protostars first become visible as they dissipate their surrounding clouds of gas and dust. Compare with Figure 9-7, which shows the formation of a star of about 1 $M_\odot$ as a dashed line up to the birth line and as a solid line from the birth line to the main sequence. (Illustration design by author)

The **birth line** in the H–R diagram in Figure 9-8 shows where contracting protostars first become visible. Protostars cross the birth line shortly before they reach the main sequence. Thus, the early evolution of a protostar is hidden from sight.

Although you cannot see protostars at visible wavelengths before they cross the birth line, you can detect them in the infrared. The dust absorbs light from the protostar and grows warm, and warm dust radiates copious infrared. Infrared observations such as those made by the Infrared Astronomy Satellite reveal many bright sources of infrared radiation that are probably protostars buried in dust clouds.

Study **Observational Evidence of Star Formation** on pages 184 and 185, and notice three important points. First, you can be sure that star formation is going on right now because you can find regions containing stars so young that they must have formed recently. Second, notice that these regions of star formation are rich in gas and dust and contain infrared protostars and stars still contracting toward the main sequence. The third thing to notice is how observations give insight into the details of star formation. The observation of jets coming from protostars is strong evidence that protostars form surrounded by disks of gas and dust.

All of these observations confirm the theoretical model of contracting protostars. Although star formation still holds many mysteries, the general process seems clear. In at least some cases, interstellar gas clouds are compressed by passing shock waves, and the clouds' gravity, acting unopposed, draws the matter inward to form protostars.

The planets of our solar system formed with the sun over an interval of no more than a few million years. At some point during the sun's contraction, when it was surrounded by a swirling disk of gas and dust, solid bodies condensed and grew into planets. Chapter 16 will discuss planet building, but it is important to note here that stars may form planets as they contract.

Inquire I Review I Analyze

What evidence do astronomers have that stars are forming right now? First, you should note that some extremely luminous stars can't live very long, so when you see such stars, such as the hot, blue stars in Orion, you know they must have formed in the last few million years. Star formation must be a continuous process, or you would not see any of these short-lived stars. But you have more direct evidence when you look at T Tauri stars, which lie above the main sequence and are often associated with gas and dust. In fact, you see entire associations of T Tauri stars as well as other associations of short-lived O stars. Many regions of gas and dust contain objects visible in the infrared that appear to be actual protostars. Furthermore, you can find cases where jets flowing outward from protostars create Herbig–Haro objects. In a few cases, such as the Eagle Nebula, the dense cores of gas and dust that are the forerunners of star formation are exposed by the evaporation of the nebula.

There seems to be no doubt that star formation is an ongoing process, and astronomers can even discuss how it happens. For example, what evidence do they have that protostars are often surrounded by disks of gas and dust?

■ ■ ■

Connections: Gravity pulls clouds of gas and dust together to make new stars, but this raises a natural question. If all this matter is falling inward to make a star, why does it stop? How do contracting protostars turn themselves into stable main-sequence stars? The answer is that the contracting stars begin to generate their own energy.

9-2 Fusion in Stars

ALL MAIN-SEQUENCE STARS fuse hydrogen into helium to generate energy. As you saw in Chapter 7, the sun fuses hydrogen using a chain of reactions called the proton–proton chain. Upper-main-sequence stars, being more massive than the sun, can fuse hydrogen in a more efficient way, and we must discuss that process here. Also, aging stars fuse fuels other than hydrogen; so, to complete our story of energy generation in stars, we must discuss those other fusion reactions.

The CNO Cycle

Main-sequence stars more massive than the sun fuse hydrogen into helium using the **CNO cycle,** a hydrogen-fusion process that uses carbon, nitrogen, and oxygen as stepping-stones (■ Figure 9-9). The CNO cycle begins with a carbon nucleus, transforms it first into a nitrogen nucleus, then into an oxygen nucleus, and then back to a carbon nucleus. The carbon is unchanged in the end, but along the way four hydrogen nuclei are fused to make a helium nucleus plus energy, just as in the proton–proton chain.

The CNO cycle begins with a carbon nucleus combining with a hydrogen nucleus. Because a carbon nucleus has a charge six times higher than hydrogen, the Coulomb barrier is high, and temperatures higher than 16,000,000 K are required to make the CNO cycle work. The center of the sun is not quite hot enough. In stars more massive than about 1.1 solar masses, the cores are hotter, and those stars use the CNO cycle and not the less efficient proton–proton chain.

Heavy-Element Fusion

At later stages in the life of a star, when it has exhausted its hydrogen fuel, it may fuse other nuclear fuels. As you learned previously, the ignition of these fuels requires high temperatures, because the nuclei have large positive charges. The particles must

■ Figure 9-9

The CNO cycle uses ^{12}C as a catalyst to combine four hydrogen atoms (^{1}H) to make one helium atom (^{4}He) plus energy. The carbon atom reappears at the end of the process, ready to start the cycle over.

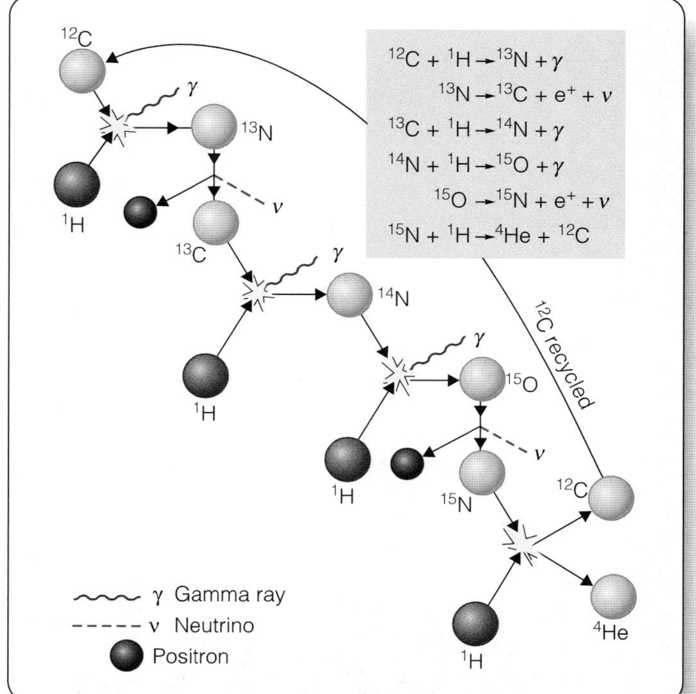

Observational Evidence of Star Formation

At visible, infrared, and radio wavelengths, astronomers find many objects in the sky that tell how stars form.

The nebula around the star S Monocerotis is bright with hot stars. Such stars live short lives of only a few million years, so they must have formed recently. Such regions of young stars are common. The entire constellation of Orion is filled with young stars and clouds of gas

Nebulae containing young stars usually contain **T Tauri** stars. These stars fluctuate irregularly in brightness, and many are bright in the infrared, suggesting they are surrounded by dust clouds and in some cases by dust disks. Doppler shifts show that gas is flowing away from many T Tauri stars. Located near the birth line in the H–R diagram, the T Tauri stars appear to be newborn stars just blowing away their dust cocoons. T Tauri stars appear to have ages ranging from 100,000 years to 100,000,000 years.

Spectra of T Tauri stars show signs of an active chromosphere as we might expect from young, rapidly rotating stars with powerful dynamos and strong magnetic fields.

Visual-wavelength image

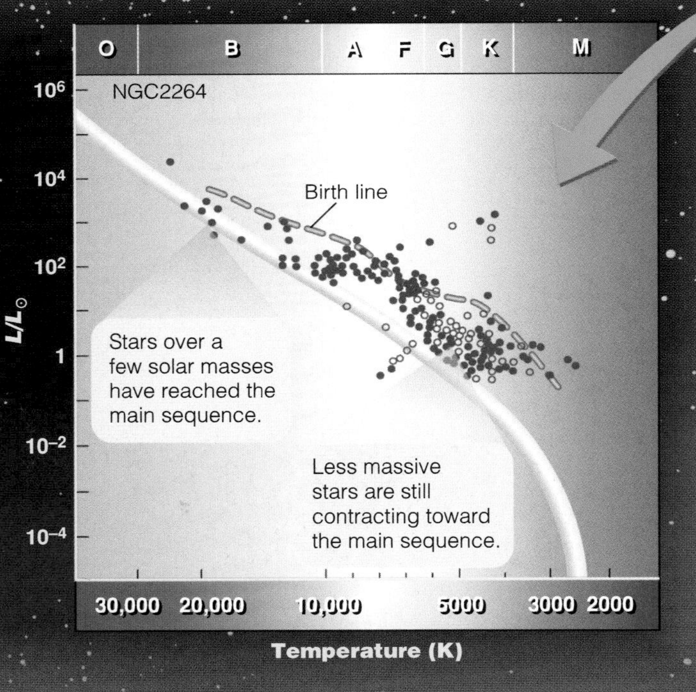

The star cluster NGC2264, imbedded in the nebula on this page, is only a few million years old. Lower-mass stars have not yet reached the main sequence, and the cluster contains many T Tauri stars (open circles), which are found above and to the right of the main sequence, near the birth line. The faintest stars in the cluster were too faint to be observed in this study.

Bok globules, named after astronomer Bart Bok, are dense, dusty clouds seen silhouetted against glowing nebulae. Only a light-year or so in diameter, they contain from 10 to 1000 solar masses, and infrared observations show that most are very cold at their centers. Only a few have warm interiors and appear to be contracting to form protostars.

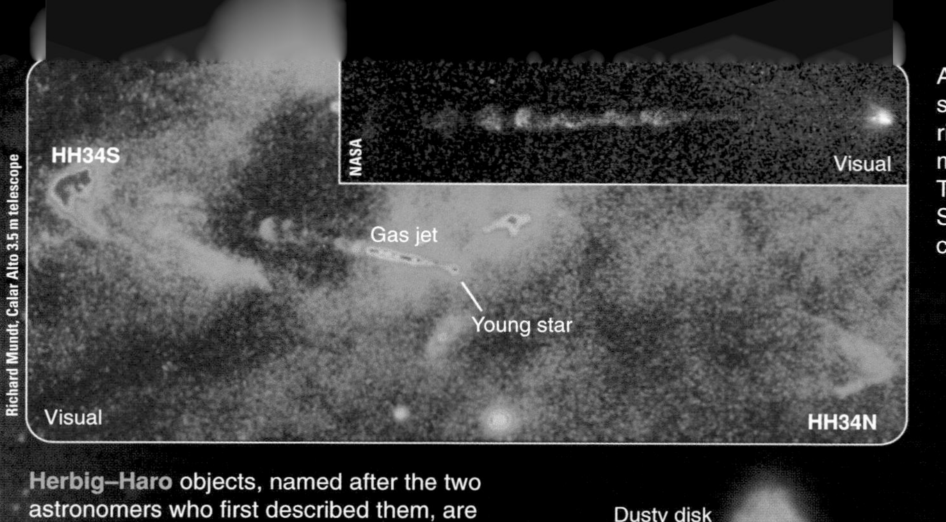

At the center of this image, a newborn star is emitting powerful jets to left and right. Where the jets strike the interstellar medium, they produce Herbig–Haro objects. The inset shows how irregular the jet is. Such jets can be over a light-year long and contain gas traveling at 100 km/s or more.

Herbig–Haro objects, named after the two astronomers who first described them, are small nebulae that fluctuate in brightness. They appear to be produced by flickering jets from newborn stars exciting the interstellar medium.

Herbig–Haro object

Dusty disk

Jet

Jet

Herbig–Haro object

Matter flowing into a protostar swirls through a thick disk and, by a process believed to involve magnetic fields, ejects high-energy jets in opposite directions. Observation of these **bipolar flows** is evidence that protostars are surrounded by disks because only disks could focus the flows into jets.

The Eagle Nebula, shown at the right in a ground-based photograph, is a region of known star formation. Turn your book a quarter turn to the right to see the eagle flying with a salmon in its talons.

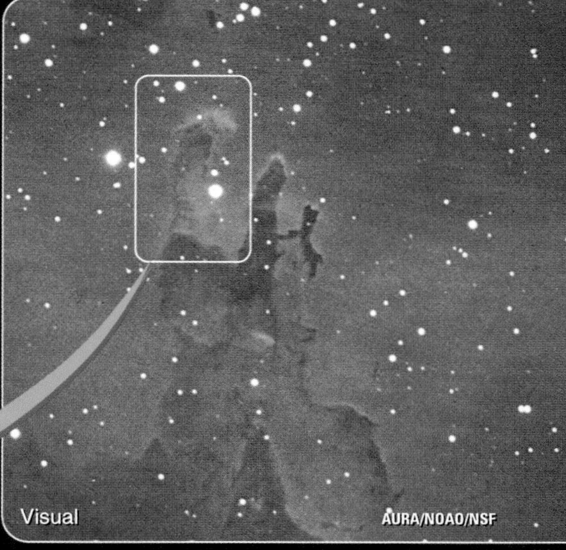

This Hubble Space Telescope image of part of the Eagle Nebula shows that radiation from bright stars out of the field to the upper right is evaporating the dust and driving away the gas to expose small globules of denser gas and dust. Infrared observations show that about 15 percent of these objects have formed protostars, and at least one seems to contain a newborn star (arrow). In part because these objects were first found in the Eagle Nebula, astronomers have enjoyed calling them EGGs — evaporating gaseous globules. They are evidently stars exposed in the act of forming.

travel at high velocities to overcome the electrostatic repulsion and force the particles close enough together to react. Helium fusion requires a temperature of at least 100 million K, and carbon fusion requires 600 million K.

You can summarize the helium-fusion process in two steps:

$$^{4}\text{He} + {}^{4}\text{He} \rightarrow {}^{8}\text{Be} + \gamma$$

$$^{8}\text{Be} + {}^{4}\text{He} \rightarrow {}^{12}\text{C} + \gamma$$

Because a helium nucleus is called an alpha particle, these reactions are commonly known as the **triple-alpha process.** Helium fusion is complicated by the fact that beryllium-8, produced in the first reaction of the process, is very unstable and may break up into two helium nuclei before it can absorb another helium nucleus. Three helium nuclei can also form carbon directly, but such a triple collision is unlikely.

At temperatures above 600,000,000 K, carbon fuses rapidly in a complex network of reactions illustrated in ■ Figure 9-10, where each arrow represents a different nuclear reaction. The process is complicated because nuclei can react by adding a proton, a neutron, or a helium nucleus or by combining directly with other nuclei. Unstable nuclei can decay by ejecting an electron, a positron, or a helium nucleus or by splitting into fragments.

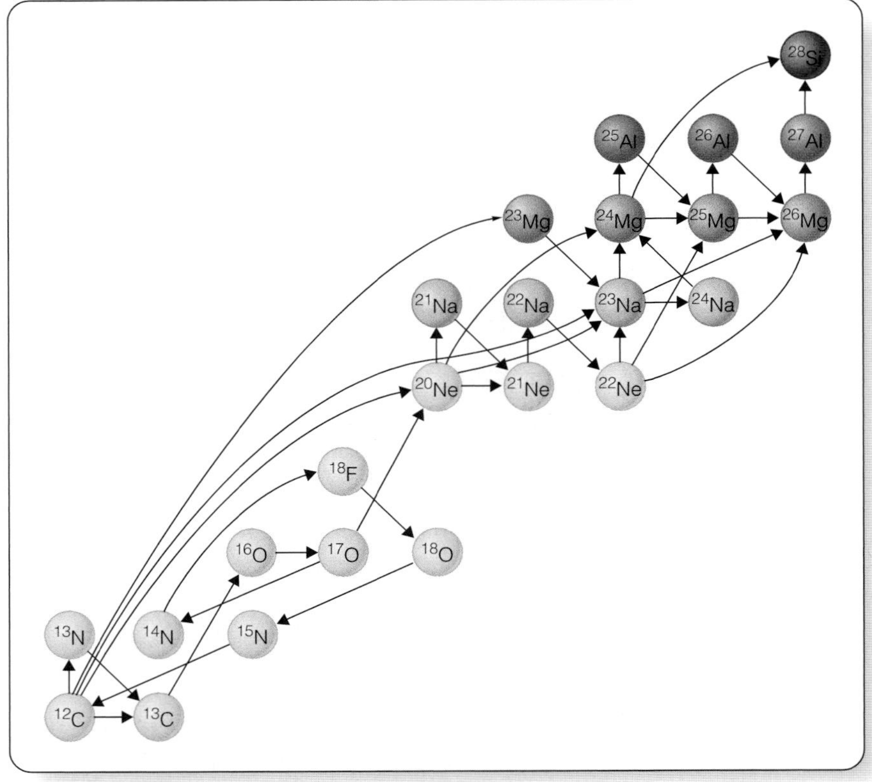

■ **Figure 9-10**

Carbon fusion involves many possible reactions that build numerous heavy atoms.

Reactions at still higher temperatures can convert magnesium, aluminum, and silicon into yet heavier atoms. These reactions involving heavy elements will be important in the study of the deaths of massive stars in the next chapter.

The Pressure–Temperature Thermostat

Nuclear reactions in stars manufacture energy and heavy atoms under the supervision of a built-in thermostat that keeps the reactions from erupting out of control. That thermostat is the relation between gas pressure and temperature.

In a star, the nuclear reactions generate just enough energy to balance the inward pull of gravity. Consider what would happen if the reactions began to produce too much energy. The star balances gravity by generating energy, so the extra energy flowing out of the star would force it to expand. The expansion would lower the central temperature and density and slow the nuclear reactions until the star regained stability. Thus the star has a built-in regulator that keeps the nuclear reactions from occurring too rapidly.

The same thermostat keeps the reactions from dying down. Suppose the nuclear reactions began making too little energy. Then the star would contract slightly, increasing the central temperature and density and increasing the nuclear energy generation.

The stability of a star depends on the relation between gas pressure and temperature. If an increase or decrease in temperature produces a corresponding change in pressure, then the thermostat functions correctly, and the star is stable. You will see in this chapter how the thermostat accounts for the mass–luminosity relation. In the next chapter, you will see what happens to a star when the thermostat breaks down completely and the nuclear fires rage unregulated.

Inquire I Review I Analyze

Why does the fusion of heavier atoms require higher temperatures than the fusion of hydrogen?

Nuclear fusion happens when atomic nuclei collide and fuse to form a new nucleus. Inside a star, the gas is ionized, so the nuclei have no orbiting electrons, and thus the nuclei have positive charges. Two objects with positive charges will repel each other in what is called the Coulomb barrier. To start fusion, the nuclei must collide at high enough velocity to overcome that barrier. Heavier atomic nuclei such as helium and carbon have higher positive charges, so their Coulomb barriers are higher. That requires that the gas be hotter so the collisions between nuclei will be violent enough to overcome the barrier.

Nuclear fusion in stars follows rules that are amazingly simple. For example, what rule sees that a star will produce exactly the energy needed?

■ ■ ■

Connections: The law of gravity and the rules of nuclear fusion determine how stars work and allow astronomers to understand what the inside of a star is like. In other words, they can describe the internal structure of a star.

9-3 Stellar Structure

THE NUCLEAR FUSION at the center of stars heats their interior, creates high gas pressure, and thus balances the inward force of gravity. If there is a single idea in modern astronomy that can be called critical, it is this concept of balance. Stars are simple, elegant power sources held together by their own gravity and supported by their nuclear fusion.

Having discussed the birth of stars and their generation of energy, you can now consider the structure of a star—that is, the variation in temperature, density, pressure, and so on from the surface of the star to its center. A star's structure depends on how it generates its energy, on four simple laws of structure, and on what it is made of.

It will be easier to think about stellar structure if you imagine that the star is divided into concentric shells like those in an onion. You can then discuss the temperature, density, pressure, and so on in each shell. Of course, these helpful shells do not really exist; stars have no separable layers.

The Laws of Mass and Energy

The first two laws of stellar structure have something in common—they are both conservation laws. They say that certain things cannot be created out of nothing or vanish into nothing. Such conservation laws are powerful tools to help you understand how nature works.

The law of **conservation of mass** is a basic law of nature, which can be applied to the structure of stars. It says that the total mass of a star must equal the sum of the masses of its shells. This is like saying the weight of a cake must equal the sum of the weight of its layers.

The law of **conservation of energy** is another basic law of nature. It says that the amount of energy flowing out of the top of a layer in the star must be equal to the amount of energy coming in at the bottom plus whatever energy is generated within the layer. That means that the energy leaving the surface of the star, its luminosity, must equal the sum of the energies generated in all the layers inside the star. This is like saying that all the new cars driving out of a factory must equal the sum of all the cars made on each of the production lines.

These two laws may seem so familiar and so obvious you hardly need to state them, but they are important clues to the structure of stars. The third law of stellar structure is familiar because you have been using a closely related law in the preceding sections.

Hydrostatic Equilibrium

When you think about a star, it is helpful to think of it as if it were made up of layers. The weight of each layer must be supported by the layer below. The deeper layers must support the weight of all of the layers above. Because the inside of a star is made up of gas, the weight pressing down on a layer must be balanced by the gas pressure in the layer. If the pressure is too low, the weight from above will compress the layer, and if the pressure is too high, the layer will expand and lift the layers above.

This balance between weight and pressure is called **hydrostatic equilibrium.** *Hydro* (from the Greek word for water) tells you the material is a fluid, the gases of a star, and *static* tells you the fluid is stable, neither expanding or contracting.

■ Figure 9-11 shows this hydrostatic balance in the imaginary layers of a star. The weight pressing down on each layer is shown by lighter red arrows, which grow larger with increasing depth because the weight grows larger. The pressure in each layer is shown by darker red arrows, which must grow larger with increasing depth to support the weight.

The law of hydrostatic equilibrium can tell you something important about the inside of a star. The pressure in a gas depends on the temperature and density of the gas. Near the surface, there is little weight pressing down, so the pressure need not be high. Deeper in the star, the pressure must be higher, and that means that the temperature and density of the gas must also be higher. Hydrostatic equilibrium tells you that stars have to be hot inside to maintain the pressure needed to support their own weight. The layers are hot because, as you have seen, energy flows outward from the core of the star where it is made by nuclear fusion.

Now you recognize hydrostatic equilibrium. It is closely related to the pressure–temperature thermostat discussed earlier.

Of course, how hydrostatic equilibrium actually works depends on what an object is made of. Nearly all stars are made of hydrogen and helium gas, with some trace of heavier elements. To understand how a star works, you must describe how that gas responds to changes in temperature and pressure. Hydrostatic equilibrium also applies to planets, including Earth, but Earth is made of rock and metal, so again, understanding how hydrostatic equilibrium supports Earth requires that Earth scientists know how rock and metal respond to changes in temperature and pressure. You will see in later chapters how the structure of a white dwarf is dramatically different from that of a star such as the sun because the white dwarf's gas behaves differently from that in most normal stars.

Although the law of hydrostatic equilibrium can tell you some things about the inner structure of stars, you need one more law to completely describe a star. You need a law that describes the flow of energy from the center to the surface.

AceAstronomy™ Log into AceAstronomy and select this chapter to see Astronomy Exercise "Hydrostatic Equilibrium." You can try, but you can't make this star violate hydrostatic equilibrium.

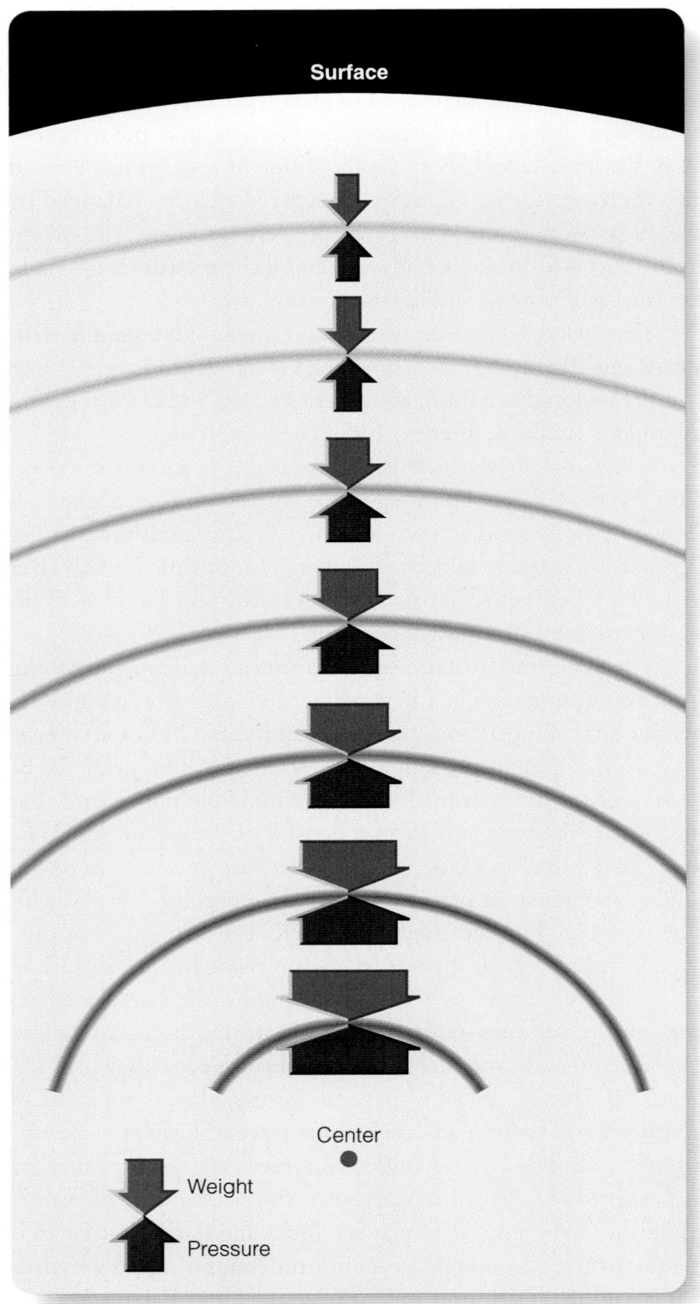

■ **Figure 9-11**

The law of hydrostatic equilibrium says the pressure in each layer of a stable star balances the weight on that layer. Thus, as the weight increases from the surface of a star to its center, the pressure also increases.

Energy Transport

The surface of a star radiates light and heat into space and would quickly cool if that energy were not replaced. Because the inside of the star is hotter than the surface, energy must flow outward from the core, where it is generated, to the surface, where it radiates away. This flow of energy through the shells determines their temperature, which, as you saw previously, determines how much weight each shell can balance. To understand the structure of a star, you must understand how energy moves from the center through the shells to the surface.

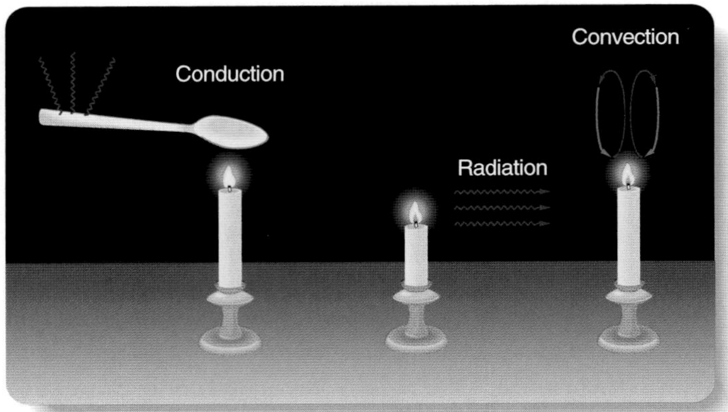

■ **Active Figure 9-12**

The three modes by which energy may be transported from the flame of a candle, as shown here, are the three modes of energy transport within a star.

AceAstronomy™ Log into AceAstronomy and select this chapter to see Active Figure "Conduction, Convection, and Radiation." Experiment with the candle flame.

The law of **energy transport** says that energy must flow from hot regions to cooler regions by conduction, convection, or radiation.

Conduction is the most familiar form of heat flow. If you hold the bowl of a spoon in a candle flame, the handle of the spoon grows warmer. Heat, in the form of motion among the molecules of the spoon, is conducted from molecule to molecule up the handle, until the molecules of metal under your fingers begin to move faster and you sense heat (■ Figure 9-12). Thus, conduction requires close contact between the molecules. Because the particles (atoms) in most stars are not in close contact, conduction is unimportant. Conduction is significant in white dwarfs, which have tremendous internal densities.

The transport of energy by radiation is another familiar experience. Put your hand beside a candle flame, and you can feel the heat. What you actually feel are infrared photons radiated by the flame (Figure 9-12). Because photons are packets of energy, your hand grows warm as it absorbs them.

Radiation is the principal means of energy transport in the sun's interior (■ Figure 9-13). Photons are absorbed and re-emitted in random directions over and over as they work their way outward. This process breaks the high-energy photons common near the sun's center into large numbers of low-energy photons in the cooler outer layers. Thus, the energy of a single high-energy photon at the center of the sun takes about 1 million years to reach the sun's surface, where it emerges as roughly 1600 photons of visible light.

The flow of energy by radiation depends on how difficult it is for the photons to move through the gas. If the gas is cool and

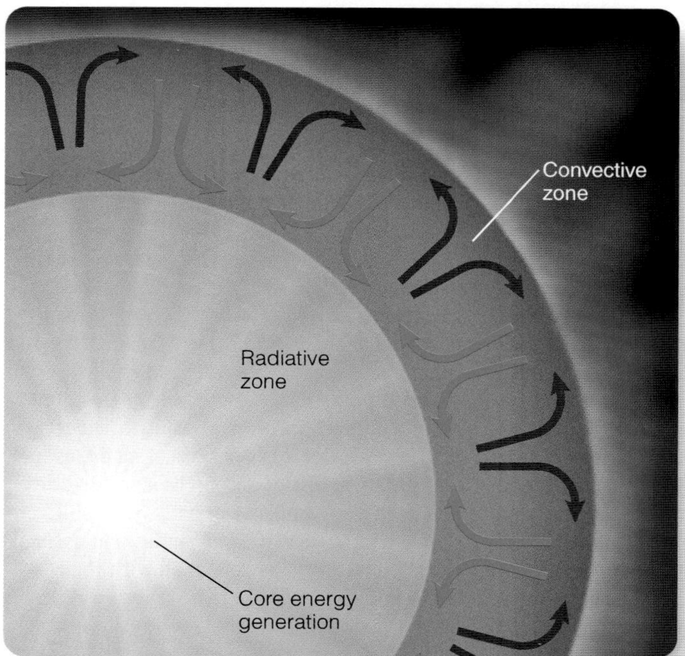

■ **Figure 9-13**

A cross section of the sun. Near the center, nuclear fusion reactions generate high temperatures. Energy flows outward through the radiative zone as photons. In the cooler, more opaque outer layers, the energy is carried outward by rising convection currents of hot gas (red) and sinking currents of cooler gas (blue).

dense, the photons are more likely to be absorbed or scattered, and so the radiation does not get through easily. Such a gas is opaque. In a hot, thin gas, the photons can get through more easily; such a gas is less opaque. The **opacity** of the gas, its resistance to the flow of radiation, depends strongly on its temperature.

If the opacity is high, radiation cannot flow through the gas easily, and it backs up like water behind a dam. When enough heat builds up, the gas begins to churn as hot gas rises upward and cool gas sinks downward. This heat-driven circulation of a fluid is convection, the third way energy can move in a star. You are familiar with convection; the rising wisp of smoke above a candle flame is carried by convection. Energy is carried upward in these convection currents as rising hot gas (red in Figure 9-13) and also as sinking cool gas (blue in Figure 9-13). In stars, energy can be carried upward in convection currents hundreds or thousands of miles in diameter.

Convection is important in stars because it carries energy and because it mixes the gas. Convection currents flowing through the layers of a star tend to homogenize the gas, giving it a uniform composition throughout the convective zone. As you might expect, this mixing affects the fuel supply of the nuclear reactions, just as the stirring of a campfire makes it burn more efficiently.

The four laws of stellar structure are summarized in ■ Table 9-1. These laws, properly understood, can tell you how stars are born, how they live, and how they die.

■ **Table 9-1 | The Four Laws of Stellar Structure**

I. Conservation of mass	Total mass equals the sum of shell masses.
II. Conservation of energy	Total luminosity equals the sum of energy generated in each shell.
III. Hydrostatic equilibrium	The weight on each layer is balanced by the pressure in that layer.
IV. Energy transport	Energy moves from hot to cool regions by conduction, radiation, or convection.

Stellar Models

The laws of stellar structure, described in general terms in the previous sections, can be written as mathematical equations. By solving those equations in a special way, astronomers can build a mathematical model of the inside of a star.

If you wanted to build a model of a star, you would have to divide the star into about 100 concentric shells, and then write down the four equations of stellar structure for each shell. You would then have 400 equations that would have 400 unknowns, namely, the temperature, density, mass, and energy flow in each shell. Solving 400 equations simultaneously is not easy, and the first such solutions, done by hand before the invention of the electronic computer, took months of work. Now a properly programmed computer can solve the equations in a few seconds and print a table of numbers that represents the conditions in each shell of the star. Such a table is a **stellar model** (Window on Science 9-2).

The table shown in ■ Figure 9-14 is a model of the sun. The bottom line, for radius equal to 0.00, represents the center of the sun, and the top line, for radius equal to 1.00, represents the surface. The other lines tell you the temperature and density in each shell, the mass inside each shell, and the fraction of the sun's luminosity flowing outward through the shell. You can use the table to understand the sun. The bottom line tells you the temperature at the center of the sun is over 15 million Kelvin. At such a high temperature the gas is highly transparent, and energy flows as radiation. Nearer the surface, the temperature is lower, the gas is more opaque, and the energy is carried by convection. This is confirmed by the evidence of convection visible on the sun's surface (Chapter 7).

Stellar models also let you look into a star's past and future. In fact, you could use models as time machines to follow the evolution of stars over billions of years. To look into a star's future, for instance, you could use a stellar model to determine how fast the star uses its fuel in each shell. As the fuel is consumed, the chemical composition of the gas changes and the amount of energy generated declines. By calculating the rate of these changes, you could predict what the star will look like at any point in the future.

Window on Science 9-2

Quantitative Thinking with Mathematical Models

One of the most powerful tools in science is the mathematical model, a group of equations carefully designed to mimic the behavior of the object that scientists want to study. Astronomers build mathematical models of stars using only four equations, but other systems are much more complicated and may require many more equations.

Many sciences use mathematical models. Medical scientists have built mathematical models of the nerves that control the human heart, and physicists have built mathematical models of the inside of an atomic nucleus. Economists have built mathematical models of certain aspects of economic systems, such as the municipal bond market, and Earth scientists have built mathematical models of Earth's atmosphere. In each case, the mathematical model allows the scientists to study something that is difficult to study in reality. The model can reveal regions we cannot observe, speed up a slow process, slow down a fast process, or allow scientists to perform experiments that would be impossible in reality. Astronomers, for example, can change the abundance of different chemical elements in a model star to see how its structure depends on its composition.

Many of these mathematical models require very large and fast computers—in some cases, supercomputers. A modern desktop computer takes only seconds to compute a stellar model; but a model of the motions of the planets of our solar system millions of years into the future would require the largest and fastest computers in the world. Most astronomers are apt computer programmers, and some are experts. A few astronomers have built their own highly specialized computers to compute specific kinds of models. Thus, the mathematical model is one of the most important tools in astronomy.

As is true for any scientific model, a mathematical model is only as reliable as the assumptions that go into its creation. In Window on Science 2-1, you saw that the celestial sphere is an adequate model of the sky for some purposes but breaks down if you extend it too far. So too with mathematical models. You can think of a mathematical model as a numerical expression of one or more theories. Although such models can be very helpful, they are always based on theories and assumptions and so must be compared with the real world at every opportunity. As always in science, there is no substitute for a careful comparison with reality.

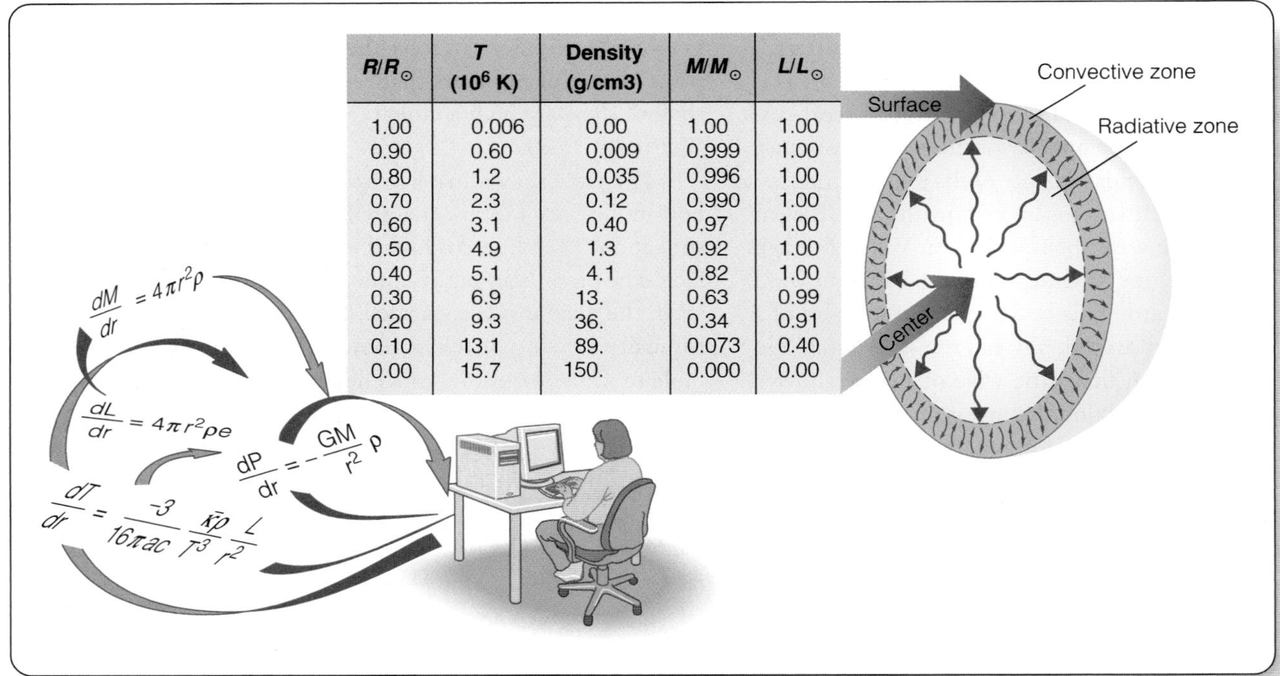

$R/R_\odot$	T (10^6 K)	Density (g/cm3)	$M/M_\odot$	$L/L_\odot$
1.00	0.006	0.00	1.00	1.00
0.90	0.60	0.009	0.999	1.00
0.80	1.2	0.035	0.996	1.00
0.70	2.3	0.12	0.990	1.00
0.60	3.1	0.40	0.97	1.00
0.50	4.9	1.3	0.92	1.00
0.40	5.1	4.1	0.82	1.00
0.30	6.9	13.	0.63	0.99
0.20	9.3	36.	0.34	0.91
0.10	13.1	89.	0.073	0.40
0.00	15.7	150.	0.000	0.00

■ Figure 9-14

A stellar model is a table of numbers that represent conditions inside a star. Such tables can be computed using the four laws of stellar structure, shown here in mathematical form. The table in this figure describes the sun. (Illustration design by author)

Although this sounds simple, it is actually a highly challenging problem involving nuclear and atomic physics, thermodynamics, and sophisticated computational methods. Only since the 1950s have electronic computers made the rapid calculation of stellar models possible, and the advance of astronomy since then has been heavily influenced by the use of such models to study the structure and evolution of stars. The summary of star formation in this chapter is based on thousands of stellar models. You will continue to rely on theoretical models as you study main-sequence stars in the next section and the deaths of stars in the next chapter.

Inquire | Review | Analyze

What would happen if the sun stopped generating energy?
Stars are supported by the outward flow of energy generated by nuclear fusion in their interiors. That energy keeps each layer of the star just hot enough so that the gas pressure can support the weight of the layers above. Each layer in the star must be in hydrostatic equilibrium; that is, the inward weight is balanced by outward pressure. If the sun stopped making energy in its interior, nothing would happen at first, but over many thousands of years the loss of energy from its surface would reduce the sun's ability to withstand its own gravity, and it would begin to contract. You wouldn't notice much for 100,000 years or so, but eventually, the sun would lose its battle with gravity.

Stars are elegant in their simplicity. Nothing more than a cloud of gas held together by gravity and warmed by nuclear fusion, a star can achieve stability, balancing its weight by generating nuclear energy. But how does the star manage to make exactly the right amount of energy to support its weight?

■ ■ ■

Connections: You have traced the birth of stars from the first instability in the interstellar medium to the final equilibrium of nuclear fusion and gravity. Now you must look carefully at these stable stars and ask two questions. First, what evidence shows that stars really work this way? And second, how long can these stars remain stable? The answers to these two questions will give you a new insight into the lives of the stars.

9-4 Main-Sequence Stars

When a contracting protostar begins to fuse hydrogen, it stops contracting and becomes a stable main-sequence star. The most massive stars are so hot they light up the remaining nearby gas in a beautiful nebula, as if announcing their birth (■ Figure 9-15), but that gas is quickly blown away, and the stars begin their long, uneventful lives as main-sequence stars. If you discount the peculiar white dwarfs, then 90 percent of all true stars are main-sequence stars.

■ **Figure 9-15**

Monster star birth regions are excited to glow by massive stars just reaching the main sequence and by the supernova explosion of the most massive of these stars. Expanding bubbles of hot gas create arcs and cavities. The Tarantula Nebula is slightly smaller than NGC604 but contains more hot stars—over 200. (Tarantula visual: ESO X ray: NASA/CXC/Penn State/L. Townsley et al.; NGC604:NASA/Hubble Heritage Team/AURA/STScI)

The Mass–Luminosity Relation

Observations of the temperature and luminosity of stars show that main-sequence stars obey a simple rule—the more massive a star is, the more luminous it is. That rule, the mass–luminosity relation discussed in the previous chapter, is the key to understanding the stability of main-sequence stars. In fact, the mass–luminosity relation is predicted by the theories of stellar structure, giving astronomers direct observational confirmation of those theories.

To understand the mass–luminosity relation, you must consider the law of hydrostatic equilibrium, which says that pressure balances weight, and the pressure–temperature thermostat, which regulates energy production. A star that is more massive than the sun has more weight pressing down on its interior, so the interior must have a high pressure to balance that weight. That means the massive star must have its pressure–temperature thermostat turned up to keep the gas in its interior hot and the pressure high. A star less massive than the sun has less weight on its interior and thus needs less internal pressure; therefore, its pressure–temperature thermostat is set lower. To sum up, massive stars are more luminous because they must support more weight by making more energy.

This tells you why the main sequence must have a lower end. Masses below about 0.08 solar mass cannot raise their central temperature high enough to begin hydrogen fusion. Called **brown dwarfs,** such objects are only about a dozen times larger than Earth and are still warm from contraction but do not generate energy by hydrogen fusion. They have contracted as far as they can and are slowly cooling off.

Brown dwarfs fall in the gap between low-mass M stars and massive planets like Jupiter. They would look red to your eyes, but they emit most of their energy in the infrared. The warmer brown dwarfs fall in spectral class L and the cooler in spectral class T (Chapter 8). Molecules can form in the atmospheres of red dwarfs, but brown dwarfs are so cool that solid particles of silicates, metals, and other minerals can condense in cloud layers. Unlike stars, brown dwarfs appear to have weather.

Because they are so small and cool, brown dwarfs are very-low-luminosity objects and are difficult to find. Nevertheless, a few hundred are known. They may be as common as M stars. The evidence shows that nature does indeed make brown dwarfs when a forming star does not have enough mass to begin hydrogen fusion. This observational detection of the lower end of the main sequence further confirms the theories of stellar structure.

Now that you know how stars form and how they maintain their stability through the mass–luminosity relation, you can predict the evolution of main-sequence stars.

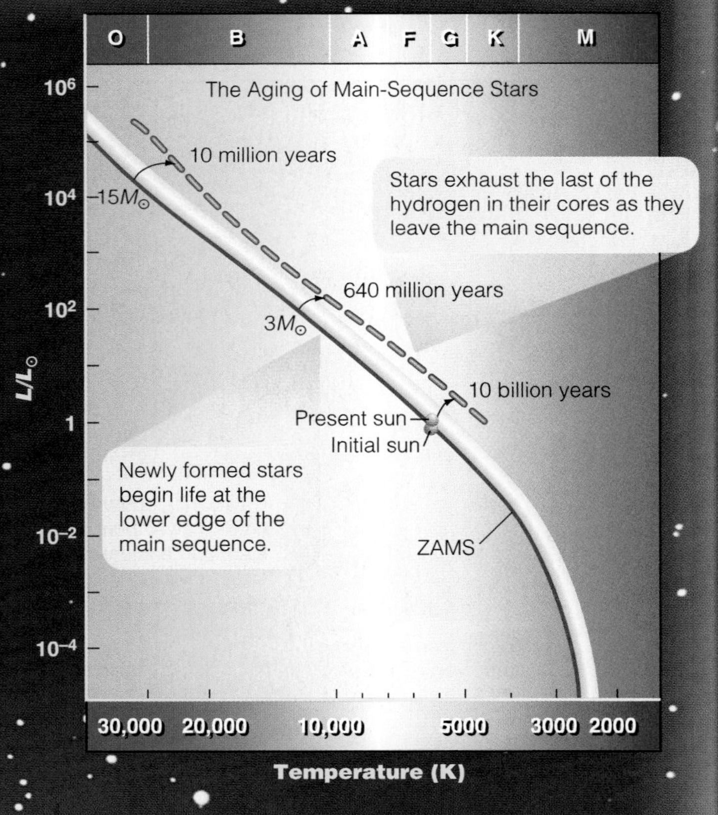

The Life of a Main-Sequence Star

While a star is on the main sequence, it is stable, so you might think its life would be totally uneventful. But a main-sequence star balances its gravity by fusing hydrogen, and as the star gradually uses up its fuel, that balance must change. Thus, even the stable main-sequence stars are changing as they consume their hydrogen fuel.

Recall that hydrogen fusion combines four nuclei into one. Thus, as a main-sequence star fuses its hydrogen, the total number of particles in its interior decreases. Each newly made helium nucleus can exert the same pressure as a hydrogen nucleus, but because the gas has fewer nuclei, its total pressure is less. This unbalances the gravity–pressure stability, and gravity squeezes the core of the star more tightly. As the core contracts, its temperature increases and the nuclear reactions run faster, releasing more energy. This additional energy flowing outward through the envelope forces the outer layers to expand. As the star becomes larger, it becomes more luminous, and eventually the expansion begins to cool the surface.

As a result of these gradual changes in main-sequence stars, the main sequence is not a sharp line across the H–R diagram but rather a band. Stars begin their stable lives fusing hydrogen on the

Figure 9-16

Contracting protostars reach stability at the lower edge of the main sequence, the zero-age main sequence (ZAMS). As a star converts hydrogen in its core into helium, it moves slowly across the main sequence, becoming slightly more luminous and slightly cooler. Once a star consumes all of the hydrogen in its core, it can no longer remain a stable main-sequence star. More massive stars age rapidly, but less-massive stars use up the hydrogen in their cores more slowly and live longer main-sequence lives.

Reasoning with Numbers | 9-1

The Life Expectancies of Stars

You can estimate the amount of time a star spends on the main sequence—its life expectancy, τ—by estimating the amount of fuel it has and the rate at which it consumes that fuel:

$$\tau = \frac{\text{fuel}}{\text{rate of consumption}}$$

The amount of fuel a star has is proportional to its mass M, and the rate at which it uses up its fuel is proportional to its luminosity L. Thus its life expectancy must be proportional to M/L. But you can simplify this equation further because, as you saw in the last chapter, the luminosity of a star depends on its mass raised to the 3.5 power ($L = M^{3.5}$). Thus the life expectancy is:

$$\tau = \frac{M}{M^{3.5}}$$

which is the same as

$$\tau = \frac{1}{M^{2.5}}$$

If you express the mass in solar masses, the lifetime will be in solar lifetimes.

Example: How long can a 4-solar-mass star live? **Solution:**

$$\tau = \frac{1}{4^{2.5}} = \frac{1}{4 \cdot 4\sqrt{4}}$$

$$= \frac{1}{32} \text{ solar lifetimes}$$

Studies of solar models show that the sun, presently 5 billion years old, can last another 5 billion years. Thus a solar lifetime is approximately 10 billion years, and a 4-solar-mass star will last for about

$$\tau = \frac{1}{32} \times (10 \times 10^9 \text{ yr})$$

$$= 310 \times 10^6 \text{ years}$$

lower edge of this band, which is known as the **zero-age main sequence (ZAMS),** but gradual changes in luminosity and surface temperature move the stars upward and slightly to the right, as shown in ■ Figure 9-16. By the time they reach the upper edge of the main sequence, they have exhausted nearly all the hydrogen in their centers. Thus you find main-sequence stars scattered throughout the band at various stages of their main-sequence lives.

The sun is a typical main-sequence star; and, as it undergoes these gradual changes, Earth will suffer. When the sun began its main-sequence life about 5 billion years ago, it was only about 70 percent as luminous as it is now, and by the time it leaves the main sequence in another 5 billion years, the sun will have twice its present luminosity. Long before that, the rising luminosity of the sun will raise Earth's average temperature, melt the polar caps, modify Earth's climate, and ultimately drive away our oceans and atmosphere. Life on Earth will probably not survive these changes in the sun, but we have a billion years or more to prepare.

The average star spends 90 percent of its life on the main sequence. This explains why 90 percent of all true stars are main-sequence stars—you are most likely to see a star during that long, stable period while it is on the main sequence. To illustrate: Imagine that you photograph, at a single instant, a crowd of 20,000 people. We all sneeze now and then, but the act of sneezing is very short compared with a human lifetime, so you would expect to find very few people in your photograph in the act of sneezing. Rather, most people in the photo would be in that much more common nonsneezing state. Our short human lifetimes are like a snapshot of the universe, so we see most stars on the main sequence, where stars spend most of their time.

The number of years a star spends on the main sequence depends on its mass (■ Table 9-2). Massive stars consume fuel rapidly and live short lives, but low-mass stars conserve their fuel and shine for billions of years. For example, a 25-solar-mass star will exhaust its hydrogen and die in only about 7 million years. The sun has enough fuel to last about 10 billion years. The red dwarfs, although they have little fuel, use it up very slowly and may be able to survive for 100 billion years or more. (**Reasoning with Numbers 9-1** explains how we can quickly estimate the life expectancies of stars from their masses.)

Nature makes more low-mass stars than high-mass stars, but this fact is not sufficient to explain the vast numbers of low-mass stars that fill the sky. An additional factor is the stellar lifetime. Because low-mass stars live long lives, there are more of them in the sky than massive stars. Look at page 167, and notice how

■ Table 9-2 | Main-Sequence Stars

Spectral Type	Mass (Sun = 1)	Luminosity (Sun = 1)	Years on Main Sequence
O5	40	405,000	1×10^6
B0	15	13,000	11×10^6
A0	3.5	80	440×10^6
F0	1.7	6.4	3×10^9
G0	1.1	1.4	8×10^9
K0	0.8	0.46	17×10^9
M0	0.5	0.08	56×10^9

Star Formation in the Orion Nebula

The visible Orion Nebula shown below is a pocket of ionized gas on the near side of a vast, dusty molecular cloud that fills much of the southern part of the constellation Orion. The molecular cloud can be mapped by radio telescopes. To scale, the cloud would be many times larger than this page. As the stars of the Trapezium were born in the cloud, their radiation has ionized the gas and pushed it away. Where the expanding nebula pushes into the larger molecular cloud, it is compressing the gas and may be triggering the formation of the protostars that can be detected at infrared wavelengths within the molecular cloud.

Side view of Orion Nebula

Hot Trapezium stars

Protostars

To Earth

Expanding ionized hydrogen

Molecular cloud

Daniel Good

A few hundred stars lie within the nebula, but only the four brightest, those in the Trapezium, are easy to see with a small telescope. A fifth star, at the narrow end of the Trapezium, may be visible on nights of good seeing.

Trapezium

The cluster of stars in the nebula is less than 2 million years old. This must mean the nebula is similarly young.

Visual-wavelength image

The near-infrared image below reveals about 50 low-mass, very cool stars that must have formed recently.

Infrared

NASA

Roughly 1000 young stars with hot chromospheres appear in this X-ray image of the Orion Nebula.

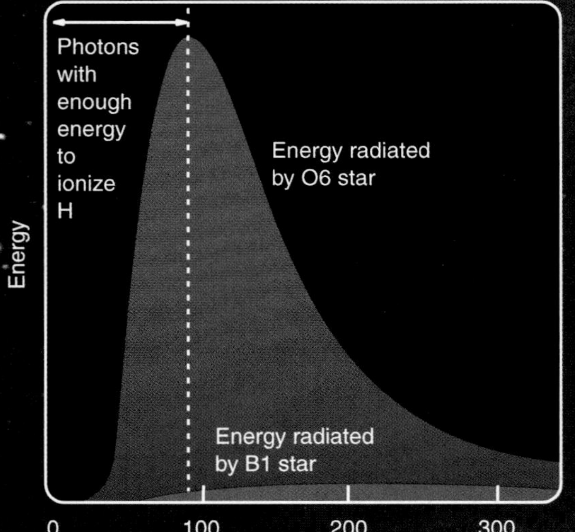

Of all the stars in the Orion Nebula, only one is hot enough to ionize the gas. Only photons with wavelengths shorter than 91.2 nm can ionize hydrogen. The second-hottest stars in the nebula are B1 stars, and they emit little of this ionizing radiation. The hottest star, however, is an O6 star 30 times the mass of the sun. At a temperature of 40,000 K, it emits plenty of photons with wavelengths short enough to ionize hydrogen. Remove that one star, and the nebula would turn off its emission.

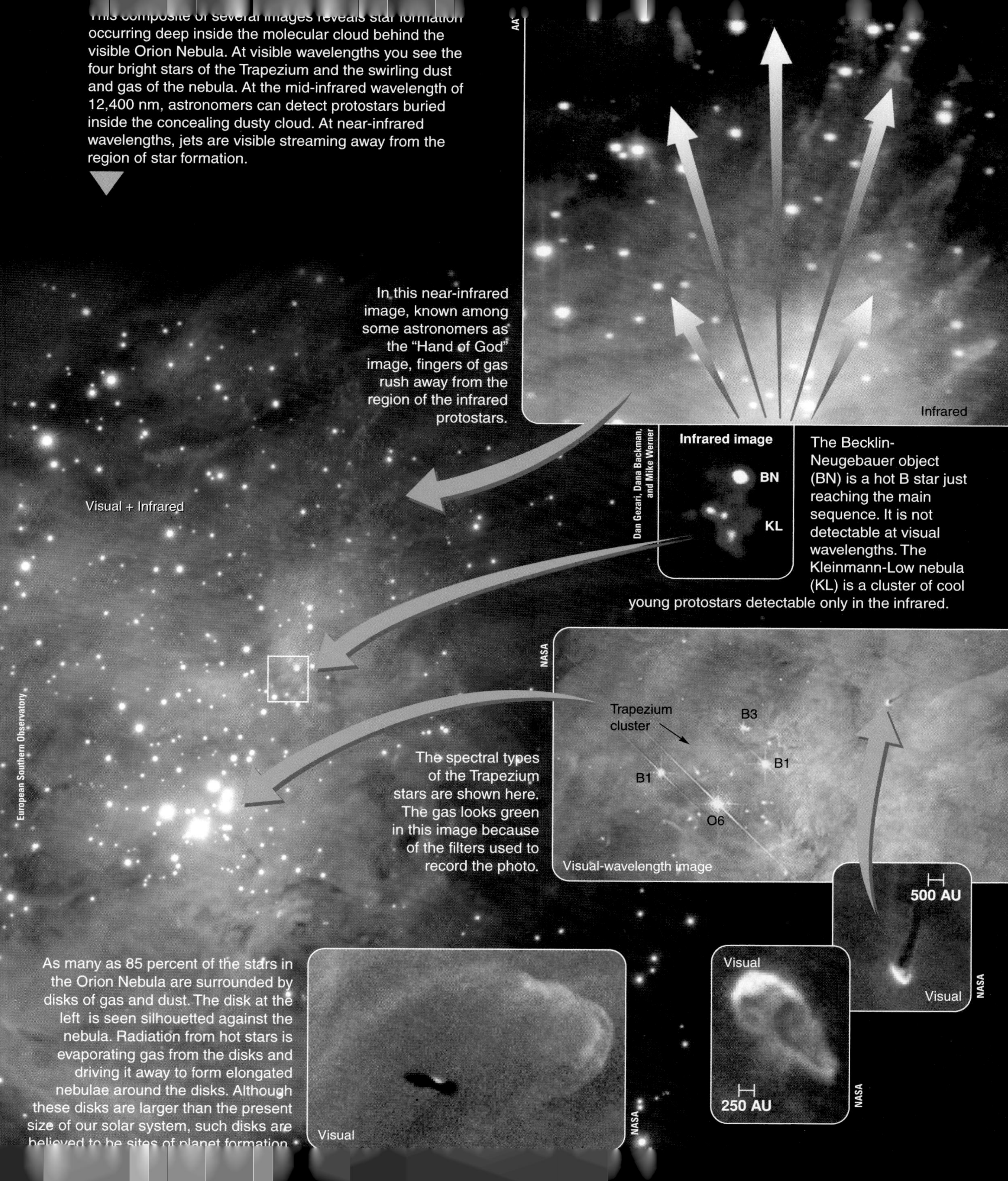

This composite of several images reveals star formation occurring deep inside the molecular cloud behind the visible Orion Nebula. At visible wavelengths you see the four bright stars of the Trapezium and the swirling dust and gas of the nebula. At the mid-infrared wavelength of 12,400 nm, astronomers can detect protostars buried inside the concealing dusty cloud. At near-infrared wavelengths, jets are visible streaming away from the region of star formation.

In this near-infrared image, known among some astronomers as the “Hand of God” image, fingers of gas rush away from the region of the infrared protostars.

The Becklin-Neugebauer object (BN) is a hot B star just reaching the main sequence. It is not detectable at visual wavelengths. The Kleinmann-Low nebula (KL) is a cluster of cool young protostars detectable only in the infrared.

The spectral types of the Trapezium stars are shown here. The gas looks green in this image because of the filters used to record the photo.

As many as 85 percent of the stars in the Orion Nebula are surrounded by disks of gas and dust. The disk at the left is seen silhouetted against the nebula. Radiation from hot stars is evaporating gas from the disks and driving it away to form elongated nebulae around the disks. Although these disks are larger than the present size of our solar system, such disks are believed to be sites of planet formation.

much more common the lower-main-sequence stars are than the massive O and B stars. The main-sequence K and M stars are so faint they are difficult to locate, but they are very common. The O and B stars are luminous and easy to locate; but, because of their fleeting lives, there are never more than a few on the main sequence at any one time.

Ace Astronomy™ Log into AceAstronomy and select this chapter to see Astronomy Exercise "Mass–Lifetime Relation." You can experiment with different mass stars to see how fast they age and how they change.

Inquire | Review | Analyze

Why is there a mass–luminosity relation?

Observational evidence shows that the luminosity of a main-sequence star depends on its mass. The more massive a star is, the more luminous it is. You can understand that if you remember that a star supports itself by being hot inside. The high temperature produces a high pressure, which balances the weight of the layers pressing inward. That is called hydrostatic equilibrium. A massive star has a stronger gravitational field, and it has more mass to support, so the pressure needed to keep it from collapsing is very high. That means the internal temperature must be high, and that means the star must generate energy rapidly. All that energy, leaking outward through the surface of the star, makes the star more luminous.

Often you can understand a process in nature by breaking it down into logical steps. For example, explain why massive stars must live shorter lifetimes than low-mass stars.

■ ■ ■

Connections: Science is a way of understanding nature, and although scientists use theories, they must always go back to reality and compare the theories with observations. In this chapter we have constructed an elaborate theory to explain how stars are born and reach stability and checked it a number of times with observations; but there is one fascinating and beautiful nest of newborn stars you should study before we proceed. The Great Nebula in Orion is our final reality check on the formation of stars.

9-5 The Orion Nebula

ON A CLEAR WINTER NIGHT, you can see with your naked eye the Great Nebula of Orion as a fuzzy wisp in Orion's sword. With binoculars or a small telescope it is striking, and through a large telescope it is breathtaking. At the center lie four brilliant blue-white stars known as the Trapezium, the brightest of a cluster of a few hundred stars. Surrounding the stars are the glowing filaments of a nebula more than 8 pc across. Like a great thundercloud illuminated from within, the churning currents of gas and dust suggest immense power. The significance of the Orion Nebula lies hidden, figuratively and literally, beyond the visible nebula. The region is ripe with star formation.

Evidence of Young Stars

You should not be surprised to find star formation in Orion. The constellation is a brilliant landmark in the winter sky because it is marked by hot, blue stars. These stars are bright in our sky not because they are nearby, but because they are tremendously luminous. These O and B stars cannot live more than a few million years, so you can conclude that they must have been born recently. Furthermore, the constellation contains large numbers of T Tauri stars, which are known to be young. Orion is rich with young stars.

The history of star formation in the constellation of Orion is written in its stars. The stars at Orion's west shoulder are about 12 million years old, whereas the stars of Orion's belt are about 8 million years old. The stars of the Trapezium at the center of the Great Nebula are no older than 2 million years. Apparently, star formation began near the west shoulder, and the massive stars that formed there triggered the formation of the stars you see in Orion's belt. That star formation may have triggered the formation of the stars you see in the Great Nebula. Like a grass fire, star formation has swept across Orion from northwest to southeast.

Study **Star Formation in the Orion Nebula** on pages 194 and 195 and notice four points. First, the nebula you see is only a small part of a vast, dusty molecular cloud. You see the nebula because the stars born within it have ionized the gas and driven it outward, breaking out of the molecular cloud. Also notice that a single hot star is almost entirely responsible for ionizing the gas. The third point to notice is that infrared observations reveal clear evidence of active star formation deeper in the molecular cloud just to the northwest of the Trapezium. Finally, notice that many stars visible in the Orion Nebula are surrounded by disks of gas and dust. Such disks do not last long and are clear evidence that the stars are very young.

In the next million years, the familiar outline of the Great Nebula will change, and a new nebula may begin to form as the protostars in the molecular cloud ionize the gas, drive it away, and become visible. Centers of star formation may develop and then dissipate as massive stars are born and force the gas to expand. If enough massive stars are born, they can blow the entire molecular cloud apart and bring the successive generations of star formation to a final conclusion. The Great Nebula in Orion and its invisible molecular cloud are a beautiful and dramatic example of the continuing cycle of star formation.

Inquire | Review | Analyze

What did Orion look like to the ancient Egyptians, to the first humans, and to the dinosaurs?

The Egyptian civilization had its beginning only a few thousand years ago, and that is not very long in terms of the history of Orion. The stars you see in the constellation are hot and young, but they are a few million years old, so the Egyptians saw the same constellation you see. (They called it Osiris.) Even the Orion Nebula hasn't changed very much in a few thousand years, and Egyptians may have admired it in the dark skies along the Nile.

Our oldest human ancestors lived about 3 million years ago, which was about the time that the youngest stars in Orion were forming. Thus, they may have looked up and seen some of the stars you see, but some stars have formed since that time. Also, the Great Nebula is excited by the Trapezium stars, and they are not more than a few million years old, so our early ancestors probably didn't see the Great Nebula.

The dinosaurs saw something quite different. The last of the dinosaurs died about 65 million years ago, long before the birth of the brightest stars in the constellation you see. The dinosaurs, had they had the brains to appreciate the view, might have seen bright stars along the Milky Way, but when you think back millions of years, you must remember that the stars are moving through space and that the sun is orbiting the center of our galaxy. The night sky above the dinosaurs contained totally different star patterns.

Modern astronomers can see evidence of star formation that was invisible only decades ago. Why are many astronomers interested in star formation observing at infrared wavelengths?

■ ■ ■

Connections: The ancient Aztecs of central Mexico told the story of how the stars, known as the Four Hundred Southerners, were scattered across the sky when they lost a cosmic battle with their brother, the great war god Huitzilopochtli. Modern astronomy tells a less colorful story of the birth of stars, but the modern story is supported by evidence and leads to further questions. If stars are born, then how do they die? That story begins in the next chapter.

Study and Review Tools

Summary

9-1 | The Birth of Stars

How are stars born?

- Stars are born from the gas and dust of the interstellar medium.
- Astronomers know there is an interstellar medium because they can see emission, reflection, and dark nebulae. Also, they can detect interstellar absorption lines in the spectra of distant stars. The dust in the interstellar medium makes distant stars look fainter and redder than expected.
- In large, dense, cool clouds of gas, hydrogen can exist as molecules rather than as atoms. These molecular clouds are sites of star formation. Such clouds can be triggered to collapse by collision with a shock wave, which compresses and fragments the gas cloud, producing a cluster of protostars.
- A contracting protostar is large and cool and would appear in the red-giant region of the H–R diagram were it not surrounded by a cocoon of dust and gas. The dust in the cocoon absorbs the protostar's light and reradiates it as infrared radiation. Many infrared sources are probably protostars.
- T Tauri stars may be stars in later stages as the surrounding dust and gas are absorbed by the protostar or driven into space. Very young star clusters contain large numbers of T Tauri stars.
- Some protostars have been found emitting bipolar flows of gas as they expel material from their surroundings. Where these flows strike existing clouds of gas, we can see nebulae called Herbig–Haro objects. The bipolar flows are evidently focused by disks of gas and dust around the protostars.
- As a protostar grows hot enough to begin hydrogen fusion at its core, it settles onto the main sequence.

9-2 | Fusion in Stars

How do stars make energy?

- Stars of the sun's mass or less make their energy by fusing hydrogen into helium in a process called the proton–proton chain.
- More massive main-sequence stars fuse hydrogen into helium in the CNO process.
- The relationship between pressure and temperature, the pressure-temperature thermostat, sees that the star generates just enough energy to support itself.

9-3 | Stellar Structure

How do stars maintain their stability?

- In a star, the weight of each layer pressing inward must be balanced by the pressure inside pressing outward. This is called hydrostatic equilibrium.
- The energy generated in the core of a star must flow outward by conduction, convection, or radiation. Conduction is not usually important inside stars.
- Much of what astronomers know about stars is based on detailed mathematical models of the interior of stars.

9-4 | Main-Sequence Stars

- The luminosity of a star depends on its mass. More massive stars have more weight to support and their pressure–temperature thermostats must make more energy. That makes them more luminous.
- The main sequence has a lower end because stars less massive than 0.08 solar masses cannot get hot enough to begin hydrogen fusion. Such objects become brown dwarfs.

How long can a star survive?

- How long a star can remain on the main sequence depends on its mass. The more massive a star is, the faster it uses up its hydrogen fuel. A 25-solar-mass star will exhaust its hydrogen and die in only about 7 million years, but the sun is expected to last for 10 billion years.

9-5 | The Orion Nebula

What evidence do astronomers have that theories of star formation are correct?

- The Great Nebula in Orion is an active region of star formation. The bright stars in the center of the nebula formed within the last few million years, and infrared telescopes detect protostars buried inside the molecular cloud that lies behind the visible nebula.

New Terms

interstellar medium (p. 174)
nebula (p. 175)
interstellar reddening (p. 175)
interstellar dust (p. 175)
emission nebula (p. 176)
HII region (p. 176)
reflection nebula (p. 176)
dark nebula (p. 177)
infrared cirrus (p. 178)
molecular cloud (p. 179)
shock wave (p. 179)
association (p. 180)
protostar (p. 180)
evolutionary track (p. 180)
birth line (p. 182)
CNO (carbon–nitrogen–oxygen) cycle (p. 183)
T Tauri star (p. 184)
Bok globule (p. 184)
Herbig–Haro object (p. 185)
bipolar flow (p. 185)
triple-alpha process (p. 186)
conservation of mass (p. 187)
conservation of energy (p. 187)
hydrostatic equilibrium (p. 187)
energy transport (p. 188)
opacity (p. 189)
stellar model (p. 189)
brown dwarf (p. 192)
zero-age main sequence (p. 193)

Review Questions

Ace Astronomy™ Assess your understanding of this chapter's topics with additional quizzing and animations at **http://astronomy.brookscole.com/sh9e**

1. What evidence can you cite that the spaces between the stars are not empty?
2. What evidence can you cite that the interstellar medium contains both gas and dust?
3. Why would an emission nebula near a hot star look red, while a reflection nebula near its star looks blue?
4. Why do astronomers rely heavily on infrared observations to study star formation?
5. What observational evidence can you cite that star formation is a continuous process?
6. How are Herbig–Haro objects related to star formation?
7. How do the proton–proton chain and the CNO cycle resemble each other? How do they differ?
8. Why does the CNO cycle require a higher temperature than the proton–proton chain?
9. How does the pressure–temperature thermostat control the nuclear reactions inside stars?
10. Step by step, explain how energy flows from the sun's core to Earth.
11. Why is there a mass–luminosity relation?
12. Why is there a lower limit to the mass of a main-sequence star?
13. Why does a star's life expectancy depend on its mass?
14. What evidence can you cite that star formation is happening right now in the Orion Nebula?
15. The image at right shows two nebulae, one pink in the background and one black in the foreground. What kind of nebulae are these, and how are they related to star formation?
16. The star at right appears to be ejecting a jet of gas. What is happening to this star?

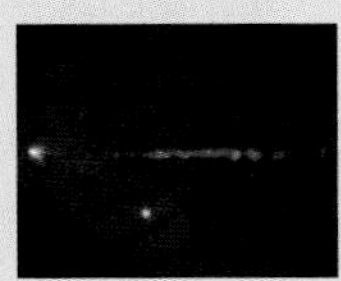

(NASA and the Hubble Heritage Team, STScI/AURA)

(STScI and NASA)

Discussion Questions

1. When you see distant streetlights through smog, they look dimmer and redder than they do normally. But when you see the same streetlights through fog or falling snow, they look dimmer but not redder. Use your knowledge of the interstellar medium to discuss the relative sizes of the particles in smog, fog, and snow compared with the wavelength of light.
2. If planets form as a natural by-product of star formation, which do you think are more common—stars or planets?

Problems

1. The interstellar medium dims starlight by about 1.9 magnitudes/1000 pc. What fraction of photons survive a trip of 1000 pc? (*Hint:* See Reasoning with Numbers 2-1.)
2. A small Bok globule has a diameter of 20 seconds of arc. If the nebula is 1000 pc from Earth, what is the diameter of the globule?
3. If a giant molecular cloud has a diameter of 30 pc and drifts relative to neighboring clouds at 20 km/s, how long will it take to travel its own diameter?
4. If the dust cocoon around a protostar emits radiation most strongly at a wavelength of 30 microns, what is the temperature of the dust? (*Hint:* See Reasoning with Numbers 6-1.)
5. The gas in a bipolar flow can travel as fast as 300 km/s. How long would it take to travel 1 light-year?
6. Circle all ^{1}H and ^{4}He nuclei in Figure 9-9. Explain how both the proton–proton chain and the CNO cycle can be summarized as $4\ ^1\text{H} \rightarrow\ ^4\text{He} + \text{energy}$.
7. In the model shown in Figure 9-14, how much of the sun's mass is hotter than 13,000,000 K?
8. If a brown dwarf has a surface temperature of 1500 K, at what wavelength will it emit the most radiation? (*Hint:* See Reasoning with Numbers 6-1.)
9. What is the life expectancy of a 16-solar-mass star?
10. If the O6 V star in the Orion Nebula is magnitude 5.4, how far away is the nebula? (*Hint:* Use spectroscopic parallax.)
11. The hottest star in the Orion Nebula has a surface temperature of 40,000 K. At what wavelength does it radiate the most energy? (*Hint:* See Reasoning with Numbers 6-1.)

Media Cluster

ACTIVE FIGURES

Ace Astronomy™ To access the resources in the Media Cluster, log into AceAstronomy at http://astronomy.brookscole.com/sh9e and select Chapter 9.

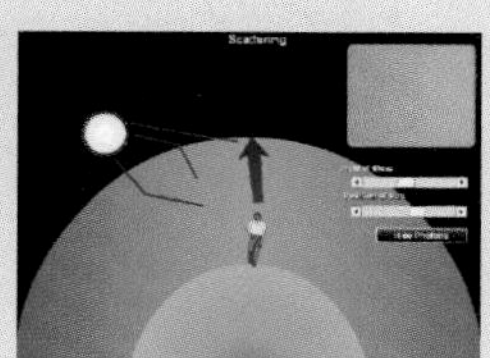

Scattering in Earth's Atmosphere
By absorption, reflection, and refraction, wavelengths of light are selected and filtered to produce the colorful world around you. In this animation, learn more about why you see blue skies and red sunsets and how the sun's position in the sky plays a part.

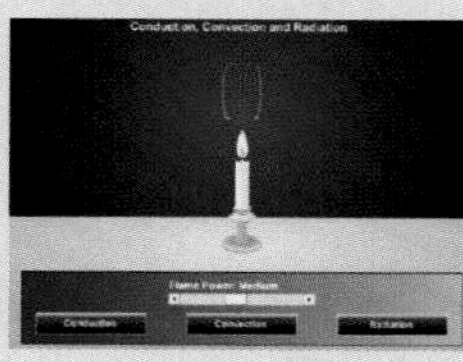

Conduction, Convection, and Radiation
Whether inside a star or on the surface of a planet, the methods of heat transfer available are the same. Make sure you understand the difference between conduction, convection, and radiation with this animation.

ASTRONOMY EXERCISES

Why Is the Sky Blue?
The sky is blue and the setting sun is red for the same reason that reflection nebulae are blue and distant stars are reddened. You can explore that process here in the topic begun in the Active Figure above.

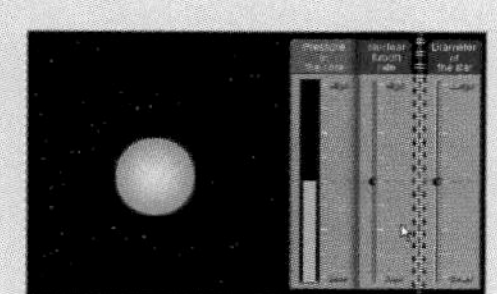

Hydrostatic Equilibrium
This animation allows you to adjust the rate of nuclear fusion and the diameter of a star in hydrostatic equilibrium. You can try, but you can't make this star violate hydrostatic equilibrium.

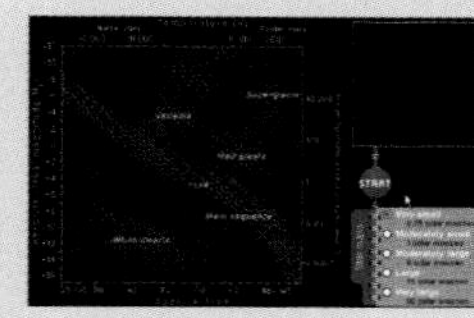

Mass–Lifetime Relation
As stars age, they undergo a number of fundamental changes, such as temperature and chemical composition. In this animation, experiment with different mass stars to see how fast they age and trace their evolution on the H–R diagram.

Critical Inquiries for the Web

1. What can you discover about disks of gas and dust detected around young stars? How are such disks observed, and of what are they composed?
2. Not only have astronomers found brown dwarfs, but they have been able to learn a surprising amount about these faint little failed stars. What can you discover about brown dwarfs on the web? How do the spectra of brown dwarfs help distinguish them from faint stars?
3. What is BM Orionis? Where is it located, and what does it do? Why might it be interesting to observe even in a small telescope?

Exploring *TheSky*

1. The following nebulae are all star-formation regions: M42, M20, M8, M17. What kind of nebulae are they? (*Hint:* To center on an object, use **Find** under the **Edit** menu. Choose **Messier Objects** and pick from the list.)
2. Locate M8 in the sky, zoom in, and identify other nebulae in the region. Study the photo of NGC6559.

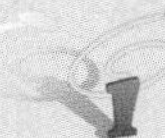

- Go to the Brooks/Cole Astronomy Resource Center (http://astronomy.brookscole.com) for critical thinking exercises, articles, and additional readings from InfoTrac College Edition, Brooks/Cole's online student library.

10 The Deaths of Stars

Natural laws have no pity.

ROBERT HEINLEIN
The Notebooks of Lazarus Long

THE STARS ARE dying. Even as you read this sentence, in galaxies all across the universe, stars are dying. Occasionally, astronomers see a star's last gasps. Sometimes what appears to be a new star appears in the sky, grows brighter, then fades after a few months, and vanishes. In this chapter, you will discover that a **nova,** what seems to be a new star in the sky, is in fact produced by a very old dying star and that a **supernova** is an astonishing violent explosion triggered by the death of certain kinds of stars. Modern astronomers find a few novae (plural of *nova*) each year, but supernovae (plural) are so rare that there are only one or two supernovae each century in our galaxy. Astronomers know that stars die as supernovae

Matter ejected repeatedly from the dying star at the center has formed the nebula known as the Cat's Eye. (NASA, ESA, HEIC, and the Hubble Heritage Team)

Guidepost

Looking Back

You now understand stars well. The preceding chapter described how stars are born and how they resist their own gravity by fusing nuclear fuels in the center and keeping their central pressure high. Stars are the lighting system of the universe, but they can last only as long as their fuel.

This Chapter

Now you are ready to consider the fate of the stars. Here we will see how stars die, and you will find the answers to five essential questions:

What happens to a star when it uses up the last of the hydrogen in its core?

What evidence shows that stars really evolve?

How will the sun die?

What happens if an evolving star is in a binary star system?

How do massive stars die?

The deaths of stars are important because life on Earth depends on the sun, but also because the deaths of massive stars create the atomic elements of which we are made. If stars didn't die, you would not exist.

Looking Ahead

In the chapters that follow, you will discover that some of the matter that was once stars becomes trapped in dead ends—white dwarfs, neutron stars, and black holes. But some matter from dying stars escapes back into the interstellar medium and is incorporated into new stars and the planets that circle them. The deaths of stars are part of a great cycle of stellar birth and death that includes our sun, our planet, and us.

AceAstronomy™ The AceAstronomy icon throughout the text indicates an opportunity for you to test yourself on key concepts and to explore animations and interactions on the AceAstronomy website at: http://astronomy.brookscole.com/sh9e

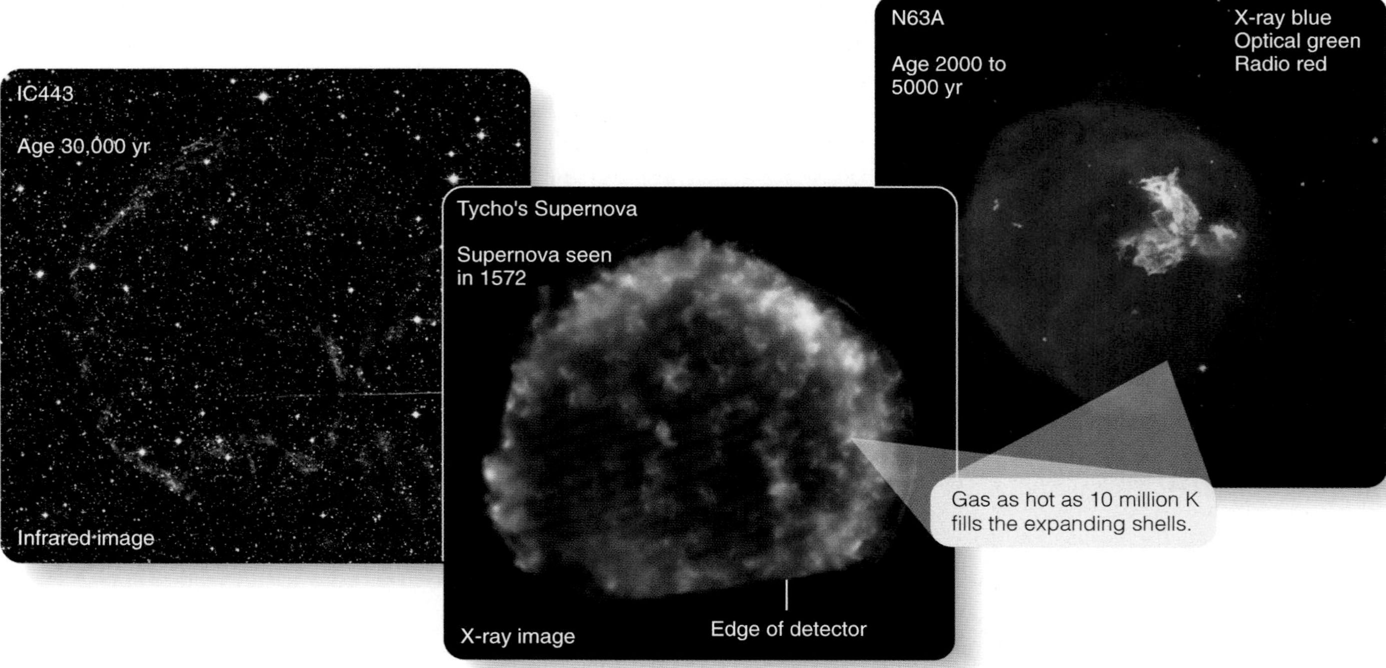

■ **Figure 10-1**

Supernova explosions are rare, but they leave behind expanding shells filled with hot, low-density gas. (IC443: 2MASS; Tycho's Supernova: NASA/CXC/SAO; N63A: NASA/CXC/Rutgers/J. Warren et al., STScI/U.Ill/Y. Chu; ATCA/U.Ill/J. Dickel et al.)

because they occasionally see them flare in other galaxies and because modern telescopes can locate the shattered remains of these violent explosions (■ Figure 10-1).

In the previous chapter, you saw that stars resist their own gravity by generating energy through nuclear fusion. The energy keeps their interiors hot, and the resulting high pressure balances gravity and prevents the star from collapsing. Stars, however, have limited fuel. When they exhaust their fuel, gravity wins, and the stars die.

The mass of a star is critical in determining its fate. Massive stars use up their nuclear fuel at a furious rate and die after only a few million years, whereas the lowest-mass stars use their fuel sparingly and may be able to live hundreds of billions of years. Massive stars can die in violent explosions that appear as supernovae, but low-mass stars die quiet deaths. To follow the evolution of stars to their graves, we will sort the stars into three categories according to their masses: low-mass red dwarfs, medium-mass sunlike stars, and massive upper-main-sequence stars. These stars lead dramatically different lives and die in different ways.

The death of a star leads invariably to one of three final states. Most stars, including stars like the sun, become white dwarfs, stars about the size of Earth with no usable fuels. The most massive stars explode and leave behind either a neutron star or a black hole, exotic objects you will study in detail in the next chapter.

The end states of stellar evolution—white dwarfs, neutron stars, and black holes—represent the final victory of gravity.

10-1 Giant Stars

A MAIN-SEQUENCE STAR GENERATES its energy by nuclear fusion reactions that combine hydrogen to make helium. The period during which the star fuses hydrogen lasts a long time, and the star remains on the main sequence for 90 percent of its total existence as an energy-generating star. When the hydrogen is exhausted, however, the star begins to evolve rapidly. It swells into a giant star and then begins to fuse helium, but it can remain in this giant stage for only about 10 percent of its total lifetime; then it must die. The giant-star stage is the first step in the death of a star.

Expansion into a Giant

The nuclear reactions in a main-sequence star's core fuse hydrogen to produce helium. Because the core is cooler than 100,000,000 K, the helium cannot fuse in nuclear reactions, so it accumulates at the star's center like ashes in a fireplace. Initially, this helium ash has little effect on the star, but as hydrogen is exhausted and the stellar core becomes almost pure helium, the star loses the ability to generate nuclear energy. Because it is the energy generated at the center that opposes gravity and supports the star, the core begins to contract as soon as the energy generation starts to die down.

Although the contracting helium core cannot generate nuclear energy, it does grow hotter because it converts gravitational energy into thermal energy (see Chapter 9). The rising temperature heats the unprocessed hydrogen just outside the core, hydrogen that was never before hot enough to fuse. When the temperature of the surrounding hydrogen becomes high enough, hydrogen fusion begins in a spherical layer or shell around the exhausted core of the star. Like a grass fire burning outward from an exhausted campfire, the hydrogen-fusion shell creeps outward, leaving helium ash behind and increasing the mass of the helium core.

The flood of energy produced by the hydrogen-fusion shell pushes toward the surface, heating the outer layers of the star and forcing them to expand dramatically (■ Figure 10-2). Stars like the sun become giant stars 10 to 100 times the diameter of the sun, and the most massive stars become supergiants some 1000 times larger than the sun. This explains the large diameters and low densities of the giant and supergiant stars. In Chapter 8, you found that giant and supergiant stars were 10 to 1000 times larger in diameter than the sun and from 10 to 10^6 times less dense. Now you understand that these stars were once normal main-sequence stars that expanded to large size and low density when hydrogen-shell fusion began.

The expansion of the envelope dramatically changes the star's location in the H–R diagram. Just as contraction heats a star, expansion cools it. As the outer layers of gas expand, energy is absorbed in lifting and expanding the gas. The loss of that energy lowers the temperature of the gas. Consequently the point that represents the star in the H–R diagram moves quickly to the right (in less than a million years for a star of 5 solar masses). A massive star moves to the right across the top of the H–R diagram and becomes a supergiant, while a medium-mass star like the sun becomes a red giant (■ Figure 10-3). As the radius of a giant star continues to increase, the enlarging surface area makes the star more luminous, moving its point upward in the H–R diagram. Aldebaran, the glowing red eye of Taurus the Bull, is such

■ Figure 10-2

When a star runs out of hydrogen at its center, the core contracts to a small size, becomes very hot, and begins nuclear fusion in a shell (blue). The outer layers of the star expand and cool. The red giant star shown here has an average density much lower than the air at Earth's surface. Here $M_{\odot}$ stands for the mass of the sun, and $R_{\odot}$ stands for the radius of the sun. (Illustration design by author)

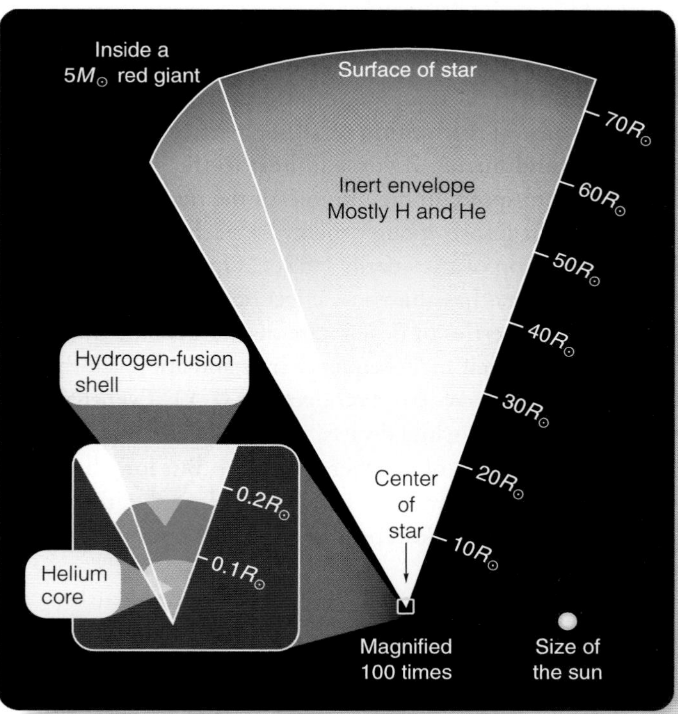

■ Figure 10-3

The evolution of a massive star moves the point that represents it in the H–R diagram to the right of the main sequence into the region of the supergiants such as Deneb and Betelgeuse. The evolution of medium-mass stars moves their points in the H–R diagram into the region occupied by giants such as those shown here.

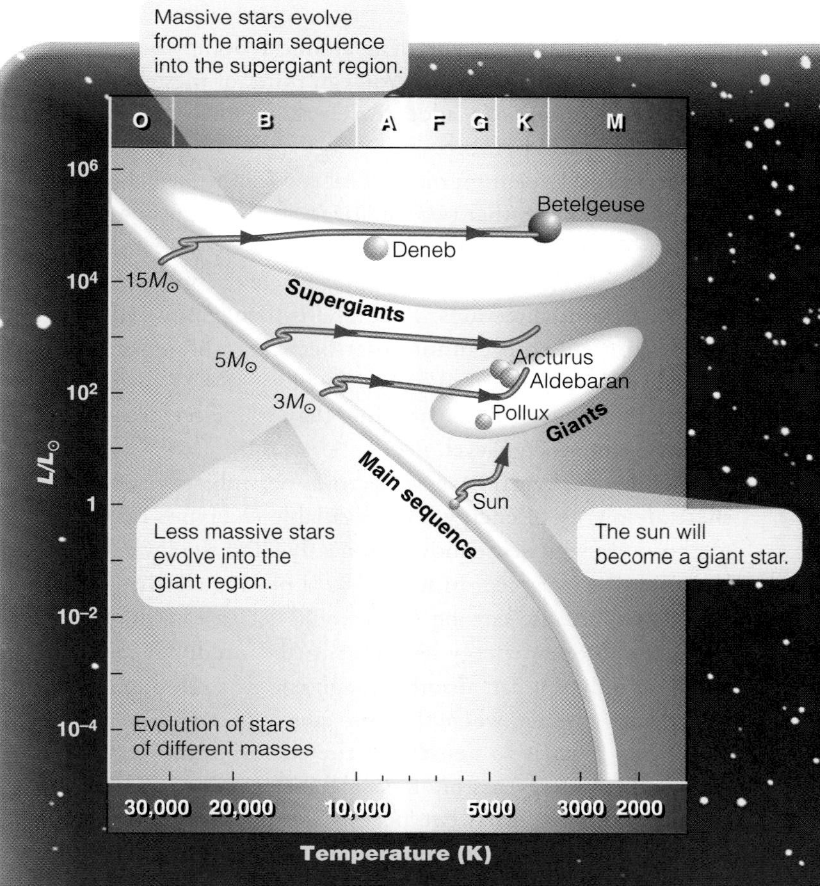

a red giant, with a diameter 25 times that of the sun but with a surface temperature much cooler than the sun.

Degenerate Matter

Although the hydrogen-fusion shell can force the envelope of the star to expand, it cannot stop the contraction of the helium core. Because the core is not hot enough to fuse helium, gravity squeezes it tighter, and it becomes very small. If you represent the helium core of a giant star with a baseball, the outer envelope of the star would be about the size of a baseball stadium. Yet the core would contain about 12 percent of the star's mass compressed to very high density. When gas is compressed to such extreme densities, it begins to behave in astonishing ways that can alter the evolution of the star. Thus, to follow the story of stellar evolution, you must consider the behavior of gas at extremely high densities.

Normally, the pressure in a gas depends on its temperature. The hotter the gas is, the faster its particles move and the more pressure it exerts; but the gas inside a star is ionized, so there are two kinds of particles: atomic nuclei and free electrons. If the gas is compressed to very high densities, for example, in the core of a giant star, the difference between these two kinds of particles becomes important.

If the density is very high, the particles of the gas are forced close together, and two laws of quantum mechanics become significant. First, quantum mechanics says that the moving electrons confined in the star's core can have only certain amounts of energy, just as the electron in an atom can occupy only certain energy levels (see Chapter 6). You can think of these permitted energies as the rungs of a ladder. An electron can occupy any rung but not the spaces between.

The second quantum mechanical law (called the Pauli exclusion principle) says that two identical electrons cannot occupy the same energy level. Because electrons spin in one direction or the other, two electrons can occupy a single energy level if they spin in opposite directions. That level is then completely filled, and a third electron cannot enter because, whichever way it spins, it will be identical to one or the other of the two electrons already in the level. Thus, no more than two electrons can occupy the same energy level.

A low-density gas has few electrons per cubic centimeter, so there are plenty of energy levels available (■ Figure 10-4). If a gas becomes very dense, however, nearly all of the lower energy levels may be occupied. In such a gas, a moving electron cannot slow down, because slowing down would decrease its energy, and there are no open energy levels for it to drop down to. It can speed up only if it can absorb enough energy to leap to the top of the energy ladder, where there are empty energy levels.

When a gas is so dense that the electrons are not free to change their energy, astronomers call it **degenerate matter.** Although it is a gas, it has two peculiar properties that can affect the star. First, the degenerate gas resists compression. To compress the gas, you must push against the moving electrons, and changing their motion means changing their energy. That requires tremendous effort, because you must boost them to the top of the energy ladder. Thus, degenerate matter, though still a gas, takes on the consistency of hardened steel.

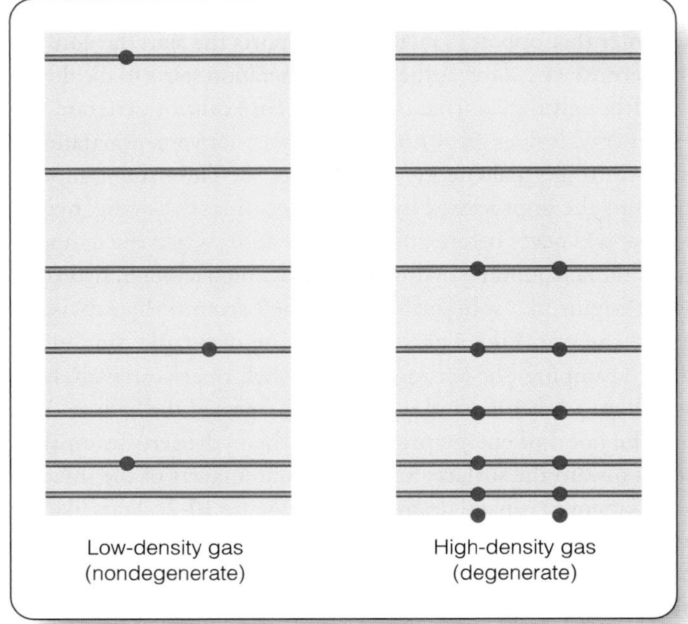

■ **Figure 10-4**

Electron energy levels are arranged like rungs on a ladder. In a low-density gas many levels are open, but in a degenerate gas all lower-energy levels are filled.

Second, the pressure of degenerate gas does not depend on temperature but rather on the speed of the electrons, which cannot be changed without tremendous effort. The temperature, however, depends on the motion of all the particles in the gas, both electrons and nuclei. If you add heat to the gas, most of that energy goes to speed up the motions of the nuclei, and only a few electrons can absorb enough energy to reach the empty energy levels at the top of the energy ladder. Thus, changing the temperature of the gas has almost no effect on the pressure.

These two properties of degenerate matter become important when stars end their main-sequence lives and approach their final collapse (**Window on Science 10-1**). Eventually, many stars collapse into white dwarfs, and you will discover that these tiny stars are made of degenerate matter. But long before that, the cores of many giant stars become so dense that they are degenerate, a situation that can produce a cosmic bomb.

Helium Fusion

Hydrogen fusion in main-sequence stars leaves behind helium ash, which cannot begin fusing because the temperature is too low. Helium nuclei have a positive charge twice that of a proton,

Causes in Nature: The Very Small and the Very Large

One of the most interesting lessons of science is that the behavior of very small things often determines the structure and behavior of very large things. In degenerate matter, the quantum-mechanical behavior of electrons helps determine the evolution of giant stars. Such links between the very small and the very large are common in astronomy, and you will see in later chapters how the nature of certain subatomic particles may determine the fate of the entire universe.

This is a second reason why astronomers study atoms. Recall that the first reason is that atoms interact with light and astronomers must understand that interaction in order to analyze the light. The second reason is the link between the very small and the very large. It would be impossible to understand the evolution of the largest stars if astronomers failed to understand how degenerate electrons behave. Similarly, astronomers studying galaxies must understand how microscopic dust in space reddens starlight, and astronomers studying planets must know how atoms form crystals.

As you study any science, look for the ways very small things affect very large things. Sociologists and psychologists know that the mass behavior of very large groups of people can depend on the behavior of a few key individuals. Biologists study the visible consequences of atomic bonds in molecules called genes, and meteorologists know that tiny changes in temperature in one part of the world can affect worldwide climate weeks or months later.

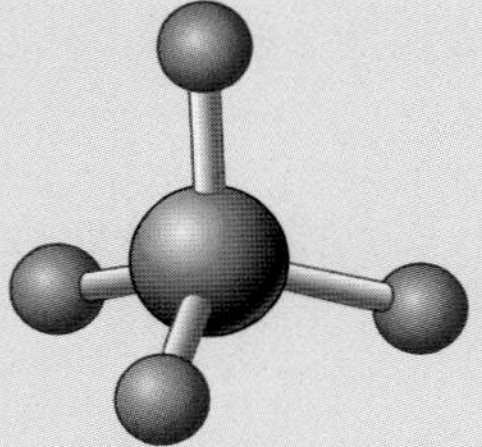

The properties of atoms and molecules determine the structure of the largest objects in the universe.

Nature is the sum of many tiny parts, and science is the study of nature. Thus, one way to do science is to search out the tiny causes that determine the way nature behaves on the grandest scales.

and so, to overcome the repulsion between nuclei, they must collide at a velocity higher than that at which hydrogen fuses. The temperature for hydrogen fusion is too cool to fuse helium. As the star becomes a giant star fusing hydrogen in a shell, the inner core of helium ash contracts and grows hotter. It may even become degenerate, but when it finally reaches a temperature of 100,000,000 K, it begins to fuse helium nuclei to make carbon (see Chapter 9).

How a star begins helium fusion depends on its mass. Stars more massive than about 3 solar masses contract rapidly, their helium-rich cores heat up, and helium fusion begins gradually. But less-massive stars evolve more slowly, and their cores contract so much the gas becomes degenerate. On Earth, a teaspoon of the gas would weigh as much as an automobile. In this degenerate matter, the pressure does not depend on temperature, and that means the pressure–temperature thermostat does not regulate energy production. When the temperature becomes hot enough, helium fusion begins to make energy and the temperature rises, but pressure does not increase because the gas is degenerate. The higher temperature increases the helium fusion even further, and the result is a runaway explosion called the **helium flash** in which, for a few minutes, the core of the star can generate more energy per second than does an entire galaxy.

Although the helium flash is sudden and powerful, it does not destroy the star. In fact, if you were observing a giant star as it experienced the helium flash, you would probably see no outward evidence of the eruption. The helium core is quite small (Figure 10-2), and all of the energy of the explosion is absorbed by the distended envelope. In addition, the helium flash is a very short-lived event in the life of a star. In a matter of minutes to hours, the core of the star becomes so hot it is no longer degenerate, the pressure–temperature thermostat brings the helium fusion under control, and the star proceeds to fuse helium steadily in its core (■ Figure 10-5).

The sun will experience a helium flash, but stars less massive than about 0.4 solar mass never get hot enough to ignite helium. Stars more massive than 3 solar masses ignite helium before their contracting cores become degenerate.

Whether a star experiences a helium flash or not, the ignition of helium in the core changes the structure of the star. The star now makes energy in its helium-fusion core and in its hydrogen-fusion shell. The energy flowing outward from the core can halt the contraction of the core, and the distended envelope of the star contracts and grows hotter. Thus, the point that represents the star in the H–R diagram moves downward and to the left toward the hot side of the H–R diagram.

Helium fusion produces carbon and oxygen atoms in the core, atoms that require much higher temperatures to fuse. Thus, as the helium fuel is used up, carbon and oxygen atoms accumulate in an inert core. Once again, the core contracts and heats up, and a helium-fusion shell ignites below the hydrogen-fusion shell. The star now makes energy in two fusion shells, it quickly expands, and its surface cools once again. The point that represents the star in the H–R diagram moves back to the right, completing a loop (Figure 10-5).

Now you can understand why giant stars are so rare (see page 167). A star spends about 90 percent of its lifetime on the main sequence and only 10 percent as a giant star. During any

as theories predict? To answer that question, let's turn to clusters of stars.

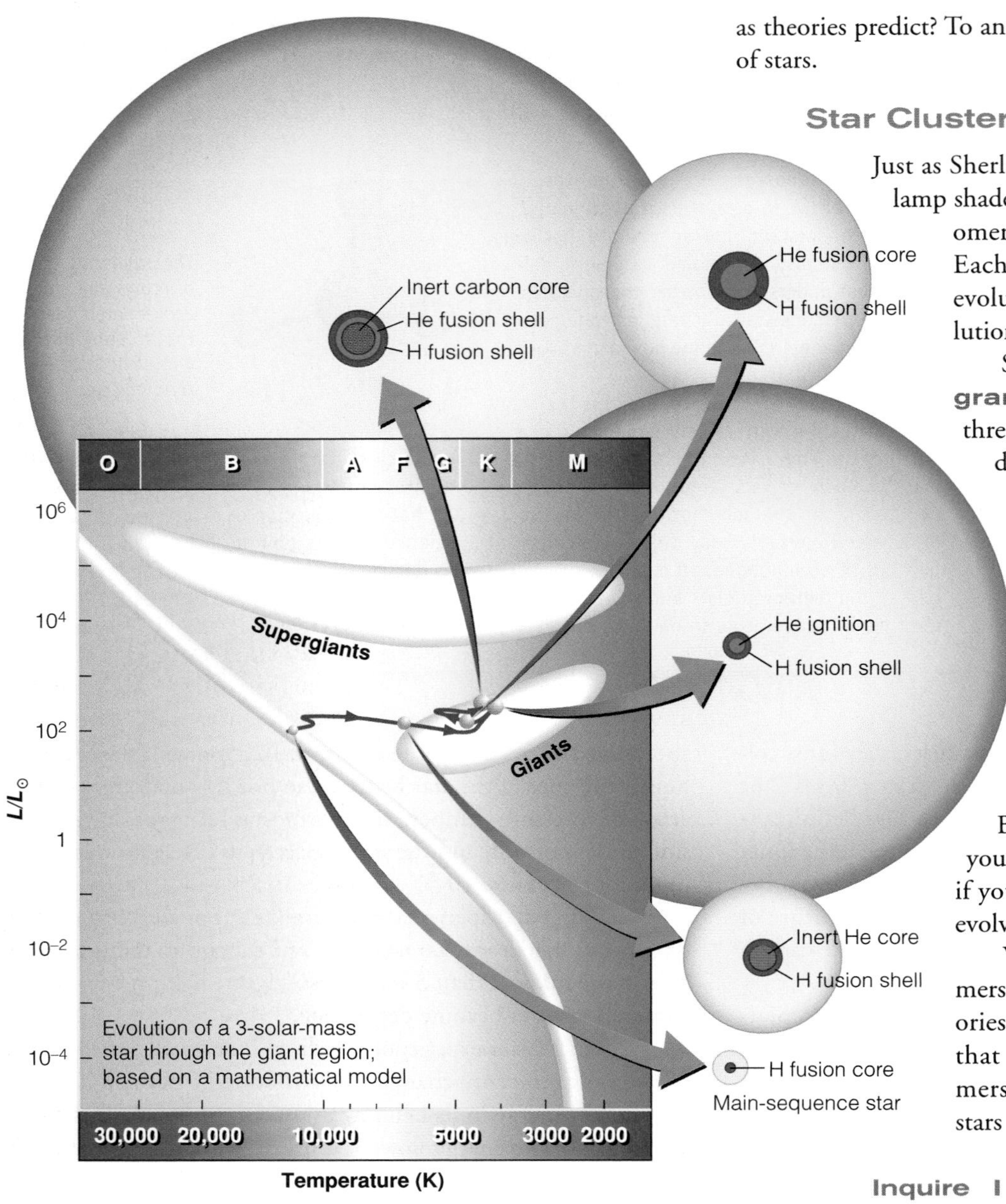

Star Clusters: Evidence of Evolution

Just as Sherlock Holmes studies peculiar dust on a lamp shade as evidence to solve a mystery, astronomers look at star clusters and say, "Ah ha!" Each star cluster freezes a moment in the evolution of the cluster and makes the evolution of the stars visible.

Study **Star Cluster H–R Diagrams** on pages 208 and 209 and notice three important points. First, there are two different kinds of star clusters, but they are similar in the way their stars evolve. We will discuss these clusters in more detail in a later chapter. Second, notice that you can estimate the age of a star cluster by observing the distribution of the points that represent its stars in the H–R diagram. Finally, notice that the shape of a star cluster's H–R diagram is governed by the evolutionary path the stars take. By comparing clusters of different ages, you can visualize how stars evolve almost as if you were watching a film of a star cluster evolving over billions of years.

Were it not for star clusters, astronomers would have little confidence in the theories of stellar evolution. Star clusters make that evolution visible and assure astronomers that they really do understand how stars are born, live, and die.

Figure 10-5

When a main-sequence star exhausts the hydrogen in its core, it evolves rapidly to the right in the H-R diagram as it expands to become a cool giant. These changes are caused at first by the ignition of a hydrogen fusion shell. When helium ignites in the core, the star contracts and heats up slightly, but when helium begins to fuse in a shell, the star again expands and cools.

particular human lifetime, only a fraction of the visible stars will be passing through the giant stage.

What happens to a star after helium fusion depends on its mass, but no matter what tricks the star plays to delay its end, it cannot survive long. It must eventually collapse and end its career as a star. We will trace the details of this process of stellar death in the remainder of this chapter, but first you must ask a critical question. What evidence shows that stars actually evolve

Inquire | Review | Analyze

How are giant stars different from main-sequence stars?

You have seen that main-sequence stars begin the death process by expanding and cooling to become giant or supergiant stars, and in our discussion you discovered that these bloated stars are fundamentally different from main-sequence stars. Stars like the sun fuse hydrogen in their cores, and the energy flows outward to support the weight of the star. The pressure–temperature thermostat regulates the rate of energy production of a stable main-sequence star. But a giant star has used up the hydrogen in its core, and it can no longer maintain that main-sequence stability. Hydrogen fusion continues in a shell, but the core of helium ash contracts while the envelope of the star expands and cools. Later, helium fusion will begin in the core of the star and will produce an inert carbon–oxygen core and a helium-fusion shell below the hydrogen-fusion shell. Thus, the internal structure of a giant star is dramatically different from the simple structure of a main-sequence star. One of the pitfalls of critical analysis is to assume that similar consequences must arise from similar causes; in this case, main-sequence stars and giants look very similar, but they are quite different inside.

Consider a 5-solar-mass star and the sun. They are both spheres of hot gas held together by gravity and supported by internal pressure. They look quite similar, and both will become giant stars, yet their internal evolution includes an important difference, the helium flash. What makes the difference?

■ ■ ■

Connections: You have now heard the story of stellar evolution in a general way, and you are ready to add some details. Let's begin by considering the low- and medium-mass stars.

10-2 The Deaths of Lower-Main-Sequence Stars

CONTRACTING STARS HEAT UP by converting gravitational energy into thermal energy. Low-mass stars have little gravitational energy, so when they contract, they can't get very hot. This limits the fuels they can ignite. In Chapter 9, you saw that protostars less massive than 0.08 solar mass cannot get hot enough to ignite hydrogen. Consequently, this section will concentrate on stars more massive than 0.08 solar mass but no more than a few times the mass of the sun.

Structural differences divide the lower-main-sequence stars into two subgroups—very-low-mass red dwarfs and medium-mass stars such as the sun. The critical difference between the two groups is the extent of interior convection. If the star is convective, fuel is constantly mixed, and the resulting evolution is drastically altered.

Red Dwarfs

Stars less massive than about 0.4 solar mass—the red dwarfs—have two advantages over more massive stars. First, they have very small masses, and thus they have very little weight to support. Their pressure–temperature thermostats are set low, and they consume their hydrogen fuel very slowly. Our discussion of the life expectancies of stars in Chapter 9 predicted that the red dwarfs could live very long lives.

The red dwarfs have a second advantage because they are totally convective. That is, they are stirred by circulating currents of hot gas rising from the interior and cool gas sinking inward. This means the stars are mixed like a pot of soup that is constantly stirred as it cooks. Hydrogen is consumed and helium accumulates uniformly throughout the star, which means the star is not limited to the fuel in its core. It can use all of its hydrogen to prolong its life on the main sequence.

Because a red dwarf is mixed by convection, it cannot develop an inert helium core surrounded by unprocessed hydrogen. Then it can never ignite a hydrogen shell and thus cannot become a giant star. Rather, nuclear fusion converts hydrogen into helium, which cannot fuse because the star cannot get hot enough. What astronomers know about stellar evolution indicates that these red dwarfs should use up nearly all of their hydrogen and live very long lives on the lower main sequence. They could survive for a hundred billion years or more. Of course, astronomers can't test this part of their theories because the universe is only about 14 billion years old, so not a single red dwarf has died of old age anywhere in the universe. Every red dwarf that has ever been born is still shining today.

Medium-Mass Stars

Stars with masses between roughly 4 solar masses and 0.4 solar mass,* including the sun, evolve in the same way. They can ignite hydrogen and helium and become giants, but they cannot get hot enough to ignite carbon, the next fuel after helium. When they reach that impasse, they collapse and become white dwarfs. There are two keys to the evolution of these sunlike stars: the lack of complete mixing and mass loss.

The interiors of medium-mass stars are not completely mixed (■ Figure 10-6). Stars of 1.1 solar masses or less have no convection near their centers, so they are not mixed at all. Stars with a mass greater than 1.1 solar masses have small zones of convection at their centers, but this mixes no more than about 12 percent of the star's mass. Thus medium-mass stars, whether they have convective cores or not, are not mixed, and the helium ash accumulates in an inert helium core surrounded by unprocessed hydrogen. When this core contracts, the unprocessed hydrogen ignites in a shell and swells the star into a giant.

In the giant stage, the core of the star contracts and the envelope expands. The star fuses helium in its core and then in a shell surrounding a core of carbon and oxygen. This core contracts and grows hotter; but, because the star has too low a mass, the core cannot get hot enough to ignite carbon fusion. Thus the carbon–oxygen core is a dead end for these medium-mass stars.

All of this discussion is based on theoretical models of stars and a general understanding of how stars evolve. Does it really happen? Astronomers need observational evidence to confirm this theoretical discussion, and the gas that is lost from these giant stars gives visible evidence that sunlike stars do indeed die in this way.

Planetary Nebulae

When a medium-mass star like the sun becomes a distended giant, it can expel its outer atmosphere to form one of the most interesting objects in astronomy, a **planetary nebula,** so called because it looks like the greenish-blue disk of a planet such as Uranus or Neptune. In fact, a planetary nebula has nothing to do with a planet. It is composed of ionized gases expelled by a dying star.

*This mass limit is uncertain, as are many of the masses quoted here. The evolution of stars is highly complex, and such parameters are difficult to specify.

r Cluster H–R grams

lusters are beautiful, but they are also o understanding the evolution of stars. the stars in a star cluster formed at the same time from the same cloud . That means the stars have the same nd the same initial chemical composition. ifferences you see among stars in a star r must arise from differences in mass, at makes stellar evolution visible.

n cluster is a collection of 10 to 1000 a region about 25 pc in diameter. Some lusters are quite small and some are large, y all have an open, transparent appearance e the stars are not crowded together.

Open Cluster
The Jewel Box

In a star cluster each star follows its orbit around the center of mass of the cluster.

Visual-wavelength image

AURA/NOAO/NSF

A **globular cluster** can contain 10^5 to 10^6 stars in a region only 10 to 30 pc in diameter. The term “globular cluster” comes from the word “globe,” although globular cluster is pronounced like “glob of butter.” These clusters are nearly spherical, and the stars are much closer together than the stars in an open cluster.

Globular Cluster
47 Tucanae

Visual-wavelength image

ATFB

Astronomers can construct an H–R diagram for a star cluster by plotting a point to represent the luminosity and temperature of each star.

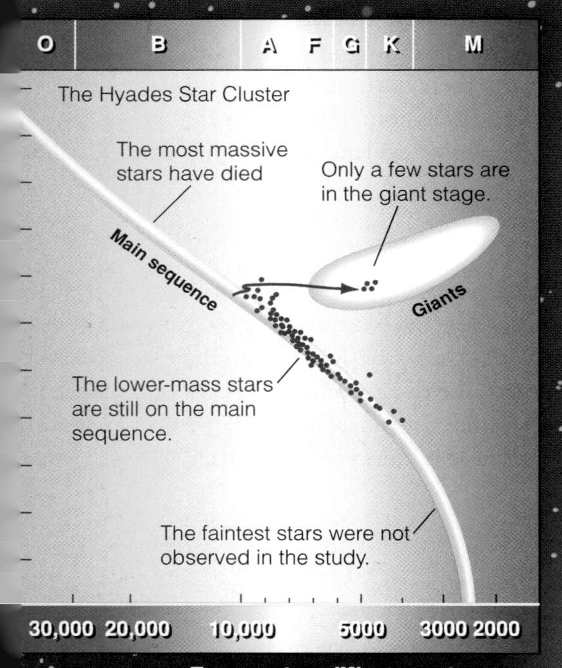

O
B
A
F
G
K
M
The Hyades Star Cluster
The most massive stars have died
Only a few stars are in the giant stage.
Main sequence
Giants
The lower-mass stars are still on the main sequence.
The faintest stars were not observed in the study.
30,000
20,000
10,000
5000
3000
2000
Temperature (K)

The H–R diagram of a star cluster can make the evolution of stars visible. The key is to remember that all of the stars in the star cluster have the same age but differ in mass. The H–R diagram of a star cluster provides a snapshot of the evolutionary state of the stars at the time you happen to be alive. The diagram here shows the 650-million-year-old star cluster called the Hyades. The upper main sequence is missing because the more massive stars have died, and our snapshot catches a few medium-mass stars leaving the main sequence to become giants.

As a star cluster ages, its main sequence grows shorter like a candle burning down. You can judge the age of a star cluster by looking at the turnoff point, the point on the main sequence where stars evolve to the right to become giants. Stars at the **turnoff point** have lived out their lives and are about to die. Consequently, the life expectancy of the stars at the turnoff point equals the age of the cluster.

AceAstronomy™ Log into AceAstronomy and select this chapter to see Active Figure “Cluster Turnoff” and notice how the shape of a cluster's H–R diagram changes with time.

heoretical models of stars, you could construct a film to show how the iagram of a star cluster changes as it ages. You can then compare theory vith observation (right) to understand how stars evolve. Note that the time step for each frame in this film increases by a factor of 10.

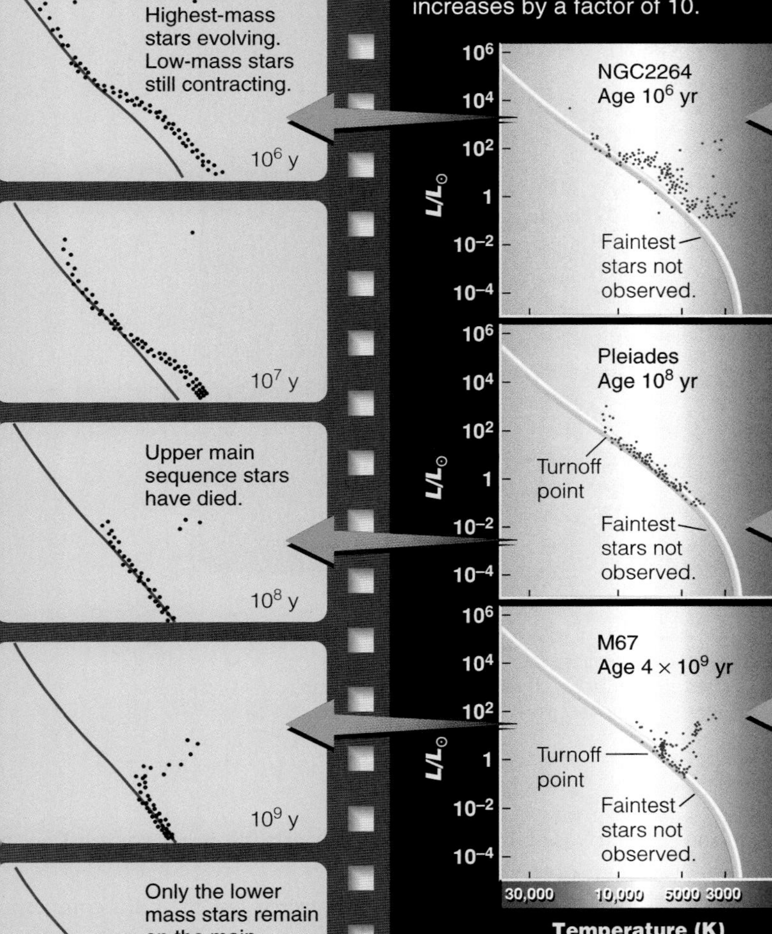

NGC 2264 is a very young cluster still embedded in the nebula from which it formed. Its lower-mass stars are still contracting, and it is rich in T Tauri stars.

Visual

The nebula around the Pleiades is produced by gas and dust through which the cluster is passing. Its original nebula dissipated long ago.

Visual

M67 is an old open cluster. In photographs, such clusters have a uniform appearance because they lack hot, bright stars. Compare with the Jewel Box on the opposite page.

Visual

ar cluster H–R diagrams, as shown below, resemble the last in the film, which tells you that globular clusters are very old.

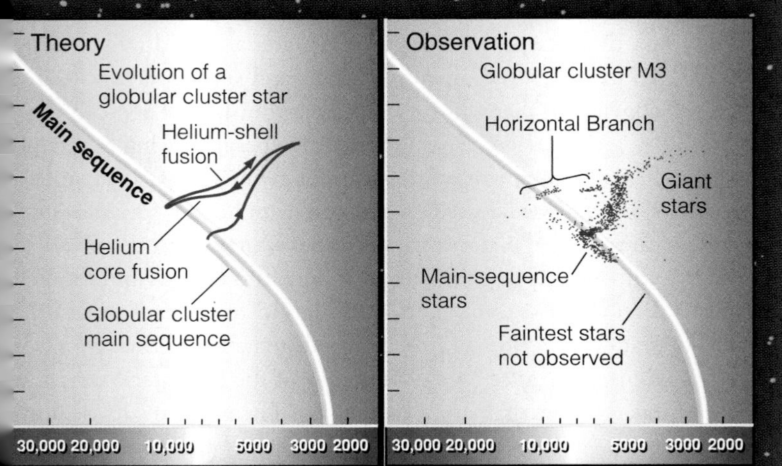

The H–R diagrams of globular clusters have very faint turnoff points showing that they are very old clusters. The best analysis suggests these clusters are about 11 billion years old.

The **horizontal branch** stars are giants fusing helium in their cores and then in shells. The shape of the horizontal branch outlines the evolution of these stars.

The main-sequence stars in globular clusters are fainter and bluer than the zero-age main sequence. Spectra reveal that globular cluster stars are poor in elements heavier than helium, and that means their gases are less opaque. That means energy can flow outward more easily, which makes the stars slightly smaller and hotter. Again the shape of star cluster H–R diagrams illustrates principles of stellar evolution

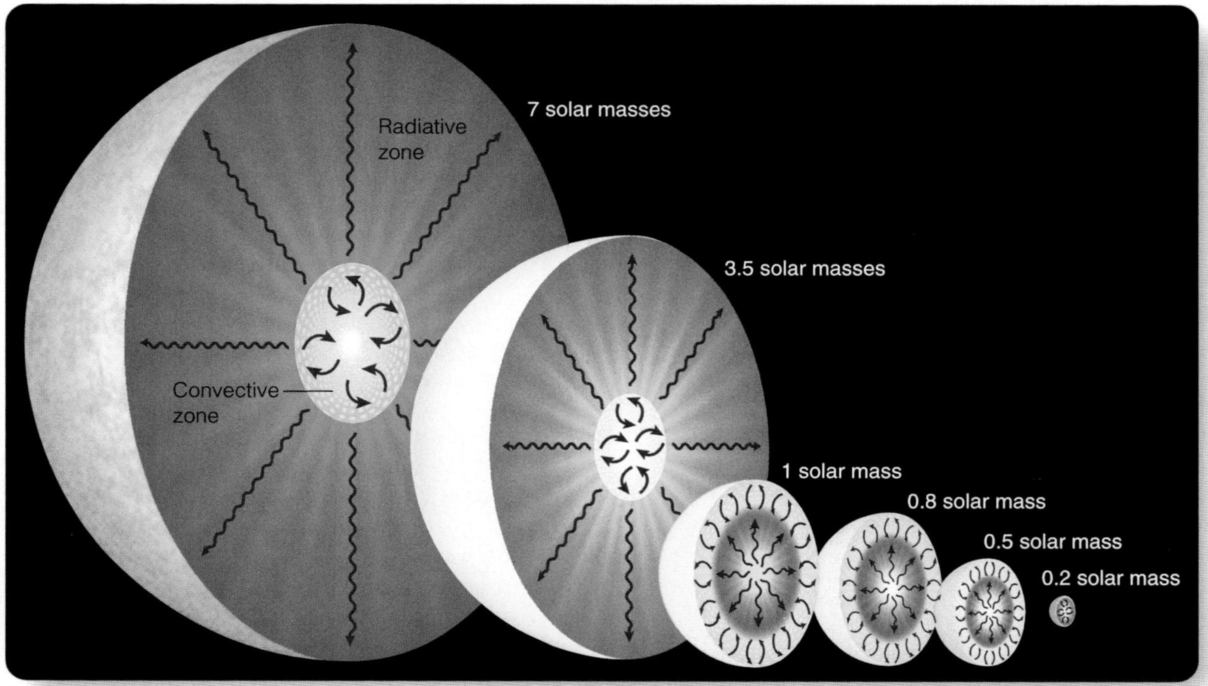

Figure 10-6

Inside stars. The more massive stars have small convective interiors and radiative envelopes. Stars like the sun have radiative interiors and convective envelopes. The lowest-mass stars are convective throughout. The "cores" of the stars where nuclear fusion occurs (not shown) are smaller than the interiors. (Illustration design by author)

Study **The Formation of Planetary Nebulae** on pages 212 and 213 and notice four things. First, you can understand what planetary nebulae are like by using simple observational methods previously discussed. Second, notice the model astronomers have developed to explain planetary nebulae. The real nebulae are more complex than the simple model of a slow wind and a fast wind, but the model gives you a way to organize the observed phenomena. The third thing to notice is how oppositely directed jets (much like bipolar flows) produce many of the asymmetries seen in planetary nebulae. Finally, notice the fate of the star itself; it must contract into a white dwarf.

The medium-mass stars die by ejecting gas into space and contracting into white dwarfs. This suggests another way you can search for evidence of the deaths of medium-mass stars: You can study white dwarfs.

White Dwarfs

Chapter 8 surveyed the stars and revealed that white dwarfs are the second most common kind of star (see page 167). Only red dwarfs are more abundant. Now you can recognize these white dwarfs as the remains of medium-mass stars that fused hydrogen and helium, failed to ignite carbon, drove away their outer layers to form planetary nebulae, and collapsed and cooled to form white dwarfs. The billions of white dwarfs in our galaxy must be the remains of medium-mass stars.

The first white dwarf discovered was the faint companion to Sirius. In that visual binary system, the bright star is Sirius A. The white dwarf, Sirius B, is 10,000 times fainter than Sirius A. The orbital motions of the stars (shown in Figure 8-12) tell you that the white dwarf's mass is about 1 solar mass, and its blue-white color tells you that its surface is hot, about 25,000 K. Although it is very hot, it has a very low luminosity, so it must have a small surface area (see Reasoning with Numbers 8-3)—in fact, its diameter is only about twice Earth's. Dividing its mass by its volume reveals that it is very dense—over 3×10^6 g/cm^3. On Earth, a teaspoonful of Sirius B material would weigh more than 15 tons. Thus, basic observations and simple physics lead to the conclusion that white dwarfs are astonishingly dense.

A normal star is supported by energy flowing outward from its core, but a white dwarf cannot generate energy by nuclear fusion. It has exhausted its hydrogen and helium fuel and produced carbon. As the star collapses into a white dwarf, it converts gravitational energy into thermal energy, and its interior becomes very hot; but it cannot get hot enough to ignite carbon fusion. The star contracts until it becomes degenerate. Although a tremendous amount of energy flows out of the hot interior, it is not the energy flow that supports the star. The white dwarf is supported against its own gravity by the inability of its degenerate electrons to pack into a smaller volume.

The interior of a white dwarf is mostly carbon and oxygen ions immersed in a whirling storm of degenerate electrons. Theory predicts that as the star cools these ions will lock together to form a crystal lattice, so there may be some truth in thinking of aging white dwarfs as great crystals of carbon and oxygen. Near the surface, where the pressure is lower, a layer of ionized gases makes up a hot atmosphere, which, because of the strong gravity, is pulled down into a shallow layer. If Earth's atmosphere were equally shallow, people on the top floors of skyscrapers would have to wear space suits.

Clearly, a white dwarf is not a true star. It generates no nuclear energy, is almost totally degenerate, and, except for a thin layer at its surface, contains no gas. Instead of calling a white dwarf a "star," you can call it a "compact object." The next chapter, discusses two other compact objects—neutron stars and black holes.

A white dwarf's future is bleak. As it radiates energy into space, its temperature gradually falls, but it cannot shrink any smaller because its degenerate electrons cannot get closer together. This degenerate matter is a very good thermal conductor, so heat flows to the surface and escapes into space, and the white dwarf gets fainter and cooler, moving downward and to the right in the H–R diagram. Because the white dwarf contains a tremendous amount of heat, it needs billions of years to radiate that heat through its small surface area. Eventually, such objects may become cold and dark, so-called **black dwarfs.** Our galaxy is not old enough to contain black dwarfs. The coolest white dwarfs in our galaxy are about the temperature of the sun.

Perhaps the most interesting thing astronomers have learned about white dwarfs has come from mathematical models. The equations predict that if you added mass to a white dwarf, its radius would *shrink* because added mass would increase its gravity and squeeze it tighter. If you added enough to raise its total mass to about 1.4 solar masses, its radius would shrink to zero (■ Figure 10-7). This is called the **Chandrasekhar limit** after Subrahmanyan Chandrasekhar, the astronomer who discovered it. It seems to imply that a star more massive than 1.4 solar masses could not become a white dwarf unless it shed mass in some way.

Stars do lose mass. Young stars have strong stellar winds, and aging giants also lose mass (■ Figure 10-8). This suggests that stars more massive than the Chandrasekhar limit can eventually die as white dwarfs if they reduce their mass. Theoretical models show that an 8-solar-mass star should be able to reduce its mass

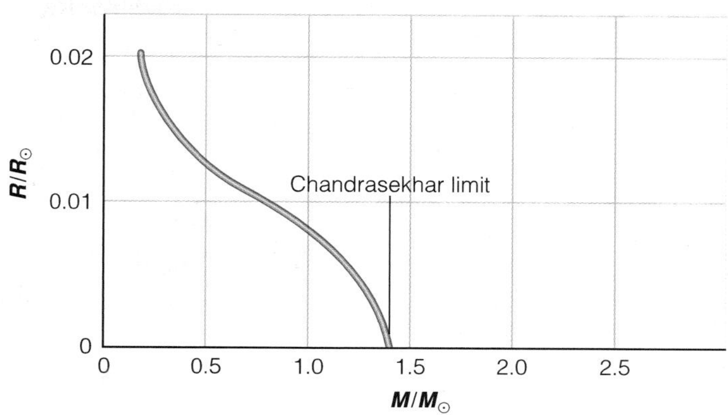

0.02
0.01
0
$R/R_\odot$
Chandrasekhar limit
0
0.5
1.0
1.5
2.0
2.5
$M/M_\odot$

■ **Figure 10-7**

The more massive a white dwarf is, the smaller its radius. Stars more massive than the Chandrasekhar limit of 1.4 $M_\odot$ cannot be white dwarfs.

■ **Figure 10-8**

Stars can lose mass if they are very hot, very large, or both. The red supergiant V838 Mon has lost mass in the past, as revealed by light emitted when it erupted temporarily in January 2002. A young massive star such as WR124 and the hot, blue stars that make up the constellation Orion constantly lose mass into space. Warmed dust in these gas clouds can make them glow in the infrared. (V838 Mon: NASA and The Hubble Heritage Team AURA/STScI; WR124: NASA; Orion: NASA/IPAC courtesy Deborah Levine)

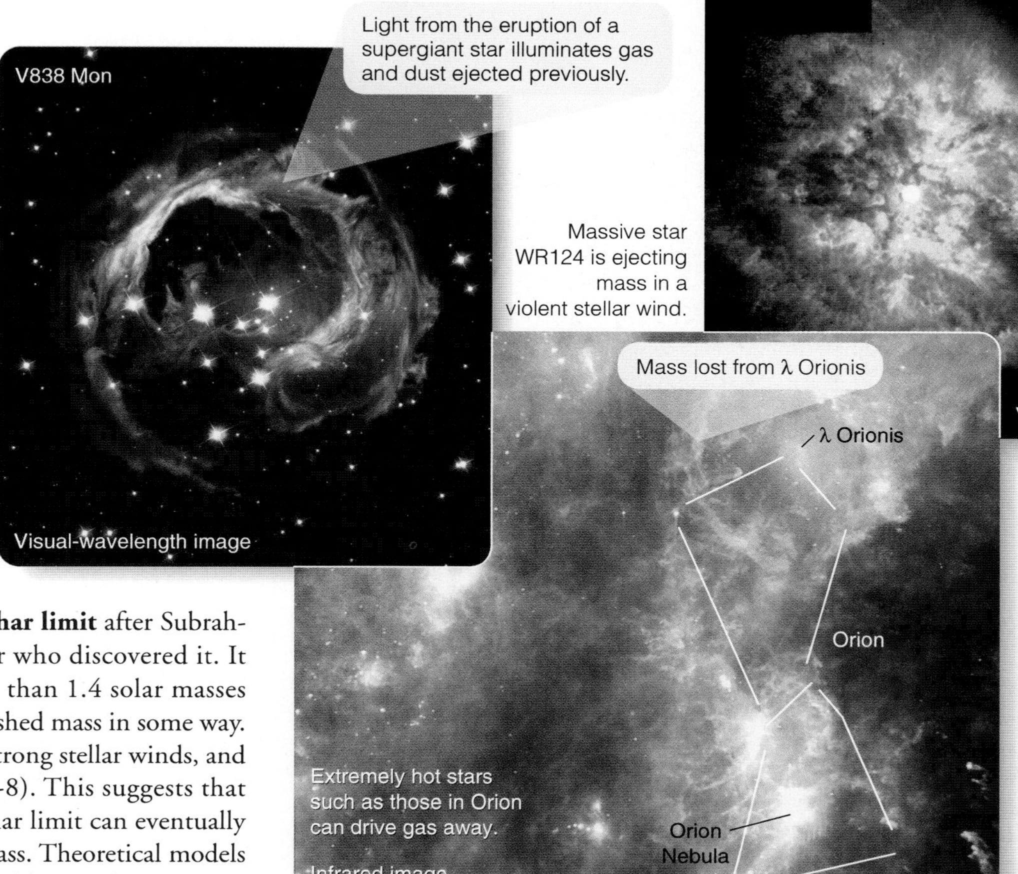

V838 Mon
Light from the eruption of a supergiant star illuminates gas and dust ejected previously.
Visual-wavelength image
Massive star WR124 is ejecting mass in a violent stellar wind.
Visual
Mass lost from λ Orionis
λ Orionis
Orion
Orion Nebula
Extremely hot stars such as those in Orion can drive gas away.
Infrared image

The Formation of Planetary Nebulae

Simple observations tell astronomers what planetary nebulae are like. Their angular size and their distance indicate that their radii range from 0.2 to 3 light-years. The presence of emission lines in their spectra assures astronomers they are excited, low-density gas. Doppler shifts show they are expanding at 10 to 20 km/s. If you divide radius by velocity, you find planetary nebulae are no more than 10,000 years old. Older planetary nebulae evidently become mixed into the interstellar medium.

NGC6369 Visual

Hubble Heritage Team, STScI/AURA/NASA

This nearly spherical planetary nebula has a low-luminosity outer envelope and a highly excited inner region.

Astronomers find about 1500 planetary nebulae in the sky. Because planetary nebulae are short-lived formations, you can conclude that they must be a common part of stellar evolution. Medium-mass stars up to a mass of about 8 solar masses are destined to die by forming planetary nebulae.

Visual

Hubble Heritage Team, AURA/STScI/NASA

The Ring Nebula in the constellation Lyra is visible even in small telescopes. Note the hot blue star at its center and the radial texture in the gas, suggesting outward motion.

The process that produces planetary nebulae involves two stellar winds. First, as an aging giant, the star gradually blows away its outer layers in a slow breeze of low-excitation gas that is not easily visible. Once the hot interior of the star is exposed, it ejects a high-speed wind that overtakes and compresses the gas of the slow wind like a snowplow, while ultraviolet radiation from the hot remains of the central star excites the gases to glow like a giant neon sign.

Slow stellar wind from a red giant

The gases of the slow wind are not easily detectable.

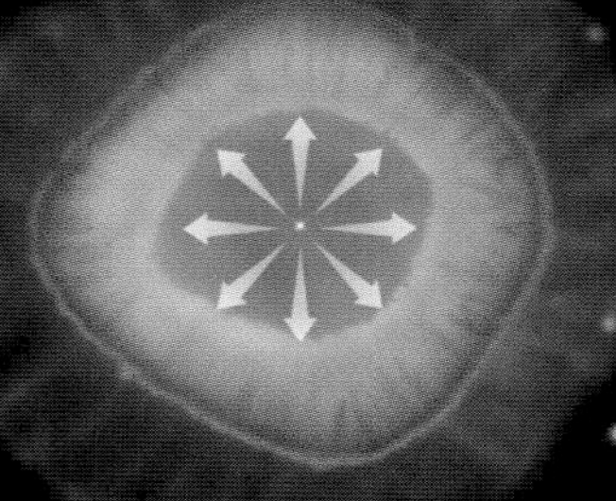

Fast wind from exposed interior

You see a planetary nebula where the fast wind compresses the slow wind.

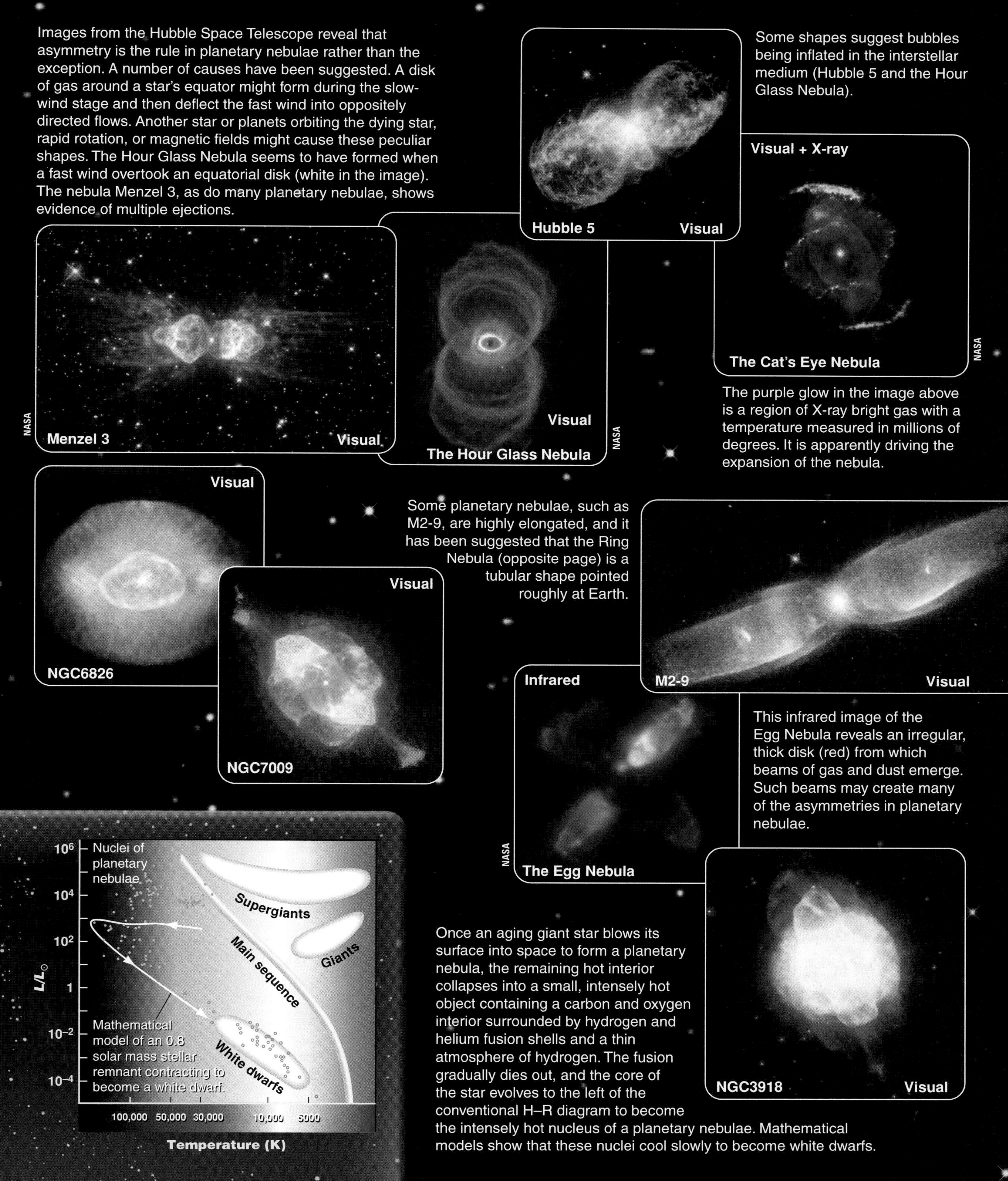

Images from the Hubble Space Telescope reveal that asymmetry is the rule in planetary nebulae rather than the exception. A number of causes have been suggested. A disk of gas around a star's equator might form during the slow-wind stage and then deflect the fast wind into oppositely directed flows. Another star or planets orbiting the dying star, rapid rotation, or magnetic fields might cause these peculiar shapes. The Hour Glass Nebula seems to have formed when a fast wind overtook an equatorial disk (white in the image). The nebula Menzel 3, as do many planetary nebulae, shows evidence of multiple ejections.

Some shapes suggest bubbles being inflated in the interstellar medium (Hubble 5 and the Hour Glass Nebula).

The purple glow in the image above is a region of X-ray bright gas with a temperature measured in millions of degrees. It is apparently driving the expansion of the nebula.

Some planetary nebulae, such as M2-9, are highly elongated, and it has been suggested that the Ring Nebula (opposite page) is a tubular shape pointed roughly at Earth.

This infrared image of the Egg Nebula reveals an irregular, thick disk (red) from which beams of gas and dust emerge. Such beams may create many of the asymmetries in planetary nebulae.

Once an aging giant star blows its surface into space to form a planetary nebula, the remaining hot interior collapses into a small, intensely hot object containing a carbon and oxygen interior surrounded by hydrogen and helium fusion shells and a thin atmosphere of hydrogen. The fusion gradually dies out, and the core of the star evolves to the left of the conventional H–R diagram to become the intensely hot nucleus of a planetary nebulae. Mathematical models show that these nuclei cool slowly to become white dwarfs.

to 1.4 solar masses before it collapses, and slightly more massive stars may also be able to get under the limit. Thus, a wide range of medium-mass stars eventually die as white dwarfs.

Inquire | Review | Analyze

What produced the energy that makes a planetary nebula glow? The low-density gas in a planetary nebula produces an emission spectrum, and you can trace the energy that it radiates as light back to the life history of the star. The gas of the nebula is excited by ultraviolet radiation from the hot central object in the planetary nebula. That object is hot because it was once the interior of the star that ejected its surface to create the planetary nebula. Furthermore, the inside of the star was hot because nuclear fusion reactions in and around the core fused hydrogen into helium and then helium into carbon. The light you see coming from a planetary nebula is energy that was originally locked inside the nuclei of hydrogen and helium atoms.

The goal of science is to tell the story of natural objects. You have traced the life stories of medium mass stars to their ultimate deaths as white dwarfs. At that point, what has become of the hydrogen the star once contained in its core?

■ ■ ■

Connections: Medium-mass stars die by producing beautiful planetary nebulae and tiny white dwarfs. The next logical step is to discuss the deaths of more massive stars, but before that, you must explore the peculiar and sometimes violent evolution of stars in binary systems.

10-3 The Evolution of Binary Systems

STARS IN BINARY SYSTEMS can evolve independently of each other if they orbit at a large distance from each other. In this situation, one of the stars can swell into a giant and collapse without disturbing its companion. But some binary stars are as close to each other as 0.1 AU, and when one of those stars begins to swell into a giant, its companion can suffer in peculiar ways.

These interacting binary stars are interesting in their own right. The stars share a complicated history and can evolve to experience strange and violent phenomena. But such systems are also important because they can help astronomers understand the ultimate fate of stars and certain observed phenomena, such as the temporary appearance of new stars in the sky. In the next chapter we will use interacting binary stars as tools to help find black holes.

Mass Transfer

Binary stars can sometimes interact by transferring mass from one star to the other. Of course, the gravitational field of each star holds its mass together, but the gravitational fields of the two stars, combined with the rotation of the binary system, define a dumbbell-shaped volume around the pair of stars called the **Roche lobes.** The surface of this volume is called the **Roche surface.** Matter inside a star's Roche lobe is gravitationally bound to the star. The size of the Roche lobes depends on the mass of the stars and on the distance between the stars. If the stars are far apart, the lobes are very large, and the stars easily control their own mass.

The **Lagrangian points** are places in the orbital plane of a binary star system where a bit of matter can reach stability. For astronomers, the most important of these points is the **inner Lagrangian point** where the two Roche lobes meet (■ Figure 10-9). If matter can leave a star and reach the inner Lagrangian point, it can flow into the other star. Thus, the inner Lagrangian point is the connection through which the stars can transfer matter.

In general, there are only two ways matter can escape from a star and reach the inner Lagrangian point. First, if a star has a strong stellar wind, some of the gas blowing away from the star can pass through the inner Lagrangian point and be captured by the other star. Second, if an evolving star expands so far that it fills its Roche lobe, which can occur if the stars are close together and the lobes are small, then matter can overflow through the inner Lagrangian point onto the other star. Mass transfer driven by a stellar wind tends to be slow, but mass can be transferred rapidly by an expanding star.

Evolution with Mass Transfer

Mass transfer between stars can affect the evolution of the stars in surprising ways. In fact, this explains a problem that puzzled astronomers for many years.

In some binary systems, the less massive star has become a giant, while the more massive star is still on the main sequence. If higher-mass stars evolve faster than lower-mass stars, how do the lower-mass stars in such binaries manage to leave the main

■ Figure 10-9

A pair of binary stars control the region of space located inside the Roche surface. The Lagrangian points are locations of stability, with the inner Lagrangian point making a connection through which the two stars can transfer matter.

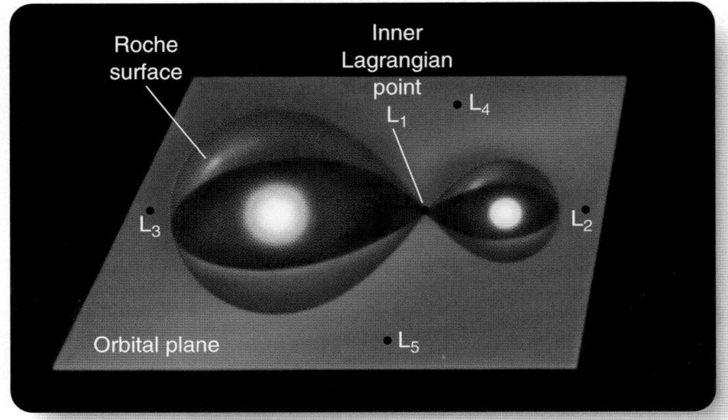

sequence first? This is called the Algol paradox, after the binary system Algol (Figure 8-19).

Mass transfer explains how this could happen. Imagine a binary system that contains a 5-solar-mass star and a 1-solar-mass companion. The two stars formed at the same time, so you might expect the higher-mass star to evolve faster and leave the main sequence first. When it expands into a giant, however, it can fill its Roche lobe and transfer matter to the low-mass companion. The higher-mass star could lose mass and evolve into a lower-mass star, and the companion could gain mass and become a high-mass star still on the main sequence. Thus, you might find a system such as Algol containing a 5-solar-mass main-sequence star and a 1-solar-mass giant.

The first four frames of ■ Figure 10-10 show mass transfer producing a system like Algol. The last frame shows an additional stage in which the giant star has collapsed to form a white dwarf, and the more massive companion has expanded and is transferring matter back to the white dwarf. Such systems can become the site of tremendous explosions. To see how this can happen, we must consider how mass falls into a star.

■ **Figure 10-10**

A pair of stars orbiting close to each other can exchange mass and modify their evolution.

The Evolution of a Binary System

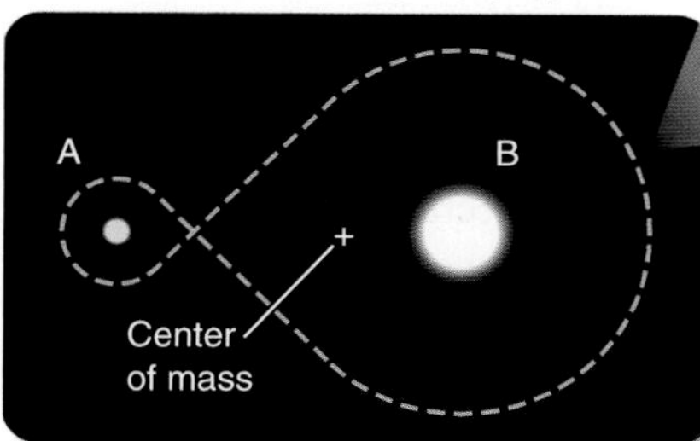
A
B
Center of mass
Star B is more massive than Star A.

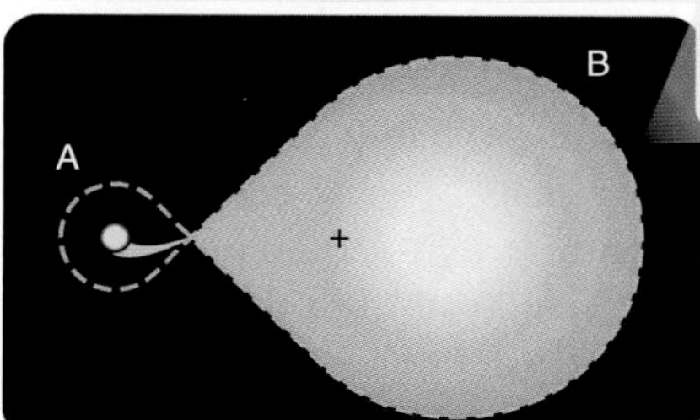
A
B
Star B becomes a giant and loses mass to Star A.

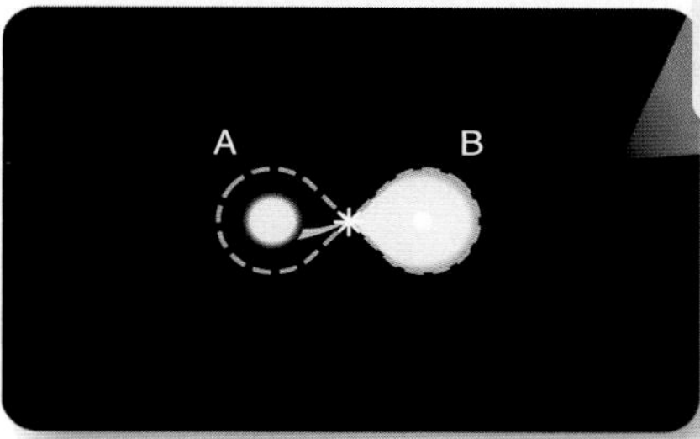
A
B
Star B loses mass, and Star A gains mass.

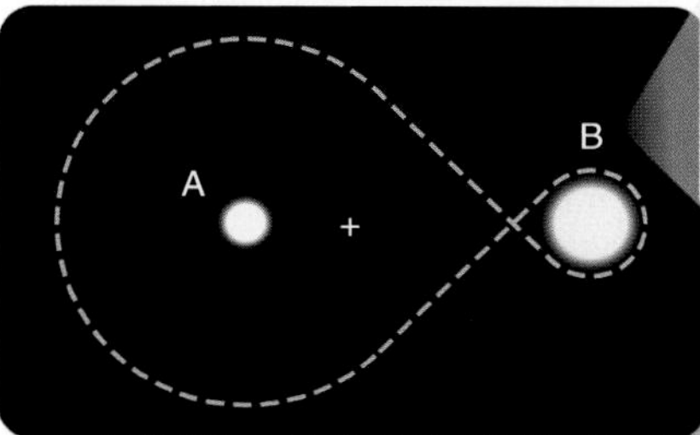
A
B
Star A is a massive main-sequence star with a lower-mass giant companion—an Algol system.

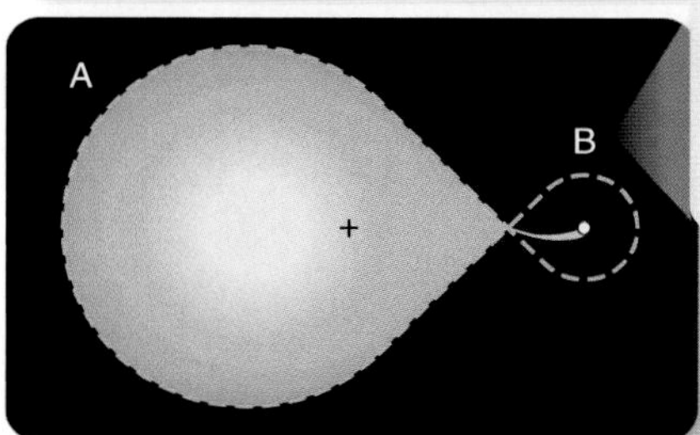
A
B
Star A has now become a giant and loses mass back to the white dwarf that remains of Star B.

Accretion Disks

Matter flowing from one star to another cannot fall directly into the star. Rather, because of conservation of angular momentum, it must flow into a whirling disk around the star.

Angular momentum refers to the tendency of a rotating object to continue rotating. All rotating objects possess some angular momentum, and in the absence of external forces, an object maintains (conserves) its total angular momentum. An ice skater takes advantage of conservation of angular momentum by starting a spin slowly with arms extended and then drawing them in. As her mass becomes concentrated closer to her axis of rotation, she spins faster (■ Figure 10-11). The same effect causes the slowly circulating water in a bathtub to spin in a whirlpool as it approaches the drain.

■ **Figure 10-11**

A skater demonstrates conservation of angular momentum when she spins faster by drawing her arms and legs closer to her axis of rotation.

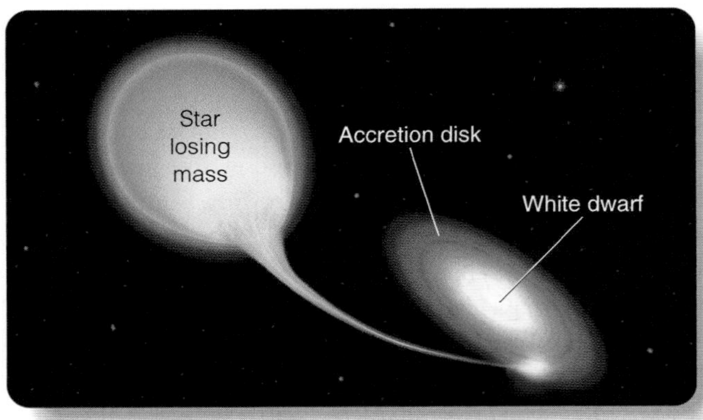

■ **Figure 10-12**

Matter falling into a compact object forms a whirling accretion disk. Friction and tidal forces can make the disk very hot.

Mass transferred through the inner Lagrangian point in a binary system toward a white dwarf must conserve its angular momentum. Thus it must flow into a rapidly rotating whirlpool called an **accretion disk** around the white dwarf (■ Figure 10-12).

Two important things happen in an accretion disk. First, the gas in the disk grows very hot due to friction and tidal forces, the same kind of gravitational forces that make Earth's oceans ebb and flow (Chapter 3). The disk also acts as a brake, ridding the gas of its angular momentum and allowing it to fall into the white dwarf. The temperature of the gas in the inner parts of the accretion disk can exceed a million Kelvin, causing the gas to emit X rays. In addition, the matter falling inward from the accretion disk can cause a violent explosion if it accumulates on a white dwarf.

Novae

At the beginning of this chapter you saw that the word *nova* refers to a new star that appears in the sky for a while and then fades away. Modern astronomers know that the nova is not a new star but an old star flaring up. After a nova fades, astronomers can photograph the spectrum of the remaining faint point of light. Invariably, they find a short-period spectroscopic binary containing a normal star and a white dwarf. A nova is evidently an explosion involving a white dwarf.

Observational evidence can tell you how nova explosions occur. As the explosion begins, spectra show blueshifted absorption lines, which tells you the gas is dense and coming toward you at a few thousand kilometers per second. After a few days, the spectral lines change to emission lines, which tells you the gas has thinned, but the blueshifts remain, showing that a cloud of debris has been ejected into space.

You can understand nova explosions as the result of mass transfer from a normal star through the inner Lagrangian point into an accretion disk around the white dwarf. As the matter loses its angular momentum in the accretion disk, it settles inward onto the surface of the white dwarf and forms a layer of unused nuclear fuel—mostly hydrogen. As the layer deepens, it becomes denser and hotter until the hydrogen fuses in a sudden explosion that blows the surface off of the white dwarf. Although the expanding cloud of debris contains less than 0.0001 solar mass, it is hot and its expanding surface area makes it very luminous. Nova explosions can become 100,000 times more luminous than the sun. As the debris cloud expands, cools, and thins over a period of weeks and months, the nova fades from view.

The explosion of its surface hardly disturbs the white dwarf and its companion star. Mass transfer quickly resumes, and a new layer of fuel begins to accumulate. How fast the fuel builds up depends on the rate of mass transfer. Accordingly, you can expect novae to repeat each time an explosive layer accumulates. Many novae take thousands of years to build an explosive layer, but some take only decades (■ Figure 10-13).

The End of Earth

Astronomy is about us. Although this chapter has been discussing the deaths of medium-mass stars, white dwarfs, and novae explosions, it has also been discussing the future of our planet. The sun is a medium-mass star and must eventually die by becoming a giant, possibly producing a planetary nebula, and collapsing into a white dwarf. That will spell the end of Earth.

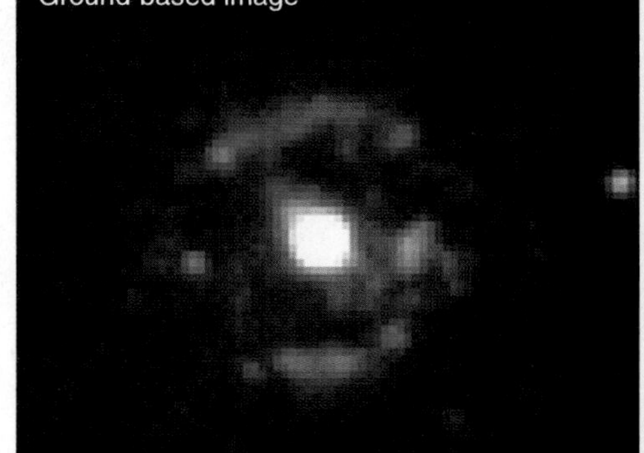

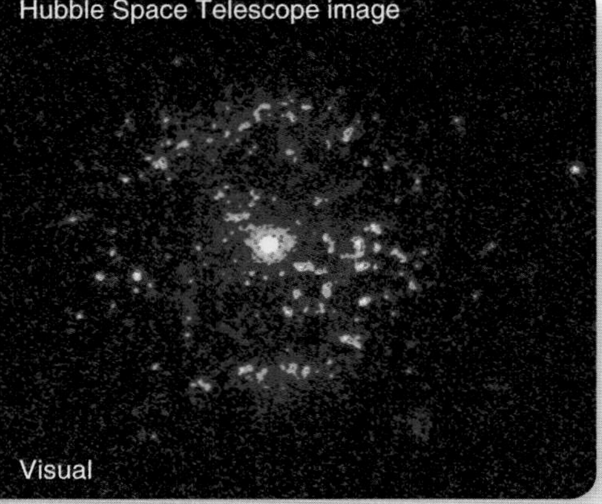

■ **Figure 10-13**

Nova T Pyxidis erupts about every two decades, expelling shells of gas into space. The shells of gas are visible from ground-based telescopes, but the Hubble Space Telescope reveals much more detail. The shell consists of knots of excited gas that presumably form when a new shell collides with a previous shell. (M. Shara and R. Williams, STScI; R. Gilmozzi, ESO; and NASA)

Mathematical models of the sun suggest that it may survive for 5 billion years or so, but it is already growing more luminous as it fuses hydrogen into helium. In a few billion years, it will exhaust hydrogen in its core and swell into a giant star about 100 times its present radius. That giant sun will be about as large as the orbit of Earth, so that will mark the end of our world. Whether the expanding sun becomes large enough to totally engulf Earth, its growing luminosity will certainly evaporate our oceans, drive away our atmosphere, and even vaporize much of Earth's crust.

While it is a giant star, the sun will lose mass into space. This mass loss is a relatively gentle process, so any cinder that might remain of Earth would not be disturbed. Of course, when the sun finally collapses into a white dwarf, the solar system will become a much colder place, but not much of Earth will remain by then.

If the collapsing sun becomes hot enough, it will ionize the expelled gas and light up a planetary nebula. Models show that a collapsing star of less than 0.55 solar mass can take as long as a million years to heat up hot enough to ionize its nebula. By that time, the expelled gases are long gone. Models of the sun are not precise enough to indicate how much mass will be left once the sun has ejected its outer layers. If it leaves behind too little mass, it may heat too slowly. Also, some research suggests that a star needs a binary companion to speed up its spin and make it eject a planetary nebula. This is an area of active research, and there are no firm conclusions. Are you disappointed that the sun may not form a planetary nebula? At least that potential embarrassment lies a few billion years in the future.

There is no danger that the sun will explode as a nova; it has no binary companion. And, as you will see, the sun is not massive enough to die the violent death of the massive stars.

The most important lesson of astronomy is that we are part of the universe and not just observers. The atoms we are made of are destined to return to the interstellar medium in just a few billion years. That's a long time, and it is possible that the human race will migrate to other planetary systems. That might save the human race, but our planet is star dust.

Inquire | Review | Analyze

How does modern astronomy explain the Algol paradox?

When you encounter a paradox in nature, it is usually a warning that you don't understand things as well as you thought you did. The Algol paradox is a good example. The binary star Algol contains a lower-mass giant star and a more massive main-sequence star. Because the two stars must have formed together, they are the same age; and the more massive star should have evolved first and left the main sequence. But in binary systems such as Algol, the lower-mass star has left the main sequence first, or so it seems.

You can understand this paradox if you add mass transfer to your theory of stellar evolution. The first star to leave the main sequence and become a giant must be the more massive star; but if the stars are close together, the star that is initially more massive can fill its Roche lobe and transfer mass back to its companion. The companion can grow more massive and the giant can grow less massive. Thus, the lower-mass giant star you see in the system may have originally been more massive and may have evolved away from the main sequence, leaving its companion behind to increase in mass as it decreased.

Mass transfer can explain the Algol paradox, and it can also explain the violent explosions called novae. How does mass transfer explain why novae can explode over and over in the same binary system?

■ ■ ■

Connections: The deaths of low- and medium-mass stars can lead to fascinating objects such as planetary nebulae and white dwarfs, and mass transfer can produce violent explosions called novae. But what are the explosions astronomers call supernovae? To answer that question you must consider the way massive stars die.

10-4 The Deaths of Massive Stars

YOU HAVE SEEN that low- and medium-mass stars die relatively quietly as they exhaust their hydrogen and helium and then eject their surface layers to form planetary nebulae. In contrast, massive stars live spectacular lives (■ Figure 10-14) and destroy themselves in violent explosions.

Nuclear Fusion in Massive Stars

Stars on the upper main sequence have too much mass to die as white dwarfs, but their evolution begins much like that of their lower-mass cousins. They consume the hydrogen in their cores and ignite hydrogen shells; as a result, they expand and become giants or, for the most massive stars, supergiants. Their cores contract and fuse helium first in the core and then in a shell, producing a carbon–oxygen core.

Unlike medium-mass stars, the massive stars are able to ignite carbon fusion at a temperature of about 1 billion Kelvin. Carbon fusion produces more oxygen and neon; but, as soon as the carbon is exhausted in the core, the core contracts and carbon ignites in a shell. This pattern of core ignition and shell ignition continues with fuel after fuel, and the star develops a layered structure as shown in Figure 10-14, with a hydrogen-fusion shell above a helium-fusion shell above a carbon-fusion shell above . . . After carbon fuses, oxygen, neon, and magnesium fuse to make silicon and sulfur, and then the silicon fuses to make iron.

The fusion of these nuclear fuels goes faster and faster as the massive star evolves rapidly. Recall that massive stars must consume their fuels rapidly to support their great weight, but other factors also cause the heavier fuels like carbon, oxygen, and silicon to fuse at increasing speed. For one thing, the amount of energy released per fusion reaction decreases as the mass of the fusing

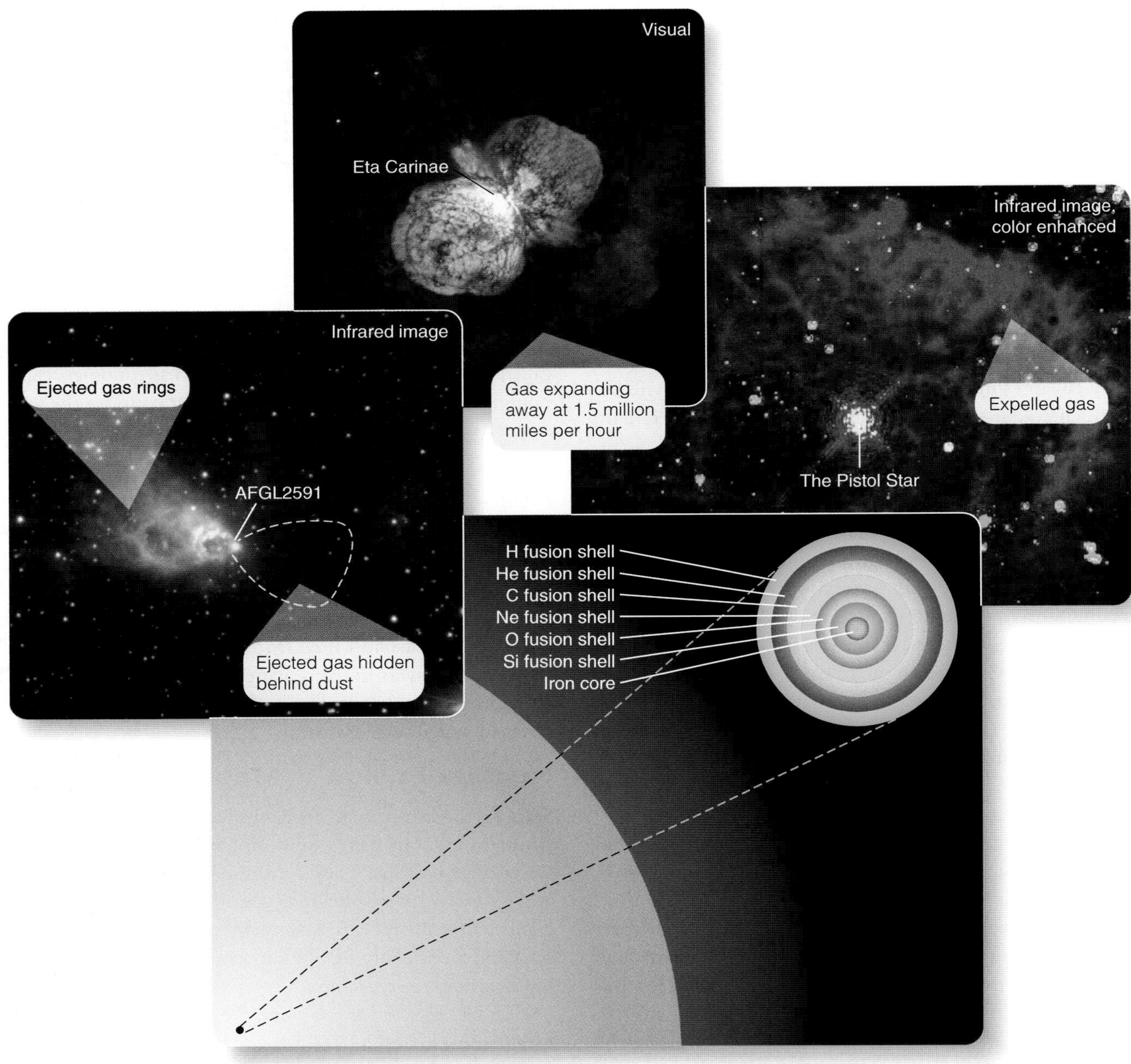

■ Active Figure 10-14

Massive stars live fast and die young. The three shown here are among the most massive stars known, containing 100 solar masses or more. They are rapidly ejecting gas into space. The centers of these massive stars develop Earth-size cores (magnified 100,000 times in this figure) composed of concentric layers of gases undergoing nuclear fusion. The iron core at the center leads eventually to a star-destroying explosion. (AFGL2591: Gemini Observatory/NSF/C. Aspin; Eta Carinae and the Pistol star: NASA)

AceAstronomy™ **Log into AceAstronomy and select this chapter to see the Active Figure called "Inside Stars." Compare the interior of a sunlike star with that of a massive star.**

atom increases. To support its weight, a star must fuse oxygen much faster than it fused hydrogen. Also, there are fewer nuclei in the core of the star by the time heavy nuclei begin to fuse. Four hydrogens made a helium nucleus, and three heliums made a carbon, so there are 12 times fewer nuclei of carbon available for fusion than there were hydrogen. This means the heavy elements are used up and fusion goes very quickly in massive stars (■ Table 10-1). Hydrogen fusion can last 7 million years in a 25-solar-mass star, but that same star will fuse its oxygen in 6 months and its silicon in a day.

■ Table 10-1 | Heavy-Element Fusion in a 25-$M_\odot$ Star

Fuel	Time	Percentage of Lifetime
H	7,000,000 years	93.3
He	500,000 years	6.7
C	600 years	0.008
O	0.5 years	0.000007
Si	1 day	0.00000004

Supernova Explosions of Massive Stars

Theoretical models of evolving stars combined with nuclear physics allow astronomers to describe what happens inside a massive star when the last nuclear fuels are exhausted. It begins with iron nuclei and ends in cosmic violence.

Silicon fusion produces iron, the most tightly bound of all atomic nuclei (see Figure 7-11). Nuclear fusion is able to release energy by combining less tightly bound nuclei into a more tightly bound nucleus, but iron is the limit. Once the gas in the core of the star has been converted to iron, there are no nuclear reactions that can combine iron nuclei and release energy. Thus, the iron core is a dead end in the evolution of a massive star.

As a star develops an iron core, energy production begins to decline, and the core contracts. For nuclei less massive than iron, such contraction heats the gas and ignites new fusion fuels, but nuclear reactions involving iron tend to remove energy from the core in two ways. First, the iron nuclei begin capturing electrons and breaking into smaller nuclei. The degenerate electrons helped support the core, so the loss of the electrons allows the core to contract even faster. Second, temperatures are so high that the average photon is a high-energy gamma ray, and these gamma rays are absorbed by atomic nuclei, which break into smaller fragments. The removal of the gamma rays further cools the core and allows it to contract more. With the loss of the electrons and the gamma rays, the core of the star collapses inward in less than a tenth of a second.

This collapse happens so rapidly that the most powerful computers are unable to predict the details. Thus models of supernova explosions contain many approximations. Nevertheless, the models predict that the collapsing core of the star must quickly become a neutron star or a black hole, the subjects of the next chapter, while the envelope of the star is blasted apart in a supernova explosion.

Although the core of the star cannot generate energy by nuclear fusion, it can draw on the tremendous energy stored in its gravitational field. As the core contracts, the temperature shoots up, but it is not enough to stop the collapse.

To understand how the inward collapse of the core can produce an outward explosion, you can think about a traffic jam. The collapse of the innermost part of the degenerate core allows the rest of the core to fall inward, and this creates a tremendous traffic jam as all of the nuclei fall toward the center. It is as if every car owner in Indiana suddenly tried to drive into downtown Indianapolis. There would be a traffic jam not only downtown but also in the suburbs; and as more cars arrived, the traffic jam would spread outward. Similarly, as the inner core falls inward, a shock wave (a "traffic jam") develops and begins to move outward.

The shock wave moves outward through the star aided by two additional sources of energy. First, when the iron nuclei in the core are disrupted, they produce a flood of neutrinos. In fact, for a short time the core produces more energy per second than all of the stars in all of the visible galaxies in the universe, and 99 percent of that energy is in the form of neutrinos. This flood of neutrinos carries large amounts of energy out of the core, allowing the core to collapse further, and helps heat the gas outside the core and accelerate the outward-bound shock wave. Also, the torrent of energy flowing out of the core triggers tremendous turbulence, and intensely hot gas rushes outward from the interior (■ Figure 10-15). Again, this rising hot gas carries energy out into the envelope and helps drive the shock wave outward. Within a few hours, the shock wave bursts outward through the surface of the star and blasts it apart.

The supernova seen from Earth is the brightening of the star as its distended envelope is blasted outward by the shock wave. As months pass, the cloud of gas expands, thins, and fades, but the rate at which it fades tells astronomers more about the death throes of the star. Essentially all of the iron in the core of the star is destroyed when the core collapses, but the violence in the outer layers can produce densities and temperatures high enough to trigger nuclear fusion reactions that produce as much as half a solar mass of radioactive nickel-56. The nickel gradually decays to form radioactive cobalt, which decays to form normal iron. The rate at which the supernova fades matches the rate at which these radioactive elements decay. Thus the destruction of iron in the core of the star is matched by the production of iron through nuclear reactions in the expanding outer layers.

The presence of nuclear fusion in the outer layers of the supernova testifies to the violence of the explosion. A typical supernova is equivalent to the explosion of 10^{28} megatons of TNT—about 3 million solar masses of high explosive.

Collapsing massive stars can trigger violent supernova explosions. There is, however, more than one kind of supernova.

Types of Supernovae

In studying supernovae in other galaxies, astronomers have noticed that there are a number of different types. **Type II supernovae** have spectra containing hydrogen lines and appear to be produced by the collapse and explosion of a massive star, the

The Exploding Core of a Supernova

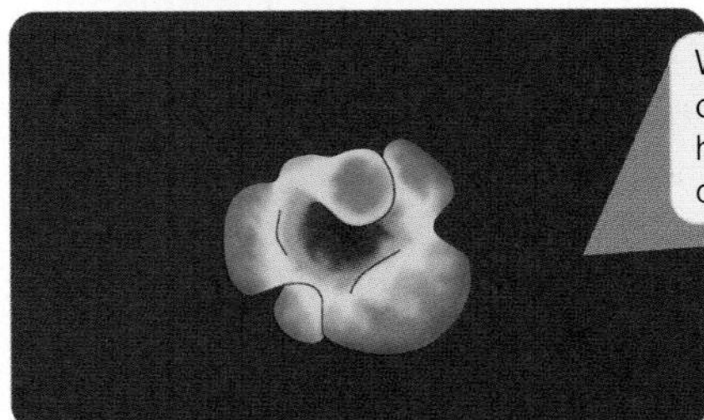

Only 0.4 s after

core remains spherical.

Within 0.025 s,
convection begins as
hot gas (blue) rushes
outward in great plumes.

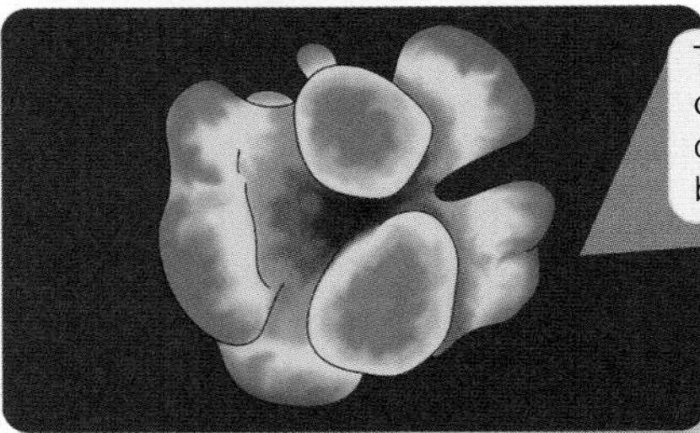

The asymmetric

blowing it apart in hours.

■ Figure 10-15

As the iron core of a massive star begins to collapse, intensely hot gas triggers violent convection. Even as the outer parts of the core continue to fall inward, the turbulence blasts outward and reaches the surface of the star within hours, creating a supernova eruption. This diagram is based on mathematical models and shows only the exploding core of the star. On this scale a diagram showing the entire star would be over 100 km in diameter. (Adapted from models by Dupuy and Warren)

process discussed in the previous section. **Type I supernovae** have no hydrogen lines in their spectra, and astronomers have found at least two ways a supernova could occur and lack hydrogen—type Ia and type Ib supernovae.

A type Ia supernova is believed to occur when a white dwarf gaining mass in a binary star system exceeds the Chandrasekhar limit and collapses. The collapse of a white dwarf is different from the collapse of a massive star because the core of the white dwarf contains usable fuel. As the collapse begins, the temperature and density shoot up and the carbon–oxygen core begins to fuse in violent nuclear reactions. In a flicker of a stellar lifetime, the carbon–oxygen interior is entirely consumed and the outermost layers are blasted away in a violent explosion that at its brightest is about six times more luminous than a type II supernova. The white dwarf is entirely destroyed. No neutron star or black hole is left behind. No hydrogen lines are seen in the spectrum of a type Ia supernova explosion because white dwarfs contain very little hydrogen.

The less common type Ib supernova is believed to occur when a massive star in a binary system loses its hydrogen-rich outer layers to its companion star. The remains of the massive star could develop an iron core and collapse, as described in the previous section, producing a supernova explosion that lacked hydrogen lines in its spectrum. A type Ib supernova is just a type II supernova in which the massive star has lost its atmosphere. Hydrogen lines appear in the spectra of type II supernovae explosions because those massive stars have not lost their atmospheres.

To summarize, a type II supernova is caused by the collapse of a massive star. A type Ia supernova is caused by the collapse of a white dwarf. A type Ib supernova is caused by the collapse of a massive star that has lost its outermost envelope of hydrogen. These different types of supernovae can be recognized from their spectra.

Astronomers working with the largest and fastest computers are using modern theory to try to understand supernova explosions. The companion to theory is observation, so you should ask what observational evidence supports this story of supernova explosions.

Observations of Supernovae

In AD 1054, Chinese astronomers saw a "guest star" appear in the constellation known in the West as Taurus the Bull. The star quickly became so bright it was visible in the daytime, and then, after a month, it slowly faded, taking almost two years to vanish from sight. When modern astronomers turned their telescopes to the location of the "guest star," they found a peculiar nebula now known as the Crab Nebula for its many-legged shape. In fact, the legs of the Crab Nebula are filaments of gas that are moving away from the site of the explosion at about 1400 km/s. Comparing the radius of the nebula, 1.35 pc, with its velocity of expansion reveals that the nebula began expanding nine or ten centuries ago, just when the "guest star" made its visit. The Crab Nebula is clearly the remains of the supernova seen in AD 1054 (■ Figure 10-16).

The blue glow of the Crab Nebula is produced by **synchrotron radiation**. This form of electromagnetic radiation is produced by rapidly moving electrons spiraling through magnetic fields and is common in the nebula produced by supernovae. In the case of the Crab Nebula, the electrons travel so fast they emit visual wavelengths; but, in most such nebulae, the electrons move slower and the synchrotron radiation is at radio wavelengths. This presents a puzzle. The high-speed electrons in the Crab Nebula should have radiated most of their energy away by now and slowed down to produce only radio-wavelength photons. The nebula must contain a powerful energy source, and in the next chapter you will discover that the Crab Nebula contains a neutron star—the last remains of the star that died in AD 1054.

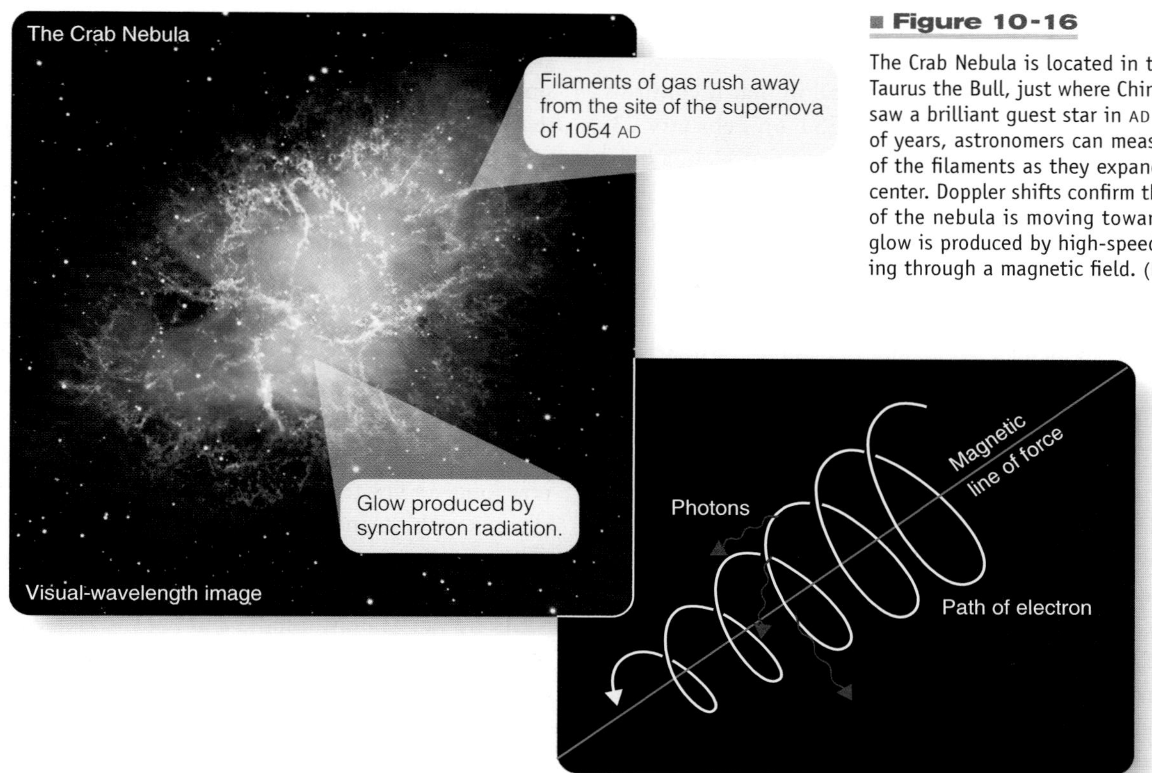

■ **Figure 10-16**

The Crab Nebula is located in the constellation Taurus the Bull, just where Chinese astronomers saw a brilliant guest star in AD 1054. Over tens of years, astronomers can measure the motions of the filaments as they expand away from the center. Doppler shifts confirm that the near side of the nebula is moving toward us. The foggy glow is produced by high-speed electrons moving through a magnetic field. (ESO)

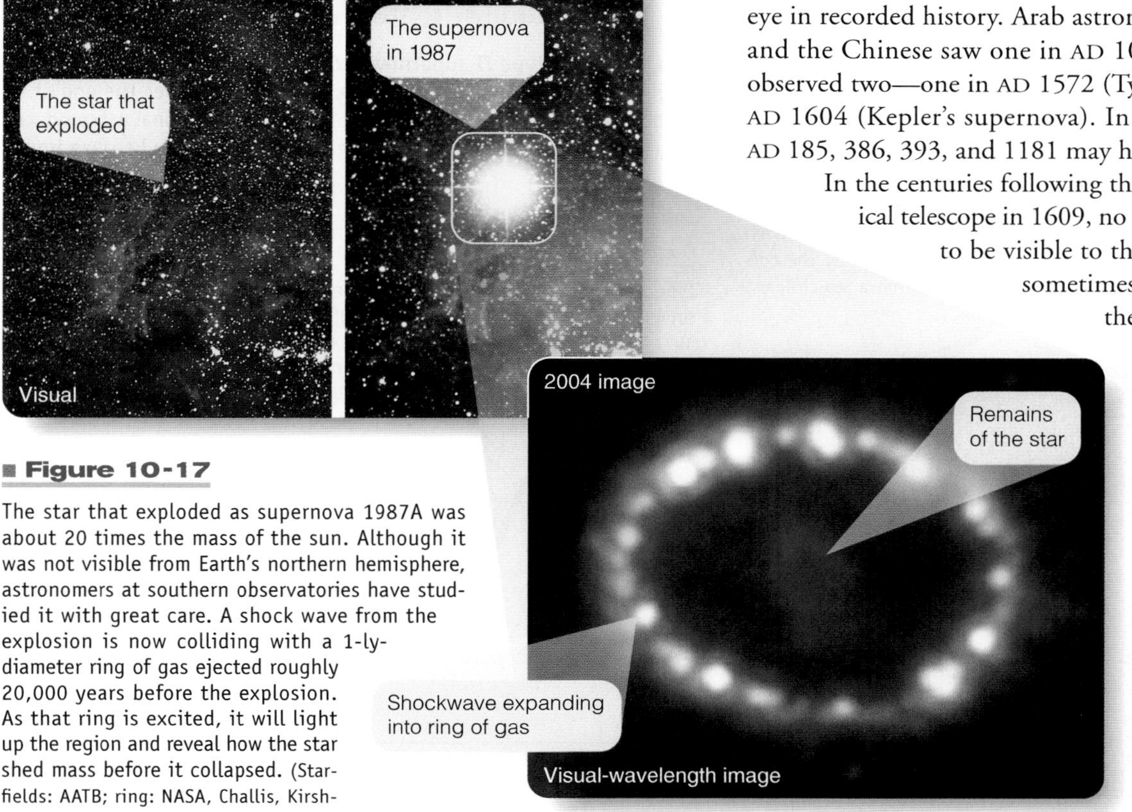

■ **Figure 10-17**

The star that exploded as supernova 1987A was about 20 times the mass of the sun. Although it was not visible from Earth's northern hemisphere, astronomers at southern observatories have studied it with great care. A shock wave from the explosion is now colliding with a 1-ly-diameter ring of gas ejected roughly 20,000 years before the explosion. As that ring is excited, it will light up the region and reveal how the star shed mass before it collapsed. (Starfields: AATB; ring: NASA, Challis, Kirshner, and Sugerman)

Supernovae are rare. Only a few have been seen with the naked eye in recorded history. Arab astronomers saw one in AD 1006, and the Chinese saw one in AD 1054. European astronomers observed two—one in AD 1572 (Tycho's supernova) and one in AD 1604 (Kepler's supernova). In addition, the guest stars of AD 185, 386, 393, and 1181 may have been supernovae.

In the centuries following the invention of the astronomical telescope in 1609, no supernova was bright enough to be visible to the naked eye. Supernovae are sometimes seen in distant galaxies, but they are faint and hard to study. Then, in the early morning hours of February 24, 1987, astronomers around the world were startled by the discovery of a naked-eye supernova still growing brighter in the southern sky (■ Figure 10-17). The supernova, known officially as SN1987A, is only 53,000 pc away in the Large Magellanic Cloud, a small satellite galaxy to our own Milky Way Galaxy. This first naked-eye supernova in 383 years has given astronomers a ringside seat for the most spectacular event in stellar evolution.

One observation of SN1987A is critical in that it confirms the theory of core collapse. At 2:35 AM EST on February 23, 1987, nearly 4 hours before the supernova was first seen, a blast of neutrinos swept through Earth, including perhaps 20 trillion that passed harmlessly through each human body on Earth. Instruments buried in a salt mine under Lake Erie and in a lead mine in Japan, though designed for another purpose, recorded 19 neutrinos in less than 15 seconds. Neutrinos are so difficult to detect that the 19 neutrinos actually detected mean that some 10^{17} neutrinos must have passed through the detectors in those 15 seconds. Furthermore, the neutrinos were arriving from the direction of the supernova. Thus, astronomers conclude that the burst of neutrinos was released when the iron core collapsed, and the supernova was seen hours later when the shock wave blasted the star's surface into space.

Most supernovae are seen in distant galaxies (■ Figure 10-18), and careful observations allow astronomers to compare types. Type Ia supernovae, caused by the collapse of white dwarfs, are more luminous at maximum brightness and decline rapidly at first and then more slowly. Type II supernovae, produced by the collapse of massive stars, are not as bright at maximum. They decline in a more irregular way. Recall that type Ia supernovae have no hydrogen lines in their spectra, but type II supernovae do.

SN1987A was a type II supernova, although its light curve is not typical (■ Figure 10-19). It was produced by the explosion of a hot, blue supergiant rather than the usual cool, red supergiant. Evidently, the star was a red supergiant a few thousand years ago but had contracted and heated up slightly, becoming smaller, hotter, and bluer before it exploded. Scientists believe that most type II supernovae are caused by the collapse of red supergiants.

Although the supernova explosion fades to obscurity in a year or two, an expanding shell of gas marks the site of the explosion. The gas, originally expelled at 10,000 to 20,000 km/s, may carry away a fifth of the mass of the star. The collision of that expanding gas with the surrounding interstellar medium can sweep up even more gas and excite it to produce a **supernova remnant,** the nebulous remains of a supernova explosion (■ Figure 10-20).

Supernova remnants look quite delicate and do not survive very long—a few tens of thousands of years—before they gradually mix with the interstellar medium and vanish. The Crab Nebula is a young remnant, only about 950 years old and about 8.8 ly in diameter. Older remnants can be larger. Some supernova remnants are visible only at radio and X-ray wavelengths (**Window on Science 10-2**). They have become too tenuous to emit detectable light, but the collision of the expanding hot gas with the interstellar medium can generate radio and X-ray radiation. You saw in Chapter 9 that the compression of the interstellar medium by expanding supernova remnants can also trigger star formation.

Inquire I Review I Analyze

What causes a type II supernova explosion?

A type II supernova occurs when a massive star reaches the end of its usable fuel and develops an iron core. The iron is the final ash produced by nuclear fusion, and it cannot fuse to produce energy because iron is

Figure 10-18

This supernova (arrow) was seen outside the central disk of the galaxy NGC4526. From its spectrum and from the way its brightness faded with time, astronomers recognize it as a type Ia supernova produced by the collapse and total destruction of a white dwarf. (High-Z Supernova Search Team, HST, NASA)

■ Figure 10-19

Type I supernovae decline rapidly at first and then more slowly, but type II supernovae pause for about 100 days before beginning a steep decline. Supernova 1987A was odd in that it did not rise directly to maximum brightness. These light curves have been adjusted to the same maximum brightness. Generally, type II supernovae are about 2 magnitudes fainter than type I.

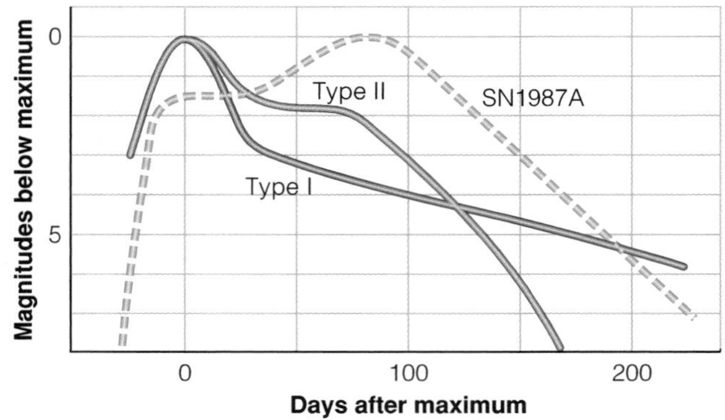

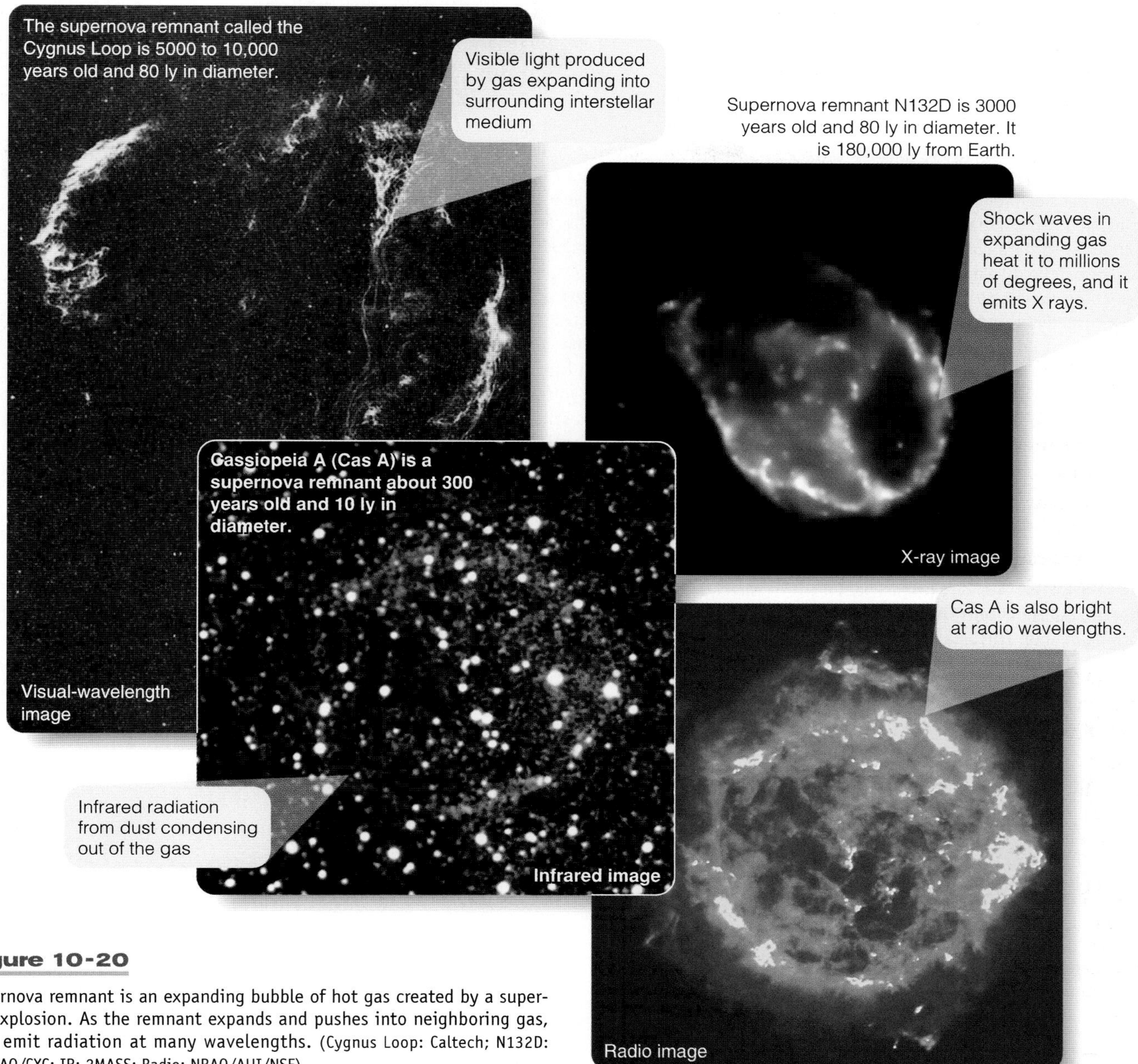

The supernova remnant called the Cygnus Loop is 5000 to 10,000 years old and 80 ly in diameter.
Visible light produced by gas expanding into surrounding interstellar medium
Visual-wavelength image
Supernova remnant N132D is 3000 years old and 80 ly in diameter. It is 180,000 ly from Earth.
Shock waves in expanding gas heat it to millions of degrees, and it emits X rays.
X-ray image
Cassiopeia A (Cas A) is a supernova remnant about 300 years old and 10 ly in diameter.
Infrared radiation from dust condensing out of the gas
Infrared image
Cas A is also bright at radio wavelengths.
Radio image

Figure 10-20

A supernova remnant is an expanding bubble of hot gas created by a supernova explosion. As the remnant expands and pushes into neighboring gas, it can emit radiation at many wavelengths. (Cygnus Loop: Caltech; N132D: NASA/SAO/CXC; IR: 2MASS; Radio: NRAO/AUI/NSF)

the most tightly bound nucleus. When energy generation begins to fall, the star contracts; but because iron can't ignite, there is no new energy source to stop the contraction. In a fraction of a second, the core of the star falls inward and a shock wave moves outward. Aided by a flood of neutrinos and sudden turbulence, the shock wave blasts the star apart, and it brightens as its surface gases expand into space.

Type II supernova explosions are easy to recognize because their spectra contain hydrogen lines. Use what you know about type Ia supernova explosions to explain why their spectra do not contain visible hydrogen lines.

■ ■ ■

Connections: The study of the deaths of stars has led you to discover astonishing objects of unbelievable density, temperature, and violence—all consequences of the victory of gravity over matter. But you have not considered the strangest circumstance of all. What happens when degenerate matter can't support the weight of a dying star—that is, when the mass of the compact object exceeds the Chandrasekhar limit? The answer is in the title of the next chapter.

Window on Science | 10-2

Astronomical Images, False Color, and Reality

Astronomers use computers to add false color to images and reveal things our eyes cannot see. Radio, infrared, ultraviolet, and X-ray radiation are not visible to human eyes, so data collected at these wavelengths can be displayed in images that are given colors visible to our eyes. The choice of colors is usually arbitrary, but the variation in color across an image does have meaning. A radio astronomer might choose shades of red through blue to display a radio map of a supernova remnant; and, although the choice of colors was arbitrary, you would be able to identify the regions of strongest radio emission as the reddest regions on the map.

Astronomers also use false color to enhance visible-wavelength photographs. By converting the image to digital form and assigning colors to different levels of intensity, the astronomer can bring out subtle details that might not otherwise be visible. This is a common technique in other fields as well. Physicians often apply false color to medical X rays and CAT scans. Biologists use false color to analyze microscope photographs, and geologists use false color to study images of Earth recorded by Earth satellites.

Even a simple photograph of an astronomical object such as the Crab Nebula is a kind of false-color image. The nebula is too faint to see with the human eye, and even if you flew in a spaceship to within a short distance of the nebula, it would still be too faint to see without special equipment. You could not see its shape or its colors. The photograph you see is a time exposure made using special filters, and thus it reveals a part of nature that is invisible to human eyes. Even a black-and-white photograph of a nebula shows an aspect of the universe you can never see with unaided eyes. In that sense, all astronomical photographs are false-color images. Their virtue is not that they are colorful, but that they are meaningful.

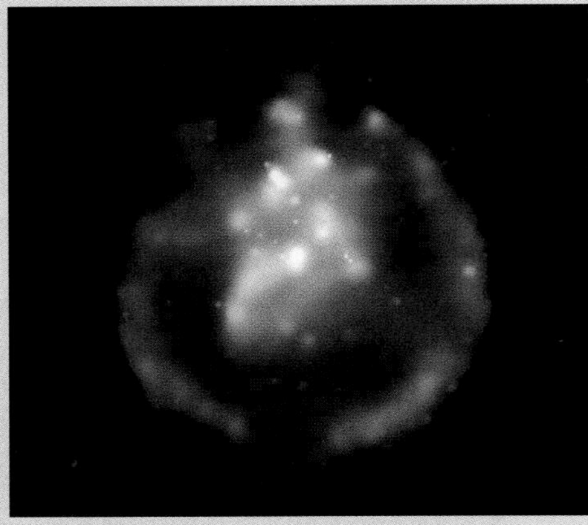

A false-color image of supernova remnant SNR 0103-72.6. (NASA/CXC/Rutgers/J. Hughes)

Study and Review Tools

Summary

10-1 | Giant Stars

What happens to a star when it uses up the last of the hydrogen in its core?

- As a star uses up the last of its hydrogen, the nuclear reactions die down and the core contracts and heats up. Hydrogen fusion begins in a spherical layer around the core—a hydrogen-fusion shell.
- The contraction of the star's core ignites helium, first in the core and later in a shell. If the star is massive enough, it can eventually fuse carbon and other elements.
- If a star's mass lies between about 0.4 and 3 solar masses, its helium core becomes degenerate before the helium ignites. In degenerate gas, pressure does not depend on temperature, so there is no pressure–temperature thermostat to control the reactions. As a result, the core explodes in a helium flash. All of the energy produced is absorbed by the star.
- Energy from the hydrogen-fusion shell swells the star into a cool giant.

What evidence shows that stars really evolve?

- You can see evidence of stellar evolution in the H–R diagrams of star clusters. The stars begin their evolution at about the same time but evolve at different rates, depending on their masses. The most massive leave the main sequence first and are followed later by progressively less massive stars. This makes the evolution of stars visible in the H–R diagram.
- You can estimate the age of a star cluster from the turnoff point in its H–R diagram.

10-2 | The Deaths of Lower-Main-Sequence Stars

How will the sun die?

- Red dwarfs less massive than about 0.4 solar mass are completely mixed and will have very little hydrogen left when they die. They cannot ignite a hydrogen-fusion shell, so they cannot become giant stars. They will remain on the main sequence for many times the present age of the universe.

- Medium-mass stars between about 0.4 and 4 solar masses, including the sun, become cool giants and fuse helium but cannot fuse carbon.
- Medium-mass stars produce planetary nebulae and become white dwarfs.

10-3 | The Evolution of Binary Systems

What happens if an evolving star is in a binary star system?

- Close binary stars evolve in complex ways because they can transfer mass from one star to the other. This explains why some binary systems contain a main-sequence star more massive than its giant companion—the Algol paradox.
- Mass is transferred from an evolving star through an accretion disk around the receiving star. Accretion disks can become hot enough to emit light and even X rays.
- Mass transferred onto the surface of a white dwarf can build up a layer of fuel that erupts in a nova explosion. A white dwarf can erupt repeatedly so long as mass transfer continues to form new layers of fuel.

10-4 | The Deaths of Massive Stars

How do massive stars die?

- Stars more massive than about 8 solar masses cannot lose mass fast enough to reduce their mass low enough to die by ejecting a planetary nebula and collapsing into a white dwarf. Such massive stars must die more violent deaths.
- The massive stars on the upper main sequence fuse nuclear fuels up to iron but cannot generate further nuclear energy because iron is the most tightly bound of all atomic nuclei. When a massive star forms an iron core, the core collapses and triggers a supernova explosion known as a Type II supernova.
- A type Ia supernova can occur when mass transferred onto a white dwarf pushes it over the Chandrasekhar limit and it collapses suddenly, fusing all of its carbon at once.
- A Type Ib supernova occurs when a massive star in a binary system loses its outer layers of hydrogen before it explodes.
- The spectra of type II supernovae contain hydrogen lines, but the spectra of type Ia and type Ib supernovae do not.
- A supernova expels an expanding shell of gas, which we see as a supernova remnant. The supernova of 1054 AD produced a supernova remnant known as the Crab Nebula. The supernova 1987A is only a few years old, but its expanding gases will eventually form a supernova remnant.

New Terms

nova (p. 200)
supernova (p. 200)
degenerate matter (p. 204)
helium flash (p. 205)
planetary nebula (p. 207)
open cluster (p. 208)
globular cluster (p. 208)
turnoff point (p. 208)
horizontal branch (p. 209)
black dwarf (p. 211)
Chandrasekhar limit (p. 211)
Roche lobe (p. 214)
Roche surface (p. 214)
Lagrangian points (p. 214)
inner Lagrangian point (p. 214)
angular momentum (p. 215)
accretion disk (p. 216)
type II supernova (p. 219)
type I supernova (p. 220)
synchrotron radiation (p. 220)
supernova remnant (p. 222)

Review Questions

Ace Astronomy™ Assess your understanding of this chapter's topics with additional quizzing and animations at **http://astronomy.brookscole.com/sh9e**

1. Why does helium fusion require a higher temperature than hydrogen fusion?
2. How can the contraction of an inert helium core trigger the ignition of a hydrogen-fusion shell?
3. Why does the expansion of a star's envelope make it cooler and more luminous?
4. Why is degenerate matter so difficult to compress?
5. How does the presence of degenerate matter in a star trigger the helium flash?
6. How can star clusters confirm astronomers' theories of stellar evolution?
7. Why don't red dwarfs become giant stars?
8. What causes an aging giant star to produce a planetary nebula?
9. Why can't a white dwarf contract as it cools? What is its fate?
10. Why can't a white dwarf have a mass greater than 1.4 solar masses?
11. How can a star of as much as 8 solar masses form a white dwarf when it dies?
12. How can you explain the Algol paradox?
13. How can the inward collapse of the core of a massive star produce an outward explosion?
14. What is the difference between type I and type II supernovae?
15. What is the difference between a supernova explosion and a nova explosion?
16. The star cluster in the photo at the right contains many hot, blue, luminous stars. Sketch its H–R diagram and discuss its probable age.

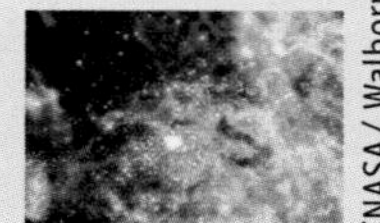
(NASA/ Walborn, Maíz-Apellániz, and Barbá)

17. What processes caused a medium-mass star to produce the nebula at the right? The nebula is now about 0.1 ly in diameter and still expanding. What will happen to it?

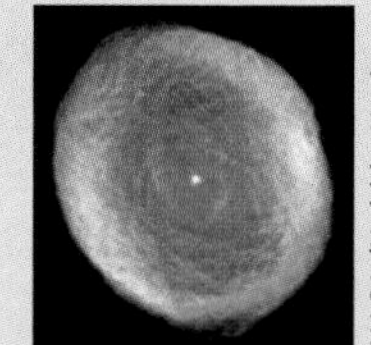
(NASA/Hubble Heritage Team/STScI/AURA)

18. The image at right combines X-ray (blue), visible (green), and radio (red) images. Observations show the sphere is expanding at a high speed and is filled with very hot gas. What kind of object produced this nebula? Roughly how old do you think it must be?

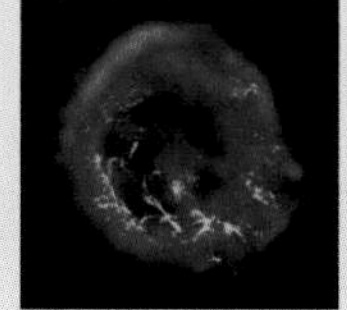
(NASA/CAC/SAO/ CSIRO/ATNF/ATCA)

Discussion Questions

1. How do you know the helium flash occurs if it cannot be observed? Can you accept something as real if you can never observe it?
2. False-color radio images and time-exposure photographs of astronomical images show aspects of nature you can never see with unaided eyes. Can you think of common images in newspapers or on television that reveal phenomena you cannot see?

Problems

1. About how long will a 0.4 solar mass star spend on the main sequence? (*Hint:* See Reasoning with Numbers 9-1.)
2. If the stars at the turnoff point in a star cluster have masses of about 4 solar masses, how old is the cluster?

3. About how far apart are the stars in an open cluster? in a globular cluster? (*Hint:* What share of the cluster's volume belongs to a single star?)
4. The Ring Nebula in Lyrae is a planetary nebula with an angular diameter of 76 seconds of arc and a distance of 5000 ly. What is its linear diameter? (*Hint:* See Reasoning with Numbers 3-1.)
5. If the Ring Nebula is expanding at a velocity of 15 km/s, typical of planetary nebulae, how old is it?
6. Suppose a planetary nebula is 1 pc in radius. If the Doppler shifts in its spectrum show it is expanding at 30 km/s, how old is it? (*Hints:* 1 pc equals 3×10^{13} km, and 1 year equals 3.15×10^{7} seconds.)
7. If a star the size of the sun expands to form a giant 20 times larger in radius, by what factor will its average density decrease? (*Hint:* The volume of a sphere is $(\frac{4}{3})\pi r^3$.)
8. If a star the size of the sun collapses to form a white dwarf the size of Earth, by what factor will its density increase? (*Hints:* The volume of a sphere is $(\frac{4}{3})\pi r^3$. See Appendix A for the radii of the sun and Earth.)
9. The Crab Nebula is now 1.35 pc in radius and is expanding at 1400 km/s. About when did the supernova occur? (*Hint:* 1 pc equals 3×10^{13} km.)
10. If the Cygnus Loop is 25 pc in diameter and is 10,000 years old, with what average velocity has it been expanding? (*Hints:* 1 pc equals 3×10^{13} km, and 1 year equals 3.15×10^{7} seconds.)
11. Observations show that the gas ejected from SN1987A is moving at about 10,000 km/s. How long will it take to travel one astronomical unit? one parsec? (*Hints:* 1 AU equals 1.5×10^{8} km, and 1 pc equals 3×10^{13} km.)

Media Cluster

ACTIVE FIGURES

Ace Astronomy™ To access the resources in the Media Cluster, log into AceAstronomy at **http://astronomy.brookscole.com/sh9e** and select Chapter 10.

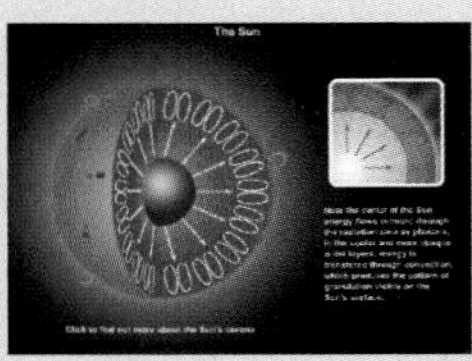

Inside Stars

You can use the laws of physics to study the interior of the sun and compare it to other stars in the universe. This cutaway animation of the sun lets you explore the various structures of stars, understand their inner workings, and view photos and movies of the real thing.

Future of the Sun

The sun's evolution profoundly affects the planets and moons in our solar system. Here you can select the vantage point of Earth, Venus, Mars, or Callisto and get a quick view of the effects of the sun's evolution over 10 billion years.

VIRTUAL ASTRONOMY LABS

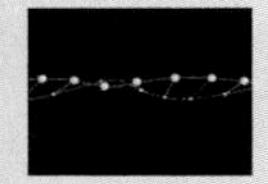

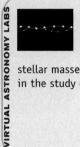

Lab 12: Binary Stars

This lab investigates some things that can be learned from different types of binary stars. Newton's form of Kepler's third law is used to determine stellar masses. You also see how the Doppler effect can be used in the study of binary systems.

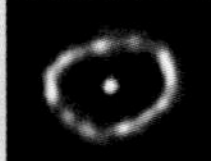

Lab 13: Stellar Explosions, Novae, and Supernovae

This lab investigates two kinds of explosions that highly evolved stars can produce, novae and type Ia supernovae, and how the latter can be used to estimate the distance of faraway galaxies.

Critical Inquiries for the Web

1. As seen in on pages 212 and 213, there is an incredible diversity of appearance for planetary nebulae. Browse the web for images and information on these dying stars and discuss why you see a range of shapes of planetary nebulae.
2. Naked-eye supernovae in our galaxy are rare, but astronomers have noted supernovae in other galaxies for years. Look for summaries of observations of recent supernovae. Are similar numbers of type I and type II being seen? Compare the number of supernovae seen during the last few years with that of two decades ago. Why are so many more supernovae being found in recent years than in the past?
3. What is U Scorpii? Why might you call it a time bomb?

Exploring *TheSky*

1. Locate the planetary nebulae M57, M97, and M27. How does their shape distinguish them from the star formation nebulae such as M42 and M8? (*Hint:* To find an object, use **Find** under **Edit.** Choose **Messier Objects,** and pick from the list.)
2. The Crab Nebula is M1. Locate it, zoom in, measure its angular size in seconds of arc, and compute its diameter, assuming it is about 6000 ly from Earth.
3. Locate the supernova remnant called the Cygnus Loop just south of ε Cygni. How big is this object in angular diameter compared to the full moon? (*Hint:* Under the **View** menu, choose **Labels** and **Setup.** Check **Bayer Designation,** go to Cygnus, and zoom in on ε Cygni until the Cygnus Loop appears.)

- Go to the Brooks/Cole Astronomy Resource Center **(http://astronomy.brookscole.com)** for critical thinking exercises, articles, and additional readings from InfoTrac College Edition, Brooks/Cole's online student library.

11 Neutron Stars and Black Holes

Almost anything is easier to get into than out of.

AGNES ALLEN

X-ray image

GRAVITY ALWAYS WINS. However a star lives, it must eventually die in one of three final states—white dwarfs, neutron stars, or black holes. These objects, often called compact objects, are small monuments to the power of gravity. Almost all of the energy available has been squeezed out of compact objects, and you find them in their final, high-density states. In this chapter, you must compare evidence and hypothesis with great care. Theory predicts the existence of these objects; but by their nature, they are difficult to detect. Is theory right? Do these objects really exist? To confirm the theories, astronomers have searched for real objects that could be identified as having the properties predicted by theory. That is, they have tried to find real neutron stars and real black holes. No medieval knight ever rode off on a more

The supernova remnant SNR 0540–69.3 is 40 ly in diameter and contains a neutron star illuminating the gas cloud at the center (white). About the size of a mountain on Earth, the neutron star emits 30,000 times more energy than the sun.
(NASA/CXC/SAO)

Guidepost

Looking back

Through the last two chapters you have traced the story of stars from their birth as clouds of gas in the interstellar medium to their final collapse. By now you are asking a simple question, "What is left?" The answer, of course, depends on the mass of the star. Stars like the sun leave behind white dwarfs, but more massive stars leave behind the strangest beasts in the cosmic zoo.

This Chapter

Now you are ready to meet neutron stars and black holes, and your exploration will answer four essential questions:

Why would you expect neutron stars to exist?

How do you know neutron stars really exist?

How does theory predict the existence of black holes?

How can you be sure that black holes really exist?

This chapter will show you clear examples of how astronomers combine observations and theory to understand nature.

Looking Ahead

This chapter ends the story of individual stars, but it does not end the story of stars. In the next chapter, you will begin exploring the giant communities in which stars live—the galaxies.

Ace Astronomy™ The AceAstronomy icon throughout the text indicates an opportunity for you to test yourself on key concepts and to explore animations and interactions on the AceAstronomy website at: http://astronomy.brookscole.com/sh9e

difficult quest, but modern astronomers have found objects that do seem to be neutron stars and black holes. You can judge their success by critically analyzing the evidence to see if it does confirm the theoretical predictions.

11-1 Neutron Stars

A **NEUTRON STAR** is a star of a little over 1 solar mass compressed to a radius of about 10 km. Its density is so high that the matter is stable only as a fluid of neutrons. Theory predicts that such an object would spin a number of times a second, be nearly as hot at its surface as the inside of the sun, and have a magnetic field a trillion times stronger than Earth's. Two questions should occur to you immediately. First, how could any theory predict such a wondrously unbelievable star? And second, do such neutron stars really exist?

Theoretical Prediction of Neutron Stars

The neutron was discovered in the laboratory in 1932, and its properties suggested something fantastic. Neutrons spin in much the way that electrons do, which means that neutrons must obey the Pauli exclusion principle. In that case, if neutrons are packed together tightly enough, they can become degenerate just as electrons do. White dwarfs are supported by degenerate electrons, and, in 1932, the Russian physicist Lev Landau predicted that neutron stars might exist supported by degenerate neutrons. Of course, the inside of a neutron star would have to be much denser than the inside of a white dwarf.

Only two years later, in 1934, Walter Baade and Fritz Zwicky provided a pedigree for neutron stars. They suggested that some of the most luminous novae in the historical record were not true novae but were caused by the collapse of a massive star in an explosion they called a supernova. What was left of the core of the star, they proposed, was a small, high-density neutron star.

Atomic physics provides an explanation of how the collapsing core of a massive star could form a neutron star. If the collapsing core is more massive than the Chandrasekhar limit of 1.4 solar masses, then it cannot reach stability as a white dwarf. The weight is too great to be supported by degenerate electrons. The collapse of the core continues, and the atomic nuclei are broken apart by gamma rays. Almost instantly, the increasing density forces the freed protons to combine with electrons and become neutrons. In a fraction of a second, the collapsing core becomes a contracting ball of neutrons. As you saw in the previous chapter, the envelope of the star is blasted away in a supernova explosion. The core of the star is left behind as a neutron star.

Theoretical calculations predict that a neutron star will be only 10 or so kilometers in radius (■ Figure 11-1) and will have a density of about 10^{14} g/cm^3. On Earth, a sugar-cube-sized lump of this material would weigh 100 million tons. This is roughly the density of the atomic nucleus, and you can think of a neutron star as matter with all of the empty space squeezed out of it.

How massive can a neutron star be? That is a critical question, and a difficult one to answer because scientists don't know the strength of pure neutron material. They can't make such matter in the laboratory, so its properties must be predicted theoretically. The most widely accepted calculations suggest that a neutron star cannot be more massive than 2 to 3 solar masses. If a neutron star were more massive than that, the degenerate neutrons would not be able to support the weight, and the object would collapse (presumably into a black hole).

Simple physics, the physics you have used in previous chapters to understand normal stars, predicts that neutron stars should be hot, spin rapidly, and have strong magnetic fields. You have seen that contraction heats the gas in a star. As the gas particles fall inward, they pick up speed; and, when they collide, their high speeds become thermal energy. The sudden collapse of the core of a massive star to a radius of 10 km should heat it to millions of degrees. Furthermore, neutron stars should cool slowly because the heat can escape only from the surface, and neutron stars are so small they have little surface from which to radiate. Thus, basic theory predicts that neutron stars should be very hot.

■ **Figure 11-1**

A tennis ball and a road map illustrate the relative size of a neutron star. Such an object, containing slightly more than the mass of the sun, would fit with room to spare inside the beltway around Washington, D.C. (Photo by author)

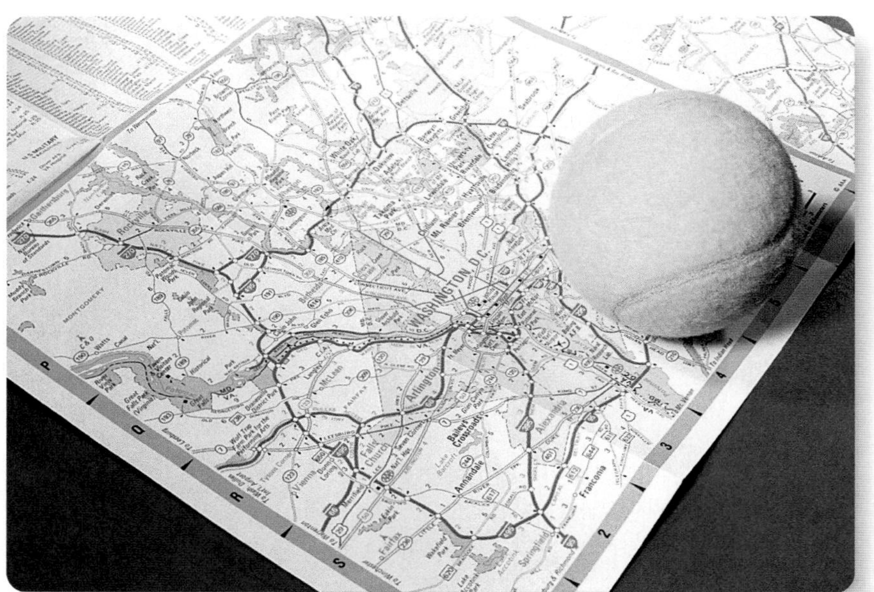

The conservation of angular momentum predicts that neutron stars should spin rapidly. All stars rotate to some extent because they form from swirling clouds of interstellar matter. As such a star collapses, it must rotate faster because it conserves angular momentum. Recall that you see this happen when ice skaters spin slowly with their arms extended and then speed up as they pull their arms closer to their bodies (see Figure 10-11). In the same way, a collapsing star must spin faster as it pulls its matter closer to its axis of rotation. If the sun collapsed to a radius of 10 km, its period of rotation would decrease from 25 days to about 0.001 second. You might expect the collapsed core of a massive star to rotate 10 or 100 times a second.

Basic theory also predicts that a neutron star should have a powerful magnetic field. Whatever magnetic field a star has is frozen into the star. The gas of the star is ionized, and that means the magnetic field cannot move easily through the gas. When the star collapses, the magnetic field is carried along and squeezed into a smaller area, which could make the field a billion times stronger. Because some stars have magnetic fields over 1000 times stronger than the sun's, you might expect a neutron star to have a magnetic field as much as a trillion times stronger than the sun's. For comparison, that is about 10 million times stronger than any magnetic field ever produced in the laboratory.

Basic physics predicted the properties of neutron stars, but it also predicted that such objects should be difficult to observe. Neutron stars are very hot, so from our understanding of black body radiation you can predict they will radiate most of their energy in the X-ray part of the spectrum, radiation that could not be observed in the 1940s and 1950s because astronomers could not put their telescopes above Earth's atmosphere. Also, the small surface areas of neutron stars mean that they will be faint objects. Thus astronomers of the mid-20th century were not surprised that none of the newly predicted neutron stars were found. Neutron stars were, at that point, entirely theoretical objects. Progress came not from theory but from observation.

The Discovery of Pulsars

In November 1967, Jocelyn Bell, a graduate student at Cambridge University in England, found a peculiar pattern on the paper chart from a radio telescope. Unlike other radio signals from celestial bodies, this was a series of regular pulses (■ Figure 11-2). At first she and the leader of the project, Anthony Hewish, thought the signal was interference, but they found it day after day in the same place in the sky. Clearly, it was celestial in origin.

Another possibility, that it came from a distant civilization, led them to consider naming it LGM for Little Green Men. But within a few weeks, the team found three more objects in other parts of the sky pulsing with different periods. The objects were clearly natural, and the team dropped the name LGM in favor of **pulsar**—a contraction of *pulsing star.* The pulsing radio source Bell had observed with her radio telescope was the first known pulsar.

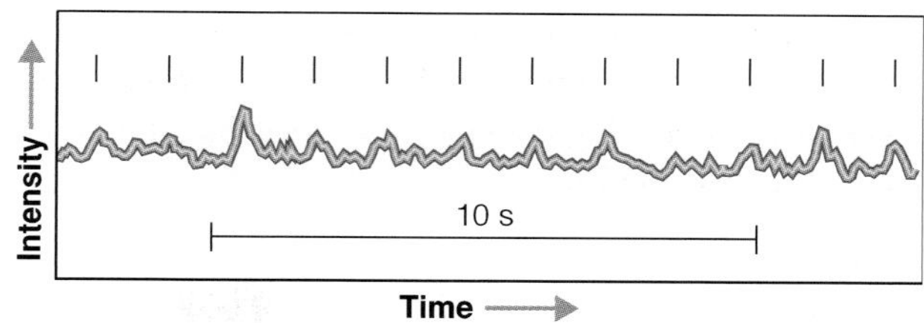

■ Figure 11-2

The 1967 detection of regularly spaced pulses in the output of a radio telescope led to the discovery of pulsars. This record of the radio signal from the first pulsar, CP1919, contains regularly spaced pulses (marked by ticks). The period is 1.33730119 seconds.

As more pulsars were found, astronomers argued over their nature. Periods ranged from 0.033 to 3.75 seconds and were nearly as exact as an atomic clock. Months of observation showed that many of the periods were slowly growing longer by a few billionths of a second per day. Thus, whatever produced the regular pulses had to be highly precise, nearly as exact as an atomic clock, but it had to gradually slow down.

It was easy to eliminate possibilities. Pulsars could not be stars. A normal star, even a small white dwarf, is too big to pulse that fast. Nor could a star with a hot spot on its surface spin fast enough to produce the pulses. Even a small white dwarf would fly apart if it spun 30 times a second.

The pulses themselves gave the astronomers a clue. The pulses lasted only about 0.001 seconds. This places an upper limit on the size of the object producing the pulse. If a white dwarf blinked on and then off in that interval, an observer would not see a 0.001-second pulse. The near side of the white dwarf would be about 6000 km closer to Earth, and light from the near side would arrive 0.022 seconds before the light from the bulk of the white dwarf. Thus its short blink would be smeared out into a longer pulse. This is an important principle in astronomy—an object cannot change its brightness appreciably in an interval shorter than the time light takes to cross its diameter. If pulses from pulsars are no longer than 0.001 second, then the objects cannot be larger than 300 km (190 miles) in diameter.

Only a neutron star is small enough to be a pulsar. In fact, a neutron star is so small, it can't vibrate slowly enough, but it can spin as fast as 1000 times a second without flying apart. The missing link between pulsars and neutron stars was found in late 1968, when astronomers discovered a pulsar at the heart of the Crab Nebula. The Crab Nebula is a supernova remnant, and theory predicts that some supernovae leave behind a neutron star (■ Figure 11-3).

The short pulses and the discovery of the pulsar in the Crab Nebula are strong hints that pulsars are neutron stars. By combining theory and observation, astronomers can devise a model of a pulsar.

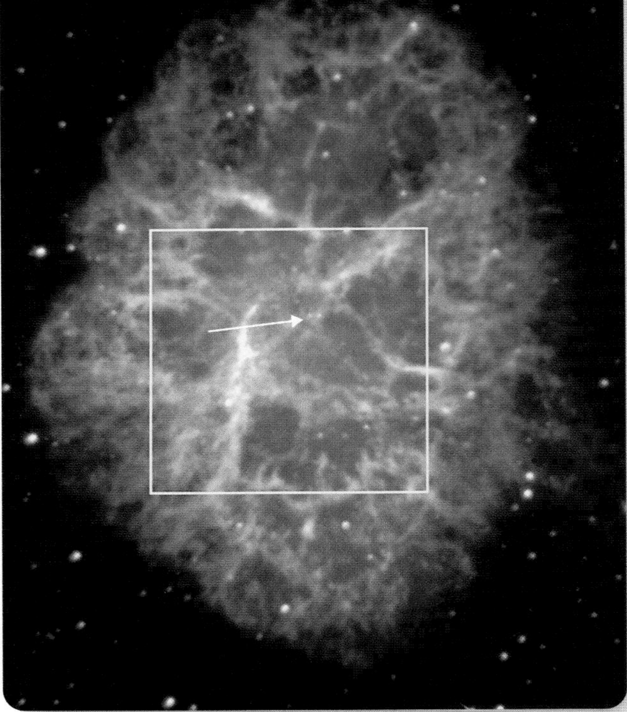

■ Figure 11-3

The pulsar at the center of the Crab Nebula (arrow) is detectable in visual-wavelength photographs. The star just to the right of the pulsar lies much closer to Earth and is not in the Crab Nebula. The white box outlines the area imaged on page 235. (Caltech)

A Model Pulsar

Scientists often work by building a model of a natural phenomenon—not a physical model made of plastic and glue, but an intellectual conception of how nature works in a specific instance. The model may be limited and incomplete, but it helps them organize their understudying of pulsars.

The Lighthouse Model of a Pulsar is shown on pages 234 and 235. Notice four important points. First, a pulsar does not pulse but rather emits beams of radiation that sweep around the sky as the neutron star rotates. Second, the mechanism that produces the beams involves extremely high energies and is not fully understood. The third thing to notice is that astronomers tend to see only those pulsars whose beams sweep over Earth. Finally, notice how modern space telescopes observing at various wavelengths can help to confirm and refine the model.

Neutron stars are not simple objects, and modern astronomers need both general relativity and quantum mechanics to try to understand them. Nevertheless, the life story of pulsars can be understood.

The Evolution of Pulsars

When a pulsar first forms, it is spinning fast, perhaps nearly 100 times a second. The energy it radiates into space comes from its energy of rotation, so as it blasts beams of radiation outward, its rotation slows. The average pulsar is apparently only a few million years old, and the oldest is about 10 million years old. Presumably, older neutron stars rotate too slowly to generate detectable radio beams.

From this you can expect that a young neutron star should emit powerful beams of radiation. The Crab Nebula gives us an example of such a system. Only about 950 years old, the Crab pulsar is so powerful it emits photons of radio, infrared, visible, X-ray, and gamma-ray wavelengths (■ Figure 11-4). Careful measurements of its brightness with high-speed instruments show

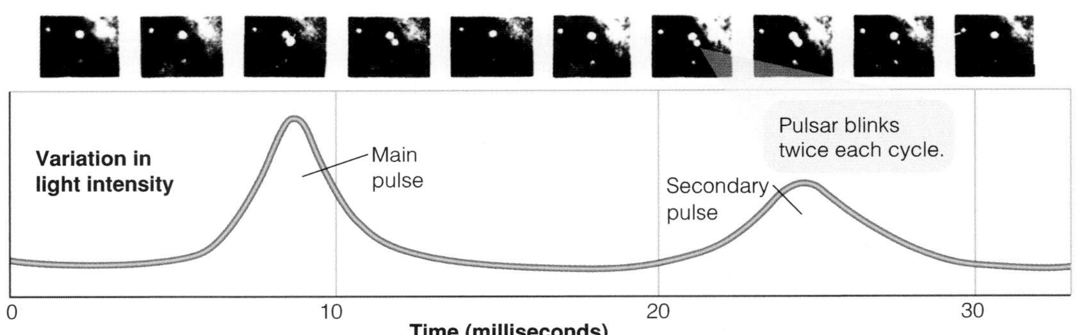

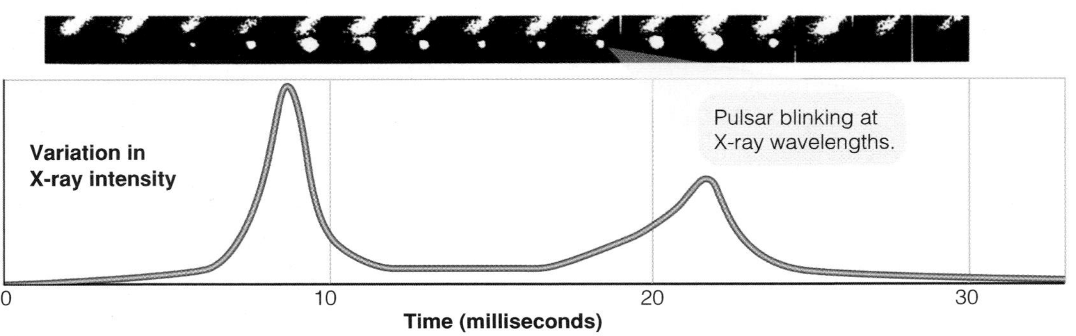

■ Figure 11-4

High-speed images of the Crab Nebula pulsar show it pulsing at visual wavelengths and at X-ray wavelengths. (© AURA, Inc., NOAO, KPNO) The period of pulsation is 33 milliseconds, and each cycle includes two pulses as its two beams of unequal intensity sweep over Earth. (Courtesy F. R. Harnden, Jr., from The Astrophysical Journal, *published by the University of Chicago Press; © 1984 The American Astronomical Society)*

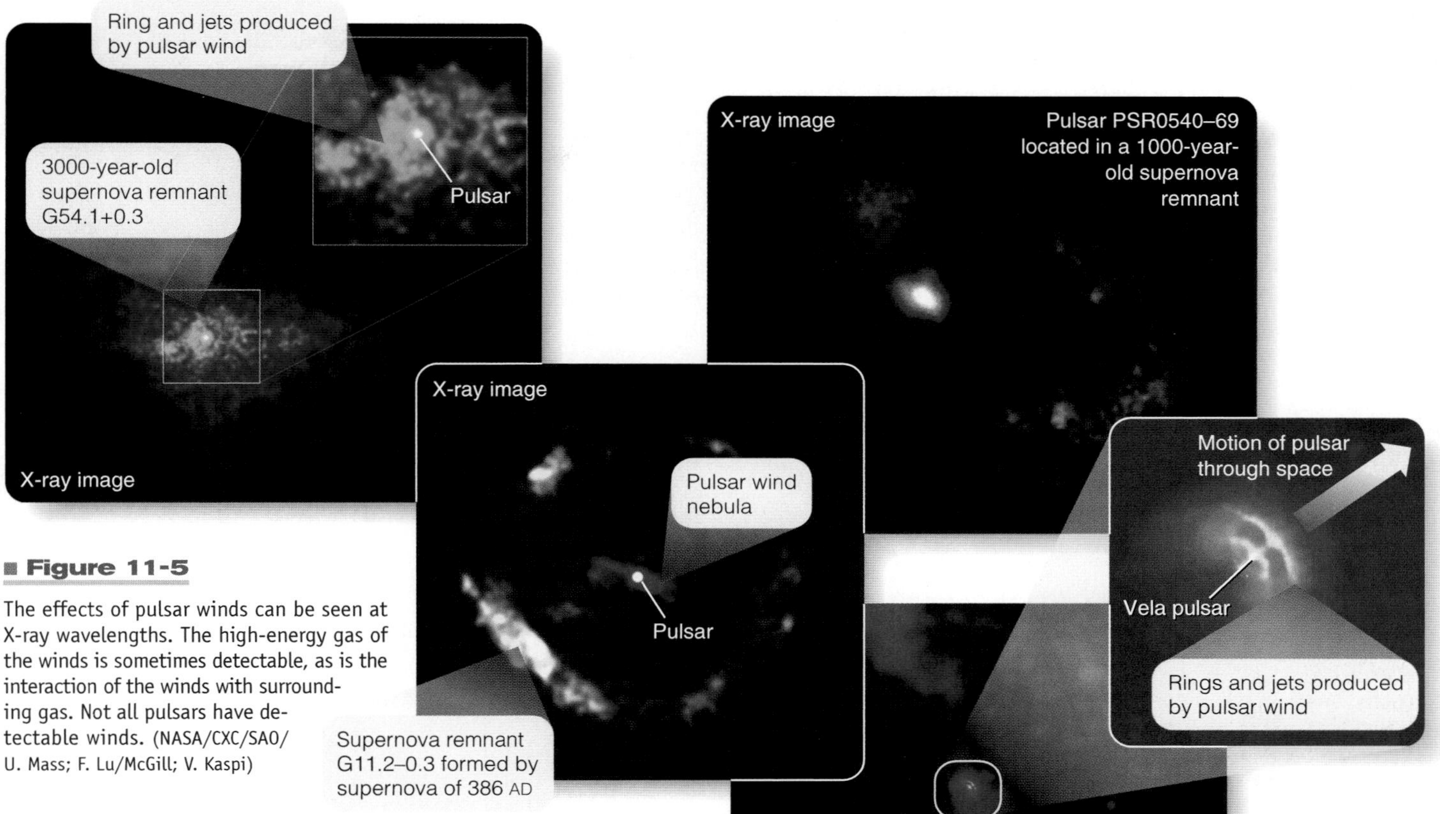

■ **Figure 11-5**

The effects of pulsar winds can be seen at X-ray wavelengths. The high-energy gas of the winds is sometimes detectable, as is the interaction of the winds with surrounding gas. Not all pulsars have detectable winds. (NASA/CXC/SAO/U. Mass; F. Lu/McGill; V. Kaspi)

that it blinks twice for every rotation. One beam sweeps almost directly over Earth, and astronomers detect a strong pulse. Half a rotation later, the edge of the other beam sweeps past, and astronomers detect a weaker pulse.

You would expect only the most energetic pulsars to produce short-wavelength photons and thus pulse at visible wavelengths. The Crab Nebula pulsar is young and powerful, and it produces visible pulses, and so does a pulsar called the Vela pulsar (located in the Southern Hemisphere constellation Vela). The Vela pulsar is fast, pulsing about 11 times a second and, like the Crab Nebula pulsar, is located inside a supernova remnant. Its age is estimated at about 20,000 to 30,000 years, young in terms of the average pulsar. Thus you can suspect that pulsars are capable of producing optical pulses only when they are young.

The energy in the beams is only a small part of the energy emitted by a pulsar. Roughly 99.9 percent of the energy flowing away from a pulsar is carried as a **pulsar wind** of high-speed atomic particles. This can produce small, high-energy nebulae near a young pulsar (■ Figure 11-5).

You might expect to find all pulsars inside supernova remnants, but the statistics must be examined with care. Not every supernova remnant contains a pulsar, and not every pulsar is located inside a supernova remnant. Many supernova remnants probably contain pulsars whose beams never sweep over Earth, and it is difficult to detect such pulsars. Also, some pulsars move through space at high velocity (■ Figure 11-6), which suggests

■ **Figure 11-6**

Many neutron stars have high velocities through space. Here the neutron star known as RXJ185635−3754 was photographed on three different dates as it rushed past background stars. (NASA and F. M. Walter)

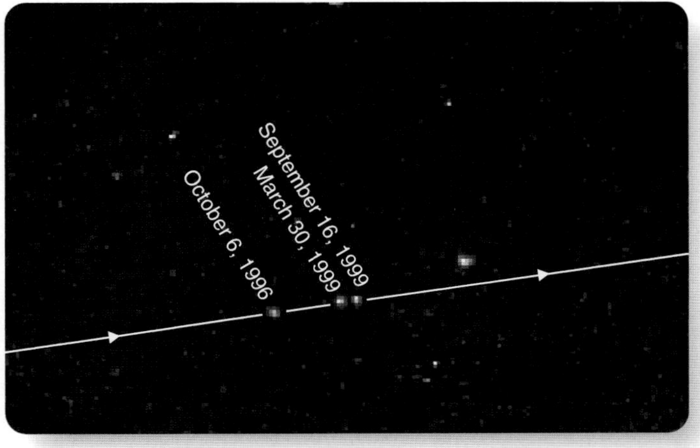

The Lighthouse Model of a Pulsar

Astronomers think of pulsars not as pulsing objects, but rather as objects emitting beams. As they spin, the beams sweep around the sky; when a beam sweeps over Earth, observers detect a pulse of radiation. Understanding the details of this lighthouse model is a challenge, but the implications are clear. Although a neutron star is only a few kilometers in radius, it can produce powerful beams. Also, observers tend to notice only those pulsars whose beams happen to sweep over Earth.

In this artist's conception, gas trapped in the neutron star's magnetic field is excited to emit light and outline the otherwise invisible magnetic field.

Beams of electromagnetic radiation would probably be invisible unless they excited local gas to glow.

How a neutron star can emit beams is one of the challenging problems of modern astronomy, but astronomers have a general idea. A neutron star contains a powerful magnetic field and spins very rapidly. The spinning magnetic field generates a tremendously powerful electric field, and the field causes the production of electron–positron pairs. As these charged particles are accelerated through the magnetic field, they emit photons in the direction of their motion, which produce powerful beams of electromagnetic radiation emerging from the magnetic poles.

What color should an artist use to paint a neutron star? With a temperature of a million degrees, the surface emits most of its electromagnetic radiation at X-ray wavelengths. Nevertheless, it would probably look blue-white to your eyes.

Neutron Star Rotation with Beams

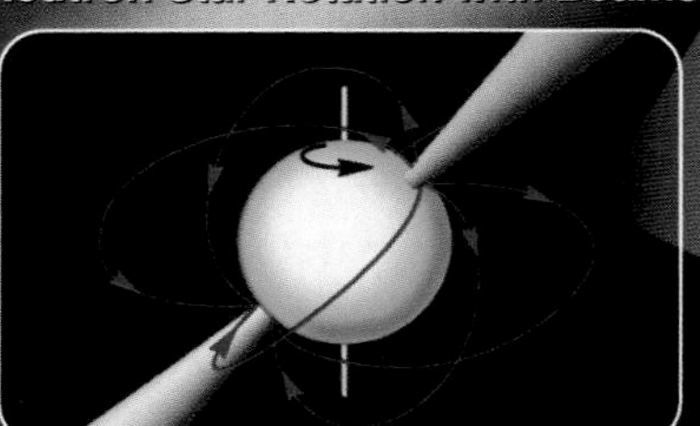

As in the case of Earth, the magnetic axis of a neutron star could be inclined to its rotational axis.

The rotation of the neutron star will sweep its beams around like beams from a lighthouse.

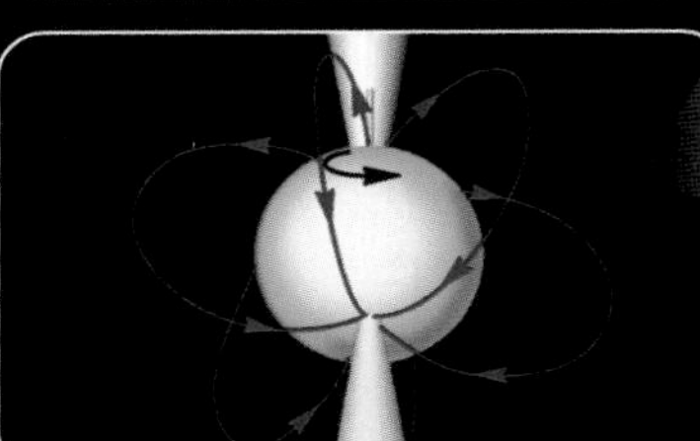

While a beam points roughly toward Earth, observers detect a pulse.

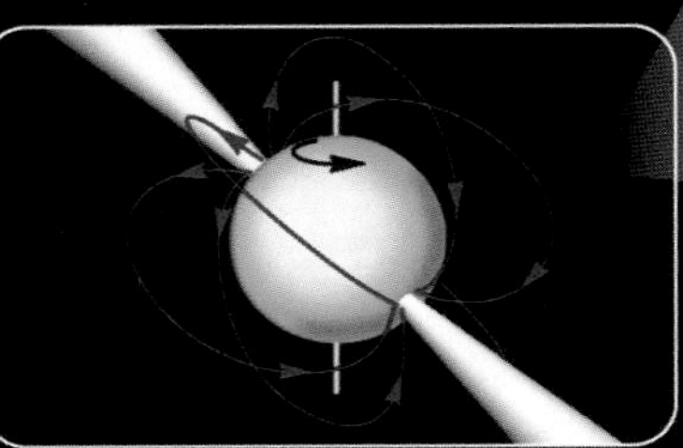

While neither beam is pointed toward Earth, observers detect no energy.

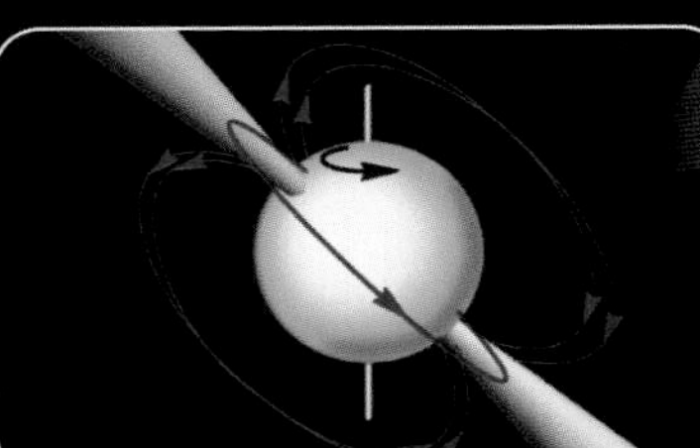

Beams may not be as exactly symmetric as in this model.

Log into AceAstronomy and select this chapter to see the Active Figure called "Neutron Star." Adjust the inclination of the neutron star's magnetic field to produce pulses.

NASA

Crab pulsar

Visual-wavelength image

The hazy glow of the Crab Nebula is produced by synchrotron radiation. In the nearly 10 centuries since the supernova, the high-speed electrons should have radiated their energy away, and the synchrotron radiation should have faded. Evidently, the Crab pulsar powers the nebula. The Hubble Space Telescope image above shows the region boxed in Figure 11-3. Circular wisps excited by the pulsar change and flicker from day to day.

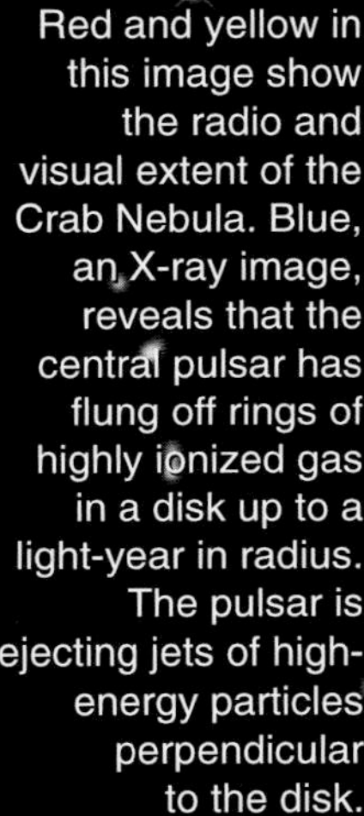

Red and yellow in this image show the radio and visual extent of the Crab Nebula. Blue, an X-ray image, reveals that the central pulsar has flung off rings of highly ionized gas in a disk up to a light-year in radius. The pulsar is ejecting jets of high-energy particles perpendicular to the disk.

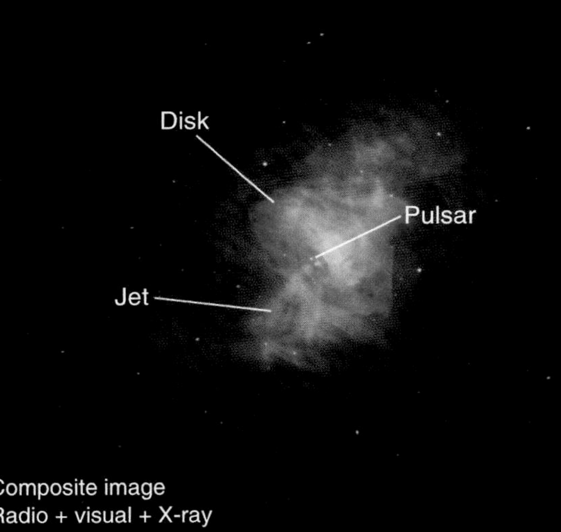

NASA

Composite image
Radio + visual + X-ray

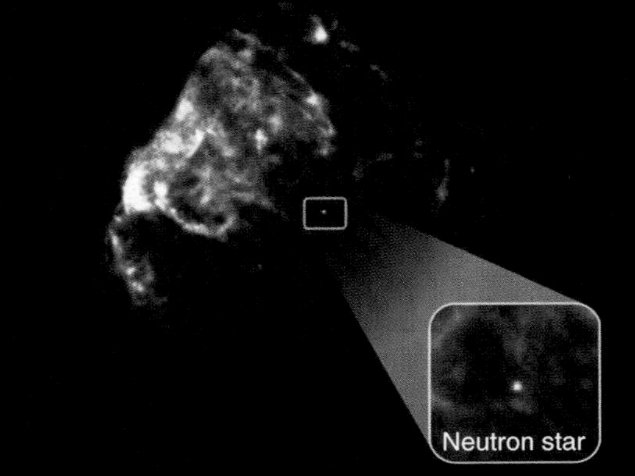

X-ray image of Puppis supernova remnant

NASA

Hubble Space Telescope visual-wavelength image

Neutron star

NASA

If a pulsar's beams do not sweep over Earth, observers detect no pulses, and the neutron star is difficult to find. A few such objects are known, however. The Puppis A supernova remnant is about 4000 years old and contains a point source of X rays believed to be a neutron star. The isolated neutron star in the right-hand image has a temperature of 700,000 K.

that supernova explosions can occur asymmetrically, perhaps because of the violent turbulence in the exploding core. Such a violent, off-center explosion could give the neutron star a high velocity through space. Also, some supernovae probably occur in binary systems and fling the two stars apart at high velocity. In any case, pulsars are known to have such high velocities that many probably escape the disk of our galaxy. You should not be surprised that many neutron stars quickly leave their supernova remnant behind. And finally, you must remember that a pulsar can remain detectable for 10 million years or so, but a supernova remnant cannot survive more than about 50,000 years before it is mixed into the interstellar medium. Thus, you should not be surprised that most pulsars are not in supernova remnants or that most supernova remnants do not contain pulsars.

The explosion of Supernova 1987A in February 1987 apparently formed a neutron star. You can draw this conclusion because a burst of neutrinos was detected passing through Earth, and theory predicts that the collapse of a massive star's core into a neutron star would produce such a burst of neutrinos. At first the neutron star would be hidden at the center of the expanding shells of gas ejected into space, but as the gas expands and thins astronomers might be able to see it. Or, if its beams don't sweep over Earth, they might be able to detect it from its X-ray and gamma-ray emission. Although, years after the explosion, no neutron star has been detected, astronomers continue to watch the site, hoping to see the youngest pulsar known.

One reason pulsars are so fascinating is the extreme conditions found in spinning neutron stars. To see natural processes of even greater violence, you have only to look at pulsars in binary systems.

Binary Pulsars

Over a thousand pulsars are now known, and some are located in binary systems. These pulsars are of special interest because astronomers can learn more about the neutron star by studying the orbital motions of the binary. Also, in some cases, mass can flow from the companion star onto the neutron star, and that produces high temperatures and X rays.

The first binary pulsar was discovered in 1974 when astronomers Joseph Taylor and Russell Hulse noticed that the pulse period of the pulsar PSR1913+16 was changing. The period first grew longer and then grew shorter in a cycle that took 7.75 hours. Thinking of the Doppler shifts seen in spectroscopic binaries, the radio astronomers realized that the pulsar had to be in a binary system with an orbital period of 7.75 hours. When the orbital motion of the pulsar carries it away from Earth, observers see the pulse period slightly lengthened, just as the wavelength of light emitted by a receding source is lengthened. That is, observers see a redshift. Then, when the pulsar rounds its orbit and approaches Earth, they see the pulse period slightly shortened—a blueshift. From these changing Doppler shifts, the astronomers could calculate the radial velocity of the pulsar around its orbit just as if it were a spectroscopic binary star (Chapter 8). The resulting graph of radial velocity versus time could be analyzed to find the shape of the pulsar's orbit (■ Figure 11-7). When Taylor and Hulse analyzed PSR1913+16, they discovered that the binary system consisted of two neutron stars separated by a distance roughly equal to the radius of our sun.

Yet another surprise was hidden in the motion of PSR1913+16. In 1916, Einstein's general theory of relativity described gravity as a curvature of space-time. Einstein realized that any rapid change in a gravitational field should spread outward at the speed of light as **gravitational radiation.** Gravity waves have not been detected, but Taylor and Hulse were able to show that the orbital period of the binary pulsar was slowly growing shorter because the stars are gradually spiraling toward each other. They are radiating orbital energy away as gravitational radiation. Taylor and Hulse won the Nobel prize in 1993 for their work with binary pulsars.

■ Figure 11-7

The radial velocity of pulsar PSR1913+16 can be found from the Doppler shifts in its pulsation. Analysis of the radial velocity curve allows astronomers to determine the pulsar's orbit. Here the center of mass does not appear to be at a focus of the elliptical orbit because the orbit is inclined. (Adapted from data by Joseph Taylor and Russell Hulse)

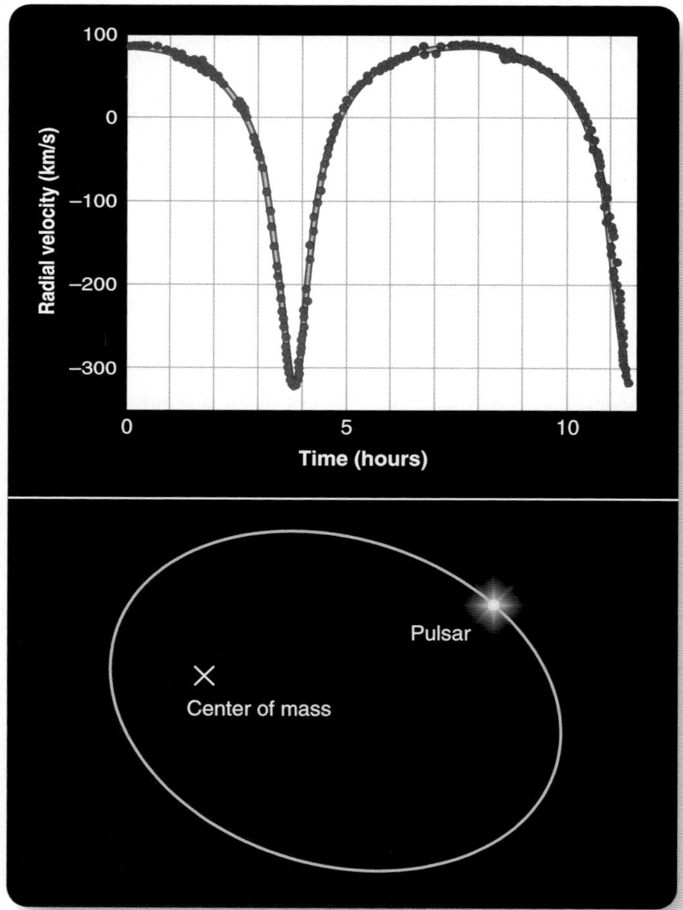

Dozens of binary pulsars have been found; by analyzing the Doppler shifts in their pulse periods, astronomers can estimate the mass of the neutron stars. Typical masses are about 1.35 solar masses, in good agreement with models of neutron stars.

Binary pulsars are relatively quiet systems now, but they may once have been violently active. The typical companion star to a binary pulsar was probably once a giant star losing mass to the neutron star, and that matter falling into the neutron star would have liberated tremendous energy. A single marshmallow dropped onto the surface of a neutron star from a distance of 1 AU would hit with the impact of a 3-megaton nuclear bomb. Even a small amount of matter flowing into a neutron star can generate high temperatures and release X rays and gamma rays.

As an example of such an active system, examine Hercules X-1. It emits pulses of X rays with a period of about 1.2 seconds, but every 1.7 days the pulses vanish for a few hours (■ Figure 11-8a). Astronomers can understand this system by comparing it to an eclipsing binary star. Hercules X-1 seems to contain a 2-solar-mass star and a neutron star that orbit each other with a period of 1.7 days. Matter flowing from the normal star into an accretion disk around the neutron star can reach temperatures of millions of degrees and emit a powerful X-ray glow. Interactions with the neutron star's magnetic field can produce beams of X rays that sweep around with the rotating neutron star (Figure 11-8b). Earth receives a pulse of X rays every time a beam points this way. The X rays shut off completely every 1.7 days when the neutron star is eclipsed behind the normal star. Hercules X-1 is a complex system with many different high-energy processes going on simultaneously, and this quick analysis only serves to illustrate how complex and powerful such binary systems are during mass transfer.

The X-ray source 4U 1820−30 illustrates another way neutron stars can interact with normal stars. In this system, a neutron star orbits a white dwarf with a period of only 11 minutes (■ Figure 11-9a and b). The separation between the two objects is only about a third the distance between Earth and the moon. To explain how such a very close pairing could originate, theorists suggest that a neutron star collided with a giant star and went into an orbit *inside* the star. (Recall the low density of the outer envelope of giant stars.) The neutron star would have gradually eaten away the giant star's envelope from the inside, leaving the white dwarf behind. Matter still flows from the white dwarf into an accretion disk and then down to the surface of the neutron star (Figure 11-9c), where it accumulates until it ignites to produce periodic bursts of X rays. Objects called **X-ray bursters** are thought to be such binary systems involving mass transferred to a neutron star. Notice the similarity between this mechanism and that responsible for novae.

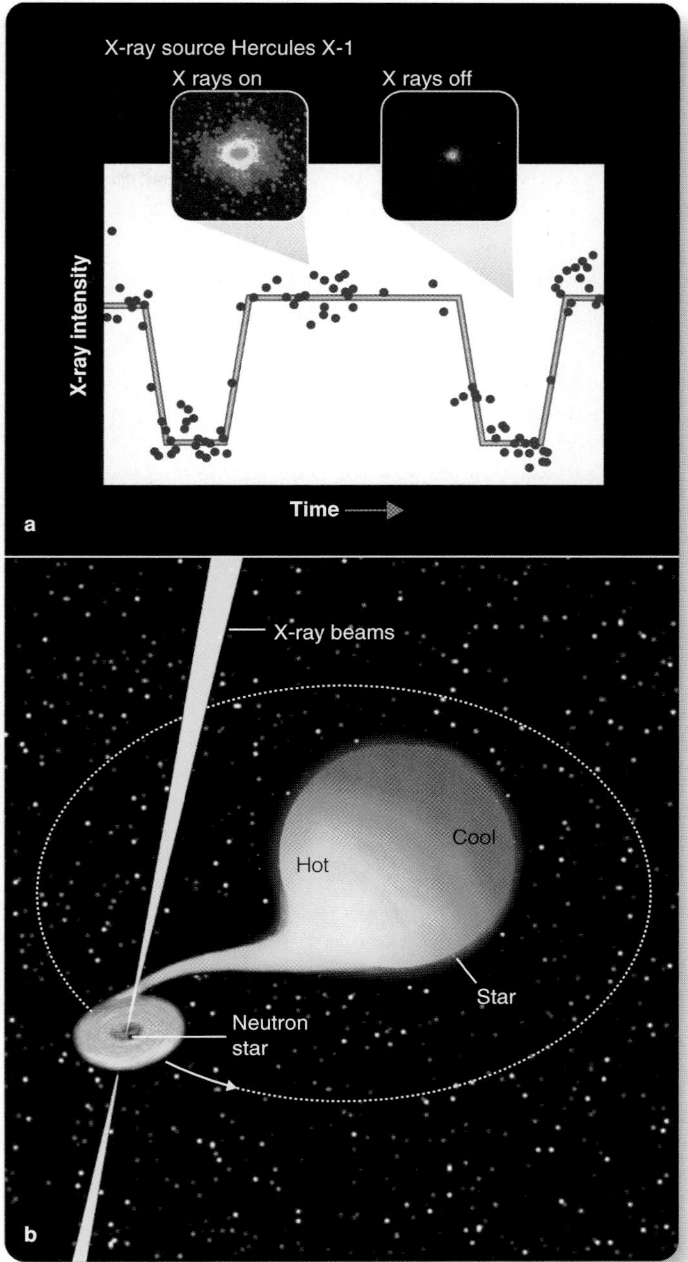

■ **Figure 11-8**

Sometimes the X-ray pulses from Hercules X-1 are on, and sometimes they are off. A graph of X-ray intensity versus time looks like the light curve of an eclipsing binary. (Insets: J. Trümper, Max-Planck Institute) (b) In Hercules X-1 matter flows from a star into an accretion disk around a neutron star producing X rays, which heat the near side of the star to 20,000 K compared with only 7000 K on the far side. X rays turn off when the neutron star is eclipsed behind the star.

The Fastest Pulsars

This discussion of pulsars suggests that newborn pulsars should blink rapidly, and old pulsars should blink slowly, but the handful that blink the fastest may be quite old. One of the fastest known pulsars is cataloged as PSR1937+21 in the constellation Vulpecula. It pulses 642 times a second and is slowing down only slightly. The energy stored in the rotation of a neutron star at this

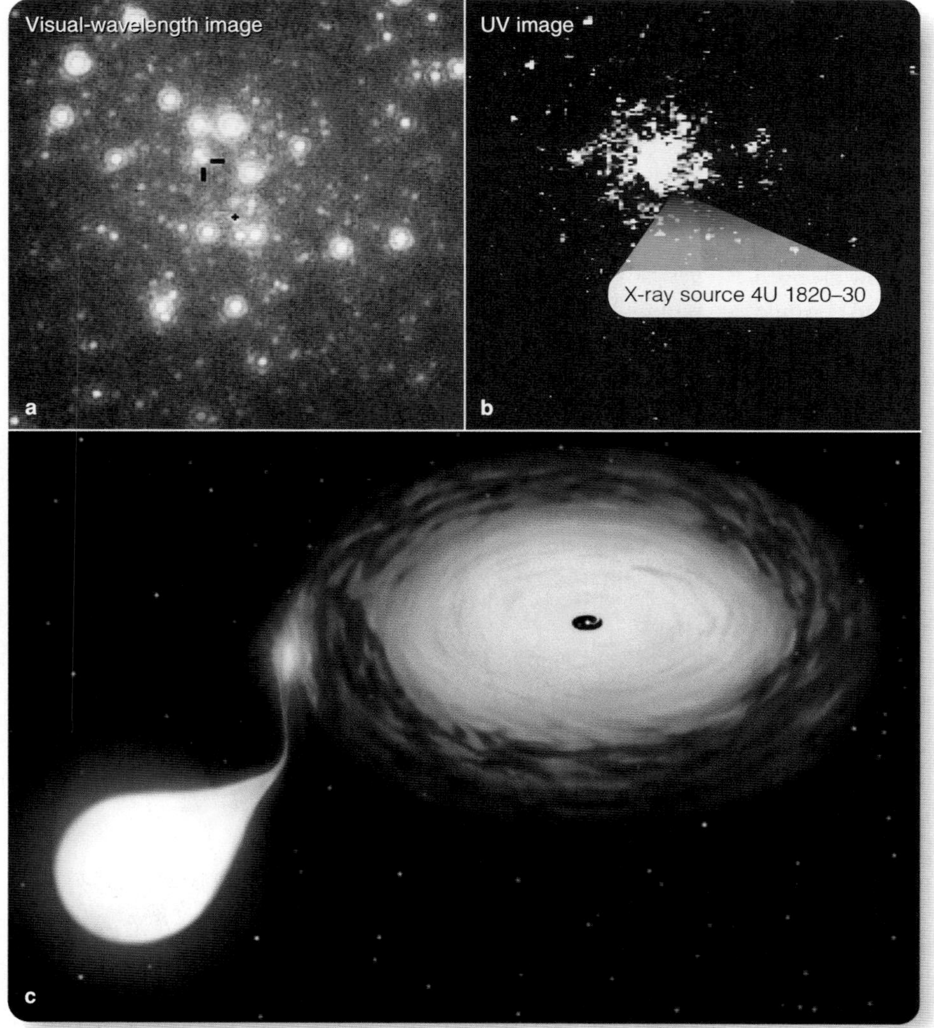

■ **Figure 11-9**

(a) At visible wavelengths, the center of star cluster NGC6624 is crowded with stars. (b) In the ultraviolet, one object stands out, an X-ray source consisting of a neutron star orbiting a white dwarf. (c) An artist's conception shows matter flowing from the white dwarf into an accretion disk around the neutron star. (a and b, Ivan King and NASA/ESA)

rate is equal to the total energy of a supernova explosion, so it seemed difficult at first to explain this pulsar. It now appears that PSR1937+21 is an old neutron star that has gained mass and rotational energy from a companion in a binary system. Like water hitting a mill wheel, the matter falling on the neutron star has spun it up to 642 rotations per second. With its weak magnetic field, it slows down very slowly and will continue to spin for a very long time.

A number of other very fast pulsars have been found. They are known generally as **millisecond pulsars** because their pulse periods are almost as short as a millisecond (0.001 s). This produces some fascinating physics because the pulse period of a pulsar equals the rotation period of the neutron star. If a neutron star 10 km in radius spins 642 times a second, as does PSR1937+21, then the period is 0.0016 seconds, and the equator of the neutron star must be traveling about 40,000 km/s. That is fast enough to flatten the neutron star into an ellipsoidal shape and is nearly fast enough to break it up.

All scientists should be made honorary citizens of Missouri, the "Show Me" state, because scientists demand evidence. The hypothesis that the millisecond pulsars are spun up by mass transfer from a companion star is quite reasonable, but astronomers demand evidence, and evidence has been found. For example, the pulsar PSRJ1740−5340 has a period of 42 milliseconds and is orbiting with a bloated red star that is losing mass to the neutron star. This appears to be a pulsar in the act of being spun up to high speed. For another example, consider the X-ray source XTEJ1751−305, a pulsar with a period of only 2.3 milliseconds. X-ray observations show that it is in the act of gaining mass from a companion star. The orbital period is only 42 minutes, and the mass of the companion star is only 0.014 solar mass. The evidence suggests this neutron star has devoured all but the last morsel of its binary partner.

Although some millisecond pulsars have binary companions, some are solitary neutron stars. A pulsar known as the Black Widow may explain how a fast pulsar can lack a companion. The Black Widow has a period of 1.6 milliseconds, meaning it is spinning 622 times per second, and it orbits with a low-mass companion. Presumably the neutron star was spun up by mass flowing from the companion, but spectra show that the blast of radiation and high-energy particles from the neutron star is now boiling away the surface of the companion. The Black Widow has eaten its fill and is now evaporating the remains of its companion. It will soon be a solitary millisecond pulsar (■ Figure 11-10).

"Show me," say scientists; and, in the case of neutron stars, the evidence seems very strong. Of course, you can never prove a theory is absolutely true (**Window on Science 11-1**), but the evidence for neutron stars is so strong that astronomers have great confidence that they really do exist. Other theories that describe how they emit beams of radiation and how they form and evolve are less certain, but continuing observations at many wavelengths are expanding the understanding of these last embers of massive stars. In fact, observations of one pulsar have turned up objects no one predicted.

The Impossibility of Proof in Science

No scientific theory or hypothesis can be proved correct. You can test a theory over and over by performing experiments or making observations, but you can never prove that the theory is absolutely true. It is always possible that you have misunderstood the theory or the evidence, and the next observation you make might disprove the theory.

For example, you might propose the theory that the sun is mostly iron. You might test the theory by looking at the iron lines in the solar spectrum, and the strength of the lines would suggest your theory is right. Although your observation has confirmed your theory, it has not proven the theory is right. You might confirm the theory many times before you realized that at the temperature of the sun iron absorbs photons much more efficiently than hydrogen. Although the hydrogen lines are weak in the sun's spectrum, they tell you that most of the atoms in the sun are hydrogen and not iron.

The nature of scientific thinking can lead to two kinds of mistakes. Sometimes nonscientists will say, "You scientists just want to tear everything down—you don't believe anything." Scientists test a theory over and over to test its worth. If a theory survives many tests, scientists begin to have confidence it is true.

The second error nonscientists make is to say, "You scientists are never sure of anything." Again, the scientist knows that no theory can be proved correct. That the sun will rise tomorrow is very likely, and scientists have great confidence in that theory. But in the end it is still a theory.

People will say of an idea they dislike, "That is only a theory," as if a theory were simply a random guess. In fact, a theory can be a well-tested truth in which all scientists have great confidence. Yet you can never prove that any theory is absolutely true.

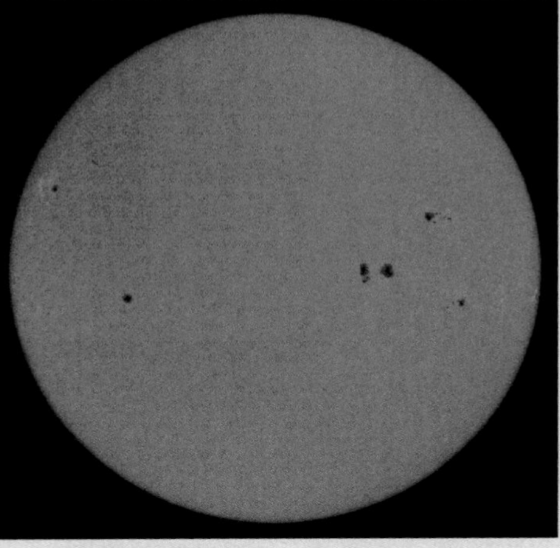

It is only a theory, but astronomers have tremendous confidence that the sun is made almost entirely of hydrogen and helium. (SOHO/MDI)

Pulsar Planets

Finding planets orbiting stars other than the sun is very difficult, and only about a hundred are known. Oddly, the first such planets were found orbiting a neutron star.

Because a pulsar's period is so precise, astronomers can detect tiny variations by comparison with atomic clocks. When astronomers checked pulsar PSR1257+12, they found variations in the period of pulsation much like those caused by the orbital motion of the binary pulsar (■ Figure 11-11a). However, in the case of PSR1257+12, the variations were much smaller; and, when they were interpreted as Doppler shifts, it became evident that the pulsar was being orbited by at least two objects with planetlike masses of 4.3 and 3.9 Earth masses. The gravitational tugs of the planets make the pulsar wobble about the center of mass of the system by no more than 800 km, and that produces the tiny changes in period (Figure 11-11b).

Astronomers greeted this discovery with both enthusiasm and skepticism. As usual, they looked for ways to test the hypothesis. Simple gravitational theory predicts that the planets should interact and slightly modify each other's orbit. When the data were analyzed, that interaction was found, further confirming the hypothesis that the variations in the period of the pulsar are caused by planets. In fact, further data revealed the presence of a third planet of about the mass of Earth's moon, and a fourth planet of about 100 Earth masses is now believed to follow a

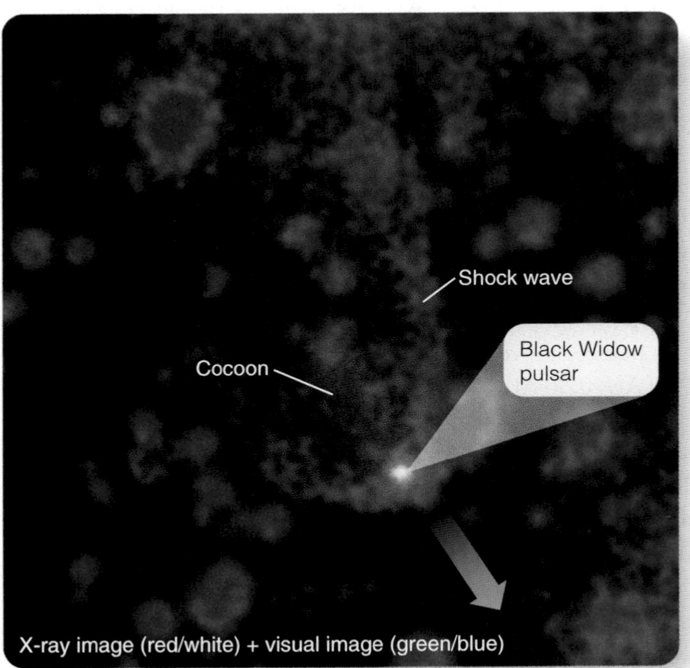

■ Figure 11-10

The Black Widow pulsar and its companion star are moving rapidly through space, creating a shock wave like the bow wave of a speedboat. The shock wave confines high-energy particles shed by the pulsar into an elongated cocoon (red). (X-ray: NASA/CXC/ASTRON/B. Stappers et al.; Optical: AAO/J. Bland-Hawthorn & H. Jones)

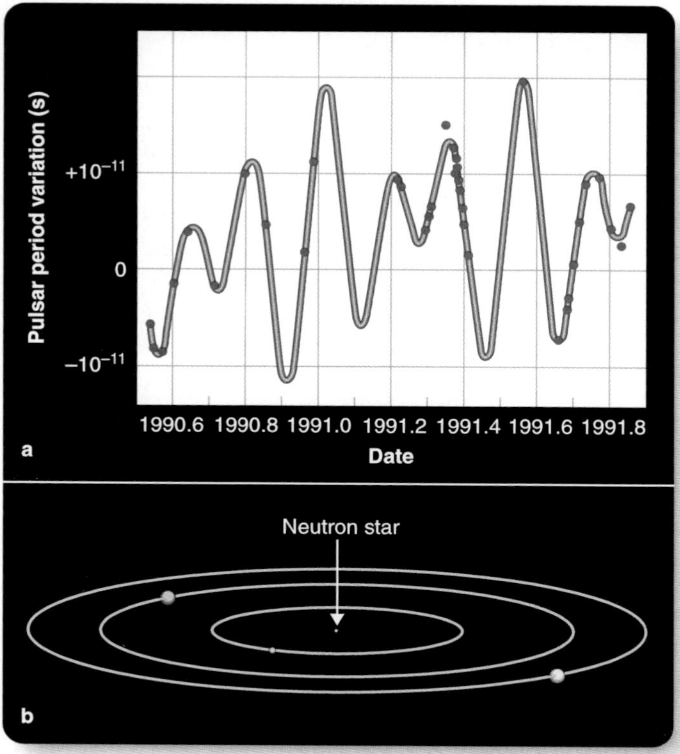

Figure 11-11

(a) The dots in this graph are observations showing that the period of pulsar PSR1257+12 varies from its average value by a fraction of a billionth of a second. The blue line shows the variation that would be produced by planets orbiting the pulsar. (b) As the planets orbit the pulsar, they cause it to wobble by less than 800 km, a distance that is invisibly small in this diagram. (Adapted from data by Alexander Wolszczan)

much larger orbit. This illustrates the astonishing precision of studies based on pulsar timing. At least one other pulsar is known where small shifts in pulse timing imply the presence of an orbiting planet.

Astronomers wonder how a neutron star can have planets. The planets are very close to the pulsar. The inner three orbit at 0.19 AU, 0.36 AU, and 0.47 AU—closer to the pulsar than Venus is to the sun. Any planets that orbit a star would be lost or vaporized when the star exploded. Furthermore, a star about to explode as a supernova would be a large giant or a supergiant, and planets only a few AU distant would be inside such a large star and could not survive. It seems more likely that these planets are the remains of a stellar companion that was devoured by the neutron star. In fact, the pulsar is very fast (162 pulses per second), suggesting that it was spun up in a binary system.

Another planet has been found orbiting in a binary system containing a neutron star and a white dwarf. Because this system is located in a very old star cluster and contains a white dwarf, astronomers suspect that the planet may be very old. Planets probably orbit other neutron stars, and small shifts in the timing of the pulses may eventually reveal their presence.

You can imagine what these worlds might be like. Formed from the remains of dying stars, they might have chemical compositions richer in heavy elements than Earth. You can imagine visiting these worlds, landing on their surfaces, and hiking across their valleys and mountains. Above you, the neutron star would glitter in the sky, a tiny point of light.

Inquire | Review | Analyze

How can a neutron star be found at X-ray wavelengths?

First, you should remember that a neutron star is very hot because of the heat released when it contracts to a radius of 10 km. It could easily have a surface temperature of 1,000,000 K, and Wien's law (Reasoning with Numbers 6-1) tells you that such an object will radiate most intensely at a very short wavelength typical of X rays. However, you know that the total luminosity of a star depends on its surface temperature and its surface area, and a neutron star is so small it can't radiate much energy. X-ray telescopes have found such neutron stars, but they are not easy to locate.

There is, however, a second way a neutron star can radiate X rays. If a normal star in a binary system loses mass to a neutron star companion, the inflowing matter will hit with so much energy that it will be heated to very high temperatures. It may form a very hot accretion disk that can radiate intense X rays easily detectable by X-ray telescopes orbiting above Earth's atmosphere.

If you discovered a pulsar, what observations would you make to determine whether it was young or old, single or a member of a binary system, alone or accompanied by planets?

■ ■ ■

Connections: Perhaps the strangest planets in the universe are those orbiting pulsars. But however strange a pulsar planet may be, you can imagine going to one. Our next topic, in contrast, seems beyond the reach of even our imaginations.

11-2 Black Holes

YOU HAVE NOW STUDIED white dwarfs and neutron stars, two of the three end states of dying stars. Now we turn to the third—black holes.

Although the physics of black holes is difficult to discuss without sophisticated mathematics, simple logic is sufficient to predict that they should exist. The problem is to consider their predicted properties and try to confirm that they exist. What objects observed in the heavens could be real black holes? More difficult than the search for neutron stars, the quest for black holes has nevertheless met with success.

To begin our discussion of black holes, you must consider a simple question. How fast must an object travel to escape from the surface of a celestial body? The answer will lead to black holes.

Escape Velocity

Suppose you threw a baseball straight up. How fast must you throw it if it is not to come down? Of course, gravity would pull back on the ball, slowing it, but if the ball were traveling fast enough to start with, it would never come to a stop and fall back. Such a ball would escape from Earth. The escape velocity (see Chapter 4) is the initial velocity an object needs to escape from a celestial body (■ Figure 11-12).

Whether you are discussing a baseball leaving Earth or a photon leaving a collapsing star, the escape velocity depends on two things: the mass of the celestial body and the distance from the center of mass to the escaping object. If the celestial body has a large mass, its gravity is strong and you need a high velocity to escape; but if you begin your journey farther from the center of mass, the velocity needed is less. For example, to escape from Earth, a spaceship would have to leave Earth's surface at 11 km/s (25,000 mph), but if you could launch spaceships from the top of a tower 1000 miles high, the escape velocity would be only 8.8 km/s (20,000 mph). If you could make an object massive enough or small enough, its escape velocity could be greater than the speed of light. Relativity tells us that nothing can travel faster than the speed of light, and even photons, which have no mass, would be unable to escape. Such a small, massive object could never be seen because light could not leave it.

■ Figure 11-12

Escape velocity, the velocity needed to escape from a celestial body, depends on mass. The escape velocity at the surface of a very small body would be so low you could jump into space. Earth's escape velocity is much larger, about 11 km/s (25,000 mph).

Long before Einstein and relativity, Rev. John Mitchell, a British gentleman astronomer, realized that Newton's laws of gravity and motion had peculiar consequences. In 1783, he pointed out that an object 500 times the radius of the sun but of the same density would have an escape velocity greater than the speed of light. Then, "all light emitted from such a body would be made to return towards it." Mitchell didn't know it, but he was talking about a black hole.

AceAstronomy™ Log into AceAstronomy and select this chapter to see Astronomy Exercise "Black Hole." You can change the mass of the object and watch it become a black hole.

Schwarzschild Black Holes

If the core of a star collapses and contains more than about 3 solar masses, no force can stop it. The object cannot stop collapsing when it reaches the size of a white dwarf because degenerate electrons cannot support the weight. It cannot stop when it reaches the size of a neutron star because degenerate neutrons cannot support the weight. No force remains to stop the object from collapsing to zero radius.

As an object collapses, its density and the strength of its surface gravity increase; and if an object collapses to zero radius, its density and gravity become infinite. Mathematicians call such a point a **singularity,** but in physical terms it is difficult to imagine an object of zero radius. Some theorists believe that a singularity is impossible and that the laws of quantum physics must somehow halt the collapse at some subatomic radius. Astronomically, it seems to make little difference.

If the contracting core of a star becomes small enough, the escape velocity in the region around it is so large that no light can escape. No information about the object or about the region of space near it can be received, and astronomers refer to this region as a **black hole.** To see how a black hole can exist, you must consider general relativity.

In 1916, Albert Einstein published a mathematical theory of space and time that became known as the general theory of relativity. Einstein treated space and time as a single entity—space-time. His equations showed that gravity could be described as a curvature of space-time, and almost immediately the astronomer Karl Schwarzschild found a way to solve the equations to describe the gravitational field around a single, nonrotating, electrically neutral lump of matter. That solution contained the first general relativistic description of a black hole, and nonrotating, electrically neutral black holes are now known as Schwarzschild black holes. In recent decades, theorists such as Roy P. Kerr and Stephen W. Hawking have found ways to apply the sophisticated mathematical equations of the general theory of relativity and quantum mechanics to charged, rotating black holes. For our discussion the differences are minor, and you may proceed as if all black holes were Schwarzschild black holes.

Schwarzschild's solution shows that if matter is packed into a small enough volume, then space-time curves back on itself. Objects can still follow paths that lead into the black hole, but no path leads out, so nothing can escape, not even light. Consequently, the inside of the black hole is totally beyond the view of an outside observer. The **event horizon** is the boundary between the isolated volume of space-time and the rest of the universe, and the radius of the event horizon is called the **Schwarzschild radius, R_S**—the radius within which an object must shrink to become a black hole (■ Figure 11-13).

Although Schwarzschild's work was highly mathematical, his conclusion is quite simple. The Schwarzschild radius (in meters) depends only on the mass of the object (in kilograms):

$$R_S = \frac{2GM}{c^2}$$

In this simple formula, G is the gravitational constant, M is the mass, and c is the speed of light. A bit of arithmetic shows that a 1-solar-mass black hole will have a Schwarzschild radius of 3 km, a 10-solar-mass black hole will have a Schwarzschild radius of 30 km, and so on (■ Table 11-1). Even a very massive black hole would not be very large.

■ **Table 11-1 | The Schwarzschild Radius**

	Mass ($M_\odot$)	R_S
Star	10	30 km
Star	3	9 km
Star	2	6 km
Sun	1	3 km
Earth	0.000003	0.9 cm

Every object with mass has a Schwarzschild radius, but not every object is a black hole. For example, Earth has a Schwarzschild radius of about 1 cm, but it could become a black hole only if you squeezed it inside that radius. Fortunately, Earth will not collapse spontaneously into a black hole because its mass is less than the critical mass of about 3 solar masses. Only exhausted stellar cores more massive than this can form black holes under the sole influence of their own gravity. In this chapter, we are interested in black holes that might originate from the deaths of massive stars. These would have masses larger than 3 solar masses. In later chapters, we will encounter black holes whose masses might exceed a million solar masses.

Do not think of black holes as giant vacuum cleaners that will pull in everything in the universe. A black hole is just a gravitational field, and at a reasonably large distance its gravity is no greater than that of a normal object of similar mass. If the sun were replaced by a 1-solar-mass black hole, the orbits of the planets would not change at all. The gravity of a black hole becomes extreme only when you approach close to it. ■ Figure 11-14 illustrates this by representing gravitational fields as curvature of the fabric of space-time. The universe contains many black holes. So long as you and other objects stay at a safe distance from the black holes, they have no catastrophic effects.

■ **Active Figure 11-13**

A black hole forms when an object collapses to a small size (perhaps to a singularity) and the escape velocity becomes so great light cannot escape. The boundary of the black hole is called the event horizon because any event that occurs inside is invisible to outside observers. The radius of the black hole R_S is the Schwarzschild radius.

AceAstronomy™ Log into AceAstronomy and select this chapter to see the Active Figures called "Schwarzschild Radius." Take control of this diagram.

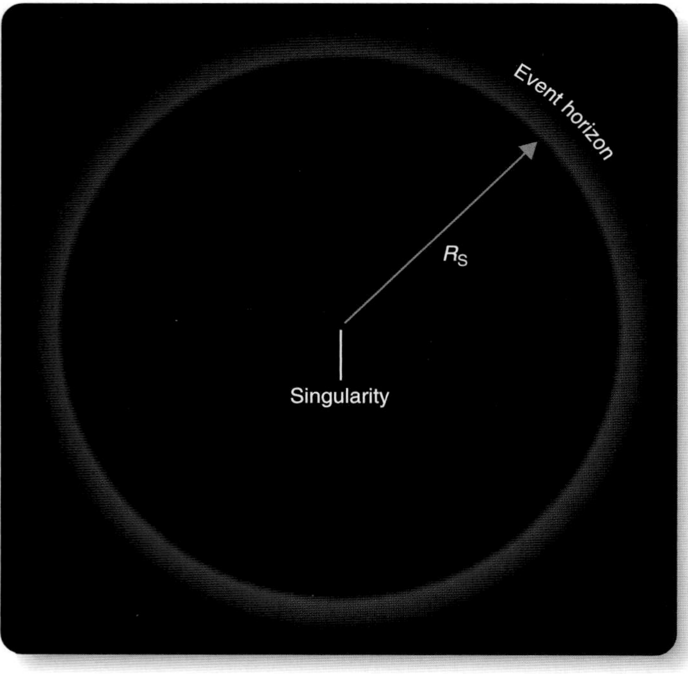

A Leap into a Black Hole

Before you can search for real black holes, you must understand what theory predicts about the appearance of a black hole. To explore that idea, you can imagine that you leap, feet first, into a Schwarzschild black hole.

If you were to leap into a black hole of a few solar masses from a distance of an astronomical unit, the gravitational pull would not be very large, and you would fall slowly at first. Of course, the longer you fell and the closer you came to the center, the faster you would travel. Your wristwatch would tell you that you fell for about 65 days before you reached the event horizon.

Gravitational field around
a 5-solar-mass star

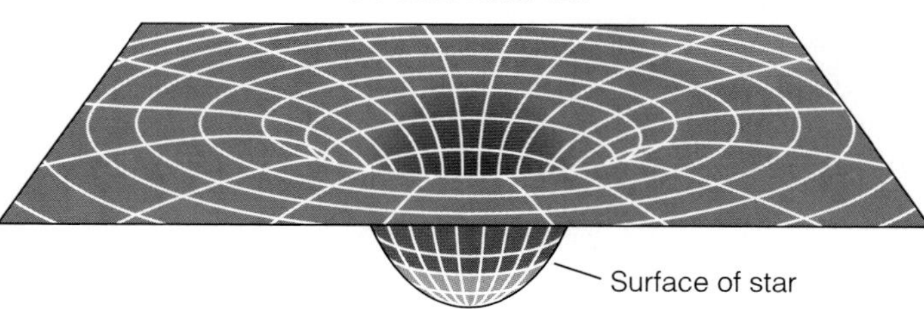

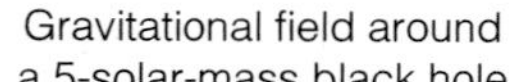

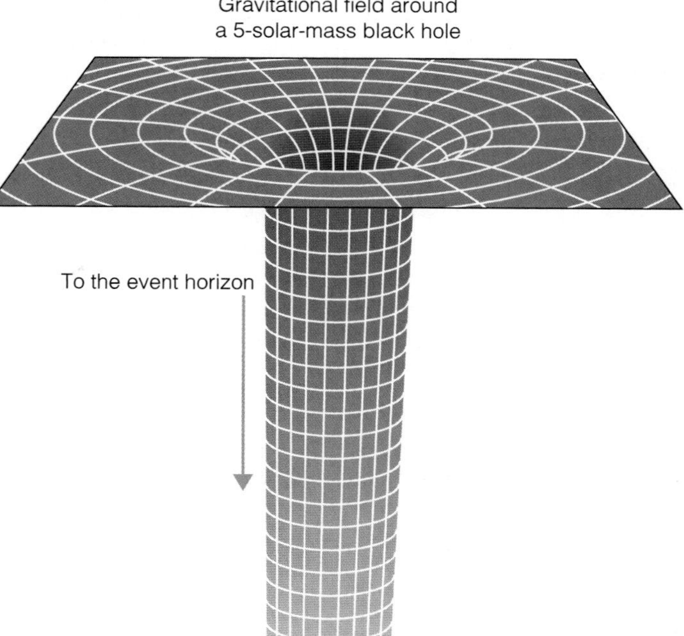

■ Figure 11-14

If you fell into the gravitational field of a star, you would hit the star's surface before you fell very far. Because a black hole is so small, you could fall much deeper into its gravitational field and eventually cross the event horizon. At a distance, the two gravitational fields are the same.

Your friends who stayed behind would see something different. They would see you falling more slowly as you came closer to the event horizon because, as explained by general relativity, clocks slow down in curved space-time. This is known as **time dilation.** In fact, your friends would never actually see you cross the event horizon. To them you would fall more and more slowly until you seemed hardly to move. Generations later, your descendants could focus their telescopes on you and see you still inching closer to the event horizon. You, however, would have sensed no slowdown and would conclude that you had crossed the event horizon after only about 65 days.

Another relativistic effect would make it difficult to see you with normal telescopes. As light travels out of a gravitational field, it loses energy, and its wavelength grows longer. This is known as the **gravitational redshift.** Although you would notice no effect as you fell toward the black hole, your friends would need to observe at longer and longer wavelengths in order to detect you.

While these relativistic effects seem merely peculiar, other effects would be quite unpleasant. Imagine again that you are falling feetfirst toward the event horizon of a black hole. You would feel your feet, which would be closer to the black hole, being pulled in more strongly than your head. This is a tidal force, and at first it would be minor. But as you fell closer, the tidal force would become very large. Another tidal force would compress you as your left side and your right side both fell toward the center of the black hole. For any black hole with a mass like that of a star, the tidal forces would crush you laterally and stretch you longitudinally long before you reached the event horizon (■ Figure 11-15). The friction from such severe distortions of your body would heat you to millions of degrees, and you would emit X rays and gamma rays. (Needless to say, this would render you inoperative as a thoughtful observer.)

Some years ago a popular book suggested that you could travel through the universe by jumping into a black hole in one place and popping out of another somewhere far across space.

Natural Checks on Fraud in Science

Fraud is actually quite rare in science. The nature of science makes fraud difficult, and the way scientists publish their research makes it almost impossible. In fact, you can think of science as a set of unwritten rules of behavior that have evolved to prevent scientists from lying to one another or to themselves, even by accident.

Suppose for a moment that you wanted to commit scientific fraud. You would have to invent data supposedly obtained from experiment or observation. You might invent X-ray data supposedly obtained by observing an X-ray binary star. Or if you were interested in theory, you might invent a fraudulent mathematical calculation of the physics going on in an X-ray binary. You might get away with it for a short time, but one of the most important rules in science is that good results must be reproducible. Other people must be able to repeat your observations, experiments, and calculations. In fact, most scientists routinely repeat other scientists' work as a way of getting started on a research topic. As soon as someone tries to repeat your fraudulent research, you will be caught. The more important a scientific result is, the sooner other scientists will repeat it, so you don't have much of a chance of getting away with scientific fraud. In this way, science is self-correcting.

Even if you could invent some convincing scientific research, you would probably have difficulty publishing it. When a scientist submits an article to a scientific journal, it is subject to peer review. That is, the editor of the journal sends the article to one or two other experts in the field for comment and suggestions. These reviewers often make helpful suggestions, but they may also point out errors that have to be fixed before the journal can publish the article. In some cases, an article may be so flawed the editor will refuse to publish it at all. If you submitted your fraudulent research on X-ray binaries, the reviewers would almost certainly notice things wrong with it, and it would never get into print.

Scientists know the rules, and they use them. If someone makes a big discovery and is interviewed by the press, scientists will begin asking, "Has this work been published in a peer-reviewed journal yet?" That is, they want to know if other experts have checked the work. Until research is published, it isn't really official, and most scientists would treat the results with care.

Fraud isn't impossible in science. Some cases have even been in the national news. Big grant money is a terrible temptation. But in science, "the truth will out." Because of the way scientists reproduce research and because of the way research is published, scientific fraud is quite rare.

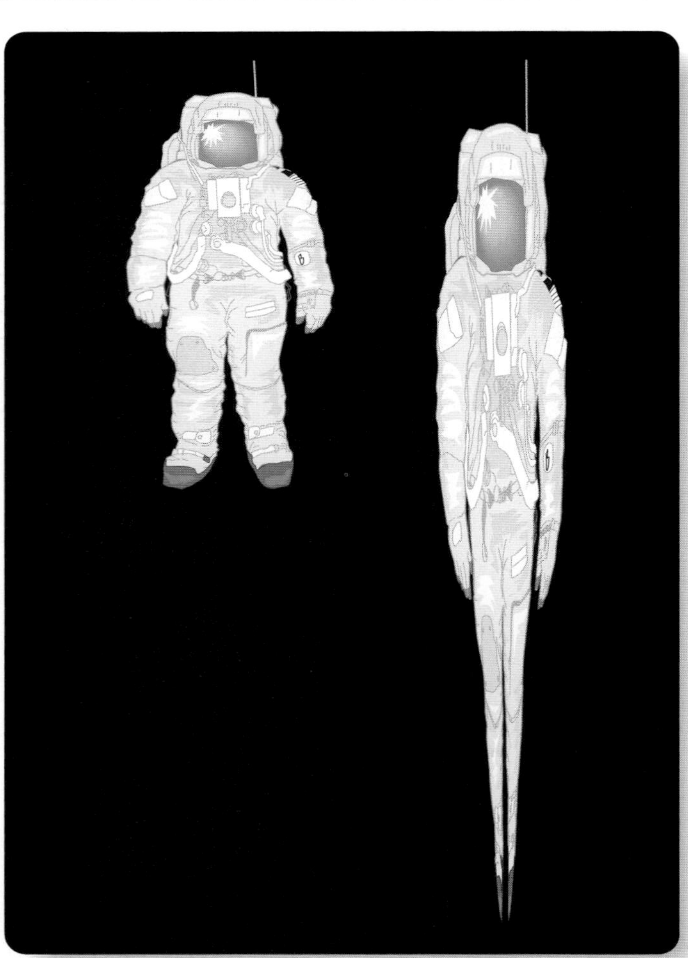

That might make for good science fiction, but tidal forces would make it an unpopular form of transportation even if it worked.

Your imaginary leap into a black hole is not entirely frivolous. You now know how to find a black hole: Look for a strong source of X rays. It may be a black hole into which matter is falling.

The Search for Black Holes

Do black holes really exist? Beginning in the 1970s, astronomers searched for observational evidence that their theories were correct. They tried to find one or more objects that were obviously black holes. That very difficult search is a good illustration of how the unwritten rules of science help us understand nature (**Window on Science 11-2**).

A black hole alone is totally invisible because nothing can escape from the event horizon. But a black hole into which matter is flowing would be a source of X rays. Of course, X rays can't escape from inside the event horizon, but X rays emitted by the heated matter flowing into the black hole could escape. An isolated black hole in space will not have much matter flowing into

Figure 11-15

Leaping feetfirst into a black hole. A person of normal proportions (left) would be distorted by tidal forces (right) long before reaching the event horizon around a typical black hole of stellar mass. Tidal forces would stretch the body lengthwise while compressing it laterally. Friction from this distortion would heat the body to high temperatures.

it, but a black hole in a binary system might receive a steady flow of matter transferred from the companion star. Thus, you can search for black holes by searching among X-ray binaries.

Some X-ray binaries such as Hercules X-1 contain a neutron star, and they will emit X rays much as would a binary containing a black hole. You can tell the difference in two ways. If the compact object emits pulses, you know it is a neutron star. Also, if the mass of the compact object is greater than 3 solar masses, the object cannot be a neutron star, and you can conclude that it must be a black hole.

The first X-ray binary suspected of harboring a black hole was Cygnus X-1, the first X-ray object discovered in Cygnus. It contains a supergiant B0 star and a compact object orbiting each other with a period of 5.6 days. Matter flows from the B0 star as a strong stellar wind, and some of that matter enters a hot accretion disk around the compact object (■ Figure 11-16). The accretion disk is about 5 times larger in diameter than the orbit of Earth's moon, and the inner few hundred kilometers of the disk has a temperature of about 2 million Kelvin—hot enough to radiate X rays. The compact object is invisible, but Doppler shifts in the spectrum reveal the motion of the B0 star around the center of mass of the binary. From the geometry of the orbit, astronomers can calculate the mass of the compact object—at least 3.8 solar masses, well above the maximum for a neutron star.

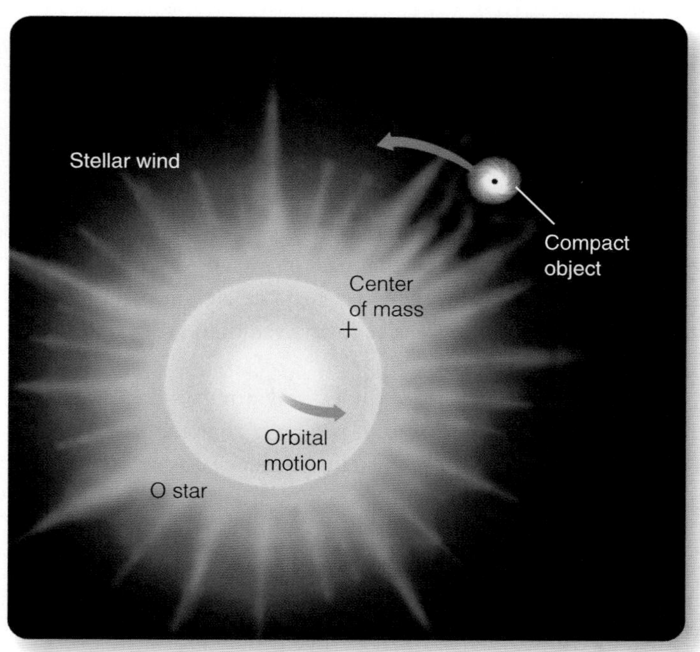

■ **Figure 11-16**

The X-ray source Cygnus X-1 is a supergiant O star and a compact object orbiting each other. Gas from the O star's stellar wind flows into the hot accretion disk, and the X rays we detect come from the disk.

To confirm that black holes existed, astronomers needed a conclusive example, an object that couldn't be anything else. Cygnus X-1 failed the test. Perhaps the B0 star was not a normal star, in which case assumptions about its mass might be far off. Also, there could be a third star in the system, and that would distort the analysis. At the time, astronomers could not conclusively show that Cygnus X-1 contained a compact object with a mass greater than 3 solar masses. More recent research has given astronomers more confidence that the compact object is indeed a black hole with a mass slightly less than 10 solar masses.

As X-ray telescopes have found more X-ray objects, the list of black hole candidates has grown to a few dozen. A few of these objects, such as the first two in ■ Table 11-2, contain massive stellar companions, either giants, supergiants, or massive main-sequence stars. This makes such systems difficult to analyze because such massive companions dominate the system. A good example of this sort is LMC X-3 (LMC refers to the Large Magellanic Cloud, a small galaxy near our own). The compact object in LMC X-3 has a mass of about 10 solar masses. One reason astronomers think the compact object is so massive is that it distorts the shape of the B main-sequence star into an egg shape; as the system rotates, the light from the B star varies because they see the side of the egg and then the end. By analyzing the light change, astronomers can determine the shape of the B star and from that find the mass of the compact object.

■ **Table 11-2 | Nine Black Hole Candidates**

Object	Location	Companion Star	Orbital Period	Mass of Compact Object
Cygnus X-1	Cygnus	O supergiant	5.6 days	$>3.8\ M_\odot$
LMC X-3	Dorado	B3 main-sequence	1.7 days	$\sim 10\ M_\odot$
A0620-00	Monocerotis	K main-sequence	7.75 hours	$10 \pm 5\ M_\odot$
V404 Cygni	Cygnus	K main-sequence	6.47 days	$12 \pm 2\ M_\odot$
J1655–40	Scorpius	F–G main-sequence	2.61 days	$6.9 \pm 1\ M_\odot$
QZ Vul	Vulpecula	K main-sequence	8 hours	$10 \pm 4\ M_\odot$
4U 1543-47	Lupus	A main-sequence	1.123 days	$2.7–7.5\ M_\odot$
V4641 Sgr	Sagittarius	B supergiant	2.81678 days	$8.7–11.7\ M_\odot$
XTEJ1118+480	Ursa Major	K main-sequence	0.170113 days	$>6\ M_\odot$

Many of the black hole candidates are binary systems in which the normal star is a lower-mass main-sequence star. Such systems do not remain X-ray sources continuously but suffer X-ray nova outbursts as matter flows rapidly into the accretion disk. A year or so after an outburst, the flow has stopped, the accretion disk has dimmed, and astronomers can detect the spectrum of the main-sequence star. The spectral type and Doppler motions of the main-sequence companion reveal the mass of the compact object with only limited uncertainty. The lower-mass stellar companion makes these X-ray nova systems easier to analyze than are systems with massive companions.

A number of examples of these X-ray nova systems are shown in Table 11-2. A0620−00 is an old nova that erupted again in 1975. It contains an ordinary main-sequence K star and a compact object that orbit each other with a period of 7.75 hours. From the orbital motion and the distortion of the K star, astronomers conclude the compact object must have a mass between 5 and 15 solar masses. Three of the best understood examples are V404 Cygni; J1655–40, also known as Nova Scorpii 1994; and QZ Vul. The compact objects in these systems seem much too massive to be anything but black holes.

The growing list of X-ray binaries with compact objects exceeding 3 solar masses has convinced astronomers that black holes really do exist. The problem now is to understand how these objects interact with matter flowing into them through accretion disks to produce X rays, gamma rays, and jets of matter.

Jets of Energy from Compact Objects

Your impression of a black hole might suggest that it is impossible to get any energy out of such an object. In later chapters, you will meet galaxies that extract vast amounts of energy from massive black holes, so you should pause here to see how a compact object can produce energy.

Whether a compact object is a black hole or a neutron star, it has a strong gravitational field. Any matter flowing into that field is accelerated inward, and because it must conserve angular momentum, it flows into an accretion disk made so hot by friction that the inner regions can emit X rays and gamma rays. Thermal and magnetic processes can cause the inner parts of the accretion disk to eject high-speed jets of matter in opposite directions along the axis of rotation. This process recalls the bipolar flows ejected by protostars, but it is much more powerful. You have seen in the Chandra X-ray image on page 235 that the Crab Nebula pulsar is ejecting jets of highly excited gas. The Vela pulsar, also a young neutron star, is ejecting a jet (Figure 11-5). Systems containing black holes can do the same. The black hole candidate J1655–40 is observed at radio wavelengths to be ejecting oppositely directed jets at 92 percent the speed of light.

One of the most powerful examples of this process is an X-ray binary called SS433. Its optical spectrum shows sets of spectral lines that are Doppler shifted by about one-fourth the speed of light, with one set shifted to the red and one set shifted to the blue. The object is both receding and approaching at a fantastic speed. Furthermore, the two sets of lines shift back and forth across each other with a period of 164 days.

Apparently, SS433 is a binary system in which a compact object (probably a black hole) pulls matter from its companion star and forms an extremely hot accretion disk. Jets of high-temperature gas blast away in beams aimed in opposite directions. As the disk precesses, it sweeps these beams around the sky once every 164 days, and you see light from gas trapped in both beams. One beam produces a redshift, and the other produces a blueshift (■ Figure 11-17). SS433 is a prototype that illustrates how the gravitational field around a compact object can produce powerful beams of radiation and matter. You will meet this phenomenon again when you study peculiar galaxies.

Gamma-Ray Bursts

The Cold War had an odd connection to neutron stars and black holes. In 1963, a nuclear test ban treaty was signed, and by 1968, the United States was able to put a series of Vela satellites in orbit to watch for nuclear tests that were violations of the treaty. A nuclear detonation emits gamma rays, so the Vela satellites were designed to watch for bursts of gamma rays coming from Earth. The experts were startled when the satellites began detecting about one gamma-ray burst a day coming from space. When those data were finally declassified in 1973, astronomers realized that the bursts might be coming from neutron stars or black holes. These objects are now known as **gamma-ray bursters.**

The Compton Gamma Ray Observatory reached orbit in 1991 and immediately began reporting gamma-ray bursts at the rate of a few a day. The intensity of the gamma rays rises to a maximum in seconds and then fades away quickly; a burst is usually over in seconds or minutes. Futhermore, the Compton Observatory discovered that the gamma-ray bursts were coming from all over the sky and not from any particular region. This helped astronomers sort out the different theories.

Some theories proposed that the gamma-ray bursts were being produced among the stars in our galaxy, but the Compton Observatory observation that the bursts were coming from all over the sky eliminated that possibility. If the gamma-ray bursts were produced among stars in our galaxy, you would expect to see them most often from along the Milky Way. The observation of bursts occurring all over the sky meant that bursts were probably coming from distant galaxies.

Gamma-ray bursts are hard to study because they occur without warning and fade so quickly, but in 1997, the Italian–Dutch satellite BeppoSAX began giving Earth's astronomers immediate notice when a gamma-ray burst occurred, along with an accurate position in the sky. When astronomers quickly looked at those positions at visible wavelengths, they could see the fading glow

tron stars. One of these objects produced a burst of gamma rays that reached Earth on August 27, 1998, and temporarily ionized Earth's upper atmosphere, disrupting radio communication worldwide.

Bursts of higher-energy gamma rays are spread uniformly across the sky and do not appear to repeat, so they can't be explained by magnetars. Two theories have been proposed. One suggestion is that gamma-ray bursts occur when two neutron stars orbiting each other lose energy by gravitational radiation and fall together. Theoretical models show that the merger of two neutron stars would produce a tremendous eruption lasting only seconds as the neutron stars ripped each other apart with tides and vanished into a newborn black hole. Such an event could produce bursts of gamma rays, possibly in beams.

The second theory proposes that gamma-ray bursts are produced when a star more massive than 25 solar masses exhausts its nuclear fuels. The interior of the star would collapse into a black hole and leave behind gas falling in through a hot accretion disk. Energy from the sudden formation of the accretion disk could blow the outer layers of the star apart in a supernova explosion and eject beams of high-energy particles along the axis of rotation. Interactions between the beams and surrounding gas could convert some of the energy into gamma rays. These have been called **hypernovae** or **collapsars** (■ Figure 11-18)

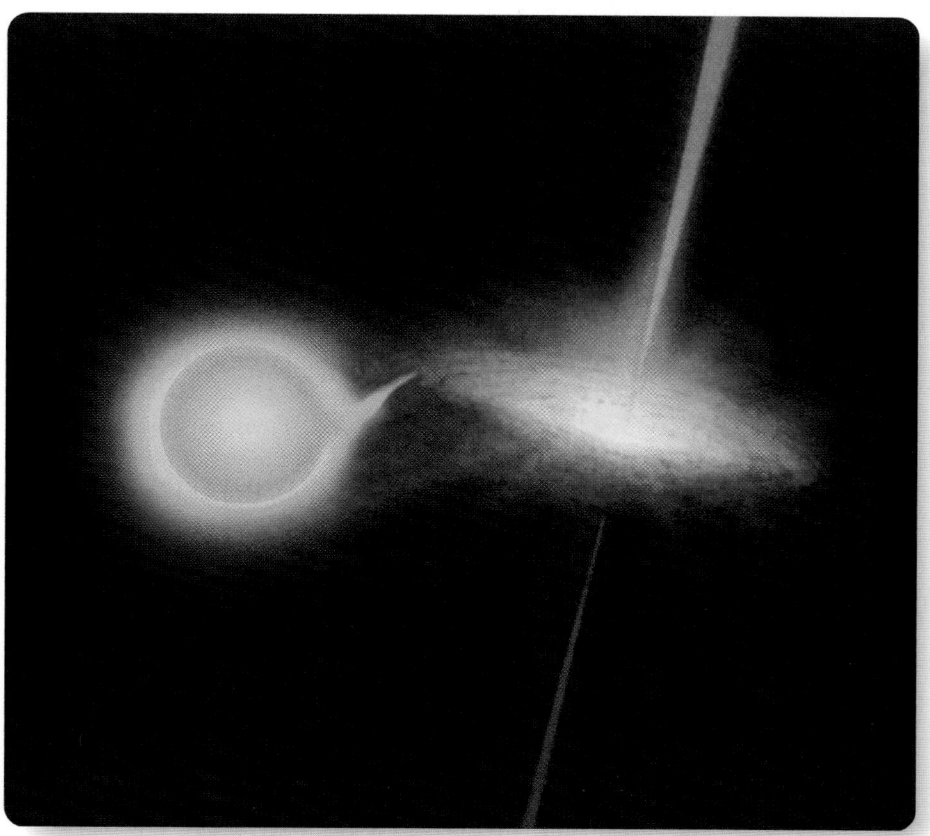

■ Active Figure 11-17

In this artist's impression, matter from a normal star at left flows into an accretion disk around a compact object at right. Processes in the spinning disk eject gas and radiation in jets perpendicular to the disk. The jet inclined slightly toward us has a blueshift, and the jet inclined slightly away from us has a redshift. (Adapted from NASA)

AceAstronomy™ Log into AceAstronomy and select this chapter to see the Active Figures called "End States of Stars." Compare the final states of high-, medium-, and low-mass stars.

of the explosion. The spectra of these glowing clouds reveal that most gamma-ray bursts are in very distant galaxies.

Given the intensity of a gamma-ray burst and its distance, astronomers can calculate its luminosity. If the gamma-ray bursts occur at very great distances, out among the galaxies, then they must be very powerful—at least as powerful as the most violent supernova explosions. Or perhaps the eruption blasts gamma rays out in beams. In that case, the explosion doesn't have to be quite as violent and could still be detected at very great distances if a beam were pointed at Earth.

Some bursts of lower-energy (soft) gamma rays repeat and are therefore known as **soft gamma-ray repeaters (SGRs).** Only a few are known, and they lie along the Milky Way. Astronomers suspect that they occur on young neutron stars that have magnetic fields 100 times stronger than the average neutron star. Dubbed **magnetars,** these objects can produce bursts of gamma rays when shifts in the magnetic field break the crust of the neu-

Earth's astronomers set up a fast alert system in which satellites could almost instantly report the position of gamma-ray bursts for observation by ground-based telescopes. The results showed that at least some gamma-ray bursts leave behind a fading glow like that from a supernova explosion in a distant galaxy. The turning point came on March 29, 2003, when astronomers rushed to study a powerful gamma-ray burst at a spot in the constellation Leo. The fading glow visible in ground-based telescopes had a complex spectrum that gradually organized itself into the spectrum of a hypernova.

The evidence seems conclusive that at least some gamma-ray bursts are produced by hypernovae. That does not rule out merging neutron stars, but no conclusive observations support that theory. We may eventually find that gamma-ray bursts can be produced in more than one way.

Incidentally, if a gamma-ray burst occurred only 1600 ly from Earth, the distance to a nearby binary pulsar, the gamma-ray burst would shower Earth with radiation equivalent to a

A Hypernova Explosion

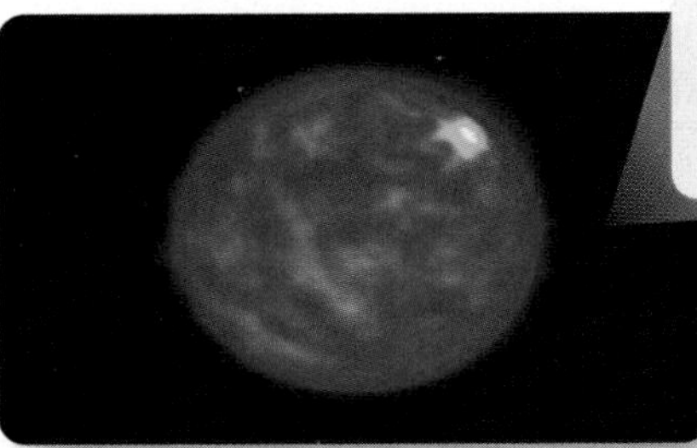

The collapsing core of a massive star drives its energy along the axis of rotation because. . .

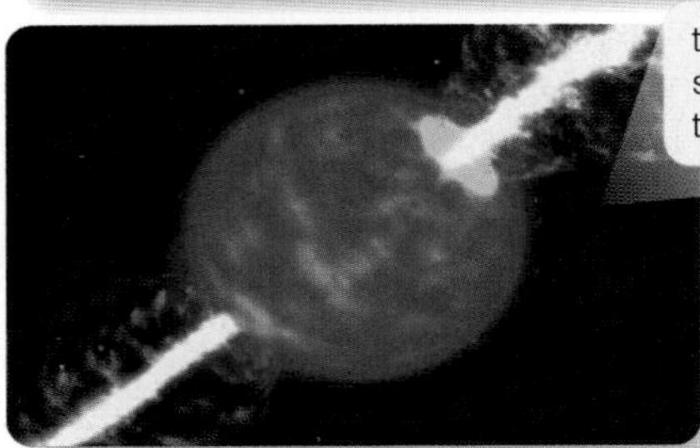

the rotation of the star slows the collapse of the equatorial regions.

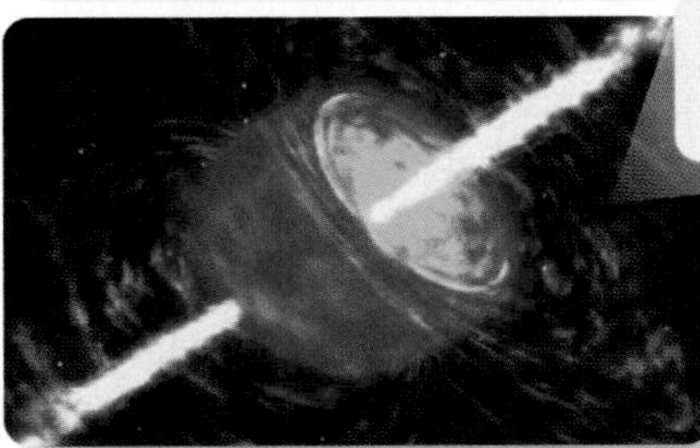

Within seconds, the remaining parts of the star fall in.

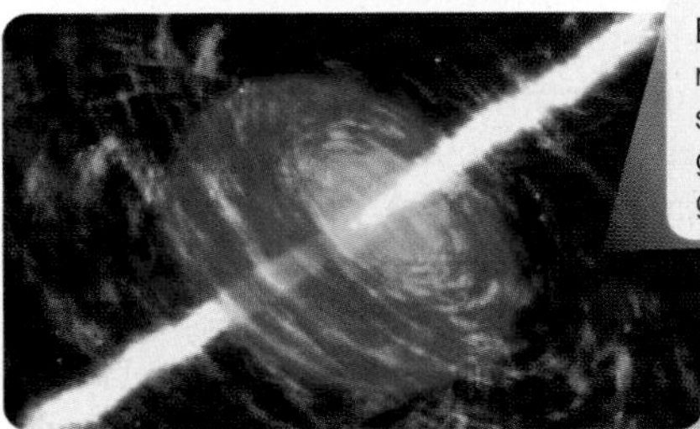

Beams of gas and radiation strike surrounding gas and generate beams of gamma rays.

The gamma-ray burst fades in seconds, and a hot accretion disk is left around the black hole.

■ Active Figure 11-18

The collapse of the cores of extremely massive stars can produce hypernova explosions, which are believed to be the source of at least some gamma-ray bursts. (NASA/Skyworks Digital)

AceAstronomy™ Log into AceAstronomy and click on Movie Gallery to see a movie of a "Hypernova."

10,000-megaton nuclear blast. (The largest bombs ever made were a few megatons.) The gamma rays could create enough nitric oxide in the atmosphere to produce intense acid rain and would destroy the ozone layer and expose life on Earth to deadly levels of solar ultraviolet radiation. Gamma-ray bursts may occur relatively near the Earth as often as every few 100 million years and could be one of the causes of the mass extinctions that show up in the fossil record.

It may seem odd that such rare events as merging neutron stars and hypernovae are blamed for something so common that gamma-ray telescopes observe one or more every day. But you must remember that these events are so powerful they can be detected over very great distances. There may be 30,000 neutron star binaries in each galaxy, so mergers must occur now and then in any galaxy. Massive stars explode as hypernovae only rarely in any one galaxy, but astronomers can detect these explosions among a vast number of galaxies. Thus the 1000 or so gamma-ray bursts that the Compton Observatory detected each year appear to be coming from the final deaths of stars far across space.

Inquire | Review | Analyze

How can a black hole emit X rays?

Once a bit of matter falling into a black hole crosses the event horizon, no light or other electromagnetic radiation it emits can escape from the black hole. The matter becomes lost to view. But if it emitted radiation before it crossed the event horizon, that radiation could escape, and you could detect it. Furthermore, the powerful gravitational field near a black hole stretches and distorts in-falling matter, and internal friction heats the matter to millions of degrees. Wien's law (Reasoning with Numbers 6-1) tells you that matter at such a high temperature should emit X rays. Any X rays emitted before the matter crosses the event horizon will escape, and thus you can look for black holes by looking for X-ray sources. Of course, an isolated black hole will probably not have much matter falling in, but black holes in binary systems may have large amounts of matter flowing in from the companion star. Thus, you can search for black holes by looking for X-ray binaries.

The search for black holes has succeeded in finding a few strong candidates, but the problem is being sure a particular binary system contains a black hole and not a neutron star. What observations would you make of an X-ray binary system to distinguish between a black hole and a neutron star?

■ ■ ■

Connections: Compact objects emitting X rays and producing precessing jets of radiation and gas may not be as unusual as they seem. Many stars collapse to form black holes or neutron stars in binary systems, but these are objects of only a few solar masses. You will see similar phenomena many times more powerful when we explore the galaxies in Chapters 12, 13, and 14.

11-1 | Neutron Stars

Why would you expect neutron stars to exist?

- When a supernova explodes, the core collapses to very small size. Theory predicts that the collapsing core cannot support itself as a white dwarf if its mass is greater than 1.4 solar masses, the Chandrasekhar limit. If its mass lies between 1.4 solar masses and about 3 solar masses, it can halt its contraction and form a neutron star.
- A neutron star is supported by the pressure of the degenerate gas of neutrons. Theory predicts that a neutron star should be about 10 km in radius, spin very fast because it conserves angular momentum as it contracts, and have a powerful magnetic field.

How do you know neutron stars really exist?

- Pulsars, rapidly pulsing radio sources, were discovered in 1967 and were eventually understood to be spinning neutron stars. The discovery of a pulsar in the supernova remnant called the Crab Nebula was a key link in the story.
- Pulsars are evidently spinning neutron stars that emit beams of radiation from their magnetic poles. As they spin, they sweep the beams around the sky; if the beams sweep over Earth, astronomers detect pulses.
- A spinning neutron star slows as it radiates its energy into space. Most of the energy emitted by a pulsar is carried away as a pulsar wind.
- Dozens of pulsars have been found in binary systems, which allows astronomers to estimate the masses of the neutron stars. Such masses are consistent with the predicted masses of neutron stars.
- In some binary systems, mass flows into a hot accretion disk around the neutron star and causes the emission of X rays.
- The fastest pulsars, the millisecond pulsars, appear to be old pulsars that have been spun up to high speed by mass flowing from binary companions.
- Planets have been found orbiting at least one neutron star. They may be the remains of a companion star that was mostly devoured by the neutron star.

11-2 | Black Holes

How does theory predict the existence of black holes?

- If the collapsing core of a supernova has a mass greater than 3 solar masses, then it must contract to a very small size—perhaps to a singularity, an object of zero radius. Near such an object, gravity is so strong that not even light can escape, and the region is called a black hole.
- The outer boundary of a black hole is the event horizon; no event inside is detectable. The radius of the event horizon is the Schwarzschild radius, amounting to only a few kilometers for a black hole of stellar mass.

How can you be sure that black holes really exist?

- If you were to leap into a black hole, your friends who stayed behind would see two relativistic effects. They would see your clock slow relative to their own clock because of time dilation. Also, they would see your light redshifted to longer wavelengths.
- You would not notice these effects, but you would feel powerful tidal forces that would deform and heat your mass until you grew hot enough to emit X rays. Any X rays you emitted before crossing the event horizon could escape.
- To search for black holes, you must look for binary star systems in which mass flows into a compact object and emits X rays. If the mass of the compact object is greater than about 3 solar masses, then the object is presumably a black hole. A number of such objects have been located.
- Black holes and neutron stars at the center of accretion disks can eject powerful beams of radiation and gas. Such beams have been detected.
- Gamma-ray bursters appear to be related to violent events involving neutron stars and black holes. Many bursts appear to arise during hypernovae, the collapse of the most massive stars to form black holes.

New Terms

neutron star (p. 230)
pulsar (p. 231)
pulsar wind (p. 233)
gravitational radiation (p. 236)
X-ray burster (p. 237)
millisecond pulsar (p. 238)
singularity (p. 241)
black hole (p. 241)
event horizon (p. 242)
Schwarzschild radius (R_S) (p. 242)
time dilation (p. 243)
gravitational redshift (p. 243)
gamma-ray burster (p. 246)
soft gamma-ray repeater (SGR) (p. 247)
magnetar (p. 247)
hypernova (p. 247)
collapsar (p. 247)

Review Questions

Ace Astronomy™ Assess your understanding of this chapter's topics with additional quizzing and animations at **http://astronomy.brookscole.com/sh9e**

1. How are neutron stars and white dwarfs similar? How do they differ?
2. Why is there an upper limit to the mass of neutron stars?
3. Why do you expect neutron stars to spin rapidly?
4. If neutron stars are hot, why aren't they very luminous?
5. Why do you expect neutron stars to have a powerful magnetic field?
6. Why did astronomers conclude that pulsars could not be pulsating stars?
7. What does the short length of pulsar pulses tell you?
8. How does the lighthouse model explain pulsars?
9. What evidence do you have that pulsars are neutron stars?
10. Why would astronomers at first assume that the first millisecond pulsar was young?
11. How can a neutron star in a binary system generate X rays?
12. If the sun has a Schwarzschild radius, why isn't it a black hole?
13. How can a black hole emit X rays?
14. What evidence can you cite that black holes really exist?
15. How can mass transfer into a compact object produce jets of high-speed gas? X-ray bursts? gamma-ray bursts?
16. Discuss the possible causes of gamma-ray bursters.
17. The X-ray image at the right shows the supernova remnant G11.2–0.3 and its central pulsar in X rays. The blue nebula near the pulsar is caused by the pulsar wind. How old do you think this system is? Discuss the appearance of this system a million years from now.

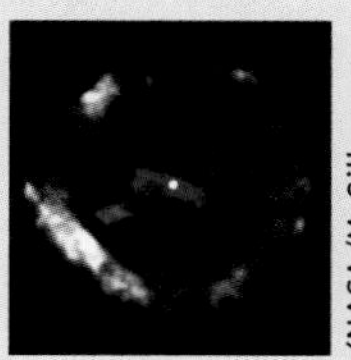
(NASA/McGill, V. Kaspi et al.)

18. What is happening in the artist's impression at the right? How would you distinguish between a neutron star and a black hole in such a system?

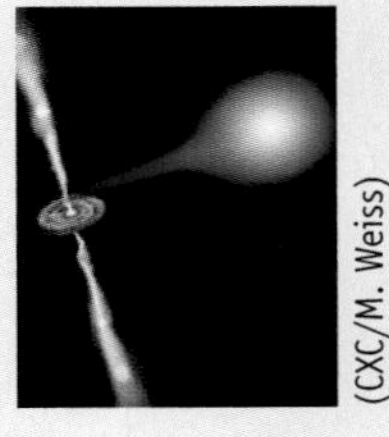
(CXC/M. Weiss)

Discussion Questions

1. In your opinion, has the existence of neutron stars been sufficiently tested to be called a theory, or should it be called a hypothesis? What about the existence of black holes?
2. Why wouldn't an accretion disk orbiting a giant star get as hot as an accretion disk orbiting a compact object?

Problems

1. If a neutron star has a diameter of 10 km and rotates 642 times a second, what is the speed of the surface at the neutron star's equator in terms of the speed of light? (*Hint:* The circumference of a circle is $2\pi r$.)
2. A neutron star and a white dwarf have been found orbiting each other with a period of 11 minutes. If their masses are typical, what is the average distance between them? (*Hint:* See Reasoning with Numbers 8-4.)
3. If Earth's moon were replaced by a typical neutron star, what would the angular diameter of the neutron star be, as seen from Earth? (*Hint:* See Reasoning with Numbers 3-1.)
4. What is the Schwarzschild radius of Jupiter (mass $= 2 \times 10^{27}$ kg)? of a human adult (mass = 75 kg)? (*Hint:* See Appendix A for the values of G and c.)
5. If the inner accretion disk around a black hole has a temperature of 10^6 K, at what wavelength will it radiate the most energy? (*Hint:* See Reasoning with Numbers 6-1.)
6. What is the orbital period of a bit of matter in an accretion disk 2×10^5 km from a 10-$M_\odot$ black hole? (*Hint:* See Reasoning with Numbers 8-4.)
7. If a 20-$M_\odot$ star and a neutron star orbit each other every 13.1 days, then what is the average distance between them? (*Hint:* See Reasoning with Numbers 8-4.)
8. What is the orbital velocity at a distance of 7400 meters from the center of a 5-solar-mass black hole? What kind of particles could orbit at this distance? (*Hint:* See Reasoning with Numbers 4-1.)
9. Compare the orbit in Problem 8 with an orbit having the same velocity around a 2-solar-mass neutron star. Why is this orbit impossible? (*Hint:* See Reasoning with Numbers 4-1.)

Media Cluster

Ace Astronomy™ To access the resources in the Media Cluster, log into AceAstronomy at **http://astronomy.brookscole.com/sh9e** and select Chapter 11.

ACTIVE FIGURES

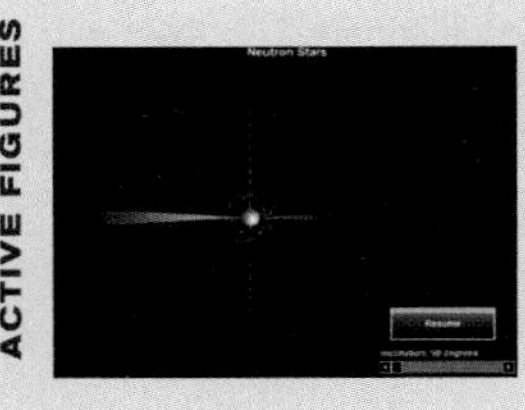

Neutron Star
Some neutron stars are pulsars; some are not. This animation shows you how the alignment of the star's axis of rotation and its polar axis creates the two effects. You can adjust the inclination of the neutron star's magnetic field to produce pulses.

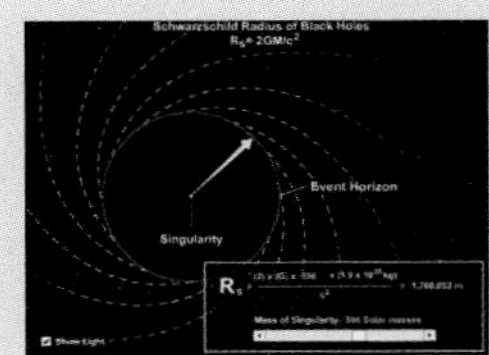

Schwarzschild Radius
Nothing can approach closer than the event horizon of a black hole and still come back out. Its size—the Schwarzschild radius—depends only on the black hole's mass. In this animation you take control of your own black hole to explore this relationship.

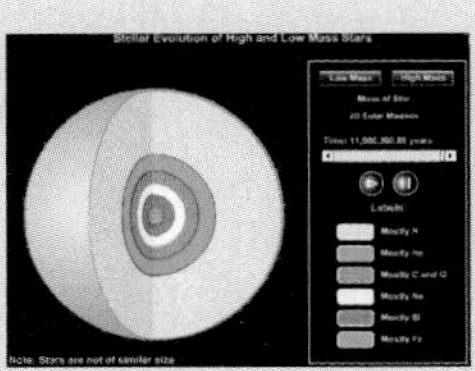

End States of Stars
In this animation you can contrast the lifespans and stages of nuclear fusion in high- and low-mass stars.

ASTRONOMY EXERCISE

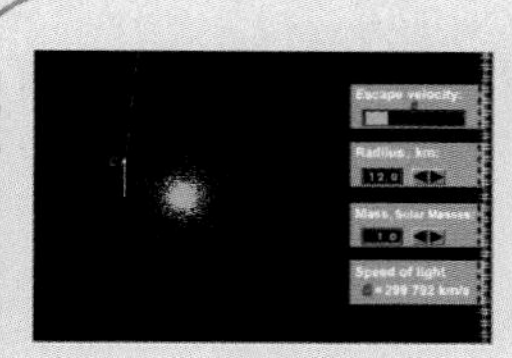

Black Hole
You can change the mass and radius of the object in this animation. See if you can discover what it takes to make a black hole.

VIRTUAL ASTRONOMY LABS

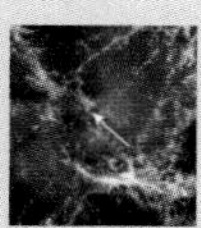

Lab 14: Neutron Stars and Pulsars
This lab looks at the extraordinary properties of neutron stars, the dense balls of neutrons that may remain after some stars have exploded. Later the lab examines pulsars, neutron stars that appear to emit radiation in rapid pulses.

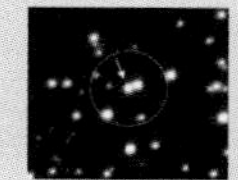

Lab 15: General Relativity and Black Holes
This lab explores the properties of black holes. It includes an exercise illustrating Einstein's general theory of relativity and an exercise on binary quasars. The lab concludes with a discussion and an exercise on black hole detection.

Critical Inquiries for the Web

1. Imagine that you are on a mission to explore one of the pulsar planets noted in the chapter. What would you find there? Look for information about pulsars and the known pulsar planets on the web and describe what you might encounter on such a mission.
2. What would you experience if you were to pilot a spacecraft near a black hole? Visit black-hole-related Internet sites to determine what the gravitational effects and general environment would be. Also use the Internet to find the limits of human tolerance to strong gravitational forces. (*Hint:* Look for information about astronaut training and find out how many g's a human can withstand.) Use these sources to give a brief account about what your voyage would be like.
3. Search for information about gamma-ray bursters. What is the latest news in this developing story?

Exploring *TheSky*

1. The Crab Nebula pulsar is located in the Crab Nebula, also known as M1. Locate it, zoom in, and compare its shape and size with Figure 11-3. (*TheSky* does not show the pulsar.)

Go to the Brooks/Cole Astronomy Resource Center **(http://astronomy.brookscole.com)** for critical thinking exercises, articles, and additional readings from InfoTrac College Edition, Brooks/Cole's online student library.

12 The Milky Way Galaxy

A hypothesis or theory
is clear, decisive, and positive,
but it is believed by no one
but the man who created it.
Experimental findings,
on the other hand,
are messy, inexact things,
which are believed by
everyone except the man
who did that work.

HARLOW SHAPLEY
Through Rugged Ways to the Stars

Visual-wavelength image

YOU ARE FANTASTICALLY wealthy. You live inside one of the largest star systems in the universe. Our Milky Way Galaxy is over 75,000 ly in diameter and contains over 100 billion stars. If Earth is the only inhabited planet in the galaxy, then the galaxy belongs to all of us; and, sharing equally, each person on Earth owns 50 stars plus assorted planets, moons, comets, and so on. Even if we must share with a few other inhabited planets, you are rich beyond any dream. Of course, you lack transportation to visit your dominions, but you can still admire our galaxy. It is, after all, ours. ▌ It isn't obvious that we live in a galaxy. We are inside, and we see nearby stars scattered all over the sky, while the more distant clouds of stars in our galaxy make a faint band of light circling the sky (■ Figure 12-1a). The ancient Greeks named that band

The Milky Way, our galaxy seen from inside, rises behind a telescope dome and the highly polished surface of a submillimeter telescope at the La Silla European Southern Observatory in Chile. (ESO and Nico Housen)

Guidepost

Looking Back

In previous chapters, you learned how stars are born from clouds of gas and dust, how they evolve as they consume their nuclear fuels, and how they die. You discovered that most stars live long, quiet lives but that the most massive stars die in violent supernova explosions.

This Chapter

Now you are ready to see stars in their vast communities called galaxies. This chapter discusses our home galaxy, the Milky Way galaxy, and attempts to answer four essential questions:

How do we know we live in a galaxy?

How did our galaxy form and evolve?

What are the spiral arms?

What lies at the very center?

In this chapter you will see more examples of how scientists use evidence and theory to understand nature. If in some cases the evidence seems contradictory and the theories incomplete, do not be disappointed. The adventure of discovery is not yet over.

Looking Ahead

In the chapters that follow, you will meet some of the billions of galaxies that fill the depths of the universe. Understanding the Milky Way Galaxy is only a step in understanding the universe as a whole.

AceAstronomy™ The AceAstronomy icon throughout the text indicates an opportunity for you to test yourself on key concepts and to explore animations and interactions on the AceAstronomy website at: http://astronomy.brookscole.com/sh9e

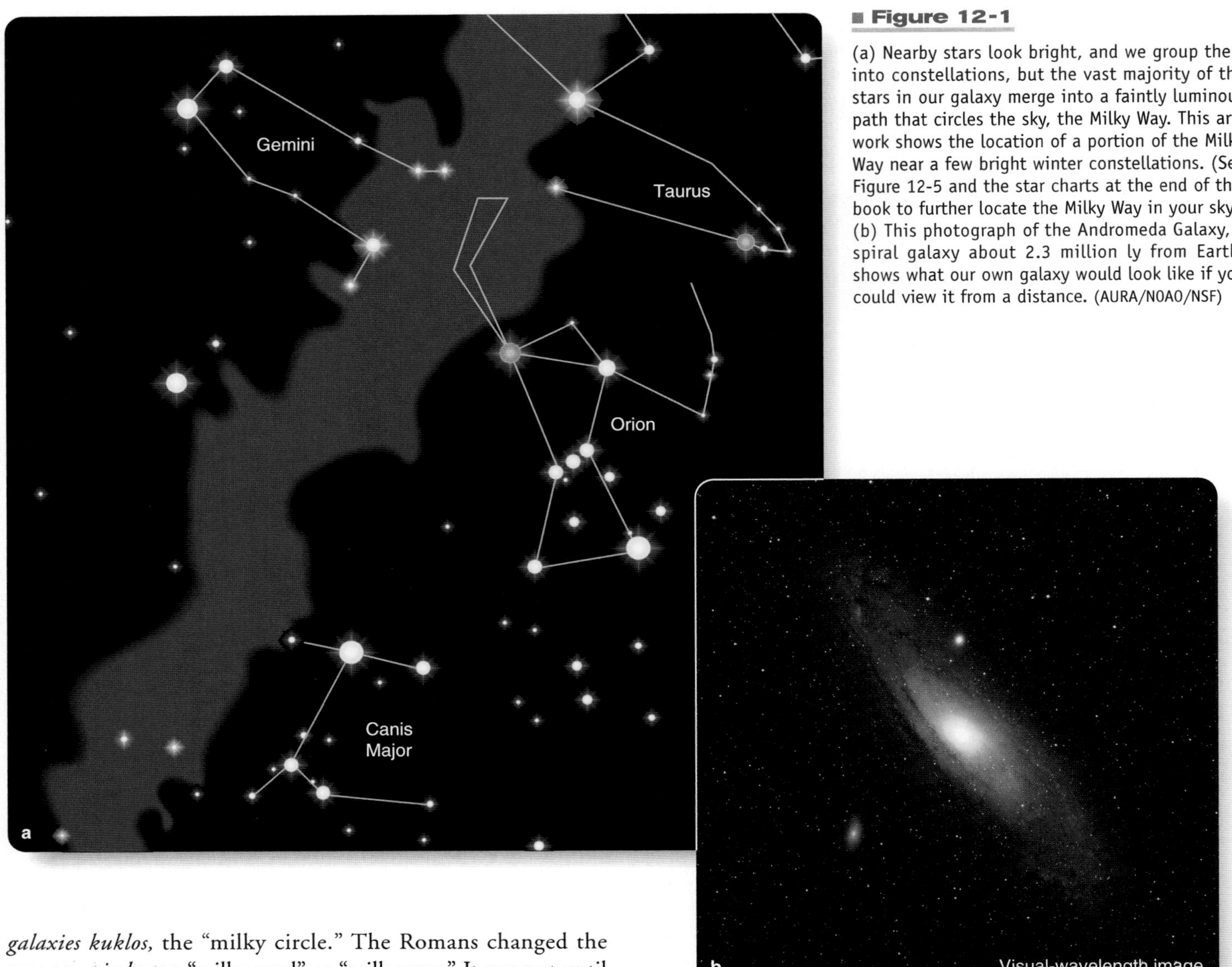

■ Figure 12-1

(a) Nearby stars look bright, and we group them into constellations, but the vast majority of the stars in our galaxy merge into a faintly luminous path that circles the sky, the Milky Way. This artwork shows the location of a portion of the Milky Way near a few bright winter constellations. (See Figure 12-5 and the star charts at the end of this book to further locate the Milky Way in your sky.) (b) This photograph of the Andromeda Galaxy, a spiral galaxy about 2.3 million ly from Earth, shows what our own galaxy would look like if you could view it from a distance. (AURA/NOAO/NSF)

galaxies kuklos, the "milky circle." The Romans changed the name to *via lactea,* "milky road" or "milky way." It was not until early in the 20th century that astronomers understood that they lived inside a great wheel of stars and that the universe is filled with other such star systems. Drawing on the Greek word for milk, they called them galaxies.

Almost every celestial object visible to your naked eyes is part of our Milky Way Galaxy. The only exception visible from Earth's northern hemisphere is the Andromeda Galaxy (cataloged as M31), just visible to your unaided eyes as a faint patch of light in the constellation Andromeda.* Our galaxy probably looks much like the Andromeda Galaxy (Figure 12-1b).

Our goal is to understand our galaxy—how it formed and evolved and why it takes the form it does. As always, we begin by gathering evidence for comparison with theories. In this case, our evidence includes the size, mass, and shape of our galaxy and the distribution of stars within it.

*Consult the star charts at the end of this book to locate the Milky Way and the Andromeda Galaxy (labeled M31 on the chart).

12-1 The Discovery of the Galaxy

IT SEEMS ODD to say astronomers discovered something that is all around us; but, until the early 20th century, no one knew what the Milky Way was.

The Great Star System

Galileo's telescope revealed that the glowing Milky Way was made up of stars, and later astronomers realized that the sun must be located in a great wheel-shaped cloud of stars, which they called the star system. If the star system were spherical, for example, you would see the stars scattered more or less uniformly over the sky. Only a wheel shape could produce the band of the Milky

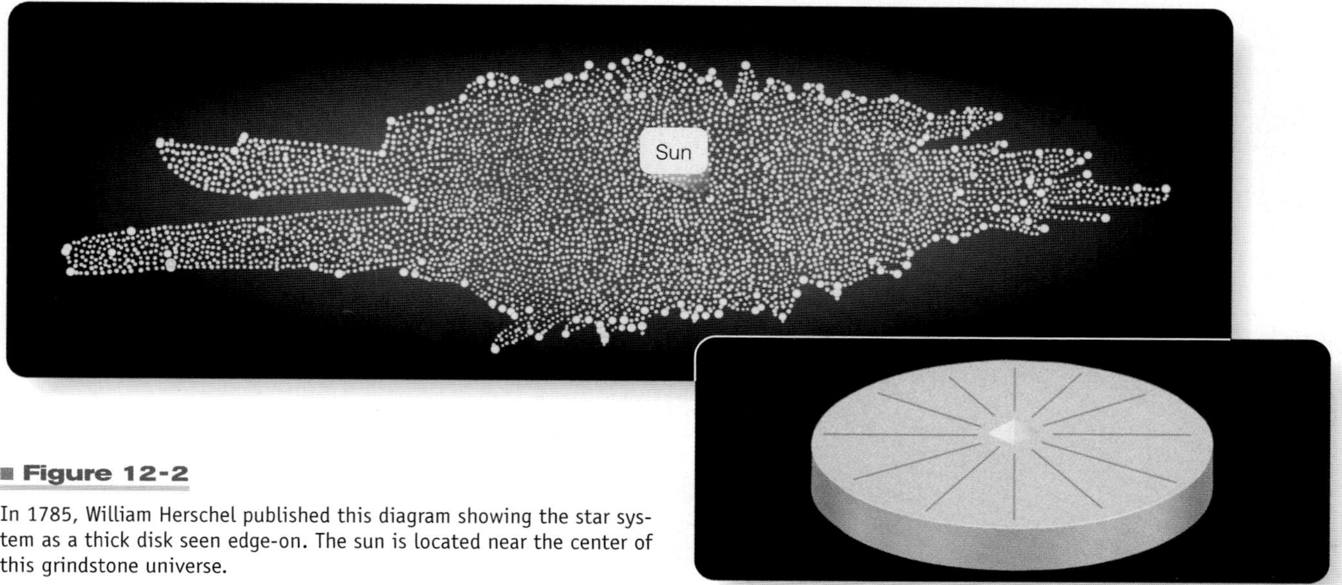

■ **Figure 12-2**

In 1785, William Herschel published this diagram showing the star system as a thick disk seen edge-on. The sun is located near the center of this grindstone universe.

Way encircling the sky. In 1750, Thomas Wright (1711–1786), drawing on the technology of the time, referred to the wheel-shaped star system as the grindstone universe by analogy with the thick disks of stone used in mills.

The English astronomer Sir William Herschel (1738–1822) and his sister Caroline Herschel (1750–1848), a talented astronomer herself, attempted to gauge the true shape of the star system by counting stars in 683 different directions in the sky. Where they saw more stars, they assumed, the star system extended further into space. The result of their research was an irregular disk shape with the sun located near the center, confirming the grindstone model of the universe (■ Figure 12-2).

In some spots on the sky, the Herschels saw very few stars, and these "holes in the sky" produced great irregularities along the edge of their diagram, as shown in Figure 12-2. Modern astronomers know that these empty spots are caused by dense clouds of gas and dust that block our view of more distant stars, but the Herschels, like many astronomers, did not understand the importance of gas and dust in space. Because they could not count stars very far into space, they concluded the sun was near the center of the grindstone universe.

The Herschels were not able to measure the size of the star system, but later astronomers were able to estimate distances to stars in statistical ways, and they concluded that the star system was only about 15,000 ly in diameter.

As the 20th century began, astronomers believed that the sun was located near the center of a rather small, wheel-shaped star system. How the human race realized the truth about their location in the universe is an adventure that begins with a woman studying stars that pulsate and leads to a man studying star clusters. We will follow that story in the next two sections because it is an important historical moment in human history, and because it illustrates how scientists build on the work of their predecessors and step by step refine their ideas about the natural world we live in.

Cepheid Variable Stars

Although poets think of stars as unchanging, astronomers have known for centuries that some stars change in brightness. Of course, novae and supernovae appear and fade, but many stars change periodically, growing brighter, then fainter, then brighter. Some of these **variable stars** are eclipsing binaries, but some are stars that pulsate like beating hearts. The period of pulsation is the time it takes a star to complete a cycle from bright to faint to bright (■ Figure 12-3).

Although there are a number of different kinds of variable stars, there is one type that is important to our story of the discovery of the Milky Way. The **Cepheid variable stars** are named after the first such star discovered, δ Cephei. Many Cepheid variables are known today; they have periods from 1 to 60 days and lie in a region of the H–R diagram known as the **instability strip** (Figure 12-3). Cepheids are clearly giant stars. A related kind of variable star, an **RR Lyrae variable star,** named after the variable star RR in the constellation Lyra, has a period of about half a day and lies at the bottom of the instability strip.

Stars in the instability strip are unstable and pulsate because a layer of partially ionized helium in the star's atmosphere can absorb and release energy like a spring. In hotter stars, the layer is too high to make the star unstable; and, in cooler stars, the layer is too low. Stars in the instability strip happen to have this layer in just the right place to make them pulsate as energy flows out of their interiors. As stars evolve into the instability strip, they become unstable; they stop pulsating when they evolve out of the strip. Massive stars are very luminous, and they cross the

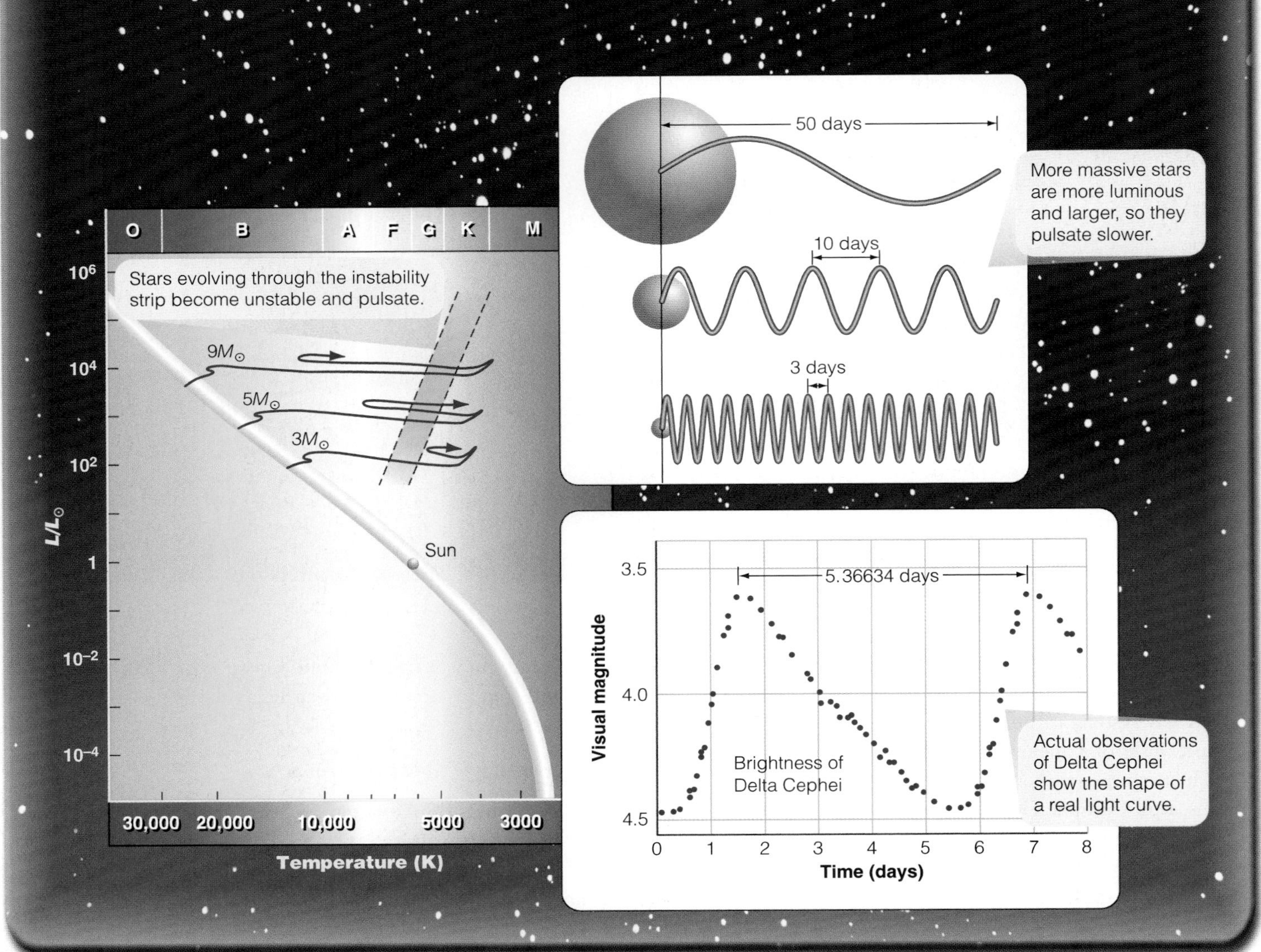

■ **Figure 12-3**

The more massive a star is, the more luminous it is. Massive stars become larger when they leave the main sequence, and those larger stars pulsate with longer periods when they pass through the instability strip. Because both luminosity and period depend on mass, there is a relationship between period of pulsation and luminosity.

instability strip higher in the H–R diagram. Because these massive stars are larger, they pulsate slower, just as large bells vibrate slower and have deeper tones. Lower-mass stars are less luminous, cross the instability strip lower in the H–R diagram, and, being smaller, pulsate faster. This explains why the long-period Cepheids are more luminous than the short-period Cepheids.

In the early part of the 20th century, astronomers knew none of this. In 1912, Henrietta Leavitt (1868–1921) photographed a star cloud in the southern sky known as the Small Magellanic Cloud. She found many variable stars there, and she noticed that the brightest had the longest periods. She didn't know the distance to the cloud, so she couldn't calculate absolute magnitudes, but because all of the variables were at nearly the same distance she concluded that there was a relationship between period and luminosity. This **period–luminosity relation** is shown in ■ Figure 12-4, where the magnitudes are given as modern absolute magnitudes.

Leavitt's study of variable stars was the key that astronomers needed to unlock the secret of the Milky Way.

AceAstronomy™ Log into AceAstronomy and select this chapter to see Astronomy Exercise "Cepheid Variable." You can make your own observations and discover the relationship between period of pulsation and luminosity.

The Size of the Milky Way

By the beginning of the 20th century, astronomers studying the distribution of the stars in the sky believed that they lived near the center of a disk-shaped cloud of stars that was not very large.

Figure 12-4

The period–luminosity diagram is a graph of the brightnesses of variable stars versus their periods of pulsation. You could plot brightness as luminosity; but, because the diagram is used in distance calculations, it is more convenient with absolute magnitude on the vertical axis. Modern astronomers know that there are two types of Cepheids, something that astronomers in the early 20th century could not recognize in their limited data.

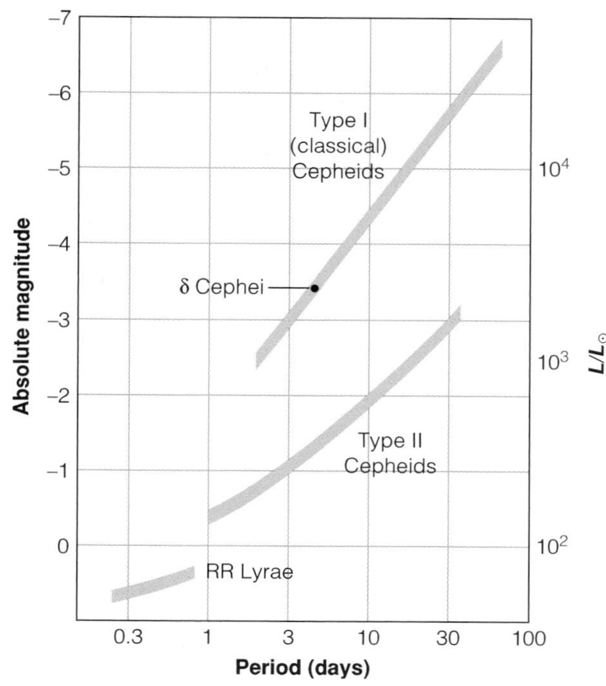

Many believed that the star system was isolated in an otherwise empty universe.

A young astronomer named Harlow Shapley (1885–1972) began the discovery of the true nature of the Milky Way when he noticed that different kinds of star clusters have different distributions in the sky. In Chapter 10, we discussed two types of star clusters, open clusters and globular clusters. Open clusters are concentrated along the Milky Way, and globular clusters are scattered all over the sky. But Shapley noticed that the globular clusters were strongly concentrated toward the constellations Sagittarius and Scorpius (■ Figure 12-5).

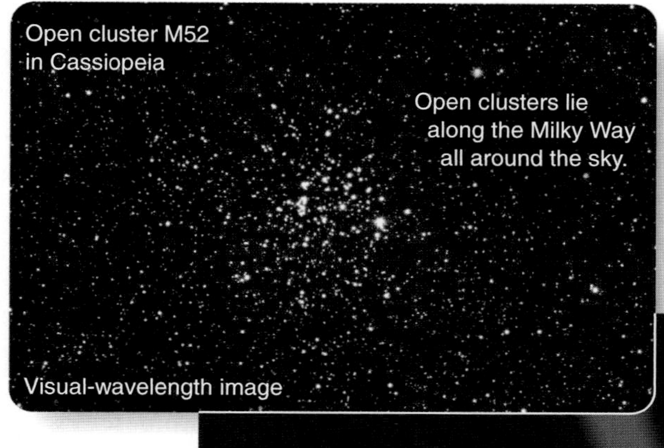

Figure 12-5

Nearly half of cataloged globular clusters (red dots) are located in or near Sagittarius and Scorpius. A few of the brighter globular clusters, labeled with their catalog designations, are visible in binoculars or small telescopes. Constellations are shown as they appear above the southern horizon on a summer night as seen from latitude 40° N, typical for most of the United States. (M52: NOAO/AURA/NSF; M19: Doug Williams, N. A. Sharp/NOAO/AURA/NSF)

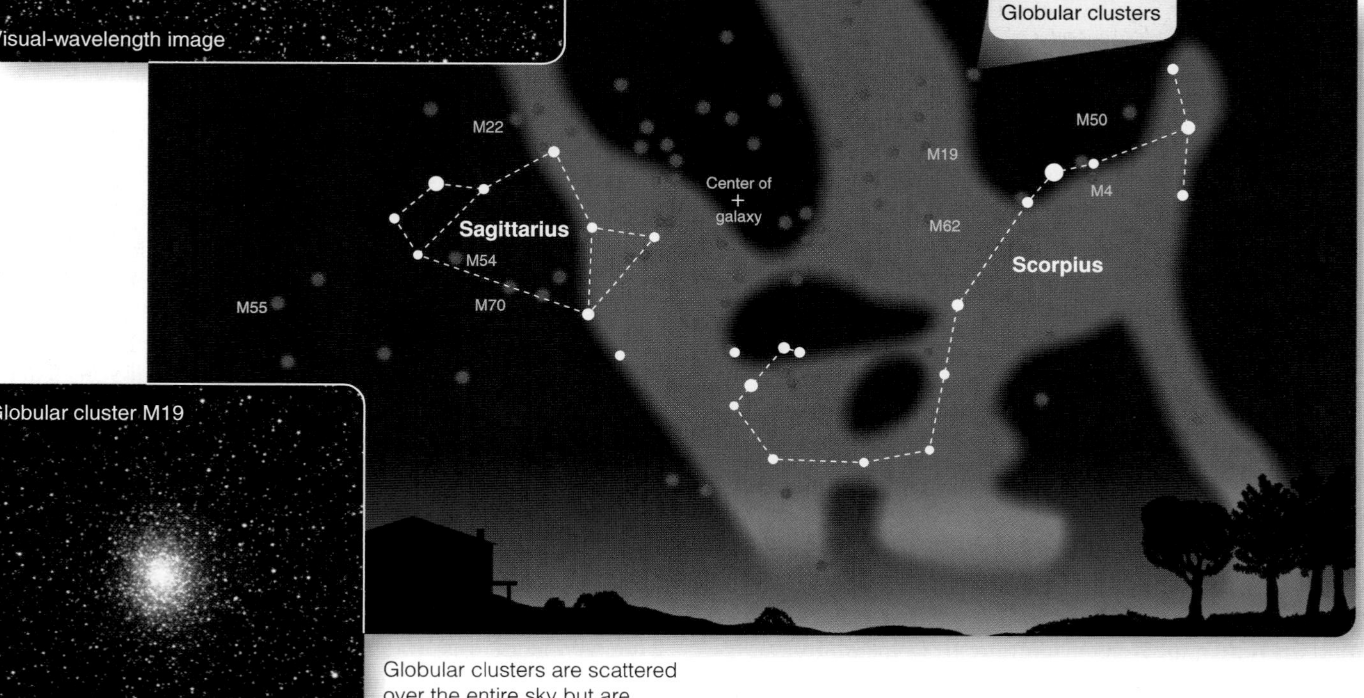

Window on Science 12-1

The Calibration of One Parameter to Find Another

Astronomers often say that Shapley "calibrated" the Cepheids for the determination of distance, meaning that he did all the detailed background work so that the Cepheids could be used to find distances. Other astronomers could use the Cepheids without repeating the detailed calibration.

Calibration is actually very common in science. Chemists, for instance, have carefully calibrated the colors of certain compounds against acidity. They can quickly measure the acidity of a solution by dipping into it a slip of paper containing the indicator compound and looking at the color. They don't have to repeat the careful calibration every time they measure acidity.

Engineers in steel mills have calibrated the color of molten steel against its temperature. They can use a hand-held device that measures the color of a ladle of molten steel and then looks up the temperature from a calibrated table. They don't have to repeat the calibration every time. Astronomers have made the same kind of color–temperature calibration for stars.

As you read about any science, notice how calibrations are used to simplify common measurements. But notice, too, how important it is to get the calibration right. An error in calibration can throw off every measurement made with that calibration. Some of the biggest errors in science have been errors of calibration.

Shapley assumed that this great cloud of globular clusters was controlled by the combined gravitational field of the entire star system, in which case he could study the size and extent of the star system by studying the globular clusters. To do that, he needed to measure the distances to as many globular clusters as possible.

Globular clusters are much too far away to have measurable parallaxes, but they do contain variable stars. Shapley knew of Henrietta Leavitt's work on these stars, but she knew only their apparent magnitudes. Cepheids and RR Lyrae stars are relatively rare, and there are none near enough to have measurable parallaxes. Consequently, Leavitt could not find the absolute magnitude of any of the variable stars she discovered, but only their apparent magnitudes. Shapley knew that he could find the distance to the globular clusters if he could find the absolute magnitude of the variable stars in the clusters, so he turned his attention to variable stars.

All stars are moving through space, and over periods of a few years these motions, known as **proper motions,** can be detected as small shifts in the positions of the stars in the sky. The more distant stars have undetectably small proper motions, but the nearer stars have larger proper motions. Clearly, proper motions contain clues to distance. Although no Cepheids are close enough to have measurable parallaxes, Shapley found 11 with measured proper motions. Through a statistical process, he was able to find their average distance and thus their average absolute magnitude. That meant he could replace Leavitt's apparent magnitudes with absolute magnitudes on the period–luminosity diagram (as shown in Figure 12-4). That is, he knew how bright the variable stars really were.

Finally, he could find the distance to the globular clusters. He could identify the variable stars he found in the clusters, and he could find their apparent magnitude from his photographs. The comparison of apparent and absolute magnitude gave him the distance to the star cluster using the magnitude–distance relationship given in Reasoning with Numbers 8-2. Notice how Shapley proceeded step by step in his research; astronomers say he **calibrated** the variable stars for distance determination (**Window on Science 12-1**).

Shapley later wrote that it was late at night when he plotted the direction and distance to the globular clusters and found that, just as he had supposed, they formed a great swarm. But it was not centered on the sun. The center of the swarm of clusters lay many thousands of light-years in the direction of Sagittarius. Evidently the center of the star system was in Sagittarius (■ Figure 12-6). He found the only other person in the building, a cleaning lady, and the two stood looking at his graph as he explained that they were the only two people on Earth who understood that humanity lives, not at the center of a small star system, but in the suburbs of a vast wheel of stars, a galaxy.

It is interesting to note that Shapley made two mistakes. He assumed that space was empty. Later, astronomers realized that the interstellar medium dims the more distant stars and makes them look farther away than they really are. Also, Shapley didn't know that there are two types of Cepheids. Those he saw in his clusters were the fainter type, but those he used for calibration were brighter. That meant he overestimated the luminosity of his cluster stars and overestimated the distance to the clusters. Consequently his estimate for the diameter of our galaxy was bigger than the modern value. The point is not that he made a few mistakes, but that he got the main point right. We live in the suburbs of a very big wheel of stars.

Why did astronomers before Shapley think they lived near the center of a small star system? Space is filled with gas and dust that dims the view of distant stars. When you look toward the band of the Milky Way, you can see only the neighborhood near the sun. Most of the star system is invisible and, like travelers in a fog, you seem to be at the center of a small region. Shapley was able to see the globular clusters at greater distances because they lie outside the plane of the star system and are not dimmed very much by the gas and dust (■ Figure 12-7).

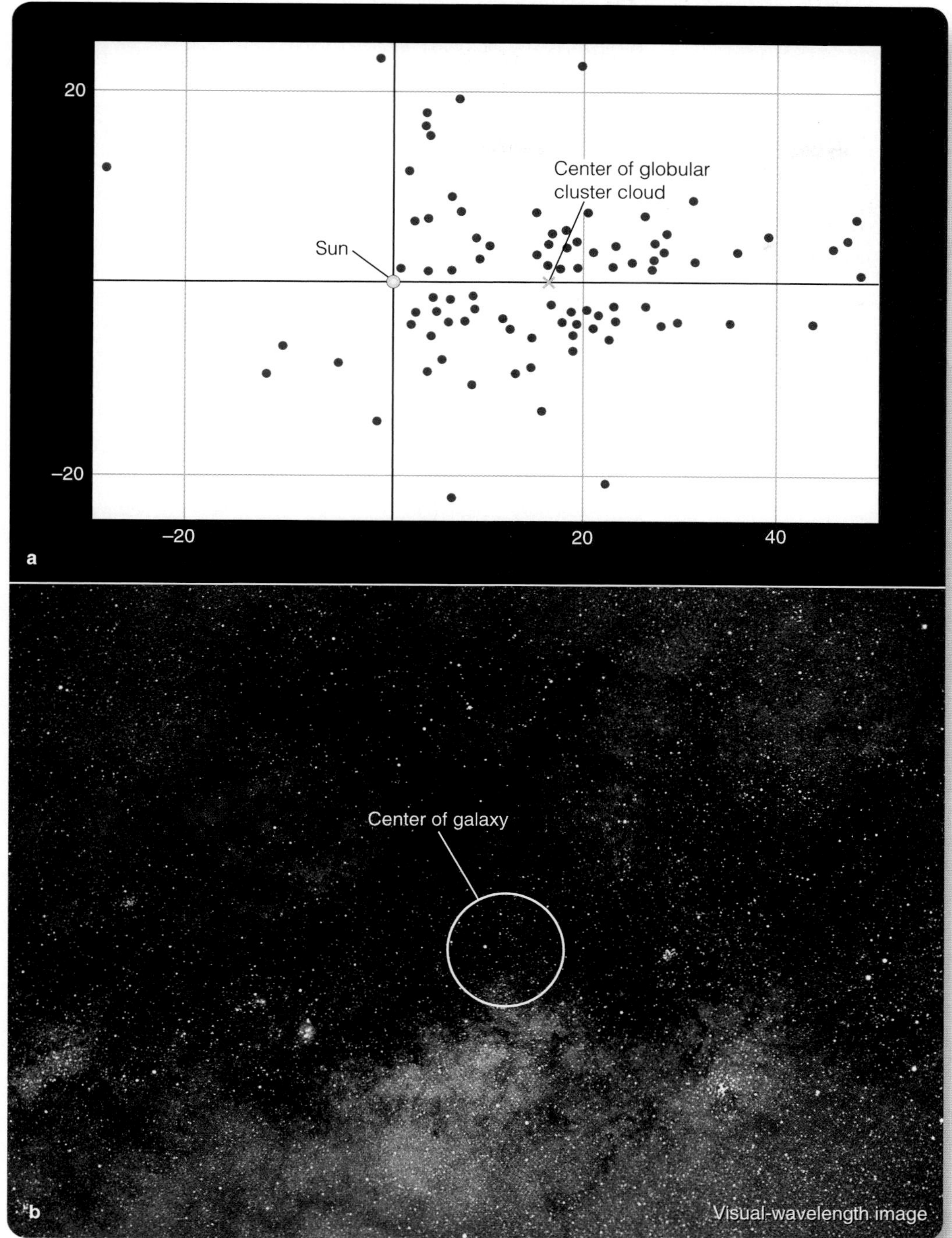

■ **Figure 12-6**

(a) Shapley's study of globular clusters showed that they were not centered on the sun, at the origin of this graph, but rather formed a great cloud centered far away in the direction of Sagittarius. Distances on this graph are given in thousands of parsecs. (b) Looking toward Sagittarius, you see nothing to suggest this is the center of the galaxy. Gas and dust block your view. Only the distribution of globular clusters told Shapley the sun lay far from the center of the star system. (Daniel Good)

Building on Shapley's work, other astronomers began to suspect that some of the faint patches of light visible through telescopes were other galaxies like our own. Shapley and astronomer Heber Curtis met in 1920 in a famous debate known as the **Shapley–Curtis debate.** Curtis claimed the faint objects were other galaxies, and Shapley argued they were not other galaxies but were part of our star system. It isn't clear who won the debate; but, in 1923, Edwin Hubble photographed individual stars in the Andromeda Galaxy, and in 1924 he identified Cepheids there. Clearly the faint patches were other star systems like our own.

How big is our galaxy? Modern studies say the disk is at least 75,000 ly in diameter, but there is more to our galaxy than just the disk. To get a true impression, you must conduct a careful analysis.

An Analysis of the Galaxy

Our galaxy, like many others, contains two primary components—a disk and a sphere. ■ Figure 12-8 shows these compo-

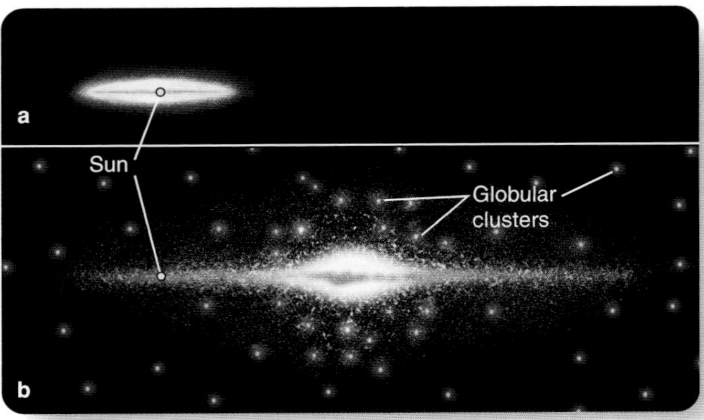

■ **Figure 12-7**

(a) Early studies of the star system concluded it was a small cloud of stars and was centered on the sun. (b) Shapley's study of globular clusters revealed that the galaxy was much larger and that the sun did not lie at the center.

nents and other features we will discuss in this section as we analyze the structure of our galaxy.

The **disk component** consists of all matter confined to the plane of rotation—that is, everything in the disk itself. This includes stars, open star clusters, and nearly all of the galaxy's gas and dust. Because the disk contains lots of gas and dust, it is the site of most of the star formation in the galaxy, and so the disk is illuminated by the brilliant, blue, massive stars. Consequently, the disk of the galaxy tends to be blue.

The dimensions of the disk are uncertain for a number of reasons. The thickness is uncertain because the disk does not have sharp boundaries. Stars become less crowded as you move away from the plane. Also, the thickness depends on what kind of object you study—O stars lie within a narrow disk only about 300 ly thick, but sunlike stars are more widely spread. The diameter of the disk and the position of the sun are also difficult to determine. Gas and dust block the view in the plane of the galaxy so you cannot see to the center or to the edge, and the outer edge of the disk is not well defined. The best studies suggest the sun is about 8.5 kpc from the center, where 1 kpc is a **kiloparsec,** or 1000 pc. Earth seems to be about two-thirds of the way from the center to the edge, so the diameter of our galaxy appears to be about 25 kpc. This is often approximated as 75,000 ly, which is a bit smaller than 25 kpc, but this illustrates the uncertainty in this number. It isn't known to better than 10 percent. This is the diameter of the luminous part of our galaxy, the part you would see from a distance. You will learn later that strong evidence suggests that our galaxy is much larger than this but that the outer parts are not luminous.

One way to explore our galaxy is to use radio telescopes to map the distribution of un-ionized hydrogen. Fortunately for astronomers, neutral hydrogen in space is capable of radiating photons with a wavelength of 21 cm. This happens because the

■ **Figure 12-8**

An artist's conception of the Milky Way Galaxy seen face-on and edge-on. Notice the position of the sun. Hot blue stars in the disk of the galaxy light up the spiral arms. Only the inner part of the halo is shown here.

spinning proton and electron act like tiny magnets, as shown in ■ Figure 12-9. When the spinning electron in the hydrogen atom reverses its spin, the magnetic fields reverse, and the atom is left with a tiny amount of excess energy, which it radiates as a photon with a wavelength of 21 cm. Radio astronomers can map our entire galaxy at a wavelength of 21 cm because these long-wavelength photons are unaffected by the microscopic dust grains that scatter photons of light. Thus, radio telescopes can "see" our entire galaxy, while optical telescopes cannot.

Observations made at other wavelengths can also let you see through the dust clouds. Infrared photons have wavelengths long enough to be unaffected by the dust. Thus, a map of the sky at long infrared wavelengths reveals the disk of our galaxy (■ Figure 12-10).

The most striking features of the disk component are the **spiral arms**—long curves of bright stars, star clusters, gas, and

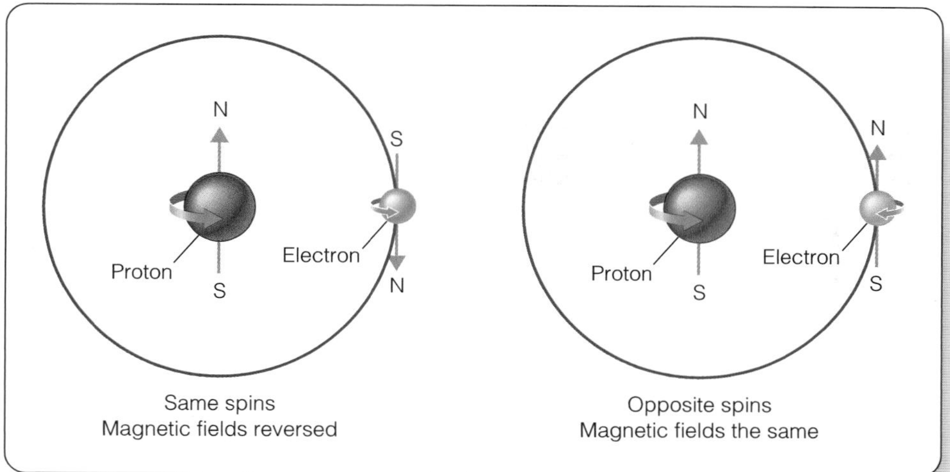

■ Figure 12-9

Both the proton and the electron in a neutral hydrogen atom spin and consequently have small magnetic fields. Because they have opposite electrostatic charges, they have opposite magnetic fields when they spin in the same direction. When they spin in opposite directions, their magnetic fields are aligned. As explained in the text, this allows cold, neutral hydrogen in space to emit radio photons with a wavelength of 21 cm.

dust. Such spiral arms are easily visible in other galaxies, and you will see later that our own galaxy has a spiral pattern.

The second component of our galaxy is the **spherical component,** which includes all matter in our galaxy scattered in a roughly spherical distribution around the center. This includes a large halo and the nuclear bulge.

The **halo** is a spherical cloud of thinly scattered stars and globular clusters. It contains only about 2 percent as many stars as the disk of the galaxy and has very little gas and dust. Thus, no new stars are forming in the halo. In fact, the vast majority of the halo stars are old, cool giants and dim lower-main-sequence stars. Detailed studies, however, suggest that most halo stars are old white dwarfs that are much too dim to be easily detected. Nevertheless, astronomers can map the halo of our galaxy by studying the giant stars. The halos of other galaxies are too faint to detect.

The **nuclear bulge** is the dense cloud of stars that surrounds the center of our galaxy. It has a radius of about 2 kpc and is slightly flattened. It is hard to observe because thick dust in the disk scatters radiation of visible wavelengths, but observations at longer wavelengths can penetrate the dust. The bulge seems to contain little gas and dust, and there is thus little star formation. Most of the stars are old, cool stars like those in the halo.

The vast numbers of stars in the disk, halo, and nuclear bulge lead to a basic question: How massive is the galaxy?

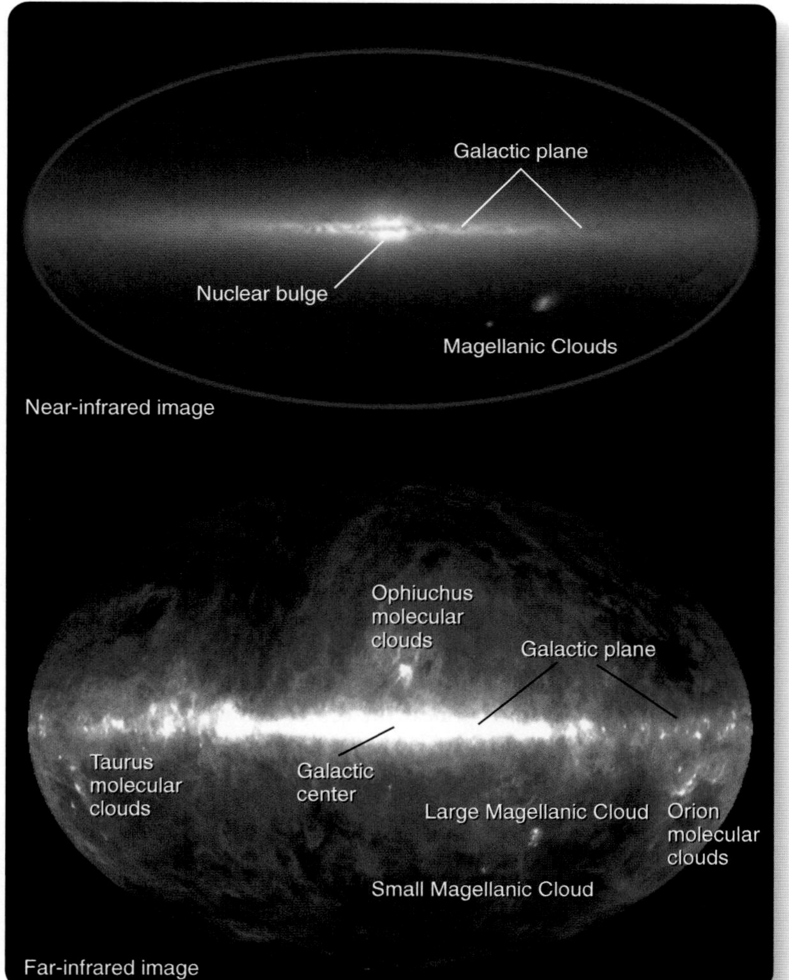

■ Active Figure 12-10

In these infrared images, the entire sky has been projected onto ovals with the center of the galaxy at the center of each oval. The Milky Way extends from left to right. In the near-infrared, the nuclear bulge is prominent, and dust clouds block our view along the Milky Way. At longer wavelengths, the dust emits significant black body radiation and glows brightly. (Near-IR: 2Mass; Far-IR: DIRBE image courtesy Henry Freudenreich)

AceAstronomy™ **Log into AceAstronomy and select this chapter to see the Active Figures called "Explorable Milky Way." Explore the Milky Way Galaxy at different wavelengths.**

The Mass of the Galaxy

When you needed to find the masses of stars, you studied the orbital motions of the pairs of stars in binary systems. To find the mass of the galaxy, you must look at the orbital motions of the stars within the galaxy. Every star in our galaxy follows an orbit around the center of mass of the galaxy. In the disk, the stars follow parallel, circular orbits, and astronomers say the disk of the galaxy rotates. That rotation can reveal the mass of the galaxy. Consequently, any discussion of the mass of the galaxy is also a discussion of the rotation of the galaxy and the orbits of the stars within the galaxy.

You can find the orbits of stars by finding how they move. Of course, you could use the Doppler effect to find a star's radial velocity (see Reasoning with Numbers 6-2). In addition, if you could measure the distance to and proper motion of a star, you could find the velocity of the star perpendicular to the radial direction. Combining all of this information, you could then find the shape of the star's orbit.

The orbital motions of the stars in the halo are strikingly different from those in the disk (■ Figure 12-11). In the halo, each star and globular cluster follows its own randomly tipped elliptical orbit. These orbits carry the stars and clusters far out into the spherical halo, where they move slowly, but when they fall back into the inner part of the galaxy, their velocities increase. Thus, motions in the halo do not resemble a general rotation but are more like the random motions of a swarm of bees. In contrast, the stars in the disk of the galaxy move in the same direction in nearly circular orbits that lie in the plane of the galaxy. The sun is a disk star and follows a nearly circular orbit around the galaxy that never carries it out of the disk.

■ Figure 12-11

(a) Stars in the galactic disk have nearly circular orbits that lie in the plane of the galaxy. (b) Stars in the halo have randomly oriented, elliptical orbits.

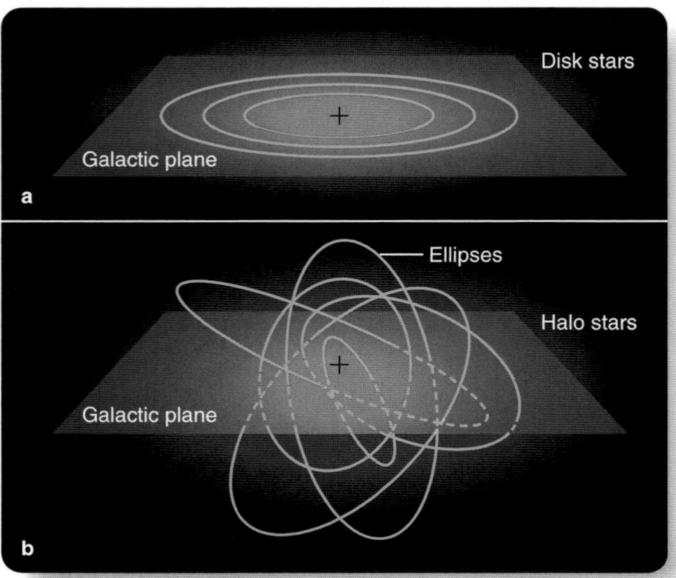

You can use the orbital motion of the sun to find the mass of the galaxy inside the sun's orbit. By observing the radial velocity of other galaxies in various directions around the sky, astronomers can tell that the sun moves about 220 km/s in the direction of Cygnus, carrying Earth and the other planets of our solar system along with it. Because its orbit is a circle with a radius of 8.5 kpc, you can divide the circumference of the orbit by the velocity and find that the sun completes a single orbit in about 240 million years. Obviously, we notice no change in a time as short as a human lifetime.

If you think of the sun and the center of mass of our galaxy as two objects orbiting each other, you can find the mass of the galaxy (see Reasoning with Numbers 8-4). You need to divide the separation in AU cubed by the period in years squared to find the mass in solar masses. The radius of the sun's orbit is about 8.5 kpc, and the orbital period is 240 million years. Converting kiloparsecs to AU and doing the arithmetic tells you the galaxy must have a mass of about 100 billion solar masses.

This estimate is uncertain for a number of reasons. First, you don't know the radius of the sun's orbit with great certainty. Astronomers estimate the radius as 8.5 kpc, but they could be wrong by 10 percent or more, and this radius gets cubed in the calculation, so it makes a big difference. Second, this estimate of the mass includes only the mass inside the sun's orbit. Mass spread uniformly outside the sun's orbit will not affect its orbital motion. Thus 100 billion solar masses is a lower limit for the mass of the galaxy, but no one knows exactly how much to increase the estimate to include the rest of the galaxy.

This estimate is also uncertain because it depends on knowledge of the rotation of the galaxy, and that rotation is very complex. The motion of the stars near the sun shows that the disk does not rotate as a solid body. Each star follows its own orbit, and stars in some regions have shorter or longer orbital periods than the sun. This is called **differential rotation** and is different from the rotation of a solid body—a carousel, for instance. Three wooden horses side-by-side on a carousel will stay together as the carousel turns, but three stars lined up in the galaxy will draw apart because they travel at different orbital velocities and have different orbital periods.

A graph of the orbital velocity of stars at various orbital radii in the galaxy is called a **rotation curve** (■ Figure 12-12). If all of the mass in the galaxy were concentrated at its center, then orbital velocity would be high near the center and would decline as you moved outward. This kind of motion has been called *Keplerian motion* because it follows Kepler's third law, as in the case of our solar system, where nearly all of the mass is in the sun. Of course, the galaxy's mass is not all concentrated at its center, but if most of the mass is inside the orbit of the sun, then you would expect to see orbital velocities decline at greater distances. Many observations confirm, however, that velocities do not decline and may actually increase at greater distance; this observation shows that these larger orbits are enclosing more mass. In other words, sig-

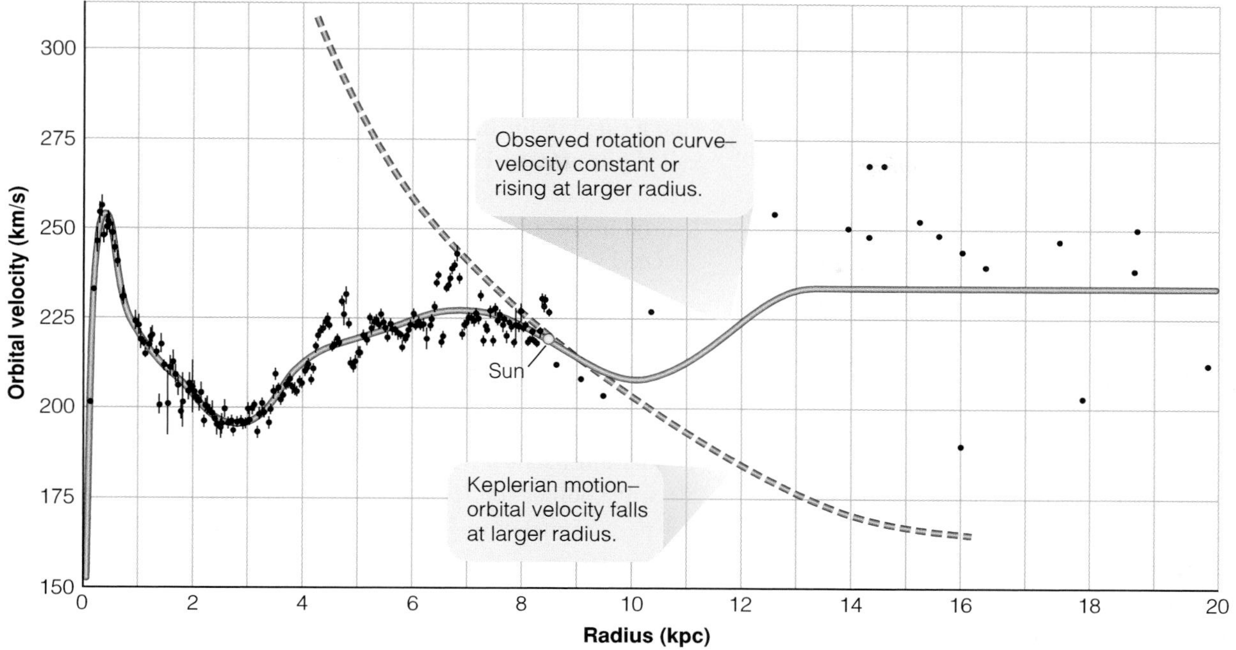

■ **Figure 12-12**

The rotation curve of our galaxy is plotted here as orbital velocity versus radius. Data points show measurements made by radio telescopes. Observations outside the orbit of the sun are much more uncertain, and the data points scatter widely. Orbital velocities do not decline outside the orbit of the sun, as you would expect if most of the mass of the galaxy were concentrated toward the center (Keplerian motion). Rather, the curve is approximately flat at great distances, suggesting that the galaxy contains significant mass outside the orbit of the sun. (Adapted from a diagram by Françoise Combes)

nificant amounts of matter in our galaxy lie in its outer parts. Although it is difficult to determine a precise edge to the visible galaxy, it seems clear that large amounts of matter are located beyond the traditional edge of the galaxy—the edge you would see if you journeyed into space and looked back at our galaxy.

The evidence is clear that extra mass lies in an extended halo sometimes called a **dark halo** or **galactic corona.** It may extend up to 10 times farther than the edge of the visible disk and could contain up to two trillion solar masses. Some small fraction of this mass is made up of low-luminosity stars and white dwarfs, but most of the matter is invisible. Astronomers call it **dark matter** and conclude that it must be some as yet unknown form of matter. The following three chapters will return to the problem of the dark matter. It is one of the fundamental problems of modern astronomy.

Inquire | Review | Analyze

If gas and dust block our view, how do we know how big our galaxy is?

The gas and dust in our galaxy block our view only in the plane of the galaxy. When you look away from the plane of the galaxy, you look out of the gas and dust and can see to great distances. The globular clusters are scattered through the halo, with a strong concentration in the direction of Sagittarius. When you look above or below the plane of the galaxy, you can see those clusters, and careful observations reveal variable stars in the clusters. By using a modern calibration of the period–luminosity diagram, astronomers can find the distance to those clusters. If you assume that the distribution of the clusters is controlled by the gravitation of the galaxy as a whole, then you can find the distance to the center of the galaxy by finding the center of the distribution of globular clusters.

In the critical analysis of ideas, it is important to ask, "How do we know?" and that is especially true in science. For example, how do we know the mass of our galaxy?

■ ■ ■

Connections: The disk component and spherical component of our galaxy are quite different from each other. In the next section, you will discover that they also differ in chemical composition; that clue, combined with your knowledge of stellar evolution, suggests a model for the formation of the galaxy.

12-2 The Origin of the Milky Way

JUST AS PALEONTOLOGISTS RECONSTRUCT the history of life on Earth from the fossil record, astronomers try to reconstruct our galaxy's past from the fossil it left behind as it formed and evolved. That fossil is the spherical component of the galaxy. The stars in the halo formed when the galaxy was young. The chemical composition and the distribution of these stars can provide clues to how our galaxy formed.

The Age of the Milky Way

To begin, you should ask yourself how old our galaxy is. That question is easy to answer because you already know how to find the age of star clusters. But there are uncertainties that make that easy answer hard to interpret.

The oldest open clusters have ages of 9 or 10 billion years. These ages are determined by analyzing the turnoff point in the cluster H–R diagram (see Chapter 10), but three things make these ages uncertain. First, finding the age of a star cluster becomes more difficult for older clusters because they change more slowly. Also, the location of their turnoff points and giant regions depends on their chemical composition, which can differ slightly from cluster to cluster. Finally, open clusters are not strongly bound by their gravity, so there may have been older open clusters that dissipated as their stars wandered away. In any case, from open clusters, you can get a rough age for the disk of our galaxy of at least 9 billion years.

Globular clusters have faint turnoff points in their H–R diagrams and are clearly old, but finding these ages is difficult. Clusters differ slightly in chemical composition, which must be accounted for in calculating the stellar models from which ages are determined. Also, to find the age of a cluster, astronomers must know the distance to the cluster. Precise parallaxes from the Hipparcos satellite have allowed astronomers to more precisely calibrate the Cepheid variable stars, and careful studies with the newest large telescopes have refined the H–R diagrams. Globular cluster ages seem to average about 11 billion years. There is some uncertainty in the measurements of ages, but some clusters do seem to be older than others. Given the spread in globular cluster ages, the halo of our galaxy must be at least 13 billion years old.

The ages of star clusters tell you that the disk is younger than the halo. You can combine these ages with subtle differences in the chemical compositions of stars to tell the story of our galaxy.

Stellar Populations

In the 1940s, astronomers realized that there were two types of stars in the galaxy. The type they were accustomed to studying was that located in the disk, such as the stars near the sun. These they called **population I** stars. The second type, called **population II** stars, are usually found in the halo, in globular clusters, or in the central bulge. In other words, the two stellar populations are associated with the two components of the galaxy.

The stars of the two populations are very similar. They burn nuclear fuels and evolve in nearly identical ways. They differ only in their abundance of atoms heavier than helium, atoms that astronomers refer to collectively as **metals.** (Note that this is not the way the word *metal* is commonly used by nonastronomers.) Population I stars are metal rich, containing 2 to 3 percent metals, while population II stars are metal poor, containing only about 0.1 percent metals or less. The metal content of the star defines its population.

Population I stars belong to the disk component of the galaxy and are sometimes called *disk population stars.* They have circular orbits in the plane of the galaxy and are relatively young stars that formed within the last few billion years. The sun is a population I star.

Population II stars belong to the spherical component of the galaxy and are sometimes called the *halo population stars.* These stars have randomly tipped, elliptical orbits and are old stars. The metal-poor globular clusters are part of the halo population.

You can now understand the two kinds of Cepheid variables shown in the period–luminosity diagram (Figure 12-4). Type I Cepheids are population I stars with metal abundance similar to that of the sun. The type II Cepheids are population II stars. The difference in metal abundance affects the ease with which radiation flows through the gas, and that changes the balance between gravity and energy flowing outward through the star. A few percent difference in metal abundance may seem like a small detail, but it makes a big difference to a star.

Since the discovery of stellar populations, astronomers have realized that there is a gradation between populations (■ Table 12-1). The most metal-rich stars are called *extreme population I stars.* Slightly less metal-rich population I stars are called

■ Table 12-1 | Stellar Populations

	Population I		Population II	
	Extreme	**Intermediate**	**Intermediate**	**Extreme**
Location	Spiral arms	Disk	Nuclear bulge	Halo
Metals (%)	3	1.6	0.8	Less than 0.8
Shape of orbit	Circular	Slightly elliptical	Moderately elliptical	Highly elliptical
Average age (yr)	100 million and younger	0.2–10 billion	2–10 billion	10–14 billion

Window on Science 12-2

Understanding Nature as Processes

One of the most common organizing themes in science is the concept of a process—a sequence of events that leads to some result or condition. Scientists try to discover the natural processes that shape the universe. Telling the stories of these processes explains why the universe is the way it is. At the same time, recognizing these processes and learning to tell their stories makes science easier to understand.

For example, astronomers try to assemble the sequence of events that led to the formation of the chemical elements. They know that the first stars formed from nearly pure hydrogen and helium and that those stars made certain heavier elements in their cores. Supernova explosions made other elements and mixed the elements back into the interstellar medium. New stars formed from this gas, and generation after generation of stars added to the heavier elements that we are made of. Telling that story explains much of the science of the stars.

As you study any science, be alert for processes as organizing themes. Science, at first glance, seems to be nothing but facts, but one of the ways to organize the facts is to fit them into the story of a process. If you study biology and learn the process by which plant cells know whether to become roots, stems, or leaves, you will have organized a great deal of factual information into a story that not only is easy to remember but also makes sense.

When you see a process in science, ask yourself a few basic questions. What conditions prevailed at the beginning of the process? What sequence of steps occurred? Can some steps occur simultaneously, or must one step occur before another can occur? What is the final state that this process produces? Most important, what evidence is there that nature really works this way?

Not everything in science is a process, but identifying a process will help you remember and understand a great many details and principles.

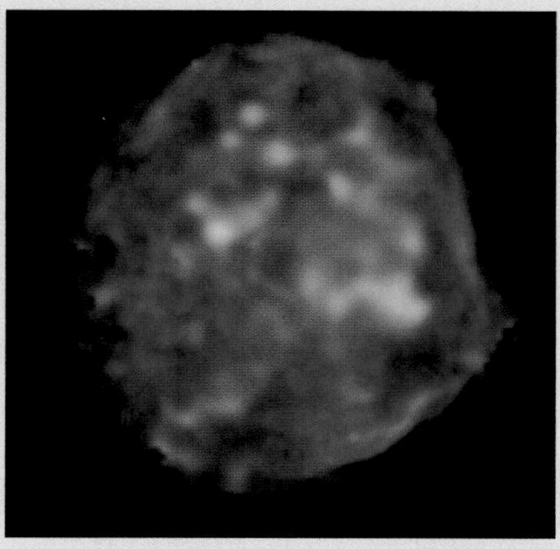

Supernova remnant N49B contains knots rich in magnesium (blue-green), which was created deep inside the star by the process that built the chemical elements. (NASA/CXC/PSU/S. Park et al.)

intermediate population I stars. The sun is such a star. Stars even less metal rich, such as stars in the nuclear bulge, are *intermediate population II stars.* The most metal-poor stars are those in the halo and in globular clusters. These are *extreme population II stars.*

Why do the disk and halo stars have different metal abundances? The two types of star must have formed at different stages in the life of the galaxy, at stages when the chemical composition of the galaxy differed. This is a clue to the history of our galaxy, but to use the clue, you must discuss the cycle of element building.

The Element-Building Cycle

The atoms of which you are made were created in a process that spanned a number of generations of stars. Natural processes are all around you, and you must learn to recognize and understand them if you are to understand nature (**Window on Science 12-2**). The process that built the chemical elements may be one of the most important processes in the history of our galaxy.

You saw in Chapter 10 how elements heavier than helium but lighter than iron are built up by nuclear reactions inside evolving stars. Atoms more massive than iron are made by short-lived nuclear reactions that occur during a supernova explosion. This explains why lower-mass atoms like carbon, nitrogen, and oxygen are so common and why atoms more massive than iron—such as gold, silver, platinum, and uranium—are so rare. ■ Figure 12-13a shows the abundance of the chemical elements, but notice that the graph has an exponential scale. To get a feeling for the true abundance of the elements, you should draw this graph using a linear scale as in Figure 12-13b. Then you see how rare the elements heavier than helium are.

Most of the matter in stars is hydrogen and helium, and the metals (including carbon, nitrogen, oxygen, and so on) were cooked up inside stars. When the galaxy first formed, there should have been no metals because stars had not yet manufactured any. The gas from which the galaxy condensed must have been almost pure hydrogen (80 percent) and helium (20 percent). (Where the hydrogen and helium came from is a mystery we will save until Chapter 15.)

The first stars to form from this gas were metal poor, and now, 13 billion years later, their spectra still show few metal lines. Of course, they may have manufactured some atoms heavier than helium, but because the stars' interiors are not mixed, those heavy atoms stay trapped at the centers of the stars where they were produced and do not affect the spectra (■ Figure 12-14). The population II stars in the halo are the survivors of an earlier generation of stars to form in the galaxy.

Most of the first stars evolved and died, and the most massive became supernovae and enriched the interstellar gas with metals. Succeeding generations of stars formed from gas clouds that were more enriched, and each generation added to the enrichment through the supernovae of massive stars. By the time

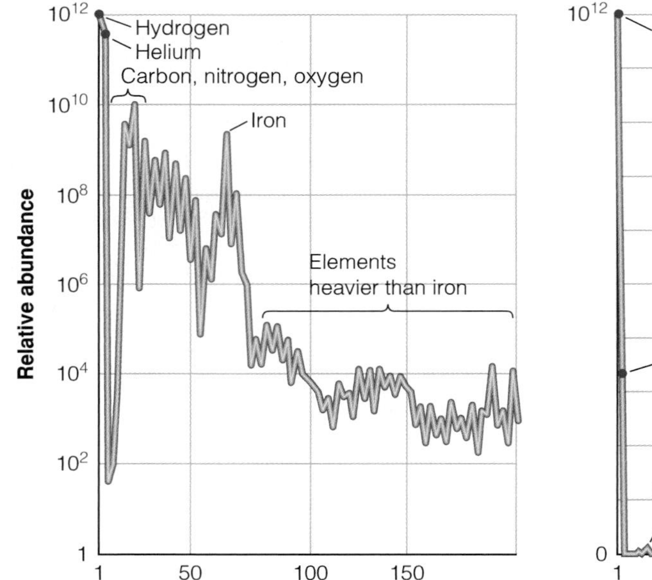

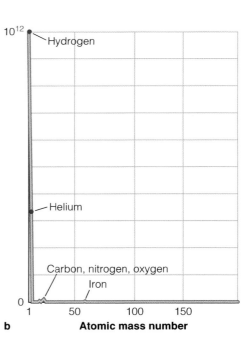

■ Figure 12-13

The abundance of the elements in the universe. (a) When the elements are plotted on an exponential scale, you see that elements heavier than iron are about a million times less common than iron and that all elements heavier than helium (the metals) are quite rare. (b) The same data plotted on a linear scale provide a more realistic impression of how rare the metals are. Carbon, nitrogen, and oxygen make small peaks near atomic mass 15, and iron is just visible in the graph.

the sun formed, 5 billion years ago, the element-building process had added about 1.6 percent metals. Since then, the metal abundance has increased further, and stars forming now incorporate 2 to 3 percent metals and become extreme population I stars. Thus metal abundance varies between populations because of the production of heavy atoms in successive generations of stars.

The lack of metals in the spherical component of the galaxy tells you it is very old, a fossil left behind by the galaxy when it was young and drastically different from its present disk shape.

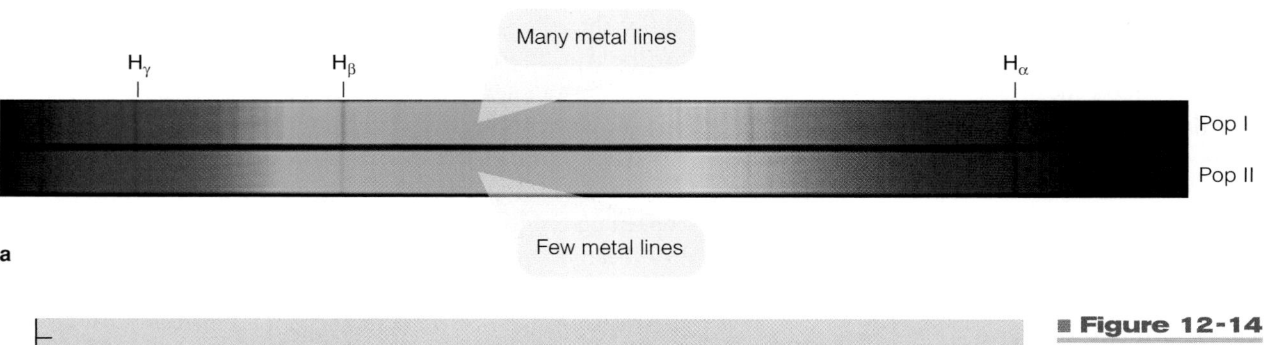

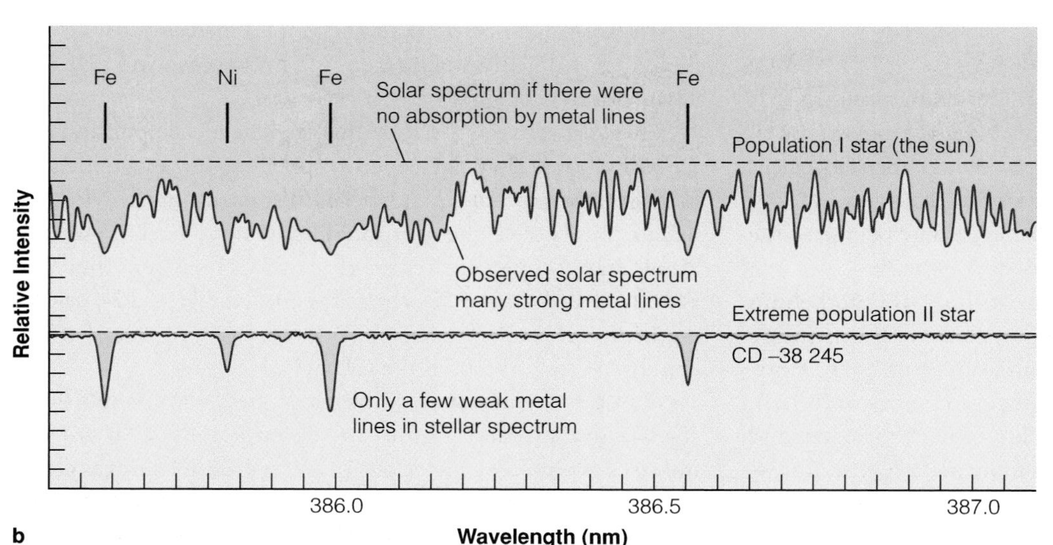

■ Figure 12-14

(a) The difference between population I stars and population II stars is dramatic. Examine the upper spectrum here and notice the hundreds of faint spectral lines. The lower spectrum has fewer and weaker lines. (AURA/NOAO/NSF) (b) A graph of such spectra reveals overlapping absorption lines of metals completely blanketing the population I spectrum. The lower spectrum is that of an extremely metal-poor star with only a few, weak metal lines of iron (Fe) and nickel (Ni). This population II star has about 10,000 times less metal content than the sun. (Adapted from an ESO illustration)

Thus, the study of element building and stellar populations leads to the fundamental question, "How did our galaxy form?"

The History of the Milky Way Galaxy

In the 1950s, astronomers began to develop a hypothesis to explain the formation of our galaxy. Recent observations, however, are forcing a reevaluation of that traditional hypothesis.

The traditional hypothesis says that the galaxy formed from a single large cloud of gas over 13 billion years ago. As gravity pulled the gas together, the cloud began to fragment into smaller clouds, and because the gas was turbulent, the smaller clouds had random velocities. Stars and star clusters that formed from these fragments went into randomly shaped and randomly tipped orbits. Of course, these first stars were metal poor because no stars had existed earlier to enrich the gas with metals. Thus, the contraction of the large gas cloud produced the spherical component of the galaxy.

The second stage of this hypothesis accounts for the disk component. The contracting gas cloud was roughly spherical at first, but the turbulent motions canceled out, as do eddies in recently stirred coffee, leaving the cloud with uniform rotation. A rotating, low-density cloud of gas cannot remain spherical. A star is spherical because its high internal pressure balances its gravity, but in a low-density cloud, the pressure cannot support the weight. Like a blob of pizza dough spun in the air, the cloud must flatten into a disk (■ Figure 12-15).

This contraction into a disk took billions of years, with the metal abundance gradually increasing as generations of stars were born from the gradually flattening gas cloud. The stars and globular clusters of the halo were left behind by the cloud when it was spherical, and subsequent generations of stars formed in flatter distributions. The gas distribution in the galaxy now is so flat that the youngest stars are confined to a disk only about 100 parsecs thick. These stars are metal rich and have nearly circular orbits.

This traditional hypothesis accounts for many of the Milky Way's properties. Advances in technology, however, have improved astronomical observation; and, beginning in the 1980s, contradictions arose. For example, many halo stars have metal abundances similar to those of globular clusters and may have escaped from such clusters, but some stars are even more metal poor. One star has 200,000 times less metal content than the sun. These stars may be older than the globular clusters. Also, some of the oldest stars in our galaxy are in the central bulge, not in the halo. Furthermore, not all globular clusters have the same age, and the younger clusters seem to be in the outer halo. The traditional hypothesis says that the halo formed first and that the clusters within it should have a uniform age.

Ages are a key problem for the traditional hypothesis. The globular clusters are about 10 to 13 billion years old, but the disk seems much younger. The oldest open star clusters in the disk are only 9 to 10 billion years old. Also, white dwarfs tend to accumulate over the eons, and there aren't as many white dwarfs in the disk as there should be if the disk were 10 to 13 billion years old. A successful hypothesis should account for these ages.

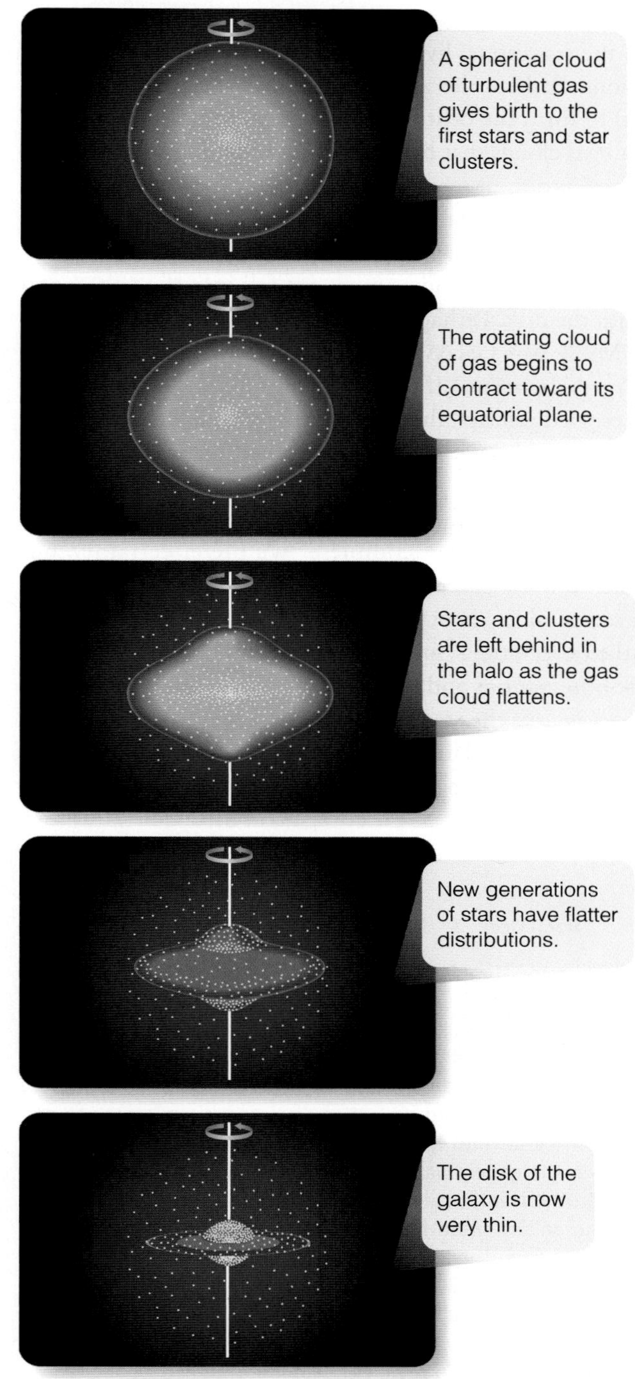

Origin of the Halo and Disk
A spherical cloud of turbulent gas gives birth to the first stars and star clusters.
The rotating cloud of gas begins to contract toward its equatorial plane.
Stars and clusters are left behind in the halo as the gas cloud flattens.
New generations of stars have flatter distributions.
The disk of the galaxy is now very thin.

■ **Figure 12-15**

The traditional theory for the origin of our galaxy begins with a spherical gas cloud that flattens into a disk.

Another problem is that the oldest stars are metal poor but not metal free. There must have been at least a few massive stars to create these metals before the formation of the oldest stars seen in the halo.

Can the traditional hypothesis be modified to explain these observations? Perhaps the galaxy began with the contraction of a gas cloud to form the central bulge, and the halo later accumulated from gas clouds that had been slightly enriched in metals by an early generation of massive stars. This would explain the age of the central bulge and the metals in the oldest stars.

The disk could have formed later as the gas that was already in the galaxy flattened and as more gas fell into the galaxy and settled into the disk. Perhaps entire galaxies were captured by the growing Milky Way Galaxy. (You will see in the next chapter that such mergers do occur.) If our galaxy absorbed a few small but partially evolved galaxies, then some of the globular clusters in the halo may be hitchhikers. This would explain the range of globular cluster ages.

The problem of the formation of our galaxy is frustrating because the theories are incomplete. We prefer certainty, but astronomers are still gathering observations and testing hypotheses. You see an older theory that has proven to be inadequate to explain the observations, and you see astronomers attempting to refine the observations and devise new theories. The metal abundances and ages of the stars in our galaxy seem to be important clues, but metal abundance and age do not tell the whole story. Astronomer Bernard Pagel was thinking of this when he said, "Cats and dogs may have the same age and metallicity, but they are still cats and dogs."

Inquire | Review | Analyze

Why do metal-poor stars have the most elliptical orbits?

Of course, the metal abundance of a star cannot affect its orbit, so any analysis must not confuse cause and effect with the relationship between these two factors. Both chemical composition and orbital shape depend on a third factor—age. The oldest stars are metal poor because they formed before there had been many supernova explosions to create and scatter metals into the interstellar medium. Those stars formed long ago when the galaxy was young and motions were not organized into a disk. In the traditional hypothesis, the first stars formed before the collapsing galaxy could form a disk; some astronomers think the oldest stars date from the assembly of the halo from smaller clouds of gas and stars. In either case, those first stars tended to take up randomly shaped orbits, and many of those orbits are quite elliptical. Thus, today the most metal-poor stars follow the most elliptical orbits.

Nevertheless, even the oldest stars in our galaxy contain some metals. They are metal poor, not metal free. How does that affect the theories?

■ ■ ■

Connections: The origin of our galaxy is an astronomical work in progress, but that is not the only mystery hidden in the Milky Way. The next two sections discuss two special problems that astronomers are trying to understand—the nature of spiral arms and the secret of the galactic nucleus.

12-3 Spiral Arms and Star Formation

THE MOST STRIKING FEATURE of galaxies like the Milky Way is the system of spiral arms that wind outward through the disk. These arms contain swarms of hot, blue stars, clouds of dust and gas, and young star clusters. These young objects hint that the spiral arms involve star formation. Trying to understand the spiral arms, you face two problems. First, how can you be sure our galaxy has spiral arms when our view is obscured by gas and dust? Second, why doesn't the differential rotation of the galaxy destroy the arms? The solution to both problems involves star formation.

Tracing the Spiral Arms

Studies of other galaxies show that spiral arms contain hot, blue stars. Thus, one way to study the spiral arms of our own galaxy is to locate these stars. Fortunately, this is not difficult, since O and B stars are often found in associations and, being very bright, are easy to detect across great distances. Unfortunately, at these great distances their parallax is too small to measure, so their distances must be found by other means, usually by spectroscopic parallax (see Chapter 8).

O and B associations in the sky are not located randomly but reveal parts of three spiral arms near the sun, which have been named for the prominent constellations through which they pass (■ Figure 12-16). If you could penetrate the gas and dust, you could locate other O and B associations and trace the spiral arms farther, but like a traveler in a fog, you see only the region near by.

Objects used to map spiral arms are called **spiral tracers.** O and B associations are good spiral tracers because they are bright and easy to see at great distances. Other tracers include young open clusters, clouds of hydrogen ionized by hot stars (emission nebulae), and certain higher-mass variable stars.

Notice that all spiral tracers are young objects. O stars, for example, live for only a few million years. If their orbital velocity is about 250 km/s, they cannot have moved more than about 500 pc since they formed. This is less than the width of a spiral arm. Because they don't live long enough to move away from the spiral arms, they must have formed there.

The youth of spiral tracers provides an important clue about spiral arms. Somehow they are associated with star formation. Before you can follow this clue, however, you must extend the map of spiral arms to show the entire galaxy.

AceAstronomy™ Log into AceAstronomy and select this chapter to see Astronomy Exercise "Milky Way Galaxy" and see a 3-D model of your home star system.

■ Figure 12-16

Many of the galaxies in the sky are disk shaped, and most of those galaxies have spiral arms. You can suspect that our own disk-shaped galaxy also has spiral arms. Images of other galaxies show that spiral arms are marked by hot, luminous stars that must be very young; and this should make you suspect that spiral arms are related to star formation. Gas and dust block our view of most of the disk of our galaxy, but nearby are young O and B stars falling along bands that appear to be segments of spiral arms. (Images: NASA, Hubble Heritage Team)

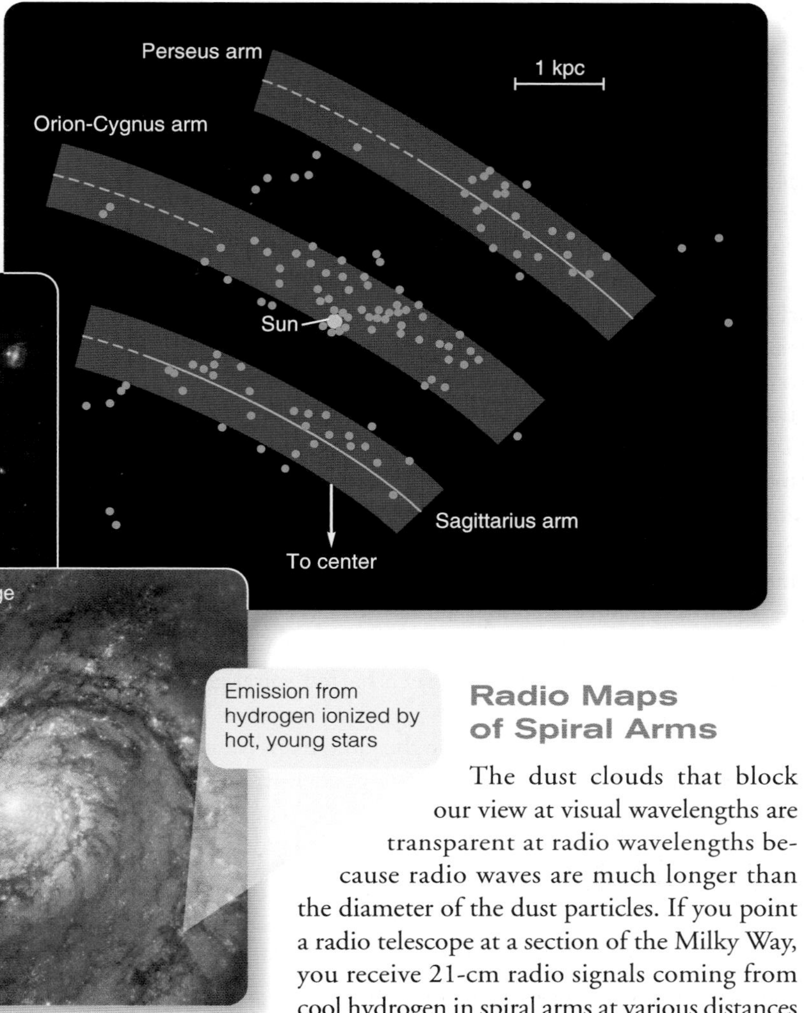

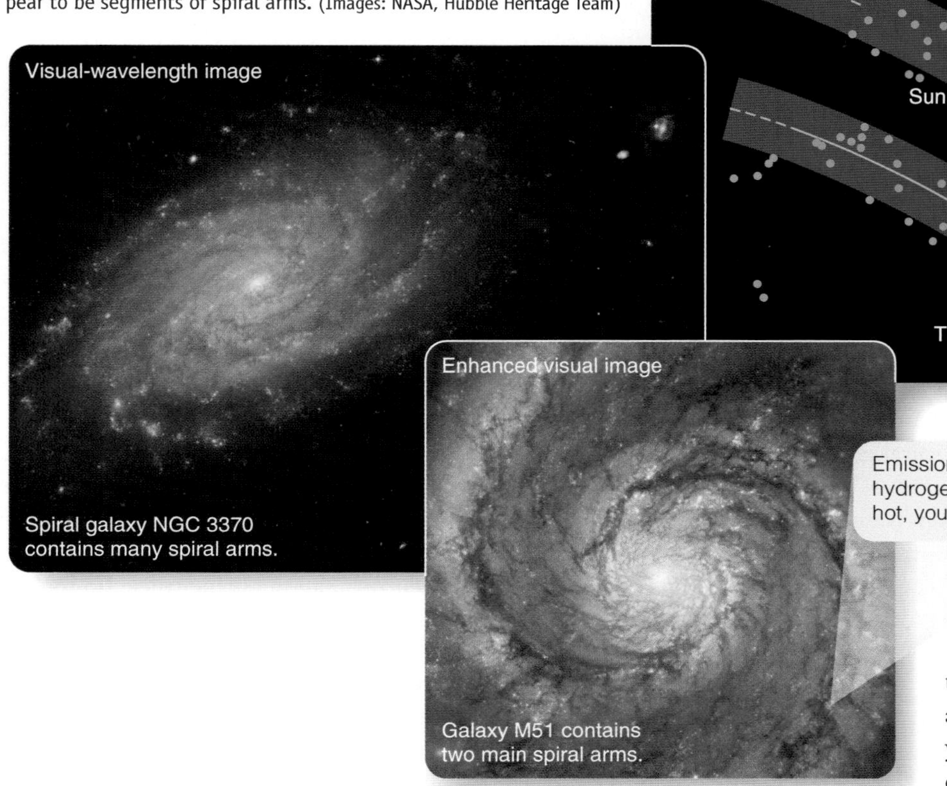

Radio Maps of Spiral Arms

The dust clouds that block our view at visual wavelengths are transparent at radio wavelengths because radio waves are much longer than the diameter of the dust particles. If you point a radio telescope at a section of the Milky Way, you receive 21-cm radio signals coming from cool hydrogen in spiral arms at various distances across the galaxy. Fortunately, the signals can be unscrambled by measuring the Doppler shifts of the 21-cm radiation, except in the direction toward the nucleus. There, the orbital motions of gas clouds are perpendicular to the line of sight, and all of the radial velocities are zero. Thus, the radio map shown in ■ Figure 12-17 reveals spiral arms throughout the disk of the galaxy, except in the wedge-shaped region toward the center.

The analysis of the 21-cm radial velocity data requires that astronomers estimate the orbital velocity of the clouds at different distances from the center of the galaxy. Because these velocities are not precisely known and because turbulent motions in the gas distort the radial

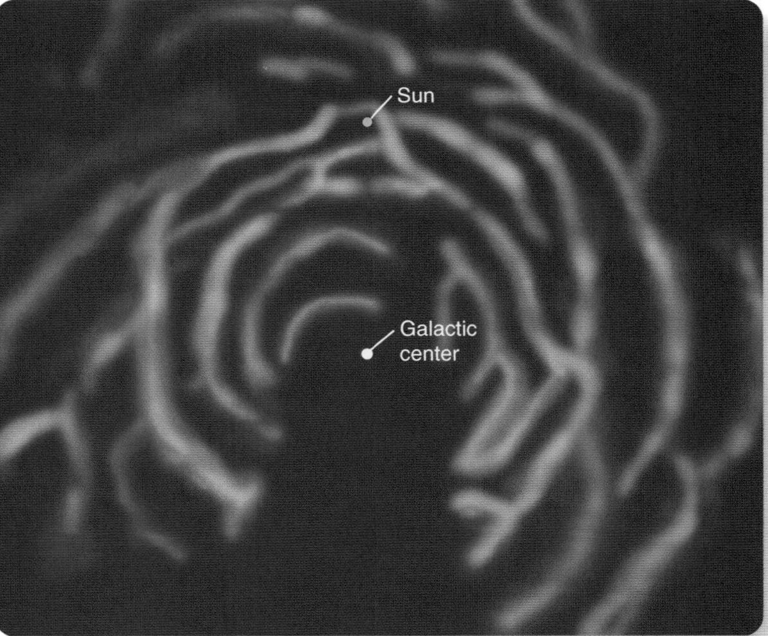

■ Figure 12-17

This map of our galaxy was made using 21-cm radio observations, and it confirms that our galaxy has spiral arms. Branches and spurs in the spiral arms are visible. The conical region around and beyond the center of the galaxy is the region where 21-cm radiation from gas clouds at different distances cannot be unscrambled. (Adapted from a radio map by Gart Westerhout)

velocities, you can't trace a perfect spiral pattern in Figure 12-17. You can be confident our galaxy has spiral arms, but you can't see the overall pattern.

Radio astronomers can also use the strong radio emission of carbon monoxide (CO) to locate giant molecular clouds in the plane of the galaxy. These clouds seem to lie along spiral structures, suggesting again that spiral arms involve star formation.

Radio maps combined with optical data allow astronomers to deduce the spiral pattern of our galaxy. The segments near the sun are part of a spiral pattern that continues throughout the disk. But the spiral arms are rather irregular and are interrupted by branches, spurs, and gaps. The stars in Orion, for example, appear to be a detached segment of a spiral arm, a spur. There are significant sources of error in the radio mapping method, but many of the irregularities along the arms seem real, and photographs of nearby spiral galaxies show similar features. Studies comparing all of the available data on our galaxy's spiral pattern with patterns seen in other galaxies do not necessarily agree, although the newest models suggest that the nuclear bulge in our galaxy is elongated into a bar (■ Figure 12-18). You will see in the next chapter that such barred spiral galaxies are common.

The most important feature in the radio maps is easy to overlook—spiral arms are regions of higher gas density. Spiral tracers showed that the arms contain young objects and thus indicated active star formation. Now radio maps confirm this suspicion by showing that the material needed to make stars is abundant in spiral arms.

The Density Wave Theory

Having mapped the spiral pattern, you can ask, "What are spiral arms?" You can be sure that they are not physically connected structures such as bands of a magnetic field holding the gas in place. Like a kite string caught on a spinning hubcap, such arms would be wound up and pulled apart by differential rotation within a few tens of millions of years. Yet spiral arms are common in disk-shaped galaxies and must be reasonably permanent features.

The most popular theory since the 1950s is called the **density wave theory,** which proposes that spiral arms are waves of compression, rather like sound waves, that move around the galaxy, triggering star formation. Because these waves move slowly, orbiting gas clouds overtake the spiral arms from behind and create a moving traffic jam within the arms.

You can imagine the density wave as a traffic jam behind a truck moving slowly along a highway. Seen from an airplane overhead, the jam seems a permanent, though slow-moving, feature. But individual cars move up through the jam behind the truck, await their turn, and pass the truck. So too do clouds of gas overtake the spiral density wave, become compressed in the "traffic jam," and eventually move out in front of the arm, leaving the slower-moving density wave behind.

Mathematical models of this process have been very successful at generating spiral patterns that look like this and other spiral galaxies. In each case, the density wave takes on a regular two-armed spiral pattern winding outward from the nuclear bulge to the edge of the disk. In addition, the theory predicts that the spiral arms will be stable over a period of a billion

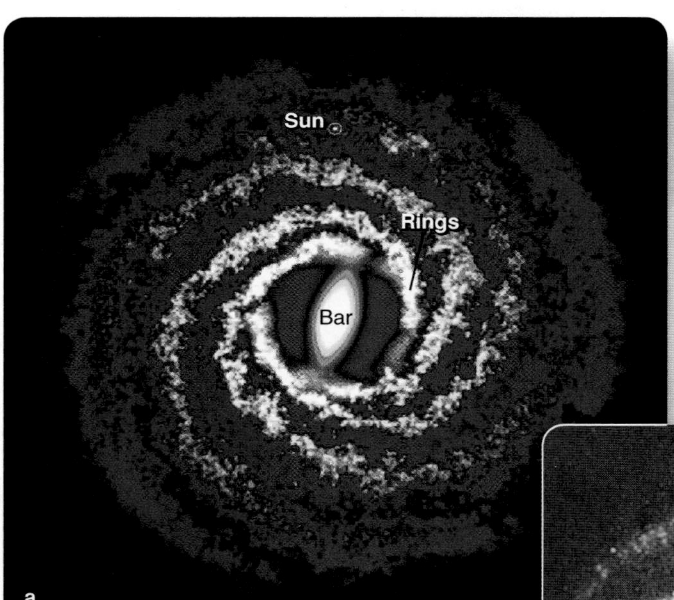

■ Figure 12-18

(a) The reconstruction of our galaxy is based on far-infrared observations of dust. The model includes a central bar surrounded by a ring from which spring four spiral arms. (Courtesy Henry Freudenreich) (b) This painting, based on radio and optical studies of the galaxy, includes a bar-shaped nucleus and a number of irregular arms with branches and spurs. (Painting by M. A. Seeds, based on a study by G. De Vaucouleurs and W. D. Pence)

years. The differential rotation does not wind them up because they are not physically connected structures.

Of course, star formation will occur where the gas clouds are compressed. Stars pass through the spiral arms unaffected, like bullets passing through a wisp of fog, but large clouds of gas slam into the spiral density wave from behind and are suddenly compressed (■ Figure 12-19). You saw in Chapter 9 that sudden compression could trigger the formation of stars in a gas cloud. Thus, new star clusters should form along the spiral arms.

The brightest stars, the O and B stars, live such short lives that they never travel far from their birthplace and are found only along the arms. Their presence is what makes the spiral arms glow so brightly (■ Figure 12-20). Lower-mass stars, like the sun, live longer and have time to move out of the arms and continue their journey around the galaxy. The sun may have formed in a star cluster about 5 billion years ago when a gas cloud smashed into a spiral arm. Since that time, the sun has escaped from its cluster and made about 20 trips around the galaxy, passing through spiral arms many times.

The density wave theory is very successful in explaining the properties of spiral galaxies, but it has two problems. First, what stimulates the formation of the spiral pattern? Of course, some process must generate the spiral arms in the first place, but the pattern must be restimulated or it would die away over a period of a billion years or so. Mathematical models show that the galaxy is naturally unstable to certain disturbances, just as a guitar string is unstable to certain vibrations. Any sudden disturbance—the rumble of a passing truck, for example—can set the string vibrating at its natural frequencies. Similarly, minor fluctuations in the galaxy's disk or gravitational interactions with passing galaxies might generate a density wave.

■ Figure 12-19

According to the density wave theory, star formation occurs as gas clouds pass through spiral arms.

The Spiral Density Wave

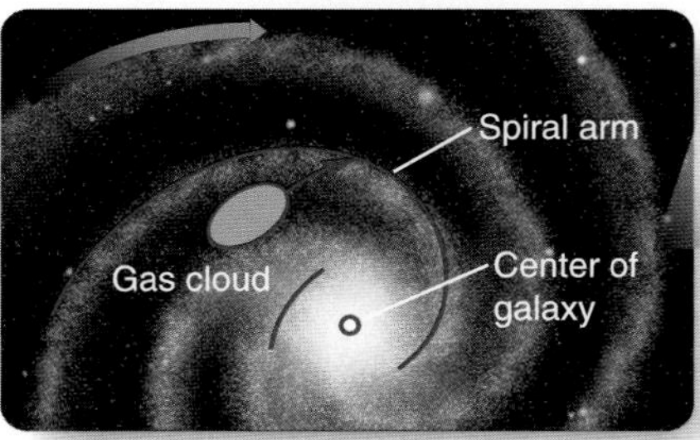

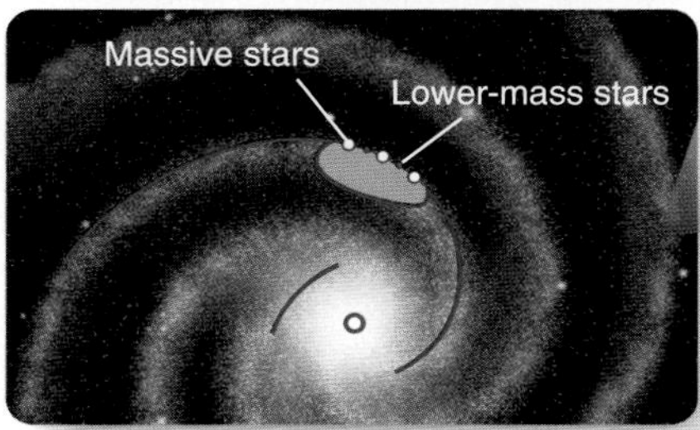

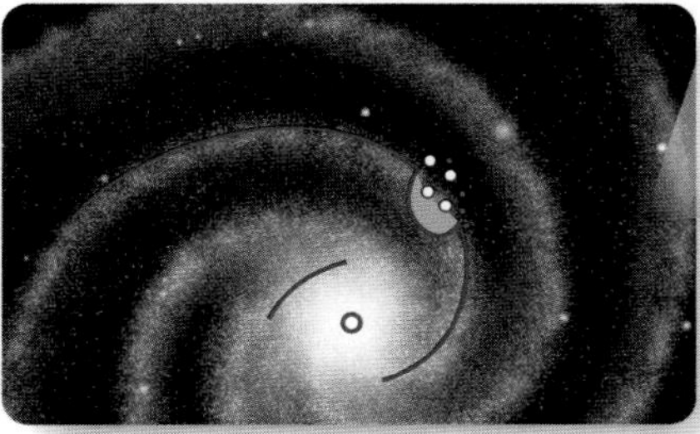

Massive stars are highly luminous and light up the spiral arm.

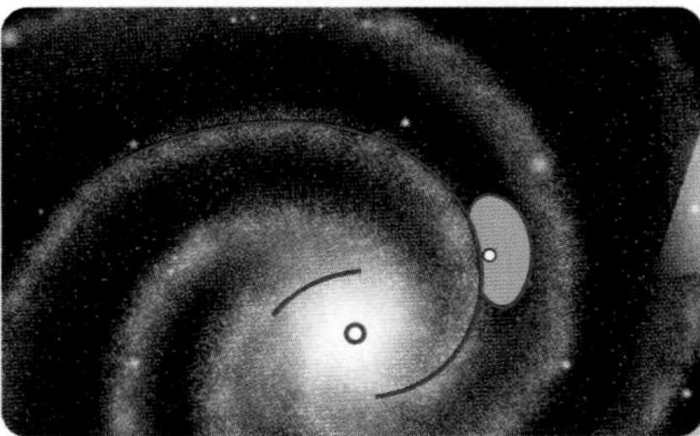

The most massive stars die quickly.

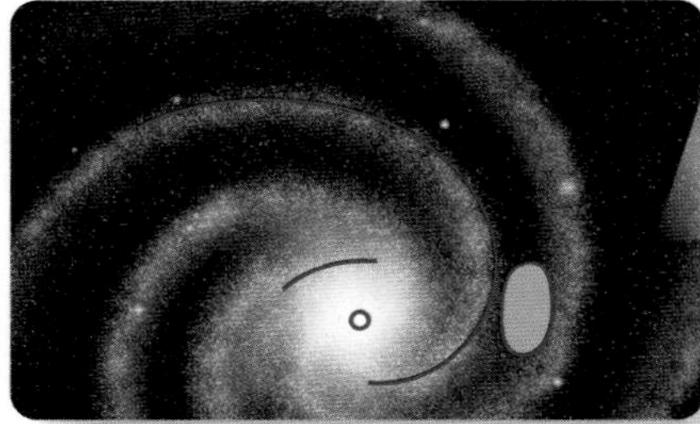

Low-mass stars live long lives but are not highly luminous.

■ Figure 12-20

This chance alignment of a small galaxy in front of a larger, more distant galaxy illustrates the nature of spiral arms. The small galaxy's spiral arms are bright with O and B stars where they are silhouetted against the darkness of space; but, seen against the glare of the distant galaxy, the arms are revealed to contain large amounts of dusty gas. (NASA and The Hubble Heritage Team)

The second problem with the density wave theory is that it does not account for the branches and spurs in the spiral arms of our own and other galaxies. Computer models of density waves produce regular, two-armed spiral patterns. Some galaxies, called grand-design galaxies, do indeed have symmetric two-armed patterns, but others do not. Other galaxies have a great many short spiral segments, giving them a fluffy appearance. These galaxies have been termed **flocculent,** meaning "woolly." Our galaxy is probably an intermediate galaxy. How can you explain these variations if the density wave theory always produces two-armed, grand-design, spiral arms? Perhaps the solution lies in a process that sustains star formation once it begins.

Star Formation in Spiral Arms

Star formation is a critical process in the creation of the spiral patterns. It makes spiral arms visible and may shape the spiral pattern itself.

The spiral density wave creates spiral arms by the gravitational attraction of the stars and gas flowing through the arms. Even if there were no star formation at all, rotating disk galaxies could form spiral arms. But without star formation to make young, hot, luminous stars, the spiral arms would be difficult to see. It is the star formation that lights up the spiral arms and makes them so prominent. Thus, star formation helps determine what you see when you look at a spiral galaxy.

But star formation can also control the shape of spiral patterns if the birth of stars in a cloud of gas can renew itself and continue making more new stars. Consider a newly formed star cluster with a single massive star. The intense radiation from that hot star can compress nearby parts of the gas cloud and trigger further star formation (■ Figure 12-21). Also, massive stars evolve so quickly that their lifetimes are only an instant in the history of a galaxy. When they explode as supernovae, the expanding gases can compress neighboring clouds of gas and trigger more star formation. Examples of such **self-sustaining star formation** have been found. The Orion complex consisting of the Great Nebula in Orion and the star formation buried deep in the dark interstellar clouds behind the nebula is such a region.

Self-sustaining star formation can produce growing clumps of new stars, and the differential rotation of the galaxy can drag the inner edge ahead and let the outer edge lag behind to produce a cloud of star formation shaped like a segment of a spiral arm, a spur. Mathematical models of galaxies filled with such segments of spiral arms do have a spiral appearance, but they lack the bold, two-armed spiral that astronomers refer to as the grand-design spiral pattern (■ Figure 12-22). Rather, astronomers suspect that self-sustaining star formation can produce the branches and spurs so prominent in flocculent galaxies, but only the spiral density wave can generate the beautiful two-armed spiral patterns.

This discussion of star formation in spiral structure illustrates the importance of natural processes. The spiral density wave creates graceful arms, but it is the star formation in the arms that makes them stand out so prominently. Self-sustaining star formation can act in some galaxies to modify the spiral arms and produce branches and spurs. In some galaxies, it can make the spiral pattern flocculent. By searching out and understanding the details of such natural processes, you can begin to understand the overall structure and evolution of the universe around us.

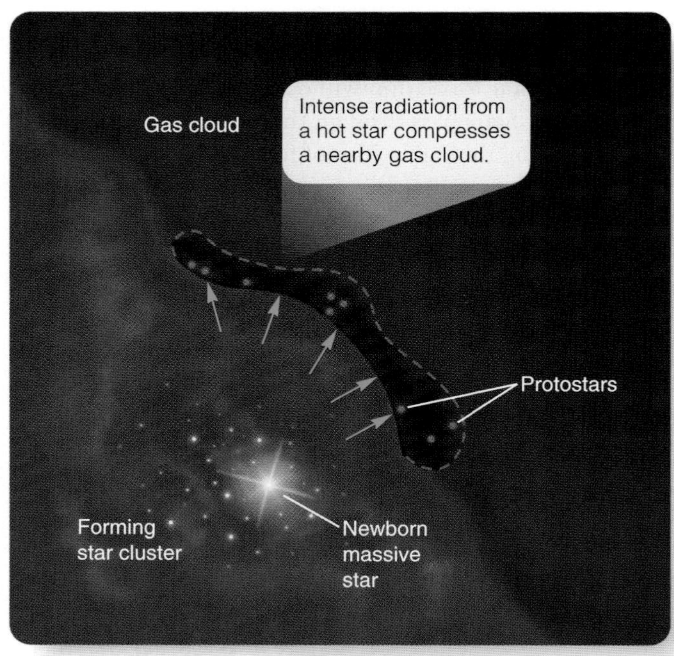

■ Figure 12-21

Most stars in a star cluster are too small and too cool to affect nearby gas clouds; but, once a massive star forms, it becomes so hot and so luminous its radiation can push gas away and compress a nearby gas cloud. In the densest regions of the cloud, new stars can begin forming.

Inquire | Review | Analyze

Why can't you use solar-type stars as spiral tracers?

Sometimes the timing of events is the critical factor in an analysis. In this case, you must think about the evolution of stars and their orbital periods around the galaxy. Stars like the sun live about 10 billion years, but the sun's orbital period around the galaxy is 240 million years. The sun almost certainly formed when a gas cloud passed through a spiral arm, but since then the sun has circled the galaxy many times and has passed through spiral arms often. Thus, the sun's present location has nothing to do with the location of any spiral arms. An O star, however, lives only a few million years. It is born in a spiral arm and lives out its entire lifetime before it can leave the spiral arm. Short-lived stars such as O stars are found only in spiral arms. G stars, like the sun, are found all around the galaxy.

The spiral arms of our galaxy must make it beautiful in photographs taken from a distance, but you are trapped inside it. How do you know that the spiral arms you can trace out near the sun actually extend across the disk of our galaxy?

■ ■ ■

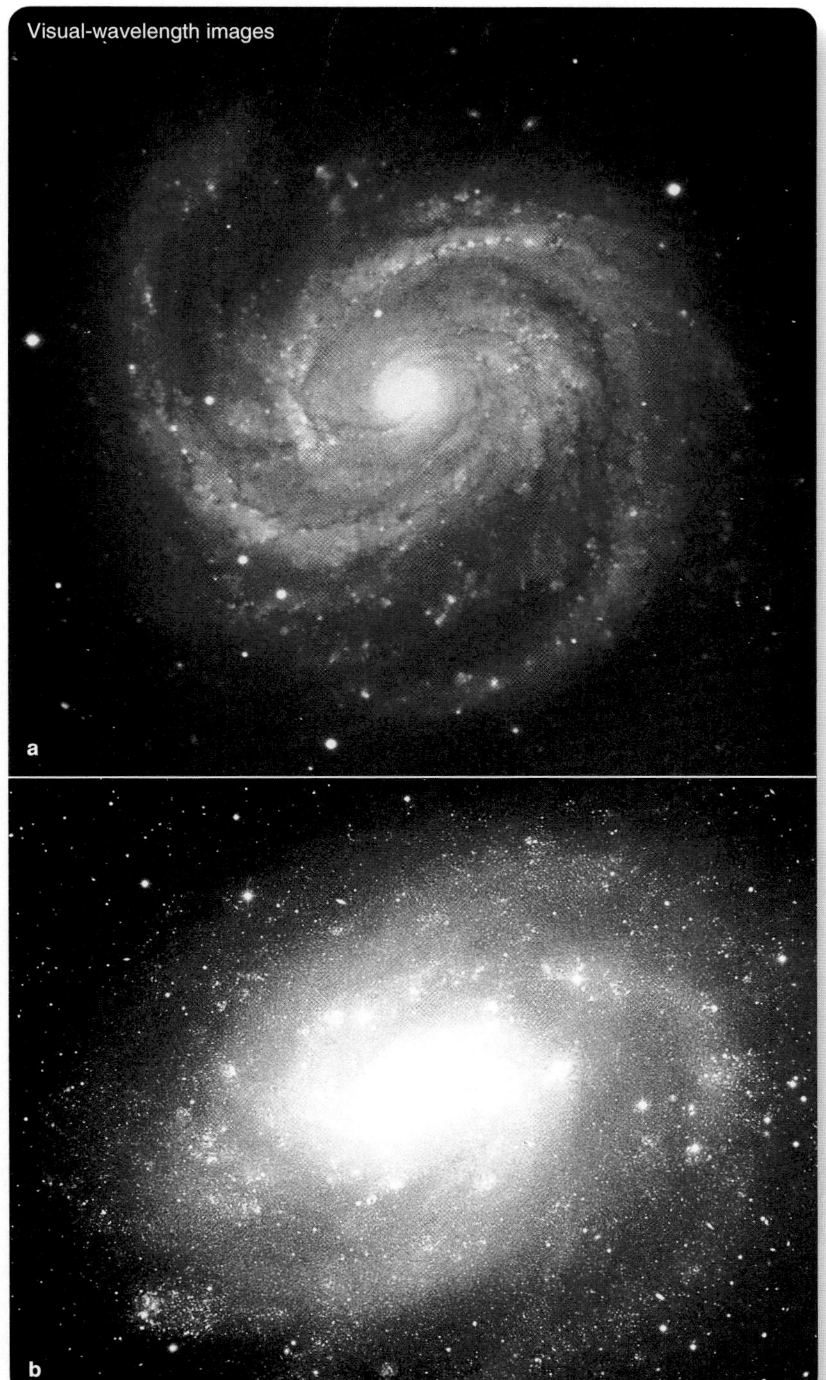

Figure 12-22

(a) Some galaxies are dominated by two spiral arms; but, even in these galaxies, minor spurs and branches are common. The spiral density wave can generate the two-armed, grand-design pattern, but self-sustained star formation may be responsible for the irregularities. (b) Many spiral galaxies do not appear to have two dominant spiral arms. Spurs and branches suggest that star formation is proceeding rapidly in such galaxies. (Anglo-Australian Telescope Board)

Connections: Spiral arms are a puzzle of great beauty, but the center of our galaxy hides a puzzle of great power.

12-4 The Nucleus

THE MOST MYSTERIOUS REGION of our galaxy is its very center, the nucleus. At visual wavelengths, this region is totally hidden by gas and dust that dim the light by 30 magnitudes (Figure 12-6b). If a trillion (10^{12}) photons of light left the center of the galaxy on a journey to Earth, only one would make it through the gas and dust. The longer-wavelength infrared photons are scattered much less often: One in every ten makes it to Earth. Consequently, visual wavelength observations reveal nothing about the nucleus. Rather, astronomers must observe at longer wavelengths such as in the infrared and radio parts of the spectrum.

The Center of the Galaxy

Harlow Shapley's study of globular clusters placed the center of our galaxy in Sagittarius, and the first radio maps of the sky showed a powerful radio source located in Sagittarius. Infrared surveys detected an intense flood of radiation coming from the same source, and higher-resolution radio maps revealed a complex collection of radio sources, with one, **Sagittarius A***, (abbreviated Sgr A* and usually pronounced "sadge A-star"), lying at the expected location of the galactic core.

Observations show that Sgr A* is only a few astronomical units in diameter but is a powerful source of radio energy. The tremendous amount of infrared radiation coming from the central area appears to be produced by crowded stars and by dust warmed by those stars. But what could be as small as Sgr A* and produce so much energy?

Study **Sagittarius A*** on pages 274 and 275 and notice three important points. First, observations at visual wavelengths are useless in the study of Sgr A*. Only at longer wavelengths can astronomers image the complex features near the center of our galaxy. A second thing to notice is the evidence of star formation in the gas clouds near the center. Not only do astronomers see clouds of gas forming new stars, but they also see supernova remnants produced by the deaths of massive stars, and those massive stars must have been born recently. Third, notice the evidence that Sgr A* is a supermassive black hole into which gas is flowing.

Sagittarius A*

The constellation of Sagittarius is so filled with stars and with gas and dust you can see nothing at visual wavelengths of the center of our galaxy. At infrared wavelengths, as shown at bottom left, you see the glow of the cool stars, gas, and dust at the center of our galaxy.

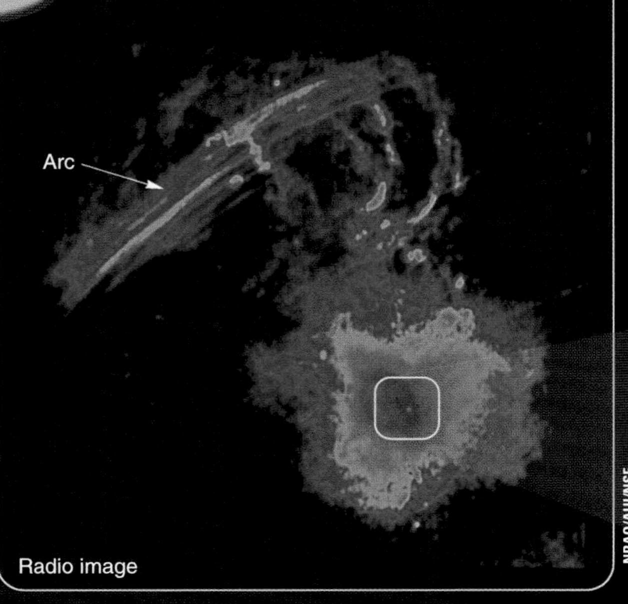

The radio map above shows Sgr A and the Arc filaments, 50 parsecs long. The image was made with the VLA radio telescope. The contents of the white box are shown on the opposite page.

The image below is a wide-field radio image of the center of our galaxy. Many of the features are supernova remnants (SNR), and a few are clouds of star formation. Peculiar features such as threads, the Arc, and the Snake may be gas trapped in magnetic fields. At the center lies Sagittarius A, the center of our galaxy.

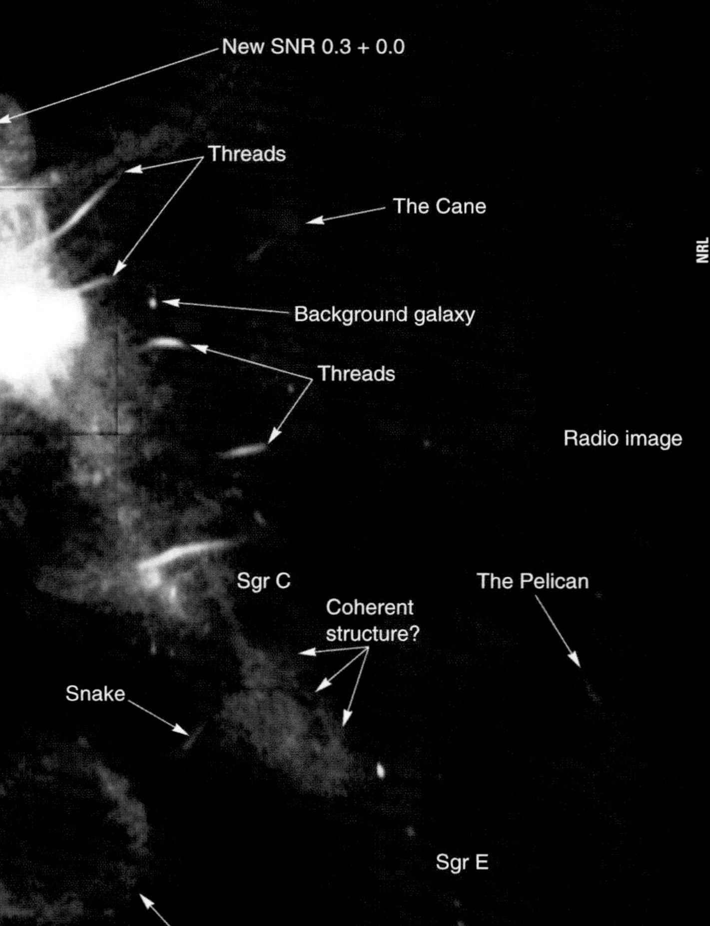

Infrared photons with wavelengths longer than 4 microns (4000 nm) come almost entirely from warm interstellar dust. The radiation at these wavelengths coming from Sagittarius is intense, and that indicates that the region contains lots of dust and is crowded with stars that warm the dust.

Infrared image

2MASS

This high-resolution radio image of Sgr A (the white boxed area on the opposite page) reveals a spiral swirl of gas around an intense radio source known as Sgr A*, the presumed central object in our galaxy. About 3 pc across, this spiral lies in a low-density cavity inside a larger disk of neutral gas. The arms of the spiral are thought to be streams of matter flowing into Sgr A* from the inner edge of the larger disk (drawing at right).

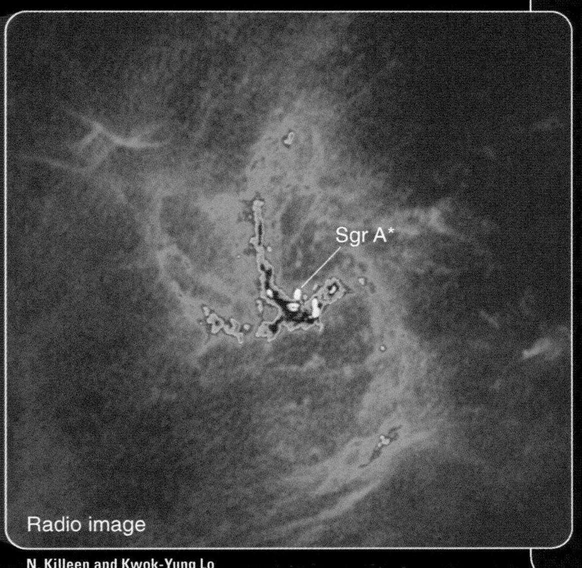

N. Killeen and Kwok-Yung Lo

The Chandra X-ray Observatory has imaged Sgr A* and detected over 2000 other X-ray sources in the area.

NASA/CXC/MIT/F.K. Baganoff et al.

Evidence of a Black Hole at the Center of Our Galaxy

Since the middle 1990s, astronomers have been able to use large infrared telescopes and active optics to follow the motions of stars orbiting around Sgr A*. A few of those orbits are shown here. The size and period of the orbit allows astronomers to calculate the mass of Sgr A* using Kepler's third law. The orbital period of the star SO-2, for example, is 15.2 years and the semimajor axis of its orbit is 950 AU. The combined motions of the observed stars suggest that Sgr A* has a mass of 2.6 million solar masses.

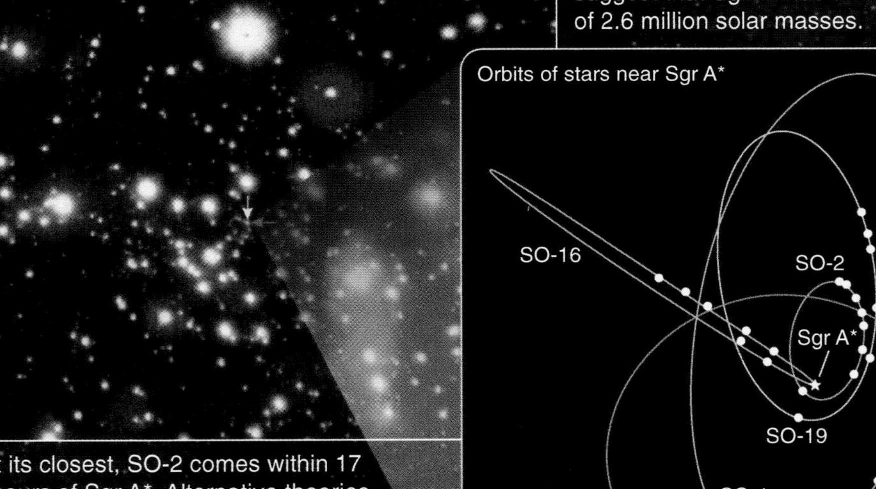

ESO

At its closest, SO-2 comes within 17 light-hours of Sgr A*. Alternative theories that Sgr A* is a cluster of stars, of neutron stars, or of stellar black holes are eliminated. Only a single black hole could contain so much mass in so small a region.

Our solar system is half a light-day in diameter.

The evidence of a massive black hole at the center of our galaxy seems conclusive. It is much too massive to be the remains of a dead star, however, and astronomers conclude that it probably formed as the galaxy first took shape.

◀ A black hole with a mass of 2.6 million solar masses would have an event horizon smaller than the smallest dot in this diagram. A slow dribble of only 0.0002 solar masses of gas per year flowing into the black hole could produce the observed energy. A sudden increase as when a star falls in could produce a violent eruption.

Astronomers continue to test the hypothesis that the center of our galaxy contains a supermassive black hole. Such an object is sufficient to explain the observations, but is it necessary? Is there some other way to explain what is observed? For example, astronomers have suggested that gas flowing inward could trigger tremendous bursts of star formation. Such theories have been considered and tested against the evidence, but none appears to be adequate to explain the observations. So far, the only theory that seems adequate is that our galaxy is home to a supermassive black hole. That is, a central black hole appears to be not only adequate but necessary to explain the observations.

Meanwhile, observations are allowing astronomers to improve their models. For instance, Sgr A* is not as bright in X rays as it should be if it has a hot accretion disk with matter constantly flowing into the black hole. Observations of X-ray and infrared flares lasting only a few hours suggest that mountain-size blobs of matter may occasionally fall into the black hole and be ripped apart and heated by tides. But the black hole may be mostly dormant and lack a fully developed hot accretion disk because little matter is flowing into it at the present time.

Such a supermassive black hole could not be the remains of a single dead star. It contains much too much mass. It may have formed when the galaxy first formed, or it may have accumulated over billions of years as matter drifted inward in the galaxy. To understand this strange object, you must compare our Milky Way Galaxy with other galaxies.

Inquire | Review | Analyze

Why do astronomers think the center of our galaxy contains a large mass?

The best way to measure the mass of an astronomical object is to watch something orbit around it. You can then use Kepler's third law to find the mass inside the orbit. Because of gas and dust, astronomers can't see to the center of our galaxy at visual wavelengths, but infrared observations can detect individual stars orbiting Sgr A*. In the case of the star S2, for example, astronomers can see the shape of its orbit and determine its orbital period. The stars that orbit Sgr A* have small orbits and very short periods, and a detailed calculation shows that Sgr A* contains roughly 2.6 million solar masses.

In addition to these orbital motions, the large amount of infrared radiation at wavelengths longer than 4 microns (4000 nm) also implies that vast numbers of stars are crowded into the center. Why?

■ ■ ■

Connections: Observations of the Milky Way Galaxy suggest it is a place of great beauty and great power. Is it normal? Are all galaxies like ours? To answer that question you must survey other galaxies, which we begin in the next chapter.

Study and Review Tools

Summary

12-1 | The Discovery of the Galaxy

How do we know we live in a galaxy?

- The hazy band of the Milky Way is our wheel-shaped galaxy seen from within, but its size and shape are not obvious. William and Caroline Herschel counted stars over the sky to show that it seemed to be shaped like a grindstone with the sun near the center, but they could not see very far into space because of gas and dust.
- In the early 20th century, Harlow Shapley calibrated Cepheid variable stars to find the distance to globular clusters and demonstrated that our galaxy is much larger than what we can see and that the sun is not at the center.
- Modern observations suggest that our galaxy contains a disk component about 75,000 ly in diameter and the sun is two-thirds of the way from the center to the visible edge. The nuclear bulge around the center and an extensive halo containing old stars and little gas and dust make up the spherical component.
- The mass of the galaxy can be found from its rotation curve. Kepler's third law tells astronomers the galaxy contains over 100 billion solar masses, and the rising rotation curve at great distance from the center shows that the halo must contain much more mass. Because that mass is not emitting detectable electromagnetic radiation, astronomers call it dark matter.

12-2 | The Origin of the Milky Way

How did our galaxy form and evolve?

- The oldest star clusters indicate that the disk of our galaxy is younger than the halo, and the oldest globular clusters appear to be about 13 billion years old. So our galaxy must have formed about 13 billion years ago.
- Stellar populations are an important clue to the formation of our galaxy. The first stars to forming our galaxy, termed population II stars, were poor in elements heavier than helium—elements that astronomers call metals. As generations of stars manufactured metals and spread them back into the interstellar medium, the metal abundance of more recent

generations increased. Population I stars, including the sun, are richer in metals.

- Because the halo is made up of population II stars and the disk is made up of population I stars, astronomers conclude that the halo formed first and the disk later. A theory that the galaxy formed from a single, roughly spherical cloud of gas and gradually flatted into a disk has been amended to include mergers with other galaxies and infalling gas contributing to the disk.

12-3 | Spiral Arms and Star Formation

What are the spiral arms?

- The most massive stars live such short lives they don't have time to move from their place of birth. Because they are scattered along the spiral arms, astronomers conclude that the spiral arms are sites of star formation.
- The spiral arms can be traced through the sun's neighborhood by using spiral tracers such as O and B stars, but to extend the map over the entire galaxy, astronomers must use radio telescopes to see through the gas and dust.
- The spiral density wave suggests that the spiral arms are regions of compression that move around the disk. When an orbiting gas cloud overtakes the compression wave, the gas cloud forms stars. Another process, self-sustaining star formation, may act to modify the arms with branches and spurs as the birth of massive stars triggers the formation of more stars by compressing neighboring gas clouds.

12-4 | The Nucleus

What lies at the very center?

- The nucleus of the galaxy is invisible at visual wavelengths, but radio, infrared, and X-ray radiation can penetrate the gas and dust. These wavelengths reveal crowded central stars and warmed dust.
- The very center of the Milky Way galaxy is marked by a radio source, Sagittarius A*. The core must be less than a few astronomical units in diameter, but the motion of stars around the center shows that it must contain roughly 2.6 million solar masses. A massive black hole is the only object that could contain so much mass in such a small space.

New Terms

variable star (p. 255)
Cepheid variable star (p. 255)
instability strip (p. 255)
RR Lyrae variable star (p. 255)
period–luminosity relation (p. 256)
proper motion (p. 258)
calibration (p. 258)
Shapley–Curtis debate (p. 259)
disk component (p. 260)
kiloparsec (kpc) (p. 260)
spiral arm (p. 260)
spherical component (p. 261)
halo (p. 261)
nuclear bulge (p. 261)
differential rotation (p. 262)
rotation curve (p. 262)
dark halo (p. 263)
galactic corona (p. 263)
dark matter (p. 263)
population I (p. 264)
population II (p. 264)
metals (p. 264)
spiral tracer (p. 268)
density wave theory (p. 270)
flocculent (p. 272)
self-sustaining star formation (p. 272)
Sagittarius A* (p. 273)

Review Questions

Ace Astronomy™ Assess your understanding of this chapter's topics with additional quizzing and animations at **http://astronomy.brookscole.com/sh9e**

1. Why isn't it possible to tell from the appearance of the Milky Way that the center of our galaxy is in Sagittarius?
2. Why is there a period–luminosity relation?
3. How can astronomers use variable stars to find distance?
4. Why is it difficult to specify the thickness or diameter of the disk of our galaxy?
5. Why didn't astronomers before Shapley realize how large the galaxy is?
6. How do astronomers know how old our galaxy is?
7. Why do astronomers conclude that metal-poor stars are older than metal-rich stars?
8. How can astronomers find the mass of the galaxy?
9. What evidence is there that our galaxy has an extended corona of dark matter?
10. How do the orbits of stars around the Milky Way Galaxy help explain its origin?
11. What evidence contradicts the traditional theory for the origin of our galaxy?
12. Why do spiral tracers have to be short-lived?
13. What evidence is there that the density wave theory is not fully adequate to explain spiral arms in our galaxy?
14. What is the evidence that the center of our galaxy contains a massive black hole?
15. Why must astronomers use infrared telescopes to observe the motions of stars around Sgr A*?
16. Why does the galaxy shown at the right have so much dust in its disk? How big do you suppose the halo of this galaxy really is?

(NASA/Hubble Heritage Team/STScI/AURA)

17. Why are the spiral arms in the galaxy at the right blue? What color would the halo be if it were bright enough to see in this photo?

(NASA/Hubble Heritage Team and A. Riess, STScI)

Discussion Questions

1. How would this chapter be different if interstellar matter didn't absorb starlight?
2. Are there any observations you could make with the Hubble Space Telescope that would allow you to better understand the nature of Sgr A*?

Problems

1. Make a scale sketch of our galaxy in cross section. Include the disk, sun, nucleus, halo, and some globular clusters. Try to draw the globular clusters to scale size.
2. Because of dust, you can see only about 5 kpc into the disk of the galaxy. What percentage of the galactic disk can you see? (*Hint:* Consider the area of the entire disk and the area you can see.)
3. If the fastest passenger aircraft can fly 1600 km/hr (1000 mph), how many years would it take to reach the sun? the galactic center? (*Hint:* 1 pc $= 3 \times 10^{13}$ km.)
4. If the RR Lyrae stars in a globular cluster have apparent magnitudes of 14, how far away is the cluster? (*Hint:* See Reasoning with Numbers 8-2.)

5. If interstellar dust makes an RR Lyrae star look 1 magnitude fainter than it should, by how much will astronomers overestimate its distance? (*Hint:* See Reasoning with Numbers 8-2.)
6. If a globular cluster is 10 minutes of arc in diameter and 8.5 kpc away, what is its diameter? (*Hint:* Use the small-angle formula from Reasoning with Numbers 3-1.)
7. If you assume that a globular cluster 4 minutes of arc in diameter is actually 25 pc in diameter, how far away is it? (*Hint:* Use the small-angle formula from Reasoning with Numbers 3-1.)
8. If the sun is 5 billion years old, how many times has it orbited the galaxy?
9. If the true distance to the center of our galaxy is found to be 7 kpc and the orbital velocity of the sun is 220 km/s, what is the minimum mass of the galaxy? (*Hints:* Find the orbital period of the sun and then see Reasoning with Numbers 8-4.)
10. Infrared radiation from the center of our galaxy with a wavelength of about 2×10^{-6} m (2000 nm) comes mainly from cool stars. Use this wavelength as λ_{max} and find the temperature of the stars. (*Hint:* See Reasoning with Numbers 6-1.)

Media Cluster

ACTIVE FIGURE

Ace Astronomy™ To access the resources in the Media Cluster, log into AceAstronomy at http://astronomy.brookscole.com/sh9e and select Chapter 12.

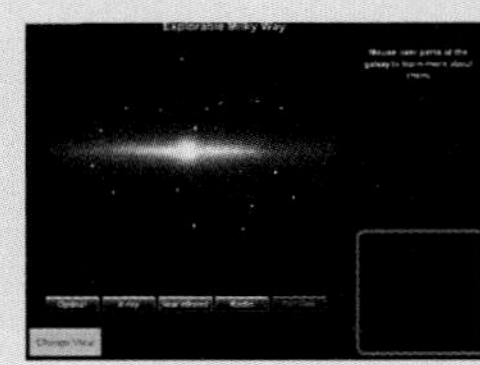

Explorable Milky Way
Viewing at different wavelengths has helped Earth astronomers to better see and understand the structures in surrounding galaxies and their own Milky Way. In this animation, explore the Milky Way Galaxy at different wavelengths.

ASTRONOMY EXERCISES

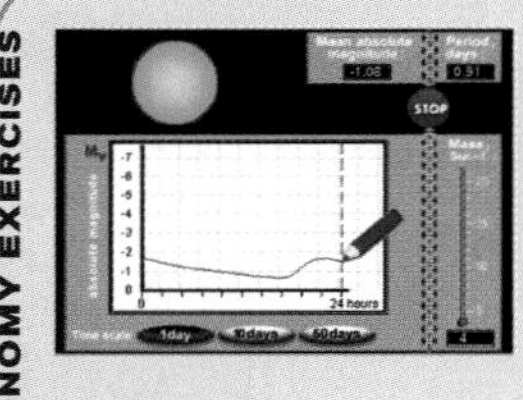

Cepheid Variable
Make your own observations and discover the relationship between period of pulsation and luminosity with this animation.

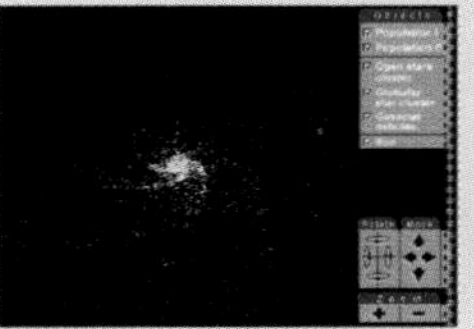

Milky Way Galaxy
What would your home galaxy look like if you could see it from the outside? Find out with this 3-D model of the Milky Way that you can spin around to view from any angle.

Critical Inquiries for the Web

1. Henrietta Leavitt discovered the period–luminosity relation for Cepheids while working on the staff at Harvard College Observatory under Edward Pickering. She was one of several women "computers" on staff there a century ago. Search the web for information on Leavitt, her colleagues, and their work at Harvard. List three of the women employed by Pickering and note their contributions to astronomy. What was life like for a woman in astronomy at the beginning of the 20th century?
2. How is our view of other galaxies affected by Earth's position in the Milky Way? Search the Internet or other sources for sky maps that trace the path of the Milky Way through the constellations. Which constellations does the Milky Way run through? Choose one of these constellations and another constellation that lies far from the band of the Milky Way. Consult the web for lists of galaxies visible in these two constellations. In which one have more galaxies been found? Why is this so?
3. Massive stars like Eta Carinae can play a significant role in the formation of stars along a spiral arm. Look for information on the web related to Eta Carinae, which is one of the most carefully studied stars in the heavens. Briefly summarize the observed behavior of this star and discuss why it is a good candidate for contributing to self-sustaining star formation in our galaxy.

Exploring *TheSky*

1. Locate Sagittarius and examine the shape of the Milky Way there and the profusion of globular clusters. (*Hint:* To turn on Messier object labels, use **Labels** and **Setup** under the **View** menu.)
2. Locate the following globular clusters: M3, M4, M5, M10, M12, M13, M15, M22, M55, M92. Where are they located in the sky? (*Hint:* Use **Find** under the **Edit** menu.)
3. Compare the distribution of globular clusters with that of open clusters. (*Hint:* Use **Filters** under the **View** menu to turn off everything but globular clusters, the Milky Way, the Galactic Equator, and Constellation Boundaries. Use the thumbwheel at the bottom of the sky window to rotate the sky. Now repeat with globular clusters off and open clusters on.)

- Go to the Brooks/Cole Astronomy Resource Center **(http://astronomy.brookscole.com)** for critical thinking exercises, articles, and additional readings from InfoTrac College Edition, Brooks/Cole's online student library.

13 Galaxies

The ability to theorize
is highly personal;
it involves art,
imagination, logic,
and something more.

EDWIN HUBBLE
The Realm of the Nebulae

Visual-wavelength image

SCIENCE FICTION HEROES flit effortlessly between the stars, but almost none voyage between the galaxies. As you leave our home galaxy, the Milky Way, behind, you voyage out into the vast depths of the universe, out among the galaxies, space so deep it is unexplored even in fiction. ▌ Less than a century ago, astronomers did not understand that there were galaxies. Nineteenth-century telescopes revealed faint nebulae scattered among the stars, and some were spiral. Astronomers argued about the nature of these nebulae, but it was not until the 1920s that astronomers understood that some were other galaxies much like our own, and it was not until recent decades that astronomical telescopes could reveal the tremendous beauty and intricacy of the galaxies (■ Figure 13-1). ▌ In this chapter, you will try to understand

The Sombrero Galaxy is about 28 million light-years from Earth. Its nuclear bulge and inner halo are bright, and dust marks the outer edge of its disk.
(NASA/Hubble Heritage Team/STScI/AURA)

Guidepost

Looking Back

In the preceding chapter, you learned about our Milky Way Galaxy, an important object to Earthlings but only one of many billions of galaxies visible in the sky. You can no more understand galaxies by understanding a single example, the Milky Way, than you could understand humanity by understanding a single person.

This Chapter

This chapter will expand your horizon to discuss the different kinds of galaxies and their complex histories. Here you can expect answers to five essential questions:

What do galaxies look like?

How do astronomers find the distance to galaxies?

How do galaxies differ in size, luminosity, and mass?

Do other galaxies contain supermassive black holes and dark matter as does our own galaxy?

Why are there different kinds of galaxies?

These are perfectly reasonable questions, but the answers will lead you to an astonishing insight into how galaxies are born and how they evolve.

Looking Ahead

You can carry the lessons of this chapter into the next, which will discuss the violently active galaxies powered by their supermassive black holes, and on into Chapter 15, which will discuss the evolution of the universe as a whole and the influence of dark matter.

Ace Astronomy™ The AceAstronomy icon throughout the text indicates an opportunity for you to test yourself on key concepts and to explore animations and interactions on the AceAstronomy website at: http://astronomy.brookscole.com/sh9e

Searching for Clues through Classification in Science

Classification is one of the most basic and most powerful of scientific tools. It is often a way to begin studying a topic, and it can lead to dramatic insights. Charles Darwin, for example, sailed around the world with a scientific expedition aboard the ship *HMS Beagle*. Everywhere he went, he studied the living things he saw and tried to classify them. For example, he studied the tortoises, mockingbirds, and finches he saw on the Galápagos Islands. His classifications of these and other animals eventually led him to think about evolution and natural selection. Scientists in many fields depend on carefully designed systems of classification.

Classification is an everyday mode of thought. You use classifications when you order lunch, buy shoes, catch a bus, and so on. In science, classifications reveal the relationships between different kinds of objects and at the same time save the scientist valuable time by bringing order out of seeming disorder. An economist can classify different kinds of businesses and does not have to analyze each of the millions of businesses as if it were unique. Classifications of minerals, plants, psychological learning styles, modes of transportation, or sandwiches bring order to the world and help you deal with information.

Astronomers use classifications of galaxies, stars, moons, telescopes, and more. Whenever you encounter a scientific discussion, look for the classifications on which it is based. Classifications are the orderly framework on which much of science is built.

The careful classification of living things can reveal surprising relationships. The birds, including this flamingo, are descended from dinosaurs. (M. Seeds)

NGC4414 60 million ly
Young blue stars
illuminate spiral arms.
Visual-wavelength image

M83 12 million ly
Dust clouds glow red in this infrared image.
Infrared image

ESO510-G13
150 million ly
Dusty disk of galaxy warped by interaction with another galaxy
New stars forming in dust clouds.
Visual-wavelength image

■ Figure 13-1

A century ago, photos of galaxies looked like spiral clouds of haze. Modern images of these relatively nearby galaxies reveal newborn stars and clouds of gas and dust. (NGC4414 and ESO510-G13: Hubble Heritage Team, AURA/STScI/NASA; M83: ESO)

Selection Effects in Science

Many different kinds of science depend on selecting objects to study. Scientists studying insects in the rain forest, for example, must choose which ones to catch. They can't catch every insect they see, so they might decide to catch and study any insect that is red. If they are not careful, a selection effect could bias their data and lead them to incorrect conclusions without their ever knowing it. The reason selection effects are dangerous in science is that they can have powerful influences without being evident. Only by carefully designing a research project can scientists avoid selection effects.

For example, suppose you decide to measure the speed of cars on a highway. There are too many cars to measure every one, so you reduce the workload and measure the speed of only red cars. It is quite possible that this selection criterion will mislead you because people who buy red cars may be more likely to be younger and drive faster. Should you measure only brown cars? No, because you might suspect that only older, more sedate people would buy a brown car. Should you measure the speed of any car in front of a truck? Perhaps you should pick any car following a truck? Again, you may be selecting cars that are traveling a bit faster or a bit slower than normal. Only by very carefully designing your experiment can you be certain that the cars you measure are traveling at representative speeds.

Astronomers face the danger of selection effects quite often. Very luminous stars are easier to see at great distances than faint stars. Spiral galaxies are brighter, bluer, and more noticeable than elliptical galaxies. What astronomers see through a telescope depends on what they notice, and that is powerfully influenced by selection effects.

Scientists engaged in observation must spend considerable time designing their experiments. They must be careful to observe an unbiased sample if they expect to make logical deductions from their results. The scientists in the rain forest, for example, should not catch and study only the red insects. Often, the most brightly colored insects are poisonous (or at least taste bad) to predators. Catching only brightly colored insects could produce a highly biased sample of the insect population. Careful scientists must plan their work with great care and avoid any possible selection effects.

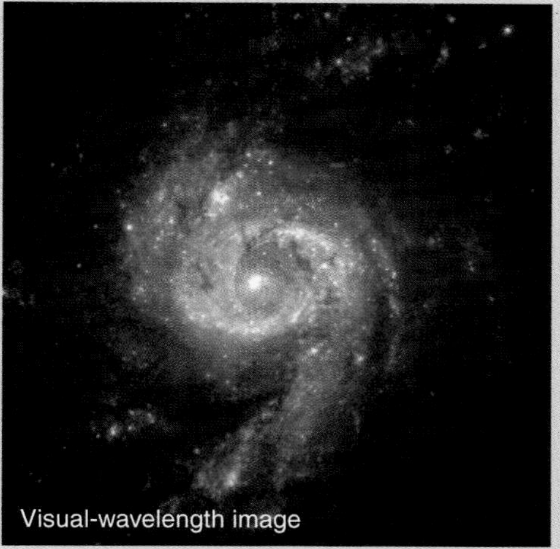

Things that are bright and beautiful, such as spiral galaxies, may attract a disproportionate amount of attention. Scientists must be aware of such selection effects. (Hubble Heritage Team/STScI/AURA/NASA)

how galaxies form and evolve, and you will discover that the amount of gas and dust in a galaxy is a critical clue. You will also discover that interactions between galaxies can influence their structure and evolution.

Before you can build theories, however, you must gather some basic data concerning galaxies. You must classify the different kinds of galaxies and discover their basic properties—diameter, luminosity, and mass. Once you know what galaxies are like, once you can characterize them by describing their typical properties, you will be ready to theorize about their origin and evolution.

13-1 The Family of Galaxies

ASTRONOMY BOOKS OFTEN PICTURE disk galaxies with spiral arms because they are beautiful pinwheels in the sky; but some galaxies are nearly featureless clouds of stars, and others are distorted muddles of gas and dust. You must begin by sorting out this jumble of different kinds of galaxies.

The Shapes of Galaxies

Astronomers classify galaxies according to their shape using a system developed in the 1920s by Edwin Hubble (namesake of the Hubble Space Telescope). Creating a system of classification is a fundamental technique in science (**Window on Science 13-1**).

Study **Galaxy Classification** on pages 284 and 285 and notice two important points. First, the amount of gas and dust in a galaxy strongly influences its appearance. Galaxies rich in gas and dust have active star formation and contain hot, bright stars. Such galaxies tend to be bluer and contain emission nebulae. Galaxies that are poor in gas and dust contain few or none of these highly luminous stars, so those galaxies look redder and have a much more uniform look.

Second, notice the wide range in the shapes of galaxies. Some look like spherical or football-shaped clouds of stars, but others are thin disks. Still others are so irregular they look like "scrambled galaxy."

You might also wonder what proportion of the galaxies are elliptical, spiral, and irregular, but that is a difficult question to answer. In catalogs of galaxies, about 70 percent are spiral, but that is the result of a selection effect (**Window on Science 13-2**). Spiral galaxies contain hot, bright stars and are thus very luminous and easy to see. Most ellipticals are fainter and harder to notice. Small galaxies, such as dwarf ellipticals and dwarf irregulars, may be very common, but they are hard to detect. From

Galaxy Classification

The most commonly used system of galaxy classification is based on the appearance of galaxies in photographs made at visual wavelengths.

AURA/NOAO/NSF

Visual-wavelength image

M87 is a giant elliptical galaxy classified E1. It is a number of times larger in diameter than our own galaxy and is surrounded by a swarm of over 500 globular clusters.

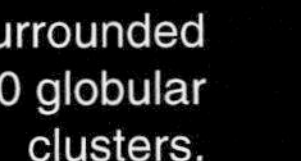

Elliptical galaxies are round or elliptical, contain no visible gas and dust, and lack hot, bright stars. They are classified with a numerical index ranging from 1 to 7; E0s are round, and E7s are highly elliptical. The index is calculated from the largest and smallest diameter of the galaxy used in the following formula and rounded to the nearest integer.

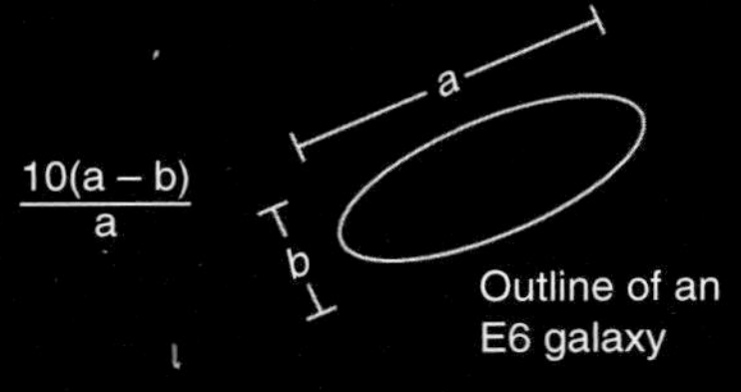

Anglo-Australian Telescope Board

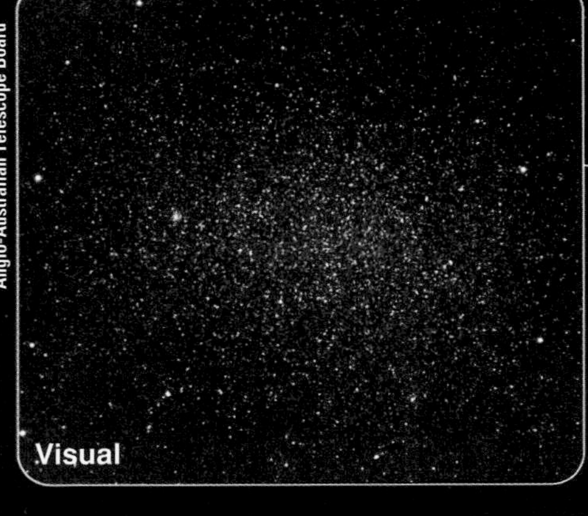

The Leo 1 dwarf elliptical galaxy is not many times bigger than a globular cluster.

Spiral galaxies contain a disk and spiral arms. Their halo stars are not visible, but presumably all spiral galaxies have halos. Spirals contain gas and dust and hot, bright O and B stars. The presence of short-lived O and B stars alerts us that star formation is occurring in these galaxies. Sa galaxies have larger nuclei, less gas and dust, and fewer hot, bright stars. Sc galaxies have small nuclei, lots of gas and dust, and many hot, bright stars. Sb galaxies are intermediate.

Anglo-Australian Telescope Board

Sb

NGC 3627 Visual

Ace Astronomy™

Log into AceAstronomy and select this chapter to see Active Figure "Galaxy Types" and review the classification of galaxies.

Anglo-Australian Telescope Board

Roughly 2/3 of all spiral galaxies are **barred spiral galaxies** classified SBa, SBb, and SBc. They have an elongated nucleus with spiral arms springing from the ends of the bar. Our own galaxy is a barred spiral.

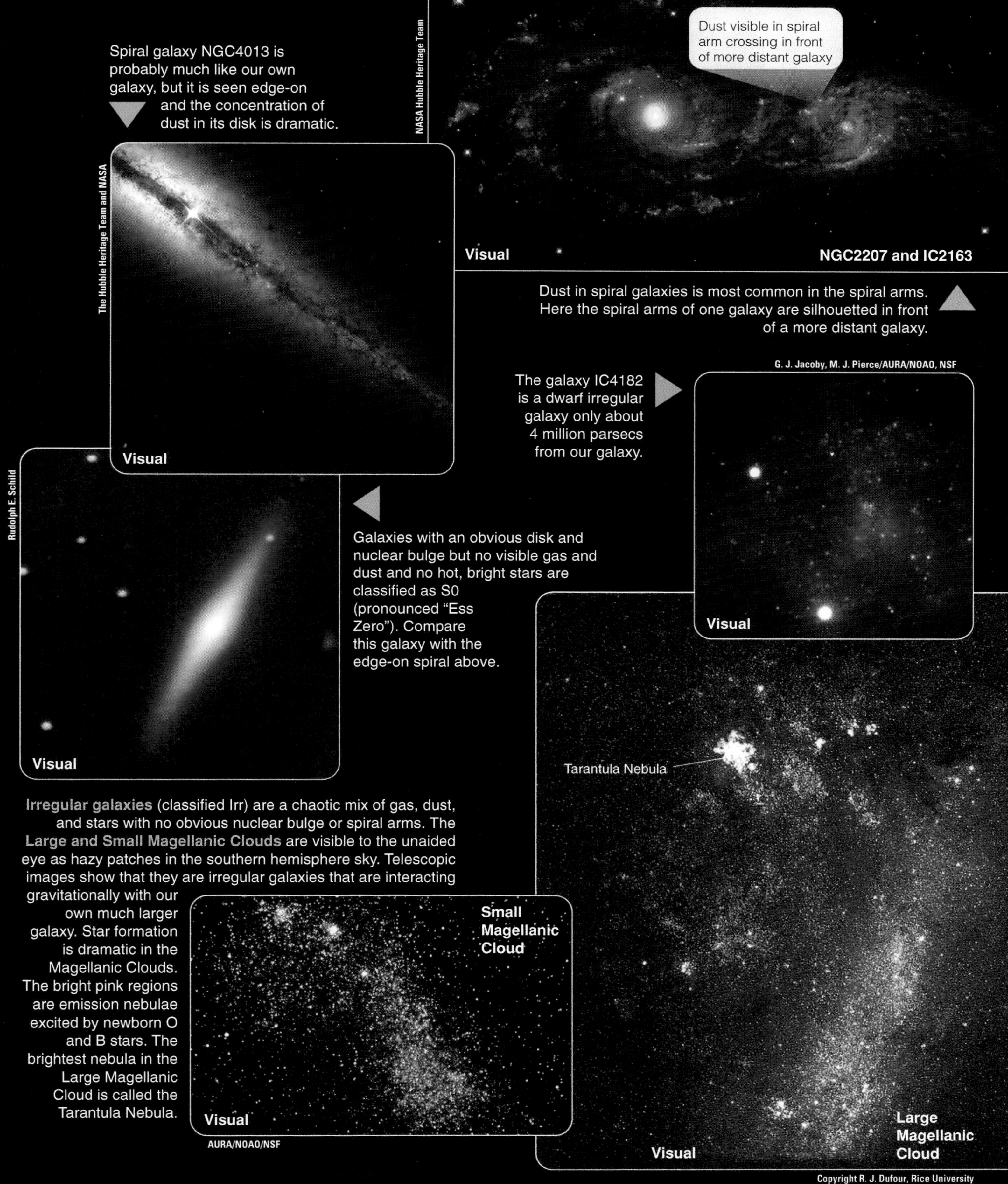

Spiral galaxy NGC4013 is probably much like our own galaxy, but it is seen edge-on and the concentration of dust in its disk is dramatic.
NASA Hubble Heritage Team
Dust visible in spiral arm crossing in front of more distant galaxy
The Hubble Heritage Team and NASA
Visual
NGC2207 and IC2163
Dust in spiral galaxies is most common in the spiral arms. Here the spiral arms of one galaxy are silhouetted in front of a more distant galaxy.
G. J. Jacoby, M. J. Pierce/AURA/NOAO, NSF
The galaxy IC4182 is a dwarf irregular galaxy only about 4 million parsecs from our galaxy.
Rudolph E. Schild
Visual
Galaxies with an obvious disk and nuclear bulge but no visible gas and dust and no hot, bright stars are classified as S0 (pronounced "Ess Zero"). Compare this galaxy with the edge-on spiral above.
Visual
Visual
Tarantula Nebula
Irregular galaxies (classified Irr) are a chaotic mix of gas, dust, and stars with no obvious nuclear bulge or spiral arms. The Large and Small Magellanic Clouds are visible to the unaided eye as hazy patches in the southern hemisphere sky. Telescopic images show that they are irregular galaxies that are interacting gravitationally with our own much larger galaxy. Star formation is dramatic in the Magellanic Clouds. The bright pink regions are emission nebulae excited by newborn O and B stars. The brightest nebula in the Large Magellanic Cloud is called the Tarantula Nebula.
Small Magellanic Cloud
Visual
AURA/NOAO/NSF
Visual
Large Magellanic Cloud
Copyright R. J. Dufour, Rice University

■ Figure 13-2

Nearly all of the objects in this deep image of an apparently empty spot on the sky are galaxies. The largest objects here are relatively nearby, but most are billions of light-years from Earth. Presumably, the entire sky is carpeted with galaxies like this. (NASA/ESA/GOODS Team and M. Giavalisco, STScI)

careful studies astronomers can conclude that ellipticals are more common than spirals, and that irregulars make up only about 25 percent of all galaxies. Among spiral galaxies, about two-thirds are barred spirals.

Why are there different kinds of galaxies? Later in this chapter we will try to understand what determines the stellar content and shape of a galaxy. The basic classification system will be an important tool.

How Many Galaxies?

Like leaves on the forest floor, galaxies carpet the sky. Pick any spot on the sky away from the dust and gas of the Milky Way, and you are looking deep into space. Photons that have traveled many billions of years enter your eye, but they are too few to register on your retina. Only the largest telescopes can gather enough light to detect the most distant galaxies.

In 1995, astronomers picked a seemingly empty spot on the sky near the Big Dipper and used the Hubble Space Telescope to record a time exposure that lasted an astonishing 10 days. This became known as a Hubble *deep* field; it was deep in that it recorded very faint objects. The image revealed that the "empty spot" on the sky was filled with galaxies.

More recent deep fields have probed even fainter objects. A research effort called GOODS (Great Observatories Origins Deep Survey) has used the Hubble Space Telescope, the Chandra X-Ray Observatory, the Spitzer Space Telescope, and the XXM-Newton X-Ray Telescope, plus the largest ground-based telescopes, to study two selected areas in the northern and southern sky.

The GOODS images reveal tremendous numbers of galaxies (■ Figure 13-2). There is no reason to believe that the two regions of the sky chosen for study are unusual, so it seems likely that the entire sky is carpeted with galaxies. At least 100 billion would be visible with today's telescopes. Surely there are other galaxies too distant or too faint to see.

Inquire | Review | Analyze

What color are galaxies?

What you know about stars and stellar populations allows you to analyze this question. Different kinds of galaxies have different colors, depending mostly on how much gas and dust they contain. If a galaxy contains large amounts of gas and dust, it probably contains lots of young stars, and a few of those young stars will be massive, hot, luminous O and B stars. They will give the galaxy a distinct blue tint. In contrast, a galaxy that contains little gas and dust will probably contain few hot stars. Its most luminous stars will be red giants, and they will give the galaxy a red tint. Because the light from a galaxy is a blend of the light from billions of stars, the colors are only tints. Nevertheless, the most luminous stars in a galaxy determine its color. From this you can conclude that elliptical galaxies tend to be red and the disks of spiral galaxies tend to be blue.

Of course, the halo of a spiral galaxy would be both dimmer and redder than the disk. Why?

■ ■ ■

Connections: If you want to create a theory to explain the shapes of galaxies, you need answers to some basic questions. How big are galaxies? How luminous are they? Are they all the same mass, or are some more massive than others? To answer those questions, you must discover the basic properties of galaxies. We begin that task in the next section.

13-2 Measuring the Properties of Galaxies

LOOKING BEYOND THE EDGE of our Milky Way Galaxy, astronomers find many billions of galaxies. Great clusters of galaxies, some containing thousands of galaxies, fill space as far as they can see. What are the properties of these star systems? What are the diameters, luminosities, and masses of galaxies? Just as in your study of stellar characteristics (Chapter 8), the first step in your study of galaxies is to find out how far away they are. Once you know a galaxy's distance, its size and luminosity are relatively easy to find. Later in this section, you will see that finding mass is more difficult.

Distance

The distances to galaxies are so large that it is not convenient to express them in light-years, parsecs, or even kiloparsecs. Instead, astronomers use the unit **megaparsec (Mpc),** or 1 million pc. One Mpc equals 3.26 million ly, or approximately 2×10^{19} miles.

To find the distance to a galaxy, you must search among its stars for a familiar object of known luminosity. Such objects are called **distance indicators.**

Because their period is related to their luminosity (see Figure 12-4), Cepheid variable stars are reliable distance indicators. If you know the period of the star's variation, you can use the period–luminosity diagram to learn its absolute magnitude. By comparing its absolute and apparent magnitudes, you can find its distance (see Reasoning with Numbers 8-2). ■ Figure 13-3 shows a galaxy in which the Hubble Space Telescope detected Cepheids.

Even with the Hubble Space Telescope, Cepheids are not visible in galaxies much beyond 80 million ly (25 Mpc), so astronomers must search for less common but brighter distance indicators and calibrate them using nearby galaxies containing visible Cepheids. For example, by studying nearby galaxies with distances known from Cepheids, astronomers have found that the brightest globular clusters have absolute magnitudes of about −10. If they find globular clusters in a more distant galaxy, they can assume that the brightest have similar absolute magnitudes and thus calculate the distance.

Planetary nebulae have proven to be very useful distance indicators, which is surprising considering how faint their central star is. The central stars of planetary nebulae seem faint because they are very small; and, in spite of their high temperature, their small surface area makes them faint. They also seem faint because they are very hot and radiate most of their energy in the ultraviolet. However, the planetary nebula absorbs this ultraviolet radiation and reradiates it as visible emission lines, and astronomers have been able to calibrate the brightest planetary nebulae by studying nearby galaxies, such as the Andromeda Galaxy. Like brightly colored paper lanterns illuminated from within, planetary nebulae in other galaxies can be identified out to roughly 25 Mpc by their emission at the wavelength of 500.7 nm, the wavelength of an emission line of oxygen; that identification allows the calculation of the distance to the galaxy.

When a supernova explodes in a distant galaxy, astronomers rush to observe it. Studies show that type Ia supernovae, those caused by the collapse of a white dwarf, all reach about the same absolute magnitude at maximum. By searching for Cepheids and other distance indicators in nearby galaxies where type Ia supernovae have occurred, astronomers have been able to calibrate these supernovae. Look at the image of the small galaxy IC4182 on page 285. It is an important link in this calibration because it is nearby, and contains Cepheids, and astronomers saw a type Ia supernova explode there in 1937.

Over a century ago, laboratory scientists referred to a "standard candle" as a source of a known amount of light. Astronomers still use the term for distance indicators of known luminosity. Cepheids, for instance, are good standard candles for finding distance. Recently an astronomer referred to type Ia supernovae as "standard bombs." However violent they are, they have been calibrated and can be used as standard candles to find distance.

Because type Ia supernovae are much brighter than Cepheids, they can be seen at great distances. The drawback is that supernovae are rare, and none may occur during your lifetime in a galaxy you might be studying. You will see in Chapter 15 how these distance indicators have reshaped our understanding of the history of the universe.

At the greatest distances, astronomers must calibrate the total luminosity of the galaxies themselves. For example, studies of nearby galaxies show that an average galaxy like our Milky Way Galaxy has a luminosity about 16 billion times the sun's. If astronomers see a similar galaxy far away, they can measure its apparent magnitude and calculate its distance. Of course, it is important to recognize the different types of galaxies, and that is difficult to do at great distances (■ Figure 13-4). Averaging the distances to the brightest galaxies in a cluster can reduce the uncertainty in this method.

Notice how astronomers use calibration (Window on Science 12-1) to build a distance scale reaching from the nearest galaxies to the most distant visible galaxies. Often astronomers refer to this as the distance pyramid or the distance ladder because each step depends on the steps below it. Of course, the foundation of the distance scale rests on the Cepheid variable stars and astronomers' understanding of the luminosities of the stars in the H–R diagram, which ultimately rest on measurements of the parallax of stars.

The most distant visible galaxies are roughly 10 billion ly (3000 Mpc) away, and at such distances you see an effect akin to

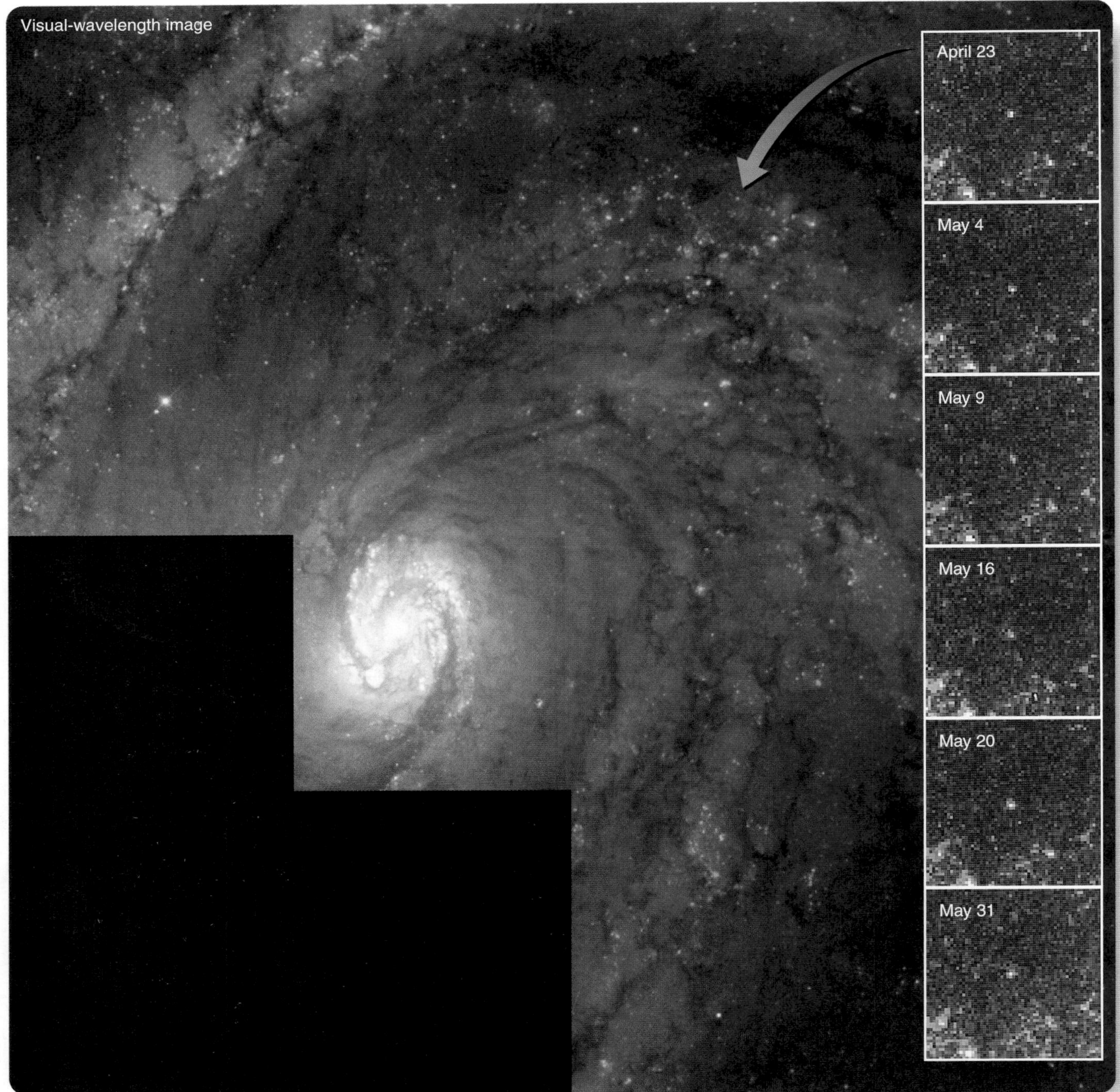

Figure 13-3

The vast majority of spiral galaxies are too distant for Earth-based telescopes to detect Cepheid variable stars. The Hubble Space Telescope, however, can locate Cepheids in some of these galaxies, as it has in the bright spiral galaxy M100. From a series of images taken on different dates, astronomers can locate Cepheids (inset), determine the period of pulsation, and measure the average apparent brightness. They can then deduce the distance to the galaxy—51 million ly for M100. (J. Trauger, JPL; Wendy Freedman, Observatories of the Carnegie Institution of Washington; and NASA)

time travel. When you look at a galaxy millions of light-years away, you do not see it as it is now but as it was millions of years ago when its light began the journey toward Earth. Thus, when you look at a distant galaxy, you look back into the past by an amount called the **look-back time,** a time in years equal to the distance to the galaxy in light-years.

You may have experienced look-back time if you have ever made a long-distance phone call carried by satellite or watched a

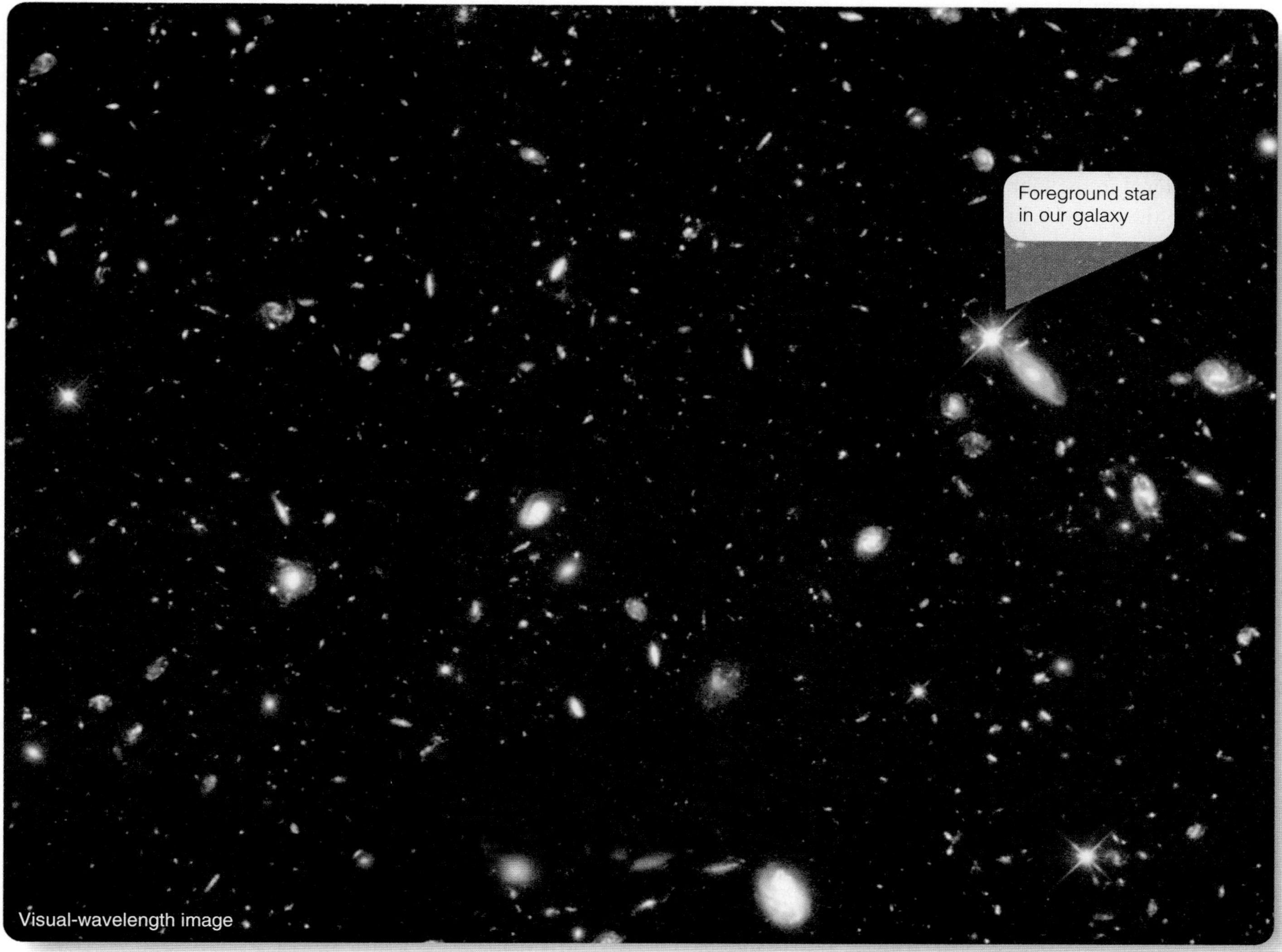

Figure 13-4

Distant galaxies can be distinguished from stars by their shape. In this segment of the northern Hubble Deep Field, only four stars are visible; they are sharp points of light with diffraction spikes produced by the telescope optics. A few galaxies are obviously spirals, but most are only elongated shapes and cannot be classified at this distance. (R. Williams and the Hubble Deep Field Team, STScI, NASA)

TV newscaster interview someone on the other side of the world via satellite. A half-second delay occurs as a radio signal carries a question 23,000 miles out to a satellite, then back to Earth, and then carries the answer out to the satellite and again back to Earth. That half-second look-back delay can make people hesitate on long distance phone calls and produces a seemingly awkward delay in intercontinental TV interviews.

The look-back time to nearby objects is usually not significant. The look-back time across a football field is a tiny fraction of a second. The look-back time to the moon is 1.3 seconds, to the sun only 8 minutes, and to the nearest star about 4 years. The Andromeda Galaxy has a look-back time of about 2 million years, a mere eye blink in the lifetime of a galaxy. But when you look at more distant galaxies, the look-back time becomes an appreciable part of the age of the universe. You will see evidence in Chapter 15 that the universe began about 14 billion years ago. Thus, when you look at the most distant visible galaxies, you are looking back 10 billion years to a time when the universe may have been significantly different. This effect will be important in our discussions in this and the next chapter.

The Hubble Law

Although astronomers find it difficult to measure the distance to a galaxy, they often estimate such distances using a simple relationship.

Early in the 20th century, astronomers noticed that the spectral lines in galaxy spectra were shifted slightly toward longer wavelengths—what astronomers call *redshifts.* Interpreted as a

Reasoning with Numbers | 13-1

The Hubble Law

The apparent velocity of recession of a galaxy, V_r, in kilometers per second is equal to the Hubble constant, H, multiplied times the distance to the galaxy, d, in megaparsecs.

$$V_r = Hd$$

Astronomers use this as a way to estimate distance from a galaxy's apparent velocity of recession.

Example: If a galaxy has a radial velocity of 700 km/s, and H is 70 km/s/Mpc,* then the distance to the galaxy equals the velocity divided by the Hubble constant, which is

$$d = \frac{700}{70 \text{ Mpc}} = 10 \text{ Mpc}$$

The galaxy is about 10 megaparsecs from Earth.

*H has the units of a velocity divided by a distance. These are usually written as km/s/Mpc, meaning km/s per Mpc.

consequence of the Doppler shift, these redshifts implied that the galaxies had large radial velocities and were receding from Earth. (Review the Doppler effect in Chapter 6.)

In 1929, the American astronomer Edwin Hubble published a graph that plotted the apparent velocity of recession versus distance for a number of galaxies. The points in the graph fell along a straight line (■ Figure 13-5). This relation between apparent velocity of recession and distance is known as the **Hubble law,** and the slope of the line is known as the **Hubble constant.**

The Hubble law is important in astronomy for two reasons. It is taken as evidence that the universe is expanding, a subject discussed in detail in Chapter 15. This chapter discusses the Hubble law because astronomers use it to estimate the distance to galaxies. As shown in **Reasoning with Numbers 13-1**, the distance to a galaxy can be found by dividing its apparent velocity of recession by the Hubble constant. This is a very useful calculation, because it is usually possible to obtain a spectrum of a galaxy and measure its redshift even if it is too far away to have observable distance indicators. Obviously, knowing the Hubble constant is important.

■ Figure 13-5

Edwin Hubble's first diagram of the apparent velocities of recession and distances of galaxies did not probe very deeply into space. It did show, however, that the galaxies are receding.

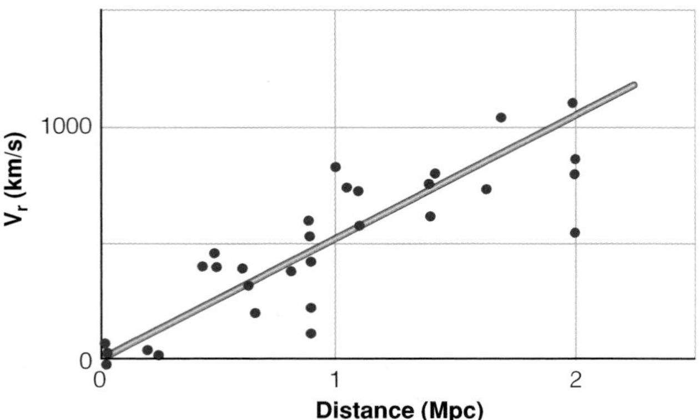

Edwin Hubble's original measurement of H was too large because of errors in his measurements of the distances to galaxies. Later astronomers have struggled to measure this important constant. The most precise measurements of the Hubble constant yield a value of about 70 km/s/Mpc with an uncertainty of about 5 percent.

Ace Astronomy™ Log into AceAstronomy and select this chapter to see Astronomy Exercise "Hubble Relations." You can make your own observations to find the Hubble Constant.

Diameter and Luminosity

The distance to a galaxy is the key to finding its diameter and its luminosity. With even a modest telescope and a CCD camera, you could photograph a galaxy and measure its angular diameter in seconds of arc. If you knew the distance to the galaxy, you could use the small angle formula (Reasoning with Numbers 3-1) to find its linear diameter. Also, if you measured the apparent magnitude of the galaxy, you could use the distance to find its absolute magnitude and from that its luminosity, as shown in Reasoning with Numbers 8-2.

The results of such observations show that galaxies differ dramatically in size and luminosity. Irregular galaxies tend to be small, 1 to 25 percent the size of our galaxy, and of low luminosity. Although they are common, they are easy to overlook. Our Milky Way Galaxy is large and luminous compared with most spiral galaxies, though astronomers know of a few spiral galaxies that are even larger and more luminous—nearly four times larger in diameter and about 10 times more luminous. Elliptical galaxies cover a wide range of diameters and luminosities. The largest, called giant ellipticals, are five times the size of our Milky Way. But many elliptical galaxies are very small, dwarf ellipticals, only 1 percent the diameter of our galaxy.

To put galaxies in perspective, you can use an analogy. If our galaxy were an 18-wheeler, the smallest dwarf galaxies would be the size of pocket-size toy cars, and the largest giant ellipticals would be the size of jumbo jets. Even among spiral galaxies that looked similar, the sizes could range from that of a go-cart to that of a small yacht.

Clearly, the diameter and luminosity of a galaxy do not determine its type. Some small galaxies are irregular, and some are elliptical. Some large galaxies are spiral, and some are elliptical. Other factors must influence the origin and evolution of galaxies.

Of the three basic parameters that describe a galaxy, you have found two—diameter and luminosity. The third, as was the case for stars, is more difficult to discover.

Mass

Although the mass of a galaxy is difficult to determine, it is an important quantity. It tells you how much matter the galaxy contains, which provides clues to the galaxy's origin and evolution. This section will examine two fundamental ways to find the masses of galaxies.

One way to find the mass of a galaxy is to watch it rotate. You know the stars in the outer parts of the galaxy are in orbit, so you can use Kepler's third law to find the mass. All you need to know is the size of the galaxy and the orbital period with which its outer stars circle the galaxy.

Recall from the previous section that it is easy to find the size of the galaxy if you know the distance. You just measure its angular size and use the small-angle formula (Reasoning with Numbers 3-1). That gives you the size of the orbits of the outer stars; that is, it tells you the semimajor axis of their orbits a.

You can't observe the orbital period, P, directly because humans don't live long enough to see a galaxy rotate, but you can find the orbital velocity of the stars using the Doppler effect. Once you know the orbital velocity, you can find the orbital period by dividing the circumference of the orbit by the velocity.

Once you know a in astronomical units and P in years, you can use Kepler's third law to find the mass of the galaxy (Reasoning with Numbers 8-4). Notice that this method depends on knowing the distance to the galaxy and being able to measure the rate at which it rotates.

Measuring the orbital velocity of the stars in a galaxy requires a special use of a spectrograph. Light enters a spectrograph through a narrow slit. If you focused the image of the galaxy on the slit of a spectrograph, you would see a bright spectrum formed by the bright nucleus of the galaxy, but you would also see fainter emission lines produced by ionized gas in the disk of the galaxy. Because the galaxy rotates, one side moves away from Earth and one side moves toward Earth, so the emission lines would be redshifted on one side of the galaxy and blueshifted on the other side. You could measure those changes in wavelength, use the Doppler formula to find the velocities, and plot a diagram showing the velocity of rotation at different distances from the center of the galaxy—a diagram called a *rotation curve.* The artwork in ■ Figure 13-6a shows the process of creating a rotation curve, and Figure 13-6b shows a real galaxy, its spectrum, and its rotation curve.

Measuring the mass of a galaxy from its rotation curve is called the **rotation curve method,** and it is the best way to find the masses of galaxies, but it works only for the nearer galaxies, whose rotation curves can be observed. More distant galaxies appear so small astronomers cannot measure the radial velocity at different points along the galaxy. Also, recent studies of our own galaxy and others show that the outer parts of the rotation curve do not decline to lower velocities (Figure 13-6b). This shows that the outermost visible parts of some galaxies do not travel more slowly and reveals that the galaxies contain large amounts of mass outside this radius, perhaps in extended galactic coronas like the one that seems to surround our own galaxy (see Chapter 12). Because the rotation curve method can be applied only to nearby galaxies and because it cannot determine the masses of galactic coronas, you must look at the second way to find the masses of galaxies.

The **cluster method** of finding galactic mass depends on the motions of galaxies within a cluster. If you measure the radial velocities of many galaxies in a cluster, you find that some velocities are larger than others because of the orbital motions of the individual galaxies in the cluster. Given the range of velocities and the size of the cluster, you can ask how massive a cluster of this size must be to hold itself together with this range of velocities. Dividing the total mass of the cluster by the number of galaxies in the cluster yields the average mass of the galaxies. This method contains the built-in assumption that the cluster is not flying apart. If it is, your result is too large. Because it seems likely that most clusters are held together by their own gravity, the method is probably valid.

A related way of measuring a galaxy's mass is called the **velocity dispersion method.** It is really a version of the cluster method. Instead of observing the motions of galaxies in a cluster, astronomers observe the motions of matter within a galaxy. In the spectra of some galaxies, broad lines indicate that stars and gas are moving at high velocities. If astronomers assume the galaxy is bound by its own gravity, they can ask how massive it must be to hold this moving matter within the galaxy. This method, like the one before, assumes that the system is not coming apart.

The masses of galaxies cover a wide range. The smallest contain about 10^{-6} as much mass as the Milky Way, and the largest contain as much as 50 times more than the Milky Way. The structure and evolution of a star is determined by its mass, but this is clearly not so for galaxies. We must search further for an explanation of the different types of galaxies.

Supermassive Black Holes in Galaxies

Rotation curves show the motions of the outer parts of a galaxy, but it is possible to detect the Doppler shifts of stars orbiting close to the centers of galaxies. Although these motions are not usually shown on rotation curves, they reveal something astonishing.

Measurements show that the stars near the centers of most galaxies are orbiting very rapidly. To hold stars in such small, short-

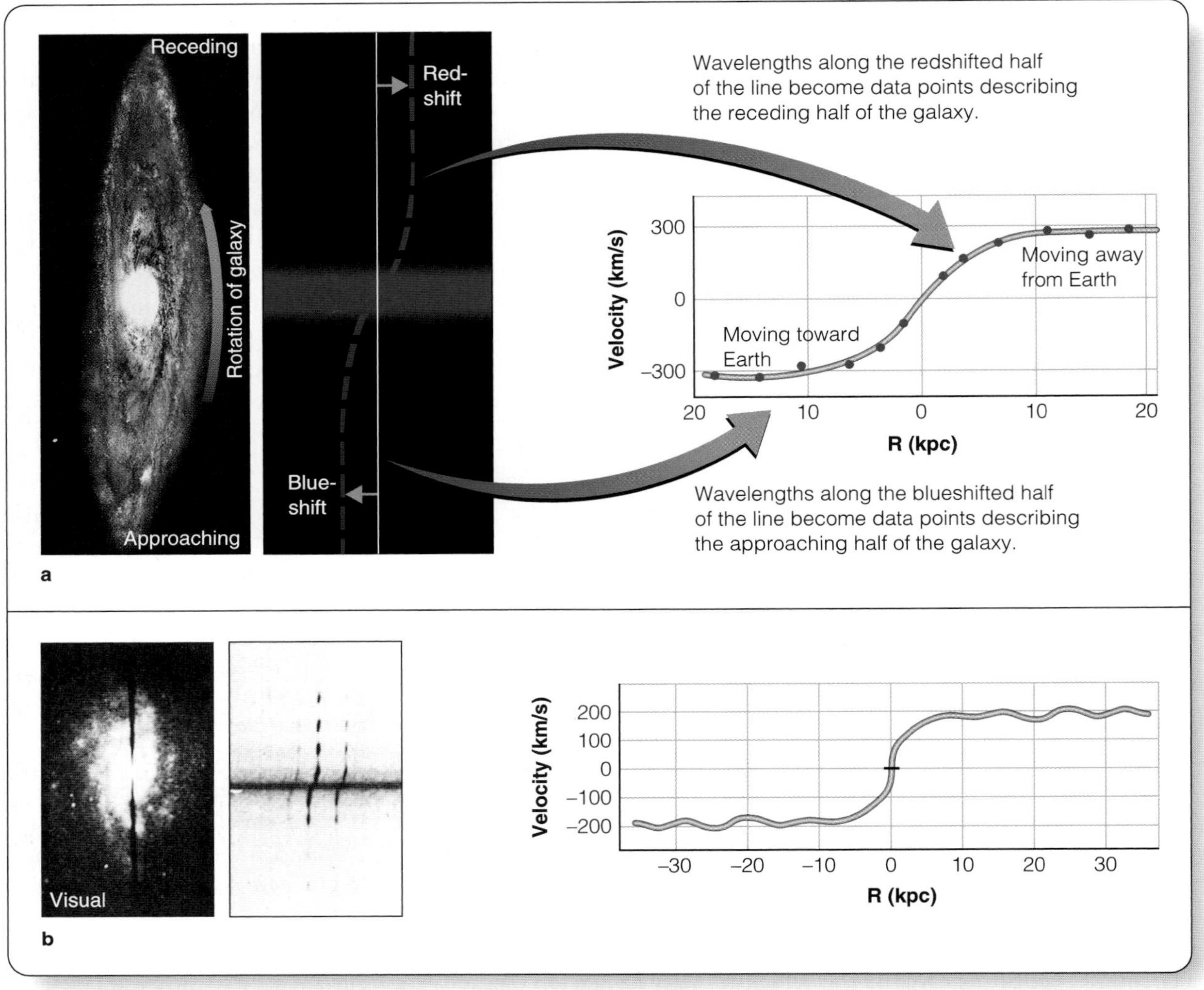

Figure 13-6

(a) In the artwork in the upper half of this diagram, the astronomer has placed the image of the galaxy over a narrow slit so that light from the galaxy can enter the spectrograph and produce a spectrum. A very short segment of the spectrum shows an emission line redshifted on the receding side of the rotating galaxy and blueshifted on the approaching side. Converting these Doppler shifts into velocities, the astronomer can plot the galaxy's rotation curve (right). (b) Real data are shown in the bottom half of this diagram. Galaxy NGC2998 is shown over the spectrograph slit, and the segment of the spectrum includes three emission lines. (Courtesy Vera Rubin)

period orbits, the centers of galaxies must contain masses of a million to a few billion solar masses. Yet no object is visible, so most astronomers believe that the nuclei of galaxies contain supermassive black holes. You saw in Chapter 12 that the Milky Way contains a supermassive black hole at its center. Evidently that is typical of galaxies.

Such supermassive black holes cannot be the remains of a dead star. Rather, these black holes must have formed as the galaxy formed, or they must have accumulated over billions of years as matter sank into the centers of the galaxies. Measurements show that the masses of these supermassive black holes are related to the mass of the nuclear bulges. A galaxy with a large nuclear bulge has a supermassive black hole whose mass is greater than that in a galaxy with a small nuclear bulge. This suggests that the supermassive black holes formed long ago as the galaxies formed. Of course, matter has continued to drain into the black holes, but they do not appear to have grown dramatically since they formed.

A billion-solar-mass black hole sounds like a lot of mass, but note that it is roughly 1 percent of the mass of a galaxy. The 2.6-million-solar-mass black hole at the center of the Milky Way Galaxy contains only a thousandth of a percent of the mass of the galaxy. In the next chapter you will discover that these supermassive black holes can produce titanic eruptions, but they represent only a small fraction of the mass of a galaxy.

Dark Matter in Galaxies

Given the size and luminosity of a galaxy, astronomers can make a rough guess as to the amount of matter it should contain. Astronomers know how much light stars produce, and they know about how much matter there is between the stars, so it is quite possible to estimate very roughly the mass of a galaxy from its luminosity. But when astronomers measure the masses of galaxies, they often find that the measured masses are much larger than expected from the luminosities of the galaxies. You discovered this effect in Chapter 12 when you studied the rotation curve of our own galaxy and concluded that it must contain large amounts of dark matter. This seems to be true of most galaxies. Measured masses of galaxies amount to 10 to 100 times more mass than you can see.

X-ray observations reveal more evidence of dark matter. X-ray images of galaxy clusters show that many of them are filled with very hot, low-density gas. The amount of gas present is much too small to account for the dark matter. Rather, the gas is important because it is very hot and its rapidly moving atoms have not leaked away. Evidently the gas is held in the cluster by a very strong gravitational field. To provide enough gravity to hold such hot gas, the cluster must contain much more matter than you can see. The detectable galaxies in the Coma cluster, for instance, amount to only a small fraction of the total mass of the cluster (■ Figure 13-7).

Dark matter is not an insignificant issue. Observations of galaxies and clusters of galaxies reveal that 90 to 95 percent of the matter in the universe is dark matter. The universe you see—the kind of matter that you and the stars are made of—has been compared to the foam on an invisible ocean.

Gravitational Lensing and Dark Matter

When you solve a math problem and get an answer that doesn't seem right, you check your work for a mistake. Astronomers using orbital motion to measure the mass of galaxies find evidence of large amounts of dark matter, so they look for ways to check their work. Fortunately, there is another way to detect dark matter that does not depend on observing orbital motion. You can follow a light beam.

Albert Einstein described gravity as a curvature of space. The presence of mass actually distorts space, and that is what we feel as gravity. He predicted that a light beam traveling through a gravitational field would be deflected by the curvature of space much as a golf ball is deflected as it rolls over a curved putting green. That effect has been observed and is a strong confirmation of Einstein's theories.

Gravitational lensing occurs when light from a distant object passes a nearby massive object and is deflected by the gravitational field. The gravitational field of the nearby object acts as a lens to deflect the passing light.

Astronomers have used gravitational lensing to detect dark matter. When light from very distant galaxies passes through a cluster of galaxies on its way to Earth, it can be deflected by the strong

■ Figure 13-7

(a) The Coma cluster of galaxies contains at least 1000 galaxies and is especially rich in E and S0 galaxies. Two giant galaxies lie near its center. Only the central area of the cluster is shown in this image. If the cluster were visible in the sky, it would span eight times the diameter of the full moon. (L. Thompson © NOAO) (b) In false colors, this X-ray image of the Coma cluster shows it filled and surrounded by hot gas. The box outlines the area of the cluster shown in part a. (Simon D. M. White, Ulrich G. Briel, and J. Patrick Henry)

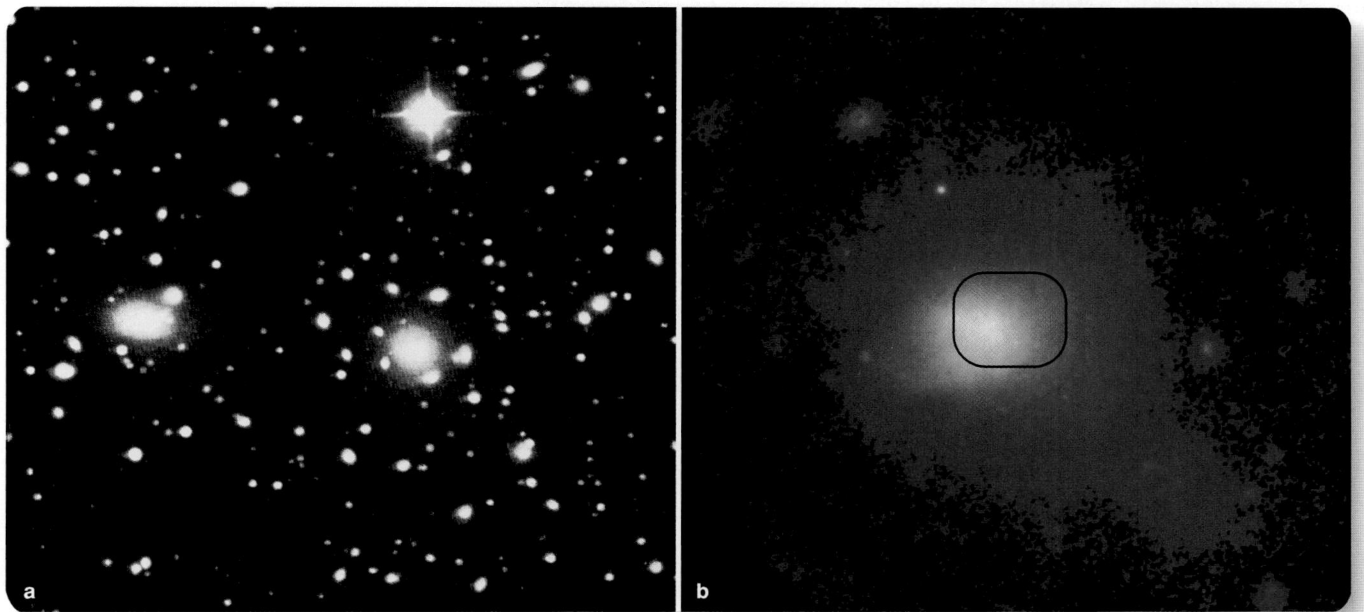

gravitational field. That distorts the images of the distant galaxies into curving arcs, and the amount of the distortion depends on the mass of the cluster of galaxies (■ Figure 13-8). Observations of gravitational lensing made with very large telescopes reveal that clusters of galaxies contain far more matter than what we can see. That is, they must contain large amounts of dark matter.

Dark matter is difficult to detect, and it is even harder to explain. Some astronomers have suggested that dark matter consists of low-luminosity white dwarfs and brown dwarfs scattered through the halos of galaxies. Both observation and theory support the idea that galaxies have massive extended halos, and searches for white dwarfs and brown dwarfs in the halo of our galaxy have been successful. Nevertheless, the searches have not turned up enough of these low-luminosity objects to make up all of the dark matter.

Recall from Chapter 12 that the dark matter can't be hidden in vast numbers of black holes and neutron stars, because astronomers don't see the X rays these objects would emit. Remember, you need 10 to 100 times more dark matter than visible matter, and that many black holes should produce X rays that are easy to detect.

Because observations imply that the dark matter can't be hidden in normal objects, some theorists have suggested that the dark matter is made up of unexpected forms of matter. Until recently neutrinos were thought to be massless, but studies now suggest they have a very small mass. Thus they may represent part of the dark matter.

Dark matter remains one of the fundamental unresolved problems of modern astronomy. You will return to this problem in Chapter 15 when we try to understand how dark matter affects the nature of the universe, its past, and its future.

Inquire | Review | Analyze

Why do you have to know the distance to a galaxy to find its mass? Often the critical analysis of a problem is complicated by interrelationships between factors. Usually a bit of care is enough to reveal those connections. To find the mass of a galaxy, you find the size of the orbits and the orbital periods of stars at the galaxy's outer edge and then use Kepler's third law to find the mass inside the orbits. Measuring the orbital velocity of the stars as the galaxy rotates is easy if you can obtain a rotation curve from a spectrum. But you must also know the radius of the star's orbits in meters or astronomical units. That is where the distance to the galaxy comes in. You can observe the radius of the stellar orbits at the outer edge of the galaxy in seconds of arc and then use the distance to the galaxy in the small angle formula to convert to meters. Once you know velocity and orbital radius in meters, you can find the mass.

Distance is critical in astronomy. For example, what would happen to your estimate of the diameter and luminosity of galaxies if astronomers discovered that the Cepheid variable stars were slighly more luminous than had been believed?

■ ■ ■

■ Figure 13-8

The mass of galaxy cluster 0024+ 1654 bends the light of a much more distant galaxy to produce arcs that are actually distorted images of the distant galaxy. This gravitational lens effect reveals that many galaxy clusters contain much more than is visible as galaxies. Thus galaxy clusters must contain large amounts of dark matter. (W. N. Colley and E. Turner, Princeton, and J. A. Tyson, Bell Labs, and NASA)

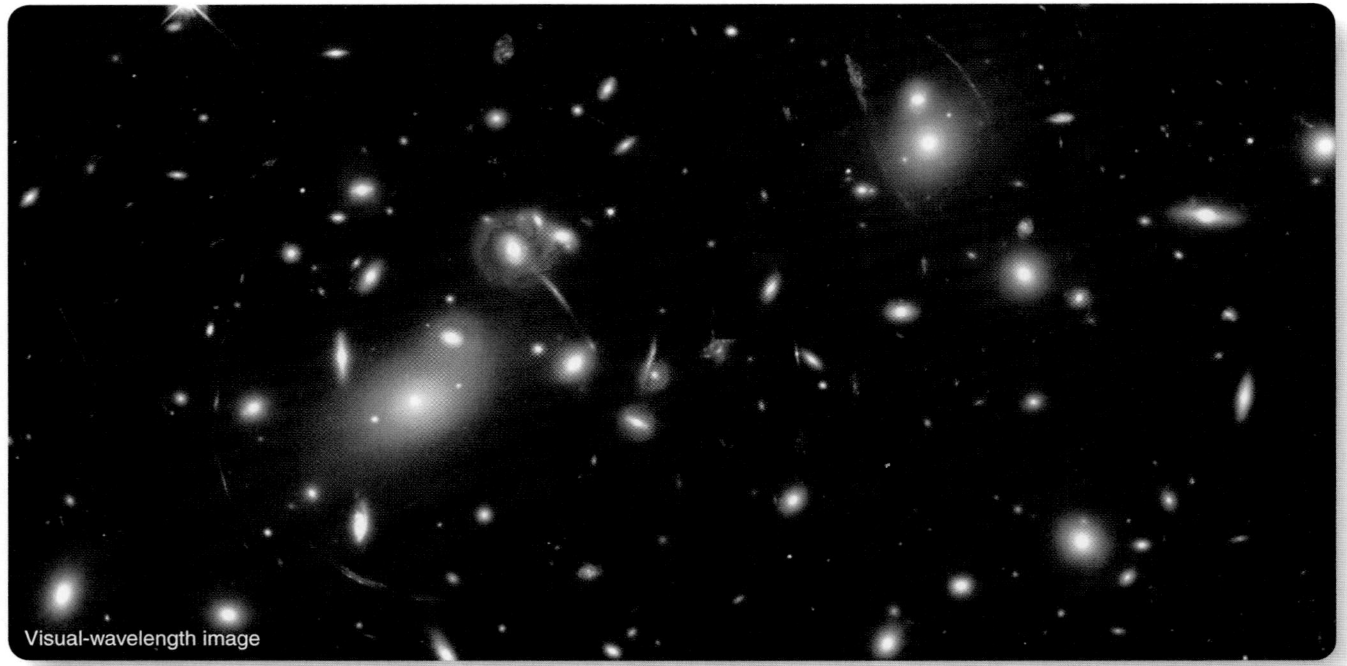

Connections: You have gathered basic data to help you characterize galaxies. That is, you can describe their general characteristics. Now you turn to the more difficult and more interesting question. How did galaxies get to be the way they are?

13-3 The Evolution of Galaxies

YOUR GOAL IN THIS CHAPTER has been to build a theory to explain the evolution of galaxies. In Chapter 12, you developed a theory that described the origin of our own Milky Way Galaxy; presumably, other galaxies formed similarly. But why did some galaxies become spiral, some elliptical, and some irregular? Clues to that mystery lie in the clustering of galaxies.

Clusters of Galaxies

The distribution of galaxies is not entirely random. Galaxies tend to occur in clusters ranging from a few to thousands. Deep photographs made with the largest telescopes reveal clusters of galaxies scattered out to the limits of detectability. This clustering of the galaxies can help you understand their evolution.

For this discussion, you can sort clusters into two groups: rich and poor. **Rich galaxy clusters** contain over a thousand galaxies, mostly elliptical, scattered through a spherical volume about 3 Mpc (10^7 ly) in diameter. Such a cluster is very crowded, with the galaxies more concentrated toward the center. Rich clusters often contain one or more giant elliptical galaxies at their centers.

The Coma cluster (located in the constellation Coma Berenices) is an example of a rich cluster (Figure 13-7). It lies over 100 Mpc from Earth and contains at least 1000 galaxies, mostly E and S0 galaxies. Its galaxies are highly crowded around a central giant elliptical galaxy and a large S0.

One of the nearest clusters, the Virgo cluster, contains over 2500 galaxies and is, by the above definition, a rich cluster. It does contain a giant elliptical galaxy, M87, near its center, but it is not very crowded and contains mostly spiral galaxies.

Poor galaxy clusters contain fewer than 1000 galaxies and are irregularly shaped. Instead of being crowded toward the center, poor clusters tend to have subcondensations, small groupings within the clusters.

Our own **Local Group,** which contains the Milky Way, is a good example of a poor cluster (■ Figure 13-9a). It contains a few dozen members scattered irregularly through a volume slightly over 1 Mpc in diameter. Of the brighter galaxies, 14 are elliptical, 3 are spiral, and 4 are irregular.

The total number of galaxies in the Local Group is uncertain because some lie in the plane of the Milky Way Galaxy and are difficult to detect. For example, a small dwarf galaxy, known as the Sagittarius Dwarf, has been found on the far side of our own galaxy, where it is almost totally hidden behind the star clouds of Sagittarius (Figure 13-9b). Even closer to the center of the Milky Way Galaxy is the Canis Major Dwarf Galaxy (Figure 13-9c). The galaxy was found by mapping the distribution of red supergiants detected by the 2MASS infrared survey. There are, almost certainly, other small galaxies in our Local Group hidden behind the stars, gas, and dust of our Milky Way Galaxy.

We have classified galaxy clusters into rich and poor clusters to reveal a fascinating and suggestive clue to the evolution of galaxies. In general, rich clusters tend to contain 80 to 90 percent E and S0 galaxies and few spirals. Poor clusters contain a larger percentage of spirals; and, among isolated galaxies, those that are not in clusters, 80 to 90 percent are spirals. This suggests that a galaxy's environment is important in determining its structure and has led astronomers to suspect that the secrets to galaxy evolution lie in collisions between galaxies.

Colliding Galaxies

Astronomers are finding more and more evidence to show that galaxies collide, interact, and merge. In fact, collisions among galaxies may dominate their evolution.

You should not be surprised that galaxies collide with each other. The average separation between galaxies is only about 20 times their diameter. Like two blindfolded elephants blundering about under a circus tent, galaxies should bump into each other once in a while. Stars, on the other hand, almost never collide. In the region of our galaxy near the sun, the average separation between stars is about 10^7 times their diameter. Thus a collision between two stars is about as likely as a collision between two blindfolded gnats flitting about in a baseball stadium.

Study **Interacting Galaxies** on pages 296 and 297 and notice four important points. First, interacting galaxies can distort each other with tides and may even trigger the formation of spiral arms. In fact, the flocculent galaxies, which lack strong two-armed patterns, tend to be isolated and thus probably have not had recent interactions with other galaxies. The second thing to notice is how interactions can trigger star formation. Galaxies that contain lots of gas and dust but are not making many stars usually have few neighbors and have consequently not suffered interactions recently. Third, notice how shells of stars, counter rotation, and multiple nuclei provide evidence of mergers between galaxies. Finally, notice the beautiful ring galaxies—bull's-eyes left behind by high-speed collisions.

Evidence of galaxy mergers is all around us. Our Milky Way Galaxy is a cannibal galaxy, snacking on the Magellanic Clouds as they orbit around it, and its tides are pulling the Sagittarius Dwarf Galaxy apart. The Canis Major Dwarf galaxy has been almost completely digested as tides have pulled stars away to form great streamers wrapped around the Milky Way galaxy. Almost certainly, our galaxy has dined on other small galaxies.

You have seen the evidence that collisions and mergers between galaxies are common and can produce dramatic changes

Interacting Galaxies

When two galaxies collide, they can pass through each other with the stars so small and so far apart that they never collide. Gas clouds and magnetic fields do collide, but the biggest effects may be tidal. Even when two galaxies just pass near each other, tides can cause dramatic effects, such as long streamers called tidal tails. In some cases, two galaxies can merge and form a single galaxy.

Tidal Distortion

Small galaxy passing near a massive galaxy.

Gravity of a second galaxy represented as a single massive object

When a galaxy swings past a massive object such as another galaxy, tides are severe. Stars near the massive object try to move in smaller, faster orbits while stars further from the massive object follow larger slower orbits. Such tides can distort a galaxy or even rip it apart.

Galaxy interactions can stimulate the formation of spiral arms

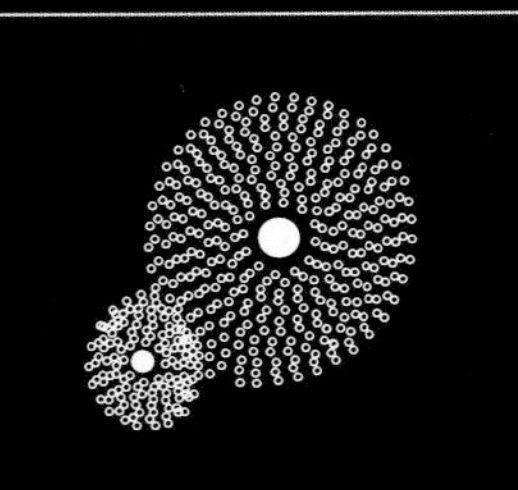

In this computer model, two uniform disk galaxies pass near each other.

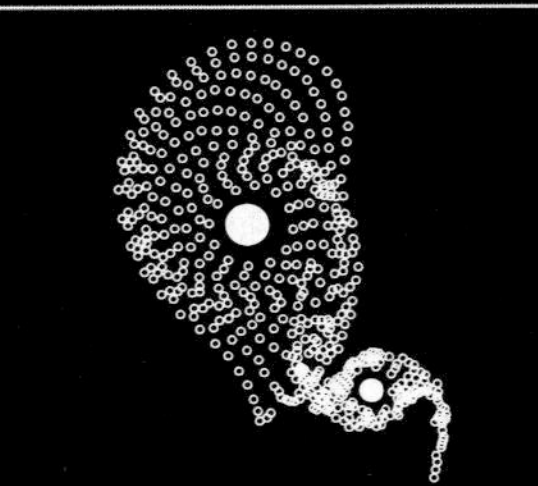

The small galaxy passes behind the larger galaxy so they do not actually collide.

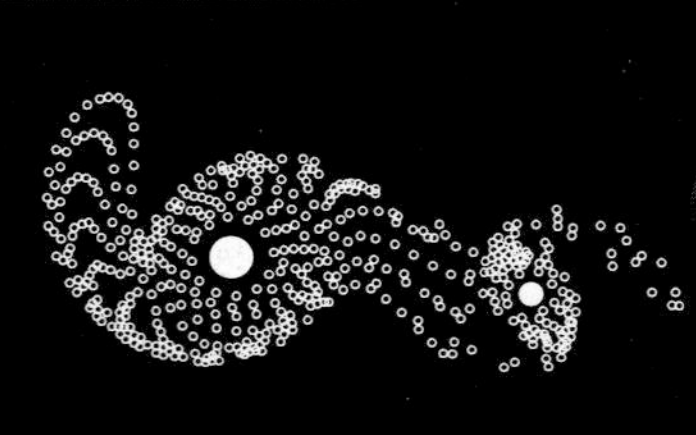

Tidal forces deform the galaxies and trigger the formation of spiral arms.

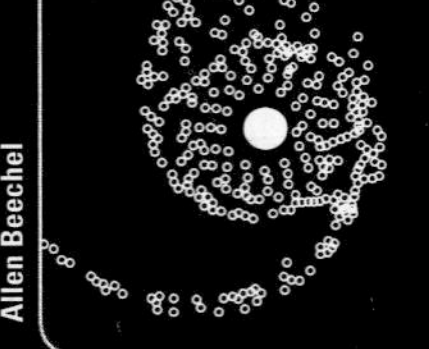

The upper arm of the large galaxy passes in front of the small galaxy.

Allen Beechel

A photo of the well-known Whirlpool Galaxy resembles the computer model.

Visual

NOAO

Visual false-color image

The Mice are a pair of galaxies whipping around each other and being distorted.

NOAO

Computer model of the Mice

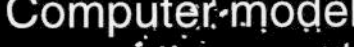

Allen Beechel

The merger of galaxies is called galactic cannibalism. Models show that merging galaxies spiral around their common center of mass while tides rip stars away and form shells.

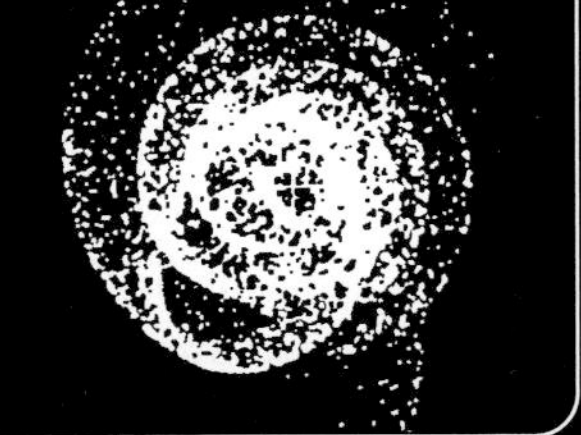
Computer model

François Schweizer and Alan Toomre

Shells of stars

Visual enhanced image

NOAO

Such shells have been found around elliptical galaxies such as NGC5128. It is peculiar in many ways and even has a belt of dusty gas. The shells revealed in this enhanced image are evidence that the giant galaxy has cannibalized at least one smaller galaxy. The giant galaxy itself may be the result of the merger of two large galaxies.

spiral and irregular galaxies can't be young. The galaxy classes tell you something important, but a single galaxy does not change from one class to another any more than a cat can change into a dog.

Another old idea held that galaxies that form from rapidly rotating clouds of gas would have lots of angular momentum and would contract slowly to form disk-shaped spiral galaxies. Clouds of gas that rotated less rapidly would contract faster, form stars quickly, use up all the gas and dust, and become elliptical galaxies. The evidence clearly shows that galaxies didn't form and evolve in isolation. Collisions and mergers dominate the history of the galaxies.

The ellipticals appear to be the product of galaxy mergers, which triggered star formation that used up the gas and dust. In fact, astronomers see star formation being stimulated to high levels in many galaxies (■ Figure 13-10). **Starburst galaxies** are very luminous in the infrared because a collision has triggered a burst of star formation that is heating the dust. The warm dust reradiates the energy in the infrared. The Antennae (page 297) contain over 15 billion solar masses of hydrogen gas and will become a starburst galaxy as the merger triggers rapid star formation. Supernovae in a starburst galaxy may eventually blow away any remaining gas and dust that doesn't get used up making stars. A few collisions and mergers could leave a galaxy with no gas and dust from which to make new stars. Astronomers now suspect that most ellipticals are formed by the merger of at least two or three galaxies. The dwarf ellipticals, too small to be formed by mergers, may be fragments left over by the merger of larger galaxies.

In contrast, spirals seem never to have suffered major collisions. Their thin disks are delicate and would be destroyed by tidal forces in a collision with a massive galaxy. Also, they retain plenty of gas and dust and continue making stars. Our own Milky Way Galaxy has evidently never merged with another large galaxy, but it has clearly cannibalized smaller galaxies. Astronomers have found streams of stars in the halo of our galaxy that are too metal rich for their location. Another stream contains globular clusters with similar ages. These streams are evidently the remains of smaller galaxies that were absorbed.

The nearby Andromeda Galaxy appears to be doing the same thing. Its halo contains streams of stars and star clusters that it has stripped from its swarm of small satellite galaxies.

Barred spiral galaxies are common, but mathematical models show that the bars do not last long. It may take tidal interactions with other galaxies to regenerate the bars.

Other processes can alter galaxies. The S0 galaxies may have lost much of their gas and dust in a burst of star formation, but

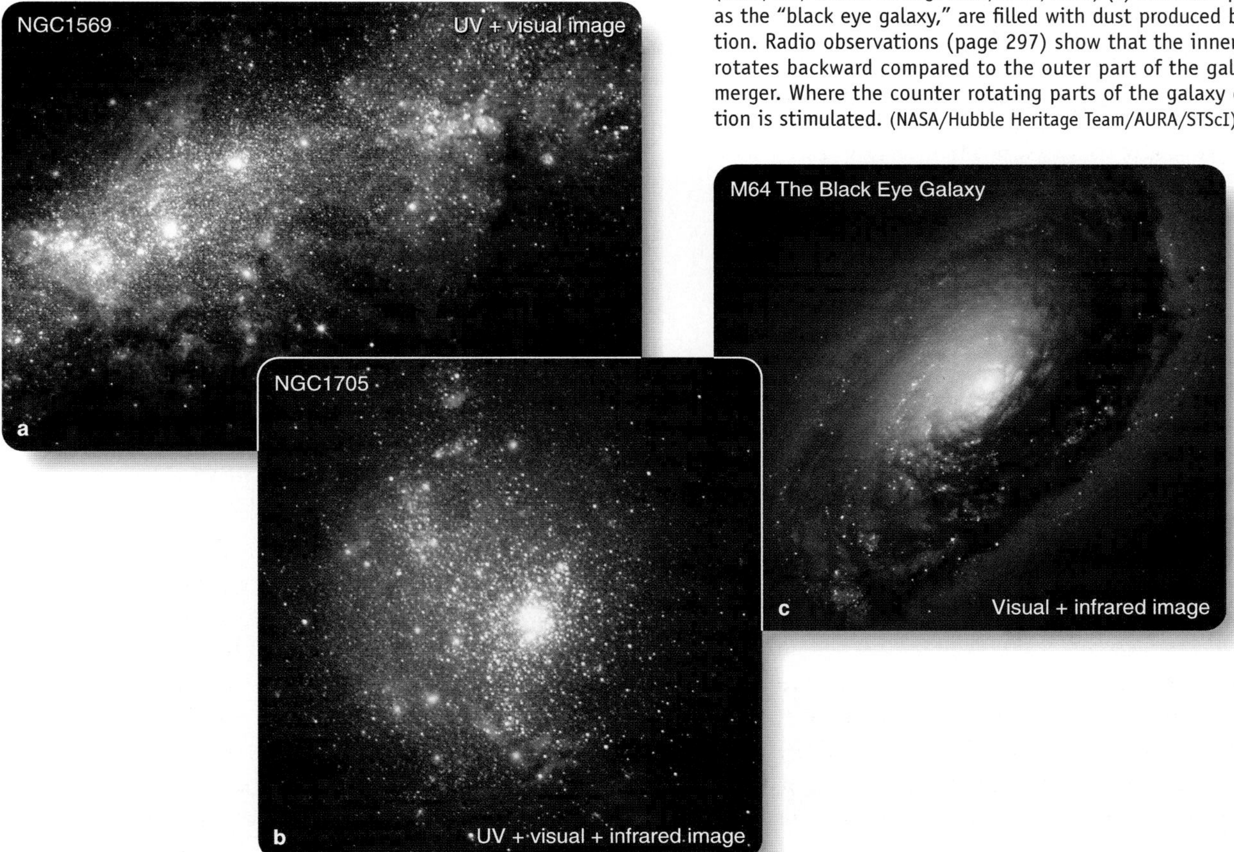

■ Figure 13-10

Rapid Star formation: (a) NGC1569 is a starburst galaxy filled with clouds of young stars and supernovae. At least some starbursts are triggered by interactions between galaxies. (ESA/NASA/P. Anders) (b) The dwarf irregular galaxy NGC1705 began a burst of star formation about 25 million years ago. (NASA/ESA/ Hubble Heritage Team/AURA/STScI) (c) The inner parts of M64, known as the "black eye galaxy," are filled with dust produced by rapid star formation. Radio observations (page 297) show that the inner part of the galaxy rotates backward compared to the outer part of the galaxy, a product of a merger. Where the counter rotating parts of the galaxy collide, star formation is stimulated. (NASA/Hubble Heritage Team/AURA/STScI)

they still managed to remain disk shaped. Also, galaxies moving through the gas trapped in dense clusters of galaxies may have their own gas and dust blown away. For example, X-ray observations show that the Coma cluster contains thin, hot gas between the galaxies (Figure 13-7). A galaxy orbiting through that gas would encounter a tremendous wind blowing its gas and dust away (■ Figure 13-11). This could explain some of the dwarf ellipticals, which are too small to be made of merged spirals. In contrast, the irregular galaxies may be small fragments of galaxies ripped apart by collisions.

The evolution of galaxies is not a simple process. Gas clouds are falling into some galaxies and adding mass. Other galaxies such as dwarf ellipticals and irregular galaxies may be small galaxies distorted as they passed near a larger galaxy, or they may be the splashed fragments from a collision of larger galaxies.

Astronomers are just beginning to understand the exciting and complex story of the galaxies, but it is already clear that galaxy evolution is much like a pie-throwing contest and just about as neat.

The Furthest Galaxies

Observations with the largest and most sophisticated telescopes are taking astronomers back to the age of galaxy formation. At great distances the look-back time is so large that they see the universe as it was soon after the galaxies began to form. There were more spirals then and fewer ellipticals. On the whole, galaxies long ago were more compact and more irregular than they are now. The observations show that galaxies were closer together then; about a third of all distant galaxies are in close pairs, but only 7 percent of nearby galaxies are in pairs. The observational evidence clearly supports the hypothesis that galaxies have evolved by merger.

At great distances, astronomers see tremendous numbers of small, blue, irregularly shaped galaxies that have been called blue dwarfs. The blue color indicates that they are rapidly forming stars, but the role of these blue dwarfs in the formation of galaxies is not clear. Blue dwarf galaxies are not found at small look-back times, so they no longer exist in the present universe. They may be the clouds of gas and stars that were long ago absorbed in the formation of larger galaxies, or they may have used up their gas and dust and have faded to obscurity today.

At the limits of the largest telescopes, astronomers see faint, red galaxies (■ Figure 13-12). They look red because of their great redshifts. The floods of ultraviolet light emitted by these star-forming galaxies have been shifted into the far red part of the spectrum. Their look-back times are so great, they appear as they were when the universe was only a billion years old. These galaxies may be among the first to begin shining after the beginning of the universe, a story told in Chapter 15.

Inquire | Review | Analyze

How did elliptical galaxies get that way?

A growing body of evidence suggests that elliptical galaxies have been subject to collisions in their past and that spiral galaxies have not. During collisions, a galaxy can be driven to use up its gas and dust in a burst of star formation or may be stripped of gas and dust by an encounter. Thus, elliptical galaxies now contain little star-making material. The beautiful disk typical of spiral galaxies is very orderly, with all the stars

■ Figure 13-11

The distorted galaxy C153 is orbiting through the thin gas in its home cluster of galaxies at 4.5 million miles per hour. At that speed it feels a tremendous wind stripping gas out of the galaxy in a trail 200,000 ly long. Such galaxies could quickly loose almost all of their gas and dust. (NASA, W. Keel, F. Owen, M. Ledlow, and D. Wang)

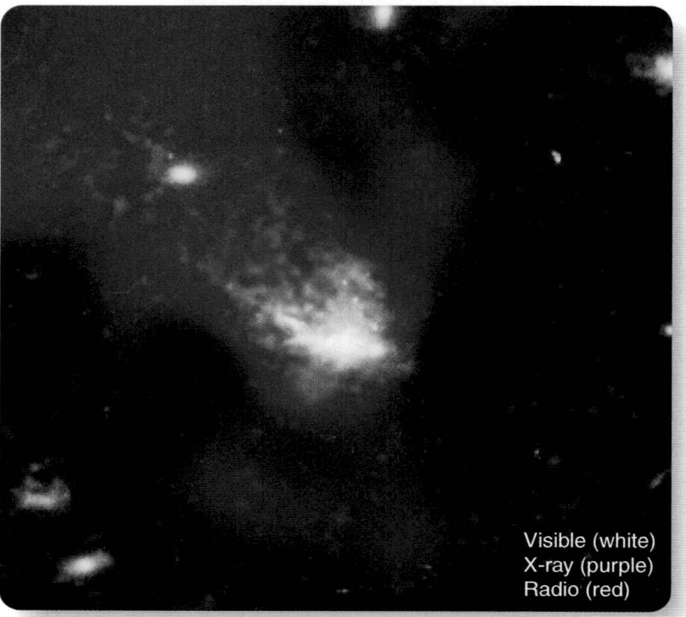

■ Figure 13-12

Arrows point to three very distant red galaxies. The look-back time to these galaxies is so large they appear as they were when the universe was only a billion years old. Such highly redshifted galaxies are believed to be among the first to form stars and begin shining after the beginning of the universe about 14 billion years ago. (NASA, H.-J. Yan, R. Windhorst, and S. Cohen, Arizona State University)

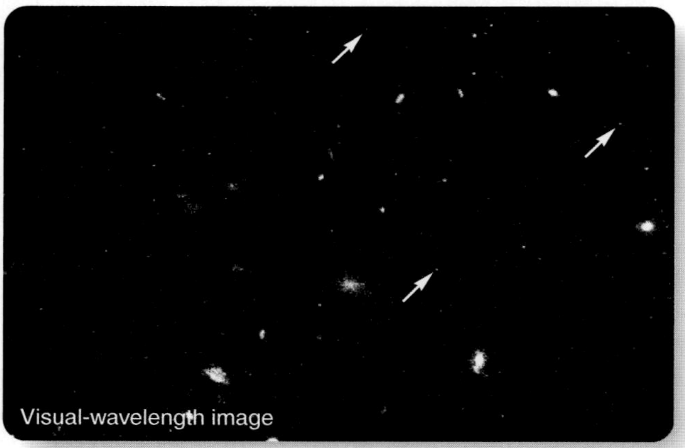

following similar orbits. When galaxies collide, the stellar orbits get scrambled, and an orderly disk galaxy could be converted into the chaotic swarm of stars typical of elliptical galaxies. Thus, it seems likely that elliptical galaxies have had much more complex histories than spiral galaxies have had.

What evidence supports the story in the preceding paragraph?

■ ■ ■

Connections: Before you can consider the universe as a whole, you must examine the galaxies from a different perspective. Some galaxies are suffering tremendous eruptions in their centers. In the next chapter, you will discover that these peculiar galaxies are closely related to collisions and mergers between galaxies.

Study and Review Tools

Summary

13-1 | The Family of Galaxies

What do galaxies look like?

- Galaxies can be divided into three classes—elliptical, spiral, and irregular—with subclasses specifying the galaxy's shape.
- Elliptical galaxies contain little gas and dust and cannot make new stars. Consequently, they lack hot, blue stars and have a reddish tint.
- Spiral galaxies contain more gas and dust and support active star formation, especially along the spiral arms. Some of the newborn stars are massive, hot, and blue, and that gives the spiral arms a blue color.
- The halo and nuclear bulge of a spiral galaxy usually lack gas and dust and contain little star formation. The halos and nuclear bulges have a reddish tint because they lack hot, blue stars.
- Irregular galaxies have no obvious shape but contain gas and dust and support star formation.

13-2 | Measuring the Properties of Galaxies

How do astronomers find the distance to galaxies?

- Astronomers find the distance to galaxies using distance indicators sometimes called standard candles, objects of known luminosity. The most accurate distance indicators are the Cepheid variable stars. Others include globular clusters, planetary nebulae, and supernovae explosions.
- Astronomers can estimate the distance to a galaxy by dividing its apparent rate of recession (its radial velocity) by the Hubble constant. Apparent rate of recession is related to distance by the Hubble law.

How do galaxies differ in size, luminosity, and mass?

- Some dwarf ellipticals and dwarf irregular galaxies are only a few percent the size and luminosity of our galaxy, but some giant elliptical galaxies are five times larger than the Milky Way galaxy.
- Astronomers measure the masses of galaxies in two basic ways—the rotation curve method and the velocity dispersion method. The rotation curve method is more accurate but can be applied only to nearby galaxies.

Do other galaxies contain supermassive black holes and dark matter as does our own galaxy?

- Stars near the centers of galaxies are orbiting at high velocities, which suggests the presence of supermassive black holes in the centers of most galaxies.
- The mass of a galaxy's supermassive black hole is proportional to the mass of its nuclear bulge. That suggests that the supermassive black holes must have formed when the galaxy formed.
- Observations of individual galaxies show that galaxies contain 10 to 100 times more dark matter than visible matter.
- The hot gas held inside some clusters of galaxies and the gravitational lensing caused by the mass of galaxy clusters add further evidence of dark matter.

13-3 | The Evolution of Galaxies

Why are there different kinds of galaxies?

- Rich clusters of galaxies contain fewer spirals than poor clusters of galaxies, and that is evidence that galaxies evolve by collisions and mergers.
- When galaxies collide, tides twist and distort their shapes, and the compression of gas clouds triggers star formation.
- Large galaxies can absorb smaller galaxies in what is called galactic cannibalism. There is clear evidence that our own Milky Way galaxy is devouring some of the small galaxies that orbit nearby and that our galaxy has consumed other small galaxies in the past.
- Shells of stars, counter-rotating parts of galaxies, streams of stars in the halos of galaxies, and multiple nuclei are evidence that galaxies can merge.
- The merger of two larger galaxies can scramble star orbits and drive bursts of star formation to use up gas and dust. Most larger ellipticals have evidently been produced by past mergers.
- Spiral galaxies have thin, delicate disks and appear not to have suffered mergers with large galaxies.
- A galaxy moving through the gas in a cluster of galaxies can be stripped of its own gas and dust and may become an S0 galaxy.
- At great distance and great look-back times, astronomers see that galaxies were smaller, more irregular, and closer together. Also, there were more spirals and fewer ellipticals long ago.
- At the largest distances are small irregular clouds of stars that may be the objects that fell together to begin forming galaxies when the universe was very young.

New Terms

elliptical galaxy (p. 284)
spiral galaxy (p. 284)
barred spiral galaxy (p. 284)
irregular galaxy (p. 285)

Large Magellanic Cloud (p. 285)
Small Magellanic Cloud (p. 285)
megaparsec (Mpc) (p. 287)
distance indicator (p. 287)
look-back time (p. 288)
Hubble law (p. 290)
Hubble constant (*H*) (p. 290)
rotation curve method (p. 291)
cluster method (p. 291)
velocity dispersion method (p. 291)
gravitational lensing (p. 293)
rich galaxy cluster (p. 295)
poor galaxy cluster (p. 295)
Local Group (p. 295)
tidal tail (p. 296)
galactic cannibalism (p. 296)
ring galaxy (p. 297)
starburst galaxy (p. 299)

Review Questions

Ace Astronomy™ Assess your understanding of this chapter's topics with additional quizzing and animations at **http://astronomy.brookscole.com/sh9e**

1. Why didn't astronomers at the beginning of the 20th century recognize galaxies for what they are?
2. How can a classification system aid a scientist?
3. What is the difference between an E0 galaxy and an E1 galaxy?
4. What is the difference between an Sa and an Sb galaxy? between an SBb and an Sb?
5. Why can't galaxies evolve from elliptical to spiral? Why can't they evolve from spiral to elliptical?
6. How do selection effects make it difficult to decide how common elliptical and spiral galaxies are?
7. Why are Cepheid variable stars good distance indicators? What about planetary nebulae?
8. Why is it difficult to measure the Hubble constant?
9. How is the rotation curve method related to binary stars and Kepler's third law?
10. What evidence is there that galaxies contain dark matter? supermassive black holes?
11. What evidence is there that galaxies collide and merge?
12. Why are the shells visible around some elliptical galaxies significant?
13. Ring galaxies often have nearby companions. What does that suggest?
14. Propose an explanation for the lack of gas, dust, and young stars in elliptical galaxies.
15. How do deep images by the Hubble Space Telescope confirm the hypothesis about galaxy evolution?
16. This image of M33, the Pinwheel Galaxy, has emission by ionized hydrogen enhanced as bright pink. Discuss the location of these clouds of gas and explain how that provides important evidence toward understanding spiral arms.

(P. Massey, Lowell, N. King, STScI, S. Holmes, Charleston, G. Jacoby, WIYN/AURA/NSF)

17. In the image at right you see two interacting galaxies; one is nearly face on and the other is nearly edge on. Discuss the shapes of these galaxies and describe what is happening.

(WIYN/NOAO/NSF)

Discussion Questions

1. Why do astronomers believe that galaxy collisions are likely, but star collisions are not?
2. Should an orbiting infrared telescope find irregular galaxies bright or faint in the far infrared? Why? What about elliptical galaxies?

Problems

1. If a galaxy contains a type I (classical) Cepheid with a period of 30 days and an apparent magnitude of 20, what is the distance to the galaxy?
2. If you find a galaxy that contains globular clusters that are 2 seconds of arc in diameter, how far away is the galaxy? (*Hints:* Assume that a globular cluster is 25 pc in diameter and see Reasoning with Numbers 3-1.)
3. If a galaxy contains a supernova that at its brightest has an apparent magnitude of 17, how far away is the galaxy? (*Hints:* Assume that the absolute magnitude of the supernova is -19 and see Reasoning with Numbers 8-2.)
4. If you find a galaxy that is the same size and mass as our Milky Way Galaxy, what orbital velocity would a small satellite galaxy have if it orbited 50 kpc from the center of the larger galaxy? (*Hint:* See Reasoning with Numbers 4-1.)
5. Find the orbital period of the satellite galaxy described in Problem 4. (*Hint:* See Reasoning with Numbers 8-4.)
6. If a galaxy has a radial velocity of 2000 km/s and the Hubble constant is 70 km/s/Mpc, how far away is the galaxy? (*Hint:* Use the Hubble law.)
7. If you find a galaxy that is 20 minutes of arc in diameter, and you measure its distance to be 1 Mpc, what is its diameter? (*Hint:* See Reasoning with Numbers 3-1.)
8. You have found a galaxy in which the outer stars have orbital velocities of 150 km/s. If the radius of the galaxy is 4 kpc, what is the orbital period of the outer stars? (*Hints:* 1 pc = 3.08×10^{13} km, and 1 yr = 3.15×10^{7} s.)
9. A galaxy has been found that is 5 kpc in radius and whose outer stars orbit the center with a period of 200 million years. What is the mass of the galaxy? On what assumptions does this result depend? (*Hint:* See Reasoning with Numbers 8-4.)

Media Cluster

ASTRONOMY EXERCISE

Ace Astronomy™ To access the resources in the Media Cluster, log into AceAstronomy at **http://astronomy.brookscole.com/sh9e** and select Chapter 13.

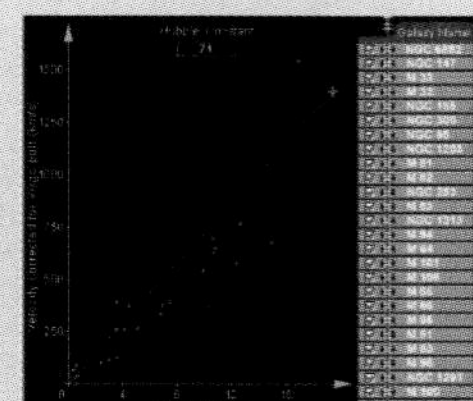

Hubble Relation
This animation lets you plot the red-shifts of several galaxies and make your own observations to find the Hubble constant.

VIRTUAL ASTRONOMY LABS

Lab 16: Astronomical Distance Scales
In this lab you explore some methods for determining distances in astronomy.

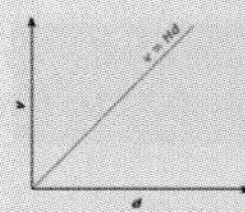

Lab 19: The Hubble Law
This lab discusses the Hubble law, shows how it can be used to estimate distances to remote objects, and illustrates that the law is a consequence of the expansion of space itself.

Critical Inquiries for the Web

1. How far out into the universe can you see Cepheid variables? Research sources on the Internet to find other galaxies whose distances have been found through observation of Cepheids. List the galaxies in which Cepheids have been identified and the distances determined from these data.
2. How does the Milky Way stack up against the other galaxies in the Local Group? Look for information on the other galaxies in our cluster and rank the top six members in order of total mass.
3. In the early 1900s the nature of the "spiral nebulae" was not well understood. In 1920 a "great debate" was held between Harlow Shapley, who held that these objects were relatively nearby swirling clouds of gas, and Heber Curtis, who saw them as distant "island universes." Use the Internet to find information about the debate, outline the lines of evidence used by the two participants to present their views, and explain who was right and who was wrong.
4. Locate a web page dedicated to the Messier Catalog—a list of galaxies, clusters, and nebulae that is often used as a list of targets for small telescopes. Be sure that your destination includes images of the objects. For each of the galaxies in the Messier list, determine its Hubble classification. You may be given this information at the site, but examine the images to see if the features of these galaxies conform to a particular Hubble type.

Exploring *TheSky*

1. Locate the Andromeda Galaxy, also known as M31, and its companion galaxies. Zoom in on it and estimate its angular size compared to the full moon. (*Hint:* Use **Find** under the **Edit** menu.)
2. Take a survey of galaxies and see how many are spiral and how many are elliptical. Is there any selection effect in your method? (*Hint:* Use **Filters** under the **View** menu to turn off everything but **Galaxies** and **Mixed Deep Sky** objects. Make sure the Messier labels are switched on using the **Labels** and **Setup** under the **View** menu.)
3. Locate the Sombrero Galaxy (M104). Study the photographs and discuss this galaxy's special properties. Zoom in on it and estimate its angular size compared to the moon.
4. Study the distribution of galaxies and notice how they cluster together. Can you find the Virgo cluster? Zoom in until more galaxies appear and then scroll north to find the Coma Cluster. Zoom in on Leo to find the cluster there. What other clusters can you find?
5. Describe the galaxy located near the south celestial pole.

Go to the Brooks/Cole Astronomy Resource Center **(http://astronomy.brookscole.com)** for critical thinking exercises, articles, and additional readings from InfoTrac College Edition, Brooks/Cole's online student library.

14 Galaxies with Active Nuclei

Virtually all scientists have the bad habit of displaying feats of virtuosity in problems in which they can make some progress and leave until the end the really difficult central problems.

M. S. LONGAIR
High Energy Astrophysics

Visual (blue) plus radio (red)

DO YOU LIKE fireworks? A barrage of skyrockets in the dark and a few loud bangs make a great show, but compared to astronomical eruptions, an Earthly fireworks display is a whimper. You are about to meet some of the most energetic events in the universe. Nova and supernova explosions are tiny compared to the energy pouring out of the nuclei of certain galaxies. You will discover that the origin of these outbursts is related to the formation and history of galaxies. You are joining an effort to understand these peculiar galaxies that began in the 1950s when radio astronomers first noticed that some galaxies were sources of unusually strong radio waves. Those galaxies were called **radio galaxies.** By the 1970s, astronomers had put spacecraft in orbit that could observe in many other parts of the electromagnetic spectrum, and

The galaxy Fornax A has erupted to expel two lobes visible at radio wavelengths. (Image courtesy of NRAO/AUI)

Guidepost

Looking Back

In the last few chapters, you have explored our own and other galaxies. In earlier chapters, you studied stars, and discovered a process that ejects oppositely directed jets of gas and radiation from such diverse objects as protostars and X-ray binaries. Now you are ready to put those ideas together to study some of the most powerful objects in the universe.

This Chapter

Some galaxies have nuclei that are both small and powerful. Astronomers see some in the act of ejecting jets much more powerful than those from protostars or from X-ray binaries. As you study these active galaxies, you will be combining many of the ideas you have discovered so far to answer five essential questions:

What evidence shows that some galactic nuclei are active?

What is the energy source for this activity?

What triggers the nucleus of a galaxy into activity?

What are the most distant active galaxies?

What can active galaxies tell us about the history of the universe?

Looking Ahead

In the next chapter, you will study the universe as a whole. The active galaxies are the last background data you need before you try to understand the birth and evolution of the entire universe and the galaxies that fill it.

Ace Astronomy™ The AceAstronomy icon throughout the text indicates an opportunity for you to test yourself on key concepts and to explore animations and interactions on the AceAstronomy website at: http://astronomy.brookscole.com/sh9e

they discovered that the radio galaxies were bright at many other wavelengths. Consequently these galaxies are now called **active galaxies.**

The flood of energy pouring out of active galaxies originates in the nuclei of the galaxies, which are referred to as **active galactic nuclei (AGN).** You are joining the research team at a time when the evidence has become overwhelming that these AGN are produced by matter plunging into supermassive black holes. But don't be disappointed that you are joining the team late; there are plenty of mysteries left to be solved.

14-1 Active Galactic Nuclei

ONLY A FEW PERCENT OF GALAXIES are active, so astronomers must search carefully to find examples. Let's begin with some galaxies that look normal at first glance and then meet some of their more twisted cousins.

Seyfert Galaxies

In 1943, Mount Wilson astronomer Carl K. Seyfert published his study of spiral galaxies. Observing at visual wavelengths, Seyfert found that some spiral galaxies have small, highly luminous nuclei (■ Figure 14-1). Today, these **Seyfert galaxies** are recognized by the peculiar spectra of these bright nuclei.

The light from a normal galaxy comes from stars, and its spectrum is the combined spectra of many millions of stars. Consequently, the spectra of normal galaxies contain a few of the main absorption lines found in the spectra of stars. But the spectrum of the Seyfert galaxy nuclei contains broad emission lines of highly ionized atoms. Emission lines suggest a hot, low-density gas, and ionized atoms suggest that the gas is very excited. The width of the spectral lines suggests large Doppler shifts produced by high velocities in the nuclei; gas approaching Earth would produce blueshifted spectral lines and gas going away would produce redshifted lines. The combined light would contain broad spectral lines. The velocities at the center of Seyfert galaxies are roughly 10,000 km/s, about 30 times greater than velocities at the center of normal galaxies.

■ Figure 14-1

Seyfert galaxies are spirals with small, highly luminous nuclei. Modern observations reveal bursts of star formation and ejected gas in some Seyfert galaxies. (AATB; Hubble Heritage Team, AURA/STScI, NASA; NASA, Andrew S. Wilson; Patrick L. Shopbell, Chris Simpson, Thaisa Storchi-Bergmann, and F. K. B. Barbosa; and Martin J. Ward)

Short-exposure photos of NGC1566 reveal a small, bright nucleus.
Visual-wavelength image
Seyfert galaxy NGC7742 has a bright ring of star formation around its active nucleus.
Gas flowing away from nucleus.
Cone of gas being ejected from nucleus
Visual + near-infrared image

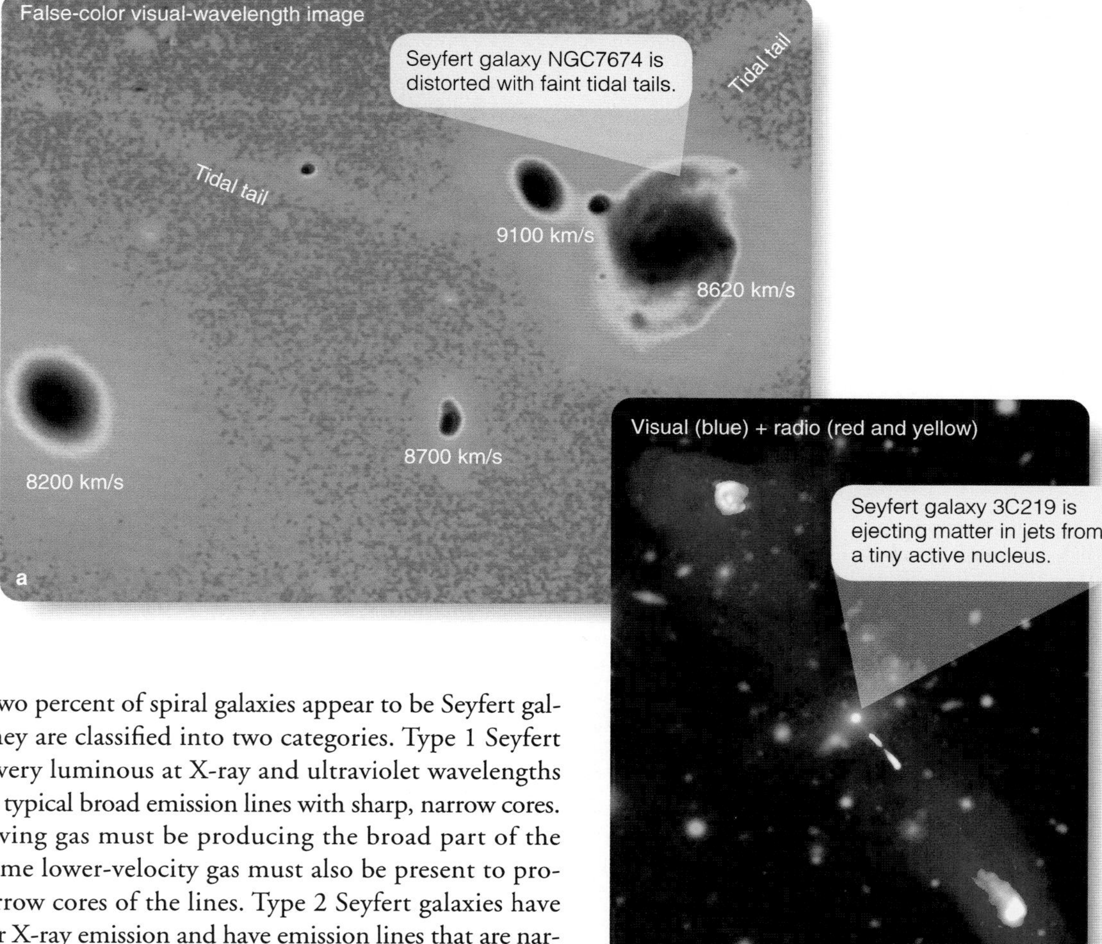

■ **Figure 14-2**

(a) The Seyfert galaxy NGC7674 appears to be distorted by interacting with its companion galaxies. You can be sure they are a group because they all have similar apparent velocities of recession and must be at about the same distance. (John W. Mackenty, Institute for Astronomy, University of Hawaii) (b) Seyfert galaxy 3C219 is the brightest galaxy in a small, compact cluster of galaxies. Compare its radio image with that of the double-lobed radio galaxies discussed in the next section. (NRAO/AUI/NSF)

About two percent of spiral galaxies appear to be Seyfert galaxies, and they are classified into two categories. Type 1 Seyfert galaxies are very luminous at X-ray and ultraviolet wavelengths and have the typical broad emission lines with sharp, narrow cores. Rapidly moving gas must be producing the broad part of the lines, but some lower-velocity gas must also be present to produce the narrow cores of the lines. Type 2 Seyfert galaxies have much weaker X-ray emission and have emission lines that are narrower than those of type 1 Seyfert galaxies but still broader than spectral lines produced by a normal galaxy.

In addition, astronomers discovered that the brilliant nuclei of Seyfert galaxies fluctuate rapidly, especially at X-ray wavelengths. A Seyfert nucleus can change its X-ray brightness by a significant amount in only minutes. As you saw in Chapter 11, an astronomical body cannot change its brightness in a time shorter than the time it takes light to cross its diameter. If the Seyfert nucleus can change in a few minutes, then it cannot be larger in diameter than a few light-minutes. In spite of their small size, the cores of Seyfert galaxies produce tremendous amounts of energy. The brightest emit a hundred times more energy than the entire Milky Way. Something in the centers of these galaxies not much bigger than Earth's orbit produces a galaxy's worth of energy.

The shapes of Seyferts can give you clues to their energy source. They are three times more common in interacting pairs of galaxies than in isolated galaxies. Also, about 25 percent have peculiar shapes suggesting tidal interactions with other galaxies (■ Figure 14-2a). This statistical evidence (**Window on Science 14-1**) hints that Seyfert galaxies may have been triggered into activity by collisions or interactions with companions. Some Seyferts are expelling matter in oppositely directed jets (Figure 17-2b), a geometry you have seen on smaller scales when matter flows into neutron stars and black holes and forms an accretion disk.

All of this evidence leads modern astronomers to conclude that the cores of Seyfert galaxies contain supermassive black holes—black holes with masses as high as a billion solar masses The gas in the centers of Seyfert galaxies is traveling so fast it would escape from a normal galaxy. Only very large central masses could exert enough gravity to hold the gas inside the nuclei, and that suggests a supermassive black hole. Encounters with other galaxies could throw matter into the black hole; and, as you have seen in Chapter 11, large amounts of energy can be liberated by

It Wouldn't Stand Up in Court: Statistical Evidence

Notice that some scientific evidence is statistical. For example, we might note that Seyfert galaxies are three times more likely to have a nearby companion than a normal galaxy is. This is statistical evidence because you can't be certain that any single Seyfert galaxy will prove to have a companion. Yet the probability is higher than if it were a normal galaxy, and that leads you to suspect that interactions between galaxies are involved. Such statistical evidence can tell you something in general about the cause of Seyfert eruptions.

Statistical evidence is common in science, but it is inadmissible in most courts of law. The American legal system is based on the principle of reasonable doubt, so most judges would not allow statistical evidence of guilt. For example, plaintiffs have had great trouble suing tobacco companies for causing their cases of lung cancer, even though there is a clear statistical link between smoking and lung cancer. The statistics say something in general about smoking but cannot be used to prove that any specific case of lung cancer was caused by smoking. There are other causes of lung cancer, so there is always a reasonable doubt as to the cause of any specific case.

You can use statistical evidence in science if you do not demand that the statistics demonstrate anything conclusively about any single example. To continue our example of Seyfert galaxies, we don't demand that the statistics predict that any specific galaxy has a companion. Rather, we use the statistical evidence to gain a general insight into the cause of active galactic nuclei. That is, we are trying to understand galaxies as a whole, not to convict any single galaxy.

Of course, if you surveyed only a few galaxies, your statistics might not be very good, and a critic might be justified in making the common complaint, "Oh, that's only statistics." For example, if you surveyed only four galaxies and found that three had companions, your statistics would not be very reliable. But if you surveyed 1000 galaxies, your statistics could be very good indeed, and your conclusions could be highly significant.

Thus, scientists can use statistical evidence if it passes two tests. It cannot be used to draw conclusions about specific cases, and it must be based on large enough samples so the statistics are significant. With these restrictions, statistical evidence can be a powerful scientific tool.

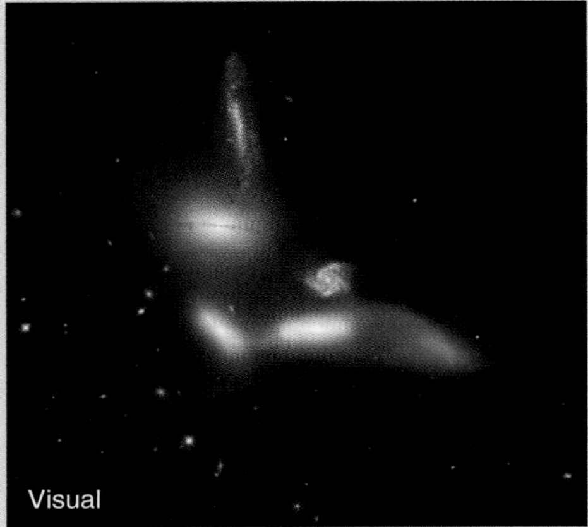

Guilt by association is statistical evidence and can be applied to galaxies. (NASA)

matter flowing into a black hole. We will expand this theory later in this chapter, but first we must search for more evidence of supermassive black holes by turning our attention to galaxies that emit powerful radio signals.

Double-Lobed Radio Sources

Beginning in the 1950s, radio astronomers found that some sources of radio energy in the sky consisted of pairs of radio-bright regions. When optical telescopes studied the locations of these **double-lobed radio sources,** they revealed galaxies located between the two lobes. Apparently, the galaxies were producing the radio lobes.

As you study **Cosmic Jets and Radio Lobes** on pages 310 and 311, notice three important points. First, the geometry suggests that radio lobes are inflated by jets of excited gas emerging from the central galaxy. This has been called the **double-exhaust model.** Second, notice how the presence of synchrotron radiation and hot spots is evidence in support of the model. The third thing you should notice is how matter falling into a central black hole can produce these jets. You have seen similar jets produced by rotating disks in the case of protostars and neutron stars; although the details are not understood, it is presumably the same process that produces all of these jets. The complex shapes of jets shown in the high-resolution radio maps of galaxies such as 3C449 (the 449th source in the *Third Cambridge Catalog of Radio Sources*) can be understood through the motions of the galactic nuclei.

Study the image of NGC5128 on page 310 for a further clue. In visual-wavelength images, the galaxy looks like a round elliptical galaxy surrounded by a belt of dust and gas. Doppler shifts reveal that the spherical cloud of stars is rotating about an axis perpendicular to the axis of rotation of the dust disk. That is, the galaxy appears to have two different axes of rotation. You will remember from the previous chapter that different parts of a galaxy rotating in different directions is evidence that the galaxy was produced by a merger. Furthermore, bright blue stars have been identified in the dust ring; those stars must have formed recently, and you recall that mergers can trigger star formation. NGC5128 appears to be the product of the merger of two galaxies. That is an important discovery because, as you will see later, interactions and mergers between galaxies can trigger eruptions.

Drawing on all of the evidence available and guided by their understanding of the behavior of light and matter, astronomers have constructed a convincing scientific argument (**Window**

Scientific Arguments

An argument can be a shouting match, but another definition of the word is "a discourse intended to persuade." Scientists construct arguments as part of the business of science not because they want to persuade others they are right, but because they want to test their ideas. For example, you can construct a scientific argument to show that the radio lobes that flank some active galaxies have their origin in jets coming from the cores of the galaxies. If you do your best and your argument is not persuasive, you begin to suspect you are on the wrong track. But if your argument seems convincing, you gain confidence that you are beginning to understand radio lobes.

A scientific argument is a logical presentation of evidence and theory with interpretations and explanations that help us understand some aspect of nature. Geologists might construct an argument to explain volcanism on midocean islands, and the argument could include almost anything—maps, for example, or mineral samples compared with seismic evidence and mathematical models of subsurface magma motions. Such an argument might involve the physics of radioactive decay or observations of the shapes of bird beaks on the islands. The scientists would be free to include any evidence or theory that helps persuade, but they must observe one fundamental rule of scientific argument: They must be totally honest. The purpose of a scientific argument is to test your understanding, not to win votes or sell soap. Dishonesty in a scientific argument is self-deluding, and scientists consider such dishonesty the worst possible behavior.

On the other hand, scientists are human, and they often behave no better than other humans. They sometimes defend their own theories and attack opposing theories more vigorously than the evidence warrants. They may unintentionally overlook contradictory evidence or unconsciously misinterpret data. Of course, if a scientist did these things intentionally, you would be shocked because he or she would be violating the most serious rules of honesty in science. In the struggle to understand how nature works, however, it is not surprising to see that human passions rise and that scientific debate becomes heated. That is why science is so exciting—it is a struggle to understand nature. Nonetheless, no matter how scientists behave in the excitement of a controversy, they all aspire to that ideal of the scientific argument—the cool, logical presentation of evidence that leads to a new understanding of nature.

Much of this book or any other scientific book consists of scientific arguments—logical presentations of evidence. As you read about any science, look for the arguments and practice organizing and presenting them to test your own understanding. The "Inquire Review Analyze" that ends each chapter section in this book is intended to illustrate how arguments are constructed. Use the same plan when you develop your own arguments.

Geologists preparing a logical argument to explain volcanism must be scientifically honest and compare theories with all known evidence.

on Science 14-2) that active galactic nuclei are powered by supermassive black holes. Now we must try to understand what these supermassive energy machines are like.

Exploring Supermassive Black Holes

In the case of active galaxies, "the really difficult, central problem" is the source of the energy. The evidence seems conclusive that the energy is produced when matter flows into supermassive black holes. You have seen in earlier chapters how matter flowing into a stellar-mass black hole forms a hot accretion disk and can emit jets in opposite directions. It is not surprising that the hot accretion disk around a supermassive black hole can emit much more powerful jets.

Can you be sure the energy sources at the centers of active galaxies are really supermassive black holes? Let's review the evidence, and you be the judge. The evidence takes the form of observations of size and observations of motion.

Earlier in this chapter, you saw evidence that the cores of some active galaxies are very small. The cores of Seyfert galaxies fluctuate in brightness over very short time periods, and that must mean they are very small—so small that astronomers know of nothing except a supermassive black hole that could produce so much energy in such a small region.

The motion of stars and gas near the centers of galaxies is important because it can tell you the amount of mass located there. Consider the giant elliptical galaxy M87 (page 284). It has a very small, bright nucleus and a visible jet of matter 1800 pc long racing out of its core (■ Figure 14-3). Radio observations show that the nucleus must be no more than a light-week in diameter. This evidence is suggestive, but observations of motions around the center are decisive. A high-resolution image shows that the core lies at the center of a spinning disk, and the jet lies along the axis of the disk. Only 60 ly from the center, the gas in the disk is orbiting at 750 km/s. Substituting radius and velocity into the equation for circular velocity (see Reasoning with Numbers 4-1)

Cosmic Jets and Radio Lobes

Many radio sources consist of two bright lobes — double-lobed radio sources — with a galaxy, often a peculiar or distorted galaxy, located between them. Evidence suggests these active galaxies are emitting jets of high-speed gas that inflate the lobes as cavities in the intergalactic medium. Where the jets impact the far side of the cavities, they create hot spots.

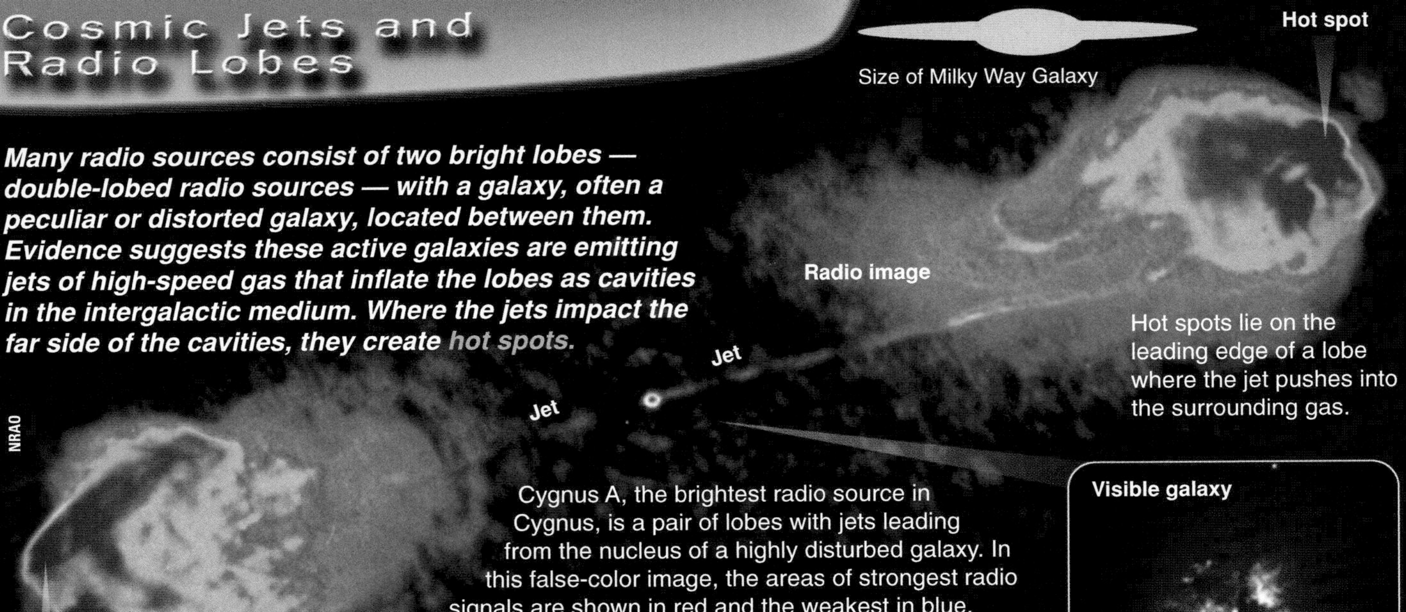

Size of Milky Way Galaxy

Hot spot

Radio image

Jet

Jet

Hot spot

NRAO

Hot spots lie on the leading edge of a lobe where the jet pushes into the surrounding gas.

Cygnus A, the brightest radio source in Cygnus, is a pair of lobes with jets leading from the nucleus of a highly disturbed galaxy. In this false-color image, the areas of strongest radio signals are shown in red and the weakest in blue. Because the radio energy detected is synchrotron radiation, astronomers conclude that the jets and lobes contain very-high-speed electrons, usually called relativistic electrons, spiraling through magnetic fields about 1000 times weaker than Earth's field. The total energy in a radio lobe is about 10^{53} J — what you would get if you turned the mass of a million suns directly into energy.

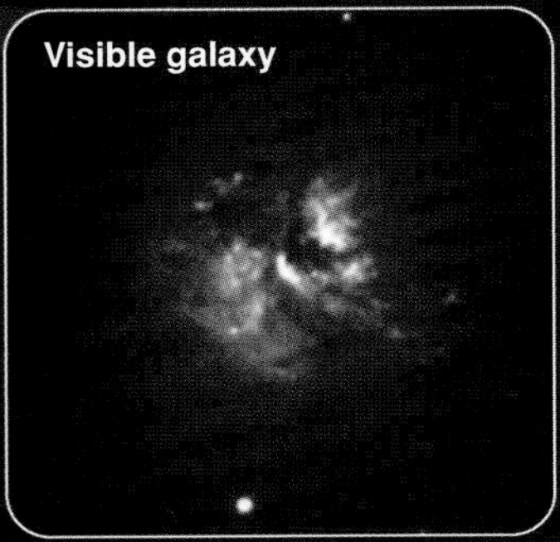

Visible galaxy

Used with permission, Fosbury R. A. E., Vernet, J., Villar-Martin, M., Cohen, M. H., Ogle, P. M., Tran, H. D. & Hook, R. N. 1998, Optical continuum structure of Cygnus A. "KNAW colloquium on: The most distant radio galaxies," Amsterdam, 15–17 October 1997, Roettgering H, Best P and Lehnert M eds, Reidel, astro-ph/9803310

Radio galaxy NGC5128 lies between two radio lobes, and the combined radio and X-ray image at the right shows a high-energy jet at the very center of the galaxy pointing to the upper left into the northern radio lobe.

NASA/CXC/NRAO/VLA

Nucleus

Radio = Red
X-ray = Blue

Infrared

E. Schreier, STScI, NASA

An infrared image of the center of the galaxy NGC5128 reveals a small, bright disk of hot gas surrounding the nucleus.

If the outer radio lobes of Centaurus A were visible to your eyes, they would look 10 times larger than the full moon.

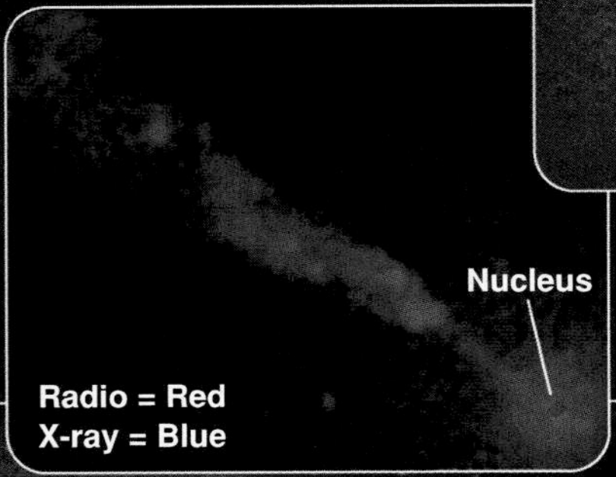

X-ray: NASA/CXC/M. Karovska et al., Radio: NRAO/VLA/Schiminovich, et al., Radio: NRAO/VLA/J. Condon et al., Optical: Digitized Sky Survey U.K. Schmidt Image/STScI.

Radio (green)

X-ray (blue)

Visual + radio + X-ray

NGC5128 appears to be a giant elliptical galaxy experiencing a collision with a dusty spiral galaxy. It is known to radio astronomers as Centaurus A, an inner pair of radio lobes and a larger outer pair.

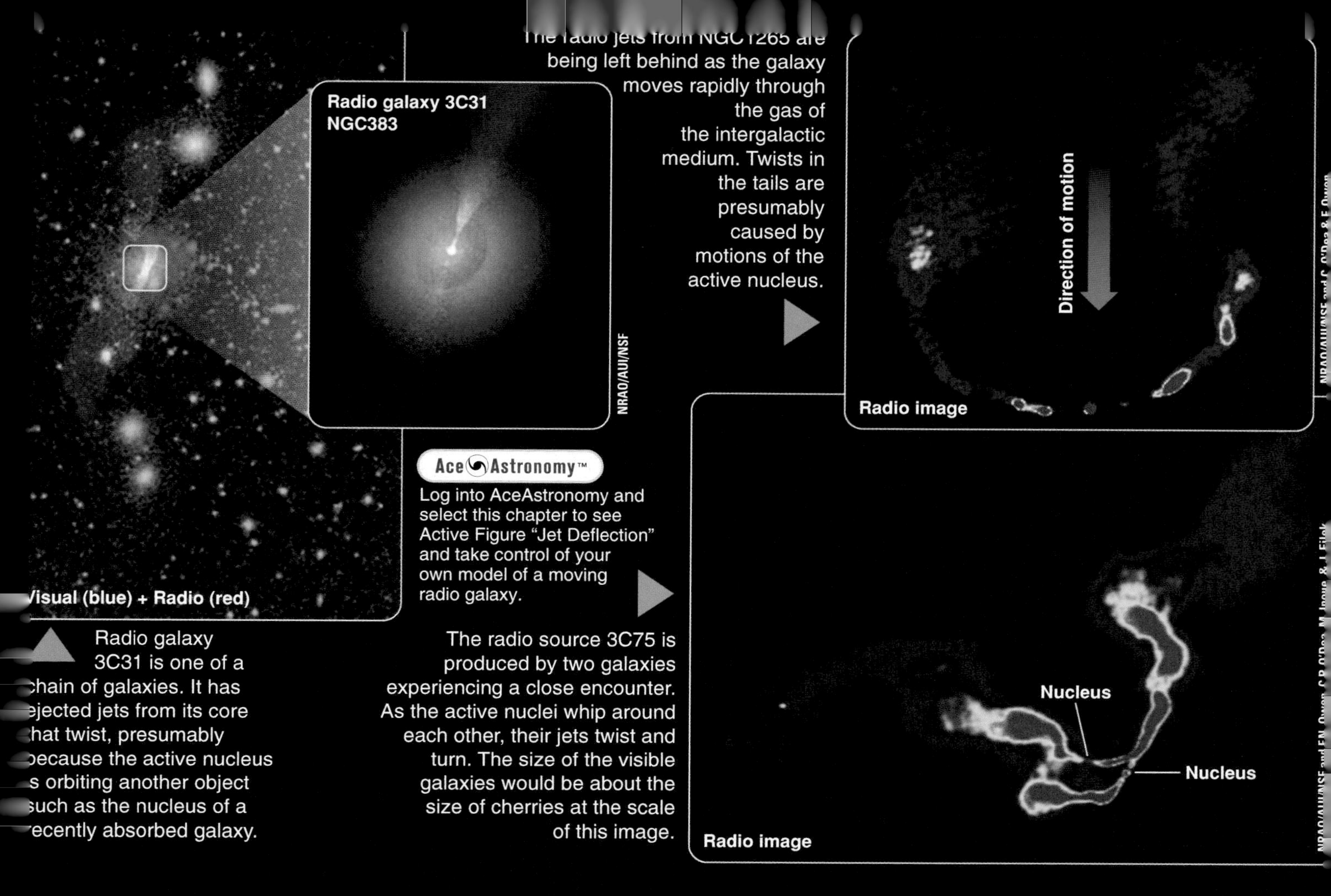

The radio jets from NGC1265 are being left behind as the galaxy moves rapidly through the gas of the intergalactic medium. Twists in the tails are presumably caused by motions of the active nucleus.

Radio galaxy 3C31 is one of a chain of galaxies. It has ejected jets from its core that twist, presumably because the active nucleus is orbiting another object such as the nucleus of a recently absorbed galaxy.

Ace Astronomy™

Log into AceAstronomy and select this chapter to see Active Figure "Jet Deflection" and take control of your own model of a moving radio galaxy.

The radio source 3C75 is produced by two galaxies experiencing a close encounter. As the active nuclei whip around each other, their jets twist and turn. The size of the visible galaxies would be about the size of cherries at the scale of this image.

High-energy jets appear to be caused by matter flowing into a supermassive black hole in the core of an active galaxy. Conservation of angular momentum forces the matter to form a whirling accretion disk around the black hole. How that produces a jet is not entirely understood, but it appears to involve magnetic fields that are drawn into the accretion disk and tightly wrapped to eject high-temperature gas. The twisted magnetic field confines the jets in a narrow beam and causes synchrotron radiation.

Jets from active galaxies may have velocities from thousands of kilometers per second up to a large fraction of the speed of light. Compare this with the jets in bipolar flows, where the velocities are only a few hundred kilometers per second. Active-galaxy jets can be millions of light-years long. Bipolar-flow jets are typically a few light-years long. The energy is different, but the geometry is the same.

Black hole

Magnetic field lines

Accretion disk

Excited matter traveling at very high speeds tends to emit photons in the direction of travel. Consequently, a jet pointed roughly toward Earth will look brighter than a jet pointed more or less away. This may explain why some radio galaxies have one jet brighter than the other, as in Cygnus A shown at the top of the opposite page. It may also explain why some radio galaxies appear to have only one jet. The other jet may point generally away from Earth and be too faint to detect, as in the case of NGC5128, also shown on the opposite page.

Adapted from a diagram by Ann Field, NASA, STScI

■ Figure 14-3

The jet at the center of giant elliptical galaxy M87 is only a few percent the diameter of the entire galaxy. The jet is emerging at nearly half the speed of light. Hubble Space Telescope images reveal a small, rapidly spinning disk at the center of the galaxy with the jet emerging parallel to the axis of the disk. Very-high-resolution radio observations show that the jet originates in a tiny active region at the very center of the disk. (M87: AURA/NOAO/NSF; jet: NASA/STScI; radio image: NRAO/AUI/NSF and J. Biretta)

tells you that the central mass must be roughly 2.4 billion solar masses. Only a supermassive black hole could pack so much mass into such a small region.

For another example, look at ■ Figure 14-4. NGC4261 is a distorted galaxy ejecting jets into a pair of radio lobes. High-resolution images made by the Hubble spacecraft reveal a dusty disk surrounding the bright central core. The axis of the disk points along the jets into the radio lobes. In the case of NGC4261, you can see the energy source in the act of producing jets and radio lobes.

In astronomy, evidence means observations, and the observations seem conclusive. Active galactic nuclei are powered by supermassive black holes. That probably leaves you with a few questions, and astronomers are struggling to answer those same

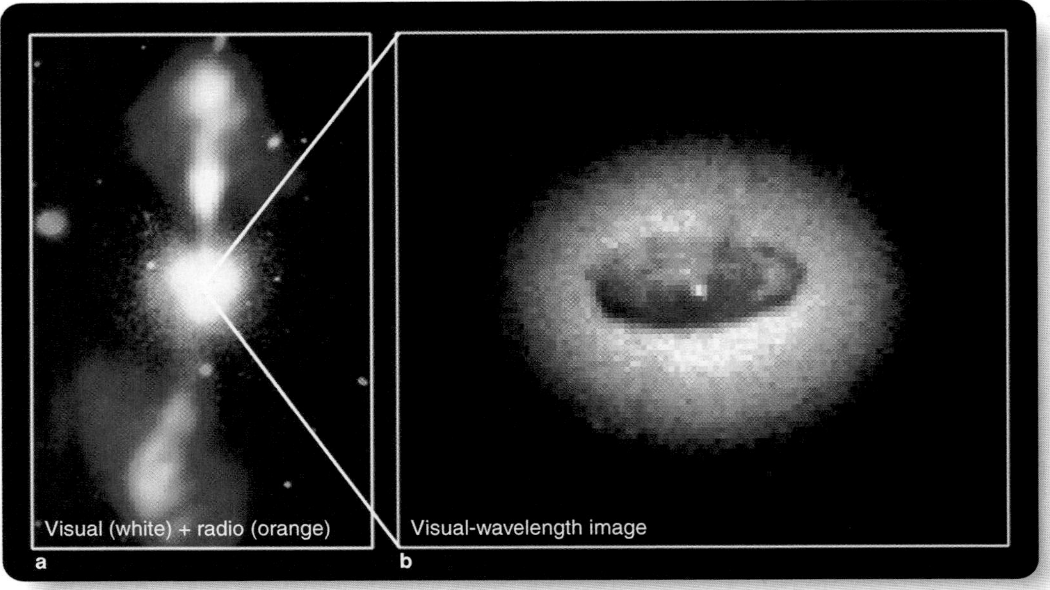

■ Figure 14-4

(a) Galaxy NGC4261 looks like a distorted fuzzy blob in an Earth-based photograph at visible wavelengths (white), but radio telescopes reveal jets leading into radio lobes (orange). (b) A Hubble Space Telescope image of the core of the galaxy reveals a bright central spot surrounded by a disk with its axis pointing at the radio lobes. (L. Ferrarese, Johns Hopkins University, and NASA)

questions. How do supermassive black holes and their accretion disks produce eruptions? Where did the supermassive black holes come from?

The Search for a Unified Model

When a field of research is young, scientists often find many seemingly different phenomena such as double-lobed radio galaxies, Seyfert galaxies, cosmic jets, and so on. As the research matures, the scientists begin seeing similarities and eventually are able to unify the different phenomena as different aspects of a single process. This is the real goal of science, to organize evidence and theory in logical arguments that explain how nature works (Window on Science 14-2). Astronomers studying active galaxies are now struggling to find a **unified model** of active galaxy cores.

It seems clear that the cores of active galaxies contain supermassive black holes surrounded by accretion disks that are extremely hot near the black hole but cooler farther out. Theory predicts that the central cavity in the disk where the black hole lurks is very narrow but that the disk there is "puffed up" and thick. This means the black hole may be hidden deep inside this narrow central well. The hot inner disk seems to be the source of the jets often seen coming out of active galaxy cores, but the process by which jets are generated is not understood. The outer part of the disk, according to calculations, is a fat, dense doughnut (a torus) of dusty gas (■ Figure 14-5).

What you see when you view the core of an active galaxy must depend on how this disk is tipped with respect to your line of sight. You should note, at this point, that the accretion disk may be tipped at a steep angle to the plane of a galaxy, so just because you see a galaxy face-on doesn't mean you are looking at the accretion disk face-on. Thus the important factor is not the inclination of the galaxy but the inclination of the accretion disk.

For example, in 1968 astronomers realized that an object they had thought was an irregular variable star in the constellation Lacerta was actually the core of an active galaxy. The visible spectrum is featureless, but the spectrum of surrounding nebulosity is that of a giant elliptical galaxy. The object, and others like it known as **BL Lac objects** or **blazars,** are 10,000 times more luminous than the Milky Way Galaxy and fluctuate in only hours. They are believed to be the cores of active galaxies in which you happen to be looking directly into the brilliant jets. In this case, the accretion disk is apparently face-on, and you are looking directly into the central cavity around the black hole—down the dragon's throat.

If the accretion disk is tipped slightly, you may not be looking directly into the jet, but you may be able to see some of the intensely hot gas in the central cavity. This region emits broad spectral lines because the gas is orbiting at high velocities and the high Doppler shifts smear out the lines. Seyfert 1 galaxies may be explained by this phenomenon.

If you view the accretion disk from the edge, you cannot see the central area at all because the thick dusty torus blocks your view. Instead you see radiation emitted by gas lying above and below the central disk. Because this gas is farther from the center, it orbits more slowly and has smaller Doppler shifts. Thus you see narrower spectral lines. This might account for the Seyfert 2 galaxies.

This unified model is far from complete. The actual structure of accretion disks is poorly understood, as is the process by which the disks produce jets. Furthermore, the spiral Seyfert galaxies

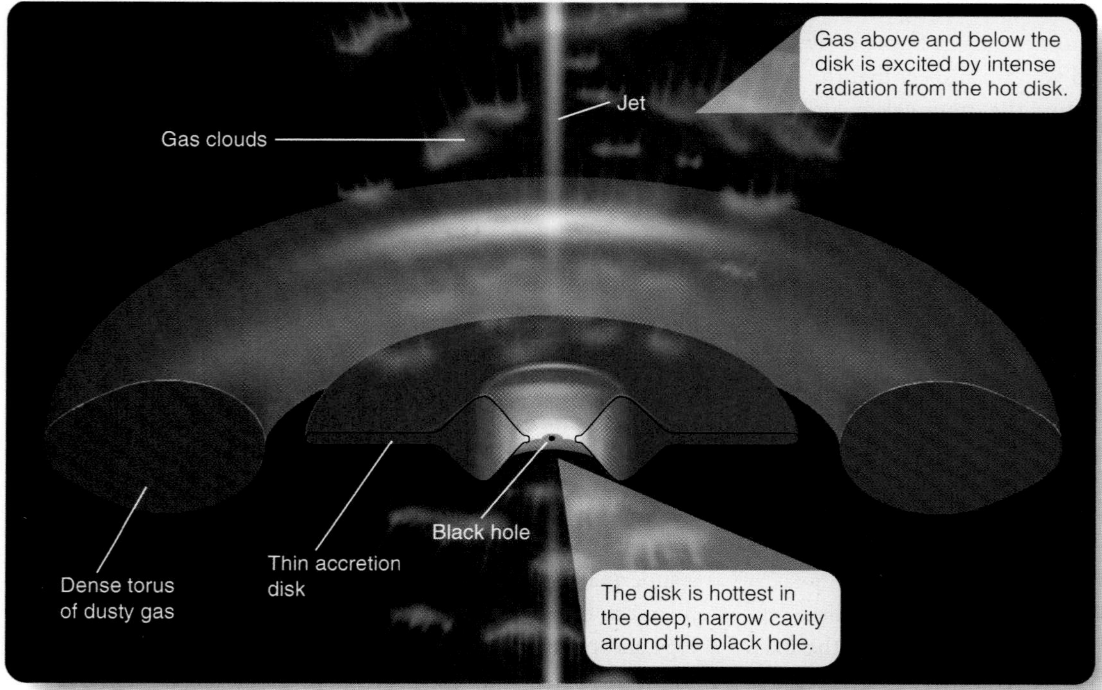
Gas above and below the disk is excited by intense radiation from the hot disk.
Jet
Gas clouds
Black hole
Thin accretion disk
Dense torus of dusty gas
The disk is hottest in the deep, narrow cavity around the black hole.

■ **Figure 14-5**

A sketch of the accretion disk around a supermassive black hole. Matter flowing inward passes first through a large, opaque torus; then into a thinner, hotter disk; and finally into a hot, narrow cavity around the black hole. This diagram is not to scale. The central cavity may be only 0.01 pc in radius, while the outer torus may be 1000 pc in radius.

are clearly different from the giant elliptical galaxies that have double radio lobes. Also, our discussion of unification has ignored important factors such as differences in the rate at which mass flows inward and the influence of magnetic fields. Unification does not explain all of the differences among active galaxies. Rather, it is a model that provides some clues to what is happening in the cores of active galaxies.

The Origin of Supermassive Black Holes

Naturally you are wondering where these supermassive black holes came from, and that question leads us to understand why they erupt and gives us an insight into the birth of galaxies.

In the previous chapter, you learned that most galaxies seem to contain a supermassive black hole at their center. Even our own Milky Way Galaxy contains a central black hole. But only a few percent of galaxies have active galactic nuclei. That must mean that most of the supermassive black holes are dormant; presumably they are not being fed large amounts of matter. A slow trickle of matter flowing into the supermassive black hole at the center of our galaxy could explain the mild activity seen there. But it would take a larger meal to trigger an eruption such as those seen in active galactic nuclei.

What could trigger a supermassive black hole to erupt? The answer is something that you studied back in Chapter 4—tides. You have seen how tides twist interacting galaxies and rip matter away into tidal tails, but mathematical models show that those same interactions can throw matter inward. A sudden flood of matter flowing into the accretion disk around a supermassive black hole would trigger it into eruption. This explains why active galaxies are often distorted; they have been twisted by tidal forces as they interacted or merged with another galaxy. Some active galaxies have nearby companions, and you can suspect that the companions are guilty of tidally distorting the other galaxy and triggering an eruption.

Tidal forces are implicated again when you imagine what happens to a lump of matter passing close to a supermassive black hole. ■ Figure 14-6 shows how a passing star would be ripped apart and, at least partially, consumed by the black hole. Inflowing gas, dust, and an occasional star would be an energy feast for a supermassive black hole.

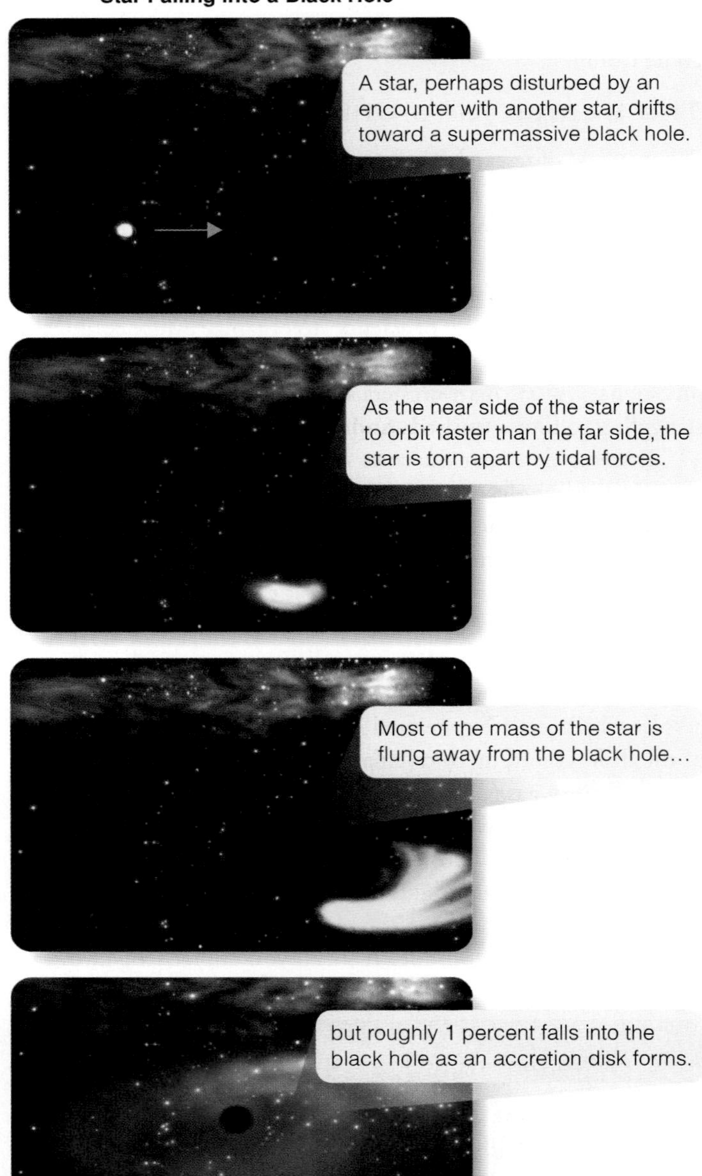

Star Falling into a Black Hole
A star, perhaps disturbed by an encounter with another star, drifts toward a supermassive black hole.
As the near side of the star tries to orbit faster than the far side, the star is torn apart by tidal forces.
Most of the mass of the star is flung away from the black hole...
but roughly 1 percent falls into the black hole as an accretion disk forms.

■ Figure 14-6

Orbiting X-ray telescopes detected an X-ray flare in the galaxy RX J1242-11. Equaling the energy of a supernova explosion, the flare was evidently caused when a star wandered too close to the 100-million-solar-mass black hole at the center of the galaxy. When tidal forces ripped the star apart, some of the mass fell into the black hole, and the rest was flung away. That sudden meal for the black hole was enough to trigger an outburst. (ESA)

A few dozen supermassive black holes have measured masses, and their masses are correlated with the masses of the host galaxies' nuclear bulges. In each case, the mass of the black hole is about 0.5 percent the mass of the surrounding nuclear bulge. But there is no relationship between the masses of the black holes and the masses of the disks of galaxies. This bit of statistics provides an exciting insight into how galaxies form.

Apparently galaxies form their nuclear bulges and disks separately. As the nuclear bulge forms, a small fraction of the mass loses orbital momentum and sinks to the middle where it forms a supermassive black hole. All of that matter flowing together to form the black hole would release a tremendous amount of energy and trigger a violent eruption. Long ago, when galaxies were actively forming, the birth of the nuclear bulges must have triggered active galactic nuclei. The formation of a nuclear bulge was evidently a violent process.

Recall that when we discussed the formation of the Milky Way Galaxy, we suspected that the disk of gas and dust formed later as matter settled into the galaxy. By that stage, the nuclear bulge and central black hole were formed, and the gradual development of the disk didn't trigger a violent eruption. "Disk formation is wimpy," said one astronomer.

Now you can see how supermassive black holes were formed and how they erupt. Many are triggered into eruption by interactions or mergers with other galaxies. Some eruptions, however, may have been triggered by the formation of the nuclear bulge. That would have happened long ago, and astronomers would only see it in galaxies at great look-back times. Are such eruptions seen? That question introduces you to one of the biggest adventures of modern astronomy—the subject of the next section.

Inquire | Review | Analyze

Construct an argument to show that double-lobed radio galaxies and Seyfert galaxies have similar energy sources.

Observations show that double-lobed radio sources are produced by two jets flowing out of a galaxy's core and inflating the radio lobes. The evidence of small size and very-high-speed motions in the core shows that these galaxies must contain large amounts of mass in small regions. Seyfert galaxies also have small nuclei, as shown by their rapid fluctuations, and the high velocities in their cores imply that the cores must contain very large amounts of mass. Consequently both kinds of objects are suspected to contain supermassive black holes at their centers.

The preceding paragraph is a quick summary of a logical argument citing evidence and theory to reach a conclusion. Construct an argument to propose that galaxy interactions are necessary to trigger active galactic nuclei.

■ ■ ■

Connections: Observational evidence and basic theory have led you to the unified model of active galactic nuclei. Although the model seems successful, there are still difficulties, and astronomers continue to struggle to understand the details. Nevertheless, the unified model may help you understand the mysterious quasars.

14-2 Quasars

THE LARGEST TELESCOPES DETECT MULTITUDES of faint points of light with peculiar emission spectra, objects called **quasars** (also known as *quasi-stellar objects,* or *QSOs*). Although astronomers now recognize quasars as some of the most distant visible objects in the universe, they were a mystery when they were first identified. The discovery of these objects in the 1960s and the struggle to understand them comprise a classic example of how scientists explore hypotheses, gather evidence, and build confidence in new ways to understand nature. Let's begin by telling that story.

The Discovery of Quasars

In the early 1960s, radio interferometers (see Chapter 5) showed that a number of radio sources are much smaller than normal radio galaxies. Photographs of the location of these radio sources did not reveal a galaxy, not even a faint wisp, but rather a single starlike point of light. The first of these objects so identified was 3C48, and later the source 3C273 was added. Though these objects emitted radio signals like those from radio galaxies, they were obviously not normal radio galaxies. Even the most distant photographable galaxies look fuzzy, but these objects looked like stars. Their spectra, however, were totally unlike stellar spectra, so the objects were called quasi-stellar objects (■ Figure 14-7).

For a few years, the spectra of quasars were a mystery. A few unidentifiable emission lines were superimposed on a continuous spectrum. But in 1963, Maarten Schmidt at Hale Observatories tried redshifting the hydrogen Balmer lines to see if they could be made to agree with the lines in 3C273's spectrum. At a redshift of 15.8 percent, three lines clicked into place (■ Figure 14-8). Other quasar spectra quickly yielded to this approach, revealing even larger redshifts.

The redshift z is the change in wavelength $\Delta\lambda$ divided by the unshifted wavelength λ_0:

$$\text{redshift} = z = \frac{\Delta\lambda}{\lambda_0}$$

Quasar spectra contain bright emission lines. The brightest are the Balmer lines of hydrogen, but other spectral lines also appear. The redshifts of quasars can be so large that spectral lines can be shifted completely out of the visible spectrum, and lines in the ultraviolet can be shifted into the visible spectrum. ■ Figure 14-9 shows the spectra of four quasars with relatively low redshifts in which the H_α Balmer line normally seen in the red part of the spectrum is shifted into the infrared. Quasars have been found with even larger redshifts.

To understand the significance of these large redshifts and the large velocities of recession they imply, you must refer to the Hubble law. Recall that the Hubble law states that galaxies have apparent velocities of recession proportional to their distances, and thus the distance to a quasar is equal to its apparent velocity of recession divided by the Hubble constant. The large redshifts of the quasars imply that they must be at great distances.

The redshift of 3C273 is 0.158, and the redshift of 3C48 is 0.37. These are large redshifts but not as large as the largest then known for galaxies, about 1.0. Soon, however, quasars were found with redshifts much larger than that of any known galaxy. Some quasars are evidently so far away that galaxies at those distances are very difficult to detect. Yet the quasars are easily photographed. This leads to the conclusion that quasars must have 10 to 1000 times the luminosity of a large galaxy. Quasars must be ultraluminous.

Soon after quasars were discovered, astronomers detected fluctuations in brightness over times as short as a few days. Recall

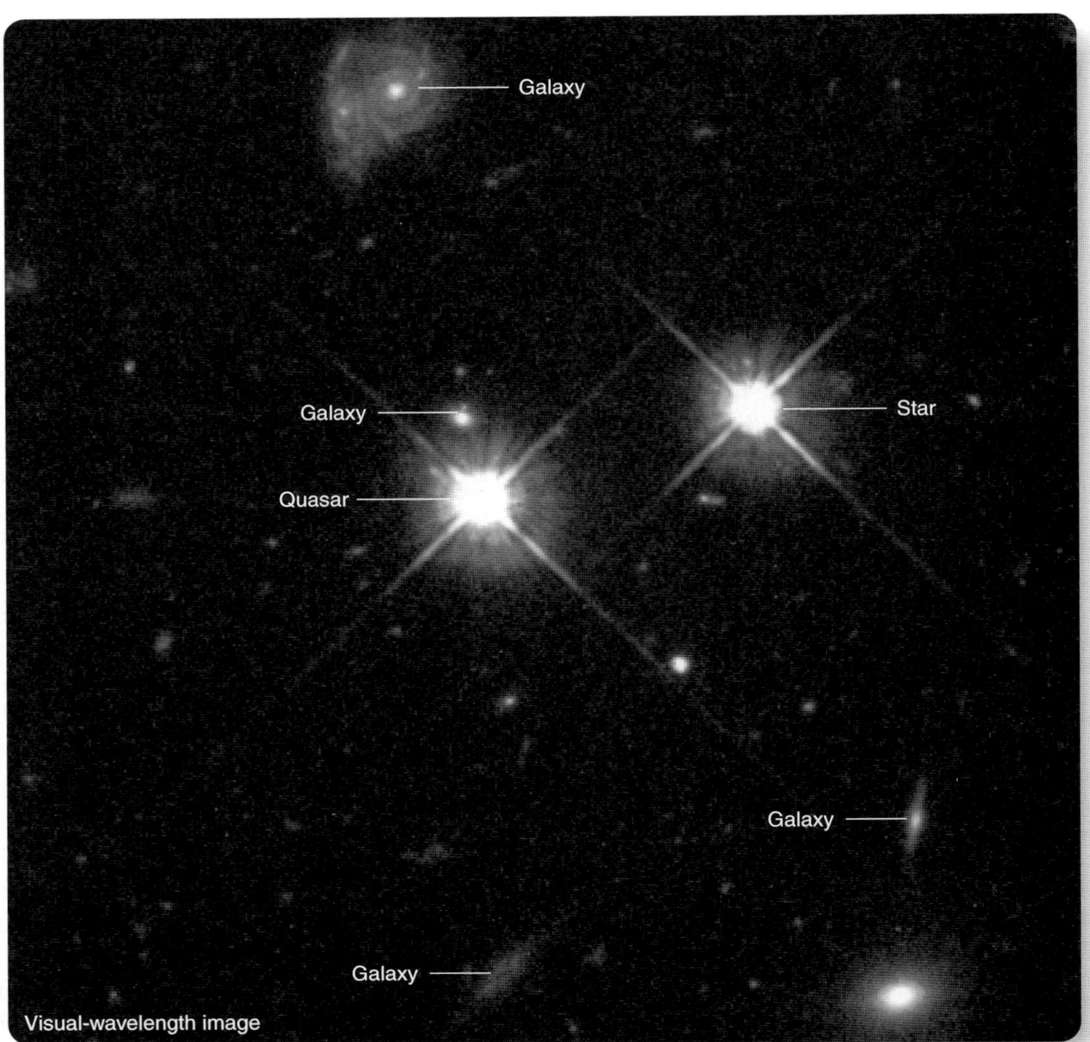

■ Figure 14-7

Quasars have starlike images clearly different from the images of distant galaxies. The spectra of quasars are unlike the spectra of stars or galaxies. (C. Steidel, Caltech, NASA)

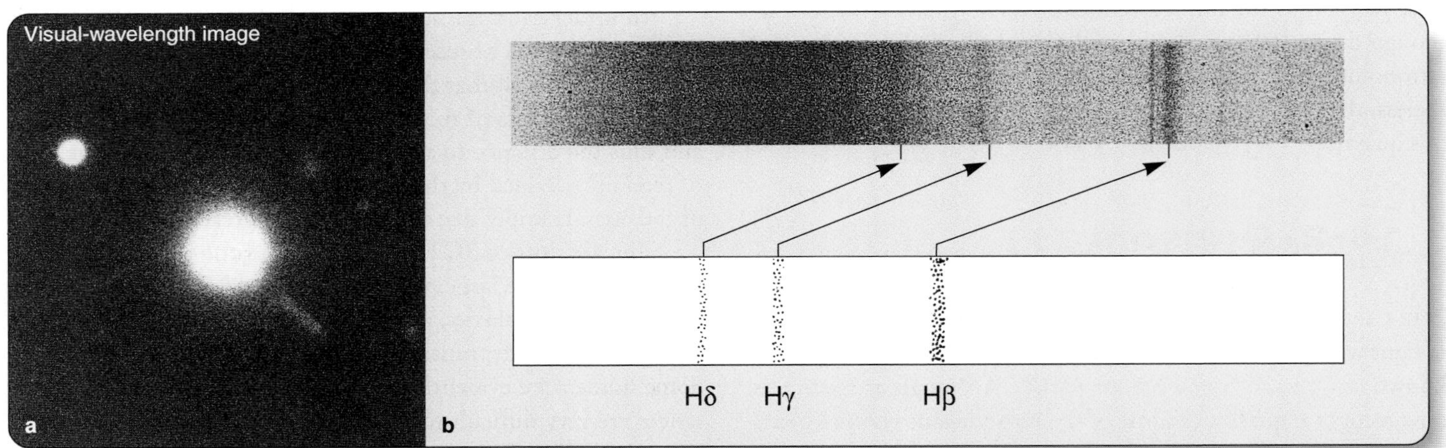

■ Figure 14-8

(a) This image of 3C273 shows the bright quasar at the center surrounded by faint fuzz. Note the jet protruding to lower right. (b) The spectrum of 3C273 (top) contains three hydrogen Balmer lines redshifted by 15.8 percent. The drawing shows the unshifted positions of the lines. (Courtesy Maarten Schmidt)

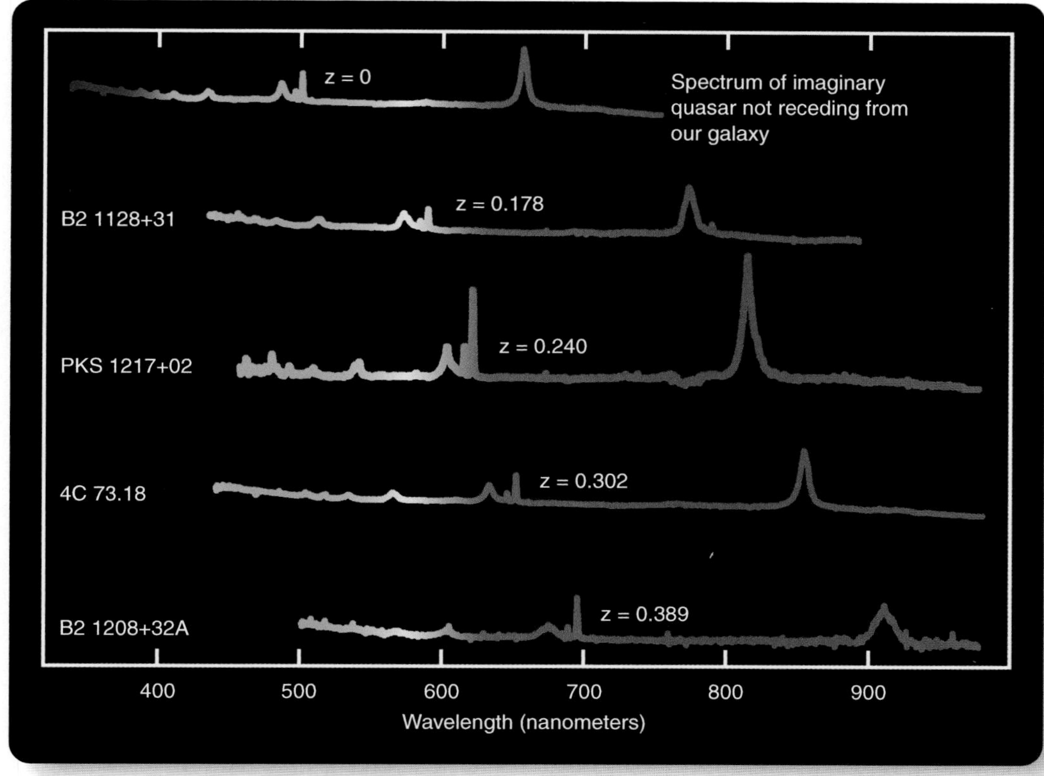

■ **Figure 14-9**

Spectra of four quasars are compared here with an idealized spectrum for an imaginary quasar that has no redshift. The first three Balmer lines of hydrogen are visible, plus lines of other atoms, but the redshifts of the quasars move these lines to longer wavelengths. Nevertheless, the relative spacing of the spectral lines is unchanged, and astronomers can recognize the lines even with a large redshift. (C. Pilachowski, M. Corbin, AURA/NOAO/NSF)

from our discussion of pulsars (Chapter 11) that an object cannot change its brightness appreciably in less time than it takes light to cross its diameter. The rapid fluctuations in quasars showed that they are small objects, not more than a few light-days or light-weeks in diameter.

Thus, by the late 1960s astronomers faced a problem: How could quasars be ultraluminous but also be very small? What could make 10 to 1000 times more energy than a galaxy in a region as small as our solar system? Since that time, evidence has accumulated that quasars are the active cores of very distant galaxies, and the rest of this chapter will discuss that evidence. The distance to quasars is so important to our discussion and so much larger than the distances considered before that you need to pay special attention to how astronomers estimate the distances to these objects.

The Distance to Quasars

By now you may have detected a slipup in the logic of the previous section. Some quasars have redshifts greater than 1, and the largest known are over 6. If you substitute those redshifts into the Doppler formula you studied in Chapter 6, you get velocities greater than the velocity of light, and that is supposed to be impossible. We need to straighten this problem out before we can go on discussing quasars.

Let's begin by eliminating a common but incorrect explanation. For over a generation, astronomy textbooks have listed a version of the Doppler formula called the relativistic Doppler formula. Einstein derived it to describe the Doppler effect for objects traveling at very high velocities. Tradition aside, that is the wrong formula because the redshifts of the galaxies and the quasars are not really Doppler shifts. In the next chapter you will discover a much more sophisticated way of understanding these redshifts; but, for the moment, we must give up that traditional formula.

Modern astronomers understand that the redshifts of the galaxies are caused by the expansion of the universe, and that the Hubble law allows us to find the distance from its apparent velocity of recession. For relatively nearby galaxies, astronomers commonly convert redshifts into velocities, but as the next chapter will show these are not really velocities through space. That is why you must be careful to call them *apparent* velocities of recession.

Unfortunately the exact mathematical relationship between the redshift of a very distant galaxy or quasar and its apparent velocity of recession is not yet known. That means you can't calculate the precise distance to quasar from its redshift, but the very large redshifts of the quasars assure you that the quasars are very far away. Some quasars are over 10 billion light-years away and, because of their large look-back times, they appear as they were when the universe was only 10 percent of its present age.

Perhaps you have another question about quasar distances at this point; "How can you be sure quasars really are that far away?" Astronomers faced with explaining how a small object could produce so much energy asked themselves the same question. In the 1970s, using some of the biggest telescopes on Earth, astronomers were able to photograph faint objects near some quasars. The spectra of those objects look like the spectra of normal galaxies, and they have the same redshift as the quasar. That implied that the quasar was located in a galaxy that was a member of a very distant cluster of galaxies. In the early 1980s, astronomers were able to photograph faint nebulosity surrounding some quasars, which was called quasar fuzz. The spectra of quasar fuzz looked like the spectra of normal but very distant galaxies.

A discovery made in 1979 provides a dramatic illustration of the tremendous distances to quasars. The object cataloged as 0957+561 lies just a few degrees west of the bowl of the Big Dipper and consists of two quasars separated by only 6 seconds of arc. The spectra of these quasars are identical and even have the same redshift of 1.4136. Quasar spectra do resemble each other, but in detail they are as different as fingerprints. When two quasars so close together were discovered to have the same redshift and identical spectra, astronomers concluded that they were two separate images of the same quasar.

The two images are formed by a gravitational lens. You met gravitational lenses in Chapter 13. In this case, the gravitational field of a galaxy located between Earth and the quasar bends the light from the quasar and forms multiple images. Astronomers know of many dozens of quasars that are being distorted by grav-

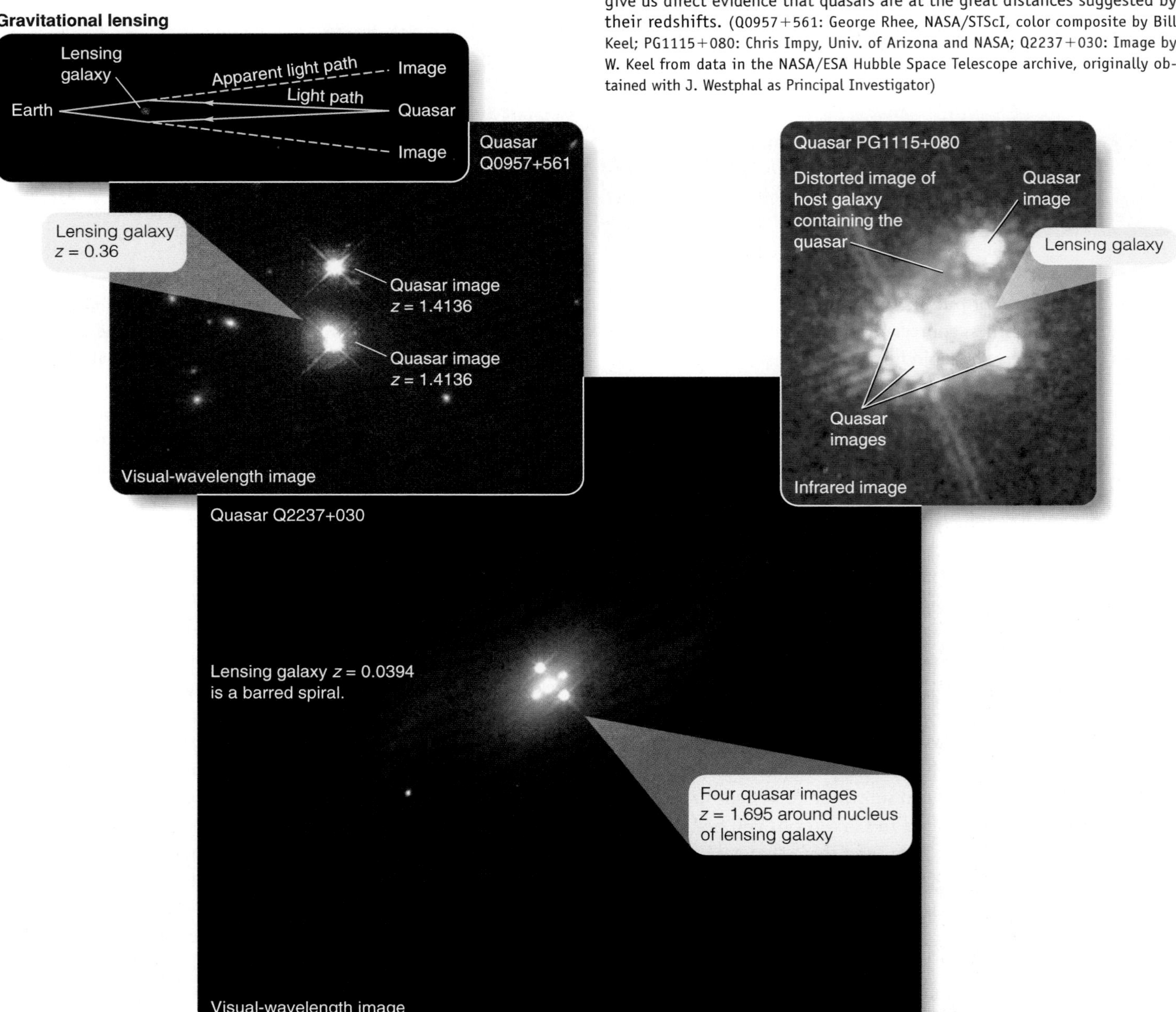

■ Figure 14-10

Gravitational lensing can occur when a relatively nearby lensing galaxy is aligned with a distant quasar. This can produce multiple images of the quasar. In the case of the quasar at upper right, the host galaxy is visible as a faint, distorted ring around the lensing galaxy. Gravitational lenses give us direct evidence that quasars are at the great distances suggested by their redshifts. (Q0957+561: George Rhee, NASA/STScI, color composite by Bill Keel; PG1115+080: Chris Impy, Univ. of Arizona and NASA; Q2237+030: Image by W. Keel from data in the NASA/ESA Hubble Space Telescope archive, originally obtained with J. Westphal as Principal Investigator)

itational lenses (■ Figure 14-10); and, in most cases, the lensing galaxy is so far away it is difficult to detect. This provides strong evidence that quasars really are distant objects as their large redshifts imply.

For more evidence that quasars are very distant, look again at Figure 14-7. Just above the image of the quasar lies the small, oval image of an elliptical galaxy, which, evidenced by its redshift, is about 7 billion light-years away. The spectrum of the quasar contains absorption lines with the same redshift as the elliptical galaxy. Evidently the lines are produced by the thin gas in the elliptical galaxy. From this you can conclude that the quasar, though very bright, is actually farther away than the distant elliptical galaxy.

The quasars are very distant and must be ultraluminous and quite small. To understand quasars, you must examine their observed characteristics and compare that evidence with the unified model of active galaxies.

Building a Model Quasar

The most satisfying part of science is combining evidence and theory to understand some aspect of nature. You have seen the evidence that quasars are very small but ultraluminous, and you are familiar with active galaxies, so now it is time to see how quasars fit into the picture. The best way to proceed is to build a model of a quasar.

Quasars seem similar to active galaxies, and some quasars even have jets and radio lobes (■ Figure 14-11). You can use the unified model of a hot accretion disk orbiting a supermassive black hole at the center of a galaxy. Matter flowing inward to feed the black hole can produce eruptions, jets, and radio lobes.

This model could explain the different kinds of radiation Earth receives from quasars (see Figure 14-5). A small percentage of quasars are strong radio sources, and the radio radiation may come from synchrotron radiation produced in the high-energy gas and magnetic fields in the jets. Much of the light from a quasar is spread in a continuous spectrum, and that is the light that fluctuates quickly. Thus, it must come from a small region. Our model could produce this light in the innermost region surrounding the black hole. Because this is a very small region, roughly the size of our solar system, rapid fluctuations can occur, probably because of random fluctuations in the flow of matter into the accretion disk. The emission lines in quasar spectra don't fluctuate rapidly, which suggests they are produced in a larger region. In our model, emission lines are emitted by clouds of gas surrounding the core in a region many light-years in diameter and excited by the intense synchrotron radiation streaming out of the central cavity.

Just as in the unified model of active galactic nuclei, the orientation of the accretion disk is important (see Figure 14-5). If the disk faces Earth so that one of its jets points directly toward you, you may see one kind of quasar. But if the disk is tipped slightly so the jet is not pointing straight at Earth, you may see a slightly different kind of quasar. Astronomers are now using this unified model to sort out the different kinds of quasars and active galaxies so they can understand how they are related. For example, using infrared radiation to penetrate dust, astronomers observed the core

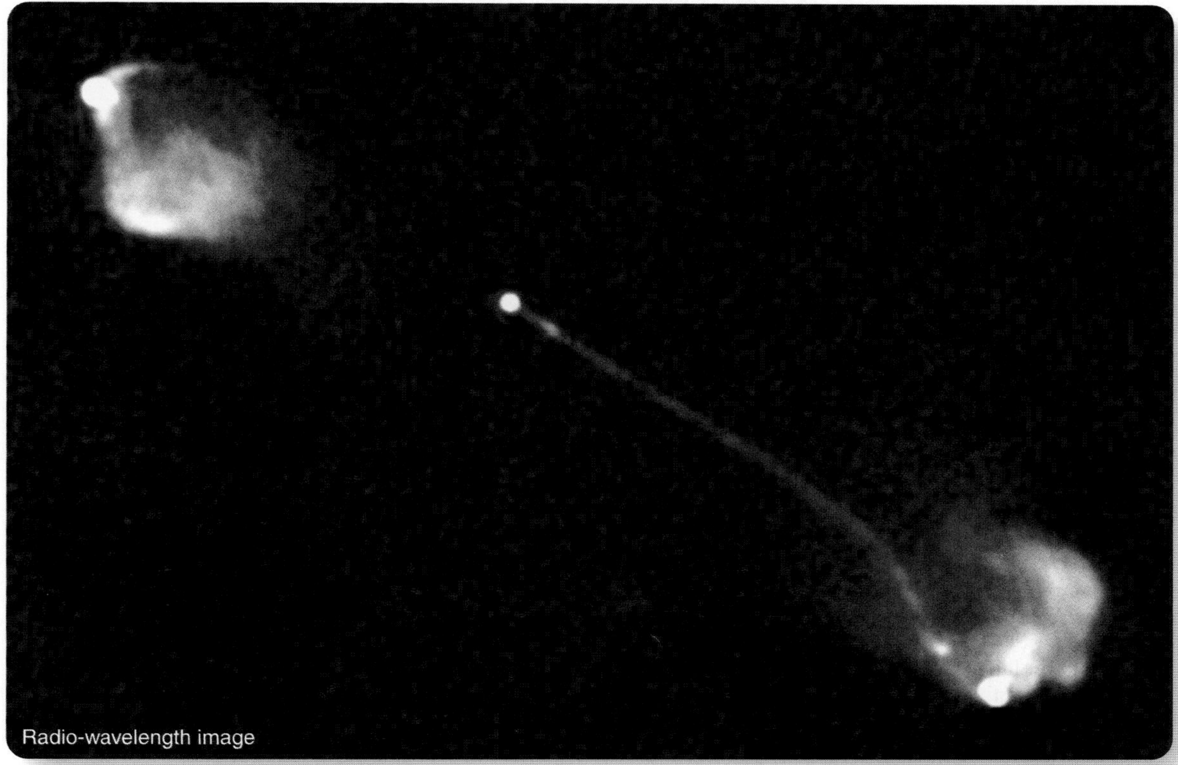

■ **Figure 14-11**

This radio image of quasar 3C175 reveals that it is ejecting a jet and is flanked by radio lobes. Presumably we see only one jet because it is directed more or less toward Earth, and the other jet is invisible because it is directed more or less away from Earth. The presence of jets and radio lobes suggests that quasars are distant active galaxies. (NRAO/AUI/NSF)

of the double-lobed radio galaxy Cygnus A (see page 310) and found an object much like a quasar. Astronomers have begun to refer to such objects as buried quasars.

Quasars Through Time

When you look at a photo of a quasar, the look-back time is large, and you are seeing the universe as it was long ago. The light journeying from quasars carries information about how the quasars formed, evolved, and died out. Here you will use that information to try to tell the story of the quasars through time.

In the next chapter, you will see evidence that the universe began about 14 billion years ago. The first clouds of gas began forming stars and falling together to form galaxies, and astronomers suspect that some of that matter formed supermassive black holes at the centers of star clouds that became the nuclear bulges of galaxies. The abundance of matter flooding into these black holes could have triggered outbursts that are seen as quasars.

You should also note that galaxies were closer together when the universe was young and had not expanded very much. Because they were closer together, the forming galaxies collided more often, and you have seen how collisions between galaxies could throw matter into supermassive black holes and trigger eruptions. Recall that quasars are often located in host galaxies that are distorted as if they were interacting with other galaxies (■ Figure 14-12).

The evidence suggests that when the universe was young, forming galaxies created supermassive black holes and, aided by interactions with other galaxies, large amounts of matter fell into those black holes, triggering quasar eruptions.

Quasars are most common with redshifts of about 2 and less common with redshifts above 2.7. The largest quasar redshifts are over 6, and such high-redshift quasars are quite rare. Evidently, if you looked at quasars with redshifts of 2 or so, you would be looking back to an age when galaxies were actively forming, colliding, and merging. In that age of quasars, they were about 1000 times more common than they are now. If you looked back to higher redshifts, you would see fewer quasars because you would be looking back to an age when the universe was so young it had not yet begun to form many galaxies. Nevertheless, even during the age of quasars, quasar eruptions must have been unusual. At any one time, only a few galaxies had quasars erupting in their cores.

Then where are all the dead quasars? An astronomer commented recently, "There is no way to get rid of supermassive black holes, so all of the galaxies that had short-lived quasars still have those black holes." Why don't astronomers see those dead quasars today? Actually, they have found quite a few.

Many dead quasars are not truly dead—only sleeping. Astronomers have discovered that most galaxies contain supermassive black holes, and those black holes may have suffered quasar eruptions when the universe was younger, when galaxies were closer together, and when dust and gas were more plentiful. Quasar eruptions became less common as galaxies became more stable and as the abundance of gas and dust in the centers of galaxies was exhausted. The dormant black holes sleeping at the centers of galaxies today can be triggered to become active galactic nuclei by collisions. Such AGN are much less energetic than quasars, evidently because there is now much less gas and dust available to feed the supermassive black holes.

Most galaxies contain sleeping supermassive black holes into which matter slowly trickles. Our own Milky Way Galaxy is a good example. It could have been a quasar long ago, but today, its supermassive black hole is resting.

Inquire | Review | Analyze

Why are most quasars so far away?

To analyze this question you must combine two factors. First, quasars are the active cores of galaxies, and only a small percentage of galaxies

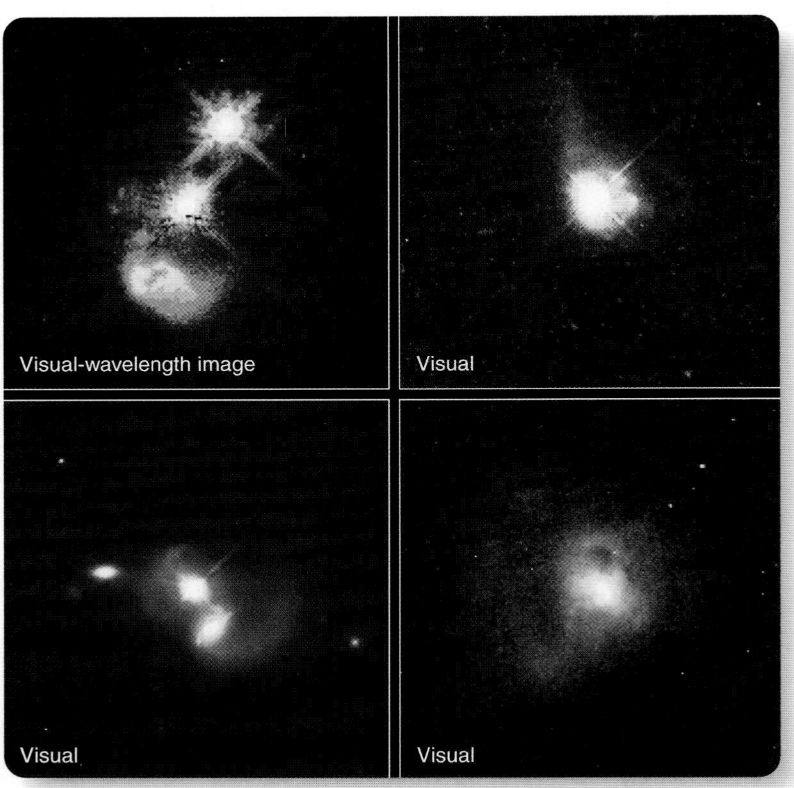

■ Figure 14-12

The bright object at the center of each of these images is a quasar. Fainter objects near the quasar are galaxies distorted by collisions. Compare the upper left ring-shaped galaxy with the ring galaxy on page 297 and compare the tail in the lower left image with the Antennae galaxies on page 297. (J. Bahcall, Institute for Advanced Study, Mike Disney, University of Wales, NASA)

contain active cores. In order to sample a large number of galaxies in your search, you must extend your search to great distances. Most of the galaxies lie far away because the amount of space searched increases rapidly with distance. Just as most seats in a baseball stadium are far from home plate, most galaxies are far from our Milky Way Galaxy. Thus, most of the galaxies that might contain quasars lie at great distances.

But a second factor is much more important. The farther you look into space, the farther back in time you look. It seems that there was a time in the distant past when quasars were more common, and thus you see most of those quasars at large look-back times, meaning at large distances. For these two reasons, most quasars lie at great distances.

If our description of quasars is correct, then quasars must be triggered into eruption. What observational evidence supports that argument?

■ ■ ■

Connections: Our study of active galaxies has led us far out in space and back in time to quasars. The light now arriving from the most distant quasars left them when the universe was only a fraction of its present age. The quasars stand at the very threshold of the study of the history of the universe itself, the subject of the next chapter.

Study and Review Tools

Summary

14-1 | Active Galactic Nuclei

What evidence shows that some galactic nuclei are active?

- Some galaxies have peculiar properties. Seyfert galaxies, for example, are spirals with small, highly luminous cores.
- Spectra of the nuclei of Seyfert galaxies show that they contain highly excited gas.
- Double-lobed radio galaxies emit radio energy from areas on either side of the galaxies. These lobes appear to be inflated by jets ejected from the nuclei of the galaxies.
- Some giant elliptical galaxies have small, energetic cores, with, in some cases, jets of matter rushing outward.

What is the energy source for this activity?

- The evidence is very strong that active galactic nuclei contain supermassive black holes into which matter is flowing through hot accretion disks. This can eject jets in opposite directions.
- Relativistic jets of excited gas from the core of an active galaxy can push into the intergalactic medium and inflate radio lobes like balloons.
- Hot spots in the radio lobes are the points where the gas hits the intergalactic medium.
- Astronomers can see instances where galaxies moving through the intergalactic medium leave behind trails of hot gas from their jets. In other cases, the motions of the nucleus can produce twisted jets.
- The mass of a central black hole can be found by observing the velocity of stars orbiting the black hole or the rotational speed of its accretion disk. Masses range from millions to billions of solar masses.
- Such massive black holes cannot be stellar black holes. That is, they cannot have been formed by dying stars but must have formed as the nuclear bulges of the galaxies began to contract.

What triggers the nucleus of a galaxy into activity?

- Most galaxies appear to contain supermassive black holes at their centers, but they are dormant because large amounts of matter are not flowing inward. Only when a supermassive black hole is fed does it erupt.
- Matter flowing into a hot accretion disk around a black hole can eject high-energy jets of gas and radiation perpendicular to the disk. This process is not well understood, but it is also observed coming from accretion disks around protostars, around neutron stars, and around stellar-mass black holes.
- Interactions between galaxies can throw matter into the center, feed the black hole, and trigger eruptions. This explains why active galaxies are often distorted or have nearby companions.
- According to the unified model, what you see depends on the tilt of the accretion disk. If you see into the core, you see broad spectral lines and rapid fluctuations. If you see the disk edge on, you see only narrow spectral lines.
- If the jet from the black hole points directly at you, you see a blazar.

14-2 | Quasars

What are the most distant active galaxies?

- The quasars appear to be very distant, highly luminous active galaxies.
- Astronomers know they are very far away because they have very high redshifts.
- To be visible at such great distances, they must be ultraluminous.
- Because quasars can change their brightness quickly, you can conclude they must be small—only a few times larger than our solar system.
- Observations of the spectra of hazy objects near quasars and the spectra of quasar fuzz show that quasars are the active cores of very distant galaxies.

What can active galaxies tell us about the history of the universe?

- Because quasars lie at great distances, astronomers see them as they were long ago—perhaps 10 billion years ago—when the universe was young and just forming galaxies.
- The best images show that the host galaxies of quasars are distorted, and that suggests that they have been involved in mergers or collisions, which could have triggered eruptions of their supermassive black holes.
- It is also possible that at least some quasars are erupting as matter falls together to create a supermassive black hole as a galaxy begins to form. That is, some quasars may be caused by the formation of the first galaxies when the universe was young.

New Terms

radio galaxy (p. 304)
active galaxy (p. 306)
active galactic nuclei (AGN) (p. 306)
Seyfert galaxy (p. 306)
double-lobed radio source (p. 308)
double-exhaust model (p. 308)
hot spot (p. 310)
unified model (p. 313)
BL Lac object (p. 313)
blazar (p. 313)
quasar (p. 315)

Review Questions

Ace Astronomy™ Assess your understanding of this chapter's topics with additional quizzing and animations at **http://astronomy.brookscole.com/sh9e**

1. What is the difference between the terms *radio galaxy* and *active galaxy?*
2. What evidence shows that the energy source in a double-lobed radio galaxy lies at the center of the galaxy?
3. How does the peculiar rotation of NGC5128 help explain the origin of this active galaxy?
4. What statistical evidence suggests that Seyfert galaxies have suffered recent interactions with other galaxies?
5. How does the unified model explain the two kinds of Seyfert galaxies?
6. What observations are necessary to identify the presence of a supermassive black hole at the center of a galaxy?
7. How does the unified model implicate collisions and mergers in triggering active galaxies?
8. Why were quasars first noticed as being peculiar?
9. How do the large redshifts of quasars lead astronomers to conclude they must be very distant?
10. What evidence shows that quasars are ultraluminous but must be very small?
11. How do gravitational lenses provide evidence that quasars are distant?
12. What evidence is there that quasars occur in distant galaxies?
13. How can our model quasar explain the different radiation received from quasars?
14. What evidence is there that quasars must be triggered by collisions and mergers?
15. Why are there few quasars at low redshifts and at high redshifts but many at redshifts of about 2?
16. The image at right combines visual (blue) with radio (red) to show the galaxy radio astronomers call Fornax A. Explain the features of this image. Is it significant that the object is a distorted elliptical galaxy in a cluster?

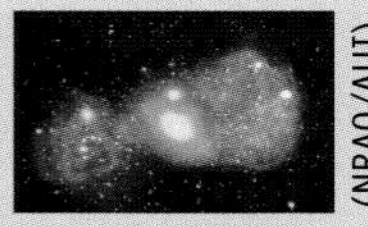
(NRAO/AUI)

17. Explain the features of this radio image of the galaxy IC708.

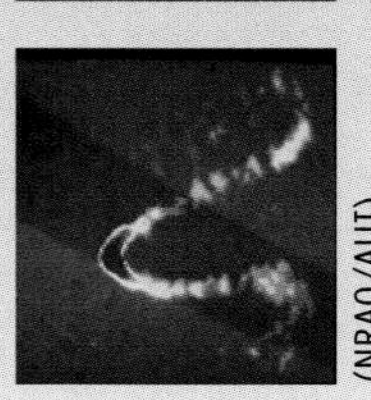
(NRAO/AUI)

18. A radio image of quasar 3C334 is shown at the right. Why do you see only one jet? What does it mean that this looks so much like a radio image of a double-lobed radio galaxy?

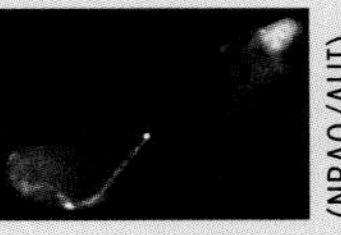
(NRAO/AUI)

Discussion Questions

1. Why do quasars, active galaxies, SS433, and protostars have similar geometry?
2. By custom, astronomers refer to the unified *model* of AGN and not to the unified *hypothesis* or unified *theory*. In your opinion, which of the words seems best?
3. Do you think that our galaxy has ever been an active galaxy? Could it have hosted a quasar when it was young?
4. If a quasar is triggered in a galaxy's core, what would it look like to people living in the outer disk of the galaxy? Could life continue in that galaxy? (Begin by deciding how bright a quasar would look seen from the outer disk of the galaxy, considering both distance and dust.)

Problems

1. The total energy stored in a radio lobe is about 10^{53} J. How many solar masses would have to be converted to energy to produce this energy? *Hints:* Use $E = mc^2$. One solar mass equals 2×10^{30} kg.)
2. If the jet in NGC5128 is traveling at 5000 km/s and is 40 kpc long, how long will it take for gas to travel from the core of the galaxy to the end of the jet? (*Hint:* 1 pc equals 3×10^{13} km.)
3. Cygnus A is roughly 225 Mpc away, and its jet is about 50 seconds of arc long. What is the length of the jet in parsecs? (*Hint:* See Reasoning with Numbers 3-1.)
4. Use the small-angle formula to find the linear diameter of a radio source with an angular diameter of 0.0015 second of arc and a distance of 3.25 Mpc.
5. If the active core of a galaxy contains a black hole of 10^6 solar masses, what will the orbital period be for matter orbiting the black hole at a distance of 0.33 AU? (*Hint:* See Reasoning with Numbers 8-4.)
6. If a quasar is 1000 times more luminous than an entire galaxy, what is the absolute magnitude of such a quasar? (*Hint:* The absolute magnitude of a bright galaxy is about -21.)
7. If the quasar in Problem 6 were located at the center of our galaxy, what would its apparent magnitude be? (*Hints:* See Reasoning with Numbers 8-2 and ignore dimming by dust clouds.)
8. If the Hubble constant is 70 and a quasar has an apparent velocity or recession of 45,000 km/s, how far away is it?
9. The hydrogen Balmer line H_β has a wavelength of 486.1 nm. It is shifted to 563.9 nm in the spectrum of 3C273. What is the redshift of this quasar? (*Hint:* What is $\Delta\lambda$?)

Media Cluster

ACTIVE FIGURES

Ace Astronomy™ To access the resources in the Media Cluster, log into AceAstronomy at http://astronomy.brookscole.com/sh9e and select Chapter 14.

Jet Deflection
This animation lets you vary the velocity of a galaxy producing jets and observe how the jets bend in the "intergalactic wind." When the galaxy's velocity is higher, the bending is greater.

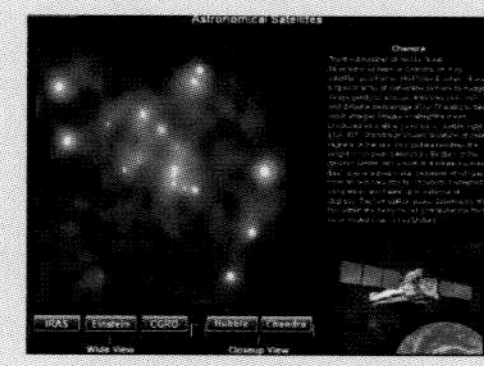

M31 at Many Wavelengths
In this animation you can compare how five astronomical satellites, including Hubble and Chandra, view the Andromeda Galaxy.

VIRTUAL ASTRONOMY LABS

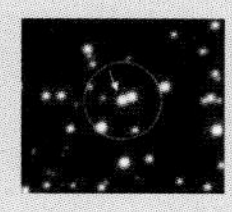

Lab 15: General Relativity and Black Holes
This lab explores the properties of black holes. It includes an exercise illustrating Einstein's general theory of relativity and an exercise on binary quasars. The lab concludes with a discussion and an exercise on black hole detection.

Lab 18: Active Galactic Nuclei
This lab looks at the extraordinary properties of neutron stars, the dense balls of neutrons that may remain after some stars have exploded. Later the lab examines pulsars, neutron stars that appear to emit radiation in rapid pulses.

Critical Inquiries for the Web

1. What object currently holds the distinction as the farthest known galaxy? Search the web for information on this distant object and find out its redshift and distance. What is the look-back time for this object?
2. Gravitational lenses were first predicted by Einstein in 1936 but were not observed until recently. Search the web for instances of gravitational lensing of galaxies and quasars. For a particular case, discuss how the lens effect allows astronomers to determine information about the lensing and/or lensed objects that might not have been available without the alignment.

- Go to the Brooks/Cole Astronomy Resource Center **(http://astronomy.brookscole.com)** for critical thinking exercises, articles, and additional readings from InfoTrac College Edition, Brooks/Cole's online student library.

15 Cosmology in the 21st Century

The Universe, as has been observed before, is an unsettlingly big place, a fact which for the sake of a quiet life most people tend to ignore.

DOUGLAS ADAMS
The Restaurant at the End of the Universe

WHAT IS THE biggest number you can name? A billion? How about a billion billion? That is only 10^{18}. How about 10^{100}? That very large number is called a googol, a number that is at least a billion billion times larger than the total number of atoms in all of the galaxies visible in the universe. Can you name a number bigger than a googol? Try a billion googols, or better yet, a googol of googols. You can go even bigger than that—10 raised to a googol. That very big number is called a googolplex. No matter how large a number you name, you can immediately name an even bigger number. That is what infinity means—big without limit. ▌ If the universe is infinite in size, then you can name any distance you want, and the universe is bigger than that. Try a googol to the googol light-years. If the universe is infinite, there is more universe beyond that distance.

In this computer model of the evolution of the universe, dark matter has drawn normal matter together to form great clouds of galaxies spread across space. (Courtesy MPA and Joerg Colberg)

Guidepost

Looking Back

Since Chapter 1 you have been on an outward journey through the universe. You have studied the appearance of the night sky seen from our planet, the birth and death of stars, and the interaction of galaxies. Now you have reached the limit of your journey in space and in time—the study of the universe as a whole.

This Chapter

The ideas in this chapter are the biggest and the most difficult in all of science. Can you imagine an edge to the universe or the first instant of time? Perhaps you have heard that the universe began with an event called the big bang. Now it is time to try to imagine what it was like.

As you explore cosmology, you will find answers to four essential questions:

Does the universe have an edge in space or time?

What evidence exists that the universe began with a big bang?

How can the universe expand if it has no edge?

Why is the universe the way it is?

These questions are deceptively simple, but they will lead you into a new understanding of what we are and where we are.

Looking Ahead

Once you have finished this chapter, you will have modern insight into the nature of the universe, and it will be time to focus on your place in that universe—the subject of the rest of this book.

AceAstronomy™ The AceAstronomy icon throughout the text indicates an opportunity for you to test yourself on key concepts and to explore animations and interactions on the AceAstronomy website at: http://astronomy.brookscole.com/sh9e

Is the universe infinite or finite? In this chapter, you will try to answer that question and others, not by playing games with big numbers, but through **cosmology,** the study of the universe as a whole.

Studying the entire universe is difficult because you have expectations about the universe that are in some cases wrong. This chapter is carefully organized to begin with your common expectations, test them against observations, compare them with theories, and build a modern understanding of cosmology. Like a mountaineer climbing a high peak, you are leaving a safely prepared base camp and must proceed carefully step by step to reach your goal.

15-1 Introduction to the Universe

EVERYONE KNOWS what a gold mine looks like, but few people have actually visited a gold mine. If you explored a real gold mine, you might be surprised at what you find. Similarly, you have an impression of the universe as a vast depth filled with galaxies (■ Figure 15-1), but as you begin exploring the universe, you must sort out your expectations so they do not mislead you.

The Edge–Center Problem

In your daily life, you are accustomed to boundaries. Rooms have walls, athletic fields have boundary lines, countries have borders, and oceans have shores. It is natural to think of the universe having an edge, but that idea leads us astray.

If the universe had an edge, imagine going to that edge. What would you find there: a wall of cardboard? a great empty space? nothing? This is not an edge to the distribution of matter, but an edge to space itself, so presumably you could not reach beyond the edge and feel around.

An edge to the universe seems to violate common sense, and modern cosmologists assume that the universe cannot have an edge. That is, the universe must be unbounded.

If the universe has no edge, then it cannot have a center. You find the centers of things—galaxies, globular clusters, oceans, pizzas—by reference to their edge. If the universe cannot have an edge, then it cannot have a center.

You must think carefully about the universe and avoid using the ideas of an edge or a center. Those ideas are misleading. Of course, if the universe is infinite, then it has no edge and no center. You have no edge–center problem if you think of the universe as infinite. Could the universe be finite—could it contain a limited volume? You will see later in this chapter how the universe could be finite but have no edge or center. Before you go further with that idea, you must deal with another of your expectations, the beginning of the universe.

■ **Figure 15-1**

The entire sky is filled with galaxies. Some lie in clusters of thousands, and others are isolated in nearly empty voids between the clusters. In this image of a typical spot on the sky, bright objects with spikes caused by diffraction in the telescope are nearby stars. All other objects are galaxies ranging from the nearby face-on spiral at upper right to the most distant galaxies visible only in the infrared, shown as red in this composite image. (R. Williams, STScI HDF-South Team, NASA)

The Necessity of a Beginning

Of course you have noticed that the night sky is dark. However, reasonable assumptions about the geometry of the universe can lead you to the conclusion that the night sky should glow as brightly as a star's surface. This conflict between observation and theory is called **Olbers's paradox** after Heinrich Olbers, a Viennese physician and astronomer who discussed the problem in 1826.

However, Olbers's paradox is not Olbers's, and it isn't a paradox. The problem of the dark night sky was first discussed by Thomas Digges in 1576 and was further analyzed by astronomers such as Johannes Kepler in 1610 and Edmund Halley in 1721. Olbers gets the credit through an accident of scholarship on the part of modern cosmologists who did not know of previous discussions. What's more, Olbers's paradox is not a paradox. You will be able to understand why the night sky is dark by revising your assumption about the nature of the universe.

To begin, let's state the so-called paradox. Suppose you assume that the universe is infinite and uniformly filled with stars. (The aggregation of stars into galaxies makes no difference to your argument.) If you look in any direction, your line of sight must eventually reach the surface of a star. Look at ■ Figure 15-2, which uses the analogy of lines of sight in a forest (**Window on Science 15-1**). When you are deep in a forest, every line of sight terminates on a tree trunk. In this example, every line of sight out

Thinking about an Abstract Idea: Reasoning by Analogy

"The economy is overheating and it may seize up," an economist might say. Economists like to talk in analogies because economics is often abstract, and one of the best ways to think about abstract problems is to find a more approachable analogy. Rather than discuss the details of the national economy, you might make conclusions about how the economy works by thinking about how a gasoline engine works. Of course, if your analogy is not a good one, it can mislead you.

Much of astronomy is abstract, and cosmology is the most abstract subject in astronomy. Furthermore, cosmology is highly mathematical, and unless you are prepared to learn some of the most difficult mathematics known, you must use analogies, such as lines of sight in a forest.

Reasoning by analogy is a powerful technique. An analogy can reveal unexpected insights and lead you to further discoveries. But carrying an analogy too far can be misleading. You might compare the human brain to a computer, and that would help you understand how data flow in and are processed and how new data flow out. But your analogy is flawed. Data in computers are stored in specific locations, but memories are stored in the brain in a distributed form. No single brain cell holds a specific memory. If you carry the analogy too far, it can mislead you. Whenever you reason by analogy, you should be alert for potential problems.

As you study any science, be alert for analogies. They are tremendously helpful, but you must always take care not to carry them too far.

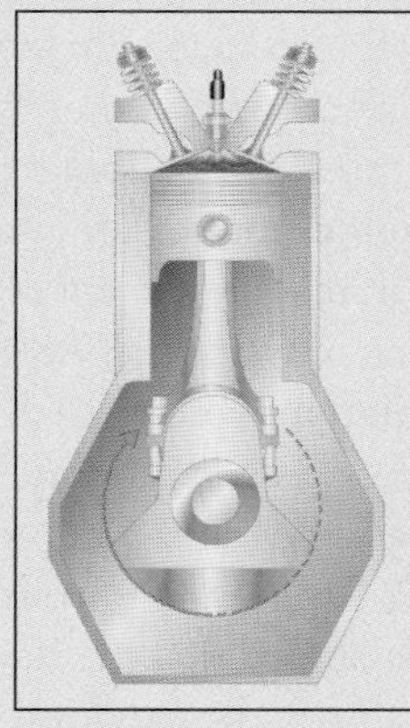

How is the economy like a gasoline engine? Thinking with analogies is a powerful tool when you respect the limits.

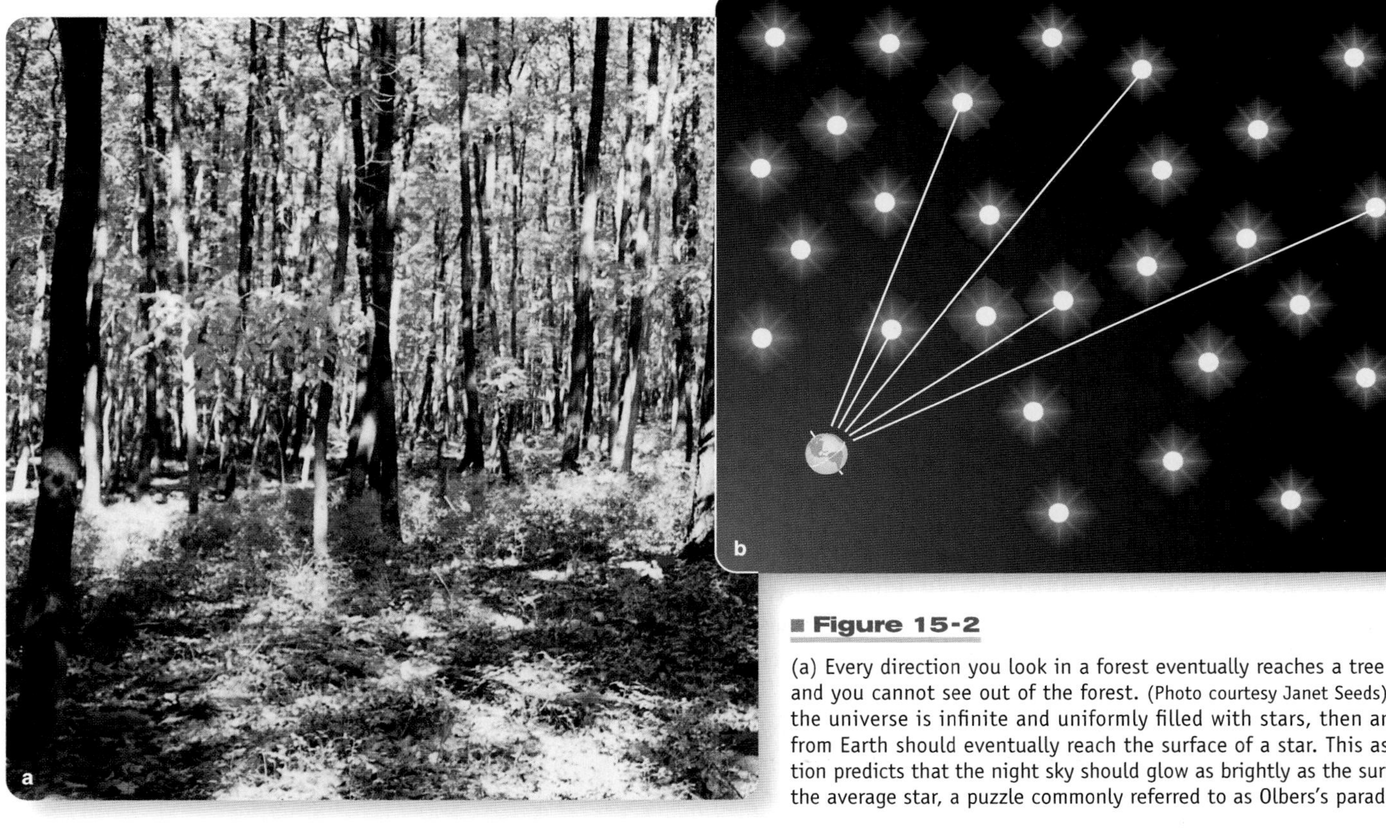

Figure 15-2

(a) Every direction you look in a forest eventually reaches a tree trunk, and you cannot see out of the forest. (Photo courtesy Janet Seeds) (b) If the universe is infinite and uniformly filled with stars, then any line from Earth should eventually reach the surface of a star. This assumption predicts that the night sky should glow as brightly as the surface of the average star, a puzzle commonly referred to as Olbers's paradox.

into space should terminate on the surface of a star, and the entire sky should be as bright as the surface of an average star. It should not get dark at night.

Of course, the more distant stars would be fainter than nearby stars because of the inverse square law. However, the farther you look into space the larger the volume you see, and the two effects cancel out. Then given your assumptions, every spot on the sky must be occupied by the surface of a star, and the night sky should not be dark.

Can you imagine the entire sky glowing with the brightness of the surface of the sun? The glare would be overpowering. In fact, the radiation would rapidly heat Earth and all other celestial objects to the average temperature of the surface of the stars, a few thousand degrees. Thus, you can pose Olbers's paradox in another way: Why is the universe so cold?

Olbers assumed that the sky was dark because clouds of matter in space absorb the radiation from distant stars. But this interstellar medium would heat up to the average surface temperature of the stars, and the gas and dust clouds would glow as brightly as the stars.

It is interesting to note that Copernicus and other astronomers of his time saw no paradox in the dark night sky because they believed in a finite universe. Once they looked beyond the sphere of stars, there was nothing to see except, perhaps, the dark floor of heaven. Soon after the time of Galileo and Newton, western astronomers began to think of an infinite universe, and that led to the paradox.

Today, cosmologists believe they understand why the sky is dark and the universe is cold. Olbers's paradox makes the incorrect prediction that the night sky should be bright because it is based on an incorrect assumption. The universe is not eternal. That is, it is not infinitely old.

This solution to Olbers's paradox was first suggested by Edgar Allan Poe in 1848. He proposed that the night sky was dark because the universe was not infinitely old but had been created at some time in the past. The more distant stars are so far away that light from them has not reached Earth yet. That is, if you look far enough, the look-back time is greater than the age of the universe, and you look back to a time before stars began to shine. Thus, the night sky is dark because the universe is not infinitely old.

This is a powerful idea because it clearly illustrates the difference between the universe and the observable universe. The universe is everything that exists, and it could be infinite. But the **observable universe** is the part that you can see. You will learn later that the universe is about 14 billion years old. In that case, the observable universe has a radius of about 14 billion light-years. Do not confuse the observable universe, which is finite, with the universe as a whole, which could be infinite.

When we described Olbers's paradox, we assumed *without noticing* that the universe was infinitely old. This illustrates the importance of assumptions in cosmology and serves as a warning that our commonsense expectations are not dependable. All of astronomy is reasonably unreasonable—that is, reasonable assumptions often lead to unreasonable results. That is especially true in cosmology, so you must proceed with care.

Our expectation that the universe be eternal and unchanging is clearly wrong. Olbers's paradox makes a beginning a necessity. In the next section we will discover that we should not expect the universe to be static; observations show that it is expanding.

Cosmic Expansion

In the early 20th century, astronomers began photographing the spectra of galaxies, and they noticed that the spectral lines in galaxy spectra had redshifts. In 1929, Edwin P. Hubble published his discovery that the size of the redshift was proportional to the

■ Active Figure 15-3

These galaxy spectra extend from the near-ultraviolet at left to the blue part of the visible spectrum at right. The two dark absorption lines of once-ionized calcium are prominent in the near-ultraviolet. The redshifts in galaxy spectra are expressed here as velocities of recession. Note that the apparent velocity of recession is proportional to distance. (Caltech)

AceAstronomy™ Log into AceAstronomy and select this chapter to see the Active Figure called "Cosmic Redshift." Graph your own observations of redshift and distance.

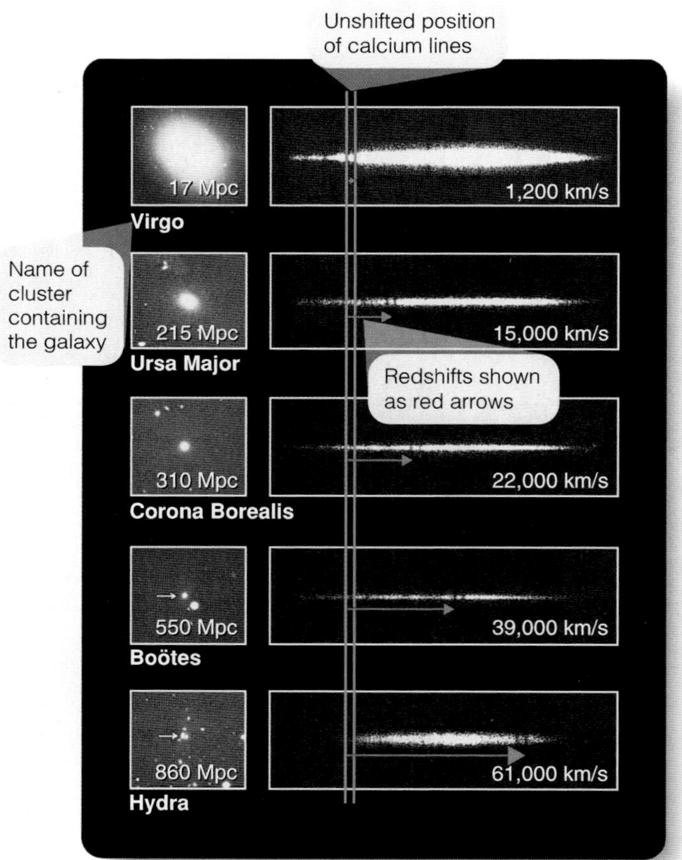

distance to the galaxy. (You met this as the Hubble law in Chapter 13, where you used it to estimate the distances to galaxies.) When interpreted using the Doppler effect, these redshifts imply that the galaxies are receding from each other and that the universe is expanding.

The redshifts of the galaxies are dramatic. Nearby galaxies have small redshifts, but when you look at more distant galaxies, the redshifts are quite large. ■ Figure 15-3 shows spectra of galaxies in galaxy clusters at various distances. Notice that the more distant galaxies look smaller in the photographs. The Virgo cluster is relatively nearby, and its redshift is small. The Hydra cluster is very distant, and its redshift is so large that the two dark lines formed by once-ionized calcium are shifted from the near-ultraviolet well into the visible part of the spectrum.

The important point about Hubble's discovery is that the redshifts are proportional to distance. That shows that the galaxies are receding from each other and the universe is expanding uniformly. But notice that the expansion does not imply that Earth is at the center of the universe. To see why, look at ■ Figure 15-4, which shows an analogy in which you imagine baking raisin bread. As the dough rises, the raisins are pushed away from each other uniformly at velocities that are proportional to distance. Two raisins that were originally close to each other are pushed apart slowly, but two raisins that were far apart, having more dough between them, are pushed apart faster. If bacterial astronomers lived on a raisin in your raisin bread, they could observe the redshifts of the other raisins and derive a bacterial Hubble law. They could conclude that their universe was expanding uniformly. But it does not matter which raisin the bacterial astronomers lived on. There is no center to the expansion, and they would get the same Hubble law no matter what raisin they lived on. Similarly, there is no center to the expansion of the universe and astronomers in any galaxy will see the same law of expansion.

When you look at Figure 15-4, you can see the edge of the loaf of raisin bread and you can identify a center. This happens because the raisin bread analogy breaks down when you look at the crust of the bread. Remember that the universe cannot have an edge or a center, so there can be no center to the expansion. We must think and speak very carefully as we discuss cosmology (**Window on Science 15-2**).

The Necessity of a Big Bang

The discovery of the expansion of the universe led astronomers to an astonishing conclusion. The universe must have begun with an event of cosmic fury.

Imagine that you have a videotape of the expanding universe, and you run it backward. You would see the galaxies moving toward each other. There is no center to the expansion of the universe, so you would not see the galaxies approaching a single spot; but, rather, you would see the distances between all galaxies decreasing, and eventually the galaxies would begin to merge. If you run the videotape far enough back, you would see the matter and energy of the universe compressed into a high-density, high-temperature state. The expanding universe must have begun from this moment of extreme conditions, which modern astronomers call the **big bang.**

You must not allow yourself to think of the big bang as an explosion. You have concluded that the universe cannot have a center, so the big bang cannot have occurred at a single point. The entire universe must have been filled with the high-density, high-temperature state called the big bang. Your challenge in this chapter is to understand how that could have happened without a center. That problem will be resolved when you refine your understanding of space and time.

For now, you can solve an easier problem. How long ago did the universe begin? You can estimate the age of the universe with a simple calculation. If you must drive to a city 100 miles away and you can travel 50 miles per hour, you divide distance by rate of travel and learn the travel time—in this example, 2 hours. To find the age of the universe, you can divide the distance between two galaxies by the velocity with which they are separating and find out how long they have taken to reach their present separation. The details of this simple calculation are given in **Reasoning with Numbers 15-1.**

The Hubble time gives an estimate of the age of the universe. You will fine tune this estimate later in this chapter, but

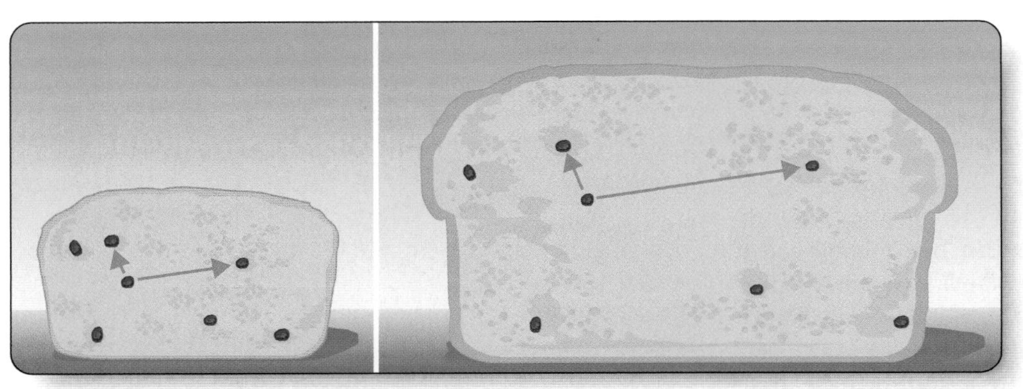

■ Active Figure 15-4

The raisin bread analogy for the expansion of the universe. As the dough rises, raisins are pushed apart with velocities proportional to distance. The expansion has no center. A colony of bacteria living on any raisin will find that the redshifts of the other raisins are proportional to their distances.

AceAstronomy™ Log into AceAstronomy and select this chapter to see the Active Figure called "Raisin Bread." You can make observations from any raisin as the dough rises.

Window on Science | 15-2

Why Scientists Speak Carefully: Words Lead Thoughts

There are certain words you should never say, not even as a joke. You should never call your friend "fool," for example, because you might begin to think of your friend as a fool. Words lead thoughts. It works in advertising and politics, and it can work in science, too. Using words carelessly can lead you to think carelessly.

In science, there are certain ways to say things, not because scientists are sticklers for good grammar, but because if they say things wrongly they begin to think things wrongly. For example, a biologist would never let you say a beehive knows it must store food. "No, don't say it that way," the biologist would object. "The hive is just a collection of individual creatures and it can't 'know' anything. Even the individual bees don't really 'know' things. The instinctive behavior of individual bees causes food to accumulate in the hive. That's the way to say it."

All scientists are careful of language because careless words can mislead. You would never refer to the center of the universe, for example, but you must also be careful not to say things like "galaxies flying away from the big bang." Those words imply a center to the expansion of the universe, and you know the expansion must be centerless. Rather, you should say "galaxies flying away from each other."

Whatever science you study, notice the customary ways of using words. It is not just a matter of convention. It is a matter of careful thought.

You should not say that a bee "knows" how to make honey. How you speak guides how you think.

for the moment you can conclude that basic observations of the recession of the galaxies require that the universe began with a big bang about 14 billion years ago.

Your instinct is to think of the big bang as a historical event, like the Gettysburg Address—something that happened long ago and happened in a particular place. But the big bang isn't really over. The look-back time makes it possible for you to observe the big bang directly. The look-back time to nearby galaxies is only a few million years, and the look-back time to more distant galaxies is a large fraction of the age of the universe (■ Figure 15-5). Suppose you look between the distant galaxies, back to the time of the big bang. You should be able to detect the hot gas that filled the universe long ago.

Although your imagination tries to visualize the big bang as a localized event, you must keep firmly in mind that the big bang did not occur at a single place but filled the entire volume of the universe. The matter of which you are made was part of that big bang, so you are inside the remains of that event, and the universe continues to expand around you. You cannot point to any particular place and say "The big bang occurred over there." The big bang occurred everywhere, and whatever direction you look, at great distances you look back to the age when the universe was filled with hot gas (■ Figure 15-6).

The radiation that comes from this great distance has a tremendous redshift. The most distant visible objects are faint galaxies and quasars, with redshifts of about 10. In contrast, the ra-

Reasoning with Numbers | 15-1

The Age of the Universe

Dividing the distance to a galaxy by the apparent velocity with which it recedes tells you the age of the universe, and the Hubble constant simplifies your task further. The Hubble constant H has the units km/s per Mpc, which is a velocity divided by a distance. If you calculate 1/H, you have a distance divided by velocity. To make the division give you an age, you must convert megaparsecs into kilometers, and then the distances will cancel out and leave you with an age in seconds. To get years, you must divide by the number of seconds in a year. If you make these simple changes in units, the age of the universe in years is approximately 10^{12} divided by H in its normal units, km/s/Mpc.

$$\tau_u = \frac{1}{H} \times 10^{12} \text{ years}$$

This estimate of the age of the universe is known as the **Hubble time.** For example, if H is 70 km/s/Mpc, then the age of the universe is roughly $10^{12}/70$, which equals 14 billion years.

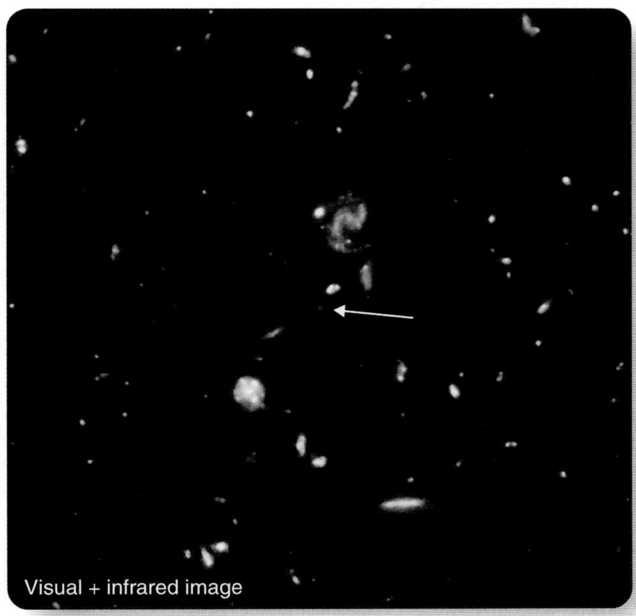

■ Figure 15-5

This faint galaxy (arrow) is believed to be so distant that its light has been traveling for most of the history of the universe. Here you see it as it was about 1 billion years after the big bang. The galaxy is bright in the infrared but is not detectable at visual wavelengths. (Ken Lanzetta and Amos Yahil, Stony Brook and NASA)

diation from the hot gas of the big bang has a redshift of about 1100. Thus the light emitted by the big bang gases arrives at Earth as infrared radiation and short radio waves. You can't see it with your eyes, but you should be able to detect it with infrared and radio telescopes. Unlike the Gettysburg Address, the big bang isn't over, and you could see it happening if you could detect the radiation it emitted. That amazing discovery is the subject of the next section.

AceAstronomy™ Log into AceAstronomy and select this chapter to see Astronomy Exercise "Age of the Universe" and take control of the Hubble constant.

The Cosmic Background Radiation

If radiation is now arriving from the big bang, then it should be detectable. The story of that discovery begins in the mid-1960s when two Bell Laboratories physicists, Arno Penzias and Robert Wilson, were using a horn antenna to measure the radio bright-

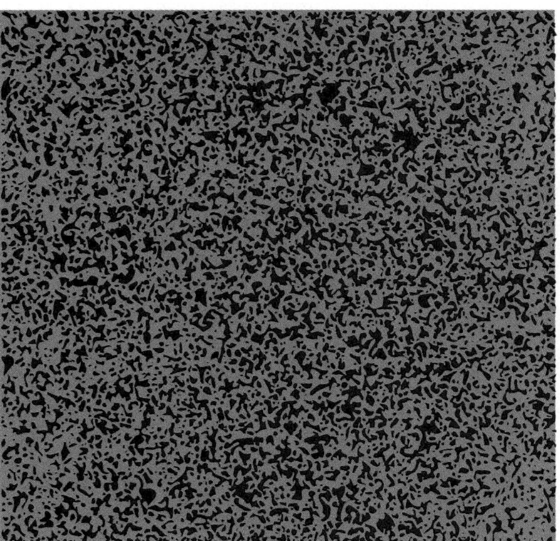

a A region of the universe during the big bang

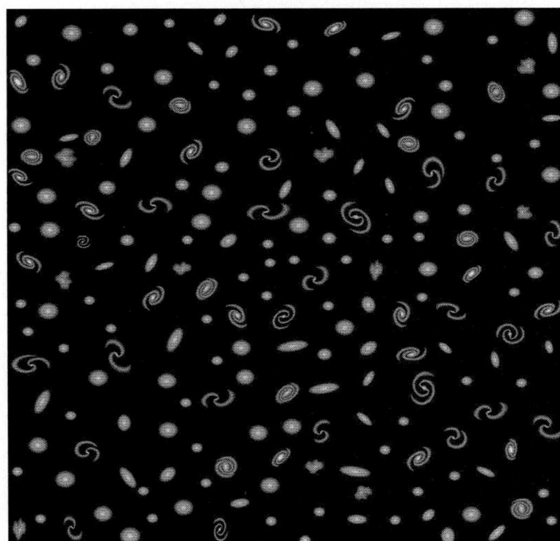

b A region of the universe now

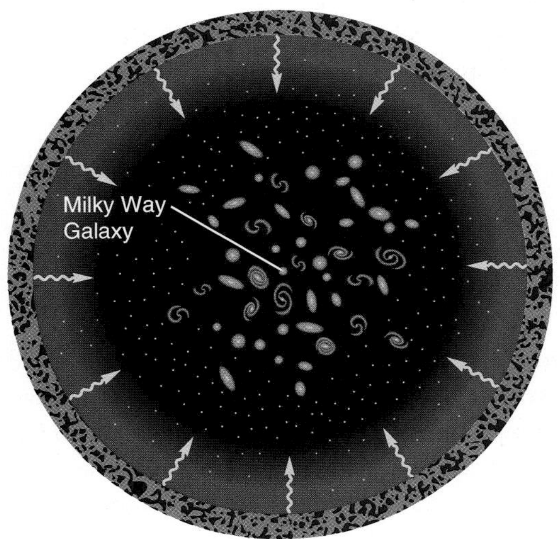

c The present universe as it appears from our galaxy

■ Figure 15-6

Three views of a small region of the universe centered on our galaxy. (a) During the big bang, the region is filled with hot gas and radiation. (b) Later, the gas has formed galaxies, but you can't see the universe this way because the look-back time distorts what you see. (c) Near you, you see galaxies, but farther away you see young galaxies (dots), and at a great distance you see radiation (arrows) coming from the hot gas of the big bang.

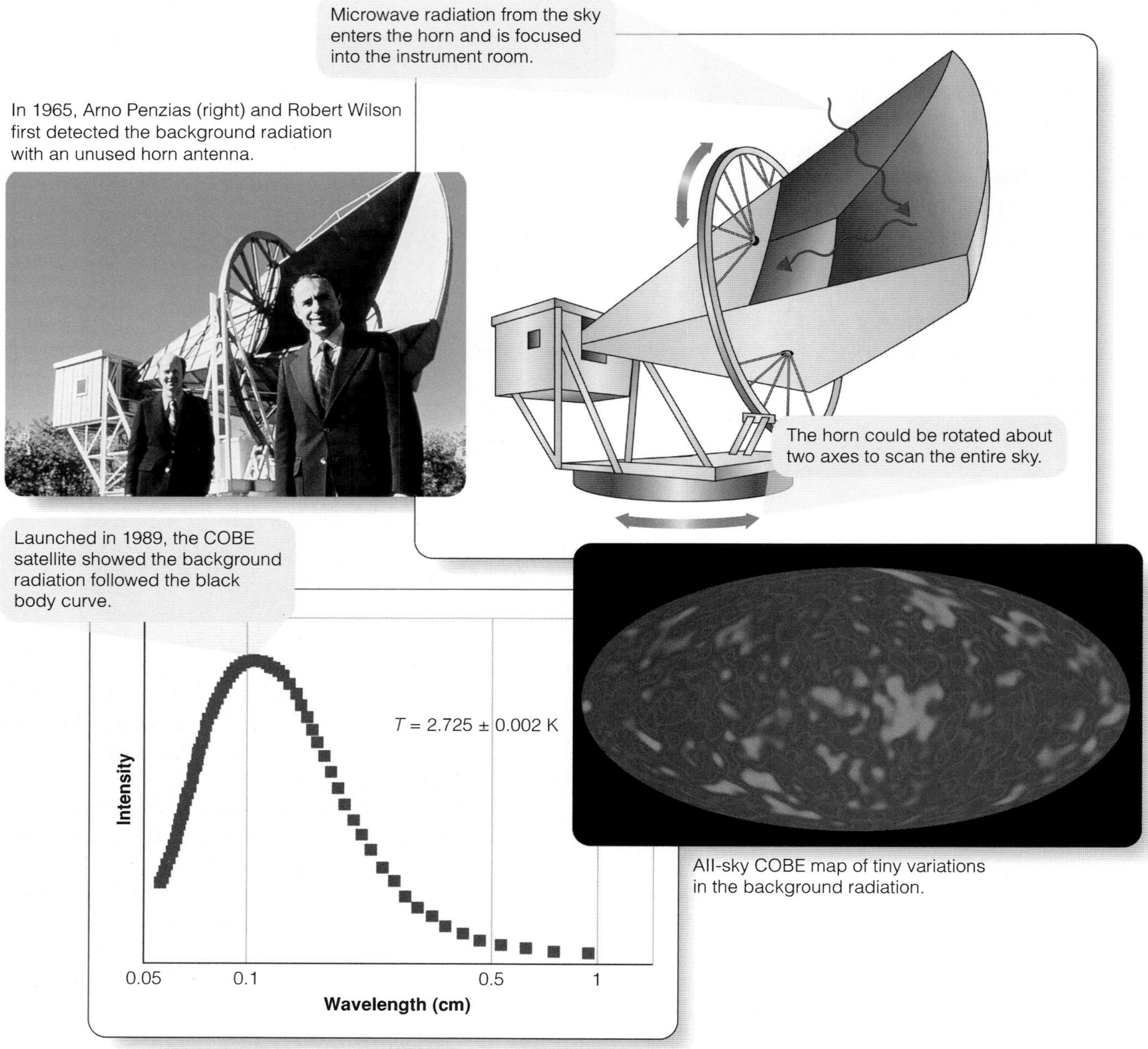

Microwave radiation from the sky enters the horn and is focused into the instrument room.
In 1965, Arno Penzias (right) and Robert Wilson first detected the background radiation with an unused horn antenna.
The horn could be rotated about two axes to scan the entire sky.
Launched in 1989, the COBE satellite showed the background radiation followed the black body curve.
T = 2.725 ± 0.002 K
Intensity
0.05
0.1
0.5
1
Wavelength (cm)
All-sky COBE map of tiny variations in the background radiation.

■ Figure 15-7

When the cosmic microwave background radiation was first detected in 1963, technology did not allow measurements at many wavelengths. Not until infrared detectors could be put in orbit was it conclusively shown that the background radiation, as predicted by theory, followed the black body curve. (Photo, AT&T Archives; COBE map, NASA)

ness of the sky (■ Figure 15-7). Their measurements showed a peculiar noise in the system, which they at first attributed to droppings from pigeons living inside the antenna. Perhaps they would have enjoyed cleaning the antenna more if they had known they would win the 1978 Nobel Prize for physics for the discovery they were about to make.

When the antenna was cleaned, they measured the brightness of the sky at radio wavelengths and again found the low-level radio noise. The pigeons were innocent, but what was causing the signal?

The explanation for the noise goes back to 1948, when George Gamow predicted that the early big bang would be very hot and would emit copious black body radiation. A year later, physicists Ralph Alpher and Robert Herman pointed out that the large redshift of the big bang would lengthen the wavelengths of the radiation into the far-infrared and radio part of the spectrum. There was no way to detect this radiation until the mid-1960s, when Robert Dicke at Princeton concluded the radiation should be just strong enough to detect with newly developed techniques. Dicke and his team began building a receiver.

When Penzias and Wilson heard of Dicke's work, they recognized the noise they had detected as radiation from the big bang, the **cosmic microwave background radiation.**

The detection of the background radiation was tremendously exciting, but astronomers wanted confirmation. Theory predicted that the radiation should look like black body radiation coming from a very cool source, but the critical observations in the far-infrared could not be made from the ground. It was not until January 1990 that satellite measurements confirmed that the background radiation was black body radiation with an apparent temperature of 2.725 ± 0.002 K—in good agreement with the theory.

It may seem strange that the hot gas of the big bang seems to have a temperature of only 2.7 K, but recall the tremendous redshift. Observers on Earth see light that has a redshift of about 1100—that is, the wavelengths of the photons are about 1100 times longer than when they were emitted. The gas clouds that emitted the photons had a temperature of about 3000 K, and they emitted black body radiation with a λ_{max} of about 1000 nm (Chapter 6). Although this is in the near-infrared, the gas would also have emitted enough visible light to glow orange-red. But the redshift has made the wavelengths about 1100 times longer, so λ_{max} is about 1 million nm (equivalent to 1 mm). Thus, the hot gas of the big bang seems to be 1100 times cooler, about 2.7 K.

The cosmic microwave background radiation is evidence that a big bang really did occur. In fact, the evidence is so strong that an alternative theory of the universe was abandoned. The **steady-state theory,** popular with some astronomers during the 1950s and 1960s, held that the universe was eternal and unchanging. If the universe had no beginning, then it could never have had a big bang. According to the theory, new matter continuously appeared from nothing to maintain the density of the expanding, steady-state universe. The discovery of the cosmic microwave background radiation in 1965 led astronomers to abandon the steady-state theory within a few years. You can be sure there was a big bang because, using radio and infrared telescopes, you can observe it happening.

The Story of the Big Bang

Simple observations of the darkness of the night sky and the redshifts of the galaxies tell you that the universe must have begun with a big bang. In fact, you have direct observational evidence of the big bang in the form of the cosmic microwave background radiation. Theorists can combine these observations with modern physics to tell the story of how the big bang occurred. As you review this story, remember that the big bang did not occur at a specific place. It filled the entire volume of the universe from the first moment.

Cosmologists cannot begin their history of the big bang at time zero because no one understands the physics of matter and energy under such extreme conditions, but they can come close. If you could visit the universe when it was very young, only 10 millionths of a second old, for instance, you would find it filled with high-energy photons having a temperature well over 1 trillion (10^{12}) K and a density greater than 5×10^{13} g/cm^3, nearly the density of an atomic nucleus. When astronomers say the photons had a given temperature, they mean that the photons were the same as black body radiation emitted by an object of that temperature. Thus, the photons in the early universe were gamma rays of very short wavelength and therefore very high energy. When astronomers say that the radiation had a certain density, they refer to Einstein's equation $E = m_0c^2$. You can express a given amount of energy in the form of radiation as if it were matter of a given density.

If photons have enough energy, two photons can combine and convert their energy into a pair of particles—a particle of normal matter and a particle of **antimatter.** When an antimatter particle meets its matching particle of normal matter—when an antiproton meets a normal proton, for example—the two particles annihilate each other and convert their mass into energy in the form of two gamma rays. In the early universe, the photons had enough energy to produce proton–antiproton pairs or neutron–antineutron pairs. But when the particles collided with their antiparticles, they converted their mass back into energy. Thus, the early universe was a dynamic soup of energy flickering from photons into particles and back again.

While all this went on, the expansion of the universe cooled the gamma rays to lower energy. By the time the universe was 0.0001 second old, its temperature had fallen to 10^{12} K. By this time, the average energy of the gamma rays had fallen below the energy equivalent to the mass of a proton or a neutron, so the gamma rays could no longer produce such heavy particles. The particles combined with their antiparticles and quickly converted most of the mass into photons.

It would seem from this that all the protons and neutrons should have been annihilated with their antiparticles, but for reasons that are poorly understood a small excess of normal particles existed. For every billion protons annihilated by antiprotons, one survived with no antiparticles to destroy it. Consequently, you live in a world of normal matter, and antimatter is very rare.

Although the gamma rays did not have enough energy to produce protons and neutrons, they could produce electron–positron pairs, which are about 1800 times less massive than protons and neutrons. This continued until the universe was about 4 seconds old, at which time the expansion had cooled the gamma rays to the point where they could no longer create electron–positron pairs. The protons, neutrons, and electrons of which our universe is made were produced during the first 4 seconds of its history.

This soup of hot gas and radiation continued to cool and eventually began to form atomic nuclei. High-energy gamma rays can break up a nucleus, so the formation of such nuclei could not occur until the universe had cooled somewhat. By the time the universe was about 2 minutes old, protons and neutrons could link to form deuterium, the nucleus of a heavy hydrogen atom; by the end of the next minute, further reactions began converting deuterium into helium. But no heavier atoms could be built because no stable nuclei exist with atomic weights of 5 or 8 (in units of the hydrogen atom). Cosmic element building during the big bang had to proceed rapidly, step by step, like someone hopping up a flight of stairs (■ Figure 15-8). The lack of stable nuclei at atomic weights of 5 and 8 meant there were missing steps in the stairway, and the step-by-step reactions had great difficulty jumping over these gaps. A tiny amount of lithium was produced but nothing heavier.

By the time the universe was 30 minutes old, it had cooled sufficiently that nuclear reactions had stopped. About 25 percent of the mass was helium nuclei, and the rest was in the form of protons—hydrogen nuclei. This is the cosmic abundance seen today in the oldest stars. (The heavier elements, remember, were built later by nucleosynthesis inside many generations of massive stars.) The cosmic abundance of helium was fixed during the first minutes of the universe.

At first, the universe was dominated by radiation. The gamma rays interacted continuously with the matter, and they cooled together as the universe expanded. The gas was ionized because it was too hot for the nuclei to capture electrons to form neutral atoms, and the free electrons made the gas very opaque. A photon could not travel very far before it collided with an electron and was deflected. Thus, radiation and matter were locked together.

■ Figure 15-8

Cosmic element building. During the first few minutes of the big bang, temperatures and densities were high, and nuclear reactions built heavier elements. Because there are no stable nuclei with atomic weights of 5 or 8, the process built very few atoms heavier than helium.

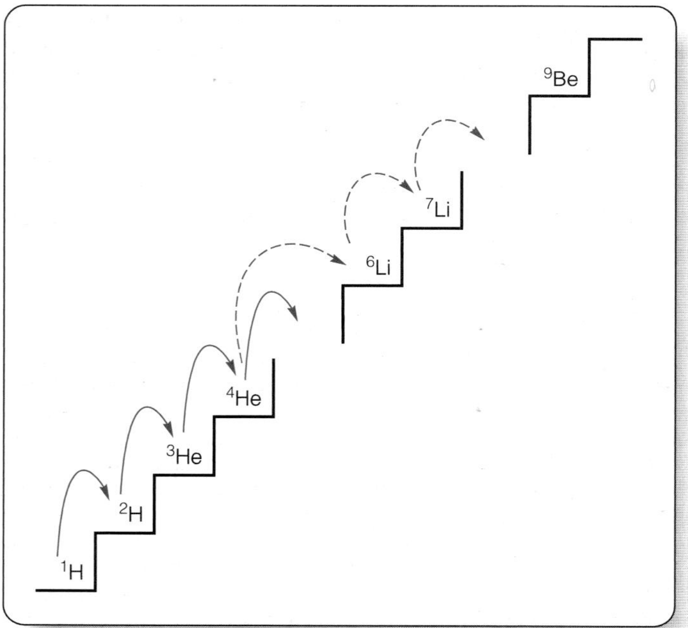

As the young universe expanded, it went through three important changes. First, when the universe reached an age of about 50,000 years, the density of the energy present as photons became less than the density of the gas. Before this, matter could not clump together because the intense sea of photons smoothed the gas out. Once the density of the radiation fell below that of matter, the matter could begin to draw together under the influence of gravity and form the clouds that eventually became galaxies.

As the universe continued to expand, the ionized gas became less dense, which meant that the free electrons were spread farther apart. As the universe reached the age of a few hundred thousand years, the second important change began. The free electrons were spread so far apart that the photons could travel for thousands of parsecs before getting scattered off an electron. That is, the universe began to grow more transparent. At about the same time, the third change happened. As the falling temperature of the universe reached 3000 K, protons were able to capture and hold free electrons to form neutral hydrogen, a process called **recombination.** As the free electrons were gobbled up, the gas finally became transparent, and the photons could travel through the gas without being absorbed or deflected (■ Figure 15-9). Although the gas continued to cool, the photons no longer interacted with the gas, and consequently the photons retained the black body temperature that the gas had at recombination. Those photons, with a black body temperature of 3000 K, are what is observed today as the cosmic microwave background radiation. Remember that the large redshift makes that gas appear to have a temperature of about 2.7 K.

Recombination left the gas of the big bang neutral and transparent. At first the universe was filled with the glow of the hot gas, but as the universe expanded and cooled, the glow faded into the infrared, and the universe entered what cosmologists call the **dark age,** a period lasting hundreds of millions of years during which the universe expanded in darkness.

The dark age ended as the first stars were born. The gas from which these first stars formed contained almost no metals and was consequently highly transparent. Models show that, because the first stars formed from this metal-poor gas, they were very massive, very luminous, and very short lived. That first violent burst of star formation produced enough ultraviolet light to begin ionizing the gas, and today's astronomers, looking back to the most distant visible quasars and galaxies, can see traces of that **reionization** of the universe (■ Figure 15-10).

■ Figure 15-11 summarizes the story of the big bang, from the formation of helium in the first three minutes, through energy–matter equality, recombination, and finally reionization. Reionization marks the end of the dark ages and the beginning of the age of stars and galaxies in which you live today.

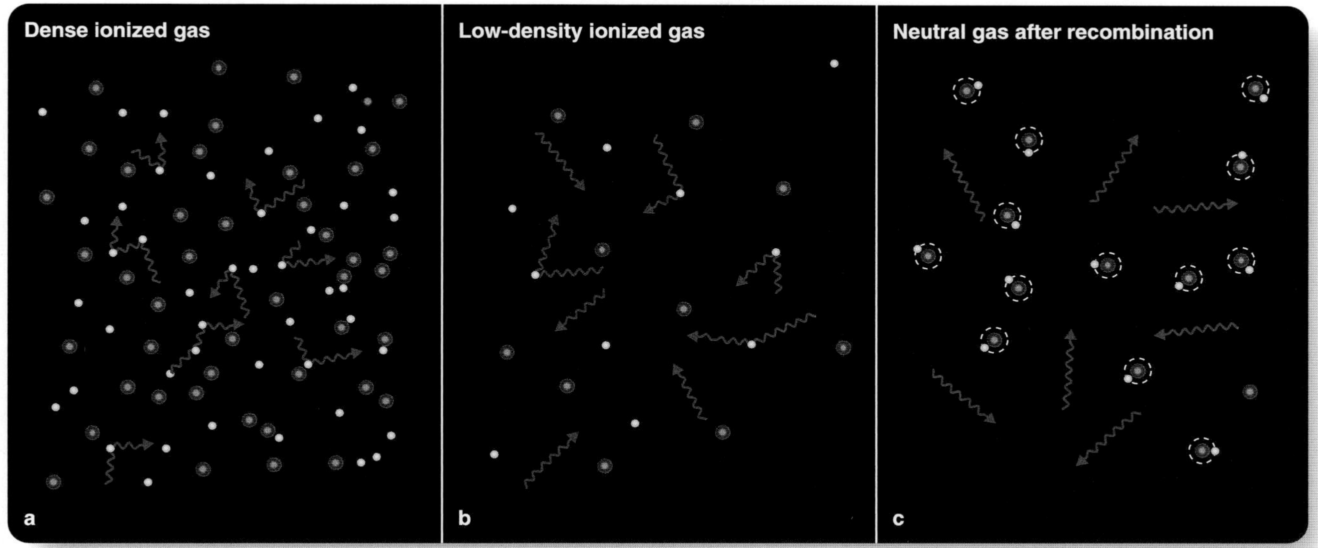

Figure 15-9

Photons scatter from electrons (blue) easily but hardly at all from the much more massive protons (red). (a) When the universe was very dense and ionized, photons could not travel very far before they scattered off an electron. This made the gas opaque. (b) As the universe expanded, the electrons were spread further apart, and the photons could travel farther; this made the gas more transparent. (c) After recombination, most electrons were locked to protons to form neutral atoms, and the gas was highly transparent.

Inquire I Review I Analyze

How do you know there was a big bang?

The cosmic microwave background radiation consists of photons emitted by the hot gas of the big bang, so when astronomers detect those photons, they are "seeing" the big bang. Of course, all scientific evidence must be interpreted, so you must understand how the big bang could produce radiation all around you before you can accept the background radiation as evidence. First, you must recognize that the big bang event filled all of the universe with hot, dense gas. The big bang didn't happen in a single place; it happened everywhere. At recombination, the expansion of the universe reached the stage where the matter became transparent and the radiation was free to travel through space. Today that radiation from the age of recombination arrives from all over the sky. It is all around you because you are part of the big bang event, and if you look out into space to great distance, you look back in time and see the hot gas in any direction you look. You can't see the radiation as light because of the large redshift that has lengthened the wavelengths by a factor of 1100 or so, but

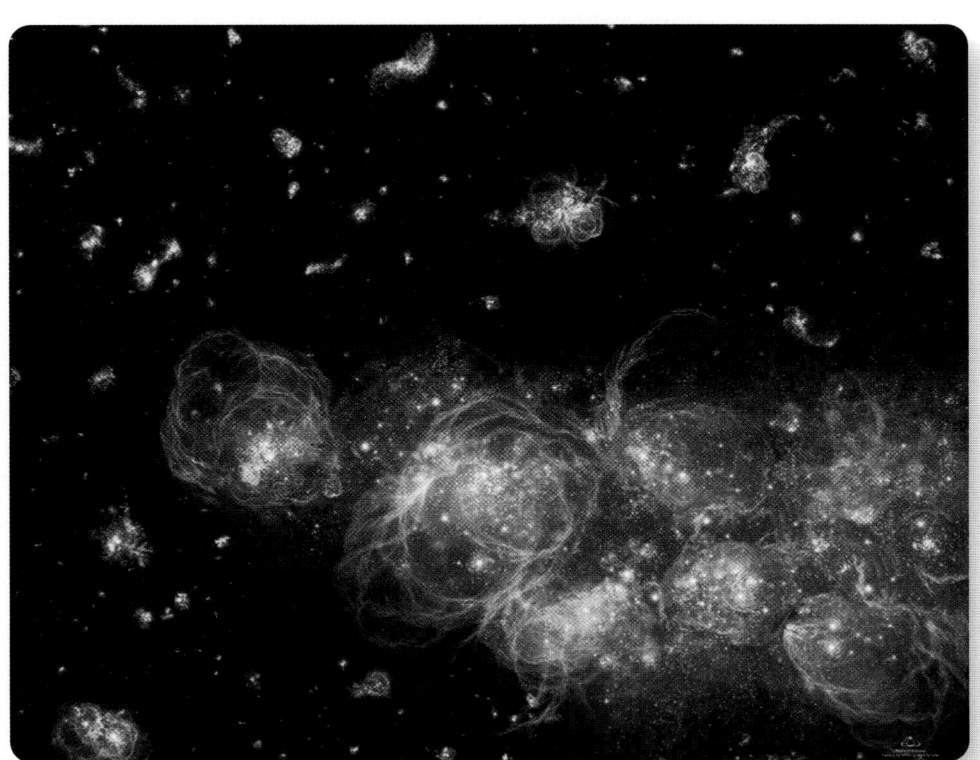

Figure 15-10

In this artist's conception of reionization, the first stars produce floods of ultraviolet photons that ionize the gas in expanding bubbles. Such a storm of star formation ended an age when the universe had expanded in darkness. Spectra of the most distant quasars reveal that those first galaxies were surrounded by neutral gas that had not yet been fully ionized. Thus the look-back time allows modern astronomers to observe the age of reionization. (K. Lanzetta, SUNY, A. Schaller for STScI, and NASA)

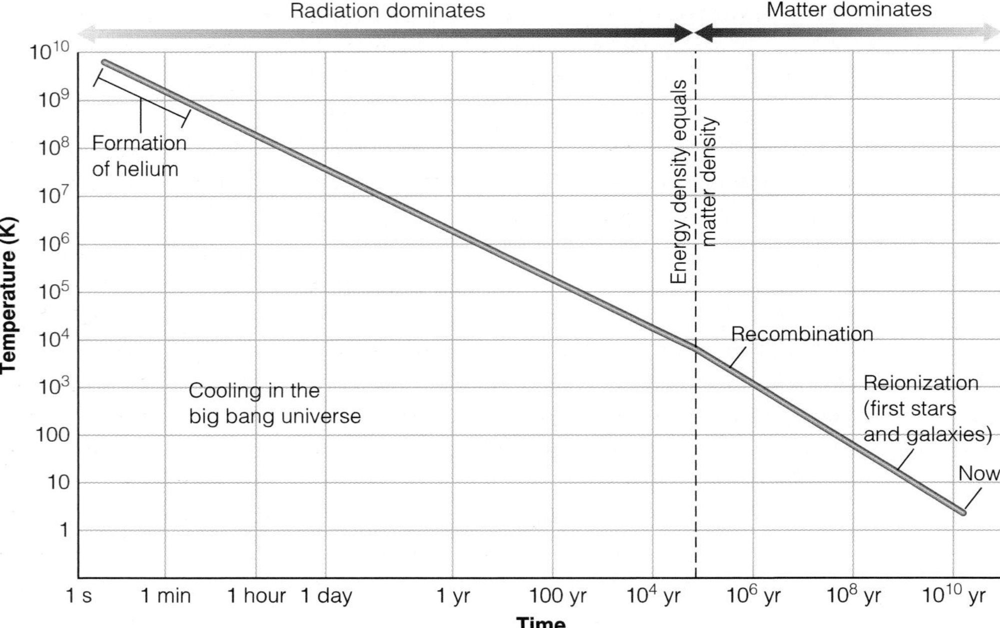

■ Figure 15-11

During the first few minutes of the big bang, some hydrogen was fused to produce helium, but the universe quickly became too cool for such fusion reactions to occur. The rate of cooling increased as matter began to dominate over radiation. Recombination freed the radiation from the influence of the gas, and reionization was caused by the birth of the first stars.

you could detect the radiation as photons with infrared and short radio wavelengths.

With this interpretation, the cosmic microwave background radiation is powerful evidence that there was a big bang. That tells you how the universe began. Why do you think the universe cannot have a center?

■ ■ ■

Connections: So far in this chapter you have compared your simple expectations with basic observations, and you have discovered some fundamental properties of the universe. Now it is time to stretch your imagination and refine your expectations. The universe is much more interesting than you have imagined to this point.

15-2 The Shape of Space and Time

BY NOW YOU ARE PROBABLY WONDERING how the universe can expand if it does not have an edge or a center. Solving that problem is the key to understanding modern cosmology, but your everyday expectations about how the world works are not a help. To solve the puzzle, you must put your reasonable expectations on hold and look carefully at how space and time behave on cosmic scales.

Looking at the Universe

Let's begin by carefully examining the general appearance of the universe. Not only do you see no edge to the universe and no center, there is no preferred direction.

The gas and dust in our own galaxy blocks your view of the distant universe along the Milky Way, but once you correct for that, the universe looks the same over the entire sky. Astronomers refer to this as **isotropy,** the property of being the same in all directions. No matter what direction you look you see the same kinds of galaxies and galaxy clusters. The universe seems isotropic.

Furthermore, the cosmic microwave background radiation is isotropic. Astronomers must correct for Doppler shifts caused by the motion of Earth around the sun, the sun around our galaxy, and our galaxy around the center of mass of the local group, but those are only local variations. Once those local variations are taken into account, the background radiation is almost perfectly uniform over the entire sky.

It certainly appears that the universe is the same everywhere. Of course, there are local variations; some points in space are inside galaxies, and some are isolated far from any galaxy, but overall, in its general properties, one place in the universe is the same as any other place. That is, there are no special places in the universe. You cannot point at any particular location in the universe and say, "That location is different from any other place in the universe." Astronomers refer to this as **homogeneity,** the property of being uniform. On the largest scales, the universe appears to be homogeneous.

Astronomers observe that the universe is isotropic and homogeneous on the largest scales, and that leads to an assumption that is critically important in cosmology. The **cosmological principle** says that any observer in any galaxy sees the same general features of the universe. Of course some observers live in spiral galaxies and some live in elliptical galaxies, but those are only local variations. According to the cosmological principle, when you look at the general features of the universe, you should see the same kind of universe no matter where your planet is located.

Evolutionary changes are not included in the cosmological principle. If the universe is expanding and the galaxies are evolving, observers living at different times may see galaxies at different stages. The cosmological principle says that once observers correct for evolutionary changes, they should see the same general features.

The cosmological principle is actually an extension of the Copernican principle. Copernicus said that Earth is not in a special place; it is just one of a number of planets orbiting the sun. The cosmological principle says that there are no special places in the universe. Local irregularities aside, one place is just like another. Your location in the universe is typical of all other locations.

If you accept the cosmological principle, then you may not imagine that the universe has an edge or a center. Such locations would not be the same as all other locations. Observations confirm this conclusion. No matter what direction you look in the sky, no matter where you look in space, the universe is generally the same. You see no edge and no center.

The Shape of the Expanding Universe

How can the universe expand if it has no edge and no center? You are now ready to resolve that central puzzle of cosmology. To do so you must understand the nature of space and time. Remember that you are a small creature, and you live on a small world. The distances and times you experience are quite small. It should not surprise you that space and time are actually more complicated than your expectations suggest.

Study **The Nature of Space-Time** on pages 338 and 339 and notice three important ideas. First, the redshifts of the galaxies are not produced by the Doppler effect. The galaxies do not move at high velocities through space. Space is expanding and carrying the galaxies away from each other and stretching photons to longer and redder wavelengths. Second, notice that the curvature of space-time allows the universe to have no edge or center even if it is finite. Finally, notice that you could make measurements in our three-dimensional universe to determine the curvature of space-time.

It seems to violate common sense that the universe can expand without having a center or an edge, but you must not depend on your limited imagination as a three-dimensional creature. You must speak carefully and thoughtfully when you discuss cosmology so as not to mislead yourself (Window on Science 15-2). You must not try to visualize the expansion of the universe as an outer edge moving into previously unoccupied space. Open, flat, or closed, the universe has no edge, so it does not need additional room to expand. The universe contains all of the volume that exists, and the expansion is a change in the nature of space-time that causes that volume to increase.

Model Universes

Modern cosmologists have been able to use general relativity to construct mathematical models of the universe under different assumptions. For example, they can describe the nature of the universe if it is open, closed, or flat. These models have had a strong influence on the development of cosmology.

The general curvature of the entire universe is determined by its density. If the average density of the universe is equal to the **critical density,** or 9×10^{-30} g/cm^3, space-time will be flat. If the average density of the universe is less than the critical density, the universe is negatively curved and open. If the average density is greater than the critical density, the universe is positively curved and closed.

You learned earlier that an open universe and a flat universe are both infinite. If their behavior is ruled entirely by their matter density, both kinds of universes will expand forever. The gravitation of the matter in the universe, present as a curvature of space-time, will cause the expansion to slow, but, if the universe is open, it will never come to a stop.

The behavior of these cosmological models is shown in ■ Figure 15-12. The parameter R, which comes out of the mathematical models, is a measure of the extent to which the universe has expanded. As the universe expands, R gets larger. Of course, R can't be the radius of the universe because there is no center or edge. It might be tempting to think that R is something like the

■ Figure 15-12

Open universe models expand without end, and the corresponding curves fall in the region shaded orange. Closed models expand and then contract again (red curve). A flat universe (dotted line) marks the boundary between open and closed universes. The estimated age of the universe depends on the rate at which the expansion is slowing down.

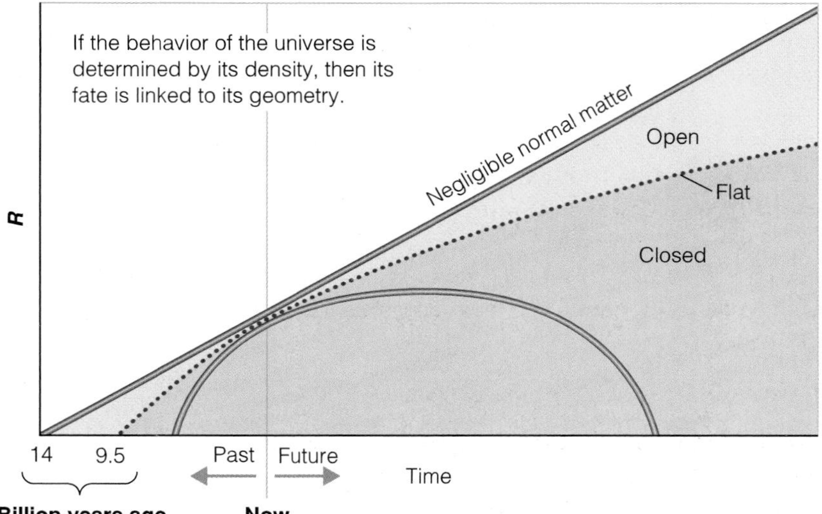

The Nature of Space-Time

In 1929, Edwin Hubble discovered that the redshifts of the galaxies are proportional to distance — a relationship now known as the Hubble law. It was taken as dramatic evidence that the universe is expanding — that the galaxies are moving away from each other. That leads you to consider not only the nature of the expansion but the nature of space and time.

Distance is the separation between two points in space, and time is the separation between two events. Albert Einstein showed how to relate space to time and treat the whole as space-time. You can think of space-time as a canvas on which the universe is painted, a canvas that can stretch.

Astronomers often express galaxy redshifts as apparent velocities of recession, but these redshifts are not Doppler shifts. They are caused by the expansion of space-time.

For decades textbooks have described the cosmological redshifts using Einstein's relativistic Doppler formula, but that formula applies to motion through space and not to the behavior of space itself. The true relation between redshift and apparent velocity of recession is not accurately known.

Notice that this description of the expansion of the universe means the galaxies are not really moving any more than the raisins are swimming through the bread dough. Except for orbital motion among neighboring galaxies, the galaxies are motionless in space-time, and it is space-time that is stretching like a rubber sheet and carrying the galaxies away from each other.

Could the universe be finite and still have no edge?

How can it expand if it has no edge?

How can the universe expand if there is no space outside for it to expand into?

To answer these questions you must use Einstein's theory of general relativity, which predicts that the presence of mass can curve space-time, a curvature you experience as gravity. Furthermore, the theory predicts that even empty space-time can have a curvature, so what you see in the universe depends on the overall curvature of space-time. Finally, the theory shows that space-time could have the property of expansion. That is, the expansion of space-time is not caused by external

What are the cosmological redshifts?

A distant galaxy emits a short-wavelength photon toward our galaxy.

Grid shows expansion of space-time.

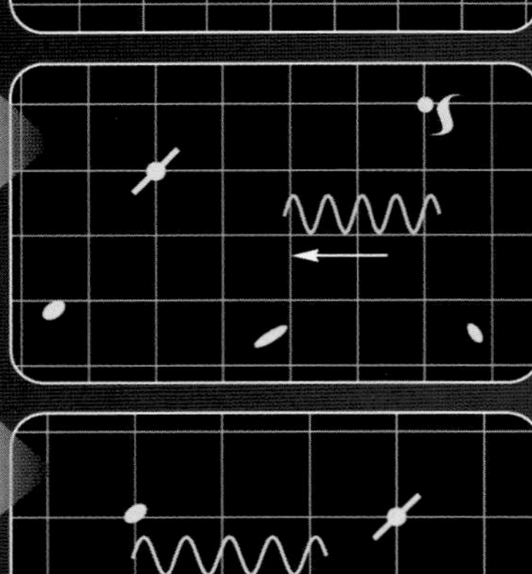

The expansion of space-time stretches the photon to longer wavelength as it travels.

The farther the photon has to travel, the more it is stretched.

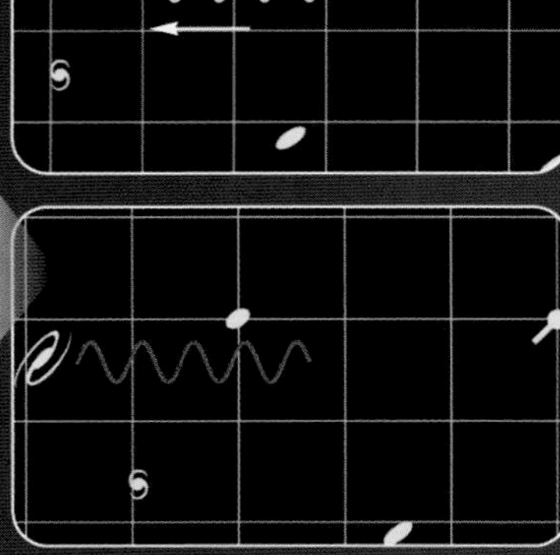

When the photon arrives at our galaxy, you see it with a longer wavelength — a redshift that is proportional to distance.

The stretching of space-time not only moves the galaxies away from each other, but it lengthens the wavelength of photons traveling through space-time.

To think about space-time curvature, you can use an analogy of a two-dimensional ant on an orange. If he is truly two-dimensional, he will not be able to understand up and down. He can only travel forward and back, right and left. Then as he walks over the surface of his spherical universe, he will eventually realize he has been everywhere because his universe is covered by his footprints. He will conclude that he lives in a finite universe that has no edge and no center.

Similarly, our universe could be curved back on itself so that it could be finite but you would never find an edge. To find out how our universe is really curved, astronomers must study very distant objects.

Notice that the center of the orange cannot be the center of the ant's two-dimensional universe because the center of the orange does not lie on the surface. That is, the center of the orange is not part of the ant's universe.

Don't forget: These analogies are only two-dimensional. When you think about our universe, you must think of a three-dimensional universe. It is as difficult for you to imagine curvature in our three-dimensional universe as it is for ants to imagine curvature in their two-dimensional universes.

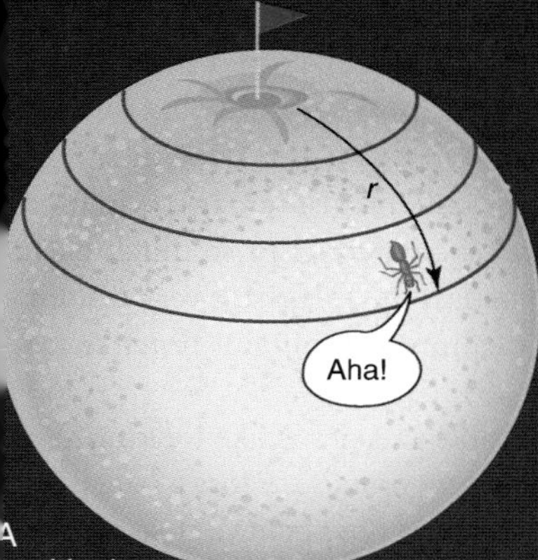

There are three possible ways our universe might be curved. It could have positive curvature, analogous to the surface of a sphere. Models of such universes are called **closed universes** because they are finite. Of course, ours could be a **flat universe** with zero curvature. Another possibility is that our universe has negative curvature, analogous to a two-dimensional universe shaped like a saddle or a potato chip. Models of these universes are called **open universes.**

A positively curved universe, also called a closed universe, is finite and has no edge. In the two-dimensional analogy, ants on an orange would notice that the areas of large circles were less than πr^2.

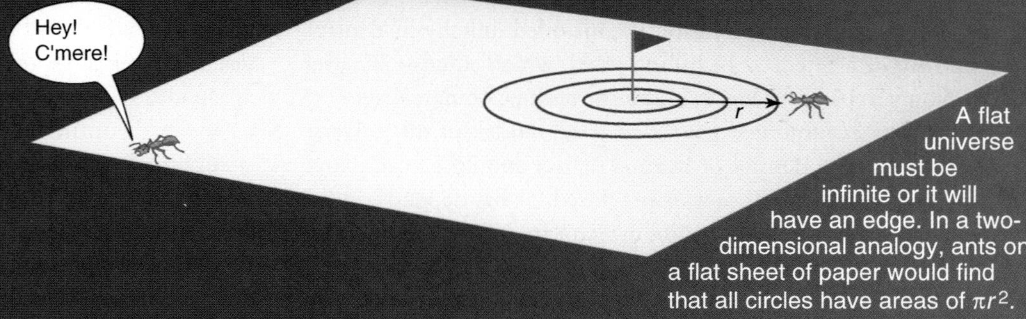

A flat universe must be infinite or it will have an edge. In a two-dimensional analogy, ants on a flat sheet of paper would find that all circles have areas of πr^2.

Although our ant might be unable to sense a third dimension, it could still measure the curvature of its universe by drawing circles. On a flat universe, the area of a circle would always be πr^2 no matter how big the circle was. But on a spherical surface, large circles would contain less than πr^2. On a saddle-shaped surface, large circles would contain more than πr^2. In the same way, you could detect the curvature of our three-dimensional universe, but you would have to make measurements over very large distances.

Whether our universe is closed, flat, or open, it cannot have an edge or a center.

A negatively curved universe, also called an open universe, must be infinite or it will have an edge. In a two-dimensional analogy, ants on a saddle shape will discover that large circles have areas greater than πr^2.

average distance between galaxies, but remember that galaxies did not exist when the universe was young. For this discussion, you must be content to think of R as an indicator of expansion.

According to the models, if the universe is flat, it will barely slow to a stop after an infinite time. Thus the models predict that an open or a flat universe will expand forever, and the galaxies will eventually become black, cold, solitary islands in a universe of darkness.

These models predict a different fate for a closed universe. In a closed universe, the gravitational field, present as curved space-time, is sufficient to slow the expansion to a stop and make the universe contract. Eventually, the contraction will compress all matter and energy back into the high-energy, high-density state from which the universe began. This end to the universe, a big bang in reverse, has been called the big crunch. Nothing in the universe could avoid being destroyed by this "crunch."

Some theorists have suggested that the big crunch will spring back to produce a new big bang and a new expanding universe. This theory, called the **oscillating universe theory,** predicts that the universe undergoes alternate stages of expansion and contraction. Theoretical work, however, suggests that successive bounces of an oscillating universe would be smaller until the oscillation ran down, and the theory is no longer taken seriously.

Notice in Figure 15-12 that you must know the geometry of the universe before you can determine the age. Earlier we calculated the Hubble time $\frac{1}{H}$ as an estimate of the age of the universe. Figure 15-12 shows that the Hubble time is the age of a model universe that contains so little matter its expansion is hardly slowed at all. If the density of a model universe is higher, the expansion will be slowed, and that means the true age will be less than $\frac{1}{H}$. If the universe is flat, its true age will be $\frac{2}{3}$ of $\frac{1}{H}$. If H equals 70 km/s/Mpc, then $\frac{1}{H}$ equals 14 billion years. But if the universe is flat and its evolution is controlled only by its density, then its true age is $\frac{2}{3}$ of 14 billion years, which equals a bit over 9 billion years. This is less than the age of the globular clusters, so the value of H combined with theoretical models of the universe suggests that the density of normal matter should be quite low.

Whether the universe is open, closed, or flat depends on the density of matter, but it is quite difficult to measure the density of the universe. Astronomers could count galaxies in a given volume, multiply by the average mass of a galaxy, and divide by the volume; but the average mass of a galaxy is poorly known, and many small galaxies are too faint to see even if they are nearby. Studies of light coming from distant galaxies suggest that large amounts of normal matter lie in great, nonluminous clouds between the galaxies, so any estimates must include that matter. Also, you must include the mass equivalent to all of the energy in the universe because mass and energy are related. When all of the mass in the universe is added up, it provides a density of a bit less than 5 percent of the critical density needed to close the universe.

Many astronomers will admit that, for theoretical reasons discussed later, they prefer to believe the universe is flat. Science, however, is not ruled by preferences or by theory, but rather by evidence. Observations of normal matter do not reveal enough density to make the universe flat, but an important factor is missing. Clear evidence exists that there is a large amount of dark matter in the universe. Could there be enough dark matter to make the universe flat?

Dark Matter in Cosmology

There is more to the universe than meets the eye. In Chapter 13, you learned that galaxies have much stronger gravitational fields than you would expect based on the amount of matter you see. Even when you add in the nonluminous gas and dust that you expect to find in galaxies, their gravitational fields are much stronger than you would expect. You concluded that galaxies and clusters of galaxies must contain dark matter.

This is not a small issue. Judging by their gravitational fields, galaxies and clusters of galaxies contain as much as 10 times more matter than you expect from what you see. To correct for this dark matter, you would not just add a small percentage. You have to multiply by a *factor* as large as 10.

In the last few chapters, you have seen a number of different examples of evidence for dark matter. Our galaxy rotates faster in its outer regions than expected. Galaxies in clusters move faster than expected, and the clusters hold onto very hot gas, so they must be much more massive than what you see. Look at ■ Figure 15-13. Gravitational lensing is the most dramatic evidence of dark matter. The galaxy cluster shown contains so much dark matter it warps space-time and focuses the images of very distant galaxies into short arcs. Clearly the universe contains much more dark matter than normal matter.

Looking at the universe of visible galaxies is like looking at a ham sandwich and seeing only the mayonnaise. Most of the universe is invisible, and most astronomers now believe that the invisible matter is not the normal matter of which you and the stars are made. To follow that line of evidence, you must look for the atoms made during the big bang.

During the first few minutes of the big bang, nuclear reactions converted some protons into helium and a small amount into other elements. How much of these elements was created depends critically on the density of the material. Deuterium, for example, is an isotope of hydrogen in which the nucleus contains a proton and a neutron. The amount of deuterium produced in the big bang depends strongly on the density of normal matter. If there were a lot of normal particles such as protons and neutrons, then they would have collided with the deuterium nuclei and converted them into helium. If the density of normal particles was less, more deuterium would survive.

Lithium of atomic mass 7, known as lithium-7, is another nucleus that could have been made in small amounts during the big bang. Figure 15-8 shows that there is a gap between helium and lithium; there is no stable nucleus with atomic mass 5, so regular nuclear reactions during the big bang could not convert he-

■ Figure 15-13

Gravitational lensing shows that galaxy clusters contain much more mass than what is visible. The yellowish galaxies in this image are members of a relatively nearby cluster of galaxies. Most of the objects in this image are blue or red images of very distant galaxies focused by the gravitational field of the cluster. Some of these imaged galaxies may be over 13 billion light-years away. That you can see them at all is evidence that the galaxy cluster contains large amounts of dark matter. (NASA, Benites, Broadhurst, Ford, Clampin, Hartig, Illingworth, ACS Science Team and ESA)

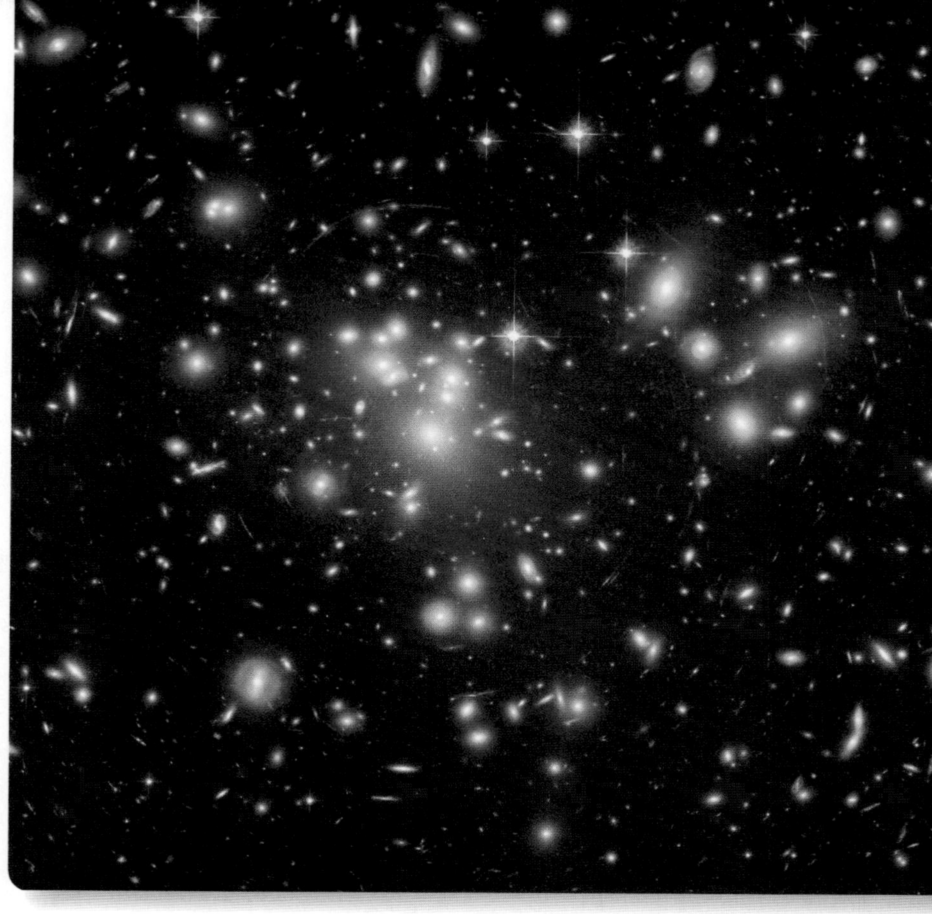

lium into lithium. If, however, the density of normal matter such as protons and neutrons was high enough, a few nuclear reactions could have leaped the gap and produced a few lithium atoms.

Deuterium is so easily converted into helium that none can be made in stars. In fact, stars destroy what deuterium they have by converting it into helium. Lithium too is destroyed in stars. Using the largest telescopes, astronomers have been able to measure the chemical abundance of gas clouds near quasars. The look-back times to these gas clouds is so great that you see them as they were before stars could have altered the abundance of the elements. As shown in ■ Figure 15-14, the observed amount of deuterium sets a lower limit, and the observed abundance of lithium-7 sets an upper limit on the density of normal matter in the universe. The normal matter that you and I are made of cannot make up much more than 4 percent of the critical density. Yet observations show that galaxies and galaxy clusters contain large amounts of dark matter. The protons and neutrons that make up normal matter belong to a family of subatomic particles called *baryons,* so cosmologists believe that the dark matter must be **nonbaryonic matter.** Only a small amount of the mass in the universe can be baryonic; the dark matter must be nonbaryonic.

This nonbaryonic matter must be made up of particles that don't often interact with normal matter. If the nonbaryonic particles don't interact very often, you would not notice them except for their gravitational influences. Particles called axions and neutralinos, for example, are predicted by some theories but have never been detected in the laboratory. If they exist, their mass could exert gravitational forces and they could be part of the dark matter. Data from a large particle detector show that related particles called WIMPs, weakly interacting massive particles, may indeed exist, but the result is still being tested. To what extent WIMPs contribute to the dark matter is unknown.

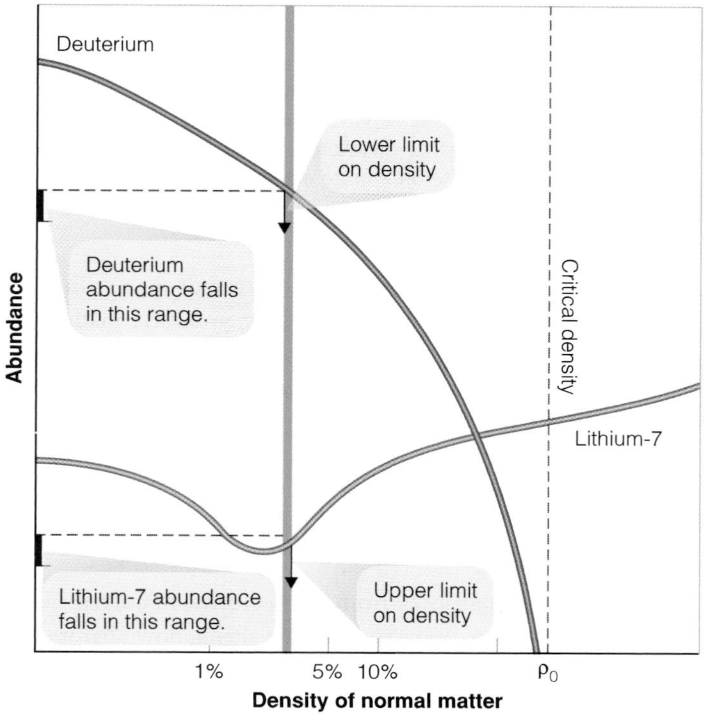

Deuterium
Lower limit on density
Deuterium abundance falls in this range.
Critical density
Lithium-7
Lithium-7 abundance falls in this range.
Upper limit on density
Abundance
1%
5% 10%
ρ_0
Density of normal matter

■ Figure 15-14

This diagram compares observation with theory. Theory predicts how much deuterium and lithium-7 you would observe for different densities of normal matter (red and blue curves). The observed density of deuterium falls in a narrow range shown at upper left and sets a lower limit on the possible density of normal matter. The observed density of lithium-7 sets an upper limit. This means the true density of normal matter must fall in a narrow range represented by the green column. Certainly, the density of normal matter is much less than the critical density.

For some years, astronomers thought that neutrinos could be an important part of the dark matter. You saw in Chapter 7 how observations made by underground detectors suggested that neutrinos oscillate, and that is important because, according to quantum mechanics, if neutrinos oscillate, they cannot have zero mass. Although the mass of neutrinos has not been measured accurately, it is clearly very small. Even a small mass might be important because there are so many neutrinos in the universe—10^8 neutrinos for every normal particle.

Theoretical models of galaxy formation in the early universe show that the dark matter can't be mostly neutrinos. Dark matter composed of neutrinos and similar particles that travel at or near the speed of light is called **hot dark matter.** Such fast-moving particles do not clump together easily and could not have stimulated the formation of objects as small as galaxies and clusters of galaxies. The most successful models of galaxy formation require that the dark matter be made up of **cold dark matter,** meaning the particles move slowly and can clump into smaller structures. WIMPS, for example, are massive and slow moving and could be part of the cold dark matter.

Nonbaryonic dark matter does not interact significantly with normal matter or with photons, which is why you can't see it. But that means that dark matter was not affected by the intense radiation that dominated the universe when it was very young. That radiation prevented normal matter from contracting to begin forming galaxies. But the dark matter was immune, and it could contract to form clouds. Once the density of the radiation fell low enough, normal matter could begin falling into the clouds of dark matter to form the first galaxies. Models with cold, nonbaryonic dark matter are most successful at forming galaxies and clusters of galaxies early in the history of the universe.

Now that you understand more about dark matter, you can evaluate the possibility that the halos of galaxies contain large numbers of very-low-mass stars, brown dwarfs, or planets too faint to see. These **MACHOs,** for Massive Compact Halo Objects, can be detected if they pass between our telescopes on Earth and a distant star. The gravitational field of the MACHO acts as a gravitational lens, focusing the light of the star and making it grow brighter for a period of a few weeks as the MACHO passes between the star and Earth. Extensive searches have detected such events (■ Figure 15-15) but not in large numbers. Furthermore, searches have turned up large numbers of white dwarfs in the halo of our galaxy. However common these halo objects are, they are made of baryonic matter, and the abundance of deuterium and lithium tells us that the dark matter is nonbaryonic. It does not appear that MACHOs and white dwarfs can make up a significant part of the dark matter.

Although there is good evidence that dark matter exists, it has proven difficult to identify. No form of dark matter has been found that is abundant enough to provide the critical density needed to make the universe flat. In fact, all forms of dark matter appear to add up to less than 30 percent of the critical density. As you will see later in this chapter, there is more to the universe than meets the eye, and more even than the dark matter.

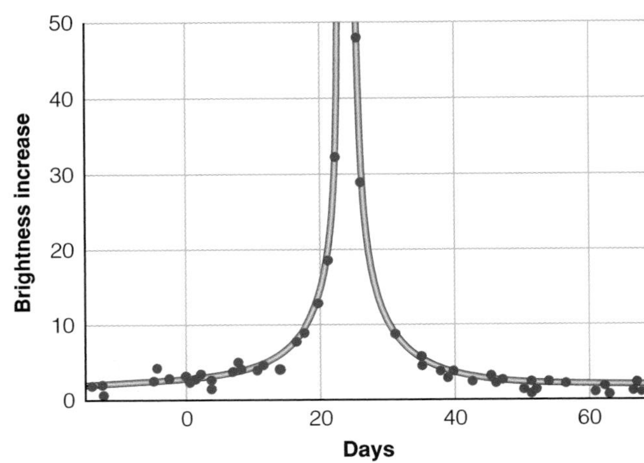

■ **Figure 15-15**

Measurements of the brightness of a distant star (dots) increased and then decreased exactly as predicted by general relativity (curve) for a star's passage behind a gravitational lens caused by a MACHO. In this example, the star brightened by a factor of over 40.

Inquire | Review | Analyze

Why do astronomers think that dark matter can't be baryonic? Isotopes like deuterium and lithium-7 were produced in the first minutes of the big bang, and the abundance of those elements depends strongly on the density of protons and neutrons. Because these particles belong to the family of particles called baryons, astronomers refer to normal matter as baryonic. Measurements of the abundance of deuterium and lithium-7 show that the universe cannot contain more baryons than about 4 percent of the critical density. Yet observations of galaxies and galaxy clusters show that dark matter must make up almost 30 percent of the critical density. Consequently, astronomers conclude that the dark matter must be made up of nonbaryonic particles.

Finding the dark matter is important because the density of matter in the universe determines its curvature. How does curvature allow you to avoid an edge or a center in a finite model of the universe?

■ ■ ■

Connections: By adding the curvature of space-time and effects of dark matter to your cosmology, you have made your theories much more sophisticated. You are now ready to explore the most recent and most exciting advances in cosmology.

15-3 21st-Century Cosmology

IF YOU ARE A LITTLE DIZZY from the weirdness of curved space-time and dark matter, make sure you are sitting down before you read much further. As the 21st century began, cosmologists made some startling discoveries, and all around the world astronomers looked at each other and said, "What? What!" The most amazing thing about these amazing discoveries is that they fit so well with some of the things you have seen earlier in this chapter. To get a running start on these new discoveries, let's go back a couple decades.

Inflation

In 1980, astronomers faced a problem. The big bang models of the universe could not explain two important features of the universe. To solve those two problems, a new theory of the big bang was created, and that theory has been startlingly successful. To introduce the new theory, let's begin with the two problems.

One of the problems is called the **flatness problem.** The universe seems to be balanced near the boundary between an open and a closed universe. That is, it seems nearly flat. Given the vast range of possibilities, from zero to infinite, it seems peculiar that the density of the universe is within a factor of 10 of the critical density that would make it flat. If dark matter is as common as it seems, the density may be as close as a factor of three to being perfectly flat.

Even a small departure from critical density when the universe was young would be magnified by subsequent expansion. To be so near critical density now, the density of the universe during its first moments must have been within 1 part in 10^{49} of the critical density. So the flatness problem is: Why is the universe so nearly flat?

The second problem with the big bang theory is the isotropy of the primordial background radiation. When astronomers correct for the motion of our galaxy, they see the same background radiation in all directions to at least 1 part in 1000. Yet when they look at background radiation coming from two points in the sky separated by more than a degree, they look at two parts of the big bang that were not causally connected when the radiation was emitted. That is, when recombination occurred and the gas of the big bang became transparent to the radiation, the universe was not old enough for any signal to have traveled from one of these regions to the other. Thus, the two spots astronomers look at did not have time to exchange heat and even out their temperatures. Then how did every part of the entire big bang universe get to be so nearly the same temperature by the time of recombination? This is called the **horizon problem** because the two spots are said to lie beyond their respective light–travel horizons.

The key to these two problems and to others involving subatomic physics may lie with a theory called the **inflationary universe** because it predicts a sudden expansion when the universe was very young, an expansion even more extreme than that predicted by the big bang theory.

To understand the inflationary universe, you must recall that physicists know of only four forces—gravity, the electromagnetic force, the strong force, and the weak force (Chapter 7). You are familiar with gravity, and the electromagnetic force is responsible for making magnets stick to refrigerator doors and cat hair stick to wool sweaters charged with static electricity. The strong force holds atomic nuclei together, and the weak force is involved in certain kinds of radioactive decay.

For many years, theorists have tried to unify these forces; that is, they have tried to describe the forces with a single mathematical law. A century ago, James Clerk Maxwell showed that the electric force and the magnetic force were really the same effect, and physicists now count them as a single electromagnetic force. In the 1960s, theorists succeeded in unifying the electromagnetic force and the weak force in what they called the electroweak force, effective only for processes at very high energy. At lower energies, the electromagnetic force and the weak force behave differently. Now theorists have found ways of unifying the electroweak force and the strong force at even higher energies. These new theories are called **grand unified theories,** or **GUTs.**

According to the inflationary universe, the universe expanded and cooled until about 10^{-35} second after the big bang, when it became so cool that the forces of nature began to disconnect from each other and behave in different ways. This released tremendous energy, which suddenly inflated the universe by a factor between 10^{20} and 10^{30} (■ Figure 15-16). At that time the part of the universe that you can see now, the entire observable universe, was no larger than the volume of an atom, but it suddenly inflated to the volume of a cherry pit and then continued its slower expansion to its present extent.

■ Figure 15-16

When the universe was very young and hot (top), the four forces of nature were indistinguishable. As the universe began to expand and cool, the forces separated and triggered a sudden inflation in the size of the universe.

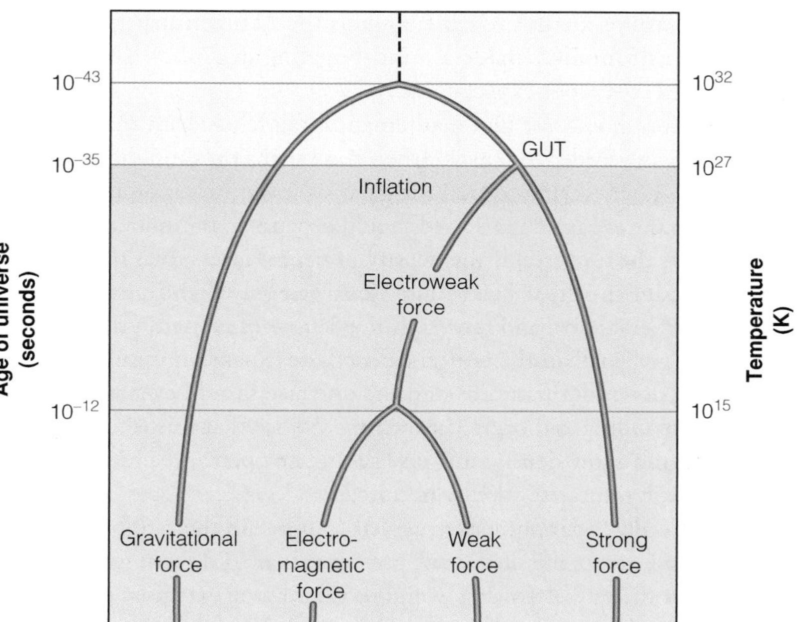

That sudden inflation can solve the flatness problem and the horizon problem. The sudden inflation of the universe would have forced whatever curvature it had toward zero, just as inflating a balloon makes a small spot on its surface flatter. Thus, you now see the universe nearly flat because of that sudden inflation long ago. In addition, because the part of the universe you can see was once no larger in volume than an atom, it had plenty of time to equalize its temperature before the inflation occurred. Now you see the same temperature for the background radiation in all directions.

The inflationary universe is based, in part, on quantum mechanics, and a slightly different aspect of quantum mechanics may explain why there was a big bang at all. Theorists believe that a universe totally empty of matter could be unstable and decay spontaneously by creating pairs of particles until it was filled with the hot, dense state called the big bang. This theoretical discovery has led some cosmologists to believe that the universe could have been created by a chance fluctuation in space-time. In the words of physicist Frank Wilczyk, "The reason there is something instead of nothing is that 'nothing' is unstable."

The inflationary theory predicts that the universe is almost perfectly flat. That is, the true density must equal the critical density. A theory can never be used as evidence, but the beauty of the inflationary theory has given many cosmologists confidence that the universe must be flat. Observations, however, seem to indicate that the universe does not contain enough matter (baryonic plus dark) to be flat, so we must search further to understand the nature of the universe.

The Acceleration of the Universe

Ever since Hubble discovered the expansion of the universe, astronomers have known what to expect. They have also known that expectations are not reliable; and, as the 21st century approached, astronomers made a mind-boggling discovery that contradicted everyone's expectations.

Both common sense and mathematical models suggest that as the galaxies recede from each other, the expansion should be slowed by gravity trying to pull the galaxies toward each other. How much the expansion is slowed should depend on the amount of matter in the universe. If the density of matter is less than the critical density, the expansion should be slowed only slightly, and the universe should expand forever. If the density of matter in the universe is greater than the critical density, the expansion should be slowing down dramatically, and the universe should eventually stop expanding and begin contracting. Notice that this is the same as saying a low-density universe should be open and a high enough density universe should be closed.

For decades, astronomers struggled to measure the Hubble constant and detect the slowing of the expansion. A direct measurement of the rate of slowing would reveal the true curvature of the universe. This was one of the key projects for the Hubble Space Telescope, and two teams of astronomers spent years making the measurements.

Detecting a change in the rate of expansion is a difficult project because it requires accurate measurements of the distances to very remote galaxies, and both teams used the same technique. They calibrated type Ia supernovae as distance indicators. A type Ia supernova occurs when a white dwarf gains matter from a companion star, exceeds the Chandrasekhar limit, and collapses in a supernova explosion. Because all such white dwarfs should collapse at the same mass limit, they should all produce explosions of the same luminosity, and that makes them good distance indicators.

In Chapter 13, you read that astronomers refer to distance indicators such as Cepheid variable stars as *standard candles*, objects of known luminosity. Type Ia supernovae have been described as *standard bombs*. They are titanic explosions, but they all reach the same peak brightness, and that makes them good distance indicators.

The two teams calibrated type Ia supernovae by locating such supernovae occurring in nearby galaxies whose distance was known from Cepheid variables and other reliable distance indicators. Once the peak luminosity of type Ia supernovae had been determined, they could be used to find the distance to more distant galaxies.

Both teams announced their results in 1998. The expansion of the universe is not slowing down. It is speeding up! That is, the expansion of the universe is accelerating (■ Figure 15-17).

The announcement that the expansion of the universe is accelerating was exciting, but astronomers immediately began testing it. It depends critically on the calibration of type Ia supernovae as distance indicators, and some astronomers suggested that the calibration might be wrong (Window on Science 12-1). Some unknown kind of dust might make very distant supernovae too faint without making them too red. No reddening had been detected, so normal dust seemed to be ruled out. Or perhaps the very distant supernovae were too faint because astronomers see them at great look-back times, and some unknown process made those early supernovae a bit less luminous than more recent supernovae. Those problems with the calibration seem to be ruled out by a supernova that occurred in 1997. Known as SN1997ff, it was recognized in 2001 as a type Ia supernova with the very large redshift of 1.7, implying a distance of 10 billion light-years. The observations of the maximum brightness are uncertain, but they seem to rule out problems with the calibration of type Ia supernovae. The universe really does seem to be expanding faster and faster. How can that be?

Dark Energy and Acceleration

If the expansion of the universe is accelerating, then there must be a force of repulsion in the universe, and astronomers are struggling to understand what it could be. One possibility leads back to 1916.

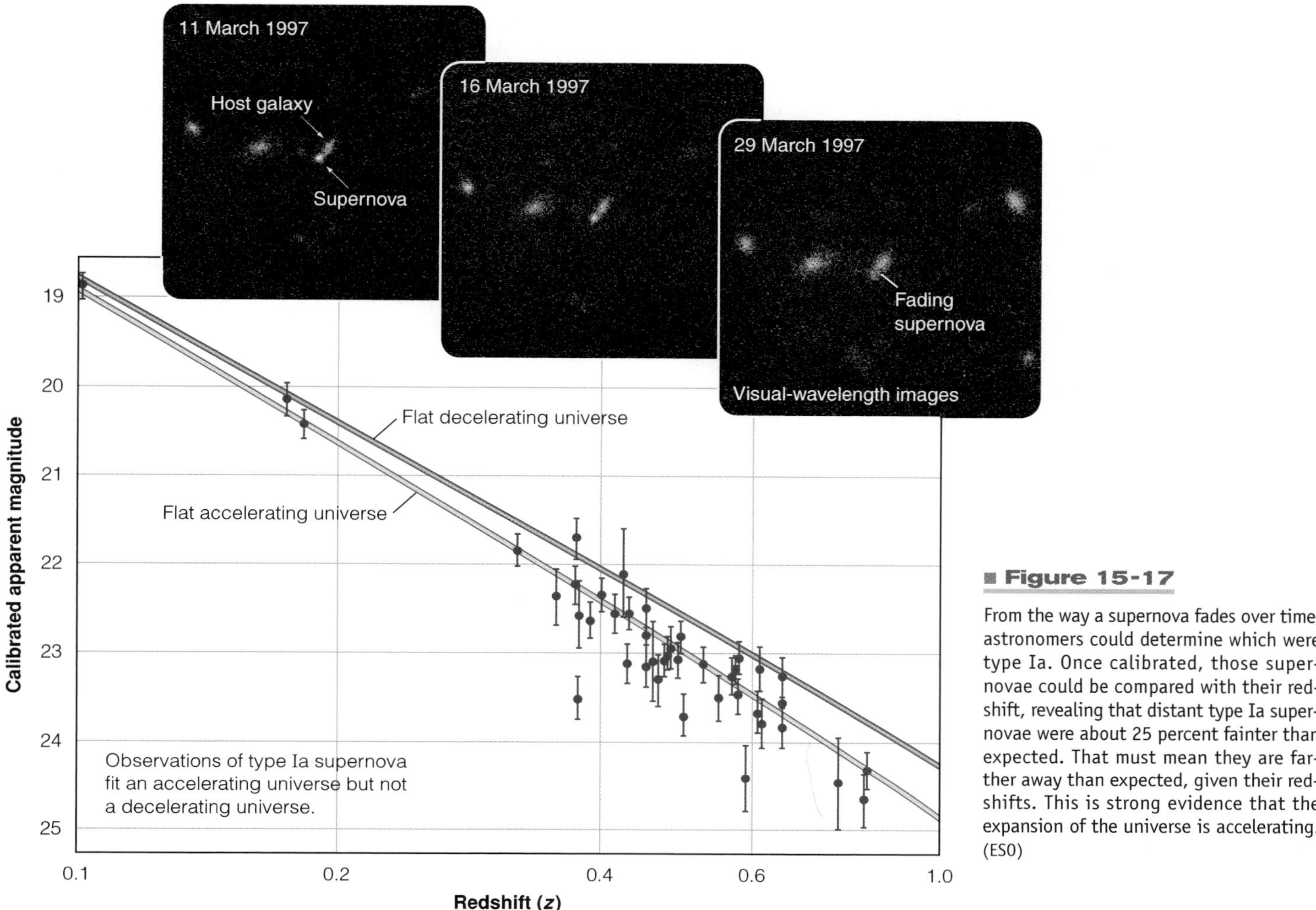

■ Figure 15-17

From the way a supernova fades over time, astronomers could determine which were type Ia. Once calibrated, those supernovae could be compared with their redshift, revealing that distant type Ia supernovae were about 25 percent fainter than expected. That must mean they are farther away than expected, given their redshifts. This is strong evidence that the expansion of the universe is accelerating. (ESO)

When Albert Einstein published his theory of general relativity in 1916, he recognized that his equations describing space and time implied that space had to contract or expand. The galaxies could not float unmoving in space because their gravity would pull them together. The only solutions seemed to be a universe that was contracting under the influence of gravity or a universe in which the galaxies were rushing away from each other so rapidly that gravity could not pull them together. In 1916, astronomers did not yet know that the universe was expanding, so Einstein made what he later said was a mistake.

To balance the attractive force of gravity, Einstein added a constant to his equations called the **cosmological constant,** represented by an upper-case lambda (Λ). The constant plays the mathematical role of a force of repulsion that balances the gravitation in the universe so it does not have to contract or expand. Thirteen years later, in 1929, Edwin Hubble announced that the universe was expanding, and Einstein said introducing the cosmological constant was his biggest blunder. Modern astronomers aren't so sure.

One explanation for the acceleration of the universe is that the cosmological constant is not zero but rather represents a force that drives a continuing acceleration in the expansion of the universe. Because the cosmological constant remains constant with time, the universe would have to have experienced this force throughout its history.

Another solution is to suppose that totally empty space, the vacuum, contains energy, which drives the acceleration. This is an interesting possibility because, for years, theoretical physicists have discussed an energy inherent in empty space. Astronomers have begun referring to this energy of the vacuum as **quintessence.** Unlike the cosmological constant, quintessence need not remain constant over time, so it may be more successful in explaining the observed distribution of galaxies and galaxy clusters.

Whichever explanation is right, the cosmological constant or quintessence, the acceleration alerts you to a form of energy spread throughout space. Astronomers refer to this energy as **dark energy,** the energy that drives the acceleration of the universe but does not contribute to the formation of starlight or the cosmic microwave background radiation.

You will recall that acceleration and dark energy were first discovered when astronomers found that supernovae just a few billion light-years away were slightly fainter than expected. The acceleration of the expansion made those supernovae a bit farther way than expected, and so they looked fainter. Since then, astronomers have continued to find even more distant type Ia supernovae, some as distant as 12 billion light-years. The more distant of those supernovae are not too faint; they are too bright! That reveals even more about dark energy.

The very distant supernovae are a bit too bright because they are not as far away as expected, and that confirms a theoretical prediction based on dark energy. When the universe was young, galaxies were closer together, and gravity could overpower dark energy and slow the expansion. That makes the very distant supernovae a bit too bright. As the universe expanded, galaxies moved farther apart, gravity became weaker than dark energy, and acceleration began. That makes the less distant supernovae a bit too faint. Sometime about 6 billion years ago, the universe shifted gears from deceleration to acceleration, and supernovae make that visible.

Furthermore, dark energy can help you understand the curvature of the universe. The theory of inflation makes the specific prediction that the universe is flat. Dark energy seems to confirm that prediction. According to Einstein's most famous equation, $E = mc^2$, you know that the dark energy is equivalent to a mass spread through space. Baryonic matter plus dark matter makes up about a third of the critical density, and dark energy appears to make up two-thirds. That is, the total density of the universe seems to equal the critical density, which means that the geometry of the universe is flat.

The Age and Fate of the Universe

Acceleration helps with another problem. If the Hubble constant equals 70 km/s/Mpc, then the Hubble time is about 14 billion years. If the universe is flat, then the age of the universe is two-thirds of the Hubble time, which equals about 9 billion years. That is younger than the globular clusters, and that is a problem that has been hard to explain. However, if the expansion of the universe has been accelerating, then it must have been expanding slower in the past, and that means it can be older than two-thirds of the Hubble time. The latest estimates suggest that the true age of the universe is about 14 billion years, and that solves the age problem.

For many years cosmologists have enjoyed saying, "Geometry is destiny." By that they meant that the destiny of the universe is determined by its geometry. An open universe must expand forever, and a closed universe must fall back. But that is true only if the universe is ruled by gravity. If acceleration dominates gravity, then geometry is not destiny, and even a closed universe might expand forever.

The ultimate fate of the universe depends on the nature of dark energy. If dark energy is described by the cosmological constant, then the force driving acceleration does not change with time, and our flat universe will expand forever with the galaxies getting farther and farther apart and using up their gas and dust and the stars dying until each galaxy is isolated, burnt out, dark, and alone. If, however, dark energy is described by quintessence, then it may be increasing with time and the universe may accelerate faster and faster as space pulls the galaxies away from each other and eventually pulls the galaxies apart and then pulls the stars apart and finally rips the individual atoms apart. This has been called the **big rip.** Don't worry. Even if a big rip is in our future, nothing will be happening for at least 30 billion years.

Maybe there will be no big rip. Critically important observations made by the Chandra X-Ray Observatory have been used to measure the amount of hot gas and dark matter in over two dozen galaxy clusters. Because distance is involved in the calculation, the X-ray astronomers could compare the amount of hot gas and dark matter and solve for distance. The most distant of these clusters was 8 billion light-years away. This result is critical for two reasons. First, the redshifts and distances of these galaxies confirm that the universe was slowing down early on but that it shifted gears about 6 billion years ago and is now accelerating. Notice that this method does not depend on type Ia supernovae at all, so it is independent confirmation that acceleration is real. The results are also critical because they are almost good enough to rule out quintessence. If dark energy is described by the cosmological constant and not by quintessence, then there will be no big rip (■ Figure 15-18).

Has a friend ever told you something so amazing you had to check it? Perhaps you called someone else to confirm the news. That's how astronomers feel about acceleration. It is so unexpected it demands confirmation. Type Ia supernovae revealed it, and the Chandra X-ray observations confirm it, but is there any other way to confirm that the universe is flat and accelerating?

One way to measure the curvature of space-time is to draw circles on it and measure the areas of the circles. But that is only practical in a simple cartoon analogy. In reality, astronomers must measure the size of things at great look-back times, and that measurement is yielding yet more exciting results, which are described in the next section.

The Origin of Structure and the Curvature of the Universe

For a number of years, astronomers have studied the distribution of galaxies in space, what they refer to as **large-scale structure.** Those studies have led to astonishing insights that are changing our understanding of the universe.

When you look at galaxies in the sky you see them in clusters ranging from a few galaxies to thousands, and those clusters appear to be grouped into **superclusters** (■ Figure 15-19). The Local Supercluster, in which we live, is a roughly disk-shaped swarm of galaxy clusters 50 to 75 Mpc in diameter. By measuring the redshifts and positions of over 100,000 galaxies in great slices across the sky, astronomers have been able to create maps

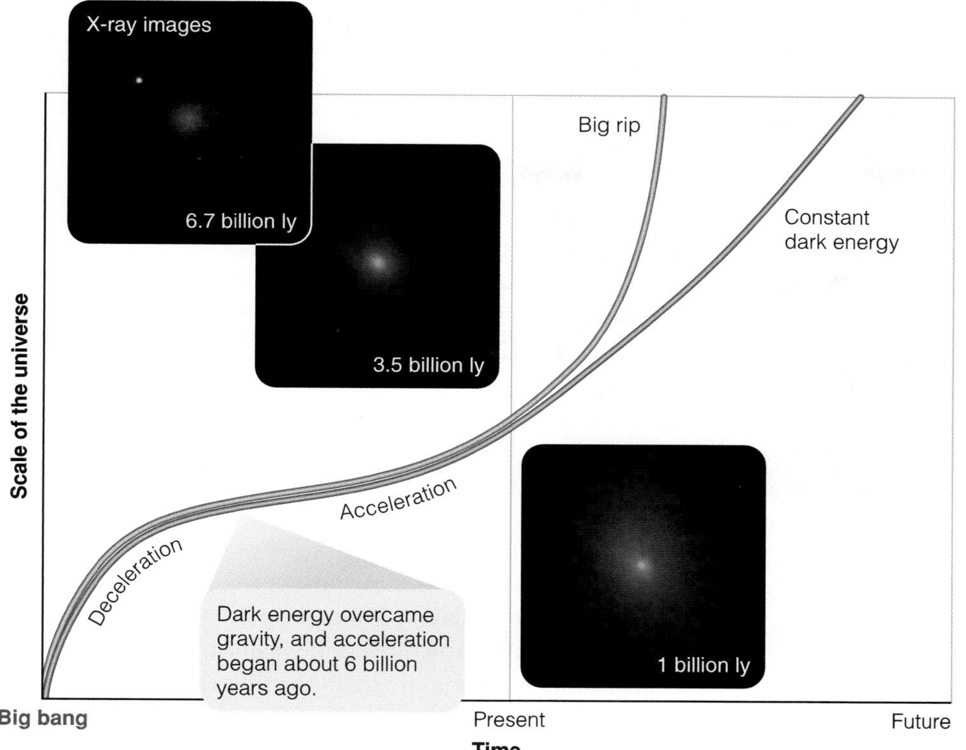

■ Figure 15-18

X-ray observations of hot gas in galaxy clusters confirm that in its early history the universe was decelerating because gravity was stronger than the dark energy. As expansion weakened the influence of gravity, dark energy began to force an acceleration. The evidence is not conclusive, but it most directly supports the cosmological constant and weighs against quintessence, which means the universe may not face a big rip. This diagram is only schematic, and the two curves are drawn separated for clarity; at the present the two curves have not diverged from each other. (NASA/CXC/IoA/S. Allen et al.)

revealing that the superclusters are distributed in long, narrow filaments and thin walls that outline great voids nearly empty of galaxies (■ Figure 15-20). These filaments and walls of superclusters stretching half a billion light-years are the largest structures in the universe. No larger objects exist.

These vast structures are a problem for cosmologists because the cosmic microwave background radiation is very uniform, and that means the gas of the big bang must have been extremely uniform at the time of recombination. Yet the look-back time to the farthest galaxies is 93 percent of the way back to the big bang.

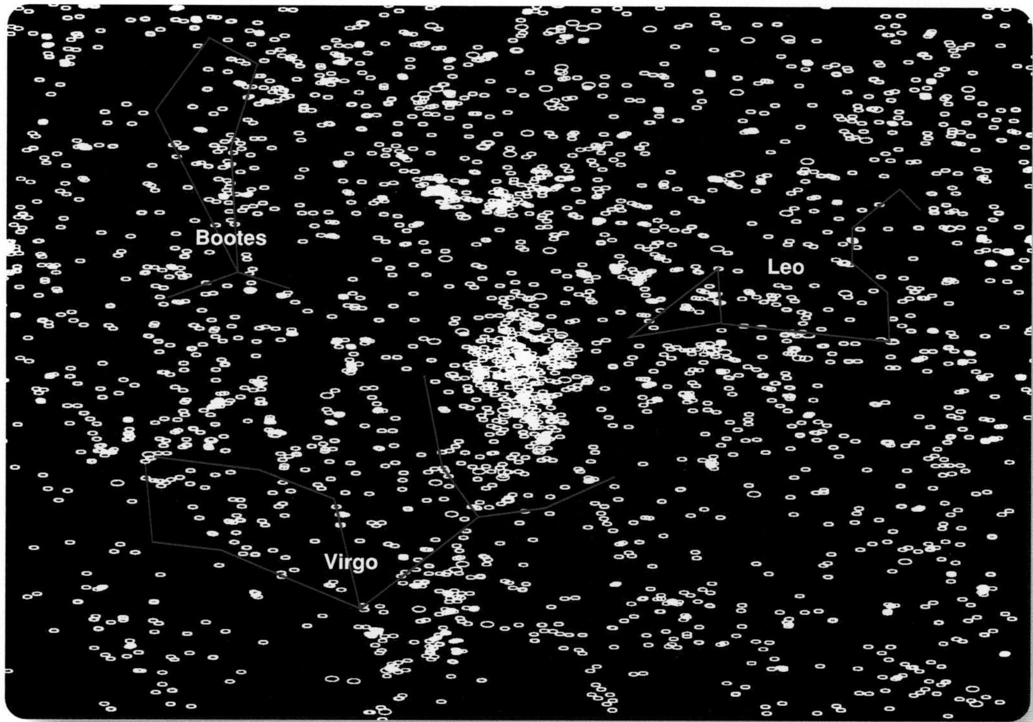

■ Figure 15-19

The distribution of brighter galaxies in the sky reveals the great Virgo cluster (center), containing over 1000 galaxies only about 17 Mpc away. Other clusters fill the sky, such as the more distant Coma cluster just above the Virgo cluster in this diagram. The Virgo cluster is linked with others to form the Local Supercluster.

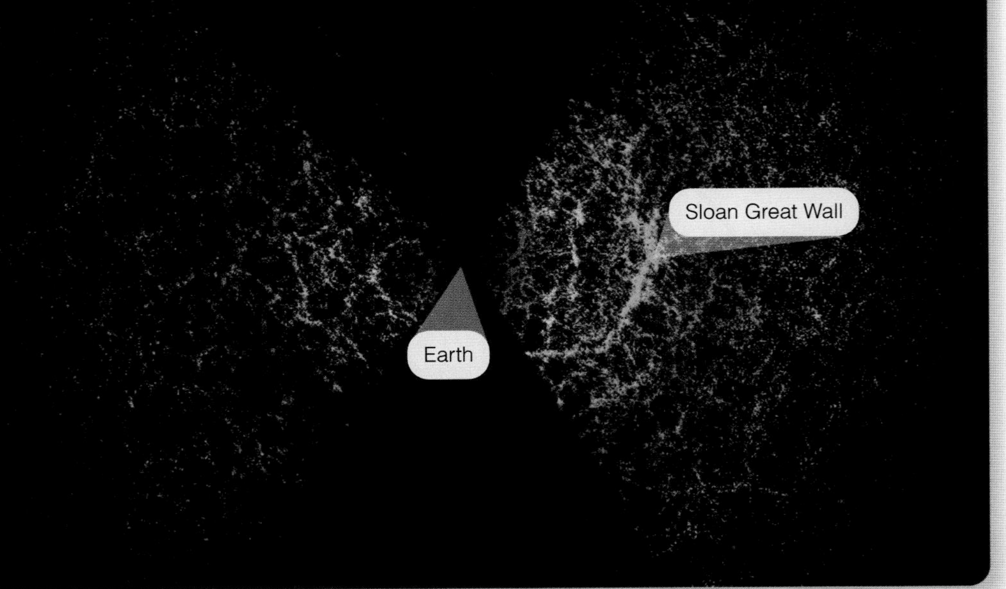
Sloan Great Wall
Earth

■ **Figure 15-20**

Nearly 70,000 galaxies are plotted in this double slice of the universe extending outward in the plane of Earth's equator. The nearest galaxies are shown in red and the more distant in green and blue. The galaxies form filaments and walls enclosing empty voids. The Sloan Great Wall is almost 1.4 billion light-years long and is the largest known structure in the universe. The most distant galaxies in this diagram are roughly 3 billion light-years from Earth. (Sloan Digital Sky Survey)

Growth of Structure in the Universe

Soon after the big bang, radiation and hot gas are almost uniformly spread through the universe.

Cold dark matter, immune to the influence of light, can contract to form clouds...

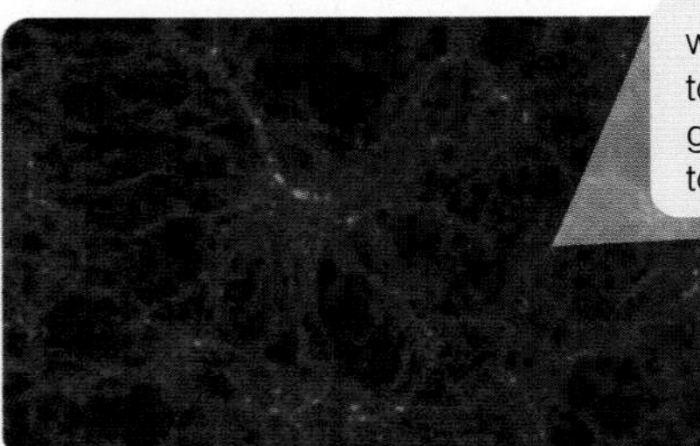
which pull in normal gas to form superclusters of galaxies. Gravity continues to pull clusters together.

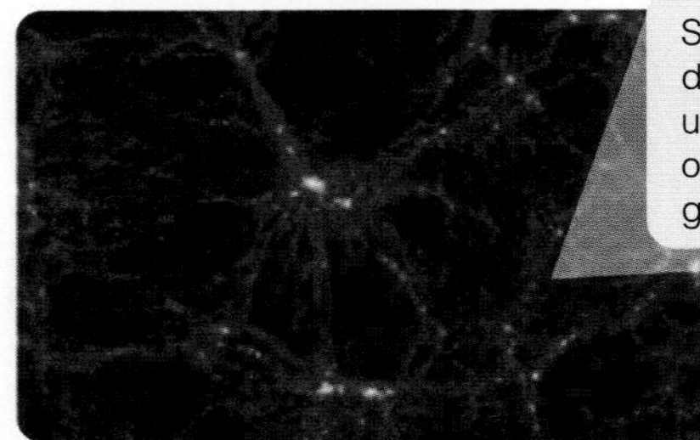
Statistical tests show the distribution in this model universe resembles the observed distribution of galaxies.

How did the uniform gas at the time of recombination coagulate so quickly to form galaxies? Worse, how did that highly uniform gas form such large structures as the filaments, voids, and walls?

Baryonic matter is so rare in the universe that it does not have enough gravity to pull itself together rapidly. Astronomers believe the nonbaryonic dark matter must provide the gravity, and models have attempted to describe how dark matter can form structure. As you have seen earlier in this chapter, hot dark matter does not clump into small enough structures, so the dark matter must be cold dark matter (■ Figure 15-21).

But what started the clumping of matter? Theorists say that subatomic quantum fluctuations in space would have been stretched at the moment of inflation to very large but very subtle variations in gravitational fields that could have stimulated the formation of clusters, filaments, and walls. The structure you see in Figure 15-20 may be the ghostly traces of quantum fluctuations in the infant universe.

The Two-Degree-Field (2DF) Redshift Survey mapped the position and redshift of 250,000 galaxies and 30,000 quasars. As expected, the galaxies are spread in filaments and walls, but a careful statistical analysis of the clumpyness of the distribution confirms that the universe is accelerating. This is an important result because it is another confirmation that is independent of the brightness of type Ia supernovae. When a theory is confirmed by observations of many different types, scientists have much more confidence that it is a true description of nature.

The inflationary theory of the universe makes very specific predictions about the sizes of the fluctuations an observer on Earth should see in the cosmic microwave background radiation.

■ **Figure 15-21**

This computer model traces the formation of structure in the universe from soon after the big bang to the present. (Adapted from a model by Kauffmann, Colberg, Diaferio, & White: Max-Planck Institut für Astronomie)

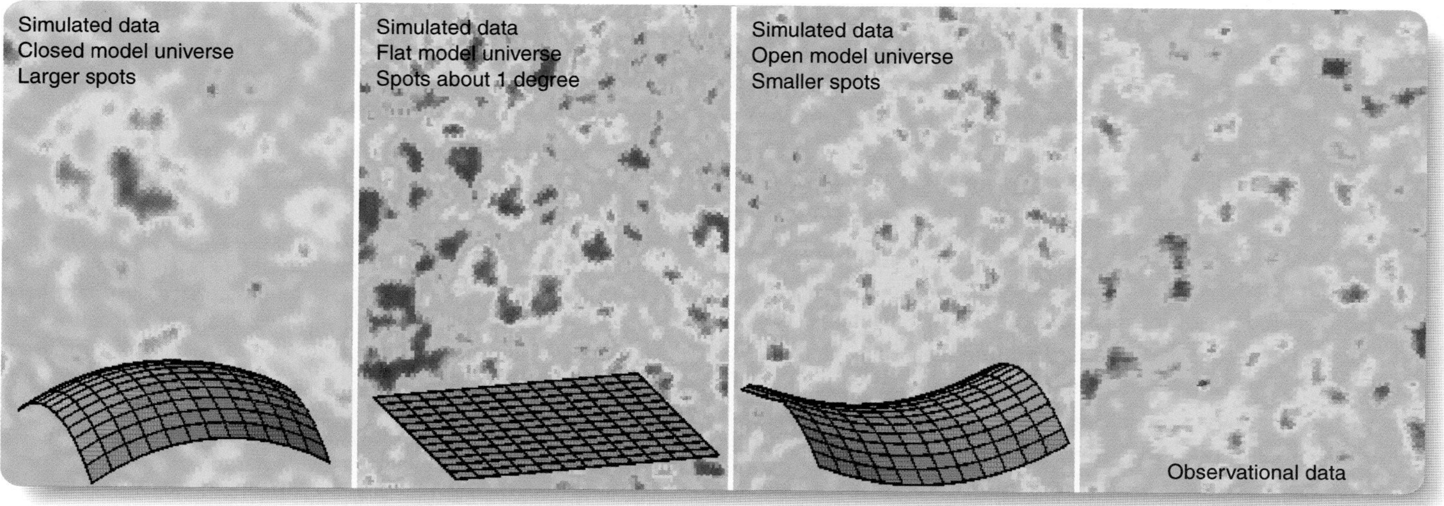

■ Figure 15-22

You can see the difference yourself. Compare the observations of the irregularities in the background radiation at right with the three simulations at left. The observed size of the irregularities fits best with cosmological models having flat geometry. Detailed mathematical analysis confirms your visual impression. The universe is flat. (Courtesy of the BOOMERANG Collaboration.)

Observations made by the COBE satellite in 1992 detected the largest variations, but detecting the smaller variations is critical for testing the theory. A half-dozen teams of astronomers have built specialized telescopes to make these observations. Some have flown under balloons high in the atmosphere, and others have observed from the ice of Antarctica. The most extensive observations have been made by the Wilkinson Microwave Anisotropy Probe (WMAP), a robotic infrared telescope that observed from space.

The background radiation is very isotropic—it looks almost exactly the same in all directions. However, when the average intensity is subtracted from each spot on the sky, small irregularities are evident. That is, some spots on the sky look a tiny bit hotter and brighter than other spots. Those irregularities contain lots of information.

If irregularities in the cosmic microwave background radiation were produced by inflation, then their size must depend on the velocity of the pressure waves (sound) that could travel through the gas at the time of recombination. Theory predicts most of the irregularities should be about 1 degree in diameter if the universe is flat. If the universe were open, the most common irregularities would be smaller. Careful measurements of the size of the irregularities in the cosmic background radiation show that the observations fit the theory very well for a flat universe (■ Figure 15-22). Not only is the theory of inflation confirmed, an exciting result itself, but these data confirm that the universe is flat, which indirectly confirms the existence of dark energy and the acceleration of the universe.

Cosmologists can analyze the irregularities using sophisticated mathematics to find out how commonly the spots of different sizes recur. The mathematics show that spots about 1 degree in diameter are the most common, but spots of other sizes recur as well, and it is possible to plot a spectrum such as ■ Figure 15-23 to show how common different size irregularities are.

■ Figure 15-23

This graph shows how commonly irregularities of different sizes occur in the cosmic microwave background radiation. Irregularities of about 1 degree in diameter are most common. Models of the universe that are open or closed are ruled out. The data fit a flat model of the universe very well. Crosses on data points show the uncertainty in the measurements.

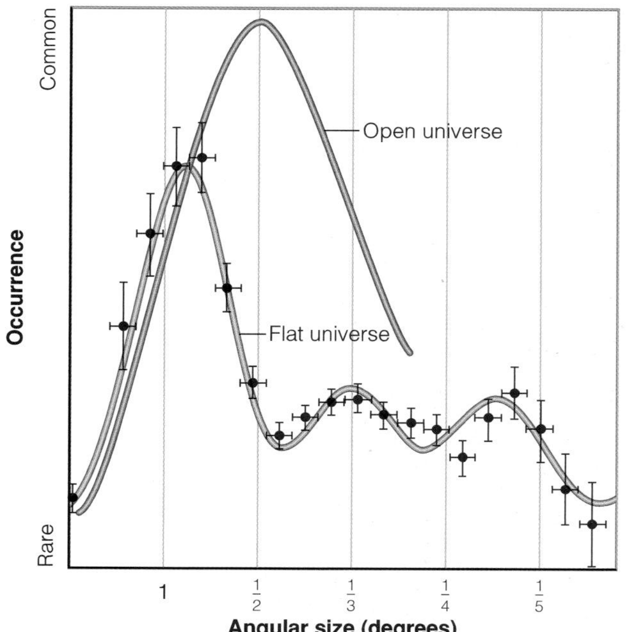

The results from the WMAP observations tell cosmologists a great deal about the universe. The universe is flat, accelerating, and will expand forever. The age of the universe derived from the data is 13.7 billion years. Furthermore, the smaller peaks in the curve reveal that the universe contains 4 percent baryonic (normal) matter, 23 percent dark matter, and 73 percent dark energy. The Hubble constant is confirmed to be 71 km/s/Mpc. The inflationary theory is confirmed, and the data support the cosmological constant version of dark energy, although quintessence is not ruled out. Hot dark matter is ruled out. The dark matter needs to be cold dark matter to clump together so rapidly after the big bang. In fact, the first stars began to produce light when the universe was only about 200 million years old. This is much earlier than most astronomers had expected.

WMAP and other studies of the cosmic microwave background radiation have revolutionized cosmology. At last, astronomers have accurate observations against which to test theories. The basic constants are known to a percent or so.

On reviewing these results, one cosmologist announced that "Cosmology is solved!" but that may be premature. Scientists don't understand dark matter or the dark energy that drives the acceleration, so over 90 percent of the universe is not understood. Hearing this, another astronomer suggested a better phrase was "cosmology in crisis." Certainly there are further mysteries to be explored, but cosmologists are growing more confident that they can describe the overall properties of the universe.

Inquire | Review | Analyze

How does the inflation theory solve the flatness problem?
The flatness problem can be stated as a question: Why is the universe so nearly flat? After all, the density of matter in the universe could be anything from zero to infinite, but the observed density of baryonic and dark matter adds up to about 27 percent of the critical density that would produce a flat universe. Furthermore, the density must have been astonishingly close to the critical density when the universe was very young. Otherwise it would not be so close now. The inflationary theory solves this problem by proposing that the universe underwent a moment of rapid inflation when it was a tiny fraction of a second old. That inflation drove the universe toward flatness just as the inflation of a balloon makes a spot on the balloon nearly flat.

Understanding theory in cosmology is critically important, but science depends on evidence. What evidence indicates that the expansion of the universe is accelerating?

■ ■ ■

Connections: Although we have traced the origin of the universe, the origin of the elements, and the birth and death of stars, we have left out one important class of objects—planets. How do Earth and the other planets fit into this grand scheme of origins? That is the subject of the following chapters.

Study and Review Tools

Summary

15-1 | Introduction to the Universe

Does the universe have an edge in space or time?

- Astronomers conclude that it is impossible for the universe to have an edge because this would introduce logical inconsistencies. That is, an edge to the universe does not make sense.
- If the universe has no edge, then it cannot have a center.
- The darkness of the night sky leads to the conclusion that the universe is not infinitely old. If it were infinite in extent and age, then every spot on the sky would glow as brightly as the surface of a star. This problem, known incorrectly as Olbers's paradox, forces us to conclude that the universe had a beginning.

What evidence indicates that the universe began with a big bang?

- Edwin Hubble's 1929 discovery that the redshift of a galaxy is proportional to its distance is known as the Hubble law. Tracing this expansion backward in time would bring you to an initial high-density, high-temperature state commonly called the big bang.
- Although the expanding universe began from this big bang, it has no center. The galaxies do not recede from a single point. They recede from each other.
- The cosmic microwave background radiation is black body radiation with a temperature of about 2.73 K uniformly spread over the entire sky. It is the light from the big bang freed from the gas at the moment of recombination and redshifted by a factor of 1100.
- The background radiation is strong evidence that the universe began with a big bang.
- During the first three minutes of the big bang, nuclear fusion converted some of the hydrogen into helium but was unable to make many other heavy atoms because no stable nuclei exist with weights of 5 or 8. Today hydrogen and helium are common in the universe, but heavier atoms are rare.

15-2 | The Shape of Space and Time

How can the universe expand if it has no edge?

- The geometry of space and time is critical to understanding cosmology. The universe appears to be isotropic and homogeneous. That is, in its major features, the universe looks the same in all directions and in all locations.
- Isotropy and homogeneity lead to the cosmological principle, the assumption that there are no special places in the universe. Except for local differences, every place is the same.

- General relativity explains that the cosmic redshift is caused by the stretching of photons as they travel through expanding space-time.
- If the universe is infinite (open), then it has no edge and no center, and space-time can expand and carry the galaxies away from each other.
- If the universe is finite (closed) but cannot have an edge, then it must be curved back on itself. Then space-time can expand and carry the galaxies away from each other.
- Whether the universe is open, closed, or flat depends on its density. If the density of the universe is less than the critical density, it is open. If the density is more than the critical density, it is closed. If the density equals the critical density, the universe is flat—balanced between open and closed.
- The amount of deuterium and lithium-7shows that the normal, baryonic matter can make up only about 4 percent of the critical density. Dark matter appears to make up less than 30 percent.

15-3 | 21st-Century Cosmology

Why is the universe the way it is?

- The inflationary theory proposes that the universe expanded dramatically when it was a tiny fraction of a second old.
- The flatness problem is explained by inflation because the sudden inflation forced the universe to become flat, just as a spot on an inflating balloon becomes flatter as the balloon inflates.
- The horizon problem is explained by inflation because the universe was so small before inflation that energy could move and equalize the temperature everywhere.
- Observations of type Ia supernovae reveal that the expansion of the universe is speeding up because of energy present in empty space as dark energy.
- Dark energy added to dark matter and baryonic matter makes the observed density of the universe equal to the critical density and confirms the prediction made by inflation that the universe is flat.
- The nature of dark energy is unknown. It may be described by Einstein's cosmological constant or it may change with time and be better called quintessence.
- Corrected for inflation, the observed value of the Hubble constant implies that the universe is about 14 billion years old. The future fate of the universe depends on the nature of dark energy.
- The sudden inflation of the universe is believed to have magnified tiny quantum mechanical fluctuations in space-time. These very large but very weak differences in gravity worked with dark matter to draw together baryonic matter and create the clusters of galaxies, filaments, and voids that astronomers call structure.
- Statistical observations of the large-scale structure of the universe confirm that it is flat and contains 4 percent baryonic matter, 23 percent dark matter, and 73 percent dark energy.

New Terms

cosmology (p. 326)
Olbers's paradox (p. 326)
observable universe (p. 328)
big bang (p. 329)
Hubble time (p. 330)
cosmic microwave background radiation (p. 333)
steady-state theory (p. 333)
antimatter (p. 333)
recombination (p. 334)
dark age (p. 334)
reionization (p. 334)
isotropy (p. 336)
homogeneity (p. 336)
cosmological principle (p. 336)
critical density (p. 337)
closed universe (p. 339)
flat universe (p. 339)
open universe (p. 339)
oscillating universe theory (p. 340)
nonbaryonic matter (p. 341)
hot dark matter (p. 342)
cold dark matter (p. 342)
MACHOs (p. 342)
flatness problem (p. 343)
horizon problem (p. 343)
inflationary universe (p. 343)
grand unified theories (GUTs) (p. 343)
cosmological constant (p. 345)
quintessence (p. 345)
dark energy (p. 345)
big rip (p. 346)
large-scale structure (p. 346)
supercluster (p. 346)

Review Questions

Ace Astronomy™ Assess your understanding of this chapter's topics with additional quizzing and animations at **http://astronomy.brookscole.com/sh9e**

1. How does the darkness of the night sky tell you something important about the universe?
2. How can Earth be located at the center of the observable universe if you accept the Copernican principle?
3. Why can't an open universe have a center? Why can't a closed universe have a center?
4. What evidence indicates that the universe is expanding? that it began with a big bang?
5. Why couldn't atomic nuclei exist when the universe was younger than 2 minutes?
6. Why is it difficult to determine the present density of the universe?
7. How does the inflationary universe theory resolve the flatness problem? the horizon problem?
8. If the Hubble constant is really 100 km/s/Mpc, much of what astronomers understand about the evolution of stars and star clusters must be wrong. Explain why. (*Hint:* What would the age of the universe be?)
9. Why must the universe have been very uniform during its first million years?
10. What is the difference between hot dark matter and cold dark matter? What difference does it make to cosmology?
11. What evidence shows that the expansion of the universe is accelerating?
12. What evidence is there that the universe is flat?
13. Explain why some of the galaxies in this photo have elongated, slightly curved images. What do such observations tell you about the universe?

14. The image at the right shows irregularities in the background radiation. Why isn't the background radiation perfectly uniform? What does the size of these irregularities tell you?

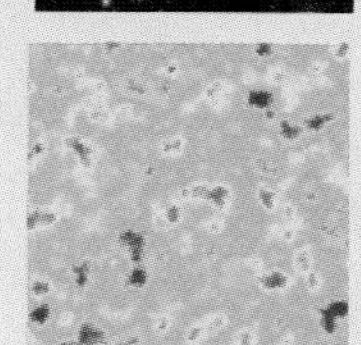

Discussion Questions

1. Do you think Copernicus would have accepted the cosmological principle? Why or why not?
2. If you reject any model of the universe that has an edge in space because you can't comprehend such a thing, shouldn't you also reject any model of the universe that has a beginning or an ending? Are those just edges in time, or is there a difference?

Problems

1. Use the data in Figure 15-3 to plot a velocity–distance diagram. Find H and estimate the Hubble time.
2. If a galaxy is 8 Mpc away from us and recedes at 456 km/s, what is H? What is the Hubble time? How old would the universe be if it were flat and there were no acceleration?

3. At what wavelength of maximum intensity did the gas of the big bang radiate at the time of recombination? By what factor is that different from the wavelength of maximum of the cosmic microwave background radiation?
4. If the average distance between galaxies is 2 Mpc, and the average mass of a galaxy is 10^{11} solar masses, what is the average density of matter in the universe? (*Hints:* The volume of a sphere is $\frac{4}{3}\pi r^3$, and the mass of the sun is 2×10^{33} g.)
5. Figure 15-12 is based on a Hubble constant of 70 km/s/Mpc. How would you have to change the diagram to fit a Hubble constant of 50 km/s/Mpc?
6. Hubble's first estimate of the Hubble constant was 530 km/s/Mpc. If his distances were too small by a factor of 7, what answer should he have obtained?
7. What was the maximum age of the universe predicted by Hubble's first estimate of the Hubble constant?
8. If the value of the Hubble constant were found to be 60 km/s/Mpc, what would the Hubble time be? How old would the universe be if it were flat and there were no acceleration? How would acceleration change your answer?

Media Cluster

ACTIVE FIGURES

Ace Astronomy™ To access the resources in the Media Cluster, log into AceAstronomy at http://astronomy.brookscole.com/sh9e and select Chapter 15.

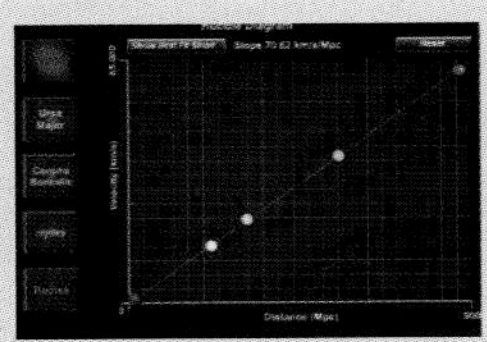

Cosmic Redshift
Graph your own observations of redshift and distance.

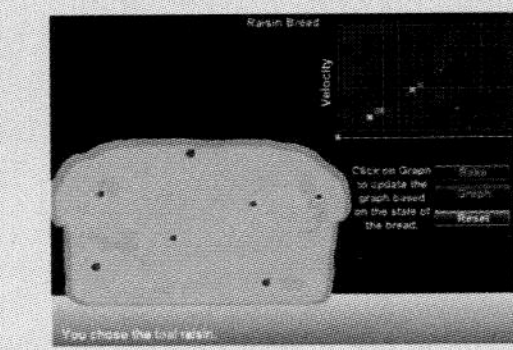

Raisin Bread
You can make observations from any raisin as the dough rises.

ASTRONOMY EXERCISE

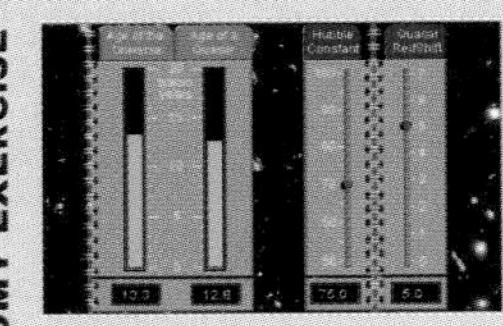

Age of the Universe
Graph your own observations of redshift and distance.

VIRTUAL ASTRONOMY LABS

Lab 17: Evidence for Dark Matter
This lab investigates two phenomena that provide possible evidence for the existence of dark matter—galactic rotation curves and gravitational lensing by unseen objects.

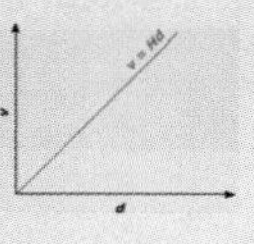

Lab 18: The Hubble Law
This lab discusses the Hubble law, shows how it can be used to estimate distances to remote objects, and illustrates that the law is a consequence of the expansion of space itself.

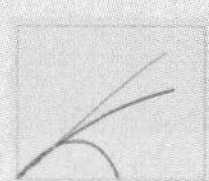

Lab 20: Fate of the Universe
In this lab we investigate what modern cosmology has learned about the history and the ultimate fate of the universe.

Critical Inquiries for the Web

1. What is the latest news concerning acceleration and the flatness of the universe? Search the web for the most recent observations.
2. The steady-state theory was once a rival cosmology of the big bang. Search for websites that provide information on steady-state cosmology. (Be careful to locate legitimate sites that discuss the theory, rather than sites where individuals use steady-state ideas as part of nonscientific arguments on cosmology.) What were the key predictions of steady-state cosmology? How has recent evidence led to its decline?
3. Search for the web pages of large surveys of galaxies, such as the 2dF Deep Field Survey and the Sloan Digital Sky Survey. What are they discovering about the largest and most distant objects in the universe?

Exploring *TheSky*

1. Search for galaxy clusters. (*Hint:* Use the **View** menu to turn off everything but galaxies, deep sky objects, constellations, and labels. Zoom in until the field of view is 40° or smaller and the fainter galaxies appear. Search for clusters of galaxies in Ursa Major, Canes Venatici, Lynx, Virgo, and Coma Bereneces.)
2. Locate small clusters of galaxies and compare the brightness of the galaxies with those in large clusters. Can you tell that the smaller clusters tend to be farther away, or is the range of galaxy luminosities too great? (*Hint:* Click on a galaxy to find its magnitude.)
3. Can you find galaxies and clusters of galaxies along the Milky Way? Turn on the Milky Way and search for galaxy clusters within its outline.

Go to the Brooks/Cole Astronomy Resource Center **(http://astronomy.brookscole.com)** for critical thinking exercises, articles, and additional readings from InfoTrac College Edition, Brooks/Cole's online student library.

16 The Origin of the Solar System

What place is this?

Where are we now?

CARL SANDBURG
Grass

MICROSCOPIC CREATURES LIVE in the roots of your eyelashes. Don't worry. Everyone has them, and they are harmless.* They hatch, fight for survival, mate, lay eggs, and die in the tiny spaces around the roots of your eyelashes without doing you any harm. Some live in renowned places—the eyelashes of a glamorous movie star—but the tiny beasts are not self-aware; they never stop to say, "Where are we?" Humans are more intelligent; we have the ability and the responsibility to wonder where we are in the universe and how we came here. ▌ You should study the solar system for many reasons. You need

** Demodex folliculorum* has been found in 97 percent of individuals and is a characteristic of healthy skin.

In this artist's impression, the planet Jupiter forms at the center of a great swirl of gas and dust. (NASA)

Guidepost

Looking Back

You have become an expert on the universe. You have studied the appearance, origin, structure, and evolution of stars, galaxies, and the universe itself. But so far, our discussion has left out one important class of objects—planets. Now it is time for you to correct that omission.

This Chapter

In this chapter you will look back on what you have learned and find your place. You live on a planet. What does that mean? Where do you fit? Most of all, you want to know how the solar system formed. As you explore our solar system in space and time, you will find answers to four essential questions:

Is our solar system unique?

What are the general properties of our solar system?

How old is our solar system?

How do planets form?

Looking Ahead

One reason you should learn about the origin of the solar system is that you live here, but there is another reason. In the next three chapters you will explore in more detail the planets, comets, and asteroids. By studying the origin of the solar system first, you give yourself a framework for understanding these fascinating worlds.

AceAstronomy™ The AceAstronomy icon throughout the text indicates an opportunity for you to test yourself on key concepts and to explore animations and interactions on the AceAstronomy website at: http://astronomy.brookscole.com/sh9e

to understand Earth as a planet because six billion of us are living on Earth and changing it in ways we don't understand. You should also study the solar system because, as you are about to discover, there are more planets in the universe than stars. Above all, you should study the solar system because it is your home in the universe. Because humans are an intelligent species, you have the right and the responsibility to wonder what you are. Our kind have inhabited this solar system for at least a million years, but only within a lifetime have we begun to understand what a solar system is. Like sleeping passengers on a train, we waken, look out at the passing scenery, and mutter: "What place is this? Where are we now?"

16-1 The Great Chain of Origins

YOU ARE LINKED through a great chain of beginnings that leads backward through time to the first instant when the universe began roughly 14 billion years ago. The gradual discovery of the links in that chain is one of the most exciting adventures of the human intellect. In earlier chapters, you studied some of that story: the origin of the universe in the big bang, the formation of galaxies, the origin of stars, and the growth of the chemical elements. Here you will explore further to consider the formation of planets.

A Review of the Origin of Matter

Look at your thumb. The matter in your thumb came into existence within moments of the beginning of the universe. Astronomers have strong evidence that the universe began in an event called the big bang (Chapter 15), and by the time the universe was 3 minutes old, the protons, neutrons, and electrons in your thumb had come into existence. You are made of very old matter.

Although those particles formed quickly, they were not linked together as they are today. Most of the matter was hydrogen, and about 25 percent was helium. Very few heavier atoms were made in the big bang. Although helium is very rare in our bodies, your thumb contains many of those ancient hydrogen atoms, unchanged since the universe began.

Within a few hundred million years after the big bang, matter began to collect to form galaxies containing billions of stars. Astronomers understand that nuclear reactions inside stars combine low-mass atoms such as hydrogen to make heavier atoms (Chapter 9). Generation after generation of stars cooked the original particles, linking them together to build atoms such as carbon, nitrogen, and oxygen (Chapter 10). Look at your thumb. Even the calcium in the bone and the iron in the blood were assembled inside stars.

The atoms heavier than iron were created by rapid nuclear reactions that can only occur when a massive star explodes (Chapter 10). Gold and silver are rare in your body, but iodine is critical in your thyroid gland, and there are no doubt a few of those iodine atoms circulating through your thumb at this very moment, thanks to the violent deaths of massive stars.

Our galaxy contains at least 100 billion stars, of which our sun is one. The sun formed from a cloud of gas and dust about 5 billion years ago, and the atoms in your thumb were part of that cloud. How the sun took shape, how the cloud gave birth to the planets, how those atoms in your thumb found their way onto Earth and into you is the story of this chapter. As you explore the origin of our solar system, keep in mind the great chain of origins that created the atoms. As the geologist Preston Cloud remarked, "Stars have died that we might live."

The Origin of Planets

Over roughly the last two centuries, astronomers have proposed two kinds of theories for the origin of the planets. Catastrophic theories proposed that the planets formed from some improbable cataclysm such as the collision of the sun and another star. Evolutionary theories proposed that the planets formed gradually and naturally with the sun. Since about 1940, evidence has accumulated to support the evolutionary theories. In fact, nearly all astronomers now accept that planets form naturally as a by-product of star formation (**Window on Science 16-1**).

In earlier chapters, you saw how stars can form from the gravitational contraction of gas and dust clouds. In at least some cases, that contraction must be triggered by compression of the gas cloud, perhaps by a nearby supernova explosion.

As stars form in these contracting clouds, they remain surrounded by cocoons of dust and gas, and the rotation of the cloud causes that dust and gas to form a spinning disk around the protostar. When the center of the star grows hot enough to ignite nuclear reactions, its surface quickly heats up, becomes more luminous, and blows away the gas and dust cocoon.

You have seen clear evidence that disks of gas and dust around young stars are common. Infrared observations of T Tauri stars, for instance, show that some are surrounded by gas clouds rich in dust, and spectra show that these stars are blowing away their nebulae at speeds up to 200 km/s. The presence of bipolar flows from young stars confirms that such systems contain gas and dust distributed in a disk-shaped cloud (see Chapter 9).

Our own planetary system probably formed in such a disk-shaped cloud around the sun. When the sun became luminous enough, the remaining gas and dust were blown away into space, leaving the planets orbiting the sun. This is known as the **solar nebula theory** because the planets form from the nebula around the proto-sun (■ Figure 16-1). If the theory is right, then planets do form as a by-product of star formation, and thus most stars should have planetary systems.

Window on Science 16-1

Two Kinds of Theories: Evolution and Catastrophe

Many theories in science can be classified as either evolutionary, in that they involve gradual processes, or catastrophic, in that they depend on specific, unlikely events. Scientists have generally preferred evolutionary theories. Nevertheless, catastrophic events do occur.

Something in people prefers catastrophic theories, perhaps because they like to see spectacular violence from a safe distance, which may explain the success of movies that include lots of car crashes and explosions. Also, cataclysmic theories resonate with Old Testament accounts of catastrophic events and special acts of creation. Thus, we have an understandable interest in catastrophic theories.

Nevertheless, most scientific theories are evolutionary. Such theories do not depend on unlikely events or special acts of biblical creation, and thus they are open to scientific investigation. For example, geologists much prefer theories of mountain building that are evolutionary, with the mountains being pushed up slowly as centuries pass. All the evidence of erosion and the folding of rock layers shows that the process is gradual. Because most such natural processes are evolutionary, scientists sometimes find it difficult to accept any theory that depends on catastrophic events.

You will see in this and later chapters that catastrophes do occur. Earth, for example, is bombarded by debris from space, and some of those impacts are very large. As you study astronomy or any other natural science, notice that most theories are evolutionary but that you must allow for the possibility of unpredictable catastrophic events.

Mountains evolve to great heights by rising slowly, not catastrophically. (Janet Seeds)

Disks around Other Suns

The first question that might occur to you is, "Do astronomers see any of these disks around other stars?" The evidence is clear. Disks around young stars are common.

Both visible- and radio-wavelength observations detect dense disks of gas orbiting young stars. For example, at least 50 percent of the stars in the Orion Nebula are encircled by dense disks of gas and dust (■ Figure 16-2). A young star is visible at the center of each disk, and astronomers estimate that the disks contain at least a few times Earth's mass in a region a few times larger in diameter than our solar system. The Orion star-forming region is only a few million years old, so it does not seem likely that planets could have formed in these disks yet. Furthermore, the intense radiation from the hot stars in the area is evaporating the disks so fast planets may never have a chance to grow large. The important point for astronomers is that so many of these young stars have disks. Evidently, disks of gas and dust are a common feature of star formation.

The Hubble Space Telescope can detect dense disks of gas and dust around young stars in a slightly different way. The disks show up by the shadows they cast in the nebulae that surround the newborn stars (■ Figure 16-3).

The Solar Nebula Hypothesis

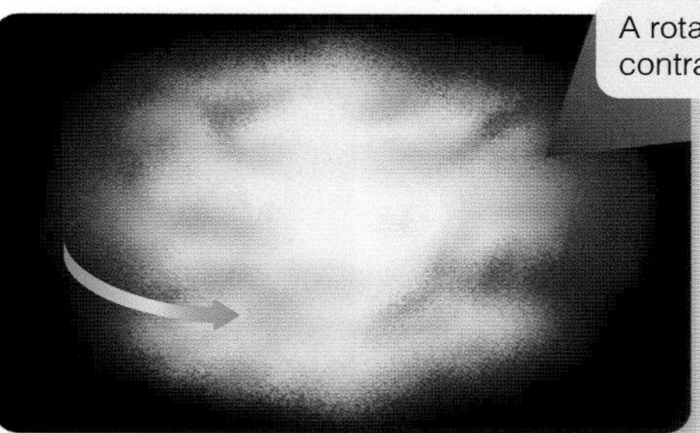

A rotating cloud of gas contracts and flattens...

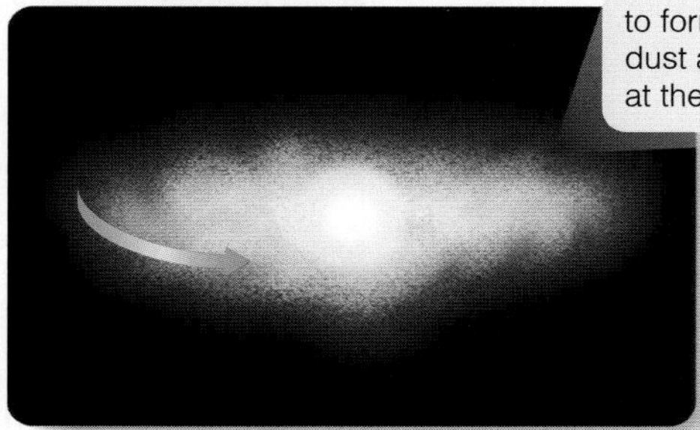

to form a thin disk of gas and dust around the forming sun at the center.

Planets grow from gas and dust in the disk and are left behind when the disk clears.

■ Figure 16-1

The solar nebula hypothesis proposes that the planets formed along with the sun.

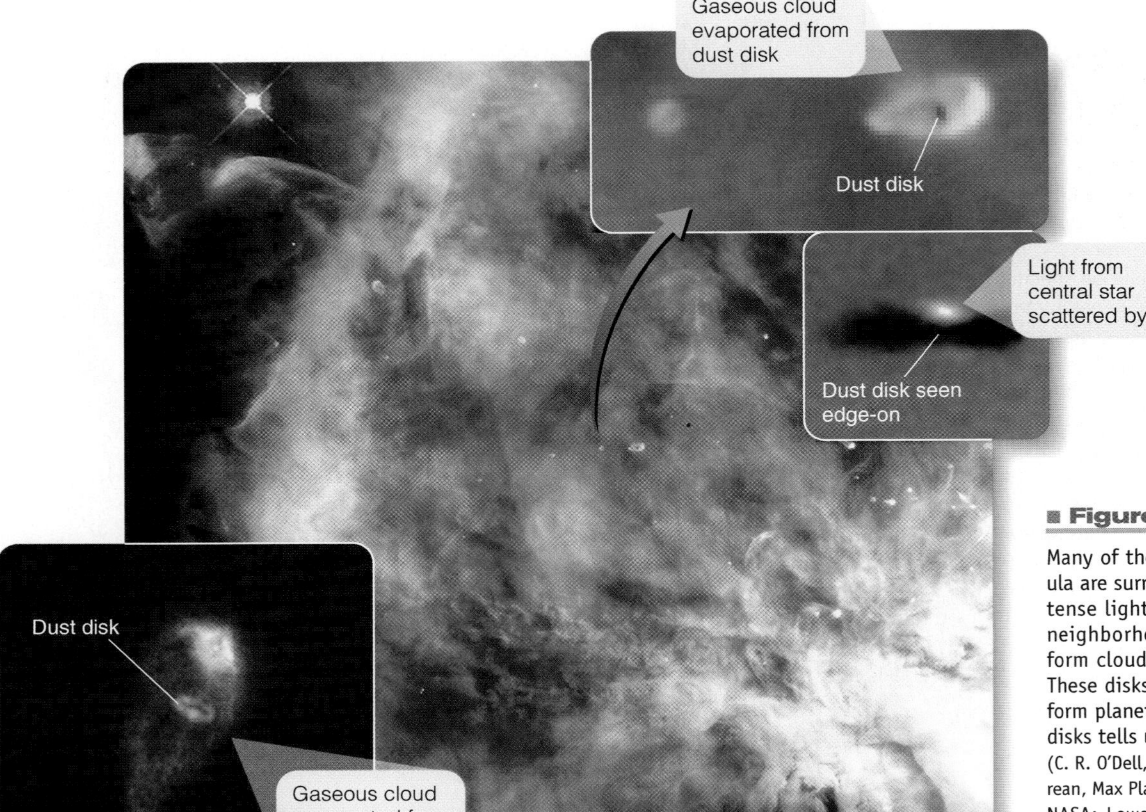

■ Figure 16-2

Many of the young stars in the Orion Nebula are surrounded by disks of dust, but intense light from the brightest star in the neighborhood is evaporating the dust to form clouds of gas around the dust disks. These disks may evaporate before they can form planets, but the large number of such disks tells us that these disks are common. (C. R. O'Dell, Rice, NASA; Dark disk: M. McCaughrean, Max Planck Inst. for Astronomy, C. R. O'Dell, NASA; Lower left insert: J. Bally, H. Throop, C. R. O'Dell, NASA)

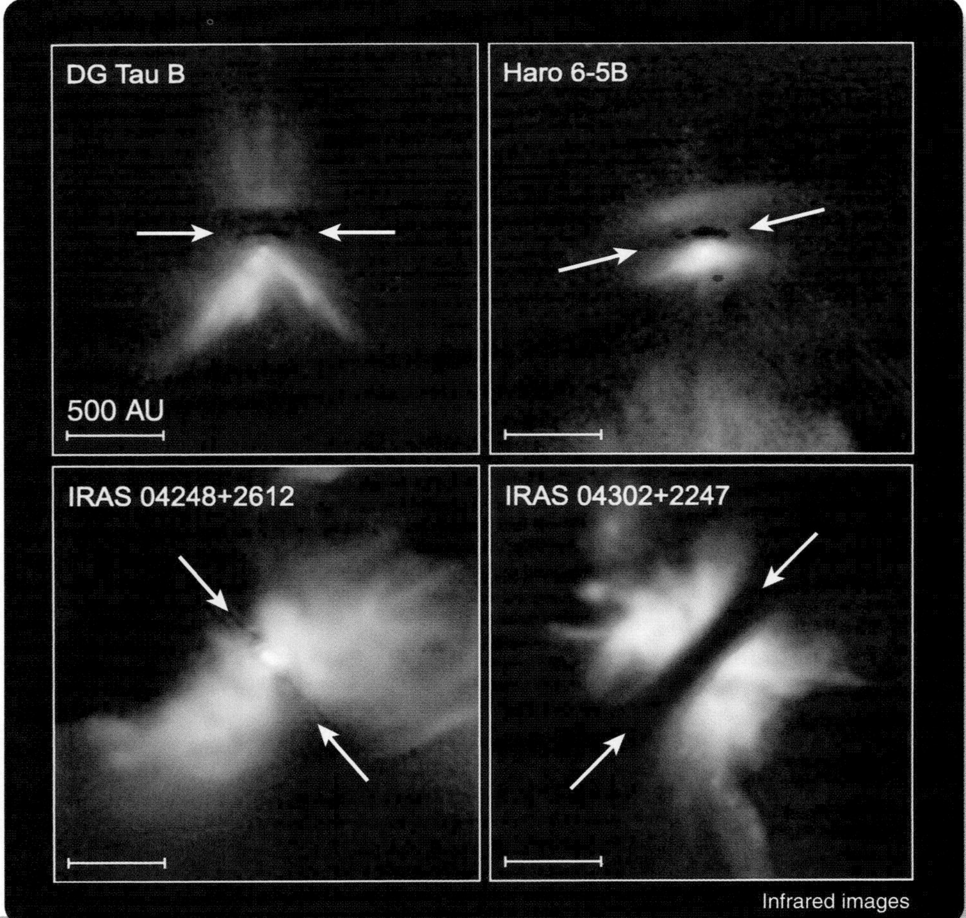

■ Figure 16-3

Dust disks around young stars are evident in these Hubble Space Telescope infrared images. The stars are so young that material is still falling inward and being illuminated by the light from the star. Dark bands across the nebulae (arrows) are caused by dense disks of gas and dust that orbit the stars. Note that the star at lower left is double. (D. Padgett, IPAC/Caltech; W. Brandner, IPAC; K. Stapelfeldt, JPL; and NASA)

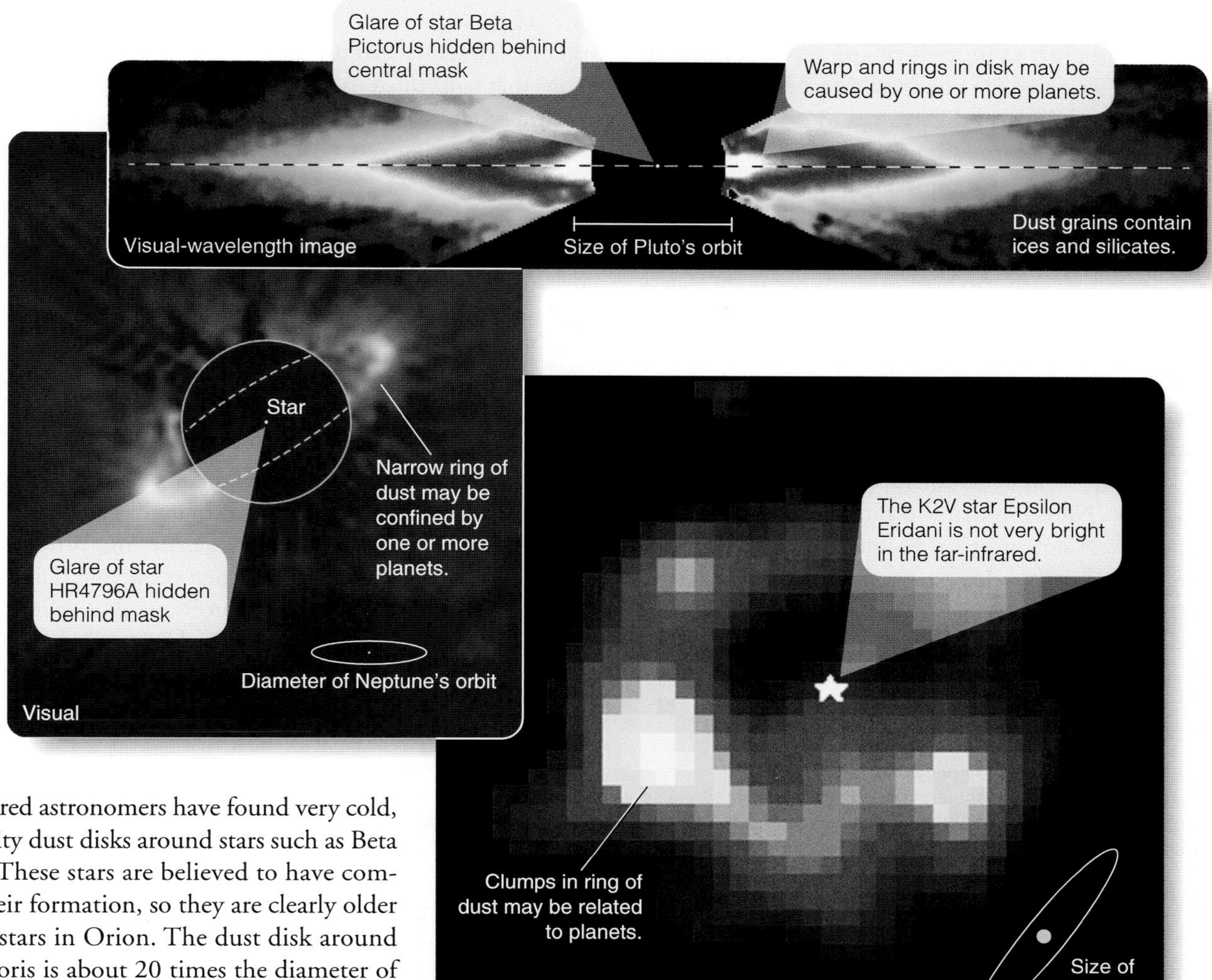

Glare of star Beta Pictorus hidden behind central mask
Warp and rings in disk may be caused by one or more planets.
Visual-wavelength image
Size of Pluto's orbit
Dust grains contain ices and silicates.
Star
Narrow ring of dust may be confined by one or more planets.
Glare of star HR4796A hidden behind mask
Diameter of Neptune's orbit
Visual
The K2V star Epsilon Eridani is not very bright in the far-infrared.
Clumps in ring of dust may be related to planets.
Size of Pluto's orbit
Far-infrared image

Infrared astronomers have found very cold, low-density dust disks around stars such as Beta Pictoris. These stars are believed to have completed their formation, so they are clearly older than the stars in Orion. The dust disk around Beta Pictoris is about 20 times the diameter of our solar system, and, like the other known low-density disks, has an inner zone with even lower density. These inner regions may be places where planets have formed (■ Figure 16-4). Such tenuous dust disks are sometimes called "debris disks" because they are understood to be debris released in collisions among small bodies such as comets and asteroids. Our own solar system contains such dust in its inner regions, and astronomers believe it has an extensive debris disk of cold dust extending far beyond the orbits of the planets.

Notice the difference between the two kinds of disks that astronomers have found. The low-density dust disks such as the one around β Pictoris are produced by dust from collisions among comets and asteroids. Such disks are evidence that planetary systems have already formed. The dense disks of gas and dust such as those seen round the stars in Orion are sites where planets could be forming right now.

The observational evidence gives astronomers confidence that disks of gas and dust around other stars are common, and the evidence of debris disks suggests that planets have formed in some of these disks. Of course, you are wondering if you can't see these planets directly. That's not easy, but astronomers are making progress.

■ Active Figure 16-4

Dust disks have been detected orbiting a number of stars; but, in the visible part of the spectrum, the dust is at least 1000 times fainter than the stars. In the far infrared, the stars are not bright, and the dust can be detected. Warps, rings, and clumps in these disks suggest the gravitational influence of planets. (Beta Pic: C. Burrows and J. Krist, STScI and NASA; HR4769A: A. Weinberger, E. Becklin, UCLA, G. Schneider, U. of Arizona, and NASA; Eps Eri: Joint Astronomy Center.)

Ace Astronomy™ Log into AceAstronomy and select this chapter to see the Active Figure called "Circumstellar Disks." Notice that what you observe in a disk depends on wavelength.

Planets Orbiting Other Suns

A planet orbiting another star is called an **extrasolar planet.** Such a planet would be quite faint and difficult to detect so close to the glare of its star. But there are ways to find these planets. All you have to do is imagine walking a dog.

You will remember that Earth and its moon orbit around their common center of mass, and two stars in a binary system orbit around their center of mass. When a planet orbits a star, the star moves very slightly as it orbits the center of mass of the planet–star system. Think of someone walking a dog on a leash; the dog runs around pulling on the leash, and even if it was an invisible dog, you could plot its path by watching how its owner was jerked back and forth. Astronomers can detect a planet orbiting another star by watching how the star moves as the planet tugs on it.

The first planet detected this way orbits the star 51 Pegasi. As the planet circles the star, the star wobbles slightly, and this very small motion of the star is detectable as Doppler shifts in the star's spectrum (■ Figure 16-5). From the motion of the star and estimates of the star's mass, astronomers can deduce that the planet has half the mass of Jupiter and orbits only 0.05 AU from the star. Half the mass of Jupiter amounts to 160 Earth masses, so this is a large planet. Note also that it orbits very close to its star.

Astronomers were not surprised by the announcement that a planet orbits 51 Pegasi; for years astronomers had assumed that many stars had planets. Nevertheless, they greeted the discovery with typical skepticism (**Window on Science 16-2**). That skepticism led to careful tests of the data and further observations that confirmed the discovery. In fact, over 100 planets have been discovered in this way, including at least three planets orbiting the star Upsilon Andromidae—a true planetary system (Figure 16-5).

Another way to search for planets is to look for changes in the brightness of the star when the orbiting planet crosses in front of the star. The decrease in light is very small, but it is detectable, and astronomers have used this technique to detect a few planets. Very large search programs are underway, however, and more planets will be found by this technique.

The planets discovered so far tend to be massive and have short periods because lower-mass planets or longer-period planets are harder to detect. Low-mass planets don't tug on their stars very much, and present-day spectrographs can't detect the very small velocity changes that these gentle tugs produce. Planets with longer periods are harder to detect because Earth's astronomers have not been making high-precision observations for many years. Jupiter takes 11 years to circle the sun once, so it will take

■ Figure 16-5

Just as someone walking a lively dog is tugged around, the star 51 Pegasi is tugged around by the planet that orbits it every 4.2 days. The wobble is detectable in precision observations of its Doppler shift. Someone walking three dogs is pulled about in a more complicated pattern, and you can see something similar in the Doppler shifts of Upsilon Andromedae, which is orbited by three planets.

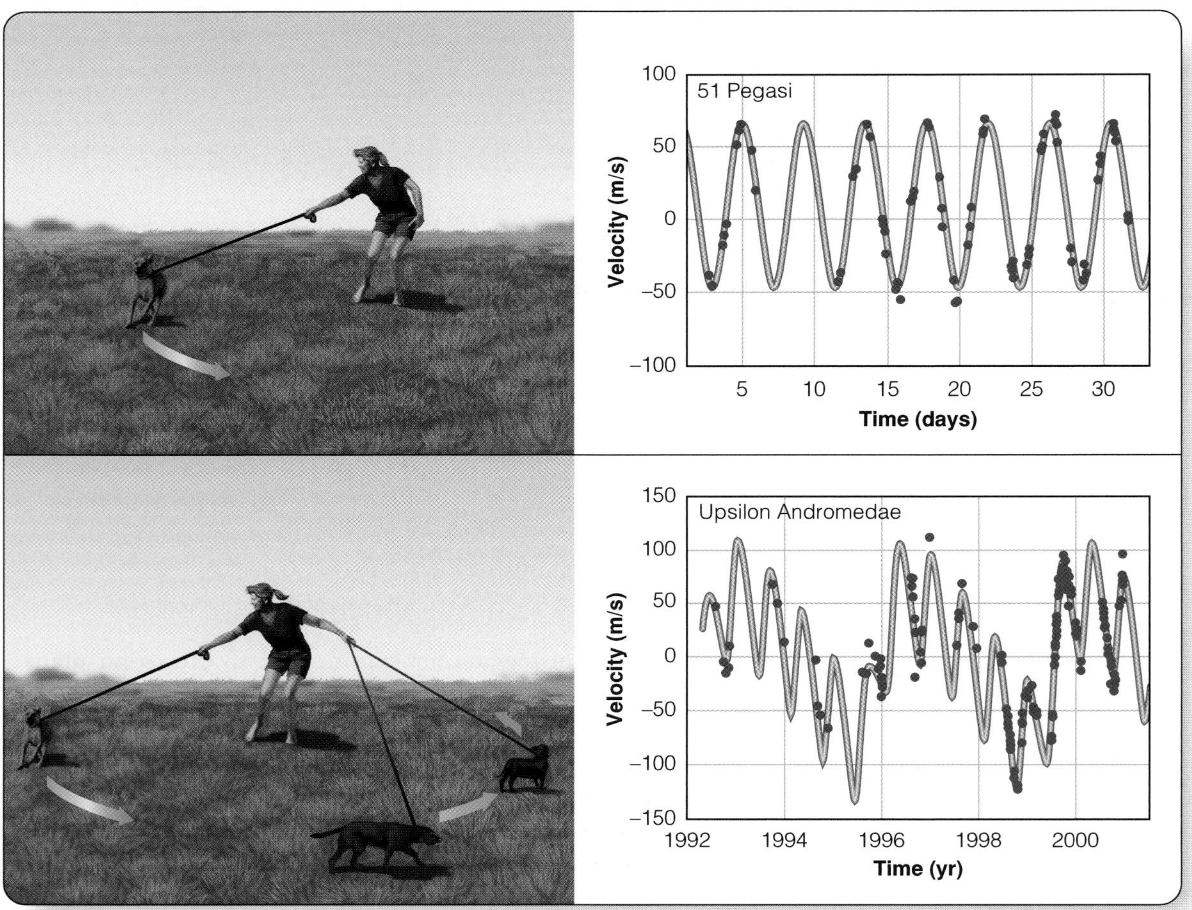

Window on Science 16-2

Scientists: Courteous Skeptics

"Scientists are just a bunch of skeptics who don't believe in anything." That is a common complaint about scientists, but it misinterprets the fundamental characteristic of the scientist. Yes, scientists are skeptical about new ideas and discoveries, but they do hold strong beliefs about how nature works. Ultimately, scientists try to tell stories about how nature works, and skepticism is their guide.

Some people think that "telling a story" is telling a fib. The stories that scientists tell are exactly the opposite; perhaps you can call them antifibs, because they are as true as scientists can make them. The stories are explanations that have been tested over and over to make them accurate descriptions of nature. Skepticism is the tool scientists use to test every aspect of an explanation.

When planets were discovered orbiting 51 Pegasi, astronomers were skeptical—not because they thought the observations were wrong, but because that is how science works. Every observation is tested, and every discovery is confirmed. Only an idea that survives many tests begins to be accepted as a scientific truth.

To nonscientists, this makes scientists seem like irritable skeptics, but among scientists it is not bad manners to say, "Really, how do you know that?" or "Why do you think that?" or "Show me the evidence!"

Scientists try to tell stories about how nature works, stories that are sometimes called theories. A story by John Steinbeck can be a brilliant work of art invented to help you understand some aspect of your life, and another story, such as a TV script, can be little more than chewing gum for the mind. But scientists' stories are different in a critical way. They are true. To create such stories, scientists cannot create facts to fit their story. Rather, every link in a scientific story must be based on evidence and logic. Of course, evidence can be misunderstood, and logical errors happen. To test every aspect of their story, scientists must be continuously skeptical. That is the only way to discover those scientific truths that help you understand how nature works.

Skepticism is not a refusal to hold beliefs. Scientists often believe sincerely in their theories once those theories have been tested over and over. Rather, skepticism is the tool scientists use to find those natural principles worthy of belief.

years for astronomers to see the longer-period wobbles produced by planets lying farther from their stars. Consequently you should not be surprised that most of the first planets discovered are massive and have short periods.

The new planets may seem odd for another reason. In our own solar system, the large planets lie farther from the sun. How could big planets form so near their stars? Theorists find that planets that form in an especially dense disk of matter could spiral inward. Thus it is possible for a few planets to become the massive, short-period planets that they can detect most easily.

A few of the newly discovered extrasolar planets have elliptical orbits, and that seems odd compared with our solar system in which the planetary orbits are nearly circular. Theorists point out, however, that planets may interact in some young planetary systems and can be thrown into elliptical orbits. This is probably rare among planetary systems, but astronomers find these extreme systems more easily because they tend to produce big wobbles.

The preceding paragraphs should reassure you that massive planets in small orbits or in elliptical orbits are not outrageous. At present astronomers know only one planetary system well, our own. As they find more extrasolar planets, astronomers will get a better understanding of what is normal and what is rare.

The discovery of extrasolar planets gives astronomers added confidence in the solar nebula theory. The theory predicts that planets are common, and astronomers are finding them orbiting many stars.

AceAstronomy™ Log into AceAstronomy and select this chapter to see Astronomy Exercise "Extrasolar Planets" and study the variables that help astronomers detect extrasolar planets.

Inquire | Review | Analyze

Why does the solar nebula theory imply planets are common?
Often, the implications of a theory are more important in its analysis than its own conjecture about nature. If the solar nebula theory is correct, the planets of our solar system formed from the disk of gas and dust that surrounded the sun as it condensed from the interstellar medium. If that is true, then it should be a common process. Most stars should form with disks of gas and dust around them, and so you would expect them to have planets.

Without knowing any more about the theory, the implication that planets are common suggests a way you could test the theory. If you look at a large number of stars and find lots of planets, the solar nebula theory is confirmed. But if you search and find few planets, it is contradicted. What observational evidence is there so far on this issue?

■ ■ ■

Connections: So far over 100 other stars are known to have planets. You can expect more such worlds to be discovered, and that raises the question of just how planets form. To answer that question, let's begin by surveying our own solar system to discover the distinguishing characteristics of planetary systems. Those are the characteristics that any satisfactory theory must explain.

16-2 A Survey of the Solar System

TO TEST THEIR THEORIES, astronomers must search the present solar system for evidence of its past. This section surveys the solar system and compiles a list of its most significant characteristics, potential clues to how it formed.

Let's begin with the most general view of the solar system. It is, in fact, almost entirely empty space. Imagine that you reduce the solar system until Earth is the size of a grain of table salt, about 0.3 mm (0.01 in.) in diameter. The moon is a speck of pepper about 1 cm (0.4 in.) away, and the sun is the size of a small plum 4 m (13 ft) from Earth. Mercury, Venus, and Mars are grains of salt. Jupiter is an apple seed 20 m (66 ft) from the sun, and Saturn is a smaller seed over 36 m (120 ft) away. Uranus and Neptune are slightly larger than average salt grains, and Pluto, the farthest planet (on average), is a speck of pepper over 150 m (500 ft) from the central plum. You would need a powerful microscope to detect the asteroids. From the largest planets to the smallest asteroids the objects in our solar system are only scattered, tiny bits of matter—the last remains of the solar nebula.

Revolution and Rotation*

The planets revolve around the sun in orbits that lie close to a common plane. The orbit of Mercury, the planet closest to the sun, is tipped 7° to Earth's orbit, and Pluto's orbit is tipped 17.2°. The rest of the planets' orbital planes are inclined by no more than 3.4°. Thus, the solar system is basically disk-shaped.

The rotation of the sun and planets on their axes also seems related to this disk shape. The sun rotates with its equator inclined only 7.25° to Earth's orbit, and most of the other planets' equators are tipped less than 30°. The rotations of Venus, Uranus, and Pluto are peculiar, however. Venus rotates backward compared with the other planets, and both Uranus and Pluto rotate on their sides (with their equators almost perpendicular to their orbits). You will learn about these planets in detail in Chapters 17 and 18, but later in this chapter we will try to understand how they could have acquired their peculiar rotations.

Apparently, the preferred direction of motion in the solar system—counterclockwise as seen from the north—is also related to its disk shape. All the planets revolve counterclockwise around the sun; and, with the exception of Venus, Uranus, and Pluto, they rotate counterclockwise on their axes. Furthermore, nearly all of the moons in the solar system, including Earth's moon, orbit around their planets counterclockwise. With only a few exceptions, most of which are understood, revolution and rotation in the solar system follow a disk theme.

Two Kinds of Planets

Perhaps the most striking clue to the origin of the solar system comes from the division of the planets into two categories: the small Earthlike worlds and the giant Jupiterlike worlds. The difference is so dramatic we can hardly keep from shouting, "Aha, this must mean something!"

Study **Terrestrial and Jovian Planets** on pages 364 and 365 and notice three important points. First, the two kinds of planets are distinguished by their location. The four inner planets are quite different from the next four outward. Also, notice that the planets differ dramatically in mass and density. Jupiter's family of planets is much more massive but much less dense than Earth's family. Finally, notice that craters are common all through the solar system. The only reason you don't see craters on Jupiter and its family is that those planets do not have solid surfaces.

The division of the planets into two families is a clue to how our solar system formed, but you can learn little from the individual planets because they have evolved since they formed. For further clues you must look at smaller objects that have remained largely unchanged since the birth of the solar system.

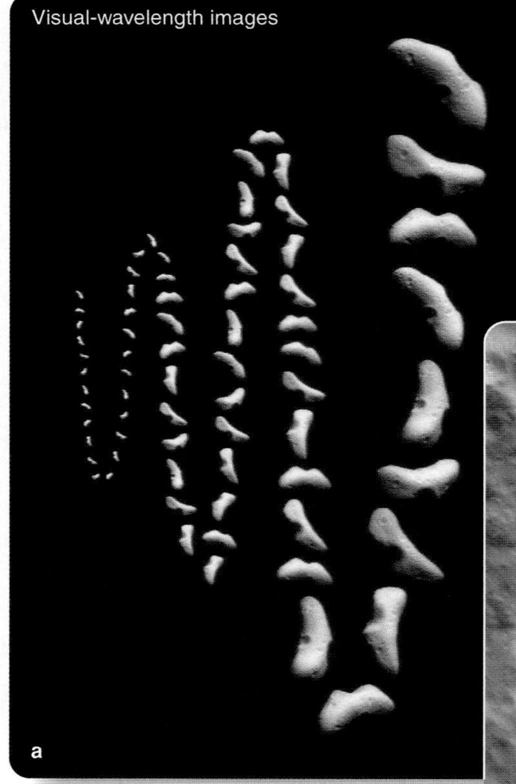

■ Figure 16-6

(a) Over a period of three weeks the NEAR spacecraft approached the asteroid Eros and recorded a series of images, arranged here in an entertaining pattern showing the irregular shape and 5-hour rotation of the asteroid. Eros is 34 km (21 mi) long. (b) This close-up of the surface of Eros shows an area about 11 km (7 mi) from top to bottom. (Johns Hopkins University, Applied Physics Laboratory, NASA)

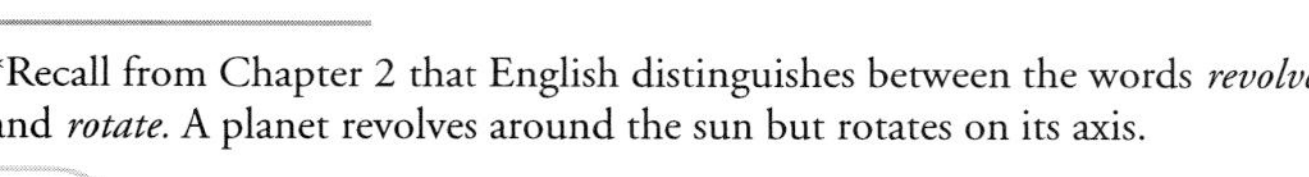

*Recall from Chapter 2 that English distinguishes between the words *revolve* and *rotate*. A planet revolves around the sun but rotates on its axis.

Space Debris

The sun and planets are not the only remains of the solar nebula. The solar system is littered with three kinds of space debris: asteroids, comets, and meteoroids. Although these objects represent a tiny fraction of the mass of the system, they are a rich source of information about the origin of the planets.

The **asteroids,** sometimes called minor planets, are small rocky worlds, most of which orbit the sun in a belt between the orbits of Mars and Jupiter. Roughly 20,000 asteroids have been identified, of which over 2000 follow orbits that bring them into the inner solar system, where they can occasionally collide with a planet. Earth has been struck many times in its history. Some asteroids are located in Jupiter's orbit, some have been found beyond the orbit of Saturn, and a growing number are being discovered among and even beyond the outermost planets.

About 200 asteroids are larger than 100 km (60 mi) in diameter, and tens of thousands are bigger than 10 km (6 mi). There are probably a million or more that are larger than a kilometer and billions that are even smaller. Because even the largest are only a few hundred kilometers in diameter, Earth-based telescopes can detect no details on their surfaces, and the Hubble Space Telescope can image only the largest features.

Astronomers do, however, have clear evidence that asteroids are irregularly shaped, cratered worlds. A number of spacecraft have visited asteroids and sent back photos. For instance, the NEAR spacecraft rendezvoused with the asteroid Eros, went into orbit, and studied the asteroid in detail. Like most asteroids, Eros is an irregular, rocky body pocked by craters (■ Figure 16-6). These observations will be discussed in detail in Chapter 19, but in this quick survey of the solar system you can note that all evidence suggests the asteroids have suffered many impacts from collisions with other asteroids.

Older theories proposed that the asteroids are the remains of a planet that broke up, but modern astronomers recognize the asteroids as the debris left over by a planet that failed to form at a distance of 2.8 AU from the sun. When we discuss the formation of planets later in this chapter, we must be prepared to explain why material in the solar nebula failed to form a planet at a distance of 2.8 AU.

In contrast to asteroids, the brightest **comets** are impressively beautiful objects (■ Figure 16-7a). Most comets are faint, however, and are difficult to locate even at their brightest. A comet may take months to sweep through the inner solar system, during which time it appears as a glowing head with an extended tail of gas and dust.

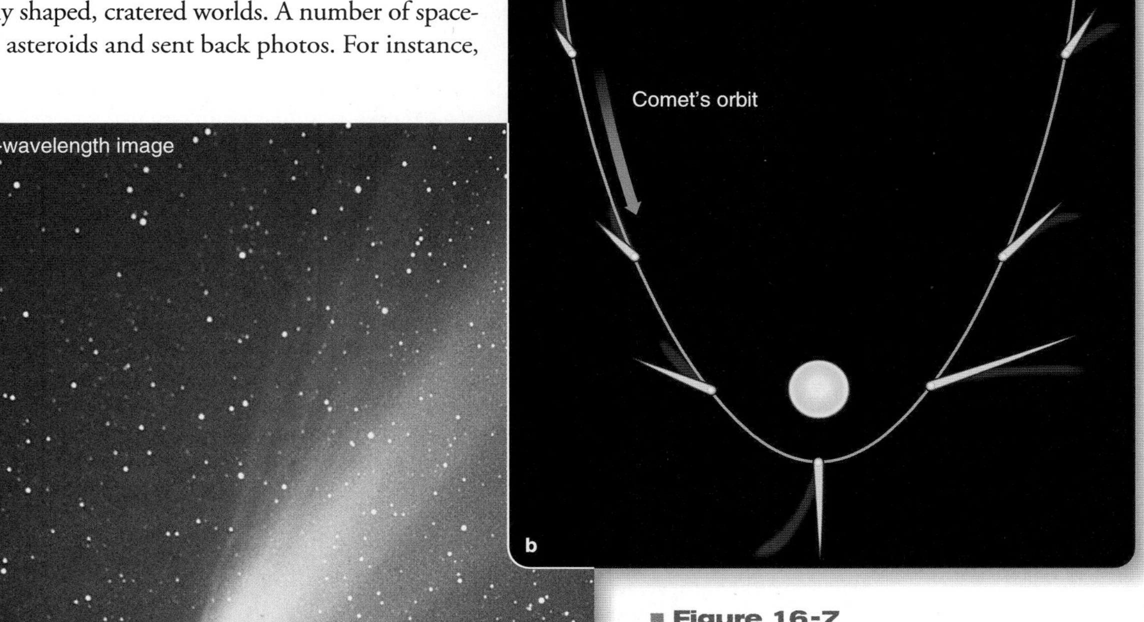

■ **Figure 16-7**

(a) A comet may remain visible in the evening or morning sky for weeks as it moves through the inner solar system. Comet West was in the sky during March 1976. (Celestron International) (b) A comet in a long, elliptical orbit becomes visible when the sun's heat vaporizes its ices and pushes the gas and dust away in a tail.

Terrestrial and Jovian Planets

The distinction between the terrestrial planets and the Jovian planets is dramatic. The inner four planets, Mercury, Venus, Earth, and Mars, are terrestrial planets, meaning they are small, dense, rocky worlds with little or no atmosphere. The outer four planets, Jupiter, Saturn, Uranus, and Neptune, are Jovian planets, meaning they are large, low-density worlds with thick atmospheres and liquid interiors. Pluto does not fit this scheme, being small but low density; you will see in a later chapter that it is a very special world and might not be considered a planet at all.

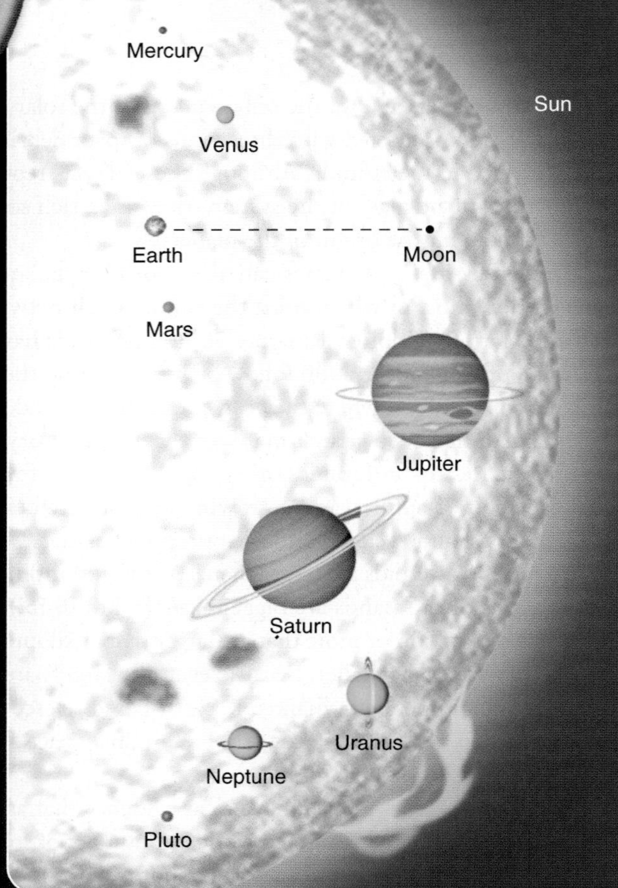

The planets and the sun to scale. Saturn's rings would just reach from Earth to the moon.

Planetary orbits to scale. The terrestrial planets lie quite close to the sun, whereas the Jovian planets are spread far from the sun outside the asteroid belt. The elliptical shape of Pluto's orbit is visible here.

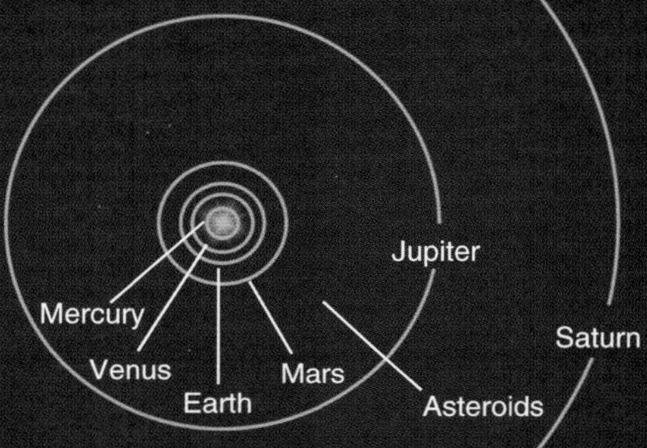

Saturn

Uranus

Of the terrestrial planets, Earth is most massive, but the Jovian planets are much more massive. Jupiter is over 300 Earth masses, and Saturn is nearly 100 Earth masses. Uranus and Neptune are 15 and 17 Earth masses.

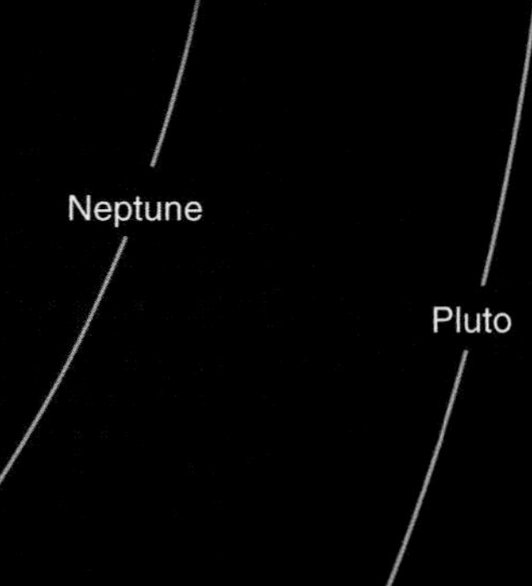

Mercury is only 40 percent larger than Earth's moon, and its weak gravity cannot retain a permanent atmosphere. Like the moon, it is covered with craters from meteorite impacts.

Mercury

Earth's moon

NASA

UCO/Lick Observatory

Craters are common on all of the surfaces in the solar system that are strong enough to retain them. Earth has about 150 impact craters, but many more have been erased by erosion. Besides the planets, the asteroids and nearly all of the moons in the solar system are scarred by craters. Ranging from microscopic to hundreds of kilometers in diameter, these craters have been produced over the ages by meteorite impacts. When astronomers see a rocky or icy surface that contains few craters, they know that the surface is young.

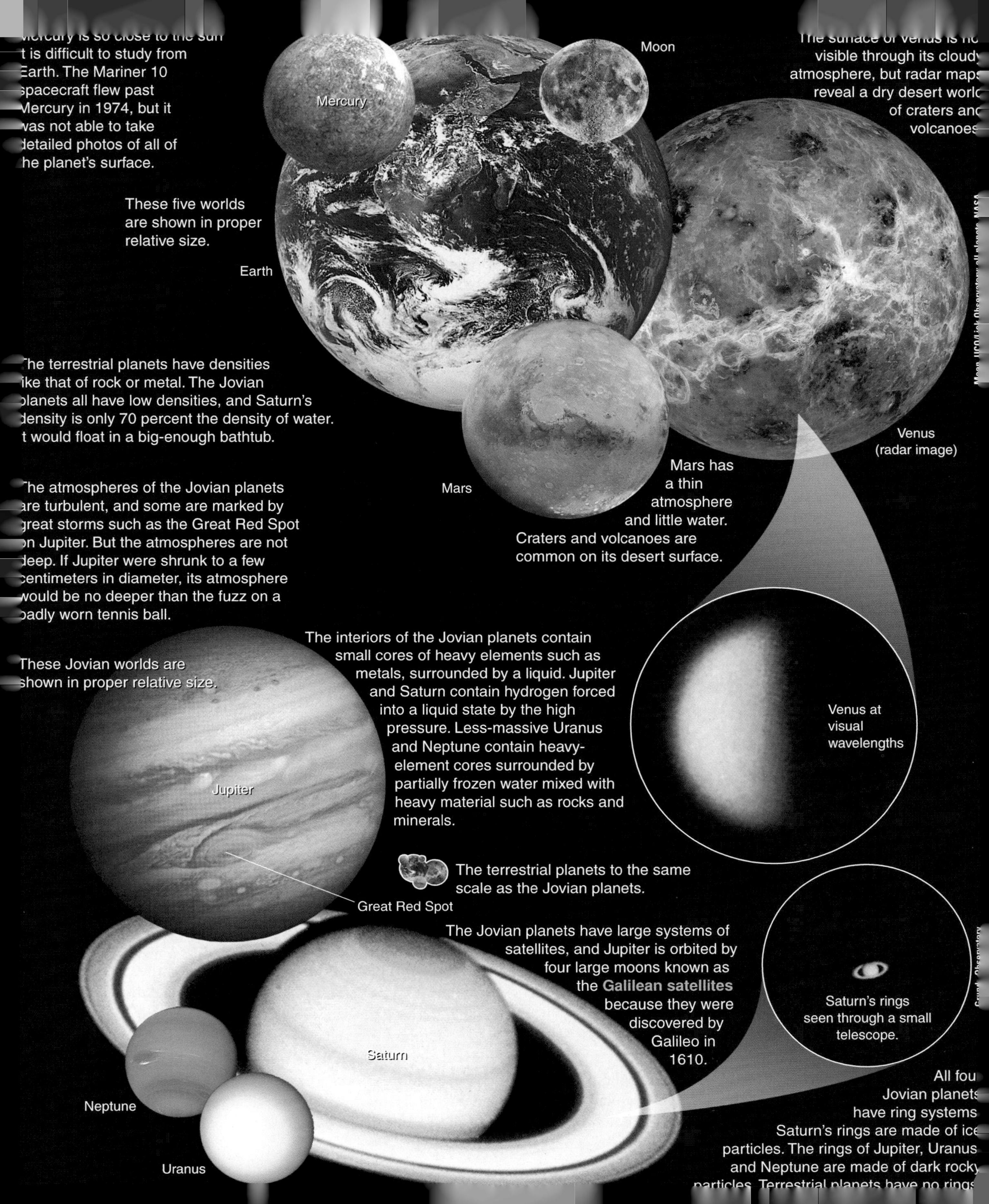
Mercury is so close to the sun it is difficult to study from Earth. The Mariner 10 spacecraft flew past Mercury in 1974, but it was not able to take detailed photos of all of the planet's surface.
Moon
The surface of Venus is not visible through its cloudy atmosphere, but radar maps reveal a dry desert world of craters and volcanoes.
Mercury
These five worlds are shown in proper relative size.
Earth
The terrestrial planets have densities like that of rock or metal. The Jovian planets all have low densities, and Saturn's density is only 70 percent the density of water. It would float in a big-enough bathtub.
Venus (radar image)
Mars has a thin atmosphere and little water. Craters and volcanoes are common on its desert surface.
Mars
The atmospheres of the Jovian planets are turbulent, and some are marked by great storms such as the Great Red Spot on Jupiter. But the atmospheres are not deep. If Jupiter were shrunk to a few centimeters in diameter, its atmosphere would be no deeper than the fuzz on a badly worn tennis ball.
The interiors of the Jovian planets contain small cores of heavy elements such as metals, surrounded by a liquid. Jupiter and Saturn contain hydrogen forced into a liquid state by the high pressure. Less-massive Uranus and Neptune contain heavy-element cores surrounded by partially frozen water mixed with heavy material such as rocks and minerals.
These Jovian worlds are shown in proper relative size.
Venus at visual wavelengths
Jupiter
The terrestrial planets to the same scale as the Jovian planets.
Great Red Spot
The Jovian planets have large systems of satellites, and Jupiter is orbited by four large moons known as the Galilean satellites because they were discovered by Galileo in 1610.
Saturn's rings seen through a small telescope.
Saturn
Neptune
Uranus
All four Jovian planets have ring systems. Saturn's rings are made of ice particles. The rings of Jupiter, Uranus and Neptune are made of dark rocky particles. Terrestrial planets have no rings.

The beautiful tail of a comet can be longer than an astronomical unit, but it is produced by an icy nucleus only a few tens of kilometers in diameter. In its long, elliptical orbit, the nucleus remains frozen and inactive while it is far from the sun. As its orbit carries the nucleus into the inner solar system, the sun's heat begins to vaporize the ices, releasing gas and dust. The pressure of sunlight and the solar wind pushes the gas and dust away, forming a long tail. The motion of the nucleus along its orbit, the pressure of sunlight, and the outward flow of the solar wind can create comet tails that are long and straight or gently curved, but in either case the tails of comets always point generally away from the sun (Figure 16-7b).

For decades astronomers described comet nuclei as dirty snowballs, meaning that they were icy bodies with a little bit of embedded rock and dust. Starting with the passage of Comet Halley in 1986, astronomers have found growing evidence that comet nuclei are not made of dirty ice but rather of icy dirt. That is, the nuclei are at least 50 percent rock and dust. One astronomer has suggested the dirty snowball model be replaced with the icy mud-ball model.

The nuclei of comets are icy bodies left over from the origin of the planets. Thus, you can conclude that at least some parts of the solar nebula were rich in ices. You will see later in this chapter how important ices were in the formation of the Jovian planets, and we will discuss comets in more detail in Chapter 19.

Unlike the stately comets, **meteors** flash across the sky in momentary streaks of light (■ Figure 16-8). They are commonly called "shooting stars." Of course, they are not stars but small bits of rock and metal falling into Earth's atmosphere and bursting into incandescent vapor about 80 km (50 mi) above the ground because of friction with the air. This vapor condenses to form dust, which settles slowly to Earth, adding about 40,000 tons per year to the planet's mass.

Technically, the word *meteor* refers to the streak of light in the sky. In space, before its fiery plunge, the object is called a **meteoroid,** and any part of it that survives its fiery passage to Earth's surface is called a **meteorite.** Most meteoroids are specks of dust, grains of sand, or tiny pebbles. Almost all the meteors you see in the sky are produced by meteoroids that weigh less than 1 g. Only rarely is one massive enough and strong enough to survive its plunge and reach Earth's surface.

Thousands of meteorites have been found, and Chapter 19 will discuss their particular forms. Meteorites are mentioned here for one specific clue they can give you concerning the solar nebula: Meteorites can tell you the age of the solar system.

The Age of the Solar System

If the solar nebula theory is correct, the planets should be about the same age as the sun. The most accurate way to find the age of a celestial body is to bring a sample into the laboratory and determine its age by analyzing the radioactive elements it contains.

■ **Figure 16-8**

A meteor is the streak of glowing gases produced by a bit of material falling into the earth's atmosphere. Friction with the air vaporizes the material about 80 km (50 mi) above Earth's surface. (Daniel Good)

When a rock solidifies, it incorporates known percentages of the chemical elements. A few of these elements are radioactive and can decay into another element, called the *daughter element.* For example, the isotope of uranium ^{238}U can decay into an isotope of lead, ^{206}Pb. The **half-life** of a radioactive element is the time it takes for half of the atoms to decay. The half-life of ^{238}U is 4.5 billion years. Thus, the abundance of a radioactive element gradually decreases as it decays (■ Figure 16-9), and the abundance of the daughter element gradually increases. If you know the abundance of the elements in the original rock, you can measure the present abundance and find the age of the rock. For example, if you studied a rock and found that only 50 percent of the ^{238}U remained and the rest had become ^{206}Pb, you could conclude that one half-life must have passed and the rock was 4.5 billion years old.

Uranium isn't the only radioactive element used in radioactive dating. Potassium (^{40}K) decays with a half-life of 1.3 billion years to form calcium (^{40}Ca) and argon (^{40}Ar). Rubidium (^{87}R) decays to strontium (^{87}Sr) with a half-life of 47 billion years. Any of these elements can be used as a radioactive clock to find the age of mineral samples.

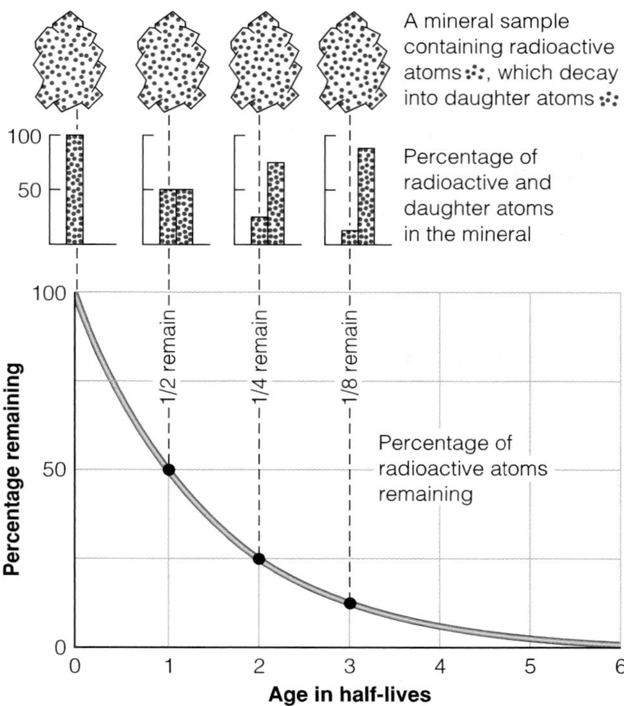

■ Active Figure 16-9

The radioactive atoms in a mineral sample (red) decay into daughter atoms (blue). Half the radioactive atoms are left after one half-life, a fourth after two half-lives, an eighth after three half-lives, and so on.

Ace Astronomy™ Log into AceAstronomy and select this chapter to see the Active Figure called "Radioactive Decay." See how radioactive atoms decay into daughter atoms.

Of course, to find a radioactive age, you need a sample in the laboratory, and the only celestial bodies from which there are samples are Earth, the moon, Mars, and meteorites.

The oldest Earth rocks so far discovered and dated are tiny zircon crystals from Australia, 4.3 billion years old. That does not mean that Earth formed 4.3 billion years ago. The surface of Earth is active, and the crust is continually destroyed and re-formed from material welling up from beneath the crust (see Chapter 17). Thus, the age of these oldest rocks tells us only that Earth is *at least* 4.3 billion years old.

One of the most exciting goals of the Apollo lunar landings was bringing lunar rocks back to Earth's laboratories, where they could be dated. Because the moon's surface is not being recycled like Earth's, some parts of it might have survived unaltered since early in the history of the solar system. Dating the rocks showed the oldest to be 4.48 billion years old. Thus, the solar system must be *at least* 4.48 billion years old.

Although no one has yet been to Mars, a few meteorites found on Earth have been identified by their chemical composition as having come from Mars. Most of these have ages of only 1.3 billion years, but one has an age of about 4.6 billion years. Thus, Mars must be at least that old.

Another important source for determining the age of the solar system is meteorites. Radioactive dating of meteorites yields a range of ages, with the oldest about 4.56 billion years old. This figure is widely accepted as the age of the solar system.

Meteorites also contain tiny grains made mostly of silicon and carbon, and the composition of those grains shows that they are from the interstellar medium. They cannot be dated radioactively, but they must have formed long before our solar system and were incorporated in the nebula that contracted to form our sun and our planetary system. These tiny particles are messengers from beyond the beginning of our solar system.

One last celestial body deserves mention: the sun. Astronomers estimate the age of the sun to be about 5 billion years, but this is not a radioactive date because they cannot obtain a sample of solar material. Instead they estimate the age of the sun from mathematical models of the sun's interior. This yields an age of about 5 billion years plus or minus 1.5 billion years, a number that is in agreement with the age of the solar system derived from the age of meteorites.

Apparently, all the bodies of the solar system formed at about the same time some 4.6 billion years ago. You can add this as the final item to your list of characteristic properties of the solar system (■ Table 16-1).

Inquire I Review I Analyze

In what ways is the solar system a disk?

First, the general shape of the solar system is that of a disk. The planets follow orbits that lie in nearly the same plane. The orbit of Mercury is inclined 7° to the plane of Earth's orbit, and the orbit of Pluto is inclined a bit more than 17°. Thus, the planets follow orbits confined to a thin disk with the sun at its center.

Second, the motions of the sun and planets also follow this disk theme. The sun and most of the planets rotate in the same direction, counterclockwise as seen from the north, with their equators near the plane of the solar system. Also, all of the planets revolve around the sun in that same direction. Thus, our solar system seems to prefer motion in the same direction, which further reflects a disk theme.

■ Table 16-1 I Characteristic Properties of the Solar System

1. Disk shape of the solar system
   - Orbits in nearly the same plane
   - Common direction of rotation and revolution
2. Two planetary types
   - Terrestrial—inner planets; high density
   - Jovian—outer planets; low density
3. Planetary ring systems and large satellite systems for Jupiter, Saturn, Uranus, and Neptune
4. Space debris—asteroids, comets, and meteors
   - Composition
   - Orbits
5. Common ages of about 4.6 billion years for Earth, the moon, Mars, meteorites, and the sun

One of the basic characteristics of our solar system is its disk shape, but another dramatic characteristic is the division of the planets into two groups. What are the distinguishing differences between the terrestrial and Jovian planets?

■ ■ ■

Connections: You have now completed a survey of the solar system and gathered the preliminary evidence you can use to evaluate the solar nebula hypothesis. Now it is time to ask how the planets formed from the solar nebula and took on the characteristics they have today.

16-3 The Story of Planet Building

THE CHALLENGE FOR MODERN PLANETARY ASTRONOMERS is to compare the characteristics of the solar system with the solar nebula theory and tell the story of how the planets formed. Thus they must re-create the solar system's early past (**Window on Science 16-3**).

The Chemical Composition of the Solar Nebula

Everything astronomers know about the solar system and star formation suggests that the solar nebula was a fragment of an interstellar gas cloud. Such a cloud would have been mostly hydrogen with some helium and tiny traces of the heavier elements.

This is precisely what you see in the composition of the sun (see Table 6-2). Analysis of the solar spectrum shows that the sun is mostly hydrogen, with a quarter of its mass being helium and only about 2 percent being heavier elements. Of course, nuclear reactions have fused some hydrogen into helium, but this happens in the sun's core and has not affected its surface composition. Thus, the composition revealed in its spectrum is essentially the composition of the gases from which it formed.

This must have been the composition of the solar nebula, and you can see that composition reflected in the chemical compositions of the planets. One factor, however, can dramatically alter the composition of a planet. You must recognize that planets grow by two processes. Planets begin growing by the sticking together of solid bits of matter. Only when a planet has grown to a mass of about 15 Earth masses does it have enough gravitation to begin capturing gas directly from the solar nebula in a process called **gravitational collapse.** The Jovian planets began forming by the aggregation of bits of rock and ice. Once they accumulated enough mass, they began growing by gravitational collapse. That is, they could capture large amounts of gas, mostly hydrogen and helium, directly from the solar nebula. Jupiter and Saturn grew rapidly and captured the most hydrogen and helium, whereas Uranus and Neptune grew more slowly and never became massive enough to capture gas as rapidly as Jupiter and Saturn. Thus, the Jovian worlds grew to be low-density, hydrogen-rich planets, with Jupiter and Saturn the largest and least dense.

The terrestrial planets, in contrast, contain very little hydrogen and helium. These small planets have such low masses they were unable to keep the hydrogen and helium from leaking into space. Indeed, because of their small masses, they were probably unable to capture very much of these lightweight gases from the solar nebula. The terrestrial planets are dense worlds because they are composed of the heavier elements from the solar nebula.

Although you can see how the chemical composition of the solar nebula is reflected in the present composition of the sun and planets, a very important question remains: How did gas and dust in the solar nebula come together to form the solid matter of the planets? You must answer that question in two stages. First, you must understand how the gas formed billions of small solid particles, and then you must explain how these particles built the planets.

The Condensation of Solids

The key to understanding the process that converted the nebular gas into solid matter is the variation in density among solar system objects. You have already noted that the four inner planets are high-density, terrestrial bodies, whereas the outermost planets are low-density, giant planets (except for Pluto, which is low in density but not a giant). This division is due to the different ways gases condensed into solids in the inner and outer regions of the solar nebula.

Even among the four terrestrial planets, you find a pattern of subtle differences in density. Merely listing the observed densities of the terrestrial planets does not reveal the pattern, because Earth and Venus, being more massive, have stronger gravity and have squeezed their interiors to higher densities. You must look at the **uncompressed densities**—the densities the planets would have if their gravity did not compress them. These densities (■ Table 16-2) show that in general the closer a planet is to the sun, the higher its uncompressed density.

This density variation probably originated when the solar system first formed solid grains. The kind of matter that condensed in a particular region would depend on the temperature of the

■ **Table 16-2 | Observed and Uncompressed Densities**

Planet	Observed Density (g/cm^3)	Uncompressed Density (g/cm^3)
Mercury	5.44	5.30
Venus	5.24	3.96
Earth	5.50	4.07
Mars	3.94	3.73
(Moon)	3.36	3.35

Window on Science 16-3

Reconstructing the Past through Science

Scientists often face problems in which they must reconstruct the past. Some of these problems are obvious, such as those faced by an archaeologist excavating the ruins of a burial tomb, but other such problems are less obvious. In each case, success requires the interplay of hypotheses and evidence to re-create a past that no longer exists.

The reconstruction of the past is obvious when astronomers use the chemical abundance of stars to reconstruct the story of the formation of our galaxy, but a biologist studying a centipede is also reconstructing the past. How did this creature come to have a segmented body with so many legs? How did it develop the metabolism that allows it to move quickly and hunt prey? Although the problem might at first seem to be one of mere anatomy, the scientist must reconstruct an environment that no longer exists.

The astronomer's problem is not just to understand what the planets are like, but to understand how they got that way. That means planetary astronomers must look at the evidence they can see today and reconstruct a history of the solar system, a past that is quite different from the present. If you had a time machine, it would be a fantastic adventure to go back and watch the planets form. Time machines may be impossible, but scientists can use the grand interplay of evidence and theory, the distinguishing characteristic of science, to journey back billions of years and reconstruct a past that no longer exists.

One way science can enrich and inform our lives is by recreating a world that no longer exists.

gas there. In the inner regions, the temperature may have been 1500 K or so. The only materials that can form grains at this temperature are compounds with high melting points, such as metal oxides and pure metals, which are very dense. Farther out in the nebula it was cooler, and silicates (rocky material) could condense. These are less dense than metal oxides and metals. In the cold outer regions, ices of water, methane, and ammonia could condense. These are low-density materials.

The sequence in which the different materials condense from the gas as you move away from the sun is called the **condensation sequence** (■ Table 16-3). It suggests that the planets, forming at different distances from the sun, accumulated from different kinds of materials. Thus, the inner planets formed from high-density metal oxides and metals, and the outer planets formed from low-density ices.

You must also remember that the solar nebula did not remain the same temperature throughout the formation of the planets but may have grown progressively cooler. Thus, a particular region of the nebula may have begun by producing solid particles of metals and metal oxides but, after cooling, began producing particles of silicates. Allowing for the cooling of the nebula makes this theory much more complex, but it also makes the processes by which the planets formed much more understandable.

■ Table 16-3 | The Condensation Sequence

Temperature (K)	Condensate	Planet (Estimated Temperature of Formation; K)
1500	Metal oxides	Mercury (1400)
1300	Metallic iron and nickel	
1200	Silicates	
1000	Feldspars	Venus (900)
680	Troilite (FeS)	Earth (600) Mars (450)
175	H_2O ice	Jovian (175)
150	Ammonia–water ice	
120	Methane–water ice	
65	Argon–neon ice	Pluto (65)

The Formation of Planetesimals

In the development of a planet, three groups of processes operate. First, grains of solid matter grow larger, eventually reaching diameters ranging from a few centimeters to kilometers. The larger of these objects, called **planetesimals,** are believed to be the bodies that the second group of processes collects into planets. Finally, a third set of processes clears away the solar nebula. The study of planet building is the study of these three groups of processes.

According to the solar nebula theory, planetary development in the solar nebula began with the growth of dust grains. These specks of matter, whatever their composition, grew from microscopic size by two processes: condensation and accretion.

A particle grows by **condensation** when it adds matter one atom at a time from a surrounding gas. Thus, snowflakes grow by condensation in Earth's atmosphere. In the solar nebula, dust grains were continuously bombarded by atoms of gas, and some of these stuck to the grains. A microscopic grain capturing a layer of gas atoms over its surface increases its mass by a much larger fraction than a gigantic boulder capturing a layer of gas atoms over its surface. Thus, condensation can increase the mass of a small grain rapidly; but, as the grain grows larger, condensation becomes less effective.

The second process is **accretion,** the sticking together of solid particles. You may have seen accretion in action if you have walked through a snowstorm with big, fluffy flakes. If you caught one of those "flakes" on your mitten and looked closely, you saw that it was made up of many tiny, individual flakes that had collided as they fell and accreted to form larger particles. In the solar nebula, the dust grains were, on the average, no more than a few centimeters apart, so they collided frequently. Their mutual gravitation was too small to hold them to each other, but other effects may have helped. Static electricity generated by their passage through the gas could have held them together, as could compounds of carbon that might have formed a sticky surface on the grains. Ice grains might have stuck together better than some other types. Of course, some collisions might have broken up clumps of grains; on the whole, however, accretion must have increased grain size. If it had not, the planets would not have formed.

There is no clear distinction between a very large grain and a very small planetesimal, but you can consider an object a planetesimal when its diameter becomes a kilometer or so (■ Figure 16-10). Objects larger than a centimeter were subject to new processes that tended to concentrate them. One important effect may have been that the growing planetesimals collapsed into the plane of the solar nebula. Dust grains could not fall into the plane because the turbulent motions of the gas kept them stirred up, but the larger objects had more mass, and the gas motions could not have prevented them from settling into the plane of the spinning nebula. This would have concentrated the solid particles into a thin plane about 0.01 AU thick and would have made further planetary growth more rapid.

This collapse of the planetesimals into the plane is analogous to the flattening of a forming galaxy. However, an entirely new process may have become important once the plane of planetesimals formed. Computer models show that the rotating disk of particles should have been gravitationally unstable and would have broken up into small clouds (■ Figure 16-11). This would further concentrate the planetesimals and help them coalesce into objects up to 100 km (60 mi) in diameter. Thus, the theory proposes that the nebula became filled with trillions of solid particles ranging in size from pebbles to tiny planets. As the largest began to exceed 100 km in diameter, new processes began to alter them, and a new stage in planet building began, the growth of protoplanets.

The Growth of Protoplanets

The coalescing of planetesimals eventually formed **protoplanets,** massive objects destined to become planets. As these larger bodies grew, new processes began making them grow faster and altered their physical structure.

If planetesimals collided at orbital velocities, it is unlikely that they would have stuck together. The average orbital velocity in the solar system is about 10 km/s (22,000 mph). Head-on collisions at this velocity would have vaporized the material. However, the planetesimals were moving in the same direction in the nebular plane and thus avoided head-on collisions. Instead, they

■ Figure 16-10

What did the planetesimals look like? You can get a clue from this photo of the 5-km-wide nucleus of Comet Wild2. Whether rocky or icy, the planetesimals must have been small, irregular bodies, pocked by craters from collisions with other planetesimals. (NASA)

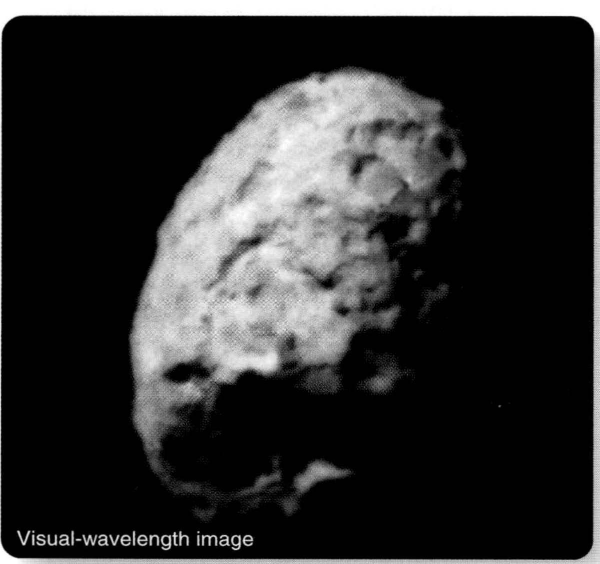

■ Figure 16-11

Gravitational instabilities in the rotating disk of planetesimals may have forced them to collect in clumps, accelerating their growth.

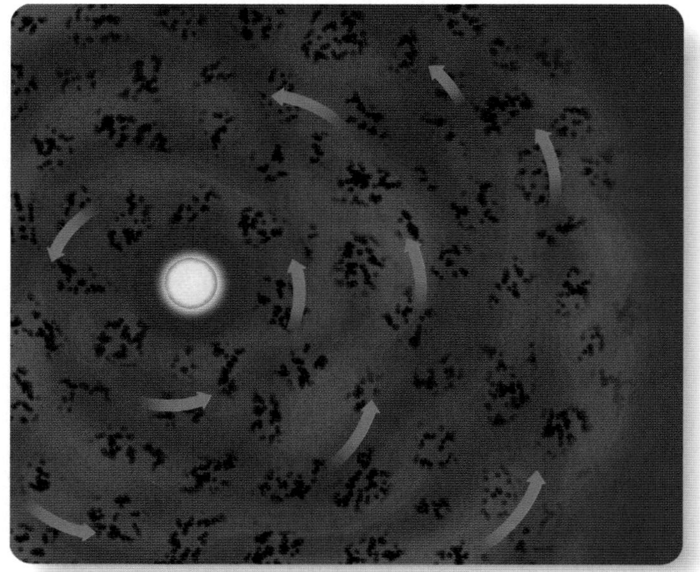

merely rubbed shoulders at low relative velocities. Such collisions would have been more likely to fuse them than to shatter them.

In addition, some adhesive effects probably helped. Sticky coatings and electrostatic charges on the surfaces of the smaller planetesimals probably aided formation of larger bodies. Collisions would have fragmented some of the surface rock, but if the planetesimals were large enough, their gravity would have held on to some fragments, forming a layer of soil composed entirely of crushed rock. Such a layer on the larger planetesimals may have been effective in trapping smaller bodies.

The largest planetesimals would grow the fastest because they had the strongest gravitational field. Not only could they hold on to a cushioning layer to trap fragments, but their stronger gravity could also attract additional material. These planetesimals probably grew quickly to protoplanetary dimensions, sweeping up more and more material. When massive enough, they trapped some of the original nebular gas to form primitive atmospheres. At some point, they crossed the boundary between planetesimals and protoplanets.

To trace the growth of a protoplanet, you can think of the formation of Earth. In its simplest form, the theory of protoplanet growth supposes that all the planetesimals had about the same chemical composition. The planetesimals accumulated gradually to form a planet-size ball of material that was of homogeneous composition throughout. Once the planet formed, heat began to accumulate in its interior from the decay of short-lived radioactive elements, and this heat eventually melted the planet and allowed it to differentiate. **Differentiation** is the separation of material according to density. When the planet melted, the heavy metals such as iron and nickel settled to the core, while the lighter silicates floated to the surface to form a low-density crust. The story of planet formation from planetesimals of similar composition is shown in the left half of ■ Figure 16-12.

This process depends on the presence of short-lived radioactive elements in the solar nebula. Astronomers know such elements were present because very old rock found in meteorites contains daughter isotopes such as magnesium-26. That isotope is produced by the decay of aluminum-26 in a reaction that has a half-life of only 0.74 million years. The aluminum-26 and similar short-lived radioactive isotopes are gone now, but they must have been produced in a supernova explosion that occurred shortly before the formation of the solar nebula. In fact, many astronomers suspect that the supernova explosion compressed nearby gas and triggered the formation of stars, one of which became the sun. Thus our solar system may exist because of a supernova explosion that occurred about 4.6 billion years ago.

Two Models of Planet Building
Planetesimals contain both rock and metal.
The first planetesimals contain mostly metals.
A planet grows slowly from the uniform particles.
Later the planetesimals contain mostly rock.
The resulting planet is of uniform composition.
A rock mantle forms around the iron core.
Heat from radioactive decay causes differentiation.
Heat from rapid formation can melt the planet.
The resulting planet has a metal core and low-density crust.
The resulting planet has a metal core and low-density crust.

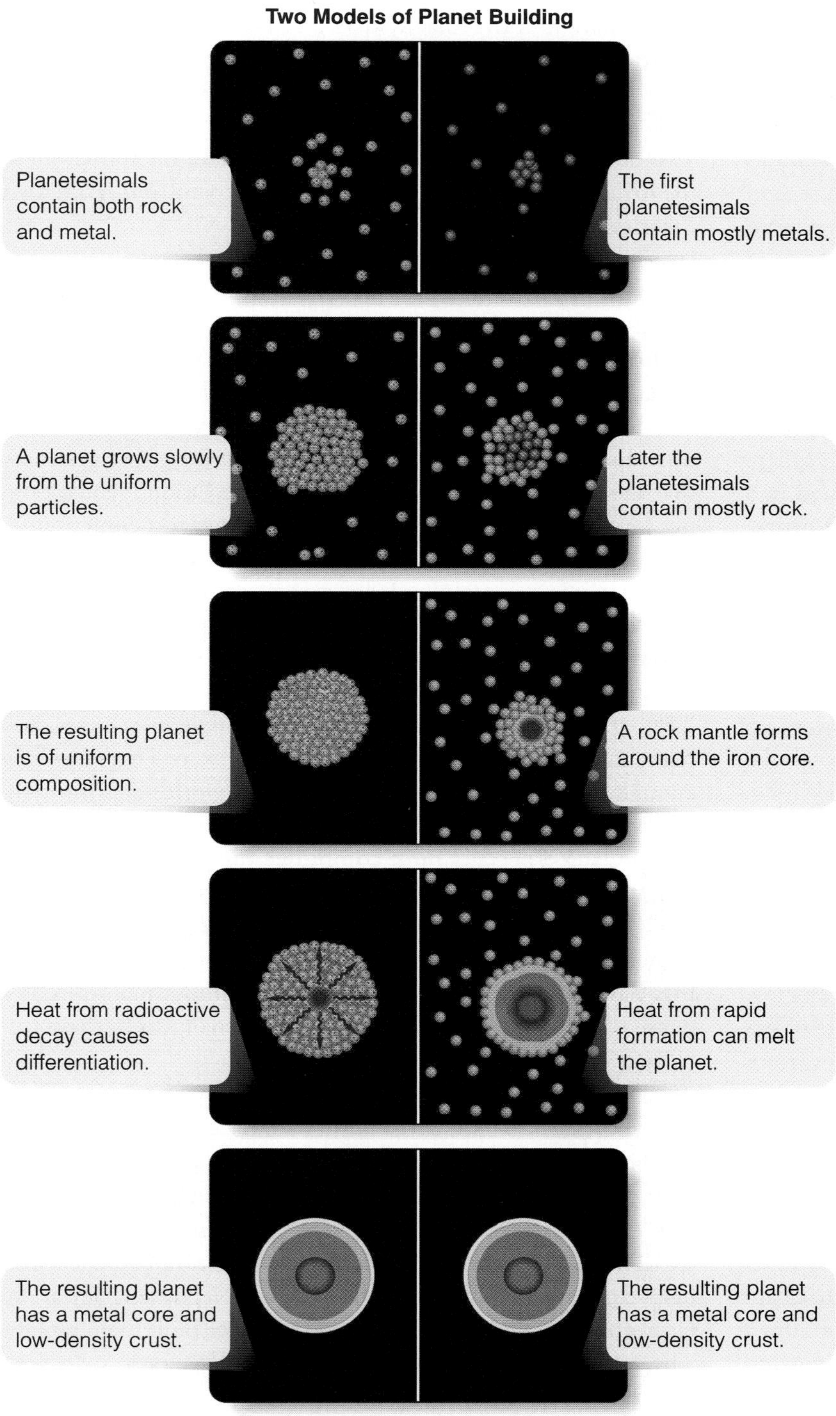

■ Figure 16-12

If the temperature of the solar nebula changed during planet building, the composition of the planetesimals may have changed. The simple model at left assumes no change occurred, but the model at the right incorporates a change from metallic to rocky planetesimals.

If planets formed in this gradual way and were then melted by radioactive decay, then Earth's present atmosphere was not its first. The first atmosphere consisted of small amounts of gases trapped from the solar nebula—mostly hydrogen and helium. They were later driven off by the heat, aided perhaps by outbursts from the infant sun, and new gases that baked from the rocks formed a secondary atmosphere. This creation of a planetary atmosphere from the planet's interior is called **outgassing.**

This simple theory of planet formation can be improved in two ways. First, it seems likely that the solar nebula cooled during the formation of the planets, so they did not accumulate from planetesimals of common composition. As planet building began, the first particles to condense in the inner solar system were metals and metal oxides, so the protoplanets may have begun by accreting metallic cores. Later, as the nebula cooled, more silicates could form, and the protoplanets added silicate mantles. A second improvement in the theory proposes that the planets grew so rapidly that the heat released by the in-falling particles, the **heat of formation,** did not have time to escape. This heat rapidly accumulated and melted the protoplanets as they formed. If this is true, then the planets would have differentiated as they formed. This improved story of planet formation is shown in the right half of Figure 16-12.

This improved theory suggests that Earth's first atmosphere was not captured from the solar nebula. If Earth formed in a molten state, then there was never a time when it had a primitive atmosphere of hydrogen and helium accumulated from the solar nebula. Rather, the gases of the atmosphere were released by the molten rock as the protoplanet grew. Those gases would not have included much water, however, so some astronomers now think that Earth's water and much of its present atmosphere accumulated late in the formation of the planets as Earth swept up volatile-rich planetesimals forming in the cooling solar nebula.

According to the solar nebula theory, the Jovian planets began growing by the same processes that built the terrestrial planets. But the Jovian planets grew faster. In the inner solar nebula, only metals and silicates could form solids, so the terrestrial planets grew slowly. The outer solar nebula not only contained solid bits of metals and silicates, but it was a blizzard of ices. The Jovian planets grew rapidly and quickly became massive enough to grow by gravitational collapse as they drew in large amounts of gas from the solar nebula. The terrestrial planets could not grow from ices, so they developed slowly and never became massive enough to grow by gravitational collapse.

If the planets formed in this way, the Jovian planets must have grown to their present size in about 10 million years. How do astronomers know this? Some of the youngest stars known are called T Tauri stars, and many of those are surrounded by disks of gas. Nearly all of those T Tauri stars that have gas disks are younger than 10 million years; older T Tauri stars have blown their gas disks away. T Tauri stars become stars like the sun, so you can conclude that the solar nebula was probably blown away by the sun within about 10 million years. That means that the Jovian planets must have been complete within 10 million years or so. The terrestrial planets grew from solids and not from the gas, so they could have continued to grow by accretion from solid debris left behind when the gas was blown away. The terrestrial planets were nearly complete within 10 million years but could have continued to grow for another 20 million years or so. All planet growth must have ended and the last of the solar nebula was blown away by 100 million years.

The solar nebula theory has been very successful in explaining the formation of the solar system. But there are some problems, and the Jovian planets are the troublemakers.

The Jovian Problem

Two observations are making it hard to explain the formation of the Jovian planets, forcing astronomers to expand and revise their theories of planet formation.

The first of the two observations is the discovery of lots of extrasolar planets. Those planets are all massive enough to be Jovian worlds. Also, the planets that have been observed crossing in front of their stars blocked enough light to show that they are as big as Jovian planets. All of this tells us that Jovian planets are common.

The second observation is the discovery that the gas and dust disks around newborn stars don't last long. Images show that the disks around the young stars in the Orion Nebula (Figure 16-2) are being evaporated by the intense ultraviolet radiation from hot stars within the nebula. It seems that nearly all stars form in clusters containing massive stars, so this must happen to most disks. Even if a disk did not evaporate quickly, the gravitational influence of the crowded stars should quickly strip away the outer parts of the disk. These observations seem to mean that disks don't last longer than about 7 million years at most, and many must evaporate within 100,000 years or so. That's not long enough to grow a Jovian planet. Yet Jovian worlds are common.

Mathematical models of the solar nebula have been computed using especially built computers running programs that take weeks to finish a calculation. The results show that the rotating solar nebula may have become unstable and formed outer planets by direct gravitational collapse. That is, massive planets may have been able to form from the gas without first forming a dense core by accretion. Jupiters and Saturns form in these calculated models within a few hundred years.

If the Jovian planets formed in this way, they could have formed quickly before the solar nebula disappeared. Terrestrial planets grow from solid bits of material in the inner nebula, so they could have continued to grow by accretion even after the disk was cleared of its gas and smaller particles of dust.

This new insight into the formation of the outer planets may help explain the formation of Uranus and Neptune. They are so far from the sun that accretion could not have built them rapidly. The gas and dust of the solar nebula must have been sparse out there, and Uranus and Neptune orbit so slowly they would not have swept up material very rapidly. The conventional view is that they grew by accretion so slowly that they never became quite massive enough to begin growing by gravitational collapse. In fact, it is hard to understand how they could have reached their present mass if they grew by accretion so far from the sun. Theoretical calculations show that they might have formed closer to the sun in the region of Jupiter and Saturn and then could have been shifted outward by gravitational interactions. In any case, the formation of Uranus and Neptune is part of the Jovian problem.

The solar nebula theory proposes that the planets formed by accreting a core and then, if they became massive enough, by gravitational collapse. The new models suggest that the outer planets may have skipped the core accretion phase.

Whatever the details of the story are, there must have been a solar nebula from which the planets formed. That is enough to allow you to explain the distinguishing characteristics of the solar system.

Explaining the Characteristics of the Solar System

Table 16-1 contains the list of distinguishing characteristics of the solar system. Any theory of the origin of the solar system should explain these characteristics.

The disk shape of the solar system is inherited from the solar nebula. The sun and planets revolve and rotate in the same direction because they formed from the same rotating gas cloud. The orbits of the planets lie in the same plane because the rotating solar nebula collapsed into a disk, and the planets formed in that disk.

The solar nebula hypothesis is evolutionary in that it calls on continuing processes to gradually build the planets. To explain the rotation of Venus, Uranus, and Pluto, however, you may need to consider catastrophic events. Uranus rotates on its side. This might have been caused by an off-center collision with a massive planetesimal. In Chapter 18 you will see evidence that Pluto's highly inclined rotation may be the result of a similar collision. Two theories have been proposed to explain the backward rotation of Venus. Theoretical models suggest that the sun can produce tides in the thick atmosphere of Venus and eventually reverse its rotation—an evolutionary theory. It is also possible that the rotation of Venus was altered by an off-center impact late in the planet's formation, and that is a catastrophic theory.

The second item in Table 16-1, the division of the planets into terrestrial and Jovian worlds, can be understood through the condensation sequence. The terrestrial planets formed in the inner part of the solar nebula, where the temperature was high and most compounds remained gaseous. Only compounds such as the metals and silicates could condense to form solid particles. Planets must begin forming by the sticking together of solid particles, because a small blob of matter does not have enough gravitation to capture gas. Thus, the planets that began growing in the inner solar system had to form mainly from metals and silicates. They became the small, dense terrestrial planets.

In contrast, the Jovian planets formed in the outer solar nebula, where the lower temperature allowed the gas to form large amounts of ices, perhaps three times more ices than silicates. Thus, the Jovian planets grew rapidly and became massive.

The heat of formation (the energy released by in-falling matter) was tremendous for these massive planets. Jupiter must have grown hot enough to glow with a luminosity of about 1 percent that of the present sun. However, because it never got hot enough to generate nuclear energy as a star would, it never generated its own energy. Jupiter is still hot inside. In fact, both Jupiter and Saturn radiate more heat than they absorb from the sun, so they are evidently still cooling.

The Jovian planets are low-density worlds because they grew in the outer solar nebula where there were large amounts of low-density ices in addition to bits of metal and silicates. Also, Jupiter and Saturn are so massive they have been able to grow by drawing in the cool gas directly from the solar nebula. The terrestrial planets could not do this because they never became massive enough and because the gas in the inner nebula was hotter and more difficult to capture.

When planets were first discovered orbiting other stars, some astronomers wondered if the solar nebula theory was quite right. The newly discovered planets are massive, so they are presumably Jovian, but they are close to their stars, where a terrestrial planet should be. Also, some of the new planets follow elliptical orbits. Models of planet formation resolved the problem. If a Jovian planet forms in a disk rich in planetesimals, it may sweep up mass so fast it migrates inward toward its star. Also, interactions among massive planets could throw a planet into an elliptical orbit. Thus the newly discovered planets probably do not pose a problem for the solar nebula explanation of terrestrial and Jovian planets.

A glance at the solar system suggests that you should expect to find a planet between Mars and Jupiter at the present location of the asteroid belt. Mathematical models suggest that there are asteroids there and not a planet because Jupiter grew into such a massive planet it was able to gravitationally disturb the motion of nearby planetesimals. Thus, the bodies that should have formed a planet just inward in the solar system from Jupiter were broken up, thrown into the sun, or ejected from the solar system. The asteroids you see today are the last remains of the rocky planetesimals.

The comets, in contrast, are evidently the last of the icy planetesimals. Some may have formed in the outer solar nebula beyond Neptune and Pluto, but many probably formed among the Jovian planets where ices could condense easily. Mathematical models show that the massive Jovian planets could have ejected some of these icy planetesimals into the far outer solar system. In a later chapter, you will see evidence that a comet forms when one of these icy bodies falls into the inner solar system.

The large satellite systems of the Jovian worlds may contain two kinds of moons. Some moons may have formed in orbit around the forming planet in a miniature of the solar nebula. But some of the smaller moons may be captured planetesimals and asteroids. The large masses of the Jovian planets would have made it easier for them to capture satellites.

Table 16-1 shows that all four Jovian worlds have ring systems, and you can understand this by considering the large mass of these worlds and their remote location in the solar system. A large mass makes it easier for a planet to hold onto orbiting ring particles; and being farther from the sun, the ring particles are not as easily swept away by the pressure of sunlight and the solar wind. It is hardly surprising, then, that the terrestrial planets, low-mass worlds located near the sun, have no planetary rings.

The last entry in Table 16-1 is the common ages of solar system bodies. The solar nebula hypothesis has no difficulty explaining that characteristic. If the hypothesis is correct, then the planets formed at the same time as the sun and thus should have roughly the same age. Although scientists have mineral samples for radioactive dating only from Earth, the moon, Mars, and meteorites, so far the ages seem to agree.

Your general understanding of the origin of the solar system also explains the origin of asteroids, meteors, and comets. They appear to be the last of the debris left behind by the solar nebula. These objects are such important sources of information about the history of our solar system that we will discuss them in detail in Chapter 19. But for now let's ask: What happened to the solar nebula?

Clearing the Nebula

The sun probably formed, along with many other stars, in a swirling nebula. Observations of young stars in Orion suggest that radiation from nearby hot stars would have evaporated the disk of gas and dust around the sun and that the gravitational influence of nearby stars would have pulled gas away. Even without these external effects, four internal processes would have gradually destroyed the solar nebula.

The most important of these internal processes was **radiation pressure.** When the sun became a luminous object, light streaming from its surface pushed against the particles of the solar nebula. Large bits of matter such as planetesimals and planets were not affected, but low-mass specks of dust and individual gas atoms were pushed outward and eventually driven from the system.

The second effect that helped clear the nebula was the solar wind, the flow of ionized hydrogen and other atoms away from the sun's upper atmosphere. This flow is a steady breeze that rushes past Earth at about 400 km/s (250 mi/s). When the sun was young, it may have had an even stronger solar wind, and irregular fluctuations in its luminosity, like those observed in young stars such as T Tauri stars, may have produced surges in the wind that helped push dust and gas out of the nebula.

The third effect for clearing the nebula was the sweeping up of space debris by the planets. All of the old, solid surfaces in the solar system are heavily cratered by meteorite impacts (■ Figure 16-13). Earth's moon, Mercury, Venus, Mars, and most of the moons in the solar system are covered with craters. A few of these craters have been formed recently by the steady rain of meteor-

■ Figure 16-13

Every old, solid surface in the solar system is scarred by craters. (a) Earth's moon is scarred by craters ranging from basins hundreds of kilometers in diameter down to microscopic pits. (b) The surface of Mercury as photographed by a passing spacecraft shows vast numbers of overlapping craters. (NASA)

ites that falls on all the planets in the solar system, but most of the craters you see appear to have been formed roughly 4 billion years ago in what is called the **heavy bombardment,** as the last of the debris in the solar nebula was swept up by the planets. You will find more evidence of this bombardment in later chapters. Thus, many of the last remaining solid objects in the solar nebula were swept up by the planets soon after they formed.

The fourth effect was the ejection of material from the solar system by close encounters with planets. If a small object such as a planetesimal passes close to a planet, it can gain energy from the planet's gravitational field and be thrown out of the solar system. Ejection is most probable in encounters with massive planets, so the Jovian planets were probably very efficient at ejecting the icy planetesimals that formed in their region of the nebula.

Attacked by the radiation and gravity of nearby stars and racked by internal processes, the solar nebula could not have survived very long. Once the gas and dust were gone, the planets could no longer gain mass, and planet building ended.

Inquire | Review | Analyze

Why are there two kinds of planets in our solar system?
Planets begin forming from solid bits of matter, not from gas. Consequently, the kind of planet that forms at a given distance from the sun depends on the kind of compounds that can condense out of the gas to form solid particles. In the inner parts of the solar nebula, the temperature was so high that most of the gas could not condense to form solids. Only metals and silicates could form solid grains, and thus the innermost planets grew from this dense material. Much of the mass of the solar nebula consisted of hydrogen, helium, water vapor, and other gases, and they were present in the inner solar nebula but couldn't form solid grains. The small terrestrial planets couldn't grow from these gases, so the terrestrial planets are small and dense.

In the outer solar nebula, the composition of the gas was the same, but it was cold enough for water vapor to condense to form ice grains. Because hydrogen and oxygen are so abundant, there was lots of ice available. Thus, the outer planets grew from solid bits of metal and silicate combined with large amounts of ice. The outer planets grew so rapidly that they became massive enough to capture gas directly, and thus they became the hydrogen- and helium-rich Jovian worlds.

The condensation sequence combined with the solar nebula hypothesis gives you a way to understand the formation of the planets. Why do some astronomers argue that the formation of the Jovian planets is a problem that needs further explanation?

■ ■ ■

Connections: The great chain of origins leads from the first instant of the big bang through the birth of galaxies, the formation of stars, the origin of our solar system, and finally to you. One of the most elegant and complex links in that chain is planet building. As you explore the solar system in detail in the following chapters, you must stay alert for further clues to the birth of the planets.

Study and Review Tools

Summary

16-1 | The Great Chain of Origins

Is our solar system unique?

- The matter in our solar system was formed in the big bang, and the atoms heavier than helium were cooked up in a few generations of stars.
- The solar nebula hypothesis proposes that the planets formed in a disk of gas and dust around the protostar that became the sun. Observations show that these disks are common.
- Hot disks of gas and dust such as those in the Orion Nebula are detected in early stages of star formation and are believed to be the kind of disk in which planets could form.
- Cold dust disks, also known as debris disks, appear to be produced by dust released by comets and collisions among asteroids. Such disks may be home to planets that have already formed.
- Planets orbiting other stars have been detected by the way they tug their stars about, creating small Doppler shifts in the stars' spectra. Planets have also been detected as they cross in front of their star. This suggests that planets are common.

16-2 | A Survey of the Solar System

What are the general properties of the solar system?

- The solar system is disk shaped in the orbital revolution of the planets and their moons and in the rotation of the planets on their axes.
- The planets are divided into two types. The inner four planets are terrestrial—small rocky Earthlike worlds. The next four outward are Jovian planets that are large and low in density.
- All four of the Jovian worlds have ring systems and large families of moons. The terrestrial planets have no rings and few moons.
- Most of the asteroids, small, irregular rocky bodies, are located between the orbits of Mars and Jupiter.
- Comets are icy bodies that fall into the inner solar system along long elliptical orbits. As the ices vaporize and release dust, the comet develops a tail that points generally away from the sun.
- Meteoroids that fall into Earth's atmosphere are vaporized by friction, and you see them as meteors. Larger and stronger meteoroids may survive to reach the ground where they are called meteorites.

How old is our solar system?

- The age of an object can be found by radioactive dating.
- The oldest rocks from Earth, the moon, and Mars have ages approaching 4.6 billion years. The oldest objects in our solar system are the meteorites, which have ages of 4.6 billion years. This is taken to be the age of the solar system.

16-3 | The Story of Planet Building

How do planets form?

- Condensation in the solar nebula converted some of the gas into solid bits of matter, which accreted to form billions of planetesimals.
- Planets begin growing by accreting solid material.
- Once a planet approaches 15 Earth masses, it can begin growing by gravitational collapse as it pulls in gas from the solar nebula.
- According to the condensation sequence, the inner part of the solar nebula was so hot that only metals and rocky minerals could form solid grains. The dense terrestrial planets grew from those solid particles and did not include many ices or low-density gases.
- The outer solar nebula was cold enough for metals, rocky minerals, and large amounts of ices to form solid particles. The Jovian planets grew rapidly and incorporated large amounts of low-density ices as gases.
- Evidence that the condensation sequence was important in the solar nebula can be found in the densities of the terrestrial planets compared to the Jovian planets. Even the uncompressed densities among the terrestrial planets show that the inner planets are the most dense.
- The terrestrial planets may have formed slowly from the accretion of planetesimals of similar composition and then differentiated later when radioactive decay heated the planet's interiors.
- It is also possible that the planets formed so rapidly that the heat of formation melted the planets and they differentiated as they formed. In that case, Earth's first atmosphere was not captured from the nebula but was baked out of Earth's interior.
- Disks of gas and dust may not last long enough to form Jovian planets by accretion and gravitational collapse. Some models suggest the Jovian planets could have formed directly and rapidly by gravitational collapse.
- In addition to intense light from hot nearby stars and the gravitational influence of passing stars, the solar nebula was eventually cleared away by radiation pressure, the solar wind, the sweeping up of debris by the planets, and by ejection.
- All of the old surfaces in the solar system were heavily cratered by an early bombardment of the debris that filled the solar system when it was young.

New Terms

solar nebula theory (p. 356)
extrasolar planet (p. 359)
asteroid (p. 363)
comet (p. 363)
terrestrial planet (p. 364)
Jovian planet (p. 364)
Galilean satellites (p. 365)
meteor (p. 366)
meteoroid (p. 366)
meteorite (p. 366)
half-life (p. 366)
gravitational collapse (p. 368)
uncompressed density (p. 368)
condensation sequence (p. 369)
planetesimal (p. 369)
condensation (p. 369)
accretion (p. 370)
protoplanet (p. 370)
differentiation (p. 371)
outgassing (p. 372)
heat of formation (p. 372)
radiation pressure (p. 374)
heavy bombardment (p. 375)

Review Questions

Ace Astronomy™ Assess your understanding of this chapter's topics with additional quizzing and animations at **http://astronomy.brookscole.com/sh9e**

1. What produced the helium now present in the sun's atmosphere? in Jupiter's atmosphere? in the sun's core?
2. What produced the iron in Earth's core and the heavier elements like gold and silver in Earth's crust?
3. What evidence indicates that disks of gas and dust are common around young stars?
4. According to the solar nebula theory, why is the sun's equator nearly in the plane of Earth's orbit?
5. Why does the solar nebula theory predict that planetary systems are common?
6. Why do astronomers think the solar system formed about 4.6 billion years ago?
7. If you visited another planetary system, would you be surprised to find planets older than Earth? Why or why not?
8. Why is almost every solid surface in our solar system scarred by craters?
9. What is the difference between condensation and accretion?
10. Why don't terrestrial planets have rings and large satellite systems like the Jovian planets?
11. How does the solar nebula theory help explain the composition of asteroids and comets?
12. How does the solar nebula theory explain the dramatic density difference between the terrestrial and Jovian planets?
13. If you visited some other planetary system in the act of building planets, would you expect to see the condensation sequence at work, or was it unique to our solar system?
14. Why do you expect to find that planets are differentiated?
15. What processes cleared the nebula away and ended planet building?
16. What do you see in the image at the right that indicates the planet formed far from the sun?
17. Why do astronomers conclude that the surface of Mercury, shown at right, is old? When did the majority of these craters form?

(NASA)

(NASA)

Discussion Questions

1. In your opinion, if planets formed with one of the first stars to form in our galaxy, how would those planets differ from the planets in our solar system? (*Hint:* To what population would the star belong?)
2. If the solar nebula hypothesis is correct, then there are probably more planets in the universe than stars. Do you agree? Why or why not?

Problems

1. The nearest star is about 4.2 ly away. If you looked back at the solar system from that distance, what would the maximum angular separation be between Jupiter and the sun? (*Hint:* 1 ly equals 63,000 AU.)
2. The brightest planet in the sky is Venus, which is sometimes as bright as apparent magnitude -4 when it is at a distance of about 1 AU. How many times fainter would it look from a distance of 1 parsec (206,265 AU)? What would its apparent magnitude be? (*Hints:* Remember the inverse square law, Chapter 8; see Chapter 2.)
3. What is the smallest-diameter crater you can identify in the photograph of Mercury on page 364? (*Hint:* See Appendix A to find the diameter of Mercury in kilometers.)
4. A sample of a meteorite has been analyzed, and the result shows that out of every 1000 nuclei of ^{40}K originally in the meteorite, only 125 have not decayed. How old is the meteorite? (*Hint:* See Figure 16-9.)
5. In Table 16-2, which object's density differs least from its uncompressed density? Why?
6. What composition might you expect for a planet that formed in a region of the solar nebula where the temperature was about 100 K?
7. Suppose that Earth grew to its present size in 10 million years through the accretion of particles averaging 100 g each. On the average, how many particles did Earth capture per second? (*Hint:* See Appendix A to find Earth's mass.)
8. If you stood on Earth during its formation as described in Problem 7 and watched a region covering 100 m^2, how many impacts would you expect to see in an hour? (*Hints:* Assume that Earth had its present radius. The surface area of a sphere is $4\pi r^2$.)
9. The velocity of the solar wind is roughly 400 km/s. How long does it take to travel from the sun to Pluto?

Media Cluster

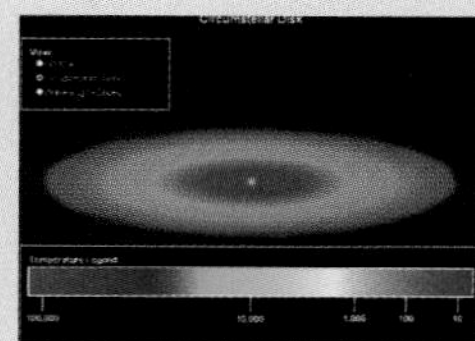

To access the resources in the Media Cluster, log into AceAstronomy at **http://astronomy.brookscole.com/sh9e** and select Chapter 16.

ACTIVE FIGURES

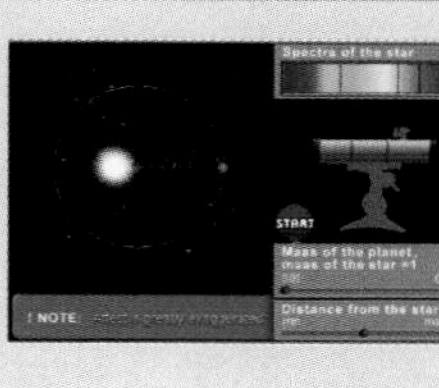

Circumstellar Disk
In this animation, you can observe the disk surrounding a star in different kinds of light. Notice that what you observe depends on wavelength.

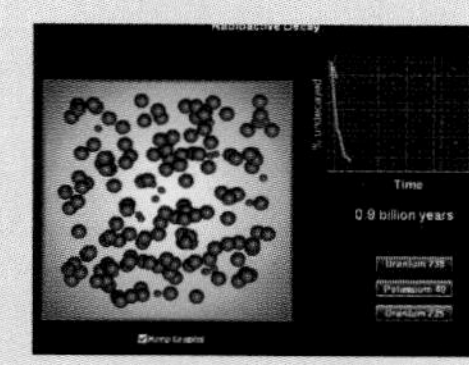

Radioactive Decay
The characteristic decay patterns of different radioactive elements provide an invaluable dating tool. In this animation, see how radioactive atoms decay into daughter atoms.

ASTRONOMY EXERCISE

Extrasolar Planets
This simulation presents a hypothetical star system with one planet orbiting the star. Study what variables help astronomers to detect extrasolar planets.

VIRTUAL ASTRONOMY LAB

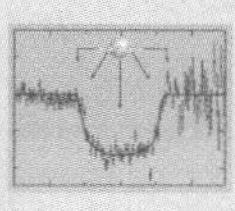

Lab 8: Extrasolar Planets
This lab examines how indirect methods such as Doppler shift techniques are used to discover and study extrasolar planets. At the end of the lab you will build your own solar system.

Critical Inquiries for the Web

1. How does our solar system compare with the others that have been found? Search the Internet for sites that give information about planetary systems around other stars. What kinds of planets have been detected by these searches so far? Discuss the selection effects (see Window on Science 13-2) that must be considered when interpreting these data.
2. The process of protoplanetary accretion is still not well understood. Search the web for current research in this field. From the results of your search, outline the basic steps in the formation of a protoplanet through accretion. What specific factors are important in these models of planet building? Do these models produce planetary systems similar to the ones we know to exist?
3. How is radioactive dating carried out on meteorites and rocks from surfaces of various bodies in the solar system? Look for websites on the details of radioactive dating, and summarize the methods used to uncover the abundances of radioactive elements in a particular sample. (*Hint:* Try looking for information on how a particular meteorite—for example, the Martian meteorite ALH84001—was studied, what age range was determined, and what radioactive elements were used to arrive at the age.)

Exploring *TheSky*

1. Look at the solar system from space. Notice how thin the disk of the solar system is and how inclined the orbits of Pluto and Mercury are. (*Hint:* Under the **View** menu, choose **3D Solar System Mode,** and then zoom in or out. Tip the solar system up and down to see it edge-on.)
2. Look at the solar system from space and notice how small the orbits of the inner planets are compared to the orbits of the outer planets. They make two distinct groups. (*Hint:* Use **3D Solar System Mode.**)
3. Watch the comets orbiting around the sun. Can you locate the comet C/198 M5 (Linear)? What is its orbit like? (*Hint:* Use **3D Solar System Mode,** and set the time step to 30 days [**30d**].)

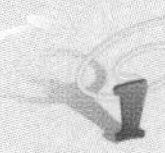

Go to the Brooks/Cole Astronomy Resource Center **(http://astronomy.brookscole.com)** for critical thinking exercises, articles, and additional readings from InfoTrac College Edition, Brooks/Cole's online student library.

17 Comparative Planetology of the Terrestrial Planets

That's one small step
for a man . . . one giant leap
for mankind.

NEIL ARMSTRONG, ON THE MOON

Beautiful, beautiful.
Magnificent desolation . . .

EDWIN E. ALDRIN, JR., ON THE MOON

IF YOU HAD been the first person to step onto the surface of the moon, what would you have said? Neil Armstrong responded to the historic significance of the first human step on the surface of another world. Buzz Aldrin was second, and he responded to the moon itself. It *is* desolate, and it *is* magnificent. But it is not unusual. Most planets in the universe probably look like Earth's moon, and astronauts may someday walk on such worlds and compare them with our moon. The comparison of one planet with another is called **comparative planetology,** and it is one of the best ways to analyze the worlds in our solar system. You will learn much more by comparing planets than you could by studying them individually.

Two human astronauts landed on the moon in 1969, and two robotic astronauts landed on Mars in 2004. The exploration of the Earthlike planets is a continuing adventure. (JSC/NASA)

Guidepost

Looking Back

In the preceding chapter, you learned how our solar system formed as a by-product of the formation of the sun. You saw how distance from the sun determined the general character of each planet. Now you are ready to visit the planets and get to know them as individuals.

This Chapter

As you study the individual planets, you must avoid the pitfall of learning facts instead of understanding principles. One way to focus on principles is to compare the planets with each other and search for similarities and contrasts. Like people, the terrestrial planets are more alike than they are different, but it is the differences that are most memorable.

As you explore, you will be searching for answers to five essential questions:

How can comparison help you understand the terrestrial planets?

What are the main features of Earth when you view it as a planet?

How does size affect a planet?

How does distance from the sun affect a planet and its atmosphere?

How does a planet's atmosphere depend on the size of the planet?

Looking Ahead

Once you are familiar with the family of the terrestrial planets, you will be ready to meet a more peculiar group of characters. The next chapter will introduce you to the worlds of the outer solar system, and the chapter after that will bring the minor characters in our solar system to center stage.

Ace Astronomy™ The AceAstronomy icon throughout the text indicates an opportunity for you to test yourself on key concepts and to explore animations and interactions on the AceAstronomy website at: http://astronomy.brookscole.com/sh9e

17-1 A Travel Guide to the Terrestrial Planets

IF YOU VISIT the city of Granada in Spain, you will probably consult a travel guide, and if it is a good guide, it will do more than tell you where to find museums and restrooms. It will tell you to look at the palace called the Alhambra and compare it with buildings in Morocco. In this chapter you are going to visit five Earthlike worlds, and this preliminary section will be your guide to important features and comparisons.

Five Worlds

You are about to visit Mercury, Venus, Earth, Earth's moon, and Mars. It may surprise you that the moon is on your itinerary. It is, after all, just a natural satellite orbiting Earth and isn't one of the planets. But the moon is a fascinating world of its own, and it makes a striking comparison with the other worlds on your list. Just as you shouldn't tour Spain without seeing Granada, you should not tour the planets without seeing the moon.

■ Figure 17-1 compares the five worlds you are about to study. The first feature you should notice is diameter. The moon is small, and Mercury is hardly much bigger. Earth and Venus are large and quite similar in size, but Mars is a medium-sized world. You will discover that size is a critical factor in determining a

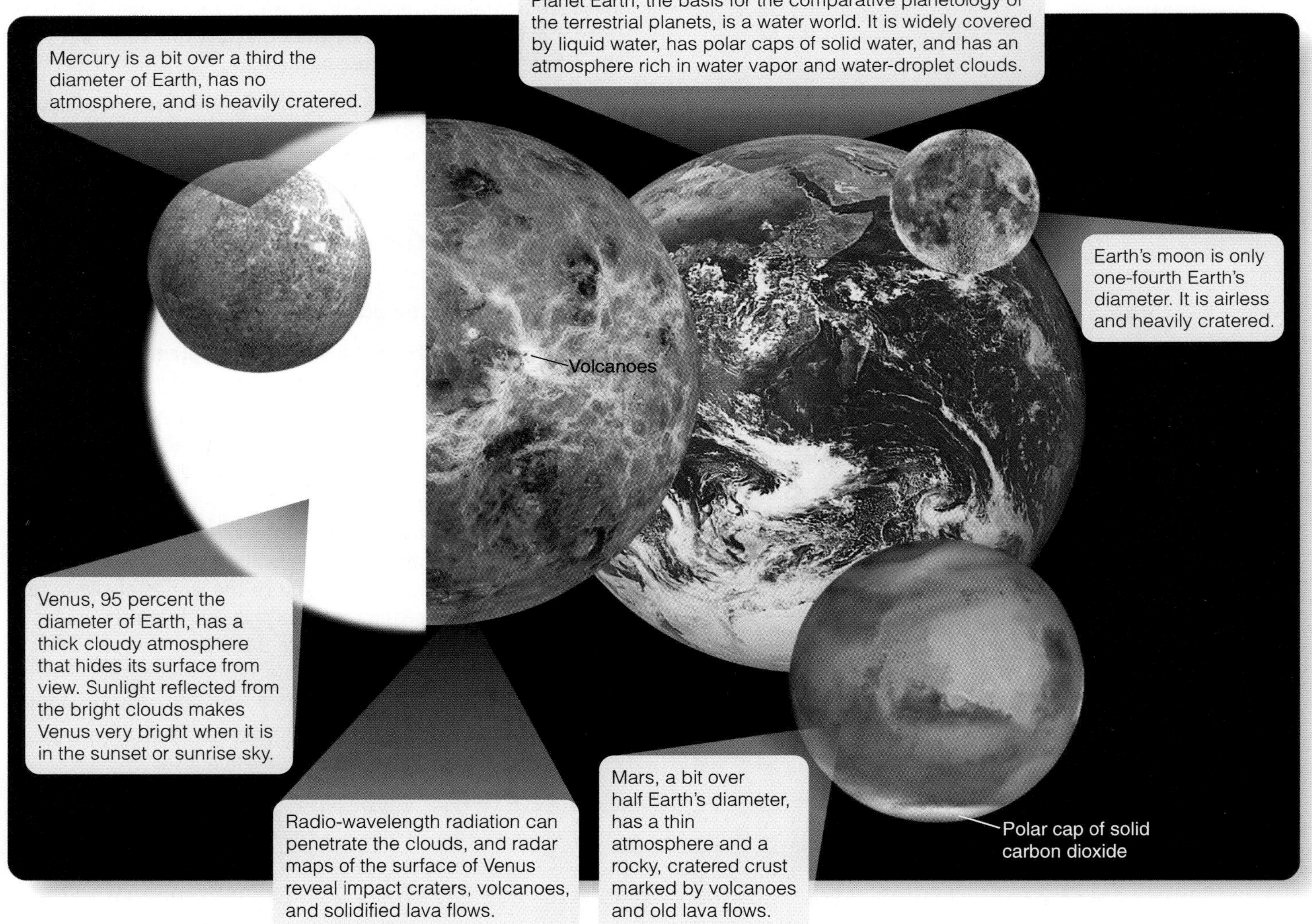

Planet Earth, the basis for the comparative planetology of the terrestrial planets, is a water world. It is widely covered by liquid water, has polar caps of solid water, and has an atmosphere rich in water vapor and water-droplet clouds.
Mercury is a bit over a third the diameter of Earth, has no atmosphere, and is heavily cratered.
Earth's moon is only one-fourth Earth's diameter. It is airless and heavily cratered.
Volcanoes
Venus, 95 percent the diameter of Earth, has a thick cloudy atmosphere that hides its surface from view. Sunlight reflected from the bright clouds makes Venus very bright when it is in the sunset or sunrise sky.
Radio-wavelength radiation can penetrate the clouds, and radar maps of the surface of Venus reveal impact craters, volcanoes, and solidified lava flows.
Mars, a bit over half Earth's diameter, has a thin atmosphere and a rocky, cratered crust marked by volcanoes and old lava flows.
Polar cap of solid carbon dioxide

■ Figure 17-1

Planets in comparison. Earth and Venus are similar in size, but their atmospheres and surfaces are very different. The moon and Mercury are much smaller, and Mars is intermediate in size. (Moon: UCO/Lick Observatory; All planets: NASA)

world's personality. Small worlds tend to be cold and dead, but larger worlds can be hot and active.

Core, Mantle, and Crust

Notice that these terrestrial worlds are made up of rock and metal. They all have rocky, low-density crusts, while much of the metal in these worlds has sunk to their centers to form dense cores. You will discover that Earth has a molten iron core, but that the moon lacks much iron at all. Between the crust and the core, a **mantle** is composed of dense rock.

When the planets formed, their surfaces were cratered by the heavy bombardment of debris in the young solar system. You will see lots of craters on these worlds, especially on Mercury and the moon, but notice that cratered surfaces are old. If a lava flow covered up the craters after the end of the heavy bombardment, few new craters could be formed because most of the debris in the solar system was gone. When you see smooth plains on the moon or on Mars, you can guess that they are younger than the cratered areas.

On your travels among the Earthlike worlds, look for signs of heat flowing up from the interior. In the preceding chapter you saw evidence that the planets probably form hot, and radioactive elements trapped in their interiors decay and generate more heat. That heat, flowing upward through the cooler crust, can make a world active with volcanoes and lava flows; but watch for the small worlds that cooled fast and have little heat flowing outward now. You will find that Mercury and the moon, being small, are inactive, and that Earth and Venus, being larger, are geologically active.

Atmospheres

When you look at Mercury and the moon in Figure 17-1, you can see their craters and plains and mountains, but the surface of Venus is completely hidden by a cloudy atmosphere even thicker than Earth's. Mars, the medium planet, has a medium atmosphere of thin gases.

As you hike the sandy deserts of Mars, you can ponder two questions. First, why do some worlds have atmospheres while some do not? You will discover that both size and temperature are important. The second question is more complex. Where did these atmospheres come from? To answer that question, you will have to study the geological history of these worlds.

Inquire | Review | Analyze

Why do you expect the inner planets to be high-density worlds?
In Chapter 16, you saw how the inner planets formed from hot inner parts of the solar nebula. No ice could form there, so the inner planets had to grow from the solid bits of rock and metal that condensed from the hot gas. So you expect the inner planets to be made mostly of rock and metal, which are dense materials.

As you visit the terrestrial planets, they will seem familiar, but you will find craters almost everywhere. What made all of those craters?

■ ■ ■

Connections: Some people are satisfied reading travel guides, but most people want to visit Granada rather than read about it. Of course, you can't really visit Venus or Mars, but one of the wonders of science is that it can take your imagination where your body cannot go. So grab your carry-on luggage. They are calling your flight, and the first world you will visit may surprise you.

17-2 Earth: Planet of Extremes

EARTH IS THE BASIS for your comparative study of the terrestrial planets, so you should visit it as if you didn't live here. You need to see it as a planet, and it is indeed a planet of extremes. Earth's interior is molten and generates a strong magnetic field. Heat flowing outward powers volcanism and earthquakes and breaks the crust into moving sections that bump against each other and push up mountains. Almost 75 percent of Earth's surface is covered by liquid water with an average depth of 3 km. Even Earth's atmosphere is an unusual mixture of nitrogen and oxygen. Once you understand Earth's complex properties, you will be ready to compare it with other unearthly worlds.

Four Stages of Planetary Development

As Earth formed in the inner solar nebula, it passed through four developmental stages (■ Figure 17-2), stages that other worlds also experienced to varying degrees.

The first stage of planetary evolution is *differentiation,* the separation of material according to density. Earth now has a dense metallic core, a thick rocky mantle, and a lower-density crust, so you know it differentiated. Some of that differentiation may have occurred very early as the heat of formation released by in-falling matter melted the growing Earth. Some of the differentiation, however, may have occurred later as radioactive decay released heat and further melted Earth, allowing the denser metals to sink to the core.

The second stage, *cratering,* could not begin until a solid surface formed. The heavy bombardment of the early solar system cratered Earth just as it did our moon. As the debris in the solar nebula cleared away, the rate of cratering impacts fell rapidly to its present low rate.

The third stage, *flooding,* began as the decay of radioactive elements heated Earth's interior. Rock melted in the upper mantle where the pressure was lower than the deep interior, and some of the molten rock welled up through fissures in the crust and flooded the deeper basins. Later, as the atmosphere cooled, water fell as rain and flooded the basins to form the first oceans. Note that on Earth flooding includes both lava and water.

The fourth stage, *slow surface evolution,* has continued for at least the past 3.5 billion years or more. Earth's surface is constantly changing as sections of crust slide over each other, push up mountains, and shift continents. In addition, moving air and water erode the surface and wear away geological features. Almost all trace of the first billion years of Earth's geology has been destroyed by the active crust and erosion.

Terrestrial planets pass through these four stages, but differences in mass, temperature, and composition emphasize some stages over others and produce surprisingly different worlds.

■ Figure 17-2

The four stages of planetary development are illustrated for Earth.

Four Stages of Planetary Development

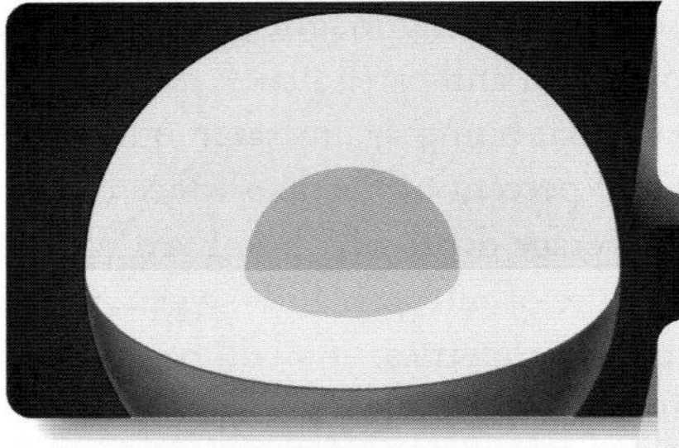

Differentiation produces a dense core, thick mantle, and low-density crust.

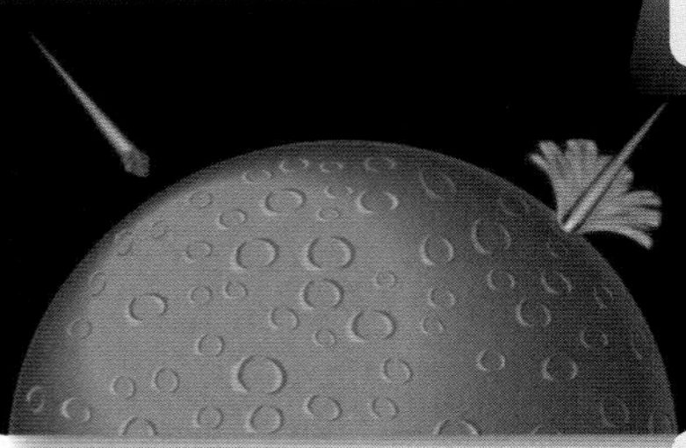

The young Earth was heavily bombarded in the debris-filled early solar system.

Flooding by molten rock and later by water can fill lowlands.

Slow surface evolution continues due to geological processes, including erosion.

Earth's Interior

From what you know of the formation of Earth, you would expect it to have differentiated; but in science, evidence rules. What does the evidence reveal about earth's interior?

Earth's mass divided by its volume tells you its average density, about 5.50 g/cm^3. (See **Celestial Profile 2**.) But the density of Earth's rocky crust is only about half that. Clearly, a large part of Earth's interior must be made of material denser than rock.

Although the deepest wells extend only a few kilometers down, Earth scientists have clear proof that Earth did differentiate. Each time an earthquake occurs, seismic waves travel through the interior and register on seismographs all over the world. Analysis of these waves shows that Earth's interior is divided into a metallic core, a dense rocky mantle, and a thin, low-density crust.

The core has a density of 14 g/cm^3, denser than lead, and is believed to be composed of iron and nickel at a temperature of roughly 6000 K. The core of Earth is as hot as the surface of the sun, but the high pressure keeps the metal a solid near the middle of the core and a liquid in its outer parts. Two kinds of seismic waves indicate that the outer core is liquid. The ***P* waves** travel like sound waves, and they can penetrate a liquid; but ***S* waves** travel as a side-to-side vibration of particles like the jiggle in a bowl of jelly. *S* waves can't travel through a liquid, so Earth scientists can deduce the size of the liquid core by observing where *S* waves get through and where they don't (■ Figure 17-3). The outer boundary of the core lies just over halfway to the surface.

■ Active Figure 17-3

P and *S* waves give you clues to Earth's interior. That no direct *S* waves reach the far side shows that Earth's core is liquid. The size of the *S* wave shadow tells you the size of the outer core.

AceAstronomy™ **Log into AceAstronomy and select this chapter to see the Active Figure called "Seismic Waves." Compare the observations at different seismic stations around the world.**

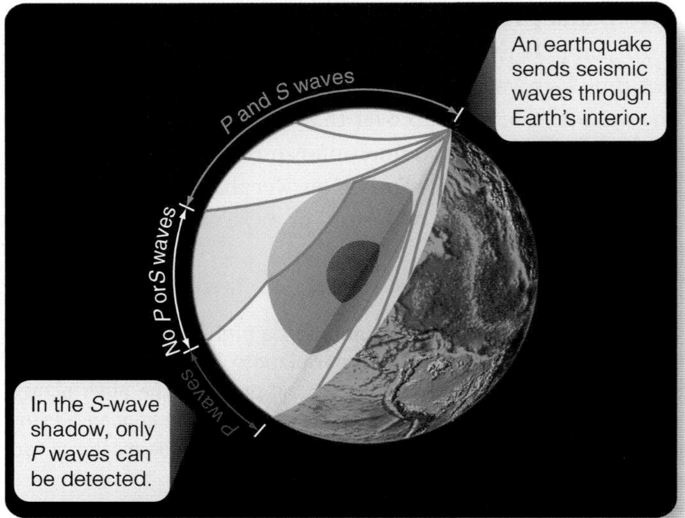

No P or S waves

In the S-wave shadow, only P waves can be detected.

Earth's magnetism gives you further clues about the core. Convection currents stir the molten liquid, and, because it is a very good conductor of electricity and is rotating as Earth rotates, it generates a magnetic field through the dynamo effect—the same process that creates the sun's magnetic field (Chapter 7). From traces of magnetic field retained by rocks that formed long ago, geologists conclude that Earth's magnetic field reverses itself every 700,000 years or so. Periodic reversals, though poorly understood, are a characteristic of the dynamo effect. Thus, the presence of a magnetic field is evidence that part of Earth's core is a liquid metal.

As shown in Celestial Profile Two, Earth's mantle is a deep layer of dense rock that lies between the molten core and the solid crust. The mantle material is a **plastic,** a material with the properties of a solid but capable of flowing under pressure. Asphalt used in paving roads is a common plastic; it shatters if struck with a sledgehammer, but it bends under the weight of a truck. Just below Earth's crust, where the pressure is less than at great depths, the mantle is most plastic.

The Earth's rocky crust is made up of low-density rocks and floats on the denser mantle. The crust is thickest under the continents, up to 60 km thick, and thinnest under the oceans, where it is only about 10 km thick.

AceAstronomy™ Log into AceAstronomy and select this chapter to see Astronomy Exercise "Convection and Magnetic Fields" and take control of a planetary interior.

Earth's Active Crust

Earth's crust is composed of lower-density rock that floats on the mantle. The image of a rock floating may seem odd, but recall that the rock of the mantle is very dense. Also, just below the crust, the mantle rock tends to be highly plastic, so great sections of low-density crust do indeed float on the mantle like great lily pads floating on a pond.

The motion of the crust and the erosive action of water make Earth's crust highly active. Look at **The Active Earth** on pages 384 and 385 and notice three important points. First, the motion of crustal plates produces much of the geological activity you see on Earth. Earthquakes, volcanism, and mountain building are all linked to motions in the crust. Second, notice how the motions of the plates push the continents around and constantly renew Earth's surface. Third, notice that most of the geological features you know—mountain ranges, the Grand Canyon, and even the outline of the continents—are recent products of Earth's active surface. The oldest rocks on Earth, small crystals from Australia, are 4.3 billion years old. Most of the crust is much younger than that.

AceAstronomy™ Log into AceAstronomy and select this chapter to see Astronomy Exercise "Convection and Plate Tectonics" and compare the interiors of the four terrestrial planets.

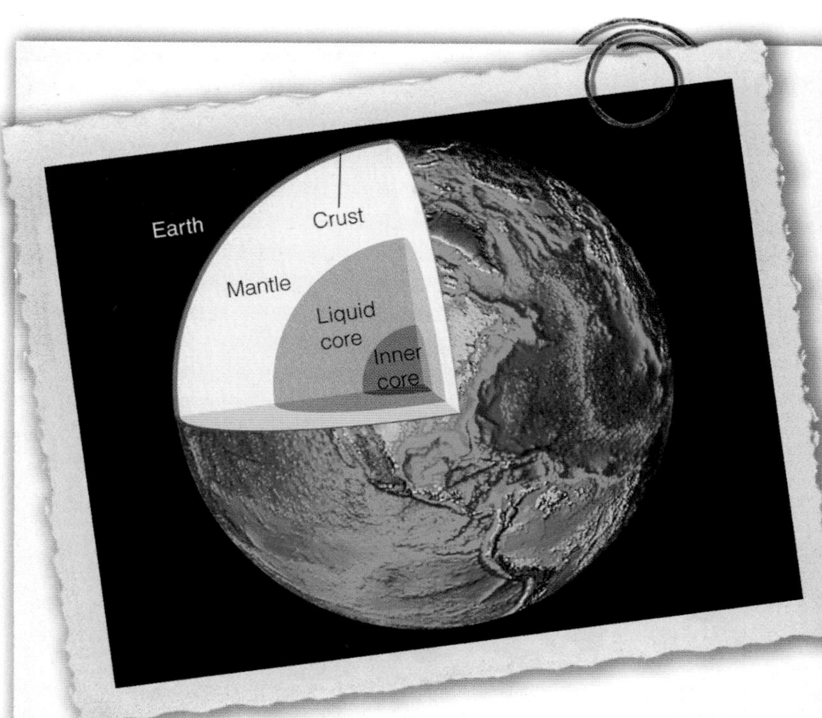

Earth's surface is marked by high continents and low sea floors, but the crust is only 10 to 60 km thick. Below that lies a deep mantle and an iron core. (NGDC)

Celestial Profile 2: Earth

Motion:

Average distance from the sun	1.00 AU (1.495979 $\times$ 10^8 km)
Eccentricity of orbit	0.0167
Maximum distance from the sun	1.0167 AU (1.5210 $\times$ 10^8 km)
Minimum distance from the sun	0.9833 AU (1.4710 $\times$ 10^8 km)
Inclination of orbit to ecliptic	0°
Average orbital velocity	29.79 km/s
Orbital period	1.00 y (365.25 days)
Period of rotation (with respect to the sun)	$24^h00^m00^s$
Inclination of equator to orbit	23°27′

Characteristics:

Equatorial diameter	12,756 km
Mass	5.976 $\times$ 10^{24} kg
Average density	5.497 g/cm^3 (4.07 g/cm^3 uncompressed)
Surface gravity	1.0 Earth gravity
Escape velocity	11.2 km/s
Surface temperature	−50° to 50°C (−60° to 120°F)
Average albedo	0.39
Oblateness	0.0034

Personality Point:

Earth comes, through Old English *eorthe* and Greek *Eraze,* from the Hebrew *erez,* which means ground. *Terra* comes from the Roman goddess of fertility and growth, *Terra Mater,* Mother Earth.

The Active Earth

Our world is an astonishingly active planet. Not only is it rich in water and therefore subject to rapid erosion, but its crust is divided into moving sections called plates. Where plates spread apart, lava wells up to form new crust; where plates push against each other, they crumple the crust to form mountains. Where one plate slides over another, you see volcanism. This process is called plate tectonics, referring to the Greek word for "builder." (An architect is literally an arch builder.)

Midocean rise
Red Sea
Midocean rise
National Geophysical Data Center

A typical view of planet Earth

William K. Hartmann

Janet Seeds

Mountains are common on Earth, but they erode away rapidly because of the abundant water.

Evidence of plate tectonics was first found in ocean floors, where plates spread apart and magma rises to form **midocean rises** made of rock called **basalt,** a rock typical of solidified lava. Radioactive dating shows that the basalt is younger near the midocean rise. Also, the ocean floor carries less sediment near the midocean rise. As Earth's magnetic field reverses back and forth, it is recorded in the magnetic fields frozen into the basalt. This produces a magnetic pattern in the basalt that shows that the sea floor is spreading away from the midocean rise.

Subduction zone
Midocean rise
Pacific Ocean
Atlantic Ocean
S. America
Africa
Plate motion
Plate motion
Plate motion
Ocean floor
Melting
Mantle

A **subduction zone** is a deep trench where one plate slides under another. Melting releases low-density magma that rises to form volcanoes such as those along the northwest coast of North America, including Mt. St. Helens.

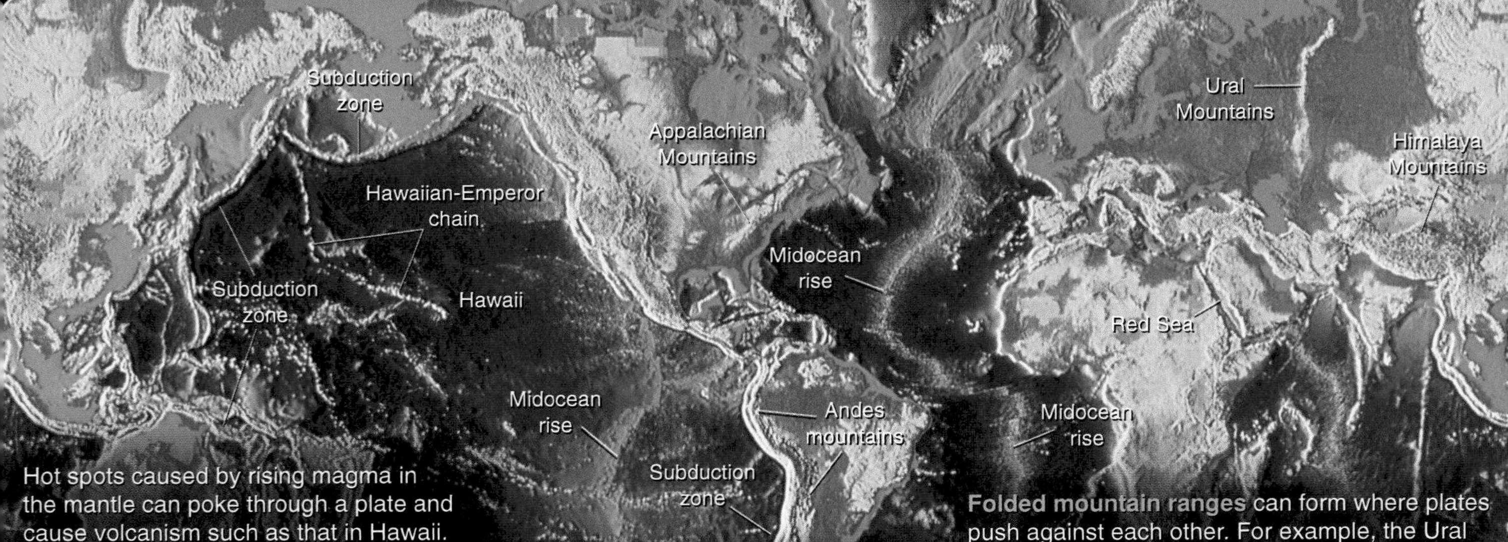

Hot spots caused by rising magma in the mantle can poke through a plate and cause volcanism such as that in Hawaii. As the Pacific plate has moved north-westward, the hot spot has punched through to form a chain of volcanic islands, now mostly worn below sea level.

Ace Astronomy™ Log into AceAstronomy and select this chapter to see Active Figure "Hot Spot Volcanoes." Notice how the moving plate can produce a chain of volcanic peaks, mostly under water in the case of Earth.

Folded mountain ranges can form where plates push against each other. For example, the Ural Mountains lie between Europe and Asia, and the Himalaya Mountains are formed by India pushing north into Asia. The Appalachian Mountains are the remains of a mountain range pushed up when North America was pushed against Africa.

The floor of the Atlantic Ocean is not being subducted. It is locked to the continents and is pushing North and South America away from Europe and Africa at about 3 cm per year, a motion called *continental drift*. Radio astronomers can measure this motion by timing pulsars from European and from American radio telescopes. Roughly 200 million years ago, North and South America were joined to Europe and Africa. Evidence of that lies in similar fossils and similar rocks and minerals found in the matching parts of the continents. Notice how North and South America fit against Europe and Africa like a puzzle.

Continental Drift

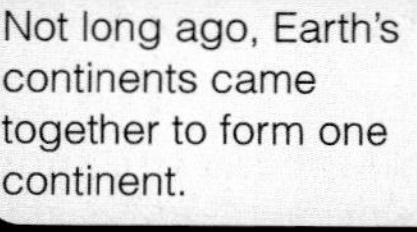

Not long ago, Earth's continents came together to form one continent.

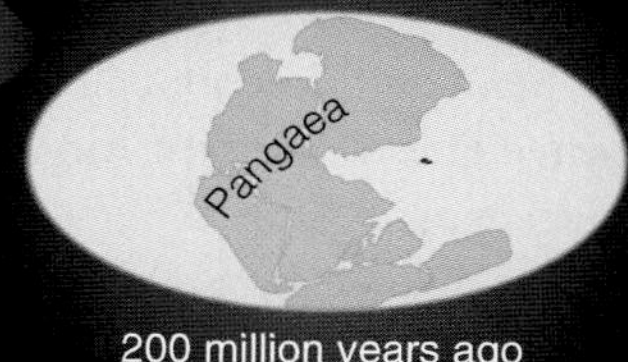

Pangaea broke into a northern and a southern continent.

Notice India moving north toward Asia.

The continents are still drifting on the highly plastic upper mantle.

Hawaii

The floor of the Pacific Ocean is sliding into subduction zones in many places around its perimeter. This pushes up mountains such as the Andes and triggers earthquakes and active volcanism all around the Pacific in what is called the Ring of Fire. In places such as southern California, the plates slide past each other, causing frequent earthquakes.

Yellow lines on this globe mark plate boundaries. Red dots mark earthquakes since 1980. Earthquakes within the plate, such as those at Hawaii, are related to volcanism over hot spots in the mantle.

A **rift valley** forms where continental plates begin to pull apart. The Red Sea has formed where Africa has begun to pull away from the Arabian peninsula.

Plate tectonics pushes up mountain ranges and causes bulges in the crust, and water erosion wears the rock away. The Colorado River began cutting the Grand Canyon only about 10 million years ago when the Colorado plateau warped upward under the pressure of moving plates. That sounds like a long time ago, but it is only 0.01 billion years. A mile down, at the bottom of the canyon, lie rocks 0.57 billion years old, the roots of an earlier mountain range that stood as high as the Himalayas. It was pushed up, worn away to nothing, and covered with sediment long ago. Many of the geological features we know on Earth have been produced by very recent events.

Earth's Atmosphere

When you think about Earth's atmosphere, you should consider three questions: How did it form? How has it evolved? How are we changing it? Answering these questions will help you understand other planets as well as our own.

Our planet's first atmosphere, its **primeval atmosphere,** was once thought to contain gases from the solar nebula, such as hydrogen and methane. Modern studies, however, indicate that the planets formed hot, so gases such as carbon dioxide, nitrogen, and water vapor would have been cooked out of the rock and metal as Earth grew. In addition, the final stages of planet building may have seen planets accreting planetesimals rich in easily vaporized materials such as water, ammonia, and carbon dioxide. Thus the primeval atmosphere must have been rich in carbon dioxide, nitrogen, and water vapor. The atmosphere you breath today is a **secondary atmosphere** produced later in Earth's history by outgassing and by green plants that produced oxygen.

Soon after Earth formed, it began to cool, and once it cooled enough, oceans began to form, and carbon dioxide began to dissolve in the water. Carbon dioxide is highly soluble in water—which explains the easy manufacture of carbonated beverages. As the oceans removed carbon dioxide from the atmosphere, it reacted with dissolved compounds in the ocean water to form silicon dioxide, limestone, and other mineral sediments. Thus, the oceans transferred the carbon dioxide from the atmosphere to the seafloor and left air rich in nitrogen.

This removal of carbon dioxide is critical to Earth because an atmosphere rich in carbon dioxide can trap heat by a process called the **greenhouse effect.** When visible-wavelength sunlight shines through the glass roof of a greenhouse, it heats the benches and plants inside. The warmed interior radiates heat in the form of infrared radiation, which can't get out through the glass. Heat is trapped in the greenhouse, and the temperature climbs until the glass itself grows warm enough to radiate heat away as fast as sunlight enters (■ Figure 17-4a). (Of course, a greenhouse also retains its heat because the walls prevent the warm air from mixing with the cooler air outside.) This is the same process that heats a car when it is parked in the sun with the windows rolled up.

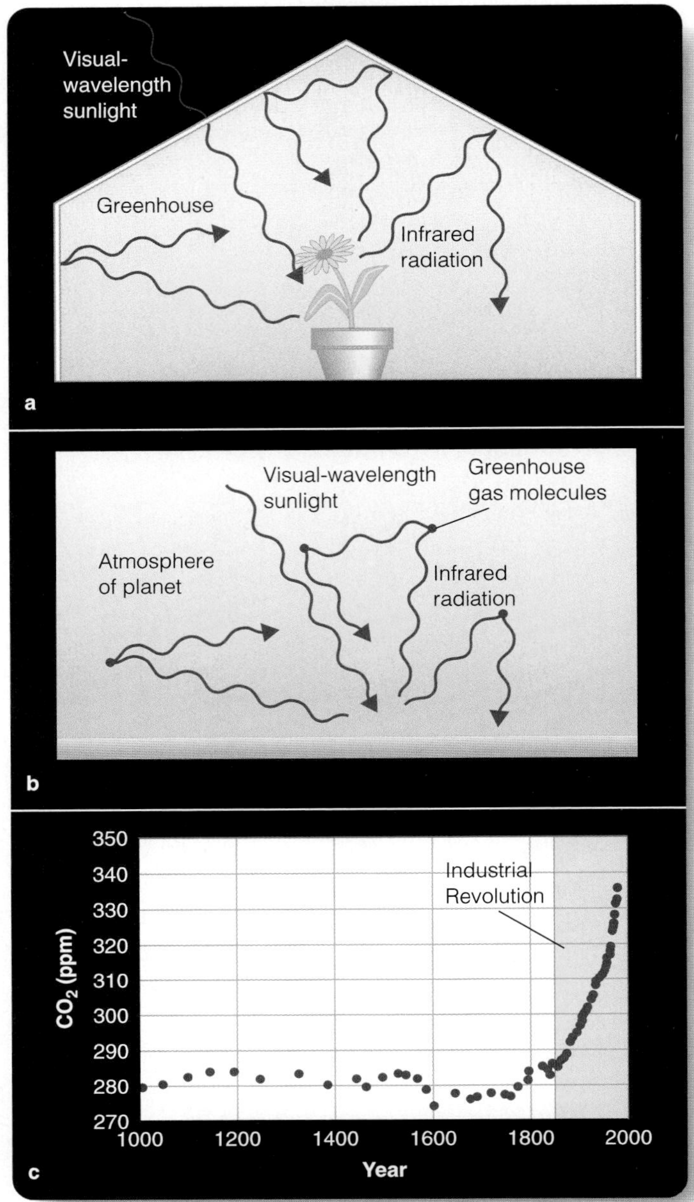

■ **Figure 17-4**

The greenhouse effect: (a) Visual-wavelength sunlight can enter a greenhouse and heat its contents, but the longer-wavelength infrared cannot get out. (b) The same process can heat a planet's surface if its atmosphere contains greenhouse gases, such as CO_2. (c) The concentration of CO_2 in Earth's atmosphere as measured in Antarctic ice cores has remained roughly constant until the beginning of the Industrial Revolution. (Graph adapted from a figure by Etheridge, Steele, Langenfelds, Francey, Barnola, and Morgan.)

Like the glass roof of a greenhouse, a planet's atmosphere can allow sunlight to enter and warm the surface. Carbon dioxide and other greenhouse gases such as water vapor and methane are opaque to infrared radiation, so an atmosphere containing enough of these greenhouse gases can trap heat and raise the temperature of a planet's surface (Figure 17-4b).

Without the greenhouse effect, Earth would be about 40 K colder. That amounts to 72°F, so Earth would be uninhabitable without the greenhouse effect. The problem is that human civilization may be increasing the amount of greenhouse gases in the atmosphere.

For roughly 4 billion years, Earth's oceans and plant life have been absorbing carbon dioxide and burying it in the form of carbonates such as limestone and in carbon-rich deposits of coal, oil, and natural gas. But in the last century or so, human civilization has begun digging up those fuels, burning them for energy, and releasing the carbon back into the atmosphere as carbon dioxide (Figure 17-4c). This process is steadily increasing the carbon dioxide concentration in our atmosphere and warming Earth's climate in what is called **global warming.**

Although the temperature increase is small now, it is a serious issue for coming decades. Climate change may alter humanity's ability to grow food, of course, but there are other consequences. For example, a small rise in Earth's temperature will begin melting the polar caps, and the released water will raise sea level. A rise of just a few meters would have catastrophic consequences for coastal areas worldwide. A rise of 8 meters would flood the White House, and a rise of a few dozen meters would flood major land areas, including much of China and all of Florida. Global warming is a controversial subject because the climate is complex, but also because solving the problem has severe political and economic costs. The problem is noted here as an example of the greenhouse effect. When we discuss Venus, you will see a planet dominated by the greenhouse effect.

AceAstronomy™ Log into AceAstronomy and select this chapter to see Astronomy Exercise "Primary Atmospheres." You can move your planet around to see how its atmosphere is affected.

AceAstronomy™ Log into AceAstronomy and select this chapter to see Astronomy Exercise "The Greenhouse Effect." Take control of the CO_2 in your experimental planet's atmosphere.

Oxygen in Earth's Atmosphere

Earth's early atmosphere may have contained some methane (CH_4) and ammonia (NH_3), but ultraviolet photons would have penetrated deep into the atmosphere and broken those molecules up to form carbon, nitrogen, and hydrogen. The hydrogen would escape to space, the carbon would form more carbon dioxide, and the nitrogen would be left in the atmosphere. This no longer happens because the atmosphere now contains free oxygen, which forms a layer of ozone (O_3) at altitudes of 15 to 30 km.

The ozone layer now protects the lower atmosphere from high-energy photons, but when Earth was young, its atmosphere had no free oxygen. Oxygen is very reactive and quickly forms oxides in the soil, so plant life is needed to keep a steady supply of oxygen in the atmosphere. Although life seems to have begun on Earth within a billion years of its formation, life did not generate free oxygen until the onset of photosynthesis about 3.3 billion years ago. Photosynthesis makes energy for the plant by absorbing carbon dioxide and releasing free oxygen. Apparently, the development of large, shallow seas along the continental margins half a billion years ago allowed ocean plants to manufacture oxygen faster than chemical reactions could remove it. Atmospheric oxygen then increased rapidly, and it is still increasing at the rate of about 1 percent every 36 million years.

The ozone layer exists because our atmosphere contains free oxygen. This layer protects you from harmful ultraviolet photons, but certain compounds called chlorofluorocarbons (CFCs), which are used in refrigeration and industry, can destroy ozone when they leak into the atmosphere. Since the late 1970s, the ozone concentration has been falling, and the intensity of harmful ultraviolet radiation at Earth's surface has been observed to be increasing year by year. While this poses an immediate problem for public health on Earth, it is also of interest astronomically. When we discuss Mars, you will see the effects of an atmosphere without ozone.

Inquire | Review | Analyze

What evidence indicates that Earth has a molten metallic core? In science, the critical analysis of ideas must eventually return to evidence. In this case, the evidence is indirect because you can never visit Earth's core. Seismic waves from distant earthquakes pass through Earth, but a certain kind of wave, the *S* waves, cannot pass through the core. Because the *S* waves cannot propagate through a liquid, Earth's core seems to be a liquid. Earth's magnetic field gives further evidence of a metallic core. The theory for the generation of magnetic fields, the dynamo effect, requires a liquid, convective, conducting core. If the core were not a liquid, it would not be able to generate a magnetic field. Thus, two different kinds of evidence tell you that our planet has a liquid core.

Every theory must be supported by evidence. What evidence can you cite to support the theory of plate tectonics?

■ ■ ■

Connections: As you learn more about other worlds and their atmospheres, you also learn more about your own. Ironically, the next world you will study has no atmosphere at all.

17-3 The Moon

THE EARTH'S MOON is the only other world that humans have visited; and, although it is small and airless, it has a fascinating geological history.

Lunar Geology

When you look at the moon in the evening sky, you can see dark and light areas on its disk (**Celestial Profile 3**), which many people refer to as "the Man in the Moon." These dark areas are the lunar lowlands, which, drawing on the Latin word for *seas,* earlier astronomers named **maria** (singular **mare**). The bright areas on the lunar disk are the lunar highlands, rugged mountainous areas.

A small telescope reveals that the surface of the moon is marked by craters. These craters are quite dramatic near the **terminator,** the boundary between daylight and darkness on the moon where shadows are long. The lunar highlands are heavily marked by craters, and the lunar lowlands contain few craters.

The key to understanding the history of the moon is this distribution of craters. Study **Impact Cratering** on pages 388 and 389 and notice three important points. First, impact craters have certain distinguishing characteristics, such as their shape and the way they eject material across the lunar surface. Second, notice the range of sizes of impact craters from giant

Impact Cratering

The craters that cover the moon and many other bodies in the solar system were produced by the high-speed impact of meteorites of all sizes. Meteorites striking the moon travel 10 to 60 km/s and can hit with the energy of many nuclear bombs.

A meteorite striking the moon's surface can deliver tremendous energy and can produce an impact crater 10 or more times larger in diameter than the meteorite. The vertical scale is exaggerated here for clarity.

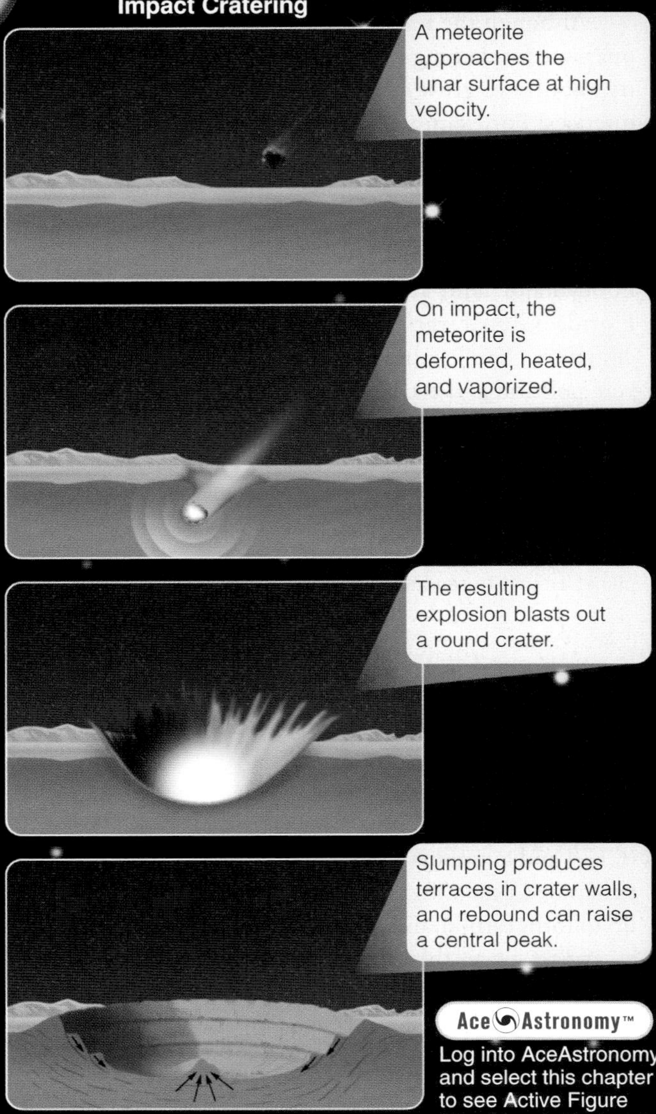

Impact Cratering

A meteorite approaches the lunar surface at high velocity.

On impact, the meteorite is deformed, heated, and vaporized.

The resulting explosion blasts out a round crater.

Slumping produces terraces in crater walls, and rebound can raise a central peak.

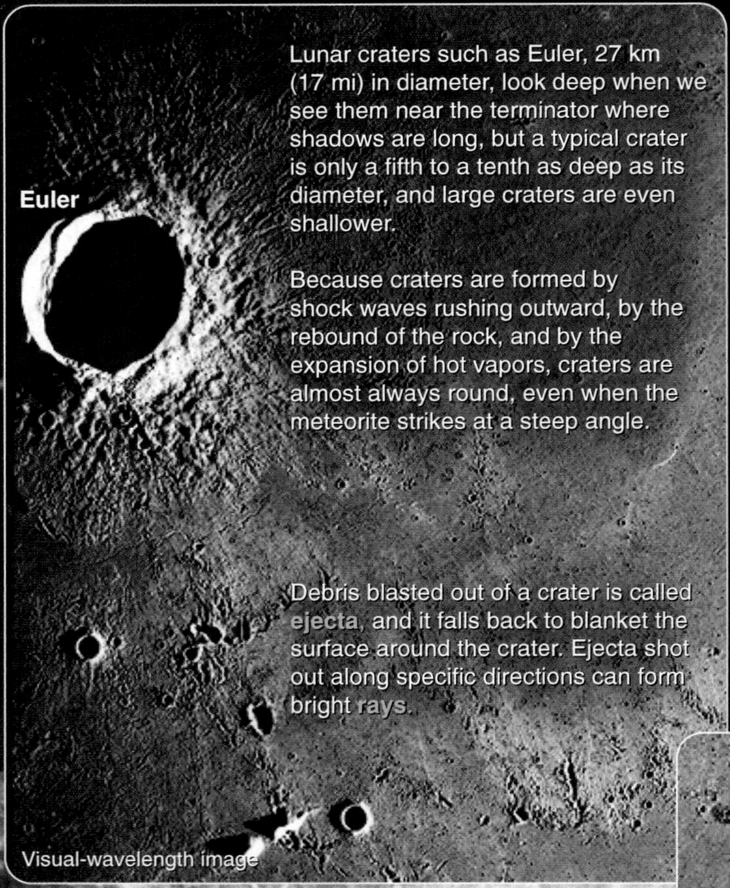

Euler

Lunar craters such as Euler, 27 km (17 mi) in diameter, look deep when we see them near the terminator where shadows are long, but a typical crater is only a fifth to a tenth as deep as its diameter, and large craters are even shallower.

Because craters are formed by shock waves rushing outward, by the rebound of the rock, and by the expansion of hot vapors, craters are almost always round, even when the meteorite strikes at a steep angle.

Debris blasted out of a crater is called **ejecta**, and it falls back to blanket the surface around the crater. Ejecta shot out along specific directions can form bright **rays**.

Visual-wavelength image

NASA

AceAstronomy™

Log into AceAstronomy and select this chapter to see Active Figure "The Moon's Craters." Notice that the structure of the craters depends on their size.

Bright ejecta blankets and rays gradually darken as sunlight darkens minerals and small meteorites stir the dusty surface. Bright rays are signs of youth. Rays from the crater Tycho, perhaps only 100 million years old, extend halfway around the moon.

Rays

Tycho

Visual

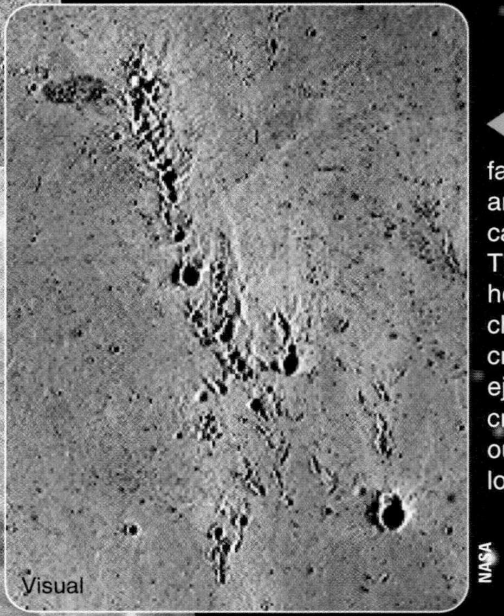

NASA

Rock ejected from distant impacts can fall back to the surface and form smaller craters called **secondary craters.** The chain of craters here is a 45-km-long chain of secondary craters produced by ejecta from the large crater Copernicus 200 km out of the frame to the lower right.

Visual

NASA

Plum Crater, 40 m (130 ft) in diameter, was visited by Apollo 16 astronauts. Note the many smaller craters visible. Lunar craters range from giant impact basins to tiny pits in rocks struck by **micrometeorites**, meteorites of microscopic size.

Visual-wavelength images

Mare Orientale

Solidified lava

The energy of an impact can melt rock, some of which falls back into the crater and solidifies. When the moon was young, craters could also be flooded by lava welling up from below the crust.

NASA

In larger craters, the deformation of the rock can form one or more inner rings concentric with the outer rim. The largest of these craters are called **multiringed basins.** In Mare Orientale on the west edge of the visible moon, the outermost ring is almost 900 km (550 mi) in diameter.

A few meteorites found on Earth have been identified chemically as fragments of the moon's surface blasted into space by cratering impacts. The fragmented nature of these meteorites indicates that the moon's surface has been battered by impact craters.

Most of the craters on the moon were produced long ago when the solar system was filled with debris from planet building. As that debris was swept up, the cratering rate fell rapidly.

Rate of Crater Formation

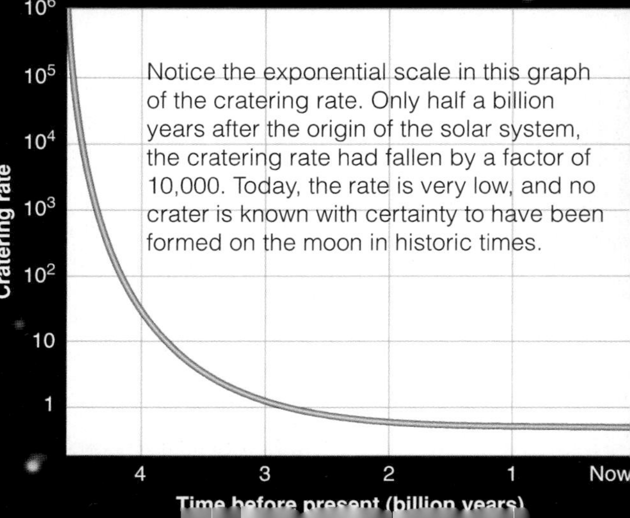

Notice the exponential scale in this graph of the cratering rate. Only half a billion years after the origin of the solar system, the cratering rate had fallen by a factor of 10,000. Today, the rate is very low, and no crater is known with certainty to have been formed on the moon in historic times.

Window on Science 17-1

Understanding Planets: Follow the Energy

One of the best ways to think about a scientific problem is to follow the energy. According to the principle of cause and effect, every effect must have a cause, and every cause must involve energy. Energy moves from regions of high concentration to regions of low concentration and, in doing so, produces changes. For example, coal burns to make steam in a power plant, and the steam passes through a turbine and then escapes into the air. In flowing from the burning coal to the atmosphere, the heat spins the turbine and makes electricity.

It is surprising how commonly scientists use energy as a key to understanding nature. A biologist might ask where certain birds get the energy to fly thousands of miles, and an economist might ask where the economic energy to support the creation of new investments comes from. Energy is everywhere, and when it moves, whether it is birds, money, or molten magma, it causes change. Energy is the cause in cause and effect.

In earlier chapters, the flow of energy from the inside of a star to its surface helped you understand how stars, including the sun, work. You saw that the outward flow of energy supports the star against its own weight, drives convection currents that produce magnetic fields, and causes surface activity such as spots, prominences, and flares. You were able to understand stars because you could follow the flow of energy outward from their interiors.

You can think of a planet by following the energy. The heat in the interior of a planet may be left over from the formation of the planet, or it may be heat generated by radioactive decay, but it must flow outward toward the cooler surface, where it is radiated into space. In flowing outward, the heat can cause convection currents in the mantle, magnetic fields, plate motions, quakes, faults, volcanism, mountain building, and more.

Heat flows out of Earth's interior and generates geological activity such as that at Yellowstone National Park. (M. Seeds)

When you think about any world, be it a small asteroid or a giant planet, think of it as a source of heat that flows outward through the planet's surface into space. If you can follow that energy flow, you can understand a great deal about the world. A planetary astronomer once said, "The most interesting thing about any planet is how its heat gets out."

■ Figure 17-5

The lunar maria are dark smooth lowlands filled with solidified lava. Apollo 11 landed in Mare Tranquillitatis, and the horizon was straight and level. The highlands are bright and cratered, and when Apollo 17 landed at Taurus-Litrow, the astronauts found the horizon mountainous and the terrain rugged. (Moon: UCO/Lick Observatory; Apollo images: NASA)

basins to microscopic pits. Finally, notice that most of the craters you see on the moon are old; they were formed long ago when the solar system was young.

Twelve Apollo astronauts visited both the highlands and the lowlands on Earth's moon between 1969 and 1972 (■ Figure 17-5). Most of the rocks they found were typical of hardened lava, and some were **vesicular basalt,** which contains holes formed by bubbles in the molten rock. These bubbles form when the rock flows out onto the surface, and the lower pressure allows gases dissolved in the rock to form bubbles. The same thing happens when you open a bottle of carbonated beverage and bubbles form. The presence of vesicular basalts shows that much of the surface of Earth's moon has been covered by successive lava flows, and the flat plains of the lunar lowlands, the maria, are actually ancient lava flows. The highlands, in contrast, are composed of low-density rock containing minerals that would be among the first to solidify and float to the top of molten rock. For example, the highlands are rich in **anorthosite,** a light-colored rock that contributes to the highlands' bright contrast with the dark lowlands. Many of the rocks all over Earth's moon are **breccias,** rocks made up of fragments of broken rock cemented together under pressure. The breccias show how extensively the lunar surface has been shattered by meteorites. Nowhere did the astronauts find what they could call bedrock; the entire surface of Earth's moon is fractured by meteorite impacts.

As the astronauts bobbed across the lunar surface, their boots kicked up the powdery dust that covers everything on Earth's moon. This lunar dust is produced by the continuous bombardment of the lunar surface by micrometeorites, which slowly grind exposed rocks into fine gray grit about the consistency of talcum powder.

The color of moon rocks is a dark gray, but Earth's moon looks quite bright in the night sky. In fact, the **albedo** of Earth's moon, the fraction of the light that it reflects, is only 0.06. In other words, it reflects only 6 percent of the light that hits it. Earth, thanks mostly to its bright clouds, has an average albedo of 0.39. Thus, our moon looks bright in the night sky only in contrast. It is in reality a cold, airless, gray world.

AceAstronomy™ Log into AceAstronomy and select this chapter to see Astronomy Exercise "The Moon's Craters." Experiment to form different kinds of craters.

The History of Earth's Moon

The four-stage history of Earth's moon is dominated by a single fact—it is small, only one-fourth the diameter of Earth. The escape velocity is low, so it has been unable to hold any atmosphere, and it cooled rapidly as its internal heat flowed outward into space. As you will see when we study other worlds, a planet's geology is largely driven by energy in the form of heat flowing outward from its interior (**Window on Science 17-1**). Small worlds have less heat and lose it more rapidly, so the moon's small size has been critical in its history.

Earth's moon is about a fourth the diameter of Earth. Its low density indicates that it contains little iron, but the size of its iron core and the amount of remaining heat are unknown. (NASA)

Celestial Profile 3: The Moon

Motion:

Average distance from Earth	384,400 km (center to center)
Eccentricity of orbit	0.055
Maximum distance from Earth	405,500 km
Minimum distance from Earth	363,300 km
Inclination of orbit to ecliptic	5°9′
Average orbital velocity	1.022 km/s
Orbital period (sidereal)	27.321661 days
Orbital period (synodic)	29.5305882 days
Inclination of equator to orbit	6°41′

Characteristics:

Equatorial diameter	3476 km
Mass	7.35×10^{22} kg (0.0123 $M_{\oplus}$)
Average density	3.36 g/cm^3 (3.35 g/cm^3 uncompressed)
Surface gravity	0.167 Earth gravity
Escape velocity	2.38 km/s (0.21 $V_{\oplus}$)
Surface temperature	−170° to 130°C (−274° to 266°F)
Average albedo	0.07

Personality Point:

Lunar superstitions are very common. *Lunatic* and *lunacy* come from *luna,* the moon. Someone who is *moonstruck* is supposed to be at least a bit nutty. Because the moon affects the ocean tides, many superstitions link the moon to water, to weather, and to women's cycle of fertility. In fact moonlight is supposed to be harmful to unborn children; but, on the plus side, moonlight rituals can remove warts.

The Apollo astronauts found that all moon rocks are igneous, meaning they solidified from molten rock.
Rocks exposed on the lunar surface become pitted by micrometeorites.
Vesicular basalt contains bubbles frozen into the rock when it was molten.
A breccia is formed by rock fragments bonded together by heat and pressure.

■ Figure 17-6

Rocks returned from the moon show that the moon formed in a molten state, that it was heavily fractured by cratering when it was young, and that it is now affected mainly by micrometeorites grinding away at surface rock. (NASA)

The Apollo moon rocks show that the moon must have formed in a molten state (■ Figure 17-6). Planetary geologists now refer to the newborn moon as a sea of magma. Denser materials sank to form a small core; and, as the magma ocean cooled, low-density minerals floated to the top to form a low-density crust. In this way the moon must have differentiated into core, mantle, and crust. The moon has a low density and is poor in iron. The moon's core is not large and must contain little metallic iron. The radioactive ages of the moon rocks tell you that the surface solidified about 4.3 billion years ago.

The second stage, cratering, began as soon as the crust solidified, and the older highlands show that the cratering was intense during the first 0.5 billion years—during the heavy bombard-

Formation of the Imbrium Basin

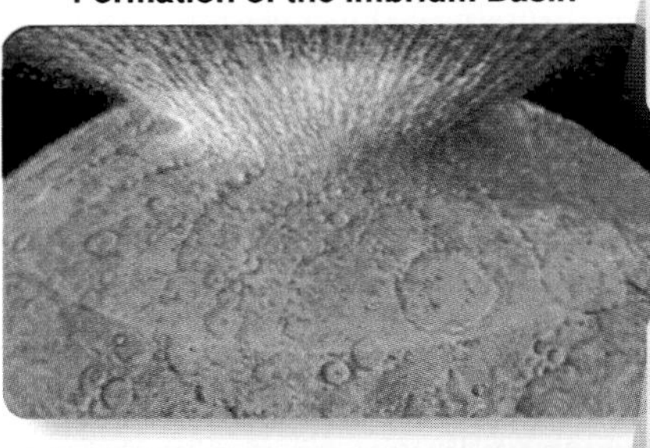

Near the end of the heavy bombardment, a giant impact creates a vast crater basin.

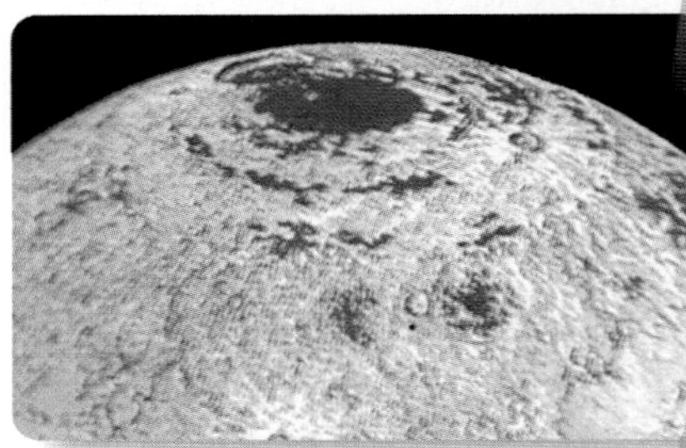

Faulting in the crust produces rings of mountains, and lava flows fill the lowest regions.

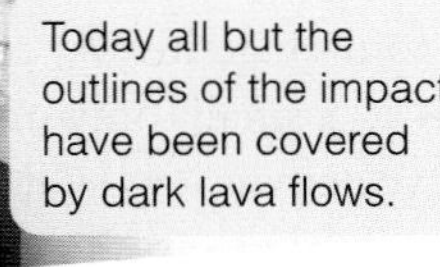

Today all but the outlines of the impact have been covered by dark lava flows.

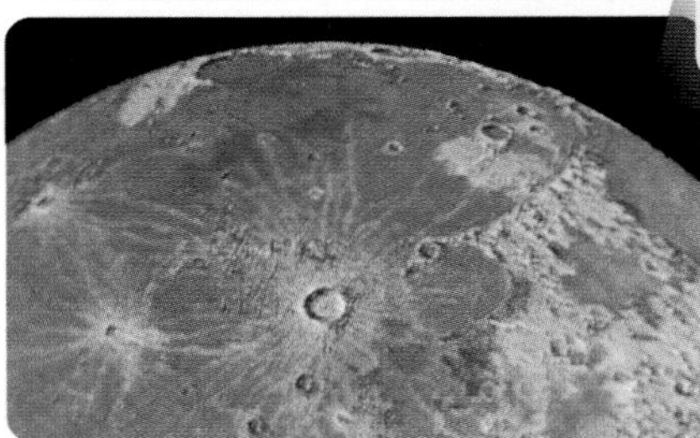

■ Figure 17-7

Mare Imbrium on the moon has a generally round outline, the consequence of its formation by a giant impact 4 billion years ago. (Courtesy Don Davis)

ment at the end of planet building. The moon's crust was shattered, and the largest impacts formed giant multiringed crater basins hundreds of kilometers in diameter (■ Figure 17-7). The basin that became Mare Imbrium, for instance, was blasted out by the impact of an object about the size of Rhode Island. This Imbrium event occurred about 4 billion years ago and blanketed 16 percent of our moon with ejecta. Between 4.1 and 3.9 billion years ago, the cratering rate fell rapidly to the current rate.

The tremendous impacts that formed the lunar basins cracked the crust as deep as 10 kilometers and led to the third stage—flooding. Though Earth's moon cooled rapidly after its formation, some process, perhaps radioactive decay, heated the subsurface material; and part of it melted, producing lava that followed the cracks up into the giant basins (■ Figure 17-8). The basins were flooded by successive lava flows of dark basalts from 3.8 to 3.2 billion years ago, thus forming the maria.

■ Figure 17-8

(a) Major impacts broke the lunar crust and produced large basins. Lava flooded up through the fractures and flooded the basins to produce maria. (Adapted from a diagram by William Hartmann) (b) In these maps color marks elevation, with red the highest regions and purple the lowest. Maria on the near side are generally circular, low, lava-filled basins. Note the circular outline of the labeled maria. The crust on the far side is thicker, and there is less flooding. Even the Aitken Basin contains little lava flooding. (NASA/Clementine)

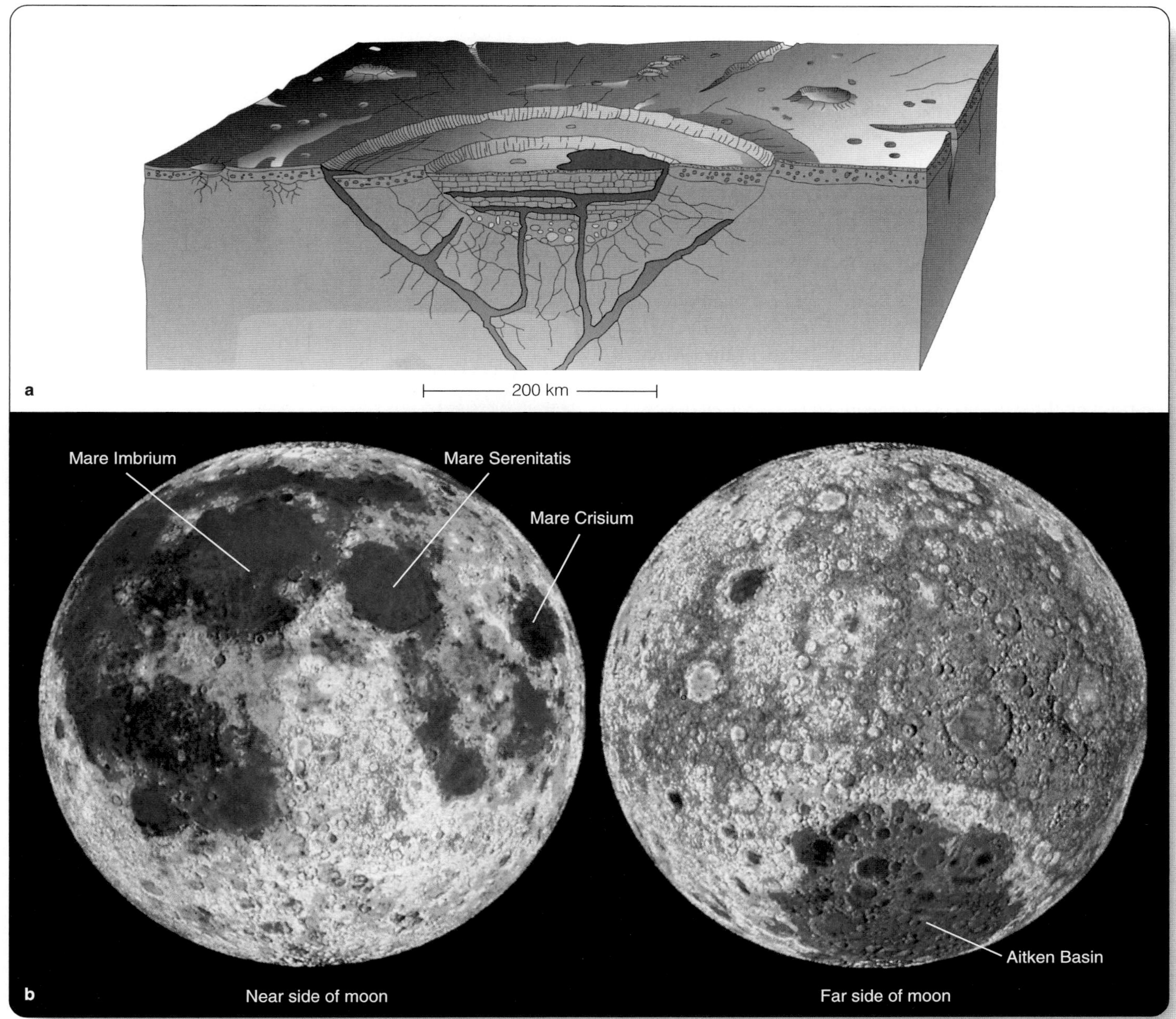

Studies of our moon show that its crust is thinner on the side toward Earth, perhaps due to tidal effects. Consequently, while lava flooded the basins on the Earthward side, it was unable to rise through the thicker crust to flood the lowlands on the far side. The largest impact basin in the solar system is the South Pole–Aitkin Basin (Figure 17-8b). It is about 2500 kilometers (1500 miles) in diameter and as deep at 13 kilometers (8 miles) in places, but flooding has never filled it with smooth lava flows.

The fourth stage, slow surface evolution, is limited because our moon lacks water and has cooled rapidly. Flooding on Earth included water, but the moon has never had an atmosphere and thus has never had liquid water. With no air and no water, erosion is limited to the constant bombardment of micrometeorites and rare larger impacts. Indeed, a few meteorites found on Earth have been identified as moon rocks ejected from our moon by major impacts within the last few million years. As our moon lost its internal heat, volcanism died down and the moon became geologically dead. Its crust never divided into moving plates—there are no folded mountain ranges—and it is now a one-plate planet, frozen between stages 3 and 4.

Although you can now tell the story of the lunar surface, we have neglected one important point. We have not said how Earth's moon formed. We will explore that idea in the next section.

The Origin of Earth's Moon

Over the last two centuries, astronomers have developed three different hypotheses for the origin of Earth's moon, but these traditional ideas have failed to survive comparison with the evidence. A theory proposed in the 1980s may hold the answer. Let's begin by testing the three unsuccessful theories against the evidence.

The **fission hypothesis** proposed that our moon broke from a rapidly spinning proto-Earth. If this happened after the proto-Earth differentiated, our moon would have formed from iron-poor material. However, moon rocks differ chemically from those of Earth. Also, if the proto-Earth had spun fast enough to break up, the Earth–moon system should now contain much more angular momentum than it does.

The **condensation hypothesis** suggested that Earth and its moon condensed from the same cloud of matter in the solar nebula. This idea doesn't work, however, because Earth and its moon have different densities and compositions. The moon, for example, is very poor in volatiles—materials such as water and sodium, which are easily vaporized.

The **capture hypothesis** suggested that our moon formed elsewhere in the solar nebula and was later captured by Earth. However, if our moon passed near enough to Earth to be captured, Earth's gravity would have ripped it to fragments.

Until the mid-1980s, astronomers had no acceptable hypothesis for the origin of Earth's moon, but at that point a hybrid theory offered hope. The **large-impact hypothesis** supposes that our moon formed when a planetesimal at least as large as Mars smashed into the proto-Earth and ejected debris into a disk around Earth, where it formed our moon (■ Figure 17-9).

This would explain a number of phenomena. If the collision occurred off center, it would have spun the Earth–moon system

■ Figure 17-9

Sometime before the solar system was 50 million years old, a collision produced Earth and the moon in its inclined orbit.

The Large-Impact Hypothesis

A protoplanet nearly the size of Earth differentiates to form an iron core.

Another body that has also formed an iron core strikes the larger body and merges, trapping most of the iron inside.

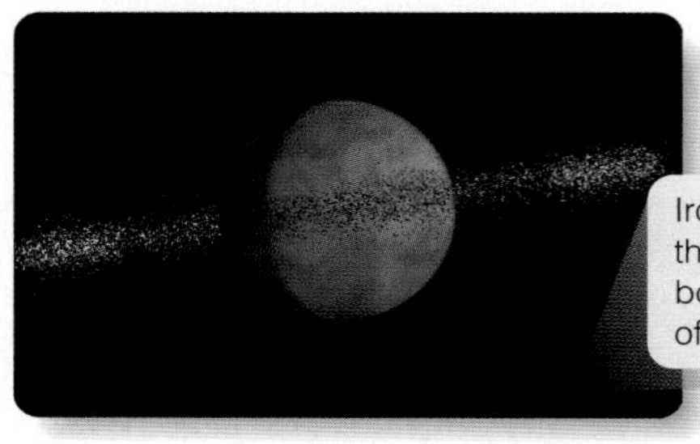

Iron-poor rock from the mantles of the two bodies forms a ring of debris.

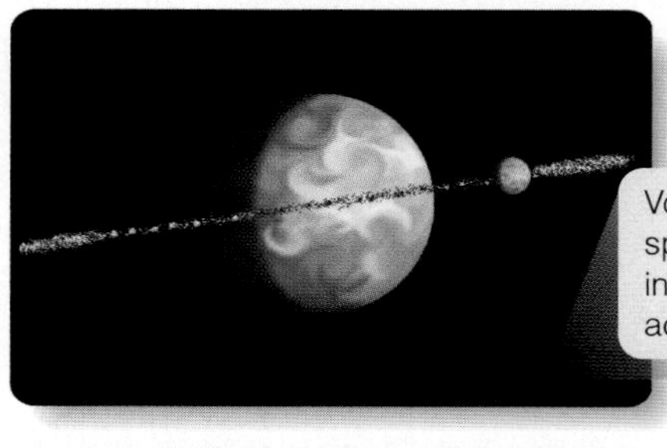

Volatiles are lost to space as the particles in the ring begin to accrete larger bodies.

Eventually the moon forms from the iron-poor and volatile-poor matter in the disk.

rapidly and would thus explain the present angular momentum. If the proto-Earth and impactor had already differentiated, the ejected material would have been mostly iron-poor mantle and crust. Furthermore, the material would have lost its volatile components while it was in space, so the moon would have formed poor in both iron and volatiles. Such an impact would have melted the proto-Earth, and the material falling together to form the moon would have been heated hot enough to melt. This fits the evidence that the moon formed as a sea of magma.

The large-impact hypothesis survives comparison with the known evidence and is now a widely considered hypothesis.

Inquire | Review | Analyze

Why are the maria nearly free of craters?

Timing is everything, someone has said, and in this case your analysis must carefully consider the sequence of events. The moon's crust formed and was heavily cratered before about 4 billion years ago. The impacts of the heavy bombardment cratered the entire lunar surface and created some very large crater basins. Later, after the end of that heavy cratering, lava welled up and filled the lowlands of the largest crater basins with basalt to form the maria. Any craters in the basins were covered over, and there were few impacts to form new craters. Thus, the maria are nearly free of craters, but the ancient highlands are heavily cratered.

Explain how timing is important in explaining the formation of an iron-poor moon in the large-impact hypothesis.

■ ■ ■

Connections: The evolution of Earth's moon has been restricted by its small size. Another world in our solar system has also suffered from its diminutive size. Mercury, like Earth's moon, is a small, rocky world, cratered by impacts and flooded by ancient lava flows that are driven by its internal heat.

17-4 Mercury

MERCURY IS INTERMEDIATE IN SIZE between Earth and its moon (**Celestial Profile 4** and ■ Figure 17-1). Like Earth's moon, Mercury cooled too quickly to develop plate tectonics. Thus, it is a cratered, dead world.

Mariner 10 at Mercury

Because Mercury's orbit keeps it near the sun, its surface temperature in direct sunlight can reach 430°C (800°F), hot enough to melt lead. So close to the sun, Mercury is difficult to observe from Earth, and little was known about it until 1974–1975, when the Mariner spacecraft flew past Mercury three times and revealed a planet whose surface is heavily cratered, much like that of Earth's moon. Analysis shows that large areas have been flooded by lava and then cratered.

The largest impact feature on Mercury is the Caloris Basin, a ringed area 1300 km (800 miles) in diameter (■ Figure 17-10), which looks much like Mare Orientale, the ringed basin on Earth's

Mercury is a bit over a third the diameter of Earth, but its high density must mean it has a large iron core. The amount of heat it retains is unknown. (NASA)

Celestial Profile 4: Mercury

Motion:

Average distance from the sun	0.387 AU (5.79×10^7 km)
Eccentricity of orbit	0.2056
Maximum distance from the sun	0.467 AU (6.97×10^7 km)
Minimum distance from the sun	0.306 AU (4.59×10^7 km)
Inclination of orbit to ecliptic	7°00′16″
Average orbital velocity	47.9 km/s
Orbital period	0.24085 y (87.969 days)
Period of rotation	58.646 days (direct)
Inclination of equator to orbit	0°

Characteristics:

Equatorial diameter	4878 km (0.382 $D_\oplus$)
Mass	3.31×10^{23} kg (0.0558 $M_\oplus$)
Average density	5.44 g/cm^3 (5.4 g/cm^3 uncompressed)
Surface gravity	0.38 Earth gravity
Escape velocity	4.3 km/s (0.38 $V_\oplus$)
Surface temperature	−170°C to 430°C (−280°F to 800°F)
Average albedo	0.1
Oblateness	0

Personality Point:

Mercury lies very close to the sun and completes an orbit in only 88 days. For this reason, the ancients named the planet after Mercury, the fleet-footed messenger of the gods. The name is also applied to the element mercury, which is also known as quicksilver because it is a heavy, quick-flowing silvery liquid at room temperatures.

■ Figure 17-10

Mercury is an airless, cratered world. The Caloris multi-ringed basin was half in shadow and half in sunlight when the Mariner 10 spacecraft flew past the planet. Lobate scarps are distributed all around the planet. (NASA, Inset: Lowell Observatory)

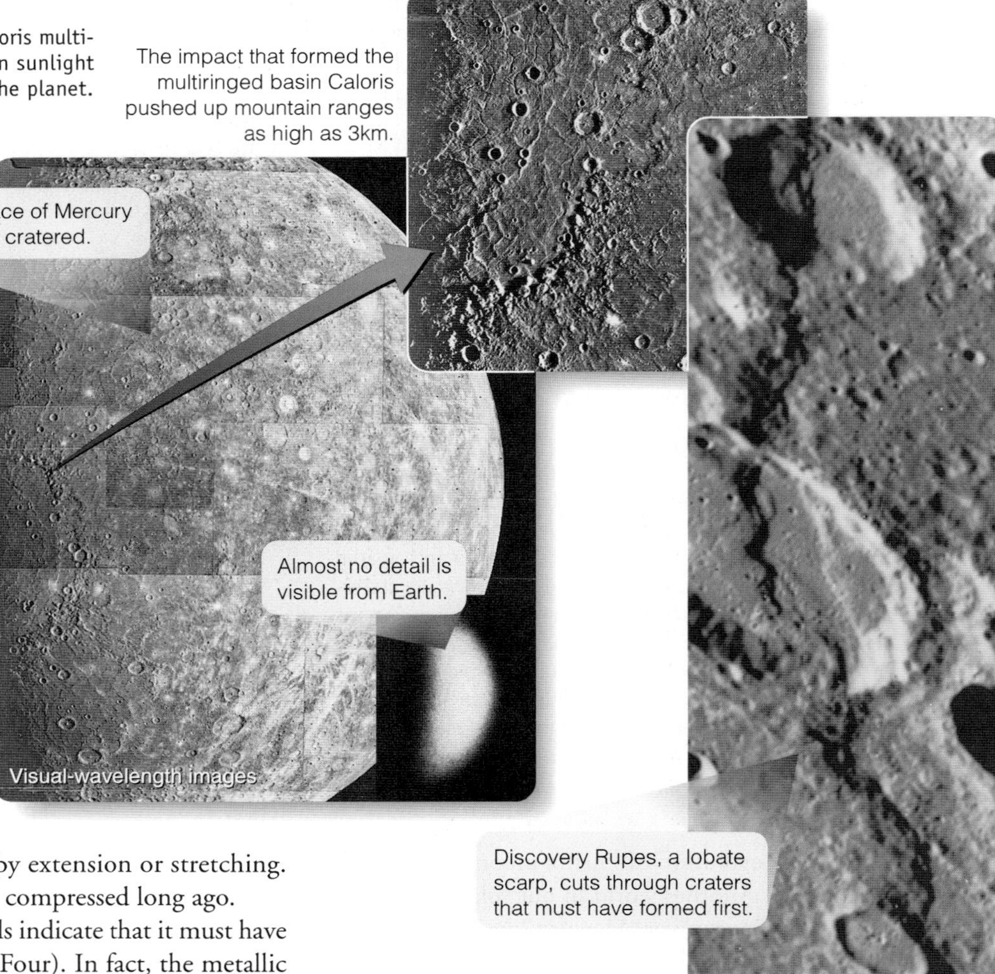
The impact that formed the multiringed basin Caloris pushed up mountain ranges as high as 3km.
The surface of Mercury is heavily cratered.
Almost no detail is visible from Earth.
Visual-wavelength images
Discovery Rupes, a lobate scarp, cuts through craters that must have formed first.

moon. Both features consist of concentric rings of cliffs formed by a large impact.

Though Mercury looks moonlike, it does have one characteristic feature Earth's moon lacks. Mariner 10 photos revealed long curving ridges called **lobate scarps** up to 3 km (1.9 miles) high and 500 km (310 miles) long as shown in Figure 17-10. The scarps even cut through craters, indicating that they formed after most of the heavy bombardment. The lobate scarps are the kind of faults that form by compression. But there are no faults on Mercury that could have formed by extension or stretching. This suggests that the entire crust was compressed long ago.

Mercury is quite dense, and models indicate that it must have a large metallic core (Celestial Profile Four). In fact, the metallic core occupies about 70 percent of the radius of the planet. In a sense, Mercury is a metal planet with a thin rock mantle.

A History of Mercury

The accumulated facts about Mercury don't really help you understand the planet until you have a unifying hypothesis. Like a story, it must make sense and bring the known facts together in a logical argument that explains how Mercury got to be the way it is (**Window on Science 17-2**).

Mercury is small, and that fact has determined much of its history. Not only is it too small to retain an atmosphere, but it has also lost much of its internal heat and thus is not geologically active.

In the first stage, Mercury differentiated to form a metallic core and a rocky mantle. Mariner 10 discovered a magnetic field about 10^{-4} times that of Earth—further evidence of a metallic core. In Chapter 16, you saw that the condensation sequence could explain the abundance of metals in Mercury, but detailed calculations show that Mercury contains even more iron than you would expect. Drawing on the large-impact hypothesis for the origin of Earth's moon, scientists have proposed that Mercury suffered a major impact soon after it differentiated, an impact so large it shattered the rocky mantle and drove much of it away. The remaining iron and rock then re-formed the present Mercury with its unusually large iron core. Such catastrophic events are rare in nature, but they do occur, so astronomers must be prepared to consider such hypotheses. (See Window on Science 16-1.)

In the second and third stages of planet formation, cratering battered the crust, and lava flows welled up to fill the lowlands, just as they did on our moon. As the small world lost internal heat, its large metal core contracted and its crust was compressed, breaking to form the lobate scarps much as the peel of a drying apple wrinkles.

Lacking an atmosphere to erode it, Mercury has changed little since the last lava hardened, and it is now a one-plate planet much like Earth's moon.

Inquire | Review | Analyze

Why don't Earth and Earth's moon have lobate scarps?
At first glance, you might suppose that any world with a metallic interior should have lobate scarps, but other factors are also important. Earth has a fairly large metallic core; but, being a large world, it has

How Hypotheses and Theories Unify the Details

Like any technical subject, science includes a mass of details, facts, figures, measurements, and observations. It is easy to be overwhelmed by the flood of details, but one of the most important characteristics of science comes to your rescue. The goal of science is not to discover more details, but to explain the details with a unifying hypothesis or theory. A good theory is like a basket that makes it easier for you to carry a large assortment of details.

This is true of all the sciences. When a psychologist begins studying the way the human eye and brain respond to moving points of light, the data are a sea of detailed measurements and observations. Once the psychologist forms a hypothesis about the way the eye and brain interact, the details fall into place as parts of a logical story. If you understand the hypothesis, the details all fit together and make sense, and thus you can remember the details without blindly memorizing tables of facts and figures. The goal of science is understanding, not memorization.

Scientists are in the storytelling business. The stories are often called hypotheses or theories, but they are really just stories to explain how nature works. The difference between scientific stories and works of fiction lies in the use of facts. Scientific stories are constructed to fit all known facts and are then tested over and over against new facts obtained by observation and experiment.

When you try to tell the story of each planet in our solar system, you pull together all the hypotheses and theories and try to make them into a logical history of how the planet got to be the way it is. Of course, your stories will be incomplete because scientists don't understand all the factors in planetary geology. Nevertheless, your story of each planet will draw together the known facts and details and attempt to make them a logical whole.

Memorizing a list of facts can give you a false feeling of security, just as when you memorize the names of things without understanding them (Window on Science 2-2). Rather than memorizing facts, you should search for the unifying hypothesis that pulls the details together into a single story. Your goal in studying science is understanding nature, not just remembering facts.

not cooled very much, so it presumably hasn't shrunk much. Of course, the geological activity on Earth's surface would erase such scarps if they did form. Earth's moon is not geologically active, but it does not contain a significant metallic core. Although Earth's moon has lost much of its internal heat, its interior is mostly rock and didn't shrink as much as metal would have.

You can, in a general way, understand lobate scarps, but how do you know the lobate scarps formed after most of the heavy bombardment was over?

■ ■ ■

Connections: Mercury, like Earth's moon, is now an inactive world. To find planets where the heat flowing outward from the interior still drives geological activity, you must look at larger worlds such as Venus and Mars.

17-5 Venus

YOU MIGHT EXPECT VENUS to be much like Earth. It is 95 percent Earth's diameter (**Celestial Profile 5**), has a similar average density, and is only 30 percent closer to the sun. Unfortunately, the surface of Venus is perpetually hidden below thick clouds, and only in the last few decades have planetary scientists discovered that Venus is a deadly hot desert world of volcanoes, lava flows, and impact craters lying at the bottom of a deep ocean of hot gases.

The Atmosphere of Venus

In composition, temperature, and density, the atmosphere of Venus is more Hades than Heaven. The air is unbreathable, very hot, and almost 100 times denser than Earth's air. How do astronomers know this? Because U.S. and Soviet space probes have descended into the atmosphere and in some cases landed on the surface.

In composition, the atmosphere of Venus is roughly 96 percent carbon dioxide. The rest is mostly nitrogen, with some argon, sulfur dioxide, and small amounts of sulfuric acid, hydrochloric acid, and hydrofluoric acid. There is only a tiny amount of water vapor. On the whole, the composition is deadly unpleasant, and spectra show that the impenetrable clouds that hide the surface are made up of droplets of sulfuric acid and microscopic crystals of sulfur (■ Figure 17-11).

This unbreathable atmosphere is 90 times denser than Earth's atmosphere. The air you breathe is 1000 times less dense than water, but on Venus the air is only 10 times less dense than water. If you could survive the unpleasant composition and intense heat, you could strap wings on your arms and fly.

The surface temperature on Venus (Celestial Profile 5) is hot enough to melt lead, and you can understand that because the thick atmosphere creates a severe greenhouse effect. Sunlight filters down through the clouds and warms the surface, but heat cannot escape because the carbon dioxide gas is opaque to infrared. Traces of sulfur dioxide and water vapor also help trap the infrared, but it is the overwhelming abundance of carbon dioxide that makes the greenhouse effect on Venus much more important than on Earth.

The Surface of Venus

Although the thick clouds on Venus are opaque to visible light, they are transparent to radio waves, so astronomers have been able

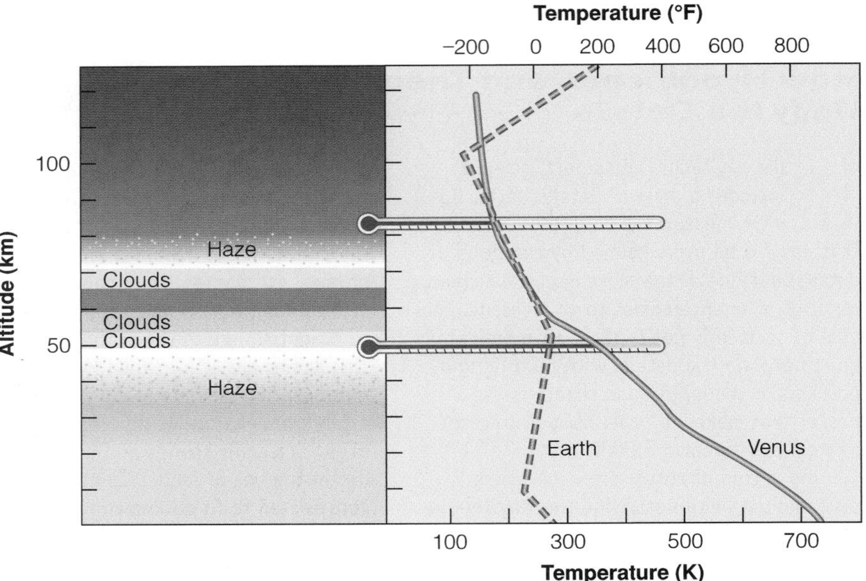

■ Active Figure 17-11

On Venus, three main cloud layers composed mostly of sulfuric acid droplets reflect much of the sunlight away. What reaches the surface is deeply reddened, like a day-long sunset. If you could insert thermometers into the atmosphere, you would find that the lower atmosphere of Venus is much hotter than that of the Earth.

Ace Astronomy™ **Log into Ace Astronomy and select this chapter to see the Active Figure called "Planetary Atmosphere." Experiment with the temperature throughout the atmosphere.**

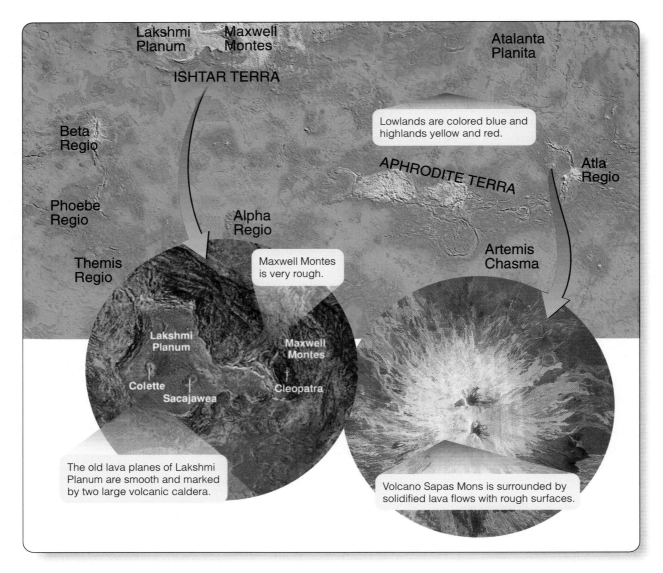

Lakshmi Planum
Maxwell Montes
ISHTAR TERRA
Atalanta Planita
Beta Regio
Lowlands are colored blue and highlands yellow and red.
APHRODITE TERRA
Atla Regio
Phoebe Regio
Alpha Regio
Themis Regio
Maxwell Montes is very rough.
Artemis Chasma
Lakshmi Planum
Maxwell Montes
Colette
Sacajawea
Cleopatra
The old lava planes of Lakshmi Planum are smooth and marked by two large volcanic caldera.
Volcano Sapas Mons is surrounded by solidified lava flows with rough surfaces.

to map Venus using radar. As early as 1965, Earth-based radio telescopes made low-resolution maps, but later both U.S. and Soviet spacecraft orbited Venus and mapped its surface by radar. Maps made in the early 1990s by the Magellan spacecraft reveal objects as small as 100 m in diameter.

Radar maps of Venus are reproduced using arbitrary colors. In some maps lowlands are colored blue, but there are no oceans on Venus. In some maps, the scientists have chosen to give Venus an overall orange glow because sunlight filtering down through the clouds bathes the landscape in a perpetual sunset glow. Some radar maps are shown in gray, the natural color of the rocks. Look carefully at radar maps of Venus and recall that its surface is a deadly dry desert.

By international agreement, names on Venus are all female, with three exceptions—Maxwell, a high mountain, and Alpha Regio and Beta Regio, two high volcanic peaks—which were named before the international naming convention was adopted.

Radar maps show that Venus is similar to Earth in one way, but strangely different in other ways. Nearly 75 percent of Earth's surface is covered by low-lying, basaltic seafloors, and 85 percent of Venus is covered by basaltic lowlands. But there is no liquid water on Venus, so its lowlands are not seafloors, and the remaining highlands are not the well-defined continents you see on Earth. Whereas Earth is dominated by plate tectonics, something different is happening on Venus.

The highland area Ishtar Terra, named for the Babylonian goddess of love, is about the size of Australia (■ Figure 17-12). At its eastern edge, the mountain called Maxwell Montes thrusts up 12 km (Everest is 8.8 km high), with the impact crater Cleopatra on its lower slopes. Bounded by mountain ranges in the north and west, the center of Ishtar Terra is occupied by Lakshmi Planum, a great plateau about 4 km above the surrounding plains. The collapsed calderas Colette and Sacajawea suggest that Lakshmi Planum is a great lava plain. The mountains bounding Ishtar Terra, including Maxwell, resemble folded mountain ranges, which suggests that limited horizontal motion in the crust as well as volcanism has helped form the highlands.

To understand how important volcanism is on Venus, you need to return to Earth and study volcanoes in general. As usual, the best way to learn about other worlds is by comparing them with each other and with Earth. Examine **Volcanoes** on pages 400 and 401 and notice three important ideas. First, there are two main types of volcanoes on Earth. Second, notice that you can recognize the volcanoes on Venus by their shape. The third

■ Figure 17-12 ◀

Notice how three radar maps show different things. The main radar map here shows elevation over most of the surface of Venus, omitting the polar areas. The detailed map of Maxwell Montes and Lakshmi Planus are colored according to roughness, with orange the roughest terrain. The map of volcano Sapas Mons also shows roughness but is given an orange color to mimic the color of sunlight at the surface. (Maxwell and Lakshmi Planum map: USGS; Other maps: NASA)

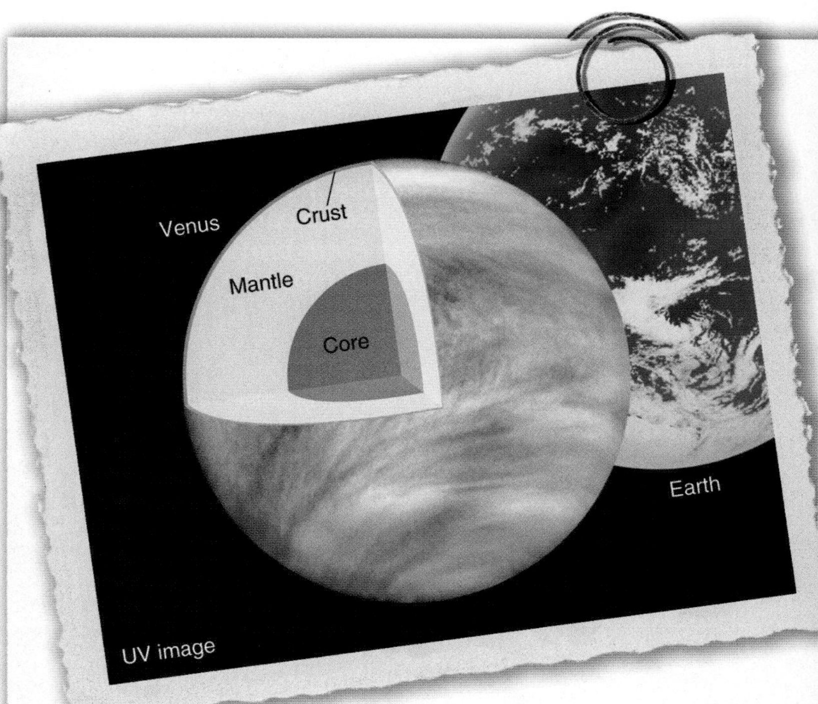

Venus is only 5 percent smaller than Earth, but its atmosphere is perpetually cloudy and its surface is hot enough to melt lead. It is believed to have a hot core about the size of Earth's.

Celestial Profile 5: Venus

Motion:

Average distance from the sun	0.7233 AU (1.082×10^8 km)
Eccentricity of orbit	0.0068
Maximum distance from the sun	0.7282 AU (1.089×10^8 km)
Minimum distance from the sun	0.7184 AU (1.075×10^8 km)
Inclination of orbit to ecliptic	3°23′40″
Average orbital velocity	35.03 km/s
Orbital period	0.61515 y (224.68 days)
Period of rotation	243.01 days (retrograde)
Inclination of equator to orbit	177°

Characteristics:

Equatorial diameter	12,104 km (0.95 $D_{\oplus}$)
Mass	4.870×10^{24} kg (0.815 $M_{\oplus}$)
Average density	5.24 g/cm^3 (4.2 g/cm^3 uncompressed)
Surface gravity	0.903 Earth gravity
Escape velocity	10.3 km/s (0.92 $V_{\oplus}$)
Surface temperature	745 K (472°C, or 882°F)
Albedo (cloud tops)	0.76
Oblateness	0

Personality Point:

Venus is named for the Roman goddess of love, perhaps because the planet often shines so beautifully in the evening or dawn sky. In contrast, the ancient Maya identified Venus as their war god Kukulkan and sacrificed human victims to the planet when it rose in the dawn sky.

Volcanoes

Molten rock (magma) is less dense than its surroundings and tends to rise. Where it bursts through Earth's crust, you see volcanism. The two main types of volcanoes on Earth provide good examples for comparison with those on Venus and Mars.

On Earth, **composite volcanoes** form above subduction zones where the descending crust melts and the magma rises to the surface. This forms chains of volcanoes along the subduction zone, such as the Andes along the west coast of South America.

Magma rising above subduction zones is not very fluid, and it produces explosive volcanoes with sides as steep as 30°.

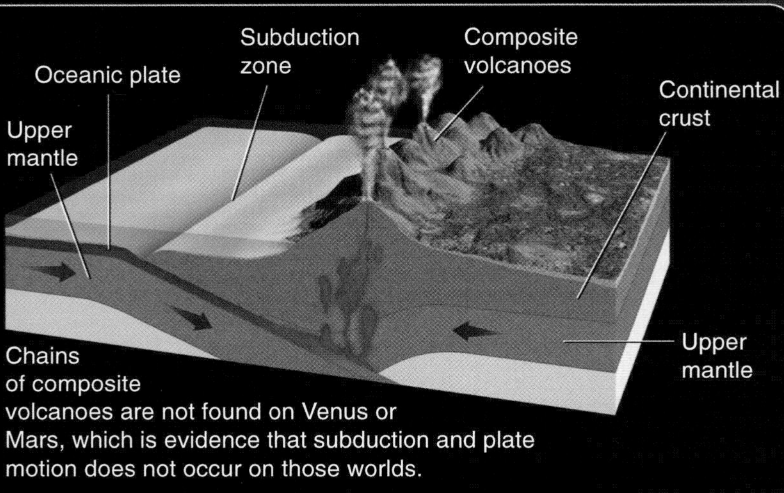

Chains of composite volcanoes are not found on Venus or Mars, which is evidence that subduction and plate motion does not occur on those worlds.

Based on *Physical Geology*, 4th edition, James S. Monroe and Reed Wicander, Wadsworth Publishing Company. Used with permission.

Mount St. Helens exploded northward on May 18, 1980, killing 63 people and destroying 600 km^2 (230 mi^2) of forest with a blast of winds and suspended rock fragments that moved as fast as 480 km/hr (300 mph) and had temperatures as hot as 350°C (660°F). Note the steep slope of this composite volcano.

USGS

Shield volcano

Lava flow

Magma chamber

Oceanic crust

A **shield volcano** is formed by highly fluid lava (basalt) that flows easily and creates low-profile volcanic peaks with slopes of 3° to 10°. The volcanoes of Hawaii are shield volcanoes that occur over a hot spot in the middle of the Pacific plate.

Magma collects in a chamber in the crust and finds its way to the surface through cracks.

Magma forces its way upward through cracks in the upper mantle and causes small, deep earthquakes.

A hot spot is formed by a rising convection current of magma moving upward through the hot, deformable (plastic) rock of the mantle.

The Cascade Range composite volcanoes are produced by an oceanic plate being subducted below North America and partially melting.

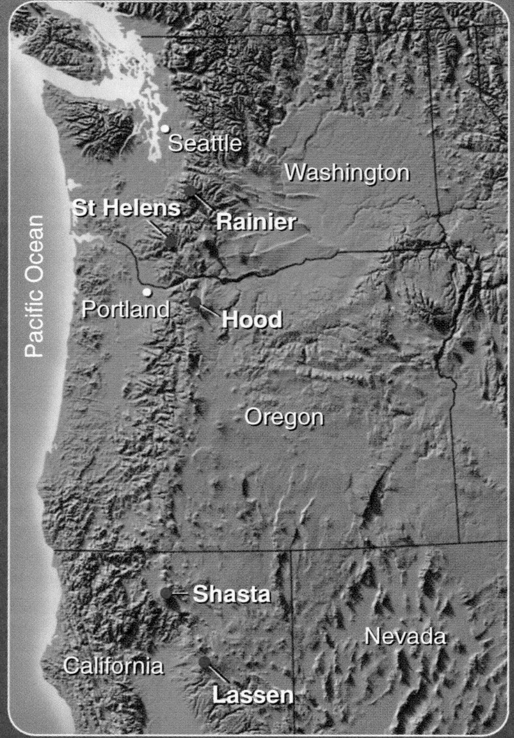

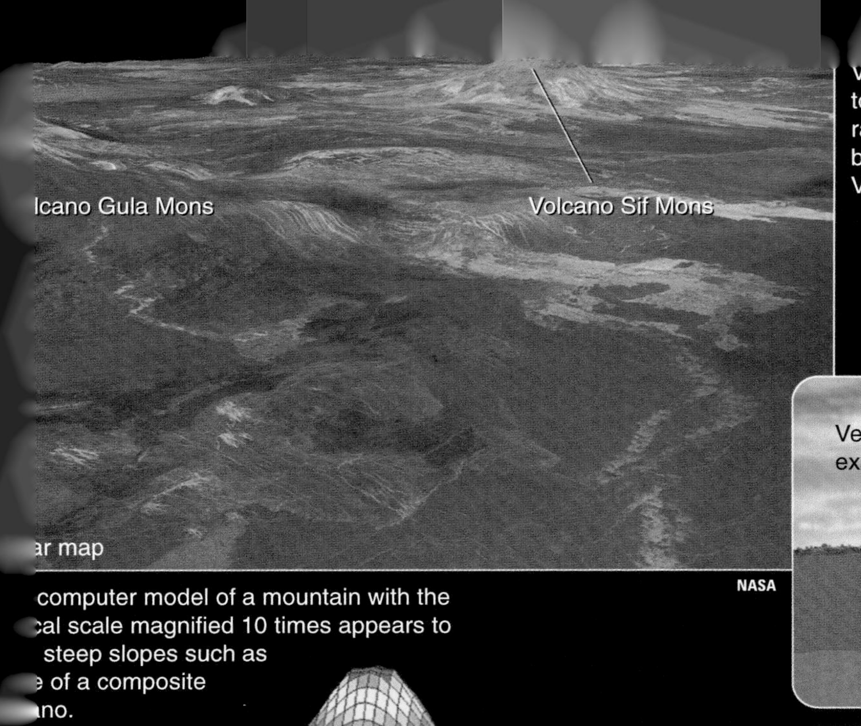

NASA

computer model of a mountain with the
cal scale magnified 10 times appears to
steep slopes such as
e of a composite
ano.

Volcanoes on Venus are shield volcanoes. They ... to be steep sided in some images created from Magellan radar maps, but that is because the vertical scale has been exaggerated to enhance detail. The volcanoes of Venus are actually shallow-sloped shield volcanoes.

Volcanism over a hot spot results in repeated eruptions that build up a shield volcano of many layers. Such volcanoes can grow very large.

Vertical scale exaggerated

Hot spot

Mike Seeds

e profile of the computer model shows
nountain has very shallow slopes
al of shield volcanoes.

Old volcanic island eroded below sea level

Plate motion

Hot spot

If the crustal plate is moving, magma generated by the hot spot can repeatedly penetrate the crust to build a chain of volcanoes. Only the volcanoes over the hot spot are active. Older volcanoes slowly erode away. Such volcanoes cannot grow large because the moving plate carries them away from the hot spot.

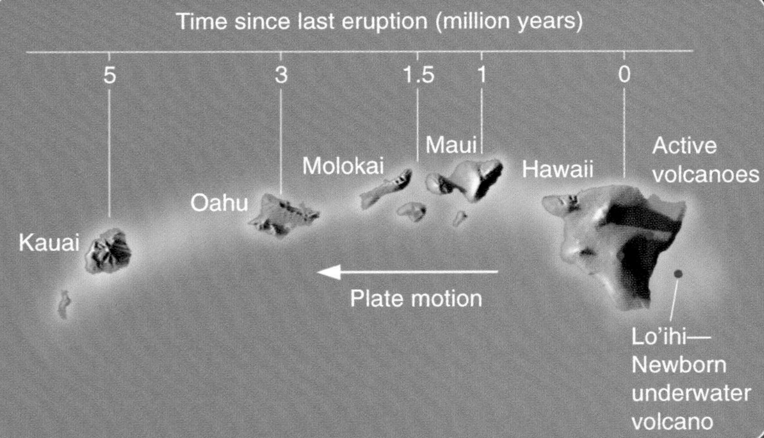

Log into AceAstronomy and select this chapter to see the Active Figure "Hotspot Volcanoes" and compare volcanism on Earth with that on Venus.

The volcanoes that make up the Hawaiian Islands have been produced by a hot spot poking upward through the middle of the moving Pacific plate.

NASA

The plate moves about 9 cm/yr and carries older volcanic islands northwest, away from the hot spot. The volcanoes cannot grow extremely large because they are carried away from the hot spot. New islands form to the southeast over the hot spot.

Olympus Mons is the largest volcano on Mars. It is a shield volcano 25 km (16 mi) high and 700 km (440 mi) in diameter at its base. Its vast size is evidence that the crustal plate must have remained stationary over the hot spot. This is evidence that Mars has not had plate tectonics.

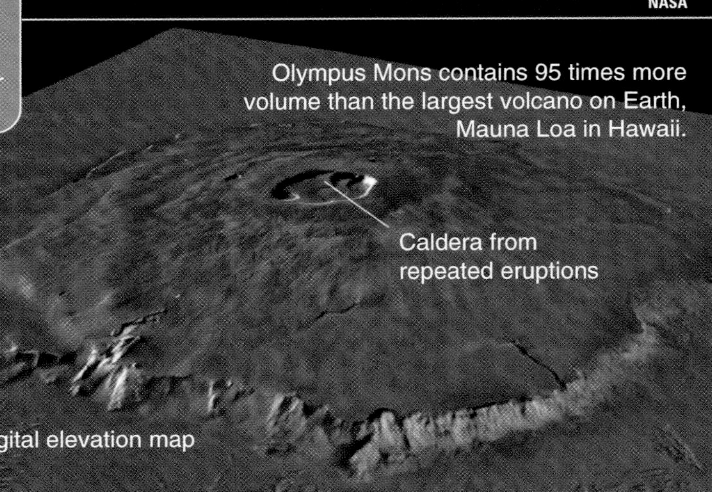

Olympus Mons contains 95 times more volume than the largest volcano on Earth, Mauna Loa in Hawaii.

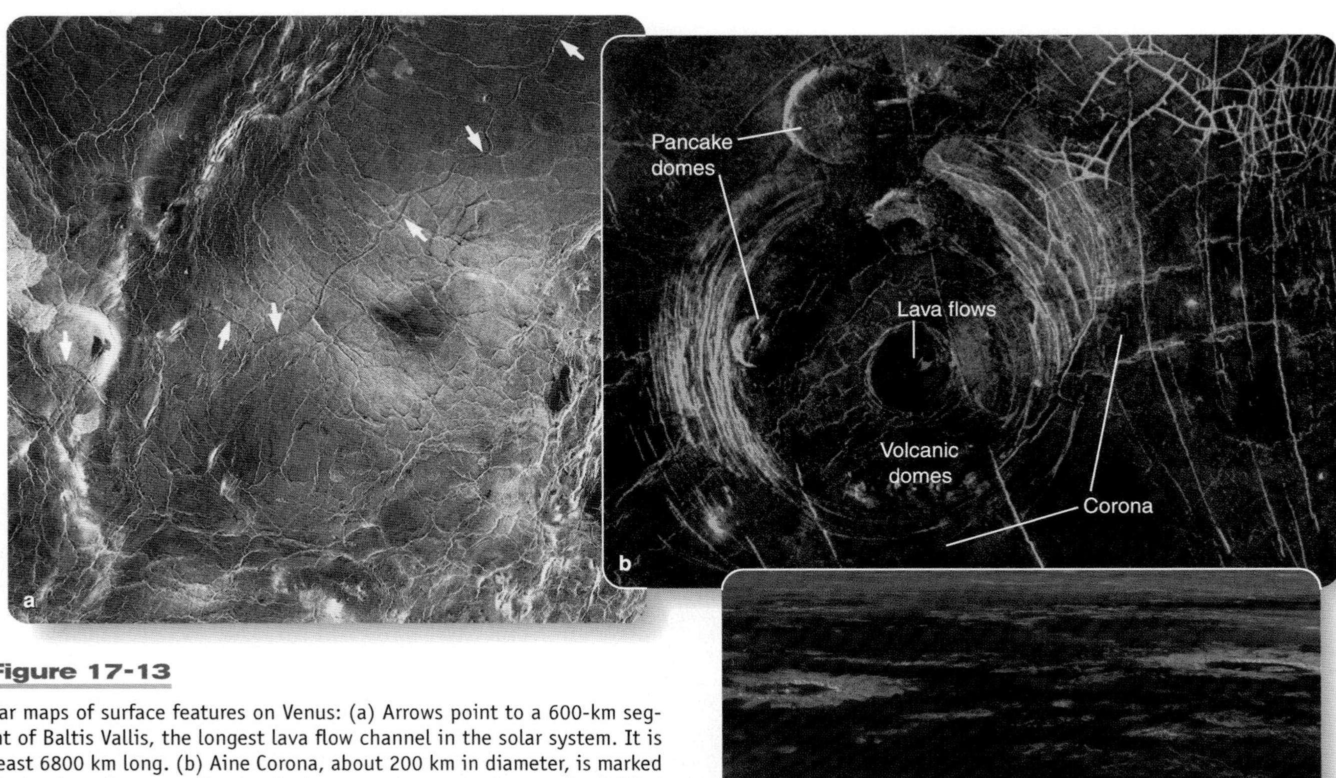

■ **Figure 17-13**

Radar maps of surface features on Venus: (a) Arrows point to a 600-km segment of Baltis Vallis, the longest lava flow channel in the solar system. It is at least 6800 km long. (b) Aine Corona, about 200 km in diameter, is marked by faults, lava flows, small volcanic domes, and pancake domes of solidified lava. (c) Impact crater Howe is 37 km in diameter. Craters in the background are 47 km and 63 km in diameter. (NASA)

idea to notice is the large size of volcanoes on Venus. The shield volcanoes found on Venus are evidence that Venus has not been dominated by plate tectonics as on Earth.

Many features on Venus testify to its volcanic history. Long, narrow lava channels meander for thousands of kilometers (■ Figure 17-13a). Radar maps reveal many smaller volcanoes, faults, and sunken regions produced when magma below the surface drained away. Other volcanic features include the **coronae,** circular bulges up to 2100 km in diameter bordered by fractures, volcanoes, and lava flows (Figure 17-13b). These appear to be produced by rising convection currents of molten magma that push up under the crust. When the magma withdraws, the crust sinks back but the circular fractures mark the edge of the corona.

Radar images show that Venus is marked by numerous craters (Figure 17-13c). The atmosphere protects the surface from smaller meteorites that would produce craters smaller than 3 km in diameter. Larger meteorites penetrate the atmosphere and have formed about 10 percent as many craters on Venus as on the maria of Earth's moon. The number of craters shows that the crust is not as ancient as the lunar maria, but not as young as Earth's active surface. The average age of the surface of Venus is roughly half a billion years. Clearly, plate tectonics cannot be renewing the surface as rapidly as it does on Earth.

No astronaut has ever stood on Venus, but Soviet spacecraft landed on the surface and survived the heat and pressure for a few hours. A few of those spacecraft have analyzed the rock and snapped a few photographs (■ Figure 17-14). The surface rocks on Venus appear to be dark gray basalts much like those in Earth's ocean floors. This confirms the evidence that volcanism is important.

The History of Venus

To tell the story of Venus you must draw together all the evidence and find hypotheses to explain two things: the thick carbon dioxide atmosphere and the peculiar geology.

Calculations show that Venus and Earth have outgassed about the same amount of carbon dioxide, but Earth's oceans have dissolved it and converted it to sediments such as limestone. If all of Earth's carbon were dug up and converted back to carbon dioxide, our atmosphere would be about as dense as the air on Venus. This suggests that the main difference between Earth and Venus is the lack of water on Venus. Venus may have had small oceans when it was young; but, being closer to the sun, it was warmer, and the carbon dioxide in the atmosphere created a greenhouse effect that made the planet even warmer. That pro-

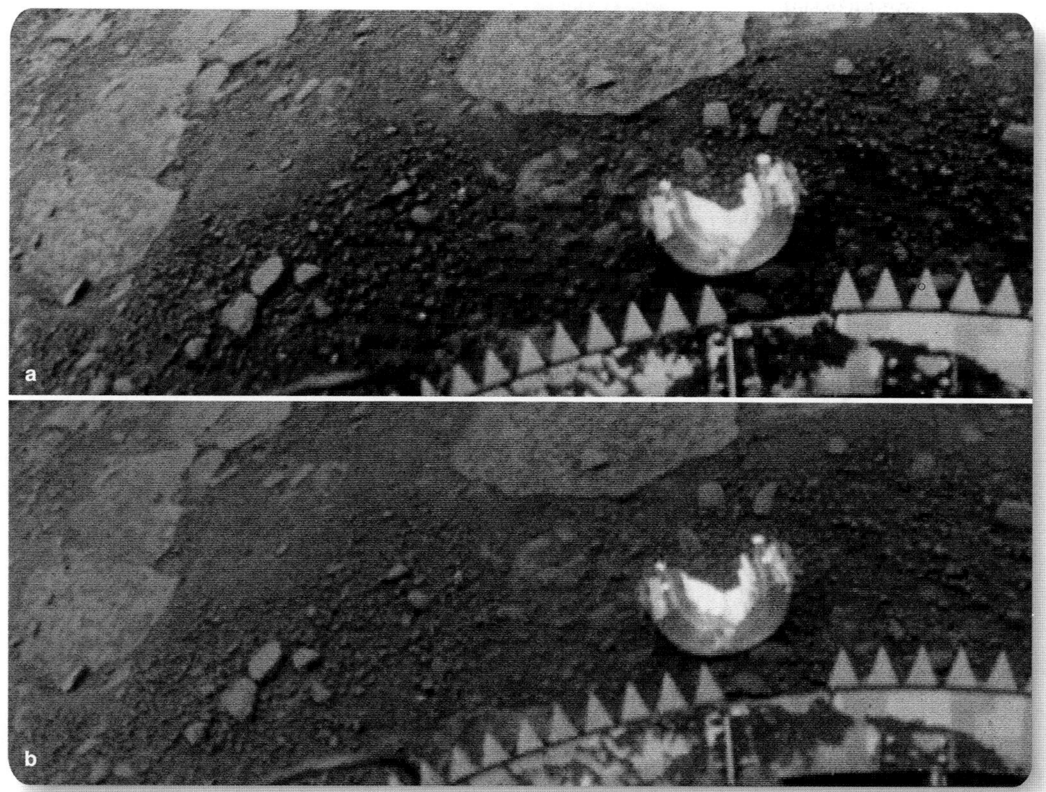

■ **Figure 17-14**

The color of the surface of Venus appears orange in the original Venera 13 visual-wavelength photograph (a), but analysis shows that the color is produced by the thick atmosphere reddening sunlight into a day-long sunset glow. The corrected image (b) reveals slabs of gray rock and dark soil typical of iron oxides at high temperatures. An instrument cover and the base of the spacecraft are visible at lower right. (C. M. Pieters through the Brown/Vernadsky Institute to Institute Agreement and the U.S.S.R. Academy of Science)

cess could have dried up any oceans that did exist and reduced the ability of the planet to purge its atmosphere of carbon dioxide. As more carbon dioxide was outgassed, the greenhouse effect grew even more severe. Thus, Venus seems to have been trapped in a runaway greenhouse effect.

The intense heat at the surface may have affected the geology of Venus by making the crust drier and more flexible so that it was unable to break into moving plates as on Earth. There is no sign of plate tectonics on Venus, but rather evidence that convection currents below the crust are deforming the crust to create coronae and push up mountains such as Maxwell. Detailed measurements of the strength of gravity over the mountains on Venus show that some must be held up not by deep roots like mountains on Earth but by rising currents of magma. Other mountains, like those around Ishtar Terra, appear to be folded mountains caused by limited horizontal motions in the crust, driven perhaps by convection in the mantle.

The small number of craters on the surface of Venus hints that the entire crust has been replaced within the last half-billion years or so. This may have occurred in a planetwide overturning as the old crust broke up and sank and lava flows created a new crust. Such drama may not be necessary, however. Models of the climate on Venus show that an outburst of volcanism could increase the greenhouse effect and drive the surface temperature up by as much as 100°C. This could soften the crust, increase the volcanism, and push the planet into a resurfacing episode. This could happen periodically on Venus, or the planet may have had a geology more like Earth's until a single resurfacing not too long ago. In either case, unearthly Venus may eventually tell us more about how our own world works.

Inquire | Review | Analyze

What evidence indicates that Venus does not have plate tectonics? On Earth, plate tectonics is identifiable by the worldwide network of faults, subduction zones, volcanism, and folded mountain chains that outline the plates. Although some of these features are visible on Venus, they do not occur in a planetwide network that outlines plates. Volcanism is widespread, but folded mountain ranges occur in only a few places, such as near Lakshmi Planum and Maxwell Montes. Rather than being dominated by the motion of rigid crustal plates, Venus may have a crust made more flexible by the heat, a crust dominated by rising plumes of molten rock that strain the crust to produce coronae and that break through to form volcanoes and lava flows.

At first glance, comparative planetology suggested that Earth and Venus should be sister worlds, but it seems they can be no more than distant cousins. You can blame the thick atmosphere of Venus for altering its geology, but that raises the question: Why isn't Earth's atmosphere similar?

■ ■ ■

Connections: Venus is certainly an unearthly world, but the familiar principles of comparative planetology help you understand it. As you turn your attention to Mars, you will see the same principles in action, but you will discover that Mars is a world that lacks what Venus has in abundance—internal heat and air.

17-6 Mars

MERCURY AND EARTH'S MOON ARE SMALL. Venus and Earth are, for terrestrial planets, large. But Mars, with a diameter 53 percent that of Earth's, is intermediate (**Celestial Profile 6**). In some ways, however, Mars is much like Earth. It rotates on its axis in 24 hours 37 minutes, and its year is 1.88 Earth years long. Because its axis of rotation is tipped, it has seasons as summer and winter come and go. Nevertheless, Mars is a forbidding world, with an ancient surface marked by craters and volcanoes and a thin, cold atmosphere.

The Atmosphere of Mars

The Martian air contains 95 percent carbon dioxide, 3 percent nitrogen, and 2 percent argon. In composition, that is much like the air on Venus, but the Martian atmosphere is very thin, less than 1 percent as dense as Earth's atmosphere. It is also never warmer than an autumn afternoon and can become as cold as −140°C (−220°F).

There is very little water in the Martian atmosphere, and the polar caps are composed of frozen carbon dioxide (dry ice) with an unknown amount of frozen water trapped below. As summer comes to a hemisphere, the carbon dioxide in its polar cap turns from solid to vapor and adds carbon dioxide to the atmosphere, while winter in the opposite hemisphere is freezing carbon dioxide out of the atmosphere and adding it to that polar cap.

Liquid water cannot survive on the surface of Mars because the air pressure is too low. Any liquid water would immediately boil away; and if you stepped out of a spaceship on Mars without your space suit, your body heat would make your blood boil. Whatever water is present on Mars must be frozen below the polar caps or frozen as permafrost in the soil.

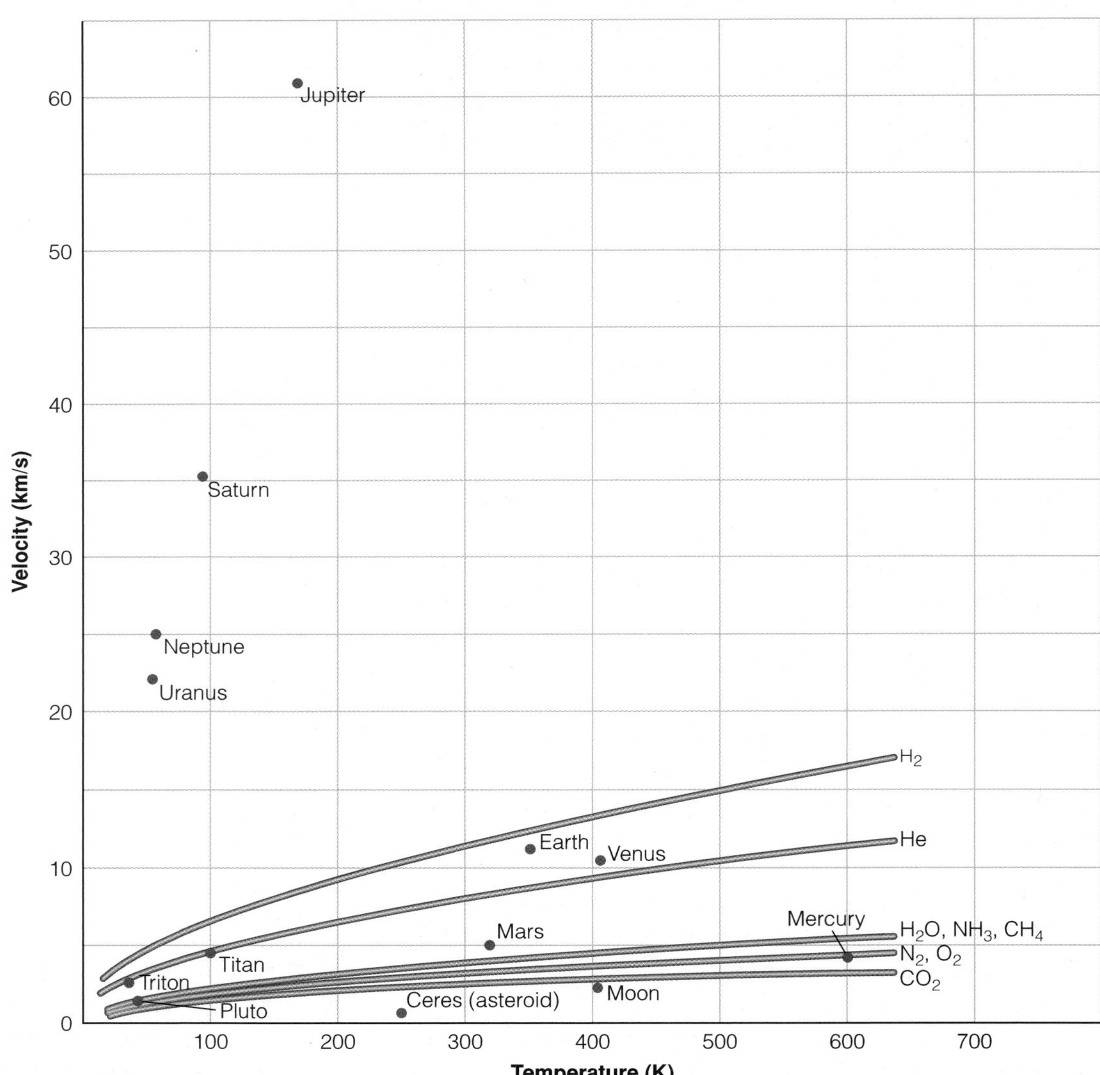

Figure 17-15

The loss of planetary gases. Dots represent the escape velocity and temperature of various solar system bodies. The lines represent the typical highest velocities of molecules of various masses. The Jovian planets have high escape velocities and can hold on to even the lowest-mass molecules. Mars can hold only the more massive molecules, and the moon has such a low escape velocity that even the most massive molecules can escape.

Although the present atmosphere of Mars is very thin, you will see evidence that the climate once permitted liquid water to flow over the surface, so Mars must have once had a thicker atmosphere. As a terrestrial planet, it should have outgassed significant amounts of carbon dioxide, nitrogen, and water vapor; but because it was small, it could not hold onto its gases. The escape velocity on Mars is only 5 km/s, less than half of Earth's, so it was easier for rapidly moving gas molecules to escape into space. Another factor is the temperature of a planet. If Mars had been colder, the gas molecules in its atmosphere would have been traveling more slowly and would not have escaped as easily. You can see this in ■ Figure 17-15, which plots the escape velocity of each planet versus the temperature of the region from which molecules would escape. For Earth, the temperature is that of the upper atmosphere. For Mercury, the temperature is that of the hot rocky surface. Clearly, small worlds cannot keep atmospheric gases easily.

A further problem is that Mars has no ozone layer to protect its atmosphere from ultraviolet radiation, which can break molecules up into smaller fragments, which escape more easily. Water, for example, can be broken up into hydrogen and oxygen. Thus, Mars was large enough to have a substantial atmosphere when it was young and may have had water falling as rain and collecting in rivers and lakes. But it gradually lost much of its atmosphere and is now a cold, dry world.

Exploring the Surface of Mars

If you ever visit another world, Mars may be your best choice. You will need a heated, slightly pressurized spacesuit with air, water, and food, but Mars is not as inhospitable as the moon. It is also more interesting, with weather, complex geology, and signs that water once flowed over its surface. You might even hope to find traces of ancient life hidden in the rocks.

Spacecraft have been visiting Mars for almost 40 years, but the pace has picked up recently. A small armada of spacecraft has gone into orbit around Mars to photograph and analyze its surface, and five spacecraft have landed. Two Viking landers touched down in 1976, and three rovers have landed in recent years.

Rovers have an advantage because they are wheeled robots that can be directed from Earth to travel from feature to feature and make detailed measurements. Pathfinder and its rover Sojourner landed in 1997. Rovers Spirit and Opportunity landed in January 2004 and carried sophisticated instruments to explore the rocky surface.

As you examine this data, you will see evidence that the surface of Mars is an old, dusty, volcanic surface, cratered by impacts and rich with hints that water was once present. (■ Figure 17-16).

Data recorded by orbiting satellites show that the southern hemisphere of Mars is a highland region heavily marked by craters, which reveals that the surface is 2 to 3 billion years old. The northern hemisphere is mostly a much younger lowland plane with

Mars is only half the diameter of Earth and probably retains some internal heat, but the size and composition of its core are not well known.

Celestial Profile 6: Mars

Motion:

Average distance from the sun	1.5237 AU (2.279 × 10^8 km)
Eccentricity of orbit	0.0934
Maximum distance from the sun	1.6660 AU (2.492 × 10^8 km)
Minimum distance from the sun	1.3814 AU (2.066 × 10^8 km)
Inclination of orbit to ecliptic	1°51′09″
Average orbital velocity	24.13 km/s
Orbital period	1.8808 y (686.95 days)
Period of rotation	$24^h37^m22.6^s$
Inclination of equator to orbit	23°59′

Characteristics:

Equatorial diameter	6796 km (0.53 $D_\oplus$)
Mass	0.6424 × 10^{24} kg (0.1075 $M_\oplus$)
Average density	3.94 g/cm^3 (3.3 g/cm^3 uncompressed)
Surface gravity	0.379 Earth gravity
Escape velocity	5.0 km/s (0.45 $V_\oplus$)
Surface temperature	−140° to 20°C (−220° to 68°F)
Average albedo	0.16
Oblateness	0.009

Personality Point:

Mars is named for the god of war. Minerva was the goddess of defensive war, but Bullfinch's *Mythology* refers to Mars's "savage love of violence and bloodshed." You can see how the planet glows blood red in the evening sky because of iron oxides in its soil.

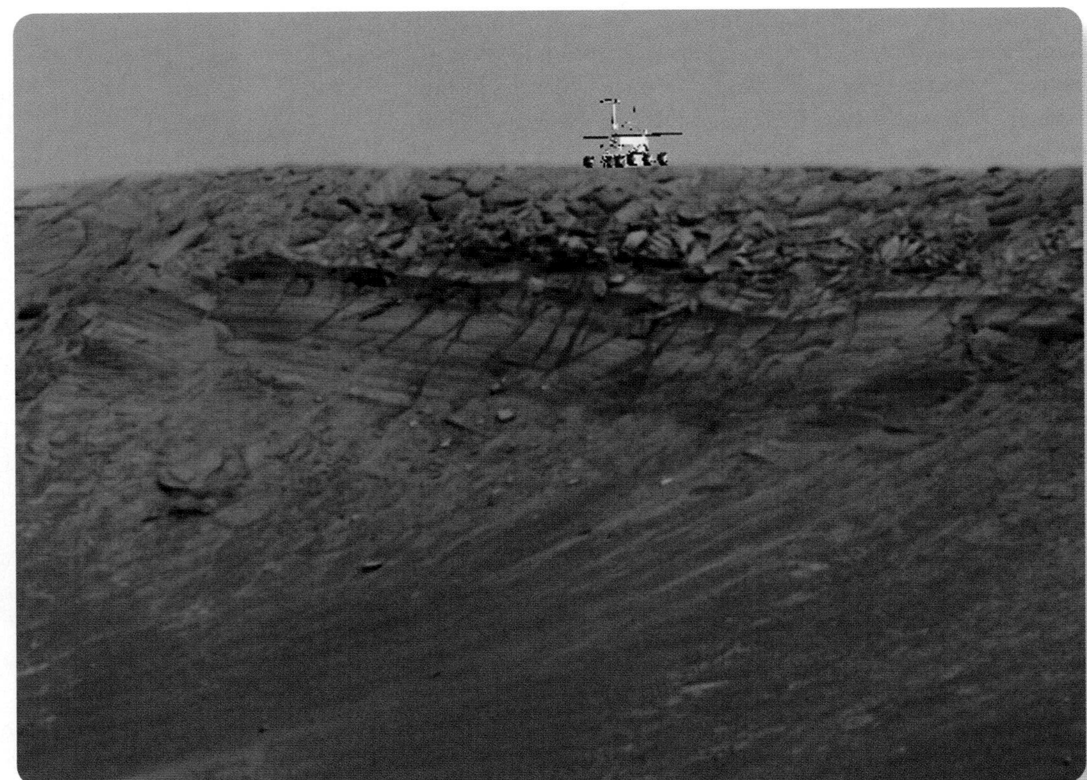

■ **Figure 17-16**

If you had been standing beside the small crater now known as Endurance, you might have seen the Rover Opportunity roll up to the edge and pause to await further instructions from Earth. This true-color image of the layered rock in the crater wall was taken by Opportunity as it circled the crater looking for a safe way down to the bottom. NASA engineers added a digital image of the rover to recreate the scene. (NASA/JPL/Cornell)

few craters (■ Figure 17-17). This lowland plane may have been smoothed by lava flows, but growing evidence suggests that it was once filled with an ocean, a controversial theory we will discuss later.

Volcanism on Mars is dramatically evident in the Tharsis region, a highland region of volcanoes and lava flows bulging 10 km (6 mi) above the surrounding surface. A similar uplifted volcanic plain, the Elysium region, is more heavily cratered and eroded

■ **Figure 17-17**

These globes of Mars are color coded to show elevation. The northern lowlands lie about 4 km below the southern highlands. Volcanoes are very high (white), and the giant impact basins, Hellas and Argyre, are low. Note the depth of the canyon Valles Marineris. (NASA)

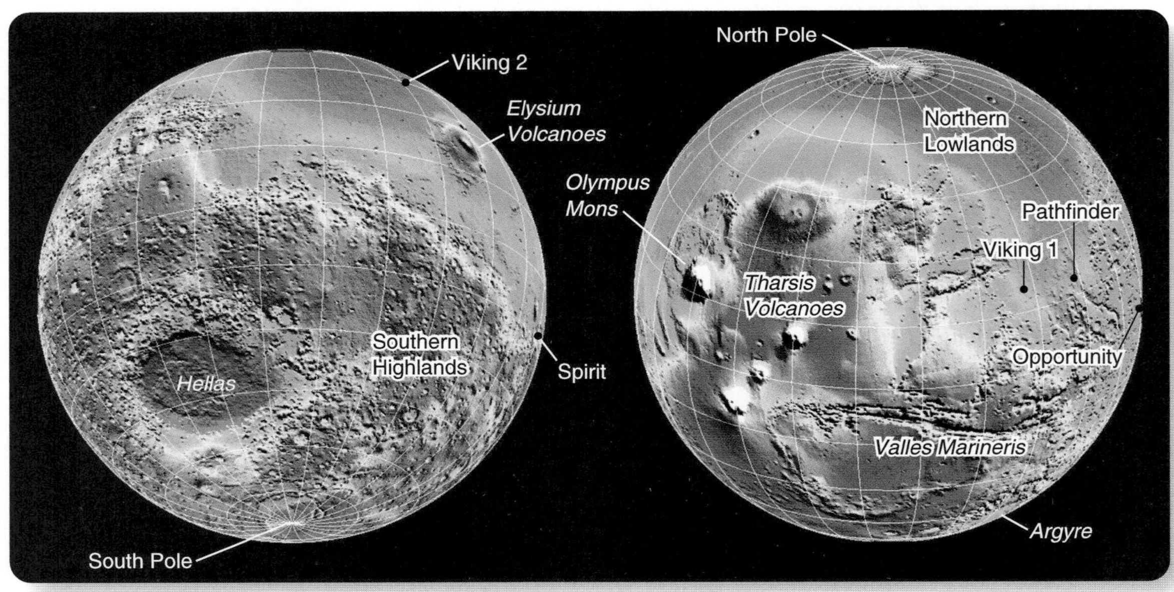

and appears to be older than the Tharsis bulge. The number of impact craters suggests that the Elysium region was active about 1.5 billion years ago and that the Tharsis region has been active within the last few hundred million years. There is no reason to think the volcanoes there are completely dead.

All of the volcanoes on Mars are shield volcanoes, which are produced by hot spots penetrating upward through the crust. Shield volcanoes are not related to plate tectonics and are not evidence of plate motion on Mars.

In fact, the largest volcano on Mars, Olympus Mons, provides evidence that plate tectonics has not occurred on Mars. Olympus Mons is 600 km (370 mi) in diameter at its base and rises 21 km (13 mi) high. The largest volcano on Earth is Mauna Loa in Hawaii, rising only 10 km (6 mi) above its base on the seafloor. Mauna Loa is so heavy that it has sunk into Earth's crust, producing an undersea moat around its base. In contrast, Olympus Mons, two times higher, has no moat and is supported entirely by the Martian crust (■ Figure 17-18). Evidently, the crust of Mars is much stronger than Earth's.

When the crust of a planet is strained, it may break, producing faults and rift valleys. Near the Tharsis region is a great valley, Valles Marineris (Figure 17-18), named after the Mariner spacecraft that first photographed it. The valley is a block of crust that has dropped downward along parallel faults. Erosion and landslides have further modified the valley into a great canyon nearly 4000 km (2500 mi) long, stretching almost 19 percent of the way around the planet. It is as deep as 6 km (4 mi) and as wide as 200 km (120 mi).

Where Valles Marineris begins, near the Tharsis region, it is marked by numerous faults, visible in Figure 17-18. This suggests that the valley is related to the uplifted volcanic plain. The number of craters in the valley indicates that it is 1 to 2 billion years old, placing its origin sometime before the end of volcanism in the Tharsis region.

Before you can tell the story of Mars, you must consider a difficult issue—water. How much water has Mars had, how much has been lost, and how much remains?

Searching for Water on Mars

You would hardly expect water on Mars. It is a deadly cold, deadly dry desert world (■ Figure 17-19). However, observations from orbiting spacecraft have revealed land forms that suggest that there

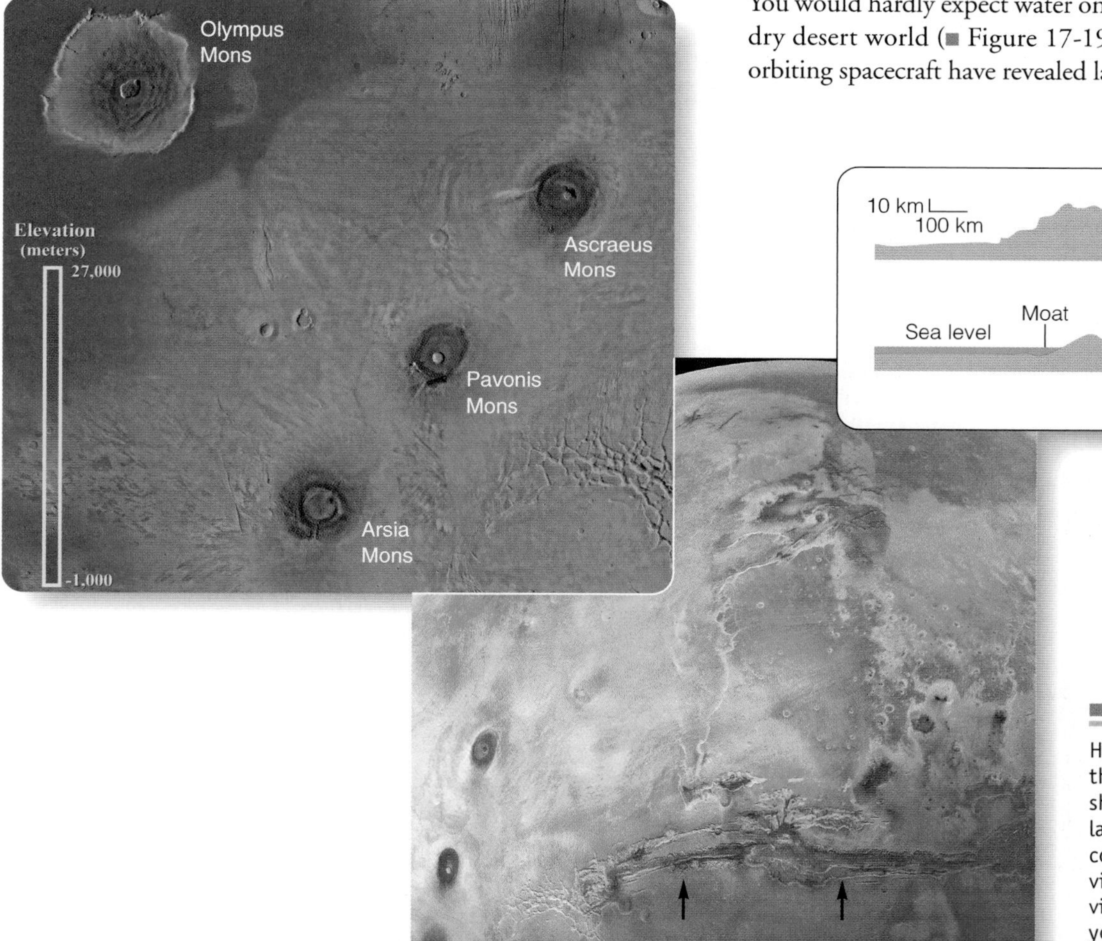

■ **Figure 17-18**

High volcanoes and deep canyons mark the surface of Mars. Olympus Mons, a shield volcano, is much larger than the largest volcano on Earth. In this false-color image, three other volcanoes are visible. Those three volcanoes are also visible in the photo along with the canyon Valles Marineris, which stretches as far as the distance from New York to Los Angeles. (Four volcanoes: © Calvin J. Hamilton, Columbia, Maryland; Photo: NASA/USGS)

Figure 17-19

Mars is a red desert planet, as shown in this true-color photo made by the Spirit rover. Rock layers are fractured by meteorite impacts, and dust is blown around the planet by winds. Dust suspended in the atmosphere colors the sky red. Yet evidence suggests that Mars had water on its surface a few billion years ago. (NASA)

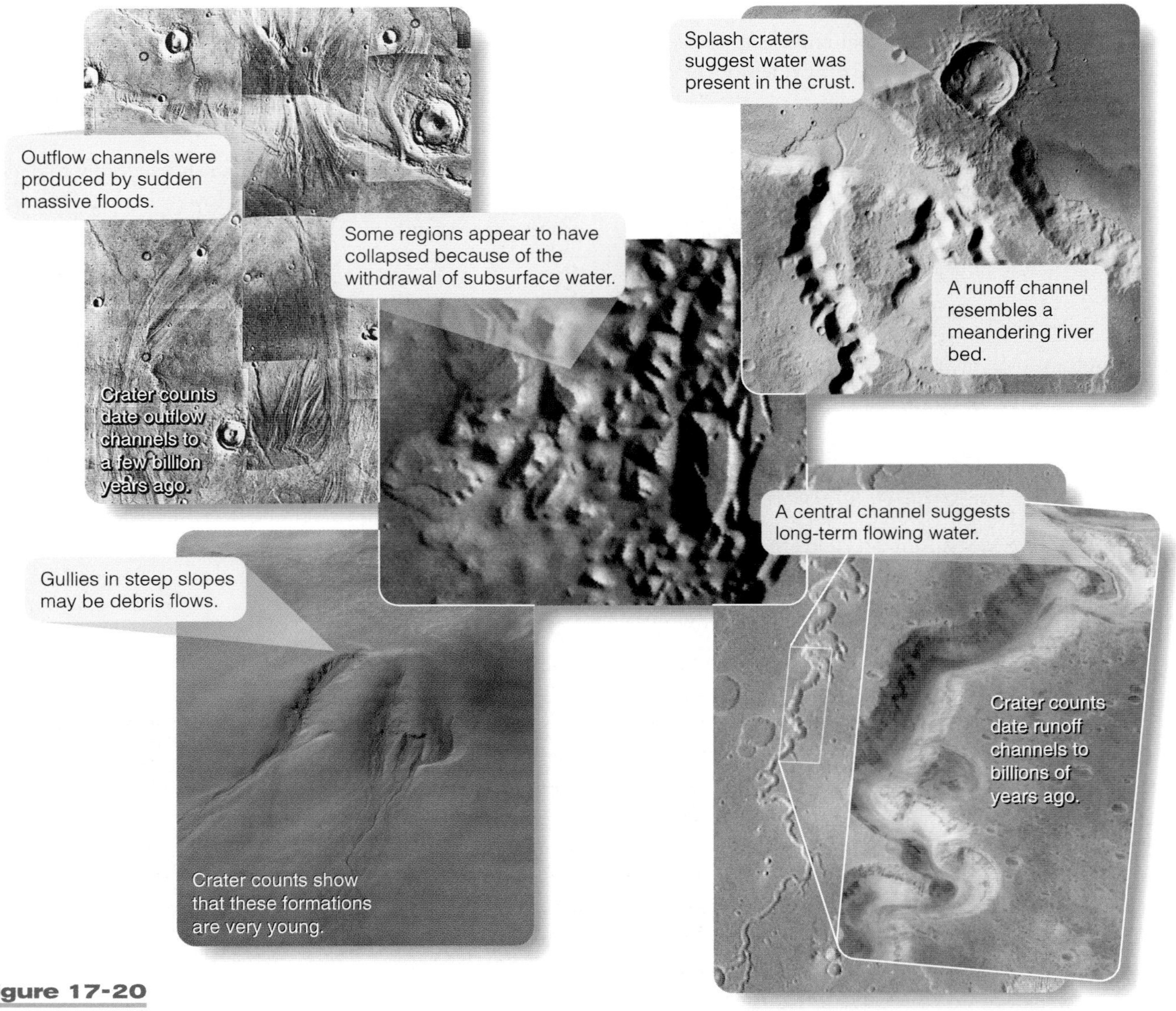

Outflow channels were produced by sudden massive floods.

Crater counts date outflow channels to a few billion years ago.

Splash craters suggest water was present in the crust.

Some regions appear to have collapsed because of the withdrawal of subsurface water.

A runoff channel resembles a meandering river bed.

A central channel suggests long-term flowing water.

Crater counts date runoff channels to billions of years ago.

Gullies in steep slopes may be debris flows.

Crater counts show that these formations are very young.

Figure 17-20

These visual-wavelength images made by the Viking orbiters and Mars Global Surveyor show some of the features that suggest liquid water on Mars. Outflow channels and runoff channels are old, but some gullies may be quite recent. (Malin Space Science Systems and NASA)

was once water on Mars, and rovers on the surface have turned up further traces of water.

In 1976, the two Viking spacecraft reached orbit around Mars and photographed its surface. Those photos revealed two water-related features. **Outflow channels** appear to have been cut by massive floods carrying as much as 10,000 times the water flowing down the Mississippi River. In a matter of hours or days, such floods swept away geological features and left scarred land such as that shown in ■ Figure 17-20. In contrast, **valley networks** look like meandering river beds with sand bars, deltas, and tributaries typical of streams that flowed for extended periods of time. The number of craters on top of these features tell us that they formed 1 to 3 billion years ago.

Images made from orbit also show regions of jumbled terrain suggesting that subsurface ice may have melted and drained away. Gullies leading down slope suggest water seeping from underground sources. A number of theories have been proposed to explain how the gullies formed, but they seem to involve water.

The atmosphere of Mars contains a tantalizing clue. Recall that deuterium is an isotope of hydrogen that contains a proton and a neutron as a nucleus. Because deuterium is twice as heavy as normal hydrogen, it escapes from a planet's atmosphere more slowly. Observations show that deuterium is 5.5 times more abundant relative to hydrogen on Mars than on Earth. That suggests that Mars once had about 20 times more water than it does now. That water, containing some deuterium, was broken into hydrogen and oxygen. The oxygen presumably formed oxides in the soil, and the hydrogen, some of it actually deuterium, escaped into space. Because deuterium is heavier and escapes more slowly than normal hydrogen, the abundance of deuterium relative to hydrogen gradually increased. This observation tells us that Mars must have once had more water.

The terrain at the edges of the northern lowlands has been compared to shorelines, and there is a growing suspicion that the northern lowlands were at least once filled with an ocean. Study Figure 17-17, where the lowlands have been color-coded blue, and notice the major outflow channels leading from the highlands into the lowlands northwest of the Viking-1 landing site and southeast of the Pathfinder landing site. Growing evidence suggests that the lowlands were filled by a major ocean 2.5 to 3.5 billion years ago.

The Mars Odyssey spacecraft, which reached orbit around Mars in February 2002, may have found the remains of that ancient water. Two instruments on Mars Odyssey are capable of detecting water within the top meter of the crust, and both agree that water ice makes up a significant fraction of the crust from latitude 60° north and from latitude −60° south. It isn't possible to determine how deep the ice goes, but the amount detected so far is enough to fill Lake Michigan twice.

Rovers Spirit and Opportunity were targeted to land in areas suspected of having had water on their surfaces. Flow features were visible at the Spirit landing site, and hematite, a mineral that forms in water, was detected from orbit at the Opportunity landing site. Both rovers reported exciting discoveries, but Opportunity found more evidence of past water. Using its analytic instruments, Opportunity discovered small spherical concretions (dubbed blueberries) that must have formed in water. In other rocks, Opportunity found layers of sediments with ripple marks and crossed layers showing they were deposited in moving water (■ Figure 17-21).

If liquid water once flowed on Mars, then its climate must have been different in the past. In the 1970s the Viking orbiters photographed layered terrain near the polar caps (■ Figure 17-22). These layers are believed to be laid down year after year as dust accumulates on the polar caps and is then left behind when the polar caps vaporize in the spring. Over periods of thousands of years, deep layers can develop. What is significant is that the Viking orbiters photographed newer layers superimposed on older layers, showing that the climate on Mars has changed repeatedly. Mars Global Surveyor has found similar layered features widespread on Mars. If these regions show traces of climate cycles, it supports studies that show that the climate on Mars may vary because of cyclic changes in the rotation and orbital revolution of the planet. Recall from Chapter 2 that Earth is affected by such cycles.

Mars has water but it is hidden. The deuterium in its atmosphere tells you it once had more water, but much of that has been lost. The climate has changed time after time, but the atmosphere has gradually grown thinner. The oceans and lakes are gone. The last of the water on Mars is frozen in the crust.

When humans reach Mars, they will not need to dig far to find ice. They can melt the ice to get water, and they can use solar power to break the water into hydrogen and oxygen. Hydrogen is fuel, and oxygen is the breath of life, so the water on Mars may prove to be buried treasure.

■ Figure 17-21

(a) Rover Opportunity photographed these hematite concretions in a rock near its landing site. The spheres appear to have grown as minerals collected around small crystals in the presence of water. Similar concretions are found on Earth. (b) The layers in this rock were deposited as sand and silt in rapidly flowing water. From the way the layers curve and cross each other, geologists can estimate that the water was at least ten centimeters deep. A few "blueberries" and one small pebble are also visible in this image. (NASA/JPL/Cornell/USGS)

Even more exciting is the realization that Mars once had bodies of liquid water on its surface. It is a desert world now, but someday an astronaut may scramble down an ancient Martian stream bed, turn over a rock, and find a fossil.

The History of Mars

Did Mars ever have plate tectonics? Where did the water go? These fundamental questions challenge you to assemble the evidence and hypotheses for Mars and tell the story of its evolution (**Window on Science 17-3**).

The four-stage history of Mars is a case of arrested development. The planet began by differentiating into a crust, mantle, and core. Studies of its rotation reveal that it has a dense core. The Mars Global Surveyor spacecraft detected no planetwide magnetic field, but it did find eight regions of the crust with fields a bit over 1 percent as strong as Earth's. Apparently, the young Mars had a molten iron core and generated a magnetic field, which became frozen into regions of the crust. The core must have cooled quickly and shut off the dynamo effect that was producing the planetwide field. Nevertheless, the magnetic regions of the crust remain behind like fossils.

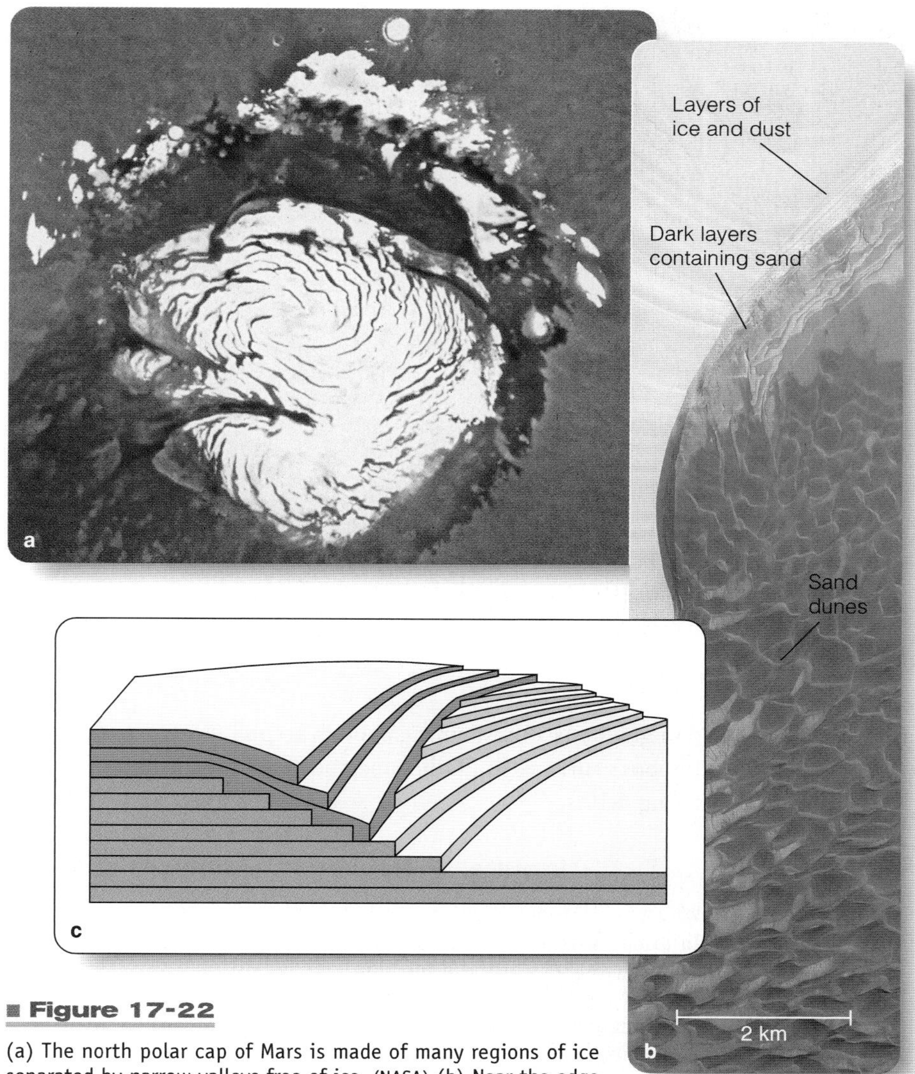

■ Figure 17-22

(a) The north polar cap of Mars is made of many regions of ice separated by narrow valleys free of ice. (NASA) (b) Near the edge of the polar cap, a set of light-colored layers of dust overlies a set of darker layers. Sand eroding from the lower layers is forming sand dunes. (NASA/JPL/Malin Space Science Systems) (c) In places, different sets of layers are superimposed, suggesting periodic changes in the Martian climate. (Adapted from a diagram by J. A. Cutts, K. R. Balasius, G. A. Briggs, M. H. Carr, R. Greeley, and H. Masursky)

The crust of Mars is now quite thick, as shown by the mass of Olympus Mons, but it was thinner in the past. Cratering may have broken or at least weakened the crust, triggering lava flows that flooded some basins. Mantle convection may have pushed up the Tharsis and Elysium volcanic regions and broken the crust to form Valles Marineris, but moving crustal plates seem never to have formed. There are no folded mountain ranges on Mars and no sign of plate boundaries. The small planet cooled rapidly, and the crust grew thick and immobile.

The large size of volcanoes on Mars is evidence that the crust does not move. On Earth, volcanoes like those that formed the Hawaiian Islands occur over rising currents of hot material in the mantle. Because the plate moves, the hot material heats the crust in a string of locations and forms a chain of volcanoes instead of a single large feature. The Hawaiian Islands are merely the most recent of a series of volcanic islands called the Hawaiian–Emperor island chain (see page 384), which stretches nearly 3800 km (2300 mi) across the Pacific Ocean floor. A lack of plate motion on Mars would have allowed a rising current of magma to heat the crust repeatedly in the same place and build a very large volcanic cone.

At some point in the history of Mars, water was abundant enough to flow over the surface in great floods. Some flood features lead into the northern lowlands, and an ocean may have existed there. The climate on Mars has changed as atmospheric gases and water were lost to space and as water was frozen into the soil as permafrost. Water from this permafrost may have flowed downhill and cut drainage channels. Under current conditions, however, liquid water cannot survive long on the Martian surface, so most of the visible flow features date from an earlier age and the water that remains is frozen in the crust.

For Mars, the fourth stage of planetary development has been one of slow decline. Volcanoes may still occasionally erupt, but the little planet has lost much of its internal heat, and most volcanism occurred long ago. The atmosphere is thin, and the surface is a forbiddingly dry, cold desert.

The Moons of Mars

Unlike Mercury or Venus, Mars has moons. Small and irregular in shape, Phobos ($28 \times 23 \times 20$ km in diameter) and Deimos ($16 \times 12 \times 10$ km) are captured asteroids.

Photographs reveal a unique set of narrow, parallel grooves on Phobos. Averaging 150 m (500 ft) wide and 25 m (80 ft) deep, the grooves run from Stickney, the largest crater, to an oddly featureless region on the opposite side of the satellite. One theory suggests that the grooves are deep fractures produced by the impact that formed the crater (■ Figure 17-23a).

The Mars Global Surveyor spacecraft reveals information about the dust on Phobos's surface. Observations made with the spacecraft's infrared spectrometer show that the moon's surface cools quickly from −4°C to −112°C (from 25°F to −170°F) as it passes from sunlight into the shadow of Mars. Solid rock would retain heat and cool more slowly, so the dust must be at least a

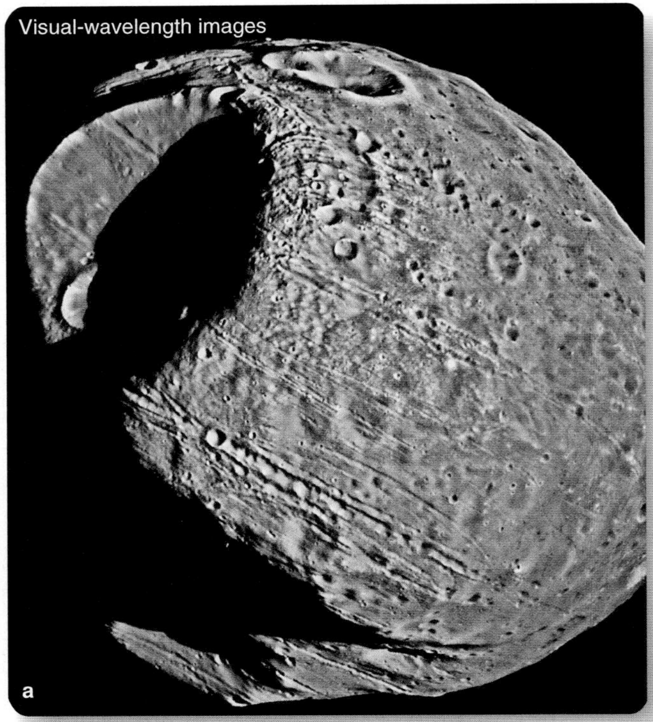

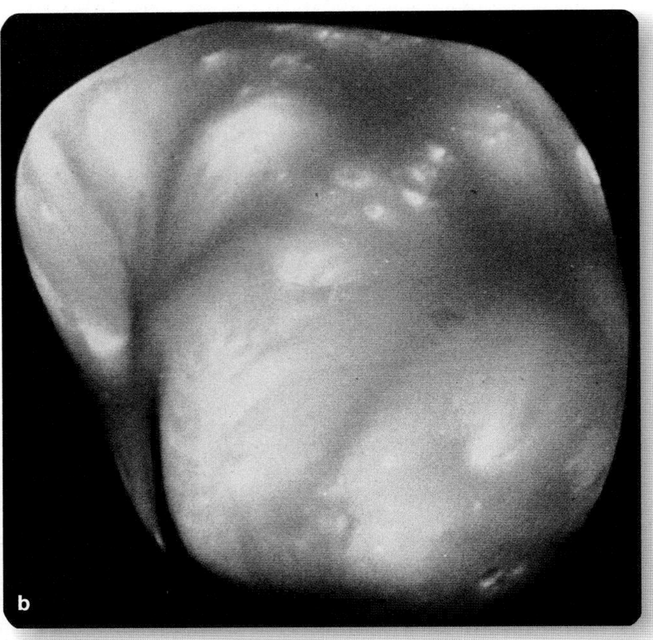

Figure 17-23

The moons of Mars are too small to pull themselves into spherical shape. (a) Phobos has an impact crater 10 km in diameter at one end with grooves radiating away. (Damon Simonelli and Joseph Ververka, Cornell University/NASA) (b) Deimos, smaller than Phobos, looks more uniform because of dusty soil covering the smaller features. (NASA)

meter deep and very fine. In photos made by the spacecraft camera, the dust blankets the terrain, but the photos also show boulders a few meters in diameter thought to be ejected by impacts.

Deimos not only has no grooves, but it also looks smoother because of a thicker layer of dust on its surface. This material partially fills craters and covers minor surface irregularities (Figure 17-23b). It seems certain that Deimos experienced collisions in its past, so fractures may be hidden below the debris.

The debris on the surfaces of the moons raises an interesting question: How can the weak gravity of small bodies hold any fragments from meteorite impacts? The escape velocity on Phobos is only about 12 m/s (40 ft/s). An athletic astronaut who could jump 2 m (6 ft) high on Earth could jump 2.8 km (1.7 miles) on Phobos. Certainly most of the fragments from an impact should escape, but the slowest particles could fall back in the weak gravity and accumulate on the surface.

Because Deimos is smaller than Phobos, its escape velocity is smaller, so it seems surprising that it has more debris on its surface. This may be related to Phobos's orbit close to Mars. The Martian gravity is almost strong enough to pull loose material off of Phobos's surface, so Phobos may be able to retain less of its cratering debris.

Inquire | Review | Analyze

Why would you be surprised to find volcanism on Phobos or Deimos? In discussing Earth's moon, Mercury, Venus, Earth, and Mars, you have seen illustrations of the principle that the larger a world is, the more slowly it loses its internal heat. It is the flow of that heat from the interior through the surface into space that drives geological activity such as volcanism and plate motion. A small world, like Earth's moon, cools quickly and remains geologically active for a shorter time than does a larger world like Earth. Phobos and Deimos are not just small; they are tiny. However they formed, any interior heat would have leaked away very quickly; with no energy flowing outward, there can be no volcanism.

Some futurists suggest that the first human missions to Mars will be not to land on the surface but to build a colony on Phobos or Deimos. These plans speculate that there may be water deep inside the moons that colonists could use. What would happen to water released in the sunlight on the surface of such small worlds?

■ ■ ■

Connections: Venus is nearly the same size as Earth, but it is unlikely that humans will ever colonize its surface. Smaller and colder Mars, however, may someday be home to large colonies. In the next chapter, we turn our attention to Jupiter-like planets, giant worlds so totally unlike Earth that you will need entirely new principles of comparative planetology to describe them.

Window on Science 17-3

The Present Is the Key to the Past

Geologists are fond of saying "The present is the key to the past." By that they mean that you can learn about the history of Earth by looking at the present condition of Earth's surface. The position and composition of various rock layers in the Grand Canyon, for example, tell you that the western United States was once at the floor of an ocean. This principle of geology was astonishing when it was first formulated in the late 1700s, and it continues to be relevant today as you try to understand other worlds, such as Venus and Mars.

In the late 18th century, naturalists first recognized that the present gave them clues to the history of Earth. This was astonishing because most people assumed that Earth had no history. That is, they assumed either that Earth had been created in its present state as described in the Old Testament or that Earth was eternal. In either case, people commonly assumed that the hills and mountains they saw around them had always existed more or less as they were. The 18th-century naturalists began to see evidence that the hills and mountains were not eternal but were the result of past processes and were slowly changing. That gave birth to the idea that Earth had a history.

As the naturalists of the 18th century made the first attempts to thoughtfully and logically explain the nature of Earth by looking at the evidence, they were inventing modern geology as a way of understanding Earth. What Copernicus, Kepler, and Newton did for the heavens in the 1600s, the first geologists did for Earth beginning in the late 1700s. Of course, the invention of geology as the study of Earth led directly to the modern attempts to understand the geology of other worlds.

Geologists and astronomers share a common goal: They are attempting to reconstruct the past (Window on Science 16-3). Whether you study Earth, Venus, or Mars, you are looking at the present evidence and trying to reconstruct the past history of the planet by drawing on observations and logic to test each step in the story. How did Venus get to be covered with lava, and how did Mars lose its atmosphere? The final goal of planetary astronomy is to draw together all of the available evidence (the present) to tell the story (the past) of how the planet got to be the way it is. Those first geologists of the late 1700s would be fascinated by the stories planetary astronomers tell today.

Layers of rock in the Martian crater Terby hint at a time when the crater was filled with a lake. (NASA/JPL/Mallin Space Science Systems)

Study and Review Tools

Summary

17-1 | A Travel Guide to the Terrestrial Planets

How can comparison help you understand the terrestrial planets?

- Earth's moon is included in this study because it is a complex world and makes a striking comparison with Earth.
- The terrestrial worlds differ mainly in size, but they all have low-density crusts, mantles of dense rock, and metallic cores.
- Comparative planetology warns you to expect that cratered surfaces are old, that heat flowing out of a planet drives geological activity, and that the nature of a planet's atmosphere depends of the size of the planet and its temperature.

17-2 | Earth: Planet of Extremes

What are the main features of Earth when you view it as a planet?

- Earth has passed through four stages as it has evolved: (1) differentiation; (2) cratering; (3) flooding by lava and water; (4) slow surface evolution.

- Seismology shows that Earth has differentiated to form a liquid metallic core, which is very hot and generates a magnetic field.
- Earth is dominated by plate tectonics that breaks the crust into moving plates. It is driven by heat flowing upward from the interior.
- Earth's first atmosphere was probably mostly carbon dioxide, but that gas was dissolved in seawater, and plant life has created oxygen.

17-3 | The Moon

How does size affect a planet?

- Earth's moon formed in a molten state and differentiated, but it contains little metal and has a low density.
- Because it is a small world, Earth's moon has lost most of its internal heat and is no longer geologically active. Its old highlands are heavily cratered, but the lowlands are filled by lava flows that formed smooth maria soon after the end of the heavy bombardment.
- The surface is fractured by impacts and covered by a regolith of powdered rock.
- The large-impact hypothesis suggests the moon formed when an impact between two planetesimals formed Earth with a disk of material around it. The moon formed from that disk.

17-4 | Mercury

- Mercury is smaller than Earth but larger than Earth's moon. It is airless and has an old, heavily cratered surface.
- Mercury has a much higher density than Earth's moon and must have a large metallic core. It may have suffered a major impact when it was young that drove off low-density crustal rock and left it with a metallic core that is large in proportion to the diameter of the planet.
- Lobate scarps are long curving cliffs formed by compression on Mercury when its large metallic core solidified and contracted.

17-5 | Venus

How does distance from the sun affect a planet and its atmosphere?

- Although Venus is almost as big as Earth, it has a thick, cloudy atmosphere of carbon dioxide that hides the surface from sight. It can be studied by radar mapping.
- The carbon dioxide atmosphere drives an intense greenhouse effect and makes Venus's surface hot enough to melt lead.
- Venus is slightly closer to the sun than Earth and was too warm for liquid water oceans to dissolve carbon dioxide from the atmosphere.
- The hot crust of Venus is dominated by volcanism and not by plate tectonics.

17-6 | Mars

How does a planet's atmosphere depend on the size of the planet?

- Mars is about half the size of Earth; it has a thin atmosphere and has lost much of its internal heat.
- The loss of atmospheric gases depends on the size of a planet and its temperature. Mars is cold, but it is small and has a lower escape velocity, and many of its lighter gases have leaked away.
- Some water may have leaked away as ultraviolet radiation from the sun broke it into hydrogen and oxygen, but some water is frozen in the polar caps and in the soil.
- Outflow channels and valley networks visible from orbit seem to have been cut by sudden floods or by longer-term drainage, but water cannot exist as a liquid on Mars now because of its low temperature and low atmospheric pressure.
- The southern hemisphere of Mars is old cratered terrain, but some large volcanoes lie in the north. The large size of these volcanoes suggests that the crust has not been divided into moving plates.
- Some volcanism may still occur on Mars, but because it is small, it has cooled and is not very active geologically.
- Rovers have found clear signs that the Martian climate was different in the past and that liquid water flowed over the surface in at least some places. The northern lowlands may even have held an ocean at one time.
- The two moons of Mars are probably captured asteroids. They are small, airless, and cratered and lost their internal heat long ago.

New Terms

comparative planetology (p. 378)
mantle (p. 381)
P wave (p. 382)
S wave (p. 382)
plastic (p. 383)
plate tectonics (p. 384)
midocean rise (p. 384)
basalt (p. 384)
subduction zone (p. 384)
folded mountain range (p. 384)
rift valley (p. 385)
primeval atmosphere (p. 386)
secondary atmosphere (p. 386)
greenhouse effect (p. 386)
global warming (p. 386)
mare (maria) (p. 387)
terminator (p. 387)
ejecta (p. 388)
rays (p. 388)
secondary crater (p. 388)
micrometeorite (p. 389)
multiringed basin (p. 389)
vesicular basalt (p. 391)
anorthosite (p. 391)
breccia (p. 391)
albedo (p. 391)
fission hypothesis (p. 394)
condensation hypothesis (p. 394)
capture hypothesis (p. 394)
large-impact hypothesis (p. 394)
lobate scarp (p. 396)
composite volcano (p. 400)
shield volcano (p. 400)
coronae (p. 402)
outflow channel (p. 409)
valley network (p. 409)

Review Questions

Ace Astronomy™ Assess your understanding of this chapter's topics with additional quizzing and animations at **http://astronomy.brookscole.com/sh9e**

1. What are the four stages in the development of a terrestrial planet?
2. Why would you expect planets to have differentiated?
3. How does plate tectonics create and destroy Earth's crust?
4. Why do astronomers suspect that Earth's primeval atmosphere was rich in carbon dioxide?
5. Why doesn't Earth have as many craters as its moon or Venus?
6. What kind of erosion is now active on Earth's moon?
7. Discuss the evidence and hypotheses concerning the origin of Earth's moon.

8. Why doesn't Earth or its moon have lobate scarps?
9. How did Earth avoid the greenhouse effect that made Venus so hot?
10. What evidence indicates that plate tectonics does not occur on Venus? on Mars?
11. What evidence suggests that Venus has been resurfaced within the last half-billion years?
12. Why is the atmosphere of Venus rich in carbon dioxide? Why is the atmosphere of Mars rich in carbon dioxide?
13. What evidence indicates that the climate on Mars has changed?
14. Why do astronomers conclude that the crust on Mars must be thick?
15. What evidence indicates that there has been liquid water on Mars?
16. In this photo, Astronaut Alan Bean works at the Apollo 12 lander. Describe the surface you see. What kind of terrain did they land on for the second mission to the moon?

(NASA)

17. Olympus Mons on Mars is a very large volcano. In this image you can see superimposed caldera at the top. What do those multiple caldera and immense size indicate about the geology of Mars?

(NASA)

18. Volcano Sif Mons on Venus is shown in this radar image. What kind of volcano is it, and why is it orange in this image? What color would the rock be if you could see it with your own eyes?

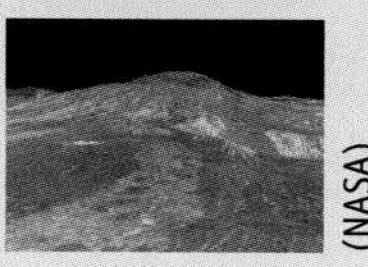

(NASA)

Discussion Questions

1. If you visited a planet in another solar system and discovered oxygen in its atmosphere, what might you guess about its surface?
2. If liquid water is rare on the surface of planets, then most terrestrial planets must have CO_2-rich atmospheres. Why?

Problems

1. If the Atlantic seafloor is spreading at 30 mm/year and is now 6400 km wide, how long ago were the continents in contact?
2. Earth is 4 times larger in diameter than its moon. How many times larger is it in surface area? in volume?
3. The smallest detail visible through Earth-based telescopes is about 1 second of arc in diameter. What size is this on Earth's moon? (*Hint:* See Reasoning with Numbers 3-1.)
4. The trenches where the seafloor slips downward are 1 km or less wide. Could Earth-based telescopes resolve such features on Earth's moon? Why are astronomers sure such features are not present on Earth's moon?
5. How long would it take radio signals to travel from Earth to Venus and back if Venus were at its farthest point from Earth? Why are such observations impractical?
6. Repeat Problem 5 for Mercury.
7. Imagine that a spacecraft has landed on Mercury, and it has transmitted radio signals to Earth at a wavelength of 10 cm. If you see Mercury at its greatest angular distance west of the sun, to what wavelength must you tune your radio telescope to detect the signals? (*Hints:* See Celestial Profile Four to find Mercury's orbital velocity, and then see Reasoning with Numbers 6-2.)
8. The smallest feature visible through an Earth-based telescope has an angular diameter of about 1 second of arc. If a crater on Mars is just visible when Mars is nearest to Earth, what is its linear diameter? (*Hint:* See Reasoning with Numbers 3-1.)
9. What is the maximum angular diameter of Phobos as seen from Earth? What surface features should you be able to see using Earth-based telescopes? (*Hint:* See Reasoning with Numbers 3-1.)
10. Phobos orbits Mars at a distance of 9380 km from the center of the planet and has a period of 0.3189 days. Calculate the mass of Mars. (*Hint:* See Reasoning with Numbers 4-1.)

Media Cluster

Ace Astronomy™ To access the resources in the Media Cluster, log into AceAstronomy at http://astronomy.brookscole.com/sh9e and select Chapter 17.

ACTIVE FIGURES

Seismic Waves
Use this animation to create an earthquake and see how tracking its seismic waves provides evidence about what lies beneath Earth's surface.

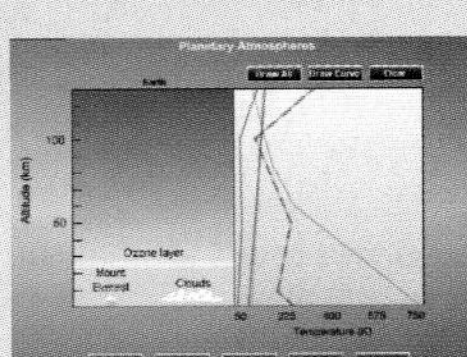

Planetary Atmospheres
Experiment with the temperature throughout the atmospheres of several planets with this animation.

The Moon's Craters
The craters that cover the moon were produced by the high-speed impact of meteorites of all sizes. Explore three types of lunar craters in this animation.

Hot Spot Volcanoes
If the crustal plate over a hot spot is moving, repeated eruptions can result in an island chain like the Hawaiian Islands. Set a series of eruptions in motion in this animation.

ASTRONOMY EXERCISES

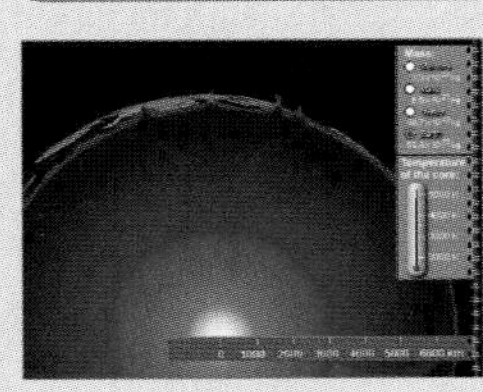

Convection and Plate Tectonics
Study the convection currents of the four Earthlike planets in this animation and see if you can determine how this relates to the amount of plate tectonic activity on each planet.

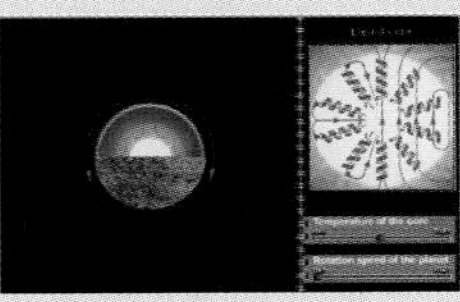

Convection and Magnetic Fields
A planet's core temperature and rotation speed are factors in the strength of its magnetic field. See the effect of these variables in this animation.

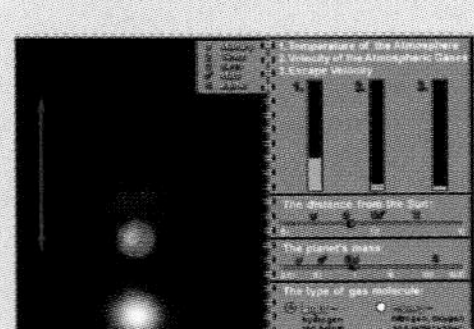

Primary Atmospheres
This animation lets you change the distance to the sun and mass of a hypothetical planet and study how these relate to the planet's retention of its atmosphere.

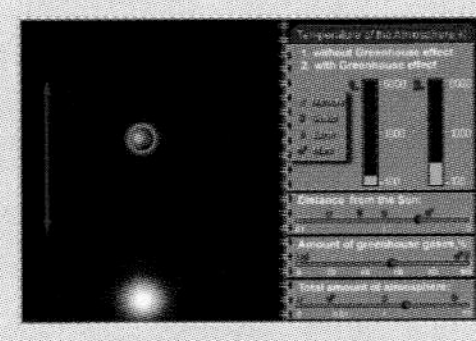

Greenhouse Effect
See how distance to the sun and amount of carbon dioxide in a planet's atmosphere affect its temperature in this animation.

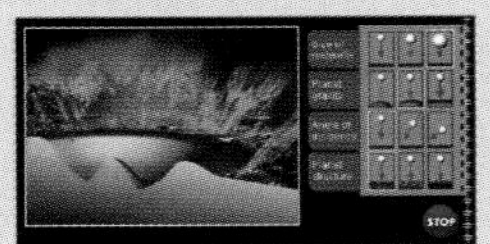

Cratering
Turn up your speakers for this one, where can you design collisions between a meteorite and a hypothetical planet.

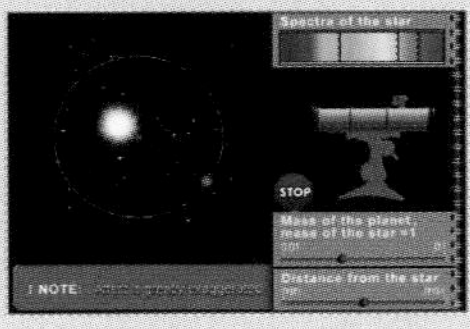

Extrasolar Planets
This simulation presents a hypothetical star system with one planet orbiting a star. Study what variables help astronomers to detect extrasolar planets.

VIRTUAL ASTRONOMY LABS

Lab 5: Planetary Geology
By thoroughly studying Earth's geology, you can leverage your knowledge to understand conditions on other planets. This lab includes an exercise in determining the interior characteristics of a planet by observing the travel times of its seismic waves.

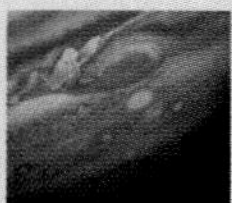

Lab 7: Planetary Atmospheres and Their Retention
This lab investigates the retention of atmospheres. You will explore the factors that govern the loss of atmospheric gases, and you will see why certain bodies can retain some gases but not others.

Critical Inquiries for the Web

1. Who decides how planetary features are named? Surface features on Venus are (mostly) named after female figures from history and mythology, whereas figures from the arts and music are used to name features on Mercury. Look for information on planetary nomenclature and summarize the way different types of features on Venus are assigned names.
2. "Martians" have fascinated humans for the last century or more. There are many online sources that chronicle the representation of life on Mars throughout history and in literature. Read about the Martians as represented by a particular literary work or nonfiction account and discuss to what extent it is (or is not) based on realistic views of the nature of Mars both in terms of our current understanding and the views of that period.
3. What would it be like to walk on the lunar surface? Apollo astronauts visited six different locations on our moon, exploring a variety of lunar terrain. Describe the horizons and general relief of the landing locations of the different missions by exploring websites that provide lunar surface photography from the missions. What differences do you see between images from landings in highlands and maria?

Exploring *TheSky*

1. Locate Mercury and Venus and zoom in until you can see their phases. Explain their phases by discussing their locations in their orbits.
2. Repeat the activity above for Mars.
3. Watch Venus and Mars circle the sky and discuss the way their phases change with their positions in their orbits.

Go to the Brooks/Cole Astronomy Resource Center **(http://astronomy.brookscole.com)** for critical thinking exercises, articles, and additional readings from InfoTrac College Edition, Brooks/Cole's online student library.

18 Comparative Planetology of the Outer Planets

There wasn't a breath

in that land of death . . .

ROBERT SERVICE
The Cremation of Sam McGee

THE SULFURIC ACID fogs of Venus seem totally alien, but compared with the planets of the outer solar system, Venus is a tropical oasis. Of the five planets beyond the asteroid belt, four have no solid surface, and one, Pluto, is so far from the sun that most of its atmosphere lies frozen on its surface. As you begin your study of these strange worlds, you will discover new principles of comparative planetology. ▌ The Jovian worlds can be studied from Earth, but much of what astronomers know has been radioed back to Earth from space probes. Pioneer 10 and Pioneer 11 explored Jupiter and Saturn in the mid-1970s, but their instruments were not highly sensitive. The Voyager 1 and Voyager 2 spacecraft were launched in 1977 on a mission to visit all four Jovian worlds, which concluded when Voyager 2 flew past Neptune in 1989. The Galileo

In 2004, the spacecraft Cassini reached orbit around the giant planet Saturn and released a probe to descend into the smoggy orange atmosphere of Saturn's moon Titan. (JPL/NASA)

Guidepost

Looking Back

In the last two chapters, you watched our solar system form and explored the terrestrial planets. Along the way you learned some important principles of comparative planetology. Now it is time to explore the outer solar system, where the planets are bigger and the temperatures are colder. Nevertheless, the principles of comparative planetology will be faithful guides.

This Chapter

You can imagine walking on Mars, and it isn't too hard to imagine a visit to the volcanic surface of Venus. The Jovian planets are a challenge to your imagination; they are so unearthly they would be unbelievable if you didn't have direct observational evidence to tell you what they are like. As you explore, you will find answers to five essential questions:

What are very massive planets like?

What evidence indicates that some moons have been active?

How are planetary rings formed and maintained?

How do the less massive Jovian worlds differ from their more massive cousins?

Is Pluto a planet or something more?

Looking Ahead

The planets of the solar system formed from small bodies in the solar nebula; to some extent, the planets are still affected by those small objects. In the next chapter, you will adapt your understanding of comparative planetology to study the last remains of the solar nebula.

AceAstronomy™ The AceAstronomy icon throughout the text indicates an opportunity for you to test yourself on key concepts and to explore animations and interactions on the AceAstronomy website at: http://astronomy.brookscole.com/sh9e

spacecraft orbited Jupiter from 1995 to 2003, and the Cassini probe to Saturn went into orbit in 2004. Throughout this discussion of the Jovian worlds, you will find images and data returned by these robotic explorers (**Window on Science 18-1**).

Today's astronomers have a tremendous library of photos, measurements, and facts about the Jovian planets; and although no spacecraft has visited Pluto, Earth's astronomers are beginning to understand its role. Our task is to discover the relationships that explain how these worlds got to be the way they are.

18-1 A Travel Guide to the Outer Planets

IF YOU TRAVEL MUCH you know that some cities make you feel at home and some do not. You are about to visit five worlds that are truly unearthly. You will not feel welcome, so this travel guide will warn you what to expect.

The Outer Planets

The outermost planets in our solar system are Jupiter, Saturn, Uranus, Neptune, and Pluto. The first four are often called the "Jovian planets," meaning they are like Jupiter. In fact, they are all individuals with their own separate personalities.

■ Figure 18-1 compares the five outer worlds, and one striking feature is diameter. Jupiter is the largest of the Jovian worlds, over 11 times the diameter of Earth. Saturn is a bit smaller, but Uranus and Neptune are quite a bit smaller than Jupiter. Pluto, in contrast, is hardly visible in this diagram. It is a tiny world and doesn't seem to fit with its companions in the outer solar system. You will discover it is truly an oddball, and some astronomers argue that it is not a planet at all.

The other feature you will notice immediately when you look at Figure 18-1 is Saturn's rings. They are bright and beautiful and composed of billions of ice particles, each orbiting around the planet. Jupiter, Uranus, and Neptune also have rings, but they are not easily detected from Earth and are not visible in this

■ Figure 18-1

The worlds of the outer solar system consist of the four Jovian planets, which are large, low-density worlds rich in hydrogen, and tiny, icy Pluto. (NASA/JPL/Space Science Institute/University of Arizona)

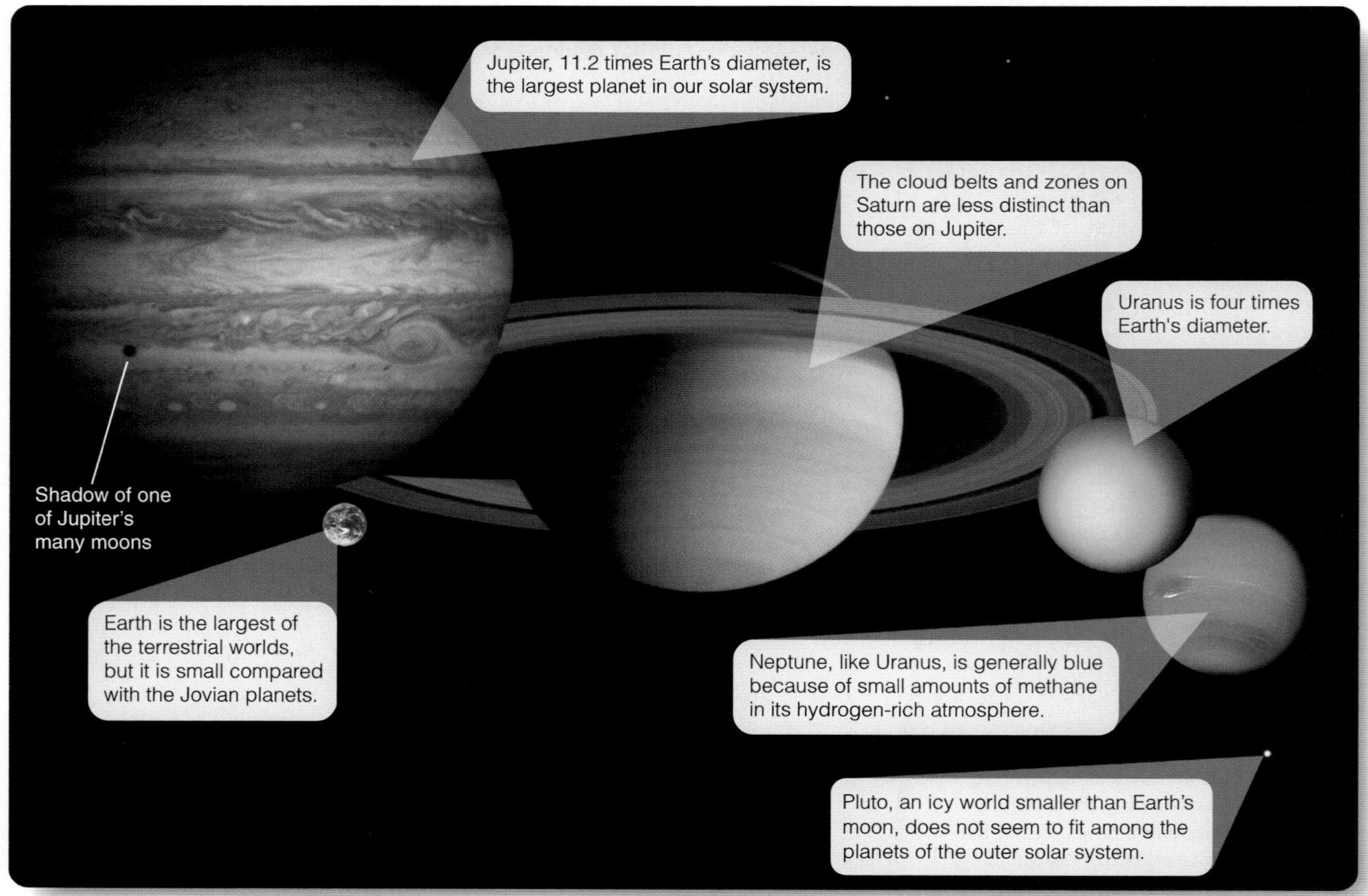

Jupiter, 11.2 times Earth's diameter, is the largest planet in our solar system.
The cloud belts and zones on Saturn are less distinct than those on Jupiter.
Uranus is four times Earth's diameter.
Shadow of one of Jupiter's many moons
Earth is the largest of the terrestrial worlds, but it is small compared with the Jovian planets.
Neptune, like Uranus, is generally blue because of small amounts of methane in its hydrogen-rich atmosphere.
Pluto, an icy world smaller than Earth's moon, does not seem to fit among the planets of the outer solar system.

Science, Technology, and Engineering

Sending spacecraft to another world is expensive, and it may seem pointless when that world seems totally hostile to human life. What practical value is there in sending a space probe to Jupiter? To resolve that question, you need to consider the distinction between science, technology, and engineering.

Science is nothing more than the logical study of nature, and the goal of science is understanding. Although much scientific knowledge proves to have tremendous practical value, the only goal of science is a better understanding of how nature works. Technology, in contrast, is the practical application of scientific knowledge to solve a specific problem. People trying to find a faster way to paint automobiles might use all the tools and techniques of science, but if their goal has some practical outcome, you should more properly call it technology rather than science. Engineering is the most practical form of technology. An engineer is likely to use well-understood technology to find a practical solution to a problem.

Of course, there are situations in which science and technology blur together. For example, humanity has a practical and tragic need to solve the AIDS problem; the world needs a cure and a method of prevention. Unfortunately, scientists don't understand the AIDS virus itself or viruses in general well enough to design a simple solution, and thus much of the work involves going back to basic science and trying to better understand how viruses interact with the human body. Is this technology or science? It is hard to decide.

You might describe science that has no known practical value as basic science or basic research. The exploration of worlds such as Jupiter would be called basic science, and it is easy to argue that basic science is not worth the effort and expense because it has no known practical use. Of course, the problem is that you have no way of knowing what knowledge will be of use until you acquire that knowledge. In the middle of the 19th century, Queen Victoria is supposed to have asked physicist Michael Faraday what good his experiments with electricity and magnetism were. He answered, "Madam, what good is a baby?" Of course, Faraday's experiments were the beginning of the electronic age. Many of the practical uses of scientific knowledge that fill the world—transistors, vaccines, plastics—began as basic research. Basic scientific research provides the raw materials that technology and engineering use to solve problems.

Exploring other worlds is of cultural value. It helps us as humans understand ourselves. (NASA/JPL/Space Science Institute)

Basic scientific research has yet one more important use that is so valuable it seems an insult to refer to it as merely practical. Science is the study of nature, and as we learn more about how nature works, we learn more about what our existence in this universe means for us. The seemingly impractical knowledge gained from space probes to other worlds tells us about our planet and our own role in the scheme of nature. Science tells us where we are and what we are, and that knowledge is beyond value.

figure. Nevertheless, as you visit these worlds you will be able to compare four different sets of planetary rings.

Atmospheres and Interiors

The four Jovian worlds have hydrogen-rich atmospheres filled with clouds. On Jupiter and Saturn, you can see that the clouds form belts and zones that circle the planets like the stripes on a child's ball. This form of atmospheric circulation is called **belt–zone circulation.** You will find traces of belts and zones on Uranus and Neptune, but they are not very distinct.

The atmospheres of the Jovian planets are not very deep. Jupiter's atmosphere makes up only about one percent of its radius. Below that atmosphere, Jupiter and Saturn are composed of liquid hydrogen, so the older term for these planets, the gas giants, should probably be changed to the liquid giants. Only near their centers do these worlds have solid cores of dense material with the composition of rock and metal.

Uranus and Neptune are sometimes called the ice giants because they are rich in water, both as a liquid and as a solid. They too have denser material in their central cores.

On your visits to these worlds, notice that they are low-density worlds that are rich in hydrogen. Jupiter and Saturn are mostly liquid hydrogen, and even Uranus and Neptune contain a much larger proportion of hydrogen than does Earth. Recall from Chapter 16 that these worlds are hydrogen-rich, low-density worlds because they formed in the outer solar nebula where water vapor could freeze to form tremendous amounts of tiny ice particles. Once these planets grew massive enough, they could draw in the hydrogen gas directly by gravitational collapse.

Satellite Systems

You can't really land your spaceship on the Jovian worlds, but you might be able to land on one of their moons. All of the Jovian worlds have large satellite systems, and even little Pluto has a moon.

As you visit the moons of the Jovian worlds, look for two processes. In many cases, the moons interact gravitationally and have adjusted their orbits. They also have fascinating gravitational interactions with planetary ring systems.

The second thing to notice is that a few moons are geologically active, while others show signs of past activity. You have learned that cratered surfaces are old, so when you see a section of a moon's surface that has few craters, you can assume that the moon must have been geologically active since the end of the heavy bombardment. Of course, geological activity depends on heat, so be alert for sources of internal heat as you visit these moons.

Inquire I Review I Analyze

Why do you expect the outer planets to be low-density worlds?
In Chapter 16, you discovered that the inner planets could not incorporate ice when they formed because it was too hot near the sun; but, in the outer solar nebula, the growing planets could accumulate lots of ice. Eventually they grew massive enough to grow by gravitational collapse, and that pulled in hydrogen and helium gas. That makes the outer planets low-density worlds.

The outer planets may be unearthly, but they are understandable. For example, why do you expect the outer planets to have rings and moons?

■ ■ ■

Connections: Earth was the basis of comparison for your study of the terrestrial planets, but Jupiter makes a better basis of comparison for the Jovian worlds. Jupiter is big and complex and has a large family of moons. So let's begin with the king of the solar system.

18-2 Jupiter

JUPITER IS THE MOST MASSIVE of the Jovian planets, containing 71 percent of all the planetary matter in our solar system. This high mass accentuates some processes that are less obvious or nearly absent on the other Jovian worlds. Just as you used Earth as the basis for comparison in your study of the terrestrial planets, we will examine Jupiter in detail so you can use it as a standard in your comparative study of the other Jovian planets.

The Interior

Density is mass divided by volume, and the density of Jupiter is only a third greater than that of water (**Celestial Profile 7**). For comparison, Earth is over 4 times denser than Jupiter. Theoretical models of Jupiter based on this density conclude that it is composed mostly of hydrogen and helium. In the interior, the pressure compresses the hydrogen into a liquid. At the center, a so-called rocky core contains heavier elements, such as iron, nickel, silicon, and so on. With a temperature 4 times hotter than the surface of the sun and a pressure 50 million times Earth's air pressure at sea level, this material is unlike any rock on Earth. The term *rocky core* refers to the chemical composition and not to the consistency of the material.

Careful measurements of the heat flowing out of Jupiter reveal that it emits about twice as much energy as it absorbs from the sun. This energy appears to be heat left over from the formation of the planet. In Chapter 16 you saw that Jupiter should have grown very hot when it formed, and some of this heat remains trapped in its interior.

Although Jupiter is mostly liquid hydrogen, there is no ocean surface on which you might imagine sailing a Jovian boat. The base of the atmosphere is so hot and the pressure is so high that there is no real distinction between liquid and gas, and if you jumped into Jupiter carrying a rubber raft you would be disappointed. As you fell deeper and deeper through the atmosphere, you would see the gas density increasing around you until you were sinking through a liquid, but you would never splash into a liquid surface.

Under high pressure, liquid hydrogen becomes **liquid metallic hydrogen**—a material that is a very good conductor of electricity. Laboratory measurements of the properties of this material show that the transition from liquid hydrogen to liquid metallic hydrogen must occur at a relatively shallow depth in Jupiter, about 10 percent of the way from the surface to the center. This mass of conducting liquid, stirred by convection currents and spun by the planet's rapid rotation, drives the dynamo effect and generates a powerful magnetic field. Jupiter's field is over 10 times stronger than Earth's.

One consequence of this magnetic field is the creation of auroras. Just as in the case of Earth, interactions between Jupiter's magnetic field and the solar wind generate powerful electric currents that flow down into the atmosphere in two rings around the planet's magnetic poles. These are visible at ultraviolet wavelengths as rings of aurora larger in diameter than Earth (■ Figure 18-2a).

A planet's magnetic field deflects the solar wind and dominates a volume of space around the planet called the **magnetosphere.** Jupiter's magnetosphere is 100 times larger than Earth's (Figure 18-2b). If you could see it in the sky, it would be 6 times larger than the full moon.

The magnetic field around Jupiter traps charged particles from the solar wind in large, doughnut-shaped radiation belts that surround the planet just as Earth is surrounded by its Van Allen radiation belts. Because Jupiter's magnetic field is stronger, its trapped radiation is a billion times more intense than Earth's. The spacecraft that have flown through these regions received over 4000 times the radiation that would have been lethal for a human.

■ Active Figure 18-2 ▶

(a) Auroras on Jupiter are confined to rings around the north magnetic pole and the south magnetic pole, as shown in these ultraviolet images. Earthly auroras follow the same pattern. The small comet-shaped spots are caused by powerful electrical currents flowing from Jupiter's moon Io. (John Clarke, University of Michigan, and NASA) (b) Jupiter's large conducting core and rapid rotation create a powerful magnetic field that holds back the solar wind and dominates a region called the magnetosphere. High-energy particles trapped in the magnetic field form giant radiation belts.

Ace Astronomy™ **Log into Ace Astronomy and select this chapter to see the Active Figure called "Explorable Jupiter." Notice the dramatic differences between images made at different wavelengths.**

However unearthly Jupiter may be, it obeys the same rules of nature that govern Earth. Thus, it has a magnetic field and auroras just as Earth does. Jupiter's atmosphere, however, seems even more unearthly than its interior.

Ace Astronomy™ Log into AceAstronomy and select this chapter to see Astronomy Exercise "Aurora." See how auroras are produced.

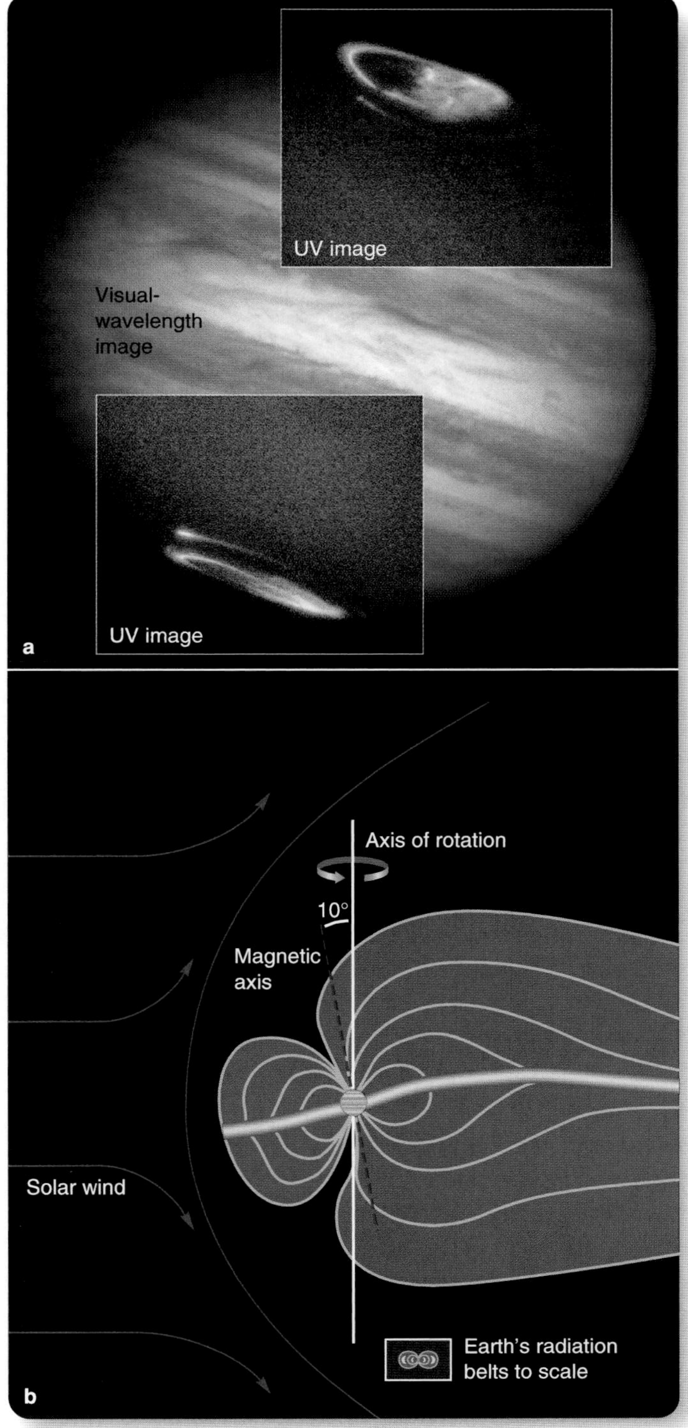

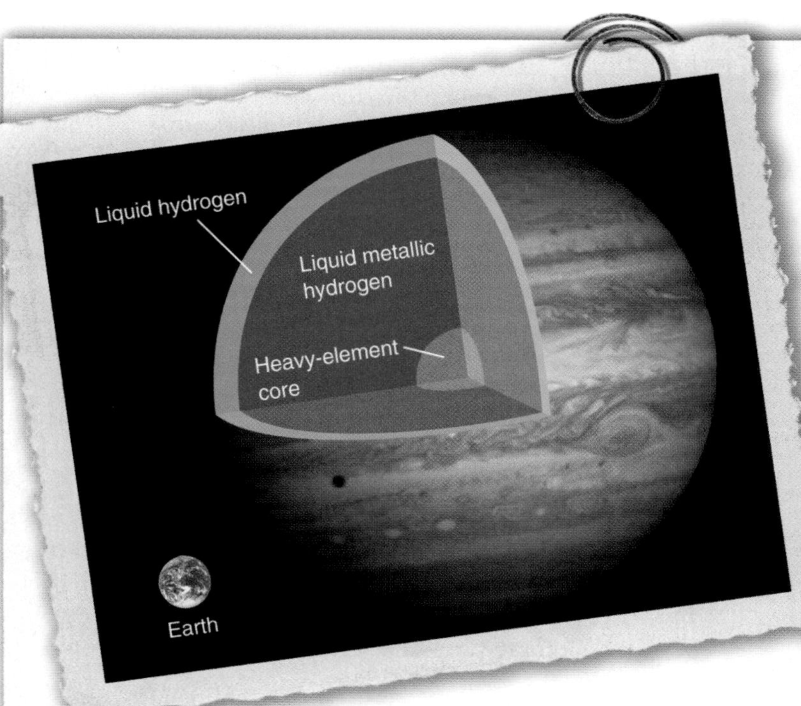

Jupiter is mostly a liquid hydrogen planet with a small core of heavy elements, not much bigger than Earth. (NASA/JPL/University of Arizona)

Celestial Profile 7: Jupiter

Motion:

Average distance from the sun	5.2028 AU (7.783 × 10^8 km)
Eccentricity of orbit	0.0484
Maximum distance from the sun	5.455 AU (8.160 × 10^8 km)
Minimum distance from the sun	4.951 AU (7.406 × 10^8 km)
Inclination of orbit to ecliptic	1°18′29″
Average orbital velocity	13.06 km/s
Orbital period	11.867 y (4334.3 days)
Period of rotation	$9^h55^m30^s$
Inclination of equator to orbit	3°5′

Characteristics:

Equatorial diameter	142,900 km (11.20 $D_{\oplus}$)
Mass	1.899 × 10^{27} kg (317.83 $M_{\oplus}$)
Average density	1.34 g/cm^3
Gravity at base of clouds	2.54 Earth gravities
Escape velocity	61 km/s (5.4 $V_{\oplus}$)
Temperature at cloud tops	−130°C (−200°F)
Albedo	0.51
Oblateness	0.0637

Personality Point:

Jupiter is named for the Roman king of the gods, and it is the largest planet in our solar system. It can be very bright in the night sky, and its cloud belts and four largest moons can be seen through even a small telescope. Its moons are even visible with a good pair of binoculars mounted on a tripod or braced against a wall.

Ace Astronomy™ Log into AceAstronomy and select this chapter to see Astronomy Exercise "Convection and Magnetic Fields." How does convection in a planet generate a magnetic field?

Ace Astronomy™ Log into AceAstronomy and select this chapter to see Astronomy Exercise "Convection and Turbulence." Not all convection is smooth. See what happens.

Jupiter's Complex Atmosphere

Jupiter is almost entirely a liquid planet, but its outermost layer, hardly more than a skin, consists of a turbulent layer of gases and clouds. The processes you will find there are repeated in slightly different ways on the other Jovian worlds.

Study **Jupiter's Atmosphere** on pages 426 and 427 and notice four important ideas. First, the atmosphere is hydrogen rich, and the clouds are composed of hydrogen-rich molecules including ammonia (NH_3), ammonium hydrosulfide (NH_4SH), and water (H_20). Notice how the cloud layers lie at certain temperatures within the atmosphere. Also, notice how the belt–zone circulation is related to the high- and low-pressure areas you find on Earth. Finally, notice that the major spots on Jupiter, although they are only circulating storms, can remain stable for decades or even centuries.

Astronomers around the world watched the atmosphere of Jupiter in the summer of 1994 as fragments of a comet slammed into the planet at high velocity. Over a period of six days, 20-some objects 0.5 km or so in diameter hit with the energy of millions of megatons of TNT, each impact creating a fireball of hot gases and leaving behind dark smudges that remained visible for months afterward (■ Figure 18-3). From models of the complex physics of the impacts, two things are clear. First, the violent impacts had little effect on the long-term circulation patterns in Jupiter's

■ Figure 18-3

In 1994, fragments of Comet Shoemaker-Levy 9 struck Jupiter. Although these were the first comet impacts ever seen by Earth's inhabitants, such events are probably common occurrences in planetary systems. (Composite and visual images: NASA; IR sequence: Mike Skrutskie; IR image: University of Hawaii)

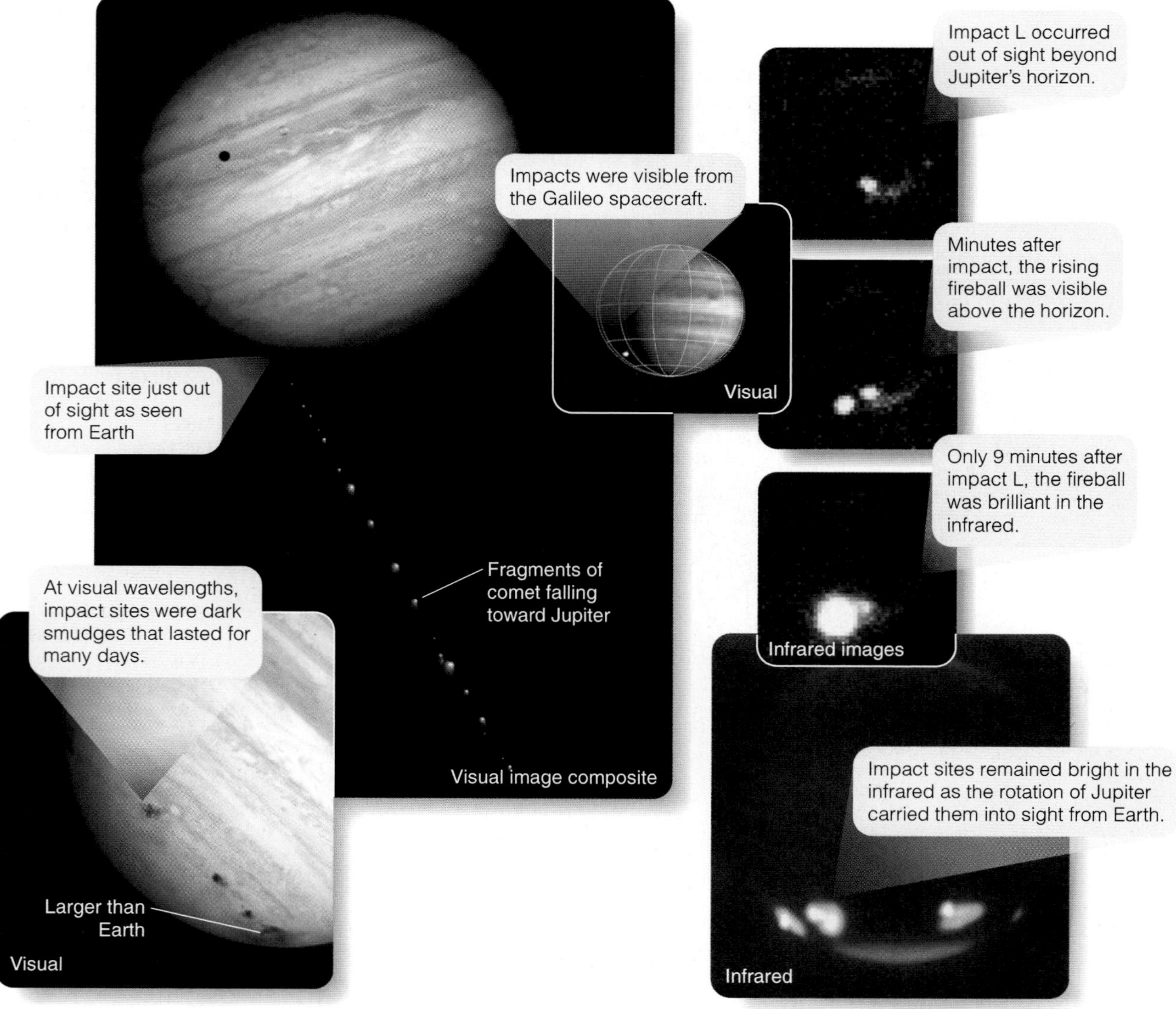
Impacts were visible from the Galileo spacecraft.
Visual
Impact L occurred out of sight beyond Jupiter's horizon.
Minutes after impact, the rising fireball was visible above the horizon.
Only 9 minutes after impact L, the fireball was brilliant in the infrared.
Infrared images
Impact site just out of sight as seen from Earth
Fragments of comet falling toward Jupiter
Visual image composite
At visual wavelengths, impact sites were dark smudges that lasted for many days.
Larger than Earth
Visual
Impact sites remained bright in the infrared as the rotation of Jupiter carried them into sight from Earth.
Infrared

atmosphere. And second, the fragments of the comet were very fragile and didn't penetrate down to the cloud layers before they exploded. In the long run, such impacts on Jupiter probably occur roughly once every century or two. Major impacts on Earth occur less often because Earth is smaller, but they are inevitable. All of the planets in the solar system are occasionally struck by these objects.

One of the interesting features of Jupiter does not lie within its cloudy atmosphere but orbits above its equator. Jupiter, like the other three Jovian planets, has a ring.

Jupiter's Ring

Astronomers have known for centuries that Saturn has rings, but Jupiter's ring was not discovered until 1979, when the Voyager 1 spacecraft sent back photos. Less than 1 percent as bright as Saturn's rings, the ghostly ring around Jupiter is a puzzle. What is it made of? Why is it there? A few simple observations help you understand.

Saturn's rings are made of bright ice chunks, but the particles in Jupiter's ring are very dark and reddish. You can guess that the ring is rocky rather than icy.

Astronomers conclude that the ring particles are mostly microscopic. Photos of the ring show that it is very bright when illuminated from behind (■ Figure 18-4a)—it is scattering light forward. Efficient **forward scattering** occurs when particles have diameters roughly the same as the wavelength of light, a few millionths of a meter. Large particles do not scatter light forward, so a ring filled with basketball-size particles would look dark when illuminated from behind. Forward scattering tells you that the ring is made of particles about the size of those in cigarette smoke.

The location of these particles near the planet is understandable. They orbit inside the **Roche limit,** the distance from a planet within which a moon cannot hold itself together by its own gravity. If a moon stays far from its planet, then the moon's gravity will be much greater than the tidal forces caused by the planet, and the moon will be able to hold itself together. If, however, the planet's moon comes inside the Roche limit, the tidal forces overcome its gravity and pull the moon apart. The International Space Station can orbit inside Earth's Roche limit because it is held together by bolts and welds, but a moon held together by its own gravity cannot survive inside the Roche limit. If a planet and its moon have similar densities, the Roche limit is 2.44 planetary radii. Jupiter's rings (and those of Saturn, Uranus, and Neptune) lie inside this limit.

Now you can understand the dust near Jupiter. If a dust speck gets knocked loose from a larger rock inside the Roche limit, the rock's gravity cannot hold the dust speck. And the billions of dust specks in the ring can't pull themselves together to make a new moon because of tidal forces inside the Roche limit.

You can be sure that the ring particles are not old. The pressure of sunlight and the powerful magnetic field alter the orbits of the particles, and they gradually spiral into the planet. Images show faint ring material extending down toward the cloud tops; this is evidently dust specks spiraling into the planet. Dust is also lost from the ring as electromagnetic effects force it out of the plane of the ring to form a low-density halo above and below the ring (Figure 18-4b). Another reason the ring particles can't be old is that the intense radiation around Jupiter will grind the dust specks down to nothing in a century or so. Thus the

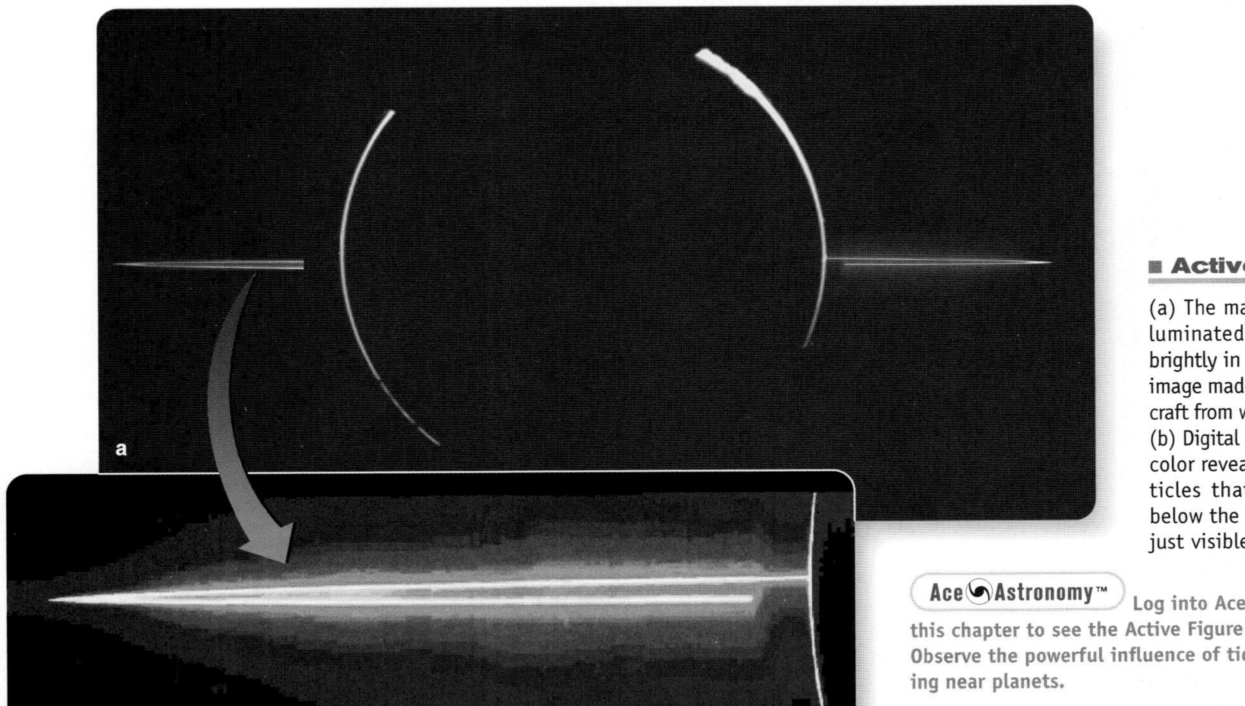

■ Active Figure 18-4

(a) The main ring of Jupiter, illuminated from behind, glows brightly in this visual-wavelength image made by the Galileo spacecraft from within Jupiter's shadow. (b) Digital enhancement and false color reveal the halo of ring particles that extends above and below the main ring. The halo is just visible in part a. (NASA)

AceAstronomy™ Log into AceAstronomy and select this chapter to see the Active Figure called "Roche Limit." Observe the powerful influence of tides on objects passing near planets.

Jupiter's Atmosphere

You probably won't ever visit Jupiter's atmosphere. Its cloud layers are deathly cold, and the deeper layers that are warmer have a crushingly high pressure. There is no free oxygen to breathe; the gases are roughly three-quarters hydrogen and a quarter helium, plus traces of water vapor, methane, ammonia, and similar molecules. Traces of sulfur and molecules containing sulfur probably make it smell bad. Of course, Jupiter has no surface, so there isn't even a place to stand. Jupiter is a nice planet to look at, but it's not a place to visit.

Belts are dark bands of clouds.

Zones are bright bands of clouds.

Shadow of Europa

Jupiter's moon Europa

NASA/JPL/Univ. of AZ

The only spacecraft to enter Jupiter's atmosphere was the Galileo probe. Released from the Galileo spacecraft, the probe entered Jupiter's atmosphere in December 1995. It parachuted through the upper atmosphere of clear hydrogen, released its heat shield, and then fell through Jupiter's stormy atmosphere until it was crushed by the increasing pressure.

Lightning bolts are common in Jupiter's turbulent clouds.

Hughes Aircraft Co

Jupiter's atmosphere is a very thin layer of turbulent gas above the liquid interior. It makes up only about 1 percent of the radius of the planet.

The Great Red Spot is a giant circulating storm in one of the southern zones. It has lasted at least 300 years since astronomers first noticed it after the invention of the telescope. Smaller spots are also circulating storms.

NASA/JPL

The visible clouds on Jupiter are composed of ammonia crystals, but models predict that deeper layers of clouds contain ammonia hydrosulfide crystals, and deeper still lies a cloud layer of water droplets. These compounds are normally white, so planetary scientists think the colors arise from small amounts of other molecules formed by lightning or by sunlight.

If you could put thermometers in Jupiter's atmosphere at difference levels, you would discover that the temperature rises below the uppermost clouds.

Far below the clouds, the temperature and pressure climb so high the gaseous atmosphere merges gradually with the liquid hydrogen interior and there is no surface.

Temperature (°F)

−300 −200 −100 0 100 212

200

100

0

−100

−200

Altitude (km)

Clear hydrogen atmosphere

Ammonia

Ammonia hydrosulfide

Water

To liquid interior

100 200 300 40

Temperature (K)

Log into Ace Astronomy and select this chapter to see the Active Figur "Planetary Atmospheres." N temperatures at w cloud laye

H L H H H L H L H L H L H

On Earth, the temperature difference between the poles and equator drives a wave shaped high-speed wind that organizes the high- and low-pressure areas into cyclonic circulations familiar from weather maps.

The poles and equator on Jupiter are about the same temperature, perhaps because of heat rising from the interior. Consequently, there are no wave-shaped winds, and the planet's rapid rotation stretches the high- and low-pressure areas into belts and zones that circle the planet.

On both Earth and Jupiter, winds circulate clockwise around the high-pressure areas in the northern hemisphere and counter-clockwise south of the equator.

Zones are brighter than belts because rising gas forms clouds high in the atmosphere, where sunlight is strong.

Belt

Zone

Altitude

North

Equat

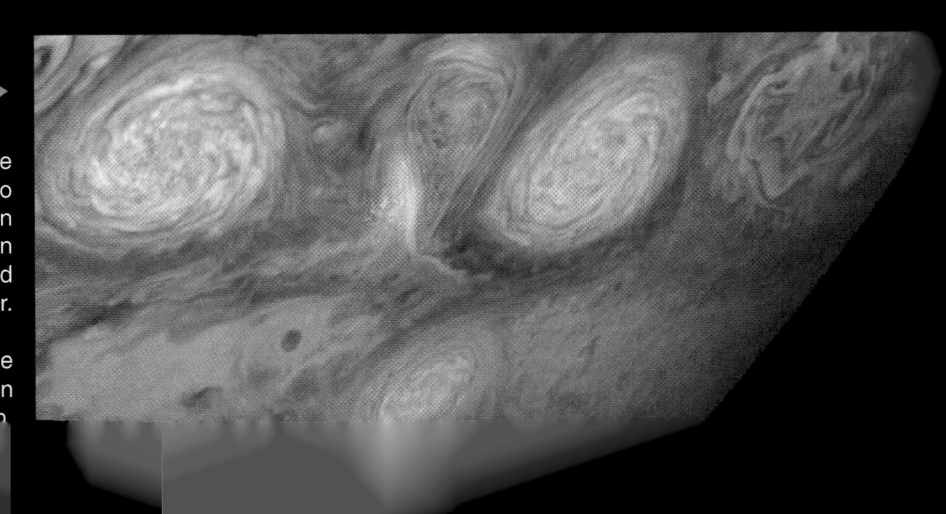

The two white ovals here are counterclockwise, high-pressure weather systems that have been visible in Jupiter's southern hemisphere since they formed in the 1930s. These two huge storms merged into one spot in February 1998. The pear-shaped circulation visible between the two storms vanished during the merger.

Features in Jupiter's atmosphere may be stable for decades or centuries, but even the Giant Red Spot may someday vanish.

■ **Figure 18-5**

The Galilean moons of Jupiter from left to right are Io, Europa, Ganymede, and Callisto. The circle shows the size of Earth's moon. (NASA)

rings you see today can't be material left over from the formation of Jupiter.

The rings of Jupiter must be continuously resupplied with new dust. Small moons that orbit near the outer edge of the rings lose dust particles as they are hit by micrometeorite impacts. Observations made by the Galileo spacecraft show that the main ring is densest at its outer edge where the small moon Adrastea orbits and that another small moon, Metis, orbits inside the ring. Galileo images reveal much fainter rings called the **gossamer rings,** extending twice as far from the planet as the main ring. These gossamer rings are most dense at the orbits of two small moons, Amalthea and Thebe, again providing evidence that the dust is being blasted into space by impacts on the inner moons.

Besides supplying the rings with particles, it is possible that the moons help confine the ring particles and keep them from spreading outward. This is an important process in planetary rings, and you will explore it in detail when we study the rings of Saturn later in this chapter.

Jupiter's Family of Moons

Jupiter has four large moons and at least two dozen smaller moons. Larger telescopes and modern techniques are rapidly finding more small moons orbiting the Jovian planets, and most of the small moons are probably captured asteroids.

In contrast, the four largest moons (■ Figure 18-5), called Galilean moons after their discoverer, Galileo, are clearly related to each other and probably formed with Jupiter. Astronomers know these worlds surprisingly well because the two Voyager spacecraft and the Galileo spacecraft have studied them in detail.

The outermost Galilean moon, Callisto, is half again as big as Earth's moon, but it has a low density of only 1.79 g/cm^3. This means it must consist roughly of 50:50 rock and ice. Observations of its gravitational field by the Galileo spacecraft reveal that it does have a dense core and a lower-density exterior, so it presumably differentiated. Also, it interacts with Jupiter's magnetic field in such a way that astronomers suspect it has a mineral-rich ocean of liquid water 100 km below its icy crust. Photos of its surface, however, show thousands of impact craters with little sign of geological activity (■ Figure 18-6).

Next inward from Callisto is Ganymede, the largest moon in the solar system. With a density of 1.9 g/cm^3, Ganymede must contain a mix of rock and ice. The Galileo spacecraft has detected a weak magnetic field, which suggests a differentiated metallic core. The icy crust is marked by old, cratered, dark areas, and younger, brighter regions of **grooved terrain** believed to be systems of faults in the brittle crust (■ Figure 18-7). Sets of grooves overlap other sets of grooves, which suggests extended episodes of geological activity.

The density of the next moon inward, Europa, is 3 g/cm^3, high enough to suggest the moon is mostly rock with a thin icy crust. The visible surface is very clean ice, contains very few craters, and has long scars suggestive of cracks in the icy crust (■ Figure 18-8). In other places, mountainlike folds cross the surface. The pattern

■ **Figure 18-6**

The dark surface of Callisto is dirty ice marked by craters in these visual-wavelength images. The youngest craters look bright because they have dug down to cleaner ice. Valhalla is the 4000-km-diameter scar of a giant impact feature, one of the largest in the solar system. Valhalla is so large and old that the icy crust has flowed back to partially heal itself, and the outer rings of Valhalla are shallow troughs marking fractures in the crust. (NASA)

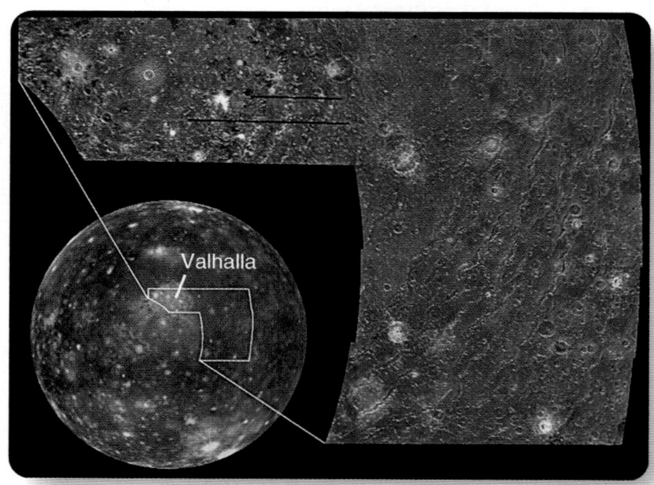

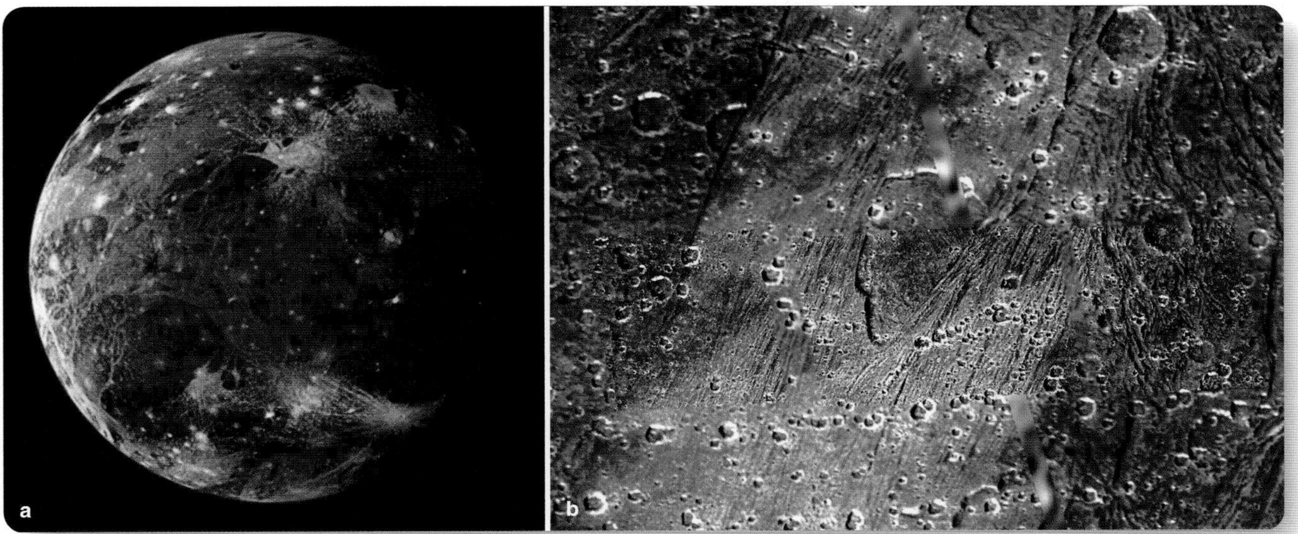

■ Figure 18-7

(a) This color-enhanced visual-wavelength image of Ganymede shows the frosty poles at top and bottom, the old dark terrain, and the brighter grooved terrain. (b) A band of bright terrain runs from lower left to upper right, and a collapsed area, a possible caldera, lies at the center in this visual-wavelength image. Caldera form where subsurface liquid has drained away, and the bright areas do contain other features associated with flooding by water. (NASA)

of folds and cracks suggests that the icy crust breaks as the moon is flexed by tides. This evidence and Europa's gravitational influence on the Galileo spacecraft reveals that a liquid-water ocean perhaps 200 km deep lies below the 100-km-thick crust. One wonders what might swim through such a lightless ocean. The lack of craters tells you Europa is an active world where craters are quickly erased.

Images from spacecraft reveal that Io, the innermost of the four Galilean moons, has over 100 volcanoes active on its surface (■ Figure 18-9). The active volcanoes throw sulfur-rich gas and ash high above the surface; the ash falls back to bury the surface at a rate of a few millimeters a year. This explains why you see no impact craters on Io—they are covered up as fast as they form. Io's density is 3.55 g/cm^3, showing that it is not ice but rather rock and metal. Its gravitational influence on the passing Galileo spacecraft shows that it has a large metallic core, a rocky mantle, and a low-density crust.

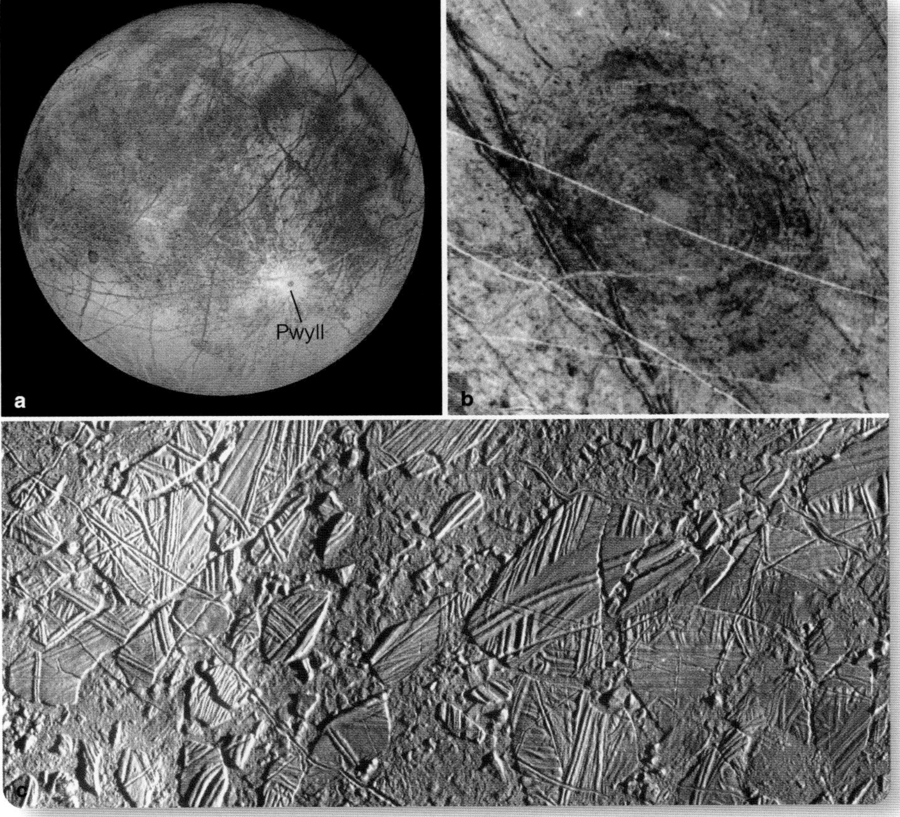

■ Figure 18-8

(a) The icy surface of Europa is shown here in natural color. Many faults are visible on its surface, but very few craters. The bright crater is Pwyll, a young impact feature. (b) This circular bull's-eye is 140 km in diameter. It is the remains of an impact by an object about the size of a mountain. Notice the younger cracks and faults that cross the older impact feature. (c) Like icebergs on an arctic ocean, blocks of crust on Europa appear to have floated apart. The blue icy surface is stained brown by mineral-rich water venting from below the crust. White areas are ejecta from the impact that formed Pwyll. (NASA)

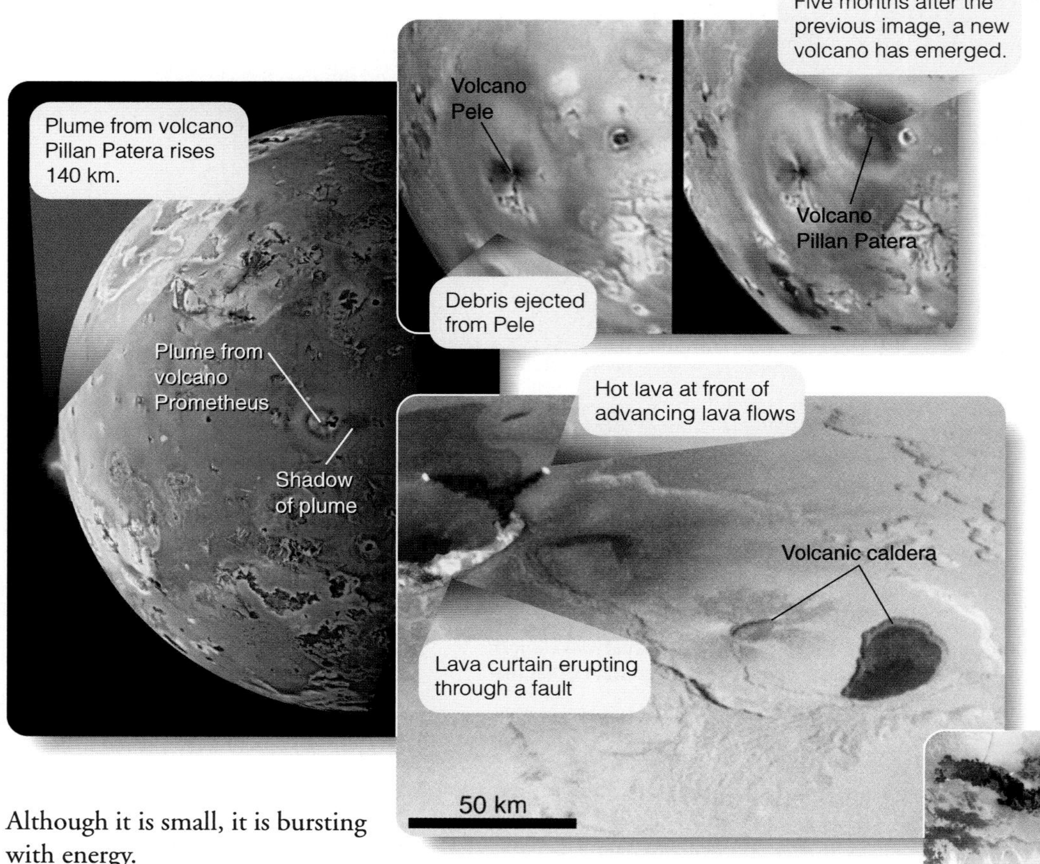

■ Figure 18-9

These color images of volcanic features on Io were produced by combining visual and near-infrared images and digitally enhancing the color. To human eyes, most of Io would look pale yellow and light orange. (NASA)

Although it is small, it is bursting with energy.

The activity you see in the Galilean moons must be driven by energy flowing outward. The violently active volcanism of Io is apparently caused by **tidal heating.** Io is too small to have remained hot from its formation, but its orbit is slightly elliptical; as it moves closer to and then farther from Jupiter, the planetary gravitational field flexes the moon with tides, and friction heats its interior. That heat flowing outward causes the volcanism. Europa is not as active as Io, but it too must have a heat source, presumably tidal heating. Ganymede is no longer active, but when it was younger it had internal heat that broke the crust to produce the grooved terrain.

In fact, the three moons are linked together in an orbital resonance. Io orbits Jupiter four times while Europa orbits twice and Ganymede orbits once. This resonance keeps Io's orbit slightly elliptical and drives the tidal heating that makes it active. Distant Callisto is not caught in this orbital resonance and appears never to have been strongly active.

A History of Jupiter

Can you put all of the evidence together and tell the story of Jupiter? Creating such a logical argument of evidence and hypotheses is the ultimate goal of planetary astronomy.

Jupiter formed far enough from the sun to incorporate large numbers of icy planetesimals, and it must have grown rapidly. Once it was a dozen times more massive than Earth, it could grow by gravitational collapse (Chapter 16), the capture of gas directly from the solar nebula. Thus, it grew rich in hydrogen and helium from the solar nebula. Its present composition is quite sunlike and resembles the composition of the solar nebula. The location of Jupiter's point in Figure 17-15 shows that its gravity is strong enough to hold on to all of its gases.

The large family of moons may be mostly captured asteroids, and Jupiter may still encounter a wandering asteroid or comet now and then. Some of these are deflected, some captured into orbit, and some, like the comet that struck Jupiter in 1994, actually fall into the planet. Dust blasted off of the inner moons by micrometeorites settles into the equatorial plane to form Jupiter's rings.

The four Galilean moons are large and seem to have formed in a disk of gas and dust around the forming planet. The innermost, Io, is densest, and the densities of the others decrease as you move away from Jupiter, which is similar to the way the densities of the planets fall as you move away from the sun. Perhaps the inner moons incorporated less ice because they formed closer to the heat of the growing planet. Nevertheless, you must recog-

Who Pays for Science?

Science is an expensive enterprise, and that raises the question of payment. Some science has direct technological applications, but some basic science is of no immediate practical value. Who pays the bill?

In Window on Science 18-1, you saw that technology is of immediate practical use; thus, business funds much of this kind of scientific research. Auto manufacturers need inexpensive, durable, quick-drying paint for their cars, and they find it worth the cost to hire chemists to study the way paint dries. Many industries have large research budgets, and some industries, such as pharmaceutical manufacturers, depend exclusively on scientific research to discover and develop new products.

If a field of research has immediate potential to help society, it is likely that government will supply funds. Much of the research on public health is funded by government institutions such as the National Institutes of Health and the National Science Foundation. The practical benefit of finding new ways to prevent disease, for example, is well worth the tax dollars.

Basic science, however, has no immediate practical use. That doesn't mean it is useless, but it does mean that the practical-minded stockholders of a company will not approve major investments in such research. Digging up dinosaur bones, for instance, is very poorly funded because no industry can make a profit from the discovery of a new dinosaur. Astronomy is another field of science that has few direct applications, and thus very little astronomical research is funded by industry.

The value of basic research is twofold. Discoveries that have no known practical use today may be critically important years from now. Thus, society needs to continue basic research to protect its own future. But basic research, such as studying Saturn's rings or digging up dinosaur bones, is also of cultural value because it tells us what we are. Each of us benefits in intangible ways from such research, and thus society needs to fund basic research for the same reason it funds art galleries and national parks—to make our lives richer and more fulfilling.

Because there is no immediate financial return from this kind of research, it falls to government institutions and private foundations to pay the bill. The Keck Foundation has built two giant telescopes with no expectation of financial return, and the National Science Foundation has funded thousands of astronomy research projects for the benefit of society. Debates rage as to how much money is enough and how much is too much, but, ultimately, funding basic scientific research is a public responsibility that society must balance against other needs. There isn't anyone else to pick up the tab.

Sending the Cassini spacecraft to Saturn cost each American $0.56 per year over the life of the project. (NASA)

nize that tidal heating has been important, and the intense heating of the inner moons could have driven off much of their ices. Thus two processes may be responsible for the compositions of the Galilean moons.

Inquire | Review | Analyze

Why is Jupiter so big?

You can analyze this question by constructing a logical argument that relates the formation of Jupiter to the solar nebula theory. Jupiter is rich in hydrogen and helium, but Earth is relatively poor in these elements. While the solar nebula existed, Earth grew by the accretion of solid, rocky planetesimals, but it never became massive enough to capture gas directly from the solar nebula. That is, it never grew by gravitational collapse. Jupiter, however, grew so rapidly from icy planetesimals in the outer solar nebula that it was eventually able to grow by gravitational collapse. By the time the solar nebula cleared away and ended planet building, Jupiter had captured large amounts of hydrogen and helium and was quite massive.

Often the present nature of a world can be traced back to the way it formed. Can you create a logical argument to explain the nature of the Galilean moons?

■ ■ ■

Connections: You have studied Jupiter in detail because it is the basis of comparison for the other Jovian worlds. Now you can turn your attention to the most beautiful Jovian world, Saturn.

18-3 Saturn

SATURN HAS PLAYED SECOND FIDDLE to its own rings since Galileo first viewed it through a telescope in 1610. The rings are dramatic, strikingly beautiful, and easily seen through even a small telescope; but Saturn itself, only slightly smaller than Jupiter, is a fascinating planet.

Although Saturn lies roughly 10 AU from the sun, astronomers know a surprising amount about it. The beautiful rings are easily visible through the telescopes of modern amateur astronomers, and large Earth-based telescopes have explored the planet's atmosphere, rings, and moons. The two Voyager spacecraft flew past Saturn in 1979, transmitting back to Earth detailed measurements and images, and the Cassini spacecraft went into orbit around Saturn in 2004 on an extended exploration of the planet, its rings, and its moons (**Window on Science 18-2**).

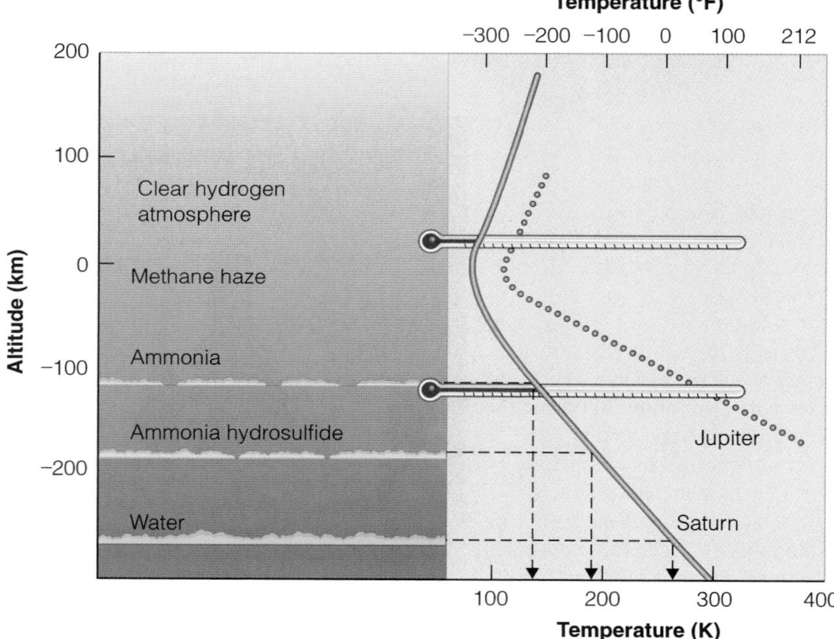

■ Active Figure 18-10

Because Saturn is farther from the sun, its atmosphere is colder than Jupiter's (dotted line). The cloud layers on Saturn form at the same temperature as do the cloud layers on Jupiter, but that puts them deeper in Saturn's hazy atmosphere, where they are not as easy to see as the clouds on Jupiter. (See page 427.)

AceAstronomy™ Log into AceAstronomy and select this chapter to see the Active Figure called "Planetary Atmospheres." See how temperature varies with height.

Saturn the Planet

Seen from Earth, Saturn shows only faint evidence of belt–zone circulation, but images from Voyager, Cassini, and the Hubble Space Telescope show that belts and zones are present and that the associated winds blow up to three times faster than on Jupiter. The belts and zones on Saturn are less visible because they occur deeper in the cold atmosphere below a layer of methane haze (■ Figure 18-10).

Saturn is less dense than water (it would float), and that suggests that it is, like Jupiter, rich in hydrogen and helium. In fact, its density is so low that it must have a relatively small core of heavy elements (**Celestial Profile 8**).

The shape of a Jovian planet can tell you about the interior. All of the Jovian planets, being mostly liquid and rotating rapidly, are slightly flattened. A planet's **oblateness** is the fraction by which its equatorial diameter exceeds its polar diameter. As photographs show, Saturn is the most oblate of the planets, and that evidence tells you that it is mostly liquid. A world with a large rocky core and mantle would not be flattened much by rotation, but an all-liquid planet would flatten significantly. Thus the oblateness of a Jovian planet, combined with its average density, can help astronomers model the interior.

Models of Saturn predict that it must have a small core of heavy elements and less liquid metallic hydrogen than Jupiter because Saturn's internal pressure is lower. Perhaps this is why

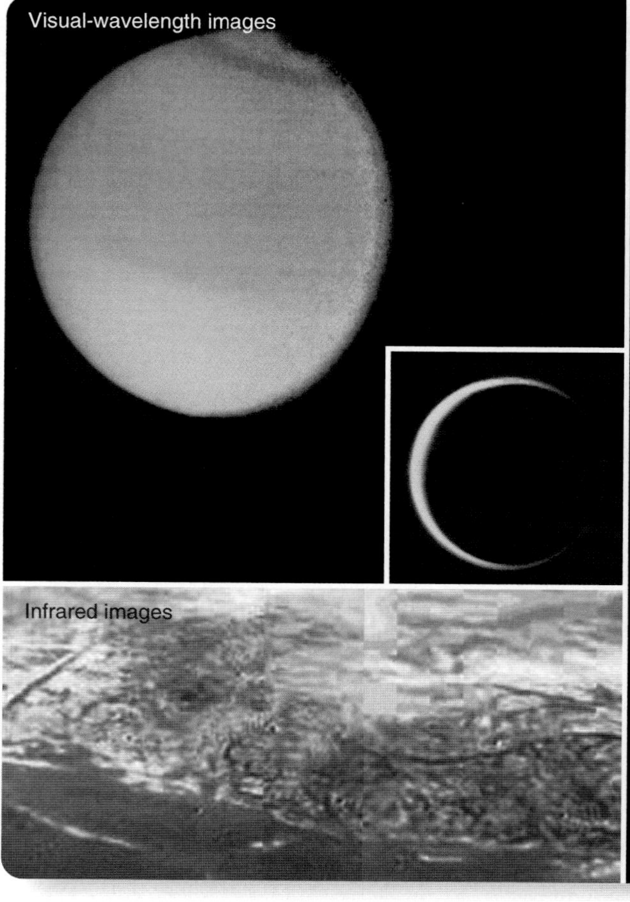

■ Figure 18-11

Titan's smoggy atmosphere has a temperature of −180°C (−290°F). The Huygens probe descended on January 14, 2005, and sent back images from high above the surface showing water and methane ice with drainage channels leading into dark lowlands, possibly filled with organic liquids. Once down, the probe radioed back this true-color image of pebble-sized chunks of ices under the glare of an orange sky. (ESA/NASA/JPL/University of Arizona)

Saturn's magnetic field is 20 times weaker than Jupiter's. Like Jupiter, Saturn has a hot interior and radiates more energy than it receives from the sun.

Saturn may look a bit bland, but it makes up for that in the splendor of its rings.

Saturn's Rings

The beautiful rings of Saturn are formed from billions of ice particles orbiting the planet in the plane of its equator. In addition to being beautiful, the rings are fascinating for what they can tell us about the worlds of the outer solar system.

Study **The Ice Rings of Saturn** on pages 434 and 435 and notice four things. First, the rings are made up of billions of dirty ice particles. Second, notice that the rings can't be as old as Saturn. The rings must be replenished now and then by impacts on Saturn's icy moons. Third, where ring particles orbit with periods that are simply related to the orbital period of one of Saturn's inner moons, a resonance can create a gap in the rings. The fourth thing to notice is a related phenomenon: Small moons near the rings can shepherd the ring particles and confine them. In these ways, the rings of Saturn, and the rings of the other Jovian moons, are created by and controlled by the planet's moons. Without the moons, there would be no rings.

Observations made by the Cassini spacecraft show that the ring particles have compositions that resemble that of Saturn's distant icy moon Phoebe. A large impact on Phoebe may be part of the complex history of Saturn's rings.

The Moons of Saturn

Saturn has over 30 known moons, many of which are small and all of which contain mixtures of ice and rock. Most of these moons were discovered from Earth, but astronomers know them well because the two Voyager spacecraft flew past Saturn in 1980 and 1981 and because the Cassini spacecraft went into orbit around Saturn in 2004 to begin an extended study of the planet, its rings, and its moons.

The largest of Saturn's moons is Titan, a few percent larger than the planet Mercury. Its density suggests that it must contain a rocky core under a thick mantle of ices. Unlike Mercury, Titan has an atmosphere, because it is so cold its gas molecules do not travel fast enough to escape (see Figure 17-15). The atmosphere is mostly nitrogen with traces of argon and methane, and it contains thick clouds that hide its surface at visible wavelengths. The Cassini spacecraft imaged Titan in the infrared and released the Huygens probe that parachuted to the surface (■ Figure 18-11).

Models of the atmosphere suggest that the clouds are formed by sunlight reacting with methane to produce particles of organic molecules that could fall to the surface. (Note that an organic molecule is one that has a carbon backbone; it does not have to originate from living things.) Models predict that the surface ices are rich in solid methane and that methane and ethane are stable as liquids and could form lakes or ponds.

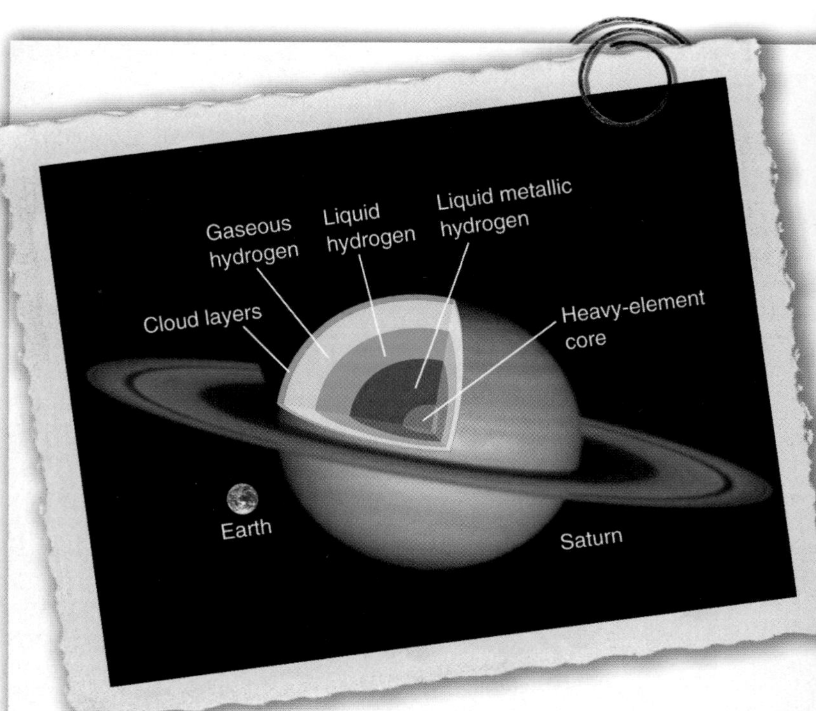

Density, oblateness, and gravity measurements made by planetary probes allow planetary astronomers to model Saturn's interior. (NASA/STScI)

Celestial Profile 8: Saturn

Motion:

Average distance from the sun	9.5388 AU (14.27 × 10^8 km)
Eccentricity of orbit	0.0560
Maximum distance from the sun	10.07 AU (15.07 × 10^8 km)
Minimum distance from the sun	9.005 AU (13.47 × 10^8 km)
Inclination of orbit to ecliptic	2°29′17″
Average orbital velocity	9.64 km/s
Orbital period	29.461 y (10,760 days)
Period of rotation	$10^h39^m25^s$
Inclination of equator to orbit	26°24′

Characteristics:

Equatorial diameter	129,660 km (9.42 $D_\oplus$)
Mass	5.69 × 10^{26} kg (95.147 $M_\oplus$)
Average density	0.69 g/cm^3
Gravity at base of clouds	1.16 Earth gravities
Escape velocity	35.6 km/s (3.2 $V_\oplus$)
Temperature at cloud tops	−180°C (−292°F)
Albedo	0.61
Oblateness	0.102

Personality Point:

The Greek god Cronus was forced to flee when his son Zeus took power. Cronus fled to Italy where the Romans called him Saturn, protector of the sowing of seed. He was celebrated in a week-long Saturnalia at the time of the winter solstice in late December. Early Christians took over the holiday to celebrate Christmas.

The Ice Rings of Saturn

The brilliant rings of Saturn are made up of billions of ice particles ranging from microscopic specks to chunks bigger than a house. Each particle orbits Saturn in its own circular orbit. Much of what astronomers know about the rings was learned when the Voyager 1 spacecraft flew past Saturn in 1980, followed by the Voyager 2 spacecraft in 1981. The Cassini Spacecraft reached orbit around Saturn in 2004. From Earth, astronomers see three rings labeled A, B, and C. Voyager and Cassini images reveal over a thousand ringlets within the rings.

Saturn's rings can't be leftover material from the formation of Saturn. The rings are made of ice particles, and the planet would have been so hot when it formed that it would have vaporized and driven away any icy material. Rather, the rings must be debris from collisions between passing comets and Saturn's icy moons. Such impacts should occur every 10 million years or so, and they would scatter ice throughout Saturn's system of moons. The ice would quickly settle into the equatorial plane, and some would become trapped in rings. Although the ice may waste away due to meteorite impacts and damage from radiation in Saturn's magnetosphere, new impacts could replenish the rings with fresh ice. The bright, beautiful rings you see today may be only a temporary enhancement caused by an impact that occurred since the extinction of the dinosaurs.

Encke's division

Cassini's division

A ring

B ring

C ring
The Crepe ring

Earth to scale

As in the case of Jupiter's ring, Saturn's rings lie inside the planet's Roché limit where the ring particles cannot pull themselves together to form a moon.

Because it is so dark, the C ring has been called the Crepe ring, referring to the black, semitransparent cloth associated with funerals.

Visual-wavelength image

NASA

An astronaut could swim through the rings. Although the particles orbit Saturn at high velocity, all particles at the same distance from the planet orbit at about the same speed, so they collide gently at low velocities. If you could visit the rings, you could push your way from one icy particle to the next. This artwork is based on a model of particle sizes in the A ring.

The C ring contains boulder-size chunks of ice, whereas most particles in the A and B ring are more like golf balls, down to dust-size ice crystals. Further, C ring particles are less than half as bright as particles in the A and B rings. Cassini observations show that the C ring particles contain less ice and more minerals.

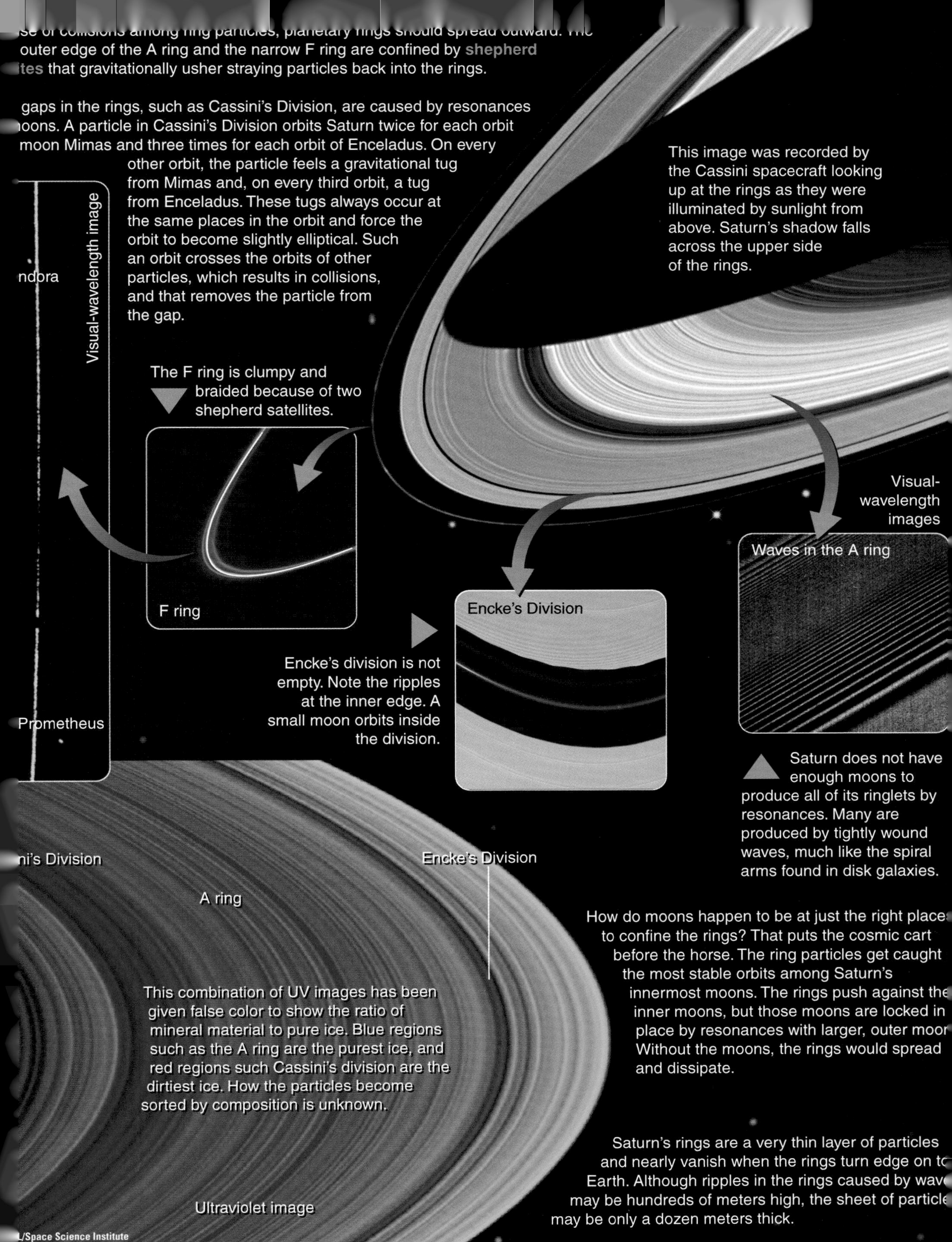
outer edge of the A ring and the narrow F ring are confined by shepherd
tes that gravitationally usher straying particles back into the rings.
gaps in the rings, such as Cassini's Division, are caused by resonances
oons. A particle in Cassini's Division orbits Saturn twice for each orbit
moon Mimas and three times for each orbit of Enceladus. On every
other orbit, the particle feels a gravitational tug from Mimas and, on every third orbit, a tug from Enceladus. These tugs always occur at the same places in the orbit and force the orbit to become slightly elliptical. Such an orbit crosses the orbits of other particles, which results in collisions, and that removes the particle from the gap.
This image was recorded by the Cassini spacecraft looking up at the rings as they were illuminated by sunlight from above. Saturn's shadow falls across the upper side of the rings.
Visual-wavelength image
ndora
The F ring is clumpy and braided because of two shepherd satellites.
F ring
Visual-wavelength images
Waves in the A ring
Encke's Division
Encke's division is not empty. Note the ripples at the inner edge. A small moon orbits inside the division.
Prometheus
Saturn does not have enough moons to produce all of its ringlets by resonances. Many are produced by tightly wound waves, much like the spiral arms found in disk galaxies.
ni's Division
Encke's Division
A ring
How do moons happen to be at just the right places to confine the rings? That puts the cosmic cart before the horse. The ring particles get caught the most stable orbits among Saturn's innermost moons. The rings push against the inner moons, but those moons are locked in place by resonances with larger, outer moon Without the moons, the rings would spread and dissipate.
This combination of UV images has been given false color to show the ratio of mineral material to pure ice. Blue regions such as the A ring are the purest ice, and red regions such Cassini's division are the dirtiest ice. How the particles become sorted by composition is unknown.
Saturn's rings are a very thin layer of particles and nearly vanish when the rings turn edge on to Earth. Although ripples in the rings caused by wave may be hundreds of meters high, the sheet of particle may be only a dozen meters thick.
Ultraviolet image
L/Space Science Institute

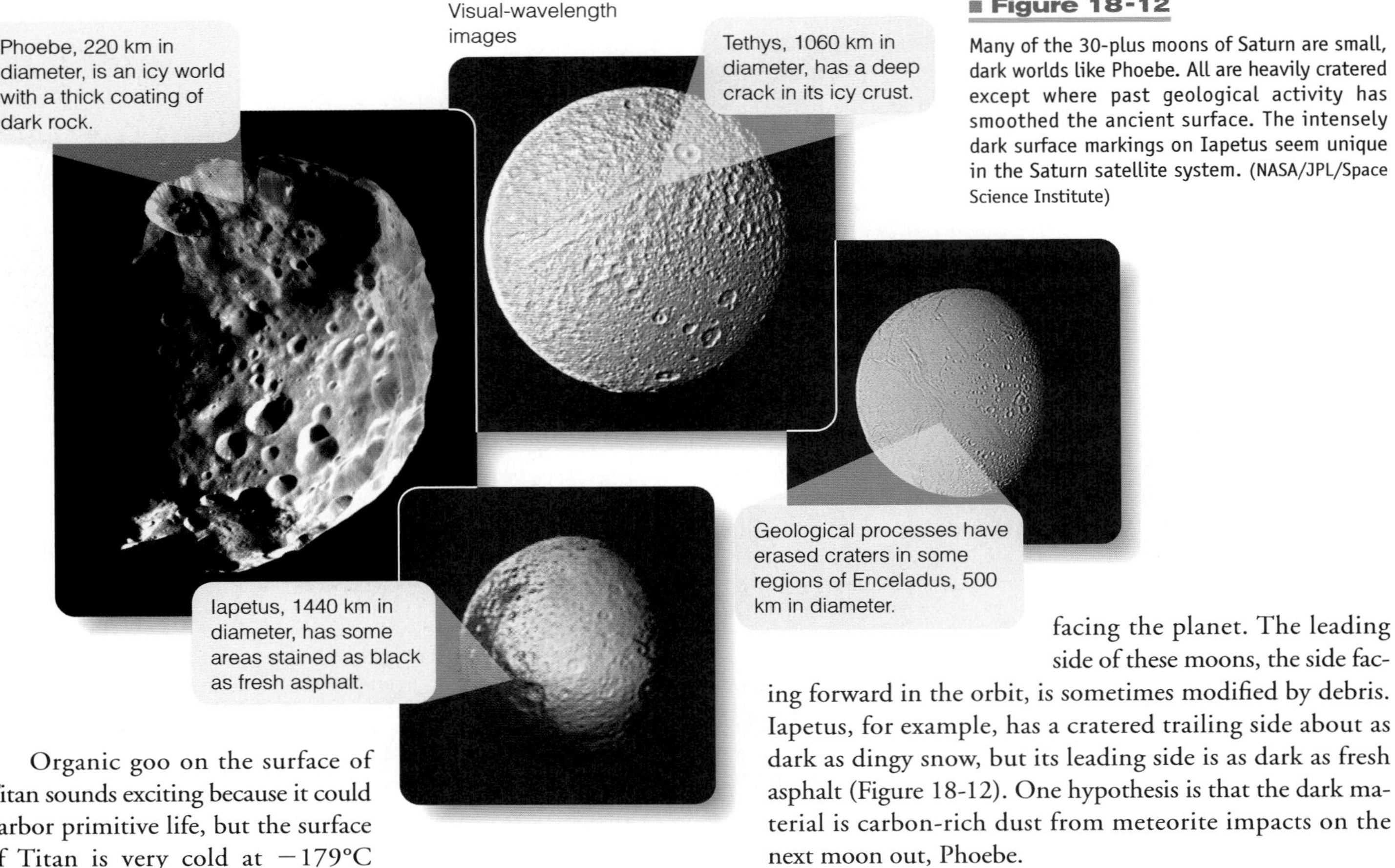

Visual-wavelength images
Phoebe, 220 km in diameter, is an icy world with a thick coating of dark rock.
Tethys, 1060 km in diameter, has a deep crack in its icy crust.
Geological processes have erased craters in some regions of Enceladus, 500 km in diameter.
Iapetus, 1440 km in diameter, has some areas stained as black as fresh asphalt.

■ **Figure 18-12**

Many of the 30-plus moons of Saturn are small, dark worlds like Phoebe. All are heavily cratered except where past geological activity has smoothed the ancient surface. The intensely dark surface markings on Iapetus seem unique in the Saturn satellite system. (NASA/JPL/Space Science Institute)

Organic goo on the surface of Titan sounds exciting because it could harbor primitive life, but the surface of Titan is very cold at −179°C (−290°F). Oddly, Titan is cooled by an inverse greenhouse effect as the organic haze blocks much of the incoming sunlight but allows infrared radiation to escape to space. Clearly, Titan is a peculiar world.

The Cassini spacecraft dropped an instrument package called the Huygens Probe into the atmosphere of Titan in January 2005. Further analysis of that data will tell astronomers more about the peculiar chemistry of Titan.

The remaining moons of Saturn are small and icy, have no atmospheres, and are heavily cratered. Most have ancient surfaces. Tethys, for example, is less than a third the diameter of Earth's moon, and its heavily cratered crust seems quite old (■ Figure 18-12). A valley 3 km deep trails three-fourths of the way around the satellite. Such cracks are found on a few other satellites and appear to have formed long ago, when the interiors of the icy moons froze and expanded.

The moon Enceladus, a bit smaller than Tethys, nevertheless shows signs of recent activity (Figure 18-12). Some parts of its surface contain 1000 times fewer craters than other regions, showing that these lightly cratered regions must be younger. At some point in its history, this moon had internal heat that produced geological activity and resurfaced some areas of its crust. The energy source is unknown, but tidal heating is likely.

Like nearly all moons in the solar system, Saturn's moons are tidally locked to their planet, rotating to keep the same side facing the planet. The leading side of these moons, the side facing forward in the orbit, is sometimes modified by debris. Iapetus, for example, has a cratered trailing side about as dark as dingy snow, but its leading side is as dark as fresh asphalt (Figure 18-12). One hypothesis is that the dark material is carbon-rich dust from meteorite impacts on the next moon out, Phoebe.

The History of Saturn

Can you put all of the evidence and hypotheses together and describe the history of Saturn? Doing so is a real test of your understanding.

Saturn formed in the outer solar nebula, where ice particles were stable and may have contained more trapped gases. The protoplanet grew rapidly and became massive enough to attract hydrogen and helium by gravitational collapse. The heavier elements sank to the middle to form a small core, and the hydrogen formed a liquid mantle containing liquid metallic hydrogen. The outward flow of heat from the interior is believed to drive convection inside the planet that produces its magnetic field. Because Saturn is smaller than Jupiter, the internal pressure is less, the planet contains less liquid metallic hyrogen, and its magnetic field is weaker.

The rings can't be primordial. That is, they can't be material left over from the formation of the planet. Such ices would have been vaporized and driven away by the heat of the protoplanet. Rather, you can suppose that the ring material is debris from occasional collisions between comets and Saturn's icy moons. Such giant impacts are not unheard of in the solar system. Jupiter was hit by a comet in 1994.

Some of Saturn's moons are probably captured asteroids that wandered too close, but some of the moons probably formed with Saturn. Many have ancient surfaces. The giant moon Titan may have formed with Saturn, or it may be a very large icy planetesimal captured into orbit around Saturn. You will see more evidence for this capture hypothesis when we explore farther from the sun.

Inquire | Review | Analyze

Why do the belts and zones on Saturn look so indistinct?
In a Jovian planet, the light-colored zones form in high-pressure regions where rising gas cools and condenses to form icy crystals of ammonia, which are visible as bright clouds. Saturn is twice as far from the sun as Jupiter, so sunlight is weaker and the atmosphere is colder. The rising gas currents don't have to rise as high to reach temperatures cold enough to form clouds. Because the clouds form deeper in the hazy atmosphere, they are not as brightly illuminated by sunlight and look dimmer. A layer of methane haze above the clouds makes the belts and zones look even less distinct.

You have used some simple physics to construct a logical argument that explains the hazy cloud features on Saturn. Can you explain why Saturn's rings have gaps and ringlets?

■ ■ ■

Connections: In any contest, Saturn would win the ribbon for most beautiful planet. But beyond Saturn's orbit are three more planets that could compete for the ribbon for most striking personality.

18-4 Uranus

NOW THAT YOU ARE FAMILIAR with the gas giants in our solar system, you will be able to appreciate how weird the ice giants, Uranus and Neptune, are. Uranus, especially, seems to have forgotten how to behave like a planet.

Uranus the Planet

Uranus is only a third the diameter of Jupiter, only a twentieth as massive, and, being four times farther from the sun, its atmosphere is over 100 degrees colder than Jupiter's (**Celestial Profile 9**).

Because Uranus is smaller than Jupiter, its internal pressure is lower, and it does not contain liquid metallic hydrogen. Models of Uranus based in part on its density and oblateness suggest that it has a small core of heavy elements and a deep mantle of partly frozen water containing rocky material and dissolved ammonia and methane. Circulation in this electrically conducting mantle may generate the planet's peculiar magnetic field, which is highly inclined to its axis of rotation. Above this mantle lies the deep hydrogen and helium of the planet's atmosphere.

Uranus rotates on its side, with its equator inclined 98° to its orbit. With an orbital period of 84 years, each of its four seasons lasts 21 years and is extreme, with the sun passing near each of its celestial poles (■ Figure 18-13). This peculiar rotation may

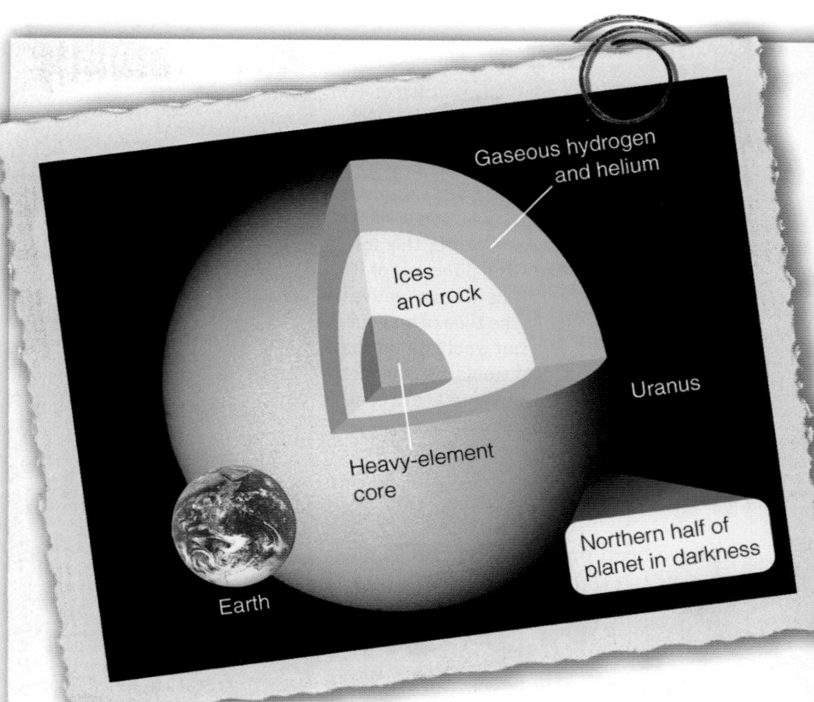

Uranus rotates on its side, and, when Voyager 2 flew past in 1986, the planet's south pole was pointed almost directly at the sun. (NASA)

Celestial Profile 9: Uranus

Motion:

Average distance from the sun	19.18 AU (28.69×10^8 km)
Eccentricity of orbit	0.0461
Maximum distance from the sun	20.1 AU (30.0×10^8 km)
Minimum distance from the sun	18.3 AU (27.4×10^8 km)
Inclination of orbit to ecliptic	0°46′23″
Average orbital velocity	6.81 km/s
Orbital period	84.013 y (30,685 days)
Period of rotation	17^h14^m
Inclination of equator to orbit	97°55′

Characteristics:

Equatorial diameter	51,118 km (4.01 $D_{\oplus}$)
Mass	8.69×10^{25} kg (14.54 $M_{\oplus}$)
Average density	1.29 g/cm^3
Gravity	0.919 Earth gravity
Escape velocity	22 km/s (1.96 $V_{\oplus}$)
Temperature above cloud tops	−220°C (−364°F)
Albedo	0.35

Personality Point:

Uranus was discovered in 1781 by the English amateur astronomer William Herschel. He named it *Georgium Sidus,* George's Star, after the English King George III. European astronomers, especially the French, refused to accept a planet named after an English king. They called it Herschel. Years later, German astronomer J. E. Bode suggested *Uranus,* the oldest of the Greek gods.

■ Figure 18-13

Uranus rotates on an axis that is tipped 97.9° from the perpendicular to its orbit, so its seasons are extreme. When one of its poles is pointed nearly at the sun (a solstice), a citizen of Uranus would see the sun near a celestial pole, and it would never rise or set. As it orbits the sun, the planet maintains the direction of its axis in space, and thus the sun moves from pole to pole. At the time of an equinox on Uranus, the sun would be on the celestial equator and would rise and set with each rotation of the planet. Compare with similar diagrams for Earth on page 29.

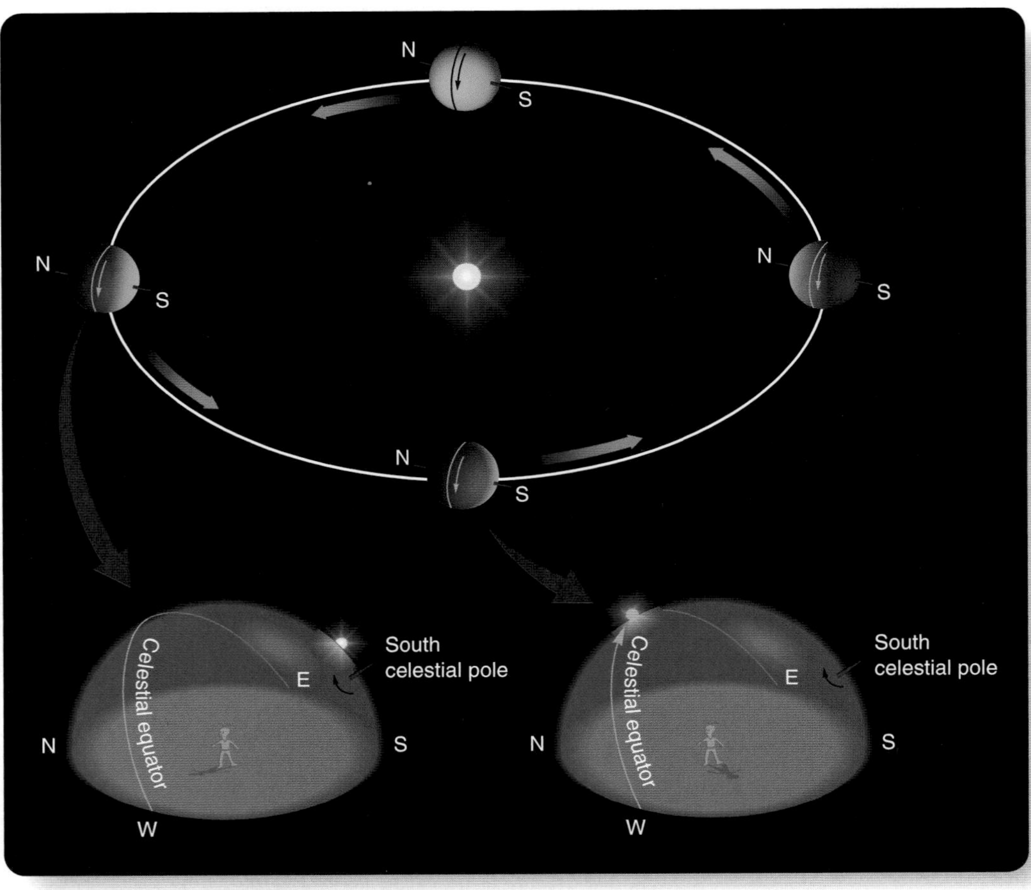

have been produced when Uranus collided with a very large planetesimal late in its formation. When Voyager 2 flew past in 1986, the planet's south pole was pointed almost directly at the sun.

Voyager 2 photos show a nearly featureless blue-green world (■ Figure 18-14a). The atmosphere is mostly hydrogen and helium (12 percent), but traces of methane absorb red light and thus make the atmosphere look blue. There is no belt–zone circulation visible in the Voyager photographs, although extreme computer enhancement does reveal a few clouds and bands around the south pole.

In the decades since Voyager 2 flew past Uranus, spring has come to the northern hemisphere of Uranus and fall to the southern hemisphere. Images made by the Hubble Space Telescope and modern Earth-based telescopes reveal changing clouds and cloud bands in both hemispheres (Figure 18-14b). Further studies may reveal the secrets of Uranus's cycle of seasons.

Laboratory experiments show that at the temperature and pressure inside the atmospheres of Uranus and Neptune, methane can decompose, and the released carbon can form crystals of diamond perhaps the size of pebbles. A continuous hailstorm of diamonds falling into a planet's interior would release energy and could help warm the planet. This process has not been observed, but it serves to warn

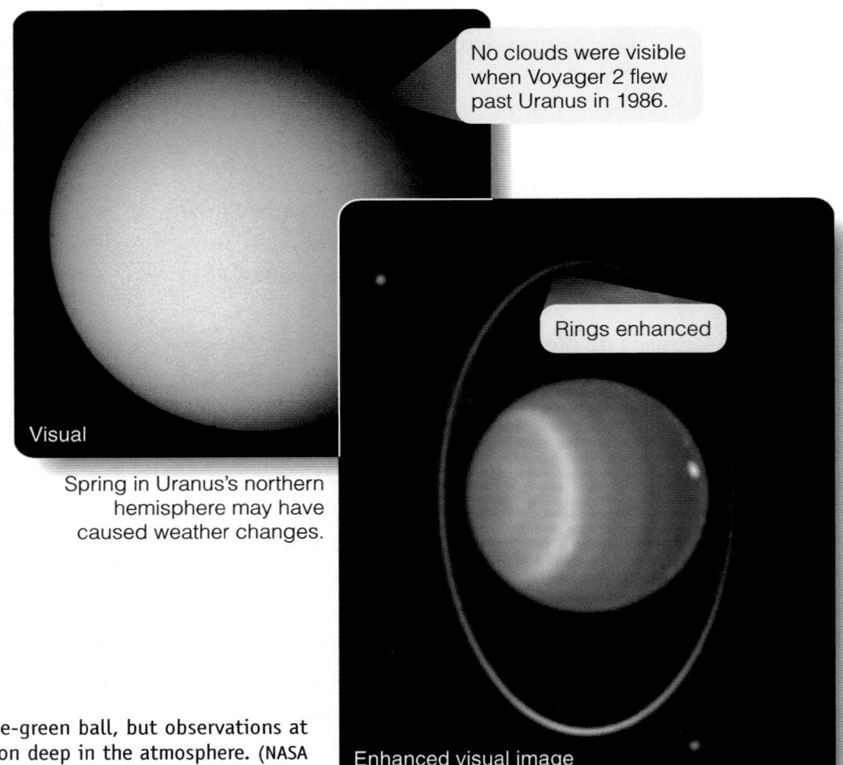

■ Figure 18-14

If you visited Uranus, it would probably look like a bland blue-green ball, but observations at certain wavelengths would reveal traces of belt–zone circulation deep in the atmosphere. (NASA and Erich Karkoschka, University of Arizona)

you that other worlds are truly un-Earthly and may harbor processes you can hardly imagine.

Observations show that Uranus is radiating about the same amount of energy that it receives from the sun. It has little heat flowing out of its interior. This may account for its limited atmospheric activity.

The Rings of Uranus

The rings of Uranus are similar to those of Neptune, so you can discuss the two ring systems together. They are dark, faint, and confined by shepherd satellites. They are not easily visible from Earth, and the first hint that these planets had rings came from **occultations,** the passage of the planet in front of a star. Nevertheless, most of what astronomers know about these ring systems comes from the observations of the Voyager 2 spacecraft.

Study **The Rings of Uranus and Neptune** on pages 440 and 441 and notice three important points. First, the particles that make up the rings are very dark and quite unlike the bright icy particles that orbit Saturn. Second, notice the importance of small moons imbedded among the rings. The moons shape and confine the rings into narrow hoops. Finally, notice how the ring particles must be occasionally resupplied by impacts on the moons.

As you read about planetary rings, notice their close relationship with moons. Because of collisions among ring particles, planetary rings tend to spread outward almost like an expanding gas. If a planet had no moons, its rings would spread out into a more and more tenuous sheet until they were gone. The spreading rings can be anchored by interacting gravitationally with small moons, which get pushed outward very slowly. Through orbital resonances, those small moons can be anchored by larger, more distant moons that are so massive they do not get pushed outward by any significant amount. In this way, a system of moons can confine and preserve a system of planetary rings.

Planetary rings are beautiful blossoms created when debris falls into a system of small moons and becomes trapped in the most stable orbits among the moons. Without moons, there could be no rings.

The Uranian Moons

Until recently, astronomers could see only five moons orbiting Uranus. Voyager 2 discovered 10 small moons in 1986, and more have been found in images recorded by new, giant telescopes on Earth. Each of the Jovian planets probably has many small, undiscovered moons.

The five major satellites were photographed by Voyager 2, and you can analyze their surface features. Oberon and Titania, the two outer moons, are about 45 percent the diameter of Earth's moon, and they have old, cratered surfaces with faults hundreds of kilometers long. Some regions of these icy moons have been resurfaced by liquid water "lava" that covered old craters and froze. In contrast, Umbrial, the next moon inward, is about a third the diameter of Earth's moon and shows no sign of activity on its ancient, cratered surface. Ariel, the fourth moon inward, is slightly smaller than Umbrial, but its cratered crust is marked by broad, smooth-floored valleys that may have been cut by flowing ice (■ Figure 18-15a).

Miranda, the innermost moon, is only 14 percent the diameter of Earth's moon, but its surface is marked by oval patterns of grooves called **ovoids** (Figure 18-15b). These are believed to have been caused by internal heat that produced convection in

■ Figure 18-15

Geological activity on moons of Uranus. (a) Ariel has an old cratered surface, but some regions are marked by broad, shallow valleys. These are evidence of past activity. (b) The face of Miranda is marred by ovoids, which are believed to have formed when internal heating caused slow convection in the ice of the moon's mantle. Note the 5-km-high cliff at the lower edge of the moon. (NASA)

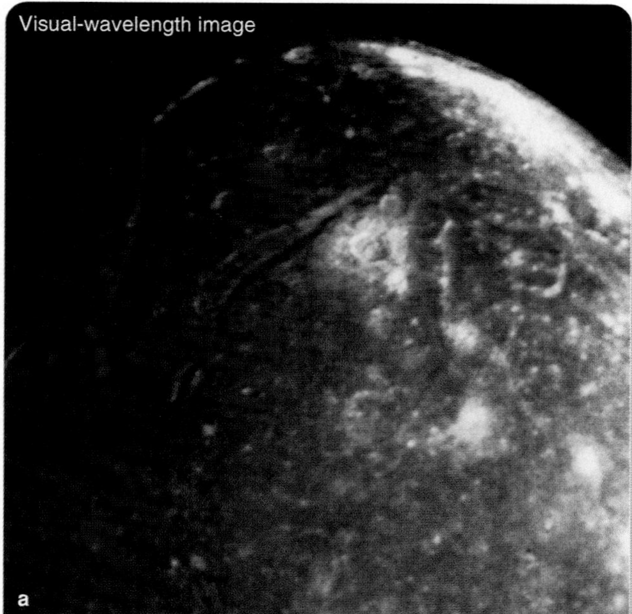

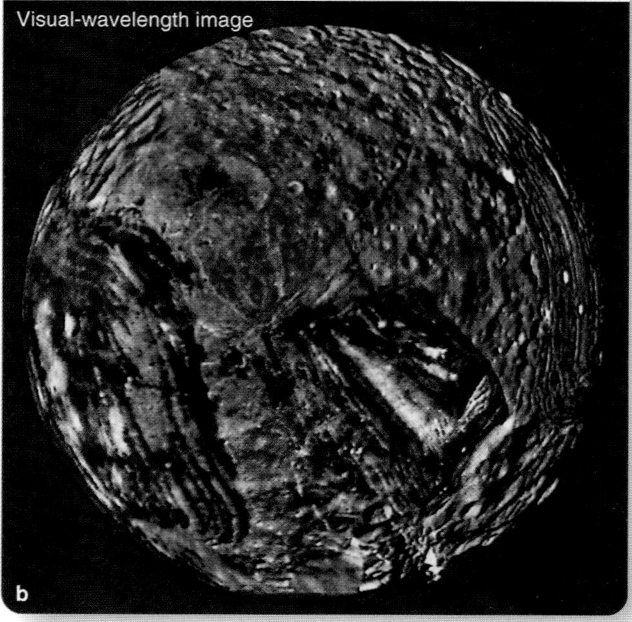

The Rings of Uranus and Neptune

Both Uranus and Neptune have rings made up of very dark particles orbiting in faint, narrow rings that are confined by small moons. Although they are similar, the two ring systems illustrate different processes that shape planetary rings.

The rings of Uranus were discovered in 1977, when Uranus crossed in front of a star. During this occultation, astronomers saw the star dim a number of times before and again after the planet crossed over the star. The dips in brightness were caused by rings circling Uranus.

Ace Astronomy™

Log into AceAstronomy and select this chapter to see the Active Figure "Uranus's Ring Detection" and animate this diagram.

Notice the eccentricity of the ε ring. It lies at different distances on opposite sides of the planet.

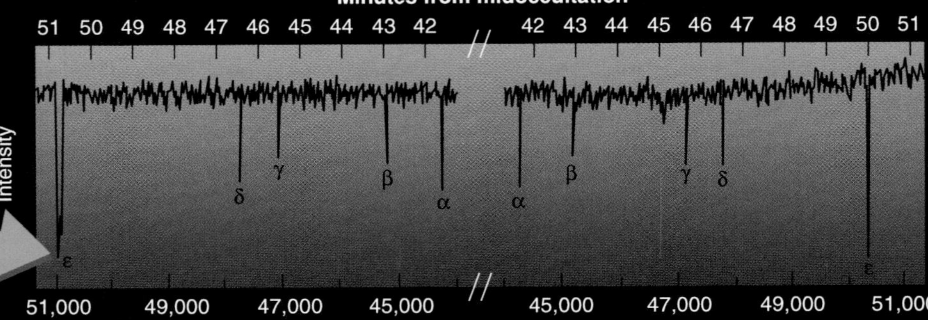

More rings were discovered by Voyager 2. The rings are identified in different ways depending on when and how they were discovered.

ε
λ
η
γ
δ
β
α
6
5
4
1986 U2R

The albedo of the ring particles is only about 0.015, darker than lumps of coal. If the ring particles are made of methane-rich ices, radiation from the planet's radiation belts could break the methane down to release carbon and darken the ices. The same process may darken the icy surface of Uranian moons.

The narrowness of the rings suggests they are shepherded by small moons. Voyager 2 found Ophelia and Cordelia shepherding the ε ring. Other small moons must be shepherding the other narrow rings. Such moons must be structurally strong to hold themselves together inside the planet's Roche limit.

Ophelia

Cordelia

The eccentricity of the ε ring is apparently caused by the eccentric orbits of Ophelia and Cordelia.

Ring particles don't last forever as they collide with each other and are exposed to radiation. The rings of Uranus may need to be resupplied with fresh particles occasionally as impacts on icy moons scatter icy debris.

When the Voyager 2 spacecraft looked back at the rings illuminated from behind by the sun, the rings were not bright. That is, the rings are not bright in forward-scattered light. That means they must not contain much dust. The nine main rings contain particles no smaller than meter-sized boulders.

Uranus

Collisions among ring particles produce dust, which is thinly scattered inward from the λ ring, but the high, tenuous atmosphere of Uranus is slowing the dust particles and making them fall into the planet. The rings of Uranus actually contain very little dust.

The moons of Neptune suggest some cataclysmic encounter long ago that put Nereid into a long-period elliptical orbit and put Triton into a retrograde orbit. You have seen evidence of major impacts throughout the solar system, so such interactions may have been fairly common. Certainly, impacts on the satellites could provide the debris that is trapped among the smaller moons to form the rings.

Inquire | Review | Analyze

Why is Neptune blue?

To analyze this question, you must be careful not to be misled by the words you use. When you look *at* something, you really turn your eyes toward it and receive light *from* the object. The light Earth receives from Neptune is sunlight that is scattered from various layers of Neptune and journeys to your eyes. Because sunlight contains a distribution of photons of all visible wavelengths, it looks white to you; but sunlight entering Neptune's atmosphere must pass through hydrogen gas that contains a small amount of methane, which is a good absorber of longer wavelengths. As a result, red photons are more likely to be absorbed than blue photons, and that makes the light bluer. Furthermore, when the light is scattered in deeper layers, the shorter-wavelength photons are most likely to be scattered, and thus the light that finally emerges from the atmosphere and reaches your telescope is poor in longer wavelengths. It looks blue.

This discussion shows how a careful, step-by-step analysis of a natural process can help you better understand how nature works. For example, can you explain why the clouds on Neptune look white?

■ ■ ■

Connections: The Voyager 2 spacecraft left the solar system after visiting Neptune. It will not pass near any other worlds unless, millions of years from now, it enters some distant solar system. Nevertheless, you have one more world to analyze.

18-6 Pluto

PLUTO IS A SMALL ICY WORLD clearly different from the Jovian worlds yet also different from the terrestrial planets of rock and metal. Since its discovery in 1930, it has been a mystery on the edge of the solar system.

Is Pluto a Planet?

You may have learned in school that there are nine planets in our solar system, but astronomers aren't so sure. Pluto isn't Jovian, and it isn't terrestrial. Is it a planet? To some extent, that controversy is just an argument over the meaning of the word "planet," but it is also revealing. Let's begin by analyzing Pluto as a planet, and you will be surprised where that takes us.

Pluto is very difficult to observe from Earth. Only a bit larger than 0.1 second of arc in diameter, the little planet is only 65 percent the diameter of Earth's moon and shows little surface detail. The best photos by the Hubble Space Telescope reveal large areas of light and dark terrain (**Celestial Profile 11**), promising

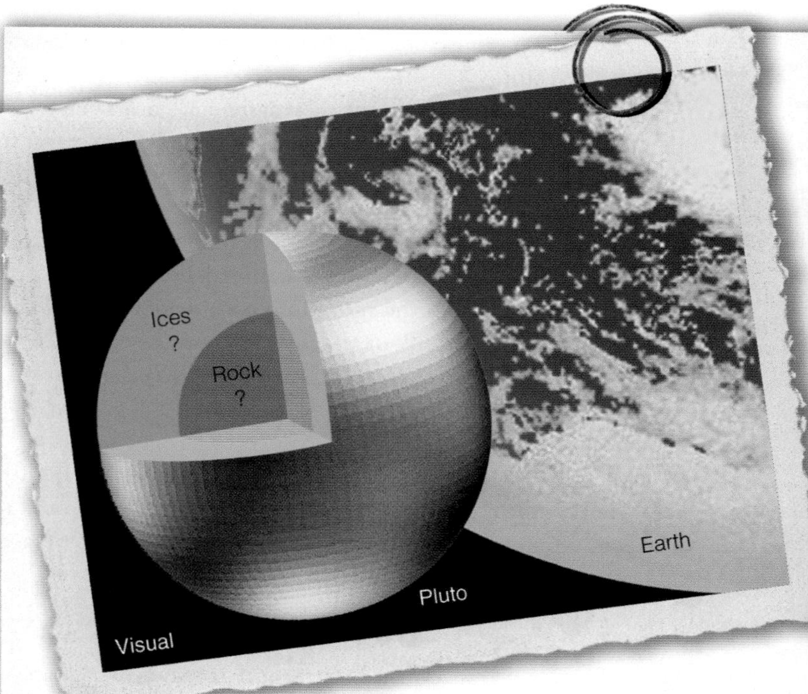

Pluto is a small, low-density world composed of ices and rock. No spacecraft has ever visited it, so the best image available shows only bright and dark regions. (NASA)

Celestial Profile 11: Pluto

Motion:

Average distance from the sun	39.44 AU (59.00 × 10^8 km)
Eccentricity of orbit	0.2484
Maximum distance from the sun	49.24 AU (73.66 × 10^8 km)
Minimum distance from the sun	29.64 AU (44.34 × 10^8 km)
Inclination of orbit to ecliptic	17°9′3″
Average orbital velocity	4.73 km/s
Orbital period	247.7 y (90,465 days)
Period of rotation	$6^d9^h21^m$
Inclination of equator to orbit	122°

Characteristics:

Equatorial diameter	2370 km (0.19 $D_\oplus$)
Mass	1.2 × 10^{22} kg (0.002 $M_\oplus$)
Average density	2.0 g/cm^3
Surface gravity	0.06 Earth gravity
Escape velocity	1.2 km/s (0.11 $V_\oplus$)
Surface temperature	−230°C (−382°F)
Albedo	0.5

Personality Point:

Pluto was discovered in 1930 by Clyde Thombaugh. As a young man, he built his own telescope on his family's wheat farm in Kansas. With no college education, he got a job at Lowell Observatory on the basis of his drawings of planets and examined millions of star images on photographic plates before he finally found Pluto.

that Pluto will reveal itself to be an interesting world when spacecraft finally visit it.

Most planetary orbits in our solar system are nearly circular, but Pluto's is quite elliptical (see Figure 1-7). On the average, it is the most distant planet, but it can come closer to the sun than Neptune. In fact, from January 21, 1979, to March 14, 1999, Pluto was closer to the sun than Neptune is. The planets will never collide, however, because Pluto's orbit is inclined 17°.

Orbiting so far from the sun, Pluto is cold enough to freeze most compounds you think of as gases, and spectroscopic observations have found evidence of nitrogen ice. It has a thin atmosphere of nitrogen and carbon monoxide with small amounts of methane.

In 1978, Pluto's moon Charon was discovered in a highly inclined orbit with a period of 6.387 days and an average orbital radius of 19,640 km. Pluto and Charon are tidally locked to face each other, so Pluto rotates at a highly inclined angle (■ Figure 18-18). Kepler's third law says the mass of the system is 6.5×10^{-9} solar masses or about 0.2 Earth masses. Most of this mass is Pluto, which is about 12 times more massive than Charon.

The mass of a body is important in astronomy because mass divided by volume is density. The density of Pluto is about 2 g/cm^3, and the density of Charon is just a bit less. Pluto and Charon must contain about 35 percent ice and 65 percent rock.

So is Pluto a planet? Before you vote, you should know that since 1992 new, large telescopes have discovered roughly a thousand icy bodies orbiting beyond Neptune in what is known as the **Kuiper belt,** which is named after the Dutch-American astronomer Gerard Kuiper (KI-Per). There are probably 70,000 or more objects in the belt. They appear to be icy bodies left over from the outer solar nebula.

None of the Kuiper belt objects are as big as Pluto, but some are close. Two large objects named Sedna and 2004DW are about 63 percent the size of Pluto. Another object called Quaoarh (kwah-o-wahr) is 50 percent the diameter of Pluto. As more and more of these objects are found, astronomers may eventually find one that is even bigger than Pluto. A few of these objects have moons of their own, or perhaps you should say that they are binaries with two icy bodies orbiting each other. In that way, they resemble Pluto and its big moon Charon.

This leaves astronomers arguing over the meaning of the word *planet.* Are there thousands of small, icy planets orbiting beyond Neptune, or are the Kuiper belt objects too small to be planets? Is Pluto a planet, or is it the king of the Kuiper belt? However you prefer to vote, you can be sure the history of the outer solar system is more interesting than anyone thought when Pluto was first discovered.

Pluto and the Plutinos

No, this section is not about a 1950s rock and roll band. It is about the history of Pluto, and it will take you back billions of years to watch the outer planets form.

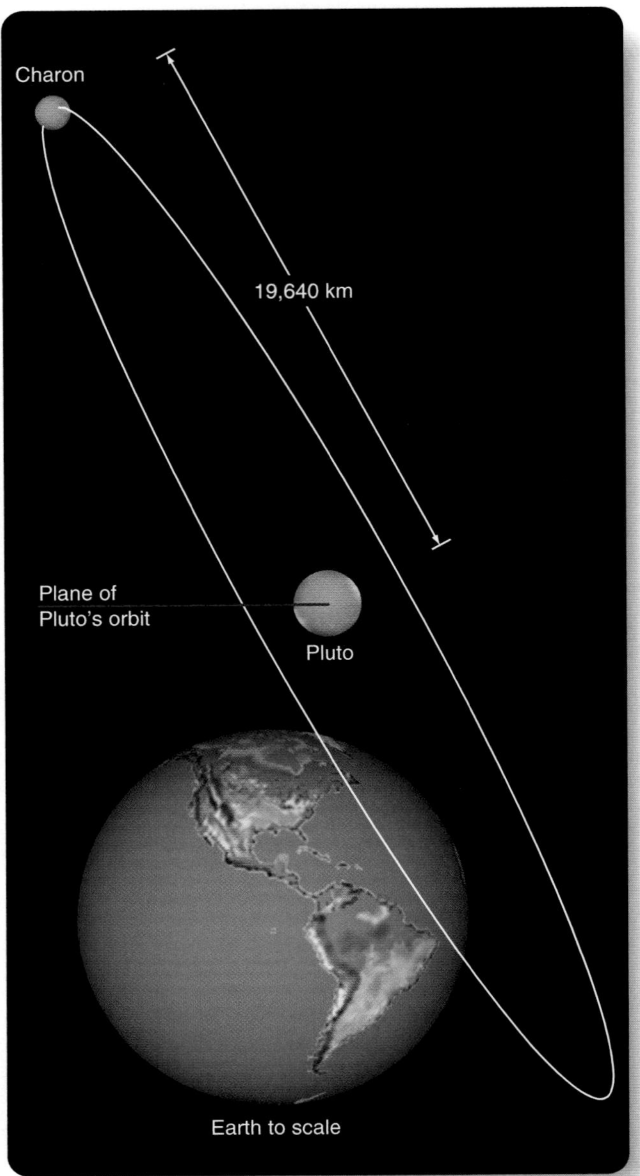

■ **Figure 18-18**

The circular orbit of Charon is seen here at an angle. The orbit is only a few times bigger than Earth and is tipped 118° to the plane of Pluto's orbit.

Let's begin by eliminating an old idea. As soon as Pluto was discovered, astronomers realized that it was much smaller than its Jovian neighbors. Some suggested that it was a moon of Neptune that had escaped. After all, their orbits actually cross, although they will never collide because Pluto's orbit is so highly inclined. That escaped-moon theory has been totally abandoned. It is hard to understand how Pluto could have escaped from Neptune and reached its present orbit. And how did it get a moon, if it was once a moon itself? Modern astronomers have a much better theory, and it links Pluto to the Kuiper belt objects.

As you have seen, some Kuiper belt objects are almost as big as Pluto, and some have orbits similar to Pluto's. Over a dozen are known that are caught in a 3:2 resonance with Neptune. That is, they orbit the sun twice while Neptune orbits three times. You learned about orbital resonances when you studied Jupiter's Galilean moons. Because Pluto is also caught in the same 3:2 resonance, these Kuiper belt objects have been named **Plutinos.** Many other Kuiper belt objects are caught in other resonances with Neptune.

If Pluto is related to the Plutinos, then Pluto is related to the Kuiper belt objects. They formed in the outer solar nebula, but how did they get caught in resonances with Neptune? Sophisticated models of the formation of the planets suggest that Uranus and Neptune may have formed closer to the sun where the solar nebula was denser. Sometime later, gravitational interactions with Jupiter and Saturn could have gradually shifted the two ice giants outward; and, as Neptune migrated outward, its orbital resonances could have swept up icy bodies like strange fishing nets pushed in front of a fishing boat.

Some icy Kuiper belt objects may have been captured as moons, and astronomers have wondered if moons such as Triton and Charon could have been Kuiper belt objects. Could Pluto have started life as a Kuiper belt object and gotten swept up by a Neptune resonance?

Now you can vote: Is Pluto a planet? Whatever the final outcome, Pluto and the Plutinos have rocked modern astronomers and forced them to reconsider their ideas on the origin of the outer worlds. As more Kuiper belt objects are discovered beyond the orbit of Neptune, astronomers will see more clearly how the outer solar nebula formed planets.

Inquire | Review | Analyze

What evidence indicates that cataclysmic impacts have occurred in our solar system?

You would expect objects that formed together from the solar nebula to rotate and revolve in the plane of the solar system, so the high inclination of Pluto and the orbit of Charon suggest a past collision or, at least, a close gravitational interaction with a passing body. And the Pluto–Charon system is not the only evidence for major impacts in the past. The peculiar orbits of Neptune's moons Triton and Nereid, the fractured and cratered state of the Jovian satellites, and the peculiar rotation of Uranus all hint that impacts and encounters with large planetesimals or the heads of comets have been important in the history of these worlds. Furthermore, the existence of planetary rings suggests that impacts have scattered small particles and replenished the ring systems. Even in the inner solar system, the backward rotation of Venus, the density of Mercury, and the formation of Earth's moon are possible consequences of major impacts when the solar system was young.

Certainly, you wouldn't imagine that planets bounce around like billiard balls, but you must recognize that the planets are not totally isolated now and may have been victims (or beneficiaries) of large impacts as they formed long ago. For example, how does the size and composition of Pluto and Charon compared with the moons of the outer Jovian worlds suggest an unexpected population of icy worlds in the outer solar system?

■ ■ ■

Connections: Like clues at the scene of a crime, the evidence in the outer solar system alerts you that all of the planets in our solar system have been and still are sitting ducks for impacts by the smaller objects that hurtle along their orbits around the sun. What are these objects and what can they tell you about the origin and evolution of our solar system? You will explore that question in the next chapter.

Study and Review Tools

Summary

18-1 | A Travel Guide to the Outer Planets

What are very massive planets like?

- The Jovian planets, Jupiter, Saturn, Uranus, and Neptune, are large, low-density worlds rich in hydrogen and helium. Pluto, small and icy, is not a Jovian world.
- The atmospheres of the Jovian planets are marked by belt–zone circulation that produces cloud belts parallel to their equators.
- Jupiter and Saturn are composed mostly of liquid hydrogen and are sometimes called liquid giants. Uranus and Neptune contain water as liquid and as ice and are sometimes called the ice giants.
- All of the Jovian worlds have large systems of satellites and rings that have had complex histories.

18-2 | Jupiter

- Jupiter must be very hot inside because heat is flowing out of it at a high rate.
- Jupiter has a core of heavy elements and a deep mantle of liquid metallic hydrogen in which the planet's magnetic field is generated.
- The magnetic field around Jupiter traps high-energy particles from the sun to form intense radiation belts.
- Jupiter's atmosphere contains three layers of clouds formed of hydrogen-rich molecules such as ammonia and water.
- Zones are high-pressure regions of rising gas, and belts are lower-pressure areas of sinking gas.
- The fragmented head of a comet hit Jupiter in 1994. Such impacts are probably common on the Jovian planets.
- Spots in Jupiter's atmosphere, including the Great Red Spot, are circulating weather patterns.

- Jupiter's ring is composed of dark particles but is bright in forward scattering, which means the particles are very small. They are probably dust from meteorite impacts on moons.
- Jupiter's ring, like all of the rings in the solar system, lies inside its planet's Roche limit.

What evidence indicates that some moons have been active?

- Many of Jupiter's moons are small, rocky bodies that are probably captured asteroids. They are too small to retain heat and are not active.
- The four Galilean satellites show signs of activity. Grooves on Ganymede, smooth ice and cracks on Europa, and active volcanoes on Io show that tidal heating has made these moons active.

18-3 | Saturn

- Saturn is less dense than water and contains a small core and less metallic hydrogen than Jupiter.
- The cloud layers on Saturn occur at the same temperature as those on Jupiter, but, because Saturn is further from the sun and colder, the cloud layers are deeper in the hydrogen atmosphere below a layer of methane haze.

How are planetary rings formed and maintained?

- Saturn's rings are composed of icy particles ranging in size from boulders to dust. In some regions the ice is cleaner than in other regions.
- Grooves in the rings can be produced by resonances with moons or by waves that propagate through the rings.
- Narrow rings and sharp edges can be confined by shepherd satellites.
- The rings cannot be primordial and must be replenished now and then with ice from comets colliding with moons.
- Saturn's moons are icy and heavily cratered.
- Saturn's largest moon, Titan, has a cold, cloudy nitrogen atmosphere. It may have methane ice and liquid ethane on its surface.
- Sunlight entering Titan's atmosphere can convert methane into organic molecules to form haze and particles that settle to the surface.

18-4 | Uranus

How do less-massive Jovian worlds differ from their more-massive cousins?

- Uranus is much less massive than Jupiter, and its internal pressure cannot produce liquid hydrogen. It has a heavy-element core and a mantle of slushy ice and rock below a hydrogen-rich atmosphere.
- Little heat flows out of Uranus, so it cannot be very hot inside.
- The atmosphere is almost featureless at visual wavelengths with a pale blue color caused by traces of methane, which absorbs red light.
- Images in the infrared or at selected wavelengths can be enhanced to show traces of belt–zone circulation.
- The rings of Uranus are narrow hoops confined by shepherd satellites. The particles appear to be ice with traces of methane darkened by the radiation belt.
- Impacts on the icy moons probably refresh the rings with new particles.
- The larger moons of Uranus are icy and heavily cratered, with signs on some of past geological activity.
- Uranus rotates on its side, perhaps because of a major impact during its early history.

18-5 | Neptune

- Neptune is an ice giant like Uranus with no liquid hydrogen.
- The atmosphere of Neptune, marked by traces of belt–zone circulation, is rich in hydrogen and colored blue by traces of methane.
- The rings of Neptune are made of icy particles in narrow hoops and contain arcs produced by the gravitational influence of one or more moons.
- Neptune's satellite system is odd in that distant Nereid follows an elliptical orbit and Triton orbits backwards.
- Triton is icy with a thin atmosphere and frosty polar caps. Smooth areas suggest past geological activity, and dark smudges mark the location of active nitrogen geysers.

18-6 | Pluto

Is Pluto a planet or something more?

- Pluto is a small, icy world with a large moon locked in a highly inclined orbit.
- Pluto may be related to a family of icy bodies in the Kuiper belt orbiting beyond Neptune. Some of these objects are nearly as large as Pluto, and some follow similar orbits.
- Some models suggest Uranus and Neptune formed closer to the sun and migrated outward pushing thousands of icy bodies outward in orbital resonances to form the Kuiper belt.
- Pluto may be the largest of the Kuiper belt objects found so far.

New Terms

belt–zone circulation (p. 421)
liquid metallic hydrogen (p. 422)
magnetosphere (p. 422)
forward scattering (p. 425)
Roche limit (p. 425)
gossamer rings (p. 428)
grooved terrain (p. 428)
tidal heating (p. 430)
oblateness (p. 432)
shepherd satellite (p. 435)
occultation (p. 439)
ovoid (p. 439)
Kuiper belt (p. 446)
Plutino (p. 447)

Review Questions

Ace Astronomy™ Assess your understanding of this chapter's topics with additional quizzing and animations at http://astronomy.brookscole.com/sh9e

1. Why is Jupiter so much richer in hydrogen and helium than Earth?
2. How can Jupiter have a liquid interior and not have a liquid surface?
3. How does the dynamo effect account for the magnetic fields of Jupiter and Saturn?
4. Why are the belts and zones on Saturn less distinct than those on Jupiter?
5. Why do astronomers conclude that neither Jupiter's ring nor Saturn's rings can be left over from the formation of the planets?
6. How can a moon produce a gap in a planetary ring system?
7. Explain how geological activity on Jupiter's moons varies with distance from the planet.
8. What makes Saturn's F ring and the rings of Uranus and Neptune so narrow?
9. Why is the atmospheric activity of Uranus less than that of Saturn and Neptune?
10. Why do astronomers suspect that Enceladus has been geologically active more recently than some other moons?
11. What are the seasons on Uranus like?
12. Why are Uranus and Neptune blue?
13. What evidence is there that Triton has been geologically active recently?
14. How do astronomers account for the origin of Pluto?
15. What evidence indicates that catastrophic impacts have occurred in the solar system's past?

16. This photo shows a segment of the surface of Jupiter's moon Callisto. Why is the surface dark? Why are some craters dark and some white? What does this image tell you about the history of Callisto?

(NASA/JPL)

17. The Cassini spacecraft recorded this photo of Saturn's A ring and Encke's division. What do you see in this photo that tells you about processes that confine and shape planetary rings?

(NASA/JPL/Space Science Institute)

18. Two images of Uranus show it as it would look to the eye and through a red filter that enhances methane clouds in the northern hemisphere. Why didn't Voyager 2 photograph the northern hemisphere? What do the visible atmospheric features tell you about circulation on Uranus?

(NASA and Heidi Hammel)

Discussion Questions

1. Some astronomers argue that Jupiter and Saturn are unusual, while other astronomers argue that all solar systems should contain one or two such giant planets. What do you think? Support your argument with evidence.
2. Why don't the terrestrial planets have rings? If you were to search for a ring among the terrestrial planets, where would you look first?

Problems

1. What is the maximum angular diameter of Jupiter as seen from Earth? Repeat this calculation for Saturn and Pluto. (*Hints:* See Celestial Profiles Seven, Eight, and Eleven, and also Reasoning with Numbers 3-1.)
2. What is the angular diameter of Jupiter as seen from Callisto? from Io? (*Hint:* See Reasoning with Numbers 3-1.)
3. Measure the photograph in Celestial Profile Eight and calculate the oblateness of Saturn.
4. If you observe light reflected from Saturn's rings, you should see a redshift at one edge of the rings and a blueshift at the other edge. If you observe a spectral line and see a difference in wavelength of 0.056 nm, and the unshifted wavelength (observed in the laboratory) is 500 nm, what is the orbital velocity of particles at the outer edge of the rings? (*Hint:* See Reasoning with Numbers 6-2.)
5. One way to recognize a distant planet is by its motion along its orbit. If Uranus circles the sun in 84 years, how many seconds of arc will it move in 24 hours? (*Hint:* Ignore the motion of Earth.)
6. If the ε ring is 50 km wide and the orbital velocity of Uranus is 6.81 km/s, how long a blink should you expect to see when the ring crosses in front of the star?
7. What is the angular diameter of Pluto as seen from the surface of Charon? (*Hint:* See Figure 18-18.)
8. If Pluto has a surface temperature of 50 K, at what wavelength will it radiate the most energy? (*Hint:* See Reasoning with Numbers 6-1.)
9. How long did it take radio commands to travel from Earth to Voyager 2 as it passed Neptune?
10. Use the orbital radius and orbital period of Charon to calculate the mass of the Pluto–Charon system. (*Hints:* Express the orbital radius in meters and the period in seconds. Then see Reasoning with Numbers 4-1.)

Media Cluster

Ace Astronomy™ To access the resources in the Media Cluster, log into AceAstronomy at **http://astronomy.brookscole.com/sh9e** and select Chapter 18.

ACTIVE FIGURES

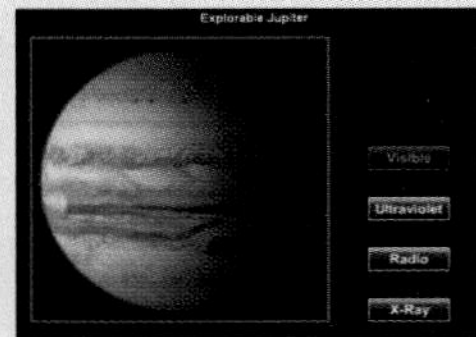

Explorable Jupiter
Study Jupiter at visible, ultraviolet, radio, and X-ray wavelengths in this animation. What features can you see at the other wavelengths that are not visible to the human eye?

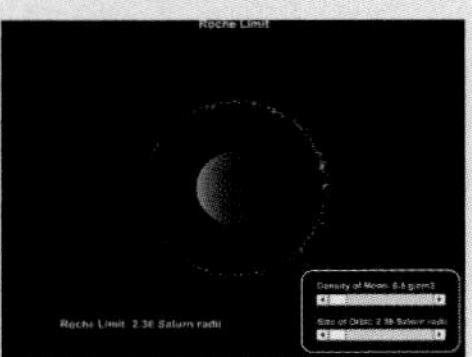

Roche Limit
Some moons remain intact in orbit, and some break apart to become rings like Saturn's rings. See if you can figure out why with this animation.

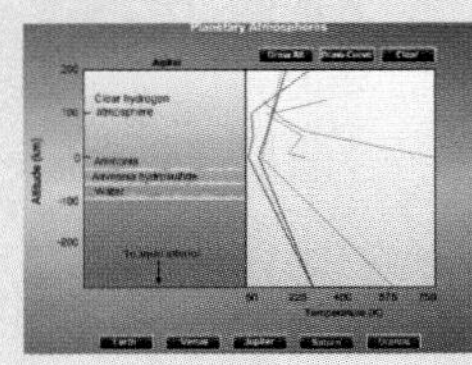

Planetary Atmospheres
Experiment with the temperature throughout the atmospheres of several planets with this animation.

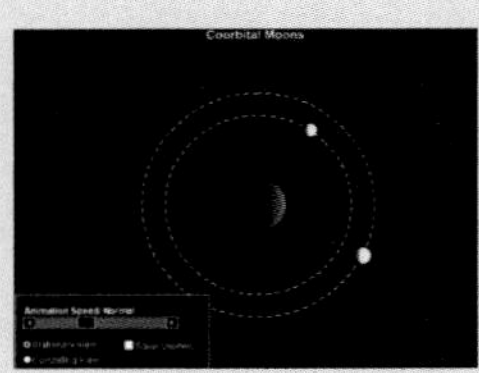

Coorbital Moons
Study the phenomenon of two of Saturn's moons repeatedly swapping orbits in this animation.

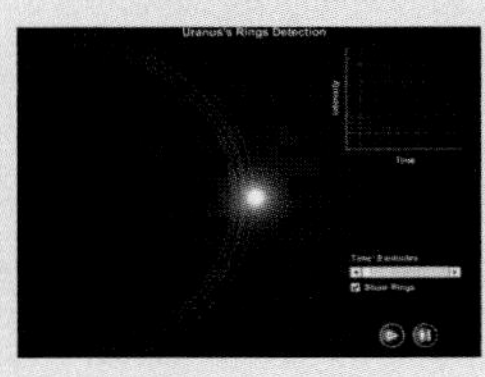

Uranus's Rings
In this animation you can replicate the observations that first provided a glimpse of the rings of Uranus.

ASTRONOMY EXERCISES

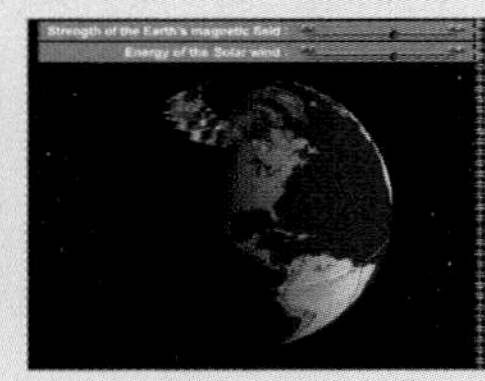

Auroras
Use this simulation to study the formation of auroras. You can vary the strength of Earth's magnetic field and the speed and number of particles in the solar wind.

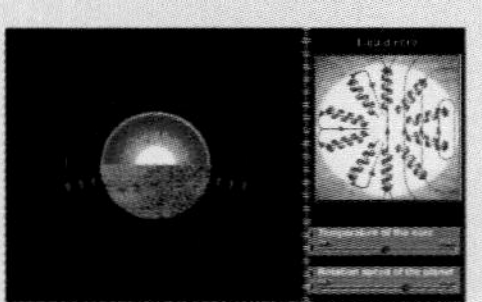

Convection and Magnetic Fields
A planet's core temperature and rotation speed are factors in the strength of its magnetic field. See the effect of these variables in this animation.

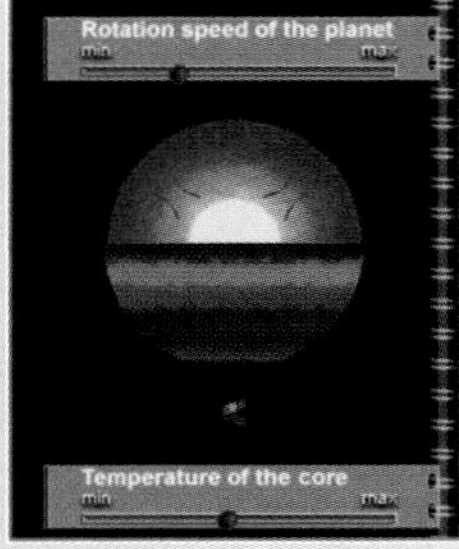

Convection and Turbulence
Core temperature and rotation speed also contribute to the amount of turbulence in a planet's atmosphere. Explore their relationship in this animation.

VIRTUAL ASTRONOMY LAB

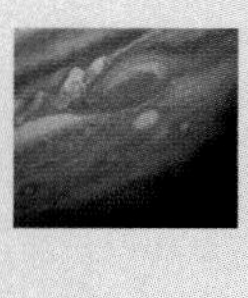

Lab 7: Planetary Atmospheres and Their Retention
This lab investigates the retention of atmospheres. You will explore the factors that govern the loss of atmospheric gases, and you will see why certain bodies can retain some gases but not others.

Critical Inquiries for the Web

1. If you lived on the surface of Pluto and looked into the sky to observe Charon, what phases would you see? (*Hint:* Be sure to consider your location on the planet when answering this question.)
2. What factors caused Voyager 2 to see a bland atmosphere when it encountered Uranus in 1986? Given these circumstances, would images of the Uranian atmosphere taken by a space probe arriving at Uranus in 2006 be similar to those taken in 1986, or would there be significant differences?
3. Imagine what you'd think if you had been the first person ever to see Saturn through a telescope. When Galileo first observed Saturn in 1610, he did not recognize that it was a ringed planet. It was many years later before the strange apparition of Saturn was finally attributed to a ring structure. Search the web for information on historical observations of Saturn, summarize the observations of Galileo and others, and determine who was first to recognize what he saw as a ring. Why do you suppose it took so long to understand that Saturn is a ringed planet?
4. Should Pluto be called a planet or not? Search the web for news and debate on this issue. What is your opinion?
5. Who was Clyde Tombaugh? What can you find out about his life after he discovered Pluto?

Exploring *TheSky*

1. Zoom in on Jupiter and observe the orbital motion of its moons. (*Hint:* Under **Tools,** choose **Time Skip** and then **Tracking Setup.** Lock on Jupiter. Then use the **Time Skip** buttons to make the moons move around their orbits.)
2. Calculate the mass of Jupiter from your own observations of the orbital period and orbital radius of Jupiter's moons. (*Hint:* See Reasoning with Numbers 4-1.)
3. Repeat Activites 1 and 2 above for Saturn.

- Go to the Brooks/Cole Astronomy Resource Center **(http://astronomy.brookscole.com)** for critical thinking exercises, articles, and additional readings from InfoTrac College Edition, Brooks/Cole's online student library.

19 Meteorites, Asteroids, and Comets

When they shall cry
"PEACE, PEACE" then cometh
sudden destruction!
COMET'S CHAOS?—
What Terrible events
will the Comet bring?

FROM A RELIGIOUS PAMPHLET PREDICTING THE END OF THE WORLD BECAUSE OF THE APPEARANCE OF COMET KOHOUTEK, 1973

YOU ARE NOT afraid of comets, of course; but, not long ago, people viewed them with terror. In 1910, Comet Halley was spectacular. On the night of May 19, Earth actually passed through the tail of the comet—and millions of people panicked. The spectrographic discovery of cyanide gas in the tails of comets led many to believe that life on Earth would end. Householders in Chicago stuffed rags around doors and windows to keep out the gas, and bottled oxygen was sold out. Con artists in Texas sold comet pills and inhalers to ward off the noxious fumes. An Oklahoma newspaper reported (in what was apparently a hoax) that a religious sect tried to sacrifice a virgin to the comet. Throughout history, bright comets have been seen as portents of doom. Even the more recent appearance of bright comets has generated predictions of the

The Stardust spacecraft flew through the dust and gas spewing from the nucleus of Comet Wild 2 in January 2004. (NASA/JPL)

Guidepost

Looking Back

In Chapter 16 you began your study of planetary astronomy by asking how our solar system formed. In the two chapters that followed, you surveyed the planets and found only limited evidence of the origin of the solar system. The planets are big, and their surfaces have been altered as heat has flowed outward from their interiors. Now you are ready to study smaller, better-preserved objects that can tell you more about the age of planet building.

This Chapter

Compared with planets, the comets and asteroids are unevolved objects. You will find them much as they were when they formed 4.6 billion years ago. The fragments of these objects that fall to Earth, the meteors and meteorites, will give you a close look at these ancient planetesimals. As you explore, you will find answers to four essential questions:

Where do meteors and meteorites come from?

What are the asteroids?

Where do comets come from?

What happens when an asteroid or comet hits Earth?

Looking Ahead

As you finish this chapter, you will have an astronomer's insight into your place in nature. You live on the surface of a planet. There are other planets. Are they inhabited too? That is the subject of the next chapter.

Ace Astronomy™ The AceAstronomy icon throughout the text indicates an opportunity for you to test yourself on key concepts and to explore animations and interactions on the AceAstronomy website at: http://astronomy.brookscole.com/sh9e

end of the world. Comet Kohoutek in 1973, Comet Halley in 1986, and Comet Hale–Bopp in 1997 caused concern among the superstitious. A bright comet moving slowly through the night sky is so out of the ordinary (■ Figure 19-1) that you should not be surprised if it generates some instinctive alarm.

In fact, comets are graceful and beautiful visitors to our skies. Astronomers think of comets as messengers from the age of planet building. By studying comets, you can learn about the conditions in the solar nebula from which planets formed. But comets are only the icy remains of the solar nebula; the asteroids are the rocky debris left over from planet building. Because you cannot easily visit comets and asteroids, let's begin by discussing the fragments of those bodies that fall into our atmosphere—the meteorites.

19-1 Meteorites

YOU LEARNED ABOUT METEORITES in Chapter 16 when we discussed the age of the solar system. There you saw that the solar system is filled with small particles called meteoroids, which can fall into Earth's atmosphere at speeds of 10 to 40 km/s. Friction with the air heats the meteoroids to glowing, and you see them vaporize as meteors streaking across the night sky. If a meteoroid is big enough and strong enough, it can survive its plunge through the atmosphere and reach Earth's surface. Once the object strikes Earth's surface, it is called a meteorite. The largest of these objects can blast out craters on Earth's surface, but such impacts are rare. The vast majority of meteorites are too small to form craters. These meteorites fall all over Earth, and their value is in what they can reveal about the origin of the planets.

Inside Meteorites

Meteorites can be divided into three broad categories. *Iron* meteorites are solid chunks of iron and nickel. *Stony* meteorites are silicate masses that resemble Earth rocks. *Stony-iron* meteorites are mixtures of iron and stone. These types of meteorites are illustrated in ■ Figure 19-2.

■ Figure 19-1

Comet Hyakutake swept through the inner solar system in 1996 and was dramatic in the northern sky. Seen here from Kitt Peak National Observatory, the comet passed close to the north celestial pole (behind the observatory dome in this photo). Notice the Big Dipper below and to the left of the head of the comet. (Courtesy Tod Lauer)

Iron meteorites are very heavy for their size and have a dark, irregular surface.
Stony meteorites tend to have a fusion crust caused by melting in Earth's atmosphere.
A stony-iron meteorite cut and polished reveals a mixture of iron and rock.
Chondrules are small, glassy spheres found in chondrites.
Cut, polished, and etched with acid, iron meteorites show a Widmanstätten pattern.
This carbonaceous chondrite contains chondrules and volatiles, including carbon, that make the rock very dark.

Figure 19-2

The three main types of meteorites, irons, stones, and stony-irons, have distinctive characteristics. (Lab photos courtesy of Russell Kempton, New England Meteoritical)

Iron meteorites are dense and heavy. They often have dark, rusted surfaces and fluted shapes caused by their passage through the atmosphere. When they are sliced open, polished, and etched with nitric acid, they reveal regular bands called **Widmanstätten patterns** (Figure 19-2). The patterns arise from crystals of nickel-iron alloys that have grown very large, indicating that the meteorite cooled from a molten state no faster than a few degrees per million years. Explaining how iron meteorites could have cooled so slowly will be a major step in analyzing their history.

Stony meteorites called **chondrites** have chemical compositions that resemble a cooled lump of matter from the sun with the volatile gases removed. Although there are many kinds of chondrites, most contain **chondrules,** rounded bits of glassy rock ranging from microscopic to as big as a pea (Figure 19-2). The origin of chondrules is unknown, but they appear to have formed in the young solar system as droplets of molten rock that cooled and hardened rapidly. Because subsequent melting of the meteorite would have destroyed chondrules, their presence in meteorites indicates that the metorites have never been melted since they formed.

Nevertheless, some kinds of chondrites show signs that they have been heated slightly. Those meteorites are poor in volatiles. The condensing solar nebula should have incorporated carbon compounds and water into the forming solids. If a meteorite were heated slightly, it could lose its volatiles. Yet, some chondrites are rich in volatiles and probably formed in the presence of water. Thus the history of chondrites illustrates the complex history of our solar system.

The **carbonaceous chondrites** generally contain both chondrules and volatile compounds, including significant amounts of carbon. Heating would have modified and driven off these fragile compounds. The carbonaceous chondrites are, along with certain kinds of chondrites, among the least modified bodies in our solar system.

Some stony meteorites contain no chondrules, and they are called **achondrites.** They also lack volatiles and appear to have been subjected to intense heat that melted chondrules and drove off volatiles, leaving behind rock with compositions similar to Earth's lavas.

In addition to iron meteorites and stony meteorites, some are called **stony-iron** meteorites. They are a mixture of iron and stone, and they appear to have formed when a mixture of molten iron and rock cooled and solidified.

The meteorites carry hints about the origin of our solar system, but they also carry an older secret. Among the smallest grains in meteorites are specks of minerals whose abundance of isotopes brands them as star dust—grains of interstellar matter that predate our solar system.

The Origin of Meteors and Meteorites

The mineral composition of the meteorites provides a clue to the origins of these objects. Somehow they have survived almost unchanged since the age of planet building in the solar nebula.

You can find evidence of the origin of meteors through one of the most pleasant observations in astronomy. You can observe a **meteor shower,** a display of meteors that are clearly related in a common origin. For example, the Perseid meteor shower occurs each year in August (■ Table 19-1), and on that night you might see as many as 40 meteors an hour if you stretched out on a lawn chair at a dark site and watched the sky. Like many natural phenomena, a meteor shower is more enjoyable when you know more about it (**Window on Science 19-1**).

On any clear, moonless night of the year you might see 5 to 15 meteors per hour, but these are not related to each other. During a meteor shower, you will see meteors that are related in that they seem to come from a single spot on the sky. The Perseid shower, for example, appears to come from a spot in the constellation Perseus. These showers are seen when Earth passes near the orbit of a comet; the meteors in meteor showers must be produced by dust and debris released by the icy head of the comet. The meteors appear to come from a single place in the sky because they are particles traveling along parallel paths through space. Like railroad tracks extending from a point on the horizon, the meteors appear to approach from a point in space (■ Figure 19-3). The orbits of comets are filled with such debris. For example, the telescope aboard the Infrared Astronomy Satellite detected the dusty orbits of a number of comets glowing in the far-infrared because of the sun-warmed dust scattered along the orbits.

Studies of meteors show that most of the meteors you see on any given night (whether or not there is a shower) are produced by tiny bits of debris from comets. These specks of matter are so small and so weak they vaporize completely in the atmosphere and never reach the ground, but from their motions astronomers

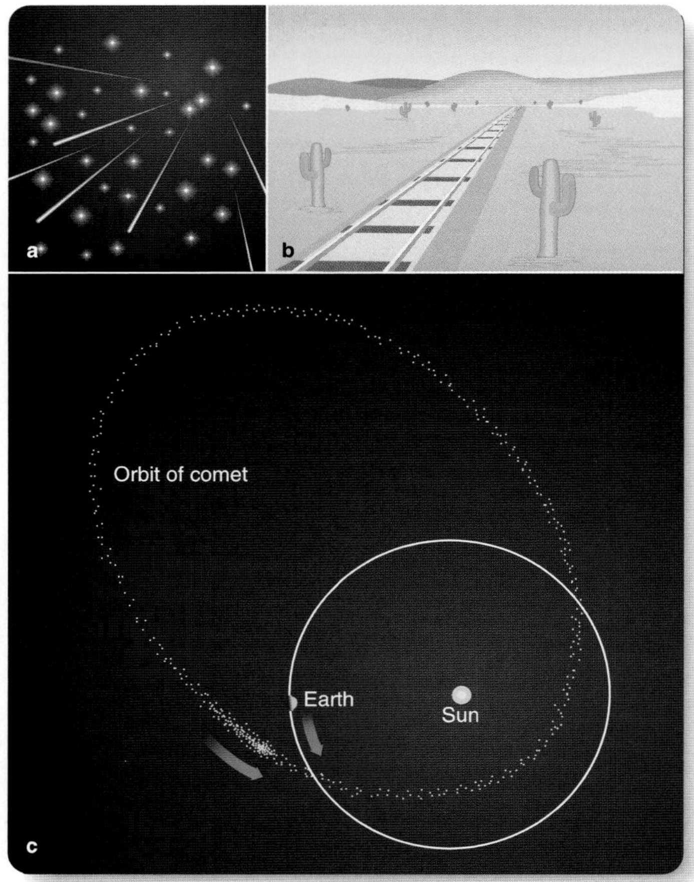

■ **Figure 19-3**

(a) Meteors in a meteor shower enter Earth's atmosphere along parallel paths, but perspective makes them appear to diverge from a radiant point in the sky. (b) Similarly, parallel railroad tracks appear to diverge from a point on the horizon. (c) Meteors in a shower are debris left behind as a comet's icy nucleus vaporizes. The rocky and metallic bits of matter spread along the comet's orbit. If Earth passes through such material, you can see a meteor shower.

■ **Table 19-1 | Meteor Showers**

Shower	Dates	Hourly Rate	Radiant* R.A.	Radiant* Dec.	Associated Comet
Quadrantids	Jan. 2–4	30	15^h24^m	50°	
Lyrids	April 20–22	8	18^h4^m	33°	1861 I
η Aquarids	May 2–7	10	22^h24^m	0°	Halley?
δ Aquarids	July 26–31	15	22^h36^m	−10°	
Perseids	Aug. 10–14	40	3^h4^m	58°	1982 III
Orionids	Oct. 18–23	15	6^h20^m	15°	Halley?
Taurids	Nov. 1–7	8	3^h40^m	17°	Encke
Leonids	Nov. 14–19	6	10^h12^m	22°	1866 I Temp
Geminids	Dec. 10–13	50	7^h28^m	32°	

*R.A. and Dec. give the celestial coordinates (right ascension and declination) of the radiant of each shower.

Enjoying the Natural World

You can enjoy a meteor shower as Mother Nature's fireworks, but your enjoyment is much greater once you begin to understand what causes meteors and why meteors in a shower follow a pattern. When you know that meteor showers help you understand the origin of our world, the evening display of shooting stars is even more exiciting. Science typically increases your enjoyment of the natural world by revealing the significance of things you might otherwise enjoy only in a casual way.

Everyone likes flowers, for example. You enjoy them in a casual way, admiring their colors and fragrances, their graceful petals and massed blooms. But botanists know that evolution has carefully designed flowers to attract insects and spread pollen. The bright colors attract insects, and the shapes of the flowers provide little runways so the insect will find it easy to land. Some color patterns even guide the insect in like landing lights at an airport, and many flowers, such as orchids and snapdragons, force the insect to crawl inside in just the right way to exchange pollen and fertilize the flower. The nectar is bug bait. Once you begin to understand what flowers are for, a visit to a garden becomes not only an adventure of color and fragrance but also an adventure in understanding the beauty of design and function, an adventure in meaning as well as an adventure of the senses.

And your understanding of a natural phenomenon helps you understand and enjoy related phenomena. For example, some flowers attract flies for pollination, and such blossoms smell like rotting meat. A few flowers depend on bats, and they open their blossoms at night. Many plants, such as pine trees, depend on the wind to spread their pollen, and those plants do not have colorful flowers at all; but flowers that depend on hummingbirds have long trumpet-shaped blossoms that just fit the hummingbirds' beaks.

The more science tells you about nature, the more enjoyable the natural world is. You can enjoy a meteor shower as nothing more than a visual treat, but the more you know about it, the more interesting and exciting it becomes. The natural world is filled with meaning, and science, as a way of discovering and understanding that meaning, gives you new opportunities to enjoy the world around you.

The beauty of flowers becomes more interesting when you know the reasons for their shapes and colors. (M. Seeds)

can deduce their orbits, and those orbits match the orbits of comets. Thus nearly all of the meteors come from comets.

Meteorites that reach Earth's surface are structurally much stronger than cometary material. They are iron and stone, and appear to be fragments of larger bodies.

Some meteorites appear to be fragments of planetesimals that were large enough to grow hot from radioactive decay, melt, and differentiate to form iron-nickel cores and rocky mantles. The molten iron cores would have been well insulated by the thick rocky mantles, and the iron would have cooled slowly to produced Widmanstätten patterns. Collisions could break such bodies up and produce different kinds of meteorites (■ Figure 19-4). Iron meteorites appear to be fragments from the iron cores. Some stony meteorites that have been strongly heated appear to have come from the mantles and surfaces of such bodies. The stony-iron meteorites apparently come from the boundary where the stony mantle meets the iron core. On the other hand, chondrites are probably fragments of smaller bodies that never melted, and the carbonaceous chondrites may have formed in small, colder bodies further from the sun.

These theories trace the origin of meteorites to planetesimal-like parent bodies, but they leave you with a mystery. The small meteorites in the solar system cannot be fragments of the planetesimals that formed the planets, because small meteorites would have been swept up by the planets in only a billion years or less. They could not have survived for 4.6 billion years.

In fact, when astronomers study the orbits of objects seen to fall to Earth as meteorites, those orbits lead back into the asteroid belt. Thus astronomers have good evidence to believe that the meteorites now in museums all over the world must have broken off asteroids within the last billion years. Although nearly all meteors are pieces of comets, the meteorites are pieces of asteroids.

Inquire | Review | Analyze

How can you say that meteors come from comets, but meteorites come from asteroids?

A selection effect can determine what you notice when you observe nature, and a very strong selection effect prevents you from finding meteorites that originated in comets. Cometary particles are physically weak, and they vaporize in our atmosphere easily. Thus, very few ever reach the ground, and you are unlikely to find them. Furthermore, even if a cometary particle reached the ground, it would be so fragile that it would weather away rapidly, and, again, you would be unlikely to find it. Asteroidal particles, however, are made from rock and metal and so are stronger. They are more likely to survive their plunge through the atmosphere and more likely to survive erosion on the ground. Meteors from the asteroid belt are rare. Almost all of the meteors you see come from comets, but not a single meteorite is known to be cometary.

The Origin of Meteorites

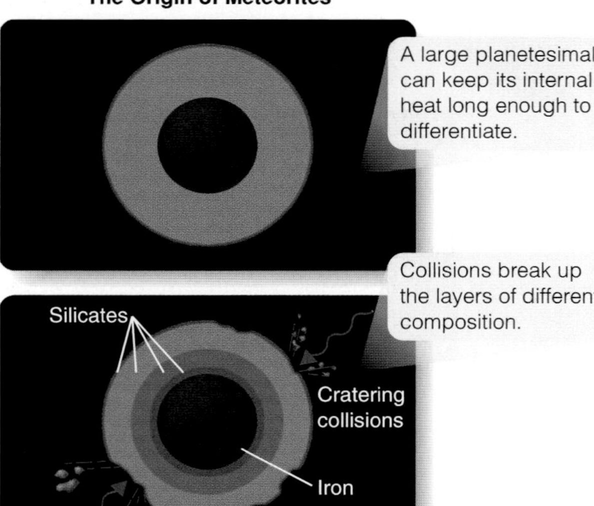

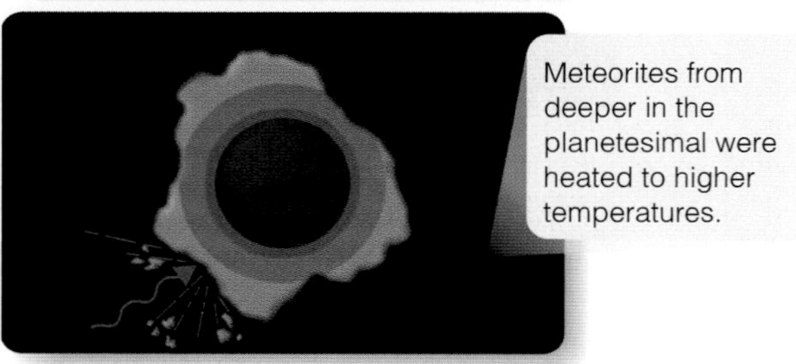

Meteorites from deeper in the planetesimal were heated to higher temperatures.

Fragments from near the core might have been melted entirely.

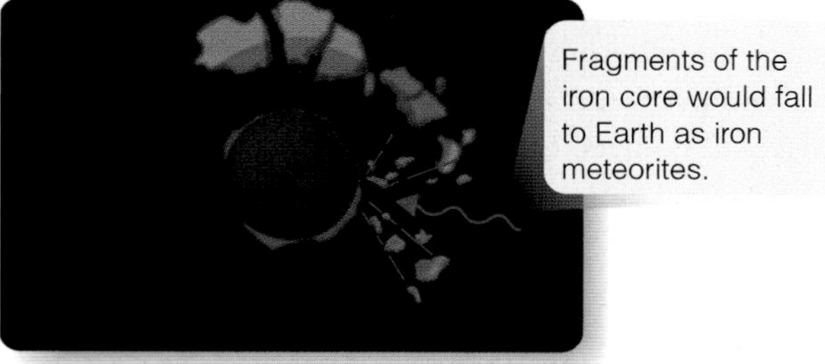

Fragments of the iron core would fall to Earth as iron meteorites.

■ **Figure 19-4**

Planetesimals formed when the solar system was forming may have melted and separated into layers of different density and composition. The fragmentation of such a body could produce many types of meteorites. (Adapted from a diagram by C. R. Chapman)

The meteorites are valuable because they give hints about the process of planet building in the solar nebula. What evidence do you find in the meteorites that tells you they were once part of larger bodies broken up by impacts?

■ ■ ■

Connections: The vast majority of meteorites must have originated in the asteroids. Your next goal is to try to understand the nature and origin of these tiny rocky worlds.

19-2 Asteroids

SPACE PIRATES LURK in the asteroid belt in old-time pulp science fiction, but astronomers have found that there isn't much in the asteroid belt for a pirate to stand on. Thousands of asteroids are known, but most are quite small; and, given its vast size, the asteroid belt between Mars and Jupiter is mostly empty.

Nevertheless, you must consider the nature of the asteroids. Chapter 16 identified the asteroids as a characteristic of the solar system and concluded that they were the last remains of material that was unable to form a planet between Mars and Jupiter. You are now ready to examine these small worlds in more detail.

Properties of Asteroids

Asteroids are distant objects too small to study in detail with Earth-based telescopes. Yet astronomers have learned a surprising amount about these little worlds, and spacecraft have provided a few close-ups.

Study **Observations of Asteroids** on pages 460 and 461 and notice four important points. First, the asteroids show clear signs of past collisions. They are irregular in shape, heavily cratered, and, in some cases, composed of fragments in floating rubble piles. Some asteroids are double objects or have moons. Note that although gravity is weak at the surface of an asteroid, they are covered with a regolith of broken rock fragments. The third thing to notice is that Vesta shows signs of past internal heating and geological activity. Finally, notice the three main types of asteroids and how they are distributed through the asteroid belt. That will be a clue to help you understand their origin.

Before you continue, you should note that not all asteroids lie in the asteroid belt; a few thousand objects larger than 1 km follow orbits that cross Earth's orbit. A number of searches are under way to locate these near-Earth objects (NEOs). For example, LONEOS (Lowell Observatory Near Earth Object Search) is searching the entire sky once a month and should be able to locate a thousand NEOs over the next 10 years. Astronomers are searching for these asteroids not only because they want to understand the asteroids better, but because such asteroids must collide with Earth occasionally. Although such collisions occur very rarely, a single impact could cause planetwide devastation. You will learn about such impacts on Earth later in this chapter.

The Origin of the Asteroids

An old theory proposed that asteroids are the remains of a planet that exploded. Planet-shattering death rays may make for exciting science-fiction movies, but in reality planets do not explode. The gravitational field of a planet holds the mass tightly, and disrupting the planet would take tremendous energy. Shattering Earth, for example, would take all of the energy generated by the sun over a period of two weeks. In addition, the total mass of the asteroids is only about one-twentieth the mass of the moon, hardly enough to be the remains of a planet.

Astronomers believe that the asteroids are the remains of material that was unable to form a planet at 2.8 AU from the sun because of the gravitational influence of Jupiter, the next planet outward. If this is true, then the asteroids are the remains of planetesimals fragmented by collisions with one another. This would explain why the C-type asteroids, which appear to be carbonaceous, are more common in the outer asteroid belt. It is cooler there, and the condensation sequence (see Chapter 16) predicts that carbonaceous material would form there more easily than in the inner belt.

As you saw in the case of Vesta, a few asteroids may have been geologically active with lava flowing on their surfaces when they were young. Perhaps they incorporated short-lived radioactive elements such as aluminum-26. Such elements could have been produced by a supernova explosion, and such an explosion might have triggered the formation of the sun and planets. But not all asteroids have been active. Ceres, at 900 km in diameter, is almost twice as big as Vesta, but it shows no spectroscopic sign of past activity.

Although there are still mysteries to solve, you can understand the compositions of the meteorites. They are fragments of planetesimals, some of which developed molten cores, differentiated, may have suffered lava flows on their surfaces, and then cooled slowly. The largest asteroids astronomers see today may be nearly unbroken planetesimals, but the rest are just the fragments produced by 4.6 billion years of collisions.

Inquire | Review | Analyze

What is the evidence that asteroids have been fragmented?

First, you might note that the solar nebula theory of the formation of the solar system predicts that planetesimals collided and either stuck together or fragmented. This is suggestive, but it is not evidence. A theory can never be used as evidence to support some other theory or hypothesis. Evidence means observations or the results of experiments, so you must turn to observations of asteroids. The spacecraft photographs of asteroids show irregularly shaped little worlds heavily scarred by impact craters. In fact, radar images show what may be pairs of bodies in contact, and images of Ida reveal its small satellite, Dactyl. Still more double asteroids have been found. Furthermore, meteorites appear to come from the asteroid belt, and a few have been linked to specific asteroids such as Vesta. All of this evidence suggests that the asteroids have suffered violent collisions in their history.

The impact fragmentation of asteroids has been important, but it has not erased all traces of the original planetesimals from which the asteroids formed. What evidence could you cite to show what those planetesimals were like?

■ ■ ■

Connections: The asteroids are only the last crumbs left behind by failed planet building near Jupiter. Farther out in the solar system are other remains—ice shards from the age of planet formation.

19-3 Comets

OF ALL THE FOSSILS LEFT BEHIND by the solar nebula, comets are the most beautiful. Asteroids are dark, rocky worlds, and meteors are flitting specks of fire, but the comets move with the grace and beauty of a great ship at sea (■ Figure 19-5). Comet Hale–Bopp, for example, was visible for weeks in 1996 and 1997. Scientifically, comets are interesting because observations of the tail and head of a comet tell about the small icy nucleus and about the ancient solar nebula from which our solar system's wealth of comets is inherited.

■ Figure 19-5

Comet C/2001 Q4 (NEAT) was discovered in 2001 by the Near Earth Asteroid Tracking (NEAT) system operated by NASA. The comet was brightest in 2004. A comet can remain bright in the sky for weeks or months as it sweeps along its orbit through the inner solar system. (T. Rector, University of Alaska Anchorage, Z. Levay and L. Frattare, Space Telescope Science Institute and WIYN/NOAO/AURA/NSF)

Observations of Asteroids

Seen from Earth, asteroids look like faint points of light moving in front of distant stars. Not many years ago they were known mostly for drifting slowly through the field of view and spoiling long time exposures. Some astronomers referred to them as "the vermin of the sky." Spacecraft have now visited asteroids, and the images radioed back to Earth show that the asteroids are small but fascinating worlds preserving signs of the processes that built the planets.

Visual-wavelength image

The Near Earth Asteroid Rendezvous (NEAR) spacecraft visited the asteroid Eros in 2000 and found it to be heavily cratered by collisions and covered by a layer of crushed rock ranging from dust to large boulders. The NEAR spacecraft eventually landed on Eros.

Eros appears to be a solid fragment of rock.

10 km

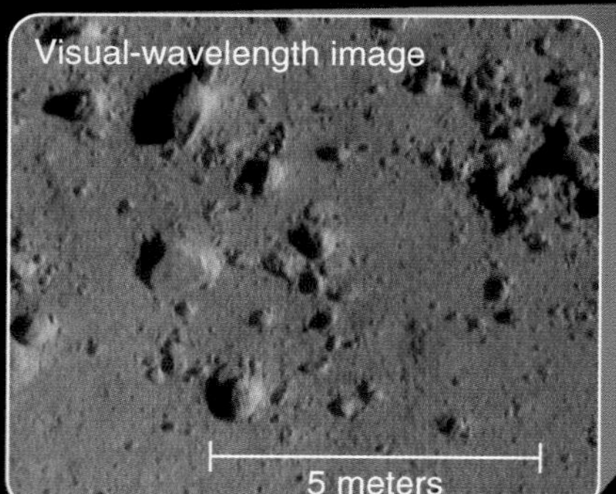

Most asteroids are too small for their gravity to pull them into a spherical shape. Impacts break them into irregularly shaped fragments.

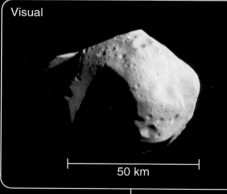

Visual

50 km

The surface of Mathilde is very dark rock.

The mass of an asteroid can be found from its gravitational influence on passing spacecraft. Mathilde has a low mass, and that makes its density so low it cannot be solid rock. Like many asteroids, Mathilde may be a rubble pile from broken fragments with large empty spaces between fragments.

If you walked across the surface of an irregularly shaped asteroid such as Eros, you would find gravity very weak; and in many places, it would not be perpendicular to the surface.

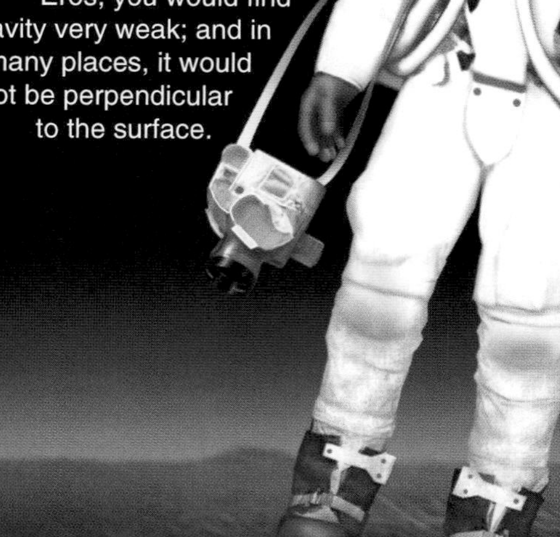

Enhanced visual image

5 km

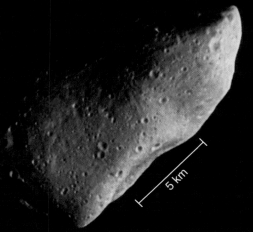

Like most asteroids, Gaspra would look gray to your eyes; but, in this enhanced image, color differences probably indicate difference in mineralogy.

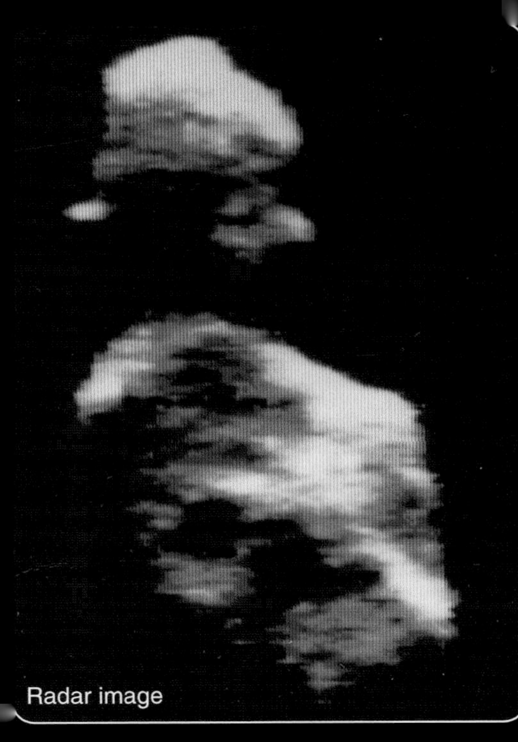

Radar image

Asteroids that pass near Earth can be imaged by radar. The asteroid Toutatis is revealed to be a double object—two objects orbiting close to each other or actually in contact.

Ida

Double asteroids are more common than was once thought, reflecting a history of collisions and fragmentation. The asteroid Ida is orbited by a moon Dactyl only about 1.5 km in diameter.

Dactyl

30 km

Enhanced visual + infrared

Occasional collisions among the asteroids release fragments, and Jupiter's gravity scatters them into the inner solar system as a continuous supply of meteorites.

Visual-wavelength image

500 km

Model

Elevation map

13-km-deep crater

Elevation

-12km +12km

The large asteroid Vesta provides evidence that some have suffered geological activity. No spacecraft has visited it, but its spectrum resembles that of solidified lava. Images made by the Hubble Space Telescope allow the creation of a model of its shape. It has a huge crater at its south pole. A family of small asteroids is evidently composed of fragments from Vesta, and a certain class of meteorites, spectroscopically identical to Vesta, are believed to be fragments from the asteroid. The meteorites appear to be solidified basalt.

Meteorite from Vesta

Vesta appears to have had internal heat at some point in its history, perhaps due to the decay of radioactive minerals. Lava flows have covered at least some of its surface.

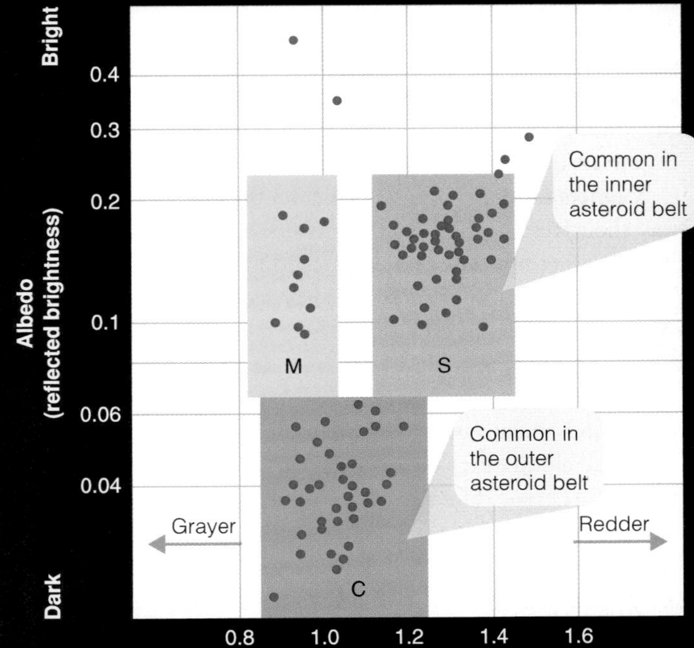

Although asteroids would look gray to your eyes, they can be classified according to their albedos (reflected brightness) and spectroscopic colors. S-types are brighter and tend to be reddish. They are the most common kind of asteroid and appear to be the source of the most common chondrites.

M-type asteroids are not too dark but are also not very red. They may be mostly iron-nickel alloys.

C-type asteroids are as dark as lumps of sooty coal and appear

Properties of Comets

As always, let's begin our study of a new kind of object by summarizing its observational properties. What do comets look like, and how do they behave? The observations are the evidence that reveal the secrets of the comets.

Study **Comet Observations** on pages 464 and 465 and notice three important properties of comets. First, comets have two kinds of tails, which tells you that the nucleus contains ices of water and other compounds plus rocky material most evident as dust. Second, notice the importance of the solar wind and solar radiation in shaping the tails of comets. Finally, notice the evidence that comet nuclei are fragile.

Astronomers can put these and other observations together to discuss the structure of comet nuclei.

AceAstronomy™ Log into AceAstronomy and select this chapter to see Astronomy Exercise "Comets." Design your own comet and see what happens.

The Geology of Comet Nuclei

The nuclei of comets are quite small and cannot be studied in detail from Earth-based telescopes. Nevertheless, astronomers are beginning to understand the geology of these peculiar worlds.

Comet nuclei contain ices of water and other volatile compounds such as carbon dioxide, carbon monoxide, methane, ammonia, and so on. These ices are the kinds of compounds that should have condensed from the outer solar nebula, and that makes astronomers think that comets are ancient samples of the gases and dust from which the planets formed.

When the nuclei of comets approach the sun, the ices absorb energy from sunlight and sublime—change from a solid directly into a gas—to produce the observed tails. As the gases break down and combine chemically, they release many compounds found in comet tails. Vast clouds of hydrogen gas observed around the heads of comets are derived from the breakup of molecules from the ices.

Five spacecraft flew past the nucleus of Comet Halley when it visited the inner solar system in 1985 and 1986. The Deep Space 1 spacecraft flew past the nucleus of Comet Borrelly in 2001, and the Stardust spacecraft flew past Comet Wild 2 (pronounced *Vilt two*) in 2004. Photos show that these comet nuclei are irregular in shape and very dark, with jets of gas and dust spewing from active regions on the nuclei (■ Figure 19-6). In general, these nuclei are darker than a lump of coal, which suggests the composition of the carbon-rich meteorites called carbonaceous chondrites.

From the gravitational influence of a nucleus on a passing spacecraft, astronomers can find the mass and density of the nucleus. Comet nuclei appear to have densities of 0.1 to 0.25 g/cm^3, much less than the density of ice. From these observations astronomers can conclude that comet nuclei are not solid balls of ice but must be fluffy mixtures of ices and dust with significant amounts of empty space.

Photographs of the comas of comets often show jets springing from the nucleus and being swept back by the pressure of sunlight and by the solar wind to form the tail (Figure 19-6). Studies of the motions of these jets as the nucleus rotates and the photographs of the nucleus of Comet Halley reveal that the jets originate from active regions that may be faults or vents. As the rotation of a cometary nucleus carries an active region into sunlight, it begins venting gas and dust, and as it rotates into darkness it shuts down.

The irregular shape of the nucleus of Comet Halley combined with its observed low density suggests that comet nuclei are not solid objects. To call them dirty snowballs or icy mud balls is misleading. Rather, they appear to be irregular bodies with large voids. Although comet nuclei seem to have porous crusts of dark material, the ice and rock are not uniformly mixed through the interior. Breaks in the crust can expose pockets of highly volatile ices and cause sudden bursts of gas production.

The nucleus of Comet Halley was observed to be 16 by 8 by 7 km, the nucleus of Comet Hale–Bopp is believed to be 80 km in diameter, and the nucleus of Comet Wild 2 was smaller. Each passage near the sun costs such an object many millions of tons of ices, so the nuclei slowly lose their ices until there is nothing left but dust and rock.

The Origin of Comets

Family relationships among the comets provide clues to their origin. Most comets have long, elliptical orbits with periods greater than 200 years. These are known as long-period comets. Their orbits are randomly inclined, with comets falling into the inner solar system from all directions. As many circle the sun clockwise as counterclockwise.

In contrast, about 100 or so of the 600 well-studied comets have orbits with periods less than 200 years. These short-period comets follow orbits that lie within 30° of the plane of the solar system, and most revolve around the sun counterclockwise—the same direction the planets orbit. Comet Halley, with a period of 76 years, is a short-period comet.

Comets cannot survive long before the heat of the sun drives away their ices and reduces them to inactive bodies of rock and dust. A comet may last only 100 to 1000 orbits around the sun. The comets seen in our skies can't have survived 4.6 billion years since the formation of the solar system, so there must be a continuous supply of new comets. Where do they come from?

In the 1950s, Dutch astronomer Jan Oort proposed that the long-period comets are objects that fall in from the **Oort cloud,** a spherical cloud of icy bodies believed to extend from 10,000 to 100,000 AU from the sun (■ Figure 19-7). Astronomers estimate that the cloud contains several trillion icy bodies. Far from the sun, they are very cold, lack comas and tails, and are invisible. The gravitational influence of occasional passing stars could per-

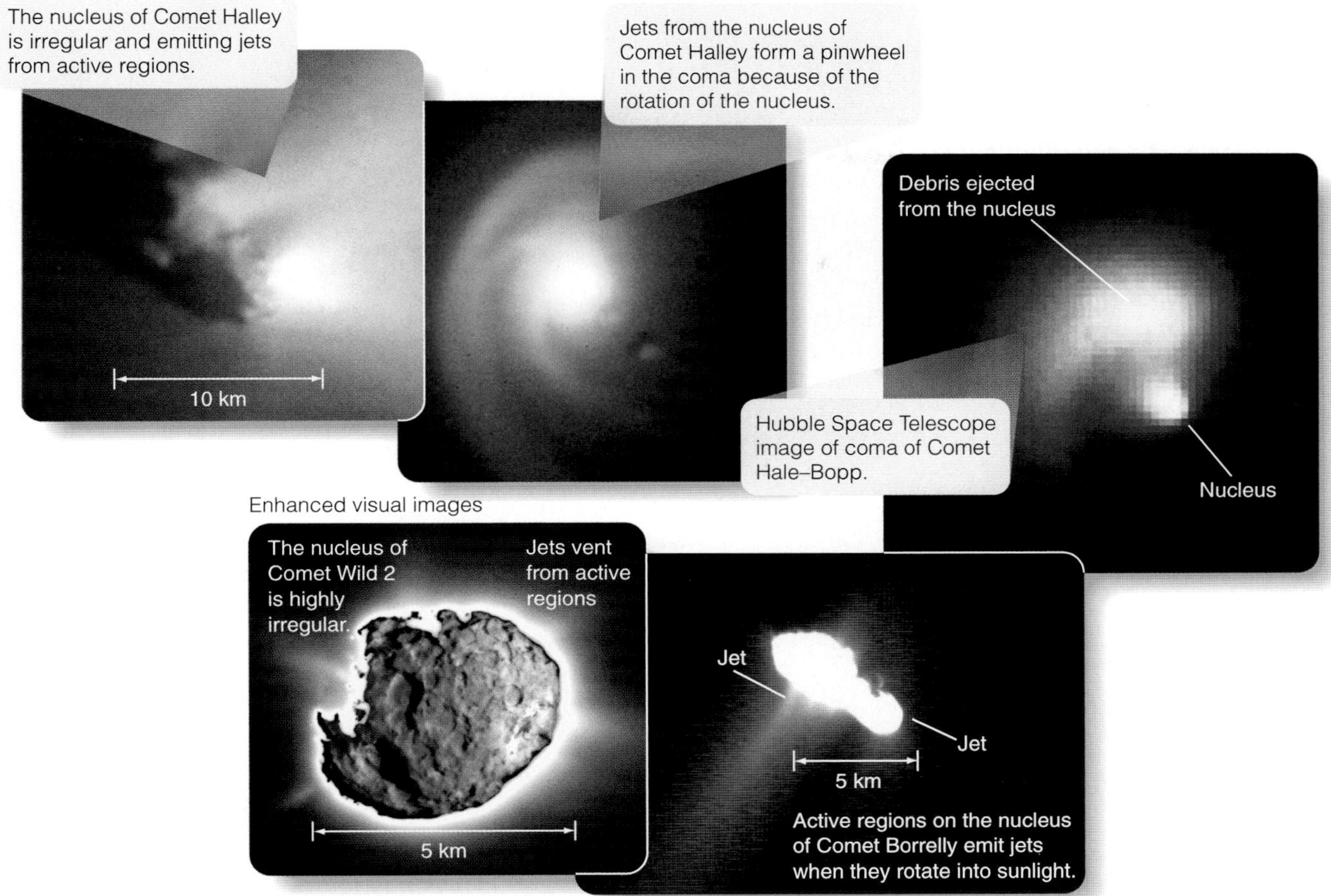
The nucleus of Comet Halley is irregular and emitting jets from active regions.
10 km
Jets from the nucleus of Comet Halley form a pinwheel in the coma because of the rotation of the nucleus.
Debris ejected from the nucleus
Hubble Space Telescope image of coma of Comet Hale–Bopp.
Nucleus
Enhanced visual images
The nucleus of Comet Wild 2 is highly irregular.
Jets vent from active regions
5 km
Jet
Jet
5 km
Active regions on the nucleus of Comet Borrelly emit jets when they rotate into sunlight.

Figure 19-6

Visual-wavelength images made by spacecraft and by the Hubble Space Telescope show how the nucleus of a comet produces jets of gases from regions where sunlight vaporizes ices. (Halley nucleus: ©1986 Max-Planck Institute; Halley coma: Steven Larson; Comet Borrelly and Comet Hale–Bopp: NASA)

turb a few of these objects to fall into the inner solar system, where the heat of the sun warms their ices and transforms them into comets. Because the Oort cloud is spherical, these long-period comets fall inward from random directions.

Some of the short-period comets, including Comet Halley, appear to have originated in the Oort cloud and probably had their orbits altered by a close encounter with Jupiter. But many of the short-period comets cannot have begun in the Oort cloud. Interactions with a planet can't capture objects from the Oort cloud into the orbits that some short-period comets occupy. There must be another source of icy bodies in our solar system. In 1951, Dutch-American astronomer Gerard P. Kuiper proposed that the formation of the solar system should have left behind a belt of small, icy planetesimals beyond the Jovian planets and in

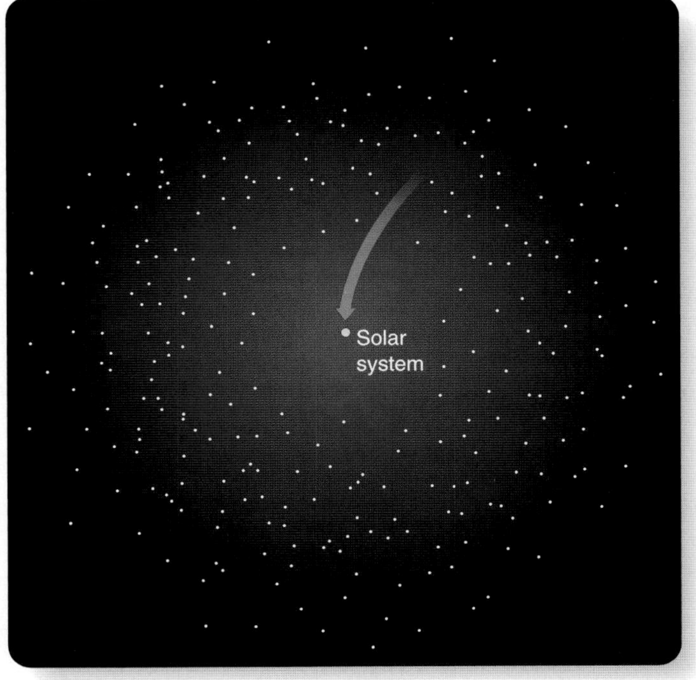
Solar system

Figure 19-7

The long-period comets appear to originate in the Oort cloud. Objects that fall into the inner solar system from this cloud arrive from all directions.

Comet Observations

Comets are beautiful to the eye and to the intellect. Their vast size and magnificent beauty spring from a small, dirty lump. Their simple structure contains details that arise from powerful processes related to the solar wind. Most of all, the luminous glow of comets tells about the origin of our solar system.

Gas tail (Type I)

Dust tail (Type II)

A **type I** or **gas tail** is produced by ionized gas carried away from the nucleus by the solar wind. The spectrum of a gas tail is an emission spectrum. The atoms are ionized by the ultraviolet light in sunlight. The wisps and kinks in gas tails are produced by the magnetic field embedded in the solar wind.

Spectra of gas tails reveal atoms and ions such as H_2O, CO_2, CO, H, OH, O, S, C, and so on. These are released by the vaporizing ices or produced by the breakdown of those molecules. Some gases, such as hydrogen cyanide (HCN), must be formed by chemical reactions.

A **type II** or **dust tail** is produced by dust from the vaporizing ices of the nucleus. The dust is pushed gently outward by the pressure of sunlight, and it reflects an absorption spectrum, the spectrum of sunlight. The dust is not affected by the magnetic field of the solar wind, so dust tails are more uniform than gas tails. Dust tails are often curved because the dust particles follow their individual orbits around the sun once they leave the nucleus.

Nucleus

The nucleus of a comet (not visible here) is a small, fragile lump of porous rock containing ices of water, carbon dioxide, ammonia, and so on. Comet nuclei can be 10 to 100 km in diameter.

Coma

The **coma** of a comet is the cloud of gas and dust that surrounds the nucleus. It can be over 1,000,000 km in diameter, bigger than the sun.

When a spacecraft named ICE passed through the gas tail of a comet, it found a magnetic field from the solar wind draped over the nucleus like seaweed draped over a fishhook.

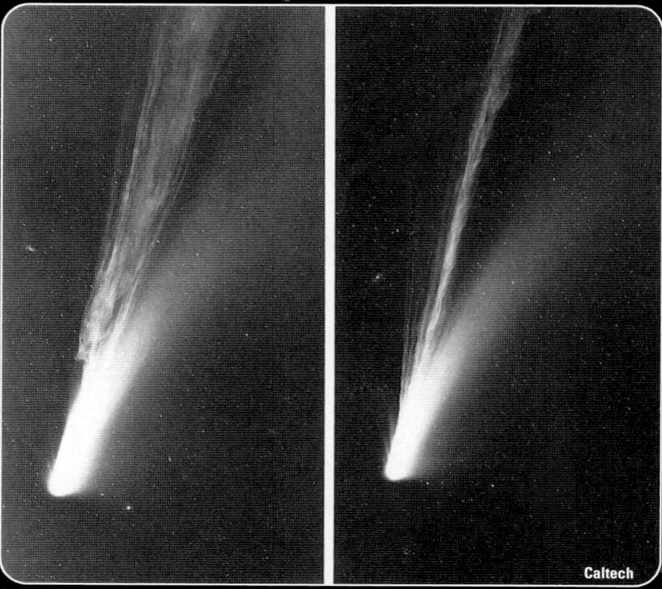

Comet Mrkos in 1957 shows how the gas tail can change from night to night due to changes in the magnetic field in the solar wind.

Visual-wavelength images

AceAstronomy™

Log into AceAstronomy and select this chapter to see Active Figure "Comet Tails." Notice how dust and gas move differently as they leave the head of a comet.

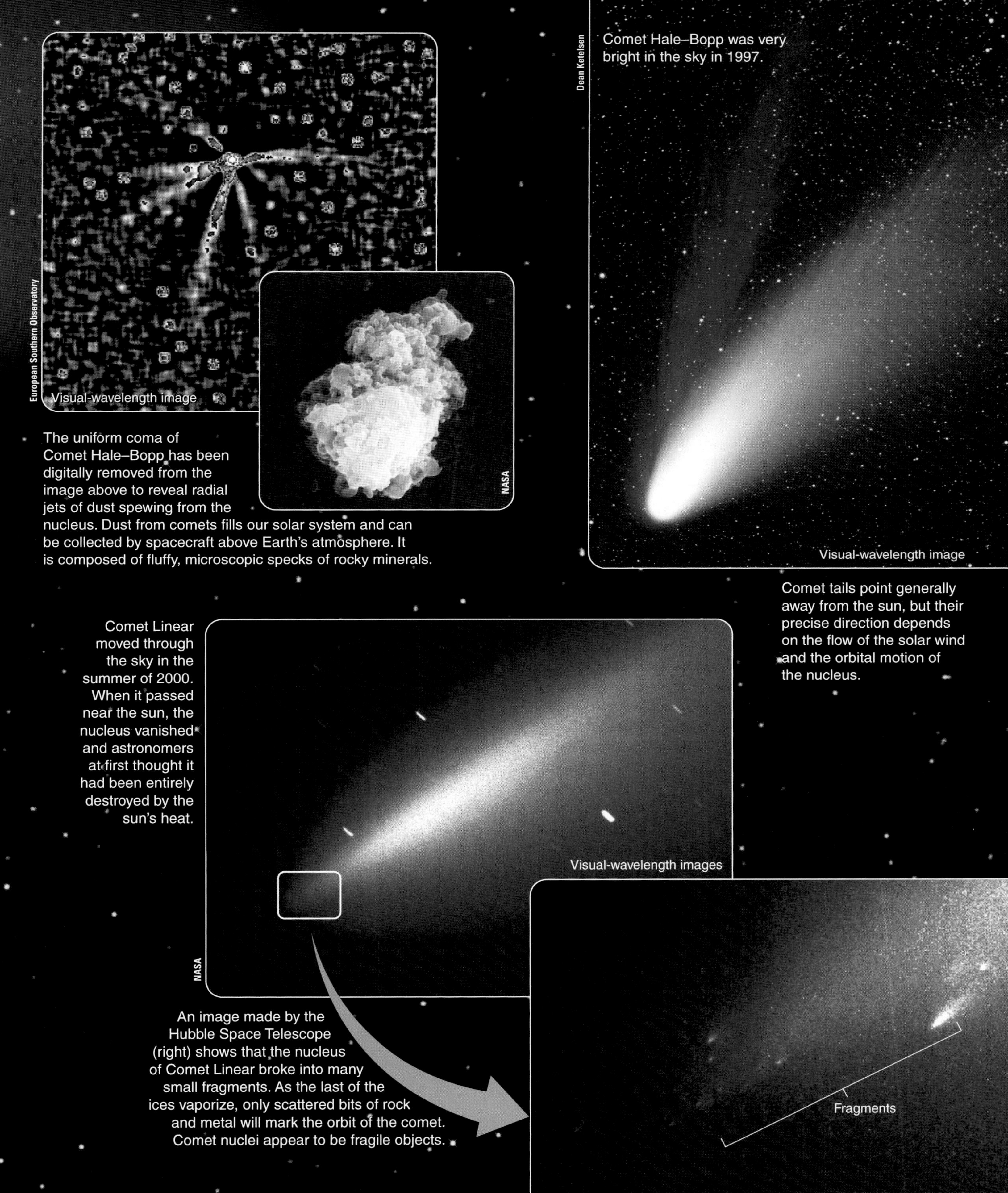

The uniform coma of Comet Hale–Bopp has been digitally removed from the image above to reveal radial jets of dust spewing from the nucleus. Dust from comets fills our solar system and can be collected by spacecraft above Earth's atmosphere. It is composed of fluffy, microscopic specks of rocky minerals.

Comet Hale–Bopp was very bright in the sky in 1997.

Comet tails point generally away from the sun, but their precise direction depends on the flow of the solar wind and the orbital motion of the nucleus.

Comet Linear moved through the sky in the summer of 2000. When it passed near the sun, the nucleus vanished and astronomers at first thought it had been entirely destroyed by the sun's heat.

An image made by the Hubble Space Telescope (right) shows that the nucleus of Comet Linear broke into many small fragments. As the last of the ices vaporize, only scattered bits of rock and metal will mark the orbit of the comet. Comet nuclei appear to be fragile objects.

the plane of the solar system. Those objects could not be detected with the telescopes and instruments available in those days, but the proposed band of icy bodies became known as the Kuiper belt. You learned about the Kuiper belt in the previous chapter in the discussion of the true nature of Pluto.

Modern astronomers have found hundreds of Kuiper-belt objects in orbits extending from Neptune at 30 AU out to about 50 AU from the sun. Few are seen further away, and those that are found out there were probably scattered outward by interactions with other belt objects. The entire Kuiper belt would be hidden behind the yellow dot in Figure 19-7.

Not only have astronomers detected the Kuiper belt around our sun, they can detect similar belts around other stars. A number of nearby stars such as Beta Pictoris (Chapter 16) are surrounded by disks of dust believed to be released by comets in the equivalent of Kuiper belts around these stars. Thus the detection of the dust implies the presence of Kuiper belts.

The Kuiper-belt objects that have been found are mostly too big to become comets, but there must be lots of smaller objects out there too. When a small Kuiper-belt object is perturbed into the inner solar system, it can interact with planets and be captured into orbit as a short-period comet.

If comets are icy planetesimals from the Oort cloud and the Kuiper belt, how did those planetesimals form? The Kuiper-belt objects appear to have formed as icy planetesimals in the outer solar nebula not much farther from the sun than the outer planets. But the objects in the Oort cloud lie much farther from the sun, and they can't have formed out there. The solar nebula would have been too tenuous at such great distances. Also, you would expect objects that formed from the nebula to be confined to a disk and not distributed in a sphere. Rather, astronomers think the objects now in the Oort cloud formed in the outer solar system among the present orbits of the Jovian planets. As the Jovian planets grew more massive, they swept up some of these planetesimals and ejected others to form the Oort cloud. If this idea is true, then even the long-period comets are icy planetesimals.

Impacts on Earth

For centuries, superstitious people have associated comets with doom, which seems silly. Comets are graceful visitors from the icy fringes of the solar system. Of course, comets and asteroids must hit planets now and then, so you might wonder just how dangerous such impacts would be.

Earthlings watched in awe during the summer of 1994 as the fragmented head of a comet (■ Figure 19-8) slammed into Jupiter and produced impacts equaling millions of megatons of TNT (Figure 18-3). Such impacts on Jupiter probably occur every century or so, and you might expect similar impacts on Earth, which is smaller and has less gravitational power, to occur much less often. Nevertheless, these impacts do occur, and you can find chains of craters on moons that seem to have been formed by fragmented comets (Figure 19-8).

Small meteorite impacts occur quite often, and a building is damaged by a falling meteorite every few years. Larger impacts

■ Figure 19-8 ▶

(a) Tides from Jupiter pulled apart the nucleus of Comet Shoemaker–Levy 9 to form a long strand of icy bodies and dust that fell back to strike Jupiter two years later. (b) A 40-km-long crater chain on Earth's moon and (c) another 140-km-long crater chain on Jupiter's moon Callisto were apparently formed by the impact of fragmented comet nuclei. The impacts that form such chains probably occur within a span of seconds. (NASA)

■ Active Figure 19-9 ▲

(a) The Barringer Meteorite Crater (near Flagstaff, Arizona) is nearly a mile in diameter and was formed about 50,000 years ago by the impact of an iron meteorite roughly 90 m in diameter. It hit with energy equivalent to that of a 3-megaton hydrogen bomb. Notice the raised and deformed rock strata all around the crater. For scale, locate the brick building on the far rim at right. (M. A. Seeds) (b) Like all larger-impact features, the Barringer Meteorite Crater has a raised rim and scattered ejecta. (USGS)

AceAstronomy™ Log into AceAstronomy and select this chapter to see the Active Figure called "Impacts on Earth." Notice how large impacts are less common than smaller ones.

are less common, and truly large impacts are rare. Earth is marked by about 150 meteorite craters that illustrate the power of an impacting object (■ Figure 19-9). A large impact could have devastating consequences.

Studies of sediments laid down all over the world 65 million years ago at the time of the extinction of the dinosaurs have found overabundances of the element iridium—common in meteorites but rare in Earth's crust. This finding suggests that the impact of a large meteorite may have altered the atmosphere and climate on Earth so dramatically that the dinosaurs and over 75 percent of the species then on Earth became extinct.

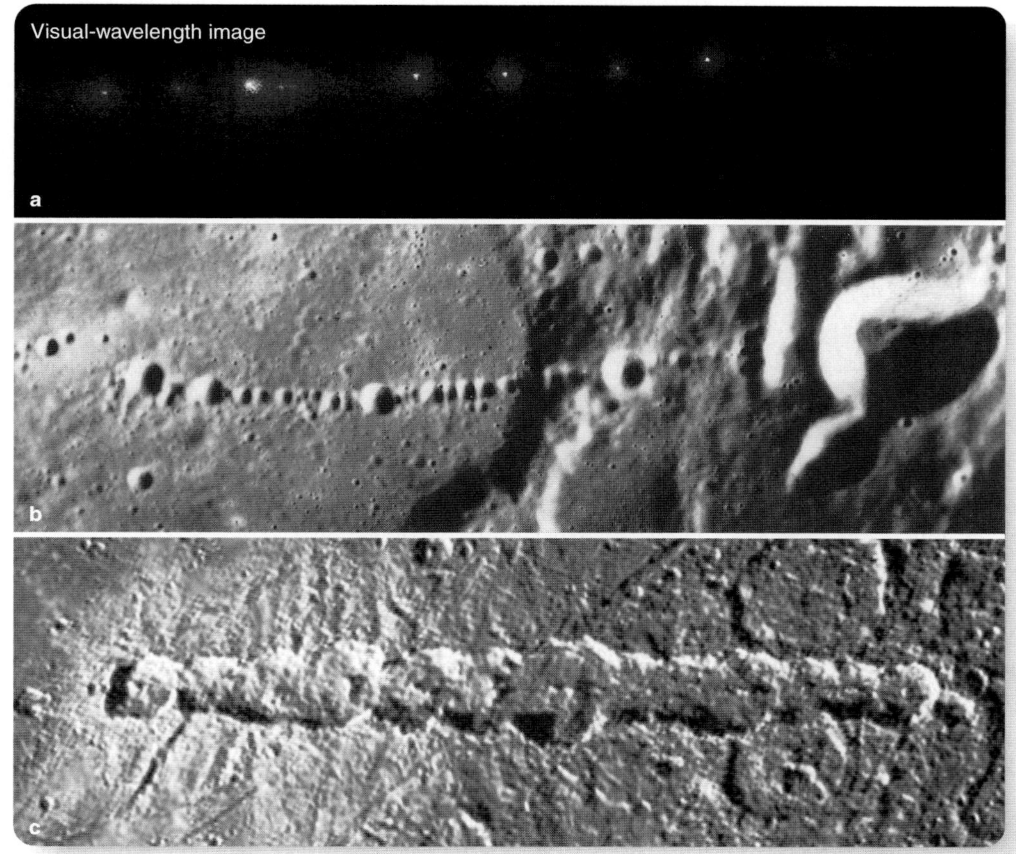

Mathematical models and observations of the impact of the comet fragments on Jupiter in 1994 have combined to create a plausible scenario of the events following a major impact on Earth. Of course, creatures living near the site of the impact would probably die in the initial shock, and an impact at sea would create tsunamis (tidal waves) many hundreds of meters high that would devastate coastal regions for many kilometers inland halfway around the world. But the worst effects would begin after the initial explosion. On land or sea, a major impact would excavate large amounts of pulverized rock, heat it to high temperatures, and loft it high above the atmosphere. As this material fell back, Earth's atmosphere would be turned into a glowing oven of red-hot rock streaming through the air as a rain of meteors, and the heat would trigger massive forest fires around the world. Soot from such fires has been detected in the layers of clay laid down at the end of the Cretaceous period. Once the firestorms cooled, the remaining dust in the atmosphere would block sunlight and produce deep darkness for a year or more, killing off most plant life. At the same time, large amounts of carbon dioxide locked in limestone deposits and released into the atmosphere by the impact would produce intense acid rain. All of these consequences make it surprising that any life could have survived such an impact.

Geologists have located a crater at least 150 km in diameter centered near the village of Chicxulub in the northern Yucatán (■ Figure 19-10). Although the crater is totally covered by sediments, mineral samples show that it contains shocked quartz typical of impact sites and that it is the right age. The impact of an object 10 to 14 km in diameter formed the crater about 65 million years ago, just when the dinosaurs and many other species died out, and many Earth scientists now believe that this is the scar of the impact that ended the Cretaceous period.

Chicxulub may not be the only impact that triggered an extinction. The biggest extinction on record occurred 250 million years ago at the end of the Permian period, when 95 percent of life in the oceans and 80 percent of life on land died out. That has been called the Great Dying. Core samples from the ocean floor northwest of Australia reveal shattered rock, meteoric fragments, and shocked quartz. Seismic and gravity data support the idea that a major impact occurred there at about the right time to cause the Great Dying.

Could such impacts happen again? Earth gets hit by small meteorites every day and by larger objects less often. Impacts by large asteroids may happen many millions of years apart, but they continue to happen. In mid-March 1998, newspaper headlines announced, "Mile-Wide Asteroid to Hit Earth in October 2028." The frightening story was true except that the news media did not emphasize the uncertainty in the orbit. Within days, astronomers found the asteroid on old photographic plates, recalculated the orbit adding the new data, and concluded that the asteroid, known as 1997XF$_{11}$, would miss Earth by 600,000 miles. There will be no impact by this asteroid in 2028, but there are plenty more asteroids that haven't been discovered. It is just a matter of time.

A solar system is a dangerous place to put an inhabited planet. Comets, asteroids, and meteoroids constantly rain down on the planets, and Earth gets hit often. Chicxulub isn't the only large impact scar on Earth. About 150 are known, including giant craters buried under sediment in Iowa and another underlying most of Chesapeake Bay.

AceAstronomy™ Log into AceAstronomy and select this chapter to see the Astronomy Exercise "Cratering." What factors determine the size of a crater?

Inquire | Review | Analyze

How do comets help explain the formation of the planets?
According to the solar nebula hypothesis, the planets formed from planetesimals that accreted in a disk-shaped nebula around the forming sun. The planetesimals that formed in the inner solar nebula were warm and could not incorporate much ice. The asteroids may be the last remains of such bodies. In the outer solar nebula, it was colder, and the planetesimals would have contained large amounts of ices. Many of these planetesimals were destroyed when they fell together to make the planets, but some may have survived. The icy bodies of the Oort cloud and the Kuiper belt may be the last surviving icy planetesimals in our solar system. When these bodies fall into the inner solar system, you see them as comets, and the gases they release indicate that many of them are rich in volatile materials such as water and carbon dioxide. These are the ices you would expect to find in the icy planetesimals. Furthermore, comets are rich in dust, and the planetesimals must have included large amounts of dust frozen into the ices when they formed. Thus, the nuclei of comets seem to be frozen samples of the ancient solar nebula.

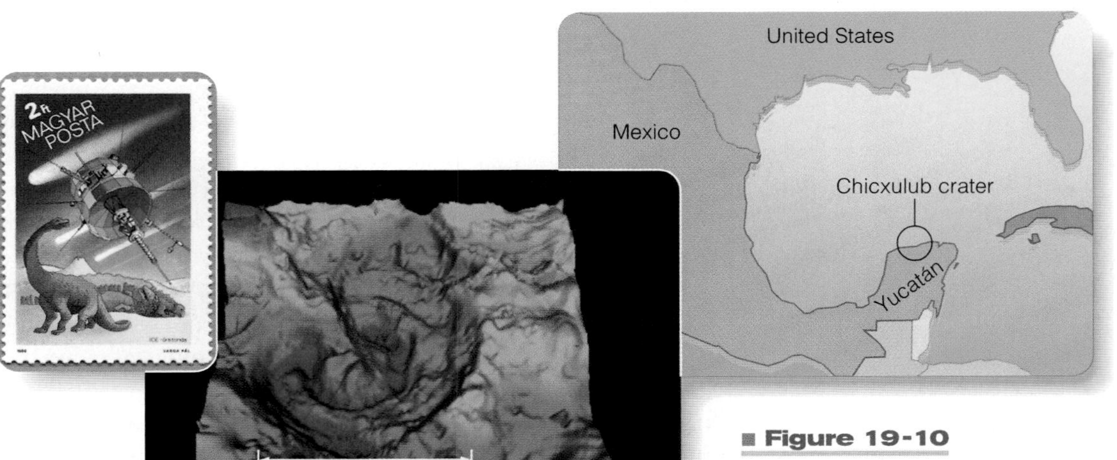

■ Figure 19-10

The theory that the impact of one or more comets altered Earth's climate and drove dinosaurs to extinction has become so popular it appeared on this Hungarian stamp. The spacecraft shown (ICE) flew through the tail of a comet in 1985. Note the dead dinosaurs in the background. The giant impact scar buried in Earth's crust near the village of Chicxulub in the northern Yucatán was formed about 65 million years ago by the impact of a large comet or asteroid. This gravity map shows the extent of the crater hidden below limestone deposited long after the impact. (Virgil L. Sharpton, University of Alaska, Fairbanks)

Nearly all of the mass of a comet is in the nucleus, but the light you see comes from the coma and the tail. What kind of spectra do you get from comets, and what does that tell you about the process that converts dirty ice into a comet?

■ ■ ■

Connections: Life on Earth seems fragile and exposed. Are we just lucky to have survived so long? Does life exist on other worlds? If so, has it, too, survived long enough to develop intelligence? That is the story of the next chapter.

Study and Review Tools

Summary

19-1 | Meteorites

Where do meteors and meteorites come from?

- Meteorites can be classified according to their mineral content.
- Iron meteorites are mostly iron and nickel; when sliced open, polished, and etched, they show Widmanstätten patterns. These tell us that the metal cooled from a molten state very slowly.
- Stony meteorites included chondrites, which contain small, glassy particles called chondrules, believed to be very ancient droplets of molten material formed in the solar nebula.
- Stony meteorites that are rich in volatiles and carbon are called carbonaceous chondrites. They are among the least modified meteorites.
- An achondrite is a stony meteorite that contains no chondrules and no volatiles. They appear to have been melted after they formed and, in some cases, resemble solidified lavas.
- Many meteorites appear to have formed as part of larger bodies that melted, differentiated, and cooled very slowly. Later they were broken up, and fragments from the core became iron meteorites, while fragments from the outer layers became stony meteorites.
- Carbonaceous chondrites appear to have formed further from the sun.
- The evidence, including the orbits of meteorites seen to fall, suggests that meteorites are fragments of asteroids.
- The vast majority of meteors, including those in meteor showers, appear to be low-density, fragile bits of debris from comets.

19-2 | Asteroids

What are the asteroids?

- Asteroids are irregular in shape and heavily cratered from collisions. Their surfaces are covered by gray, pulverized rock, and some asteroids have such low densities they appear to be fragmented rubble piles.
- Most asteroids lie in a belt between Mars and Jupiter, although some follow orbits that cross into the inner solar system. If they pass near Earth, they are called Near Earth Objects (NEOs).
- C-type asteroids are more common in the outer asteroid belt where it is cooler. They are darker and may be carbonaceous.
- S-type asteroids are the most common and appear to be the source of the most common kind of meteorites, the chondrites. S-type asteroids are more common in the inner belt.
- M-type asteroids appear to have nickel-iron compositions and may be the cores of broken asteroids.
- The asteroids formed as rocky planetesimals between Mars and Jupiter, but Jupiter prevented them from forming a planet. Collisions have broken all but the largest of the asteroids.

19-3 | Comets

Where do comets come from?

- A comet is produced by a lump of rock and ices 10 to 100 km in diameter. In long, elliptical orbits, the icy body stays frozen until it nears the sun. Then some of the ices vaporize and release dust and gas that is blown away to form a tail.
- A type I, or gas, tail is ionized gas carried away by the solar wind.
- A type II, or dust, tail is solid debris released from the nucleus and blown outward by the pressure of sunlight.
- The coma of a comet can be up to a million kilometers in diameter.
- Spacecraft flying past comets have revealed that the nucleus of a comet has a very dark, rocky crust and that jets of vapor and dust issue from active regions on the sunlit side.
- The low density of the nuclei shows that they are irregular mixtures of ices and silicates, probably containing large voids.
- Comets are believed to have formed as icy planetesimals in the outer solar system, and some were ejected to form the Oort cloud. Comets falling in from the Oort cloud become long-period comets.
- Other icy bodies formed in the outer solar system and now make up the Kuiper belt beyond Neptune. Objects from the Kuiper belt that fall into the inner solar system can become short-period comets.

What happens when an asteroid or comet hits Earth?

- A major impact on Earth can trigger extinctions because of changes to the climate.
- Material falling back into the atmosphere after a major impact can trigger global forest fires.
- Large amounts of carbon dioxide released into the atmosphere by a major impact can produce intense acid rain.
- A major impact can fill the atmosphere with dust for years and plunge the planet into darkness.
- An impact at Chicxulub in the Yucatán 65 million years ago appears to have triggered the extinction of 75 percent of the species then on Earth, including the dinosaurs. The Great Dying at the end of the Permian may also have been triggered by a major impact.

New Terms

Widmanstätten pattern (p. 455)
chondrite (p. 455)
chondrule (p. 455)
carbonaceous chondrite (p. 455)
achondrite (p. 455)
stony-iron meteorite (p. 455)
meteor shower (p. 456)
Oort cloud (p. 462)
type I, or gas, comet tail (p. 464)
type II, or dust, comet tail (p. 464)
coma (p. 464)

Review Questions

AceAstronomy™ Assess your understanding of this chapter's topics with additional quizzing and animations at **http://astronomy.brookscole.com/sh9e**

1. What do Widmanstätten patterns indicate about the history of iron meteorites?
2. What do chondrules tell you about the history of chondrites?
3. Why are there no chondrules in achondritic meteorites?
4. Why do astronomers refer to carbonaceous chondrites as "unmodified"?
5. How do observations of meteor showers reveal one of the sources of meteoroids?
6. How can most meteors be cometary if all meteorites are asteroidal?
7. Why do astronomers think the asteroids were never part of a planet?
8. What evidence indicates that the asteroids are fragmented?
9. What evidence indicates that some asteroids have differentiated?
10. What evidence indicates that some asteroids have had active surfaces?
11. How is the composition of meteorites related to the formation and evolution of asteroids?
12. What is the difference between a type I tail and a type II tail?
13. What evidence indicates that cometary nuclei are rich in ices?
14. Why do short-period comets tend to have orbits near the plane of the solar system?
15. How did the bodies in the Kuiper belt and the Oort cloud form?
16. What do you see in this image that tells you how big planetesimals were when the solar system was forming?

(Russell Kempton, New England Meteoritical)

17. Discuss the surface of the asteroid Mathilde. What do you see that tells you something about the history of the asteroids?

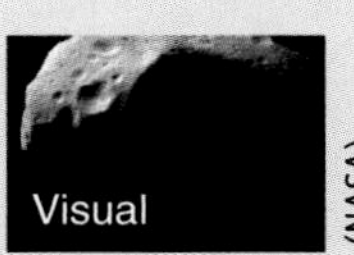

(NASA)

18. What do you see in this image of the nucleus of Comet Borrelly that tells you how comets produce their comas and tails?

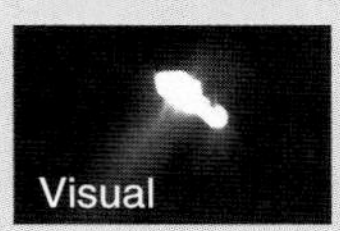

(NASA)

Discussion Questions

1. Futurists suggest humans may someday mine the asteroids for materials to build and supply space colonies. What kinds of materials could Earthlings get from asteroids? (*Hint:* What are S-, M-, and C-type asteroids made of?)
2. If cometary nuclei were heated by internal radioactive decay rather than by solar heat, how would comets differ from what is observed?
3. From what you know now, do you think the government should spend money to locate near-Earth asteroids? How serious is the risk?

Problems

1. Large meteorites are hardly slowed by Earth's atmosphere. Assuming the atmosphere is 100 km thick and that a large meteorite falls perpendicular to the surface, how long does it take to reach the ground? (*Hint:* About how fast do meteoroids travel?)
2. What is the orbital velocity of a meteoroid whose average distance from the sun is 2 AU? (*Hint:* Find the orbital period from Kepler's third law. See Table 4-1.)
3. If a single asteroid 1 km in diameter were fragmented into meteoroids 1 m in diameter, how many would it yield? (*Hint:* The volume of a sphere $= \frac{4}{3}\pi r^3$.)
4. What is the orbital period of a typical asteroid? (*Hint:* Use Kepler's third law. See Table 4-1.)
5. If half a million asteroids, each 1 km in diameter, were assembled into one body, how large would it be? (*Hint:* The volume of a sphere $= \frac{4}{3}\pi r^3$.)
6. What is the maximum angular diameter of Ceres as seen from Earth? Could Earth-based telescopes detect surface features? Could the Hubble Space Telescope? (*Hint:* See Reasoning with Numbers 3-1.)
7. If the velocity of the solar wind is about 400 km/s and the visible tail of a comet is 10^8 km long, how long does it take an atom to travel from the nucleus to the end of the visible tail?
8. If you saw Comet Halley when it was 0.7 AU from Earth and it had a visible tail 5° long, how long was the tail in kilometers? Suppose that the tail was not perpendicular to your line of sight. Is your answer too large or too small? (*Hint:* See Reasoning with Numbers 3-1.)
9. What is the orbital period of a cometary nucleus in the Oort cloud? What is its orbital velocity? (*Hints:* Use Kepler's third law. The circumference of a circular orbit $= 2\pi r$.)
10. The mass of an average comet's nucleus is about 10^{12} kg. If the Oort cloud contains 200×10^9 cometary nuclei, what is the mass of the cloud in Earth masses? (*Hint:* Mass of Earth $= 6 \times 10^{24}$ kg.)

Media Cluster

ACTIVE FIGURES

AceAstronomy™ To access the resources in the Media Cluster, log into AceAstronomy at **http://astronomy.brookscole.com/sh9e** and select Chapter 19.

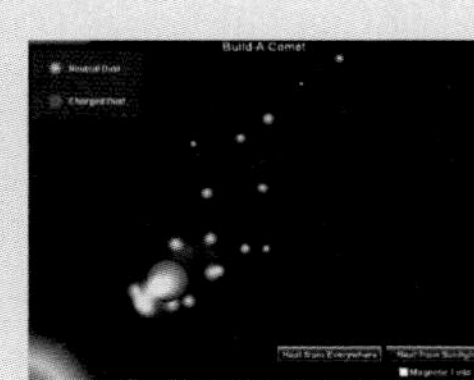

Build a Comet
This animation shows how radiation from the sun creates the tails of comets. You can compare this to "heat from everywhere," which causes particles to be shed in every direction.

Impacts on Earth
Watch asteroids pummel Earth in this animation and learn about the probability of a "killer asteroid" hitting Earth.

ASTRONOMY EXERCISES

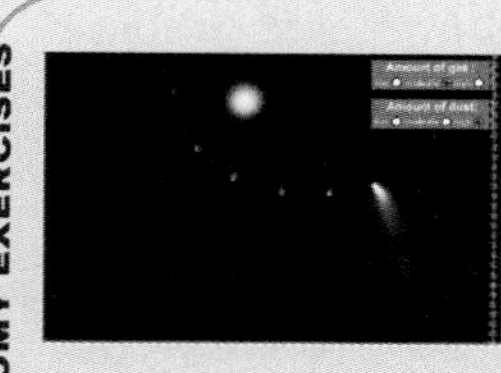

Comets
In this animation you can move a comet to different positions relative to the sun. You can also change the amount of gas and dust in the comet's nucleus and see how this affects the comet's tails.

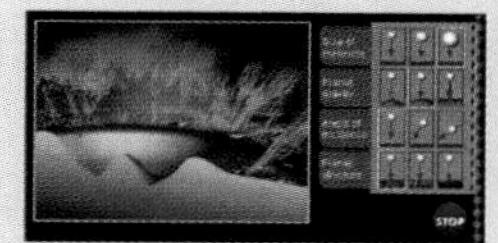

Cratering
Turn up your speakers for this one, where can you design collisions between a meteorite and a hypothetical planet.

VIRTUAL ASTRONOMY LAB

Lab 9: Asteroids and Kuiper-Belt Objects
This lab presents an overview of asteroid properties, including an exercise on how to discover asteroids. You also investigate the class of objects known as Kuiper-belt objects.

Critical Inquiries for the Web

1. Some nights are better for looking for meteors than others (see Table 19-1). These showers are known to be associated with comets, but how are these associations made? There are several sites on the Internet that provide information on meteor showers, including historical data and information on parent comets. Pick a shower whose parent comet is known and summarize how astronomers came to know that the meteors and the comet are related.
2. The chances are small that you will be killed by an asteroid impact, but if there are objects out there that astronomers are not aware of whose orbits intersect Earth, we could be in for a surprise one day. Look for information on the LONEOS project and other investigations into near-Earth asteroids. How many such objects have been discovered? What is the record for closest known passage of an asteroid to Earth?

Exploring *TheSky*

1. Of the five brightest asteroids, which has the most inclined orbit? (*Hint:* Use **Filters** in the **View** menu to turn off everything but the sun, stars, ecliptic, and minor planets. You can use **Tracking Setup** under **Time Skip** in the **Tools** menu to lock onto an object and follow it along the ecliptic as time passes.)
2. Do asteroids go through retrograde motion? (*Hint:* Use **Filters** in the **View** menu to turn off the stars and turn on the Equatorial Grid. See Activity 1 above.)
3. Of the five brightest asteroids, which has the most elliptical orbit? (*Hint:* Use **3D Solar System Mode** in the **View** menu and watch objects orbit the sun.)
4. Of the comets shown, which has the smallest orbit? (*Hint:* Use **3D Solar System Mode** in the **View** menu and watch objects orbit the sun.)

- Go to the Brooks/Cole Astronomy Resource Center **(http://astronomy.brookscole.com)** for critical thinking exercises, articles, and additional readings from InfoTrac College Edition, Brooks/Cole's online student library.

20 Life on Other Worlds

Did I solicit thee from darkness
to promote me?

JOHN MILTON
Paradise Lost

As LIVING THINGS, we have been promoted from darkness. We are made of heavy atoms that could not have formed at the beginning of the universe. Successive generations of stars fusing light elements into heavier elements have built the atoms so important to our existence. When a dark cloud of interstellar gas enriched in these heavy atoms fell together to form our sun, a small part of the cloud gave birth to the planet you inhabit. ▌ If life can originate on other worlds, and if intelligence is a natural result of the evolution of life forms, then you might expect that other intelligent races inhabit other worlds. Visits between worlds seem impossible, but perhaps humans can detect their existence by radio. ▌ Alien life forms could be quite different from Earthlings, but if they are alive, then they must share certain characteristics. Your

Every life form we know of has evolved to live somewhere on Earth. The Wekiu bug lives with the astronomers at 13,600 feet atop Hawaiian volcano Mauna Kea. It lives in the icy cinders and eats insects carried up by ocean breezes. (Kris Koenig/Coast Learning Systems)

Guidepost

Looking Back

This chapter is either unnecessary or critical, depending on your point of view. If you believe that astronomy is the study of the physical universe above the clouds, then you are done; the last 19 chapters completed your study of astronomy. But, if you believe that astronomy is the study of your position in the universe, not just your physical location but also your role as a living being in the evolution of the universe, then everything you have done so far was just preparation for this chapter.

This Chapter

You will begin by tackling a question that has troubled scientists and philosophers for centuries: What is life? You won't get an answer to the question, but just asking it will illuminate the problem and prepare you to search for life on other worlds. In fact, as you read this chapter, you will ask four essential questions:

What is life?

How did life originate on Earth?

Could life begin on other worlds?

Could Earthlings communicate with civilizations on other worlds?

We can't yet answer these questions, but often in science asking a question is more important than getting an answer.

Looking Ahead

You are different now. You have explored the universe from the phases of the moon to the big bang, from the origin of Earth to the death of the sun. Astronomy is important, not because it is about stars and galaxies, but because it is about us. It tells us what we are, and once you know astronomy, you see yourself and your world in a different way. Astronomy changes us. You are different now.

Ace Astronomy™ The AceAstronomy icon throughout the text indicates an opportunity for you to test yourself on key concepts and to explore animations and interactions on the AceAstronomy website at: http://astronomy.brookscole.com/sh9e

The Nature of Scientific Explanation

Science is a way of understanding the world around you, and the heart of that understanding is the explanations that science gives you for natural phenomena. Whether you call these explanations stories, histories, theories, or hypotheses, they are attempts to describe how nature works based on fundamental rules of evidence and intellectual honesty. While you may take these explanations as factual truth, you should understand that they are not the only explanations that satisfy the rules of logic.

A separate class of explanations involves religion, and those explanations can be quite logical. The Old Testament description of the creation of the world, for instance, does not fit scientific observations, but if you accept the existence of an omnipotent being, then the biblical explanation is internally logical and acceptable. Of course, it is not a scientific explanation, but religion is a matter of faith and not subject to the rules of evidence. Religious explanations follow their own logic, and you would be wrong to demand that they follow the rules of evidence that govern scientific explanations, just as you would be wrong to demand that scientific explanations accept certain religious beliefs on faith. Both kinds of explanations are logical, but the rules are different.

If scientific explanations are not the only logical explanations, then why do people give them such weight? First, you must notice the tremendous success of scientific explanations in producing technological advances in your daily life. Diseases such as chickenpox are more childhood irritations than life-threatening illnesses, thanks to the application of scientific explanations to modern medicine. The power of science to shape our world can lead you to think its explanations are unique. But second, the process called science depends on the use of evidence to test and perfect explanations, and the logical rigor of this process gives scientists great confidence in their conclusions.

Scientific explanations have provided tremendous insight into the workings of nature, and consequently both scientists and nonscientists tend to forget there can be other kinds of explanations. The so-called conflict between science and religion has been symbolized for centuries by the trial of Galileo. That conflict is easier to understand when you consider the nature of scientific explanations and the role of evidence in testing scientific understanding.

Galileo's telescope gave him a new way to know about the world.

goal in this chapter is to use your knowledge of science to explain the greatest of mysteries—the origin and evolution of life on Earth and on other worlds (**Window on Science 20-1**).

20-1 The Nature of Life

WHAT IS LIFE? Philosophers have struggled with that question for thousands of years, so it is unlikely that it will be answered here. But we must agree on a working model of life before we can speculate on its occurrence on other worlds. To that end, let's identify in living things two important aspects: a physical basis and a unit of controlling information.

The Physical Basis of Life

On Earth, the physical basis of life is the carbon atom (■ Figure 20-1). Because of the way this atom bonds to other atoms, it can form long, complex, stable chains that are capable of extracting, storing, and utilizing energy. Other chemical bases of life may exist. Science fiction stories and movies abound with silicon creatures, living things whose body chemistry is based on silicon rather than carbon. However, silicon forms weaker bonds than carbon does, and it cannot form double bonds as easily. Consequently, it cannot form the long, complex, stable chains that carbon can. Silicon is 135 times more common on Earth than carbon is, yet there are no silicon creatures among us. All Earth life is carbon based. Thus, the likelihood that distant planets are inhabited by silicon people seems small, but we cannot rule out life based on noncarbon chemistry.

In fact, nonchemical life might be possible. What is required is some mechanism capable of supporting the extraction and utilization of energy that has been identified as life. One could at least imagine life based on electromagnetic fields and ionized gas. No one has ever met such a creature, but science fiction writers conjure up all sorts.

Clearly, you could range far in space and time, theorizing about different bases for alien life, but to make progress you must discuss what humans know best—carbon-based life on Earth. How can a lump of carbon-rich matter live? The answer lies in the information that guides its life processes.

Information Storage and Duplication

Living cells are tiny chemical factories. They must store all of the recipes for those chemicals in a safe place, use them to fulfill the

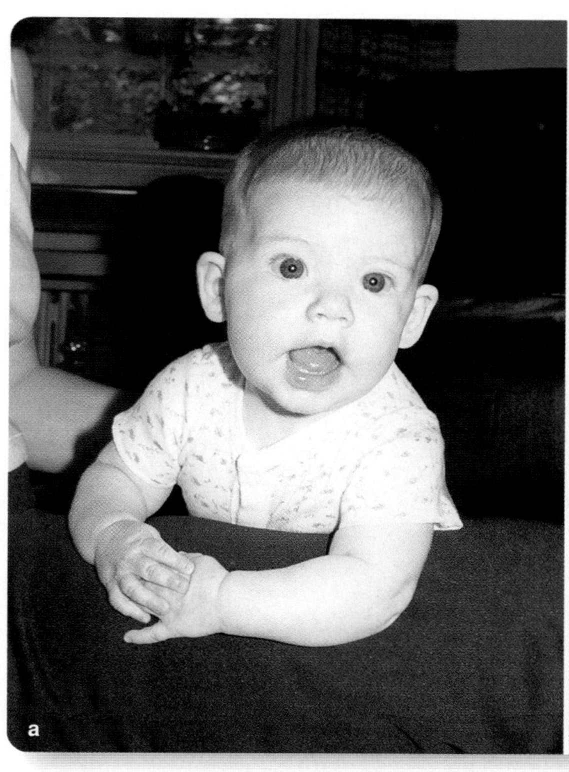

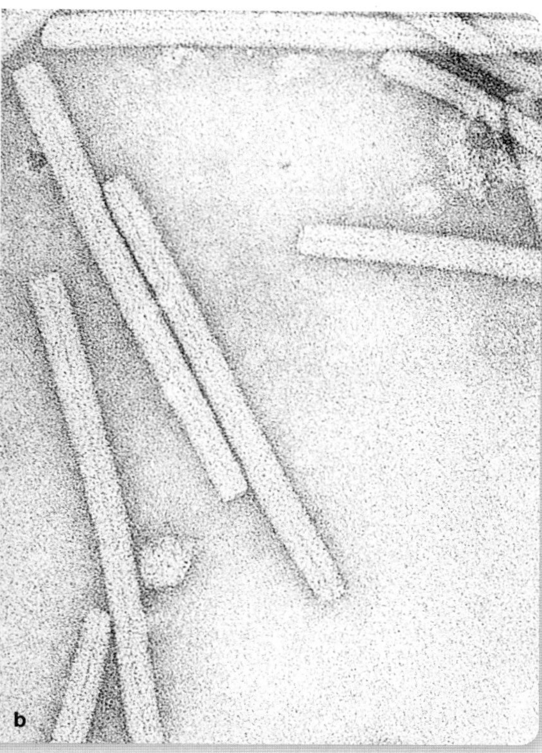

■ Figure 20-1

All living things on Earth are based on carbon chemistry. Even the long molecules that carry genetic information, DNA and RNA, have a framework defined by chains of carbon atoms. (a) Katie, a complex mammal, contains about 30 AU of DNA. (Michael Seeds) (b) Each rod of the tobacco mosaic virus contains a single spiral strand of RNA about 0.01 mm long. (L. D. Simon) All life on Earth stores its genetic information in such carbon-chain molecules.

cell's task, and hand down duplicates of the recipes to offspring. That information is encoded on long molecules.

Study **DNA: The Code of Life** on pages 476 and 477 and notice three important points. First, the chemical recipes of life are stored as templates on DNA molecules. The templates automatically guide specific chemical reactions within the cell. Second, the instructions stored in DNA are the genetic information handed down to offspring. When people say "You have your mother's eyes," they are talking about DNA codes. Finally, notice how the DNA molecule reproduces itself when a cell divides so that each new cell contains a copy of the original information.

Although the DNA molecule must preserve its coded information from damage and make accurate copies, it must also be capable of making mistakes now and then. To see why, we must consider how new DNA recipes are created.

Modifying the Information

If living things are to survive for many generations, then the information stored in their DNA must change as the environment changes. Without change in DNA, a slight warming of the climate, for example, might kill a species of plant, in turn starving the rabbits, deer, and other plant eaters, and leaving the hawks, wolves, and mountain lions with no prey. If the information stored in DNA could never change, then environmental changes would quickly drive life forms to extinction. If life is to survive in a changing world, then the information in DNA must be changeable. Living things must evolve.

Species evolve by **natural selection.** Each time an organism reproduces, its offspring receive the data stored in the DNA, but some variation is possible. For example, most of the rabbits in a litter may be normal, but it is possible for one to get a DNA recipe that gives it stronger teeth. If it has stronger teeth, it may be able to eat something other than the plant the others depend on, and if that plant is becoming scarce, the rabbit with stronger teeth has a survival advantage. It can eat other plants and so will be healthier than its littermates and have more offspring. Some of these offspring will also have stronger teeth, as the altered DNA recipes are handed down to the new generation. Thus nature selects and preserves those attributes that contribute to the survival of the species. Those creatures that are unfit die. Natural selection is merciless to the individual, but it gives the species the best possible chance to survive in a changing environment.

The only way nature can obtain new DNA patterns from which to select the best is from DNA molecules that have changed. This can happen through chance mismatching of base pairs—errors—in the reproduction of the DNA molecule. Another way this can occur is through damage to reproductive cells from exposure to radioactivity such as cosmic rays or natural radioactivity in the soil. In any case, an offspring born with altered DNA is called a **mutant.** Most mutations make no difference at all because they change segments of DNA that are not being used. Many mutations are fatal, and the individual dies long before it can have offspring of its own. But in rare cases, a mutation may give a species a new survival advantage. Then natural selection makes it

DNA: The Code of Life

The key to understanding life is information — the information that guides all of the processes in an organism. In most living things on Earth, that information is stored on a long spiral molecule called DNA (deoxyribonucleic acid).

The DNA molecule looks like a spiral ladder with rails made of phosphates and sugars. The rungs of the ladder are made of four chemical bases arranged in pairs. The bases always pair the same way. That is, base A always pairs with base T, and base G always pairs with base C.

Information is coded on the DNA molecule by the order in which the base pairs occur. To read that code, molecular biologists have to "sequence the DNA." That is, they must determine the order in which the base pairs occur along the DNA ladder.

DNA automatically combines raw materials to form important chemical compounds. The building blocks of these compounds are relatively simple **amino acids.** Segments of DNA act as templates that guide the amino acids to join together in the correct order to build specific **proteins,** chemical compounds important to the structure and function of organisms. Some proteins called **enzymes** regulate other processes. In this way, DNA recipes regulate the production of the compounds of life.

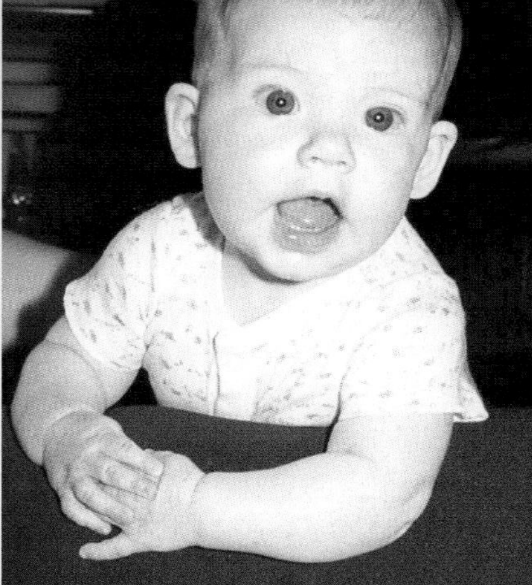

The traits you inherit from your parents, the chemical processes that animate you, and the structure of your body are all encoded in your DNA.

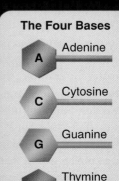

The Four Bases

A Adenine

C Cytosine

G Guanine

T Thymine

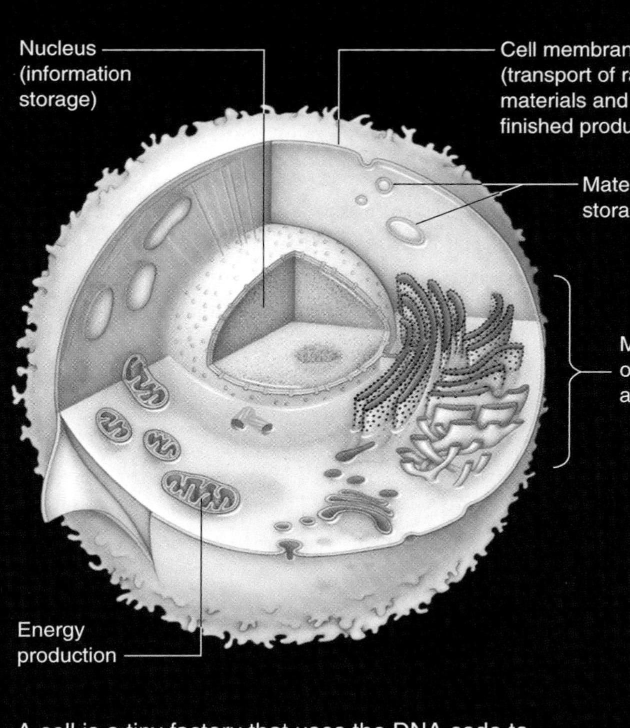

A single cell from a human being contains about 1.5 meters of DNA containing about 4.5 billion base pairs — enough to record the entire works of Shakespeare 200 times. A typical human contains a total of about 600 AU of DNA. Yet the DNA in each cell, only 1.5 meters in length, contains all of the information to create a new human. A clone is a new creature created from the DNA code found in a single cell.

Original DNA

A cell is a tiny factory that uses the DNA code to manufacture chemicals. Most of the DNA remains safe in the nucleus of a cell, and the code is copied to create a molecule of **RNA (ribonucleic acid).** Like a messenger carrying blueprints, the RNA carries the code out of the nucleus to the work site where the proteins and enzymes are made.

To divide, a cell must duplicate its DNA. The DNA ladder splits, and new bases match to the exposed bases of the ladder to build two copies of the original DNA code. Because the base pairs almost always match correctly, errors in copying are rare. One set of the DNA code goes to each of the two new cells.

Copy DNA

DNA, coiled into a tight spiral, makes up the **chromosomes** that are the genetic material in a cell. A **gene** is a segment of a chromosome that controls a certain function. When a cell divides, each of the new cells receives a copy of the chromosomes, as genetic information is handed down to new generations.

Copy DNA

Ace Astronomy™

Log into AceAstronomy and select this chapter to see the Active Figure called "DNA." Explore the structure of DNA.

Cell Reproduction by Division

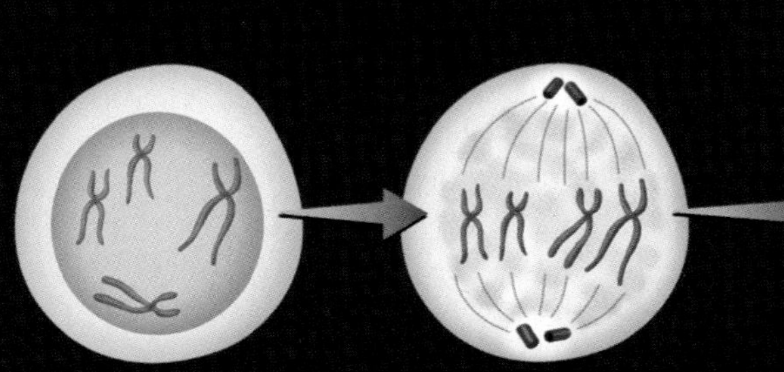

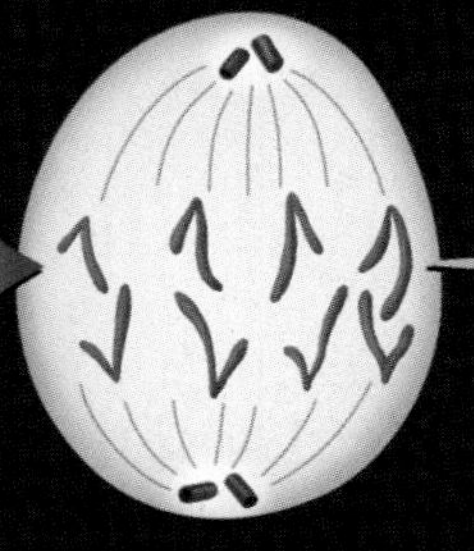

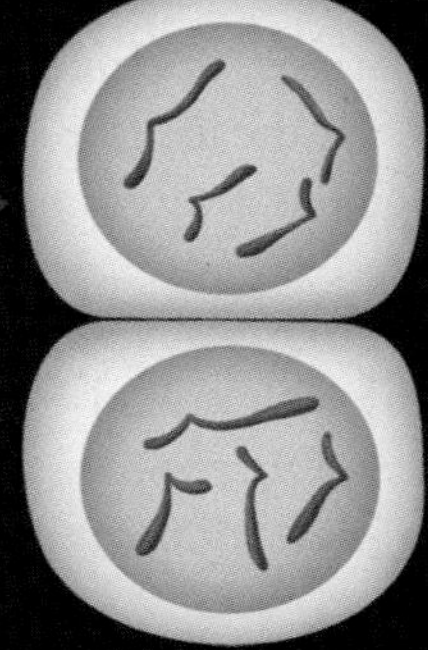

As a cell begins to divide, its DNA duplicates itself.

The duplicated chromosomes move to the middle.

The two sets of chromosomes separate, and . . .

the cell divides to produce . . .

two cells, each containing a full set of the DNA code.

likely that the new DNA message will survive and be handed down, making the species more capable of surviving.

Evolution is not random. Of course, the errors that occur in the DNA code are indeed random, but natural selection is not random. Those changes in the DNA that help a species survive are preserved in future generations. With each passing generation, the species becomes more fit to survive in its environment.

Inquire | Review | Analyze

Why can't the information in DNA be permanent?

The information stored in a creature's DNA provides the recipes that make the creature what it is. For example, the DNA in a starfish must contain all the recipes for making the various kinds of proteins needed to consume and digest food. That information must be passed on to offspring starfish, or they will be unable to survive. But the information must be changeable because Earth's environment is changeable. Ice ages come and go, mountains rise, lakes dry up, and ocean currents shift. If the environment changes in some way, one or more of the recipes may no longer work. In this example, a change in the temperature of the ocean water may kill off the specific shellfish the starfish eat. If they can't digest other shellfish, the entire species will become extinct. Natural variation in DNA means that among all the infant starfish in any generation, some of the recipes are different; if the environment changes, all of the old-style starfish may die, but a few—those with the different DNA—can carry on.

The survival of life depends on this delicate balance between reliable reproduction and the introduction of small variations in DNA information. What are some of the ways these small changes in DNA can arise?

■ ■ ■

Connections: Life is based not only on information, but also on the duplication of information. Today, that process seems so complex that it is hard to imagine how it could have begun.

20-2 The Origin of Life

IF LIFE ON EARTH is based on the storage of information in these long, complex, carbon-chain molecules, how could it have ever gotten started? Obviously, 4.5 billion chemical bases didn't just happen to drift together to form the DNA formula for a human being. The key is evolution. Once a life form begins to reproduce itself, natural selection preserves the most advantageous traits. Over long periods of time spanning thousands, perhaps millions, of generations, the life form becomes more fit to survive. This often means the life form becomes more complex. Thus life could have begun as a very simple process that gradually became more sophisticated as it was modified by evolution.

Let's begin our search on Earth, where fossils and an intimate familiarity with carbon-based life provide a glimpse of the first living matter. Once you discover how Earthly life could have begun, you can look for signs that life began on other planets in our solar system. Finally, you can speculate on the chances that other planets, orbiting other stars, have conditions that give rise to life.

The Origin of Life on Earth

The oldest fossils hint that life began in the sea. The first living things on Earth left behind a very poor fossil record. They were at first single-celled creatures, and even when they became more complex, multicellular creatures, they contained no hard parts such as bones or shells that form good fossils. What fossils can be found are clearly the remains of ocean creatures.

Although the oldest rocks contain no obvious fossils, microscopes reveal traces of ancient life. The oldest widely accepted evidence is 2.5-billion-year-old fossils of bacteria. Rock from western Australia that is nearly 3.5 billion years old contains microscopic features that may be fossils of living things (■ Figure 20-2), but that conclusion is controversial. While the experts debate the details, you can safely conclude that life was present on Earth as simple organisms at least 2.5 billion years ago and possibly a billion years before that.

A little over a half-billion years ago something happened, perhaps a change in Earth's climate, and life exploded into a wide diversity of complex forms. This marks the beginning of the **Cambrian period,** and is sometimes called the *Cambrian explosion.* The best known of the Cambrian creatures may be the trilobites (■ Figure 20-3). Although the Cambrian creatures were diverse, there are no Cambrian fossils of land plants or animals. Evidently, land surfaces were totally devoid of life until only 400 million years ago.

The fossil record shows that life began as simple organisms in the sea soon after Earth formed. Only recently in geological terms has life become complex.

The key to the origin of this life may lie in an experiment performed by Stanley Miller and Harold Urey in 1952. This **Miller experiment** sought to reproduce the conditions on Earth under which life began. In a closed glass container, the experimenters

■ Figure 20-2

Among the oldest fossils known, this microscopic filament resembles modern bacterial forms (artist's reconstruction at right). This fossil was found in the 3.5-billion-year-old chert of the Pilbara Block in northwestern Australia. (Courtesy J. William Schopf)

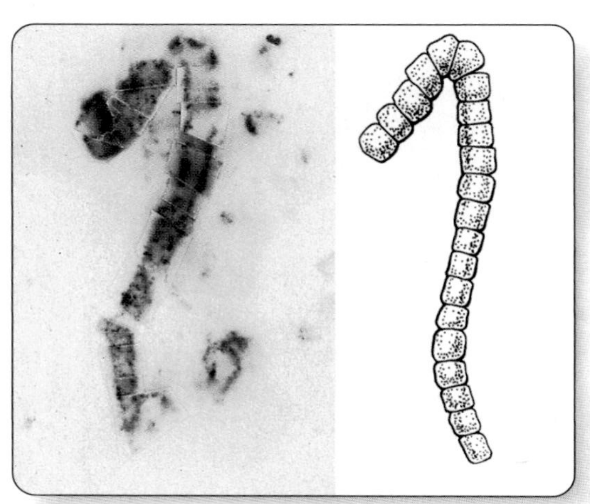

placed water (to represent the oceans), the gases hydrogen, ammonia, and methane (to represent the primitive atmosphere), and an electric arc (to represent lightning bolts). The apparatus was sterilized, sealed, and set in operation (■ Figure 20-4).

After a week, Miller and Urey stopped the experiment and analyzed the material in the flask. Among the many compounds the experiment produced, they found four amino acids (building blocks of protein), various fatty acids, and urea, a molecule common to many life processes. Evidently, the energy from the electric arc had molded the atmospheric gases into some of the basic components of living matter. Other energy sources, such as hot silica (to simulate hot lava spilling into the sea) and ultraviolet radiation (to simulate sunlight), give similar results.

Recent studies of the composition of meteorites and models of planet formation suggest that Earth's first atmosphere did not resemble the gases used in the Miller experiment. Earth's first atmosphere was probably composed of carbon dioxide, nitrogen, and water vapor. This finding, however, does not invalidate the Miller experiment. When such gases are processed in a Miller apparatus, a tarry gunk rich in organic molecules soon coats the inside of the chamber.

The Miller experiment did not create life, nor did it necessarily imitate the exact conditions on the young Earth. Rather, it is important because it shows that complex organic molecules form naturally in a wide variety of circumstances. The chemical deck is stacked to deal nature a hand of complex molecules. If you could travel back in time, you would probably find Earth's first

■ Figure 20-3

Trilobites made their first appearance in the Cambrian oceans. The smallest were almost microscopic, and the largest were bigger than dinner plates. This example, about the size of a human hand, lived 400 million years ago in an ocean floor that is now a limestone deposit in Pennsylvania. (Grundy Observatory photograph)

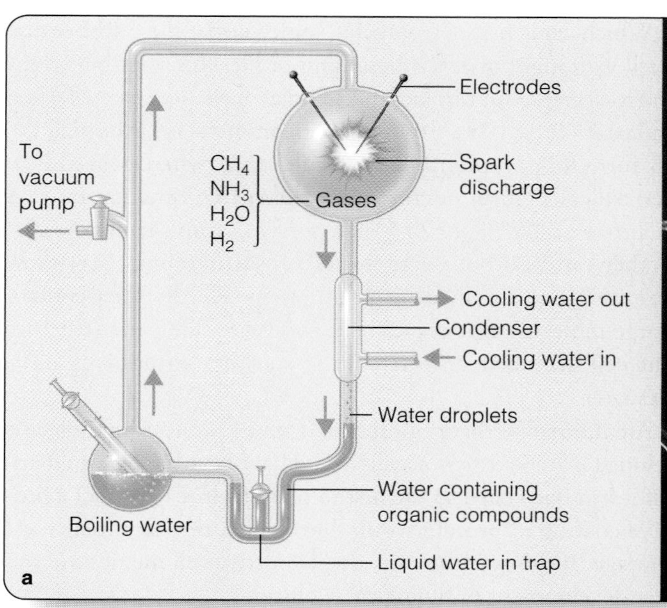

■ Figure 20-4

(a) The Miller experiment circulated gases through water in the presence of an electric arc. This simulation of primitive conditions on Earth produced amino acids, the building blocks of proteins. (b) Stanley Miller with a Miller apparatus. (Courtesy Stanley Miller)

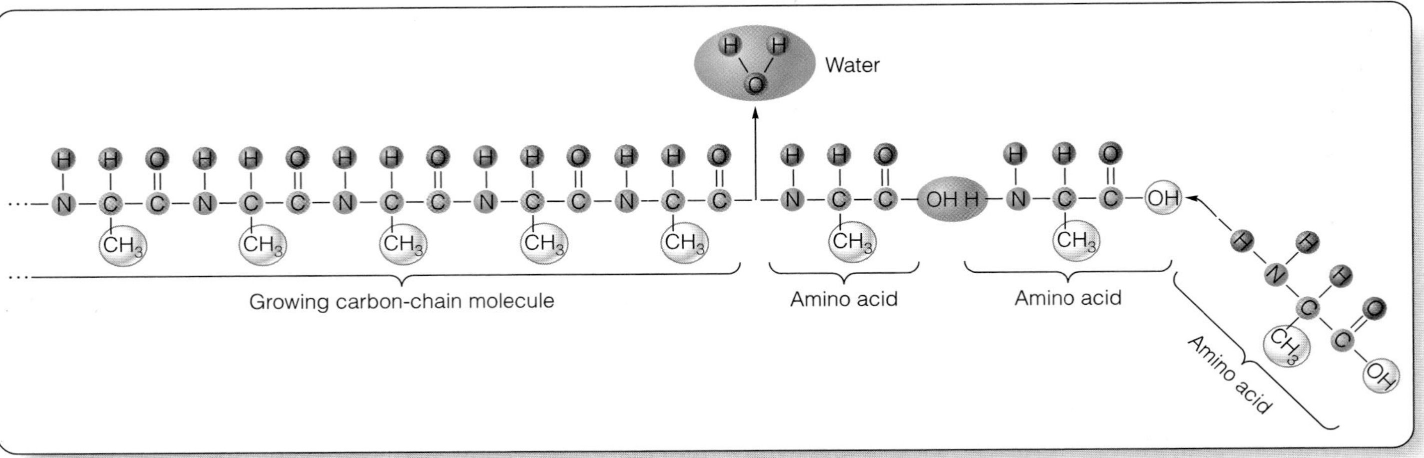

■ Figure 20-5

Amino acids can link together through the release of a water molecule to form long carbon-chain molecules. The amino acid in this hypothetical example is alanine, one of the simplest.

oceans filled with a rich mixture of organic compounds in what some have called the **primordial soup.**

The next step on the journey toward life is for the compounds dissolved in the oceans to link up and form larger molecules. Amino acids, for example, can link together to form proteins. This linkage occurs when amino acids join together end to end and release a water molecule (■ Figure 20-5). For many years, experts have assumed that this process must have happened in sun-warmed tidal pools where evaporation concentrated the broth. But recent studies suggest that the young Earth was subject to extensive volcanism and large meteorite impacts that periodically modified the climate enough to destroy any life forms exposed on the surface. Thus, the early growth of complex molecules likely took place among the hot springs along the midocean ridges. These complex molecules would not have been photosynthetic—taking energy from sunlight. Rather, they would have taken energy from the heat and chemicals emerging from the hot springs of the midocean ridge. That energy could have powered the growth of long protein chains. Deep in the oceans, they would have been safe from climate changes.

Although these proteins might have contained hundreds of amino acids, they would not have been alive. Not yet. Such molecules would not have reproduced but would have merely linked together and broken apart at random. Because some molecules are more stable than others, however, and because some molecules bond together more readily than others, this blind **chemical evolution** would have led to the concentration of the varied smaller molecules into the most stable larger forms. Eventually, somewhere in the oceans, a molecule took shape that could reproduce itself. At that point the chemical evolution of molecules became the biological evolution of living things.

An alternative theory proposes that primitive living things such as reproducing molecules did not originate on Earth but came here in meteorites or comets. Radio astronomers have found a wide variety of organic molecules in the interstellar medium, and some studies have found similar compounds inside meteorites (■ Figure 20-6). Such molecules form so readily that scientists would be surprised if they were not present in space. A few investigators, however, have speculated that living, reproducing molecules originated in space and came to Earth as a cosmic contamination. If this is true, every planet in the universe is contaminated with the seeds of life. However entertaining this theory may be, it is presently untestable, and an untestable theory is of little use in science.

Whether life originated in the oceans or in space, science still faces the problem of how life began. Experts studying the origin of life proceed on the assumption that life began as reproducing molecules in Earth's oceans.

Which came first, reproducing molecules or the cell? Because the cell is thought of as the basic unit of life, this question seems to make no sense, but in fact the cell may have originated during chemical evolution. If a dry mixture of amino acids is heated, the acids form long, proteinlike molecules that, when poured into water, collect to form microscopic spheres that function in ways similar to cells (■ Figure 20-7). They have a thin membrane surface, they can absorb material from their surroundings, they grow in size, and they can divide and bud just as cells do. They contain no large molecule that copies itself, however. Thus the structure of the cell may have originated first and the reproducing molecules later.

An alternative theory proposes that the replicating molecule developed first. Such a molecule would have been exposed to damage if it had been bare, so the first to manufacture or attract a protective coating of protein would have had a significant survival advantage. If this was the case, the protective cell membrane was a later development of biological evolution.

Figure 20-6

A sample of the Murchison meteorite, a carbonaceous chondrite that fell in 1969 near Murchison, Australia. Analysis of the interior of the meteorite revealed evidence of amino acids. Whether the first building blocks of life originated in space is unknown, but the amino acids found in meteorites illustrate how commonly amino acids and other complex molecules occur even in the absence of living things. (Courtesy Chip Clark, National Museum of Natural History)

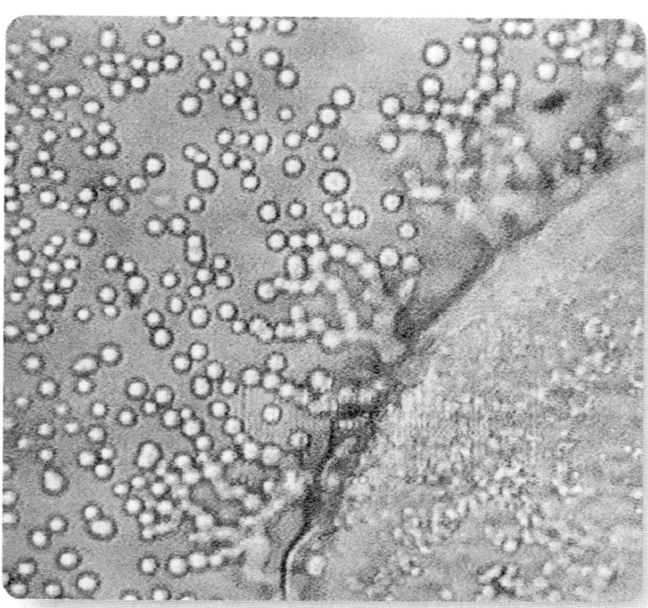

Figure 20-7

Single amino acids can be assembled into long proteinlike molecules. When such material cools in water, it can form microspheres, microscopic spheres with double-layered boundaries similar to cell membranes. Microspheres may have been an intermediate stage in the evolution of life between complex molecules and cells holding molecules reproducing genetic information. (Courtesy Sidney Fox and Randall Grubbs)

The first living things must have been single-celled organisms much like modern bacteria. Some of the oldest fossils known are **stromatolites,** structures produced by communities of photosynthesizing bacteria that grew in mats and, year by year, deposited layers of minerals that were later fossilized. One of the oldest such fossils known is 3.5 billion years old (■ Figure 20-8). If such bacteria were common when Earth was young, the early atmosphere may have contained a small amount of oxygen produced by the photosynthesis. Recent studies suggest that an oxygen abundance of only 0.1 percent would have been sufficient to provide an ozone screen that would protect organisms from the sun's ultraviolet radiation.

How evolution shaped creatures to live in the ancient oceans, to photosynthesize and respire, to become multicellular, and to reproduce sexually is a fascinating story, but it can't be explored in detail here. You can see that life could have begun through simple chemical reactions building complex molecules, and that once some DNA-like molecule formed, it protected its own survival with selfish determination. Over billions of years, the genetic information stored in living things kept those qualities that favored survival and discarded the rest. As Samuel Butler said, "The chicken is the egg's way of making another egg." In that sense, all living matter on Earth is merely the physical expression of DNA's mindless determination to continue its existence.

Perhaps this seems harsh. Human experience goes far beyond mere reproduction. *Homo sapiens* has art, poetry, music, philosophy, religion, science. Perhaps all of the great accomplishments of human intelligence represent more than mere reproduction of DNA. Nevertheless, intelligence, the ability to analyze complex situations and respond with appropriate action, must have begun as a survival mechanism. For example, a fixed escape strategy stored in the DNA is a disadvantage for a creature that frequently moves from one environment to another. A rodent that always escapes from predators by automatically climbing the nearest tree would be in serious jeopardy if it met a hungry fox in a treeless clearing. Even a faint glimmer of intelligence might allow the rodent to analyze the situation and, finding no trees, to choose running over climbing. Thus, intelligence, of which *Homo sapiens* is so proud, may have developed in ancient creatures as a way of making them more versatile.

Any discussion of the evolution of life seems to involve highly improbable coincidences until you consider how many years have passed in the history of Earth. We read the words—4.6 billion years—so easily, but it is in truth hard to grasp the meaning of such a long period of time.

Geologic Time

Humanity is a very new experiment on planet Earth. You can fit all of the evolution that leads from the primitive life forms in the oceans of the Cambrian period 600 million years ago to fishes, amphibians, reptiles, and mammals into a single chart such as

■ Active Figure 20-8

A 3.5-billion-year-old fossil stromatolite from western Australia is one of the oldest known fossils (above left). Stromatolites were formed, layer by layer, by mats of bacteria living in shallow water. Such life may have been common in shallow seas when Earth was young (right). Stromatolites are still being formed today in similar environments. (Mural by Peter Sawyer; photo courtesy Chip Clark, National Museum of Natural History)

AceAstronomy™ Log into AceAstronomy and select this chapter to see the Active Figure called "Future of the Sun." Watch Earth's temperature change as the sun ages.

■ Figure 20-9. You can take comfort in thinking that creatures like us have walked on Earth for roughly 3 million years, but when you add human history to the chart, you discover that the entire history of humanity makes up no more than a thin line at the top. In fact, if you tried to represent the entire 4.6-billion-year history of Earth on the chart, the portion describing the rise of life on the land would be an unreadably small segment.

Imagine that you could tell the story of the 4.6-billion-year history of life on Earth in a video program that ran continuously for one year. You could tune in now and then to see what was happening. In your program, Earth forms as the video begins on January 1, and through all of January and February it cools and is cratered, and the first oceans form. Search as you might, you would find no trace of life in these oceans until sometime in March or early April, when the first living things develop. The slow development of these simplest of living forms grinds on slowly through the spring and summer of your videotape. The entire 4-billion-year history of Precambrian evolution lasts until the video reaches mid-November, when the primitive ocean life explodes into the more complex Cambrian organisms such as trilobites.

While your year-long videotape plays on and on, you might amuse yourself by looking at the land instead of the oceans, but you would be disappointed. The land is a lifeless waste with no plants or animals of any kind. Not until November 28 in your video does life appear on the land; but, once it does, it evolves rapidly into a wide range of plants and animals. Dinosaurs, for example, appear about December 12th and vanish by Christmas evening as mammals and birds flourish.

Throughout the one-year run of your video you would see no humans; and, even during the last days of the year, as the mammals rise to dominate the landscape, you would see no people. But you would have to clear your calendar for New Year's Eve because you would want to be tuned in. Things begin to happen about suppertime when you would glimpse vaguely human forms moving through the grasslands. By late evening, they would begin making stone tools. The Stone Age would last till about 11:45 PM, and the first signs of civilization, towns and cities, would not ap-

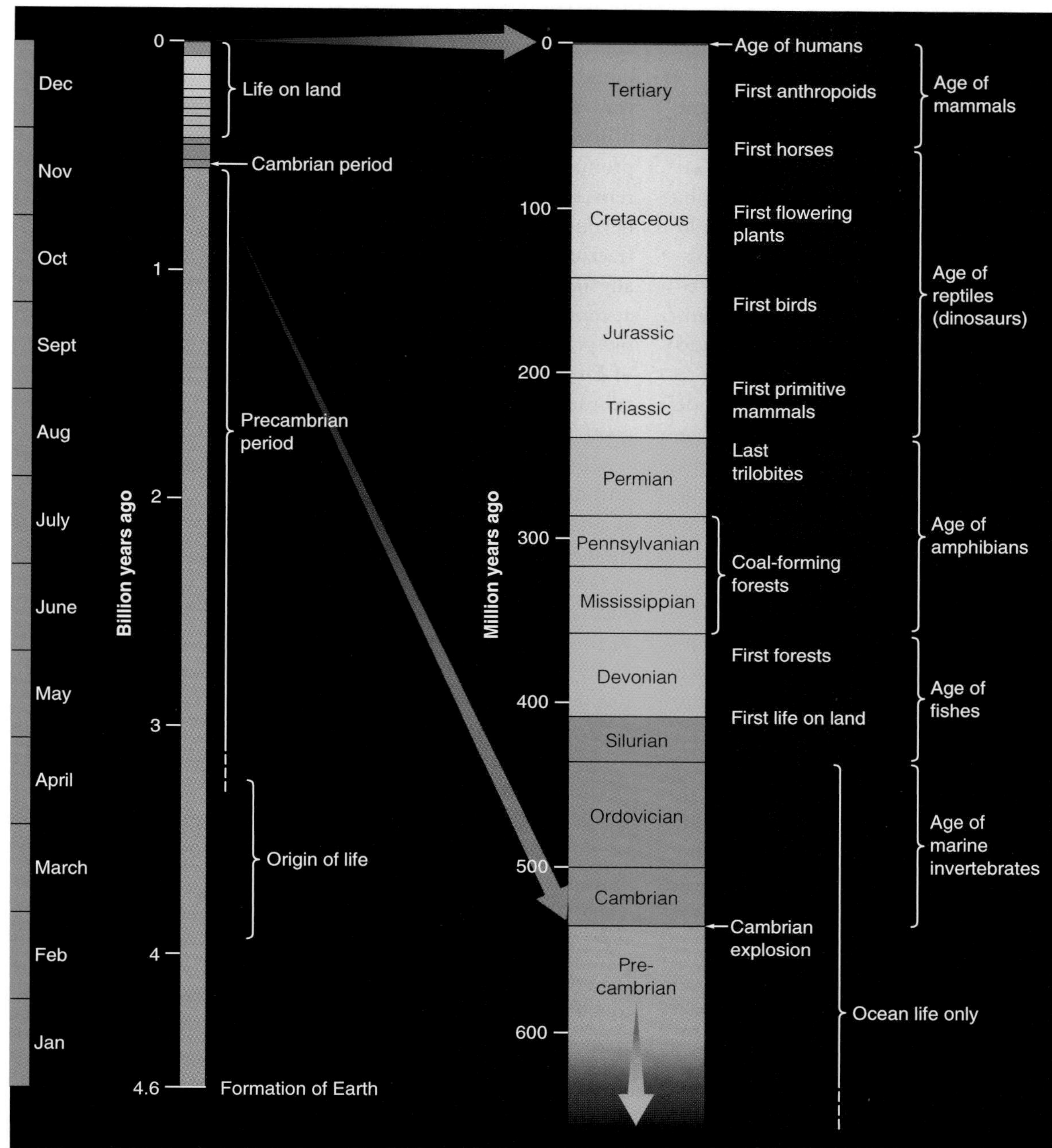

Active Figure 20-9

Complex life has developed on Earth only recently. If the entire history of Earth were represented in a time line (left), you would have to magnify the end of the line to see details such as life leaving the oceans and dinosaurs appearing. The age of humans would still be only a thin line at the top of your diagram. If the history of Earth were a year-long videotape, humans would not appear until the last hours of December 31.

AceAstronomy™ Log into AceAstronomy and select this chapter to see the Active Figure called "Earth Calendar." Animate this diagram.

pear until 11:54 PM. No fair slowing down the tape or hitting pause; you have to watch carefully as events flash across the screen. Babylon flourishes, the Pyramids rise in a moment, Troy falls, armies rush to battle, and the Christian era begins only 14 seconds before the New Year. The fall of Rome, the Middle Ages, and the Renaissance flicker past, and the Declaration of Independence is signed only 1 second before your tape ends.

By converting the history of Earth into a year-long videotape, you have placed the rise of life in perspective. Tremendous amounts of time were needed for the first simple living things to evolve in the oceans, and even more time was needed for the evolution of complex creatures that could colonize the land. As life became more complex, it evolved and diversified faster and faster, as if evolution were drawing on a growing library of solutions that had been previously invented with great effort to solve earlier problems. The burst of diversity on land led eventually to the rise of intelligent creatures like you, a process that has taken 4.6 billion years.

If life could originate on Earth and develop into intelligent creatures, perhaps the same thing could have happened on other planets. This raises three questions. First, could life originate on another world if conditions were suitable? You can't be certain, but scientists have found natural, chemical processes that could lead to living things, so your answer to this question should probably be yes. The second question is: Will life always evolve toward intelligence? Perhaps intelligence arose on Earth under unusual conditions, but you must recall that intelligence appears to be a way to make a species more versatile and thus more likely to survive. If that is true, then intelligence may develop under a wide range of conditions if enough time is available. But what of the third question: Are suitable conditions so rare that life almost never gets started? The only way to answer that is to search for life on other planets. Let's begin, in the next section, with the other planets in our solar system.

Life in Our Solar System

Although we can imagine life based on something other than carbon chemistry, scientists know of no examples to tell us how such life might arise and survive. We must limit our discussion to life as we know it and the conditions it requires. The most important requirement is the presence of liquid water, not only as part of the chemical reactions of life, but also as a medium to transport nutrients and wastes within the organism. Also, it seems that life on Earth began in the oceans and developed there for nearly 4 billion years before it was able to emerge onto the land. Certainly, any world where you hope to find life must have liquid water, and that means it must have moderate temperatures.

The water requirement automatically eliminates many worlds in our solar system. The moon is airless, and although some data suggest ice frozen in the soil at its poles, it has never had liquid water on its surface. In the vacuum of the lunar surface, liquid water would boil away rapidly. Mercury too is airless and cannot have had liquid water on its surface for long periods of time. Venus has some traces of water vapor in its atmosphere, but it is much too hot for liquid water to survive. If there were any lakes or oceans of water on its surface when it was young, they must have evaporated quickly. Even if life began there, no traces would be left now.

The inner solar system seems too hot, and the outer solar system seems too cold. The Jovian planets have deep atmospheres, and at a certain level, they have moderate temperatures where water might condense into liquid droplets. But it seems unlikely that life could begin there. The Jovian planets have no surfaces where oceans could nurture the beginning of life, and currents in the atmosphere seem destined to circulate gas and water droplets from regions of moderate temperature to other levels that are much too hot or too cold for life to survive.

A few of the satellites of the Jovian planets might have suitable conditions for life. Jupiter's moon Europa seems to have a liquid-water ocean below its icy crust (see Figure 18-8), and minerals dissolved in that water would provide a rich broth of possibilities for chemical evolution. Nevertheless, Europa is not a promising site to search for life because conditions may not have remained stable for the billions of years needed for life to evolve beyond the microscopic stage. The subsurface ocean is kept from freezing by tidal heating. If Jupiter's moons interact gravitationally and modify their orbits, Europa may have been frozen solid at some points in history. Such periods of freezing would probably prevent life from developing. Drilling through the icy crust of Europa to search for life in its ocean will be a wonderful adventure for future generations, but Europa does not seem to be a good bet to harbor any form of complex life.

Saturn's moon Titan has an atmosphere of nitrogen, argon, and methane and may have liquid methane and ethane on its surface. You saw in Chapter 18 how sunlight can convert the methane in the atmosphere to organic smog particles that settle to the surface. The chemistry of life that might crawl or swim on such a world is unknown, but life there may be unlikely because of the temperature. The surface of Titan is a deadly −179°C (−290°F). Chemical reactions occur slowly or not at all at such low temperatures, so the chemical evolution needed to begin life may never have occurred on Titan.

Mars is the most likely place for life in our solar system. The evidence, however, is not encouraging. In 1976, two robotic spacecraft, Viking 1 and Viking 2, landed at two different places on the Martian surface. The spacecraft scooped up soil samples and subjected them to tests for the presence of living organisms. For example, they gave some soil samples a dose of nutrient-rich water and watched for signs the nutrients were taken up by biological processes. The results seem negative. Although some peculiar chemical processes were detected, no evidence was found that clearly indicated the presence of any living things in the soil.

Rovers on Mars and orbiting spacecraft have found dramatic evidence that liquid water once flowed over the surface (■ Figure 20-10), but that does not mean that life did originate there. Only detailed study of the rocky soil can answer that question, and that means Earth must send a geologist to Mars or bring Mars rocks back to Earth. Nature has performed the later task for you.

Meteorite ALH84001 (■ Figure 20-11a) was found on the Antarctic ice in 1984. Years later an analysis of its chemical composition showed that it had originated on Mars. It was probably part of debris ejected into space by a large impact on Mars. Some of that debris fell to Earth, and a small number of these meteorites have been found. ALH84001 is important because a team of scientists studied it and announced in 1996 that it contained chemical and physical traces of ancient life on Mars (Figure 20-11b).

Figure 20-10

Mars rover Opportunity descended into the crater Endurance and then looked back up the crater wall. Holes mark test borings. In places, Opportunity found layering in the rocks that showed they had formed as layers of sand and silt deposited by flowing water. Water no longer flows on the surface of Mars, but signs of past life may lie hidden in these rocks. (NASA/JPL/Cornell)

Figure 20-11

(a) Meteorite ALH84001 is one of a dozen meteorites known to have originated on Mars. It was claimed that the meteorite contained chemical and physical traces of ancient life on Mars, including what appear to be fossils of microscopic organisms (b). The evidence has not been confirmed, and the validity of the claim is highly questionable. (NASA)

This discovery was received with great excitement by the press and the public. It would, indeed, be dramatic evidence if true, because it would show that life could begin on another world.

Scientists were excited too, but being professionally skeptical (see Window on Science 16-2), they began testing the results immediately. In many cases, the results did not confirm the conclusion that life once existed on Mars. Some chemical contamination from water on Earth has occurred, and some chemicals in the meteorite may have originated without the presence of life. The physical features that look like fossil bacteria may be mineral formations in the rock. Although studies of ALH84001 continue, it does not provide unchallenged evidence that life once existed on Mars.

Spacecraft now visiting Mars may help scientists understand the past history of water there and paint a more detailed picture of present conditions. Landers are planned that will eventually land on Mars, collect rocks, and return them to Earth. Nevertheless, conclusive evidence may have to wait until a geologist in a space suit can wander the dry streambeds of Mars cracking open rocks and searching for fossils.

We are left to conclude that, so far as we know, our solar system is bare of life except for Earth. Consequently our search for life in the universe takes us to other planetary systems.

Life in Other Planetary Systems

Might life exist in other solar systems? To consider this question, let us try to decide how common planets are and what conditions a planet must fulfill for life to originate and evolve to intelligence. The first question is astronomical; the second is biological. Our ability to discuss the problem of life outside our solar system is severely limited by lack of experience.

In Chapter 16 you learned that planets form as a natural byproduct of star formation and that a number of extrasolar planets have been found circling nearby stars. From this you can conclude that planetary systems are very common.

If a planet is to become a suitable home for life, it must have a stable orbit around its sun. This is simple in a solar system like our own, but in a binary system most planetary orbits are unstable. Most planets in such systems would not last long before they were swallowed up by one of the stars or ejected from the system.

Thus, single stars are the most likely to have planets suitable for life. Because our galaxy contains at least 10^{11} stars, half of which are single, there could be roughly 5×10^{10} planetary systems in which life might exist.

A few million years of suitable conditions does not seem to be enough time to originate life. On our planet it took at least 0.5 to 1 billion years for the first cells to evolve and 4.6 billion years for intelligence to evolve. Clearly, conditions on a planet must remain acceptable over a long time. This eliminates massive stars that remain stable on the main sequence for only a few million years. If it takes a few billion years for life to originate and evolve to intelligence, no star hotter than about F5 will do. This is not really a serious restriction, because upper-main-sequence stars are rare anyway.

In previous sections, you saw how life on Earth depends on water. In the past, astronomers have used that to define a **life zone** (or ecosphere) around a star, a region within which planets have temperatures that permit the existence of liquid water. A cool star has a small life zone, and a hot star has a large life zone. The life zone around the sun extends from about the orbit of Venus to the orbit of Mars.

You might use the life zone to eliminate lower-main-sequence stars as candidates for life. The M dwarfs at the bottom of the main sequence are so faint a planet would have to orbit very close to stay warm. Some astronomers have argued that such a planet would become tidally locked to its star, and one side of the planet would be in perpetual darkness. Water would tend to freeze out on the dark side and end chances for life to develop. Whether the atmosphere could distribute heat to the dark side is unknown, but most astronomers think the dinky red dwarfs are not good candidates for life.

Recent discoveries, however, are making the whole idea of a life zone seem just a little bit silly. For one thing, scientists on Earth are finding living things in astonishing places such as the bottom of icy lakes in Antarctica and far underground in solid rock. Living things have been found in boiling springs where the water is highly acidic. It appears that life is much tougher than some scientists thought. Also, spacecraft visiting other worlds have found conditions that might support life outside the so-called life zone of the sun. The biggest water ocean in our solar system is under the ice on Jupiter's moon Europa. Liquid has flooded parts of Neptune's moon Triton, and Saturn's moon Titan has a complex surface chemistry of carbon-rich molecules. We can imagine life in these environments, and they all lie outside the conventional life zone of the sun. Perhaps the life zone is most useful as a warning that life is proving to be more varied than many people expected.

One restriction on life is the slow evolution of a planet's star. You learned in Chapter 9 that main-sequence stars gradually grow more luminous as they convert their hydrogen to helium. The sun was only about 70 percent as luminous when it formed as it is today. As a star grows more luminous, it should slowly warm its planets. A planet might have liquid water on its surface long enough for life to begin but then be sterilized as its star slowly grows more luminous and drives away the planet's atmosphere and oceans. Perhaps a planet must form at just the right distance from its star to remain hospitable to life for billions of years.

The fundamental question in this discussion is simple. If conditions are right, will life begin? Early in this chapter you learned that life could begin through simple chemical reactions, so perhaps we should change our question and ask: What could prevent life from beginning? Given what we know about life, it should arise whenever conditions permit, and our galaxy should be filled with planets that are inhabited with living creatures.

Inquire | Review | Analyze

What evidence indicates that life is at least possible on other worlds? The evidence is limited almost entirely to Earth, but it is promising. Fossils show that life originated in Earth's oceans almost 4 billion years ago, and biologists have proposed relatively simple chemical processes that could have created these first reproducing molecules. The fossils show that life developed from a very slow beginning into more and more complex creatures that filled the oceans. The pace of evolution quickened dramatically about half a billion years ago, the beginning of the Cambrian period, when life began taking on complex forms; later, when life emerged onto the surface of the land, it evolved rapidly to produce the tremendous diversity you see around you. Human intelligence has been a very recent development; it is only a few million years old.

If this process occurred on Earth, then it seems reasonable that it could have occurred on other worlds as well. Thus, the evidence suggests you can expect that life might begin and evolve to intelligent forms on any world where conditions are right. What are the conditions you should expect of other worlds that host life?

■ ■ ■

Connections: It is both easy and fun to speculate about life on other worlds, but it leads to a simple question: Is there really life beyond Earth? If we can't leave Earth and visit other worlds,

UFOs and Space Aliens

If you discuss life on other worlds, then you might be tempted to use UFO sightings and supposed visits by aliens from outer space as evidence to test your hypotheses. Scientists don't do so for two reasons, both related to the reliability of these observations.

First, the reputation of UFO sightings and alien encounters does not inspire confidence that these data are reliable. Most people hear of such events in the grocery store tabloids, daytime talk shows, or sensational "specials" on viewer-hungry cable networks. You must take note of the low reputation of the media that report UFOs and space aliens. Most of these reports are simply made up for the sake of sensation, and you cannot use them as reliable evidence.

Second, the remaining UFO sightings, those not simply made up, do not survive careful examination. Most are mistakes and unconscious misinterpretations of natural events made by honest people. In short, there is no dependable evidence that Earth has been visited by aliens from space.

That's too bad. A confirmed visit by intelligent creatures from beyond our solar system would answer many questions. It would be exciting, enlightening, and, like any real adventure, a bit scary. But none of the UFO sightings is dependable, and we are left with no direct evidence of life on other worlds.

UFOs from space are fun to think about, but there is no evidence that they are real.

then the only life we can detect will be beings intelligent enough to communicate with us. Why haven't we heard from them?

20-3 Communication with Distant Civilizations

IF OTHER CIVILIZATIONS EXIST, perhaps we can communicate with them in some way. Sadly, travel between the stars is more difficult in real life than in science fiction—and may in fact be impossible. If we can't physically visit, perhaps we can communicate by radio. Again, nature places restrictions on such conversations, but the restrictions are not too severe. As you will see, the real problem lies with the life expectancy of civilizations.

Travel Between the Stars

Practically speaking, roaming among the stars is tremendously difficult because of three limitations: distance, speed, and fuel. The distances between stars are almost beyond comprehension. It does little good to explain that if you use a golf ball in New York City to represent the sun, the nearest star would be another golf ball in Chicago. It is only slightly better to note that the fastest commercial jet would take about 4 million years to reach the nearest star.

The second limitation is a speed limit—you cannot travel faster than the speed of light. Though science fiction writers invent hyperspace drives so their heroes can zip from star to star, the speed of light is a natural and unavoidable limit that you cannot exceed. This, combined with the large distances between stars, makes interstellar travel very time consuming.

The third limitation is that you can't even approach the speed of light without using a fantastic amount of fuel. Even if you ignore the problem of escaping from Earth's gravity, you must still use energy stored in fuel to accelerate to high speed and to decelerate to a stop when you reach your destination. To return to Earth, assuming you wish to, you have to repeat the process. These changes in velocity require a tremendous amount of fuel. If you flew a spaceship as big as a large yacht to a star 5 light-years (1.5 pc) away and wanted to get there in only 10 years, you would use 40,000 times as much energy as the United States consumes in a year.

Travel for a few individuals might be possible if you accept very long travel times. That would require some form of suspended animation (currently unknown) or colony ships that carry a complete, though small, society in which people are born, live, and die generation after generation. Whether the occupants of such a ship would retain the social characteristics of humans over a long voyage is questionable.

These three limitations not only make it difficult for us to leave our solar system, but also would make it difficult for aliens to visit Earth. Reputable scientists have studied "unidentified flying objects" (UFOs) and related phenomena and have never found any evidence that Earth is being visited or has ever been visited by aliens from other worlds (**Window on Science 20-2**). Thus, humans are unlikely ever to meet an alien face to face. The only way we can communicate with other civilizations is via radio.

Radio Communication

Nature places two restrictions on our ability to communicate with distant societies by radio. One has to do with simple physics, is well understood, and merely makes the communication difficult. The second has to do with the fate of technological civilizations, is still unresolved, and may severely limit the number of societies we can detect by radio.

Radio signals are electromagnetic waves that travel at the speed of light. Because even the nearest civilizations must be a few light-years away, this limits Earth's ability to carry on a conversation with distant beings. If you ask a question of a creature 10 light-years away, you will have to wait 20 years for a reply. Clearly, the give-and-take of normal conversation will be impossible.

Instead, you could simply broadcast a radio beacon of friendship to announce your presence. Such a beacon would have to consist of a pattern of pulses obviously designed by intelligent beings, to distinguish it from natural radio signals emitted by nebulae, pulsars, and so on. For example, pulses counting off the first dozen prime numbers would do. In fact, Earth is already broadcasting a recognizable beacon. Short-wavelength radio signals, such as TV and FM, have been leaking into space for the last 50 years or so. Any civilization within 50 light-years might already have detected us.

If you intentionally broadcast a signal, you can anticode it. That is, you can arrange it to make it easy to decode. You could transmit pulses to represent 1s and gaps to represent 0s. A message counting through the first few prime numbers would distinguish your signal from natural sources of radio noise. You could even transmit a picture by sending a string of 1s and 0s that can be arranged in only two ways. One way produces nonsense, but the other way produces a meaningful picture (■ Figure 20-12).

In 1974, at the dedication of the 1000-ft radio telescope at Arecibo, radio astronomers transmitted such a signal toward the globular cluster M13, which is located 26,000 ly from Earth. When the signal finally arrives, any aliens who detect it will be able to arrange its 1679 pulses in only two ways, as 23 rows of 73 or 73 rows of 23. The second arrangement will form a picture that describes life on Earth (■ Figure 20-13).

It took only minutes to transmit the Arecibo message. If more time were taken, a more detailed picture could be sent, and if astronomers were sure their radio telescope was pointed at a listening civilization, they could send a long series of pictures. With pictures we could teach aliens our language and tell them all about our life, our difficulties, and our accomplishments.

If we can think of sending such signals, aliens can think of it too. If astronomers point their radio telescopes in the right direction and listen at the right wavelength, they might hear other intelligent races calling out to one another. This raises two questions: Which stars are the best candidates, and what wavelengths are most likely? You have already answered the first question. Main-sequence G and K stars have the most favorable characteristics. But the second question is more complex.

Only certain wavelengths are useful for communication. You cannot use wavelengths longer than about 30 cm because the signal would be lost in the background radio noise from our galaxy. Nor can you go to wavelengths much shorter than 1 cm because of absorption within our atmosphere. Thus only a certain range of wavelengths, a radio window, is open for communication.

This communications window is very wide, so a radio telescope would take a long time to tune over all the wavelengths searching for intelligent signals. Nature may have provided a way to narrow the search, however. Within the communications window lie the 21-cm line of neutral hydrogen and the 18-cm line of OH. The interval between these two lines has been dubbed the

■ Figure 20-12

An anticoded message is designed for easy decoding. Here a string of 35 radio pulses, represented as 1s and 0s, can be arranged in only two ways, as 5 rows of 7 or 7 rows of 5. The second way produces a friendly message. Any number of pulses can be used so long as it is the product of two prime numbers. Then the pulses can be arranged in only two ways.

An anticoded message

1 0 1 0 0 1 1 1 1 1
0 0 1 0 1 0 0 1 0 0
0 1 0 1 0 0 1 0 1 0
0 1 0 1 0

5 rows of 7

1	0	1	0	0	1	1
1	1	1	0	0	1	0
1	0	0	1	0	0	0
1	0	1	0	0	1	0
1	0	0	1	0	1	0

7 rows of 5

1	0	1	0	0
1	1	1	1	1
0	0	1	0	1
0	0	1	0	0
0	1	0	1	0
0	1	0	1	0
0	1	0	1	0

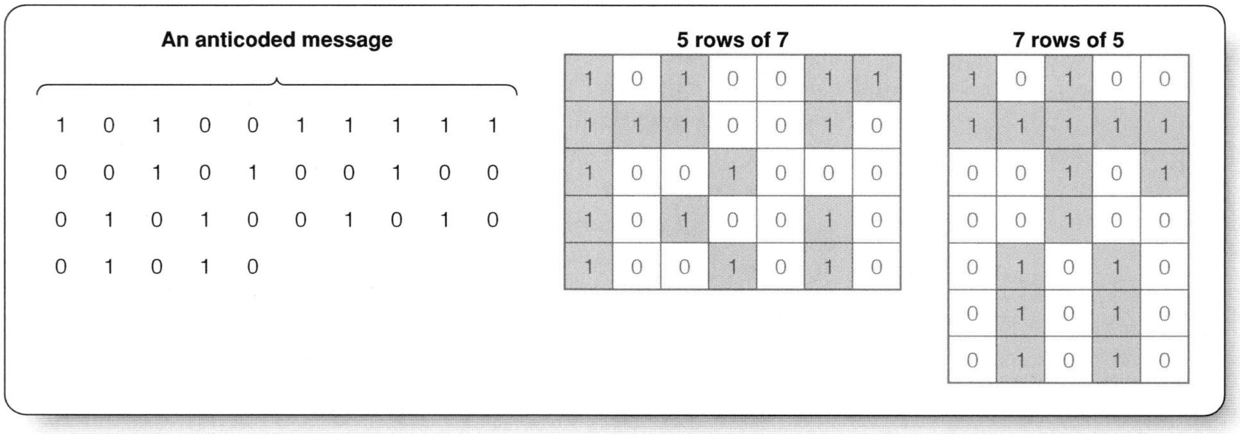

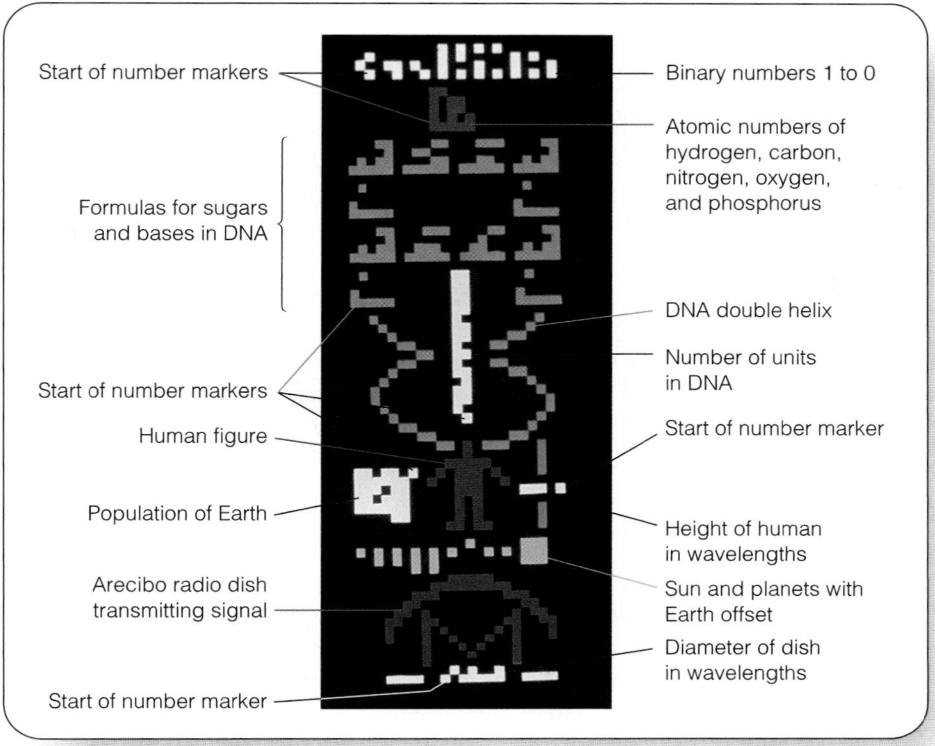

■ Active Figure 20-13

The Arecibo message to M13 begins by counting from 1 to 10 and goes on to describe our solar system and the biochemistry of life on Earth (color added for clarity). Binary numbers give the height of the human figure (1110) and the diameter of the telescope dish (100101111110) in units of the wavelength of the signal, 12.3 cm. (NASA)

AceAstronomy™ Log into AceAstronomy and select this chapter to see the Active Figure called "Interstellar Communication." Send messages and probes to nearby stars.

water hole because the combination of H and OH yields water (H_2O). Water is the fundamental solvent in our life form, so it might seem natural for similar water creatures to call out to each other at wavelengths in the water hole (■ Figure 20-14). But even silicon creatures would be familiar with the 21-cm line of hydrogen. Restricting the search to the water hole helps, but a radio telescope must still scan over millions of wavelengths just to search the entire water hole.

This is not idle speculation. A number of searches for extraterrestrial radio signals have been made, and some major searches are now under way. The field has become known as **SETI,** Search for Extra-Terrestrial Intelligence, and it has generated heated debate. Some scientists and philosophers argue that life on other worlds can't possibly exist, so it is a waste of money to search. Others argue that life on other worlds is common and could be detected with present technology. Congress funded a NASA search for a short time but then ended support in the early 1990s because political leaders feared public reaction. In fact, the annual cost of a major search is only about as much as a single Air Force attack helicopter. The controversy may spring in part from the theological and philosophical controversy that would result from the discovery of intelligent life on another world.

In spite of the controversy, searches are under way. The NASA SETI project canceled by Congress is now supported by private funds as Project Phoenix. It is located at the SETI Institute and has been using radio telescopes at major radio astronomy observatories to listen to the radio emissions of a list of candidate stars. About two billion radio-frequency bands must be examined for each star, so the project depends heavily on computers to search for signs of artificial radio transmissions in the recorded radio emissions of each star.

Astonishing amounts of computer power are needed to search for weak or unusual signals. Consequently, the Berkeley SETI team, with the support of the Planetary Society, has recruited owners of personal computers linked to the Internet to participate in a project called seti@home. Participants download a screensaver that searches data files from the Arecibo radio telescope for meaningful signals whenever the owner is not using the computer. About 2.3 million people have signed up and are providing 1000 years of computer time per day for the project. For information, locate the seti@home project at http://setiathome.ssl.berkeley.edu/.

■ Figure 20-14

Radio noise from various sources makes it difficult to detect distant signals at wavelengths longer than 30 cm or shorter than 1 cm. In this range, radio emission from H atoms and from OH marks a small wavelength range dubbed the water hole, which may be a likely place for communication.

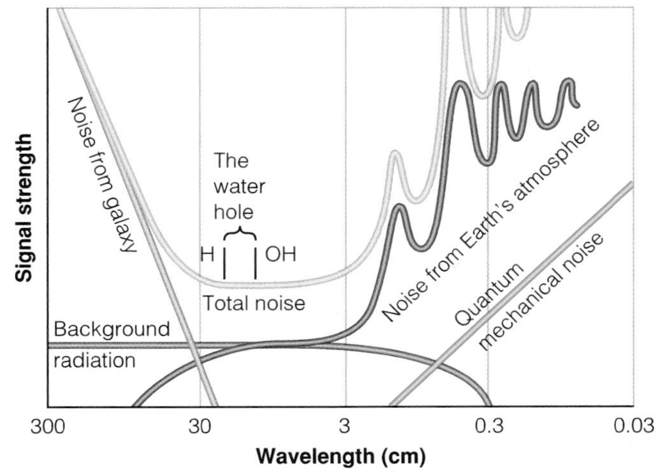

Since 1985, META, Megachannel Extra-Terrestrial Assay, has been searching the entire sky at millions of frequency bands in the water hole. Funded by the Planetary Society, the search uses radio antennas at Harvard and near Buenos Aires, Argentina. The project has instituted a second, even more efficient search called BETA. Since it began, META has detected dozens of candidate signals. That is, it has found dozens of radio signals that satisfy its most basic criteria. Unfortunately, none of these signals has proved to be continuous, and no signal has been detected when the radio telescopes were redirected at any of the candidate stars. Many of these signals are probably unusual noise from sources on Earth, but the candidate signals are still under study.

One ingenious search is called SERENDIP, Search for Extra-terrestrial Radio Emission from Nearby Developed Intelligent Populations. Rather than monopolize an entire radio telescope, SERENDIP rides piggyback on the 305-m Arecibo Telescope. Wherever the radio astronomers point the telescope, the receiver samples the signal, looking for intelligent signals over millions of frequency bands. If candidate signals are found, the receivers note the position of the radio telescope for future investigation.

Radio noise pollution is a serious problem for these searches. Year by year society generates more radio signals from sources on Earth and satellites in orbit. Poorly designed radio transmitters generate noise at unexpected frequencies, and other electronic devices, such as computers, emit radio-frequency signals. Furthermore, society, industry, and government press to use wider and wider sections of the electromagnetic spectrum. Radio astronomers struggle to hear through the radio babble, and SETI searches suffer because many of the noise signals mimic the patterns expected from other civilizations. It would be ironic if we fail to detect signals from another world because our own world has become too noisy.

What should you expect if signals are found? By international agreement, those searching for signals have agreed to share candidate signals and obtain reliable confirmation before making public announcements. False alarms would be embarrassing. The chances of success depend on the number of inhabited worlds in our galaxy, and that number is difficult to estimate.

How Many Inhabited Worlds?

The technology exists, and given enough time the searches will find other inhabited worlds, assuming there are at least a few out there. If intelligence is common, then scientists should find the signals soon—in the next few decades—but if intelligence is rare in the universe, it may be a very long time before they confirm that we are not alone.

Simple arithmetic can give you an estimate of the number of technological civilizations with which you might communicate, N_c. The first proposed formula for N_c is now known as the **Drake equation,** named after radio astronomer Frank Drake, a pioneer in the search for extraterrestrial intelligence. We will use a version of the Drake equation modified slightly to make it a bit easier to understand. The formula gives N_c, the number of communicative civilizations in a galaxy, as

$$N_c = N^* \cdot f_P \cdot n_{LZ} \cdot f_L \cdot f_I \cdot F_S$$

N^* is the number of stars in a galaxy, and f_P represents the fraction of all stars that have planets. If all single stars have planets, f_P is about 0.5. The factor n_{LZ} is the average number of planets in a solar system suitably placed in the life zone, f_L is the fraction of suitable planets on which life begins, and f_I is the fraction of planets where a life form evolved to intelligence. These factors can be roughly estimated, but the remaining factor is much more uncertain.

F_S is the fraction of a star's life during which the life form is communicative. Here we assume that a star lives about 10 billion years. If a society survives at a technological level for only 100 years, our chances of communicating with it are small. But a society that stabilizes and remains technological for a long time is much more likely to be in the communicative phase at the proper time to signal to us. If you assume that technological societies destroy themselves in about 100 years, F_S is 100 divided by 10 billion, or 10^{-8}. But if societies can remain technological for a million years, then F_S is 10^{-4}. The influence of the factors in the formula is shown in ■ Table 20-1.

■ Table 20-1 | The Number of Technological Civilizations per Galaxy

		Estimates	
	Variables	**Pessimistic**	**Optimistic**
N^*	Number of stars per galaxy	2×10^{11}	2×10^{11}
f_P	Fraction of stars with planets	0.01	0.5
n_{LZ}	Number of planets per star that lie in life zone for longer than 4 billion years	0.01	1
f_L	Fraction of suitable planets on which life begins	0.01	1
f_I	Fraction of life forms that evolve to intelligence	0.01	1
F_S	Fraction of star's life during which a technological society survives	10^{-8}	10^{-4}
N_c	Number of communicative civilizations per galaxy	2×10^{-5}	10×10^{6}

If the optimistic estimates are true, there may be a communicative civilization within a few dozen light-years of us, and we could locate it by searching through only a few thousand stars. On the other hand, if the pessimistic estimates are correct, Earth may be the only planet in our galaxy capable of communication. We may never know until we understand how technological societies function.

Ace Astronomy™ Log into AceAstronomy and select this chapter to see the Active Figure called "Drake Equation." Make your own calculation of the number of inhabited planets in our galaxy.

Ace Astronomy™ Log into AceAstronomy and select this chapter to see Astronomy Exercise "Extrasolar Planets." How common are planets orbiting other stars?

Inquire | Review | Analyze

Why does the number of inhabited worlds Earth might hear from depend on how long civilizations survive at a technological level? When astronomers turn radio telescopes toward the sky and scan millions of frequency bands, they take a snapshot of the universe at a particular time when humans are living and able to build radio telescopes. To detect other civilizations, they must be in a similar technological stage so they will be broadcasting either intentionally or accidentally. If scientists search for decades and detect no other signal, it may mean not that life is rare but rather that civilizations do not survive at a technological level for very long. Only a few decades ago, our civilization was threatened by nuclear war, and now Earth is threatened by the pollution of our environment. If nearly all civilizations in our galaxy are either on the long road up from primitive life forms in oceans or on the long road down from nuclear war or environmental collapse, there may be no one transmitting during the short interval when Earth is capable of building radio telescopes with which to listen.

Radio communication between inhabited worlds is limited because it requires very fast computers to search many frequency intervals. Why must scientists search so many frequencies when they suspect that the water hole would be a good place to listen?

■ ■ ■

Connections: Are we the only thinking race? If we are, we bear the sole responsibility to understand and admire the universe. Then we are the sole representatives of that state of matter called intelligence. The mere detection of signals from another civilization would demonstrate that we share the universe with others. Although we might never leave our solar system, such communication would end the self-centered isolation of humanity and stimulate a reevaluation of the meaning of our existence. We may never realize our full potential as humans until we communicate with nonhuman intelligent life.

Study and Review Tools

Summary

20-1 | The Nature of Life

What is life?

- The process of life extracts energy from the surroundings, maintains the organism, and modifies the surroundings to promote the organism's survival.
- Living things have a physical basis—the arrangement of matter and energy that makes life possible. Life on Earth is based on carbon chemistry.
- Living things must also have a controlling unit of information, which you can recognize as genetic information passed to each new generation.
- Genetic information for life on Earth is stored in long carbon-chain molecules such as DNA.
- The DNA molecule stores information in the form of chemical bases linked together like the rungs of a ladder. Copied by the RNA molecule, the patterns of bases act as recipes for the manufacture of proteins and enzymes.
- When a cell divides, the DNA molecules split lengthwise and duplicate themselves so that each of the new cells can receive a copy of the genetic information.
- Errors in duplication or damage to the DNA molecule can produce mutants, organisms that contain new DNA information and have new properties.
- Natural selection determines which of these new organisms are best suited to survive, and the species evolves to fit its environment.

20-2 | The Origin of Life

How did life originate on Earth?

- The oldest fossils on Earth are at least 2.5 billion years old. The fossils show that life began in the oceans.
- Life began on Earth as simple organisms like bacteria, and it evolved into more complex creatures.
- The Miller experiment shows that the building blocks of life form naturally under a wide range of circumstances.
- Chemical evolution concentrated simple molecules into larger, stronger molecules, but those molecules did not reproduce copies of themselves.
- Biological evolution begins when molecules begin reproducing.
- Not until about half a billion years ago did life become complex in what is called the Cambrian explosion.
- Life emerged from the oceans only about 0.4 billion years ago, and human intelligence developed over the last 3 million years.

Could life begin on other worlds?

- Life as we know it on Earth requires liquid water and moderate temperatures.
- No other planet in our solar system appears to harbor life at present. Most are too hot or too cold, although life might have begun on Mars before it became too cold and dry.
- Life in the oceans of Europa or Titan seems possible but unlikely.

- Because the origin of life and its evolution into intelligent creatures took so long on Earth, scientists eliminate short-lived stars such as middle- and upper-main-sequence stars as homes for life.
- Main-sequence G and K stars are thought to be likely candidates for searches for life, but M stars seem unlikely.
- The life zone around a star may be larger than scientists had expected, given the wide variety of living things being found in extreme environments on Earth.

20-3 | Communication with Distant Civilizations

Could Earthlings communicate with civilizations on other worlds?

- Because of distance, speed, and fuel, travel between the stars seems almost impossible for humans or for aliens who might visit Earth.
- Radio communication may be possible, but a conversation would be impossible because of very long travel times for radio signals.
- Broadcasting a radio beacon of pulses would distinguish the signal from naturally occurring radio emission and identify a civilization as technological.
- A signal could be anticoded so it would be easy for another civilization to decode.
- The best place in the radio spectrum for communication is in the water hole, the wavelength range from the 21-cm line of hydrogen to the emission line of OH. Even so, millions of radio wavelengths need to be tested to fully survey the water hole for a given target.
- Sophisticated searches are now underway to detect radio transmissions from civilizations on other worlds, but such SETI programs are hampered by limited computer power and radio noise pollution.
- The number of civilizations in our galaxy that are at a technological level and able to communicate may be limited by the lifetime of such civilizations.

New Terms

natural selection (p. 475)
mutant (p. 475)
DNA (deoxyribonucleic acid) (p. 476)
amino acid (p. 476)
protein (p. 476)
enzyme (p. 476)
RNA (ribonucleic acid) (p. 477)
chromosome (p. 477)
gene (p. 477)
Cambrian period (p. 478)
Miller experiment (p. 478)
primordial soup (p. 480)
chemical evolution (p. 480)
stromatolite (p. 481)
life zone (p. 486)
water hole (p. 489)
SETI (p. 489)
Drake equation (p. 490)

Review Questions

Ace Astronomy™ Assess your understanding of this chapter's topics with additional quizzing and animations at **http://astronomy.brookscole.com/sh9e**

1. If life is based on information, what is that information?
2. What would happen to a life form if the information handed down to offspring was always the same? How would that endanger the future of the life form?
3. How does the DNA molecule produce a copy of itself?
4. Give an example of natural selection acting on new DNA patterns to select the most advantageous characteristics.
5. Why do scientists believe that life on Earth began in the sea?
6. Why do scientists think that liquid water is necessary for the origin of life?
7. What is the difference between chemical evolution and biological evolution?
8. What was the significance of the Miller experiment?
9. How does intelligence make a creature more likely to survive?
10. Why are upper-main-sequence stars unlikely sites for intelligent civilizations?
11. Why do we suspect that travel between stars is nearly impossible?
12. How does the stability of technological civilizations affect the probability that Earth can communicate with them?
13. What is the water hole, and why would it be a good place to look for other civilizations?
14. The star cluster NGC2264, shown here, contains cool red giants and main-sequence stars from hot blue stars all the way down to red dwarfs. Discuss the likelihood that planets orbiting any of these stars might be home to life. Don't neglect to estimate the age of the cluster.

(T.A. Rector, B.A. Wolpa, NRAO/NOAO/AURA/NSF)

15. If you could search for life in the galaxy shown, would you look among disk stars or halo stars? Discuss the factors that influence your decision.

(ESO)

Discussion Questions

1. What would you change in the Arecibo message if humanity lived on Mars instead of Earth?
2. What do you think it would mean if decades of careful searches for radio signals for extraterrestrial intelligence turned up nothing?

Problems

1. A single human cell encloses about 1.5 m of DNA containing 4.5 billion base pairs. What is the spacing between these base pairs in nanometers? That is, how far apart are the rungs on the DNA ladder?
2. If you represent the history of the Earth by a line 1 m long, how long a segment would represent the 400 million years since life moved onto the land? How long a segment would represent the 3-million-year history of human life?
3. If a human generation, the time from birth to childbearing, is 20 years, how many generations have passed in the last million years?
4. If a star must remain on the main sequence for at least 5 billion years for life to evolve to intelligence, how massive could a star be and still harbor intelligent life on one of its planets? (*Hint:* See Reasoning with Numbers 9-1.)
5. If there are about 1.4×10^{-4} stars like the sun per cubic light-year, how many lie within 100 light-years of Earth? (*Hint:* The volume of a sphere is $\frac{4}{3}\pi r^3$.)
6. Mathematician Karl Gauss suggested planting forests and fields in a gigantic geometric proof to signal to possible Martians that intelligent life exists on Earth. If Martians had telescopes that could resolve details no smaller than 1 second of arc, how large would the smallest element of Gauss's proof have to be? (*Hint:* See Reasoning with Numbers 3-1.)
7. If you detected radio signals with an average wavelength of 20 cm and suspected that they came from a civilization on a distant planet, roughly how much of a change in wavelength should you expect to see because of the orbital motion of the distant planet? (*Hint:* See Reasoning with Numbers 6-2.)
8. Calculate the number of communicative civilizations per galaxy from your own estimates of the factors in Table 20-1.

Media Cluster

Ace Astronomy™ To access the resources in the Media Cluster, log into AceAstronomy at **http://astronomy.brookscole.com/sh9e** and select Chapter 20.

ACTIVE FIGURES

Future of the Sun
Over billions of years, the sun will gradually mature and die. This animation lets you observe the death of the sun from four different vantage points in our solar system.

Earth Calendar
This animation shows how the formation of Earth and its life forms would look if the entire history of Earth were condensed into one calendar year.

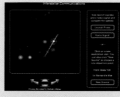

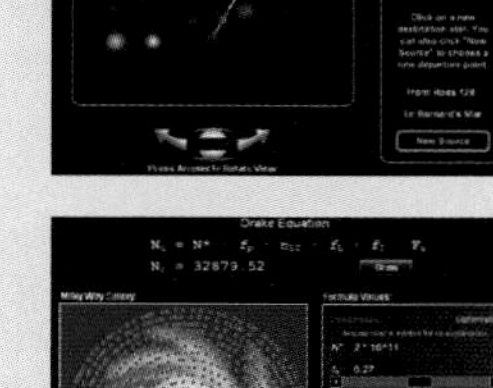

Interstellar Communication
In this animation, you can launch a space probe between distant stars or send radio signals and compare the two methods of communication.

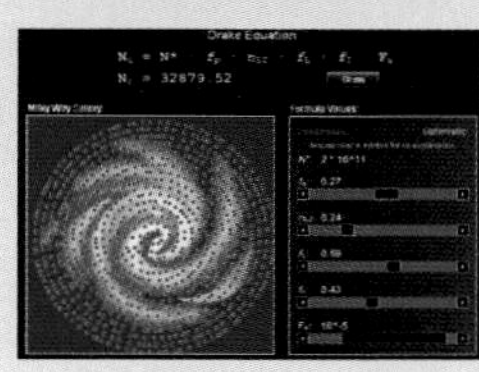

Drake Equation
Explore how different values for factors in the Drake equation affect predictions about how plentiful intelligent life is in a galaxy.

ASTRONOMY EXERCISE

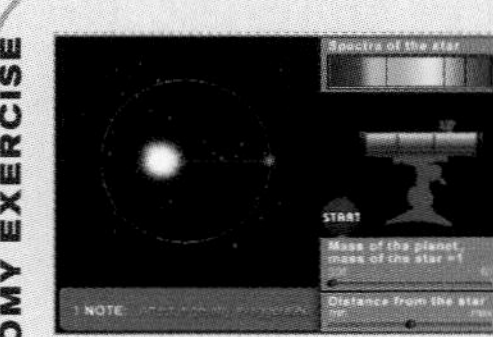

Extrasolar Planets
This simulation presents a hypothetical star system with one planet orbiting a star. Study what variables help astronomers detect extrasolar planets.

VIRTUAL ASTRONOMY LAB

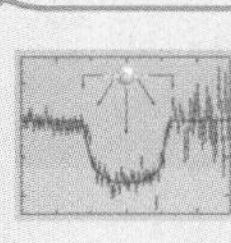

Lab 8: Extrasolar Planets
This lab examines how indirect methods such as Doppler shift techniques are used to discover and study extrasolar planets. At the end of the lab you will build your own solar system.

Critical Inquiries for the Web

1. The popular movie *Contact* focused interest on the SETI program by profiling the work of a radio astronomer dedicated to the search for extraterrestrial intelligence. Visit websites that give information about the movie, SETI programs, and radio astronomy, and discuss how realistic the movie was in capturing how such research is done.
2. Where outside the solar system would you look for habitable planets? NASA has increasingly focused its interest on this question. Look for information online about programs dedicated to detecting which planets might support life as we know it. What criteria are used to choose targets for the planned searches? What methods will be used to carry out the searches?

Go to the Brooks/Cole Astronomy Resource Center **(http://astronomy.brookscole.com)** for critical thinking exercises, articles, and additional readings from InfoTrac College Edition, Brooks/Cole's online student library.

Afterword

The aggregate of all our joys and sufferings, thousands of confident religions, ideologies and economic doctrines, every hunter and forager, every hero and coward, every creator and destroyer of civilizations, every king and peasant, every young couple in love, every hopeful child, every mother and father, every inventor and explorer, every teacher of morals, every corrupt politician, every superstar, every supreme leader, every saint and sinner in the history of our species, lived there on a mote of dust, suspended in a sunbeam.

CARL SAGAN (1934–1996)

Earth photographed by Voyager 1 from the edge of the solar system. (NASA)

Our journey is over, but before we part company, there is one last thing to discuss—the place of humanity in the universe. Astronomy gives us some comprehension of the workings of stars, galaxies, and planets, but its greatest value lies in what it teaches us about ourselves. Now that you have surveyed astronomical knowledge, you can better understand your own position in nature.

To some, the word *nature* conjures up visions of furry rabbits hopping about in a forest glade dotted with pastel wildflowers. To others, nature is the blue-green ocean depths filled with creatures swirling in a mad struggle for survival. Still others think of nature as windswept mountaintops of gray stone and glittering ice. As diverse as these images are, they are all Earthbound. Having studied astronomy, you can view nature as a beautiful mechanism composed of matter and energy interacting according to simple rules to form galaxies, stars, planets, mountaintops, ocean depths, and forest glades.

Perhaps the most important astronomical lesson is that you are a small but important part of the universe. Most of the universe is lifeless. The vast reaches between the galaxies appear to be empty of all but the thinnest gas, and the stars are much too hot to preserve the chemical bonds that seem necessary to allow life to survive and develop. Only on the surfaces of a few planets, where temperatures are moderate, could atoms link together to form living matter.

If life is special, then intelligence is precious. The universe must contain many planets devoid of life, planets where the wind has blown unfelt for billions of years. There may also be planets where life has developed but has not become complex, planets on which the wind stirs wide plains of grass and rustles dark forests. On some planets, insects, fish, birds, and animals may watch the passing days unaware of their own existence. It is intelligence, human or alien, that gives meaning to the landscape.

Science is the process by which intelligence tries to understand the universe. Science is not the invention of new devices or processes. It does not create home computers, cure the mumps, or manufacture plastic spoons—that is engineering and technology, the adaptation of scientific understanding for practical purposes. Science is understanding nature, and astronomy is understanding nature on the grandest scale. Astronomy is the science by which the universe, through its intelligent lumps of matter, tries to understand its own existence.

As the primary intelligent species on this planet, we are the custodians of a priceless gift—a planet filled with living things. This is especially true if life is rare in the universe. In fact, if Earth is the only inhabited planet, our responsibility is overwhelming. In any case, we are the only creatures who can take action to preserve the existence of life on Earth, and ironically, it is our own actions that are the most serious hazards.

The future of humanity is not secure. We are trapped on a tiny planet with limited resources and a population growing faster than our ability to produce food. In our efforts to survive, we have already driven some creatures to extinction and now threaten others. If our civilization collapses because of starvation, or if our race destroys itself somehow, the only bright spot is that the rest of the creatures on Earth will be better off for our absence.

But even if we control our population and conserve and recycle our resources, life on Earth is doomed. In a few billion years, the sun will leave the main sequence and swell into a red giant, incinerating Earth. Earth will be lifeless long before that, however. Within the next few billion years, the growing luminosity of the sun will first alter Earth's climate and then boil away its atmosphere and oceans. Our Earth is, like everything else in the universe, only temporary.

To survive, humanity must leave Earth and search for other planets. Colonizing the moon and other planets of our solar system will not save us, because they face the same fate as Earth when the sun dies. But travel to other stars is tremendously difficult and may be impossible with the limited resources we have in our small solar system. We and all the living things that depend on us for survival may be trapped.

This is a depressing prospect, but a few factors are comforting. First, everything in the universe is temporary. Stars die; galaxies die; perhaps the entire universe will die in a "big rip." That our distant future is limited only assures us that we are a part of a much larger whole. Second, we have a few billion years to prepare, and a billion years is a very long time. Only a few million years ago, our ancestors were learning to walk erect and communicate with one another. A billion years ago our ancestors were microscopic organisms living in the primeval oceans. To suppose that a billion years hence we humans will still be human, or that we will still be the dominant species on Earth, or that we will still be the sole intelligence on Earth, is the ultimate conceit.

Our responsibility is not to save our race for all eternity, but to behave as dependable custodians of our planet, preserving it, admiring it, and trying to understand it. That will call for drastic changes in our behavior toward other living things and a revolution in our attitude toward our planet's resources. Whether we can change our ways is debatable—humanity is far from perfect in its understanding, abilities, or intentions. We must not imagine, however, that we and our civilization are less than precious. We have the gift of intelligence, and that is the finest thing this planet has ever produced.

We shall not cease from exploration
And the end of all our exploring
Will be to arrive where we started
And know the place for the first time.

—T. S. Eliot, "Little Gidding"

Appendix A

Units and Astronomical Data

Introduction

The metric system is used worldwide as the system of units not only in science but also in engineering, business, sports, and daily life. Developed in 18th-century France, the metric system has gained acceptance in almost every country in the world because it simplifies computations.

A system of units is based on the three fundamental units for length, mass, and time. Other quantities such as density and force are derived from these fundamental units. In the English (or British) system of units (commonly used in the United States, Tonga, and Southern Yemen, but not in Britain) the fundamental unit of length is the foot, composed of 12 inches. The metric system is based on the decimal system of numbers, and the fundamental unit of length is the meter, composed of 100 centimeters.

To see the advantage of having a decimal-based system, try computing the volume of a bathtub that is 5′9″ long, 1′2″ deep, and 2′10″ wide. In the metric system the length of the tub is 1 m and 75 cm, or 1.75 m. The other dimensions are 0.35 m and 0.86 m, and the volume is just $1.75 \times 0.3 \times 0.86\ m^3$. To make the computation in English units, you must first convert inches to feet by dividing by 12. You can convert centimeters to meters by the simpler process of moving the decimal point. Thus, the computation is much easier if you measure the tub in meters and centimeters instead of feet and inches.

Because the metric system is a decimal system, it is easy to express quantities in larger or smaller units as is convenient. You can express distances in centimeters, meters, kilometers, and so on. The prefixes specify the relation of the unit to the meter. Just as a cent is $\frac{1}{100}$ of a dollar, so a centimeter is $\frac{1}{100}$ of a meter. A kilometer is 1000 m, and a kilogram is 1000 g. The meanings of the commonly used prefixes are given in Table A-1.

The SI Units

Any system of units based on the decimal system would be easy to use, but by international agreement, the preferred set of units, known as the *Système International d'Unités* (SI units), is based on the meter, kilogram, and second. These three fundamental units define the rest of the units as given in Table A-2.

The SI unit of force is the newton (N), named after Isaac Newton. It is the force needed to accelerate a 1 kg mass by 1 m/s^2, or the force roughly equivalent to the weight of an apple at Earth's surface. The SI unit of energy is the joule (J), the energy produced by a force of 1 N acting through a distance of 1 m. A joule is roughly the energy in the impact of an apple falling off a table.

Exceptions

Units help us in two ways. They make it possible to make calculations, and they help us to conceive of certain quantities. For

Table A-1 | Metric Prefixes

Prefix	Symbol	Factor
mega	M	10^6
kilo	k	10^3
centi	c	10^{-2}
milli	m	10^{-3}
micro	μ	10^{-6}
nano	n	10^{-9}

Table A-2 | SI Metric Units

Quantity	SI Unit	English Unit
Length	meter (m)	foot
Mass	kilogram (kg)	slug (sl)
Time	second (s)	second (s)
Force	newton (N)	pound (lb)
Energy	joule (J)	foot-pound (fp)

calculations the metric system is far superior, and it is used for calculations throughout this book.

But Americans commonly use the English system of units, so for conceptual purposes we can express quantities in English units. Instead of saying the average person would weigh 133 N on the moon, you could express the weight as 30 lb. Thus, the text commonly gives quantities in metric form followed by the English form in parentheses: The radius of the moon is 1738 km (1080 mi).

In SI units, density should be expressed as kilograms per cubic meter, but no human can enclose a cubic meter in his or her hand, so this unit does not help you grasp the significance of a given density. This book refers to density in grams per cubic centimeter. A gram is roughly the mass of a paperclip, and a cubic centimeter is the size of a small sugar cube, so you can conceive of a density of 1 g/cm^3, roughly the density of water. This is not a bothersome departure from SI units because we will not make complex calculations using density.

Conversions

To convert from one metric unit to another (from meters to kilometers, for example), you have only to look at the prefix. However, converting from metric to English or English to metric is more complicated. The conversion factors are given in Table A-3.

Example: The radius of the moon is 1738 km. What is this in miles? Table A-3 indicates that 1 mile equals 1.609 km, so

$$1738 \text{ km} \times \frac{1 \text{ mile}}{1.609 \text{ km}} = 1080 \text{ miles}$$

Temperature Scales

In astronomy, as in most other sciences, temperatures are expressed on the Kelvin scale, although the centigrade (or Celsius) scale is also used. The Fahrenheit scale commonly used in the United States is not used in scientific work.

Table A-3 | Conversion Factors

1 inch = 2.54 centimeters	1 centimeter = 0.394 inch
1 foot = 0.3048 meter	1 meter = 39.36 inches = 3.28 feet
1 mile = 1.6093 kilometers	1 kilometer = 0.6214 mile
1 slug = 14.594 kilograms	1 kilogram = 0.0685 slug
1 pound = 4.4482 newtons	1 newton = 0.2248 pound
1 foot-pound = 1.35582 joules	1 joule = 0.7376 foot-pound
1 horsepower = 745.7 joules/s	1 joule/s = 1 watt

Table A-4 | Temperature Scales

	Kelvin (K)	Centigrade (°C)	Fahrenheit (°F)
Absolute zero	0 K	−273°C	−459°F
Freezing point of water	273 K	0°C	32°F
Boiling point of water	373 K	100°C	212°F

Conversions:

$K = °C + 273$

$°C = \frac{5}{9}(°F - 32)$

$°F = \frac{9}{5}°C + 32$

Temperatures on the Kelvin scale are measured from absolute zero, the temperature of an object that contains no extractable heat. In practice, no object can be as cold as absolute zero, although laboratory apparatuses have reached temperatures less than 10^{-6} K. The scale is named after the Scottish mathematical physicist William Thomson, Lord Kelvin (1824–1907).

The centigrade scale refers temperatures to the freezing point of water (0°C) and to the boiling point of water (100°C). One degree centigrade is $\frac{1}{100}$ the temperature difference between the freezing and boiling points of water. Thus the prefix *centi.* The centigrade scale is also called the Celsius scale after its inventor, the Swedish astronomer Anders Celsius (1701–1744).

The Fahrenheit scale fixes the freezing point of water at 32°F and the boiling point at 212°F. Named after the German physicist Gabriel Daniel Fahrenheit (1686–1736), who made the first successful mercury thermometer in 1720, the Fahrenheit scale is used only in the United States.

It is easy to convert temperatures from one scale to another using the information given in Table A-4.

Powers of 10 Notation

Powers of 10 make writing very large numbers much simpler. For example, the nearest star is about 43,000,000,000,000 km from the sun. Writing this number as 4.3×10^{13} km is much easier.

Very small numbers can also be written with powers of 10. For example, the wavelength of visible light is about 0.0000005 m. In powers of 10 this becomes 5×10^{-7} m.

The powers of 10 used in this notation appear in the following list. The exponent tells you how to move the decimal point. If the exponent is positive, you move the decimal point to the right. If the exponent is negative, you move the decimal point to the left. Thus, 2×10^3 equals 2000.0, and 2×10^{-3} equals 0.002:

⋮

10^5	=	100,000
10^4	=	10,000
10^3	=	1,000
10^2	=	100
10^1	=	10
10^0	=	1
10^{-1}	=	0.1
10^{-2}	=	0.01
10^{-3}	=	0.001
10^{-4}	=	0.0001

⋮

If you use scientific notation in calculations, be sure you correctly enter numbers into your calculator. Not all calculators accept scientific notation, but those that can have a key labeled EXP, EEX, or perhaps EE that allows you to enter the exponent of 10. To enter a number such as 3×10^8, press the keys 3 EXP 8. To enter a number with a negative exponent, you must use the change-sign key, usually labeled +/− or CHS. To enter the number 5.2×10^{-3}, press the keys 5.2 EXP +/− 3. Try a few examples.

To read a number in scientific notation from a calculator you must read the exponent separately. The number 3.1×10^{25} may appear in a calculator display as 3.1 25 or on some calculators as 3.1 10^{25}. Examine your calculator to determine how such numbers are displayed.

* * *

Astronomy, and science in general, is a way of learning about nature and understanding the universe you live in. To test hypotheses about how nature works, scientists use observations of nature. The tables on the following pages contain some of the basic observations that support our best understanding of the astronomical universe. Of course, these data are expressed in the form of numbers, not because science reduces all understanding to mere numbers, but because the struggle to understand nature is so demanding you must use every tool available. Quantitative thinking, reasoning mathematically, is one of the most powerful tools ever invented by the human brain. Thus the tables beginning here are not nature reduced to mere numbers, but numbers supporting humanity's growing understanding of the natural world around us.

Table A-5 | Units Used in Astronomy

1 angstrom (Å)	$= 10^{-8}$ cm
	$= 10^{-10}$ m
1 astronomical unit (AU)	$= 1.495979 \times 10^{11}$ m
	$= 92.95582 \times 10^6$ miles
1 light-year (ly)	$= 6.3240 \times 10^4$ AU
	$= 9.46053 \times 10^{15}$ m
	$= 5.9 \times 10^{12}$ miles
1 parsec (pc)	= 206,265 AU
	$= 3.085678 \times 10^{16}$ m
	= 3.261633 ly
1 kiloparsec (kpc)	= 1000 pc
1 megaparsec (Mpc)	= 1,000,000 pc

Table A-6 | Constants

Astronomical unit (AU)	$= 1.495979 \times 10^{11}$ m
Parsec (pc)	= 206,265 AU
	$= 3.085678 \times 10^{16}$ m
	= 3.261633 ly
Light-year (ly)	$= 9.46053 \times 10^{15}$ m
Velocity of light (c)	$= 2.997925 \times 10^8$ m/s
Gravitational constant (G)	$= 6.67 \times 10^{-11}$ m^3/s^{2}kg
Mass of Earth ($M_\oplus$)	$= 5.976 \times 10^{24}$ kg
Earth equatorial radius ($R_\oplus$)	= 6378.164 km
Mass of sun ($M_\odot$)	$= 1.989 \times 10^{30}$ kg
Radius of sun ($R_\odot$)	$= 6.9599 \times 10^8$ m
Solar luminosity ($L_\odot$)	$= 3.826 \times 10^{26}$ J/s
Mass of moon	$= 7.350 \times 10^{22}$ kg
Radius of moon	= 1738 km
Mass of H atom	$= 1.67352 \times 10^{-27}$ kg

■ Table A-7 | The Nearest Stars

Name	Absolute Magnitude (M_v)	Distance (ly)	Spectral Type	Apparent Visual Magnitude (m_v)
Sun	4.83		G2	−26.8
Proxima Cen	15.45	4.28	M5	11.05
α Cen A	4.38	4.37	G2	0.1
B	5.76	4.37	K5	1.5
Barnard's Star	13.21	5.9	M5	9.5
Wolf 359	16.80	7.6	M6	13.5
Lalande 21185	10.42	8.1	M2	7.5
Sirius A	1.41	8.6	A1	−1.5
B	11.54	8.6	white dwarf	7.2
Luyten 726-8A	15.27	8.9	M5	12.5
B (UV Cet)	15.8	8.9	M6	13.0
Ross 154	13.3	9.4	M5	10.6
Ross 248	14.8	10.3	M6	12.2
ε Eri	6.13	10.7	K2	3.7
Luyten 789-6	14.6	10.8	M7	12.2
Ross 128	13.5	10.8	M5	11.1
61 CYG A	7.58	11.2	K5	5.2
B	8.39	11.2	K7	6.0
ε Ind	7.0	11.2	K5	4.7
Procyon A	2.64	11.4	F5	0.3
B	13.1	11.4	white dwarf	10.8
Σ 2398 A	11.15	11.5	M4	8.9
B	11.94	11.5	M5	9.7
Groombridge 34 A	10.32	11.6	M1	8.1
B	13.29	11.6	M6	11.0
Lacaille 9352	9.59	11.7	M2	7.4
τ Ceti	5.72	11.9	G8	3.5
BD + 5° 1668	11.98	12.2	M5	9.8
L 725-32	15.27	12.4	M5	11.5
Lacaille 8760	8.75	12.5	M0	6.7
Kapteyn's Star	10.85	12.7	M0	8.8
Kruger 60 A	11.87	12.8	M3	9.7
B	13.3	12.8	M4	11.2

■ Table A-8 | Properties of Main-Sequence Stars

Spectral Type	Absolute Visual Magnitude (M_v)	Luminosity*	Temp. (K)	λ_{max} (nm)	Mass*	Radius*	Average Density (g/cm^3)
O5	−5.8	501,000	40,000	72.4	40	17.8	0.01
B0	−4.1	20,000	28,000	100	18	7.4	0.1
B5	−1.1	790	15,000	190	6.4	3.8	0.2
A0	+0.7	79	9900	290	3.2	2.5	0.3
A5	+2.0	20	8500	340	2.1	1.7	0.6
F0	+2.6	6.3	7400	390	1.7	1.4	1.0
F5	+3.4	2.5	6600	440	1.3	1.2	1.1
G0	+4.4	1.3	6000	480	1.1	1.0	1.4
G5	+5.1	0.8	5500	520	0.9	0.9	1.6
K0	+5.9	0.4	4900	590	0.8	0.8	1.8
K5	+7.3	0.2	4100	700	0.7	0.7	2.4
M0	+9.0	0.1	3500	830	0.5	0.6	2.5
M5	+11.8	0.01	2800	1000	0.2	0.3	10.0
M8	+16	0.001	2400	1200	0.1	0.1	63

*Luminosity, mass, and radius are given in terms of the sun's luminosity, mass, and radius.

■ Table A-9 | The Brightest Stars

Star	Name	Apparent Visual Magnitude (m_v)	Spectral Type	Absolute Visual Magnitude (M_v)	Distance (ly)
α CMa A	Sirius	−1.47	A1 V	1.4	8.7
α Car	Canopus	−0.72	F0 Ib	−3.1	98
α Cen	Rigil Kentaurus	−0.01	G2 V	4.4	4.3
α Boo	Arcturus	−0.06	K2 III	−0.3	36
α Lyr	Vega	0.04	A0 V	0.5	26.5
α Aur	Capella	0.05	G8 III	−0.6	45
β Ori A	Rigel	0.14	B8 Ia	−7.1	900
α CMi A	Procyon	0.37	F5 IV	2.7	11.3
α Ori	Betelgeuse	0.41	M2 Ia	−5.6	520
α Eri	Achernar	0.51	B3 IV	−2.3	118
β Cen AB	Hadar	0.63	B1 II	−5.2	490
α Aql	Altair	0.77	A7 V	2.2	16.5
α Tau A	Aldebaran	0.86	K5 III	−0.7	68
α Cru	Acrux	0.90	B2 IV	−3.5	260
α Vir	Spica	0.91	B1 V	−3.3	220
α Sco A	Antares	0.92	M1 II	−5.1	520
α PsA	Fomalhaut	1.15	A3 V	2.0	22.6
β Gem	Pollux	1.16	K0 III	1.0	35
α Cyg	Deneb	1.26	A2 Ia	−7.1	1600
β Cru	Beta Crucis	1.28	B0.5 IV	−4.6	490

Table A-10 | Greatest Elongations of Mercury*

Evening Sky	Morning Sky
March 12, 2005**	April 26, 2005
July 9, 2005	Aug. 23, 2005**
Nov. 3, 2005	Dec. 12, 2005
Feb. 24, 2006**	April 8, 2006
June 20, 2006	Aug. 7, 2006
Oct. 16, 2006	Nov. 25, 2006**
Feb. 7, 2007	March 22, 2007
June 2, 2007	July 20, 2007
Sept. 29, 2007	Nov. 8, 2007
Jan. 22, 2008	March 3, 2008
May 14, 2008	July 1, 2008
Sept. 11, 2008	Oct. 22, 2008**
Jan. 4, 2009	Feb. 13. 2009
April 26, 2009**	June 13, 2009
Aug. 24, 2009	Oct. 6, 2009**
Dec. 18, 2009	

*Elongation is the angular distance from the sun to a planet.

**Most favorable elongations.

Table A-11 | Greatest Elongations of Venus

Evening Sky	Morning Sky
Nov. 3, 2005	March 25, 2006
June 9, 2007	Oct. 28, 2007
Jan. 14, 2009	June 5, 2009
Aug. 20, 2010	Jan. 8, 2011
March 27, 2012	Aug. 15, 2012
Nov. 1, 2013	March 22, 2014
June 6, 2015	Oct. 26, 2015
Jan. 12, 2017	June 3, 2017
Aug. 17, 2018	Jan. 6, 2019

Table A-12 | Meteor Showers

Shower	Dates	Hourly Rate	Radiant R.A.*	Radiant Dec.*	Associated Comet
Quadrantids	Jan. 2–4	30	15^h24^m	50°	
Lyrids	April 20–22	8	18^h4^m	33°	1861 I
η Aquarids	May 2–7	10	22^h24^m	0°	Halley?
δ Aquarids	July 26–31	15	22^h36^m	−10°	
Perseids	Aug. 10–14	40	3^h4^m	58°	1982 III
Orionids	Oct. 18–23	15	6^h20^m	15°	Halley?
Taurids	Nov. 1–7	8	3^h40^m	17°	Encke
Leonids	Nov. 14–19	6	10^h12^m	22°	1866 I Temp
Geminids	Dec. 10–13	50	7^h28^m	32°	

*R.A. and Dec. give the celestial coordinates (right ascension and declination) of the radiant of each shower.

Table A-13 | Properties of the Planets

PHYSICAL PROPERTIES (EARTH = ⊕)

Planet	Equatorial Radius (km)	Equatorial Radius (⊕ = 1)	Mass (⊕ = 1)	Average Density (g/cm³)	Surface Gravity (⊕ = 1)	Escape Velocity (km/s)	Sidereal Period of Rotation	Inclination of Equator to Orbit
Mercury	2439	0.382	0.0558	5.44	0.378	4.3	58.646^d	0°
Venus	6052	0.95	0.815	5.24	0.903	10.3	244.3^d	177°
Earth	6378	1.00	1.00	5.497	1.00	11.2	$23^h56^m04.1^s$	23°27′
Mars	3396	0.53	0.1075	3.94	0.379	5.0	$24^h37^m22.6^s$	25°19′
Jupiter	71,494	11.20	317.83	1.34	2.54	61	$9^h55^m30^s$	3°5′
Saturn	60,330	9.42	95.147	0.69	1.16	35.6	$10^h13^m59^s$	26°24′
Uranus	25,559	4.01	14.54	1.19	0.919	22	17^h14^m	97°55′
Neptune	24,750	3.93	17.23	1.66	1.19	25	16^h3^m	28°48′
Pluto	1185	0.18	0.0022	2.0	0.06	1.2	$6^d9^h21^m$	122°

ORBITAL PROPERTIES

Planet	Semimajor Axis (AU)	Semimajor Axis (10^6 km)	Orbital Period (y)	Orbital Period (days)	Average Orbital Velocity (km/s)	Orbital Eccentricity	Inclination to Ecliptic
Mercury	0.3871	57.9	0.24084	87.969	47.89	0.2056	7°0′16″
Venus	0.7233	108.2	0.61515	224.68	35.03	0.0068	3°23′40″
Earth	1	149.6	1	365.26	29.79	0.0167	0°
Mars	1.5237	227.9	1.8808	686.95	24.13	0.0934	1°51′9″
Jupiter	5.2028	778.3	11.867	4334.3	13.06	0.0484	1°18′29″
Saturn	9.5388	1427.0	29.461	10,760	9.64	0.0560	2°29′17″
Uranus	19.18	2869.0	84.013	30,685	6.81	0.0461	0°46′23″
Neptune	30.0611	4497.1	164.793	60,189	5.43	0.0100	1°46′27″
Pluto	39.44	5900	247.7	90,465	4.74	0.2484	17°9′3″

■ Table A-14 | Principal Satellites of the Solar System

Planet	Satellite	Radius (km)	Distance from Planet (10^3 km)	Orbital Period (days)	Orbital Eccentricity	Orbital Inclination
Earth	Moon	1738	384.4	27.322	0.055	5°8′43″
Mars	Phobos	14 × 12 × 10	9.38	0.3189	0.018	1°.0
	Deimos	8 × 6 × 5	23.5	1.262	0.002	2°.8
Jupiter	Metis	20	126	0.29	0.0	0°.0
	Adrastea	12 × 8 × 10	128	0.294	0.0	0°.0
	Amalthea	135 × 100 × 78	182	0.4982	0.003	0°.45
	Thebe	50	223	0.674	0.0	1°.3
	Io	1820	422	1.769	0.000	0°.3
	Europa	1565	671	3.551	0.000	0°.46
	Ganymede	2640	1071	7.155	0.002	0°.18
	Callisto	2420	1884	16.689	0.008	0°.25
	Leda	~8	11,110	240	0.146	26°.7
	Himalia	~85	11,470	250.6	0.158	27°.6
	Lysithea	~20	11,710	260	0.12	29°
	Elara	~30	11,740	260.1	0.207	24°.8
	Ananke	15	21,200	631	0.169	147°
	Carme	22	22,350	692	0.207	163°
	Pasiphae	35	23,300	735	0.40	147°
	Sinope	20	23,700	758	0.275	156°
Saturn	Pan	10	133.570	0.574	0.000	0°
	Atlas	20 × 15 × 15	137.7	0.601	0.002	0°.3
	Prometheus	70 × 40 × 50	139.4	0.613	0.003	0°.0
	Pandora	55 × 35 × 50	141.7	0.629	0.004	0°.05
	Epimetheus	70 × 50 × 50	151.42	0.694	0.009	0°.34
	Janus	110 × 80 × 100	151.47	0.695	0.007	0°.14
	Mimas	196	185.54	0.942	0.020	1°.5
	Enceladus	250	238.04	1.370	0.004	0°.0
	Tethys	530	294.67	1.888	0.000	1°.1
	Calypso	17 × 11 × 12	294.67	1.888	0.0	~1°.?
	Telesto	12	294.67	1.888	0.0	~1°?
	Dione	560	377	2.737	0.002	0°.0
	Helene	20 × 15 × 15	377	2.74	0.005	0°.15
	Rhea	765	527	4.518	0.001	0°.4
	Titan	2575	1222	15.94	0.029	0°.3
	Hyperion	205 × 130 × 110	1484	21.28	0.104	~0°.5
	Iapetus	720	3562	79.33	0.028	14°.72
	Phoebe	110	12,930	550.4	0.163	150°
Uranus	Cordelia	20	49.8	0.3333	~0	~0°
	Ophelia	15	53.8	0.375	~0	~0°
	Bianca	25	59.1	0.433	~0	~0°
	Cressida	30	61.8	0.462	~0	~0°
	Desdemona	30	62.7	0.475	~0	~0°
	Juliet	40	64.4	0.492	~0	~0°

(continued)

Table A-14 | *(continued)*

Planet	Satellite	Radius (km)	Distance from Planet (10^3 km)	Orbital Period (days)	Orbital Eccentricity	Orbital Inclination
Uranus *(continued)*	Portia	55	66.1	0.512	~0	~0°
	Rosalind	30	69.9	0.558	~0	~0°
	Belinda	30	75.2	0.621	~0	~0°
	S/1986 U10	20	76.2	0.638	~0	~0°
	Puck	85 ± 5	85.9	0.762	~0	~0°
	Miranda	242 ± 5	129.9	1.414	0.017	3.4°
	Ariel	580 ± 5	190.9	2.520	0.003	0°
	Umbriel	595 ± 10	266.0	4.144	0.003	0°
	Titania	805 ± 5	436.3	8.706	0.002	0°
	Oberon	775 ± 10	583.4	13.463	0.001	0°
	Caliban	40	7164	579	0.082	139.2°
	Stephano	~20	7900	676		
	Sycorax	80	12,174	1284	0.509	152.7°
	Prospero	~20	16,100	1950		
	Setebos	~20	17,600	2240	0.539	
Neptune	Naiad	30	48.2	0.296	~0	~0°
	Thalassa	40	50.0	0.312	~0	~0°
	Despina	90	52.5	0.333	~0	~0°
	Galatea	75	62.0	0.396	~0	~0°
	Larissa	95	73.6	0.554	~0	~0°
	Proteus	205	117.6	1.121	~0	~0°
	Triton	1352	354.59	5.875	0.00	160°
	Nereid	170	5588.6	360.125	0.76	27.7°
Pluto	Charon	626	19.7	6.38718	~0	122°

Table A-15 | **The Greek Alphabet**

A, α	alpha	H, η	eta	N, ν	nu	T, τ	tau
B, β	beta	Θ, θ	theta	Ξ, ξ	xi	Y, υ	upsilon
Γ, γ	gamma	I, ι	iota	O, o	omicron	Φ, ϕ	phi
Δ, δ	delta	K, κ	kappa	Π, π	pi	X, χ	chi
E, ε	epsilon	Λ, λ	lambda	P, ρ	rho	Ψ, ψ	psi
Z, ζ	zeta	M, μ	mu	Σ, σ	sigma	Ω, ω	omega

■ Table A-16 | Periodic Table of the Elements

Atomic number — 11; Symbol — Na; Atomic mass — 22.99

Atomic masses are based on carbon-12. Numbers in parentheses are mass numbers of most stable or best-known isotopes of radioactive elements.

Period	IA(1)	IIA(2)	IIIB(3)	IVB(4)	VB(5)	VIB(6)	VIIB(7)	VIII (8)	VIII (9)	VIII (10)	IB(11)	IIB(12)	IIIA(13)	IVA(14)	VA(15)	VIA(16)	VIIA(17)	Noble Gases (18)
1	1 H 1.008																	2 He 4.003
2	3 Li 6.941	4 Be 9.012											5 B 10.81	6 C 12.01	7 N 14.01	8 O 16.00	9 F 19.00	10 Ne 20.18
3	11 Na 22.99	12 Mg 24.31	← Transition Elements →										13 Al 26.98	14 Si 28.09	15 P 30.97	16 S 32.06	17 Cl 35.45	18 Ar 39.95
4	19 K 39.10	20 Ca 40.08	21 Sc 44.96	22 Ti 47.90	23 V 50.94	24 Cr 52.00	25 Mn 54.94	26 Fe 55.85	27 Co 58.93	28 Ni 58.7	29 Cu 63.55	30 Zn 65.38	31 Ga 69.72	32 Ge 72.59	33 As 74.92	34 Se 78.96	35 Br 79.90	36 Kr 83.80
5	37 Rb 85.47	38 Sr 87.62	39 Y 88.91	40 Zr 91.22	41 Nb 92.91	42 Mo 95.94	43 Tc 98.91	44 Ru 101.1	45 Rh 102.9	46 Pd 106.4	47 Ag 107.9	48 Cd 112.4	49 In 114.8	50 Sn 118.7	51 Sb 121.8	52 Te 127.6	53 I 126.9	54 Xe 131.3
6	55 Cs 132.9	56 Ba 137.3	57* La 138.9	72 Hf 178.5	73 Ta 180.9	74 W 183.9	75 Re 186.2	76 Os 190.2	77 Ir 192.2	78 Pt 195.1	79 Au 197.0	80 Hg 200.6	81 Tl 204.4	82 Pb 207.2	83 Bi 209.0	84 Po (210)	85 At (210)	86 Rn (222)
7	87 Fr (223)	88 Ra 226.0	89** Ac (227)	104 Rf (261)	105 Db (262)	106 Sg (263)	107 Bh (262)	108 Hs (265)	109 Mt (266)	110 Ds (269)	111 Uuu (272)	112 Uub (277)	113 Uub (284)	114 Uuq (285)	115 Uub (288)	116 Uuh (289)		

Inner Transition Elements

* Lanthanide Series 6	58 Ce 140.1	59 Pr 140.9	60 Nd 144.2	61 Pm (145)	62 Sm 150.4	63 Eu 152.0	64 Gd 157.3	65 Tb 158.9	66 Dy 162.5	67 Ho 164.9	68 Er 167.3	69 Tm 168.9	70 Yb 173.0	71 Lu 175.0
** Actinide Series 7	90 Th 232.0	91 Pa 231.0	92 U 238.0	93 Np 237.0	94 Pu (244)	95 Am (243)	96 Cm (247)	97 Bk (247)	98 Cf (251)	99 Es (252)	100 Fm (257)	101 Md (258)	102 No (259)	103 Lr (260)

The Elements and Their Symbols

Actinium	Ac
Aluminum	Al
Americium	Am
Antimony	Sb
Argon	Ar
Arsenic	As
Astatine	At
Barium	Ba
Berkelium	Bk
Beryllium	Be
Bismuth	Bi
Bohrium	Bh
Boron	B
Bromine	Br
Cadmium	Cd
Calcium	Ca
Californium	Cf
Carbon	C
Cerium	Ce
Cesium	Cs
Chlorine	Cl
Chromium	Cr
Cobalt	Co
Copper	Cu
Curium	Cm
Darmstadtium	Ds
Dubnium	Db
Dysprosium	Dy
Einsteinium	Es
Erbium	Er
Europium	Eu
Fermium	Fm
Fluorine	F
Francium	Fr
Gadolinium	Gd
Gallium	Ga
Germanium	Ge
Gold	Au
Hafnium	Hf
Hassium	Hs
Helium	He
Holmium	Ho
Hydrogen	H
Indium	In
Iodine	I
Iridium	Ir
Iron	Fe
Krypton	Kr
Lanthanum	La
Lawrencium	Lr
Lead	Pb
Lithium	Li
Lutetium	Lu
Magnesium	Mg
Manganese	Mn
Meitnerium	Mt
Mendelevium	Md
Mercury	Hg
Molybdenum	Mo
Neodymium	Nd
Neon	Ne
Neptunium	Np
Nickel	Ni
Niobium	Nb
Nitrogen	N
Nobelium	No
Osmium	Os
Oxygen	O
Palladium	Pd
Phosphorous	P
Platinum	Pt
Plutonium	Pu
Polonium	Po
Potassium	K
Praseodymium	Pr
Promethium	Pm
Protactinium	Pa
Radium	Ra
Radon	Rn
Rhenium	Re
Rhodium	Rh
Rubidium	Rb
Ruthenium	Ru
Rutherfordium	Rf
Samarium	Sm
Scandium	Sc
Seaborgium	Sg
Selenium	Se
Silicon	Si
Silver	Ag
Sodium	Na
Strontium	Sr
Sulfur	S
Tantalum	Ta
Technetium	Tc
Tellurium	Te
Terbium	Tb
Thallium	Tl
Thorium	Th
Thulium	Tm
Tin	Sn
Titanium	Ti
Tungsten	W
Uranium	U
Vanadium	V
Xenon	Xe
Ytterbium	Yb
Yttrium	Y
Zinc	Zn
Zirconium	Zr

Appendix B

Observing the Sky

Observing the sky with the naked eye is of no more importance to modern astronomy than picking up pretty pebbles is to modern geology. But the sky is a natural wonder unimaginably bigger than the Grand Canyon, the Rocky Mountains, or any other natural wonder that tourists visit every year. To neglect the beauty of the sky is equivalent to geologists neglecting the beauty of the minerals they study. This supplement is meant to act as a tourist's guide to the sky. You analyzed the universe in the regular chapters, but here you will admire it.

The brighter stars in the sky are visible even from the centers of cities with their air and light pollution. But in the countryside only a few miles beyond the cities, the night sky is a velvety blackness strewn with thousands of glittering stars. From a wilderness location, far from the city's glare, and especially from high mountains, the night sky is spectacular.

Using Star Charts

The constellations are a fascinating cultural heritage of our planet, but they are sometimes a bit difficult to learn because of Earth's motion. The constellations above the horizon change with the time of night and the seasons.

Because Earth rotates eastward, the sky appears to rotate around you westward. A constellation visible in the southern sky soon after sunset will appear to move westward, and in a few hours it will disappear below the horizon. Other constellations will rise in the east, so the sky changes gradually through the night.

In addition, Earth's orbital motion makes the sun appear to move eastward among the stars. Each day the sun moves about twice its own diameter eastward along the ecliptic, and each night at sunset, the constellations are about 1° farther toward the west.

Orion, for instance, is visible in the southern sky in January, but as the days pass, the sun moves closer to Orion. By March, Orion is difficult to see in the southwest sky soon after sunset. By June, the sun is so close to Orion it sets with the sun and is invisible. Not until late July is the sun far enough past Orion for the constellation to become visible rising in the eastern sky just before dawn.

Because of the rotation and orbital motion of Earth, you must use more than one star chart to map the sky. Which chart you select depends on the month and the time of night. The charts given in this appendix show the evening sky for each month.

Two sets of charts are included for two typical locations on Earth. The northern hemisphere charts show the sky as seen from a northern latitude typical of the United States and central Europe. The southern hemisphere star charts are appropriate for readers in Earth's southern hemisphere, including Australia, southern South America, and southern Africa.

To use the charts, select the appropriate chart and hold it overhead as shown in ■ Figure B-1. If you face south, turn the chart until the words "Southern Horizon" are at the bottom of the chart. If you face other directions, turn the chart appropriately.

■ **Figure B-1**

To use the star charts in this book, select the appropriate chart for the date and time. Hold it overhead, and turn it until the direction at the bottom of the chart is the same as the direction you are facing.

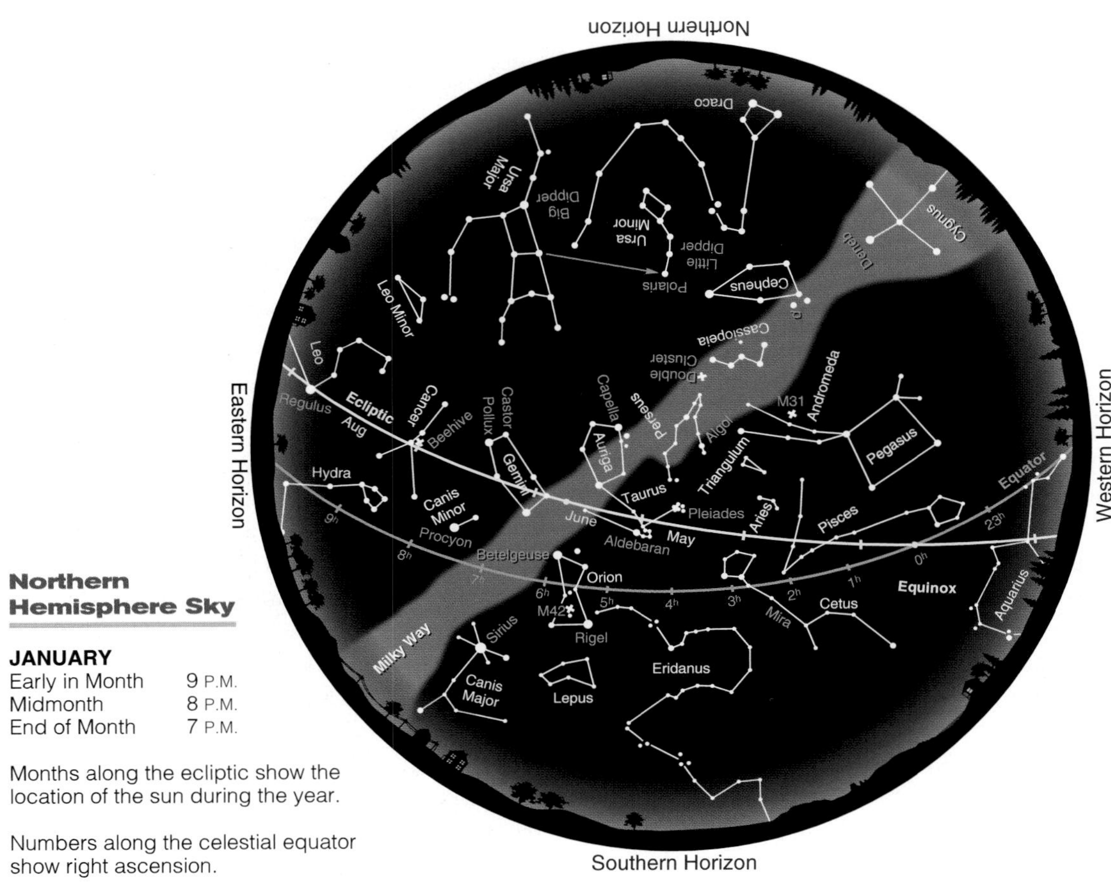

Northern Hemisphere Sky

JANUARY

Early in Month	9 P.M.
Midmonth	8 P.M.
End of Month	7 P.M.

Months along the ecliptic show the location of the sun during the year.

Numbers along the celestial equator show right ascension.

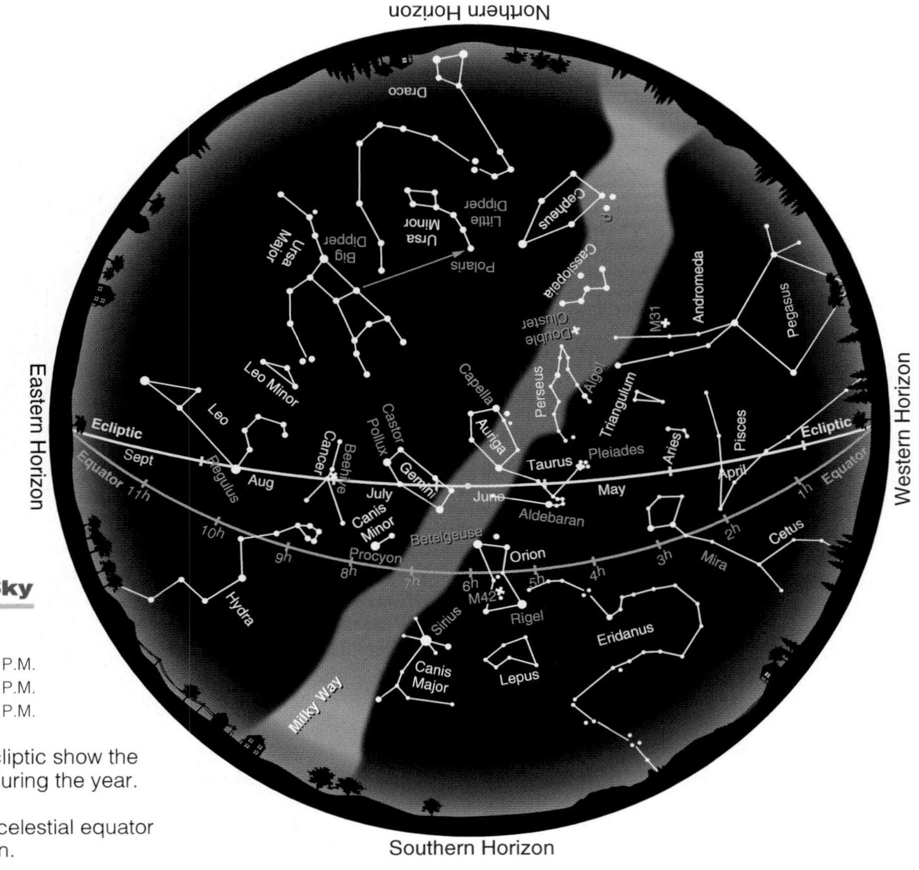

Northern Hemisphere Sky

FEBRUARY

Early in Month	9 P.M.
Midmonth	8 P.M.
End of Month	7 P.M.

Months along the ecliptic show the location of the sun during the year.

Numbers along the celestial equator show right ascension.

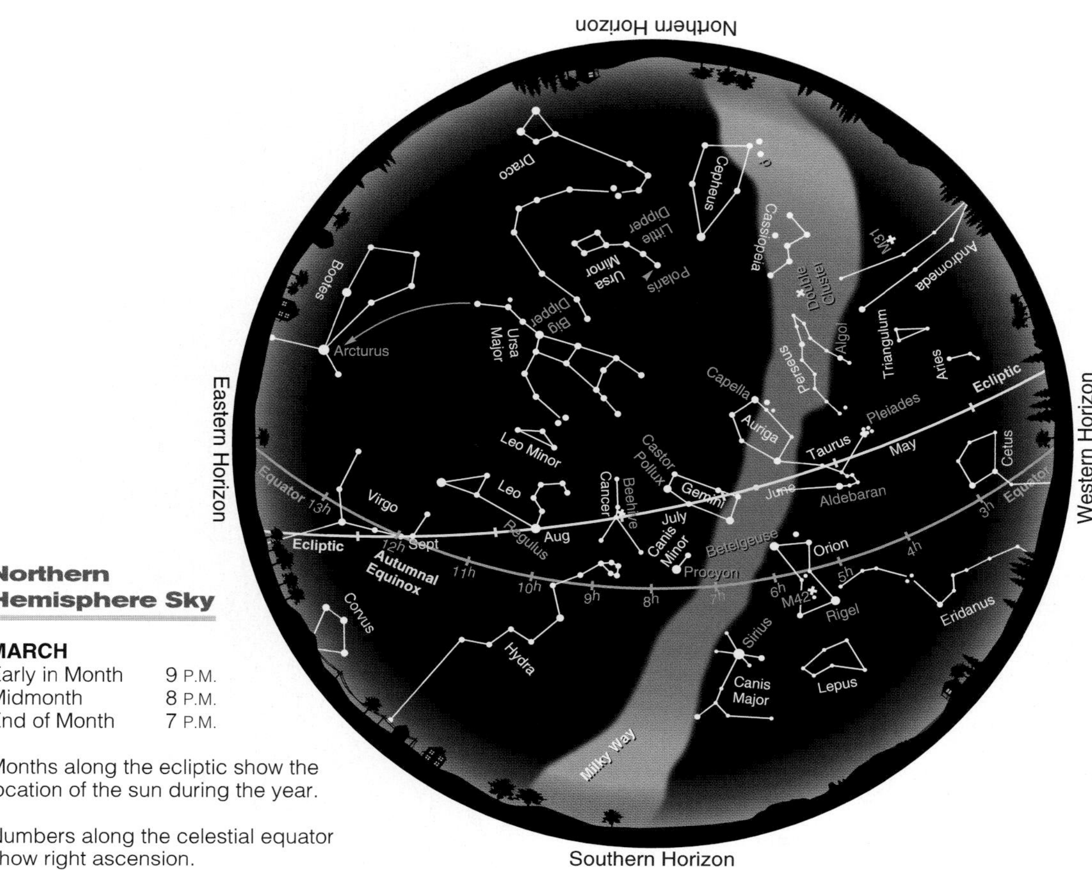

Northern Hemisphere Sky

MARCH

Early in Month	9 P.M.
Midmonth	8 P.M.
End of Month	7 P.M.

Months along the ecliptic show the location of the sun during the year.

Numbers along the celestial equator show right ascension.

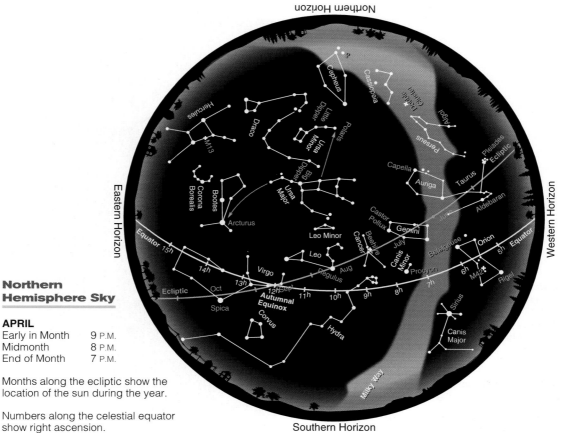

Northern Horizon

Eastern Horizon

Western Horizon

Southern Horizon

Northern Hemisphere Sky

APRIL

Early in Month	9 P.M.
Midmonth	8 P.M.
End of Month	7 P.M.

Months along the ecliptic show the location of the sun during the year.

Numbers along the celestial equator show right ascension.

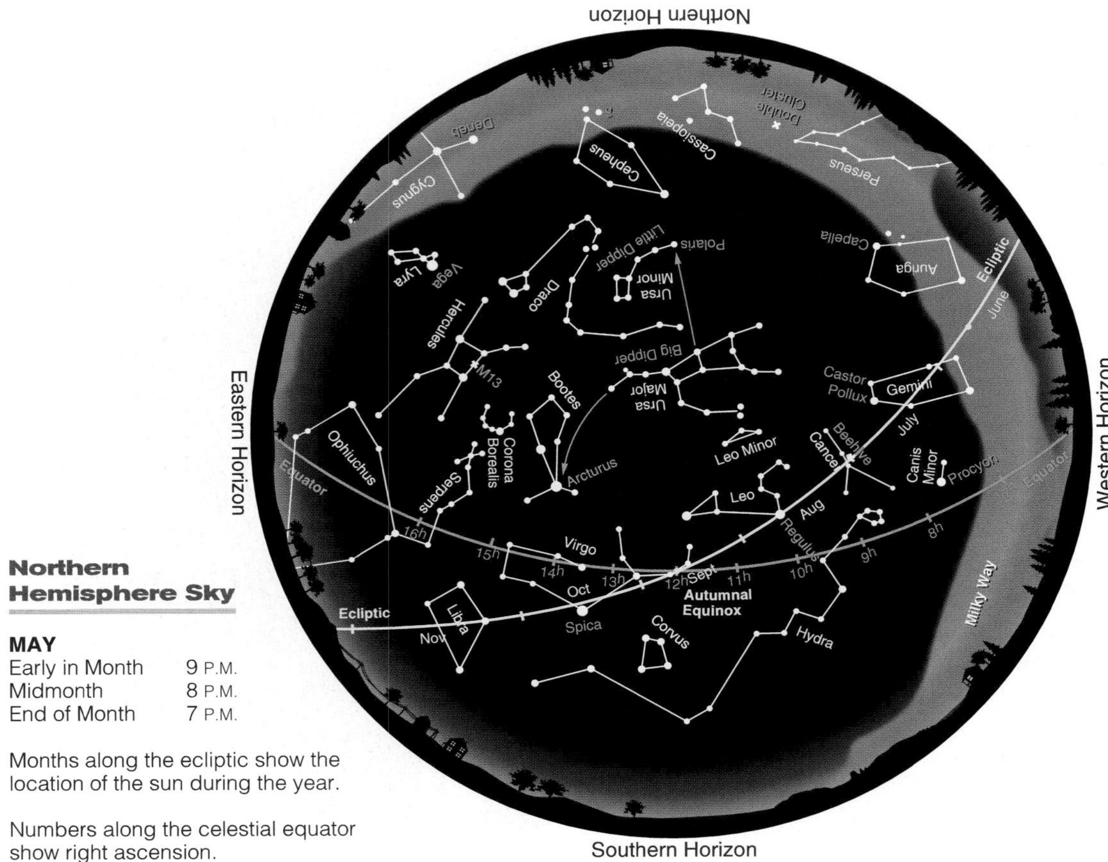

Northern Hemisphere Sky

MAY

Early in Month	9 P.M.
Midmonth	8 P.M.
End of Month	7 P.M.

Months along the ecliptic show the location of the sun during the year.

Numbers along the celestial equator show right ascension.

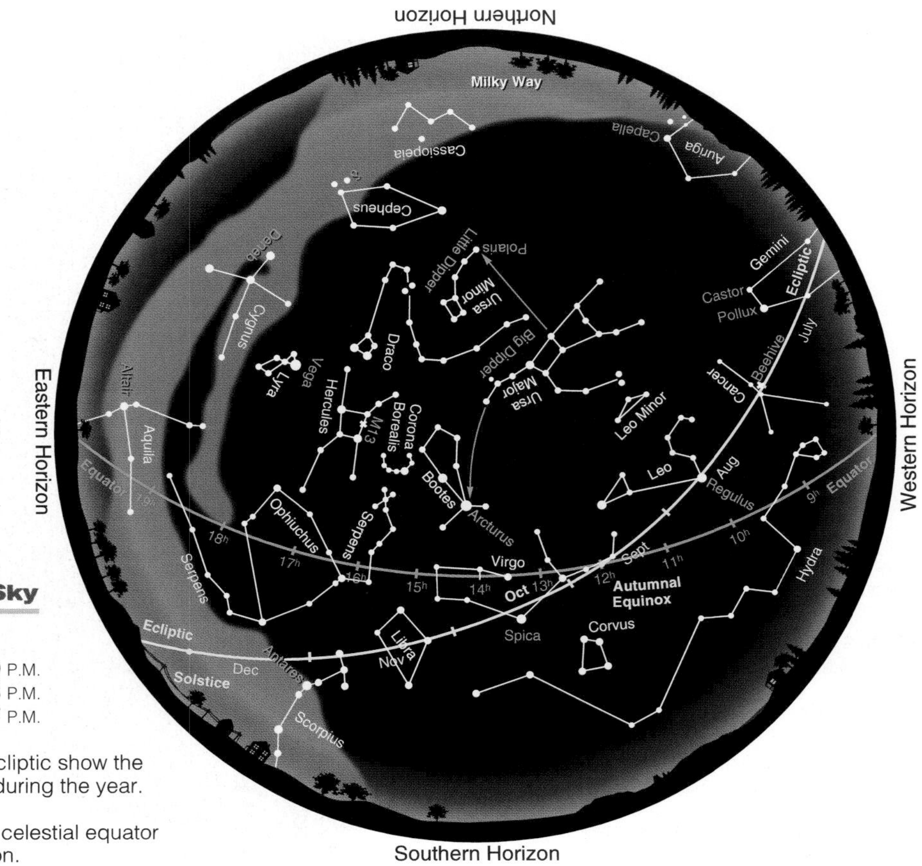

Northern Hemisphere Sky

JUNE

Early in Month	9 P.M.
Midmonth	8 P.M.
End of Month	7 P.M.

Months along the ecliptic show the location of the sun during the year.

Numbers along the celestial equator show right ascension.

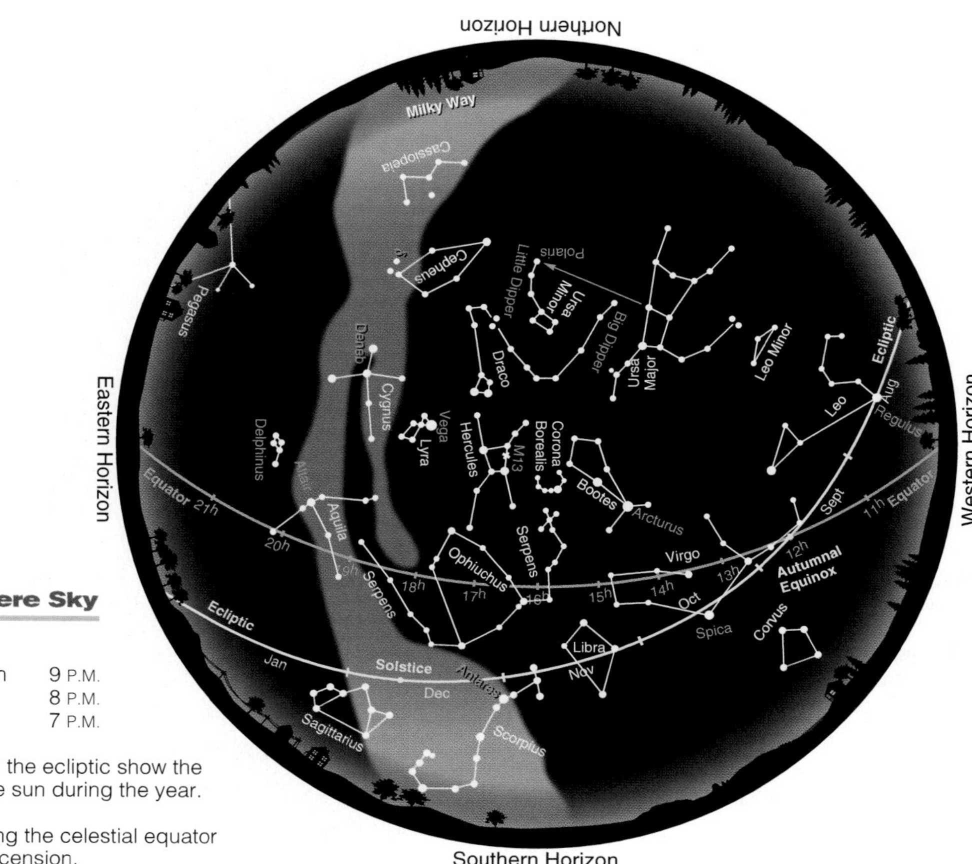

Northern Hemisphere Sky

JULY

Early in Month	9 P.M.
Midmonth	8 P.M.
End of Month	7 P.M.

Months along the ecliptic show the location of the sun during the year.

Numbers along the celestial equator show right ascension.

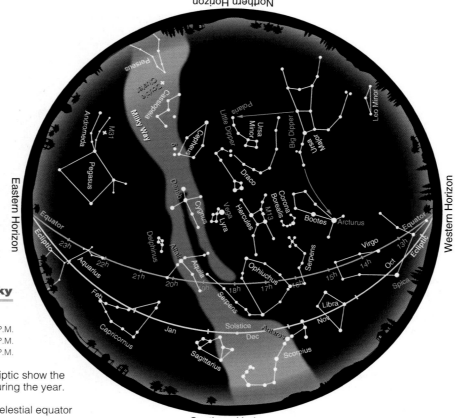

Northern Hemisphere Sky

AUGUST

Early in Month	9 P.M.
Midmonth	8 P.M.
End of Month	7 P.M.

Months along the ecliptic show the location of the sun during the year.

Numbers along the celestial equator show right ascension.

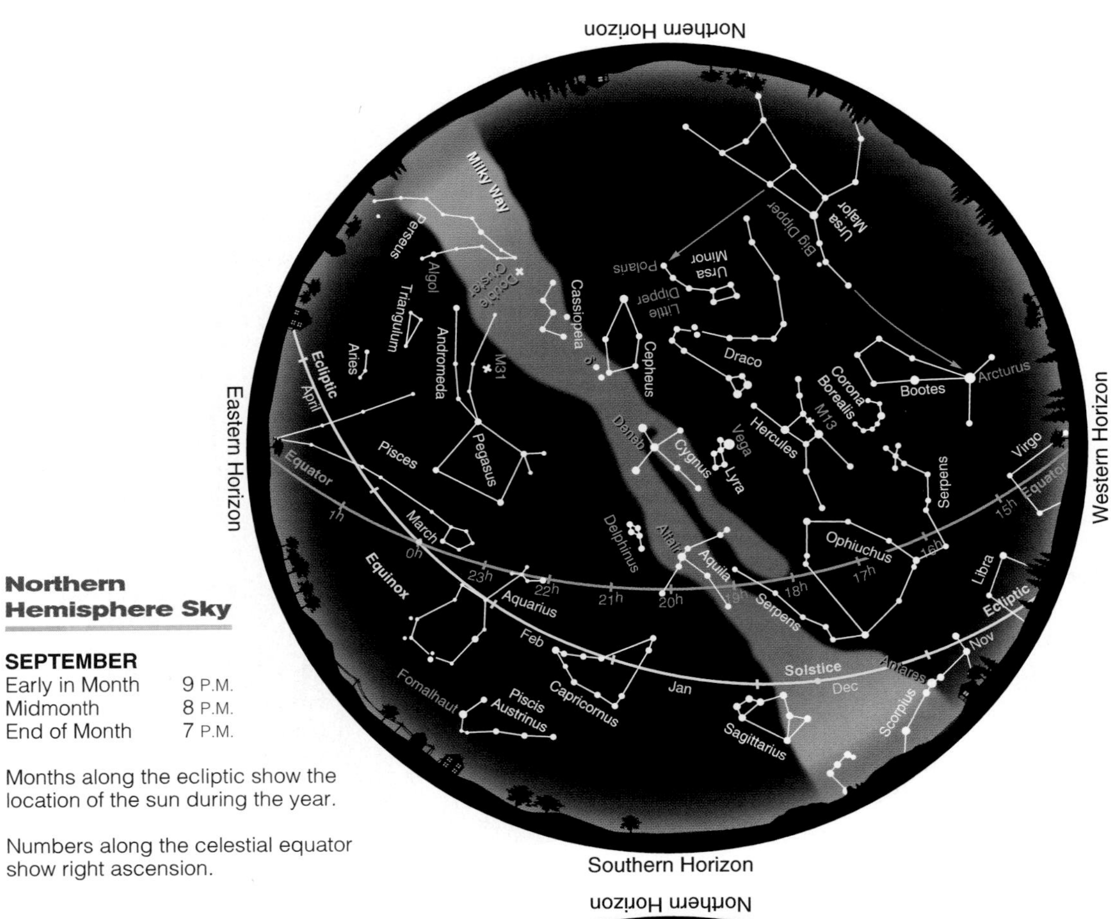

Northern Hemisphere Sky

SEPTEMBER

Early in Month	9 P.M.
Midmonth	8 P.M.
End of Month	7 P.M.

Months along the ecliptic show the location of the sun during the year.

Numbers along the celestial equator show right ascension.

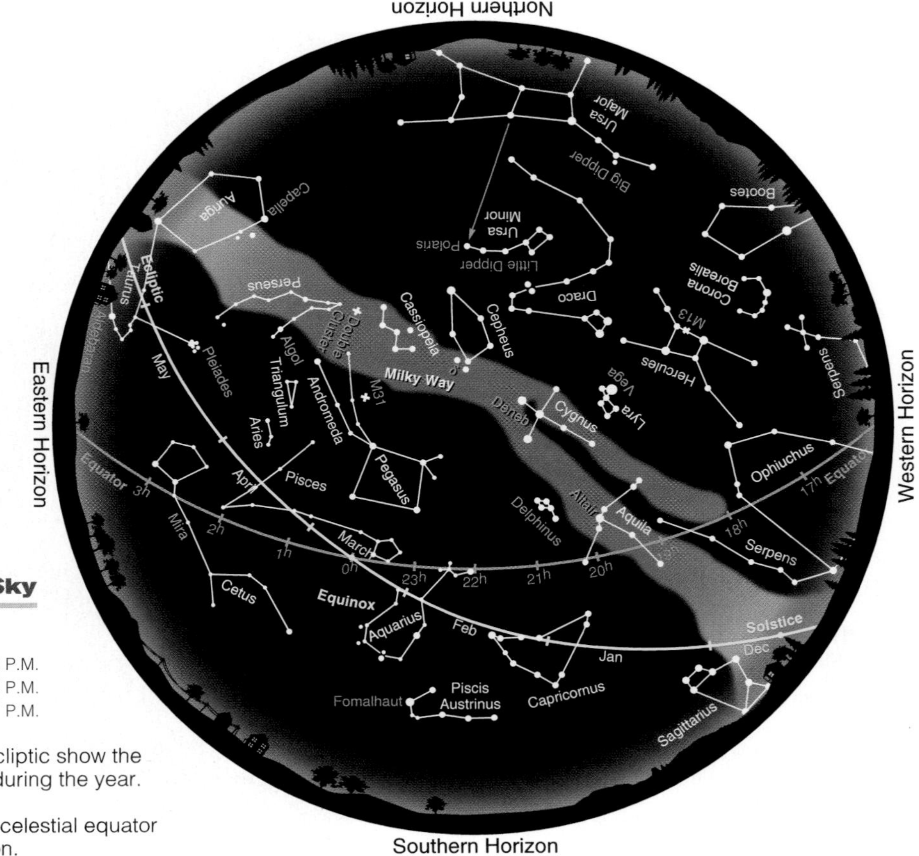

Northern Hemisphere Sky

OCTOBER

Early in Month	9 P.M.
Midmonth	8 P.M.
End of Month	7 P.M.

Months along the ecliptic show the location of the sun during the year.

Numbers along the celestial equator show right ascension.

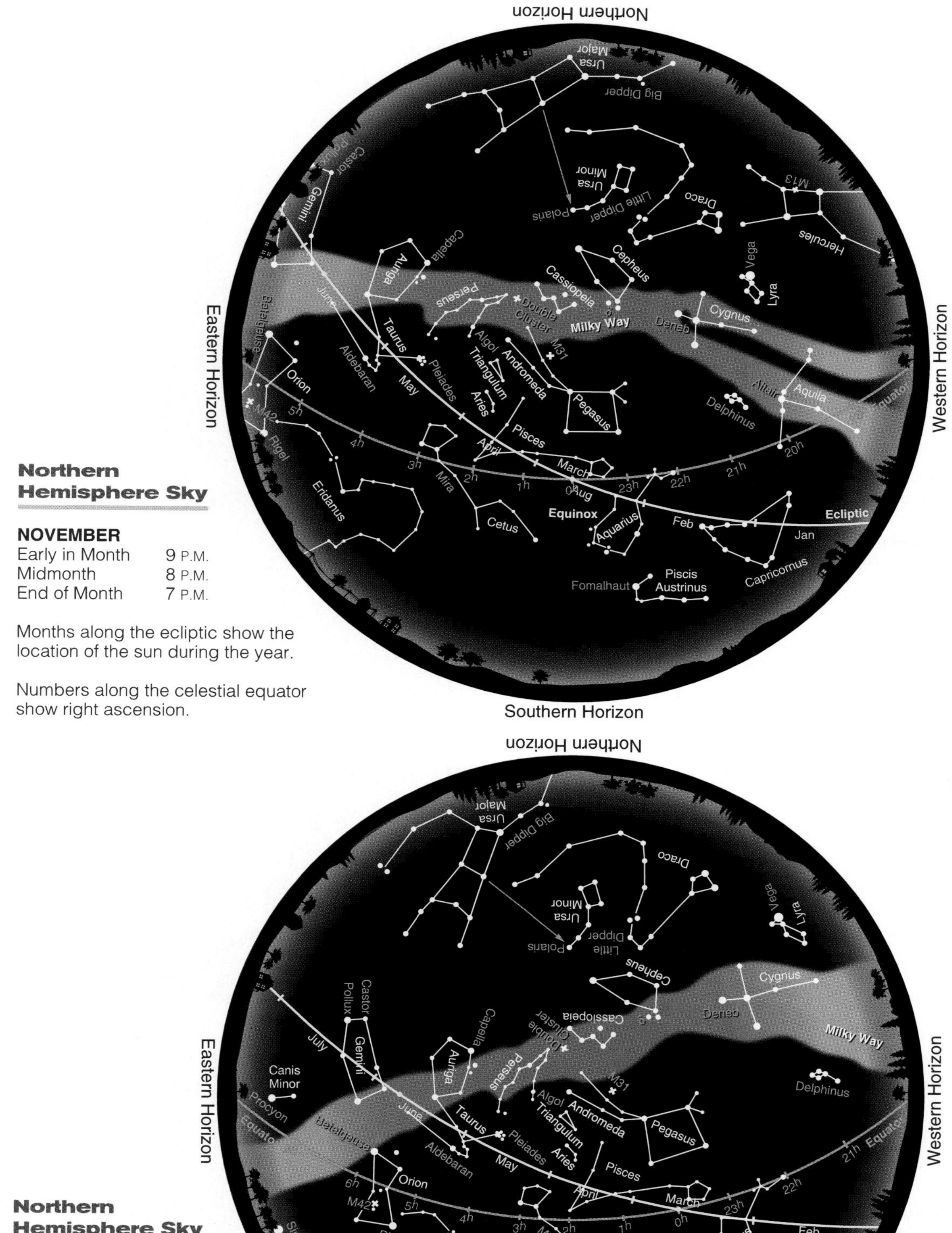

Northern Hemisphere Sky

NOVEMBER

Early in Month	9 P.M.
Midmonth	8 P.M.
End of Month	7 P.M.

Months along the ecliptic show the location of the sun during the year.

Numbers along the celestial equator show right ascension.

Northern Hemisphere Sky

DECEMBER

Early in Month	9 P.M.
Midmonth	8 P.M.
End of Month	7 P.M.

Months along the ecliptic show the location of the sun during the year.

Numbers along the celestial equator show right ascension.

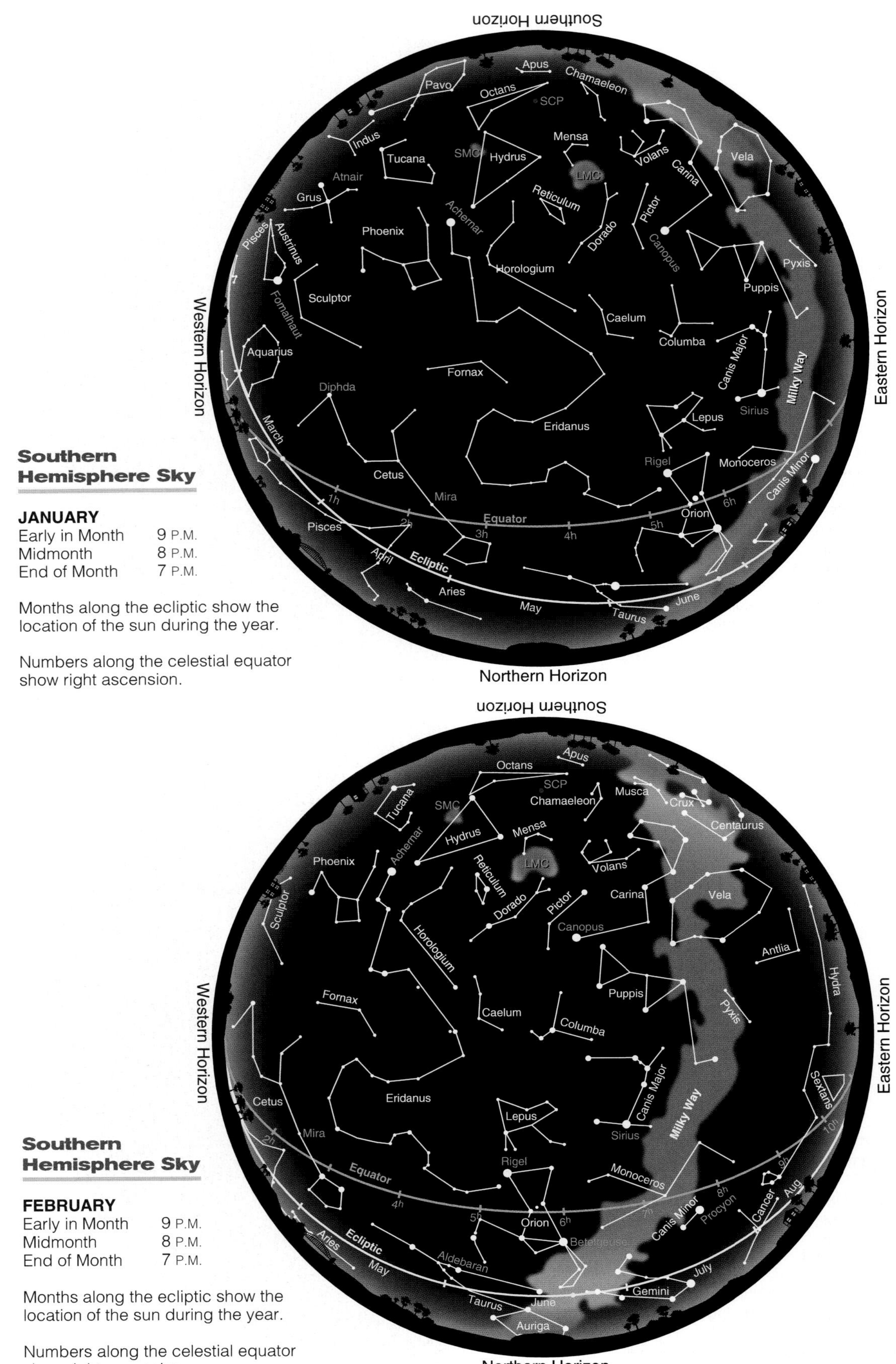

Southern Hemisphere Sky

JANUARY

Early in Month	9 P.M.
Midmonth	8 P.M.
End of Month	7 P.M.

Months along the ecliptic show the location of the sun during the year.

Numbers along the celestial equator show right ascension.

Southern Hemisphere Sky

FEBRUARY

Early in Month	9 P.M.
Midmonth	8 P.M.
End of Month	7 P.M.

Months along the ecliptic show the location of the sun during the year.

Numbers along the celestial equator show right ascension.

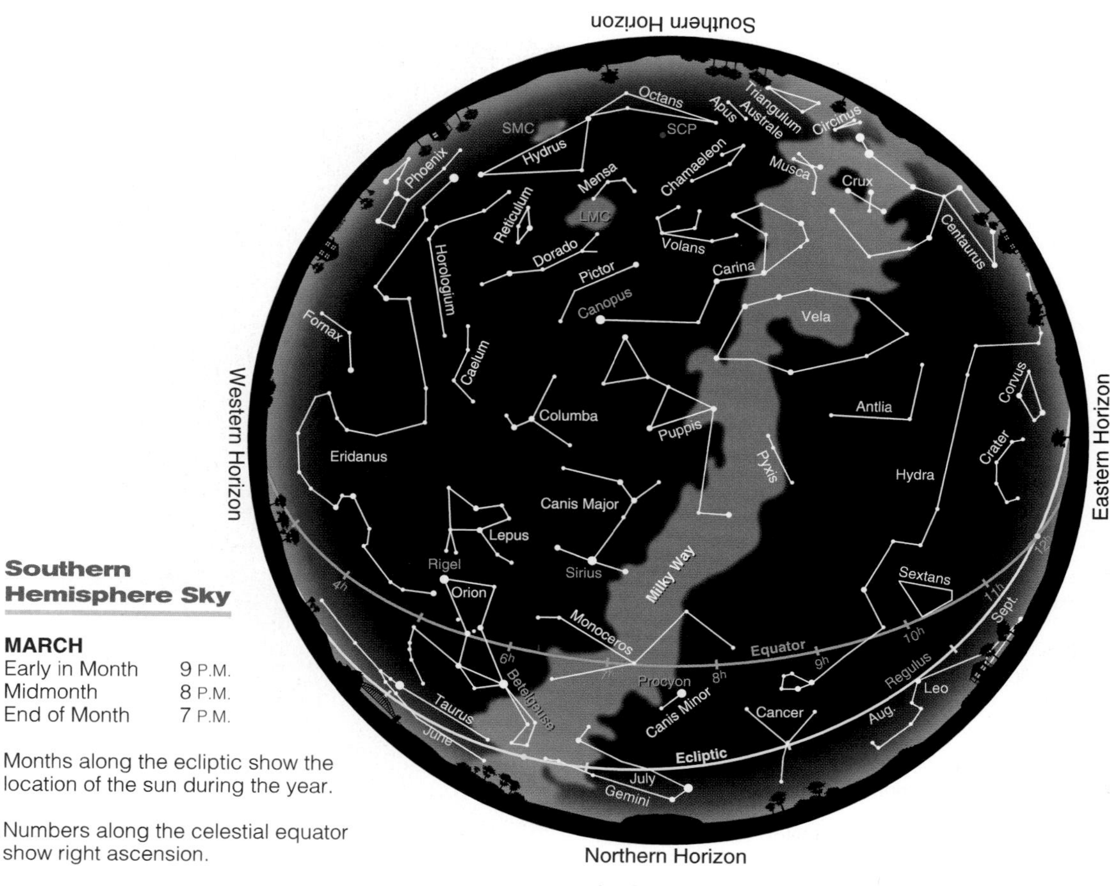

Southern Hemisphere Sky

MARCH

Early in Month 9 P.M.
Midmonth 8 P.M.
End of Month 7 P.M.

Months along the ecliptic show the location of the sun during the year.

Numbers along the celestial equator show right ascension.

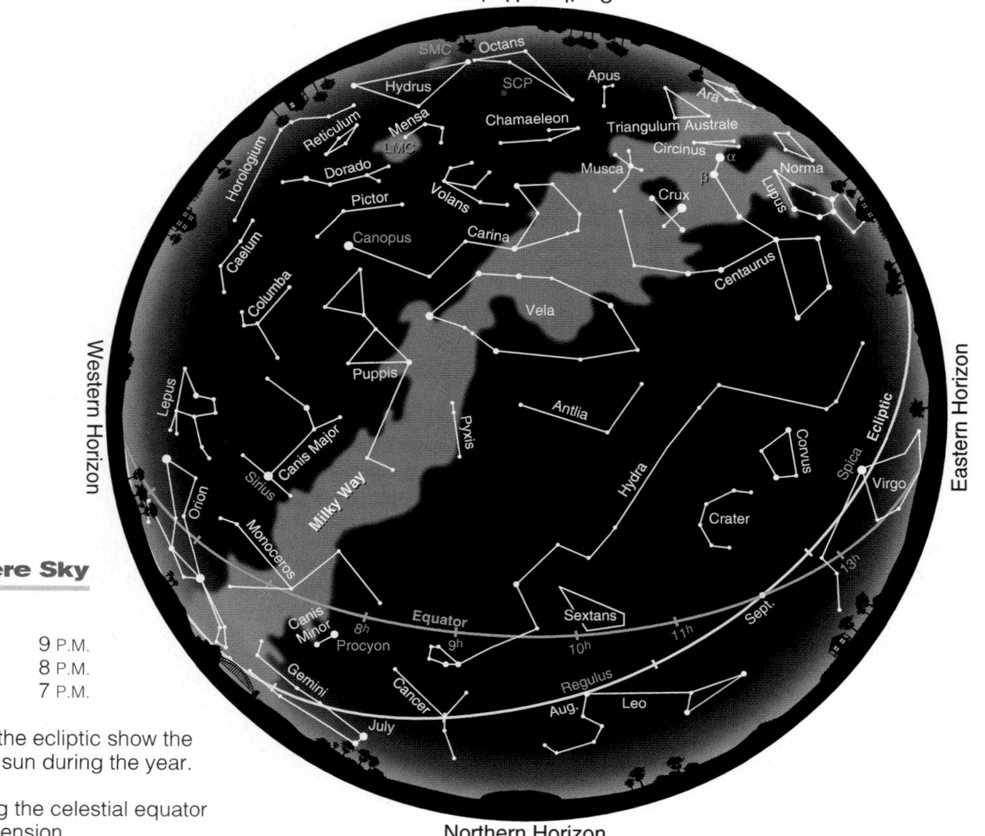

Southern Hemisphere Sky

APRIL

Early in Month 9 P.M.
Midmonth 8 P.M.
End of Month 7 P.M.

Months along the ecliptic show the location of the sun during the year.

Numbers along the celestial equator show right ascension.

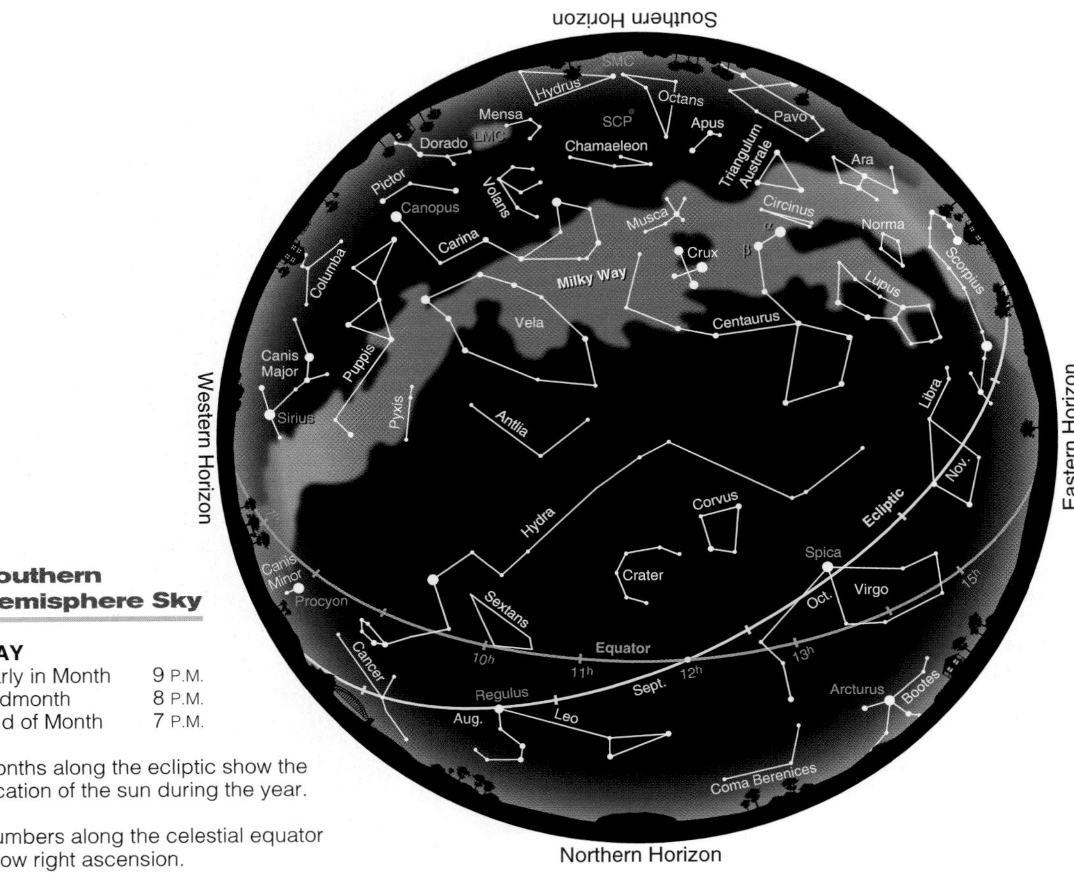

Southern Hemisphere Sky

MAY

Early in Month	9 P.M.
Midmonth	8 P.M.
End of Month	7 P.M.

Months along the ecliptic show the location of the sun during the year.

Numbers along the celestial equator show right ascension.

Southern Hemisphere Sky

JUNE

Early in Month	9 P.M.
Midmonth	8 P.M.
End of Month	7 P.M.

Months along the ecliptic show the location of the sun during the year.

Numbers along the celestial equator show right ascension.

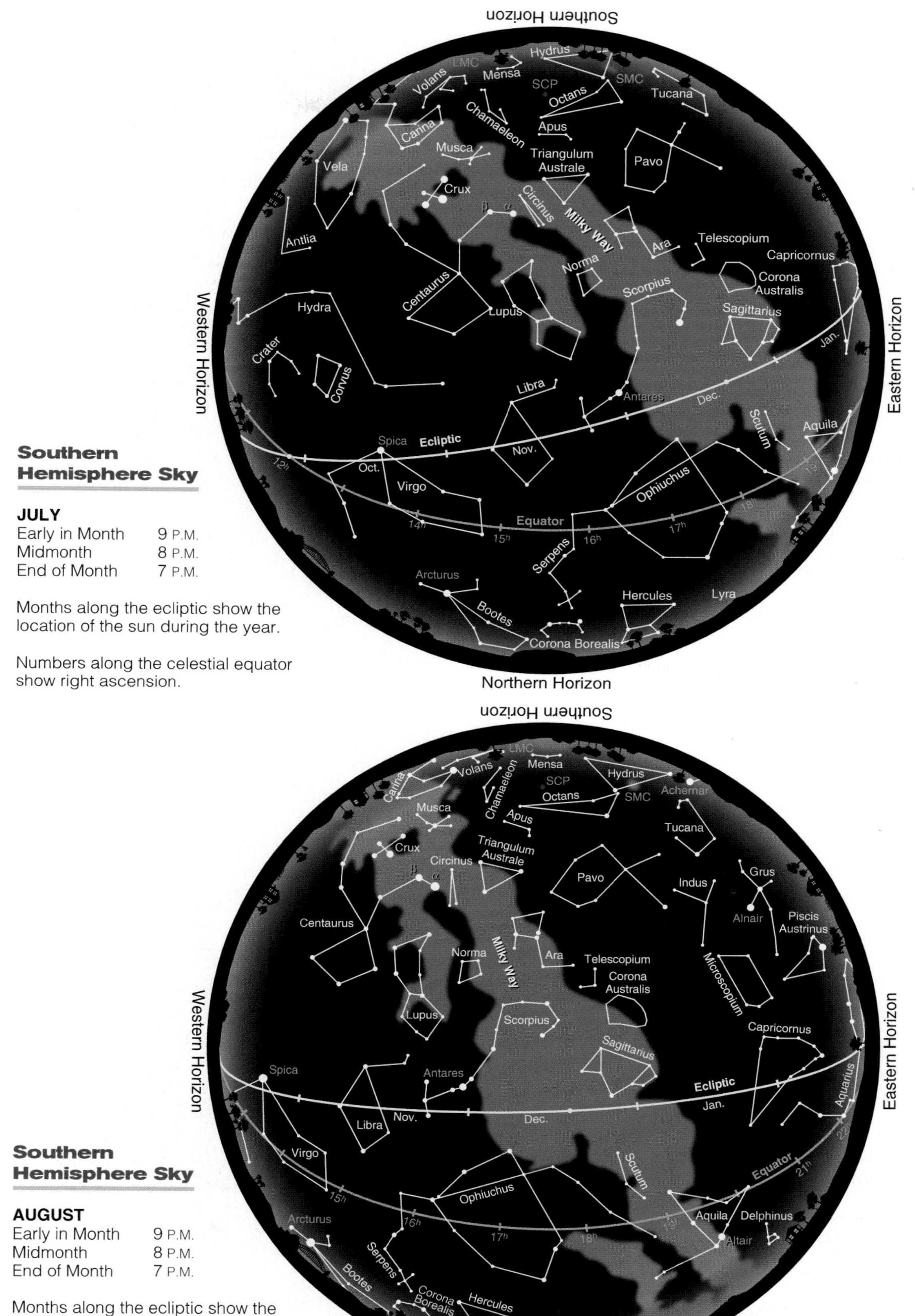

Southern Hemisphere Sky

JULY

Early in Month	9 P.M.
Midmonth	8 P.M.
End of Month	7 P.M.

Months along the ecliptic show the location of the sun during the year.

Numbers along the celestial equator show right ascension.

Southern Hemisphere Sky

AUGUST

Early in Month	9 P.M.
Midmonth	8 P.M.
End of Month	7 P.M.

Months along the ecliptic show the location of the sun during the year.

Numbers along the celestial equator show right ascension.

Southern Horizon

Musca
Chamaeleon
SCP
Mensa
LMC
Crux
Apus
Octans
Hydrus
SMC
Achernar
Circinus
Triangulum Australe
Pavo
Tucana
Phoenix
Lupus
Norma
Ara
Indus
Grus
Alnair
Sculptor
Telescopium
Corona Australis
Microscopium
Fomahaut
Libra
Scorpius
Antares
Sagittarius
Piscis Austrinus
Aquarius
Dec.
Ecliptic
Jan.
Capricornus
Feb.
Ophiuchus
Milky Way
Scutum
Serpens
17h
18h
19h
Equator
21h
22h
Pisces
Equuleus
Pegasus
Hercules
Aquila
Altair
Sagitta
Delphinus
Vulpecula
Lyra
Vega
Cygnus

Western Horizon

Eastern Horizon

Northern Horizon

Southern Hemisphere Sky

SEPTEMBER

Early in Month	9 P.M.
Midmonth	8 P.M.
End of Month	7 P.M.

Months along the ecliptic show the location of the sun during the year.

Numbers along the celestial equator show right ascension.

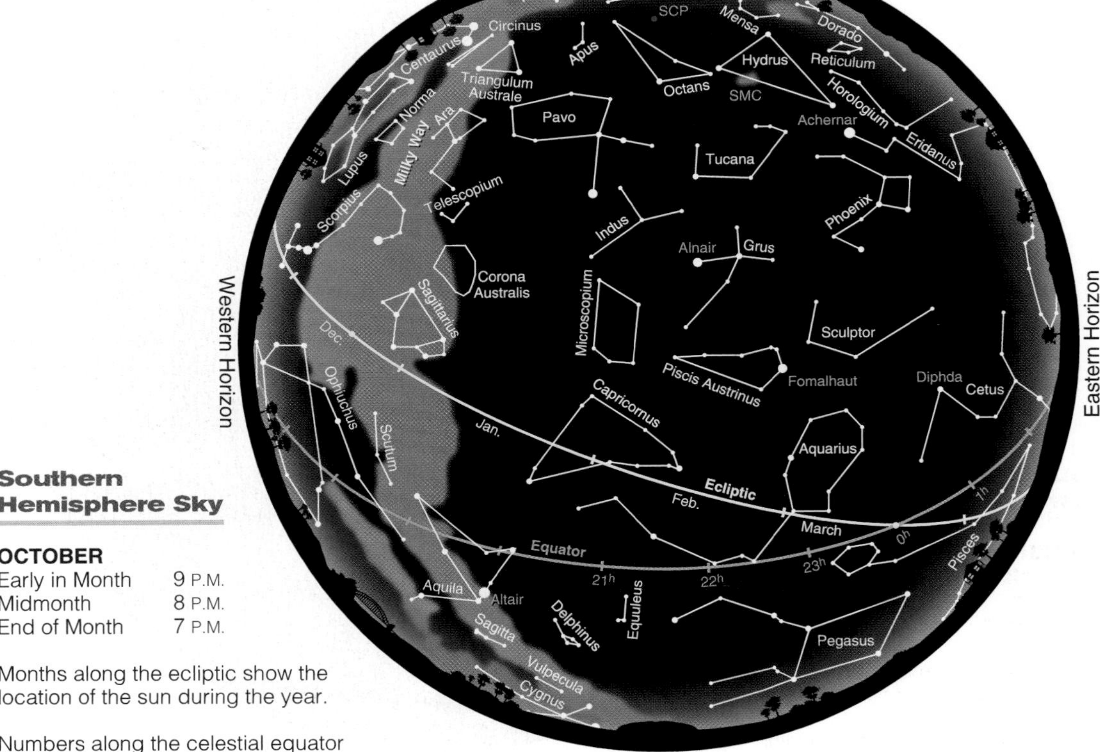

Southern Hemisphere Sky

OCTOBER

Early in Month	9 P.M.
Midmonth	8 P.M.
End of Month	7 P.M.

Months along the ecliptic show the location of the sun during the year.

Numbers along the celestial equator show right ascension.

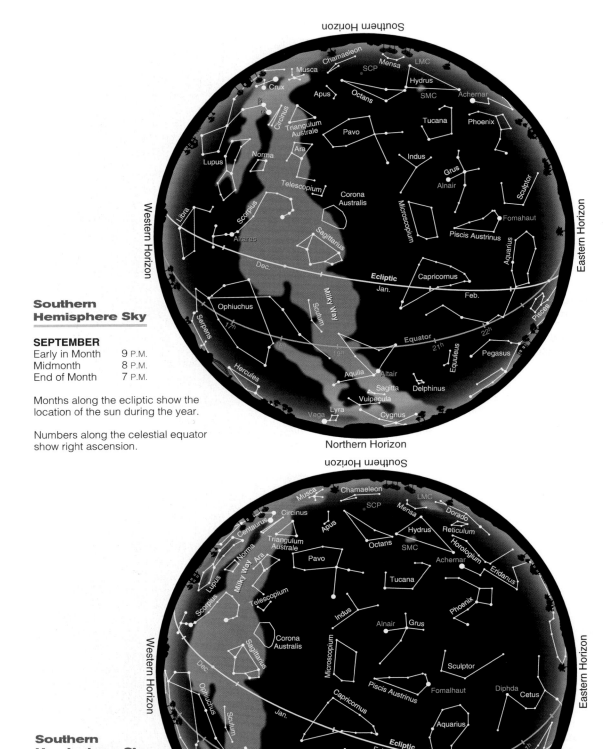

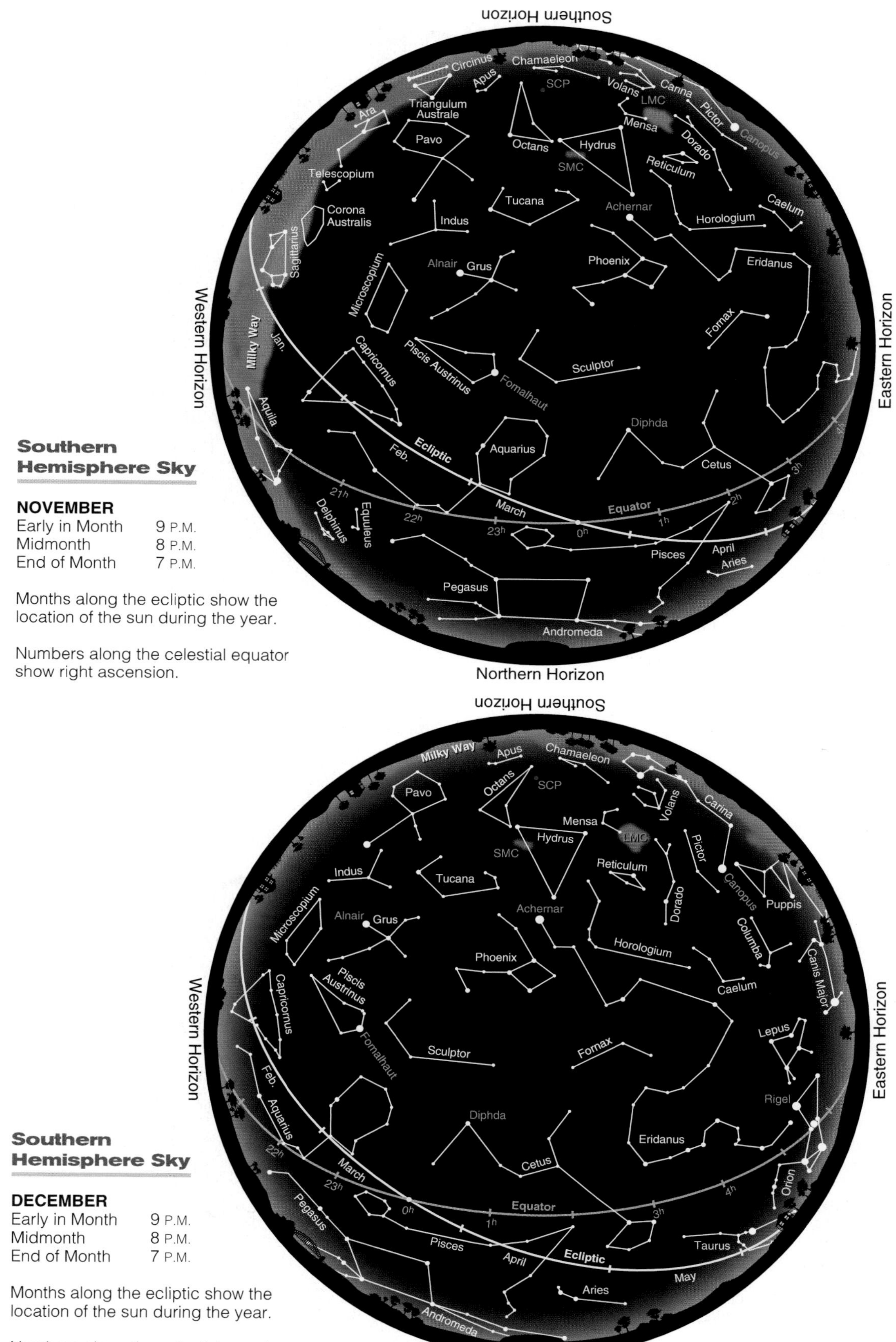

Southern Hemisphere Sky

NOVEMBER

Early in Month	9 P.M.
Midmonth	8 P.M.
End of Month	7 P.M.

Months along the ecliptic show the location of the sun during the year.

Numbers along the celestial equator show right ascension.

Southern Hemisphere Sky

DECEMBER

Early in Month	9 P.M.
Midmonth	8 P.M.
End of Month	7 P.M.

Months along the ecliptic show the location of the sun during the year.

Numbers along the celestial equator show right ascension.

Glossary

Pronunciation Guide

ā pay	ē be	ī pie	ō so	ŭ cut
ă hat	ĕ pet	ĭ pit	ŏ pot	
ä father		î pier	ô paw	
			o͞o food	

Numbers in parentheses refer to the page where the term is first discussed in the text.

absolute visual magnitude (M_v) (149) Intrinsic brightness of a star. The apparent visual magnitude the star would have if it were 10 pc away.

absolute zero (108) The theoretical lowest possible temperature at which a material contains no extractable heat energy. Zero on the Kelvin temperature scale.

absorption line (110) A dark line in a spectrum. Produced by the absence of photons absorbed by atoms or molecules.

absorption spectrum (dark-line spectrum) (110) A spectrum that contains absorption lines.

accretion (370) The sticking together of solid particles to produce a larger particle.

accretion disk (216) The whirling disk of gas that forms around a compact object such as a white dwarf, neutron star, or black hole as matter is drawn in.

achondrite (ā·kŏn′drīt) (455) Stony meteorite containing no chondrules or volatiles.

achromatic lens (82) A telescope lens composed of two lenses ground from different kinds of glass and designed to bring two selected colors to the same focus and correct for chromatic aberration.

active galactic nuclei (AGN) (306) The centers of active galaxies that are emitting large amounts of excess energy. See also **active galaxy.**

active galaxy (306) A galaxy whose center emits large amounts of excess energy, often in the form of radio emission. Active galaxies are suspected of having massive black holes in their centers into which matter is flowing.

active optics (89) Thin telescope mirrors that are controlled by computers to maintain proper shape as the telescope moves.

active region (131) A magnetic region on the solar surface that includes sunspots, prominences, flares, etc.

adaptive optics (89) A computer-controlled optical system used to partially correct for seeing in an astronomical telescope.

albedo (391) The ratio of the light reflected from an object divided by the light that hits the object. Albedo equals 0 for perfectly black and 1 for perfectly white.

alt-azimuth mounting (89) A telescope mounting that allows the telescope to move in altitude (perpendicular to the horizon) and in azimuth (parallel to the horizon). See also **equatorial mounting.**

amino acid (ŭ·mē′nō) (476) Carbon-chain molecule that is the building block of protein.

angstrom (Å) (ăng′strŭm) (79) A unit of distance. 1 Å $= 10^{-10}$ m. Commonly used to measure the wavelength of light.

angular diameter (21) The angle formed by lines extending from the observer to opposite sides of an object.

angular distance (21) The angle formed by lines extending from the observer to two locations.

angular momentum (215) A measure of the tendency of a rotating body to continue rotating. Mathematically, the product of mass, velocity, and radius.

annular eclipse (37) A solar eclipse in which the solar photosphere appears around the edge of the moon in a bright ring, or annulus. The corona, chromosphere, and prominences cannot be seen.

anorthosite (ăn·ôr′thŭ·sīt) (391) Rock of aluminum and calcium silicates found in the lunar highlands.

antimatter (333) Matter composed of antiparticles, which upon colliding with a matching particle of normal matter annihilate and convert the mass of both particles into energy. The antiproton is the antiparticle of the proton, and the positron is the antiparticle of the electron.

aphelion (ŭ·fē′le·ŭn) (29) The orbital point of greatest distance from the sun.

apogee (37) The point farthest from Earth in the orbit of a body circling Earth.

apparent visual magnitude (m_v) (16) The brightness of a star as seen by human eyes on Earth.

association (180) Group of widely scattered stars (10 to 100) moving together through space. Not gravitationally bound into clusters.

asterism (15) A named grouping of stars that is not one of the recognized constellations. Examples are the Big Dipper and the Pleiades.

asteroid (363) Small, rocky world. Most asteroids lie between Mars and Jupiter in the asteroid belt.

astronomical unit (AU) (6) Average distance from Earth to the sun; 1.5×10^8 km, or 93×10^6 mi.

atmospheric window (80) Wavelength region in which our atmosphere is transparent—at visual, infrared, and radio wavelengths.

aurora (ô·rôr′ŭ) (137) The glowing light display that results when a planet's magnetic field guides charged particles toward the north and south magnetic poles, where they strike the upper atmosphere and excite atoms to emit photons.

autumnal equinox (28) The point on the celestial sphere where the sun crosses the celestial equator going southward. Also, the time when the sun reaches this point and autumn begins in the northern hemisphere—about September 22.

Babcock model (133) A model of the sun's magnetic cycle in which the differential rotation of the sun winds up and tangles the solar magnetic field in a 22-year cycle. This is thought to be responsible for the 11-year sunspot cycle.

Balmer series (111) A series of spectral lines produced by hydrogen in the near-ultraviolet and visible parts of the spectrum. The three longest-wavelength Balmer lines are visible to the human eye.

barred spiral galaxy (284) A spiral galaxy with an elongated nucleus resembling a bar from which the arms originate.

basalt (384) Dark igneous rock characteristic of solidified lava.

belt–zone circulation (421) The atmospheric circulation typical of Jovian planets. Dark belts and bright zones encircle the planet parallel to its equator.

big bang (329) The high-density, high-temperature state from which the expanding universe of galaxies began.

big rip (346) The fate of the universe if dark energy increases with time and galaxies, stars, and even atoms are eventually ripped apart by the accelerating expansion of the universe.

binary stars (156) Pairs of stars that orbit around their common center of mass.

binding energy (105) The energy needed to pull an electron away from its atom.

bipolar flow (185) Jets of gas flowing away from a central object in opposite directions. Usually applied to protostars.

birth line (182) In the H–R diagram, the line above the main sequence where protostars first become visible.

black body radiation (108) Radiation emitted by a hypothetical perfect radiator. The spectrum is continuous, and the wavelength of maximum emission depends on the body's temperature.

black dwarf (211) The end state of a white dwarf that has cooled to low temperature.

black hole (241) A mass that has collapsed to such a small volume that its gravity prevents the escape of all radiation. Also, the volume of space from which radiation may not escape.

blazars See **BL Lac objects.**

BL Lac objects (313) Objects that resemble quasars. Thought to be highly luminous cores of distant active galaxies.

blueshift (115) A Doppler shift toward shorter wavelengths caused by a velocity of approach.

Bok globule (184) Small, dark cloud only about 1 ly in diameter that contains 10 to 1000 solar masses of gas and dust. Believed to be related to star formation.

breccia (brĕch'ē·ŭ) (391) Rock composed of fragments of earlier rocks bonded together.

bright-line spectrum See **emission spectrum.**

brown dwarf (192) A star whose mass is too low to ignite nuclear fusion. Heated by contraction.

calibration (258) The establishment of the relationship between a parameter that is easily determined and a parameter that is more difficult to determine. For example, the periods of Cepheid variables have been calibrated to reveal absolute magnitudes, which can then be used to find distance. Thus astronomers say Cepheids have been calibrated as distance indicators.

Cambrian period (kăm'brē·ŭn) (478) A geological period 0.6 to 0.5 billion years ago during which life on Earth became diverse and complex. Cambrian rocks contain the oldest easily identifiable fossils.

capture hypothesis (394) The theory that Earth's moon formed elsewhere in the solar nebula and was later captured by Earth.

carbonaceous chondrite (kär·bŭ·nā'·shŭs kŏn'-drīt) (455) Stony meteorite that contains both chondrules and volatiles. These chondrites may be the least-altered remains of the solar nebula still present in the solar system.

carbon–nitrogen–oxygen (CNO) cycle (183) A series of nuclear reactions that use carbon as a catalyst to combine four hydrogen atoms to make one helium atom plus energy. Effective in stars more massive than the sun.

Cassegrain focus (kăs'ŭ·grān) (88) The optical design in which the secondary mirror reflects light back down the tube through a hole in the center of the objective mirror.

CCD See **charge-coupled device.**

celestial equator (20) The imaginary line around the sky directly above Earth's equator.

celestial pole (north or south) (20) One of the two points on the celestial sphere directly above Earth's poles.

celestial sphere (20) An imaginary sphere of very large radius surrounding Earth and to which the planets, stars, sun, and moon seem to be attached.

center of mass (69) The balance point of a body or system of masses. The point about which a body or system of masses rotates in the absence of external forces.

Cepheid variable star (sĕ·fē'ĭd) (255) Variable star with a period of 60 days. Period of variation is related to luminosity.

Chandrasekhar limit (shăn'drä·sā'·kär) (211) The maximum mass of a white dwarf, about 1.4 solar masses. A white dwarf of greater mass cannot support itself and will collapse.

charge-coupled device (CCD) (90) An electronic device consisting of a large array of light-sensitive elements used to record very faint images.

chemical evolution (480) The chemical process that led to the growth of complex molecules on primitive Earth. This did not involve the reproduction of molecules.

chondrite (kŏn'drīt) (455) A stony meteorite that contains chondrules.

chondrule (kŏn'dro͞ol) (455) Round, glassy body found in some stony meteorites. Believed to have solidified very quickly from molten drops of silicate material.

chromatic aberration (krō·măt'ĭk) (81) A distortion found in refracting telescopes because lenses focus different colors at slightly different distances. Images are consequently surrounded by color fringes.

chromosome (477) A body within a living cell that contains genetic information responsible for the determination and transmission of hereditary traits.

chromosphere (krō'mŭ·sfîr) (36) Bright gases just above the photosphere of the sun.

circular velocity (68) The velocity an object needs to stay in orbit around another object.

circumpolar constellation (north or south) (21) A constellation so close to one of the celestial poles that it never sets or never rises as seen from a particular latitude.

closed orbit (69) An orbit that returns to the same starting point over and over. Either a circular orbit or an elliptical orbit.

closed universe (339) A model universe in which the average density is great enough to stop the expansion and make the universe contract.

cluster method (291) The method of determining the masses of galaxies based on the motions of galaxies in a cluster.

CNO cycle See **carbon–nitrogen–oxygen cycle.**

cold dark matter (342) Mass in the universe, as yet undetected except for its gravitational influence, which is made up of slow-moving particles.

collapsar (247) A star of high mass that collapses into a black hole. A possible source of gamma-ray bursts.

coma (464) The glowing head of a comet.

comet (363) One of the small, icy bodies that orbit the sun and produce tails of gas and dust when they approach the sun.

comparative planetology (378) The study of planets in relation to one another.

comparison spectrum (91) A spectrum of known spectral lines used to identify unknown wavelengths in an object's spectrum.

composite volcano (400) A volcano formed by successive lava and ash flows. They have steep sides and, on Earth, are found along subduction zones.

condensation (369) The growth of a particle by addition of material from surrounding gas, atom by atom.

condensation hypothesis (394) The theory that Earth and the moon condensed from the same cloud of material in roughly their present orbital relationship.

condensation sequence (369) The sequence in which different materials condense from the solar nebula as we move outward from the sun.

conservation of energy (187) One of the basic laws of stellar structure. The amount of energy flowing out of the top of a shell must equal the amount coming in at the bottom plus whatever energy is generated within the shell.

conservation of mass (187) One of the basic laws of stellar structure. The total mass of the star must equal the sum of the masses of the shells, and the mass must be distributed smoothly through the star.

constellation (14) One of the stellar patterns identified by name, usually of mythological gods, people, animals, or objects. Also, the region of the sky containing that star pattern.

continuous spectrum (110) A spectrum in which there are no absorption or emission lines.

convection (125) Circulation in a fluid driven by heat. Hot material rises and cool material sinks.

corona (37) On the sun, the faint outer atmosphere composed of low-density, high-temperature gas.

coronae (402) On Venus, large, round geological faults in the crust caused by the intrusion of magma below the crust.

coronal hole (137) An area of the solar surface that is dark at X-ray wavelengths. Thought to be associated with divergent magnetic fields and the source of the solar wind.

coronal mass ejection (CME) (137) Matter ejected from the sun's corona in powerful surges guided by magnetic fields.

cosmic microwave background radiation (333) Radiation from the hot clouds of the big bang explosion. The large red shift makes it appear to come from a body whose temperature is only 2.7 K.

cosmological constant (345) A constant in Einstein's equations of space and time that represents a force of repulsion.

cosmological principle (336) The assumption that any observer in any galaxy sees the same general features of the universe.

cosmology (326) The study of the nature, origin, and evolution of the universe.

Coulomb barrier (ko̅o̅·lôm) (138) The electrostatic force of repulsion between bodies of like charge. Commonly applied to atomic nuclei.

Coulomb force (105) The electrostatic force of repulsion or attraction between charged bodies.

critical density (337) The average density of the universe needed to make its curvature flat.

dark age (334) The period of time after the glow of the big bang faded into the infrared and before the birth of the first stars, during which the universe expanded in darkness.

dark energy (345) The energy believed to fill empty spaces and drive the acceleration of the expanding universe.

dark halo (263) The low-density extension of the halo of our galaxy believed to be composed of dark matter.

dark-line spectrum See **absorption spectrum.**

dark matter (263) Nonluminous matter that is detected only by its gravitational influence.

dark nebula (177) A cloud of gas and dust seen silhouetted against a brighter nebula.

deferent (dĕf′ŭr·ŭnt) (51) In the Ptolemaic theory, the large circle around Earth along which the center of the epicycle was thought to move.

degenerate matter (204) Extremely high-density matter in which pressure no longer depends on temperature due to quantum mechanical effects.

density wave theory (270) Theory proposed to account for spiral arms as compressions of the interstellar medium in the disk of the galaxy.

deoxyribonucleic acid (476) The long carbon-chain molecule that records information to govern the biological activity of the organism. DNA carries the genetic data passed to offspring.

deuterium (139) An isotope of hydrogen in which the nucleus contains a proton and a neutron.

diamond-ring effect (37) During a total solar eclipse, the momentary appearance of a spot of photosphere at the edge of the moon, producing a brilliant glare set in the silvery ring of the corona.

differential rotation (132, 262) The rotation of a body in which different parts of the body have different periods of rotation. This is true of the sun, the Jovian planets, and the disk of the galaxy.

differentiation (371) The separation of planetary material according to density.

diffraction fringe (83) Blurred fringe surrounding any image, caused by the wave properties of light. Because of this, no image detail smaller than the fringe can be seen.

disk component (260) All material confined to the plane of the galaxy.

distance indicator (287) Object whose luminosity or diameter is known. Used to find the distance to a star cluster or galaxy.

DNA See **deoxyribonucleic acid.**

Doppler effect (115) The change in the wavelength of radiation due to relative radial motion of source and observer.

double-exhaust model (308) The theory that double radio lobes are produced by pairs of jets emitted in opposite directions from the centers of active galaxies.

double-lobed radio source (308) A galaxy that emits radio energy from two regions (lobes) located on opposite sides of the galaxy.

Drake equation (490) The equation that estimates the total number of communicative civilizations in our galaxy.

dust tail (464) A comet tail composed of dust released from the nucleus and pushed away by the pressure of sunlight. Also known as a type II tail.

dynamo effect (132) The process by which a rotating, convecting body of conducting matter, such as Earth's core, can generate a magnetic field.

east point (20) One of the four cardinal directions. The point on the horizon directly east.

eccentricity, *e* (59) A number between 1 and 0 that describes the shape of an ellipse. The distance from one focus to the center of the ellipse divided by the semimajor axis.

eclipsing binary system (160) A binary star system in which the stars eclipse each other.

ecliptic (26) The apparent path of the sun around the sky.

ejecta (388) Pulverized rock scattered by meteorite impacts on a planetary surface.

electromagnetic radiation (78) Changing electric and magnetic fields that travel through space and transfer energy from one place to another; examples are light or radio waves.

electron (104) Low-mass atomic particle carrying a negative charge.

ellipse (58) A closed curve around two points called the foci such that the total distance from one focus to the curve and back to the other focus remains constant.

elliptical galaxy (284) A galaxy that is round or elliptical in outline and contains little gas and dust, no disk or spiral arms, and few hot, bright stars.

emission line (110) A bright line in a spectrum caused by the emission of photons from atoms.

emission nebula (176) A cloud of glowing gas excited by ultraviolet radiation from hot stars.

emission spectrum (bright-line spectrum) (110) A spectrum containing emission lines.

energy level (106) One of a number of states an electron may occupy in an atom, depending on its binding energy.

energy transport (188) Flow of energy from hot regions to cooler regions by one of three methods: conduction, convection, or radiation.

enzyme (476) Special protein that controls processes in an organism.

epicycle (ĕp′ŭ·sī·kŭl) (51) The small circle followed by a planet in the Ptolemaic theory. The center of the epicycle follows a larger circle (the deferent) around Earth.

equant (ē′kwŭnt) (51) In the Ptolemaic theory, the point off center in the deferent from which the center of the epicycle appears to move uniformly.

equatorial mounting (89) A telescope mounting that allows motion parallel to and perpendicular to the celestial equator.

escape velocity (69) The initial velocity an object needs to escape from the surface of a celestial body.

evening star (30) Any planet visible in the sky just after sunset.

event horizon (242) The boundary of the region of a black hole from which no radiation may escape. No event that occurs within the event horizon is visible to a distant observer.

evolutionary track (180) The path a star follows in the H–R diagram as it gradually changes its surface temperature and luminosity.

excited atom (106) An atom in which an electron has moved from a lower to a higher energy level.

extrasolar planet (359) A planet orbiting a star other than the sun.

eyepiece (80) A short-focal-length lens used to enlarge the image in a telescope. The lens nearest the eye.

false-color image (90) A representation of graphical data with added or enhanced color to reveal detail.

filament (126) A solar prominence seen from above silhouetted against the bright photosphere.

filtergram (126) A photograph (usually of the sun) taken in the light of a specific region of the spectrum—for example, an H_α filtergram.

fission hypothesis (394) The theory that the moon and Earth formed when a rapidly rotating protoplanet split into two pieces.

flare (137) A violent eruption on the sun's surface.

flatness problem (343) In cosmology, the peculiar circumstance that the early universe must have contained almost exactly the right amount of matter to make space-time flat.

flat universe (339) A model of the universe in which space-time is not curved.

flocculent (272) Describes a galaxy whose spiral arms have a woolly or fluffy appearance.

flux (149) A measure of the flow of energy out of a surface. Usually applied to light.

focal length (80) The focal length of a lens is the distance from the lens to the point where it focuses parallel rays of light.

folded mountain range (384) A long range of mountains formed by the compression of a planet's crust.

forward scattering (425) The optical property of finely divided particles to preferentially direct light in the original direction of the light's travel.

galactic cannibalism (296) The theory that large galaxies absorb smaller galaxies.

galactic corona (263) The extended, spherical distribution of low-luminosity matter believed to surround the Milky Way and other galaxies.

galaxy (8) A large system of stars, star clusters, gas, dust, and nebulae orbiting a common center of mass.

Galilean satellites (găl·ŭ·lē′ŭn) (365) The four largest satellites of Jupiter, named after their discoverer Galileo.

gamma-ray burster (246) An object very faint at visual wavelengths that produces a sudden, powerful burst of gamma rays.

gas tail (464) A comet tail composed of ionized gas atoms released from the nucleus and carried outward by the solar wind. Also called a type I comet tail.

gene (477) A unit of DNA—or sometimes RNA—information responsible for controlling an inherited physiological trait.

geocentric universe (50) A model universe with Earth at the center, such as the Ptolemaic universe.

geosynchronous satellite (68) A satellite that orbits eastward around Earth with a period of 24 hours and remains above the same spot on Earth's surface.

giant (152) Large, cool, highly luminous star in the upper right of the H–R diagram. Typically 10 to 100 times the diameter of the sun.

global warming (386) The gradual increase in the surface temperature of Earth caused by human modifications to Earth's atmosphere.

globular cluster (208) A star cluster containing 100,000 to 1 million stars in a sphere about 75 ly in diameter. Generally old, metal-poor, and found in the spherical component of the galaxy.

gossamer rings (428) Jupiter's largest and most tenuous rings of dust.

grand unified theories (GUTs) (343) Theories that attempt to unify (describe in a similar way) the electromagnetic, weak, and strong forces of nature.

granulation (125) The fine structure of bright grains covering the sun's surface.

grating (91) A piece of material in which numerous microscopic parallel lines are scribed. Light encountering a grating is dispersed to form a spectrum.

gravitational collapse (368) The process by which a forming body such as a planet gravitationally captures gas from its surroundings.

gravitational lensing (293) The process by which the gravitational field of a massive object focuses the light from a distant object to produce multiple images of the distant object or to make the distant object look brighter.

gravitational radiation (236) Disturbances in a gravitational field traveling at the velocity of light and carrying energy away from an object with a rapidly changing mass distribution.

gravitational redshift (243) The lengthening of the wavelength of a photon due to its escape from a gravitational field.

greenhouse effect (386) The process by which a carbon dioxide atmosphere traps heat and raises the temperature of a planetary surface.

grooved terrain (428) Regions of the surface of Ganymede consisting of parallel grooves. Believed to have formed by repeated fracture of the icy crust.

ground state (107) The lowest permitted electron energy level in an atom.

GUTs See **grand unified theories.**

half-life (366) The time required for half of the atoms in a radioactive sample to decay.

halo (261) The spherical region of a spiral galaxy, containing a thin scattering of stars, star clusters, and small amounts of gas.

heat (107) Energy stored in a material as agitation among its particles.

heat of formation (372) In planetology, the heat released by in-falling matter during the formation of a planetary body.

heavy bombardment (375) The intense cratering during the first 0.5 billion years in the history of the solar system.

heliocentric universe (49) A model of the universe with the sun at the center, such as the Copernican universe.

helioseismology (128) The study of the interior of the sun by the analysis of its modes of vibration.

helium flash (205) The explosive ignition of helium burning that takes place in some giant stars.

Herbig–Haro object (185) A small nebula that varies irregularly in brightness. Believed to be associated with star formation.

Hertzsprung–Russell (H–R) diagram (hĕrt′ sprŭng·rŭs·ŭl) (151) A plot of the intrinsic brightness versus the surface temperature of stars. It separates the effects of temperature and surface area on stellar luminosity. Commonly plotted as absolute magnitude versus spectral type but also as luminosity versus surface temperature or color.

homogeneity (336) The assumption that, on the large scale, matter is uniformly spread through the universe.

horizon (20) The circular boundary between the sky and Earth.

horizon problem (343) In cosmology, the circumstance that the primordial background radiation seems much more isotropic than can be explained by the standard big bang theory.

horizontal branch (209) In the H–R diagram, stars fusing helium in a shell and evolving back toward the red giant region.

hot dark matter (342) Dark matter made up of particles such as neutrinos traveling at or nearly at the speed of light.

hot spot (310) In geology, a place on Earth's crust where volcanism is caused by a rising convection cell in the mantle below. In radio astronomy, a bright spot in a radio lobe.

H–R diagram See **Hertzsprung–Russell diagram.**

H II region (176) A region of ionized hydrogen around a hot star.

Hubble constant (*H*) (290) A measure of the rate of expansion of the universe. The average value of velocity of recession divided by distance. Presently believed to be about 70 km/s/Mpc.

Hubble law (290) The linear relation between the distances to galaxies and their velocity of recession.

Hubble time (330) The age of the universe, equivalent to 1 divided by the Hubble constant. The Hubble time is the age of the universe if it has expanded since the big bang at a constant rate.

hydrostatic equilibrium (187) The balance between the weight of the material pressing downward on a layer in a star and the pressure in that layer.

hypernova (247) Produced when a very massive star collapses into a black hole. Thought to be a possible source of gamma-ray bursts.

hypothesis (59) A conjecture, subject to further tests, that accounts for a set of facts.

inflationary universe (343) A version of the big bang theory that includes a rapid expansion when the universe was very young. Derived from grand unified theories.

infrared cirrus (178) Wispy network of cold dust clouds discovered by the Infrared Astronomy Satellite.

inner Lagrangian point (214) The point of gravitational equilibrium between two orbiting stars through which matter can flow from one star to the other.

instability strip (255) The region of the H–R diagram in which stars are unstable to pulsation. A star passing through this strip becomes a variable star.

interferometry (87) The observing technique in which separated telescopes combine to produce a virtual telescope with the resolution of a much-larger-diameter telescope.

interstellar dust (174) Microscopic solid grains in the interstellar medium.

interstellar medium (174) The gas and dust distributed between the stars.

interstellar reddening (174) The process in which dust scatters blue light out of starlight and makes the stars look redder.

inverse square relation (66) A rule that the strength of an effect (such as gravity) decreases in proportion as the distance squared increases.

ion (105) An atom that has lost or gained one or more electrons.

ionization (105) The process in which atoms lose or gain electrons.

irregular galaxy (285) A galaxy with a chaotic appearance, large clouds of gas and dust, and both population I and II stars, but without spiral arms.

isotopes (105) Atoms that have the same number of protons but a different number of neutrons.

isotropy (ī·sŏt′rŭ·pē) (336) The assumption that in its general properties the universe looks the same in every direction.

joule (J) (jo͞ol) (109) A unit of energy equivalent to a force of 1 newton acting over a distance of 1 m. One joule per second equals 1 watt of power.

Jovian planet (364) Jupiterlike planet with a large diameter and low density.

Kelvin temperature scale (107) A temperature scale using Celsius degrees and based on zero at absolute zero.

kiloparsec (kpc) (260) A unit of distance equal to 1000 pc or 3260 ly.

Kirchhoff's laws (110) A set of laws that describe the absorption and emission of light by matter.

Kuiper belt (446) The collection of icy planetesimals believed to orbit in a region from just beyond Neptune out to 100 AU or more.

Lagrangian points (214) Points of gravitational stability in the orbital plane of a binary star system or of a planet and its moon.

large-impact hypothesis (394) The theory that the moon formed from debris ejected during a collision between Earth and a large planetesimal.

Large Magellanic Cloud (285) An irregular galaxy that is a satellite of our Milky Way Galaxy. It is visible in the southern sky.

large-scale structure (346) The distribution of clusters and superclusters of galaxies in filaments and walls enclosing voids.

L dwarf (114) A main-sequence star cooler than an M star.

life zone (486) A region around a star within which a planet can have temperatures that permit the existence of liquid water.

light curve (160) A graph of brightness versus time commonly used in analyzing variable stars and eclipsing binaries.

light-gathering power (82) The ability of a telescope to collect light. Proportional to the area of the telescope's objective lens or mirror.

light pollution (84) The illumination of the night sky by waste light from cities and outdoor lighting, which prevents the observation of faint objects.

light-year (ly) (7) A unit of distance. The distance light travels in 1 year.

liquid metallic hydrogen (422) A form of liquid hydrogen that is a good electrical conductor, found in the interiors of Jupiter and Saturn.

lobate scarp (lō′bāt·skärp) (396) A curved cliff such as those found on Mercury.

Local Group (295) The small cluster of a few dozen galaxies that contains our Milky Way Galaxy.

look-back time (288) The amount by which we look into the past when we look at a distant galaxy. A time equal to the distance to the galaxy in light-years.

luminosity (*L*) (149) The total amount of energy a star radiates in 1 second.

luminosity class (153) A category of stars of similar luminosity, determined by the widths of lines in their spectra.

lunar eclipse (31) The darkening of the moon when it moves through Earth's shadow.

Lyman series (lī′mĕn) (111) Spectral lines in the ultraviolet spectrum of hydrogen produced by transitions whose lowest energy level is the ground state.

MACHOs (342) Massive compact halo objects. Low-luminosity objects such as planets and brown dwarfs that contribute to the mass of the halo.

magnetar (247) A class of neutron star having very strong magnetic fields.

magnetic carpet (127) The network of small magnetic loops that covers the solar surface.

magnetosphere (măg·nē′tō·sfîr) (422) The volume of space around a planet within which the motion of charged particles is dominated by the planetary magnetic field rather than the solar wind.

magnifying power (84) The ability of a telescope to make an image larger.

magnitude scale (15) The astronomical brightness scale. The larger the number, the fainter the star.

main sequence (152) The region of the H–R diagram running from upper left to lower right, which includes roughly 90 percent of all stars.

mantle (381) The layer of dense rock and metal oxides that lies between the molten core and Earth's surface. Also, similar layers in other planets.

mare (mä′rā) (plural: **maria**) (387) One of the lunar lowlands filled by successive flows of dark lava. From the Latin for "sea."

mass (66) A measure of the amount of matter making up an object.

mass–luminosity relation (162) The more massive a star is, the more luminous it is.

Maunder butterfly diagram (môn′dŭr) (130) A graph showing the latitude of sunspots versus time. First plotted by W. W. Maunder in 1904.

Maunder minimum (môn′dŭr) (131) A period of less numerous sunspots and other solar activity between 1645 and 1715.

megaparsec (Mpc) (287) A unit of distance equal to 1,000,000 pc.

metals (264) In astronomical usage, all atoms heavier than helium.

meteor (366) A small bit of matter heated by friction to incandescent vapor as it falls into Earth's atmosphere.

meteorite (366) A meteor that survives its passage through the atmosphere and strikes the ground.

meteoroid (366) A meteor in space before it enters Earth's atmosphere.

meteor shower (456) A multitude of meteors that appear to come from the same region of the sky. Believed to be caused by comet debris.

micrometeorite (389) Meteorite of microscopic size.

midocean rise (384) One of the undersea mountain ranges that push up from the seafloor in the center of the oceans.

Milankovitch hypothesis (40) Suggestion that Earth's climate is determined by slow periodic changes in the shape of its orbit, the angle of its axis, and precession.

Milky Way (8) The hazy band of light that circles our sky. Produced by the glow of our galaxy.

Milky Way Galaxy (8) The spiral galaxy containing our sun. Visible in the night sky as the Milky Way.

Miller experiment (478) An experiment that reproduced the conditions under which life began on Earth and manufactured amino acids and other organic compounds.

millisecond pulsar (238) A pulsar with a pulse period of only a few milliseconds.

minute of arc (21) An angular measure. One-sixtieth of a degree.

model See **scientific model.**

molecular cloud (179) A dense interstellar gas cloud in which atoms are able to link together to form molecules such as H_2 and CO.

molecule (105) Two or more atoms bonded together.

morning star (30) Any planet visible in the sky just before sunrise.

multiringed basin (389) Large impact feature (crater) containing two or more concentric rims formed by fracturing of the planetary crust.

mutant (475) Offspring born with altered DNA.

nadir (20) The point on the celestial sphere directly below the observer. The opposite of the zenith.

nanometer (nm) (79) A unit of distance equaling one-billionth of a meter (10^{-9} m).

natural law (59) A theory that is almost universally accepted as true.

natural selection (475) The process by which the best traits are passed on, allowing the most able to survive.

neap tide (71) Ocean tide of low amplitude occurring at first- and third-quarter moon.

nebula (175) A glowing cloud of gas or a cloud of dust reflecting the light of nearby stars.

neutrino (139) A neutral, massless atomic particle that travels at or nearly at the speed of light.

neutron (104) An atomic particle with no charge and about the same mass as a proton.

neutron star (230) A small, highly dense star composed almost entirely of tightly packed neutrons. Radius about 10 km.

Newtonian focus (88) The optical design in which a diagonal mirror reflects light out the side of the telescope tube for easier access.

node (39) The points where an object's orbit passes through the plane of Earth's orbit.

nonbaryonic matter (341) Proposed dark matter made up of particles other than protons and neutrons (baryons).

north celestial pole (20) The point on the celestial sphere directly above Earth's North Pole.

north point (20) One of the four cardinal directions. The point on the horizon directly north.

nova (200) From the Latin, meaning "new," a sudden brightening of a star making it appear as a new star in the sky. Believed to be associated with eruptions on white dwarfs in binary systems.

nuclear bulge (261) The spherical cloud of stars that lies at the center of spiral galaxies.

nuclear fission (135) Reactions that break the nuclei of atoms into fragments.

nuclear fusion (135) Reactions that join the nuclei of atoms to form more massive nuclei.

nucleus (of an atom) (104) The central core of an atom containing protons and neutrons. Carries a net positive charge.

objective lens (80) In a refracting telescope, the long-focal-length lens that forms an image of the object viewed. The lens closest to the object.

objective mirror (80) In a reflecting telescope, the principal mirror (reflecting surface) that forms an image of the object viewed.

oblateness (432) The flattening of a spherical body. Usually caused by rotation.

observable universe (328) The part of the universe that we can see from our location in space and in time.

occultation (439) The passage of a larger body in front of a smaller body.

Olbers's paradox (ôl′bŭrs) (326) The conflict between observation and theory about why the night sky should or should not be dark.

Oort cloud (ōrt) (462) The hypothetical source of comets. A swarm of icy bodies believed to lie in a spherical shell extending to 100,000 AU from the sun.

opacity (189) The resistance of a gas to the passage of radiation.

open orbit (69) An orbit that carries an object away, never to return to its starting point.

open cluster (208) A cluster of 100 to 1000 stars with an open, transparent appearance. The stars are not tightly grouped. Usually relatively young and located in the disk of the galaxy.

open universe (339) A model of the universe in which the average density is less than the critical density needed to halt the expansion.

oscillating universe theory (340) The theory that the universe begins with a big bang, expands, is slowed by its own gravity, and then falls back to create another big bang.

outflow channel (409) Geological feature produced by the rapid motion of floodwaters. Applied to features on Mars.

outgassing (372) The release of gases from a planet's interior.

ovoid (439) The oval features found on Miranda, a satellite of Uranus.

paradigm (54) A commonly accepted set of scientific ideas and assumptions.

parallax (50) The apparent change in position of an object due to a change in the location of the observer. Astronomical parallax is measured in seconds of arc.

parsec (pc) (pär′sĕk) (147) The distance to a hypothetical star whose parallax is 1 second of arc. 1 pc = 206,265 AU = 3.26 ly.

Paschen series (pä′shŭn) (111) Spectral lines in the infrared spectrum of hydrogen produced by transitions whose lowest energy level is the third.

penumbra (pĭ·nŭm′brŭ) (31) The portion of a shadow that is only partially shaded.

perigee (37) The point closest to Earth in the orbit of a body circling Earth.

perihelion (pĕr·ŭ·hē′lē·ŭn) (29) The orbital point of closest approach to the sun.

period–luminosity relation (256) The relation between period of pulsation and intrinsic brightness among Cepheid variable stars.

permitted orbit (106) One of the energy levels in an atom that an electron may occupy.

photon (78) A quantum of electromagnetic energy. Carries an amount of energy that depends inversely on its wavelength.

photosphere (36) The bright visible surface of the sun.

planet (6) A small, nonluminous body formed by accretion in a disk around a protostar.

planetary nebula (207) An expanding shell of gas ejected from a star during the latter stages of its evolution.

planetesimal (plăn·ŭ·tĕs′ŭ·mŭl) (369) One of the small bodies that formed from the solar nebula and eventually grew into protoplanets.

plastic (383) A material with the properties of a solid but capable of flowing under pressure.

plate tectonics (384) The constant destruction and renewal of Earth's surface by the motion of sections of crust.

Plutino (447) One of the icy Kuiper belt objects that, like Pluto, are caught in a 3:2 orbital resonance with Neptune.

polar axis (89) In an equatorial telescope mounting, the axis that is parallel to Earth's axis.

poor galaxy cluster (295) An irregularly shaped cluster that contains fewer than 1000 galaxies, many spiral, and no giant ellipticals.

population I (264) Stars rich in atoms heavier than helium. Nearly always relatively young stars found in the disk of the galaxy.

population II (264) Stars poor in atoms heavier than helium. Nearly always relatively old stars found in the halo, globular clusters, or the nuclear bulge.

precession (18) The slow change in the direction of Earth's axis of rotation. One cycle takes nearly 26,000 years.

primary lens (80) In a refracting telescope, the largest lens.

primary mirror (80) In a reflecting telescope, the largest mirror.

prime focus (88) The point at which the objective mirror forms an image in a reflecting telescope.

primeval atmosphere (386) Earth's first air.

primordial soup (480) The rich solution of organic molecules in Earth's first oceans.

prominence (37) Eruption on the solar surface. Visible during total solar eclipses.

proper motion (258) The rate at which a star moves across the sky. Measured in seconds of arc per year.

protein (476) Complex molecule composed of amino acid units.

proton (104) A positively charged atomic particle contained in the nucleus of an atom. The nucleus of a hydrogen atom.

proton–proton chain (139) A series of three nuclear reactions that builds a helium atom by adding together protons. The main energy source in the sun.

protoplanet (370) Massive object resulting from the coalescence of planetesimals in the solar nebula and destined to become a planet.

protostar (180) A collapsing cloud of gas and dust destined to become a star.

pulsar (231) A source of short, precisely timed radio bursts. Believed to be spinning neutron stars.

pulsar wind (233) The breeze of high-energy particles flowing away from a spinning neutron star.

***P* wave** (382) A pressure wave. A type of seismic wave produced in Earth by the compression of the material.

quantum mechanics (105) The study of the behavior of atoms and atomic particles.

quasar (quasi-stellar object, or **QSO)** (kwā′zär) (315) Small, powerful sources of energy believed to be the active cores of very distant galaxies.

quintessence (345) The postulated energy that fills empty space and drives the acceleration of the universe.

radial velocity (V_r) (116) That component of an object's velocity directed away from or toward Earth.

radiation pressure (374) The force exerted on the surface of a body by its absorption of light. Small particles floating in the solar system can be blown outward by the pressure of the sunlight.

radio galaxy (304) A galaxy that is a strong source of radio signals.

radio interferometer (93) Two or more radio telescopes that combine their signals to achieve the resolving power of a larger telescope.

rays (388) Ejecta from meteorite impacts forming white streamers radiating from some lunar craters.

recombination (334) The stage within 300,000 years of the big bang, when the gas became transparent to radiation.

reconnection (137) On the sun, the merging of magnetic fields to release energy in the form of flares.

red dwarf (153) A faint, cool, low-mass, main-sequence star.

redshift (115) A Doppler shift toward longer wavelengths caused by a velocity of recession.

reflecting telescope (81) A telescope that uses a concave mirror to focus light into an image.

reflection nebula (176) A nebula produced by starlight reflecting off dust particles in the interstellar medium.

refracting telescope (81) A telescope that forms images by bending (refracting) light with a lens.

reionization (334) The stage in the early history of the universe when ultraviolet photons from the first stars ionized the gas filling space.

resolving power (83) The ability of a telescope to reveal fine detail. Depends on the diameter of the telescope objective.

retrograde motion (50) The apparent backward (westward) motion of planets as seen against the background of stars.

revolution (24) Orbital motion about a point located outside the orbiting body. See also **rotation.**

ribonucleic acid (477) Long carbon-chain molecules that use the information stored in DNA to manufacture complex molecules necessary to the organism.

rich galaxy cluster (295) A cluster containing over 1000 galaxies, mostly elliptical, scattered over a volume about 3 Mpc in diameter.

rift valley (385) A long, straight, deep valley produced by the separation of crustal plates.

ring galaxy (297) A galaxy that resembles a ring around a bright nucleus. Believed to be the result of a head-on collision of two galaxies.

RNA See **ribonucleic acid.**

Roche limit (rōsh) (425) The minimum distance between a planet and a satellite that holds itself together by its own gravity. If a satellite's orbit brings it within its planet's Roche limit, tidal forces will pull the satellite apart.

Roche lobe (214) The volume of space a star controls gravitationally within a binary system.

Roche surface (214) The dumbbell-shaped surface that encloses the Roche lobes around a close binary star.

rotation (24) Motion around an axis passing through the rotating body. See also **revolution.**

rotation curve (262) A graph of orbital velocity versus radius in the disk of a galaxy.

rotation curve method (291) A method of determining a galaxy's mass by observing the orbital velocity and radius of stars in the galaxy.

RR Lyrae variable star (är·är·lī′rē) (255) Variable star with period of from 12 to 24 hours. Common in some globular clusters.

Sagittarius A* (273) The powerful radio source located at the core of the Milky Way Galaxy.

Saros cycle (sĕ′rōs) (39) An 18-year, $11\frac{1}{3}$-day period after which the pattern of lunar and solar eclipses repeats.

Schmidt-Cassegrain focus (88) The optical design that uses a thin corrector plate at the entrance to the telescope tube. A popular design for small telescopes.

Schwarzschild radius (R_S) (schwôrts′ shēld) (242) The radius of the event horizon around a black hole.

scientific model (18) A tentative description of a phenomenon for use as an aid to understanding.

scientific notation (6) The system of recording very large or very small numbers by using powers of 10.

secondary atmosphere (386) The gases outgassed from a planet's interior; rich in carbon dioxide.

secondary crater (388) Impact crater formed by debris ejected from a larger impact.

secondary mirror (88) In a reflecting telescope, a mirror that directs the light from the primary mirror to a focal position.

second of arc (21) An angular measure. One-sixtieth of a minute of arc.

seeing (83) Atmospheric conditions on a given night. When the atmosphere is unsteady, producing blurred images, the seeing is said to be poor.

self-sustaining star formation (272) The process by which the birth of stars compresses the surrounding gas clouds and triggers the formation of more stars. Proposed to explain spiral arms.

semimajor axis, *a* (58) Half of the longest diameter of an ellipse.

SETI (489) The Search for Extra-Terrestrial Intelligence.

Seyfert galaxy (sē′fûrt) (306) An otherwise normal spiral galaxy with an unusually bright, small core that fluctuates in brightness. Believed to indicate the core is erupting.

Shapley–Curtis debate (259) A 1920 debate between Harlow Shapley and Heber Curtis on the nature of spiral nebulae. Curtis argued that they are other galaxies, and Shapley argued they are internal to our own galaxy.

shepherd satellite (435) A satellite that, by its gravitational field, confines particles to a planetary ring.

shield volcano (400) Wide, low-profile volcanic cone produced by highly liquid lava.

shock wave (179) A sudden change in pressure that travels as an intense sound wave.

sidereal drive (sī·dîr′ē·ŭl) (89) The motor and gears on a telescope that turn it westward to keep it pointed at a star.

sidereal period (33) The time a celestial body takes to turn once on its axis or revolve once around its orbit relative to the stars.

singularity (241) The object of zero radius into which the matter in a black hole is believed to fall.

Small Magellanic Cloud (285) An irregular galaxy that is a satellite of our Milky Way Galaxy. It is visible in the southern sky.

soft gamma-ray repeater (SGR) (247) An object that produces repeated bursts of low-energy (soft) gamma rays.

solar eclipse (35) The event that occurs when the moon passes directly between Earth and the sun, blocking our view of the sun.

solar nebula theory (356) The theory that the planets formed from the same cloud of gas and dust that formed the sun.

solar system (6) The sun and its planets, asteroids, comets, and so on.

solar wind (127) Rapidly moving atoms and ions that escape from the solar corona and blow outward through the solar system.

south celestial pole (20) The point on the celestial sphere directly above Earth's South Pole.

south point (20) One of the four cardinal directions. The point on the horizon directly south.

spectral class or type (113) A star's position in the temperature classification system O, B, A, F, G, K, M. Based on the appearance of the star's spectrum.

spectral line (104) A line in a spectrum at a specific wavelength produced by the absorption or emission of light by certain atoms.

spectral sequence (113) The arrangement of spectral classes (O, B, A, F, G, K, M) ranging from hot to cool.

spectrograph (91) A device that separates light by wavelengths to produce a spectrum.

spectroscopic binary system (158) A star system in which the stars are too close together to be visible separately. We see a single point of light, and only by taking a spectrum can we determine that there are two stars.

spectroscopic parallax (154) The method of determining a star's distance by comparing its apparent magnitude with its absolute magnitude as estimated from its spectrum.

spherical component (261) The part of the galaxy including all matter in a spherical distribution around the center (the halo and nuclear bulge).

spicule (spĭk′yōōl) (126) A small, flamelike projection in the chromosphere of the sun.

spiral arm (8, 260) Long spiral pattern of bright stars, star clusters, gas, and dust. Spiral arms extend from the center to the edge of the disk of spiral galaxies.

spiral galaxy (284) A galaxy with an obvious disk component containing gas; dust; hot, bright stars; and spiral arms.

spiral tracer (268) Object used to map the spiral arms—for example, O and B associations, open clusters, clouds of ionized hydrogen, and some types of variable stars.

spring tide (71) Ocean tide of high amplitude that occurs at full and new moon.

star (6) A globe of gas held together by its own gravity and supported by the internal pressure of its hot gases, which generate energy by nuclear fusion.

starburst galaxy (299) A galaxy undergoing a rapid burst of star formation.

steady-state theory (333) The theory (now generally abandoned) that the universe does not evolve.

stellar model (189) A table of numbers representing the conditions in various layers within a star.

stellar parallax (*p*) (147) A measure of stellar distance. See also **parallax.**

stony-iron meteorite (455) A meteorite composed of stone and iron mixed together.

stromatolite (481) A layered fossil formation caused by ancient mats of algae or bacteria, which build up mineral deposits season after season.

strong force (135) One of the four forces of nature. The strong force binds protons and neutrons together in atomic nuclei.

subduction zone (384) A region of a planetary crust where a tectonic plate slides downward.

summer solstice (28) The point on the celestial sphere where the sun is at its most northerly point. Also, the time when the sun passes this point, about June 22, and summer begins in the northern hemisphere.

sunspot (124) Relatively dark spot on the sun that contains intense magnetic fields.

supercluster (346) A cluster of galaxy clusters.

supergiant (152) Exceptionally luminous star whose diameter is 10 to 1000 times that of the sun.

supergranule (127) Very large convective features in the sun's surface.

supernova (200) The explosion of a star in which it increases its brightness by a factor of about a million.

supernova remnant (222) The expanding shell of gas marking the site of a supernova explosion.

***S* wave** (382) A shear wave. A type of seismic wave produced in Earth by the lateral motion of the material.

synchrotron radiation (sĭn′krŭ·trŏn) (220) Radiation emitted when high-speed electrons move through a magnetic field.

synodic period (sĭ·nŏd′ĭk) (33) The time a solar system body takes to orbit the sun once and return to the same orbital relationship with Earth. That is, orbital period referenced to Earth.

T dwarf (114) A very cool, low-mass star or brown dwarf located below the L stars on the main sequence.

temperature (107) A measure of the agitation among the atoms and molecules of a material. The intensity of heat.

terminator (387) The dividing line between daylight and darkness on a planet or moon.

terrestrial planet (364) An Earthlike planet—small, dense, rocky.

theory (59) A system of assumptions and principles applicable to a wide range of phenomena that have been repeatedly verified.

tidal heating (430) The heating of a planet or satellite because of friction caused by tides.

tidal tail (296) A long streamer of stars, gas, and dust torn from a galaxy during its close interaction with another passing galaxy.

time dilation (243) The slowing of moving clocks or clocks in strong gravitational fields.

transition (111) The movement of an electron from one atomic energy level to another.

triple-alpha process (186) The nuclear fusion process that combines three helium nuclei (alpha particles) to make one carbon nucleus.

T Tauri star (tôrē) (184) A young star surrounded by gas and dust. Believed to be contracting toward the main sequence.

turnoff point (208) The point in an H–R diagram at which a cluster's stars turn off of the main sequence and move toward the red-giant region, revealing the approximate age of the cluster.

type I comet tail See **gas tail.**

type I supernova (220) A supernova explosion caused by the collapse of a white dwarf.

type II comet tail See **dust tail.**

type II supernova (219) A supernova explosion caused by the collapse of a massive star.

umbra (ŭm′brŭ) (31) The region of a shadow that is totally shaded.

uncompressed density (368) The density a planet would have if its gravity did not compress it.

unified model (313) An attempt to explain the different types of active galactic nuclei using a single model viewed from different directions.

uniform circular motion (50) The classical belief that the perfect heavens could only move by the combination of uniform motion along circular orbits.

valley network (409) A system of dry drainage channels on Mars that resembles the beds of rivers and tributary streams on Earth.

variable star (255) A star whose brightness changes periodically.

velocity dispersion method (291) A method of finding a galaxy's mass by observing the range of velocities within the galaxy.

vernal equinox (28) The place on the celestial sphere where the sun crosses the celestial equator moving northward. Also, the time of year when the sun crosses this point, about March 21, and spring begins in the northern hemisphere.

vesicular basalt (vŭ·sĭk′yŭ·lŭr) (391) A porous rock formed by solidified lava with trapped bubbles.

visual binary system (156) A binary star system in which the two stars are separately visible in the telescope.

water hole (489) The interval of the radio spectrum between the 21-cm hydrogen radiation and the 18-cm OH radiation. Likely wavelengths to use in the search for extraterrestrial life.

wavelength (78) The distance between successive peaks or troughs of a wave. Usually represented by λ.

wavelength of maximum intensity (λ_{max}) (108) The wavelength at which a perfect radiator emits the maximum amount of energy. Depends only on the object's temperature.

weak force (135) One of the four forces of nature. The weak force is responsible for some forms of radioactive decay.

west point (20) One of the four cardinal directions. The point on the horizon directly west.

white dwarf (153) Dying star that has collapsed to the size of Earth and is slowly cooling off. At the lower left of the H–R diagram.

Widmanstätten pattern (wĭd′mŭn· stä·tŭn) (455) Bands in iron meteorites due to large crystals of nickel-iron alloys.

winter solstice (28) The point on the celestial sphere where the sun is farthest south. Also the time of year when the sun passes this point, about December 22, and winter begins in the northern hemisphere.

X-ray burster (237) An object that produces repeated bursts of X rays.

Zeeman effect (131) The splitting of spectral lines into multiple components when the atoms are in a magnetic field.

zenith (20) The point on the sky directly above the observer.

zero-age main sequence (ZAMS) (193) The location in the H–R diagram where stars first reach stability as hydrogen-burning stars.

zodiac (27) A band centered on the ecliptic and encircling the sky.

Answers to Even-Numbered Problems

Chapter 1
2. 3475 km; **4.** 1.05×10^8 km; **6.** about 1.2 s; **8.** about 75,000 yr; **10.** about 27

Chapter 2
2. 2800; **4.** The sun is 400,000 times brighter than the full moon.

Chapter 3
2. (a) new; (b) full; (c) first quarter; (d) third quarter; **4.** 14 days; **6.** 690 arc seconds or 11 arc minutes or about one-fifth of a degree; **8.** The eclipse would again be visible in Canada on 12 August 2026 [July 10, 1972 + 3 × ($6585\frac{1}{3}$ days)].

Chapter 4
2. Ratio of maximum to minimum distance = 6.1:1. In the Ptolemaic model, Venus is always between Earth and the sun. In the diagram of the Ptolemaic universe, the same ratio is about 1.5:1.; **4.** 32 years; **6.** 39.4 AU; **8.** The arrows are about the correct length.

Chapter 5
2. 3 m; **4.** Each of the Keck telescopes has 1.6 million times the light-gathering power of the human eye; **6.** No, the resolution of Galileo's telescopes was about 5.8 seconds of arc; **8.** 13 mm; **10.** 45 cm

Chapter 6
2. 150 nm; **4.** $2^4 = 16$; **6.** 250 nm; **8.** B, F, M, K; **10.** 0.6 nm

Chapter 7
2. 730 km; **4.** 3.6 times brighter; **6.** 400,000 yr; **8.** 9.0×10^{16} J; **10.** 0.2 kg

Chapter 8
2. 28 m
4.

m	M	d (pc)	p (seconds of arc)
7	7	10	0.1
11	1	1000	0.001
1	−2	40	0.025
4	2	25	0.040

6. about B8; **8.** Using Figure 8-8, M_v is estimated to be −7, so d = 400 pc. There is some uncertainty in this number in determining the absolute magnitude of the B0V star from Figure 8-8.; **10.** a, c, c, c, d; **12.** 4.23×10^{11} m, 6.7×10^{10} m, 2.4×10^{31} kg (12 solar masses); **14.** A 4-solar-mass main-sequence star has a luminosity of approximately 128 solar luminosities, a 9-solar-mass main-sequence star has a luminosity of approximately 2187 solar luminosities, and a 7-solar-mass main-sequence star has a luminosity of 907 solar luminosities.

Chapter 9
2. 20,000 AU or 0.097 pc; **4.** 100 K; **6.** In both reactions, the net result is that four ^{1}H nuclei are converted into one ^{4}He nucleus and energy. In the case of the proton–proton chain, six ^{1}H nuclei are put into the reaction, and the reaction returns two ^{1}H nuclei, one ^{4}He nucleus, and energy (6 ^{1}H → 2 ^{1}H + 1 ^{4}He + energy, which is equivalent to 4 ^{1}H → 1 ^{4}He + energy). The CNO cycle inputs one ^{12}C nucleus and four ^{1}H nuclei and returns one ^{12}C nucleus, one ^{4}He nucleus, and energy (1 ^{12}C + 4 ^{1}H → 1 ^{12}C + ^{4}He + energy, which is equivalent to 4 ^{1}H → 1 ^{4}He + energy); **8.** 2000 nm = 2 micrometers; **10.** Estimating the absolute magnitude of an O6V star to be −5.6 implies a distance of 1600 pc. The answer is very dependent on the estimate for the absolute magnitude.

Chapter 10
2. about 310 million years; **4.** 1.8 ly; **6.** 32,000 years; **8.** 1.3×10^6 times; **10.** 1200 km/s

Chapter 11
2. 0.00096 AU if the total mass is 2 $M_\odot$; **4.** 3 m, 1.1×10^{-25} m; **6.** 8.1 minutes; **8.** 3.0×10^8 m/s; photons

Chapter 12
2. 16%; **4.** 5000 pc; **6.** 25 pc; **8.** 20 times; **10.** 1500 K

Chapter 13
2. 2.6×10^6 pc (2.6 Mpc); **4.** 93 km/s; **6.** 28.6 Mpc; **8.** 165 million years

Chapter 14
2. 7.6 million years; **4.** 0.024 pc = 4900 AU; **6.** −28.5; **8.** 640 Mpc

Chapter 15
2. 57; 17.5 billion years; 11.7 billion years; **4.** 1.6×10^{-30} gm/cm^3; **6.** 76 km/s/Mpc; **8.** 16.6 billion years; 11 billion years; the universe could be older

Chapter 16
2. It will look $(206{,}265)^2 = 4.3 \times 10^{10}$ times fainter, apparent magnitude of about +23; **4.** about three times the half-life, or 3.9 billion years; **6.** large amounts of methane and water ices; **8.** about 130 impacts per 100 square meters

Chapter 17
2. 16 times, 64 times; **4.** No. They would be visible in photos taken from orbit around the moon; **6.** 23 minutes; **8.** about 380 km; **10.** 6.4×10^{23} kg

Chapter 18
2. From Callisto, Jupiter would appear to be 4.3°, and from Io it would appear to be 19.4°; **4.** 16.8 km/s; **6.** 7.3 s; **8.** 60,000 nm (60 microns); **10.** 1.5×10^{22} kg

Chapter 19
2. 21.1 km/s; **4.** Assuming an orbital distance of 3 AU, the period would be 5.2 years; **6.** 0.77 seconds of arc (R = 500 km, D = 1.77 AU). Earth-based telescopes can seldom resolve structures smaller than 0.5 second of arc. The Hubble Space Telescope would be able to see some very large features, since it can resolve structures as small as 0.1 second of arc; **8.** 9.2×10^6 km. If the tail is not perpendicular to your line of sight, then the tail is longer than 9.2×10^6 km; **10.** 0.033 Earth masses

Chapter 20
2. 8.9 cm; 0.67 mm; **4.** about 1.3 solar masses; **6.** 380 km; **8.** Pessimistic, 2×10^{-5}; optimistic, 10^7. Answers will vary greatly. Using intermediate values from the table, such as 2×10^{11}, 0.1, 0.1, 0.1, 0.1, 10^{-6}, yields an answer of 20.

Index

Boldface page numbers indicate definitions of key terms.

C

D

H

I

J

K

L

M

N

O

P

T

U

V